BIOREMEDIATION AND BIOECONOMY
A Circular Economy Approach

BIOREMEDIATION AND BIOECONOMY
A Circular Economy Approach

SECOND EDITION

Edited by

MAJETI NARASIMHA VARA PRASAD
Department of Plant Sciences, School of Life Sciences,
University of Hyderabad (An Institution of Eminence),
Hyderabad, Telangana, India

ELSEVIER

Elsevier
Radarweg 29, PO Box 211, 1000 AE Amsterdam, Netherlands
The Boulevard, Langford Lane, Kidlington, Oxford OX5 1GB, United Kingdom
50 Hampshire Street, 5th Floor, Cambridge, MA 02139, United States

Notices
Knowledge and best practice in this field are constantly changing. As new research and experience broaden our
understanding, changes in research methods, professional practices, or medical treatment may become
necessary.

Practitioners and researchers must always rely on their own experience and knowledge in evaluating and using
any information, methods, compounds, or experiments described herein. In using such information or methods
they should be mindful of their own safety and the safety of others, including parties for whom they have a
professional responsibility.

To the fullest extent of the law, neither the Publisher nor the authors, contributors, or editors, assume any liability
for any injury and/or damage to persons or property as a matter of products liability, negligence or otherwise, or
from any use or operation of any methods, products, instructions, or ideas contained in the material herein.

ISBN: 978-0-443-16120-9

For information on all Elsevier publications
visit our website at https://www.elsevier.com/books-and-journals

Publisher: Candice Janco
Acquisitions Editor: Jessica Mack
Editorial Project Manager: Teddy Lewis
Production Project Manager: Paul Prasad Chandramohan
Cover Designer: Mark Rogers

Typeset by STRAIVE, India

Dedication

This book is dedicated to my wife Challapalli Savithri.

Contents

I

Bioproducts from contaminated soil and water

1. Bioremediation, bioeconomy, circular economy, and circular bioeconomy—Strategies for sustainability

Majeti Narasimha Vara Prasad

2. Holistic approach to bioremediation-derived biomass and biorefineries for accelerating bioeconomy, circular bioeconomy, and carbon neutrality

Majeti Narasimha Vara Prasad

3. Production of biodiesel feedstock from trace element-contaminated lands in Ukraine

Oksana Sytar and Majeti Narasimha Vara Prasad

4. Aromatic, medicinal, and energy plantations on metalliferous/contaminated soil—Bioremediation and bioeconomy

Sai Shankar Sahu, Adarsh Kumar, Majeti Narasimha Vara Prasad, and Subodh Kumar Maiti

V

Algal bioproducts, biofuels, and biorefinery for business opportunities

---

†Deceased

21. Algae-based bioremediation bioproducts and biofuels for biobusiness

Raman Kumar, Yograj Neha, G.A. Ravishankar, and Vidyashankar Srivatsan

VI

Bioprocesses and bioengineering for boosting biobased economy

22. Building circular bio-based economy through sustainable waste management

K. Amulya, Shikha Dahiya, and S. Venkata Mohan

VII

Case studies

23. Bioremediation in Brazil: Recent evolutions and remaining challenges to boost up the bioeconomy

Geórgia Labuto, Lucélia Alcantara Barros, Marcus Leonan Costa Guimaraes, Ricardo Santos Silva, and Taciana Guarnieri Soares Guimarães

VIII
New biology

Contributors

K. Amulya National University of Ireland, Galway, Ireland

Klaus J. Appenroth Matthias Schleiden Institute-Plant Physiology, Friedrich Schiller University, Jena, Germany

B. Barbosa METRICS, Universidade Nova de Lisboa, Caparica, Portugal

Lucélia Alcantara Barros Laboratory of Integrated Sciences, Universidade Federal de São Paulo, São Paulo, Brazil

R. Chaturvedi Department of Botany, St. John's College, Dr. B.R. Ambedkar University, Agra, India

Tamara S. Chibrik[†]

J. Costa METRICS, Universidade Nova de Lisboa, Caparica; Instituto Superior de Educação e Ciências, Lisbon, Portugal

Shikha Dahiya Biochemical Process Engineering, Chemical Engineering, Department of Civil, Environmental and Natural Resources Engineering, Lulea Technical University, Luleå, Sweden

Pierre-Alexandre Deyris Laboratory of Bioinspired Chemistry and Ecological Innovations, UMR 5021, CNRS-University of Montpellier, Grabels, France

Patricia Dubois Responsable du Service Stratégie et Partenariat, Établissement Public Foncier Hauts-de-France, Euralille, France

P.J.C. Favas University of Trás-os-Montes e Alto Douro, UTAD, School of Life Sciences and the Environment, Vila Real; University of Coimbra, MARE—Marine and Environmental Sciences Centre/ARNET—Aquatic Research Network,

Department of Life Sciences, Calçada Martim de Freitas, Coimbra, Portugal

A.L. Fernando METRICS, Universidade Nova de Lisboa, Caparica, Portugal

Elena I. Filimonova Ural Federal University Named After First President of Russia B.N. Yeltsin, Yekaterinburg, Russian Federation

Margarita A. Glazyrina Ural Federal University Named After First President of Russia B.N. Yeltsin, Yekaterinburg, Russian Federation

L.A. Gomes METRICS, Universidade Nova de Lisboa, Caparica, Portugal; CNR-ITAE, Messina, Italy

S. Gopalakrishnan International Crops Research Institute for the Semi-Arid Tropics (ICRISAT), Patancheru, Telangana, India

Claude Grison Laboratory of Bioinspired Chemistry and Ecological Innovations, UMR 5021, CNRS-University of Montpellier, Grabels, France

Marcus Leonan Costa Guimaraes Laboratory of Integrated Sciences, Universidade Federal de São Paulo, São Paulo, Brazil

Taciana Guarnieri Soares Guimarães Laboratory of Integrated Sciences, Universidade Federal de São Paulo, São Paulo, Brazil

Ponlakit Jitto Faculty of Environment and Resource Studies, Mahasarakham University, Kham Riang, Kantharawichai, Maha Sarakham, Thailand

Srujana Kathi UGC-Human Resource Development Centre, Pondicherry University, Puducherry, India

---

[†]Deceased

Boda Ravi Kiran Bioengineering and Environmental Sciences Lab, Department of Energy and Environmental Engineering (DEE), CSIR-Indian Institute of Chemical Technology (CSIR-IICT), Hyderabad, Telangana, India

Adarsh Kumar Laboratory of Biotechnology, Institute of Natural Sciences and Mathematics, Ural Federal University, Ekaterinburg, Russia

Raman Kumar Applied Phycology and Food Technology Laboratory, Biotechnology Division, CSIR-Institute of Himalayan Bioresource Technology, Palampur, Himachal Pradesh; Academy of Scientific and Innovative Research (AcSIR), CSIR- Human Resource Development Centre, (CSIR-HRDC) Campus, Ghaziabad, Uttar Pradesh, India

Nichanun Kutrasaeng Science Center, Faculty of Science and Technology, Sakon Nakhon Rajabhat University, Sakon Nakhon, Thailand

Geórgia Labuto Laboratory of Integrated Sciences; Department of Chemistry, Universidade Federal de São Paulo, São Paulo, Brazil

Guillaume Lemoine Biodiversity and Ecological Engineering Referent, Etablissement Public Foncier Hauts-de-France, Euralille, France

Natalia V. Lukina Ural Federal University Named After First President of Russia B.N. Yeltsin, Yekaterinburg, Russian Federation

Subodh Kumar Maiti Department of Environmental Science and Engineering, Indian Institute of Technology (Indian School of Mines), Dhanbad, Jharkhand, India

Maria G. Maleva Ural Federal University Named After First President of Russia B.N. Yeltsin, Yekaterinburg, Russian Federation

S. Venkata Mohan Bioengineering and Environmental Sciences Lab, Department of Energy and Environmental Engineering (DEE), CSIR-Indian Institute of Chemical Technology (CSIR-IICT), Hyderabad, Telangana, India

Woranan Nakbanpote Department of Biology, Faculty of Science, Mahasarakham University, Khamriang, Kantarawichi, Mahasarakham, Thailand

Yograj Neha Applied Phycology and Food Technology Laboratory, Biotechnology Division, CSIR-Institute of Himalayan Bioresource Technology, Palampur, Himachal Pradesh, India

Tomasz K. Olszewski Department of Physical and Quantum Chemistry, Wroclaw University of Science and Technology, Wroclaw, Poland

E.G. Papazoglou Agricultural University of Athens, Athens, Greece

R. Pasumarthi International Crops Research Institute for the Semi-Arid Tropics (ICRISAT), Patancheru, Telangana, India

M.S. Paul Department of Botany, St. John's College, Dr. B.R. Ambedkar University, Agra, India

Cyril Poullain Laboratory of Bioinspired Chemistry and Ecological Innovations, UMR 5021, CNRS-University of Montpellier, Grabels, France

Majeti Narasimha Vara Prasad Department of Plant Sciences, School of Life Sciences, University of Hyderabad (An Institution of Eminence), Hyderabad, Telangana, India

J. Pratas University of Coimbra, Faculty of Sciences and Technology, Department of Earth Sciences, Coimbra, Portugal

P. Srinivasa Rao Western Colorado Research Centre, Fruita, CO, United States

G.A. Ravishankar Dayananda Sagar Institutions, Bengaluru, Karnataka, India

Sai Shankar Sahu Department of Environmental Science and Engineering, Indian Institute of Technology (Indian School of Mines), Dhanbad, Jharkhand, India

Hemen Sarma Bioremedatio Technology Group, Department of Botany, Bodoland University, Kokrajhar, Assam, India

K. Schmidt-Przewoźna   Institute of Natural Fibres and Medicinal Plants—PIB, Poznan, Poland

Abin Sebastian   Department of Botany, St George's College, Aruvithura, Kerala, India

Ricardo Santos Silva   Laboratory of Integrated Sciences, Universidade Federal de São Paulo, São Paulo, Brazil

V. Sivasubramanian   Phycospectrum Environmental Research Centre (PERC), Chennai, Tamil Nadu, India

K. Sowjanya Sree   Department of Environmental Science, Central University of Kerala, Periye, India

Vidyashankar Srivatsan   Applied Phycology and Food Technology Laboratory, Biotechnology Division, CSIR-Institute of Himalayan Bioresource Technology, Palampur, Himachal Pradesh, India

Oksana Sytar   Department of Plant Biology, Educational and Scientific Center "Institute of Biology and Medicine", Kyiv National University of Taras Shevchenko, Kyiv, Ukraine; Institute of Plant and Environmental Sciences, Slovak Agricultural University in Nitra, Nitra, Slovakia

Uraiwan Taya   Land Development Regional Office 5, Department of Land Development, Ministry of Agriculture and Cooperatives, Khon Kaen, Thailand

J.C. Tewari   Central Arid Zone Research Institute, Jodhpur, Rajasthan, India

Aliyu Ahmad Warra   Centre for Entrepreneurial Development, Federal University, Gusau, Nigeria

Preface

Bioremediation and Bioeconomy: A Circular Economy Approach provides a common platform for scientists and entrepreneurs to find sustainable solutions to the contemporary environmental issues related to pollution and contamination of inorganics and organics.

Biodiversity is the toolbox for environmental remediation. Utilizing biodiversity to remove, immobilize, contain, reduce the toxicity, and/or degrade pollutants from soil, water, or air is known as bioremediation. It can be a successful method for removing different contaminants from the abovementioned substrates on site and/or in situ, such as toxic metals, metal(loids), radioactive isotopes, solvents, salt (NaCl), polycyclic aromatic hydrocarbons (PAHs), petroleum hydrocarbons (PHCs), polychlorinated biphenyls (PCBs), and solvents (e.g., trichloroethylene [TCE]). Technologies for commercial bioremediation are underutilized globally. This book elucidates some significant examples of bioremediation via tested techniques into a common practice. There is ample proof that bioremediation can be used successfully in the field in many situations as elaborated in different chapters.

Clean biofuel production and phytoremediation solutions from contaminated lands worldwide are a reality. The goal of the European Union's H2020 Phy2Climate project is to create a link between the production of clean drop-in biofuels and the bioremediation of polluted locations. These biofuels won't pose any hazards for land use change; thus, bioremediation will clear the contaminated areas of a wide range of contaminants and make them suitable for farming while also enhancing the process' overall sustainability, legal framework, and economics. This is because circular economy matters and has been considered as a very effective strategy to address the environmental challenge that we are currently facing. In 2023, an analysis of 221 definitions conceptualizing the circular economy has been proposed by Kirchherr et al. [Resources, Conservation & Recycling 194 (2023) 107001]; Figge et al. [Ecological Economics 208 (2023) 107823].

Global system with finite resources is interwoven with the economy as an open subsystem. Both systems depend on one another. Economic activity is physically constrained by the ecological system, while the ecological system's sources and sinks are affected by the economic system. Both elements will be combined, according to the circular economy. Resources are used more sustainably when loops are closed, or when the same resources are used repeatedly. A circular economy doesn't need any additional virgin resources in its ideal condition. Thus, this book demonstrates how to increase government and business support for bioremediation. The use of Bioremediation as a "gentle remediation option" (GRO) within a more comprehensive, long-term management approach under a new paradigm of phytomanagement is elaborated in its 8 sections and 26 chapters.

Section I. Bioproducts from contaminated soil and water—six chapters

Section II. Biomass energy and biofuel from contaminated substrates—two chapters

Section III. Ornamentals and tree crops for contaminated substrates—six chapters

Section IV. Brownfield development for smart bioeconomy—five chapters

Section V. Algal bioproducts, biofuels, and biorefinery for business opportunities—two chapters

Section VI. Bioprocesses and bioengineering for boosting biobased economy—one chapter

Section VII. Case studies—two chapters

Section VIII. New biology—two chapters

Editor

Acknowledgments

It is my pleasure to thank all the contributors to this edition for their comprehensive contributions and support.

I am thankful to Candice Janco, publisher, Jessica Mack, Acquisitions Editor for excellent coordination of this fascinating project.

Thanks are due to Mr. Paul Prasad Chandramohan and his team for excellent technical help that resulted in record time of publication of this second edition.

Last but not least, I wish to thank editorial project manager, Mr. Teddy Lewis; cover designer, Mark Rogers; and typeset by Straive, India.

I acknowledge the Honorary Emeritus Professorship by the University of Hyderabad during which tenure part of this work was carried out. I thankfully acknowledge numerous colleagues, my research associates, and students for sharing knowledge, ideas, and assistance that helped shape this second edition of the book.

Bioproducts from contaminated soil and water

1

Bioremediation, bioeconomy, circular economy, and circular bioeconomy— Strategies for sustainability

Majeti Narasimha Vara Prasad

Department of Plant Sciences, School of Life Sciences, University of Hyderabad (An Institution of Eminence), Hyderabad, Telangana, India

1 Introduction

Biodiversity is the tool box for nature-based solutions (bioremediation). The following are the emerging technologies based on biodiversity and are being implemented in different situations and locations for pollution abatement.

(a) Cyanoremediation = blue green algae
(b) Genoremediation = genetic engineering
(c) Pteridoremediation = use of ferns
(d) Mycoremediation = use of fungi
(e) Phycoremediation = use of algae
(f) Rhizoremediation = root/rhizosphere
(g) Phylloremediation = use of leaves for air pollution abatement
(h) Dendroremediation = use of trees for pollution abatement

Gerhardt et al. (2017) opined that bio-phytoremediation is a proven technology and an accepted practice globally to remediate soils impacted with inorganics and organic pollutants including salt (Abdelhafeez et al., 2022). Bioeconomy is "the creation, development, production, and usage of biological goods and processes."

The following are some of the significant Assisted and Amendment Enhanced (Integrated) Sustainable Remediation Technologies (Prasad, 2021; Ahila et al., 2021; Ahmad et al., 2021; Alazaiza et al., 2022; Alpandi et al., 2021; Ammeri et al., 2022; Antony et al., 2022; Ara et al., 2021; Auchterlonie et al., 2021; Bezie et al., 2021; Bhunia et al., 2022; Gajić et al., 2022).

(a) Remediation and Combined Clean Biofuel
(b) Assisted Phytoremediation to Green Energy Production
(c) Biochar-Based Soil and Water Remediation
(d) Use of Diverse Organic Amendments in Remediation
(e) Plant Microbe-Assisted Remediation of Inorganics and Organics
(f) Advanced Technologies for Remediation of Inorganics and Organics
(g) Nanoscience in Remediation of Inorganics and Organics

Building bioeconomy using contaminated substrates has become a cutting-edge science. Decontamination of the sites (technogenic and geogenic) across the globe requires robust remediation for resource recovery and pollution abatement (Prasad and Shih, 2016). This means going beyond the 3Rs to 7Rs (Reduce, Reuse, Recycle, Recover, Redesign, Refuse, and Rethink) to multiple Rs, namely Reduce, Reclaim, Recovery, Recycle, Refuse, Remediation, Renovation, Replenish, Resilience, Restore, Reuse, Reverence to Nature, Reclamation, and Repurposing Rainwater Harvest (Prasad, 2023; Braga et al., 2021; Chatterjee et al., 2022; Cheng et al., 2022).

2 Economic development and environmental deterioration

The Environmental Kuznets curve suggests that economic development initially causes deterioration in the environment. Later due to economic growth, society begins to improve the relationship with the environment, and environmental degradation reduces. Thus, the economic growth is good for the environment. Nevertheless, critics are of the view that there is no guarantee that the economic growth will lead to an improved environment—in fact, the opposite is often the case. At the least, it requires a very targeted policy and attitudes to make sure that economic growth is compatible with an improving environment (Fig. 1).

3 Salient aspects of environmental Kuznets curve

1. It is observed that increased economic growth led to decreased pollution.
2. Economic growth increased income.
3. The link between economic growth and living standards can be weak. i.e., living standards are opposite to real GDP.
4. Improved technology saves energy and raw material.
5. Use of renewable energy gained momentum in economic growth.
6. Initially, economic development leads to shifting from farming to manufacturing. This leads to greater environmental degradation. However, increased productivity and rising real incomes see a third shift from manufacturing industrial sector to the service sector. The service sector usually has a lower environmental impact than manufacturing.
7. Economic growth and development usually see a growth in the size of government as a share of GDP. The governments are able to implement taxes and regulations in an attempt to solve environmental externalities, which harm health and living standards.

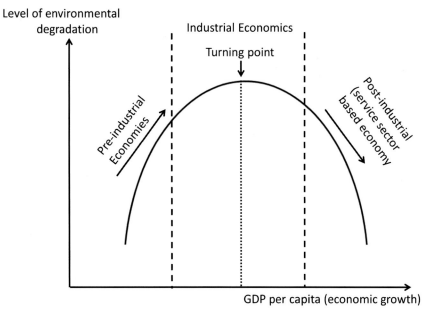

FIG. 1   Environmental Kuznets curve.

8. Rising income has a diminishing marginal utility. Having a very high salary is of little consolation to live in a deteriorated environment (e.g., congestion, pollution, and ill health). Therefore, it is obvious that a person who is earning high income will think of improving other aspects of living standards.
9. Salient aspects of environmental Kuznets curve.
10. Accelerating the shift to sustainability via carbon crops and circular economy.
11. Knowledge explosion in the field of pollution abatement involving technological and biological cycle—progressing fields of advanced research.

4 Criticisms of Kuznets environmental curve

1. There is no guarantee that economic growth will see a decline in pollutants.
2. Pollution is not simply a function of income, but many factors. For example, the effectiveness of government regulation, the development of the economy, population levels.
3. Global pollution. Many developed economies have seen a reduction in industry and growth in the service sector, but they are still importing goods from developing countries. In that sense, they are exporting environmental degradation. Pollution may reduce in the United Kingdom and United States, but countries that export to these countries are seeing higher levels of environmental degradation. One example is with regard to deforestation. Higher-income countries tend to stop the process of deforestation, but at the same time,

they still import meat and furniture from countries that are creating farmland out of forests.

4. Growth leads to greater resource use. Some economists are of the opinion that there is a degree of reduced environmental degradation post-industrialization. But, if the economy continues to expand, then inevitably some resources will continue to be used in greater measure. There is no guarantee that long-term levels of environmental degradation will continue to fall.

5. Countries with the highest GDP have highest levels of CO_2 emission. For example, per capita CO_2 emissions of United States are the highest, 17.564t, and Ethiopia 0.075t per capita in 2009.

The link between levels of income and environmental degradation is quite weak. It is possible that economic growth will be compatible with an improved environment, but it requires a very distinct set of policies and willingness to produce energy and goods in most environmentally friendly way.

There has been considerable discussion on fiscal policies aimed at abatement of pollution. One prominent question posed is "Can subsidies rather than pollution taxes break the trade-off between economic output and environmental protection?" According to Renström et al., 2021, socially responsible investing (SRI) is a long-term-oriented investment approach, which integrates ESG (i.e., Environmental, Social, Governance) factors in the research, analysis, and selection process of securities within an investment portfolio. In contrast, under the impulse of several international summits (from Kyoto in 1997 to Paris 2015), many public institutions have been implementing or proposing fiscal and regulatory policies aimed at contrasting the worsening of environmental quality and at increasing the ESG practices (Renström et al., 2021) (Figs. 2–5).

Table 1 refers to the key SDGs relevant to the bio-based industry sector, identification of the most relevant targets, and the action-oriented interpretation of these targets in the bio-based industry context (Table 2) (Sumfleth et al., 2020).

The green deal strategies aim to reduce climate change and/or to achieve climate neutrality. Climate change and pollution are closely linked. Therefore, achievement of climate neutrality pollution abatement is a must. Climate change is known to influence global weather including its adverse impact on agriculture, land use, hydrology, coastal environment, air

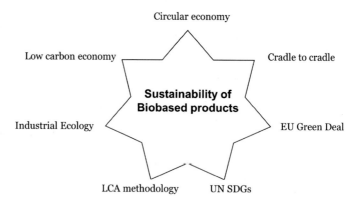

FIG. 2 Bio-based products and sutainability methodologies/strategies/approaches (Allen et al., 2020; Chintani et al., 2021; Choudhary et al., 2022; Cui et al., 2022; Davamani et al., 2021; Do et al., 2022; Ekanayake et al., 2021; Eraky et al., 2022; Gajić et al., 2022; Garate-Quispe et al., 2021; García Martín et al., 2020; Gautam and Agrawal, 2021; Goswami et al., 2022; Gravand et al., 2021; Gunwal et al., 2021).

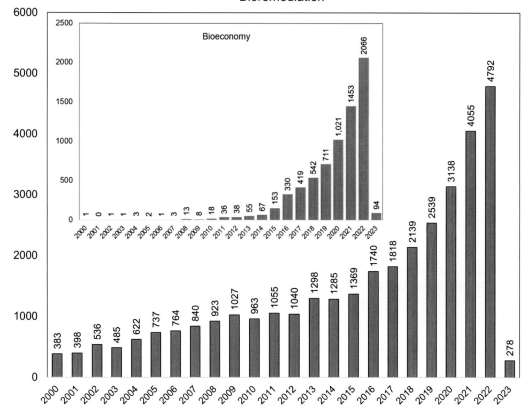

FIG. 3 Articles published on bioremediation and bioeconomy (Global). *Source: www.sciencedirect.com.*

quality, and anthropocene (Petrovič, 2021; Golobokova, 2021; Visbeck and Keiser, 2021; Bini and Rossi, 2021; Panepinto et al., 2021) (Fig. 6).

Climate change and pollution are closely linked. Climate change mitigation means efforts to reduce or prevent emission of greenhouse. Mitigation can mean using new technologies/new approaches and renewable energies, making older equipment more energy-efficient, or changing management practices or consumer behavior (Panepinto et al., 2021). The mitigation technologies are able to reduce or absorb the greenhouse gases (GHG) and, in particular, the CO_2 present in the atmosphere. The CO_2 is a persistent atmospheric gas. It seems increasingly likely that concentrations of CO_2 and other GHG in the atmosphere will overshoot the 450 ppm CO_2 target, widely seen as the upper limit of concentrations consistent with limiting the increase in global mean temperature from pre-industrial levels to around 2°C. To stay well below to the 2°C temperature thus compared with the pre-industrial level as required by the Paris Agreement, it is necessary that in the future we will obtain low (or better zero) emissions, and it is also necessary that we will absorb a quantity of CO_2 from the atmosphere, by 2070, equal to 10 Gt/y. It is proposed for massive cultivation of microalgae for CO_2 fixation as an emergent climate mitigation technology (Panepinto et al., 2021). The green deal

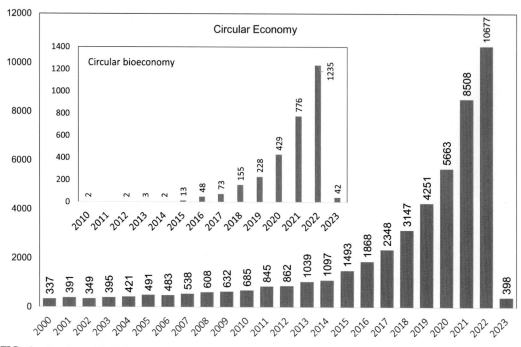

FIG. 4 Articles published on circular economy and circular bioeconomy (Global). *Source: www.sciencedirect.com.*

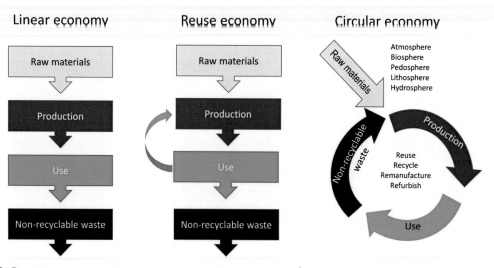

FIG. 5 Different models of economy for using the raw materials (Gunwal et al., 2021; Guo et al., 2021; Haripriyan et al., 2022; Hui et al., 2022; Irawanto, 2021).

TABLE 1 For key SDGs relevant to the bio-based industry sector, identification of the most relevant targets, and the action-oriented interpretation of these targets in the bio-based industry context.

Target (no)	Target (name)	Target goals in the bio-based industry context
	SDG 2: Zero hunger	
2.3	By 2030, double the agricultural productivity and incomes of small-scale food producers […] including through secure and equal access to land, other productive resources and inputs, knowledge, financial services, markets and opportunities for value addition and non-farm employment	Support agricultural productivity Support incomes of small-scale food producers Secure access to land Support knowledge creation Support markets Support value addition and non-farm employment
2.4	By 2030, ensure sustainable food production systems and implement resilient agricultural practices that increase productivity and production, that help maintain ecosystems, that strengthen capacity for adaptation to climate change, extreme weather, drought, flooding and other disasters and that progressively improve land and soil quality	Ensure sustainable food production systems Put in place resilient agricultural practices that increase productivity and production Put in place resilient agricultural practices that help maintain ecosystems Put in place resilient agricultural practices that strengthen capacity for adaptation to climate change, extreme weather, drought, flooding and other disasters Economic and environmental practices that progressively improve land and soil quality
2.5	By 2020, maintain the genetic diversity of seeds, cultivated plants and farmed and domesticated animals and their related wild species […] promote access to and fair and equitable sharing of benefits arising from the utilization of genetic resources and associated traditional knowledge […]	Maintain genetic diversity of seeds, cultivate plans and farmed and domesticated animals and their related wild species Promote access to fair and equitable sharing of benefits
2.A	Increase investment […] in rural infrastructure, agricultural research and extension services, technology development and plant and livestock gene banks in order to enhance agricultural productive capacity in developing countries […]	Increase rural infrastructure Increase agricultural research and extension services Increase technology development Increase plant and livestock gene banks
2.C	Adopt measures to ensure the proper functioning of food commodity markets […]	Support food commodity markets
	SDG 6: Clean water and sanitation	
6.3	By 2030, improve water quality by reducing pollution, eliminating dumping and minimizing release of hazardous chemicals and materials, halving the proportion of untreated wastewater and substantially increasing recycling and safe reuse globally	Improve water quality Pollution reduction and dumping elimination Controlling and limiting the release of hazardous chemicals and materials Reducing untreated wastewater Increase water recycling and safe reuse

Continued

TABLE 1   For key SDGs relevant to the bio-based industry sector, identification of the most relevant targets, and the action-oriented interpretation of these targets in the bio-based industry context—cont'd

6.4	By 2030, substantially increase water-use efficiency across all sectors and ensure sustainable withdrawals and supply of freshwater to address water scarcity and substantially reduce the number of people suffering from water scarcity	Increase water use efficiency Sustainable water withdrawals from all water bodies Reduce water scarcity
6.6	By 2020, protect and restore water-related ecosystems, including mountains, forests, wetlands, rivers, aquifers and lakes	Support ecosystem protection

 SDG 9: Industry, innovation, and infrastructure

9.1	Develop quality, reliable, sustainable and resilient infrastructure […] to support economic development and human well-being […]	Develop infrastructure to support economic development Develop infrastructure to support human well-being
9.4	By 2030, upgrade infrastructure and retrofit industries to make them sustainable, with increased resource-use efficiency and greater adoption of clean and environmentally sound technologies and industrial processes, with all countries taking action in accordance with their respective capabilities	Infrastructure upgrade Industry retrofit Increase resource efficiency Support clean and environmentally sound technologies and industrial processes
9.5	Enhance scientific research, upgrade the technological capabilities of industrial sectors in all countries, in particular developing countries, including, by 2030, encouraging innovation and substantially increasing the number of research and development workers per 1 million people and public and private research and development spending	Enhance scientific research Improve technological capabilities of industrial sectors Promote Innovation Increase research and development workers Increase public and private research and development spending
9.B	Support domestic technology development, research and innovation in developing countries, including by ensuring a conducive policy environment for, inter alia, industrial diversification and value addition to commodities	Support technology development Support research and innovation

 SDG 12: Responsible consumption and production

12.1	Implement the 10-year framework of programs on sustainable consumption and production […]	Sustainable consumption and production
12.2	By 2030, achieve the sustainable management and efficient use of natural resources	Achieve sustainable management of natural resources Achieve efficient use of natural resources

TABLE 1 For key SDGs relevant to the bio-based industry sector, identification of the most relevant targets, and the action-oriented interpretation of these targets in the bio-based industry context—cont'd

12.3	By 2030, halve per capita global food waste at the retail and consumer levels and reduce food losses along production and supply chains […]	Reduce food waste
12.4	By 2020, achieve the environmentally sound management of chemicals and all wastes throughout their life cycle, in accordance with agreed international frameworks, and significantly reduce their release to air, water and soil in order to minimize their adverse impacts on human health and the environment	Environmentally sound management of chemicals and all wastes Reduce chemical and waste releases to air, water and soil
12.5	By 2030, substantially reduce waste generation through prevention, reduction, recycling and reuse	Reduce waste generation
12.6	Encourage companies, especially large and transnational companies, to adopt sustainable practices and to integrate sustainability information into their reporting cycle	Support sustainable practices Integrate sustainability information
12.A	Support developing countries to strengthen their scientific and technological capacity to move toward more sustainable patterns of consumption and production	Strengthen scientific and technological capacity Support sustainable patterns of consumption and production

 SDG 13: Climate action

13.2	Integrate climate change measures into national policies, strategies and planning	Support climate change action

 SDG 14: Life below water

14.1	By 2025, prevent and significantly reduce marine pollution of all kinds […]	Reduce marine pollution
14.2	By 2020, sustainably manage and protect marine and coastal ecosystems to avoid significant adverse impacts, including by strengthening their resilience, and take action for their restoration in order to achieve healthy and productive oceans	Ecosystem protection Protect marine and coastal resilience Protect marine and coastal restoration
14.A	Increase scientific knowledge, develop research capacity and transfer marine technology […] in order to improve ocean health and to enhance the contribution of marine biodiversity to the development of developing countries […]	Increase scientific knowledge Increase research capacity Increase technology improvement to support marine protection
14.C	Enhance the conservation and sustainable use of oceans and their resources by implementing international law as reflected in UNCLOS […]	Enhance the conservation and sustainable use of oceans and their resources

Continued

I. Bioproducts from contaminated soil and water

TABLE 1 For key SDGs relevant to the bio-based industry sector, identification of the most relevant targets, and the action-oriented interpretation of these targets in the bio-based industry context—cont'd

	SDG 15: Life on land	
15.1	By 2020, ensure the conservation, restoration and sustainable use of terrestrial and inland freshwater ecosystems and their services [...]	Protect and restore Inland freshwater ecosystem and their services
15.2	By 2020, promote the implementation of sustainable management of all types of forests, halt deforestation, restore degraded forests and substantially increase afforestation and reforestation globally	Promote sustainable forest management Reduce deforestation Restore degraded forests Increase afforestation and reforestation
15.3	By 2030, combat desertification, restore degraded land and soil, including land affected by desertification, drought and floods, and strive to achieve a land degradation-neutral world	Combat desertification Restore degraded land and soil
15.4	By 2030, ensure the conservation of mountain ecosystems [...]	Protect mountain ecosystems
15.5	Take urgent and significant action to reduce the degradation of natural habitats, halt the loss of biodiversity and, by 2020, protect and prevent the extinction of threatened species	Reduce degradation of natural habitats Reduce loss of biodiversity Protect threatened species
15.9	By 2020, integrate ecosystem and biodiversity values into national and local planning, development processes, poverty reduction strategies and accounts	Integrate ecosystem and biodiversity values

Source: with permission of Allen, B., Nanni, S., Bowyer, C., Kettunen, M., Giadrossi, A., 2020. Assessing Progress Towards the SDGs: Guidance for Evaluating Biobased Solutions Projects. Institute for European Environmental Policy, AISBL.

TABLE 2 Selected topics on bioremediation and bioeconomy covered in the first edition of this work (Prasad, M.N.V. ed., 2016. Bioremediation and bioeconomy. Elsevier. pages 698.)

Making biodiesel from Ukraine's polluted fields with trace elements Energy plantations, contaminated soil, and aromatic and medicinal plants	Sytar and Prasad (2016)
Energy plantations, medicinal and aromatic plants on contaminated soil	Maiti and Kumar (2016)
Bioeconomy and Bioremediation potential for *Prosopis juliflora* (Sw) DC	Prasad and Tewari (2016)
A versatile crop that connects phytoremediation and a sustainable bioeconomy is giant reed (*Arundo donax* L.)	Fernando et al. (2016)
Bioenergy from dump site reforestation	Seshadri et al. (2016)
Gasification of bioenergy from aquatic plant biomass that has undergone phytoremediation	Kathi (2016)
Exploring the possibilities for economic development and environmental conservation through the cultivation of *Jatropha curcas* L. on contaminated land	Warra and Prasad (2016)
Possibilities of Castor bean (*Ricinus communis* L.) for phytoremediation of metallic waste with the assistance of bacteria that promote plant growth: potential for co-generation of commercial goods	Annapurna et al. (2016)
Potential of attractive plants for heavy metal phytoremediation and revenue creation	Nakbanpote et al. (2016)

TABLE 2 Selected topics on bioremediation and bioeconomy covered in the first edition of this work (Prasad, MN.V. ed., 2016. Bioremediation and bioeconomy. Elsevier. pages 698.)—cont'd

Using tree plantings on closed mines as industrial feedstock and environmental remediation	Favas et al. (2016)
Bioeconomic perspectives on the cleanup of trace elements in rice fields	Sebastian and Prasad (2016)
Cultivation of sweet sorghum using a phytoremediation strategy on soils with heavy metal contamination in order to produce bioethanol	Sathya et al. (2016)
Phytostabilizing mine overburden with mulberry and vetiver for co-product generation	Prasad et al. (2016)
Use of contaminated soil for growing dyes producing plants	Schmidt-Przewoźna and Brandys (2016)
Two case studies in Nord-Pas de Calais show how brownfield restoration can be a sensible economic growth strategy for fostering ecotourism, biodiversity, and leisure	Lemoine (2016)
In Russia's central Ural region, biological re-cultivation of mine industry deserts is promoting the development of phytocoenosis	Chibrik et al. (2016)
Phycoremediation potential for business	Sivasubramanian (2016)
Bioproducts and biofuels for the biobusiness from algae-based bioremediation	Vidyashankar and Ravishankar (2016)
Waste remediation as a means of creating a biobased economy: innovation for a sustainable future	Amulya et al. (2016)
Waste and wastewater cleanup through bioprocesses for renewable energy	Ahammad and Sreekrishnan (2016)
Bioprocesses for waste and wastewater remediation for sustainable energy	Mohanakrishna et al. (2016)
Phytomanagement of heavy metal and polycyclic aromatic hydrocarbon-contaminated locations in Assam, India's northeastern state	Sarma and Prasad (2016)
Brazilian bioremediation: scope and obstacles to advancing the bioeconomy	Labuto and Carrilho (2016)
Economic advantages of phytoremediation for soil and groundwater compared to conventional techniques	Gatliff et al. (2016)
Ecocatalysts from metal consisted soils	Grison et al. (2016)

strategies aim to reduce climate change and/or to achieve climate neutrality (Prasad and Smol, 2023). Therefore, achievement of climate neutrality pollution abatement is a must. Climate change is known to influence global weather including its adverse impact on agriculture, land use, hydrology, coastal environment, air quality, and anthropocene (Petrović, 2021; Golobokova, 2021; Visbeck and Keiser, 2021; Bini and Rossi, 2021) (Fig. 7).

Bioremediation is experiencing a knowledge boom as new domains of cutting-edge research are applied to wide range of inorganic, organic contaminants including wastewater and saline environment. Bioremediation applicability and evaluation are tremendously improved and widely accepted (Abou-Khalil et al., 2022; Ahammad and Sreekrishnan, 2016; Alazaiza et al., 2022; Ammeri et al., 2022; Amulya et al., 2016; Annapurna et al., 2016; Antony et al., 2022; Azubuike et al., 2016; Bhunia et al., 2022; Chatterjee et al., 2022; Cheng et al., 2022;

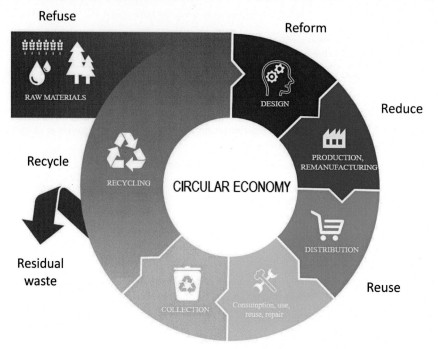

FIG. 6 Circular economy encompassing multiple "Rs", namely Refuse, Reform, Reduce, Reuse, Recycle, and finally, Residual Waste (UNEP (United Nations Environment Programme), 2021).

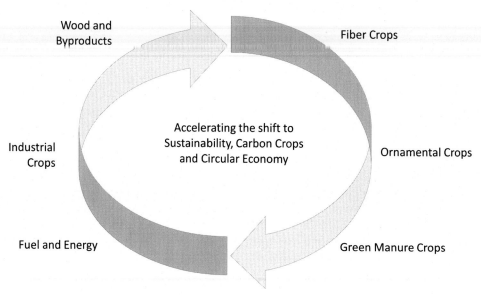

FIG. 7 Accelerating the sustainability through carbon crops and circular economy (Abid et al., 2021; Sytar and Prasad, 2016; Maiti and Kumar, 2016; Prasad, 2015; Prasad and Tewari, 2016; Fernando et al., 2016; Kathi, 2016; Warra and Prasad, 2016; Annapurna et al., 2016; Nakbanpote et al., 2016; Favas et al., 2016; Sathya et al., 2016; Prasad et al., 2016; Schmidt-Przewoźna and Brandys, 2016; Chibrik et al., 2016; Sivasubramanian, 2016; Vidyashankar and Ravishankar, 2016; Abid et al., 2021; García Martín et al., 2020; Masocha et al., 2021; Mitra et al., 2021; Mohammed et al., 2021; Nero, 2021; Singha and Pandey, 2020; Yaqoob et al., 2021).

Chibrik et al., 2016; Choudhary et al., 2022; Cui et al., 2022; Dike et al., 2022; Do et al., 2022; Eraky et al., 2022; Favas et al., 2016; Fernando et al., 2016; Gatliff et al., 2016; Goswami et al., 2022; Grison et al., 2016; Haripriyan et al., 2022; Hui et al., 2022; Jabbar et al., 2022; Karimi et al.; 2022; Kathi, 2016; Kisić et al., 2022; Kumar et al., 2022; Labuto and Carrilho, 2016; Laothamteep et al., 2022; Lemoine, 2016; Lin et al., 2022; Lopez et al., 2022; Mahmoudpour et al., 2021; Maiti and Kumar, 2016; Machhirake et al., 2021; Manikandan et al., 2022; Mishra et al., 2022; Mohanakrishna et al., 2016; Nakbanpote et al., 2016; Ng et al., 2022; Orellana et al., 2022; Ortner et al., 2022; Patel et al., 2022a, b; Prasad and Tewari, 2016; Prasad et al., 2016; Preethi et al., 2022; Qomariyah et al., 2022; Ratnasari et al., 2022; Raza et al., 2022; Roskova et al., 2022; Sajadi Bami et al., 2022; Sarma and Prasad, 2016; Sathya et al., 2016; Schmidt-Przewoźna and Brandys, 2016; Sebastian and Prasad, 2016; Seshadri et al., 2016; Singh et al., 2022; Sivasubramanian, 2016; Srivastava et al., 2022; Sytar and Prasad, 2016; Tang and Kristanti, 2022; Thakur et al., 2021; Thangavelu and Veeraragavan, 2022; Tripathi et al., 2022; Tufail et al., 2022; Vidyashankar and Ravishankar, 2016; Warra and Prasad, 2016; Yan et al., 2022) (Figs. 8–11).

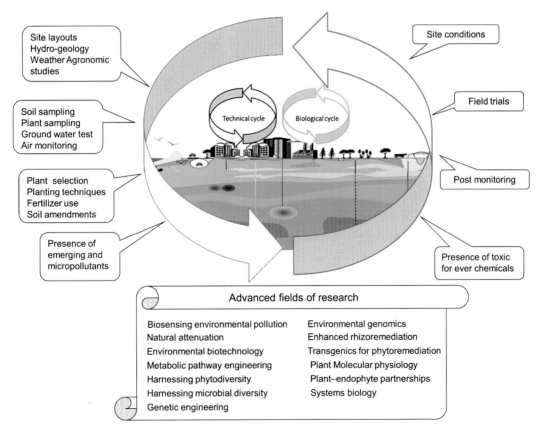

FIG. 8 Bioremediation—knowledge explosion in terms of applicability and evaluation (Prasad, 2021; Gerhardt et al., 2017; Gravand et al., 2021; Grison et al., 2016; Wan et al., 2021; Wang and Ji, 2021; Wang et al., 2021). *With permission and update from Breure, A.M., Lijzen, J.P.A., Maring, L., 2018. Soil and land management in a circular economy. Sci. Total Environ. 624, 1125–1130.*

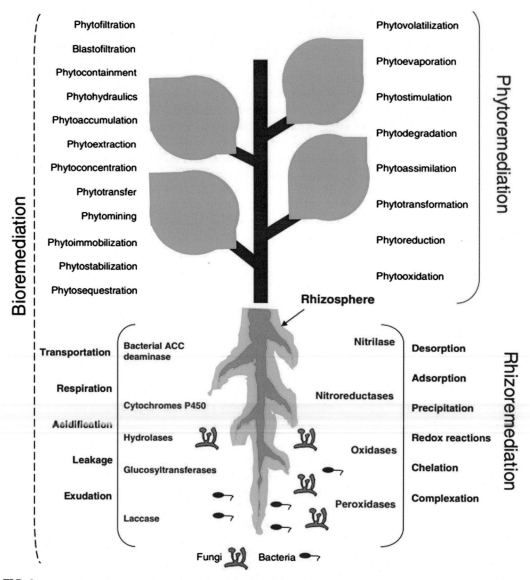

FIG. 9   Importance of soil-plant-microbial interactions in bioremediation for the cleanup of inorganics and organics (heavy metals, pesticides, solvents, explosives, crude oil, polyaromatic hydrocarbons (Ma et al., 2011). *Ma, Y., Prasad, M.N.V., Rajkumar, M., Freitas, H., 2011. Plant growth promoting rhizobacteria and endophytes accelerate phytoremediation of metalliferous soils. Biotechnol. Adv. 29, 248–258 with permission.*

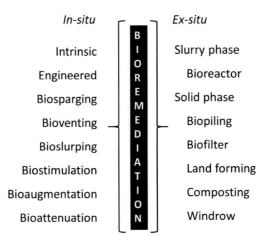

FIG. 10   Bioremediation approaches for environmental cleanup (Azubuike et al., 2016).

5   A zero-pollution ambition for a toxic-free environment

More must be done to stop pollution from being produced, and more must be done to clean up and fix pollution if we are to have an environment free of toxins. The EU must better monitor, report, prevent, and address pollution from air, water, soil, and consumer items to safeguard its population and ecosystems. The EU and Member States will need to conduct a more thorough analysis of all laws and regulations to do this. The Commission had adopted a zero-pollution action plan for air, water, and soil in 2021 to address these connected problems. It is necessary to reestablish natural functioning of groundwater and surface water. This is necessary to avoid and reduce flood damage, as well as to protect and restore biodiversity in lakes, rivers, wetlands, and estuaries. The "Farm to Fork" approach will lessen pollution from extra fertilizers. The Commission will also suggest actions to address pollution from new or particularly dangerous sources, such as microplastics and chemicals, including pharmaceuticals, and urban runoff. Addressing the cumulative effects of several contaminants is also necessary. The European Commission will apply the knowledge gained from the analysis of the existing air quality regulations. To assist local authorities in achieving cleaner air, it will also suggest strengthening the provisions on monitoring, modeling, and air quality plans. The Commission will specifically advocate for changing the air quality standards to be more in line with the suggestions of the World Health Organization.

The European Commission will develop a chemical strategy for sustainability to ensure a toxic-free environment. This will support innovation for the creation of secure and long-lasting substitutes while also assisting in better protecting people and the environment from dangerous chemicals. All parties, including business, should collaborate to improve public health, protect the environment, and boost global competitiveness. The legal system might be made stronger and more straightforward to accomplish this. To move toward a procedure of "one substance—one evaluation" and to increase openness when determining the priority of action to deal with chemicals, the Commission will examine how to employ the agencies and scientific organizations of the EU more effectively.

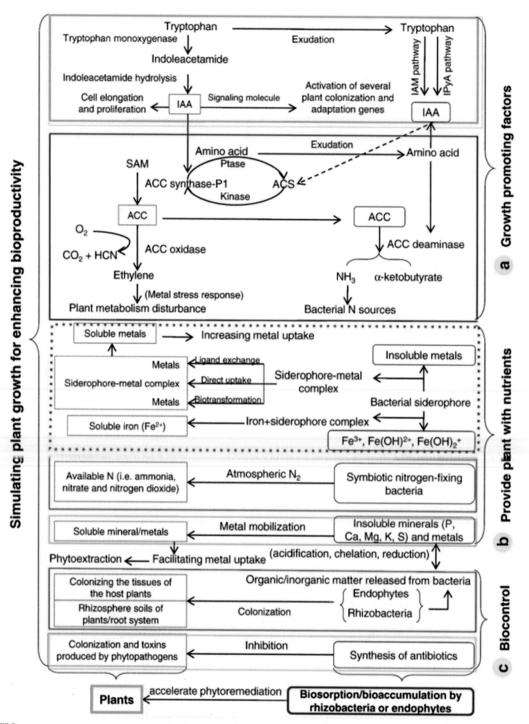

FIG. 11 Plant growth-promoting rhizobacteria and endophytes accelerate phytoremediation of metalliferous soils though modulation of (A) plant growth-promoting parameters, (B) by providing plants with nutrients, and (C) controlling disease through the production of antifungal metabolites. Abbreviations: indole-3-acetic acid (IAA), indole-3-acetamide (IAM) pathway, indole-3-pyruvate (IPyA) pathway, methionine-S-adenosylmethionine (SAM), 1-aminocyclopropane-1-carboxylate (ACC), 1-aminocyclopropane-1-carboxylate synthase (ACS), phosphatase (Ptase), ammonia (NH₃), hydrogen cyanide (HCN) (Ma et al., 2011). *Ma, Y., Prasad, M.N.V., Rajkumar, M., Freitas, H., 2011. Plant growth promoting rhizobacteria and endophytes accelerate phytoremediation of metalliferous soils. Biotechnol. Adv. 29, 248–258 with permission.*

Because of the potential harm they do to the environment, its ecosystems, and ultimately human health, a class of substances called as contaminants of emerging concern (CECs) have attracted the attention of environmental scientists and engineers. CEC regulation is still in its infancy compared with other pollutants such as organochlorines and heavy metals. As these contaminants are widespread in a range of media, including surface and groundwater, soil, air, sediment, and wastewater, it takes human intervention to remove them due to their resistance to biological processes that may occur naturally (Zhang et al., 2007; Zulfahmi et al., 2021) (Figs. 12 and 13).

6 Phytoremediation of polyfluoroalkyl and perfluoroalkyl substances (PFASs)—Forever chemicals—Successfully demonstrated PFAS project at the former Loring Airforce Base, USA

Poly-/per-fluoroalkyl substances (PFASs) are described as a broad range of anthropogenic chemicals used in food containers, non-stick cookwares, medical devices, paints, water-resistant clothing, personal protective equipment, firefighting foams, and ski waxes (Mahmoudnia et al., 2022). The contaminant group comprising *per-* and polyfluorinated alkyl substances (PFASs) does not occur naturally, rather it is exclusively of anthropogenic origin. PFASs comprise more than 4700 substances (OECD, 2018). The abbreviation "PFAS" that is used in this guidance document corresponds to the internationally uniformly used designation for the substance group. In Germany, the designation "PFC" is used for the parameter

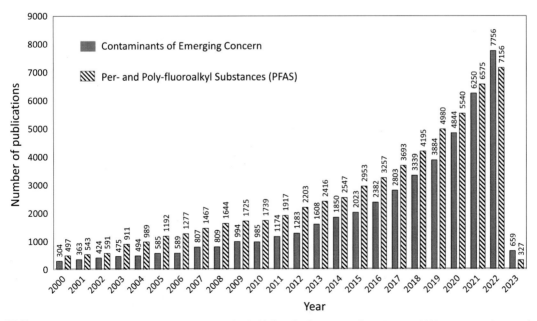

FIG. 12 Articles published in the past two and a half decades beginning from 2000 to 2023 on contaminants of emergent concern (CECs) based on https://www.sciencedirect.com/. Keywords used CEC and used "PFAS."

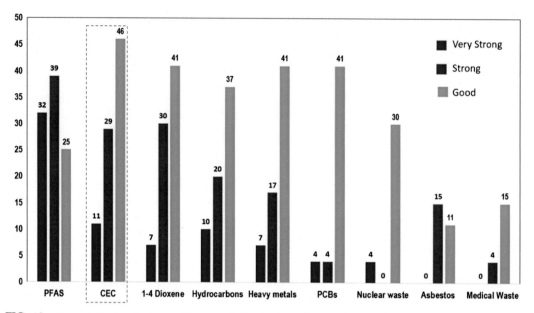

FIG. 13 Figure indicates the demand for remediation by type of contaminant in the next 3 years. Polyfluoroalkyl and perfluoroalkyl substances (PFASs) CECs are the most amenable contaminants for bioremediation among the chosen contaminants. Demand for remediation of the chosen contaminant is shown as %. *Data source: Environmental Business Journal remediation market survey 2019.*

group internationally referred to as "PFAS." PFASs are persistent, very mobile, and have a high efficacy in terms of eco- and human toxicology. The partially fluorinated, so-called polyfluorinated chemicals can be degraded to persistent, fully fluorinated (perfluorinated) chemicals; these partially fluorinated chemicals are generally referred to as "precursors." An observed complete microbial degradation of PFAS has not yet been noted in the scientific literature on this topic.

PFAS contaminants, a subset of the CEC, are becoming more and more significant (Figs. 15, 16, and 14). Guidelines for PFAS management during remediation and evaluation of PFAS-contaminated locations. Therefore, is very important to healthcare. Additionally, guidelines for PFAS-specific studies that go along with the remedial planning process need to be developed. As each individual PFAS molecule has various chemical characteristics, different remedial measures are required for each one.

Particularly in cases of widespread PFAS pollution, PFASs display high mobility and permanence, are of considerable public interest, and frequently involve a significant degree of uncertainty. No free-phase PFAS products are produced (non-aqueous phase liquids or NAPLs). They don't have microbiological life, and they mostly build up in the unsaturated soil zone and at air/water interfaces.

PFASs do not form free-phase products (non-aqueous phase liquids or NAPLs). They are not microbially mineralizable and tend to accumulate in the unsaturated soil zone and at air/water interfaces. It is conceivable for PFAS to be enriched in thick or light NAPL or at the NAPL/water interface. Long-lasting pollutant plumes can be produced by PFASs that are

FIG. 14 (A and B) A former firefighting training site at the Loring Airforce Base (2019), now the site of a collaborative phytoremediation project. (C) site preparation for growing hemp during 2020 planting season, (D–F) watering and management of hemp plot in August 2020, (G) analysis of PFAS in contaminated soil (Nason et al., 2021). *With permission from Elsevier.*

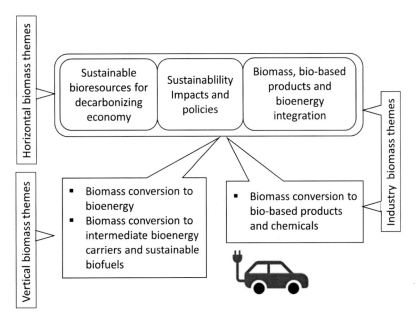

FIG. 15 Topics from biomass production, including phytoremediation solutions for contaminated lands, to processes and technologies for conversion to biofuels and bio-based products. *Credits: EUBCE 2023 and Phy2Climate Project.*

I. Bioproducts from contaminated soil and water

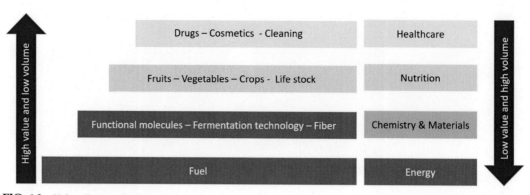

FIG. 16   Value chain in the bioeconomy, a new innovation wave utilization of contaminated substrates (primarily soil and water) for boosting bioeconomy through cascading approach, is a new innovation wave. The added value is the highest at the top of the pyramid and the lowest at the bottom. On the contrary, the volume of biomass needed for the application is the lowest at the top of the pyramid and the highest at the bottom of the pyramid (Prasad, 2015).

present in the groundwater. Depending on the redox conditions in the source and at locations remote from the point of incursion, the biotransformation of precursors can result in the creation of new perfluoroalkane carboxylic and sulfonic acids. When choosing remediation techniques, their great resistance to microbiological, chemical, and thermal deterioration is very important.

The overall goal of our work is to improve the quality of the land on the former Loring AFB, which now belongs to the Aroostook band of the Micmac Nation. Specifically, we are testing the use of fiber hemp plants for phytoremediation of *per-* and polyfluoroalkyl substances (PFASs). PFASs are a class of emerging contaminants that are highly toxic at low concentrations and are frequently found on former military bases owing to their use in firefighting foams (Hagstrom et al., 2021). They are often called "forever chemicals," as they are highly resistant to degradation. Part of the land acquired by the Micmac people was formerly used as a firefighting testing area. The US Airforce has detected concerning levels of PFAS in groundwater at this site.

7 Phy2Climate project

This is a global project being funded by the European Union's Horizon 2020 research and innovation program (Ortner et al., 2022) (Figs. 15 and 16). The overall goal of the H2020 Phy2Climate project is to create a link between the production of clean drop-in biofuels and the phytoremediation of polluted locations. The project uses a biorefinery concept with the thermo-catalytic process (TCR®) at its core to produce high-quality drop-in biofuels such as marine fuels (ISO 8217), gasoline (EN 228), and diesel (EN 590). The process of phytoremediation will decontaminate lands from a wide range of pollutants and make the restored lands available for agriculture while enhancing the process' overall sustainability, legal framework, and economics because the produced biofuels won't pose any risks associated with Land Use Change. By doing this, Phy2Climate hopes to significantly advance the Mission Innovation Challenge for Sustainable Biofuel Production, nearly all UN Sustainable

Development Goals, the EU Biodiversity Strategy for 2030, which is a component of the European Green Deal, and the new EU Soil Strategy for 2030, which was adopted in 2021.

On the one hand, it is unquestionable that there is a growing demand for land, which increases tension among the different groups of users. Land is a finite resource, and the main competitors are Feed, Food, and Fuel. From the available worldwide arable land, about 71% is dedicated to animal feed, about 18% to food, and only about 4% to biofuels (another 7% is for material use of crops). The multiple uttered food versus fuel debate is, actually, a food versus feed debate. However, the increasing demand for biofuels and bio-based products also contributes to this tension, but in a much smaller dimension (Cavelius et al., 2023).

Deforestation, soil erosion, biodiversity loss, and the depletion of essential water resources are all consequences of the rising land demand for energy crops, which results in both direct and indirect land use change (iLUC). On the other hand, a sizable portion of land is poisoned and, as a result, unfit for any use. Even worse, these regions require extensive investigation, registration as "contaminated sites," treatment, and administration, all of which add additional fuel to the fire.

Energy crop demand is driving "direct" and "indirect Land Use Change" (iLUC), which results in deforestation, soil erosion, biodiversity loss, and depletion of essential water resources. On the other side, a sizable portion of land is poisoned and thus useless for any activity. The investigation, designation as a "contaminated site," rehabilitation, and administration of such locations are all exceedingly expensive (Table 3).

TABLE 3 Bioremediation, bio-based products, and bioeconomy via Indirect Land Use Change" (iLUC) (Morese, 2021).

Salient research observation	Author(s)	Year
Case study on the bioethanol production potential from *Miscanthus* with low ILUC Risk in the Province of Lublin, Poland	Gerssen-Gondelach et al. (2014)	2014
Case study on the biodiesel production potential from rapeseed with minimal ILUC risk in Eastern Romania	Brinkman et al. (2015)	2015
In the Polish province of Lublin, Miscanthus has the ability to produce bioethanol with no danger of ILUC	Gerssen-Gondelach et al. (2016)	2016
Low ILUC risk biofuels are identified using certain methodologies, and they are certified. Netherlands ECOFYS	Peters et al. (2016)	2016
Ethanol made from Hungarian maize with low ILUC risk	Brinkman et al. (2017)	2017
Ethanol made from Hungarian maize with low ILUC risk. Bioenergy and biomass	Marnix (2017)	2017
Rapeseed biodiesel with low ILUC risk: Impacts on direct and indirect GHG emissions in Eastern Romania	Brinkman et al. (2018)	2018
Review of important global sustainability frameworks and certification programs for biomass and bioenergy, as well as their use and effects in Colombia	Ramirez-Contreras and Faaij (2018)	2018
New advancements in EU bio-based economy low iLUC policies and certification	Sumfleth et al.	2020

Continued

TABLE 3 Bioremediation, bio-based products, and bioeconomy via Indirect Land Use Change" (iLUC) (Morese, 2021)—cont'd

Salient research observation	Author(s)	Year
Low iLUC Bioenergy Production in Europe: Cost-benefit and risk analysis using Monte Carlo simulation	Mm	2021
Using a user-friendly tool to bridge modeling and certification to assess low-ILUC-risk practices for biobased materials	Balugani et al. (2022)	2022
Perennial biomass cropping and use: Shaping the policy ecosystem in European countries	Clifton-Brown et al. (2023)	2023
The potential of biofuels from first to fourth generation	Cavelius et al.	2023
China's increased purchase of bioenergy has ramifications for global land usage and sustainability	Wu et al.	2023

Thermo-catalytic reforming (TCR®) will evaluate four pilot sites with crops cultivated in various parts of Europe and South America as a feedstock for biomass thermo-chemical conversion. This cutting-edge technology can create various biofuels for sea and land transportation as well as bio-coke for the metallurgical sector. The recovered metals and metalloids will also be valued in the metal smelting process if the soil is contaminated with heavy metals (Figs. 17–20).

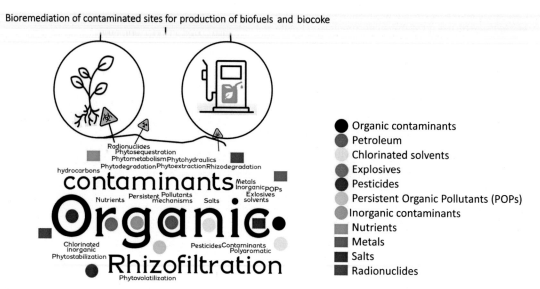

FIG. 17 Synoptic sketch of organic and inorganic contaminated sites. Application of bioremediation for production of biofuels and bio-coke and resource recovery (Polińska et al., 2021; Pranoto and Budianta, 2020; Issaka and Ashraf, 2021; Zhang et al., 2021). *Credit: Phy2Climate Project.*

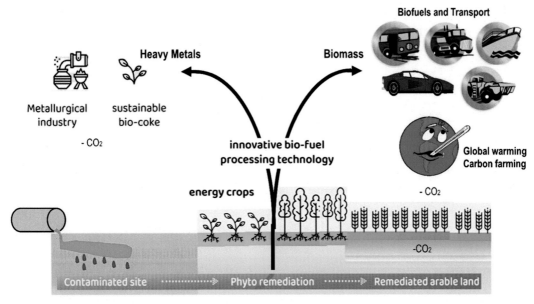

FIG. 18 Schematic sketch of Phy2Climate project envisaged for biofuels to augment sea and land transportation as well as bio-coke for the metallurgical sector. In addition, this project contributes to decarbonization (Cavelius et al., 2023; Nascimento et al., 2021). *Redrawn based on Ortner et al. (2021).*

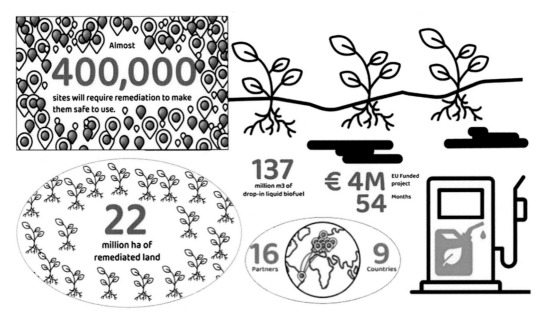

FIG. 19 Phy2Climate project scope. It is estimated that about 400,000 sites need to be remediated to make them safe for use. It is expected about 22 million ha of land to be remediated and the project consortium comprises of 16 partners in 9 countries. European funded 4 million euros for a project duration of 54 months. *Credit: Phy2Climate Project.*

I. Bioproducts from contaminated soil and water

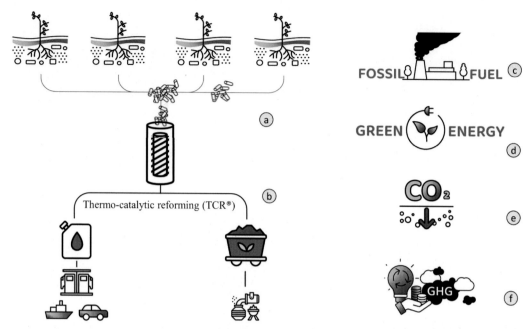

FIG. 20 Thermo-catalytic reforming (TCR®) evaluated four pilot sites with crops cultivated in various parts of Europe and South America as a feedstock for biomass thermo-chemical conversion. This cutting-edge technology is expected to produce various biofuels for sea and land transportation as well as bio-coke for the metallurgical sector. The recovered metals and metalloids will add value to the metal smelting process (Issaka and Ashraf, 2021; Li and Xu, 2022; Zhang et al., 2021). (a) Cultivation of plants/crops on contaminated sites, (b) Harvested phytomass is subjected to TCR for production of biofuel and bio-coke or biochar, (c) reduces pressure on consumption of fossil fuels, (d) green energy production, (e) carbon storage in soil, and (f) green energy for decarbonization. Such an approach would be sustainable and economic solutions to lower the pressure in land use competition and contributing to GHG reduction to replace the fossil fuels. *Credit: Phy2Climate Project.*

Acknowledgments

Thanks are due to ETA-Florence Renewable Energies, EUBCE 2023, and Phy2Climate Project for sharing knowledge.

References

Abdelhafeez, I., El-Tohamy, S., Abdel-Raheem, S., El-Dars, F., 2022. A review on green remediation techniques for hydrocarbons and heavy metals contaminated soil. Curr. Chem. Lett. 11 (1), 43–62.

Abid, R., Mahmood, S., Zahra, S., Ghaffar, S., Malik, M., Noreen, S., 2021. *Jatropha curcas* L. and *Pongamia pinnata* L. exhibited differential growth and bioaccumulation pattern irrigated with wastewater. Sains Malays. 50 (3), 559–570.

Abou-Khalil, C., Sarkar, D., Braykaa, P., Boufadel, M.C., 2022. Mobilization of per-and polyfluoroalkyl substances (PFAS) in soils: a review. Curr. Pollut. Rep. 8, 1–23.

Ahammad, S.Z., Sreekrishnan, T.R., 2016. Energy from wastewater treatment. In: Bioremediation and Bioeconomy. Elsevier, pp. 523–536.

Ahila, K.G., Ravindran, B., Muthunarayanan, V., Nguyen, D.D., Nguyen, X.C., Chang, S.W., Nguyen, V.K., Thamaraiselvi, C., 2021. Phytoremediation potential of freshwater macrophytes for treating dye-containing wastewater. Sustainability 13 (1), 329.

Ahmad, Z., Khan, S.M., Page, S., 2021. Politics of the natural vegetation to balance the hazardous level of elements in marble polluted ecosystem through phytoremediation and physiological responses. J. Hazard. Mater. 414, 125451.

Alazaiza, M.Y., Albahnasawi, A., Ahmad, Z., Bashir, M.J., Al-Wahaibi, T., Abujazar, M.S.S., Amr, S.S.A., Nassani, D.E., 2022. Potential use of algae for the bioremediation of different types of wastewater and contaminants: production of bioproducts and biofuel for green circular economy. J. Environ. Manag. 324, 116415.

Allen, B., Nanni, S., Bowyer, C., Kettunen, M., Giadrossi, A., 2020. Assessing Progress Towards the SDGs: Guidance for Evaluating Biobased Solutions Projects. Institute for European Environmental Policy, AISBL.

Alpandi, A.H., Husin, H., Sidek, A., 2021. A critical review on the development of wax inhibiting agent in facilitating remediation process of contaminated groundwater. Environ. Sci. Pollut. Res. 29, 1–11.

Ammeri, R.W., Simeone, G.D.R., Hidri, Y., Abassi, M.S., Mehri, I., Costa, S., Hassen, A., Rao, M.A., 2022. Combined bioaugmentation and biostimulation techniques in bioremediation of pentachlorophenol contaminated forest soil. Chemosphere 290, 133359.

Amulya, K., Dahiya, S., Mohan, S.V., 2016. Building a bio-based economy through waste remediation: Innovation towards sustainable future. In: Bioremediation and Bioeconomy. Elsevier, pp. 497–521.

Annapurna, D., Rajkumar, M., Prasad, M.N.V., 2016. Potential of Castor bean (*Ricinus communis* L.) for phytoremediation of metalliferous waste assisted by plant growth-promoting bacteria: possible cogeneration of economic products. In: Bioremediation and Bioeconomy. Elsevier, pp. 149–175.

Antony, S., Antony, S., Rebello, S., George, S., Biju, D.T., Reshmy, R., Madhavan, A., Binod, P., Pandey, A., Sindhu, R., Awasthi, M.K., 2022. Bioremediation of endocrine disrupting chemicals-advancements and challenges. Environ. Res. 213, 113509.

Ara, T., Nisa, W.U., Anjum, M., Riaz, L., Saleem, A.R., Hayat, M.T., 2021. Hexachloro-cyclohexane toxicity in water bodies of Pakistan: challenges and possible reclamation technologies. Water Sci. Technol. 83 (10), 2345–2362.

Auchterlonie, J., Eden, C.L., Sheridan, C., 2021. The phytoremediation potential of water hyacinth: a case study from Hartbeespoort Dam, South Africa. S. Afr. J. Chem. Eng. 37, 31–36.

Azubuike, C.C., Chikere, C.B., Okpokwasili, G.C., 2016. Bioremediation techniques-classification based on site of application: principles, advantages, limitations and prospects. World J. Microbiol. Biotechnol. 32, 1–18.

Balugani, E., Sumfleth, B., Majer, S., Marazza, D., Thrän, D., 2022. Bridging modeling and certification to evaluate low-ILUC-risk practices for biobased materials with a user-friendly tool. Sustainability 14 (4), 2030.

Bezie, Y., Taye, M., Kumar, A., 2021. Recent advancement in phytoremediation for removal of toxic compounds. In: Nanobiotechnology for Green Environment. CRC Press, pp. 195–228.

Bhunia, A., Lahiri, D., Nag, M., Upadhye, V., Pandit, S., 2022. Bacterial biofilm mediated bioremediation of hexavalent chromium: a review. Biocatal. Agric. Biotechnol. 43, 102397.

Bini, M., Rossi, V., 2021. Climate change and anthropogenic impact on coastal environments. Water 13 (9), 1182.

Braga, A.F., Borges, A.C., Vaz, L.R., Souza, T.D.D., Rosa, A.P., 2021. Phytoremediation of fluoride-contaminated water by *Landoltia punctata*. Eng. Agríc. 41, 171–180.

Brinkman, M.L.J., Pişcă, I., Wicke, B., Faaij, A., 2015. ILUC Prevention Strategies for Sustainable Biofuels: Case Study on the Biodiesel Production Potential from Rapeseed with Low ILUC Risk in Eastern Romania. Universiteit Utrecht.

Brinkman, M.L., Wicke, B., Faaij, A.P., 2017. Low-ILUC-risk ethanol from Hungarian maize. Biomass Bioenergy 99, 57–68.

Brinkman, M.L., van der Hilst, F., Faaij, A.P., Wicke, B., 2018. Low-ILUC-risk rapeseed biodiesel: potential and indirect GHG emission effects in eastern Romania. Biofuels 12, 171–186.

Cavelius, P., Engelhart-Straub, S., Mehlmer, N., Lercher, J., Awad, D., Brück, T., 2023. The potential of biofuels from first to fourth generation. PLoS Biol. 21 (3), e3002063.

Chatterjee, S., Kumari, S., Rath, S., Das, S., 2022. Prospects and scope of microbial bioremediation for the restoration of the contaminated sites. In: Microbial Biodegradation and Bioremediation. Elsevier, pp. 3–31.

Cheng, Y., Chen, L., Guo, K., Xie, J., Shu, Y., He, S., Xiao, F., 2022. Progress of uranium-contaminated soil bioremediation technology. J. Environ. Radioact. 241, 106773.

Chibrik, T.S., Lukina, N.V., Filimonova, E.I., Glazyrina, M.A., Rakov, E.A., Maleva, M.G., Prasad, M.N.V., 2016. Biological recultivation of mine industry deserts: facilitating the formation of phytocoenosis in the middle Ural region, Russia. In: Bioremediation and Bioeconomy. Elsevier, pp. 389–418.

Chintani, Y.S., Butarbutar, E.S., Nugroho, A.P., Sembiring, T., 2021. Uptake and release of chromium and nickel by vetiver grass (*Chrysopogon zizanioides* (L.) Roberty). SN. Appl. Sci. 3 (3), 1–13.

I. Bioproducts from contaminated soil and water

Choudhary, K., Vivekanand, V., Pareek, N., 2022. Recent advancements in microbial bioremediation of industrial effluents: challenges and future outlook. In: Microbial Biodegradation and Bioremediation. Elsevier, pp. 293–303.

Clifton-Brown, J., Hastings, A., von Cossel, M., Murphy-Bokern, D., McCalmont, J., Whitaker, J., Alexopoulou, E., Amaducci, S., Andronic, L., Ashman, C., Awty-Carroll, D., 2023. Perennial biomass cropping and use: shaping the policy ecosystem in European countries. GCB Bioenergy 15, 538–558.

Cui, Q., Zhang, Z., Beiyuan, J., Cui, Y., Chen, L., Chen, H., Fang, L., 2022. A critical review of uranium in the soil-plant system: distribution, bioavailability, toxicity, and bioremediation strategies. Crit. Rev. Environ. Sci. Technol. 53, 1–26.

Davamani, V., Parameshwari, C.I., Arulmani, S., John, J.E., Poornima, R., 2021. Hydroponic phytoremediation of paperboard mill wastewater by using vetiver (*Chrysopogon zizanioides*). J. Environ. Chem. Eng. 9 (4), 105528.

Dike, C.C., Khudur, L.S., Hakeem, I.G., Rani, A., Shahsavari, E., Surapaneni, A., Shah, K., Ball, A.S., 2022. Biosolids-derived biochar enhances the bioremediation of diesel-contaminated soil. J. Environ. Chem. Eng. 10 (6), 108633.

Do, C.V.T., Pham, M.H.T., Pham, T.Y.T., Dinh, C.T., Bui, T.U.T., Tran, T.D., 2022. Microalgae and bioremediation of domestic wastewater. Current opinion in green and sustainable. Chemistry 34, 100595.

Ekanayake, M.S., Udayanga, D., Wijesekara, I., Manage, P., 2021. Phytoremediation of synthetic textile dyes: biosorption and enzymatic degradation involved in efficient dye decolorization by *Eichhornia crassipes* (Mart.) Solms and *Pistia stratiotes* L. Environ. Sci. Pollut. Res. 28 (16), 20476–20486.

Eraky, M., Elsayed, M., Qyyum, M.A., Ai, P., Tawfik, A., 2022. A new cutting-edge review on the bioremediation of anaerobic digestate for environmental applications and cleaner bioenergy. Environ. Res. 213, 113708.

Favas, P.J.C., Pratas, J., Chaturvedi, R., Paul, M.S., Prasad, M.N.V., 2016. Tree crops on abandoned mines for environmental remediation and industrial feedstock. In: Bioremediation and Bioeconomy. Elsevier, pp. 219–249.

Fernando, A.L., Barbosa, B., Costa, J., Papazoglou, E.G., 2016. Giant reed (*Arundo donax* L.): a multipurpose crop bridging phytoremediation with sustainable bioeconomy. In: Bioremediation and Bioeconomy. Elsevier, pp. 77–95.

Gajić, G., Mitrović, M., Pavlović, P., 2022. Phytobial remediation by bacteria and fungi. In: Assisted Phytoremediation. Elsevier, pp. 285–344.

Garate-Quispe, J.S., de Leon, R.P., Herrera-Machaca, M., Julian-Laime, E., Nieto-Ramos, C., 2021. Growth and survivorship of *Vetiveria zizanioides* in degraded soil by gold-mining in the Peruvian Amazon. J. Degrad. Min. Lands Manag. 9 (1), 3219–3225.

García Martín, J.F., González Caro, M.D.C., López Barrera, M.D.C., Torres García, M., Barbin, D., Álvarez Mateos, P., 2020. Metal accumulation by *Jatropha curcas* L. adult plants grown on heavy metal-contaminated soil. Plan. Theory 9 (1), 418.

Gatliff, E., Linton, P.J., Riddle, D.J., Thomas, P.R., 2016. Phytoremediation of soil and groundwater: economic benefits over traditional methodologies. In: Bioremediation and Bioeconomy. Elsevier, pp. 589–608.

Gautam, M., Agrawal, M., 2021. Application potential of *Chrysopogon zizanioides* (L.) Roberty for the remediation of red mud-treated soil: an analysis via determining alterations in essential oil content and composition. Int. J. Phytoremediation 23, 1–9.

Gerhardt, K.E., Gerwing, P.D., Greenberg, B.M., 2017. Opinion: taking phytoremediation from proven technology to accepted practice. Plant Sci. 256, 170–185.

Gerssen-Gondelach, S., Wicke, B., Faaij, A., 2014. ILUC Prevention Strategies for Sustainable Biofuels: Case Study on the Bioethanol Production Potential from Miscanthus with Low ILUC Risk in the Province of Lublin, Poland. Utrecht University, Faculty of Geosciences, Copernicus Institute of Sustainable Development.

Gerssen-Gondelach, S.J., Wicke, B., Borzęcka-Walker, M., Pudełko, R., Faaij, A.P., 2016. Bioethanol potential from miscanthus with low ILUC risk in the province of Lublin, Poland. GCB Bioenergy 8 (5), 909–924.

Golobokova, L.P., 2021. Special issue editorial: air pollution estimation. Atmosphere 12 (6), 655.

Goswami, R.K., Agrawal, K., Shah, M.P., Verma, P., 2022. Bioremediation of heavy metals from wastewater: a current perspective on microalgae-based future. Lett. Appl. Microbiol. 75 (4), 701–717.

Gravand, F., Rahnavard, A., Pour, G.M., 2021. Investigation of vetiver grass capability in phytoremediation of contaminated soils with heavy metals (Pb, Cd, Mn, and Ni). Soil Sediment Contam. Int. J. 30 (2), 163–186.

Grison, C., Escande, V., Olszewski, T.K., 2016. Ecocatalysis: a new approach toward bioeconomy. In: Bioremediation and Bioeconomy. Elsevier, pp. 629–663.

Gunwal, I., Mathur, R., Agrawal, Y., Mago, P., 2021. Plants useful for phytoremediation of soil and water in India. Asian J. Plant Soil Sci. 6 (3), 1–8.

Guo, Z., Zeng, P., Xiao, X., Peng, C., 2021. Physiological, anatomical, and transcriptional responses of mulberry (*Morus alba* L.) to Cd stress in contaminated soil. Environ. Pollut. 284, 117387.

Hagstrom, A.L., Anastas, P., Boissevain, A., Borrel, A., Deziel, N.C., Fenton, S.E., Fields, C., Fortner, J.D., Franceschi-Hofmann, N., Frigon, R., Jin, L., 2021. Yale School of Public Health Symposium: an overview of the challenges and opportunities associated with per-and polyfluoroalkyl substances (PFAS). Sci. Total Environ. 778, 146192.

Haripriyan, U., Gopinath, K.P., Arun, J., Govarthanan, M., 2022. Bioremediation of organic pollutants: a mini review on current and critical strategies for wastewater treatment. Arch. Microbiol. 204 (5), 1–9.

Hui, C.Y., Guo, Y., Liu, L., Yi, J., 2022. Recent advances in bacterial biosensing and bioremediation of cadmium pollution: a mini-review. World J. Microbiol. Biotechnol. 38 (1), 1–16.

Irawanto, R., 2021. Phytoremediation model of greywater treatment in the Purwodadi Botanic Garden. IOP Conf. Ser. Earth Environ. Sci. 743 (1), 012078 (IOP Publishing).

Issaka, S., Ashraf, M.A., 2021. Phytorestoration of mine spoiled: "evaluation of natural phytoremediation process occurring at ex-tin mining catchment". In: Phytorestoration of Abandoned Mining and Oil Drilling Sites. Elsevier, pp. 219–248.

Jabbar, N.M., Alardhi, S.M., Mohammed, A.K., Salih, I.K., Albayati, T.M., 2022. Challenges in the implementation of bioremediation processes in petroleum-contaminated soils: a review. Environ. Nanotechnol. Monit. Manag. 18, 100694.

Karimi, H., Mahdavi, S., Asgari Lajayer, B., Moghiseh, E., Rajput, V.D., Minkina, T., Astatkie, T., 2022. Insights on the bioremediation technologies for pesticide-contaminated soils. Environ. Geochem. Health 44 (4), 1329–1354.

Kathi, S., 2016. Bioenergy from phytoremediated phytomass of aquatic plants via gasification. In: Bioremediation and Bioeconomy. Elsevier, pp. 111–128.

Kisić, I., Hrenović, J., Zgorelec, Ž., Durn, G., Brkić, V., Delač, D., 2022. Bioremediation of agriculture soil contaminated by organic pollutants. Energies 15 (4), 1561.

Kumar, L., Chugh, M., Kumar, S., Kumar, K., Sharma, J., Bharadvaja, N., 2022. Remediation of petrorefinery wastewater contaminants: a review on physicochemical and bioremediation strategies. Process Saf. Environ. Prot. 159, 362–375.

Labuto, G., Carrilho, E.N.V.M., 2016. Bioremediation in Brazil: scope and challenges to boost up the bioeconomy. In: Bioremediation and Bioeconomy. Elsevier, pp. 569–588.

Laothamteep, N., Naloka, K., Pinyakong, O., 2022. Bioaugmentation with zeolite-immobilized bacterial consortium OPK results in a bacterial community shift and enhances the bioremediation of crude oil-polluted marine sandy soil microcosms. Environ. Pollut. 292, 118309.

Lemoine, G., 2016. Brownfield restoration as a smart economic growth option for promoting ecotourism, biodiversity, and leisure: two case studies in Nord-Pas De Calais. In: Bioremediation and Bioeconomy. Elsevier, pp. 361–388.

Li, J., Xu, G., 2022. Circular economy towards zero waste and decarbonization. Circ. Econ. 1, 100002.

Lin, C., Cheruiyot, N.K., Bui, X.T., Ngo, H.H., 2022. Composting and its application in bioremediation of organic contaminants. Bioengineered 13 (1), 1073–1089.

Lopez, A.M.Q., Silva, A.L.D.S., Maranhão, F.C.D.A., Ferreira, L.F.R., 2022. Plant growth promoting bacteria: aspects in metal bioremediation and phytopathogen management. In: Microbial Biocontrol: Sustainable Agriculture and Phytopathogen Management. Springer, Cham, pp. 51–78.

Ma, Y., Prasad, M.N.V., Rajkumar, M., Freitas, H., 2011. Plant growth promoting rhizobacteria and endophytes accelerate phytoremediation of metalliferous soils. Biotechnol. Adv. 29, 248–258.

Machhirake, N.P., Yadav, S., Krishna, V., Kumar, S., 2021. Toxicity-removal efficiency of *Brassica juncea, Chrysopogon zizanioides* and *Pistia stratiotes* to decontaminate biomedical ash under non-chelating and chelating conditions: a pilot-scale phytoextraction study. Chemosphere, 132416.

Mahmoudnia, A., Mehrdadi, N., Baghdadi, M., Moussavi, G., 2022. Change in global PFAS cycling as a response of permafrost degradation to climate change. J. Hazard. Mater. Adv. 5, 100039.

Mahmoudpour, M., Gholami, S., Ehteshami, M., Salari, M., 2021. Evaluation of phytoremediation potential of vetiver Grass (*Chrysopogon zizanioides* (L.) Roberty) for wastewater treatment. Adv. Mater. Sci. Eng. 2021, 3059983.

Maiti, S.K., Kumar, A., 2016. Energy plantations, medicinal and aromatic plants on contaminated soil. In: Bioremediation and Bioeconomy. Elsevier, pp. 29–47.

Manikandan, A., Suresh Babu, P., Shyamalagowri, S., Kamaraj, M., Muthukumaran, P., Aravind, J., 2022. Emerging role of microalgae in heavy metal bioremediation. J. Basic Microbiol. 62 (3–4), 330–347.

Marnix, L.J., 2017. Low-ILUC-risk ethanol from Hungarian maize. Biomass Bioenergy 99, 57–68.

Masocha, B.L., Dikinya, O., Moseki, B., 2021. Improving *Jatropha curcas* L. photosynthesis-related parameters using poultry litter and its biochar. Not. Bot. Horti Agrobot. Cluj-Napoca 49 (3), 12344.

Mishra, B., Tiwari, A., Mahmoud, A.E.D., 2022. Microalgal potential for sustainable aquaculture applications: bioremediation, biocontrol, aquafeed. Clean Technol. Environ. Policy 25, 1–13.

Mitra, S., Ghose, A., Gujre, N., Senthilkumar, S., Borah, P., Paul, A., Rangan, L., 2021. A review on environmental and socioeconomic perspectives of three promising biofuel plants *Jatropha curcas, Pongamia pinnata* and *Mesua ferrea*. Biomass Biomass Bioenergy 151, 106173.

Mohammed, A.T., Jaafar, M.N.M., Othman, N., Veza, I., Mohammed, B., Oshadumi, F.A., Sanda, H.Y., 2021. Soil fertility enrichment potential of *Jatropha curcas* for sustainable agricultural production: a case study of Birnin Kebbi, Nigeria. Ann. Romanian Soc. Cell Biol. 25, 21061–21073.

Mohanakrishna, G., Srikanth, S., Pant, D., 2016. Bioprocesses for waste and wastewater remediation for sustainable energy. In: Bioremediation and Bioeconomy. Elsevier, pp. 537–565.

Morese, M.M., 2021. Cost benefit and risk analysis of low iLUC bioenergy production in Europe using Monte Carlo simulation. Energies 14 (6), 1650.

Nakbanpote, W., Meesungnoen, O., Prasad, M.N.V., 2016. Potential of ornamental plants for phytoremediation of heavy metals and income generation. In: Bioremediation and Bioeconomy. Elsevier, pp. 179–217.

Nascimento, C.W.A.D., Biondi, C.M., Silva, F.B.V.D., Lima, L.H.V., 2021. Using plants to remediate or manage metal-polluted soils: an overview on the current state of phytotechnologies. Acta Sci. Agron. 43.

Nason, S.L., Stanley, C.J., PeterPaul, C.E., Blumenthal, M.F., Zuverza-Mena, N., Silliboy, R.J., 2021. A community based PFAS phytoremediation project at the former Loring Airforce Base. iScience 24 (7), 102777.

Nero, B.F., 2021. Phytoremediation of petroleum hydrocarbon-contaminated soils with two plant species*: Jatropha curcas* and *Vetiveria zizanioides* at Ghana Manganese Company Ltd. Int. J. Phytoremediation 23 (2), 171–180.

Ng, Y.J., Lim, H.R., Khoo, K.S., Chew, K.W., Chan, D.J.C., Bilal, M., Munawaroh, H.S.H., Show, P.L., 2022. Recent advances of biosurfactant for waste and pollution bioremediation: substitutions of petroleum-based surfactants. Environ. Res. 212, 113126.

OECD, 2018. Toward a new comprehensive global database of per-polyfluoroalkyl substances (PFASs): summary report on updating the OECD 2007 list of per and polyfluoroalkyl substances (PFASs). Series on Risk Management No. 39. OECD. https://www.oecd.org/officialdocuments/publicdisplaydocumentpdf/?cote=ENV-JM-MONO (2018)7&doclanguage=en.

Orellana, R., Cumsille, A., Piña-Gangas, P., Rojas, C., Arancibia, A., Donghi, S., Stuardo, C., Cabrera, P., Arancibia, G., Cárdenas, F., Salazar, F., 2022. Economic evaluation of bioremediation of hydrocarbon-contaminated urban soils in Chile. Sustainability 14 (19), 11854.

Ortner, M., Otto, H.J., Brunbauer, L., Kick, C., Eschen, M., Sanchis, S., Audino, F., Zeremski, T., Szlek, A., Petela, K., Grassi, A., Capaccioli, S., Fermeglia, M., Vanheusden, B., Perišić, M., Young, B., Trickovic, J., Kidikas, Z., Gavrilović, O., Blázquez-Pallí, N. López Cabornero, D.; Jaggi, C.; Klein, V., 2022. Clean biofuel production and phytoremediation solutions from contaminated lands worldwide. In: 30th European Biomass Conference and Exhibition, 9–12 May 2022. (Online).

Panepinto, D., Riggio, V.A., Zanetti, M., 2021. Analysis of the emergent climate change mitigation technologies. Int. J. Environ. Res. Public Health 18, 6767.

Patel, A.K., Katiyar, R., Chen, C.W., Singhania, R.R., Awasthi, M.K., Bhatia, S., Bhaskar, T., Dong, C.D., 2022a. Antibiotic bioremediation by new generation biochar: recent updates. Bioresour. Technol. 358, 127384.

Patel, A.K., Singhania, R.R., Albarico, F.P.J.B., Pandey, A., Chen, C.W., Dong, C.D., 2022b. Organic wastes bioremediation and its changing prospects. Sci. Total Environ. 824, 153889.

Peters, D., Hähl, T., Kühner, A.K., Cuijpers, M., Stomph, T.J., van der Werf, W., Grass, M., 2016. Methodologies Identification and Certification of Low ILUC Risk Biofuels. ECOFYS Netherlands.

Petrović, F., 2021. Hydrological impacts of climate change and land use. Water 13 (6), 799.

Polińska, W., Kotowska, U., Kiejza, D., Karpińska, J., 2021. Insights into the use of phytoremediation processes for the removal of organic micropollutants from water and wastewater: a review. Water 13 (15), 2065.

Pranoto, B.S.M., Budianta, W., 2020. Phytoremediation of heavy metals contaminated soil in artisanal gold Mining at Selogiri, Wonogiri District, Central Java, Indonesia. J. Appl. Geol. 5 (2), 64–72.

Prasad, M.N.V., 2015. Phytoremediation crops and biofuels. In: Sustainable Agriculture Reviews. vol. 17, pp. 159 261.

Prasad, M.N.V. ed., 2016. Bioremediation and bioeconomy. Elsevier. pages 698.

Prasad, M.N.V. (Ed.), 2021. Handbook on Assisted and Amendments Enhanced Sustainable Remediation Technology. John Wiley & Sons.

Prasad, M.N.V., 2023. Circular economy in domestic and industrial wastewaters: challenges and opportunities. In: Water in Circular Economy. Springer, pp. 167–189.

Prasad, M.N.V., Shih, K. (Eds.), 2016. Environmental Materials and Waste: Resource Recovery and Pollution Prevention. Academic Press.

Prasad, M.N.V., Smol, M. (Eds.), 2023. Sustainable and Circular Management of Resources and Waste Towards a Green Deal. Elsevier.

Prasad, M.N.V., Tewari, J.C., 2016. *Prosopis juliflora* (Sw) DC: potential for bioremediation and bioeconomy. In: Bioremediation and Bioeconomy. Elsevier, pp. 49–76.

Prasad, M.N.V., Nakbanpote, W., Phadermrod, C., Rose, D., Suthari, S., 2016. Mulberry and vetiver for phytostabilization of mine overburden: cogeneration of economic products. In: Bioremediation and Bioeconomy. Elsevier, pp. 295–328.

Preethi, P.S., Hariharan, N.M., Vickram, S., Manian, R., Manikandan, S., Subbaiya, R., Karmegam, N., Yadav, V., Ravindran, B., Chang, S.W., Awasthi, M.K., 2022. Advances in bioremediation of emerging contaminants from industrial wastewater by oxidoreductase enzymes. Bioresour. Technol. 359, 127444.

Qomariyah, L., Kumalasari, P.I., Suprapto, S., Puspita, N.F., Agustiani, E., Altway, S., Hardi, H., 2022. Bioremediation of heavy metals in petroleum sludge through bacterial mixtures. Mater. Today Proc. 63, S140–S142.

Ramirez-Contreras, N.E., Faaij, A.P., 2018. A review of key international biomass and bioenergy sustainability frameworks and certification systems and their application and implications in Colombia. Renew. Sust. Energ. Rev. 96, 460–478.

Ratnasari, A., Syafiuddin, A., Zaidi, N.S., Kueh, A.B.H., Hadibarata, T., Prastyo, D.D., Ravikumar, R., Sathishkumar, P., 2022. Bioremediation of micropollutants using living and non-living algae-current perspectives and challenges. Environ. Pollut. 292, 118474.

Raza, N., Rizwan, M., Mujtaba, G., 2022. Bioremediation of real textile wastewater with a microalgal-bacterial consortium: an eco-friendly strategy. Biomass Convers. Biorefin., 1–13.

Renström, T.I., Spataro, L., Marsiliani, L., 2021. Can subsidies rather than pollution taxes break the trade-off between economic output and environmental protection? Energy Econ. 95, 105084.

Roskova, Z., Skarohlid, R., McGachy, L., 2022. Siderophores: an alternative bioremediation strategy? Sci. Total Environ. 819, 153144.

Sajadi Bami, M., Raeisi Estabragh, M.A., Ohadi, M., Banat, I.M., Dehghannoudeh, G., 2022. Biosurfactants aided bioremediation mechanisms: a mini-review. Soil Sediment Contam. Int. J. 31 (7), 801–817.

Sarma, H., Prasad, M.N.V., 2016. Phytomanagement of polycyclic aromatic hydrocarbons and heavy metals-contaminated sites in Assam, North Eastern State of India, for boosting bioeconomy. In: Bioremediation and Bioeconomy. Elsevier, pp. 609–626.

Sathya, A., Kanaganahalli, V., Rao, P.S., Gopalakrishnan, S., 2016. Cultivation of sweet sorghum on heavy metal-contaminated soils by phytoremediation approach for production of bioethanol. In: Bioremediation and Bioeconomy. Elsevier, pp. 271–292.

Schmidt-Przewoźna, K., Brandys, A., 2016. Utilization of contaminated lands for cultivation of dye producing plants. In: Bioremediation and Bioeconomy. Elsevier, pp. 329–359.

Sebastian, A., Prasad, M.N.V., 2016. Rice paddies for trace element cleanup: bioeconomic perspectives. In: Bioremediation and Bioeconomy. Elsevier, pp. 251–269.

Seshadri, B., Bolan, N.S., Thangarajan, R., Jena, U., Das, K.C., Wang, H., Naidu, R., 2016. Biomass energy from revegetation of landfill sites. In: Bioremediation and Bioeconomy. Elsevier, pp. 99–109.

Singh, M., Jayant, K., Bhutani, S., Mehra, A., Kaur, T., Kour, D., Suyal, D.C., Singh, S., Rai, A.K., Yadav, A.N., 2022. Bioremediation—sustainable tool for diverse contaminants management: current scenario and future aspects. J. Appl. Biol. Biotechnol. 10 (2), 48–63.

Singha, L.P., Pandey, P., 2020. Rhizobacterial community of *Jatropha curcas* associated with pyrene biodegradation by consortium of PAH-degrading bacteria. Appl. Soil Ecol. 155, 103685.

Sivasubramanian, V., 2016. Phycoremediation and business prospects. In: Bioremediation and Bioeconomy. Elsevier, pp. 421–456.

Srivastava, A., Rani, R.M., Patle, D.S., Kumar, S., 2022. Emerging bioremediation technologies for the treatment of textile wastewater containing synthetic dyes: a comprehensive review. J. Chem. Technol. Biotechnol. 97 (1), 26–41.

Sumfleth, B., Majer, S., Thrän, D., 2020. Recent developments in low iLUC policies and certification in the EU biobased economy. Sustainability 12 (19), 8147.

I. Bioproducts from contaminated soil and water

Sytar, O., Prasad, M.N.V., 2016. Production of biodiesel feedstock from the trace element contaminated lands in Ukraine. In: Bioremediation and Bioeconomy. Elsevier, pp. 3–28.

Tang, K.H.D., Kristanti, R.A., 2022. Bioremediation of perfluorochemicals: current state and the way forward. Bioprocess Biosyst. Eng. 45, 1–17.

Thakur, L.S., Varma, A.K., Goyal, H., Sircar, D., Mondal, P., 2021. Simultaneous removal of arsenic, fluoride, and manganese from synthetic wastewater by *Vetiveria zizanioides*. Environ. Sci. Pollut. Res. 28, 1–10.

Thangavelu, L., Veeraragavan, G.R., 2022. A survey on nanotechnology-based bioremediation of wastewater. Bioinorg. Chem. Appl. 2022.

Tripathi, S., Sanjeevi, R., Anuradha, J., Chauhan, D.S., Rathoure, A.K., 2022. Nano-bioremediation: nanotechnology and bioremediation. In: Research Anthology on Emerging Techniques in Environmental Remediation. IGI Global, pp. 135–149.

Tufail, M.A., Iltaf, J., Zaheer, T., Tariq, L., Amir, M.B., Fatima, R., Asbat, A., Kabeer, T., Fahad, M., Naeem, H., Shoukat, U., 2022. Recent advances in bioremediation of heavy metals and persistent organic pollutants: a review. Sci. Total Environ. 850, 157961.

UNEP (United Nations Environment Programme), 2021. From Pollution to Solution: A Global Assessment of Marine Litter and Plastic Pollution. UNEP, Nairobi, p. 146.

Vidyashankar, S., Ravishankar, G.A., 2016. Algae-based bioremediation: bioproducts and biofuels for biobusiness. In: Bioremediation and Bioeconomy. Elsevier, pp. 457–493.

Visbeck, M., Keiser, S., 2021. Climate change and its impact on the ocean. In: Transitioning to Sustainable Life below Water. MDPI, Basel, p. 87.

Wan, T., Dong, X., Yu, L., Huang, H., Li, D., Han, H., Jia, Y., Zhang, Y., Liu, Z., Zhang, Q., Tu, S., 2021. Comparative study of three *Pteris vittata*-crop intercropping modes in arsenic accumulation and phytoremediation efficiency. Environ. Technol. Innov. 24, 101923.

Wang, L., Ji, G., 2021. Glutathione and calcium biomineralization of mulberry (*Morus alba* L.) involved in the heavy metal detoxification of lead-contaminated soil. J. Soil Sci. Plant Nutr. 21 (2), 1182–1190.

Wang, W., Cui, J., Li, J., Du, J., Chang, Y., Cui, J., Liu, X., Fan, X., Yao, D., 2021. Removal effects of different emergent-aquatic-plant groups on Cu, Zn, and Cd compound pollution from simulated swine wastewater. J. Environ. Manag. 296, 113251.

Warra, A.A., Prasad, M.N.V., 2016. *Jatropha curcas* L. cultivation on constrained land: exploring the potential for economic growth and environmental protection. In: Bioremediation and Bioeconomy. Elsevier, pp. 129–147.

Yan, C., Qu, Z., Wang, J., Cao, L., Han, Q., 2022. Microalgal bioremediation of heavy metal pollution in water: recent advances, challenges, and prospects. Chemosphere 286, 131870.

Yaqoob, H., Teoh, Y.H., Sher, F., Ashraf, M.U., Amjad, S., Jamil, M.A., Jamil, M.M., Mujtaba, M.A., 2021. *Jatropha curcas* biodiesel: a lucrative recipe for Pakistan's energy sector. Processes 9, 1129.

Zhang, X.B., Peng, L.I.U., Yang, Y.S., Chen, W.R., 2007. Phytoremediation of urban waste water by model wetlands with ornamental hydrophytes. J. Environ. Sci. 19 (8), 902–909.

Zhang, J., Cao, X., Yao, Z., Lin, Q., Yan, B., Cui, X., He, Z., Yang, X., Wang, C.H., Chen, G., 2021. Phytoremediation of Cd-contaminated farmland soil via various *Sedum alfredii*-oilseed rape cropping systems: efficiency comparison and cost-benefit analysis. J. Hazard. Mater. 419, 126489.

Zulfahmi, I., Kandi, R.N., Huslina, F., Rahmawati, L., Muliari, M., Sumon, K.A., Rahman, M.M., 2021. Phytoremediation of palm oil mill effluent (POME) using water spinach (*Ipomoea aquatica* Forsk). Environ. Technol. Innov. 21, 101260.

2

Holistic approach to bioremediation-derived biomass and biorefineries for accelerating bioeconomy, circular bioeconomy, and carbon neutrality

Majeti Narasimha Vara Prasad

Department of Plant Sciences, School of Life Sciences, University of Hyderabad (An Institution of Eminence), Hyderabad, Telangana, India

1 Introduction

The contemporary world is overloaded with waste. Sound policies and affordable technologies are required for handling waste including biowaste. There is increasing demand for fuel, while food costs are steeply increasing. Therefore, there is a need to address waste management and curb pollution due to excessive usage of energy.

The ecological footprint is a measure of human demand on the Earth's resources (biotic and abiotic). It is a standardized measure of the natural capital that includes regenerable and finite resources. Within the life cycle of waste, the specific aim in terms of utilization and commercialization of a product at markets assigning value to principles of sustainable development.

Biorefineries for the valorization of biowaste to produce products and bioenergy is gaining considerable momentum. The age-old concept of landfills is discouraged due to a variety of constraints, viz. (1) land resources from a global perspective are under immense pressure, (2) production of greenhouse gases (GHGs), etc., and (3) valuable resources are buried.

Waste management has been known for several centuries as follows: dumping, burning, composting, and converting it into a useful thing (Edjabou et al., 2015; Gallardo et al., 2015). However, the concept of biowaste valorization with emphasis on the production of higher valuable materials has attained tremendous importance. The age-old concept of incineration and composting is no longer valid in all situations. In this context, the circular economic model is quite appropriate to the problem of waste management as it integrates waste into a resource-based economy and applies to all categories of waste.

The key value drivers in achieving the mentioned goals are:

(a) Environmental: lower footprint and less waste.
(b) Economic: cost of feedstock, carbon, or waste taxation.
(c) Strategic: license to operate, branding and renewable.

Thus, waste is no longer to be considered waste; it is indeed a valuable resource for the production of chemicals, materials, and energy. Fig. 1 depicts the opportunities for the recovery of resources, including the utilization of contaminated substrates (Fig. 2) (Boxes 1 and 2).

Sustainable solid waste management and waste collection, segregation, and characterization are major challenges in many developing countries (Boonrod et al., 2015). Innovative strategies for waste separation, recycling and recovery, and sustainable waste utilization and recycling methods are to be adopted, including life cycle assessment (LCA) of waste management practices (cradle to grave model) (Fig. 3) (Labunska et al., 2015; Nnaji, 2015; Ravindra et al., 2015). In developed countries, solid waste is collected in color-coded polythene bags. In recycling units, the waste is segregated by an optical sensor.

The biological sciences have undergone an innovative revolution that has given rise to the bioeconomy. It is closely related to inventions, the advancement, and the application of biological processes in human health, agricultural and livestock productivity, and biotechnology. Thus, it touches on a variety of industrial fields. Much of the global wealth created over the previous 30 years was based on a single, straightforward formula. Every 18 months, the cost of processing digital code is cut in half, while the power of digital equipment continues to increase. This change from an analog world, where music was played on phonographs using needles, pictures were developed using chemicals, and cars were essentially mechanical, started out very slowly.

FIG. 1   Waste to resource including utilization of contaminated substrates for resource recovery.

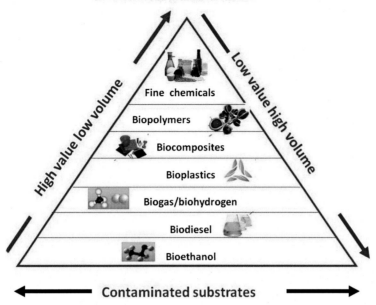

FIG. 2 Pyramid of bio-based products derived from contaminated substrates. High-value products, viz. fine chemicals, are low in quantity, and low-value products are obtained in very high quantity, e.g., biofuels.

BOX 1

Emerging domains in bioeconomy in developing world.

(a) Industrial biotechnology

Processing and productions: chemicals, plastics, enzymes

    Environmental applications: bioremediation, biosensor, methods for reducing environmental impacts

(b) Production of biofuels

Crossing and improving plants and animals, veterinary applications

(c) Human health

BOX 2

Value chain in the bioeconomy (see also Fig. 2)

    Utilization of contaminated substrates for boosting bioeconomy through cascading approach is a new innovation wave. The commercial value of the product is the highest at the top of the pyramid and lowest at the bottom. On the contrary, the volume of active principle recovery (e.g., secondary metabolite for pharma industry) is the lowest at the top of the pyramid and highest at the bottom of the pyramid (e.g., liquid biofuels).

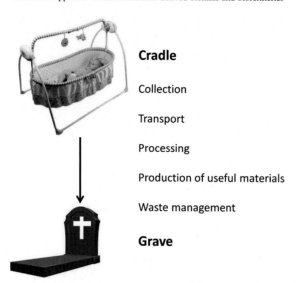

Cradle

Collection

Transport

Processing

Production of useful materials

Waste management

Grave

FIG. 3 Cradle to grave. Domestic and industrial waste. Within the life cycle of waste, the specific aim in terms of utilization and commercialization of a product at markets assigning value to principles of sustainable development.

Only 6% of the world's data was digital as of 2006. It is currently 99%. The nations and companies that recognized this wave and effectively rode it rose to wealth and power. This is the tsunami that propelled a feudal country like Korea and an African-like country like Singapore into the first world. It is a wave that most of Latin America missed.

Fortunately, because science and technology are developing at such a quick pace, there are numerous avenues, fields, and discoveries that can lift a country out of poverty and into the position of a developed country in just one generation. Korea was clever, aspirational, strategic, and diligent enough to capture the tail end of the initial wave of offshoring manufacturing and digital manufacturing. The country started competing in, and eventually partially controlling, entire industries since the effort was sufficiently concentrated, and the educational quality was high enough. Japanese steel, ship, auto, computer, TV, and phone businesses suddenly found themselves up against fierce competition. As a result, new household names such as Samsung, LG, Hyundai, and SK emerged.

Public support for regulation, intellectual property, societal attitudes, and research, development, and innovation (RDI) initiatives are likely to have an impact on the growth of the bioeconomy. Advanced knowledge of genes and intricate cellular processes, the use of renewable biomass, and the multi-sectoral integration of applied biotechnology are all necessary for its formation. The development of the bioeconomy is probably going to be influenced by public support for regulation, intellectual property, social attitudes, and research, development, and innovation initiatives. Its development is dependent on cutting-edge knowledge of genes and intricate cellular processes, as well as on the use of renewable biomass and the multi-sectoral integration of applied biotechnology (Kardung and Wesseler, 2019; Kardung et al., 2019)

2 Contemporary bioeconomy

Perhaps nothing more accurately captures the so-called "Third Industrial Revolution" than the creation of countless new polymers for 3D printers, new enzymes, prebiotics, probiotics, and molecular gastronomy, social media forecasting algorithms, new knowledge of yeast biosynthesis and its use in the field of biofuels, or the creation of artificial neural networks (autonomous chemical "perceptions") that can learn (Carus, 2017). Recent scientific breakthroughs in genome editing, genetic circuit engineering, and biological programming languages have advanced synthetic biology much more than was previously thought possible. Fig. 2 depicts the evolution of programming language and advances in synthetic biology.

Thus, a new technological revolution that could be considerably more profound in scope and impact than the new information and communications technologies (ICTs) of the past thirty years is about to begin on Earth. These are in fact no more than a few of the tools for the disruptive transformation wrought by the unraveling of the genetic code. It is now widely acknowledged that this Revolution in manufacturing and industrial production is based on "3-D Printing," "Big Data," and "Pattern Recognition" technologies, which are the wholly contemporary elements of a new industrial production process that has been adapting the infrastructure and logistics of the production and trade of goods globally. Thus, a new technological revolution that could be considerably more profound in scope and impact than the new ICTs of the past 30 years is about to begin on Earth. These are in fact no more than a few of the tools for the disruptive transformation wrought by the unraveling of the genetic code.

The fundamental idea can be stated as follows: in the decades to come, nations' ability to produce economic wealth will likely be based on their genetic makeup. Modern molecular biology tools for programming genes, together with the natural diversity and variability of genes that result from the endless combinations of them, provide the world with a virtually limitless supply for the engineering and production of new biological products. The economic paradigm is changing yet again as a result of this new revolution, with the language of the modern world switching from the digital binary code [0-1] to the genetic code [A-T-C-G] found in DNA.

Synthetic biology with emphasis on plant genetic engineering for non-food uses, novel biological crop protection strategies, fermentation, and microbial conversion as well as the production and use of enzymes. This is in addition to chemical, thermal, and mechanical aspects of the processing and use of biological materials in the industry for producing bulk chemicals, liquid and gaseous biofuels, cosmetics, drugs, and vaccines. Bioplastics, polymers, and packaging, non-wood fibers, wood products (Cao, 2017), specialty chemicals, inputs for integrated crop protection, organic fertilizers, etc. Synthetic biology offers investors an attractive opportunity: to be able to turn biowaste into a diverse range of profitable products and reclaim valuable degraded land with low-cost remediation solutions (Fig. 4).

Justus Wesseler, Wageningen University coordinated the project "Monitoring bioeconomy (Acronym = Biomonitor) from June 2018 to May 2022. The salient observations of "Biomonitor" are as follows:

The team assessed five key product categories, namely, construction materials, bioplastics, biochemicals, wood-based composites, and textiles concentrating on feedstock

I. Bioproducts from contaminated soil and water

Medical biotechnology
Bioinformatics and -omics
Agricultural biotechnology
Bioenergy

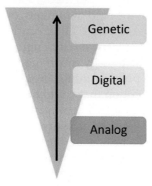

FIG. 4   Evolution of Programming Language and advances in synthetic biology.

requirements/sustainability aspects and compatibility with existing value chains. These goods are making their way into developed markets and, in many circumstances, can seamlessly integrate into existing value chains. The diversity of fossil-based items and materials that can be substituted for the spectrum of wood-based products are dealt with elaborately by Wesseler in the BIOMONITPR project (Tables 1 and 2).

TABLE 1   Products from wood-based products.

Category	Product	Main uses
Construction materials	Cross-laminated timber (CLT)	Building elements
	Laminated veneer lumber (LVL)	Building elements
	Wood foam/cellulose foam	Insulation (thermal and acoustic), packaging
Chemicals	Lignin-based adhesives	Adhesives
	Glycols	Bioplastics, biofuels, etc.
	Nanocrystalline cellulose (NCC)	Biomedical products, hydrogels, 3D-printing, etc.
	Lignin-based carbon black	Fillers for tires
	Wood-based chemicals building blocks	Bioplastics, detergents, cosmetics, resins, etc.)
	Biodiesel	Biofuels
	Bio-naphtha	Biofuels
Bioplastics	Bioplastics from tall oil	Packaging
	Bioplastics from ethylene	Packaging
	Flexible packaging material	Barrier material in packaging

TABLE 1 Products from wood-based products—cont'd

Category	Product	Main uses
Wood-based composites	Sulapac material	Packaging, daily use products, etc.
	3D molded parts	Parts for automotive, transport, and furniture industries
	Wood-plastic composites	Decking, outdoor furniture, etc.
	UPM form	Injection molding
	Bio-based polymer	Material for 3D-printing
Textile fibers	Lyocell fiber	Staple fiber for textiles
	Kuura/Ioncell fiber	Staple fiber for textiles
	Veocell	Fiber for non-wovens
	TreeToTextile (process)	Fiber for textiles

Source: "BIOMONITOR" Monitoring the Bioeconomy—Project Coordinator Justus Wesseler This project has received funding from the European Union's Horizon 2020 research and innovation program under grant agreement N° 773297.

TABLE 2 Types of biorefinery approaches reported in the literature.

Classifications	Source of feedstock	Bioconversion technology	Major bioactive molecules
Non-food feedstocks	Sugar cane bagasse Microalgal biomass Proteins	Thermo economic and environmental assessment Supercritical water gasification	Bio-ethanol Bio-fuel
Cereal biorefinery	White sorghum and corn grains, Legumes seeds Pea pods, pulses by-product	Alcoholic-alkaline treatment Alkaline/acid, solvent, and enzymatic extractions and the use of ultrafiltration membranes	Citric acid Protein, dietary fibers, minerals
Oilseed biorefinery	Oil seed crops and oil palm	Enzymatic synthesis Microbial production	Biodiesel, vegetable oil, Biosurfactants like rhamnolipids and glycolipids
Lignocellulosic biorefinery	Agricultural and crop residue (corn-cob)	Enzymatic methods, chemical treatment, acid treatment, and microbial fermentation	Bio-ethanol, Xylitol
Forest biorefinery	Walnut biomass barks, Eucalyptus, chestnut	Solvents extraction	Phenolics content, Fuels, energy, chemicals
Industrial or municipal waste-based biorefinery	Industrial and municipality waste	Microbial fuel cell (MFC) technology	Bioelectricity

Source Awasthi, M.K., Sindhu, R., Sirohi, R., Kumar, V., Ahluwalia, V., Binod, P., Juneja, A., Kumar, D., Yan, B., Sarsaiya, S., Zhang, Z., 2022. Agricultural waste biorefinery development towards circular bioeconomy. Renew. Sust. Energ. Rev. 158, 112122, with permission.

Waste can be defined as "right or useful material in a wrong place" and as such there is no waste, until it was wasted. All waste is a reusable resource. The **ecological footprint** is a measure of human demand on the Earth's resources (biotic and abiotic). It is a standardized measure of natural capital that includes regenerable and finite resources. Within the life cycle of waste, the specific aim in terms of utilization and commercialization of a product at markets assigning value to principles of sustainable development.

In order to reduce anthropogenic emissions and accomplish sustainable development goals, clean technology and recycling are essential. The trade-off between the environmental costs and advantages of recycling and clean technologies is still unclear. A life cycle evaluation is one technique to evaluate this trade-off (LCA) (Fig. 5) (Chen and Huang, 2019; Curran, 2017; D'amato et al., 2020; Fauzi et al., 2019). Bioremediation-derived biomass and possible routes of biorefinery (Badgujar and Bhanage, 2018; Katakojwala and Venkata Mohan, 2020b) are shown in Fig. 6 (Corrado and Sala, 2018; Dahiya et al., 2018).

2.1 *Ricinus communis* L. (Castor bean) for environmental cleanup and co-generation of bioproducts

Euphorbiaceae gained considerable importance for biodiesel production, e.g., castor bean and jatropha (Baioni e Silva et al., 2023; Cheban and Dibrova, 2020)). *Ricinus communis* L. (castor bean) belongs to Euphorbiaceae, is a fast-growing C_3 plant native to tropical Africa, and it is widespread throughout tropical regions. It is an erect, tropical shrub or small tree that grows up to 2–3 m and sometimes about 5 m. The stem, stalks, and leaves are reddish to purple. Leaves palmate, 6–11 lobed, serrate. It is only one species of monotypic genus *Ricinus*. It is cultivated and naturalized in India. It is an important non-edible oil seed crop in the arid and semi-arid regions.

Castor bean emerged as a prominent environmental crop-producing feedstock for biorefinery (Naik et al., 2018). Castor bean possesses the enormous adaptive potential to perform better in a diverse habitat due to its plasticity. Although, the "castor bean" was introduced as an ornamental but is being much maligned subsequently. Various adaptive traits of *R. communis* enable it to perform better in a stressed environment producing high biomass.

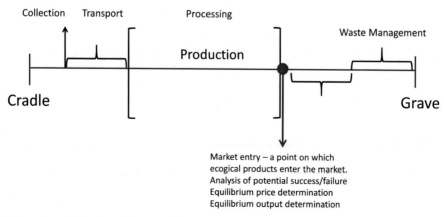

FIG. 5   Life cycle analysis of market penetration.

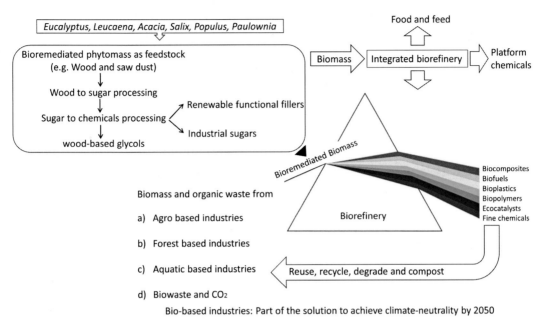

FIG. 6   Bioremediation-derived biomass and possible routes of biorefinery (Dahiya et al., 2020).

Several other studies have shown that *R. communis* performs well on co-contaminated sites and has been proposed as a candidate for environmental cleanup and co-generation of bioproducts. It has been reported that *R. communis* is tolerant to various stresses and proved to be a highly qualified candidate for phytoremediator (Alherbawi et al., 2023; Baioni e Silva et al., 2023; Kaur et al., 2023; Moncada et al., 2015; Naik et al., 2018; Román-Figueroa et al., 2023; Zhang et al., 2023).

Castor seeds contain 40%–50% oil. The oil-yielding plants are known to have better efficiency of S-metabolism. Thus their tolerance to metal through the induced synthesis of phytochelatins has been documented extensively. Besides, it is also reported as a high salt tolerant, salinity and drought tolerant and a potential ameliorator for seashore saline soil because of its well-developed and massive root system along with tap root, which penetrates soil depths to several meters, much deeper than herbaceous plants. It has an economic advantage as a cash crop in modern agriculture along with the remediation of heavy metal-contaminated soil (Fig. 7).

Contaminated lands may be unsuitable for food production due to possible accumulation of toxic substances in the food chain, but taking moderately contaminated land into use for non-food Biomass production may further relieve the problem, although obviously plant growth would be affected due to toxicity of the contaminants. Moreover, diffuse contamination affects large areas worldwide, mainly as a result of mining activity, fallout from industrial processes such as smelting, agriculture, traffic; areas elevated with contaminated dredged sediments, former landfill sites, and abandoned industrial sites. Re-using such contaminated land for non-food crops during and after soil remediation could bring them back into beneficial and sustainable use and reduce detrimental environmental, social, and economic impacts on affected communities. Research toward a better understanding and

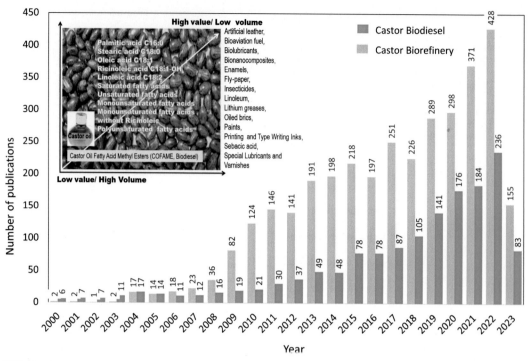

FIG. 7 Castor biodiesel and biorefinery scenario over a period of more than 2 decades. Castor seeds producing a wide variety bioresources ranging low value, high volume materials such as castor oil fatty acid methyl esters (COFAME = biodiesel) to high value products as indicated in low volume, with permission from Springer Nature (Prasad 2021)

improving plant growth in the presence of contaminants could contribute to the future use of these contaminated areas for safe and efficient Biomass production. Castor biodiesel and biorefinery scenario for more than two decades is shown.

2.2 Contaminated site cleanup and co-generation of bioproducts

Economic growth, extensive industrialization, extraction of natural resources, and lack of willpower for cleanup have resulted in environmental contamination and pollution. Large amounts of toxic waste have been dispersed in hundreds of contaminated sites spread all over India. These pollutants belong to two main classes, viz., (i) inorganic and (ii) organic. The challenge is to develop innovative and cost-effective solutions to decontaminate polluted environments. Development of cost-effective solutions to decontaminate polluted sites is a challenging task. Bioremediation is emerging as an invaluable tool for environmental cleanup. Cultivation of appropriate non-food crops and energy crops for value chain and value additions appears profitable. Bioproducts and bioenergy from contaminated substrates (soil and water), and brownfield development for smart bioeconomy are covered in this lecture with specific examples.

Prominent plants for bioremediation and biomass production in tropical countries on contaminated soil/water. Simplified value chain (a) for CLT (cross-linked timber), (b) for wood foam tiles, and (c) lignin-based adhesives (Fig. 8).

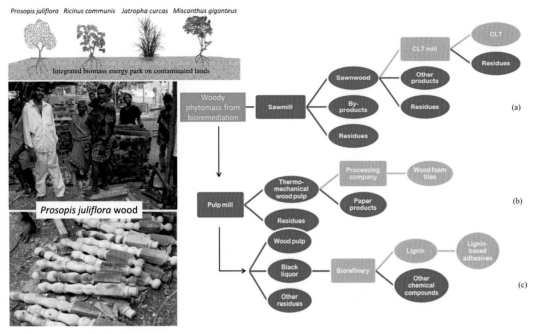

FIG. 8 Please note prominent plants for bioremediation and biomass production in tropical countries on contaminated soil/water. Simplified value chain (A) for CLT (cross-linked timber), (B) for wood foam tiles, and (C) lignin-based adhesives. *Source: "BIOMONITOR" Monitoring the Bioeconomy—Project Coordinator Justus Wesseler This project has received funding from the European Union's Horizon 2020 research and innovation programme under grant agreement N° 773297.*

2.2.1 Glycols made from woody biomass

Since the chemicals produced from woody biomass are equivalent to those produced from non-renewable resources, they are regarded as drop-in components in the value chain for applications further down the line. There is currently no full-scale production facility for the manufacture of wood-based glycols, which is still in its early stages of development. North America and Europe now have the biggest demand for bio-based glycols, while the Asia-Pacific area has the fastest-growing need for glycols. In 2024, it is anticipated that the world's capacity to produce ethylene glycol will exceed 65 million metric tons (Statista, 2021). From the almost 42 million metric tons of ethylene glycol that could be produced globally in 2019, that is a rather considerable increase (Statista, 2021). Fig. 9A depicts a simplified value chain for MEG (monoethylene glycol) (B) for bioplastics and (C) PET, PEF, and other chemicals from monoethylene glycol (MEG) lignin-based adhesives.

2.2.2 Compatibility with existing value chains for wood-based composites and textile fibers

In the interview and poll, every stakeholder who took part acknowledged that their items were already available. The value chain for this substance has therefore already been created

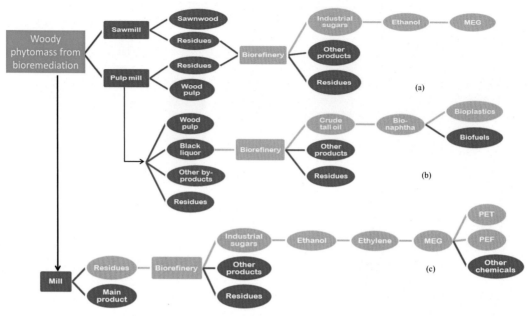

FIG. 9 Simplified value chain (A) for MEG (monoethylene glycol), (B) for bioplastics, and (C) PET, PEF and other chemicals from monoethylene glycol (MEG) lignin-based adhesives. *Source: "BIOMONITOR" Monitoring the Bioeconomy - Project Coordinator Justus Wesseler This project has received funding from the European Union's Horizon 2020 research and innovation programme under grant agreement N° 773297.*

or modified from standard value chains. According to the stakeholders, their wood-based composites are drop-in. However, when asked how much fossil-based or GHG-intensive material the wood-based composites can replace, our stakeholders' responses ranged from 10 to 19 percent to 90 percent or more, demonstrating the wide difference in substitution depending on the kind of product and its intended use. Fig. 10 displays a condensed value chain for wood-based composites.

The textile fibers made from wood are only partially compatible with the current value chains. For the activity to be economically viable, wood-based textile fiber production costs must be comparable to those of other fiber types. An integrated mill with a textile fiber production facility operating next to a pulp mill would be the most economically feasible way to achieve the goal (Fig. 10). The United Nations Sustainable Development Goals (SDG) 1–3, 6–9, and 12–15 and Anonymous (2019a,b) Green Deal agenda were identified to be influenced by bioeconomy activities (Heimann, 2019) and those are detailed in Fig. 11 (Brusseau, 2019).

It is apparent that there is no single universal definition for the circular economy. Based on the aforementioned definitions, we conclude that to achieve a circular economy is principally to slow, narrow, and close the material resource loops, all built on the foundation of renewable energy and non-toxic materials, as well as through long-lasting design, maintenance, repair, reuse, remanufacturing, refurbishing, and recycling (as illustrated in the resource

FIG. 10 Simplified value chain for (A) wood-based composites and (B) wood-based textile fibers. *Source: "BIOMONITOR" Monitoring the Bioeconomy—Project Coordinator Justus Wesseler This project has received funding from the European Union's Horizon 2020 research and innovation programme under grant agreement N° 773297.*

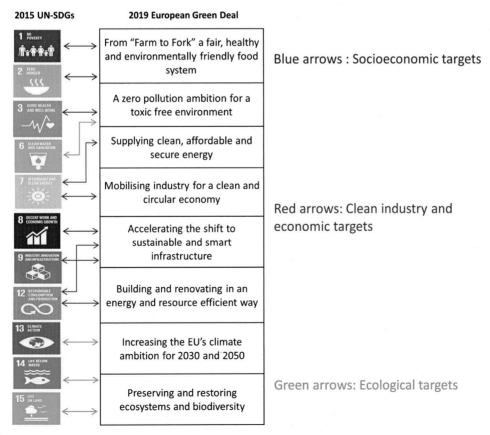

FIG. 11 UN 2015 SDG 1–3, 6–9, and 12 to 15 and EU 2019 Green Deal agenda as shown above were identified to be influenced by bioeconomy activities. Blue arrow: socioeconomic targets; green arrow: ecological targets; red arrow: clean industry and economic targets.

cycles—namely slowing, closing, and narrowing material resource loops—were first introduced by Bocken et al. (2016) and can serve as a working definition that is simple yet practical for a circular economy practitioner to interpret and implement.

Figge et al. (2023) outlined certain requirements that effective definitions must meet. Therefore, the authors proposed a definition that meets all four conditions. "Definitions of the Circular Economy: Circularity Matters," which was published in Ecological Economics. By way of illustration, we apply these requirements to a well-known definition that we discovered in the literature and point out some of the problems with current definitions.

We establish four requirements that we think make up a solid description of the circular economy. Good circular economy definitions must

(1) Talk about completing resource loops,
(2) Mention optimizing resource flows rather than reducing them,
(3) Take into account a minimum of two tiers, and
(4) Differentiate between the circular economy as an idealized, perfect type and a practical, flawed type that achieves sustainability when used in conjunction with other strategies.

The GHG emissions and resulting carbon efficiency of CO_2 utilization routes, including CO_2 to fuels or chemicals, heavily depend on the GHG intensity of key process inputs, foremost electricity, and hydrogen (Grim et al., 2020). Thus, achieving net-zero emission sectors depends on decarbonizing the energy system and appropriately integrating and linking utilization routes. Fig. 12 graphically represents this interdependency of the bio-based circular economy's sustainability as a system of systems (Tan and Lamers, 2021) (Fig. 13).

Bioremediation produces lignocellulosic biomass, which is frequently resistant and may be broken down to produce sugar chains and lignin (Katakojwala et al., 2019; Katakojwala and Venkata Mohan, 2020a). In a biorefinery setup aiming for a circular economy, lignocellulosic wastes can be treated through a variety of thermochemical and biochemical routes, including combustion, gasification, pyrolysis, pre-treatment followed by saccharification, and fermentation, to create a variety of products (Fig. 14). The most common method for using agricultural waste is to use a biochemical process to create ethanol in a biorefinery environment (Fig. 14) (Awasthi et al., 2022; Ferreira et al., 2018). Lignocellulosic biorefinery can be broadly classified under two conversion technologies: thermochemical and biochemical (Fig. 15) (Awasthi et al., 2022). Table 3 shows the performance of typical bioprocesses for lignocellulosic biorefinery. The majority of these biorefineries are in developing and progressing, techno-economic analysis (TEA) models are used to assess the process's technical and financial viability (Fig. 16, Table 4).

3 Creating a circular economy while decarbonizing

The expansion of RE is dependent on the availability of easily recyclable and reusable raw materials. A circular strategy also boosts a country's economy by making it resilient to external shocks like the COVID-19 epidemic (Anonymous, 2021; Ávila-Gutiérrez et al., 2019).

An economy that is circular values everything. By reducing, re-using, and recycling garbage, all resources are used to their fullest extent. This strategy not only lowers carbon

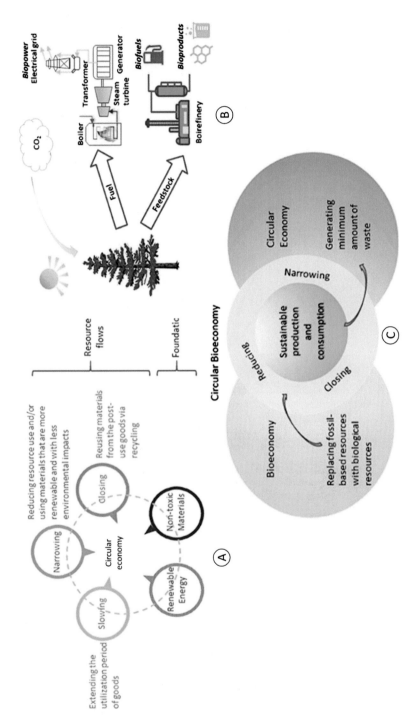

FIG. 12 (A) Components of a proposed working circular economy definition. (B) Schematic representation of a sustainable bioeconomy—renewable biological resources to renewable products (i.e., biopower, biofuels, bioproducts). (C) Schematic representation of a circular bioeconomy resulting from the intersection between circular economy and bioeconomy concepts. *Source: Tan, E.C.D., Lamers, P., 2021. Circular Bioeconomy concepts—a perspective. Front. Sustain. 2, 701509. https://doi.org/10.3389/frsus.2021.701509.*

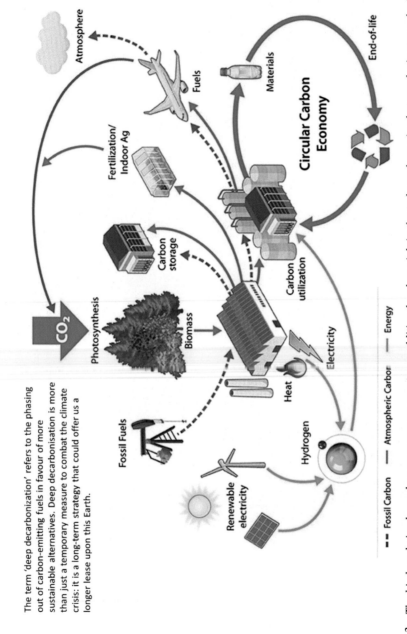

The term 'deep decarbonization' refers to the phasing out of carbon-emitting fuels in favour of more sustainable alternatives. Deep decarbonisation is more than just a temporary measure to combat the climate crisis: it is a long-term strategy that could offer us a longer lease upon this Earth.

FIG. 13 The bio-based circular carbon economy can create an additional carbon sink in the technosphere via photosynthetic atmospheric carbon sequestration in biomass feedstock, energy, and material conversion pathways plus carbon capture and (re)utilization in closed circular carbon economy loops (Baldassarre et al., 2019; Cecchin et al., 2020; CGRi, 2020) *Source: Tan, E.C.D., Lamers, P., 2021. Circular Bioeconomy concepts—a perspective. Front. Sustain. 2, 701509. https://doi.org/10.3389/frsus.2021.701509.*

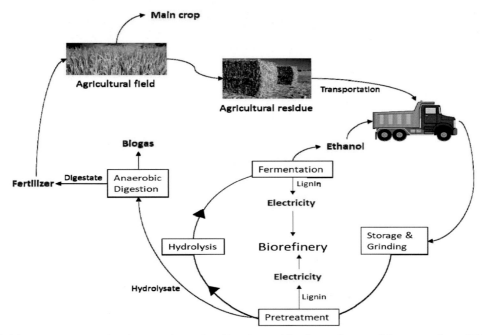

FIG. 14 Systematic process of thermochemical and biochemical for the production of diverse products. *With permission from Awasthi, M.K., Sindhu, R., Sirohi, R., Kumar, V., Ahluwalia, V., Binod, P., Juneja, A., Kumar, D., Yan, B., Sarsaiya, S., Zhang, Z., 2022. Agricultural waste biorefinery development towards circular bioeconomy. Renew. Sust. Energ. Rev. 158, 112122.*

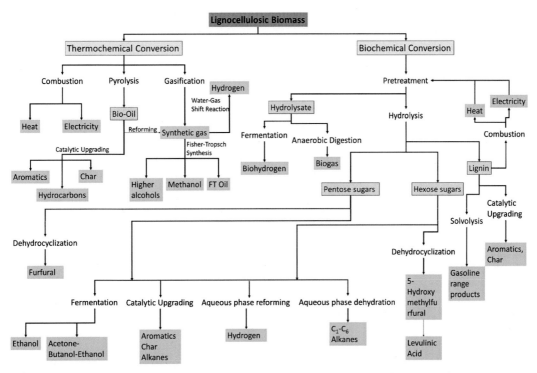

FIG. 15 Ethanol production from agricultural residue in a biorefinery setting. *With permission from Awasthi, M.K., Sindhu, R., Sirohi, R., Kumar, V., Ahluwalia, V., Binod, P., Juneja, A., Kumar, D., Yan, B., Sarsaiya, S., Zhang, Z., 2022. Agricultural waste biorefinery development towards circular bioeconomy. Renew. Sust. Energ. Rev. 158, 112122.*

TABLE 3 Performance of typical bioprocesses for agricultural waste biorefinery.

Substrate	Bioprocess	Configuration	Tem (°C)	Product	Specific yield	By-product
Olive mill solid waste	Simultaneous saccharification and fermentation (SSF)	Batch fermentation bottle	35	Ethanol	15.9 g/L ethanol	–[a]
Banana waste	SSF	Batch fermentation flask	Not clearly mentioned	Ethanol	17.1 g/L	–
Banana waste	SSF	250 mL glass bottles	35	Ethanol	48.3 g/L, reflecting 86.6% of the maximum attainable yield	–
Corn stalk	Dark fermentation + single chamber microbial electrolysis cells	5 L batch stirred anaerobic bioreactor; 64 mL cubic single chamber membrane-less MECs	36 ± 1	H_2	387.1 mL H_2/g-corn stalk	CO_2
Vinasse	Two-stage anaerobic digestion	250 mL bottles	37	H_2 & CH_4	14.8 & 274 mL/g $VS_{substrate}$	CO_2
Corn straw	Two-stage anaerobic digestion	300 mL glass bottles	37	H_2 & CH_4	24.3 & 296.0 mL/g VS	CO_2
Cow solid waste	Anaerobic digestion	H_2: 200 mL CH_4: 2 L continuously stirred anaerobic bioreactor	35	H_2 & CH_4	[b]HPR: 150.59 $\pm$ 35.70 mL/L/d (97 mL H_2 gVS^{-1}) [c]MPR: 395.41 $\pm$ 72.82 mL/L/d	CO_2
Corn-cob powder	Two-stage anaerobic digestion	Conical flasks	H_2 30, CH_4 3	Bio-hythane	22.29 mL/g TS & 141.14 mL/g TS	CO_2

Feedstock	Process	Reactor	Temperature (°C)	Product	Yield	Byproducts
Swine manure and rice bran	Dark fermentation, anaerobic digestion and microbial fuel cells	Dark fermentation 250 mL flask; AD 3-L flasks	55	CH_4	182 L CH_4 kg^{-1} COD•added	H_2, CO_2
Agricultural waste	Pyrolysis	Container	400–600	Bio-oil	Rice straw: 31.86 wt%; C. oleifera shell: 27.45 wt%	Phenols, aldehydes, and alcohols
Soybean cooking oil	Catalytic conversion	250-mL round-bottom flask	65°C	Biodiesel	98.8% conversion	–
Tequila Bagasse	Biodegradation	Culture bottle	Not given	Polyhydroxyalkanoates (PHA)	1.5 mg/mL	–
Rice husk hydrolysate, rice straw hydrolysate, dairy industry, and rice mill wastewater	Dark and photo fermentation	100 mL serum bottles	37°C	BioH_2 and poly-b-hydroxybutyrate (PHB)	1.82 ± 0.01 mol H_2/mol glucose and 19.15 ± 0.25 g/L PHB	–
Citrus peel waste	Fermentation by Actinobacillus succinogenes	Duran bottle	37°C	Succinic acid	$0.77\ g_{sa}\ g_{tsc}^{-1}$	–
Oil palm empty fruit bunch	Simultaneous saccharification and fermentation	150 mL flasks	38°C	Succinic acid	0.69 g/L/h	–
Sugarcane bagasse	Batch fermentation	BIOSTAT®B 2.5-L benchtop fermenter	28°C	Succinic acid	0.58 ± 0.01 g/g	–

[a] Production costs (ratio of operating costs and total amount produced).

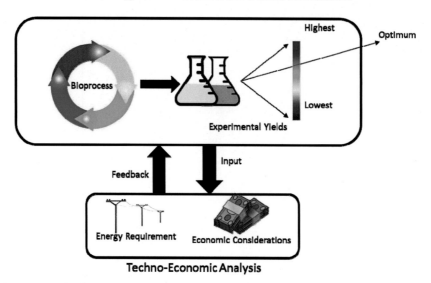

FIG. 16   Importance of techno-economic analysis to determine optimum conditions for a bioprocess. *With permission from Awasthi, M.K., Sindhu, R., Sirohi, R., Kumar, V., Ahluwalia, V., Binod, P., Juneja, A., Kumar, D., Yan, B., Sarsaiya, S., Zhang, Z., 2022. Agricultural waste biorefinery development towards circular bioeconomy. Renew. Sust. Energ. Rev. 158, 112122.*

emissions but also strengthens resistance to climate change. Additionally, it gives inventive organizations the opportunity to lower their need for raw materials, optimize their spending, and improve resource use. While reducing reliance on raw material imports, waste generation, energy use, and environmental emissions, a circular economy has the potential to hasten the path to carbon neutrality and deep decarbonization. According to the most current Circularity Gap Report, moving to a circular economy could save 23 billion tons of carbon emissions by reducing GHG emissions by 39% and easing demand on virgin resources by 28% (Anonymous, 2021). The ability to build more solar capacity and lessen reliance on fossil fuels has been made possible by the falling cost of solar energy. However, as there are more solar installations, there are also more panels that will eventually need to be replaced.

4 Conclusion

Although there are many different ways that biomass is used, the majority of the based products cited as examples in this study can use both virgin biomass and wood by-products and/or residues from industrial side streams. This also implies that, despite the fact that higher volume demands are anticipated, supply is not thought to be a limiting problem (ref).

TABLE 4 Techno-economic analysis of various biorefineries using agricultural waste as feed stocks.

Biomass ($ year)	Capacity	Process	Products	Capital investment ($ million)	Minimum selling price ($)	Remarks
Corn Stover (2017 $)	792 ton/day biomass	Integrated ethanol and xylitol/furfural process	Ethanol, xylitol/furfural, electricity	148–173	$0.7–1.65/L ethanol	Steam explosion pre-treatment, hemicellulose sugars converted to xylitol/furfural
Corn Stover (2017 $)	2000 MT/day biomass	Integrated cellulosic ethanol and HTL (valorizing lignin)	Ethanol, catechol, cresols, acetic add, furfural, formic acid, phenol, and acetaldehyde	624.5	1.06/gal ethanol	80% lignin for HTL and 20% for steam and power
Corn Stover (2015 $)	912115 ton/year biomass	Cellulosic butanol biorefinery using lignin for steam and power	Butanol, acetone, electricity	410	$1.47/L butanol	Assumed vacuum fermentation ethanol was not recovered
Corn Stover (2007 $)	2000 MT/day biomass	Cellulosic ethanol biorefinery using lignin for steam and power	Ethanol and electricity	329.4–410.5	$0.66–1.47/L ethanol	Investigated various scenarios of fermentation using various ethanologen strains
Corn Stover (2010 $)	2000 dry MT/day biomass	Cellulosic ethanol biorefinery using lignin for steam and power	Ethanol and electricity	422.5	0.57	Dilute acid pre-treatment, onsite enzyme production
Corn Stover (2007 $)	2000 MT/day biomass	Cellulosic ethanol biorefinery using lignin for steam and power	Ethanol and electricity	361–501	$0.9–1.17/L ethanol	Investigated various pre-treatments and scenarios
Grass straw (Tall fescue) (2010 $)	758 MT/day biomass	Cellulosic ethanol biorefinery using lignin for steam and power	Ethanol and electricity	91.4–114.6 ethanol	0.81–0.89/L ethanol[a]	Four pre-treatment processes: acid, alkali, hot water & steam explosion
Grass straw (Perennial ryegrass) (2012 $)	758 MT/day biomass	Cellulosic ethanol biorefinery using lignin for steam and power	Ethanol and electricity	127	0.86/L ethanol[a]	Dilute acid pre-treatment
Rice straw (2017 $)	1000 ton/year biomass	Thermochemical biorefinery	Bio-oil, bio-char	28.88	0.55/L oil	Bubbling fluidized bed reactor for pyrolysis

Continued

TABLE 4 Techno-economic analysis of various biorefineries using agricultural waste as feed stocks—cont'd

Biomass ($ year)	Capacity	Process	Products	Capital investment ($ million)	Minimum selling price ($)	Remarks
Rice straw (2014 $)	62 ton/h biomass	Thermochemical biorefinery	Ethanol, electricity cooling & heating utilities	257	—	Gasification using downdraft gasifier
Wheat straw (2017 $)	10,000 ton/year unrefined lipids (product)	Microbial process	Microbial oil (lipids), yeast protein, 2-phenylethanol, electricity	124.3[b]	$4520/ton oil[b]	Used oleaginous yeast *Metschnikowia pulcherrima*
Wheat straw (2016 $)	316 ton/day biomass	Cellulosic ethanol biorefinery using lignin for steam and power	Cellulosic ethanol biorefinery using lignin for steam and power	21.5	$1.23/L ethanol	Steam explosion
Wheat straw (2015 $)	25 ton/h biomass	Integrated ethanol and biogas production	Bio-ethanol, biogas, electricity	148–184[b]	$0.81–0.98/L ethanol[b]	Acid-impregnated steam pre-treatment; various solids loadings and pre-treatment conditions
Wheat straw (2015 $)	69–208 million L/year ethanol (product)	Cellulosic ethanol biorefinery using lignin for steam and power	Ethanol and electricity	90–200	1–1.7/L ethanol	Investigated various plant sizes and locations (affecting biomass price)
Wheat straw (2012 $)	751,000 MT/year biomass	Cellulosic butanol biorefinery using lignin for steam and power	Butanol, ethanol, acetone	193	$1.1/L butanol	Both ethanol and acetone were recovered as coproducts
Wheat straw (2012 $)	758 MT/day biomass	Cellulosic ethanol biorefinery using lignin for steam and power	Ethanol and electricity	122.6	$0.71/L ethanol[a]	Dilute acid pre-treatment
Wheat straw (2008 $)	2100 ton/day biomass	Cellulosic ethanol biorefinery using lignin for steam and power	Ethanol and electricity	389.5	0.99/L ethanol	Dilute acid pre-treatment

[a] Production costs (ratio of operating costs and total amount produced).
[b] Assuming 1.13 conversion from Euro to USD.
With permission from Awasthi, M.K., Sindhu, R., Sirohi, R., Kumar, V., Ahluwalia, V., Binod, P., Juneja, A., Kumar, D., Yan, B., Sarsaiya, S., Zhang, Z., 2022. Agricultural waste biorefinery development towards circular bioeconomy. Renew. Sust. Energ. Rev. 158, 1121–2.

Glossary

Bioenergy and carbon capture and storage BECCS The combination of Bioenergy and Carbon Capture and Storage is a risky distraction from the real solutions that require urgent action.

- Drastic emissions cuts, and a flow of finance from the global North to the global South, in line with the principle of fair shares;
- A just energy transformation and an end to fossil fuels and other dirty and harmful energy;
- Universal access to clean, democratically controlled and community-owned energy based on truly sustainable, climate-safe and locally appropriate sources like sun and wind;
- A just and climate-friendly food system based on the principles of agroecology;
- Community management of ecosystems and forests;
- No more deforestation;
- Ecosystem restoration and regrowth that respects the lives and livelihoods of communities living on the land;
- Reforestation that restores real forests and respects the lives and livelihoods of communities living on the land; we need nothing less than system change—a new model of environmental, social, political, economic, and gender justice—built on the power of the people.

Bioeconomy The production of renewable biological resources and the conversion of these resources and waste streams into value-added products, such as food, feed, bio-based products, and bioenergy.

Bioenergy Energy produced through the burning of organic material. Friends of the Earth International opposes industrial bioenergy as a form of dirty and harmful energy. Biomass: The organic material used as a fuel for bioenergy (sometimes called bio-fuel), e.g., wood, manure, sugarcane, and other crops used for fuel.

Biorefinery A biorefinery is an overall concept of a processing plant where biomass feedstocks are converted and extracted into a spectrum of valuable products and or Biorefinery is the sustainable processing of biomass into a spectrum of marketable products (food, feed, materials, chemicals) and energy (fuels, power, heat) and or Biorefinery refers to the conversion of biomass feedstock into a host of valuable chemicals and energy with minimal waste and emissions.

Bioremediation Bioremediation is a branch of biotechnology that employs the use of living organisms, like microbes and bacteria, in the removal of contaminants, pollutants, and toxins from soil, water, and other environments. Bioremediation broadly refers to any process wherein a biological system, living or dead, is employed for removing environmental pollutants from air, water, soil, flue gasses, industrial effluents, etc., in natural or artificial settings.

Blue Economy The Blue Economy is sustainable use of ocean resources for economic growth, improved livelihoods, and jobs, while preserving the health of marine and coastal ecosystem. The Blue Economy encompasses many activities that impact all of us.

Carbon capture, utilization, and storage (CCUS) Refers to a suite of technologies that can play a diverse role in meeting global energy and climate goals.

Carbon neutrality Net-zero carbon emissions to the atmosphere during the energy production process (infrastructures excluded).

Carbon sequestration Carbon sequestration is the process of capturing and storing atmospheric carbon dioxide. It is one method of reducing the amount of carbon dioxide in the atmosphere with the goal of reducing global climate change.

Carbon sinks A natural environment that has the ability to remove carbon dioxide from the atmosphere (e.g., forests and oceans).

Circular economy A circular economy (also referred to as circularity and CE) is a model of production and consumption, which involves sharing, leasing, re-using, repairing, refurbishing, and recycling existing materials and products as long as possible. CE aims to tackle global challenges such as climate change, biodiversity loss, waste, and pollution by emphasizing the design-based implementation of the three base principles of the model. The three principles required for the transformation to a circular economy are: eliminating waste and pollution, circulating products and materials, and the regeneration of nature. CE is defined in contradistinction to the traditional linear economy. The idea and concepts of circular economy (CE) have been studied extensively in academia, business, and government over the past ten years. CE has been gaining popularity since it helps to minimize emissions and

I. Bioproducts from contaminated soil and water

consumption of raw materials, open up new market prospects, and principally, increase the sustainability. The value of the resources we extract and generate should be retained in circulation through planned and integrated production chains. Wastes in a CE model are transformed into resources that may be recovered and reclaimed through recycling and reuse. The final use of a material is no longer a waste management concern, but rather a step in the design of systems and products. A circular economy promotes cleaner production, efficient resource utilization, and a reduction in the usage of virgin resources.

Circular bioeconomy (CBE) Aspires to replace the economy that is dependent on fossil fuels and create sustainable green technology in order to attain a net-zero-carbon society. It is progressing in various execution phases. Biomass carbon, which can be obtained from any organic source that can decompose, is the fundamental component of the CBE. Social, economic, and environmental variables are the main motivators.

Closed-loop Focuses on lowering the usage of raw materials and waste creation by treating effluents and returning them to reuse and/or improving the longevity of goods. It is a logistic process system that combines reverse logistics with forward logistics (procurement, manufacture, and distribution). Instead of discarding used goods, parts, and resources, closed-loop operations repurpose them to add value to other production cycles.

Cradle to cradle Design concept to put industrial ecology concepts into practice, producing goods that allow for the secure and potentially endless recyclability of materials. It focuses on creating manufactured items where reuse, adaption, and disassembly are taken into account from the beginning. It offers an economy that gets rid of garbage via recycling, remanufacturing, and refurbishing. Each cycle passage serves as a fresh cradle for a particular substance in a circular logic of creation and reuse.

Decarbonization Refers to the phasing out of carbon-emitting fuels in favor of more sustainable alternatives. Deep decarbonization is more than just a temporary measure to combat the climate crisis: it is a long-term strategy that could offer us a longer lease upon this Earth.

Energy penalty The energy penalty gives an indication of the amount of energy that needs to be spent for carbon capture in relation to the energy generated by the plant.

Green economy That aims at reducing environmental risks and ecological scarcities and that aims for sustainable development without degrading the environment. It is closely related to ecological economics, but has a more politically applied focus.

Industrial ecology The study of material and energy flows across industrial systems is the subject of one academic field. It promotes industrial systems where the participants collaborate by utilizing one another waste (residual) energy flows and waste material.

Industrial symbiosis (IS) Framework for mutually beneficial industry collaboration based on industrial ecology, sharing water resources, energy, by-products, and waste materials across all enterprises to benefit the environment and the bottom line. IS creates material flows through industrial ecosystems where waste production is minimized, energy and material consumption are maximized, and the effluents from one operation are used as raw materials in another.

Reverse logistics Process of returning used or unused goods—or parts of goods—from their usual final destination—consumer waste—to a producer in a supply chain, with the intention of regaining value or disposing of them properly. It promotes garbage collection and returns to the industry so that it can be reused or reintroduced into the production process.

References

Alherbawi, M., McKay, G., Al-Ansari, T., 2023. Development of a hybrid biorefinery for jet biofuel production. Energy Convers. Manag. 276, 116569.

Anonymous, 2019a. European Commission, 2019. Communication from the Commission to the European Parliament, the European Council, the Council, the European Economic and Social Committee and the Committee of the Regions. The European Green Deal. Document 52019DC0640, 640. 24 p. European Commission.

Anonymous, 2019b. European Commission, 2019. Annex to the Communication from the Commission to the European Parliament, the European Council, the Council, the European Economic and Social Committee and the Committee of the Regions. The European Green Deal. 4 p. European Commission.

Anonymous, 2021. Circle Economy. The Circularity Gap Report. 2021.

Ávila-Gutiérrez, M.J., Martín-Gómez, A., Aguayo-González, F., Córdoba-Roldán, A., 2019. Standardization framework for sustainability from circular economy 4.0. Sustainability 11 (22), 6490.

Awasthi, M.K., Sindhu, R., Sirohi, R., Kumar, V., Ahluwalia, V., Binod, P., Juneja, A., Kumar, D., Yan, B., Sarsaiya, S., Zhang, Z., 2022. Agricultural waste biorefinery development towards circular bioeconomy. Renew. Sust. Energ. Rev. 158, 112122.

Badgujar, K.C., Bhanage, B.M., 2018. Dedicated and waste feedstocks for biorefinery: an approach to develop a sustainable society. In: Waste Biorefinery. Elsevier, pp. 3–38.

Baioni e Silva, G., Manicardi, T., Longati, A.A., Lora, E.E., Milessi, T.S., 2023. Parametric comparison of biodiesel transesterification processes using non-edible feedstocks: castor bean and jatropha oils. Biofuels Bioprod. Biorefin. 17 (2), 297–311.

Baldassarre, B., Schepers, M., Bocken, N., Cuppen, E., Korevaar, G., Calabretta, G., 2019. Industrial symbiosis: towards a design process for eco-industrial clusters by integrating Circular Economy and Industrial Ecology perspectives. J. Clean. Prod. 216, 446–460.

Bocken, N.M.P., de Pauw, I., Bakker, C., van der Grinten, B., 2016. Productdesign and business model strategies for a circular economy. J. Ind. Prod. Eng. 33, 308–320. https://doi.org/10.1080/21681015.2016.1172124.

Boonrod, K., Towprayoon, S., Bonnet, S., Tripetchkul, S., 2015. Enhancing organic waste separation at the source behavior: a case study of the application of motivation mechanisms in communities in Thailand. Resour. Conserv. Recycl. 95, 77–90.

Brusseau, M.L., 2019. Sustainable development and other solutions to pollution and global change. In: Environmental and Pollution Science. Academic Press, pp. 585–603.

Cao, C., 2017. Sustainability and life assessment of high strength natural fibre composites in construction. In: Advanced High Strength Natural Fibre Composites in Construction. Woodhead Publishing, pp. 529–544.

Carus, M., 2017. Bio-based economy and climate change - important links, pitfalls, and opportunities. Ind. Biotechnol. 13 (2), 41–51. https://doi.org/10.1089/ind.2017.29073.mca.

Cecchin, A., Salomone, R., Deutz, P., Raggi, A., Cutaia, L., 2020. Relating industrial symbiosis and circular economy to the sustainable development debate. In: Salomone, R., Cecchin, A., Deutz, P., Raggi, A., Cutaia, L. (Eds.), Industrial Symbiosis for the Circular Economy. Strategies for Sustainability. Springer.

CGRi, 2020. Circularity Gap Reporting Initiative, Measuring and Mapping Circularity, Technical Methodology Document. https://www.circularity-gap.world/methodology. (Accessed 30 May 2020).

Cheban, I., Dibrova, A., 2020. Development of liquid biofuel market: impact assessment of the new support system in Ukraine. J. Int. Stud. 13 (1).

Chen, Z., Huang, L., 2019. Application review of LCA (life cycle assessment) in circular economy: from the perspective of PSS (product service system). Procedia CIRP 83, 210–217.

Corrado, S., Sala, S., 2018. Bio-economy contribution to circular economy. In: Designing Sustainable Technologies, Products and Policies. Springer, Cham, pp. 49–59.

Curran, M.A., 2017. Overview of goal and scope definition in life cycle assessment. In: Goal and Scope Definition in Life Cycle Assessment. Springer, Dordrecht, pp. 1–62.

D'amato, D., Gaio, M., Semenzin, E., 2020. A review of LCA assessments of forest-based bioeconomy products and processes under an ecosystem services perspective. Sci. Total Environ. 706, 135859.

Dahiya, S., Kumar, A.N., Sravan, J.S., Chatterjee, S., Sarkar, O., Mohan, S.V., 2018. Food waste biorefinery: sustainable strategy for circular bioeconomy. Bioresour. Technol. 248, 2–12.

Dahiya, S., Lakshminarayanan, S., Venkata Mohan, S., 2020. Steering acidogenesis towards selective propionic acid production using co-factors and evaluating environmental sustainability. Chem. Eng. J., 122135.

Edjabou, M.E., Jensen, M.B., Götze, R., Pivnenko, K., Petersen, C., Scheutz, C., Astrup, T.F., 2015. Municipal solid waste composition: sampling methodology, statistical analyses, and case study evaluation. Waste Manag. 36, 12–23.

Fauzi, R.T., Lavoie, P., Sorelli, L., Heidari, M.D., Amor, B., 2019. Exploring the current challenges and opportunities of life cycle sustainability assessment. Sustainability 11 (3), 636.

Ferreira, F.V., Mariano, M., Rabelo, S.C., Gouveia, R.F., Lona, L.M.F., 2018. Isolation and surface modification of cellulose nanocrystals from sugarcane bagasse waste: from a micro-to a nano-scale view. Appl. Surf. Sci. 436, 1113–1122.

Figge, F., Thorpe, A., Gutberlet, M., 2023. Definitions of the circular economy-circularity matters. Ecol, Econ., p. 208.

Gallardo, A., Carlos, M., Peris, M., Colomer, F.J., 2015. Methodology to design a municipal solid waste generation and composition map: a case study. Waste Manag. 36, 1–11.

Heimann, T., 2019. Bioeconomy and SDGs: Does the bioeconomy support the achievement of the SDGs? Earth's Future 7, 43–57.

Kardung, M., Costenoble, O., Dammer, L., Delahaye, R., Lovrić, M., van Leeuwen, M., M'Barek, R., van Meijl, H., Piotrowski, S., Ronzon, T., Verhoog, D., Verkerk, H., Vrachioli, M., Wesseler, J., Xinqi Zhu, B., 2019. Framework for measuring the size and development of the bioeconomy. In: BioMonitor Deliverable 1. Available at: www.biomonitor.eu. (Accessed 26 May 2020).

Kardung, M., Wesseler, J., 2019. EU bio-based economy strategy. In: Dries, L., Heijman, W., Jongeneel, R., Purnhagen, K., Wesseler, J. (Eds.), EU Bioeconomy Economics and Policies, Palgrave Advances in Bioeconomy: Economics and Policies. vol. II. Palgrave, Cham, pp. 277–292, https://doi.org/10.1007/978-3-030-28642-2_15.

Katakojwala, R., Kumar, A.N., Chakraborty, D., Mohan, S.V., 2019. Valorization of sugarcane waste: prospects of a biorefinery. In: Industrial and Municipal Sludge. Butterworth-Heinemann, pp. 47–60, https://doi.org/10.1016/B978-0-12-815907-1.00003-9.

Katakojwala, R., Venkata Mohan, S., 2020a. Microcrystalline cellulose production from sugarcane bagasse: sustainable process development and life cycle assessment. J. Clean. Prod. 249, 119342.

Katakojwala, R., Venkata Mohan, S., 2020b. A critical view on the environmental sustainability of biorefinery systems. Curr. Opin. Green Sustain. Chem. 27, 100392.

Kaur, R., Gera, P., Jha, M.K., Bhaskar, T., 2023. Hydrothermal treatment of pretreated castor residue for the production of bio-oil. BioEnergy Res. 16 (1), 517–527.

Grim, R.G., Huang, Z., Guarnieri, M.T., Ferrell, J.R., Tao, L., Schaidle, J.A., 2020. Transforming the carbon economy: challenges and opportunities in the convergence of low-cost electricity and reductive CO_2 utilization. Energy Environ. Sci. 13 (2), 472–494.

Labunska, I., Abdallah, M.A.E., Eulaers, I., Covaci, A., Tao, F., Wang, M., Santillo, D., Johnston, P., Harrad, S., 2015. Human dietary intake of organohalogen contaminants at e-waste recycling sites in Eastern China. Environ. Int. 74, 209–220.

Moncada, J., Cardona, C.A., Rincon, L.E., 2015. Design and analysis of a second and third generation Biorefinery: the case of castor bean and microalgae. Bioresour. Technol. 198, 836–843.

Naik, S.N., Saxena, D.K., Dole, B.R., Khare, S.K., 2018. Potential and perspective of castor biorefinery. In: Waste Biorefinery. Elsevier, pp. 623–656.

Nicolaidis Lindqvist, A., Broberg, S., Tufvesson, L., Khalil, S., Prade, T., 2019. Bio-based production systems: why environmental assessment needs to include supporting systems. Sustainability 11 (17), 4678.

Nnaji, C.C., 2015. Status of municipal solid waste generation and disposal in Nigeria. Manag. Environ. Qual. 26 (1), 53–71.

Prasad, M.N.V., 2021. Biomass Energy Sources and Conversion Technologies for Production of Biofuels. Alternative Energy Resources: The Way to a Sustainable Modern Society, pp.115–132.

Ravindra, K., Kaur, K., Mor, S., 2015. System analysis of municipal solid waste management in Chandigarh and minimization practices for cleaner emissions. J. Clean. Prod. 89, 251–256.

Román-Figueroa, C., Cea, M., Paneque, M., 2023. Industrial oilseed crops in Chile: current situation and future potential. Biofuels Bioprod. Biorefin. 17 (1), 273–290.

Statista, 2021. Production capacity of ethylene glycol worldwide from 2014 to 2024. Statista. https://www.statista.com/statistics/1067418/global-ethylene-glycol-production-capacity/.

Tan, E.C., Lamers, P., 2021. Circular bioeconomy concepts—a perspective. Front. Sustain. 2, 701509.

Zhang, Y., Wu, Z., Shi, L., Dai, L., Liu, R., Zhang, L., Lyu, B., Zhao, S., Thakur, V.K., 2023. From Castor Oil-Based Multifunctional Polyols to Waterborne Polyurethanes: Synthesis and Properties. Macromolecular Materials and Engineering., p. 2200662.

Production of biodiesel feedstock from trace element-contaminated lands in Ukraine

Oksana Sytar[a,b] and Majeti Narasimha Vara Prasad[c]

[a]Department of Plant Biology, Educational and Scientific Center "Institute of Biology and Medicine", Kyiv National University of Taras Shevchenko, Kyiv, Ukraine [b]Institute of Plant and Environmental Sciences, Slovak Agricultural University in Nitra, Nitra, Slovakia [c]Department of Plant Sciences, School of Life Sciences, University of Hyderabad (An Institution of Eminence), Hyderabad, Telangana, India

1 Introduction

The total area of Ukraine is about 60.37 million hectares, out of which agricultural land makes up 41.76 million hectares. Ukraine also has several rivers and lakes. Ukraine was suggested on cereal production within Europe as a country with a great potential to enchance production (Ryabchenko and Nonhebel, 2016). The prominent crops cultivated for grain production are wheat, oats, corn, barley, and buckwheat (Nekhay, 2012). In recent years, about 9% of the world's buckwheat production has come from Ukraine (Romanovskaja et al., 2022). Flax, hop, sugar beet, and potato are other important crops in Ukraine (Lerman et al., 1994). Soybean, rapeseed, and sunflower production has increased in recent years, boosting the agro-based economy (Fig. 1). In recent years, these three oilseeds—sunflower, rapeseed, and soybeans—occupied on average about 22% of the structure of sown areas, which is beyond scientifically substantiated limits of crop rotation saturation with these crops. Such crops are the most profitable, but they carry 3–10 times more nutrients than corn for grain (Slobodianyk et al., 2021).

After Chernobyl accident in 1986, trace element pollution and contamination in Ukraine received considerable attention. According to a recent study, crops grown near Chernobyl are still contaminated. Analysis of wheat, rye, oats, and barley grains found that concentrations of radioactive isotopes—cesium 137 and/or strontium 90—were above Ukraine's

FIG. 1 Primary feed stock for production of biodiesel in Ukraine. *Brassica* and *Helianthus* are acknowledged for their potential to polish contaminated soils. The harvested produce is ideal feed stock for bioenergy. (A) *Brassica napus* (rape seed). (B) *Helianthus annus* (sunflower). (C) *Glycine max* (soybean). (D) *Camelina sativa* (false flax).

official safe limits in almost half of samples (Labunska et al., 2021). Analysis of satellite images found a large number of burning coal waste dumps with high thermal activity in the Donetsk Coal Basin of Ukraine (Nádudvari et al., 2021). For more than 50 years, the largest Hg production facility of the former Soviet Union operated in Horlivka, a city in the Donetsk region of approximately 300,000 residents in the Donets Basin. During this period, about 30,000 metric tons of Hg were produced from ore extracted from the adjacent Mykytivka mines.

The contaminated land is unsuitable for food production. The use of such lands have extra challenges with appropriate agrotechnological interventions for establishing the sustainability and safety of an applicable production system. There are also other cases to examine, such as cost–benefit analysis, the probable entry of pollutants into plant products, and the certification of such products and their marketing, to complete the large-scale use of polluted lands (Abhilash et al., 2016; Shayanmehr et al., 2022). Contaminated lands could be used for the production of bioenergy for a little economic return. The problem of competition for fertile land can be alleviated by using contaminated land for the production of bioenergy plants.

This chapter review presents the current updated situation of heavy metals pollution in Ukraine territory and the possibility of using oilseed crops (rapeseed, soybean, and sunflower) to clean up 50% of the areas contaminated by heavy metals.

2 Monitoring heavy metal pollution in Ukraine

The concentration of heavy metals As, Cd, Cr, Cu, Hg, Pb, Zn, Sb, Co, and Ni in the soil of the European Union (EU) was estimated. Concentration of heavy metals in topsoil of the EU was calculated, and an estimated 6.24% or 137,000 km^2 of agricultural land needs local assessment and eventual remediation action (Tóth et al., 2016). It is known that problems of diffuse soil contamination is greatest in Eastern Europe when compared to other European countries. A high contamination of heavy metals is shown in the territory of Ukraine, especially in the

Chernobyl area. Unfortunately, international literature is still missing a detailed analysis of heavy metals contamination in Ukraine territory. Since 2022, it is difficult to assess heavy metals contamination due to the military conflict in Ukraine. But there is likely a constant increase in heavy metal contamination around regions with aggressive and violent conflict (Anyanwu et al., 2018). Warfare is associated with significant heavy metal contamination of the environment due to destruction of built infrastructure, consequent release of heavy metals, and direct contamination from exploded ordnance and leakage from unexploded ordnance (Bazzi et al., 2020).

In a scientific article from 2020, Czech and Ukrainian ecologists found increased levels of heavy metals in the rivers and soils of Pavlograd city, Dnipropetrovsk region (Kharytonov et al., 2020). Savosko et al. (2021a, b) established an effective predictive model developed for heavy metals inputs to soils of the industrial region of Kryvyi Rih District (Dnipropetrovsk region, Central Ukraine). It was shown that anthropogenic flows are higher in Fe and Pb inputs (60%–99%). The natural flows show significant higher levels of Ni and Cd inputs (55%–95%). For Cu, Mn, and Zn inputs, the alternate dominance of anthropogenic and natural and flows are defined (Savosko et al., 2021a). The measured content of gross forms of iron, manganese, copper, cadmium, and partially zinc in the soils of different monitoring sites in the soil on devastated lands at Kryvyi Rih Iron Mining & Metallurgical District (Ukraine) exceeded control values by 5.5–5.9 times (Savosko et al., 2021b).

Identifying non-agricultural marginal lands as a route to sustainable bioenergy provision is one of the tasks of modern ecophysiology. Non-agricultural marginal lands may be 15%–24% of urban regions (Mellor et al., 2021). This suits the Kyiv, Dnipropetrovsk, Kharkiv, Lviv, Odessa, etc. regions as all big cities with suburb areas in Ukraine are characterized by parameters of urban regions. The use of such areas for bioenergy may help to avoid food competition and possible land use conflicts via economic benefits by the development "green energy" urban areas. Bioenergy may give us environmental, economic, and social co-benefits. The research of plant's role in bioenergy has not always expressed the full range of sustainability and often focuses on the most evident environmental and economic interests, such as energy access, economic development, climate change, and sustainable production and innovation (Röder et al., 2020).

About 90% of toxic metals accumulate from the atmosphere into soil, where they migrate into groundwater, are absorbed by plants, and get into the trophic chains (Yan et al., 2020). A coefficient of load factor has been established that, in some regions of Ukraine, a level of metal pollution of soil, including mobile forms, exceed permissible level by 2–14 times (Nikolaychuk and Hrabovsky, 2000). In particular, the content of lead in soils of northern agricultural regions (Zhytomyr, Sumy, Rivne, Chernihiv, and Kyiv) exceeds the permissible level by 3–9 times (Shestapalov et al., 1996; Sytar and Prasad, 2016). A similar tendency is observed for soil in the region of Kyiv Polesye, where a concentration of metal was three times greater than the natural level (Shestapalov et al., 1996; Sytar and Prasad, 2016) (Fig. 2). A high level of Pb was observed in gray forest soils of the Vinnitsa region as well (Razanov et al., 2021).

Pb and Cd are considered the most dangerous heavy metals. Modeling of trace elements and heavy metals content in the Steppe Soils of Ukraine was done via qualitative interpretation of spatial and graphical information and modeling results. The content of Cd in the soils of the Kherson region (South Ukraine) was in the range of 0.02–0.42 mg kg^{-1}. The Cd content, which corresponds to a value of $0.20 < Cd < 0.40$ mg kg^{-1}, characterizes 48.8% of the area of agricultural land of Kherson region. During the research period, the value of Pb content in the soils of the Kherson region did not exceed the value of the maximum permissible

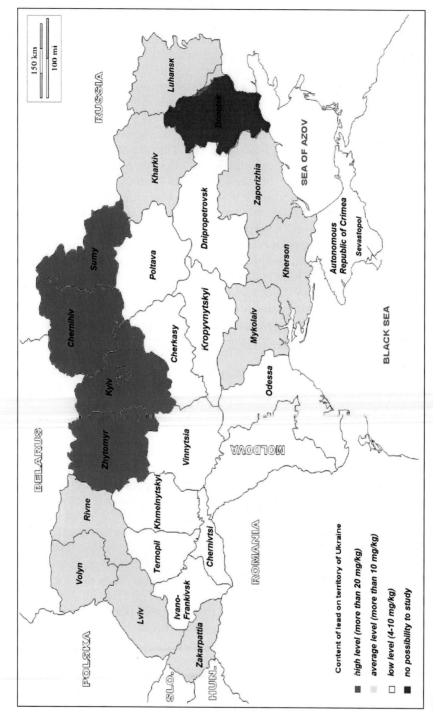

FIG. 2 The map of lead pollution in Ukraine.

concentration—6.0 mg kg^{-1} of soil. The cause of soil contamination of agricultural lands appears to be due to the unregulated use of agrochemicals (Breus and Yevtushenko, 2023).

Local background levels of heavy metals are greatly exceeded in snow close to industrial regions. Cd and Mo accumulate in forest soils in the Ukrainian Carpathians region (Shparyk and Parpan, 2004) (Fig. 3).

A similar situation has been observed with with Zn and Co pollution of soil. The maximal permissible level of Zn and Co in the air is 300 and 50 Clarke, respectively. (A Clarke is a unit of each element, and in soil, it is 50 and 8 mg kg^{-1}, respectively.) In general, for these three heavy metals (Ni, Zn, and Co), their content in the soil in many areas of Ukrainian territory is less than the maximal permissible level (Figs. 4 and 5). For example, in gray forest soils of Vinnitsa region, the levels of Zn and Co were 32.5 and 12.0 mg kg^{-1}, respectively (Razanov et al., 2021). In the Kherson region were low levels of trace elements: Zn in the soil of the region was estimated at a low level: 0.3–1.5 mg kg^{-1}, and cobalt: 0.06 g kg^{-1} (low) to 0.81 g kg^{-1} (high) (Breus and Yevtushenko, 2023).

It is important to reveal that violent conflict in Ukraine in 2022 may cayuse an increase in heavy metals pollution of already polluted areas with Pb, Ni, Cr, and Cu. Such tendency is notably critical for Ukrainian regions with high level of heavy metals level where military conflict is in an active phase: Donetsk, Zaporizhia, Kyiv, Lugansk, Sumy, Kharkiv, Kherson, Chernihiv, Mykolaiv, Zhytomyr, Dnipropetrovsk, and Odessa regions. Near 70% of planted areas for major exported plants in the areas of conflict may be under additional increased effects of heavy metals. These are wheat, corn, barley, rapeseed, and sunflower. It is suggested to use novel technologies of phytoremediation together with crop rotation of plant hyperaccumulators of heavy metals at these areas afterward (Sytar and Taran, 2022).

3 Economical background of biodiesel production in Ukraine and in the world

Ukraine is world famous for its humus-rich black soil. After the dissolution of the Soviet Union with ongoing land reform and decreasing activity in the agriculture sector, Ukraine is making full use of its agricultural potential (Schaffartzik et al., 2014).

Ukraine is often singled out for its potential as a supplier of agricultural goods to the EU region; 42 of 60 Mha of Ukraine can be used for crop production (Elbersen et al., 2009). At 0.7 ha per person (ha/cap), Ukraine has more land available than any other European country, except for Russia (0.9 ha/cap), significantly surpassing France (0.3 ha/cap) and Germany (0.1 ha/cap) (FAOSTAT, 2012), for example. The country's geographical location as a neighbor to the countries of the European Union and its access to the Black Sea is additionally advantageous to its role as a global supplier of biomass resources. The study with viability and sustainability assessment of bioenergy value chains on underutilized lands in the EU and Ukraine showed good results regarding profitability and greenhouse gas emissions for bioethanol production from willow (*Salix* sp.) in Ukraine. The miscanthus can be used for heat and power production and grass for biogas and chemicals production in Germany (Khawaja et al., 2021). A logistic scheme from producer to consumer for switchgrass and giant miscanthus biomass cultivation, including the use and supply of biomass from biomass energy crops, has been proposed. It was observed that switchgrass and giant miscanthus from

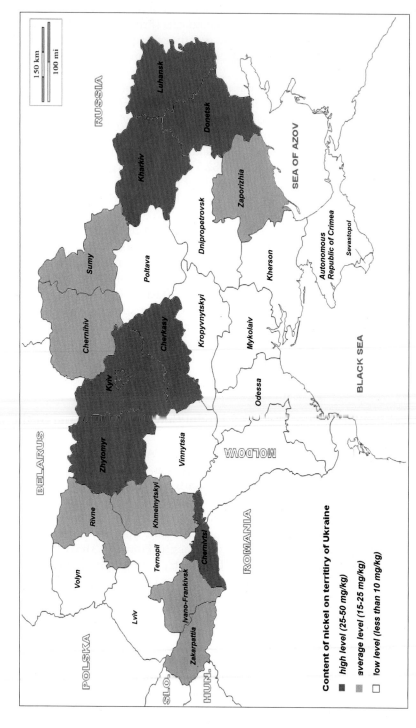

FIG. 3 The map of nickel pollution in Ukraine.

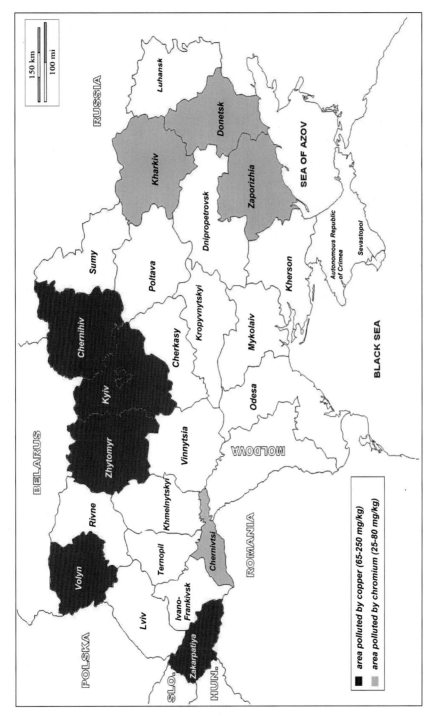

FIG. 4 The map of copper pollution in Ukraine.

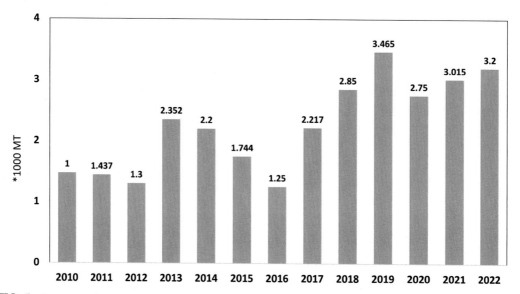

FIG. 5 Rapeseed oil production by year in Ukraine. The diagram is created from data from United States Department of Ukraine: https://www.indexmundi.com/agriculture/?country=ua&commodity=rapeseed-oilseed&graph=production.

the third to fifth year of vegetation form a high yield of dry biomass (up to 15.2 and 18.8 t/ha, respectively) with a maximum level of production profitability, up to 108.6% and 128.1%, providing high indicators of biofuel output (up to 18.2 and 24.0 t/ha) and energy (up to 313.0 and 396.0 GJ/ha) (Kulyk et al., 2020).

Rheay et al. (2021) discussed potential uses of hemp (*Cannabis sativa* L.) for paired phytoremediation and bioenergy production. The literature shows that hemp cultivation was successful at the field scale for phytoremediation and use of it in several bioenergy conversion technologies as well. Due to strict requirements, farmers grow only local industrial hemp seed that is registered in the Ukrainian Registry of Plant Varieties. The Registry contains 12 industrial hemp varieties at the time of the report writing. Farmers growing hemp have a limited number of markets to sell their product (either for seeds or fiber), taking into account that THC turnover in any form is prohibited in Ukraine by Regulation #770. Unfortunately. Total Harvested Area of hemp from 2016 to 2019 has been decreased from 2800 to 1000 ha. This same decrease was observed for seed and straw yield (Sobolev, 2020).

Ukraine is a major exporter of agricultural primary products. In 2010 and in physical units, it ranked second among the international top exporters of barley, third among the exporters of rapeseed, seventh for exports of sunflower seed, and ninth among the exporters of wheat. Rapeseed production and exports drastically increased from 2004 onward and grew by a factor of almost 20 until 2008 (FAOSTAT, 2012). From 2013 to 2017, Ukraine supported less cultivation of raw materials for production of biofuel. Although, for agricultural enterprises, since 2019, incomplete compensation for the bioethanol cost and biomass electricity equipment was covered by domestic producers. The cost of state support under such programs for 2019 is 0.9 billion UAH (Gołębiewski and Kucher, 2020).

For the full development of the market of liquid biofuels in Ukraine, there is a need to create a strong raw material base and proper infrastructure to guarantee the storage, processing, transportation, and sale of final products. However, there is a lack of a functional public policy system that can stimulate the production of bioethanol, biodiesel, and their use. The practical question of developing a financial support program for producers of this type of biofuel is important to ensure greater consumer appeal over traditional fuels. It is estimated that to meet the indicators of the National Action Plan of Ukraine for bioethanol and biodiesel production, it is crucial to merge its ability and put modern processing enterprises into operation (Gołębiewski and Kucher, 2020).

A few plants with heavy metal accumulating potential such as *Arundo donax*, *Brassica juncea*, *Jatropha curcas*, *Miscanthus* species, *Ricinus communis*, and *Salix* species have been proposed to use for cleaning polluted lands, and due to high biomass production, they can be used for the production of bioenergy. This is the dual advantage of phytoremediation as well as good economic return in terms of production of bioenergy (Tavakoli-Hashjini et al., 2020; Sameena and Puthur, 2021).

Soybean oil is used in Spain, France, Germany, Portugal, and Italy. The use of soybean oil in conventional biodiesel is, however, limited mainly because of its viscosity and winter operability. In the EU, rapeseed oil is the main biodiesel feedstock (Bušić et al., 2018).

In Ukraine, sunflower, rapeseed, soybean, and false flax are the most widespread oilseed crops that can be used for phytoremediation processes and the production of biodiesel and biogas. In Eastern Europe, the problems of diffuse soil contamination are greatest in Azerbaijan, Belarus, Moldova, Russia, and Ukraine (Denisov et al., 1997). From the total cultivation area of these crops in Ukraine, the area of heavy metal-contaminated soils is approximately 15%–20%.

At the same time, biofuels are currently not produced or consumed in relevant amounts within Ukraine. Yet in Ukraine, the potential is mainly as an alternative to liquid fossil fuel use, not to use in place of natural gas and coal in industrial applications and electricity generation. The production of biofuel requires energy inputs. But the biggest advantage that biodiesel has over gasoline and petroleum diesel is its environmental friendliness. Biodiesel burns similar to petroleum diesel with regard to regulated pollutants. On the other hand, biodiesel probably has better efficiency than gasoline. One such fuel for compression ignition engines that exhibit great potential is biodiesel. Diesel fuel can also be replaced by biodiesel made from vegetable oils. Biodiesel is now mainly being produced from soybean, rapeseed, and palm oils.

The higher heating values (HHVs) of biodiesels are relatively high. The HHVs of biodiesels (39–41 MJ/kg) are slightly lower than that of gasoline (46 MJ/kg), petrol diesel (43 MJ/kg), or petroleum (42 MJ/kg), but higher than coal (32–37 MJ/kg). Biodiesel has more than double the price of petrol diesel (Demirbas, 2007). The major economic factor to consider for input costs of biodiesel production is the feedstock, which is about 80% of the total operating cost. The high price of biodiesel is due in large part to the high price of the feedstock. Economic benefits of a biodiesel industry would include value added to the feedstock, an increased number of rural manufacturing jobs, and increased income taxes and investments in plant and equipment.

From 2013 to 2017, Ukraine did not support the cultivation of raw materials for biofuel production. However, for agricultural enterprises, since 2019, partial compensation for the cost of bioethanol and biomass electricity equipment purchased from domestic producers has been introduced. The amount of state support for 2019, under this program, is 0.9 billion UAH (CMU Cabinet of Ministers of Ukraine, 2019).

There are currently about 13 bioethanol producers in the Ukraine. Six of them work quite successfully. Last year, they produced 80,000 tons of bioethanol, which went into alternative fuel production. Expected demand in the Ukrainian market will be more than 300,000 tons of bioethanol with a 5% additive in gasoline. With an increase in bioethanol content of up to 7%, demand is expected to reach 450,000 tons with alternative fuels. In the Ukraine, molasses and corn are the most effective raw materials for bioethanol production (Gołębiewski and Kucher, 2020).

The production and utilization of biodiesel is facilitated primarily through the agricultural policy of subsidizing the cultivation of non-food crops (rapeseed, for example) (Di Lucia and Nilsson, 2007). Second, biodiesel is exempt from the oil tax. The European Union in 2005 used nearly 89% of all biodiesel production worldwide. Germany produced 1.9 billion liters, which is more than half the world total. In Germany, biodiesel is also sold at a lower price than fossil diesel fuel. Biodiesel is treated like any other vehicle fuel in the UK. Among EU countries, France, Spain, Germany, Netherlands, and Poland are currently the largest producers of biofuel. The main raw material for its production in these countries is rapeseed (Gołębiewski and Kucher, 2020). All other countries combined account for only 11% of world biodiesel consumption. By 2010, the United States is expected to become the world's largest single biodiesel market, accounting for roughly 18% of world biodiesel consumption, followed by Germany (Demirbas, 2007; Naylora and Higginsa, 2018). For Ukraine, with its very low share of renewables in total primary energy supply (TPES), the promise of reduced environmental burdens is especially appealing (Schaffartzik et al., 2014). The rise in production of biofuel was 9.7% in 2018. Indonesia and Brazil together account for almost two-thirds of the global increase in production of biofuel. The biggest producers of bioethanol and biodiesel are the United States (40% of world production) and Brazil (22.4%), which are constantly increasing their production (Gołębiewski and Kucher, 2020). Brazil and the United States are the dominant industrial players for biofuel production driven by government support (Bajpai, 2021).

Figs. 5–7 shows the yield patterns of oilseeds. Also, based on former trends, oilseed yields are projected to grow faster than cereals yields.

As a producer of rapeseed (and other crops) for the global market and thus as a producer of biofuel feedstock for other economies, Ukraine can follow the trajectory of other economies dependent on primary commodity exports (Schaffartzik et al., 2014).

4 Biodiesel produced from oilseed crops

Triacylglycerols (TAGs) produced by plants are one of the most energy-rich and abundant forms of reduced carbon available from nature. Most plant oils are derived from TAGs stored in seeds (Demirbas, 2009; Srivastava and Prasad, 2000) . During seed development, photosynthate from the mother plant is imported in the form of sugars, and the seed converts these into precursors of fatty acid biosynthesis. Given their chemical similarities, plant oils represent a logical substitute for conventional diesel, a non-renewable energy source (Bhat et al., 2022).

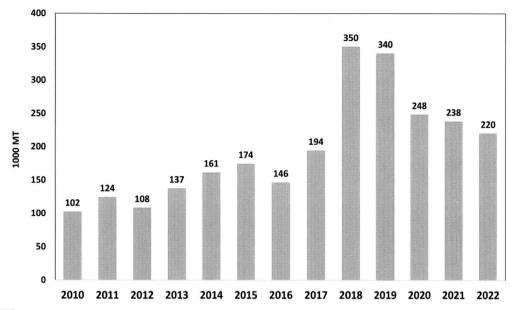

FIG. 6 Ukraine soybean oil production by year. The diagram is created from data from United States Department of Ukraine: https://www.indexmundi.com/agriculture/?country=ua&commodity=rapeseed-oilseed&graph= production.

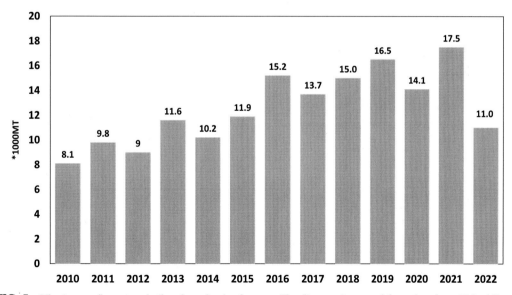

FIG. 7 Ukraine sunflower seed oilseed production by year. The diagram is created from data from United States Department of Ukraine: https://www.indexmundi.com/agriculture/?country=ua&commodity=rapeseed-oil seed&graph=production.

However, as plant oils are too viscous for use in modern diesel engines, they are converted to fatty acid esters. Most plant TAGs have a viscosity range that is much higher than that of conventional diesel: $17.3–32.9\,mm^2\,s^{-1}$ compared to $1.9–4.1\,mm^2\,s^{-1}$, respectively (Knothe and Steidley, 2005). The fatty acid methyl esters (FAMEs) found in biodiesel has a high energy density as reflected by their high heat of combustion, which is similar, if not greater, than that of conventional diesel (Knothe and Steidley, 2005). Similarly, the cetane number (a measure of diesel ignition quality) of the FAMEs found in biodiesel exceeds that of conventional diesel (Knothe and Steidley, 2005).

The higher oxygenated state compared to conventional diesel leads to lower carbon monoxide (CO) production and reduced emission of particulate matter (Graboski and McCormick, 1998). This latter air pollutant is especially problematic in European cities, motivating temporary curfews for diesel-powered vehicles. Biodiesel also contains little to no sulfur or aromatic compounds; in conventional diesel, the former contributes to the formation of sulfur oxide and sulfuric acid, while the aromatic compounds also increase particulate emissions and are considered carcinogens. In addition to the reduced CO and particulate emissions, the use of biodiesel confers additional advantages, including a higher flashpoint, faster biodegradation, and greater lubricity ((Demirbas, 2007).

The emissions reduction index from the United States National Biodiesel Board displayed that the full explosion of biodiesel as a fuel transportation decreased carbon, total hydrocarbons, polycyclic aromatic hydrocarbons, and sulfur emissions by 48%, 67%, 80%, and 100%, respectively (Ogunkunle and Ahmed, 2021). Therefore, plant oil production needs to be greatly increased for biodiesel to replace a major proportion of the current and future fuel needs of the world.

The source of biodiesel usually depends on the crops amenable to the regional climate. In South and North America, soybean oil is the most commonly biodiesel feedstock, whereas rapeseed oil is the most regular source in European Union where its share of the total area of oilseed crops is somewhat more than 80% (Bala-Litwiniak and Radomiak, 2019). Biodiesel is a monoalkyl ester of long chain fatty acids derived from renewable feedstocks, such as vegetable oil or animal fats, for use in the compression ignition engine. Biodiesel is composed of FAMEs that can be prepared from triglycerides in vegetable oils by transesterification with methanol. The biodiesel is quite similar to conventional diesel fuel in its main characteristics (Meher et al., 2006). In the United States, soybean oil is the most commonly biodiesel feedstock, whereas rapeseed (canola) oil and palm oil are the most common source for biodiesel in Europe and tropical countries, respectively (Knothe, 2002). However, any vegetable oil—corn, cottonseed, peanut, sunflower, safflower, coconut, or palm—can be used to produce biodiesel (Demirbas, 2007; Madadi et al., 2017). From a chemical point of view, oils from different sources have different fatty acid compositions.

Twenty-one fatty acids were screened in all the samples. The fatty acids commonly found in vegetable oil and fat are stearic, palmitic, oleic, and linoleic. The other fatty acids that are also present in many of the oils and fats are myristic (tetradecanoic), palmitoleic, acachidic, linolenic, and octadecatetranoic (Amat Sairin et al., 2022). There are many other fatty acids that are also found in oils with the previously mentioned common fatty acids. Erucle fatty acid is found only in three oils, crambe, camellia oil, and *Brassica carinata*.

In the oil of sunflower and soybean, the content of the fatty acid is different in the same plant species, which may be either due to the varietal or instrumental difference or in the different parts of plants. These non-edible vegetable oils (2G) can be used for biofuels production. To choose a crop for 2G oil production, further bioethanol production can be connected with country specification in crop agriculture. For example, in Bangladesh, among five major crops, rice residue alone can produce 71% of bioethanol (Mahmud et al., 2022).

Chemically, the oil/fats consist of 90%–98% triglycerides and small amount of mono- and diglycerides. Triglycerides are esters of three fatty acids and one glycerol. These contain a substantial amount of oxygen in their structures. When three fatty acids are identical, the product is simple triglycerides; when they are dissimilar, the product is mixed triglycerides fatty acids, which are fully saturated with hydrogen having no double bonds. Fatty acids with one missing hydrogen molecule and a one double bond between carbon atoms called monosaturated. The fatty acids that have more than one missing hydrogen or have more than one double bond are called polyunsaturated. Fully saturated triglycerides lead to excessive carbon deposits in engines. The fatty acids are different in relation to the chain length, degree of unsaturation, or presence of other chemical functions. Chemically, biodiesel is referred to as the mono-alkyl esters of long-chain-fatty acids derived from renewable lipid sources. Biodiesel is the name for a variety of ester-based oxygenated fuel from renewable biological sources. It can be used in compression ignition engines with little or no modifications (Demirbas, 2005, 2009).

Biodiesel is made in a chemical process called transesterification, where organically derived oils (vegetable oils, animal fats, and recycled restaurant greases) are combined with methanol and ethanol (Shimasaki, 2020). Such mixture is chemically altered to form fatty esters, such as methyl ester. Biodiesel consists of alkyl (usually methyl) esters instead of the alkanes and aromatic hydrocarbons of petroleum-derived diesel.

5 Cleanup of heavy metal from soil by using oilseed crops

The ideal characteristics of an effective plant phytoremediator are fast growth, deep and extensive roots (phreatophytes), high biomass, and easy to harvest (Sytar et al., 2021).

The application of plants for phytoremediation has two advantages: the remediation of contaminated soil and the production of biofuel (Abdelsalam et al., 2019). Which plant species to use is the most important factor that needs to be satisfied. The following information is required for effective remediation and abundant production of biofuel: (1) heavy metal tolerance of species, (2) heavy metal accumulation capacity of the plants, and (3) biofuel production capacity per unit area. Not only sunflowers, but rapeseed and soybean have also been used as principal biofuel sources. Moreover, these plants have been studied for their potential in phytoremediation to clean up contaminated soils.

Phytostabilization using plant species useful for bioenergy production has recently received increased attention. However, the water requirement of most of these species is a limitation for their use in Mediterranean climatic conditions. At the same time, the potential of the plant's biomass grown in trace element-contaminated soils is proposed for crop rotation (Bernal et al., 2021).

Integrative effects of treated wastewater with the presence of trace elements and synthetic fertilizers on productivity, energy characteristics, and elements uptake of potential energy crops in an arid agro-ecosystem was studied. It was suggested that treated municipal wastewater can be used to irrigate crops grown for bioenergy purposes, since it did not pose any harmful effect for energy crops (Al-Suhaibani et al., 2021).

The possible use of *Miscanthus* species for phytoremediation of oil-contaminated lands in Poltava region, Ukraine was studied. It was found that, by enhancing concentration of oil pollution in the experimental areas after phytoremediation of *Miscanthus giganteus*, the metabolic acidity and pH of the soil solution increases, the quantity of absorbed bases decreases, and the potency of respiration significantly increased. *Miscanthus giganteus* plants create favorable conditions for the development of microorganisms and increase their activity as a consequence of the secretion of nutrients by the roots, improving soil properties (Pysarenko and Bezsonova, 2020).

The milk thistle, grown in agricultural crop rotations on gray forest soils of the right-bank Forest-Steppe of Ukraine, Vinnitsa region, showed high efficiency of removing heavy metals from the soil. The milk thistle makes removing from the soil per 1 ha at a yield of 28.5 t/ha of Pb—160 g, Cd—32 g, Zn—839.5 g, and Co—310 g (Razanov et al., 2021). Takase et al. (2014) discovered a method of biodiesel production from non-edible *Silybum marianum* oil using methanol under ultrasonication (Takase et al., 2014). The crude oil extracted from milk thistle seeds was employed as a precursor for creating biodiesel via the cosolvent esterification-transesterification process as well (Saeed et al., 2021). The use of milk thistle for biofuel production is not been developed further due to greater use of sunflower and rapeseed. [Regarding the use of sunflower (*Helianthus annuus* L.), please see Chapter (Sytar and Prasad, 2016) for additional information.]

Oilseed sunflower genotypes exhibit excellent arsenic removal ability. Speciation discovered greater concentration of arsenic in roots than shoots. There were 37 genotypes of oilseed sunflower measured under natural field conditions for ligno-cellulosic mass production of bioenergy, growth performance, and metal uptake capacity during the remediation of As-contaminated soil (Sahito et al., 2021).

Soybean roots can accumulate the highest value of Pb—294.8 mg kg^{-1}, Zn—644.4 mg kg^{-1}, Cu—34.4 mg kg^{-1}, and Cd—10.9 mg kg^{-1} (Angelova et al., 2003). The roots of soybean were fixed and accumulated a great part of the heavy metals that had entered the soil as soybean plants formed a powerful root system with strong absorbing ability in depth as well as in width, including a great volume of soil. Soybean roots can accumulate nickel in high quantity (Malan and Farrant, 1998). Nickel uptake by plants is positively correlated with the metal's concentrations in soil solution. At the same time, soybean plants that grow in vineyard soils with high heavy metal concentrations accumulated 50% more Zn in leaves and seeds, 70% more Cu in leaves, and 90% more Cu in seeds than those plants grown in grassland soils. Zn level in seeds was greater than the permissible limit (Schwalbert et al., 2021).

The concept of differential compartmentalization already occurs in natural cultivars of sunflowers. It was shown that Cu can be phytoextracted in leaves of sunflower. At the same time, oil may be used as biofuel and sunflower plant biomass may be composted as a Cu fertilizer specialty (Kolbas et al., 2011). Another alternative for sunflower biomass after the phytoremediation process is to use it as raw material for industrial saccharification intended for bioethanol production (Edgar et al., 2021)

It was shown that oilseed sunflower accessions exhibit excellent removal ability to arsenic with partitioning of arsenic species to plant organs differed significantly. Several oilseed sunflower accessions are rich sources of lignocellulosic mass for bioenergy production. Arsenic and mercury uptake and accumulation in oilseed sunflower accessions were selected to mitigate co-contaminated soil coupled with oil and bioenergy production (Sahito et al., 2021).

The increasing demand for energy and decreasing reserves of fossil fuels have turned the focus of governments onto energy production from biomass (biofuels, biogas) (Abreu et al., 2022). Contaminated land that is unsuitable for food production and of little or no economic return could be used for the production of bioenergy. Phytoremediation, the use of vegetation for in situ restoration of contaminated soils, is generally considered a cost-effective and environmentally friendly approach (Arthur et al., 2005; Barbosa et al., 2018). However, it is increasingly recognized that the success of phytoremediation depends on its capacity to produce valuable biomass (Conesa et al., 2012).

Sunflower would be a suitable crop for the production of bioenergy on contaminated land. Forty germplasms of sunflower were screened in field conditions for phytoremediation of Pb. All experimental genotypes consummated essential concentrations of Pb, with maximum levels in shoot > root > seed, respectively. The Pb content in the oil of GP.8585 was under the level of food safety standard of China, with 59.5% oleic acid and 32.1% linoleic acid (Zehra et al., 2020). It can be used for the production of biodiesel (seed oil), as well as for biogas (straw). Sunflower oil comprises 3% of the total raw materials for biodiesel production, and it is mainly used in France and Greece (Naylora and Higginsa, 2018). Since it has been used as a crop for centuries in Ukraine, cultivation techniques and infrastructure for harvesting and processing are already available.

However, this environmentally friendly phytoremediation of heavy metal-contaminated soils is still not commercialized or used extensively on a huge range (Hauptvogl et al., 2019). Results of a 2-year phytoremediation project on soils contaminated with As, Cd, and Pb concluded that costs were lower than for most other technologies, and the benefits of phytoremediation are expected to offset the project costs in less than 7 years (Wan et al., 2016). It is expected to become a commercially viable technology in the future.

Biochar mitigates abiotic stresses by developing the soil's physical and biological parameters (Bilias et al., 2021; Radziemska et al., 2022). Biochar is able to manage production of antioxidants under trace-metal stress. The biochar application also improves the synthesis of stressed proteins and proline contents in plants, thus maintaining the osmoprotectant and osmotic potential of the plant under contamination stress. It was suggested that biochar feedstock plays a significant role in immobilization and adsorption of trace metals (Haider et al., 2022).

Treatment with 10% biochar proved equally efficient in reducing metal concentrations in shoots, but the biomass production tripled as a result of the soil fertility improvement. Thus, in addition to C sequestration, the incorporation of biochar into metal-contaminated soils could make it possible to cultivate bioenergy crops without encroaching on agricultural lands. Although additional investigations are needed, we suggest that the harvested biomass might, in turn, be used as feedstock for pyrolysis to produce both bioenergy and new biochar, which could further contribute to the reduction of CO_2 emissions (Houben et al., 2013). For example, a diverse range of physicochemically distinct biochars made from a combination of different feedstock tissues and pyrolysis temperatures from a biodiesel plant *J. curcas* were studied. It was shown that leaf biochar was rich in nutrient minerals, and trunk biochar was rich in calcium (Konaka et al., 2021).

Trace element contamination/pollution in agrocoenose; application of environmental crops, e.g., sunflower, rapeseed, soybean; and phytomanagement of trace element-contaminated soils are possible scopes for the development of phytoproducts and establishment of sustainable agrocoenose. Evaluation of Ukrainian *Camelina sativa* germplasm productivity and analysis of its amenability for efficient biodiesel production showed that that average productivity of the analyzed cultivars is 1756 kg/ha in the Ukrainian Forest-Steppe zone and 1146 kg/ha in the Steppe zone. Evaluation of potential rates of FAME- and FAEE-based biodiesel production were found to be more sustainable and feasible in the Forest-Steppe zone rather than in the Steppe zone (Blume et al., 2022). The possible ecological use of *C. sativa* plants in the various industrial areas is presented in Figs. 8 and 9 and Table 1.

Today, various techniques such as clustered regularly interspaced, short palindromic repeats technique; additional use of microorganisms; specific immobilizing agents; and optimizations of the process steps and their validation in the production of biofuel are actively in use (Table 1).

It is shown that highest seed yield, protein, oil content, and biodiesel yield production is found in Austria, Saskatchewan Canada, Southern Ethiopia, and Northeastern Poland. The climate conditions for cultivation of *C. sativa* in Ukraine may be similar to Poland region as was shown in the Blume et al. (2022) research work (Fig. 9).

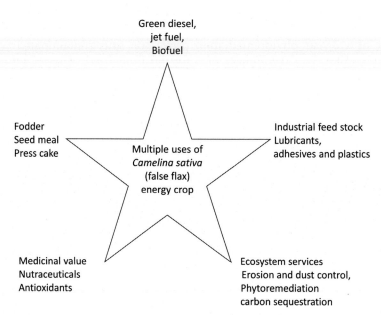

FIG. 8 Evaluation of Ukrainian *Camelina sativa* germplasm use as a crop for biodiesel production. *Source: Chaturvedi, S., Bhattacharya, A., Khare, S.K., Kaushik, G., 2017. C. sativa: An Emerging Biofuel Crop. Springer, Berlin/Heidelberg, Germany, pp. 1–38.*

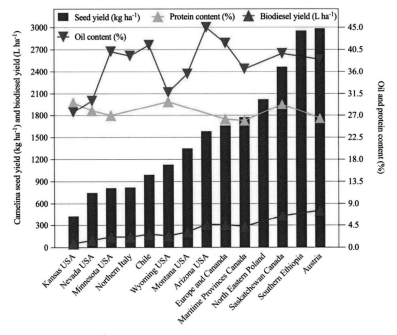

FIG. 9 Average seed yield (kg ha^{-1}), oil (%), and protein content (%) of *C. sativa* grown at various locations across the world. *Source: Neupane, D., Lohaus, R.H., Solomon, J.K.Q., Cushman, J.C., 2022. Realizing the potential of* Camelina sativa *as a bioenergy crop for a changing global climate. Plants 11, 772. https://doi.org/10.3390/plants110607.*

TABLE 1 *Camelina sativa* techniques as a bioenergy, bioremediation, industrial, and medicine use crop.

Use	Authors and year
Bioenergy Crop Foods for human consumption—*C. sativa* seeds and oil, such as chlorogenic, caffeic, sinapinic, and phytic acids; glucosinolates, which have anti-nutritive properties	Neupane et al. (2022)
Phytoremediation potential with *C. sativa* and *Enterobacter ludwigii* ZCR5 strain in association on petroleum hydrocarbon-contaminated soil	Pacwa-Płociniczak et al. (2023)
Technique of clustered regularly inter-spaced short palindromic repeats (CRISPR) for *C. sativa* plants and for phytoremediation	Bhattacharyya et al. (2022)
Zn and Cu availability and the impact of immobilizing agents in relation to potential health risk	Jakubus and Bakinowska (2022)
Optimizations of the process steps and their validation in the production of biofuel	Bhat et al. (2022)
Characterization and structural modeling of proteins involved in hyper accumulating fern's arsenic tolerance in plant and algal biomass	Deogam et al. (2022)
A sustainable foundation for second-generation biofuels and biobased chemicals is provided by low indirect land use change (ILUC)	Abreu et al. (2022)
Energy crops to bioenergy and biofuels plants	Lopez et al. (2022)
Possibilities in Europe for low indirect land use biomass for biofuels	Panoutsou et al. (2022)
Analysis of the productivity of Ukrainian *C. sativa* (False flax) germplasm and its suitability for effective biodiesel production	Blume et al. (2022)

6 Conclusions

The high concentration of heavy metal is harmful to human health through contaminated agricultural production. Most heavy metals would be accumulated in soil after a long period of time. This chapter raised the issue of heavy metal pollution in the world, more specifically in Ukraine. The most dangerous aspect of this pollution is high level lead and nickel in soils of Ukraine territory. Using heavy metal-tolerant plant species that are able to hyperaccumulate metals in plant roots and shoots is a proposed method of for cleanup of polluted soils. The widespread agriculture in Ukraine territory of soybean, rapeseed, and sunflower can be accumulators of heavy metals. But the harvest of these crops cannot be used for ecologically pure agricultural production according to international standards. Cleanup of contaminated soils will be a better use for this part of the harvest, such as biodiesel production. From an economic side, it can be profitable too. At the same time, it will also clean up the most contaminated soils in Ukraine with *C. sativa*. Cleanup of the soil by plants will be better than using an electrolysis method (where one puts the harvest into an oxide solution and uses electrical power to gather metal into a solid form).

Acknowledgments

Thanks are due to European Environment Agency and UNEP Regional Office in Europe for authorization to use their contaminated soil map.

References

Abdelsalam, I.M., Elshobary, M., Eladawy, M.M., Nagah, M., 2019. Utilization of multi-tasking non-edible plants for phytoremediation and bioenergy source—a review. Phyton 88 (2), 69.

Abhilash, P.C., Tripathi, V., Edrisi, S.A., et al., 2016. Sustainability of crop production from polluted lands. Energy Ecol. Environ. 1, 54–65. https://doi.org/10.1007/s40974-016-0007-x.

Abreu, M., Silva, L., Ribeiro, B., Ferreira, A., Alves, L., Paixão, S.M., Gouveia, L., Moura, P., Carvalheiro, F., Duarte, L.-C., Fernando, A.L., 2022. Low indirect land use change (ILUC) energy crops to bioenergy and biofuels—a review. Energies 15 (12), 4348.

Al-Suhaibani, N., Seleiman, M.F., El-Hendawy, S., Abdella, K., Alotaibi, M., Alderfasi, A., 2021. Integrative effects of treated wastewater and synthetic fertilizers on productivity, energy characteristics, and elements uptake of potential energy crops in an arid agro-ecosystem. Agronomy 11 (11), 2250. https://doi.org/10.3390/agronomy11112250.

Amat Sairin, M., Abd Aziz, S., Yoke Mun, C., Khaled, A.Y., Rokhani, F.Z., 2022. Analysis and prediction of the major fatty acids in vegetable oils using dielectric spectroscopy at 5–30 MHz. PLoS One 17 (5), e0268827. https://doi.org/10.1371/journal.pone.0268827.

Angelova, V., Ivanova, R., Ivanov, K., 2003. Accumulation of heavy metals in leguminous crops (bean, soybean, peas, lentils and gram). J. Environ. Prot. Ecol. 4, 787–795.

Anyanwu, B.O., Ezejiofor, A.N., Igweze, Z.N., Orisakwe, O.E., 2018. Heavy metal mixture exposure and effects in developing nations: an update. Toxics 6 (4), 65. https://doi.org/10.3390/toxics6040065.

Arthur, E.L., Rice, P.J., Rice, P.J., Anderson, T.A., Baladi, S.M., Henderson, K.L.D., Coats, J.R., 2005. Phytoremediation—an overview. Crit. Rev. Plant Sci. 24 (2), 109–122.

Bajpai, P., 2021. Developments in bioethanol. In: Series Title Green Energy and Technology. Springer, p. 222.

Bala-Litwiniak, A., Radomiak, H., 2019. Possibility of the utilization of waste glycerol as an addition to wood pellets. Waste Biomass Valoriz. 10, 2193–2199. https://doi.org/10.1007/s12649-018-0260-7.

Barbosa, B., Costa, J., Fernando, A.L., 2018. Production of energy crops in heavy metals contaminated land: opportunities and risks. In: Land Allocation for Biomass Crops. Springer, Cham, pp. 83–102.

Bazzi, W., Abou Fayad, A.G., Nasser, A., Haraoui, L.P., Dewachi, O., Abou-Sitta, G., Nguyen, V.K., Abara, A., Karah, N., Landecker, H., Knapp, C., McEvoy, M.M., Zaman, M.H., Higgins, P.G., Matar, G.M., 2020. Heavy metal toxicity in armed conflicts potentiates AMR in *A. baumannii* by selecting for antibiotic and heavy metal co-resistance mechanisms. Front. Microbiol. 11, 68. https://doi.org/10.3389/fmicb.2020.00068.

Bernal, M.P., Grippi, D., Clemente, R., 2021. Potential of the biomass of plants grown in trace element-contaminated soils under Mediterranean climatic conditions for bioenergy production. Agronomy 11 (9), 1750.

Bhat, R.A., Singh, D.V., Tonelli, F.M.P., Hakeem, K.R., 2022. Optimizations on steps involved on biofuel obtainment and their validation. In: Plant and Algae Biomass. Springer, Cham, pp. 107–125.

Bhattacharyya, N., Anand, U., Kumar, R., Ghorai, M., Aftab, T., Jha, N.K., Rajapaksha, A.U., Bundschuh, J., Bontempi, E., Dey, A., 2022. Phytoremediation and sequestration of soil metals using the CRISPR/Cas9 technology to modify plants: a review. Environ. Chem. Lett. 21, 1–17.

Bilias, F., Nikoli, T., Kalderis, D., Gasparatos, D., 2021. Towards a soil remediation strategy using biochar: effects on soil chemical properties and bioavailability of potentially toxic elements. Toxics 9 (8), 184. https://doi.org/10.3390/toxics9080184.

Blume, R.Y., Rakhmetov, D.B., Blume, Y.B., 2022. Evaluation of Ukrainian *Camelina sativa* germplasm productivity and analysis of its amenability for efficient biodiesel production. Ind. Crop. Prod. 187, 115477.

Breus, D., Yevtushenko, O., 2023. Agroecological assessment of suitability of the steppe soils of Ukraine for ecological farming. J. Ecol. Eng. 24 (5), 229–236. https://doi.org/10.12911/22998993/161761.

Bušić, A., Kundas, S., Morzak, G., Belskaya, H., Marđetko, N., Ivančić Šantek, M., Komes, D., Novak, S., Šantek, B., 2018. Recent trends in biodiesel and biogas production. Food Technol. Biotechnol. 56 (2), 152–173. https://doi.org/10.17113/ftb.56.02.18.5547.

CMU (Cabinet of Ministers of Ukraine), 2019. Resolution On Amendments to the Resolutions of the Cabinet of Ministers of Ukraine No. 77 of February 8, 2017 and No. 130 of January 1, 2017, No. 123, https://translate.google.com/translate?hl=ru&sl=ru&tl=uk&u=https%3A%2F%2Fza kon.rada.gov.ua%2Fgo%2F123-2019-%D0%BF, Access 10.12.2019.

Conesa, H.M., Evangelou, M.W.H., Robinson, B.H., Schulin, R., 2012. A critical view of current state of phytotechnologies to remediate soils: still a promising tool? Sci. World J., 173829 (10 pp.) https://doi.org/10.1100/2012/173829.

Demirbas, A., 2005. Biodiesel production from vegetable oils via catalytic and non-catalytic supercritical methanol transesterification methods. Prog. Energy Combust. Sci. 31 (5–6), 466–487. https://doi.org/10.1016/j.pecs.2005.09.001.

Demirbas, A., 2007. Importance of biodiesel as transportation fuel. Energy Policy 35 (9), 4661–4670. https://doi.org/10.1016/j.enpol.2007.04.003.

Demirbas, A., 2009. Progress and recent trends in biodiesel fuels. Energy Convers. Manag. 50, 14–34.

Denisov, N., Mnatsakanian, R.A., Semichaevsky, A.V., 1997. Environmental Reporting in Central and Eastern Europe: A Review of Selected. Central European University, Budapest, Hungary.

Deogam, R., Pipil, N.K., Chakraborty, N., Chatterjee, S., Purty, R.S., 2022. Characterization and structural modeling of proteins involved in arsenic tolerance of hyper accumulating fern. In: International Conference on Chemical, Bio and Environmental Engineering. Springer, Cham, pp. 415–427.

Di Lucia, L., Nilsson, L.J., 2007. Transport biofuels in the European Union: the state of play. Transp. Policy Elsevier 14 (6), 533–543. https://doi.org/10.1016/j.tranpol.2007.09.003.

Edgar, V.N., Fabián, F.L., Mario, P.C.J., Ileana, V.R., 2021. Coupling plant biomass derived from phytoremediation of potential toxic-metal-polluted soils to bioenergy production and high-value by-products—a review. Appl. Sci. 11 (7), 2982.

Elbersen, W., Wiersinga, R., Waarts, Y., 2009. Market Scan Bioenergy Ukraine. Report for the Dutch Ministry of Agriculture, Nature and Food Quality (Wageningen).

FAOSTAT, 2012. FAO Statistical Database. Food and Agriculture Organization (FAO), Rome. http://faostat.fao.org/site/573/default.aspx#ancor. (Accessed 18 June 2012).

Gołębiewski, J., Kucher, O., 2020. Development of the biofuel market in the Ukraine. Ann. PAAAE XXII (1), 104–112. https://doi.org/10.5604/01.3001.0013.7813.

Graboski, M.S., McCormick, R.L., 1998. Combustion of fat and vegetable oil derived fuels in diesel engines. Prog. Energy Combust. Sci. 24 (2), 125–164. https://doi.org/10.1016/S0360-1285(97)00034-8.

Haider, F.U., Wang, X., Farooq, M., Hussain, S., Cheema, S.A., Ain, N.U., Virk, A.L., Ejaz, M., Janyshova, U., Liqun, C., 2022. Biochar application for the remediation of trace metals in contaminated soils: implications for stress tolerance and crop production. Ecotoxicol. Environ. Saf. 230, 113165. https://doi.org/10.1016/j.ecoenv.2022.113165.

I. Bioproducts from contaminated soil and water

Hauptvogl, M., Kotrla, M., Prčík, M., Pauková, Ž., Kováčik, M., Lošák, T., 2019. Phytoremediation potential of fast-growing energy plants: challenges and perspectives—a review. Pol. J. Environ. Stud. 29 (1), 505–516.

Houben, D., Evrard, L., Sonnet, P., 2013. Beneficial effects of biochar application to contaminated soils on the bioavailability of Cd, Pb and Zn and the biomass production of rapeseed (*Brassica napus* L.). Biomass Bioenergy 57, 196–204.

Jakubus, M., Bakinowska, E., 2022. The effect of immobilizing agents on Zn and Cu availability for plants in relation to their potential health risks. Appl. Sci. 12 (13), 6538.

Kharytonov, M.M., Stankevich, S.A., Titarenko, O.V., Doležalová Weisssmannová, H., Klimkina, I.I., Frolova, L.A., 2020. Geostatistical and geospatial assessment of soil pollution with heavy metals in Pavlograd city (Ukraine). Ecol. Quest. 31 (2), 47–61. https://doi.org/10.12775/EQ.2020.013.

Khawaja, C., Janssen, R., Mergner, R., Rutz, D., Colangeli, M., Traverso, L., Morese, M.M., Hirschmugl, M., Sobe, C., Calera, A., Cifuentes, D., 2021. Viability and sustainability assessment of bioenergy value chains on underutilised lands in the EU and Ukraine. Energies 14 (6), 1566.

Knothe, G., 2002. Current perspectives on biodiesel. INFORM. 13, 900–903.

Knothe, G., Steidley, K.R., 2005. Kinematic viscosity of biodiesel fuel components and related compounds. Influence of compound structure and comparison to petrodiesel fuel components. Fuel 84, 1059–1065.

Kolbas, A., Mench, M., Herzig, R., Nehnevajova, E., Bes, C.M., 2011. Copper phytoextraction in tandem with oilseed production using commercial cultivars and mutant lines of sunflower. Int. J. Phytoremediation 13 (Suppl. 1), 55–76. https://doi.org/10.1080/15226514.2011.568536.

Konaka, T., Tadano, S., Takahashi, T., Suharsono, S., Mazereku, C., Tsujimoto, H., Masunaga, T., Yamamoto, S., Akashi, K., 2021. A diverse range of physicochemically-distinct biochars made from a combination of different feedstock tissues and pyrolysis temperatures from a biodiesel plant *Jatropha curcas*: a comparative study. Ind. Crop. Prod. 159, 113060.

Kulyk, M., Kalynychenko, O., Pryshliak, N., Pryshliak, V., 2020. Efficiency of using biomass from energy crops for sustainable bioenergy development. J. Environ. Manag. Tour. XI (5 (45)), 1040–1053.

Labunska, I., Levchuk, S., Kashparov, V., Holiaka, D., Yoschenko, L., Santillo, D., Johnston, P., 2021. Current radiological situation in areas of Ukraine contaminated by the Chornobyl accident. Part 2. Strontium-90 transfer to culinary grains and forest woods from soils of Ivankiv district. Environ. Int. 146, 106282. https://doi.org/10.1016/j.envint.2020.106282.

Lerman, Z., Csaki, C., Brooks, K.M., 1994. Land Reform and Farm Restructuring in Ukraine. World Bank, Washington, DC.

Lopez, L., Alagna, F., Bianco, L., De Bari, I., Fasano, C., Panara, F., Perrella, G., 2022. Plants: a sustainable platform for second-generation biofuels and biobased chemicals. In: Handbook of Biofuels. Academic Press, pp. 47–72.

Madadi, M., Aqleem Abbas, A., Zahoor, 2017. Green biodiesel production potential from oil seeds in Iran. Int. J. Life. Sci. Sci. Res. 3 (2), 895–904.

Mahmud, S., Redwan Haider, A.S.M., Tahmid-Shahriar, S., Salehin, S., Monjurul Hasan, A.S.M., Johansson Maria, T., 2022. Bioethanol and biodiesel blended fuels—feasibility analysis of biofuel feedstocks in Bangladesh. Energy Rep. 8, 1741–1756. https://doi.org/10.1016/j.egyr.2022.01.001.

Malan, H.L., Farrant, J.M., 1998. Effects of the metal pollutants cadmium and nickel on soybean seed development. Seed Sci. Res. 8, 445–453.

Meher, L.C., Vidya Sagar, D., Naik, S.N., 2006. Technical aspects of biodiesel production by transesterification—a review. Renew. Sust. Energ. Rev. 10 (3), 248–268.

Mellor, P., Lord, R.A., João, E., Thomas, R., Hursthouse, A., 2021. Identifying non-agricultural marginal lands as a route to sustainable bioenergy provision-a review and holistic definition. Renew. Sust. Energ. Rev. 135, 110220.

Nádudvari, A., Abramowicz, A., Fabiańska, M., Misz-Kennan, M., Ciesielczuk, J., 2021. Classification of fires in coal waste dumps based on Landsat, Aster thermal bands and thermal camera in Polish and Ukrainian mining regions. Int. J. Coal Sci. Technol. 8, 441–456. https://doi.org/10.1007/s40789-020-00375-4.

Naylora, R.L., Higginsa, M.M., 2018. The rise in global biodiesel production: implications for food security. Glob. Food Sec. 16, 75–84. https://doi.org/10.1016/j.gfs.2017.10.004.

Nekhay, O., 2012. The Agri-Food Sector in Ukraine: Current Situation and Market Outlook until 2025. European Commission Joint Research Centre Scientific and Policy Report. (74 pp.).

Neupane, D., Lohaus, R.H., Solomon, J.K., Cushman, J.C., 2022. Realizing the potential of *Camelina sativa* as a bioenergy crop for a changing global climate. Plan. Theory 11 (6), 772.

Nikolaychuk, V., Hrabovsky, O., 2000. Ecological problems of Carpathian mountains in connection with pollution of heavy metals biogeocenoses. In: Proceedings of International Conference "Problems of modern ecology", Zaporozhye, 20–22 September 2000, p. 30.

Ogunkunle, O., Ahmed, N.A., 2021. Overview of biodiesel combustion in mitigating the adverse impacts of engine emissions on the sustainable human-environment scenario. Sustainability 13 (10), 5465. https://doi.org/10.3390/su13105465.

Pacwa-Płociniczak, M., Byrski, A., Chlebek, D., Prach, M., Płociniczak, T., 2023. A deeper insight into the phytoremediation of soil polluted with petroleum hydrocarbons supported by the *Enterobacter ludwigii* ZCR5 strain. Appl. Soil Ecol. 181, 104651.

Panoutsou, C., Giarola, S., Ibrahim, D., Verzandvoort, S., Elbersen, B., Sandford, C., Malins, C., Politi, M., Vourliotakis, G., Zita, V.E., Vásáry, V., 2022. Opportunities for low indirect land use biomass for biofuels in Europe. Appl. Sci. 12 (9), 4623.

Pysarenko, P.V., Bezsonova, V.O., 2020. Potential for the utilization of biofuel plant of the second generation of *Miscanthus giganteus* for phytoremediation of oil-contaminated lands. Agrology 3 (3), 127–132.

Radziemska, M., Gusiatin, Z.M., Mazur, Z., Hammerschmiedt, T., Bęś, A., Kintl, A., Galiova, M.V., Holatko, J., Blazejczyk, A., Kumar, V., Brtnicky, M., 2022. Biochar-assisted phytostabilization for potentially toxic element immobilization. Sustainability 14 (1), 445. https://doi.org/10.3390/su14010445.

Razanov, S., Vdovenko, S., Hetman, N., Didur, I., Ogorodnichuk, G., 2021. Assessment of heavy metal pollution of gray forest soils of agricultural lands and their phytoremediation in the cultivation of milk thistle. Ukr. J. Ecol. 11 (6), 41–45. https://doi.org/10.15421/2021_221.

Rheay, H.T., Omondi, E.C., Brewer, C.E., 2021. Potential of hemp (*Cannabis sativa* L.) for paired phytoremediation and bioenergy production. GCB Bioenergy 13 (4), 525–536.

Röder, M., Mohr, A., Liu, Y., 2020. Sustainable bioenergy solutions to enable development in low-and middle-income countries beyond technology and energy access. Biomass Bioenergy 143, 105876.

Romanovskaja, D., Razukas, A., Asakaviciute, R., 2022. Influence of morphostructural elements for buckwheat (*Fagopyrum esculentum* Moench) productivity in different agricultural systems. Plants (Basel) 11 (18), 2382. https://doi.org/10.3390/plants11182382.

Ryabchenko, O., Nonhebel, S., 2016. Assessing wheat production futures in the Ukraine. Outlook Agric. 45, 165–172.

Saeed, L.I., Khalaf, A.M., Fadhil, A.B., 2021. Biodiesel production from milk thistle seed oil as nonedible oil by cosolvent esterification-transesterification process. Asia Pac. J. Chem. Eng. 16, e2647. https://doi.org/10.1002/apj.2647.

Sahito, Z.A., Zehra, A., Tang, L., Ali, Z., Bano, N., Ullah, M.A., He, Z., Yang, X., 2021. Arsenic and mercury uptake and accumulation in oilseed sunflower accessions selected to mitigate co-contaminated soil coupled with oil and bioenergy production. J. Clean. Prod. 291, 125226.

Sameena, P.P., Puthur, J.T., 2021. Heavy metal phytoremediation by bioenergy plants and associated tolerance mechanisms. Soil Sediment Contam. Int. J. 30 (3), 253–274. https://doi.org/10.1080/15320383.2020.1849017.

Savosko, V., Komarova, I., Lykholat, Y., Yevtushenko, E., Lykholat, T., 2021a. Predictive model of heavy metals inputs to soil at Kryvyi Rih District and its use in the training for specialists in the field of biology. J. Phys. Conf. Ser. 1840, 012011. https://doi.org/10.1088/1742-6596/1840/1/012011.

Savosko, V.M., Bielyk, Y.V., Lykholat, Y.V., Heilmeier, H., Grygoryuk, I.P., Khromykh, N.O., Lykholat, T.Y., 2021b. The total content of macronutrients and heavy metals in the soil on devastated lands at Kryvyi Rih Iron Mining & Metallurgical District (Ukraine). J. Geol. Geogr. Geoecology 30 (1), 153–164.

Schaffartzik, A., Plank, C., Brad, A., 2014. Ukraine and the great biofuel potential? A political material flow analysis. Ecol. Econ. 104, 12–21.

Schwalbert, R., Stefanello, L.O., Schwalbert, R.A., Tarouco, C.P., Drescher, G.L., Trentin, E., Tassinari, A., da Silva, I.B., Brunetto, G., Nicoloso, F.T., 2021. Soil tillage affects soybean growth and promotes heavy metal accumulation in seeds. Ecotoxicol. Environ. Saf. 216, 112191. https://doi.org/10.1016/j.ecoenv.2021.112191.

Shayanmehr, S., Ivanič Porhajašová, J., Babošová, M., Sabouhi, S.M., Mohammadi, H., Henneberry, S.R., Foroushani, N.S., 2022. The impacts of climate change on water resources and crop production in an arid region. Agriculture 12, 1056. https://doi.org/10.3390/agriculture12071056.

Shestapalov, V., Naboka, I., Bobyleva, O.A., 1996. Hygienic estimation of influence of heavy metals on an ecological situation of territories, damaged by Chernobyl accident. Rep. Natl. Acad. Sci. Ukr. 8, 156–163 (on Ukrainian).

Shimasaki, C., 2020. Understanding biotechnology product sectors. In: Shimasaki, C. (Ed.), Biotechnology Entrepreneurship, second ed. 2020. Academic Press, pp. 123–149 (Chapter 10) https://doi.org/10.1016/B978-0-12-815585-1.00010-3.

Shparyk, Y.S., Parpan, V.I., 2004. Heavy metal pollution and forest health in the Ukrainian Carpathians. Environ. Pollut. 130 (1), 55–630.

Slobodianyk, A., Abuselidze, G., Lymar, V., 2021. Economic efficiency of oilseed production in Ukraine. E3S Web Conf. 234, 00001. https://doi.org/10.1051/e3sconf/202123400001.

Sobolev, D., 2020. Ukraine: Industrial Hemp Report. https://apps.fas.usda.gov/newgainapi/api/Report/DownloadReportByFileName?fileName=Industrial%20Hemp%20Report_Kyiv_Ukraine_02-09-2020.

Srivastava, A., Prasad, R., 2000. Triglycerides-based diesel fuels. Renew. Sust. Energ. Rev. 4 (1), 111–133. https://doi.org/10.1016/S1364-0321(99)00013-1.

Sytar, O., Ghosh, S., Malinska, H., Zivcak, M., Brestic, M., 2021. Physiological and molecular mechanisms of metal accumulation in hyperaccumulator plants. Physiol. Plant. 173 (1), 148–166. https://doi.org/10.1111/ppl.13285.

Sytar, O., Prasad, M.N.V., 2016. Chapter 1 - Production of biodiesel feedstock from the trace element contaminated lands in Ukraine. In: Prasad, M.N.V. (Ed.), Bioremediation and Bioeconomy. Elsevier, 10.1016/B978-0-12-802830-8.00001-0.

Sytar, O., Taran, N., 2022. Effect of heavy metals on soil and crop pollution in Ukraine—a review. J. Cent. Eur. Agric. 23 (4), 881–887. https://doi.org/10.5513/JCEA01/23.4.3603.

Takase, M., Chen, Y., Liu, H., Zhao, T., Yang, L., Wu, X., 2014. Biodiesel production from non-edible *Silybum marianum* oil using heterogeneous solid base catalyst under ultrasonication. Ultrason. Sonochem. 21 (5), 1752–1762. https://doi.org/10.1016/j.ultsonch.2014.04.003.

Tavakoli-Hashjini, E., Piorr, A., Müller, K., Vicente-Vicente, J.L., 2020. Potential bioenergy production from *Miscanthus × giganteus* in Brandenburg: producing bioenergy and fostering other ecosystem services while ensuring food self-sufficiency in the Berlin-Brandenburg region. Sustainability 12 (18), 7731. https://doi.org/10.3390/su12187731.

Tóth, G., Hermann, T., Da Silva, M.R., Montanarella, L., 2016. Heavy metals in agricultural soils of the European Union with implications for food safety. Environ. Int. 88, 299–309. https://doi.org/10.1016/j.envint.2015.12.017.

Wan, X., Lei, M., Chen, T., 2016. Cost-benefit calculation of phytoremediation technology for heavy-metal-contaminated soil. Sci. Total Environ. 563–564, 796–802. https://doi.org/10.1016/j.scitotenv.2015.12.080.

Yan, A., Wang, Y., Tan, S.N., Mohd Yusof, M.L., Ghosh, S., Chen, Z., 2020. Phytoremediation: a promising approach for revegetation of heavy metal-polluted land. Front. Plant Sci. 11, 359. https://doi.org/10.3389/fpls.2020.00359.

Zehra, A., Sahito, Z.A., Tong, W., Tang, L., Hamid, Y., Khan, M.B., Ali, Z., Naqvi, B., Yang, X., 2020. Assessment of sunflower germplasm for phytoremediation of lead-polluted soil and production of seed oil and seed meal for human and animal consumption. J. Environ. Sci. (China) 87, 24–38. https://doi.org/10.1016/j.jes.2019.05.031.

Further reading

Durrett, T.P., Benning, C., Ohlrogge, J., 2008. Plant triacylglycerols as feedstocks for the production of biofuels. Plant J. 54, 593–607. https://doi.org/10.1111/j.1365-313X.2008.03442.x.

Aromatic, medicinal, and energy plantations on metalliferous/contaminated soil—Bioremediation and bioeconomy

Sai Shankar Sahu[a], Adarsh Kumar[b], Majeti Narasimha Vara Prasad[c], and Subodh Kumar Maiti[a]

[a]Department of Environmental Science and Engineering, Indian Institute of Technology (Indian School of Mines), Dhanbad, Jharkhand, India [b]Laboratory of Biotechnology, Institute of Natural Sciences and Mathematics, Ural Federal University, Ekaterinburg, Russia [c]Department of Plant Sciences, School of Life Sciences, University of Hyderabad (An Institution of Eminence), Hyderabad, Telangana, India

1 Introduction

Pollution due to heavy metal contamination is a significant environmental issue across the globe. Though numerous physical, chemical, and biological remediation technologies are available to clean up the contaminated sites, they are of high cost and low efficiency.

Bioremediation is a new emerging biological technology defined as the application of microbial consortium to reduce, degrade, and detoxify organic and inorganic contaminants. Phytoremediation, one of the most visible emerging techniques of bioremediation, is an eco-friendly, cost-effective, and esthetically pleasing technology, gaining importance to remediate or reduce metal- and metalloid-contaminated sites. The different types of phytoremediation include phytostabilization, phytoextraction, phytovolatilization, phytodegradation, and rhizofiltration. However, phytoextraction and phytostabilization are the two most important aspects of phytoremediation, which derives its scientific justification from the emerging concept of green chemistry and engineering. Both techniques involve the accumulation of metal, their uptake, and translocation from soil to plant. Phytoextraction involves the removal of metal from soil by accumulating and translocating the metal into the aerial/harvested part. After translocating, these plant materials can be incinerated or burned to recover valuable material. The residue can be disposed of appropriately under controlled conditions or, in some cases, recycled as fertilizer. In contrast to phytoextraction, phytostabilization is the immobilization of metal in soil.

Extensive research on bioremediation over the last three decades has focused on utilizing various types of native trees and grasses along with leguminous plants to not only improve the economic conditions of the local farmers but also contribute to the economic development of the country. However, few plants such as lemongrass and vetiver grass have emerged as highly effective and widely recognized plants for the efficient phytoremediation of metalliferous soils.

Moreover, the term "bioeconomy" refers to the production of food, materials, and energy using regenerative biological resources, such as crops, forests, fish, animals, and microorganisms, from the land and sea. The EU's transition to a circular economy with low carbon emissions will be accelerated with stronger growth of the bioeconomy.

Lignocellulosic biomass (LCB) is the most abundantly available raw material on earth for the production of biofuels that can help minimize reliance on nonrenewable resources. Since bioethanol is a liquid fuel, it is simple to incorporate into the current fuel distribution system. On the other hand, biochar is a carbonaceous substance that possesses a large surface area and is rich in macro- and micronutrients. It has a high metal-accumulating ability and can immobilize microorganisms effectively. Hence, a circular economy can be developed by producing biochar and bioethanol in a single system from the LCB grown from heavy metal-contaminated soil.

An underutilized species without any regular or complete exploitation is the *Cistus ladanifer*, a flowering plant commonly found in the Mediterranean region, which is mostly used in the perfume industry or as a decorative plant. It is typically connected to small family based or rural organizations. However, strong economic exploitation and well-developed value chains are still lacking. As a result, *C. ladanifer* proves to be a suitable and natural candidate for evaluation as a biomass feedstock. Finally, the potential of volarization of *C. ladanifer* has been demonstrated in the study.

2 Aromatic, medicinal, and energy plants

Metal- or metalloid-contaminated soils are potentially toxic sites generally found with sparse vegetation of invasive or natural plants. These plants often demonstrate metal stress tolerance, exhibiting the properties of exclusion, accumulation, or hyperaccumulation of metals. However, because of low biomass, these plants are not frequently used by the phytologists for remediation of contaminated sites. Aromatic plants and grasses are gaining importance in remediation because of their luxuriant biomass and metal tolerance capacity under stress conditions. Aromatic plants such as vetiver grass, lemongrass, tulsi, palmarosa, citronella, and geranium mint are ecologically feasible and viable and are widely used aromatic and medicinal plants. Of these, vetiver (*Chrysopogon zizanioides*) and lemongrass (*Cymbopogon citratus*) are the two most important grasses gaining keen attention throughout the world. Grasses and plants that have high metal tolerance, adaptability to environmental

TABLE 1 Plants and grasses with level of metal tolerance, stress tolerance, medicinal property, and energy production.

Plant/Grass	Metal tolerance	Stress tolerance	Medicinal property	Energy production
Vetiver	✓✓✓	✓✓✓	✓	✓✓✓
Lemongrass	✓✓✓	✓✓	✓✓	✓✓
Tulsi	✓	✓	✓✓✓	✓✓
Stylo	✓	✓✓		✓

✓: low; ✓✓: medium; ✓✓✓: high.

stress, and medicinal property along with the capacity to produce energy after burning are presented in Table 1.

3 Vetiver and lemongrass

Vetiver (*Chrysopogon zizanioides* (L.) Roberty) and lemongrass (*Cymbopogon citratus* (DC.) Stapf) are aromatic and perennial grass species belonging to the Poaceae family. Vetiver, commonly known as "wonder grass," is found growing throughout the world. It is considered as a miracle grass because it can tolerate and grow even in adverse conditions and on different soils, such as:

- Metal- and metalloid-contaminated soil
- Sandy soil
- Acid sulfate soil
- Sodic soil
- Mangrove soil
- Saline soil
- Lateritic soil.

The main features that promote the use of vetiver and lemongrass against other plants and grasses are listed in Table 2.

3.1 Uses and economic importance of vetiver and lemongrass

Both live and dry masses of vetiver and lemongrass have great uses and economic importance (Table 3). Live plants are widely used in soil water conservation, disaster prevention, erosion control and stabilization, protection of terrestrial and aquatic environment, and oil production, whereas dry plants can be used in industries such as cosmetics, perfumery, medicine, and handicrafts.

3.2 Economics of vetiver and lemongrass

Vetiver technology (VT) system has been established in many countries, including India, for bioremediation and resource generation. The VT system has gained popularity and proven to be cost-effective in many developing countries such as China, India, and Sri Lanka,

TABLE 2   Silent features and properties of vetiver and lemongrass.

Properties/Parameters	Vetiver grass	Lemongrass
Luxuriant growth	✓✓✓	✓✓
Noninvasive	✓	✓
Root system and biomass	✓✓✓	✓✓
Metal accumulation in root	✓✓✓	✓✓✓
Metal accumulation in shoot	✓	✓
Drought resistant	✓✓✓	✓✓
Fire-resistant	✓✓✓	✓
Flood-resistant	✓✓✓	✓
Pathogen-resistant	✓✓	✓✓✓
Soil as substrate to grow	✓✓	✓✓✓
Water as substrate to grow	✓✓	✗
Erosion control and steep slope stabilization	✓✓✓	✓✓
Explosive degradation	✓✓	✓
N and P removal	✓✓✓	✓
Dioxins and PAH removal	✓✓	✓✓
Pesticide tolerance and removal	✓✓✓	✓✓
Radionucleid removal	✓	✗
Carbon sequestration	✓✓	✓

✓: low; ✓✓: medium; ✓✓✓: high.

TABLE 3   Commercial importance of vetiver and lemongrass.

Uses	Vetiver grass	Lemongrass
Handicrafts (fans, cloth baskets, and hangers)	✓	✓
Home appliances (chairs, tables, stools, and room partitions)	✓	✗
Soaps and detergents	✓	✓
Perfumes	✓	✓
Thatching of roofs	✓	✗
Medicines	✓	✓
Veti-concrete slabs	✓	✗
Insect-repellent	✓	✓

where the labor cost is significantly lower. Even in developed countries, such as Australia, where the labor cost is high, soft engineering technology costs between 27% and 40% compared to hard engineering solutions (Diti, 2003).

India is the largest producer (300–350 tons per annum) of lemongrass oil, 80% of which is exported to the developed countries, namely, China, France, Germany, Italy, Japan, Spain, the United Kingdom, and the United States (Lal et al., 2013). According to literature, 20–25 tons of fresh lemongrass per acre can produce 70–75 kg of oil in a year, and the profit level of farmers ranged between 80% and 120%.

4 Case study: Phytostabilization of integrated sponge iron plant waste dumps by aromatic grass-legume (lemon-stylo) mixture and energy plantation (*Sesbania*)

4.1 Generation of integrated sponge iron plant waste

The principal raw materials required for sponge iron production are iron ore, noncoking coal, and dolomite. For example, production of 100 tons of sponge iron requires 154 tons (65 wt % Fe) of iron ore and 120 tons (B grade) of coal, and out of the total generated solid waste of 45 tons, 25 tons is char (dolochar). In the absence of good grade coal, sponge iron industries are using poor quality coal (F grade), which contains more than 40% by weight of ash, leading to more waste generation. Slag is also generated from the sponge iron plant or SMS (steel melting shop) plant in the process. Therefore, in an integrated sponge iron plant, large quantities of solid wastes generated in the form of dolochar, slag, and fly ash were dumped together due to the scarcity of space to cover a large area exceeding 40–50 m. During the summer, they caused severe air pollution issues, while in monsoon, the loose waste was easily carried away along with runoff to pollute the nearby water bodies.

4.2 Characteristics of waste dump materials

The waste was very homogenous, loose, and devoid of nutrients; hence, stabilization of this type of waste poses a severe challenge. Waste samples from the boilers and DRI plant of the sponge iron unit were collected for physico-chemical analysis. They are highly alkaline in nature and consist of slag, fly ash, and dolochar. The detailed characteristics of individual waste and composite waste are given in Table 4. The waste exhibited high electrical conductivity (EC), with high concentration of Na, K, and Ca, and was very fine, loose, and easily windborne from the surface of the dumps, thus worsening the quality of air. Furthermore, during the monsoon season, large amounts of waste were washed off into nearby water channels, polluting them. The representative samples of the waste were collected, air-dried, thoroughly mixed, grinded, if needed, and sieved down to micron size with standard sieves. Analyses of the physico-chemical parameters were done according to the standard procedures in seven replicates.

The properties of fly ash mainly depend on the coal used in the thermal power plant. The samples in this study had alkaline pH of 8–9. Dolochar and slag samples had pH of 12 and 11, respectively, while the composite sample's pH was approximately 9–10. The waste exhibited high EC of $3\,dSm^{-1}$ due to the presence of dolochar (nearly $5\,dSm^{-1}$) and fly ash

TABLE 4 Physico-chemical characteristics of individual wastes and composite waste generated from the integrated sponge iron unit ($n=3$).

Parameters	Fly ash from AFBC boiler	Fly ash from WHRB boiler	Dolochar	Slag	Composite waste
Color	Light gray	Dark gray	Black	Shiny black	Blackish gray
pH$_{H2O}$ (1:1) (w/v)	8.69±0.04	9.37±0.02	11.77±0.06	10.38±0.06	9.66±0.04
EC$_{H2O}$ (1:1) (w/v) dSm^{-1}	1.73±0.04	2.25±0.01	5.10±0.27	0.14±0.021	2.89±0.05
WHC (%)	71.37±0.82	63.81±0.23	94.35±0.12	33.94±0.14	65.21±2.62
LOI (%)	3.86±0.03	5.91±0.02	17.08±2.53	–	6.08±0.31
Ex. K (mg kg^{-1})	156±11.7	113.1±7.8	78±3.9	59.86±1.7	128.7±1.95
Ex. Na (mg kg^{-1})	124.2±1.38	117.3±2.07	103.5±0.69	31.24±0.69	116.5±0.69
Ex. Ca (%)	0.19±0.01	0.28±0.01	0.99±0.01	0.01±0.001	0.28±0.01
EDS analysis					
CaCO$_3$ (%)	15.69±1.23	25.5±0.01	62.32±3.11	6.09±0.37	26.2±0.09
SiO$_2$ (%)	56.82±0.68	50.64±2.36	27.55±2.16	44.28±1.23	48.61±0.69
MgO (%)	0.69±0.02	1.7±0.27	0.56±0.06	0.45±0.02	0.79±0.03
Al$_2$O$_3$ (%)	10.49±0.17	7.67±1.14	3.98±0.43	7.63±0.22	8.63±0.33
FeS$_2$ (%)	0.34±0.01	0.31±0.01	0.54±0.04	–	0.43±0.01
Fe (%)	2.34±0.04	6.41±0.02	2.42±0.02	25.09±0.29	11.86±0.12
Trace element analysis					
Cr (mg kg^{-1})	529.8±1.23	341±2.02	97.4±0.12	353.4±4.29	312±2.58

Ex: exchangeable; WHC: water-holding capacity; LOI: loss on ignition.

(1.5–2 dS m^{-1}). The composite waste consisted of fine-sized particles, mainly due to the presence of fly ash, and had a water-holding capacity of 65%, which was corroborated by the fact that fly ash and dolochar had high water-holding capacity of 63%–70%. High water-holding capacity of the dolochar particles (94%) are substantiated by SEM studies, which showed that the particles had micro-porous structure and thus high surface area to withhold maximum amount of water. The unburned carbon in the samples was determined by the loss on ignition (LOI) procedure and was found to be 3%–6% in fly ash and 17% in dolochar samples while the LOI of composite sample was 6%. High LOI in dolochar due to the presence of carbon particles was evident in SEM and EDX studies as well. Ammonium acetate extract of the samples indicated the presence of high calcium mainly in fly ash (0.3%) and dolochar samples (1%) while the composite sample had 0.3% exchangeable calcium. Rapid pyrolysis of the raw materials in the DRI leads to its porous nature with complex mineralogical compositions as found through EDX analysis and elemental mapping. They showed mineral carbon inclusion in fine dolochar samples. The major constituents of inorganic inert material present in the char are CaCO$_3$ (62%) and SiO$_2$ (27%). The contents of Al$_2$O$_3$, Fe, and MgO present in the sample were 4%, 2.5%, and 0.5%, respectively. EDX mapping showed elements in dolochar particles in the order, $Ca > C > K > O > Si > Al > Mg > S$.

4.3 Methods of phytostabilization of waste dump

A challenging task was undertaken to ecologically restore a huge waste dump covering 7 ha ground area (top flat surface area of 5 ha), aerial height of 50–70 m, and without any intermediate benching with steep slope ($>70°$). Practically no space was available to reduce the slope. The sequence of restoration operation was (1) regarding of the dump, (2) blanketing with topsoil, (3) covering the slope with coir mat, (4) sowing of grass-legume mixture on the slope, (5) construction of drainage, (6) watering arrangement, and (7) aftercare and maintenance of the site for 3 years. Using good-quality topsoil is essential for the success of any eco-restoration program (Maiti, 2013). Sources of topsoil are either stockpiled or simultaneously excavated or reused (i.e., concurrent use). In the present study, topsoil was used in a concurrent manner.

Geotextiles are used to protect slopes by preventing erosion and creating favorable soil conditions for re-vegetation, especially in the initial stage of slope restoration. Natural geotextile mat (jute mat or coir mat) is preferred since it is more effective and more environment-friendly (Shao et al., 2014). It can markedly increase the vegetation cover, reduce evapotranspiration (Mitchell et al., 2003), and help to conserve moisture, which increases the survival of young seedlings during the hot summers (Sawtschuk et al., 2012). In the present study, as there was no scope to reduce the slope of the dump by terracing, the entire slope was covered with high tensile-strength coir mat (slope length, 45–50 m).

Using grass-legume mixtures to stabilize steep slopes has now become a widely employed technique. Studies have shown that Deenanath grass or Desho grass (*Pennisetum pedicellatum* Trin.) along with forage legumes (*Stylosanthes humilis* (L.) Taub. (Caribbean Stylo)) can successfully restore a degraded site (Maiti and Saxena, 1998; Caravaca et al., 2002; Lenka et al., 2012; Araújo and Costa, 2013). Field experiments showed that forage legume (*Stylosanthes hamata*) can be used as an initial colonizer during reclamation of coal mine-degraded sites, without the application of topsoil (Maiti and Saxena, 1998). The combination of grass-legume mixture creates a nitrogen balance in the soil, and decomposition of dry plant parts creates nitrogen-rich litter and mulch for the soil. Grasses have extensive fibrous root systems that can reduce erosion by holding the loose soil particles, can tolerate adverse soil conditions, and form mulches after drying (Maiti, 2013). Generally, perennial forage-type legumes are used, but native species show greater improvement in soil fertility. Thus, during the hydroseeding of grass-legume mixture, a higher percentage of legume seeds are used, which will accelerate nutrient cycling and improve the quality of soils, biodiversity, and sustainability of the plant community (Maiti, 2013). The aims of the study are to investigate: (1) the changes in physical and chemical properties of waste dump surface due to the growth of grass-legume mixture, (2) effectiveness of use of coir mat and grass-legume seeds for stabilization of steep slope, and (3) measurement of soil temperature amelioration due to mulch and litter accumulation.

4.4 Study area

The study was carried out on waste dumps of an integrated sponge iron unit located in Raigarh district, Chhattisgarh, India, which falls between the latitudes 22°00′51″N and

FIG. 1   Aerial view of waste dump before ecological restoration.

22°02′10″N and the longitudes 83°22′35″E and 83°23′27″E, covering an area approximately seven hectares. Wastes generated during the processes were dumped on the outskirts, which was surrounded by the protecting forest. The height of the dump ranged between 40 and 50 m with steep slope (>70–80°), which was unstable, causing serious environmental problems. An aerial photograph of the original waste dump is shown in Fig. 1.

4.5 Characteristics of blanketing soil and coir mat

Topsoil blanketing of the waste dump was done to cover the waste and create a suitable substratum for the growth of vegetation and stabilization of surface. Topsoil, slightly of acidic nature (4.7), will buffer the pH of loose alkaline solid wastes. Soil was sandy loam (73%), with moderate fertility, cation exchange capacity (CEC) of 7.2 cmol(+) kg^{-1}, and base saturation of 51.3%. About 50,000 m^3 of topsoil excavated from the forest area (which was to be used for the creation of fresh dump) was transported to the restored site, spread on the waste dump, and leveled by the dozer (Fig. 2A–D).

Coir mat is a tough organic fiber having high tensile strength, rich in lignin (46%), about 1.5–2 cm thick, interweaved with coconut coir, fixed by nylon net; about 1 m wide and 30–50 m length rolls were used for stabilization of the slope (Manufacturer: Sri Venkateshwara Fiber Udyog Private Limited, Bangalore, India) (Fig. 2E and F). Coir-matting was essential to stabilize the slope and prevent erosion of materials from the steep slope during the monsoon. It was laid loosely from the berm of the slope and unrolled downwards without stretching, and stapled in each 1-m distance with a U-shaped iron nail (size: 30 cm arm length and 8 cm width between the arms).

4.6 Composition of grass-legume mixture

Grass-legume seeds consisting of *P. pedicellatum*, *S. hamata*, *Crotalaria juncea* L (Indian hemp), and *Sesbania sesban* (L.) Merr. (Common sesban), and *Hibiscus sabdariffa* L. (Indian rosella) (high

FIG. 2 (A) Leveling of waste dump; (B) source of topsoil; (C) blanketing of waste dump with topsoil; (D) leveling of topsoil on waste dump using the dozer; (E) fixing of coir mat on the dump slope; and (F) waste dump fully covered with topsoil and coir mat.

biomass-yielding, nonleguminous undershrub) were used for the re-vegetation of slopes of the waste dump. As the slope was very steep, it was initially covered with coir mat and then the topsoil was spread over it to create a substratum for germination and growth of seeds.

Three rows of tillers of *C. citratus* (after shorting) were planted along the berm (edge) of the dump (Fig. 3A and B). Additionally, one row of *Azadirachta indica* seeds was sown at 3 m intervals on the berm. All these activities were carried out before the onset of the monsoon to protect the soil and seed mixture from erosion. A seed germination test for grass-legume mixture was conducted in the laboratory/field conditions to test the viability of seeds and time taken for germination by standard test method (Maiti, 2013). Leguminous seeds have a low dormancy period, and within 6 days, 60% of the seeds are germinated while the rest germinates in 10 days. In contrast, grass seeds take a longer time to germinate. Approximately, 3350 kg of seed mixture was used and sown three times at an interval of 1 week.

FIG. 3   (A) Shorting of fresh lemongrass tillers; (B) planting of lemongrass on the berm of dump in a row; (C) growth of small grass-legume plants after a few months; (D) growth of *Sesbania sesban*; (E) growth of lemongrass; (F) growth of *Stylosanthes*; (G) constructed heli-pad on the top of the phytostabilized dump; and (H) fully restored and phytostabilized waste dump.

4.7 Technical restoration of the site

The top surface of the dump was leveled by the dozer and a 3° slope toward the central drainage system was provided (as shown in Fig. 3B and C). The acidic nature of topsoil was helpful to ameliorate the soil pH because waste material has an alkaline pH (11.9). The slope was blanketed with soil, seeded with grass-legume mixture, and covered with coir mat. The mat (dimension: $1\,m \times 30\,m$ and $50\,m$) encored at the berm and spread from top to bottom. Next, seeds of grass-legume mixed with fine soil materials (seed:soil; 1:1) were broadcasted. All these activities were carried out before monsoon. About 2000 tillers of *C. citratus* were planted along the berm of the dump in $0.5\,m \times 0.5\,m$ spacing, and toward the inner side, one row of *A. indica* was developed by seeding. The vegetated slope was irrigated in the summer months. At the initial stage (first year), a piped water supply with sprinkler distribution system was used during the winter and summer months (December-June). Plant growth monitoring and soil sampling were done after 7 months. Reinforced concrete drainage was provided to drain runoff from the top surface of the dump, while seepage from the slope was drained out through an earthen channel.

The entire slope was found to be covered, stabilized, and erosion-controlled during the monsoon with lush green growth of *Stylosanthes* legume (Fig. 3F). A distant view of the restored dump is shown in Fig. 3H. During the field survey in monsoon, it was observed that the creeping nature of *Stylosanthes* had caused it to form a 30–40 cm carpet above the coir mat, and beneath this layer, the decomposition of leaves (black in color) contributed to the soil organic matter (SOM) and initiated the formation of humus. Natural colonization of vegetation on the slope surface due to accretion of habitat by *Stylosanthes* legume was observed. The berm of the dump was stabilized with massive growth of *C. citratus*, *C. juncea*, and *A. indica*.

4.8 Amelioration of soil temperature due to mulch and litter accumulation

Soil temperature is an important parameter that controls biochemical processes and thereby influencing nutrient cycling. Mulch properties such as quantity and architecture also affect the soil microenvironment. Mulching makes soil less prone to erosion, increases infiltration and biomass, and ameliorates soil surface temperature. Different types of mulches modify soil temperatures in different ways (Fehmi and Kong, 2012). Accumulation of dry mulch (stems and branches of *C. juncea*, *S. sesban*, and *S. hamata*) in a quantity 13–15 ton ha^{-1} was observed on the waste dump, which was higher than the reference forest site (8.16 tons ha^{-1}). Higher accumulation of dry mulch was due to the massive growth of the leguminous species (*S. sesban* and *C. juncea*) and drying within 5 months. Their accumulation also had a significant influence in the reduction of rhizospheric temperature. A significant reduction in the rhizospheric temperature of 12%–17% was observed under mulch cover, while it was hardly 4% decrease on bare surfaces (area covered with minimum vegetation). The nearby reference forest site also showed 12% reduction in soil temperature under the litter (at 10 cm depth). This indicates dry mulches generated from *S. hamata*, *C. juncea*, *S. sesban*, and *Hibiscus sabdariffa* not only enhance SOC but also their dry parts ameliorate surface temperature during the summer and helps in moisture conservation.

4.9 Estimation of the cost of ecological restoration of waste dump

Topsoil is regarded as a strategic resource for successes of any ecological restoration project. Before salvaging operation of topsoil for concurrent use, inventory (quality and depth) and time of scrapping is essential. In the present project, the major challenge was to stabilize the very steep slope (>70°) with 50–60 m straight height without any bench. The slope could have been easily stabilized if dump slope could have been reduced to <28–30° by creating two additional benches of 15 m height. As there was practically no space left between the edges of the dump and forest land, a terracing option to reduce the slope angle was ruled out (Maiti and Maiti, 2015).

In the present study, topsoil was excavated (about 1 m deep) from the newly constructed dumping site and transported to the waste dump, which constituted 31% of the total cost. The running cost of heavy earth-moving machinery (HEMM) accounted for 13% of the total cost, which again depended on the efficiency of use of HEMM and supervision. The third major cost involved purchasing of coir mat at US $1.3 m^{-2}, which depends on the technical requirement and the type of coir mat. Even though, for the present work, soil was available free of cost, the cost of topsoil was at US $ 0.50 m^{-3} for restoration work, which comes to 11% of total cost. The cost of biological restoration comes close to 17%, of which aftercare and maintenance for 3 years constitutes half of the cost, and the rest was for the purchase of grass-legumes seed mixture and construction of watering facilities. The cost of grass-legume seed materials was Rs. 150 kg^{-1} (US $2.5 kg^{-1}), *C. citratus* tiller at Rs. 7.50 tiller^{-1} (US $0.125 tiller^{-1}), and *A. indica* seeds at Rs. 500 kg^{-1} (US $8.34 kg^{-1}). It is always advisable to test the real value or pure live seed count for seed lot (Maiti, 2013). The knowledge of seed dormancy and methods of seed treatment to overcome dormancy is essential to leguminous seeds. Seeds of *A. indica* do not have longer dormancy periods; therefore, for better results, the seeds should be sown immediately. The quantity of seeds to be ordered depends on the types of seeds, seed viability, methods of sowing, and time of broadcasting. It is advisable, rather than having a single broadcasting operation, to sow the seeds three or four times over an interval of 7–10 days. The average cost of eco-restoration was US $39,730 ha^{-1} (i.e., US $4 m^{-2}), out of which the total biological reclamation components constituted 17%, which includes aftercare and maintenance for 3 years (Maiti and Maiti, 2015).

5 Bioremediation and circular bioeconomy potential of lignocellulosic biomass (LCB)

A circular economy built on the use of renewable bioresources has attracted a lot of interest as civilization has evolved. Due to their high availability and low pricing, LCB resources, such as agro-industrial wastes, wood residues, grass, and municipal solid wastes, have extensive commercial applications in the manufacturing of value-added products. A variety of important biomaterials, lignin-modifying enzymes, bioenergy, useful compounds, and efficient biopolymers, including adsorbents and biofuels, can be made using different LCBs as a raw material (Shafiei Alavijeh et al., 2020). Using these lignocellulosic materials can significantly lower production costs of biomaterials and reduces generation of waste and help to mitigate the severe environmental issues caused by the enormous disposal of biomass.

Due to the significant environmental advantages and low economic costs involved, valorizing LCB for manufacturing bioethanol and biochar is of particular relevance. A significant area of study in the realm of biofuels is bioethanol, which can be made from the monosaccharides produced by the hydrolysis of cellulose and hemicellulose from LCB material (Waghmare et al., 2018). The cellulose fibers in the native biomass are covered with lignin, which is a natural polymer with a complicated 3D structure. Lignin is a polymer that possesses the robust recalcitrance properties, which is required to support structural integrity and safeguard cells (Pratto et al., 2020).

Hence, a pre-treatment step for degrading the lignin structure is required to yield more fermentable sugars from LCB. However, the most challenging and expensive phase in the manufacturing of bioethanol is pre-treatment (Huang et al., 2021). In order to break down the walls of plant cells within LCB, a variety of pre-treatment techniques, including physical, chemical, and biological ones, are currently being used. The choice of the various processes depends on the type of LCB being used due to their different physico-chemical properties and structural compositions. Pre-treatment can be used to modify the microscopic structure of the cellulose matrix and reduce the degree of crystallinity in LCB (Mohapatra et al., 2017). Moreover, pre-treatment also expands the biomass' surface area and makes the cellulose and hemicellulose components more accessible to hydrolytic enzymes. The simplified operational procedure and lower energy inputs would both benefit from the optimized pre-treatment approach. Unfortunately, a lot of fermentation inhibitors are unavoidably produced during the pre-treatment stage, and these inhibitors may prevent microbial decomposition and have an impact on the succeeding bioethanol production process (Agu et al., 2019). As a result, many detoxification techniques have been developed to get rid of these fermentation inhibitors. The method used will depend on the makeup of the various LCB sources and will mostly use commercial adsorbents and resins. Due to numerous sources and advantageous economics, biochar has been utilized widely in the detoxification process as an excellent adsorbent. Biochar has a great attraction for nonpolar compounds, high surface porosity, and an abundance of surface functional groups (Zhang et al., 2016). It is effective in removing both organic and inorganic pollutants. Moreover, the use of biochar can be advantageous for immobilizing the fermentative micro-organisms and thereby enhance the fermentation process for bioethanol production. But there have only been a few reports of using biochar to improve the production of bioethanol. However, the yields of bioethanol and biochar from different biomass materials vary (Table 5). A schematic circular economy diagram showing cultivation LCB and subsequent production of biochar and bioethanol (Fig. 4).

5.1 Bioremediation and bioeconomic potential of *Cistus ladanifer*

The perennial shrub *C. ladanifer* L. (rockrose plant), a member of the Cistaceae family, is widely distributed in the marginal areas of the Mediterranean and is known to yield labdanum, a valuable resin rich in compounds (Alves-Ferreira et al., 2023). Additionally, *C. ladanifer* has been widely employed in producing biofuels, feed, traditional medicine, feed, and cosmetics.

Due to their low cost and accessibility, rockrose plants have historically been employed as a significant resource for primary healthcare. Rockrose's aerial portions are used to make

TABLE 5   Yields of bioethanol and biochar from different biomass.

Bioethanol		Biochar	
Sources	**Yield (%)**	**Sources**	**Yield (%)**
Eucalyptus	75.31	Pine wood	25.4
Oak	53.9	Wheat straw	34.12
Corn stover	84.13	Plant residues	31.8
Switch grass	40	Organic waste	22.1
Rice straw	83.5	Rice husk	33.5
Poplar	68.3	Eucalyptus sawdust	25–35
Spruce	54.5	*Spartina alterniflora*	46.76

Source: Hou, J., et al., 2020. A critical review on bioethanol and biochar production from lignocellulosic biomass and their combined application in generation of high-value byproducts. Energy Technol. 8(5), 1–12. https://doi.org/10.1002/ente.202000025.

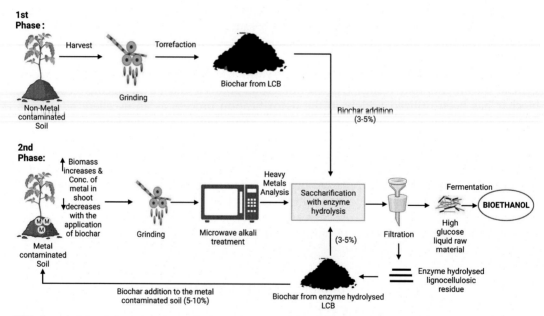

FIG. 4   1st phase: Cultivation of lignocellulosic plants in non-metal contaminated soil, after harvesting, drying, and grinding, biochar is produced by torrefaction at a temperature of 200–300°C. 2nd phase: Lignocellulosic plants are grown in metal contaminated soil and amended with biochar produced from two sources, i.e., LCB and enzyme-hydrolyzed LCB. Due to plant growth, metal concentration in shoot will decreases and soil is also remediated through phytoremediation with biochar amendment. Metal contaminated LCB is harvested and grinded followed by microwave alkali pre-treatment—saccharification with enzyme hydrolysis—filtration—fermentation of high glucose containing liquid raw material and the end product, bioethanol is produced.

extracts, get exudates, and get essential oils (Alves-Ferreira et al., 2023). The traditional use of rockrose has helped rural people thrive socioeconomically. Due to their exceptional fixative qualities, labdanum and the essential oil from *C. ladanifer* are both highly valued in the fields of perfumery, cosmetics, aromatherapy, and food flavoring (limited usage). The products that can be obtained from *C. ladanifier* are aromatic extracts such as labdanum gum, concrete, and essential oils.

5.2 Aromatic extracts from *C. ladanifer*

As a sedative, hemostatic, anti-bacterial, astringent, tonic, expectorant, and balsamic ingredient, *C. ladanifer* generates a sticky resin or aromatic exudate known as labdanum (Alves-Ferreira et al., 2023). By removing waxes, resinous matter, and oily materials and treatment with hot alkaline water (sodium carbonate), followed by acidification (sulfuric acid), crude gum of labdanum is obtained from the surface of the plant leaves and branches. Its sticky mechanical features make it suited for incense sticks or other items to burn. It has been utilized as a technical ingredient in perfumery and can be utilized in creams, masks, and makeup removers in the beauty industry (Carrion-Prieto et al., 2017). Apolar solvents are used to extract concrete, a viscous substance with a balsamic odor, from the entire plant (hexane or isopropanol, for example). The extract has waxy components and a volatile aroma once the solvent has evaporated. Essential oils are organic compounds involved in the defense mechanisms of several species and are mainly responsible for aroma. They can be synthesized by any plant organ and are kept in glandular trichomes, secretory cells, cavities, canals, or epidermic cells (Carrion-Prieto et al., 2017). Fig. 5 illustrates the commercial products that can be obtained from *C. ladanifer* using different process associated with it.

5.2.1 Bioactivity obtained from *C. ladanifer*

Moreover, several phytochemical groups have been identified in rockrose that have a range of biological activities including anti-bacterial, anti-viral, and anti-oxidant properties. These compounds' diverse functional properties provide numerous opportunities for the development of new medications and serve as excellent sources for the manufacturing of food additives, functional foods, nutritional supplements, and nutraceuticals for the pharmaceutical and natural food industries (Alves-Ferreira et al., 2023).

5.2.2 Other potential applications of *C. ladanifer*

Using the plant as an animal feed with nutritional benefits for animal health, for phytoremediation of soils contaminated by heavy metals, and as a lignocellulosic raw material for value-added products, such as the production of bioethanol, are some of the new directions regarding the valorization of rockrose.

Phytoremediation

According to reports, *C. ladanifer* can thrive and survive in soils that are highly contaminated with hazardous components as Mn, Cu, Zn, Pb, and As (El Mamoun et al., 2020). Selectivity in the uptake and translocation of metals is suggested by observations in mining

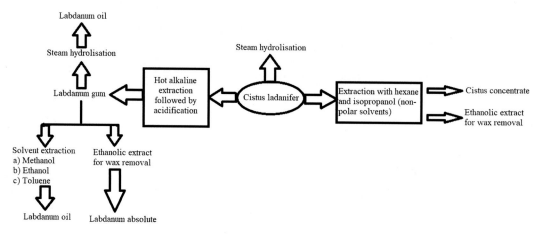

FIG. 5 Products of commercial importance from the harvested biomass of *Cistus ladanifer* applied in phytotechnologies.

sites. As a result, this species' tolerance of toxic substances in soils can increase the potential for phytoremediation in areas with high levels of contamination.

Animal feed

Interesting outcomes have been observed when *C. ladanifer* (soft stems, leaves, or extracts) has been used as a food supplement to enhance animal nutrition and to increase productivity. Inclusion of *C. ladanifer* leaves and tender stems in lambs' meals that also contain a combination of sunflower and linseed oils enhanced the amount of fatty acids found intramuscularly (Alves-Ferreira et al., 2023). The fatty acids of the food containing *C. ladanifer* gave improved health without impairing animal performance.

Value-added products from lignocellulosic material

In the context of a shift to a circular bio-based economy, the interest in using lignocellulosic residues and by-products has increased as a result of the development of industrial activity, fluctuations in the fossil fuel markets, and the need to reduce the effects of global climate change (Amoah et al., 2019). This is because the biorefinery idea has been acquiring significant relevance due to the integration of biomass conversion processes to produce energy and products with added value. They comprise biochemical, bio-based chemicals and biofuels.

C. ladanifer is used as a lignocellulosic feedstock to produce goods with added value. For the synthesis of bioethanol, lignocellulose materials must first undergo pre-treatment (for instance, with acids), followed by enzymatic hydrolysis and fermentation (Duarte et al., 2023). According to literature, diluted acid pre-treatment successfully solubilized carbohydrates from *C. ladanifer* residues with a maximum concentration of 302.2 mg of total sugars/g of dry material. Temperature, cellulase concentration, and incubation duration all had an impact on the subsequent enzymatic hydrolysis of the solid fractions produced by the acid hydrolysis pre-treatment (Bert et al., 2017).

6 Conclusion

Grass-legume mixture (*Stylosanthes-Pennisetum*) can be used as initial colonizers for stabilization of a very steep slope, after blanketing with topsoil and coir mat. Addition of fast-growing, annual, high biomass-producing species is essential to increase the SOM and moisture. Fast-growing species can form massive green cover in a very short time and play an important role in reducing erosion and conserving moisture. In the initial year, both *Stylosanthes-Pennisetum* colonized together, but from the second year onward, *Stylosanthes* covered the entire slope surface and eradicated the *Pennisetum*. Natural colonization of other herbaceous leguminous species (*Desmodium* spp., *Tephrosia*) was also observed. The short life cycle of *S. sesban* and *C. juncea* can play a role of green manure for the soil as it adds organic carbon and nitrogen to the soil after drying. This would also be economic. *S. hamata* is a very effective forage legume that uplifts the nitrogen economy of degraded soils in a short period of time. These legumes can effectively influence the nitrogen cycling in soils due to presence of root nodules, which fix atmospheric nitrogen. Aftercare and maintenance of the eco-restored site, particularly watering and protection of cattle, is essential.

However, the main elements of LCBs that could be used to produce biochar and ethanol was covered in the study. Most importantly, the study emphasized the use of LCB-derived biochar and hydrolyzed LCB-derived biochar to enhance bioethanol production and remediate heavy metal-contaminated soil while taking phytoremediation into account. Using biochar as a powerful adsorbent for the detoxification of the fermentation inhibitors in the bioethanol has the potential to reduce both resource waste and environmental pollution when bioethanol and biochar production are combined into a single self-sufficient circular economy model.

The products obtained from *C. ladanifer* extraction vary greatly. Therefore, the commercialization of *C. ladanifer* may help make better use of an underutilized endogenous resource while fostering residue management, easing environmental pressure, and fostering sustainable development by generating new markets, creation of new employment and products. Yet, in order to turn biorefineries into an industrial reality, a distribution map that highlights local potential availability and factors connected to transportation logistics must be developed. This will allow a thorough sustainability analysis to be built. When choosing biomass commercial valorization paths, it is also necessary to do a comparative techno-economic analysis of the various processes, namely, with the aid of modeling tools, as well as a life cycle evaluation in terms of environmental, social, and economic sustainability.

References

Agu, O.S., et al., 2019. Pretreatment of crop residues by application of microwave heating and alkaline solution for biofuel processing: a review. In: Renewable Resources and Biorefineries. IntechOpen, https://doi.org/10.5772/intechopen.79103.

Alves-Ferreira, J., et al., 2023. *Cistus ladanifer* as a potential feedstock for biorefineries: a review. Energies 16 (1). https://doi.org/10.3390/en16010391.

Amoah, J., et al., 2019. Co-fermentation of xylose and glucose from ionic liquid pretreated sugar cane bagasse for bioethanol production using engineered xylose assimilating yeast. Biomass Bioenergy 128 (December 2018), 105283. https://doi.org/10.1016/j.biombioe.2019.105283.

Araújo, I.C.S., Costa, M.C.G., 2013. Biomass and nutrient accumulation pattern of leguminous tree seedlings grown on mine tailings amended with organic waste. Ecol. Eng. 60, 254–260. https://doi.org/10.1016/j.ecoleng.2013.07.016.

Bert, V., et al., 2017. How to manage plant biomass originated from phytotechnologies? Gathering perceptions from end-users. Int. J. Phytoremediation 19 (10), 947–954. https://doi.org/10.1080/15226514.2017.1303814.

Caravaca, F., et al., 2002. Improvement of rhizosphere aggregate stability of afforested semiarid plant species subjected to mycorrhizal inoculation and compost addition. Geoderma 108 (1–2), 133–144. https://doi.org/10.1016/S0016-7061(02)00130-1.

Carrion-Prieto, P., et al., 2017. Valorization of *Cistus ladanifer* and *Erica arborea* shrubs for fuel: wood and bark thermal characterization. Maderas: Cienc. Tecnol. 19 (4), 443–454. https://doi.org/10.4067/S0718-221X2017005000401.

Diti, A., 2003. VGT: A Bioengineering and Phytoremediation Option for the New Millennium. APT Consulting group, Bangkok.

Duarte, B., et al., 2023. *Cistus ladanifer* metal uptake and physiological performance in biochar amended mine soils. S. Afr. J. Bot. 153, 246–257. https://doi.org/10.1016/j.sajb.2023.01.002.

El Mamoun, I., et al., 2020. Zinc, lead, and cadmium tolerance and accumulation in *Cistus libanotis*, *Cistus albidus*, and *Cistus salviifolius*: perspectives on phytoremediation. Remediation 30 (2), 73–80. https://doi.org/10.1002/rem.21638.

Fehmi, J.S., Kong, T.M., 2012. Effects of soil type, rainfall, straw mulch, and fertilizer on semi-arid vegetation establishment, growth and diversity. Ecol. Eng. 44 (May 2014), 70–77. https://doi.org/10.1016/j.ecoleng.2012.04.014.

Huang, L.Z., et al., 2021. Recent developments and applications of hemicellulose from wheat straw: a review. Front. Bioeng. Biotechnol. 9 (June), 1–14. https://doi.org/10.3389/fbioe.2021.690773.

Lal, K., et al., 2013. Productivity, essential oil yield, and heavy metal accumulation in lemon grass (*Cymbopogon flexuosus*) under varied wastewater-groundwater irrigation regimes. Ind. Crop. Prod. 45 (February 2013), 270–278. https://doi.org/10.1016/j.indcrop.2013.01.004.

Lenka, N.K., et al., 2012. Soil aggregation, carbon build up and root zone soil moisture in degraded sloping lands under selected agroforestry based rehabilitation systems in eastern India. Agric. Ecosyst. Environ. 150, 54–62. https://doi.org/10.1016/j.agee.2012.01.003.

Maiti, S.K., 2013. Ecorestoration of the Coalmine Degraded Lands. Available from: http://journal.um-surabaya.ac.id/index.php/JKM/article/view/2203.

Maiti, S.K., Maiti, D., 2015. Ecological restoration of waste dumps by topsoil blanketing, coir-matting and seeding with grass-legume mixture. Ecol. Eng. 77, 74–84. https://doi.org/10.1016/j.ecoleng.2015.01.003.

Maiti, S.K., Saxena, N.C., 1998. Biological reclamation of coalmine spoils without topsoil: an amendment study with domestic raw sewage and grass-legume mixture. Int. J. Surf. Min. Reclam. Environ. 12 (2), 87–90. https://doi.org/10.1080/09208118908944028.

Mitchell, D.J., et al., 2003. Field studies of the effects of jute geotextiles on runoff and erosion in Shropshire, UK. Soil Use Manag. 19 (2), 182–184. https://doi.org/10.1079/sum2002165.

Mohapatra, S., et al., 2017. Application of pretreatment, fermentation and molecular techniques for enhancing bioethanol production from grass biomass—a review. Renew. Sust. Energ. Rev. 78 (February), 1007–1032. https://doi.org/10.1016/j.rser.2017.05.026.

Pratto, B., et al., 2020. Experimental optimization and techno-economic analysis of bioethanol production by simultaneous saccharification and fermentation process using sugarcane straw. Bioresour. Technol. 297, 122494. https://doi.org/10.1016/j.biortech.2019.122494.

Sawtschuk, J., Gallet, S., Bioret, F., 2012. Evaluation of the most common engineering methods for maritime cliff-top vegetation restoration. Ecol. Eng. 45, 45–54. https://doi.org/10.1016/j.ecoleng.2010.12.019.

Shafiei Alavijeh, R., Karimi, K., van den Berg, C., 2020. An integrated and optimized process for cleaner production of ethanol and biodiesel from corn Stover by *Mucor indicus*. J. Clean. Prod. https://doi.org/10.1016/j.jclepro.2019.119321. Elsevier Ltd.

Shao, Q., et al., 2014. Effectiveness of geotextile mulches for slope restoration in semi-arid northern China. Catena 116, 1–9. https://doi.org/10.1016/j.catena.2013.12.006.

Waghmare, P.R., et al., 2018. Enzymatic hydrolysis of biologically pretreated sorghum husk for bioethanol production. Biofuel Res. J. 5 (3), 846–853. https://doi.org/10.18331/BRJ2018.5.3.4.

Zhang, G., et al., 2016. Effects of biochars on the availability of heavy metals to ryegrass in an alkaline contaminated soil. Environ. Pollut. 218, 513–522. https://doi.org/10.1016/j.envpol.2016.07.031.

Prosopis juliflora (Sw) DC: Potential for bioremediation and bio-based economy

Majeti Narasimha Vara Prasad[a] and J.C. Tewari[b]

[a]Department of Plant Sciences, School of Life Sciences, University of Hyderabad (An Institution of Eminence), Hyderabad, Telangana, India [b]Central Arid Zone Research Institute, Jodhpur, Rajasthan, India

1 Introduction

It is a general belief that non-native or alien plants when grow outside their natural territory could become invasive. These invasives are widely distributed in a variety of ecosystems throughout the world. Many invasive alien species support farming and forestry systems positively. However, some alien species become invasive when they are introduced deliberately or unintentionally outside their natural habitats into new areas where they express the capability to establish, invade, and outcompete native species. According to the International Union for Conservation of Nature and Natural Resources (IUCN), alien invasive species means an exotic species that becomes established in natural or semi-natural ecosystems or habitat, an agent of undesirable change which threatens the native biological diversity. Invasive species are, therefore, considered to be a serious hindrance to the conservation and profitable use of biodiversity, with significant undesirable impacts on the services provided by ecosystems. Alien invasive species are supposed to demand huge resources that would ultimately bring several undesirable changes in the ecosystem functioning. However, in the case of *P. juliflora*, though exotic, with proper management, it can be converted to be an invaluable bioresource.

2 About *P. juliflora*

P. juliflora (Sw.) DC (Velvet Mesquite), (Fabaceae, sub-family Mimosoideae) (henceforth referred to as *Prosopis*). It contains 44 species, of which 40 are native to the Americas, 3 to Asia, and 1 to Africa. *Prosopis* is conspicuously thorny with a wide, flat-topped crown. The root system of *Prosopis* consists of a deep tap root (phreatophyte), sometimes reaching unusual depth of 35 m, combined with extensive lateral roots. *Prosopis* is especially suitable for dry sites with annual rainfall between 150 and 700 mm. Tap roots contribute to a stable anchoring of the tree and expand toward groundwater reserves. They are essential during periods of drought when only deep water sources are available. The depth of the roots depends on the quality and structure of the soil and the availability of soil water; it is also determined by the density of the stand. Once the water source is reached, the roots extend horizontally in the direction of the water flow. *Prosopis* species are adapted to areas with low rainfall and long periods of drought once they are established and are able to tap groundwater or any other water source during the first years. The lateral roots play an important role during rainy seasons or periods of abundant water, for instance, in irrigated areas. The trees are also able to absorb moisture through their foliage during light rains or from dew or other atmospheric sources of moisture (Prasad and Tewari, 2016).

3 Global distribution

Africa, Argentina, Australia, Brazil (north-east), Cameroon, Caribbean, Central America, Egypt, Ethiopia, Ethiopia, Hawaii, India (Fig. 1), Kenya, Nigeria, Pakistan, Paraguay Peru, Peruvian-Ecuador, Portugal, Senegal, Spain, Sri Lanka, Sudan, Uganda, USA, and Yemen.

Prosopis has spread rapidly in the arid regions, people as well as nature have taken advantage of this alien plant. It has some biological characteristics that foster invasion; hence,

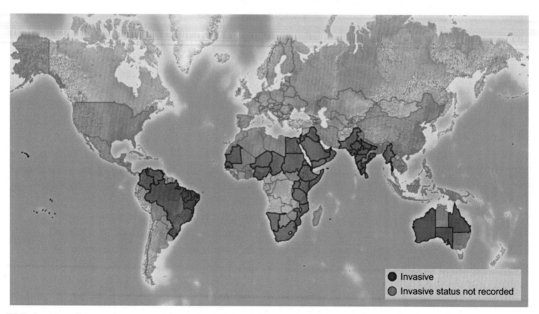

FIG. 1 Distribution *of Prosopis juliflora*. In: Invasive species compendium, Wallingford, UK: CAB International.2022 https://www.cabi.org/isc.

appropriate management practices are also needed to exploit its resource potential. In spite of its invasive nature, services provided by *Prosopis* are numerous. A large number of publications have appeared on various aspects of its bioresource potential. The notable bioeconomic aspects are 1. Phytochemistry, 2. Allelopathy 3. Antioxidant, 4. Food, 5. Biopesticide, 6. Phytoremediation 7. Bioethanol, 8. Synthesis of nanoparticles 9. Timber and fire wood, 10. Improvement of livelihood of rural community, and 11. Medicinal uses, etc. (Table 1, Fig. 2).

TABLE 1 *Prosopis juliflora* for bioremediation and bioeconomy.

	Author(s) (Year)
Characterization of a new natural fiber derived from the bark of *P. juliflora*	Saravanakumar et al. (2013)
Alkaloid-enriched extract from *P. juliflora* pods and its influence on in vitro ruminal digestion	Dos Santos et al. (2013)
Bioremediation for fueling the biobased economy	Tripathi et al. (2016)
Within 8 years of *P. juliflora* forestation reclaimed coal mine soil and stores of carbon, nitrogen, and phosphate	Ahirwal et al. (2017)
P. juliflora (mesquite) gum exudate acts as a potential excipient	Basu et al. (2017)
Useful for restoring marginal and degraded areas in India in a sustainable manner	Edrisi et al. (2018)
Using plants to manage polluted areas especially Phytomanagement of fly ash	Pandey and Bauddh (2018)
Evaluation of the antioxidant potential of mesquite grains flour in hamburger meat product	de Melo Cavalcante et al. (2019)
P. juliflora restores degraded land in North India - sustainability analysis	Edrisi et al. (2020)
Biomass is a resource for bioenergy production and environmental bioremediation	Gavrilescu (2020)
2020. Pharmacological and pharmacognostical aspect of *P. juliflora*: A review	Yadav and Rana (2020)
A case study on the remediation of bauxite residue using an integrated method using microbes and plantations	Dubey and Dubey (2021)
In the semiarid zone, mesquite (*P. juliflora*) extract was used as a phytogenic supplement for sheep completed on grass	Férrer et al. (2021)
Prosopis Species -An Invasive Species and a Potential Source of Browse for Livestock in Semi-Arid Areas of South Africa	Ravhuhali et al. (2021)
Review of catalytic activities of biosynthesized metallic nanoparticles in wastewater treatment	Saim et al. (2021)
Phytoremediation potential of invasive species growing in mining dumpsite	Singh et al. (2021)
Establishment of phytocoenoses on coal mine over burden of pranahita godavari basin, Telangana	Suthari and Prasad (2021)
Ecological restoration of fly-ash disposal areas: Challenges and opportunities	Yadav et al. (2021)

Continued

TABLE 1   *Prosopis juliflora* for bioremediation and bioeconomy—cont'd

	Author(s) (Year)
P. juliflora methyl ester as a fuel in the CI Engine	Duraisamy et al. (2022)
Utilizing biomaterials from *P. juliflora* for carbon sequestration to reduce the effects of climate change	Edrisi et al. (2022)
P. juliflora leaves extract with antifungal traits	Naik et al. (2022)
Role of Prosopis in reclamation of salt-affected soils and soil fertility improvement	Singh (2022)
Growth and yield response of cowpea (*Vigna unguiculata* L.) in sodic soil amended with marine gypsum, biochar and bioinoculants	Teja et al., 2022
The paradigm shift in *P. juliflora* use through community participation by developing value chain of value-added products from pods	Tewari et al. (2022)
Application of silicon to reduce the negative effects of soil salinity on plants: a review	Adil et al. (2023)
Coconut shell charcoal can be made in an eco-friendly way through pyrolysis *P. juliflora* (mesquite) gum exudate	Azit et al. (2023)
The function of its peroxidases in organic pollutant bioremediation	Basumatary et al. (2023)
Invasive plant species of Maharashtra state: A review	Chavre and Patil (2023)
Prosopis galactomannans are useful for production of active antifungal edible coating based on nanocomposite cast films	de Souza et al. (2023)
The primary microsymbiont of *Prosopis chilensis* in the dry soils of Eastern Morocco is *Ensifer meliloti*	El Idrissi et al. (2023)
Results from the participatory rural evaluation on social protection for inclusive development in the Afar region of Ethiopia	Fre et al. (2023)
A summary of recent developments in the removal of the antibiotic metronidazole, an emerging pollutant, from aquatic habitats	Ghosh et al. (2023)
Using sustainable grazing methods to protect biodiversity in a tropical Asian grassland	Joshi (2023)
Mechanical characterization of hybrid natural fiber composites under static and dynamic conditions for engineering applications	Kumar et al. (2023)
Potential toxicity and bioavailability of engineering nanomaterials (ENMs)	Loredo-Portales et al. (2023)
Analysis of breakthrough curves and fixed-bed column performance for the removal of phosphorus from wastewater	Lv and Li (2023)
the creation of carbon dots from waste biomass and their use	Mansi and Gaurav (2023)
Urban dust pollution tolerance indices of selected plant species for development of urban greenery in Delhi	Patel et al. (2023)
Impact of Biochar Fraction on Thermal Characteristics of Soil-Biochar Composite	Patwa et al. (2023)
Sustainable animal production and meat processing. In Lawrie's Meat Science (pp. 727–798)	Ponnampalam and Holman (2023)
Multimodality: phantom imaging for superparamagnetic graphene composites using green technology for theranostic nanosystems	Preethy et al. (2023)

TABLE 1 *Prosopis juliflora* for bioremediation and bioeconomy—cont'd

	Author(s) (Year)
Bioenergy Crop-Based Ecological Restoration of Degraded Land. In Bio-Inspired Land Remediation (pp. 1–29). Springer, Cham	Ranđelović and Pandey (2023)
Enhanced waste cooking oil biodiesel with Al2O3 and MWCNT for CI engines. Fuel, 333, p. 126429	Sathish et al. (2023)
Global principles in local traditional knowledge: A review of forage plant-livestock-herder interactions	Sharifian et al. (2023)
P. juliflora valorization via microwave-assisted pyrolysis: Optimization of reaction parameters using machine learning analysis	Suriapparao et al. (2023)
The Cytotoxicity Effect of Ethanol Extract and Alkaloid Fraction of *Mirabilis jalapa* Leaves in Hepatocarcinoma Cell Line	Suselo (2023)
Action against invasive species: Charcoal production, beekeeping, and Prosopis eradication in Kenya	Tabe-Ojong Jr (2023)
A comprehensive evaluation on thermo-chemical potential of longan waste for the production of chemicals and carbon materials	Tian et al. (2023)
Advances in the delivery systems for oral antibiotics	Wang et al. (2023)
Engineered nanomaterials in crop plants drought stress management	Weisany and Khosropour (2023)
Plant Biodiversity Conservation in Ethiopia: A Shift to Small Conservation Reserves. Springer Nature	Yaynemsa (2023)
Biomass estimation models for four priority Prosopis species: Tools required for forestry management in overexploited arid ecosystems	Zarzosa et al. (2023)

Pharma applications

Bioactive polysaccharide
Bioadhesive delivery of drug
Cast films for wound healing
Polymer for tablet formulation
Food process industry
Formulation of gels
Hydrocolloid production
Microwave assisted preparations
Low calorie fruit nectar
Spray-dried encapsulation
Tablet binder

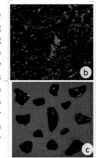

Industrial uses

Adhesive and binder
Ceramics
Coating agents
Cosmetic
Dietary fiber
Emulsifiers
Encapsulator in textiles
Lithography
Packaging films
Stabilizers
Texture modifiers
Thickeners

FIG. 2 Plant-based excipients play a key role in the formulation of efficient drug delivery mechanism. *Prosopis* gum is a popular and better used plant excipient (Basu et al., 2017).

I. Bioproducts from contaminated soil and water

4 *Prosopis* as livestock feed

The use of *Prosopis* pods as a livestock feed is of great advantage in the arid and semiarid areas of India. *Prosopis* grows naturally in drought affected areas, requiring only small quantities of water for its growth. Malnutrition is one of the multiple hazards that livestock faces under these harsh conditions. It has been observed that a mature 10-year-old tree grown under good conditions of soil and water availability is able to produce 90 kg of pods per year. A pod yield of 2000 kg/ha of *Prosopis* was estimated for the unmanaged Arizona dessert in North America and 4000–20,000 kg/ha in the arid Hawaiian savannas. In the north-east of Brazil, pod yields of 2–3 tons/ha are produced even on shallow stony soils with vegetation typical of semiarid regions and with no agricultural value. Pod production of *Prosopis* can be increased three to four times if irrigation is provided during the flowering period, which lasts 2 months. However, pod production may continue even after a drought period of 2–3 years (Prasad and Tewari, 2016).

Prosopis pods contain 20%–30% sucrose and about 15% crude protein. They can be used in various ways as animal feed without causing any adverse digestive effects when used properly. Although crushed and ground pods are suggested as an additional feed, the feasibility of these techniques is questioned due to limited availability of manpower that processes the pods, causing additional expenditure to the average livestock farmer. It is further suggested that the pods should be mixed with other feed and that continued feeding of pods to cattle as the only diet should be avoided. Pods in well-preserved silage have produced excellent livestock feed. However, the cost and availability of those additives make their use difficult. The ensiling of *Prosopis* pods together with dry grass, such as the dessert sewan grass *Lasiurus sindicus* Henr., is able to improve the nutritional value of the feed without pretreatment (Prasad and Tewari, 2016).

The inclusion of *Prosopis* pods in the diet of Indian sheep, under a feedlot system growing lambs for meat production, indicated that the incorporation of 15%–30% pods into the feed significantly lowered the feeding costs. A diet consisting of 15% pods added to other ingredients, such as wheat straw (30%), alfaalfa hay (20%), and other minor components such as wheat bran, urea, mineral mixture, and salt (3.5%), showed superior feed efficiency. In whatever way the pods are used, crushed, and ground or otherwise prepared as a mixture, whenever the seeds are destroyed in the process, it has an influence on the control of the *Prosopis* invasion.

The use of *Prosopis* pods as livestock feed provides the farmer monetary savings. It is a low-cost solution compared with other commercial animal feeds. The pods have a high concentration of proteins, and the tree easily adapts to semiarid conditions where hardly anything else is able to grow. Other products such as honey and bee wax provide additional income to the families in these areas and thus contribute to diversifying their livelihood opportunities.

Very few animals graze on the foliage of *P. juliflora* because of its unpalatable leaves and long spines. The leaves of *Prosopis cineraria*, on the other hand, are browsed on by various animals. *P. cineraria* (khejri), which is native to India, is a slow-growing tree that develops a height of 6 m only after 10–15 years, compared with a height of 12–15 m in 4–5 years in *P. juliflora*. Khejri is well adapted to low rainfall (150–500 mm annually) areas. The particular feature of the tree is that it produces its leaves, flowers, and pods during the hottest period of the year, between March and June. Thus, the tree offers great opportunities as a forage resource in the extremely arid zones during the hottest season. Although khejri produces animal feed only after 10 years, under proper management regimes, it continues production for up to two decades. Annually, the tree produces an average of 25–30 kg of dry leaf forage (Prasad and Tewari, 2016).

5 *Prosopis*—Fuel wood source

The most common use of *Prosopis* is as fuel wood, especially among the landless and poor. Farmers who are economically better off use gas, but after the growing season is over, they often engage in cutting *Prosopis* for firewood. Hotels and restaurants use *Prosopis* as fuel wood processed as charcoal for preparing food (Fig. 3).

Although people were aware of the advantage that *Prosopis* is a free fuel wood for poor, thorns cause injuries while collecting firewood. There are possibilities to genetically improve *Prosopis* through selective breeding and hybridization with other less thorny *Prosopis* species, in order to obtain a variety that would make the handling of the tree easier (Prasad and Tewari, 2016).

Prosopis was introduced into the drylands of India one-and-half centuries ago mainly for the purpose of conservation but has meanwhile become the main source of fuel wood in the rural areas, fulfilling more than 70% of the firewood requirements of rural people living in the areas (Prasad and Tewari, 2016). Today, the tree has established itself, spreading all over the arid and semiarid regions, and in some areas it has become an invasive weed.

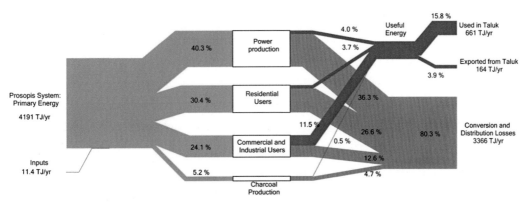

FIG. 3 Prosopis energy flow diagram, India's biofuel and biomass economies. *After Baka, J. and Bailis, R., 2014. Wasteland energy-scapes: a comparative energy flow analysis of India's biofuel and biomass economies. Ecol. Econ. 108, 8–17, with permission from Elsevier.*

I. Bioproducts from contaminated soil and water

5.1 *Prosopis* for production of timber

Prosopis timber had high demand for manufacturing various products such as furniture or boats. Large trees of more than 25 cm diameter are desirable and can be achieved through various stand management practices. There is no need to establish *Prosopis* plantations especially for this purpose. There are already large areas of invaded wastelands that can be managed to produce stem sizes that are needed in the wood manufacturing industry (Prasad and Tewari, 2016).

The quality of *Prosopis* wood is similar to several other species of the genius *Prosopis* and can easily compete with Indian rosewood (*Dalbergia latifolia* Roxb.), Indian teak (*Tectona grandis* L.), and cocobolo (*Dalbergia retusa* Hemsl.), three commonly recognized fine hardwoods, in relation to physical and mechanical properties. Tangential and radial shrinkage values are similar in *prosopis*, which indicate that the wood shrinks or swells equally in both directions. The hard surface of *Prosopis* wood makes it an ideal raw material for furniture. The wood of *Prosopis* is light brown when cut, and after drying and aging, it darkens, turning into a dark reddish golden brown color. *Prosopis* can be classified as one of the world's most precious tropical hardwoods (Prasad and Tewari, 2016).

In many South American countries, such as Argentina and Peru where *Prosopis* species are indigenous, furniture and flooring made of *Prosopis* wood are highly appreciated, and the technology to produce sawn timber has been developed during recent years to suit these species. In India, the potential for managing trees to be used as timber is still under-exploited. It has been concluded that chainsaw milling that is portable and able to cut small-diameter, crooked, and small logs is the most suitable method for the conversion of *Prosopis* wood into timber (Prasad and Tewari, 2016).

Small-scale enterprises using chainsaws to mill *Prosopis* would improve livelihoods by adding value to an already existing resource and creating new working places. At the same time, it has a controlling effect on *Prosopis* invasions (control through utilization). A critical aspect of profitable saw timber production is proper marketing. The more the timber fulfills the requirements of the potential user, the more profit can be made. As research has indicated, the most common sizes of timber that are used for the manufacture of furniture and flooring are less than 10 cm in width and 1.5 m in length. Those measures can be easily extracted from *Prosopis* trees (Prasad and Tewari, 2016).

5.2 *Prosopis* for bioethanol

Studies on lignocellulosic ethanol and on pelletization and delignification of cellulosic biomass *Prosopis juliflora* have been carried out via two-stage pretreatments involving an alkali and an acid (National Bioresource Development Board (NBDB)). *P. juliflora* resulted in 85% or higher delignification for all the three residues using 0.5% NaOH at room temperature for 24 h on the basis of holo-cellulose recovery and total reducing sugar yield per gram of initial substrate.

Enzymatic delignification of banana stem/wheat straw using laccase from *Pleurotus ostreatus* showed good results: enzymatic hydrolysis and fermentation of lignocellulose.

Two novel bifunctional cellulolytic enzymes with good activities were developed for application.

In lignocellulosic ethanol:

(i) Endoglucanase/-glucosidase chimera (EG5):
(ii) Endoglucanase/xylanase (Endo5A-GS-Xyl11D):

Also an engineered *E. coli* strain (SSY10) has been developed to ferment C5/C6 into ethanol via native pathway engineering. SSY10 produced ethanol from C5/C6 sugars at the rate of 0.7 g/L/h.

SSY10 also efficiently fermented lignocellulosic hydrolysates generated via acid treatment and via ammonia treatment to ethanol at neat theoretical maximum yield. *Paenibacillus* ICGEB2008 strain that produces cellulolytic enzymes as well as ferments sugars into ethanol and 2, 3-butanediol has been isolated. This strain could ferment glucose, xylose, cellobiose, and glycerol and produce ethanol.

5.3 *Prosopis* for honey production

Prosopis regularly produces an abundant amount of flowers that are used for forage by honeybees even during times of drought. After the introduction of *Prosopis* into Hawaii in the 1930s, the island became one of the largest producers of honey (Fig. 6).

The honey produced from *Prosopis* flowers is claimed to be of excellent quality. A substantial amount of this honey (300 metric tons) was harvested and marketed during a 5-year period in the early 1990s in the state of Gujarat alone. The rare local honeybee species *Apis florea* (Fabricius) and *Apis cerana* (Fabricius) use the nectar and pollen of *Prosopis* to produce honey, which is also used in traditional medicine (Prasad and Tewari, 2016).

A large number of beekeepers in South India earn their living from *A. cerana* bees, which are known to be good pollinators. *A. florea* (dwarf honeybee) has adjusted to the extreme climate of the arid and semiarid zones and occurs in large numbers in the arid zones of Gujarat; it is also present in smaller numbers in South India. Bees wax is used as a pharmaceutical and industrial raw material, out of which candles, creams, and balms are manufactured (Prasad and Tewari, 2016).

6 Medicinal uses

Several beneficial medicinal compounds have been reported from *P. juliflora*. Some of the prominent compounds are steroids, tannins, leucoantho cyanidin, and ellagic acid glycosides. Extracts of *P. juliflora* seeds and leaves have several in vitro pharmacological effects such as antibacterial, antifungal, and anti-inflammatory properties (Dos Santos et al., 2013; Prasad and Tewari, 2016). Flavonoids are water-soluble phytochemicals showing the antioxidant, anticancer, and anti-inflammatory activities. These prevent cells from oxidative damage and carcinogenesis. Flavonoids are also used to cure some heart-related diseases. Flavonoids occur virtually in all parts of the plant, the root, heartwood, sapwood, bark, leaf, fruit, and alkaloids and their derivatives are used as basic medicinal agents for their analgesic, antispasmodic, and bactericidal activities and properties (Dos Santos et al., 2013, Prasad and Tewari, 2016). Alkaloid-rich fractions of *P. juliflora* are active antifungal and antibacterial agents. In this study, *P. juliflora* was taken as a plant source to screen its anti-pustule active compounds. Screened compounds were tested for their anti-pustule behavior by studying the inhibition levels of Staphylococcus sp. Combined effect of bioactive compound from *P. juliflora* along with commercially available anti-pimple creams (Clinigex and Clingel) was chosen as synthetic anti-pimple cream for research studies (Prasad and Tewari, 2016).

6.1 Char coal production

Charcoal is not commonly used as fuel among rural households in Andhra Pradesh. It is more frequently used in tea shops and hotels in the urban areas. In industries, where much energy is needed, such as in the Raigado steel plant, charcoal made from *Prosopis* is used in large quantities. Dobies, the Hindi name for laundries, almost always use charcoal in the process of ironing. The irons used in India are made of heavy metal, and charcoal is put inside to heat up the iron plate. On average, an iron that is used the whole day for 8–10h consumes approximately 7kg of charcoal. On average, 1k of charcoal costs 5 Rupees. The handle of the iron is made of *Prosopis* because, according to a manufacturer, it is easier to drill a hole through *Prosopis* wood than through *A. nilotica*, the other alternative species for this purpose. Laundries are found all over India in huge numbers, which creates a high demand for charcoal. (Prasad and Tewari, 2016).

Charcoal is produced by burning *Prosopis* wood in "bhatti" formations (the local name for the kiln used). Batis are arrangements of the wood in round circles with a top layer of coal dust that is sprinkled with water. The whole process, from first forming the "bhatti" to the final loading of the charcoal onto lorries, takes around 2 weeks. The burning process itself requires between 5 and 8 days and needs to be carefully guarded to avoid an excessive flow of air into the kiln. Too much air would provoke an open fire, thus burning the wood instead of transforming it into charcoal. One bati consists usually of approximately 60 tons of fresh *Prosopis* wood, which produces 15–16 tons of charcoal. According to the craftsman, it takes approximately 4kg of fresh *Prosopis* wood to produce 1kg of charcoal. The processing site in the Nalgonda District sells 40–50 tons of charcoal monthly, mainly to industrial plants and in smaller quantities to charcoal wholesalers or restaurants (Fig. 4).

The charcoal burning process is managed and guarded by an employed family that usually lives at the burning place. This was the case in all three sites visited. Extra workers were employed, to unload the wood and arrange the bati, as well as for loading the transferred charcoal. The charcoal enterprises are owned privately and situated outside towns in areas where *Prosopis* is abundantly available. Because of their high energy content, *Prosopis* roots are preferred for charcoal production. Whenever land is cleared, for example, for real estate properties or building projects, either manually or in most cases with the help of excavators, *Prosopis* roots are dug up and sold to processing sites. The conversion of wastelands covered by *Prosopis* into eucalypt plantations also produces root material. As roots do not need to be chopped into specific sizes, they are traded for 500 Rupees per ton. Other parts of *Prosopis* wood demand the same price of 500 Rupees per ton, but the wood needs to be cut into sizes that are suitable for the construction of a bati. For cutting the wood into suitable sizes, extra workers are needed, and these wood cutters demand a payment of around 200 Rupees per ton.

FIG. 4 *Prosopis juliflora* bioremediation and bioeconomic potential. (A) *P. juliflora* phytostabilizing industrially polluted sites. (B) *P. juliflora* tree in semiarid region of Telangana. (C) Tender fruits of *P. juliflora*. (D and E) Goats and sheep feeding on tender foliage and pods of *P. juliflora*. (F and G) Wood of *P. juliflora* piled up for char coal production. (H and I) Charcoal production in traditional style. (J, K, and L) Charcoal and charcoal dust.

The investigation revealed that a medium-sized charcoal enterprise converts approximately 240 tons of *Prosopis* into 40–50 tons of charcoal monthly. The wood costs for the entrepreneur vary between 500 and 700 Rs per ton, depending if the wood needs to be cut into suitable pieces. The market price for one ton of charcoal is 5000 Rs. Calculating the monthly income of a middle-size charcoal enterprise reveals that, after the expenditure for the *Prosopis* wood, there is a monthly net income of 81,000 Rs. In addition, it creates work for several people, including wood cutters, wood collectors, and lorry drivers who transport the wood to the production ground and the charcoal to different wholesale places (Fig. 5).

7 Natural pesticide (allelopathy)

Prosopis foliage has allelopathic effects on seed germination and seedling growth of bermudagrass (*Cynodon dactylon*); three cultivars of *Zea mays* L. (R 796, Gohar, EV 1081), four cultivars of *Triticum aestivum* L. (Inqalab, Chakwal, Pak 81, Rohtas), and *Albizia lebbeck* (L.).

7.1 *Prosopis* pod and seeds are a rich source of carbohydrate

The gum of *Prosopis* is a good encapsulating material. Cardamom (*Elettaria cardamomum*) essential oil microcapsules were produced using its gum.

FIG. 5 Multiple uses of *Prosopis juliflora*, (A) Its biomass is the major feedstock for biomass-based gasifier/power plant. It is the most common tree in semiarid tropic regions. (B) Charcoal is used as a carrier for plant growth promoting bacteria, e.g., *Azospirillum*. (C) Drug de-addicting pellets produced from its extract. (D, E) High-quality animal feed from its pods. (F) and (G) *P. juliflora* provides shelter and breeding ground to migratory birds (Uppalapadu near Guntur, Andhra Pradesh, India, contributing to environmental protection and conservation. (H). Furniture products from its timber.

Prosopis seeds serve as alternate source of the galactomannans and are possible to extract into water at room temperature. Galactomannans are water-soluble neutral polysaccharides, composed of a linear mannan backbone bearing side chains of a single galactose unit. These have wide industrial applications, mainly as thickening and stabilizing agents in a range of applications. Its leaves are ground with tobacco (*Nicotiana tabacum* L.) and lime and placed on painful tooth for relief. It also well known for its antibiotic and antibacterial properties. Polyphenols and tannins of medicinal importance have been reported to be obtained from the stem bark and pods (Fig. 5–7).

8 Restoration of contaminated/degraded land, phytoremediation

Prosopis is an ideal species for stabilizing the pegmatitic tailings of mica mines in Nellore district of Andhra Pradesh, India. It is also helpful for reclamation of copper, tungsten, marble, dolomite mine tailings and is a green solution to heavy metal-contaminated soils. It is an appropriate species for rehabilitation of gypsum mine spoil in arid zone; restoration of sodic soils. It outperformed all other tree species in sand dune stabilization. *Prosopis juliflora* seedlings cultivated in hydroponics are able to bioaccumulate Ni, Cd, and Cr. Chromium (Cr) is

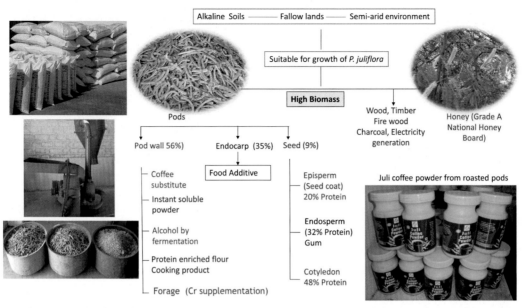

FIG. 6 Multiple uses of pods and woody biomass. Highquality honey is produced from its flowers including coffee powder-like product (Juli coffee powder).

FIG. 7 Dried wood is used for production of charcoal from which Granular Activated Carbon Cartridges are produced for purification of water with contaminants of emerging concern (Menya et al., 2023).

I. Bioproducts from contaminated soil and water

an essential mineral for ruminants in tropical regions. Its pods contain high Cr concentration (upto 150 p.p.b.), thus, supplementing Cr in fodder for animals (goats). Cr requirement for animals is >0.1 p.p.m., while the toxic level was 1000 p.p.m. (Prasad and Tewari, 2016).

Prosopis was very much helpful for reclamation of copper mine tailings, e.g., copper mines in Arizona, USA, and abandoned mine waste in Mexico (Fig. 8). It is an appropriate species for rehabilitation of gypsum mine spoil in arid zone; restoration of sodic soils. A wide variety of *Prosopis* pure stands have been observed on soils industrially polluted. It outperformed all other tree species in sand dune stabilization. Arbuscular mycorrhizal inocula isolated from its rhizosphere when inoculated to seedlings of other agroforestry and social forestry legumes accelerated the growth of these seedlings on perturbed ecosystems. It is an ideal species for afforestation and helps in the reclamation of waste lands, stabilization of technogenically contaminated soils, and regulation of soil erosion. It is an important tree for afforestion/restoration of perturbed ecosystem and grows satisfactorily without any amendments. Mycorrhizae are reported to greatly improve the growth of *Prosopis* on high-pH soils (Fig. 9).

The association between *Prosopis* and *Leptochloa fusca* was successful for the restoration of salt lands. Therefore, grass-legume-tree association needs to be tested on different sites for remediation, if necessary with biotic and abiotic amendments (Fig. 10). There are a few

FIG. 8   (A–D) In copper mines near Globe and Tucson, Arizona, USA (Arizona Ranch, Resource Management and Mine Reclamation; ASARCO Inc. Copper Operations), the ecosystem rehabilitation and mine reclamation program is primarily based on cattle. Herds of cows are impounded with electric fence on mine by providing fodder and water for varying durations. Cows not only stabilize the soil by their hoofs but also enrich soil nutrient status via urination. They also augment microbes to the soil through the dung. This process is repeated at regular intervals in cycles. Thus, cattle accelerates rhizosphere development and improves plant root association via enriching soil microorganisms and nutrients. The results obtained with cattle for rhizosphere ecodevelopment are spectacular. (E) *Prosopis juliflora* colonization on copper mine tailings, Arizona, USA.

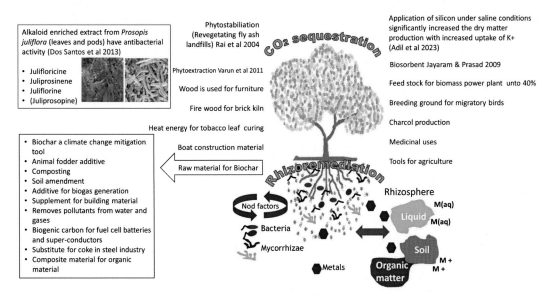

FIG. 9 Multiple uses of *Prosopis juliflora*. Alkaloid-enriched *P. juliflora* pods and leaves have antibacterial activity. *P. juliflora* biomass serves as raw material for biochar production. Above all, *P. juliflora* plays a significant role in carbon dioxide sequestration and rhizoremediation co-contaminated (inorganic and organic) substrates.

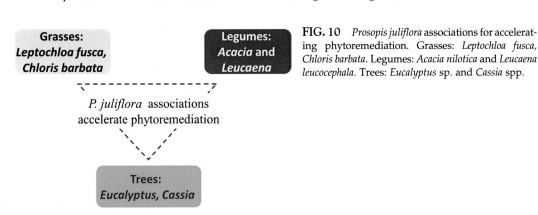

FIG. 10 *Prosopis juliflora* associations for accelerating phytoremediation. Grasses: *Leptochloa fusca, Chloris barbata*. Legumes: *Acacia nilotica* and *Leucaena leucocephala*. Trees: *Eucalyptus* sp. and *Cassia* spp.

publications reporting its fly ash landfills revegetating potential following different amendments and *Rhizobium* inoculation. However, *Prosopis* has some biological characteristics that foster invasion; hence, appropriate management practices needed to be developed for recommending it for phytoremediation (Diaz-Uribe et al., 2022). In rainfed agriculture, leaves from multipurpose trees are traditionally one of the main sources of nutrients that maintain soil fertility. It has been suggested that the use of tree legumes, such as prosopis, has potential for a minimal input farming system because *Prosopis* has the ability to grow with little or no irrigation, produces high yields, is able to fix nitrogen, and produces large amounts of high-quality protein (Prasad and Tewari, 2016) However, there is competition between trees and crops for water, nutrients, and light. Another point to consider is the number of years it takes before a constant improved soil status is reached for agricultural crop cultivation (Prasad and Tewari, 2016) (Fig. 4).

A widespread traditional practice in the semiarid tropics is to grow trees scattered or dispersed on croplands, known as "parklands." The degree of interaction between trees and crops depends on the density of trees. Trees in this system are rarely planted but occur and regenerate naturally. Farmers maintain these trees on their lands to provide fodder, fuel wood, timber, and non-wood products (Prasad and Tewari, 2016). The improvement of soil properties comes as a side product, along with the provision of shade for people and livestock.

Prosopis is potentially useful for rehabilitating degraded saline soils in dry land ecosystems where degradation is enhanced by high temperatures and irregular precipitation. *Prosopis* planted on degraded sodic soils increases soil fertility through adding to and increasing the soil organic C, nitrogen (N), available phosphorus (P) exchangeable potassium (K), calcium (Ca), and magnesium (Mg) levels. In addition, decreases in the exchangeable sodium level (Na), pH, and EC can be observed. These processes have a positive effect on the rehabilitation of sodic soils through the improvement of nutrient cycling and detoxifying sodicity. As an effect of prosopis, the crop productivity tested on wheat indicated higher germination, survival, plant growth, and grain yield (Prasad and Tewari, 2016).

Prosopis is reported to reduce EC and ESP significantly in saline-sodic soil. *Prosopis* improved the saline-sodic soil where agriculture was possible. Similarly, alkali soils were prepared suitable for forestry using *Prosopis*. The use of trees to remediate heavily metal-contaminated soils is profitable (Prasad and Tewari, 2016; Adil et al., 2023). *Prosopis* has high capacity to accumulate heavy toxic metals occurring in the soil. *Prosopis* is also able to accumulate lead (Pb) in various parts of the plant. The highest concentration of lead is found in the leaves, lower levels in the bark, and the lowest levels in the pods. The lead content in the plant can be used as an indicator to measure air pollution derived from industrial factories and vehicles. Because of the high amounts of heavy metals accumulated in *Prosopis* leaves and pods, it is recommended to keep livestock away from grazing on this species when it grows in heavy metal-polluted soils *Prosopis* colonized the industrial efflunt-laden oils produced by textile, paper products, tannery, chemical products, basic metal products, machinery parts, and transport equipment industry. The association between *P. juliflora* and *L. fusca* was successful for revegetating salt-laden lands. Therefore, grass-legume-tree association needs to be tested on different sites for remediation, if necessary with biotic and abiotic amendments. Restoration of fly ash landfills with *P. juliflora* following different amendments and Rhizobium inoculation yielded promising results. Mycorrhizae improved the growth of *P. juliflora* on high-pH soils *P. juliflora* seedlings growing in gypsum mine had high-frequency arbuscular mycorrhizal fungal infection. However, *P. juliflora* has some biological characteristics that foster invasion; hence, appropriate management practices needed to be developed for recommending it for phytoremediation. In Columbia, *Prosopis* trees improved the soil microbial community and enzymatic activities in intensive silvopastoral systems ((Prasad and Tewari, 2016).

Prosopis is also a suitable candidate for revegetating fly ash landfills. *Prosopis* has shown potential to accumulate in its tissues the metals contained in fly ash. Inoculation of *Prosopis* with fly ash-tolerant rhizobium (PJ-1) accelerates the translocation of metals from the soil to the above-ground growth. *Rhizobium* inoculation also increases the biomass of the plant in nitrogen deficient fly ash, which confirms that *Prosopis* is beneficial for revegetation and de-contamination of soil and landfills affected by fly ash (Prasad and Tewari, 2016).

Mycorrhizae are reported to greatly improve the growth of *P. juliflora* on high-pH soils. *Prosopis* was able to grow satisfactorily without amendments up to pH 9. Arbuscular mycor-rhizal inocula have been isolated from its rhizosphere (low-cost agrotechnology) and were found to accelerate the growth of other agro forestry and social forestry legumes in perturbed ecosystems. *Prosopis* was able to grow satisfactorily without amendments up to pH 9 (Fig. 4).

9 Ecosystem services: Birds breeding on *Prosopis*

Prosopis provides shelter and breeding ground to migratory birds. Classic example is Uppalapadu wetland near Guntur, Andhra Pradesh, India, thus contributing to environmen-tal protection and conservation. The Uppalapadu bird sanctuary is located in Guntur District, 7 km from Guntur town. The water tank on the edge of the village Uppalapadu is unique since they provide refuge to more than 30 migratory bird species throughout the year. Endangered species, such as Spot-Billed Pelican (*Pelecanus philippensis*), Painted Stork (*Mycteria leucocephala*) and White Ibis (*Threskiornis melanocephalus*), have shifted from Nelapattu in the Nellore District to Uppalapadu to breed on *Prosopis* trees and bushes. The water tank is surrounded by dense *Prosopis* tree and bush formations, and the small island in the middle of the tank is totally covered with *Prosopis*. According to the guard of the sanctuary, it is es-timated that the bird population of these water tanks amounts to 7000, and the number of birds stopping in the sanctuary during the peak season between September and February amounts to 20,000. Some migratory birds, such as Rosy Pastor (*Sturnus roseus*), travel hun-dreds of kilometers to come to the sanctuary. The villagers have agreed to keep this tank only for the birds while using other ponds nearby for their needs. The unique aspect of the place is that the birds can be spotted throughout the year in this small four and a half acres stretch of land (Fig. 5F and G) (Prasad and Tewari, 2016).

Acknowledgments

Thanks are due to Dr. Kurt Walter for useful discussion and constructive suggestions. The authors are thankful to Director, Central Arid Zone Research Institute (CAZRI), Jodhpur, for providing valuable information on value chain on value-added products derived from *P. juliflora*.

References

Adil, M., Shah, A.N., Khan, A.N., Younas, T., Mehmood, M.S., Mahmood, A., Asghar, R.M.A., Javed, M.S., 2023. Ame-lioration of harmful effects of soil salinity in plants through silicon application: a review. Pak. J. Bot. 55 (1), 9–18.

Ahirwal, J., Maiti, S.K., Reddy, M.S., 2017. Development of carbon, nitrogen and phosphate stocks of reclaimed coal mine soil within 8 years after forestation with Prosopis juliflora (Sw.) Dc. Catena 156, 42–50.

Azit, M.N., Ahmad, R.K., Sulaiman, S.A., 2023. Environment friendly production of coconut Shell charcoal through pyrolysis. In: Energy and Environment in the Tropics. Springer, Singapore, pp. 245–261.

Basu, S., Prasad, M.N.V., Suthari, S., Kiran, B.R., 2017. *Prosopis juliflora* (mesquite) gum exudate as a potential excip-ient. EuroBiotech J. 1 (1), 76–81.

Basumatary, D., Yadav, H.S., Yadav, M., 2023. The role of peroxidases in the bioremediation of organic pollutants. Nat. Prod. J. 13 (1), 60–77.

Chavre, B., Patil, R., 2023. Invasive plant species of Maharashtra State: a review. In: Sustainability, Agri, Food and Environmental Research, p. 11.

Diaz-Uribe, C., Walteros, L., Duran, F., Vallejo, W., Romero Bohórquez, A.R., 2022. *Prosopis juliflora* seed waste as biochar for the removal of blue methylene: a thermodynamic and kinetic study. ACS Omega 7 (47), 42916–42925.

Dos Santos, E.T., Pereira, M.L.A., Da Silva, C.F.P., Souza-Neta, L.C., Geris, R., Martins, D., Santana, A.E.G., Barbosa, L.C.A., Silva, H.G.O., Freitas, G.C., Figueiredo, M.P., 2013. Antibacterial activity of the alkaloid-enriched extract from *Prosopis juliflora* pods and its influence on in vitro ruminal digestion. Int. J. Mol. Sci. 14 (4), 8496–8516.

Dubey, K., Dubey, K.P., 2021. Remediation of bauxite residue through integrated approach of microbes and plantation: a case study. In: Handbook of Research on Microbial Remediation and Microbial Biotechnology for Sustainable Soil. IGI Global, pp. 475–489.

Duraisamy, B., Velmurugan, K., Venkatachalapathy, V.K., Thiagarajan, A., 2022. Analysis of *Prosopis juliflora* methyl ester as a fuel in the CI engine. Int. J. Res. Rev. 9 (7), 112–118.

Edrisi, S.A., El-Keblawy, A., Abhilash, P.C., 2020. Sustainability analysis of Prosopis juliflora (Sw.) DC based restoration of degraded land in North India. Land 9 (2), 59.

Edrisi, S.A., Tripathi, V., Abhilash, P.C., 2018. Towards the sustainable restoration of marginal and degraded lands in India. Trop. Ecol. 59, 397–416.

Edrisi, S.A., Tripathi, V., Dubey, P.K., Abhilash, P.C., 2022. Carbon sequestration and harnessing biomaterials from terrestrial plantations for mitigating climate change impacts. In: Biomass, Biofuels, Biochemicals. Elsevier, pp. 299–313.

El Idrissi, M.M., Ourarhi, M., Bouhnik, O., Abdelmoumen, H., 2023. *Ensifer meliloti* is the main microsymbiont of *Prosopis chilensis* in arid soils of Eastern Morocco. In: Microbial Symbionts, p. 111.

Férrer, J.P., da Cunha, M.V., dos Santos, M.V., Torres, T.R., da Silva, J.R., Véras, R.M., da Silva, D.C., da Silva, A.H., Queiroz, L., Férrer, M.T., Neto, E.L., 2021. Mesquite (Prosopis juliflora) extract as a phytogenic additive for sheep finished on pasture in the semiarid region. Chil. J. Agric. Res. 81 (1), 14–26.

Fre, Z., Woldu, G.T., Negash, Z., Araya, S.T., Tsegay, B., Teka, A.M., Kenton, N., Livingstone, J., 2023. Social protection for inclusive development in the Afar region of Ethiopia: findings from the participatory rural appraisal. In: Social Protection, Pastoralism and Resilience in Ethiopia. Routledge, pp. 41–82.

Gavrilescu, M., 2020. Biomass—A resource for environmental bioremediation and bioenergy. In: Recent Developments in Bioenergy Research. Elsevier, pp. 19–63.

Ghosh, S., Falyouna, O., Onyeaka, H., Malloum, A., Bornman, C., AlKafaas, S.S., Al-Sharify, Z.T., Ahmadi, S., Dehghani, M.H., Mahvi, A.H., Nasseri, S., 2023. Recent progress on the remediation of metronidazole antibiotic as emerging contaminant from water environments using sustainable adsorbents: a review. J. Water Process Eng. 51, 103405.

Joshi, P., 2023. Sustainable grazing practices: conserving biodiversity in an Asian tropical grassland. In: Conservation Through Sustainable Use. Routledge India, pp. 97–109.

Kumar, S.S., Vignesh, V., Prasad, V.V.S.H., Sunil, B.D.Y., Srinivas, R., Sanjay, M.R., Siengchin, S., 2023. Static and dynamic mechanical analysis of hybrid natural fibre composites for engineering applications. Biomass Convers. Bioref., 1–13.

Loredo-Portales, R., Castillo-Pérez, L.J., Alonso-Castro, A.J., Carranza-Álvarez, C., 2023. Potential toxicity and bioavailability of ENMs and their products in plant tissues. In: Physicochemical Interactions of Engineered Nanoparticles and Plants. Academic Press, pp. 277–294.

Lv, N., Li, X., 2023. Phosphorus removal from wastewater using Ca-modified attapulgite: fixed-bed column performance and breakthrough curves analysis. J. Environ. Manag. 328, 116905.

Mansi, M., Gaurav, S., 2023. Synthesis and applications of carbon dots from waste biomass. In: Carbon Dots in Analytical Chemistry. Elsevier, pp. 319–328.

de Melo Cavalcante, A.M., da Silva, O.S., da Silva Neto, G.J., de Melo, A.M., Ribeiro, N.L., 2019. Evaluation of the antioxidant potential of mesquite grains flour in hamburger meat product. J. Exp. Agric. Int. 41 (3), 1–14.

Menya, E., Jjagwe, J., Kalibbala, H.M., Storz, H., Olupot, P.W., 2023. Progress in deployment of biomass-based activated carbon in point-of-use filters for removal of emerging contaminants from water: a review. Chem. Eng. Res. Des. https://doi.org/10.1016/j.cherd.2023.02.045.

Naik, N.M., Krishnaveni, M., Mahadevswamy, M., Bheemanna, M., Nidoni, U., Kumar, V., Tejashri, K., 2022. Comparative Analysis of Supercritical Fluid and Soxhlet *Prosopis juliflora* Leaves Extract With Antifungal Traits. pp. 1–15, https://doi.org/10.21203/rs.3.rs-1676795/v1.

Pandey, V.C., Bauddh, K. (Eds.), 2018. Phytomanagement of Polluted Sites: Market Opportunities in Sustainable Phytoremediation. Elsevier.

Patel, K., Chaurasia, M., Rao, K.S., 2023. Urban dust pollution tolerance indices of selected plant species for development of urban greenery in Delhi. Environ. Monit. Assess. 195 (1), 1–21.

Patwa, D., Ravi, K., Sreedeep, S., 2023. Impact of biochar fraction on thermal characteristics of soil-biochar composite. In: Indian Geotechnical Conference. Springer, Singapore, pp. 309–317.

Ponnampalam, E.N., Holman, B.W., 2023. Sustainability II: sustainable animal production and meat processing. In: Lawrie's Meat Science. Woodhead Publishing, pp. 727–798.

Prasad, M.N.V., Tewari, J.C., 2016. *Prosopis juliflora* (Sw) DC: potential for bioremediation and bioeconomy. In: Bioremediation and Bioeconomy. Elsevier, pp. 49–76.

Preethy, K.R., Ganesan, P., Chamundeeswari, M., 2023. Multimodality: phantom imaging for superparamagnetic graphene composites using green technology for theranostic nanosystems. Appl. Phys. A 129 (1), 1–17.

Ranđelović, D., Pandey, V.C., 2023. Bioenergy crop-based ecological restoration of degraded land. In: Bio-Inspired Land Remediation. Springer, Cham, pp. 1–29.

Ravhuhali, K.E., Mudau, H.S., Moyo, B., Hawu, O., Msiza, N.H., 2021. Prosopis species—an invasive species and a potential source of browse for livestock in semi-arid areas of South Africa. Sustainability 13, 7369.

Saim, A.K., Adu, P.C.O., Amankwah, R.K., Oppong, M.N., Darteh, F.K., Mamudu, A.W., 2021. Review of catalytic activities of biosynthesized metallic nanoparticles in wastewater treatment. Environ. Technol. Rev. 10 (1), 111–130.

Saravanakumar, S.S., Kumaravel, A., Nagarajan, T., Sudhakar, P., Baskaran, R., 2013. Characterization of a novel natural cellulosic fiber from Prosopis juliflora bark. Carbohydr. Polym. 92 (2), 1928–1933.

Sathish, T., Muthukumar, K., Abdulwahab, A., Rajasimman, M., Saravanan, R., Balasankar, K., 2023. Enhanced waste cooking oil biodiesel with Al2O3 and MWCNT for CI engines. Fuel 333, 126429.

Sharifian, A., Gantuya, B., Wario, H.T., Kotowski, M.A., Barani, H., Manzano, P., Krätli, S., Babai, D., Biró, M., Sáfián, L., Erdenetsogt, J., 2023. Global principles in local traditional knowledge: a review of forage plant-livestock-herder interactions. J. Environ. Manag. 328, 116966.

Singh, G., 2022. Role of *Prosopis* in reclamation of salt-affected soils and soil fertility improvement. In: Prosopis as a Heat Tolerant Nitrogen Fixing Desert Food Legume. Academic Press, pp. 27–54.

Singh, S., Saha, L., Kumar, M., Bauddh, K., 2021. Phytoremediation potential of invasive species growing in mining dumpsite. In: Phytorestoration of Abandoned Mining and Oil Drilling Sites. Elsevier, pp. 287–305.

de Souza, G.P.G., Duarte, D.D.S., De Souza, A.V., Ribeiro Junior, P.M., De Lima, M.A.C., Britto, D.D., 2023. Active antifungal edible coating based on nanocomposite cast films from galactomannan. Int. J. Postharvest Technol. Innov. X (Y). xxxx.

Suriapparao, D.V., Reddy, B.R., Rao, C.S., Jeeru, L.R., Kumar, T.H., 2023. *Prosopis juliflora* valorization via microwave-assisted pyrolysis: optimization of reaction parameters using machine learning analysis. J. Anal. Appl. Pyrolysis 169, 105811.

Suselo, Y.H., 2023. The Cytotoxicity Effect of Ethanol Extract and Alkaloid Fraction of *Mirabilis jalapa* Leaves in Hepatocarcinoma Cell Line.

Suthari, S., Prasad, M.N.V., 2021. Establishment of phytocoenoses on coal mine over burden of Pranahita Godavari Basin, Telangana. Front. Biosci., 54.

Tabe-Ojong Jr., M.P., 2023. Action against invasive species: charcoal production, beekeeping, and Prosopis eradication in Kenya. Ecol. Econ. 203, 107614.

Teja, T.R., Alagesan, A., Avudaithai, S., Ejilane, J., Sebastian, S.P., 2022. Growth and yield response of cowpea (Vigna unguiculata L.) in sodic soil amended with marine gypsum, biochar and bioinoculants. Ecol. Environ. Conserv. 28.

Tewari, J.C., Pareek, K., Tewari, P., Sharma, A., Shiran, K., 2022. The paradigm shift in Prosopis juliflora use through community participation by developing value chain of value-added products from pods. In: Prosopis as a Heat Tolerant Nitrogen Fixing Desert Food Legume. Academic Press, pp. 213–230.

Tian, B., Wang, S., Wang, J., Feng, F., Xu, L., Ma, X., Tian, Y., 2023. A comprehensive evaluation on thermo-chemical potential of longan waste for the production of chemicals and carbon materials. Fuel 334, 126655.

Tripathi, V., Edrisi, S.A., O'Donovan, A., Gupta, V.K., Abhilash, P.C., 2016. Bioremediation for fueling the biobased economy. Trends Biotechnol. 34 (10), 775–777.

Wang, L., Fan, L., Yi, K., Jiang, Y., Filppula, A.M., Zhang, H., 2023. Advances in the delivery systems for oral anti-biotics. Biomed. Tech. 2, 49–57.

Weisany, W., Khosropour, E., 2023. Engineered nanomaterials in crop plants drought stress management. In: Engineered Nanomaterials for Sustainable Agricultural Production, Soil Improvement and Stress Management. Academic Press, pp. 183–204.

Yadav, S., Pandey, V.C., Singh, L., 2021. Ecological restoration of fly-ash disposal areas: challenges and opportunities. Land Degrad. Dev. 32 (16), 4453–4471.

Yadav, N., Rana, A.C., 2020. Pharmacological and pharmacognostical aspect of Prosopis juliflora: a review. World J. Adv. Res. Rev. 8 (1), 036–052.

Yaynemsa, K.G., 2023. Plant Biodiversity Conservation in Ethiopia: A Shift to Small Conservation Reserves. Springer Nature.

Zarzosa, P.S., Navarro-Cerrillo, R.M., Mc Cubbin, E.P., Cruz, G., Lopez, M., 2023. Biomass estimation models for four priority Prosopis species: tools required for forestry management in overexploited arid ecosystems. J. Arid Environ. 209, 104904.

Giant reed (*Arundo donax* L.)—A multi-purpose crop bridging phytoremediation with sustainable bioeconomy

A.L. Fernando[a], B. Barbosa[a], L.A. Gomes[a,b], J. Costa[a,c], and E.G. Papazoglou[d]

[a]METRICS, Universidade Nova de Lisboa, Caparica, Portugal [b]CNR-ITAE, Messina, Italy
[c]Instituto Superior de Educação e Ciências, Lisbon, Portugal [d]Agricultural University of Athens, Athens, Greece

1 Introduction

The worldwide expansion of anthropogenic activities such as the production and discard of municipal wastes, pesticides, fertilizers, emissions from vehicles and waste incinerators, and metallurgical, petrochemical, mining, and construction activities are making large-scale changes in the natural environments. These activities are also changing the rate of release of inorganic compounds such as heavy metals into the ecosphere, contributing significantly to worldwide degradation, contamination, and pollution of air, water, and soil systems where, ultimately, these compounds tend to accumulate (Barbosa et al., 2016). Alone or together, these factors and activities reduce soil and water quality, posing an imminent threat to humans, animals, and ecosystem services. Therefore, it is necessary to find solutions that promote decontamination and remediation, if possible, in a cost-effective way.

Several types of technologies have been used to remediate soil contaminated with heavy metals. The most common techniques used are (a) excavation and disposal, (b) immobilization, (c) toxicity reduction, (d) vitrification, (e) encapsulation and cover with clean soil, and (f) washing (Barbosa and Fernando, 2018). For water decontamination, common options to remove heavy metals include alkaline precipitation, ion exchange, electrochemical removal, filtration, and membrane technologies. These soil and water remediation approaches are, in most cases, very expensive and, in some cases, may lead to adverse effects on ecosystems, often requiring other methods for waste disposal (Nsanganwimana et al., 2014).

Phytoremediation, the use of plants and their associated microbes for soil, water, and air decontamination, is a cost-effective, solar-driven, and alternative or complementary technology for physicochemical approaches (Barbosa et al., 2016). Plants can be used for the extraction and stabilization of many heavy metals found in contaminated media, reducing their risks to humans and ecosystems. Much research has been conducted to determine which plant species have simultaneously the ability to produce high yields and tolerate and/or accumulate heavy metals (Barbosa et al., 2018). Energy crops have been recognized as tolerant to different abiotic stresses and contamination, and the high-yielded biomass can be used for the production of energy, paper pulp, and biomaterials (Von Cossel et al., 2019a,b). Phytoremediation capabilities have been suggested for perennial grasses (such as *Miscanthus* and switchgrass), short-rotation coppices (such as poplar trees), and annual crops (such as hemp, flax, sugar beet, and kenaf), among other energy crops (Barbosa et al., 2018; Barbosa and Fernando, 2021; Fernando et al., 2014; Gomes et al., 2023; Papazoglou and Fernando, 2017). Those phytoremediation capabilities can be explored either to remediate soils or wastewaters (WW) and the produced biomass can be used to support the growing market for renewable energy and biobased products. The re-use of WW in the irrigation of energy crops, as well as the use of marginal soils for its production, assumes increased importance in water-scarce regions such as the Mediterranean basin, contributing to reduce competition for land and water and to mitigate desertification (Barbero-Sierra et al., 2013; Barbosa et al., 2015b, 2018).

The energy crop *Arundo donax* L. is a plant with the genetic and physiological potential to remove or immobilize inorganic compounds from contaminated media, such as heavy metals contained in contaminated soils and WW (Fernando et al., 2015; Scordia et al., 2022; Zema et al., 2012). In addition, this crop can contribute to the bioeconomy, as a feedstock for bioenergy, biofuels, fiber, and other biobased products (Christou et al., 2018; Schmidt et al., 2015). Therefore, the aim of this work was to review the biomass and physiological characteristics of *A. donax*, its ecological requirements, its contribution to the bioeconomy, and case studies of its cultivation under marginal soils (namely, contaminated soils with inorganic compounds such as heavy metals) or when irrigated with WW, as well as the social, economic, and environmental risks and benefits related to this practice.

2 Giant reed (*A. donax* L.)

Giant reed (*A. donax* L.) is a herbaceous, perennial, non-food crop that produces high yields of dry biomass (Christou et al., 2018) (Fig. 1). The plant belongs to the *Poaceae* family, originating in Asia and later spreading to different subtropical wetlands and warm-temperature regions of Europe, Africa, North America, and Oceania (Christou et al., 2018). It grows spontaneously and is widely spread in ditches and riverbanks throughout the Mediterranean basin (Molari et al., 2021) (Fig. 2).

FIG. 1 Illustration of *A. donax* L. *Source: Otto Wilhelm Thomé, 1885. Flora von Deutschland, Österreich und der Schweiz, Gera, Germany.*

This robust plant with metabolism in C3 achieves photosynthetic rates comparable with those of C4 plants (Webster et al., 2016). In warm temperate regions, vegetative growth normally occurs between February and March, when new shoots emerge from the soil. The plant develops from February to October, and the flowering occurs from August to November when an inflorescence (panicle) appears in the apical leaf (Fig. 3). Cessation of growth occurs between November and February when stems start to lose moisture, leaves begin to enter into senescence, and the panicle breaks. The seeds (kernels) are sterile; hence, the species reproduces by vegetative propagation of the rhizome, a feature that gives it strength, being important for phytoremediation purposes (Fig. 4). The stems are hollow, robust, and erect, growing in dense groups derived from the same rhizome, reaching 2–4 cm in diameter and 4–6 m in height, with a thickness between 2 and 3 mm (Molari et al., 2021) (Fig. 5).

This plant tolerates a high range of ecological conditions and easily adapts to virtually all types of soil, including clay and sandy soils, gravel, and saline soils, as well as infertile or

FIG. 2   Giant reed growing in Portugal.

heavy metal-contaminated soils and it grows commonly at the margins of agricultural land, near roads, or watercourses (Christou et al., 2018; Ge et al., 2016; Lino et al., 2023) (Fig. 6). The robustness of its rhizomes allows its use in the support of sloping terrains and erosion control, which is a beneficial trait of this crop, and with significance for its production in marginal soils (Cosentino et al., 2015; Fernando et al., 2015). It prefers well-drained soils with high humidity, where it achieves greater productivity, but can survive for long periods under a dry environment (Christou et al., 2018; Curt et al., 2018). It occurs naturally in areas with 300–4000 mm of annual precipitation, with annual temperatures ranging from 9°C to 28°C and altitudes up to 2300 m (Lewandowski et al., 2003). Normally it does not tolerate low temperatures, except when dormant. During the winter dormancy, frost damage does not affect significantly giant reed, but if during the spring period there are late frosts, the plant may suffer severe physiological damages (Lino et al., 2023). Propagation is via rhizomes, stem cuttings, or from axillary buds using in vitro propagation technologies (Ge et al., 2016). Some researchers attempted to induce the development of roots on the stems to increase the efficiency of the stem cutting technique (Christou et al., 2018). The crop does not have any special requirements in terms of soil preparation. Ge et al. (2016) refer to sowing densities in the range of 5000–20,000 plants ha^{-1}. The full crop water demand can be very high in warm climates, e.g., 835 mm year^{-1} in Sicily, and 1089 mm year^{-1} in Greece (Curt et al., 2018). Therefore,

FIG. 3   Details of the *A. donax* L. panicles.

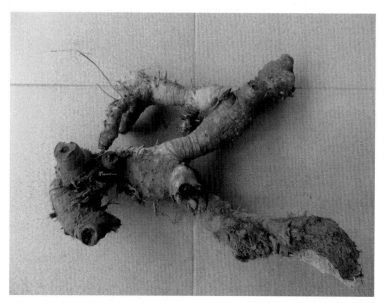

FIG. 4   Details of the *A. donax* L. rhizome and roots.

I. Bioproducts from contaminated soil and water

FIG. 5   Giant reed growing in Portugal.

FIG. 6   Giant reed growing in a harsh environment, near the sea.

under dryland conditions, irrigation is recommended to ensure crop establishment and high yields (Fernando et al., 2015; Ge et al., 2016). Borin et al. (2013) and Mantineo et al. (2009) apply ca. 450 mm by irrigation when rainfall is around 400–500 mm.

In soils with low nitrogen content, annual applications of $100 \, kg \, (N) \, ha^{-1} \, year^{-1}$ are recommended during the initial stages of crop growth (Christou et al., 2018). When the crop reaches maturity, fertilization with 40–80 kg (N) ha^{-1} is the most appropriate from an economic and environmental point of view (Christou et al., 2018; Ge et al., 2016). An annual application of ca. 150–200 kg (P, K) ha^{-1} is usually necessary to maintain the fields, but the amount to be applied should also consider the soil's fertility and plant uptake (Christou et al., 2018).

A. donax is highly resistant to most pests and it does not require the application of herbicides (Fernando et al., 2010, 2018b; Ge et al., 2016). Yet, weed control is recommended for the establishment year and pesticides are usually applied to destroy rhizomes when giant reed eradication is needed (Fernando et al., 2015). *A. donax* is not cultivated on a large scale, yet, long-term data of some pilot fields show yields of dry biomass from 17.6 to 36.5 Mg ha^{-1} year^{-1} in medium fertility soils, and in marginal soils the yields can be reduced to values in the range 5.0 to 12.3 Mg ha^{-1} year^{-1} dry matter (Christou et al., 2018). Nevertheless, *A. donax*'s invasive character may represent a drawback on its exploitation (Ge et al., 2016). Its anticipated vegetative cycle and the large production of propagating material are the most important competitive advantages when compared with native plants, especially after extreme events such as floods and fires. The most effective way to eradicate giant reed is plowing followed by broad-spectrum systemic herbicides and crop desiccants (Christou et al., 2018). Mechanical removal of the rhizomes has also been reported but this is a technique that removes both plant and soil, and given the depth and extent of the giant reed root system, may leave small rhizome fragments in the soil that may persist and would lead to major soil disturbances (Christou et al., 2018).

In the southern parts of Europe, harvest can be done either in autumn or in late winter; however, in the latter losses of around 30% (dry matter) can occur, especially if the winter is rigorous, where the wind is a very prominent factor. If the harvest takes place in autumn, it may result in very wet biomass. Nonetheless, especially in semi-arid Mediterranean climates, the weather conditions usually allow the natural drying of the biomass in the field after cutting (Christou et al., 2018). Its biomass can easily be stored, and normally no treatment is applied, but a 10%–15% of total biomass can be lost, mainly at leaf level but not with stems (El Bassam, 2010).

The biomass composition of *A. donax* can define the path used for its valorization and contribution to the bioeconomy (see Box 1), although differences can be observed due to plantation age, year, fertilization, irrigation, clones, soil characteristics, and harvest time, among others.

Its cellulose content ranges from 31% to 44%, hemicellulose content in the range of 13%–35%, and the lignin content in the range of 11%–44% dry biomass (Krička et al., 2017; Oginni and Singh, 2019; Ramos et al., 2018; Silva et al., 2015; Yang et al., 2020). The holocellulose content in *A. donax*'s fibers indicates its potential for conversion processes like anaerobic digestion and alcoholic fermentation (Lei et al., 2020; Qiu et al., 2023). When the goal is the production of second-generation ethanol, the application of pre-treatments (e.g., acid pre-treatment followed by alkaline pre-treatment, or enzymatic attack) improves the efficiency of the fibers' fermentation and reduces the costs of the ethanol production, once the fibers will be more accessible for fractionation into xylose and glucose (Lemões et al., 2018; Sidana et al., 2022).

This crop has a good anaerobic biogasification potential (ABP), and when compared to corn, rye, triticale, and sorghum, the ABP corresponds to 75%, 94%, 77%, and 120% of the

---

BOX 1

Giant reed (*Arundo donax* L.), a multi-purpose crop for the bioeconomy.

- Bioenergy, biofuels
  - ✓ Biomethane and biogas production
  - ✓ Hydrogen production
  - ✓ Second generation bioethanol
  - ✓ Heat and power
  - ✓ Biochar, Biooil, Gases (H_2, CO_2, CO, CH_4, …)
- Biobased products

- ✓ Musical instruments
- ✓ Low-cost adsorbents
- ✓ Activated carbon
- ✓ Paper pulp and resins
- ✓ Chipboards and particle boards
- ✓ Xylo-oligosaccharides and xylose
- ✓ Nanocellulose

ABP of those biomasses, respectively. Due to its high biomass productivity, the biomethane production potential *per* hectare of *A. donax* is higher than other energy crops, being able to reach 19,440 Nm^3 (CH_4) ha^{-1} (Corno et al., 2014). The utilization of this crop for biogas production can be improved with the addition of some KOH-based thermochemical pretreatment, which showed to increase methane yield by 21%, the daily rate of the production by 42%, and its concentration in biogas by 23%, when compared to the biogas produced using untreated *A. donax* biomass (Vasmara et al., 2021).

A. donax was also tested to produce hydrogen (Chai et al., 2022). In this study, the biomass of *A. donax* was pre-treated with three different ionic liquids and different loadings ($2-16 \, g \, L^{-1}$). The pre-treated biomass followed by enzymatic hydrolysis reached a sugar yield of $7.9 \, g \, L^{-1}$ and a hydrogen yield of $106.1 \, mL \, g^{-1}$. Total solids during the photo-fermentation were 68.8%, corresponding to 35.3% higher than those of untreated *A. donax*. Acetic acid was the main by-product during the hydrogen production process with ionic liquids pre-treated *A. donax* (Chen et al., 2022). Zhang et al. (2023) showed that photo-fermentative biohydrogen production can be further enhanced through pre-treatment of the lignocellulosic biomass with ultrasonication combined with ionic liquids.

The ultimate and proximate analysis of *A. donax* (Tables 1 and 2) evidence its potential to be used in thermochemical processes due to its high HHV (high heating value) (Gomes et al., 2022a; Molino et al., 2018; Ozdemir et al., 2020; Pari et al., 2016; Yang et al., 2020) and low nitrogen and sulfur content, which limit NO_x emissions, and reduce corrosion problems associated with sulfur content (Gomes et al., 2018).

TABLE 1   Ultimate analysis of *A. donax*.

C (%)	H (%)	O (%)	N (%)	S (%)	Reference
45.67	6.17	47.13	0.74	0.29	Krička et al. (2017)
42.47	6.28	50.02	0.65	0.58	Oginni and Singh (2019)
44.76	5.28	38.69	0.77	0.45	Yang et al. (2020)

TABLE 2 Proximate analysis of A. *donax*.

Fixed carbon (%)	Volatile material (%)	Ash (%)	HHV (MJ kg^{-1})	Reference
11.47	76.06	3.56	17.48	Krička et al. (2017)
18.52	79.70	1.78–2.26	18.96	Oginni and Singh (2019)
15.35	74.59	6.77		Yang et al. (2020)

The valorization through combustion processes can, however, present some problems due to the composition of the leaf and stem tissues, which presents high amounts of ash content (Table 2). Some corrosive elements are also present in significant amounts, such as K, Cl, and Si, which can cause problems in the machinery used in the thermochemical process, reducing the reactors' life span (Chai et al., 2022).

Alessina and collaborators tested the utilization of A. *donax* in gasification plants on a pilot scale in a gasifier of 20 kW$_{el}$ using a mixture of 50% wood chips and 50% A. *donax* with a biomass-specific consumption of 2.7 kg (dry) kWh^{-1}, being able to produce biochar (5% on a dry basis) and power (9515 MWh with a feedstock of 25,882 t of giant reed biomass). The gas produced showed an HHV of 5.98 MJ/Nm^3, with a composition of 20.1 (H_2), 39.5 (N_2), 11.6 (CO_2), 1.5 (CH_4), and 14.3 (CO), all in % (v/v). The process showed to be promising to tackle GHG emissions, sequestrating up to 150 kg of CO_2 for every t of A. *donax* submitted to gasification, despite the electrical efficiency being only 7.72% (Alessina et al., 2018).

Ba and collaborators promote a study to evaluate the potential of A. *donax* pyrolysis. The model established a pyrolysis temperature in the range of 600–700°C, and considered 10% of thermal losses in the pyrolysis reactor. The study concluded that in an industrial-scale pyrolysis (2000 t day^{-1}), the output could reach 28 MW power, along with 51.36 t day^{-1} biooil, 555.04 t day^{-1} vinegar, and 511.36 t day^{-1} biochar (Ba et al., 2020). When subjecting A. *donax* to pyrolysis in the range of 300–600°C, Zheng et al. (2018) observed the production of 25 different chemical products. It was also observed that the temperature is inversely proportional to biochar production, as an increase in temperature from 300°C to 600°C decreased biochar production from 43.6% to 29% (Zheng et al., 2018).

The torrefaction and low-temperature carbonization of A. *donax* was studied by Correia and collaborators, submitting the crop to a process where the temperature varied from 200°C to 350°C, from 15 to 90 min. Mass and energy yields and the biochar composition were significantly influenced by the temperature. Increasing the temperature from 250°C to 350°C and the residence time, increased the biochar heating value up to 21.4 MJ kg^{-1}. An increment in ash and fixed carbon contents was also observed in these biochars, which restricts their use as solid fuel, but other valorization options can be considered, e.g., as low-cost adsorbents, for remediation of aqueous effluents contaminated with cationic dyes or heavy metals (Correia et al., 2017).

A. *donax* fuel properties can be enhanced after hydrothermal carbonization (HTC). In the work of Nawaz and Kumar (2022), the HTC of A. *donax*, performed in a 2-L Parr reactor, produced a material with increased HHV, 25.65 MJ kg^{-1}, when compared with raw biomass, 18.43 MJ kg^{-1}, but a lower percentage of volatile matter, 54%, when compared with a raw A. *donax* (78.5%), thus with lower reactivity and more stable. Moreover, the HTC process reduced the ash content of the material, from 6.3% (raw) to 4.8% (hydrothermal carbonized material), due to the dissolution in water of some fractions of ash (Nawaz and Kumar, 2022), thus, reducing the chances to occur slagging and fouling during boiler operation due to its alkaline nature. In addition, the lignin content of the hydrothermal carbonized material was lower and the cellulose content was higher than the raw biomass.

The utilization of *A. donax* goes further than the energy end products. It is also used as raw material to manufacture musical instruments and as a reinforcing material (Zhang et al., 2021), biooil (Galletti et al., 2022), activated carbon (Üner, 2019), paper pulp and resins (Garcez et al., 2022), chipboards and particle boards (Ferrandez-Villena, et al., 2020), xylo-oligosaccharides and xylose (You et al., 2018), fractionated into cellulose and lignin, to produce nanocellulose (Pires et al., 2019b, 2022a,b), among other applications. Production of nanocellulose from giant reed biomass needs special attention. This product (nanocellulose) has a significant value in the market due to its characteristics, e.g., high specific surface area, high crystallinity, good optical properties, mechanical reinforcement, barrier properties, no toxicity, renewability, and biodegradability (Klemm et al., 2018; Malucelli et al., 2017; Pires et al., 2019a; Thomas et al., 2018). Nanocellulose-based materials (e.g., hydrogels, aerogels, or composites) show many applications: electronic devices, food packaging, paper industry, biomedicine, cosmetics, paint coatings, catalysis, fire-resistant materials, thermal insulation materials, to name a few (Klemm et al., 2018; Malucelli et al., 2017; Thomas et al., 2018).

Ahmed (2016) reviewed the potential of *A. donax* to produce activated carbon, which showed to have an excellent potential for organic and inorganic pollutants removal, especially when applied to WW treatment. Ammari (2014), on the other hand, studied the utilization of the leaves as a Cd absorbent. The WW containing Cd^{2+} showed a reduction of 92% of Cd with the incorporation of *A. donax* leaves, which were added in powder. Activated carbon produced from giant reed can be also used to store H_2. It was demonstrated by Üner (2019) that *A. donax*-derived activated carbon prepared with $ZnCl_2$ presented small mesopores and wide micropores, being capable to stock high amounts of H_2, especially under high-pressure conditions.

Due to the *A. donax* fibrous structure, the use of this crop as a construction material has also been tested, aiming to increase the renewability and sustainability of the construction materials. Karahancer et al. tested the addition of *A. donax* fibers in a hot-mix asphalt. The addition of these fibers in the mixture increased the tensile strength ratio, improving the strength of the mixture (Karahancer et al., 2016).

The potential applications of *A. donax* make this versatile crop an excellent option in biorefinery applications. *A. donax* can help the European Union (EU) and the United Nations to achieve its environmental goals, with socioeconomic improvements as collateral, being a feedstock that can contribute to reduce the consumption of fossil resources.

3 Production of *A. donax* irrigated with wastewaters: Case studies

Water supply and water quality degradation are global concerns, and in water-scarce countries, marginal-quality water will become an increasingly important component of agricultural water supplies(Barbosa et al., 2015b). Although the re-use of WW should be encouraged, it should also be controlled because its application could lead to numerous nutrients, pathogens, and potentially toxic elements being added to soils (Barbosa et al., 2015b). At the same time, the re-use of WW in the irrigation of energy crops should ensure environmental and public health.

Kausar et al. (2012) refer to the potential of *A. donax* for the removal of Cr from contaminated WW, and Bonano et al. (2013) suggest that *A. donax* can be used for the phytoextraction of Al, As, Cd, Cr, Cu, Hg, Mn, Ni, Pb, and Zn in sediments and water bodies. This crop accumulates metals preferentially in underground structures and seems to follow metal exclusion mechanisms. Yet, in this latter study, *A. donax* exhibited toxic levels of Cr in all organs for concentrations above $0.5 \, mg \, kg^{-1}$. Among all species tested by Bonano et al. (2013), *A. donax* was the one with the lowest bioaccumulation capacity.

A. donax exhibits high biomass production and tolerance to different types of WW composition. In many of the studies cited here, it was found that this plant combines high yields with a high potential for the removal of many inorganic contaminants dissolved in many different types of WW, under both natural and artificial conditions. Heavy metal contamination in water environments may threaten not only aquatic ecosystems but also human health. Phytoremediation using plants like *A. donax* to remove, detoxify, or stabilize heavy metals can be considered an important tool for cleaning polluted water, with the concomitant production of biomass that can be further valorized. In constructed wetlands, for example, heavy metals present in WW can be removed by processes such as absorption, precipitation, and plant uptake. *A. donax* can preserve the main attributes shown under natural or uncontrolled conditions (situations where it shows invasiveness behavior) as produced with WW irrigation. Tuttolomondo et al. (2015) tested *A. donax* and *Cyperus alternifolius* in a constructed wetland system receiving domestic WW, observing that *A. donax* began vegetative growth and re-growth earlier than *C. alternifolius*, leading to greater plant growth during the crop development stage. During the 2-year experiment, the average WUE (water use efficiency) for *A. donax* was greater than for *C. alternifolius* for all growth stages. *A. donax* used the water more efficiently thanks to the greater aboveground biomass production and the higher photosynthetic activity, related to high leaf area index (LAI) and total chlorophyll levels. An important consideration concerning WUE for *A. donax* is the relationship between this parameter and the availability of water in the growing medium. *A. donax* tends to increase WUE when water availability is reduced, with values ranging from 6 to $10 \, g \, dm^{-3}$, depending on the type of soil and climatic conditions (test conditions) (Tuttolomondo et al., 2015). Under constructed wetland systems, the plant's performance in removing nutrients and heavy metals is highly dependent on the water balance systems. Sun et al. (2013) tested *A. donax* among other macrophyte species and found that the plant showed excellent Fe accumulation capacity, but did not accumulate Cr, Cu, or Zn.

In other cases, depending on the conditions of the study, the plant showed high efficiency in the removal of many metals and other nutrients from WW. In a pot experiment performed in Portugal, Costa et al. (2016) tested *A. donax* irrigation with swine effluents under three different types of water regimes (950, 475, and 238 mm). The authors found that reducing the water supply reduced the yields, showing the greater dependence of *A. donax*'s biomass production on the water availability in the growing medium. In the study the irrigation with WW did not affected the yields, comparing with control irrigated with tap water. However, biomass harvested from pots irrigated with WW showed higher ash and nitrogen content, which can be a problem when biomass is used in thermochemical processes. The soil-plant system retained over 90% of the pollutant load of the WW, remediating the effluent. Moreover, the produced biomass can be economically valorized for energy or biomaterials, as the fiber content and the calorific value were not influenced by the irrigation with wastewater. Still, the higher ash and nitrogen contents in the biomass can be detrimental especially.

Elhawat et al. (2014) tested the phytoextraction potentials of two biotechnologically propagated ecotypes (one American and one Hungarian) of *A. donax* in Cu-contaminated synthetic WW (0, 1, 2, 3, 5, 10, and 26.8 mg dm^{-3}) for 6 weeks. The increment of Cu concentration in the nutrient solution slightly reduced the root, stem, and leaf biomass yield to 26.8 mg dm^{-3}. Yet Cu removal from WW ranged between 96.6% and 98.8% for the American ecotype and 97%–100% for the Hungarian ecotype. Accordingly, both populations of *A. donax* can be used to treat water bodies contaminated with Cu up to 26.8 mg dm^{-3}.

Phytoextraction and phytovolatilization of As (0, 50, 100, 300, 600, and 1000 μg dm^{-3}) contained synthetic WW was tested by Mirza et al. (2010, 2011). They observed some toxicity symptoms in *A. donax* biomass, which included the appearance of red spots on the roots, the emergence of young leaves with red color, leaf yellowing, and necrotic leaves, especially at the highest concentration. The As content recovered in plants was similar to its concentration in WW. Measurements of physiological and ultrastructural anatomical changes showed the presence of stomata on the stems of *A. donax* at 1000 μg (As) dm^{-3}, highlighting phytovolatilization as one of the mechanisms that the species uses to tolerate As, allowing them to tolerate concentrations in the range of 600–1000 μg (As) dm^{-3} without suffering any toxicity symptoms. The appearance of certain symptoms at 1000 μg (As) dm^{-3} reveals that the plant cannot tolerate this concentration.

A. donax is pointed out by Williams et al. (2008) as a suitable crop for the production of biofuels and the pulp/paper industry in Australia when irrigated with saline winery WW together with saline soils. *A. donax* can generate biomass yields of 45.2 Mg ha^{-1} (dry weight) when irrigated with 21,000 m^3 ha^{-1} of winery WW, a yield similar to the ones registered by Hidalgo and Fernandez (2001) (46 Mg ha^{-1} dry weight, in Spain), with low technical inputs under arable land and irrigated with tap water. Due to its composition, the irrigation with winery WW also allows soil fertilization. In the same study by Williams et al. (2008), *A. donax* removed large amounts of N, P, and K at rates of 528, 22, and 664 kg ha^{-1} year^{-1} with salinity up to 9 dS m^{-1}, accumulating at the same time 20.6 Mg ha^{-1} of organic carbon, highlighting the possibility to use *A. donax* in carbon-credit programs. High oven dry yields of *A. donax* together with nutrient and heavy metal removals from swine WW effluents in a closed gravel hydroponic system were registered by Mavrogianopoulos et al. (2002) in a 3-year experiment, in which extra P was added to the WW solution. After the first year of growth, annual stem production was 12–15 Mg ha^{-1} (dry weight), and during the third year, after extra P was added, the same parameter ranged from 20 to 23 Mg ha^{-1} (dry weight). The same behavior was observed in the average infiltration rate for most elements (N, K, Ca, Mg,

Fe, Mn, Zn, and Cu), which increased by 46%, and for P an increment of 169% was registered. This work shows that it is possible to combine biomass production with WW treatment and metal removal. Idris et al. (2012) evaluated the performance of *A. donax* irrigated with treated WW from a dairy processing factory, with a median electrical conductivity (EC) of $8.9\,mS\,cm^{-1}$, registering a yield of approximately $179\,Mg\,ha^{-1}\,year^{-1}$ of biomass (dry weight) during a growing season of 250 days as well as a removal percentage of 69%, 95%, and 26% of biochemical oxygen demand (BOD), suspended solids, and total N, respectively. The authors pointed out *A. donax* as a suitable crop for WW treatment, and the biomass produced as an additional opportunity for secondary income streams through its utilization. Others tested many agronomic parameters under WW irrigation practices. Barbagallo et al. (2011) tested the irrigation of *A. donax* with domestic treated WW with 66% and 100% of crop evapotranspiration (ETc), registering $40\,Mg\,ha^{-1}$ (dry weight) and $47\,Mg\,ha^{-1}$ (dry weight), respectively, as well as an energy content of $8.0\,MJ\,kg^{-1}$ in biomass. When crops were irrigated with 100% ETc, they obtained higher growth parameters such as higher culm density. The authors highlighted *A. donax* as a suitable crop for energy production in marginal lands of the Mediterranean basin, a water-scarce region. These results may also indicate good performance for fiber purposes under this sort of irrigation, but further research is needed.

In Portugal, Calheiros et al. (2012) used *A. donax* and *Sarcocornia* in the treatment of tannery WW with high and variable concentrations of complex pollutants, combined with high salinity levels, in a wetland system composed of three beds. They discovered that constructed wetland systems planted with salt-tolerant plants such as *A. donax* are promising solutions for that type of WW. By the end of the trial, the average height of the plants was around 97 cm and the root system was deep and well developed. In the second bed, *A. donax* produced $108–365\,g\,plant^{-1}$ (dry weight), and in the third bed $59–407\,g\,plant^{-1}$ (dry weight) were obtained. *A. donax* had a higher capacity than *Sarcocornia* to take up nutrients. Mass removal rates were up to $615\,kg\,ha^{-1}\,day^{-1}$, expressed in chemical oxygen demand (COD), and $363\,kg\,ha^{-1}\,day^{-1}$, expressed in 5 days biochemical oxygen demand (BOD_5), and removal efficiencies were 40%–93% for total P, 31%–89% for NH_4^+, and 41%–90% for total Kjeldahl N. The performance of *A. donax* across different constructed wetland types receiving different types of WW ranged between 50% and 90% for removing excess nutrients (N, P, and K), total suspended solids, COD, and BOD (Nsanganwimana et al., 2014). Kouki et al. (2012) tested the potential of a mixed culture of *A. donax* and *Typha latifolia* irrigated with rural WW. *A. donax* had higher height elongation (288 cm; 9-month experiment), yield ($2.4\,kg\,m^{-2}$ dry weight), and nitrogen uptake ($21.1\,mg\,kg^{-1}$ dry weight) when compared with *T. latifolia*.

Rhizofiltration is one of the phytoremediation methods that use terrestrial plants and their root system or aquatic plants as a biofilter and which can be extraordinarily effective in the sequestration of metals and radionuclides from polluted water bodies. Dürešová et al. (2014) tested a rhizofiltration system composed of *A. donax* L. for the removal of ^{109}Cd and ^{65}Zn. The authors found complete Zn and Cd removal from a solution containing $0.28\,\mu mol\,dm^{-3}\,CdCl_2$ ($78.9\,kBq\,dm^{-3}$ of $^{109}CdCl_2$) and $ZnCl_2$ ($66.6\,kBq\,dm^{-3}$ of $^{65}ZnCl_2$) in deionized water by a rhizofiltration system assembled from 20 experimental units containing juvenile plants of *A. donax* at a solution flow rate of $0.125\,cm^3\,min^{-1}$. This work confirmed that *A. donax* has the potential to be used in rhizofiltration systems for Cd and Zn removal from WW or contaminated liquids under continuous flow conditions.

Constructed wetlands can also act as a pre-treatment scheme to reduce the nutrient concentration in the effluent, being implemented between secondary treatment in WW treatment plants and land treatment systems (LTS), a strategic way to promote WW resource management in water-scarce regions. *A. donax* may be a suitable plant to use in these systems, allowing water quality improvements as well as reducing irrigation risk for LTS. In this kind of system, water treatment is achieved by using vegetation that affects hydraulic loading and nutrient uptake. In Greece, Tzanakakis et al. (2009) tested *A. donax* among other species, obtaining a total yield of 72.81 Mg ha^{-1} at the end of a 3-year experiment. At the same time, the plants accumulated 967.2 kg (N) ha^{-1} and 56.66 kg (P) ha^{-1}. Effluents were applied at rates capable of satisfying ETc in a semi-arid region of Greece, and biomass growing in LTS contributed to 35% of the nutrient recovery. These systems require additional management practices to approximate nutrient load with potential assimilation (Tzanakakis et al., 2009). Due to its capacity to control nitrate leaching, promoted by its deep and extensive root system and by maintaining a higher nitrification rate than trees, *A. donax* is a relevant candidate species for growing on poor soil irrigated by nutrient-enriched WW (Tzanakakis et al., 2011). These studies show the great adaptability of *A. donax* to different environments, as well as its potential for biomass use. The plant is well adapted to high disturbance dynamics in ecosystems, as well as to phytodepurate waters containing high amounts of heavy metals such as Cd and Ni (Papazoglou, 2007). Nevertheless, Zema et al. (2012) reported a decrease in the productivity of *A. donax* when irrigated with WW. Lower growth (reductions of 7.2% in height, 5.9% in stem diameter, and 3.1% in LAI) was recorded for plants irrigated with WW compared with plants irrigated with conventional water. Another interesting point from the study of Zema et al. (2012) concerns the mean high heating value of 18.18 MJ kg^{-1}, a value in the range referred by Angelini et al. (2005) under conventional water application and in fertile soil (16 MJ kg^{-1} in the establishing year and 18 MJ kg^{-1} from the second to the sixth year). *A. donax* irrigated with WW produced appreciable biomass and energy yields, but its behavior and response to WW seemed to be independent, where higher nutrient load and water did not lead to higher biomass yields for this crop. Another point that should be considered in this analysis refers to the fact that *A. donax* has a greater capacity to survive under WW irrigation—namely, after rhizome transplantation—when compared to other species like *Phragmites* and *Typha*. Zema et al. (2012) reported survival rates of up to 90% for *A. donax* and 75% and 65% for *Phragmites* and *Typha*, respectively. Even though in some cases this plant produces less biomass than others under similar conditions, the survival rate related to this crop may ensure higher overall biomass. Similarly, the mean low heating value was 50% lower than the maximum value measured for *Typha*, but since its biomass yield is much higher, the energy yield per unit of cultivated area (24.03 MJ m^{-2}) for *A. donax* was about 13 times the value obtained for *Typha*. *A. donax* is effective in controlling and removing pollutants such as nitrates and phosphates, as well as COD, from WW (Chang et al., 2012; Tam and Wong, 2014). In many of the studies cited here, *A. donax* had a higher capacity to control the percolation of salts and organic and inorganic contaminants. Further research activities should be developed to accurately determine the agronomic aspects of its production under WW irrigation.

4 Production of *A. donax* in contaminated soils: Case studies

The growing demand for biomass for bioenergy production and, in particular, the production of energy crops generates several conflicts concerning the use of land. Such conflicts can be solved by spatial segregation of the area for the production of energy crops, which encompasses various types of marginal lands, including heavy metal-contaminated lands (Abreu et al., 2020, 2022; Gomes et al., 2022b). The introduction of a perennial grass such as *A. donax* to poor-nutrient soils may bring many ecological advantages and provide a wide range of ecological services such as extensive vegetation cover and permanence in soils, and control of erosion dynamics, thereby contributing to carbon and water storage of soils, as well as to the reversal, control, and mitigation of desertification (Fernando et al., 2018b, c) contributing also to the remediation of the soils.

This section briefly introduces the studies that have used *A. donax* for the phytoremediation of heavy metal- and other inorganic-contaminated soils, as well as the main results concerning agronomic parameters, physiological processes, pollutant removal rates, and yields achieved in production on contaminated soils.

A. donax showed great endurance when exposed to Cd and Ni soil concentrations of up to 973.8 and 2543.3 mg kg^{-1}, respectively. Plants showed no toxicity effects on biometric parameters and physiological processes (i.e., on growth, biomass production, net photosynthesis, stomatal conductance, intercellular CO_2 concentration, stomatal resistance, chlorophyll content, and chlorophyll fluorescence) (Papazoglou et al., 2005, 2007). Plant productivity and WUE remained unaffected, indicating the extent of the tolerance to Cd and Ni of this plant species (Papazoglou, 2007). Higher Cd concentrations were determined in the lower leaves (13.7 mg kg^{-1}) and in the upper part of the stems (9.0 mg kg^{-1}), while in rhizomes Cd concentration was 9.4 mg kg^{-1}. Corresponding Ni values were 42.4 mg kg^{-1} in lower leaves, 16.0 mg kg^{-1} in upper stems, and 54.8 mg kg^{-1} in rhizomes (Papazoglou, 2009).

The studies of Guo and Miao (2010) and Miao et al. (2012) registered fast growth of giant reed as well as high yields in aerial biomass under As (254 mg kg^{-1}), Cd (76.1 mg kg^{-1}), and Pb (1552 mg kg^{-1}) contamination. The height of the plants and dry biomass reduced slightly due to the presence of heavy metals, and metal accumulation was greater at the root level. Soil amendments such as acetic acid, ethylenediaminetetraacetic acid (EDTA), and citric acid enhanced the biomass production. The concentrations of As, Cd, and Pb in giant reed shoots were significantly increased when lower levels of acetic acid, citric acid, and sepiolite were applied and higher levels of EDTA. These amendments could be considered optimum for remediation systems using *A. donax* L. Han et al. (2005) refer to high yields of biomass, to a huge growth of the root system and to a high adaptability, tolerance, and accumulation of Cd and Hg when giant reed was tested for phytoextraction of Hg (101 mg kg^{-1}) and Cd (115 mg kg^{-1}) from soils. Han and Hu (2005) tested the effects of Cu^{2+}, Pb^{2+}, Cd^{2+}, Zn^{2+}, Ni^{2+}, Hg^{2+} (100 mg kg^{-1}), and Cr^{6+} (50 mg kg^{-1}), registering a decrease in chlorophyll content (20%–56%) and effects on leaves. The plant did not tolerate this concentration of Cr^{6+} and the root system was damaged. Heavy metal concentration in soil decreased during the period of growth of the plant, probably due to the translocation of metals within the *A. donax* biomass. Río et al. (2002) tested many plant species including *A. donax* for the phytoremediation of Pb, Cu, Zn, Cd, Tl, Sb, and As (5000 mg kg^{-1}) in Doñana Natural Park, Spain. The plant accumulated 0–23 mg kg^{-1} Pb, 133–147 mg kg^{-1} Zn, 13–16 mg kg^{-1} Cu, and 7–11 mg kg^{-1} As. Under the conditions described earlier, *A. donax* did not assume a dominant status over all other heavy metal accumulator species, showing coexistence under wetland conditions, a type of environment where it normally displays invasiveness (California Invasive Plant Council, 2011).

Boularbah et al. (2006) refer to the ecological restoration of soils situated in former mining sites in Morocco, contaminated with Cd, Cu, Pb, and Zn by using many plant species in its phytoremediation. *A. donax* was used among these species, with bioaccumulation factors of 0.01 for Cu, 0.004 for Pb, and 0.04 for Zn. The total concentration of metals in the plant was $0.2 \, mg \, kg^{-1}$ for Cd, $7.1 \, mg \, kg^{-1}$ for Cu, $2.8 \, mg \, kg^{-1}$ for Pb, and $72 \, mg \, kg^{-1}$ for Zn. Sabeen et al. (2013) tested its use to treat Cd (0, 50, 100, 250, 500, 750, and $1000 \, \mu g \, dm^{-3}$) for 21 days both in hydroponics and contaminated soil environments. Roots were the plant organ with higher Cd accumulation, followed by stems and leaves. Better results were recorded in hydroponic culture than with Cd-contaminated soils: the maximum Cd content in root was $300 \, \mu g \, g^{-1}$ in the hydroponics experiment over $230 \, \mu g \, g^{-1}$ in the soil experiment. Under hydroponic conditions, translocation and bioaccumulation factors were always greater than 1, and above the reference value (1.0) for hyperaccumulation. In the soil, despite the low Cd uptake, the translocation factors were above the reference value (1.0); however, bioaccumulation values were below 1. At higher Cd exposure, plants showed some antioxidative stress as the concentration of antioxidants was increased with increasing Cd exposure. Alshaal et al. (2013) tested the phytoremediation of bauxite-derived red mud by *A. donax*. The presence of this plant species reduced the EC of red mud by 25% and that of mud-polluted soil by 6%. At the same time, *A. donax* promoted phytoextraction of Cd, Pb, Co, Ni, and Fe, with high translocation factors, especially for Ni. Its presence on the contaminated medium also improved other soil quality parameters such as pH, EC, organic carbon, microbial counts, and soil enzyme activities.

A. donax has a high potential to uptake more than one type of metal from polluted soil at significant rates, as well as to improve many properties of poor or contaminated soils. The versatility of *A. donax* allows its cultivation in soils with different stress conditions, such as salinity, pH, organic matter, and nitrogen content, or heavy metal contaminations in various concentrations and availabilities (Eid et al., 2016).

Pb and Zn EDTA extractable soil fractions were also reduced by *A. donax* in a 2-year open-air experiment aimed at assessing the *A. donax* potential for phytoextraction and soil fertility restoration, confirming the ability of this crop to grow on contaminated soils (Fiorentino et al., 2017). Compost addition gives the highest biomass production and consequently the highest metal uptakes of *A. donax* (Fagnano et al., 2015). Similarly, a study in Lisbon (Barbosa et al., 2015a) tested the adaptability and phytoremediation capacity of *A. donax* and *Miscanthus* spp. on contaminated soils (under the exposure of 450 and $900 \, mg \, kg^{-1}$ of Zn and Pb; 300 and $600 \, mg \, kg^{-1}$ of Cr), showing their suitability for phytoextraction and accumulation. In particular, the results confirm that bioaccumulation occurs mainly in the hypogeal part (i.e., rhizomes and roots), especially for Pb and Cr, while Zn is easily transported and accumulated in the aerial fractions.

Cristaldi and collaborators evaluated the association of *A. donax* with *T. harzianum* for 7 months to test its tolerance and accumulation potential when exposed to different metals. The crop was submitted to a mix containing two different doses of Ni, Cd, Cu, V, Zn, As, Pb, and Hg, being L1 as the lowest dose and L2 as the highest. It was possible to observe that the average bioaccumulation percentage occurred in L2, and for Cd, Cu, and Hg was 50%, 35%, and 45%, respectively. The addition of *T. hazianum*, however, increased the bioaccumulation percentages of Ni, Pb, and Hg, for L2, going from 26% to 38%, from 14% to 54%, and from 45% to 60%, respectively. However, the addition of microorganisms decreased the plants' capability to accumulate Cu, from 35% to 29%, As from 27% to 23%, V from 26% to 20%, and Zn from 9% to 7% (Cristaldi et al., 2020).

The potential for phytoremediation by *A. donax* depends also on the harvest period, as demonstrated by Danelli et al., who designed a field experiment in which the crop was cultivated under heavy metal-contaminated soils in a density of $2500 \, \text{plants} \, \text{ha}^{-1}$. The researchers harvested the biomass in two different periods of the year, in October and in February, for three different harvest seasons, concluding that the high yield of *A. donax* and its capability to accumulate a considerable amount of contaminants make this crop an excellent alternative for phytoextraction. The results also pointed out that the best accumulation result occurred in the third year, when the harvest was made in October, and the plant was able to extract $3.87 \, \text{kg} \, \text{ha}^{-1}$ of Zn, $2.09 \, \text{kg} \, \text{ha}^{-1}$ of Cu, and $0.007 \, \text{kg} \, \text{ha}^{-1}$ of Cd (Danelli et al., 2021).

Giant reed were tested under 4 and $8 \, \text{mg} \, \text{Cd} \, \text{kg}^{-1}$, 300 and $600 \, \text{mg} \, \text{Cr} \, \text{kg}^{-1}$, and 110 and $220 \, \text{mg} \, \text{Ni} \, \text{kg}^{-1}$ contaminated soils, in a 2 year pot experiment. Giant reed aboveground yields, in the second year harvest, and in all the levels metals of contaminated soils, reduced significantly, by 30–70%. Only soils contaminated with the low level of Ni did not affect the yields. But, the rhyzomes and roots biomass yields were not altered by the tested metals. Interestingly, the high heating value (HHV) of giant reed aerial fraction, harvested in the second year, was not affected by the different contaminations tested. Therefore, the biomass obtained from the contaminated soils can be valorized as a feedstock for energy production (Gomes et al., 2022a). Sidella et al. (2016) also studied the response of different giant reed clones to soils contaminated with lead (450 and $900 \, \text{mg} \, \text{Pb} \, \text{kg}^{-1}$). Biomass yields were not significantly affected by the lead contamination and it was possible to identify the genotypes that yielded more under the contamination stressing conditions tested. Giant reed also showed phytoextraction capacity once Pb content increased in the biomass with the increment of Pb in the soil, especially in roots and rhizomes. In the study, the genotypes that showed the capacity to phytoextract more Pb from the soils, when the plant was harvested, were also marked. Yet, giant reed needs irrigation to be more productive, especially in areas of the Globe that are more prone to desertification (Fernando et al., 2015). In this context, the work of Sidella et al. (2018) aimed to study the effect of Pb contaminated soils (450 and 900 mg Pb kg^{-1}) in the growth of *Arundo donax* under a low irrigation regime (475 mm). Under both stressed conditions, the plant growth and yields were significantly affected by the contamination, contrarily to what was observed when full irrigation was provided to surpass water stress (950 mm) (Sidella et al., 2016). In terms of phytoextraction, after two consecutive years, only 3.2‰ maximum was accumulated and harvested from the total Pb soil bioavailable fraction. Fernando et al. (2018a) also evaluated the growth of giant reed in contaminated soils and under low irrigation. Results obtained showed that giant reed biomass production was reduced due to Cr, Zn, and Pb contamination, inducing lower accumulation of metals in the biomass. Nevertheless, the crop showed favorable potential for phytoextraction and demonstrated to be suited for phytostabilization of heavy metals contamination, by prevented the leaching of heavy metals to groundwater contamination.

Yu and collaborators studied the physiological response of *A. donax* to Cd stress. It was observed that when exposed to small concentrations of Cd, the root system of the plant uses a mechanism to improve the plant's tolerance to this contaminant based on the osmosis of organic substances. These organic substances, from leaves, move to the root system to chelate Cd. Another observed point was that the bioaccumulation of Cd in all plant fractions is directly affected by its concentration in soil (Yu et al., 2018).

Another interesting study was realized by Zeng et al. in which an evaluation of the effects of intercrop of *A. donax* with woody crops in soils contaminated with As, Zn, Cd, and Pb. The selected woody crops were *B. papyrifera* or *Morus alba*. The cultivation of woody crops without intercrop system was settled as Control. The results showed a beneficial effect in terms of woody crops productivity caused by the intercrop system. Also, comparing with the Control trials, Zn and Pb uptake increased to 124% and 171%, respectively, for *A. donax* and *B. papyrifera*, while for *A. donax* and *Morus alba*, the increase in As and Pb uptakes were 150% and 75% (Zeng et al., 2018).

5 Benefits and constraints of using giant reed for phytoremediation purposes

Giant reed can be used for phytoremediation purposes of different types of contaminated soil, sediments, and water bodies. *A. donax* L. has the advantages of many annual energy crops and being perennial allows the effects of these characteristics on the ecosystems in which it is implemented to endure. The implementation of this crop on marginal land may increase the yield and profitability of the soil over time, promoting the control of diseases and pests on site by increasing biological and landscape diversity, as well as providing a source of biomass for fiber, bioenergy, and other by-products. Giant reed can also be introduced for the mitigation and reversal of desertification, because of low WUE and NUE, being able to generate commercial value for a given region. Its implementation on contaminated land may involve fewer problems for natural ecosystems and be integrated into a waste management strategy (Fernando et al., 2015). Being a perennial crop, giant reed offers additional ecological advantages, providing a wide range of ecological services such as greater vegetation cover; greater permanence in the soil, which limits soil erosion; low susceptibility to diseases; reduced need for pesticides; and, due to its extensive and deep root system, the plant can be used for leaching control of many contaminants; (Fernando et al., 2018c; Gomes et al., 2023). It is also a contributor crop for carbon sequestration (Gomes et al., 2022b).

Due to its fast growth and high cellulose content, *A. donax* L. is considered one of the most promising energy crops for marginal lands, since its culms represent an important source of cellulose for the production of paper, second-generation ethanol, biodiesel, and biopolymers (Lino et al., 2023). Its residual lignin content could be used for the production of lignin-based resin coatings and composites. When produced in heavy metals-contaminated soils or with heavy metals-rich WW, there may be an increased accumulation of those elements in the biomass. If the levels of heavy metals contained in its biomass are high, the most environmentally safe solution is the production of energy from combustion or pyro-gasification (Lievens et al., 2008a,b), followed by metal recovery from fly ash using hydrometallurgical routes (Fiorentino et al., 2013). As metal toxicity may seriously limit the microbial-driven conversion of lignin cellulose to second-generation ethanol and biopolymers, these options are usually put aside. These studies validate the logic of giant reed use with WW and in contaminated soils by offering solutions for the recovery of contaminants such as heavy metals, as well as for the provision of biomass for bioenergy and other by-product production. This is one of the main advantages of this approach and clearly meets the purposes of biorefinery and sustainable bioeconomy.

A. donax L. can also contribute to the restoration of soil ecosystems, along with biomass production, as indicated in the work of Alshaal et al. (2014), who tested the potential of *A. donax* under microbial communities in marginal soils. Giant reed showed considerable potential for recovering red mud-affected soils. Its cultivation increased the activities of most soil enzymes, especially of dehydrogenase, urease, and catalase. Regarding soil organic carbon, the authors did not find any considerable effect of its presence on soils but did find that giant reed has a special microbial community associated with its root system that helps to restore and improve many soil properties. Total fungi increased in soils after the presence of giant reed, but the total bacterial count decreased after its plantation by a range of 29%–93%. Giant reed can also be used to restore and recover soil ecosystems after exposure to natural disasters such as bushfires. *A. donax* L. shares the advantages of many hyperaccumulator plant species used in phytoremediation approaches and can be used within in situ and ex situ applications, with additional gains. Phytoremediation technology requires several growing seasons to clean a contaminated site, and owners of contaminated lands may not wish to wait several growing seasons when hyperaccumulator plants are used to clean the sites due to lower yields and no profitable revenue from it. Energy/fiber crops such as giant reed offer the possibility of generating commercial value for the biomass produced in contaminated soils, overcoming this problem from the first year.

Giant reed, like other perennial crops, can tolerate soils contaminated with one or more contaminants, showing interesting yields as well as providing quality biomass for fiber or bioenergy purposes (Fernando et al., 2015). One of the disadvantages of its use in the phytoremediation of inorganic compounds is the limited amount of these compounds that can be extracted from soil, especially when compared with the amounts of the same compounds that can be extracted by physicochemical remediation methods, and the related time needed for remediation. Nevertheless, that disadvantage is clearly outweighed by the income obtained from the use of the biomass.

The use of a wild or non-indigenous plant could provoke its spread as well as leading to several risks for ecosystem functions if used near phytotreatment sites (McIntyre, 2003). In fact, the attributes required for optimal herbaceous energy/fiber crops correspond to those of typical weeds and invasive plants: namely, rapid growth, low fertilizer inputs, high WUE and NUE, and an absence of pests and diseases. *A. donax* L. meets many of these requirements, being an herbaceous crop with high potential for invasiveness. Because of that, the implementation of this type of crop requires containment plans for their potential spread. Another disadvantage associated with its potential invasiveness is the limited information that is available to perform an appropriate risk assessment of the species (McIntyre, 2003), as well as appropriate methods for crop removal from the field after the remediation process. In fact, giant reed can reach huge population density and its fast growth can take control of the existing resources in the neighborhood, such as soil nutrients and access to light. Moreover, it can have more than one growth cycle in 1 year, and shows huge adaptability to different environments. On one hand, this could be useful in cases where there is a need for a different genetic background to approach the remediation in a particular site (for example, inexistence of diseases and plagues or other trophic relations). On the other hand, this means environmental managers have to be aware of potentially harmful impacts in situations where the crops show invasiveness and should promote its precautionary control. Because *A. donax* L. possesses a deep and extensive root system, its presence on slope terrains could also provide slope support for marginal lands, restraining erosion processes. In these sloping areas, as well as in space-restricted sites, the application of machinery and other traditional agricultural cropping techniques may not be possible (McIntyre, 2003). The application of energy and fiber crops in phytoremediation opens up the possibility of recycling all constituents contained in the phytoremediation biomass. Understanding the main pathways and main techniques to achieve that objective is fundamental to promoting phytoremediation in the remediation market, as well as for the commercialization of its by-products (McIntyre, 2003). In addition to the economic recovery when used for energy production or biomaterials, using *Arundo* in the phytoremediation of soils and WW can also provide additional benefits, such as the amount of carbon sequestered by the biomass ((Fernando et al., 2015), or the water and mineral resources saved, if irrigation uses WW (Costa et al., 2016), or improvement of soil functions, when cropped in marginal land (Fernando et al., 2018b).

6 Conclusions and recommendations

Giant reed is a good candidate for marginal and wetland soils, for incorporation into LTS and in constructed wetlands, with an enormous potential for phytoremediation. The plant is able to improve the quality of water-polluted bodies, being able to remove COD, BOD, nitrates, ammonium and phosphate ions, and heavy metals. In contaminated soils, giant reed with its associated microorganisms and fungi improves many soil properties, whether at chemical or physical (slope terrains) levels, being able to remove heavy metals and some types of radionuclides, among other inorganic compounds. This crop can be used to control water, wind, and biological erosion on marginal lands, important for the mitigation and reversal of desertification. At the same time, the plant shows good physiological and biomass responses to stress in both polluted water and soil environments, returning biomass with high-quality parameters for the production of fiber, energy, and other bioproducts. Under some specific conditions, precautionary control measures should be implemented because of its potential invasiveness.

References

Abreu, M., Reis, A., Moura, P., Fernando, A.L., Luís, A., Quental, L., Patinha, P., Gírio, F., et al., 2020. Evaluation of the potential of biomass to energy in Portugal-conclusions from the CONVERTE project. Energies 13 (4), 937.

Abreu, M., Silva, L., Ribeiro, B., Ferreira, A., Alves, L., Paixão, S.M., Gouveia, L., Moura, P., Carvalheiro, F., Duarte, L.C., et al., 2022. Low indirect land use change (ILUC) energy crops to bioenergy and biofuels—a review. Energies 15, 4348. https://doi.org/10.3390/en15124348.

Ahmed, M.J., 2016. Potential of *Arundo donax* L. stems as renewable precursors for activated carbons and utilization for wastewater treatments: review. J. Taiwan Inst. Chem. Eng. 63, 336–343. https://doi.org/10.1016/j.jtice.2016.03.030.

Alessina, G., Pedrazzi, S., Puglia, M., Morselli, N., Tartarini, P., 2018. Energy production and carbon sequestration in wet areas of Emilia Romagna region, the role of Arundo donax. Adv. Model. Anal. 55 (3), 108–113. https://doi.org/10.18280/ama_a.550302.

Alshaal, T., Szabolcsy, É., Márton, L., Czakó, M., Kátai, J., Balogh, P., Elhawat, N., Ramady, H., Fári, M., 2013. Phytoremediation of bauxite-derived red mud by giant reed. Environ. Chem. Lett. 11, 295–302.

Alshaal, T., Szabolcsy, É., Márton, L., Czakó, M., Kátai, J., Balogh, P., Elhawat, N., Ramady, H., Geröes, A., Fári, M., 2014. Restoring soil ecosystems and biomass production of a. donax L. under microbial communities-depleted soil. Bioenergy Res. 7, 268–278.

Ammari, T.G., 2014. Utilization of a natural ecosystem bio-waste; leaves of Arundo Donax reed, as a raw material of low-cost eco-biosorbent for cadmium removal from aqueous phase. Ecol. Eng. 71, 466–473. https://doi.org/10.1016/j.ecoleng.2014.07.067.

Angelini, L., Ceccarini, L., Bonari, E., 2005. Biomass yield and energy balance of giant reed (A. donax L.) cropped in Central Italy as related to different management practices. Eur. J. Agron. 22, 375–389.

Ba, Y., Liu, F., Wang, X., Yang, J., 2020. Pyrolysis of C3 energy plant (Arundo donax): thermogravimetry, mechanism, and potential evaluation. Ind. Crop. Prod. 149, 112337. https://doi.org/10.1016/j.indcrop.2020.112337.

Barbagallo, S., Cirelli, G.L., Consoli, S., Milani, M., Toscano, A., 2011. Utilizzo di acque reflue per l'irrigazione di biomasse erbacee a scope energetici. In: Convegno di Medio Termine dell'Associazione Italiana di Ingegneria Agraria, 22–24 settembre, 2011, Belgirate.

Barbero-Sierra, C., Marques, M.J., Ruíz-Perez, M., 2013. The case of urban sprawl in Spain as an active and irreversible driving force for desertification. J. Arid Environ. 90, 95–102.

Barbosa, B., Fernando, A.L., 2018. Aided phytostabilization of mine waste. In: Prasad, M.N.V., Favas, P.J.C., Maiti, S.K. (Eds.), Bio-Geotechnologies for Mine Site Rehabilitation. Elsevier Inc., Amsterdam, The Netherlands, UK, pp. 147–158.

Barbosa, B., Fernando, A.L., 2021. Ecological and ecosystem engineering for economic environmental revitalization. In: Prasad, M.N.V. (Ed.), Handbook of Ecological and Ecosystem Engineering. John Wiley & Sons Ltd, UK, pp. 25–46.

Barbosa, B., Boléo, S., Sidella, S., Costa, J., Duarte, M.P., Mendes, B., Cosentino, S.L., Fernando, A.L., 2015a. Phytoremediation of heavy metal-contaminated soils using the perennial energy crops Miscanthus spp. and Arundo donax L. Bioenergy Res. 8 (4), 1500–1511. https://doi.org/10.1007/s12155-015-9688-9.

Barbosa, B., Costa, J., Fernando, A.L., Papazoglou, E.G., 2015b. Wastewater reuse for fiber crops cultivation as a strategy to mitigate desertification. Ind. Crop. Prod. 68, 17–23.

Barbosa, B., Costa, J., Boléo, S., Duarte, M.P., Fernando, A.L., 2016. Phytoremediation of inorganic compounds. In: Ribeiro, A.B., Mateus, E.P., Couto, N. (Eds.), Electrokinetics Across Disciplines and Continents—New Strategies for Sustainable Development. Springer International Publishing, Switzerland, pp. 373–400.

Barbosa, B., Costa, J., Fernando, A.L., 2018. Production of energy crops in heavy metals contaminated land: opportunities and risks. In: Li, R., Monti, A. (Eds.), Land Allocation for Biomass. Springer International Publishing AG, pp. 83–102.

Bonano, G., Cirelli, G.L., Toscano, A., Giudice, R., Pavone, P., 2013. Heavy metal content in ash of energy crops growing in sewage-contaminated natural wetlands: potential applications in agriculture and forestry? Sci. Total Environ. 452–453, 349–354.

Borin, M., Barbera, A.C., Milani, M., Molari, G., Zimbone, S.M., Toscano, A., 2013. Biomass production and N balance of giant reed (A. donax L.) under high water and N input in Mediterranean environments. Eur. J. Agron. 51, 117–119.

Boularbah, A., Schwartz, C., Bitton, G., Aboudrar, W., Ouhammou, A., Morel, J., 2006. Heavy metal contamination from mining sites in South Morocco: 2. Assessment of metal accumulation and toxicity in plants. Chemosphere 63, 811–817.

Calheiros, C.S.C., Quitério, P.V.B., Silva, G., Crispim, L.F.C., Brix, H., Moura, S.C., Castro, P.M.L., 2012. Use of constructed wetland systems with Arundo and Sarcocornia for polishing high salinity tannery wastewater. J. Environ. Manag. 95, 66–71.

California Invasive Plant Council, 2011. A. donax (giant reed): Distribution and impact report March 2011. State Water Resources Control Board, Berkeley, CA.

Chai, Y., Bai, M., Chen, A., Peng, L., Shao, J., Shang, C., Peng, C., Zhang, J., Zhou, Y., 2022. Thermochemical conversion of heavy metal contaminated biomass: fate of the metals and their impact on products. Sci. Total Environ. 822, 153426. https://doi.org/10.1016/j.scitotenv.2022.153426.

I. Bioproducts from contaminated soil and water

Chang, J., Wu, S., Dai, Y., Liang, W., Wu, Z., 2012. Treatment performance of integrated vertical-flow constructed wetland plots for domestic wastewater. Ecol. Eng. 44, 152–159.

Chen, Z., Jiang, D., Zhang, T., Lei, T., Zhang, H., Yang, J., Shui, X., Li, F., Zhang, Y., Zhang, Q., 2022. Comparison of three ionic liquids pretreatment of *Arundo d*onax L. for enhanced photo-fermentative hydrogen production. Bioresour. Technol. 343, 126088. https://doi.org/10.1016/j.biortech.2021.126088.

Christou, M., Alexopoulou, E., Cosentino, S.L., Copani, V., Nogues, S., Sanchez, E., Monti, A., Zegada-Lizarazu, W., Pari, L., Scarfone, A., 2018. Giant reed: from production to end use. In: Alexopoulou, E. (Ed.), Perennial Grasses for Bioenergy and Bioproducts. Academic Press; Elsevier Inc., Cambridge, MA; UK, pp. 107–151.

Corno, L., Pilu, R., Adani, F., 2014. *Arundo donax* L.: a non-food crop for bioenergy and bio-compound production. Biotechnol. Adv. 32 (8), 1535–1549. https://doi.org/10.1016/j.biotechadv.2014.10.006.

Correia, R., Gonçalves, M., Nobre, C., Mendes, B., 2017. Impact of torrefaction and low-temperature carbonization on the properties of biomass wastes from *Arundo donax* L. and *Phoenix canariensis*. Bioresour. Technol. 223, 210–218. https://doi.org/10.1016/j.biortech.2016.10.046.

Cosentino, S.L., Copani, V., Scalici, G., Scordia, D., Testa, G., 2015. Soil erosion mitigation by perennial species under Mediterranean environment. Bioenergy Res. 8, 1538–1547.

Costa, J., Barbosa, B., Fernando, A.L., 2016. Wastewaters reuse for energy crops cultivation. In: Proceedings of the 7th IFIP WG 5.5/SOCOLNET Advanced Doctoral Conference on Computing, Electrical and Industrial Systems, DoCEIS 2016, Costa de Caparica, Portugal, 11–13 April 2016, first ed. 470. Springer, pp. 507–514.

Cristaldi, A., Conti, G.O., Cosentino, S.L., Mauromicale, G., Copat, C., Grasso, A., Zuccarello, P., Fiore, M., Restuccia, C., Ferrante, M., 2020. Phytoremediation potential of *Arundo donax* (giant reed) in contaminated soil by heavy metals. Environ. Res. 185, 109427. https://doi.org/10.1016/j.envres.2020.109427.

Curt, M.D., Sanz, M., Mauri, P.V., Plaza, A., Cano-Ruiz, J., Sánchez,, J., Chaya, C., Fernández, J., 2018. Effect of water regime change in a mature *Arundo donax* crop under a Xeric Mediterranean climate. Biomass Bioenergy 115, 203–209. https://doi.org/10.1016/j.biombioe.2018.04.018.

Danelli, T., Sepulcri, A., Masetti, G., Colombo, F., Sangiorgio, S., Cassani, E., Anelli, S., Adani, F., Pilu, R., 2021. *Arundo donax* L. biomass production in a polluted area: effects of two harvest timings on heavy metals uptake. Appl. Sci. 11 (3), 1147. https://doi.org/10.3390/app11031147.

Dürešová, Z., Šuňovská, A., Horník, M., Pipíška, M., Gubišová, M., Gubiš, J., Hostin, S., 2014. Rhizofiltration potential of *A. donax* for cadmium and zinc removal from contaminated wastewater. Chem. Pap. 68, 1452–1462.

Eid, E.M., Youssef, M.S.G., Shaltout, K.H., 2016. Population characteristics of giant reed (*Arundo donax* L.) in cultivated and naturalized habitats. Aquat. Bot. 129, 1–8. https://doi.org/10.1016/j.aquabot.2015.11.001.

El Bassam, N., 2010. In: Handbook of Bioenergy Crops. A Complete Reference to Species, Development and Appli cations. Earthscan, London.

Elhawat, N., Alshaal, T., Szabolesy, É., Ramady, H., Márton, L., Czakó, M., Kátai, J., Balogh, P., Sztrik, A., Molnár, M., Popp, J., Fári, M., 2014. Phytoaccumulation potentials of two biotechnologically propagated ecotypes of *A. donax* in copper-contaminated synthetic wastewater. Environ. Sci. Pollut. Res. Int. 21, 7773–7780.

Fagnano, M., Impagliazzo, A., Mori, M., Fiorentino, N., 2015. Agronomic and environmental impacts of giant reed (*Arundo donax* L.): results from a long-term field experiment in hilly areas subject to soil erosion. Bioenergy Res. 8 (1), 415–422. https://doi.org/10.1007/s12155-014-9532-7.

Fernando, A.L., Costa, J., Barbosa, B., Monti, A., Rettenmaier, N., 2018b. Environmental impact assessment of perennial crops cultivation on marginal soils in the Mediterranean Region. Biomass Bioenergy, 111, 174–186.

Fernando, A.L., Duarte, M.P., Almeida, J., Boléo, S., Mendes, B., 2010. Environmental impact assessment of energy crops cultivation in Europe. Biofuels Bioprod. Biorefin. 4, 594–604.

Fernando, A.L., Boléo, S., Barbosa, B., Costa, J., Lino, J., Tavares, C., Sidella, S., Duarte, M.P., Mendes, B., 2014. How sustainable is the production of energy crops in heavy metal contaminated soils? In: Hoffmann, C., Baxter, D., Maniatis, K., Grassi, A., Helm, P. (Eds.), Proceedings of the 22nd European Biomass Conference and Exhibition, Setting the Course for a Biobased Economy, 23–26 June 2014, Hamburg, Germany. ETA-Renewable Energies, pp. 1593–1596.

Fernando, A.L., Barbosa, B., Boléo, S., Duarte, M.P., Sidella, S., Costa, 2018a. Phytoremediation potential of heavy metal contaminated soils by the perennial energy crops miscanthus SPP. and arundo donax L. under low irrigation. In: - - (Ed.), European Biomass Conference and Exhibition Proceedings. ETA-Florence, Florence, Italy, pp. 136–139.

Fernando, A.L., Boléo, S., Barbosa, B., Costa, J., Duarte, M.P., Monti, A., 2015. Perennial grass production opportunities on marginal Mediterranean land. Bioenergy Res. 8, 1523–1537.

Fernando, A.L., Rettenmaier, N., Soldatos, P., Panoutsou, C., 2018c. Sustainability of perennial crops production for bioenergy and bioproducts. In: Alexopoulou, E. (Ed.), Perennial Grasses for Bioenergy and Bioproducts, first ed. Academic Press, Cambridge, MA, USA, pp. 245–283.

Ferrandez-Villena,, M., Ferrandez-Garcia,, A., Garcia-Ortuño, T., Ferrandez-Garcia, C.E., Garcia, M.T. F., 2020. Properties of wood particleboards containing giant reed (*Arundo donax* L.) particles. Sustainability 12 (24), 10469. https://doi.org/10.3390/su122410469.

Fiorentino, N., Fegnano, M., Adamo, P., Impagliazzo, A., Mori, M., Pepe, O., Ventorino, V., Zoina, A., 2013. Assisted phytoextraction of heavy metals: compost and Trichoderma effects on giant reed (*A. donax* L.) uptake and soil N-cycle microflora. Ital. J. Agron. 8, e29.

Fiorentino, N., Ventorino, V., Rocco, C., Cenvinzo, V., Agrelli, D., Gioia, L., Di Mola, I., Adamo, P., Pepe, O., Fagnano, M., 2017. Giant reed growth and effects on soil biological fertility in assisted phytoremediation of an industrial polluted soil. Sci. Total Environ. 575, 1375–1383. https://doi.org/10.1016/j.scitotenv.2016.09.220.

Gomes, L., Costa, J., Moreira, J., Cumbane, B., Abias, M., Santos, F., Zanetti, F., Monti, A., Fernando, A.L., 2022a. Switchgrass and giant reed energy potential when cultivated in heavy metals contaminated soils. Energies 15 (15), 5538. https://doi.org/10.3390/en15155538.

Galletti, S., Cianchetta, S., Righini, H., Roberti, R., 2022. A lignin-rich extract of giant reed (*Arundo donax* L.) as a possible tool to manage soilborne pathogens in horticulture: a preliminary study on a model pathosystem. Horticulturae 8 (7), 589. https://doi.org/10.3390/horticulturae8070589.

Garcez, L.R.Q., Gatti, T.H., Gonzalez, A.J.C., Franco, C., Ferreira, C.S., 2022. Characterization of fibers from culms and leaves of *Arundo donax* L. (*Poaceae*) for handmade paper production. J. Nat. Fibers 19 (16), 12805–12813. https://doi.org/10.1080/15440478.2022.2076005.

Ge, X., Xu, F., Vasco-Correa, J., Li, Y., 2016. Giant reed: a competitive energy crop in comparison with miscanthus. Renew. Sust. Energ. Rev. 54, 350–362. https://doi.org/10.1016/j.rser.2015.10.010.

Gomes, L.A., Costa, J., Cumbane, B., Abias, M., Pires, J.R.A., Souza, V.G.L., Santos, F., Fernando, A.L., 2023. Combating climate change with phytoremediation. Is it possible? In: Brito, P.S.D., Galvão, J.R.C.S., Monteiro, P., Panizio, R., Calado, L., Assis, A.C., Ribeiro, V.S.S. (Eds.), Proceedings of the 2nd International Conference on Water Energy Food and Sustainability (ICoWEFS 2022). ICoWEFS 2022. Springer, Cham, pp. 507–514.

Gomes, L.A., Costa, J., Santos, F., Fernando, A.L., 2022b. Environmental and socio-economic impact assessment of the switchgrass production in heavy metals contaminated soils. In: Lect. Notes Mech. Eng. Springer, pp. 410–419.

Gomes, L., Fernando, A.L., Santos, F., 2018. A Toolbox to Tackle the Technological and Environmental Constraints Associated with the Use of Biomass for Energy from Marginal Land. Proceedings of the ECOS 2018, the 31st International Conference on Efficiency, Cost, Optimization, Simulation and Environmental Impact of Energy Systems, Guimarães, Portugal, 17–22 June 2018.

Guo, Z., Miao, X., 2010. Growth changes and tissues anatomical characteristics of giant reed (*A. donax* L.) in soil contaminated with arsenic, cadmium and lead. J. Cent. S. Univ. Technol. 17, 770–777.

Han, Z., Hu, Z., 2005. Tolerance of *A. donax* to heavy metals. Chin. J. Appl. Ecol. 16, 161–165 (abstract).

Han, Z., Hu, X., Hu, Z., 2005. Phytoremediation of mercury and cadmium polluted wetland by *A. donax*. Chin. J. Appl. Ecol. 16, 945–950 (abstract).

Hidalgo, M., Fernandez, J., 2001. Biomass production of ten populations of giant reed (*A. donax* L.) under the environmental conditions of Madrid (Spain). In: Kyritsis, S., Beenackers, A.A.C., Helm, P., Grassi, A., Chiaramonti, D. (Eds.), Biomass for Energy and Industry: Proceedings of the First World Conference, 5–9 June 2000, Seville, Spain. vol. 1. James & James (Science Publishers), London, pp. 1881–1884.

Idris, S.M., Jones, P.L., Salzman, S.A., Croatto, G., Allinson, G., 2012. Evaluation of the giant reed (*A. donax*) in horizontal subsurface flow wetlands for the treatment of dairy processing factory wastewater. Environ. Sci. Pollut. Res. 19, 3525–3537.

Karahancer, S.S., Eriskin, E., Sarioglu, O., Capali, B., Saltan, M., Terzi, S., 2016. Utilization of *Arundo donax* in hot mix asphalt as a fiber. Constr. Build. Mater. 125, 981–986. https://doi.org/10.1016/j.conbuildmat.2016.08.147.

Kausar, S., Mahmood, Q., Raja, I., Khan, A., Sultan, S., Gilani, M., Shujaat, S., 2012. Potential of *A. donax* to treat chromium contamination. Ecol. Eng. 42, 256–259.

Klemm, D., Cranston, E.D., Fischer, D., Gama, M., Kedzior, S.A., Kralisch, D., Kramer, F., Kondo, T., Lindström, T., Nietzsche, S., Petzold-Welcke,, K., Rauchfuß, F., 2018. Nanocellulose as a natural source for groundbreaking applications in materials science: today's state. Mater. Today 21, 720–748. https://doi.org/10.1016/j.mattod.2018.02.001.

I. Bioproducts from contaminated soil and water

Kouki, S., Saidi, N., Rajeb, A., M'hiri, F., 2012. Potential of a polyculture of *A. donax* and *Typha latifolia* for growth and phytotreatment of wastewater pollution. Afr. J. Biotechnol. 11, 15341–15352.

Krička, T., Matin, A., Bilandžija, N., Jurišić, V., Antonović, A., Voća, N., Mateja, G., 2017. Biomass valorisation of *Arundo donax* L., Miscanthus × giganteus and *Sida hermaphrodita* for biofuel production. Int. Agrophys. 31 (4), 575–581.

Lei, Y., Xie, C., Wang, X., Fang, Z., Huang, Y., Cao, S., Liu, B., 2020. Thermophilic anaerobic digestion of *Arundo donax* cv. Lvzhou no. 1 for biogas production: structure and functional analysis of microbial communities. Bioenergy Res. 13 (3), 866–877.

Lemões, J.S., Lemons-e-Silva, C.F., Avila, S.P.F., Montero, C.R.S., Silva, S.D.A., Samios, D., Peralba, M.C.R., 2018. Chemical pretreatment of *Arundo donax* L. for second-generation ethanol production. Electron. J. Biotechnol. 31, 67–74.

Lewandowski, I., Scurlock, M.O.J., Lindvall, E., Christou, M., 2003. The development and current status of perennial rhizomatous grasses as energy crops in the US and Europe. Biomass Bioenergy 25, 335–361.

Lievens, C., Yperman, J., Vangronsveld, J., Carleer, R., 2008a. Study of the potential valorization of heavy metal contaminated biomass via phytoremediation by fast pyrolysis: part I. Influence of temperature, biomass species and solid heat carrier on the behavior of heavy metals. Fuel 87, 1894–1905.

Lievens, C., Yperman, J., Vangronsveld, J., Carleer, R., 2008b. Study of the potential valorization of heavy metal contaminated biomass via phytoremediation by fast pyrolysis: part II. Characterization of the liquid and gaseous fraction as a function of the temperature. Fuel 87, 1906–1916.

Lino, G., Espigul, P., Nogués, S., Serrat, X., 2023. Arundo donax L. growth potential under different abiotic stress. Heliyon 9 (5), e15521. doi:j.heliyon.2023.e15521.

Malucelli, L.C., Lacerda, L.G., Dziedzic, M., Filho, M.A.S.C., 2017. Preparation, properties and future perspectives of nanocrystals from agro-industrial residues: a review of recent research. Rev. Environ. Sci. Biotechnol. 16, 131–145. https://doi.org/10.1007/s11157-017-9423-4.

Mantineo, M., Agosta, D., Copani, V., Patanè, C., Cosentino, S., 2009. Biomass yield and energy balance of three perennial crops for energy use in the semi-arid Mediterranean environment. Field Crop Res. 114, 204–213.

Mavrogianopoulos, G., Vogli, V., Kyritsis, S., 2002. Use of wastewaters as a nutrient solution in a closed gravel hydroponic cultures of giant reed (*A. donax*). Bioresour. Technol. 82, 103–107.

McIntyre, T., 2003. Phytoremediation of heavy metals from soils. Adv. Biochem. Eng. Biotechnol. 78, 97–123.

Miao, Y., Xi-yuan, X., Xu-feng, M., Zhao-hui, G., Feng-yong, W., 2012. Effects of amendments on growth and metal uptake of giant reed (*A. donax* L.) grown on soil contaminated by arsenic, cadmium and lead. Trans. Nonferrous Metals Soc. China 22, 1162–1169.

Mirza, N., Mahmood, Q., Pervez, A., Ahmad, R., Farooq, R., Shah, M.M., Azim, M.R., 2010. Phytoremediation potential of *A. donax* in arsenic-contaminated synthetic wastewater. Bioresour. Technol. 101, 5815–5819.

Mirza, N., Pervez, A., Mahmood, Q., Shah, M., Shafqat, M., 2011. Ecological restoration of arsenic contaminated soil by *A. donax* L. Ecol. Eng. 37, 1949–1956.

Molari, L., Saverio, C.F., García, J.J., 2021. *Arundo donax*: a widespread plant with great potential as sustainable structural material. Constr. Build. Mater. 268, 121143.

Molino, A., Larocca, V., Valerio, V., Rimauro, J., Marino, T., Casella, P., Cerbone, A., Arcieri, G., Viola, E., 2018. Supercritical water gasification of lignin solution produced by steam explosion process on Arundo donax after alkaline extraction. Fuel 221, 513–517. https://doi.org/10.1016/j.fuel.2018.02.072.

Nawaz, A., Kumar, P., 2022. Elucidating the bioenergy potential of raw, hydrothermally carbonized and torrefied waste *Arundo donax* biomass in terms of physicochemical characterization, kinetic and thermodynamic parameters. Renew. Energ. 187, 844–856. https://doi.org/10.1016/j.renene.2022.01.102.

Nsanganwimana, F., Marchland, L., Douay, F., Mench, M., 2014. *A. donax* L., a candidate for phytomanaging water and soils contaminated by trace elements and producing plant-based feedstock. A review. Int. J. Phytoremediation 16, 982–1017.

Oginni, O., Singh, K., 2019. Pyrolysis characteristics of *Arundo donax* harvested from a reclaimed mine land. Ind. Crop. Prod. 133, 44–53.

Ozdemir, S., Yetilmezsoy, K., Nuhoglu, N.N., Dede, O.H., Turp, S.M., 2020. Effects of poultry abattoir sludge amendment on feedstock composition, energy content, and combustion emissions of giant reed (Arundo donax L.). J. King Saud Univ. Sci. 32 (1), 149–155. https://doi.org/10.1016/j.jksus.2018.04.002.

Papazoglou, E.G., 2007. *A. donax* L. stress tolerance under irrigation with heavy metals aqueous solutions. Desalination 211, 304–313.

Papazoglou, E.G., 2009. Heave metal allocation in giant reed plants irrigated with metalliferous water. Fresenius' Environ. Bull. 18, 166–174.

Papazoglou, E.G., Fernando, A.L., 2017. Preliminary studies on the growth, tolerance and phytoremediation ability of sugarbeet (*Beta vulgaris* L.) grown on heavy metal contaminated soil. Ind. Crop. Prod. 107, 463–471.

Papazoglou, E.G., Karantounias, G., Vemmos, S., Bouranis, D., 2005. Photosynthesis and growth responses of giant reed (*A. donax* L.) to the heavy metals Cd and Ni. Environ. Int. 31, 243–249.

Papazoglou, E.G., Konstantinos, G., Bouranis, D., 2007. Impact of high cadmium and nickel soil concentration on selected physiological parameters of *A. donax* L. Eur. J. Soil Biol. 43, 207–215.

Pari, L., Curt, M.D., Sánchez, J., Santangelo, E., 2016. Economic and energy analysis of different systems for giant reed (*Arundo donax* L.) harvesting in Italy and Spain. Ind. Crop. Prod. 84, 176–188. https://doi.org/10.1016/j.indcrop.2016.01.036.

Pires, J.R.A., Gomes, L.A., Souza, V.G.L., Godinho, M.H., Fernando, A.L., 2022a. Evaluation and comparison of micro/nanocellulose extracted from Arundo, Kenaf and Miscanthus. In: 30th European Biomass Conference & Exhibition Proceedings, pp. 1094–1098.

Pires, J., Souza, V., Fernando, A.L., 2019a. Valorization of energy crops as a source for nanocellulose production – current knowledge and future prospects. Ind. Crop. Prod. 140, 111642. https://doi.org/10.1016/j.indcrop.2019.111642.

Pires, J.R.A., Souza, V.G., Fernando, A.L., 2019b. Ecofriendly strategies for the production of nanocellulose from agro-industrial wastes. Eur. Biomass Conf. Exhib. Proc., 1781–1784.

Pires, J.R.A., Souza, V.G.L., Gomes, L.A., Coelhoso, I.M., Godinho, M.H., Fernando, A.L., 2022b. Micro and nanocellulose extracted from energy crops as reinforcement agents in chitosan film. Ind. Crops Prod 186, 115247. https://doi.org/10.1016/j.indcrop.2022.115247.

Qiu, Y., Lei, Y., Zhao, H., He, X., Liu, B., Huang, Y., 2023. Mesophilic anaerobic digestion of *Arundo donax* cv. Lvzhou no. 1 and *Pennisetum giganteum* for biogas production: structure and functional analysis of microbial communities. Bioenergy Res. 16, 1205–1216.

Ramos, D., El-Mansouri, N.E., Ferrando, F., Salvadó, J., 2018. All-lignocellulosic fiberboard from steam exploded *Arundo donax* L. Molecules 23 (9), 1–12.

Río, M., Font, R., Almela, C., Vélez, D., Montoro, R., Bailón, A., 2002. Heavy metals and arsenic uptake by wild vegetation in the Guadiamar rivera area after the toxic spill of the Aznalcóllar mine. J. Biotechnol. 98, 125–137.

Sabeen, M., Mahmood, Q., Irshad, M., Fareed, I., Khan, A., Ullah, F., Hussain, J., Hayat, Y., Tabassum, S., 2013. Cadmium phytoremediation by *Arundo donx* L. from contaminated soil and water. Biomed. Res. Int., 324830. https://doi.org/10.1155/2013/324830. pp. 9.

Schmidt, T., Fernando, A.L., Monti, A., Rettenmaier, N., 2015. Life cycle assessment of bioenergy and bio-based products from perennial grasses cultivated on marginal land in the Mediterranean region. BioEnergy Res. 8, 1548–1561.

Scordia, D., Papazoglou, E.G., Kotoula, D., Sanz, M., Ciria, C.S., Pérez, J., Maliarenko, O., Prysiazhniuk, O., von Cossel, M., Greiner, B.E., Lazdina, D., Makovskis, K., Lamy, I., Ciadamidaro, L., Petit-dit-Grezeriat, L., Corinzia, S.-A., Fernando, A.L., Alexopoulou, E., Cosentino, S.L., 2022. Towards identifying industrial crop types and associated agronomies to improve biomass production from marginal lands in europe. GCB Bioenergy 14 (7), 710–734.

Sidana, A., Kaur, S., Yadav, S.K., 2022. Assessment of the ability of *Meyerozyma guilliermondii* P14 to produce second-generation bioethanol from giant reed (*Arundo donax*) biomass. Biomass Convers. Biorefin. https://doi.org/10.1007/s13399-021-02211-4.

Sidella, S., Barbosa, B., Costa, J., Cosentino, S.L., Fernando, A.L., 2016. Screening of Giant Reed Clones for Phytoremediation of Lead Contaminated Soils Springer International Publishing. Perennial Biomass Crops for a Resource Constrained World, 1st. Springer International Publishing, Switzerland, pp. 191–197.

Sidella, S., Cosentino, S.L., Fernando, A.L., Costa, J., Barbosa, B., 2018. Phytoremediation of soils contaminated with lead by Arundo donax L. WASTES–Solutions, Treatments and Opportunities II. CRC Press, Boca Raton, FL, USA, pp. 383–388.

Silva, C., Lemons, F., Schirmer, M.A., Maeda, R.N., Barcelos, C.A., Pereira, N., 2015. Potential of giant reed (*Arundo donax* L.) for second generation ethanol production. Electron. J. Biotechnol. 18 (1), 10–15.

Sun, H., Wang, Z., Gao, P., Liu, P., 2013. Selection of aquatic plants for phytoremediation of heavy metal in electroplate wastewater. Acta Physiol. Plant. 35, 355–364.

Tam, N., Wong, Y., 2014. Constructed wetland with mixed mangrove and non-mangrove plants for municipal sewage treatment. In: 4th International Conference on Future Environment and Energy. IPCBEE, vol. 61. IACSIT Press, Singapore.

I. Bioproducts from contaminated soil and water

Thomas, B., Raj, M.C., Athira, B.K., Rubiyah, H.M., Joy, J., Moores, A., Drisko, G.L., Sanchez, C., 2018. Nanocellulose, a versatile green platform: from biosources to materials and their applications. Chem. Rev. 118, 11575–11625. https://doi.org/10.1021/acs.chemrev.7b00627.

Tuttolomondo, T., Licata, M., Leto, C., Leone, R., 2015. Effect of plant species on water balance in a pilot-scale horizontal subsurface flow constructed wetland planted with *A. donax* L. and *Cyperus alternifolius* L.—two-year tests in a Mediterranean environment in the West of Sicily (Italy). Ecol. Eng. 74, 79–92.

Tzanakakis, V.A., Paranychianakis, N.V., Angelakis, A.N., 2009. Nutrient removal and biomass production in land treatment systems receiving domestic effluent. Ecol. Eng. 35, 1485–1492.

Tzanakakis, V.A., Paranychianakis, N.V., Londra, P.A., Angelakis, A.N., 2011. Effluent application to the land: changes in soil properties and treatment potential. Ecol. Eng. 37, 1757–1764.

Üner, O., 2019. Hydrogen storage capacity and methylene blue adsorption performance of activated carbon produced from Arundo donax. Mater. Chem. Phys. 237, 121858. https://doi.org/10.1016/j.matchemphys.2019.121858.

Vasmara, C., Cianchetta, S., Marchetti, R., Ceotto, E., Galletti, S., 2021. Potassium hydroxyde pre-treatment enhances methane yield from giant reed (*Arundo donax* L.). Energies 14 (3), 630. https://doi.org/10.3390/en14030630.

Von Cossel, M., Lewandowski, I., Elbersen, B., Staritsky, I., Van Eupen, M., Iqbal, Y., Mantel, S., Scordia, D., Testa, G., Cosentino, S.L., Maliarenko, O., Eleftheriadis, I., Zanetti, F., Monti, A., Lazdina, D., Neimane, S., Lamy, I., Ciadamidaro, L., Sanz, M., Carrasco, J.E., Ciria, P., McCallum, I., Trindade, L.M., Van Loo, E.N., Elbersen, W., Fernando, A.L., Papazoglou, E.G., Alexopoulou, E., 2019a. Marginal agricultural land low-input systems for biomass production. Energies 12 (16), 3123. https://doi.org/10.3390/en12163123.

Von Cossel, M., Wagner, M., Lask, J., Magenau, E., Bauerle, A., Von Cossel, V., Warrach-Sagi, K., Elbersen, B., Staritsky, I., Van Eupen, M., Iqbal, Y., Jablonowski, N.D., Happe, S., Fernando, A.L., Scordia, D., Cosentino, S.L., Wulfmeyer, V., Lewandowski, I., Winkler, B., 2019b. Prospects of bioenergy cropping systems for a more social-ecologically sound bioeconomy. Agronomy 9 (10), 605. https://doi.org/10.3390/agronomy9100605.

Webster, R., Driever, S., Kromdijk, J., McGrath, J., Leakey, A.D.B., Siebke, K., Demetriades-Shah, T., Bonnage, S., Peloe, T., Lawson, T., Long, S.P., 2016. High C3 photosynthetic capacity and high intrinsic water use efficiency underlies the high productivity of the bioenergy grass Arundo donax. Sci. Rep. 6, 20694. https://doi.org/10.1038/srep20694.

Williams, C.M.J., Biswas, T.K., Schrale, G., Virtue, J.G., Heading, S., 2008. Use of saline land and wastewater for growing a potential biofuel crop (*A. donax* L.) irrigation. In: Irrigation Australia 2008, Conference May 20. South Australian Research and Development Institute, Melbourne.

Yang, J., Wang, X., Shen, B., Hu, Z., Xu, L., Yang, S., 2020. Lignin from energy plant (*Arundo donax*): pyrolysis kinetics, mechanism and pathway evaluation. Renew. Energy 161, 963–971.

You, T., Wang, R., Zhang, X., Ramaswamy, S., Xu, F., 2018. Reconstruction of Lignin and Hemicelluloses by Aqueous Ethanol Anti-Solvents to Improve the Ionic Liquid-Acid Pretreatment Performance of *Arundo donax* Linn. Biotechnology and Bioengineering 115 (1), 82–91. https://doi.org/10.1002/bit.26457.

Yu, S., Sheng, L., Zhang, C., Deng, H., 2018. Physiological response of *Arundo donax* to cadmium stress by Fourier transform infrared spectroscopy. Spectrochim. Acta A Mol. Biomol. Spectrosc. 198, 88–91. https://doi.org/10.1016/j.saa.2018.02.039.

Zema, D., Bombino, G., Andiloro, S., Zimbone, S.M., 2012. Irrigation of energy crops with urban wastewater: effects on biomass yields, soils and heating values. Agric. Water Manag. 115, 55–65.

Zeng, P., Guo, Z., Xiao, X., Peng, C., Huang, B., 2018. Intercropping *Arundo donax* with woody plants to remediate heavy metal-contaminated soil. Huan Jing Ke Xue 39 (11), 5207–5216. https://doi.org/10.13227/j.hjkx.201804136.

Zhang, D., Jiang, Q.W., Liang, D.Y., Huang, S., Liao, J., 2021. The potential application of giant reed (*Arundo donax*) in ecological remediation. Front. Environ. Sci. 9, 652367. https://doi.org/10.3389/fenvs.2021.652367.

Zhang, Q., Yang, J., Zhang, T., Shui, X., Zhang, H., Chen, Z., He, X., Lei, T., Jiang, D., Elgorban, A.M., Syed, A., Solanki, M.K., 2023. Pretreatment of *Arundo donax* L. for photo-fermentative biohydrogen production by ultrasonication and ionic liquid. Bioresour. Technol. 377, 128904. https://doi.org/10.1016/j.biortech.2023.128904.

Zheng, H., Sun, C., Hou, X., Wu, M., Yao, Y., Li, F., 2018. Pyrolysis of *Arundo donax* L. to produce pyrolytic vinegar and its effect on the growth of dinoflagellate Karenia brevis. Bioresour. Technol. 247, 273–281. https://doi.org/10.1016/j.biortech.2017.09.049.

Biomass energy and biofuel from contaminated substrates

Phytomass gasification for energy recovery from aquatic plants

Srujana Kathi[a] and Majeti Narasimha Vara Prasad[b]

[a]UGC-Human Resource Development Centre, Pondicherry University, Puducherry, India
[b]Department of Plant Sciences, School of Life Sciences, University of Hyderabad (An Institution of Eminence), Hyderabad, Telangana, India

1 Introduction

The use of biomass wastes as resources for biorefinery in a circular bioeconomy (CBE) is promising (Ubando et al., 2020; Kaur et al., 2019). The need to investigate new renewable energy sources grows as energy consumption continues to soar on a worldwide scale. While the transportation industry is expected to continue increasing in the United States and Europe, it is anticipated that growth in the rising nations of China and India would be far greater, growing at a faster rate. Rapid depletion of currently used non-renewable fossil fuels such as coal, petroleum, and natural gas is a well much debated issue. At the same time, the liberation of trapped carbon from the fossil fuels into the atmosphere in the form of carbon dioxide has led to increased discussion about global warming. The irregular distribution of fossil fuel resources geographically renders many countries severely dependent on oil imports. There is clear scientific evidence that human activities coupled with fossil fuel combustion and land-use change cause the emission of greenhouse gases (GHG), such as carbon dioxide, methane, and nitrous oxide, which are disturbing the earth's climate.

Due to the growing demand for renewable energy sources and environmentally friendly waste management techniques, the generation of biofuels and value-added goods from waste biomass is a developing subject that is of great interest (Okolie et al., 2022). Composting, vermicomposting, anaerobic digestion, and landfilling are the biochemical processes that are crucial in the valorization of biomass and solid waste for the creation

Bioremediation and Bioeconomy
https://doi.org/10.1016/B978-0-443-16120-9.00001-7

of biofuels, biosurfactants, and biopolymers (Varjani et al., 2021). Therefore, the search for further renewable energy sources seems inevitable. Biofuels are now the only fuels compatible with the current engine technologies when compared to substitute technologies such as hydrogen and fuel cells. The three main forces driving the development of bioenergy and biofuels are climate change, energy security, and rural development (Yadav et al., 2023).

People began looking into ways to use terrestrial and aquatic weeds after failing to completely remove them. Aquatic macrophytes could be kept in check by routine harvesting, allowing for-profitable usage of them. Among the many uses of these weeds, biogas generation has been found to be a sustainable choice (Deshmukh, 2012). Aquatic weeds can in fact meet the needs for biomass during the production of biogas (Mshandete, 2009). Phytoremediation has been used more frequently recently to enhance the water quality in polluted water bodies and also for minimizing the toxicity (Kathi and Khan, 2011; Kathi and Padmavathy, 2019). After phytoremediation, various methods for disposing of plants with high amounts of heavy metals have been suggested, including direct dumping, burning, high-temperature decomposition, chemical extraction, and phytomining. One of the challenges in remediation has been the disposal of plants used.

A promising and economical plant-based method called phytoremediation uses ecologically sound procedures found in wetland ecosystems to remove pollutants from wastewater, including heavy metals, herbicides, and xenobiotics (Dipu et al., 2011; Yadav et al., 2022a, b). A significant environmental co-benefit might be achieved by economically valuing the biomass created during phytoremediation in the form of bioenergy. Using specific types of biomass to partially meet energy needs is made possible by biogas technology. Up to 25% of the European Union (EU)'s transportation fuel demands might be satisfied by clean and CO_2-efficient biofuels by 2030, according to the Biomass Action Plan and the EU Strategy for Biofuels. A renewable fuel standard of 36 billion gallons (1 billion gallons of biodiesel) by 2022 was created in the United States by the Energy Independence and Security Act of 2007 (Abdullah et al., 2017). Within the EU policy, US policy, Chinese policy, and the policies of many other nations, the development of renewable energy and the reduction of GHG emissions are now established priorities. The EU has established the following goals for 2020: increasing the share of renewable energy in the EU's final energy consumption to at least 20% and increasing the share of biofuels in each Member State's road transportation sector to at least 10% (Awogbemi et al., 2021).

Gasification is one of the thermochemical processes that can be used to break down lignocellulosic biomass because of its high conversion efficiency. Any kind of lignocellulosic material can be used as feedstock in the high-temperature biomass transformation process known as gasification. Therefore, when aquatic macrophytes are used as the source of biomass, this approach may be chosen (Alauddin et al., 2010). This chapter describes aquatic weeds that are used to clean up water, analyzes biogas production from water weed phytomass, discusses thermochemical and biological processes involved in biogas production, and discusses limitations to using aquatic macrophytic phytomass for biogas production.

2 Aquatic weeds for decontamination of water

Aquatic macrophytes play a key role in the structural and functional balance of aquatic ecosystems by altering water movement regimes, providing shelter to fish and aquatic invertebrates, serving as a food source, and altering water quality by regulating oxygen balance, nutrient cycles, and accumulating heavy metals. The ability to hyperaccumulate heavy metals makes them interesting research areas, especially for the treatment of industrial effluents and sewage wastewater (Sood et al., 2012). Infestation of aquatic weeds in the water bodies leads to sedimentation and unsuitability for domestic use, interference with navigation, effects on fisheries, amplification in breeding habitat to disease-transmitting mosquitoes, blocking irrigation canals, and evapotranspiration. A lot of soil, nutrients, and heavy metals are carried away in the runoff in the lower reaches of the catchment areas. As a result, heavy aquatic weed invasion and silting occurs in different kinds of water bodies. Aquatic plants can be used for biological treatment and stabilization of contaminated water bodies (Khankhane and Varshney, 2009).

Both constructed and natural wetlands are currently being used for heavy metal removal and wastewater quality control. The diversified macrophytes growing in the wetlands enable the retention of contaminants such as metals from water passing through them. Aquatic macrophytes encompass many common weeds, enabling cost-effective treatment and remediation technologies for wastewater contaminated with inorganic and organic compounds (Prasad, 2013). The use of aquatic plants to absorb pollutants from water bodies is termed as rhizofiltration or phytofiltration (Mahendran, 2014). Plants preserve the natural structure and texture of soil using solar energy; therefore, phytoremediation is an ecologically compatible tool practically suitable to clean up a wide array of environmental contaminants (Kathi and Khan, 2011).

Water hyacinth (*Eichhornia crassipes* (Martius)) is a member of family Pontederiaceae. It is a monocotyledonous freshwater weed, native to the Brazil and Ecuador region. It is used in traditional medicine and even used to remove toxic elements from polluted water bodies (Fig. 1A–C, Singhal and Rai, 2003; Bhattacharya and Kumar, 2010). They reproduce both asexually through stolons and sexually through seeds. The seeds remain viable for up to 20 years and therefore are difficult to control (Bhattacharya and Kumar, 2010). The rapid proliferation of water hyacinth can result in detrimental effects, including exhausting oxygen in water and interfering with navigation and recreation. Wang et al. (2011), while evaluating the potential of *E. crassipes* and *Lolium perenne* L. that serve as phytoremediating plants by the floating bed on the Guxin River in Hangzhou, concluded that the growing of *E. crassipes* can reduce the turbidity of water by adsorbing a mass of silt from the river. Moreover, harvesting *E. crassipes* aids in carrying the contaminant away from the contaminated water body. The removal of water hyacinth from a polluted water body might contribute to water quality improvement. Studies on water hyacinth (*E. crassipes*), channel grass (*Vallisneria spiralis*), and water chestnut (*Trapa bispinosa*) used as phytoremediating plants for remediating industrial effluents demonstrated that the slurry of these plants produces significantly more biogas than the slurry of control plants grown in unpolluted water (Witters et al., 2012).

Pistia stratiotes L. is a hyperaccumulator, known to remove heavy metals, organic compounds, and radionuclides from the water bodies. It purifies the polluted aquatic

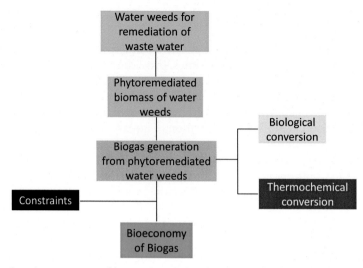

FIG. 1   Aquatic plants have a variety of functions, including phytoremediation and gasification of phytoremediated phytomass for energy recovery.

systems contaminated with harmful metals. The lower size of the plant aids in the removal of heavy metals, which is an additional advantage as compared to water hyacinth. This plant can be successfully exploited for biofuel production thereby contributing to weed management, water pollution mitigation, relieving energy problems, and protecting the aquatic ecosystem (Khan et al., 2014) (Fig. 2). Several studies have proved

FIG. 2   Irrigation canals in India are choked by *E. crassipes* (water hyacinth). (A) Waterbody infested by *E. crassipes*, (B) harvesting the clogged waterbodies, (C) irrigation canals choked with reeds and cattails, (D) weed management and biofuel production.

reed canary grass (*Phalaris arundinacea* L.) to be a noteworthy substrate for bioenergy production (Lakaniemi et al., 2011; Kacprzak et al., 2012; Kathi, 2022). There is a need to evaluate the potential biogas yield of such aquatic macrophytes as *Sacciolepis africana, Ipomoea cornea, Vossia cuspidata*, and *P. stratiotes* (Adeleye et al., 2013) (Table 1).

TABLE 1 Use of water weeds (other than duck weeds and water hyacinth, reference be made to Tables 2 and 3) in phytoremediation and co-generation of bioenergy or other economic products.

Salient findings	Author(s)	Year
Advantages and disadvantages of phytoremediation: a concise review	Farraji et al.	2016
Giant reed (*Arundo donax* L.): a multipurpose crop bridging phytoremediation with sustainable bioeconomy	Fernando et al.	2016
Phytoremediation of sediments polluted with phenanthrene and pyrene by four submerged aquatic plants	He and Chi	2016
Bioenergy from phytoremediated phytomass of aquatic plants via gasification	Kathi S	2016
Phytoremediation of textile industry effluent using aquatic macrophytes	Manjunath	2016
Waste biorefinery models towards sustainable circular bioeconomy: critical review and future perspectives	Mohan et al.	2016
Potential of ornamental plants for phytoremediation of heavy metals and income generation	Nakbanpote et al.	2016
The capacity of aquatic macrophytes for phytoremediation and their disposal with specific reference to water hyacinth	Newete and Byrne	2016
Water lettuce for removal of nitrogen and phosphate from sewage	Nivetha et al.	2016
Comprehensive review on phytotechnology: heavy metals removal by diverse aquatic plants species from wastewater	Rezania et al.	2016
Phytoremediation of toxic elements-polluted water and soils by aquatic macrophyte *Eleocharis acicularis*	Sakakibara	2016
Producing biogas from agricultural residues generated during phytoremediation process: possibility, threshold, and challenges	Tian and Zhang	2016
A review of biomass-derived heterogeneous catalyst for a sustainable biodiesel production	Abdullah et al.	2017
Phytoremediation using aquatic macrophytes	Akhtar et al.	2017
Macrophytes for the reclamation of degraded waterbodies with potential for bioenergy production	Anand et al.	2017
Heavy metal removal by aquatic plants and its disposal by using as a concrete ingredient	Chaudhuri et al.	2017
Phytoremediation of suspended solids and turbidity of palm oil mill effluent (POME) by *Ipomoea aquatica*	Farrajia et al.	2017
Phytostabilization of heavy metals by the emergent macrophyte *Vossia cuspidata* (Roxb.) Griff.: a phytoremediation approach	Galal et al.	2017

Continued

TABLE 1   Use of water weeds (other than duck weeds and water hyacinth, reference be made to Tables 2 and 3) in phytoremediation and co-generation of bioenergy or other economic products—cont'd

Salient findings	Author(s)	Year
Screening of cadmium and copper phytoremediation ability of *Tagetes erecta*, using biochemical parameters and scanning electron microscopy-energy-dispersive X-ray microanalysis	Goswami and Das	2017
Performance of aquatic plant species for phytoremediation of arsenic-contaminated water	Jasrotia et al.	2017
Phytoremediation and bioenergy production efficiency of medicinal and aromatic plants	Jisha et al.	2017
Phytoremediation of nitrogen as green chemistry for wastewater treatment system	Kinidi and Salleh	2017
Experimental and kinetics study for phytoremediation of sugar mill effluent using water lettuce (*Pistia stratiotes* L.) and its end use for biogas production	Kumar et al.	2017
Decontamination of coal mine effluent generated at the Rajrappa coal mine using phytoremediation technology	Lakra et al.	2017
December. Biological treatment of domestic wastewater by selected aquatic plants	Nagarajan et al.	2017
Wastewater phytoremediation by *Salvinia molesta*	Ng and Chan	2017
Phytoremediation for ecological restauration and industrial ecology	Pirrera and Pluchino	2017
Tolerance and hyperaccumulation of a mixture of heavy metals (Cu, Pb, Hg, and Zn) by four aquatic macrophytes	Romero-Hernández et al.	2017
Phytoremediation of industrial mines wastewater using water hyacinth	Saha et al.	2017
Uptake of cesium and cobalt radionuclides from simulated radioactive wastewater by *Ludwigia stolonifera* aquatic plant	Saleh et al.	2017
Evaluation of recovery of aquatic plants used in wastewater treatment and discharged as waste	Topal et al.	2017
Phytoremediation as an effective method to remove heavy metals from contaminated area—TG/FT-IR analysis results of the gasification of heavy metal contaminated energy crops	Werle et al.	2017
Efficiency and mechanism of the phytoremediation of decabromodiphenyl ether-contaminated sediments by aquatic macrophyte *Scirpus validus*	Zhao et al.	2017
Application of phytoremediation in the management of oil spillage	Attah-Olottah and Adeniyi	2018
Application of agro-waste derived materials as heterogeneous base catalysts for biodiesel synthesis	Basumatary et al.	2018
Characterization of biological waste stabilized by cement during immersion in aqueous media to develop disposal strategies for phytomediated radioactive waste	Bayoumi and Saleh	2018
Phytoremediation techniques for the removal of dye in wastewater	Bharathiraja et al.	2018
Removal of pesticides using aquatic plants in water resources: a review	Chander et al.	2018

TABLE 1 Use of water weeds (other than duck weeds and water hyacinth, reference be made to Tables 2 and 3) in phytoremediation and co-generation of bioenergy or other economic products—cont'd

Salient findings	Author(s)	Year
Accumulation and effects of copper on aquatic macrophytes *Potamogeton pectinatus* L.: Potential application to environmental monitoring and phytoremediation	Costa et al.	2018
Effective phytoremediation of low-level heavy metals by native macrophytes in a vanadium mining area, China	Jiang et al.	2018
Aquatic weeds as the next generation feedstock for sustainable bioenergy production	Kaur et al.	2018
Phytoremediation of copper pollution by eight aquatic plants.	Lu et al.	2018
Role of phytoremediation in reducing cadmium toxicity in soil and water	Mahajan and Kaushal	2018
Adsorption, bioaccumulation and kinetics parameters of the phytoremediation of cobalt from wastewater using *Elodea canadensis*	Mosoarca et al.	2018
Phytoremediation of heavy metals: mechanisms, methods and enhancements	Muthusaravanan et al.	2018
Simultaneous phytoremediation of chromium and phenol by *Lemna minuta* Kunth: a promising biotechnological tool	Paisio et al.	2018
October. Simultaneous phytoremediation of Ni^{2+} and bioelectricity generation in a plant-microbial fuel cell assembly using water hyacinth (*Eichhornia crassipes*)	Pamintuan et al.	2018
Case study on phytoremediation driven energy crop production using *Sida hermaphrodita*	Pogrzeba et al.	2018
Nanoparticle-plant interaction: implications in energy, environment, and agriculture	Rai et al.	2018
Synergistic phytoremediation of wastewater by two aquatic plants (*Typha angustifolia* and *Eichhornia crassipes*) and potential as biomass fuel	Sricoth et al.	2018
Innovative developments in biofuels production from organic waste materials: a review	Stephen and Periyasamy	2018
Phytoremediation potential of *Pistia stratiotes* and *Eichhornia crassipes* to remove chromium and copper	Tabinda et al.	2018
Synthesis of biomass as heterogeneous catalyst for application in biodiesel production: State of the art and fundamental review	Tang et al.	2018
November. Choice of sorbents for the pretreatment in the complex technology of phytoremediation	Ulrikh et al.	2018
Phytoremediation: halophytes as promising heavy metal hyperaccumulators	Usman et al.	2018
Mechanistic understanding and holistic approach of phytoremediation: a review on application and future prospects	Yadav et al.	2018
Phytoremediation of landfill leachate waste contaminants through floating bed technique using water hyacinth and water lettuce	Abbas et al.	2019

Continued

TABLE 1   Use of water weeds (other than duck weeds and water hyacinth, reference be made to Tables 2 and 3) in phytoremediation and co-generation of bioenergy or other economic products—cont'd

Salient findings	Author(s)	Year
Phytoremediation of heavy metals and pesticides present in water using aquatic macrophytes	Anand et al.	2019
Physiological mechanisms of aluminum (Al) toxicity tolerance in nitrogen-fixing aquatic macrophyte *Azolla microphylla* Kaulf: phytoremediation, metabolic rearrangements, and antioxidative enzyme responses	Chakraborty et al.	2019
Phytoremediation of polluted waterbodies with aquatic plants: Recent progress on heavy metal and organic pollutants	Christian and Beniah	2019
Heavy metal pollutions: state of the art and innovation in phytoremediation	DalCorso et al.	2019
Phytoremediation of effluents contaminated with heavy metals by floating aquatic macrophytes species	de Souza and Silva	2019
Remediation of uranium-contaminated sites by phytoremediation and natural attenuation	Favas et al.	2019
Phytoremediation of Lebanese polluted waters: a review of current initiatives	Ghanem et al.	2019
Uptake, accumulation and metabolization of 1-butyl-3-methylimidazolium bromide by ryegrass from water: Prospects for phytoremediation	Habibul et al.	2019
Phytoremediation potential of fast-growing energy plants: challenges and perspectives—a review	Hauptvogl et al.	2019
Phytoremediation as a promising method for the treatment of contaminated sediments	Jani et al.	2019
A sustainable biorefinery approach for efficient conversion of aquatic weeds into bioethanol and biomethane	Kaur et al.	2019
Removal of inorganic and organic contaminants from terrestrial and aquatic ecosystems through phytoremediation and biosorption	Kumar et al.	2019
Application of phytoremediation technology in decontamination of a fish culture pond fed with coal mine effluent using three aquatic macrophytes	Lakra et al.	2019
Accumulation and effects of uranium on aquatic macrophyte *Nymphaea tetragona* Georgi: Potential application to phytoremediation and environmental monitoring	Li et al.	2019
Effect of daily exposure to Pb-contaminated water on *Salvinia biloba* physiology and phytoremediation performance	Loría et al.	2019
Assessment of phytoremediation potential of *Chara vulgaris* to treat toxic pollutants of textile effluent	Mahajan et al.	2019
Recent advances in the pretreatment of lignocellulosic biomass for biofuels and value-added products	Mahmood et al.	2019
Phytoremediation of selenium-impacted water by aquatic macrophytes	Nattrass et al.	2019
Status, progress and challenges of phytoremediation-An African scenario	Odoh et al.	2019
Biofuels from agricultural wastes	Pattanaik et al.	2019
Linking phytotechnologies to bioeconomy; varietal screening of high biomass and energy crops for phytoremediation of Cr and Cu contaminated soils	PoÅ et al.	2019

TABLE 1 Use of water weeds (other than duck weeds and water hyacinth, reference be made to Tables 2 and 3) in phytoremediation and co-generation of bioenergy or other economic products—cont'd

Salient findings	Author(s)	Year
Prospects for manipulation of molecular mechanisms and transgenic approaches in aquatic macrophytes for remediation of toxic metals and metalloids in wastewaters	Prasad	2019
Bioethanol production from waste lignocelluloses: A review on microbial degradation potential	Prasad et al.	2019
Phytoremediation potential and control of *Phragmites australis* as a green phytomass: an overview	Rezania et al.	2019
Microbial phyto-power systems—a sustainable integration of phytoremediation and microbial fuel cells	Saba et al.	2019
Design of a sustainable development process between phytoremediation and production of bioethanol with *Eichhornia crassipes*	Sayago	2019
Thermo-chemical conversion of biomass and upgrading to biofuel: the thermo-catalytic reforming process—a review	Schmitt et al.	2019
Transgenic energy plants for phytoremediation of toxic metals and metalloids	Shah and Pathak	2019
Iron plaque formation in the roots of *Pistia stratiotes* L.: importance in phytoremediation of cadmium	Singha et al.	2019
Ecorestoration of polluted aquatic ecosystems through rhizofiltration	Tiwari et al.	2019
Design proposal model for improving rivers with phytoremediation method	Tokmak et al.	2019
Sustainable phytoremediation strategies for river water rejuvenation	Upadhyay et al.	2019
Phytoremediation potential of freshwater macrophytes for treating dye-containing wastewater	Ahila et al.	2020
Energy production and wastewater treatment using *Juncus, S. triqueter, P. australis, T. latifolia, and C. alternifolius* plants in sediment microbial fuel cell	Aswad et al.	2020
Phytoremediation of zinc contaminated water by marigold (*Tagetes Minuta* L). Central Asian	Awan et al.	2020
Urban pond ecosystems: preservation and management through phytoremediation. In: Fresh water pollution dynamics and remediation (pp. 263–291)	Bhat et al.	2020
Phytoremediation of heavy metals/metalloids by native herbaceous macrophytes of wetlands: current research and perspectives	Bora and Sarma	2020
Review on water hyacinth weed as a potential bio fuel crop to meet collective energy needs	Bote et al.	2020
Phytoremediation: tools & technique	Choudhary	2020
Physiological mechanisms and phytoremediation potential of the macrophyte Salvinia biloba towards a commercial formulation and an analytical standard of glyphosate	da Silva Santos et al.	2020
Phytoremediation processes of domestic and textile effluents: evaluation of the efficacy and toxicological effects in *Lemna minor* and *Daphnia magna*	de Alkimin et al.	2020

Continued

II. Biomass energy and biofuel from contaminated substrates

TABLE 1　Use of water weeds (other than duck weeds and water hyacinth, reference be made to Tables 2 and 3) in phytoremediation and co-generation of bioenergy or other economic products—cont'd

Salient findings	Author(s)	Year
Life cycle thinking applied to phytoremediation of dairy wastewater using aquatic macrophytes for treatment and biomass production	de Queiroz et al.	2020
Concept and application of phytoremediation in the fight of heavy metal toxicity	Devi and Kumar	2020
Phytoremediation of engineered nanoparticles using aquatic plants: Mechanisms and practical feasibility	Ebrahimbabaie et al.	2020
Phytoremediation using aquatic plants	Fletcher et al.	2020
Bio-methanol as a renewable fuel from waste biomass: current trends and future perspective	Gautam et al.	2020
Phytoremediation—from environment cleaning to energy generation—Current status and future perspectives	Grzegórska et al.	2020
Phytoremediation: Treating Euthrophic Lake at KotaSAS Lakeside, Kuantan by Aquatic Macrophytes	Haziq et al.	2020
Sustainable livestock wastewater treatment via phytoremediation:	Hu et al.	2020
In situ phytoremediation of uranium contaminated soils	Khan	2020
Aquatic phytoremediation strategies for chromium removal	Malaviya et al.	2020
Phytoremediation of metals by aquatic macrophytes	Manorama Thampatti et al.	2020
Efficiency of five selected aquatic plants in phytoremediation of aquaculture wastewater	Mohd Nizam et al.	2020
Absorption of lead and mercury in dominant aquatic macrophytes of Balili River and its implication to phytoremediation of water bodies	Napaldet and Buot Jr	2020
Utilisation of an aquatic plant (*Scirpus grossus*) for phytoremediation of real sago mill effluent	Nash et al.	2020
Eco-friendly transformation of waste biomass to biofuels.	Parakh et al.	2020
The evolution of botanical biofilters: developing practical phytoremediation of air pollution for the built environment. In 1st International Conference on Climate Resilient Built Environment CRBE	Pettit et al.	2020
Pteridophytes in phytoremediation. Environmental geochemistry and health	Praveen and Pandey	2020
Qualification of corroborated real phytoremediated radioactive wastes under leaching and other weathering parameters	Saleh et al.	2020a
Potential of the submerged plant Myriophyllum spicatum for treatment of aquatic environments contaminated with stable or radioactive cobalt and cesium	Saleh et al.	2020b
Trends in phytomanagement of aquatic ecosystems and evaluation of factors affecting removal of inorganic pollutants from water bodies	Shah et al.	2020
Biorefineries in circular bioeconomy: a comprehensive review	Ubando et al.	2020
Phytoremediation: a way towards sustainable Agriculture	Udawat and Singh	2020

TABLE 1 Use of water weeds (other than duck weeds and water hyacinth, reference be made to Tables 2 and 3) in phytoremediation and co-generation of bioenergy or other economic products—cont'd

Salient findings	Author(s)	Year
Discussing on "source-sink" landscape theory and phytoremediation for non-point source pollution control in China	Wang et al.	2020
Pontederia cordata, an ornamental aquatic macrophyte with great potential in phytoremediation of heavy-metal-contaminated wetlands	Xin et al.	2020
Migration of antibiotic ciprofloxacin during phytoremediation of contaminated water and identification of transformation products	Yan et al.	2020
February. Nutrients and organics removal from slaughterhouse wastewater using phytoremediation: A comparative study on different aquatic plant species	Alam et al.	2021
The phytoremediation potential of water hyacinth: A case study from Hartbeespoort Dam, South Africa	Auchterlonie et al.	2021
Trends in the development and utilization of agricultural wastes as heterogeneous catalyst for biodiesel production	Awogbemi et al.	2021
Potential for the production of biofuels from agricultural waste, livestock, and slaughterhouse waste in Golestan province, Iran	Azadbakht et al.	2021
Phytoremediation of toxic metals: a sustainable green solution for clean environment	Babu et al.	2021
Microbe-assisted phytoremediation of environmental pollutants and energy recycling in sustainable agriculture	Basit et al.	2021
A holistic approach to soil contamination and sustainable phytoremediation with energy crops in the Aegean Region of Turkey	Baştabak et al.	2021
2,4,6-Trichlorophenol (TCP) removal from aqueous solution using Canna indica L.: kinetic, isotherm and Thermodynamic studies	Enyoh and Isiuku	2021
Environmental phytoremediation and analytical technologies for heavy metal removal and assessment	Govere	2021
A review on the waste biomass derived catalysts for biodiesel production	Hamza et al.	2021
Phytoremediation of poly-and perfluoroalkyl substances: a review on aquatic plants, influencing factors, and phytotoxicity	Huang et al.	2021
Nano-Phytoremediation of heavy metals contaminated wastewater ecosystems and wetlands by constructed wetlands planted with waterlogging-tolerant mycorrhizal fungi and vetiver grass	Khan	2021a
Potential of coupling heavy metal (HM) phytoremediation by bioenergy plants and their associated HM-adapted rhizosphere microbiota (arbuscular mycorrhizal fungi and plant growth promoting microbes) for bioenergy production	Khan	2021b
Sustainable and eco-friendly approach for controlling industrial wastewater quality imparting succour in water-energy nexus system	Lata	2021
A review on disposal and utilization of phytoremediation plants containing heavy metals	Liu et al.	2021

Continued

II. Biomass energy and biofuel from contaminated substrates

TABLE 1   Use of water weeds (other than duck weeds and water hyacinth, reference be made to Tables 2 and 3) in phytoremediation and co-generation of bioenergy or other economic products—cont'd

Salient findings	Author(s)	Year
Aquatic weeds: A potential pollutant removing agent from wastewater and polluted soil and valuable biofuel feedstock	Mehariya et al.	2021
Phytoremediation of wastewater using aquatic plants: a review	Mohebi and Nazari	2021
Cultivation of S. molesta plants for phytoremediation of secondary treated domestic wastewater	Mustafa and Hayder	2021a
Recent studies on applications of aquatic weed plants in phytoremediation of wastewater	Mustafa and Hayder	2021b
April. Applications of constructed wetlands and hydroponic systems in phytoremediation of wastewater	Mustafa et al.	2021
Phytoremediation to improve eutrophic ecosystem by the floating aquatic macrophyte, water lettuce (*Pistia stratiotes* L.) at lab scale	Nahar and Hoque	2021
Phytohormonal roles in plant responses to heavy metal stress: Implications for using macrophytes in phytoremediation of aquatic ecosystems	Nguyen et al.	2021
Enhanced phytoremediation of soil heavy metal pollution and commercial utilization of harvested plant biomass:	Quarshie et al.	2021
Heavy metals and arsenic phytoremediation potential of invasive alien wetland plants *Phragmites karka* and *Arundo donax*: Water-Energy-Food (WEF) Nexus linked sustainability implications	Rai	2021
The potential of *Azolla filiculoides* for in vitro phytoremediation of wastewater	Rezooqi et al.	2021
Salvinia (*Salvinia molesta*) and Water Hyacinth (*Eichhornia crassipes*): two Pernicious Aquatic Weeds with High Potential in Phytoremediation	Tabassum-Abbasi et al.	2021
Processes and prospects on valorizing solid waste for the production of valuable products employing bio-routes: a systematic review	Varjani et al.	2021
Phytoremediation of heavy metals extracted soil and aquatic environments: current advances as well as emerging trends	Verma et al.	2021
Water contamination with atrazine: is nitric oxide able to improve *Pistia stratiotes* phytoremediation capacity?	Vieira et al.	2021
Role of bacterial consortium and synthetic surfactants in promoting the phytoremediation of crude oil-contaminated soil using *Brachiaria mutica*	Anwar-ul-Haq et al.	2022
Role of water hyacinth (*Eichhornia crassipes*) in integrated constructed wetlands: a review on its phytoremediation potential	Aqdas and Hashmi	2022
Efficacious bioremediation of heavy metals and radionuclides from wastewater employing aquatic macro-and microphytes	Das et al.	2022
A review of thermochemical conversion of waste biomass to biofuels	Jha et al.	2022
Phytoremediation: mechanisms, plant selection and enhancement by natural and synthetic agents	Kafle et al.	2022
Kinetic evaluation for removal of an anionic diazo Direct Red 28 by using phytoremediation potential of *Salvinia molesta* Mitchell	Kaushal and Mahajan	2022

TABLE 1 Use of water weeds (other than duck weeds and water hyacinth, reference be made to Tables 2 and 3) in phytoremediation and co-generation of bioenergy or other economic products—cont'd

Salient findings	Author(s)	Year
Upgrading the value of anaerobic fermentation via renewable chemicals production: a sustainable integration for circular bioeconomy	Kumar et al.	2022
Study on the application of phytoremediation of phosphate content to eutrophication in Cengklik Reservoir, Boyolali Regency	Kurniawan et al.	2022
Dried brown seaweed's phytoremediation potential for methylene blue dye removal from aquatic environments	Mansour et al.	2022
Circular economy framework for energy recovery in phytoremediation of domestic waste waters	Mustafa et al.	2022
Waste biomass valorization for the production of biofuels and value-added products: a comprehensive review of thermochemical, biological and integrated processes	Okolie et al.	2022
Salvinia (*Salvinia molesta*) and water hyacinth (*Eichhornia crassipes*): two pernicious aquatic weeds with high potential in phytoremediation	Patnaik et al.	2022
Comparison of different phytoremediation strategies for acid mine drainage (AMD)	Rahman et al.	2022
Phytoremediation and phycoremediation: a sustainable solution for wastewater treatment	Sameena et al.	2022
Systematic review of the efficiency of aquatic plants in the wastewater treatment	Seguil et al.	2022
Potential of indigenous plant species for phytoremediation of arsenic contaminated water and soil	Singh et al.	2022
Wastewater remediation and biomass production: a hybrid technology to remove pollutants	Sruthi et al.	2022
Barriers in biogas production from the organic fraction of municipal solid waste: a circular bioeconomy perspective	Yadav et al.	2022a, b
Utilization of reeds to sequester Ni and/or Cu from wastewater and to produce valuable products	Duran et al.	2023
Biogenic synthesis, molecular docking, biomedical and environmental applications of multifunctional CuO nanoparticles mediated *Phragmites australis*	Kocabas et al.	2023
Water hyacinth (*Eichhornia crassipes*) a sustainable strategy for heavy metals removal from contaminated waterbodies	Koley et al.	2023
Iron oxide doped rice biochar reduces soil-plant arsenic stress, improves nutrient values: an amendment towards sustainable development goals	Majumdar et al.	2023
Assessment of heavy metal accumulation potential of aquatic plants for bioindication and bioremediation of aquatic environment	Petrov et al.	2023
Bioenergy routes for valorizing constructed wetland vegetation: an overview	Pinho and Mateus	2023
Vermicomposting technology for organic waste management.	Sharma and Garg	2023
Metals Ti, Cr, Mn, Fe, Ni, Cu, Zn and Pb in aquatic plants of man-made water reservoir, Eastern Siberia, Russia: tracking of environment pollution	Vladimirovna Chuparina et al.	2023

3 Biofuel generation from phytoremediated biomass of water weeds

Biogas, a clean and renewable form of energy, could be a substitute for the conventional sources that cause ecological and environmental issues (Yadvika et al., 2004). Biomass constitutes lignin, hemicellulose, cellulose, mineral matter, and ash. It possesses high moisture and volatile matter constituents, low bulk density, and high calorific value. The primary rationale for exploiting aquatic plant biomass is that their primary productivity rates are significantly higher than the terrestrial bioenergy feedstocks. The conversion of the biomass of harvested aquatic macrophytes into combustible biogas proved to be an inevitable option for the control and management of environmental pollution associated with aquatic macrophytes. Potential productivity of water hyacinth and water chestnut in nutrient-enriched wastewaters has led to its selection for phytoremediation of various industrial effluents and subsequently produced biomass as a feedstock for biogas production to achieve economic success in energy harvest (Verma et al., 2007). Unlike corn-based ethanol production, which can greatly affect the world food market, non-food, crop-based, cellulosic biofuels have the advantage of not competing with food sources. Total plant biomass in order of concentration consists of cellulose, lignin, and hemicelluloses, which are cellulosic materials, in addition to ash and extractives. Ash consists mostly of metallic salts, metal oxides, and trace mineral residues left over after burning plant biomass. When considering bioenergy and biofuel production from aquatic macrophytes, lignin and cellulose assume significance as they are combustible parts of plant biomass and are consumed on burning (Witters et al., 2012).

Mitsubishi Heavy Industries, Ltd. (MHI) has been working to develop a variety of commercially viable technologies to support biomass energy. MHI successfully performed once-through operation of woody biomass gasification to methanol synthesis process at their demonstration plant with a throughput of 2 tons day^{-1} (Hishida et al., 2011). The conversion of the biomass of harvested aquatic macrophyte water hyacinth from the Niger Delta into renewable energy demonstrated an inevitable option for the control and management of environmental pollution associated with aquatic macrophytes and their usability for poverty alleviation in the Niger Delta region of Nigeria (Adeleye et al., 2013). The saccharification of cell walls for the production of reduced sugars for the conversion to value-added products or ethanol has been well described. Duckweed, the most-investigated aquatic macrophyte, can produce a theoretical ethanol yield reaching $6.42 \times 10^3 \, \mathrm{L\,ha^{-1}}$, 50% more ethanol when compared to maize, which is the main ethanol-producing feedstock in many countries (Miranda et al., 2014) (Table 2).

TABLE 2 Bioenergy potential of water hyacinth (*E. crassipes*) applied in phytoremediation edited.

Salient findings	Author(s)	Year
The ability of aquatic macrophytes to do phytoremediation and how to get rid of them, with emphasis on water hyacinth	Newete and Byrne	2016
employing water hyacinth for phytoremediation of industrial mining wastewater	Saha et al.	2017
Simultaneous phytoremediation of Ni^{2+} and bioelectricity generation in a plant-microbial fuel cell assembly using water hyacinth	Pamintuan et al.	2018
Application of water hyacinth for ammoniacal nitrogen phytoremediation	Ting et al.	2018

TABLE 2 Bioenergy potential of water hyacinth (*E crassipes*) applied in phytoremediation edited—cont'd

Salient findings	Author(s)	Year
Water hyacinth and water lettuce are used in the floating bed approach for phytoremediation of pollutants from landfill leachate waste	Abbas et al.	2019
Application of common duckweed (*Lemna minor*) in phytoremediation of environmental chemicals: current situation and outlook	Ekperusi et al.	2019
Water hyacinth combines phytoremediation with bioenergy production	Carreño-Sayago et al.	2019
Bioaccumulation and translocation of nine heavy metals by water hyacinth in the Nile Delta, Egypt: potential for phytoremediation	Eid et al.	2019
Water weeds such as *Pistia stratiotes, Eichhornia crassipes, and Oedogonium* sp. Are useful in the treatment of textile effluents	Tabinda et al.	2019
Water hyacinth is used in the phytoremediation of palm oil mill effluent (POME)	Wei et al.	2019
Review of the potential of water hyacinth weed as a biofuel crop to address global energy demands	Bote et al.	2020
Salvinia molesta, water hyacinth, water lettuce, and duckweed have the phytoremediation ability to decrease phosphorus in rice mill wastewater	Kumar and Deswal	2020
Water hyacinth's ability to clean up pollution: a case study from Hartbeespoort Dam, South Africa	Auchterlonie et al.	2021
Review of water hyacinth's uses in energy and environmental applications	Li et al.	2021
Water hyacinth and Salvinia (*Salvinia molesta*) have high potential for phytoremediation	Tabassum-Abbasi et al.	2021
Duck weed (*Lemna minor*) and water hyacinth have the ability to phytoremediate nickel-contaminated waters (II)	Mallarino et al.	2021
Review of water hyacinth's uses in energy and environmental applications	Li et al.	2021
Review of the potential for phytoremediation of the water hyacinth in integrated artificial wetlands	Aqdas and Hashmi	2022
Water hyacinth and have high potential for phytoremediation	Patnaik et al.	2022
Investigating the potential of water hyacinth for phytoremediation and the production of biofuels	Sharma et al.	2022
Use of water hyacinth for removal of heavy metals from polluted waterbodies	Koley et al.	2023

4 Mechanisms of biogas generation

Biomass fuels and residues can be converted to more valuable energy forms via a number of processes including thermal, biological, mechanical, or physical processes. Thermal conversion contributes to multiple and complex products, in very short reaction times, making use of inorganic catalysts to improve the product quality, unlike biological processing, which is usually very selective and produces a small number of discrete products using biological catalysts. Three main thermal processes—namely, pyrolysis, gasification, and combustion—are available for converting biomass to a more useful energy form (Bridgwater, 2012). Among all thermochemical conversion processes, gasification is considered one of the promising ways to convert the energy content of biomass into a clean fuel, which can be used for the synthesis of methanol and other liquid fuels (Kaewpanha et al., 2014). Steam gasification of biomass can produce a gaseous fuel with a relatively higher H_2 content having relatively low tar content (Shen et al., 2008; Fig. 3).

The biomass gasification technology package consists of a fuel- and ash-handling system and gasification system: reactor, gas cooling, and cleaning system. Torrefaction is a biomass pretreatment method carried out at approximately 200–300°C in the absence of oxygen. During this process, biomass is completely dried and partially decomposed, losing its fibrous structure. If the biomass is pretreated remotely combining with pelletization, very energy-dense fuel pellets are produced, which reduces transportation costs. Pretreatment of biomass prevents biological degradation, facilitating long-term storage. Process efficiency can be improved by dewatering and drying biomass before gasification. Plugging of feeders can be reduced by properly drying biomass, which improves air emissions. Many types of dryers are used in drying biomass, including direct and indirect fired-rotary dryers, conveyor dryers, cascade dryers, flash or pneumatic dryers, and superheated steam dryers. Selection of the appropriate dryer depends on several factors such as size and characteristics of the feedstock, requirements of operation and maintenance, energy efficiency, environmental emissions, availability of waste heat sources, available space, potential fire hazard, and capital cost.

In a biomass gasifier, biomass is heated with controlled amounts of oxygen, air, and steam so that only a relatively small portion of the fuel burns. This "partial oxidation" process provides the heat. Rather than burning, most of the carbon-containing feedstock is chemically broken apart by the gasifier's heat and pressure, setting into motion chemical reactions that produce syngas. The producer gas/syngas consists of carbon monoxide (CO) and hydrogen (H_2) along with gaseous constituents such as methane (CH_4), carbon dioxide (CO_2), and nitrogen (N_2) (Das and Najmul Hoque, 2014). The resultant mixture of hydrogen and carbon monoxide is then cooled and converted at elevated temperature or via a Fischer-Tropsch reaction. Thus, the key to gasifier design is to create conditions such that the biomass is reduced to charcoal and the charcoal is then converted at suitable temperature to produce CO and H_2.

4.1 Gasification reactions

The chemical reactions taking place in the process of gasification (Fig. 3) are the following:

1. Drying: biomass fuels usually contain up to 35% moisture. When the biomass is heated to around 100°C, the moisture gets converted into steam.
2. Pyrolysis: pyrolysis is the thermal decomposition of biomass fuels in the absence of oxygen process occurring at around 200–300°C. Biomass decomposes into solid charcoal, liquid tars, and gases.
3. Oxidation: air is introduced into a gasifier in the oxidation zone. The oxidation takes place at about 700–1400°C in which the solid carbonized fuel reacts with oxygen in the air, producing carbon dioxide and releasing 406 kJ/g mol of heat energy.

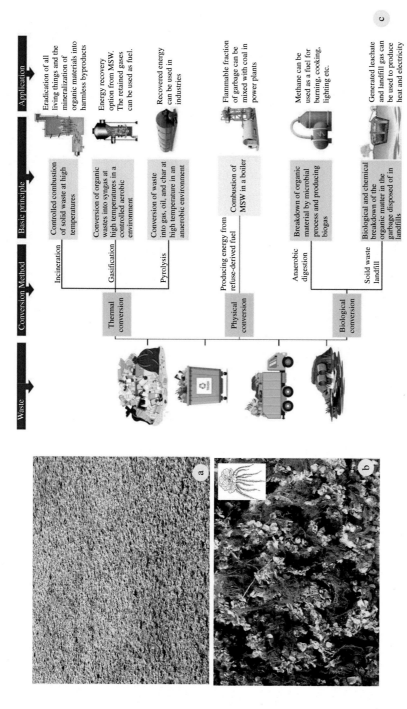

FIG. 3 (A) Abundantly grown *Pistia stratiotes* in water body for aquaculture (Water lettuce). (B) Harvested and dried biomass of *P. stratiotes*. (C) Thermochemical conversion yields biogas. The figure describes three modes of waste conversion, i.e., thermal, physical, and biological—their principles and application in energy production are briefly explained. *Source for Part C: Bhowmik, D., Chetri, S., Enerijiofi, K.E., Naha, A., Kanungo, T.D., Shah, M.P., Nath, S., 2023. Multitudinous approaches, challenges and opportunities of bioelectrochemical systems in conversion of waste to energy from wastewater treatment plants. Clean. Circ. Bioecon. 100040. Thanks are due to Elsevier Permissions.*

4. Reduction: at higher temperatures and under reducing conditions several reactions take place, which results in the formation of CO, H_2, and CH_4 (Kishore, 2008). Reduction in a gasifier is accomplished by passing carbon dioxide (CO_2) or water vapor (H_2O) across a bed of red-hot charcoal. The carbon in the hot charcoal is highly reactive with oxygen. It has high oxygen affinity that strips the oxygen off water vapor and carbon dioxide and redistributes it to as many single-bond sites as possible. The oxygen is more attracted to the bond site on the charcoal than to itself. Therefore, no free oxygen can survive in its usual diatomic O_2 form. All available oxygen will bond to available charcoal sites until all the oxygen is completed. When all the available oxygen is redistributed as single atoms, reduction stops. Through this process, CO_2 is reduced by carbon to produce two CO molecules, and H_2O is reduced by carbon to produce H_2 and CO. Both H_2 and CO are combustible fuel gases, and those fuel gases can then be piped off to do desired work elsewhere.

4.2 Types of gasifiers

Biomass gasifiers are classified according to the way air or oxygen is introduced. The design of gasifier depends on the type of fuel used and whether the gasifier is portable or stationary. Gas producers are classified according to how the air blast is introduced in the fuel column. Gasifiers are classified as updraft gasifier, downdraft gasifier, fluidized bed gasifier, cross-draft gas producer, etc.

4.2.1 Updraft or countercurrent gasifier

The oldest and simplest type of gasifier is the countercurrent or updraft gasifier. A diagram of a fluidized bed gasifier is shown in Fig. 4. An updraft gasifier has clearly defined zones for partial combustion, reduction, and pyrolysis. In this gasifier, biomass is fed at the top of the

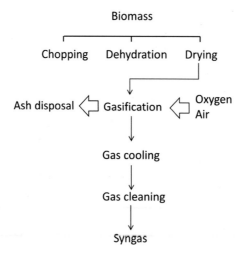

FIG. 4   Schematic diagram for syngas generation from biomass.

gasifier; air enters from the bottom and acts as countercurrent to fuel flow. The gas thus produced is collected from the top. Ashes are removed from the bottom of the gasifier. In the updraft gasifier, the hot gas passes through the fuel bed and leaves the gasifier at low temperatures. These gasifiers are best suited for applications where high-flame temperature is required and moderate amount of dust in the fuel gas is acceptable. They have high thermal efficiency and are more tolerant for fuel switching. Updraft gasifiers have outlet temperatures of 250°C and operating temperatures of 800–1200°C. They can handle moisture contents as high as 55%. High tar production requires extensive cleaning of the syngas produced, which is a noted disadvantage. Tar removal from the product gas has been a major problem in updraft gasifiers (Roos, 2010).

4.2.2 Downdraft or cocurrent gasifiers

These gasifiers are the most commonly used and easy to control, having outlet temperatures of 800°C, and operating temperatures of 800–1200°C. The resultant gas is removed at the bottom of the apparatus, so that the fuel and gas move in the same direction, as schematically shown in Fig. 5. One of the important drawbacks of downdraft gasifiers is that the feedstock must have a moisture content of about 20% or lower. Depending on the temperature of the hot zone and the residence time of the tarry vapors, a more or less complete breakdown of the tars is achieved. The tar-free gas thus produced is suitable for engine applications. Downdraft gasifiers produce syngas that typically has low tar and particulate content. They can produce as much as 20% char, but more typically char content is 2%–10% (Roos, 2010).

4.2.3 Cross-draft gasifier

Cross-draught gasifiers, schematically illustrated in Fig. 6, are an adaptation for the use of charcoal. Charcoal gasification results in very high temperatures (1500°C and more) in the oxidation zone, which can lead to material problems. Air or air/steam mixtures are introduced into the side of the gasifier near the bottom while the product gas is drawn off on the opposite side. Normally an inlet nozzle is used to bring the air into the middle of the combustion zone. They are simpler to construct and more suitable for running engines than the other types of fixed-bed gasifiers. However, they are sensitive to changes in biomass composition and moisture content.

4.2.4 Fluidized bed gasifiers

Unlike fixed-bed reactors, models with a fluidized bed have no distinct reaction zones such as drying, pyrolysis, and gasification in the reactor as the reactor is mixed and, so, close to isothermal. The design of fluidized bed gasifier is illustrated schematically in Fig. 7. Here, air is blown through a bed of solid particles at a sufficient velocity to keep them in suspension state. The bed is externally heated. The feedstock is introduced after reaching a suitable temperature. The fuel particles introduced at the bottom of the reactor can be readily mixed with the bed material and can be instantly heated up to the bed temperature, resulting in very fast pyrolysis. Gasification and tar-conversion reactions occur in the gas phase. Most systems are equipped with an internal cyclone to minimize char blowout as much as possible. Ash particles are also carried over the top of the reactor and should be removed from the gas stream if the gas is used in engine applications. Problems with feeding, instability of the bed, and fly-ash sintering in the gas channels can occur with some biomass fuels.

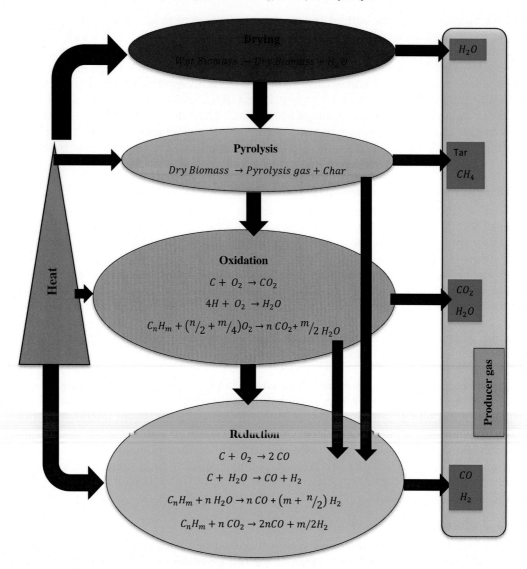

FIG. 5   The reaction mechanism for thermal gasification of biomass.

Feedstock flexibility is one of the important advantages of these gasifiers. It has the ability to deal with fluffy and fine-grained materials (sawdust, etc.) without the need of pre-processing (Figs. 8–10).

   Other biomass gasifier systems such as double fired, entrained bed, and molten bath are currently under development. Though there is a considerable overlap of the processes, each can be assumed to occupy a separate zone where fundamentally different chemical and thermal reactions take place.

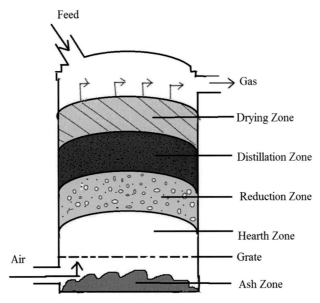

FIG. 6 Updraft gasifier.

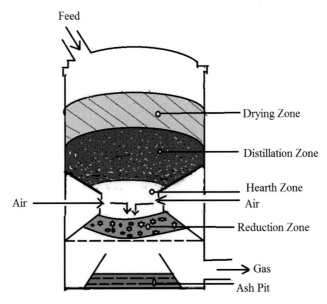

FIG. 7 Downdraft gasifier.

II. Biomass energy and biofuel from contaminated substrates

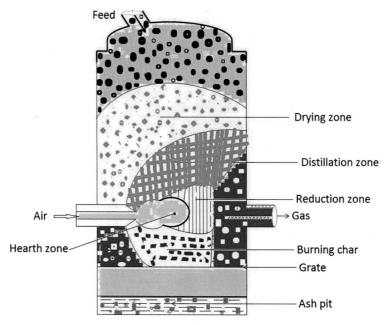

FIG. 8   Cross-draft gasifier.

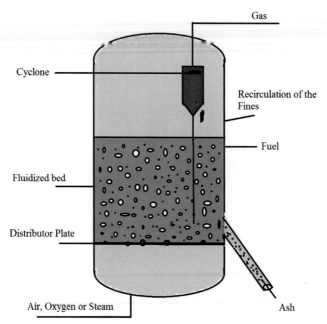

FIG. 9   Fluidized bed gasifier.

II. Biomass energy and biofuel from contaminated substrates

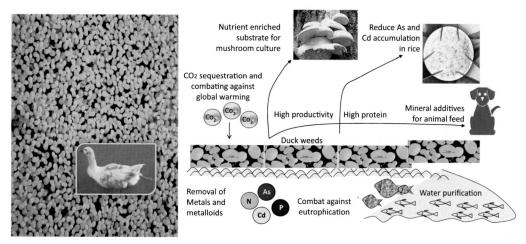

FIG. 10 Duckweeds, e.g., *Wolfiella*, are known for its high protein content and feasible for high-quality and high-quantify production. Thus, duckweeds are regarded as a promising source of protein for human food. Therefore, duckweeds are considered as human food for Western consumers (De Beukelaar et al., 2019). Also, a suitable material for CO_2 sequestration to combat global warming, to address eutrophication, and for removal of metals and metalloids from waste waters (Table 3).

TABLE 3 Duckweed—bioremediation and bioeconomy potential.

Salient findings	Author(s)	Year
Application of common duckweed (*Lemna minor*) in phytoremediation of environmental chemicals: current situation and outlook	Ekperusi et al.	2019
Evaluating and simulating *Lemna paucicostata*'s performance in phytoremediating petroleum hydrocarbons in wetlands affected by crude oil	Ekperusi, et al.	2020
The ability of *Salvinia molesta*, water hyacinth, water lettuce, and duckweed to perform phytoremediation to lessen phosphorus in rice mill wastewater	Kumar and Deswal	2020
Potential for duckweed-based clean energy production (ethanol and biogas) in Bangladesh.	Nahar and Sunny	2020
Aquatic floating duckweed (*Lemna minor*) phytoremediation of nutrients from water in the rearing of African cichlid (*Labidochromis lividus*) fingerlings	Sarkheil and Safari	2020
Duckweeds (Lemnaceae), potential energy plants, accumulate starch	Appenroth et al.	2021
Evaluation of the composite mixture's biogas potential made from duckweed biomass	Chusov et al.	2021
Duckweed plays a variety of roles in the production of bioproducts and aquatic phytoremediation	Liu et al.	2021
Evaluation of *Lemna minor* and *Eichhornia crassipes*' ability to phytoremediate nickel(II) contaminated waters	Mallarino et al.	2021
The potential for phytoremediation of microplastics in the aquatic environment is shown through long-term interactions between microplastics and the floating macrophyte *Lemna minor*	Rozman et al.	2022

5 Cleaning and upgrading biogas

The gas cleaning step is often considered an important phase in the commercialization of gasification process. In addition to the syngas, gasification produces several contaminants including tars, sulfur, and nitrogen compounds. These compounds must be abated to meet process requirements and emission standards as these gases might disrupt the subsequent catalytic conversion processes. Gas cleanup is technically complex where the output composition of the gaseous mixture is determined by the characteristics of the feedstock (Royal Society, 2008). Gasification is followed by passing the resultant gas through a cold trap and collecting in a gas bag. For the separation of carbon monoxide, methane, and carbon dioxide using helium as carrier gas, the gas compositions should be analyzed using a gas chromatograph (Kaewpanha et al., 2014). Since raw syngas produced resembles more conventional fuels like coal and oil residues, similar cleaning technologies can be adopted. There is a profitable transaction between gas cleaning and catalyst performance. Cleaning well below the specifications might be economically attractive for synthesis processes that use sensitive and expensive catalytic materials. The cleaning process includes a filter, Rectisol unit, and downstream gas polishing to remove the traces. This process comprises zinc oxide and active carbon filtering. A water-gas shift reactor will also be part of gas conditioning because the H_2/CO ratio needs adjustment. The Rectisol unit combines the removal of the bulk of the impurities and the separation of CO_2 (van der Drift and Boerrigter, 2005).

Tar removal is a major challenge to make syngas valorization technologies technically and commercially feasible. Tar tolerance levels for gas engines equal to $50\,mg\,Nm^{-3}$, gas turbines equal to $5\,mg\,Nm^{-3}$, and fuel cells equal to $1\,mg\,Nm^{-3}$ (Fjellerup et al., 2005). Tar removal methods can be divided into two categories: (1) primary methods/treatments inside the gasifier and (2) secondary methods/hot gas cleaning. For advanced syngas applications such as usage in gas engines, primary methods are not sufficient. Primary methods can be used as a tool to optimize the gas composition for the secondary cleaning step. Secondary tar reduction methods can be divided into physical/mechanical and chemical methods. Several physical methods such as cyclones, filters (baffle, fabric/ceramic), rotating particle separators, electrostatic precipitators, and scrubbers (water/organic-liquid-based) are available. Chemical methods are further sub-divided into thermal cracking, catalytic cracking, and plasma cracking (Bosmans et al., 2013). By adapting, demonstrating, and optimizing existing industrial processes as well as evolving innovative separation processes based on membranes such as pressure swing adsorption, it is possible to upgrade synthesis gas. These upgrading processes can be integrated with carbon sequestration technologies to improve GHG balance. In the conversion of synthesis gas to biofuels, it is necessary to develop, adapt, and demonstrate the Fischer-Tropsch processes as well as methanol/dimethyl ether and mixed alcohol processes, especially when biomass is used as the feedstock. As the thermochemical routes produce huge amounts of heat, the possibilities of using them in polygeneration systems should be explored (Public Consultation Draft, 2007). If the biomass gasification is integrated with combined gas steam cycles, conversion efficiencies up to 50% might be achieved (Caputo et al., 2005).

6 Factors influencing biogas yield

Gasification process is affected by a number of factors such as temperature, pressure, and height of the reactor bed, fluidization velocity, gasifying medium, moisture content of feed material, particle size, equivalence ratio, air-to-steam ratio, and presence of catalysts. These parameters are quite inter-related and each of them affects the gasification rate, product gas heating value, process efficiency, and product distribution. Large variations in biogas yields were reported based on the composition of raw materials. The gas yield can also be affected by digestion temperature, retention time, load, digestion technology (co-digestion, batch or continuous, one- or two-phase digestion), pretreatment of the raw materials, etc. (Berglund and Borjesson, 2006).

The rate of formation of combustible gases can be increased by increasing the temperature and decreasing the yield of char, thus leading to complete conversion of the fuel. The amount of steam and reaction temperature have significant influence on the gasification performance of biomass. The yield of hydrogen increases with the rise of reaction temperature and the amount of steam while excessive steam reduces the hydrogen yield (Kaewpanha et al., 2014). Hydrocarbon gases such as methane and ethylene emission increase with temperature, while the yields of higher hydrocarbons (C3–C8) decrease at temperatures above 650°C. Energy content of the syngas steadily increases up to 700°C and then decreases at still higher temperatures.

With the increase in pressure the rate of char gasification and yields of methane increases, having significant impacts at high temperatures (900–950°C). For a given reactor temperature, higher fuel beds the reaction time thereby increasing total syngas yields. Higher fluidization velocities increase the temperature of the fuel bed and lead to the production of syngas lower in energy content. The equivalence ratio means the actual fuel-to-air ratio divided by the stoichiometric fuel-to-air ratio. The higher the equivalence ratio, the higher the rate of syngas produced and vice versa. Increase in the steam-to-air ratio increases the energy content of the syngas (Sadaka, 2009). Thermal gasification is effective mostly for feedstock with low water content before the gasification process. Low nitrogen and alkali contents as well as a feedstock particle size are other desirable properties. It is faster than the anaerobic digestion but has a relatively low conversion efficiency rate of 20%–40%. New technologies such as the integrated combined gasification cycle are expected to increase the conversion efficiency up to 50%. Recently, co-gasification with coal has become popular because it facilitates the gasification of smaller quantities of plant material with lower investment and higher conversion efficiency (Elbehri et al., 2013). Some studies have found that heavy metals associated with phytoremediated phytomass affect the activity of the microorganisms that participate in biogas digestion (Cao et al., 2015).

6.1 Biological conversion of waste biomass to biofuels and bioeconomy

Biofuels can be produced from a wide variety of biomasses through thermochemical and biological conversion processes (Jha et al., 2022). Higher grade fuels can be obtained from biomass. Among them bio-methanol is being considered as an attractive liquid fuel as well as feedstock for the synthesis of enumerable valuable organic compounds currently being produced from coal, natural gas, and petroleum feedstocks (Gautam et al., 2021). Anaerobic fermentation is a well-established process for the transformation of organic waste into biogas while separation of biogas is expensive. This paper discussed an integrated upgradation opportunities of AF to produce easily marketable short-chain fatty acids (SCFAs) and medium-chain fatty acids (MCFAs) using organic municipal solid waste as a resource (Kumar et al., 2022). Anaerobic fermentation can produce biogas from organic waste, agricultural waste, and animal waste (Azadbakht et al., 2021). Yeast fermentation is a notable method for waste biomass-to-bioethanol conversion (Hossain et al., 2017).

Lignocellulosic biomass has recently attracted much attention as an alternative energy source because of its ability to provide high fuel output, while not causing environmental pollution similar to the coal or petroleum-based products (Prasad et al., 2019). Lignocellulosic biomass and other organic residues can produce biofuels such as bio-oils, biochar, syngas, biohydrogen, bioethanol, and biobutanol (Parakh et al., 2020; Mahmoud and Kathi, 2022). Agricultural wastes are a major form of lignocellulose biomass (Pattanaik et al., 2019). Catalysts derived from agricultural wastes are an affordable and readily available substitute for the ones made from imported raw materials (Awogbemi et al., 2021). Tang et al. (2018) stated that the usage of biomass as a catalyst precursor possesses several advantages such as abundant availability, less-expensive raw materials, reusability, non-toxicity, and biodegradability. Thermo-catalytic reforming is a promising conversion technology for the production of liquid bio-fuels (Schmitt et al., 2019). Liquid biofuels are getting increased attention because they have the potential to reduce the consumption of fossil fuels in transportation and industrial sectors by having similar performance efficiencies (Stephen and Periyasamy, 2018). The biodiesel production using Agro-waste-derived heterogeneous base catalysts are very active for catalyzing transesterification of oil to biodiesel with a shorter reaction time and higher conversion up to 100% as compared to commercial catalysts and other catalysts derived from other renewable biomasses such as egg shells, fish scales, and bones (Basumatary et al., 2018).

Gautam et al. (2020) reviewed various thermochemical and biochemical routes for the sustainable production of bio-methanol from waste biomass. They understood that bio-methanol is an attractive liquid fuel as well as feedstock for the synthesis of enumerable valuable organic compounds currently being produced from coal, natural gas, and petroleum feedstocks. Usage of virgin and waste biomasses to produce gaseous and liquid fuels is recommended. The selection of an appropriate catalyst depends on the amount of free fatty acids in the oil (Abdullah et al., 2017). The paper proposes the use of heterogeneous catalysts, specifically carbon-based catalysts, in the production of biodiesel. Heterogeneous catalysts are less expensive to manufacture, easier to separate, and more reusable than homogeneous catalysts (Hamza et al., 2021). A biogas-based CBE can provide a long-term way out of the organic fraction of municipal solid waste. The barriers to biogas production, as identified by barrier ranking, show that the lack of appropriate segregation facilities is the most crucial barrier, followed by waste characteristics variation, and inconsistent supply (Yadav et al., 2022a, b).

7 Bioeconomy of biogas production (see also Chapters 1 and 2 in this book for additional info)

The concept of a biorefinery has been recently developed, and it is used to process different biomass feedstocks such as lignocelluloses, algae, and various types of wastes. The bioeconomy consists of the CBE and closed-loop economy (CLE) framework, which uses biomass as an integral component to generate various products. With both CBE and CLE in place, we achieve sustainable production (Ubando et al., 2020). Bioeconomy is defined by the EU as the sustainable production and conversion of biomass for food, health, energy, and industrial products. Biofuels and the energy sector are one of the components of the newly emerging bioeconomy. It has increasingly been recognized globally that plant-based raw materials will eventually replace fossil reserves as feedstocks for industrial production, addressing both the energy and non-energy sectors including chemicals and materials (EC, 2004). Bioeconomy brings together phenomena that have so far been unrelated, such as biomass and products for mainstream consumers, business and sustainability, and ecosystem services and industrial applications. It is about sustainability, closed circles, cross-sectoral cooperation, and the use of biomass. Therefore, bioeconomy can be understood as a large, sociotechnical system that binds together technologies, markets, people, and policies defined by principles rather than the sectoral borderlines of existing industries (Manninen et al., 2014).

Considering international effects on energy balance, the biogas supply creates external economies substituting fossil fuels. At the same time, it reduces import duties due to fossil fuel trading. Moreover, biogas replaces conventional fuels, contributing to sustainable and green environment. The price of supplied energy produced by biogas competes with distorted prices in the energy market. Therefore, under competitive conditions, a decentralized economically self-sufficient biogas unit provides its energy without market distortions. In this context, it is appropriate to recognize that global markets for bioautomotive fuels are rapidly expanding with governmental support. Countries such as Brazil and the United States are supporting the usage of bioethanol, whereas Germany and Austria are encouraging biodiesel (derived from vegetable oils) for transportation purposes. From the viewpoint of economics, environment, land use, water use, chemical fertilizer use, etc., there is a strong inclination for the use of woody, grassy materials and agricultural residues, municipal wastes, and industrial wastes as a feedstock. Therefore, the production of the synthetic transportation fuels such as biomethanol, bioethanol, dimethyl ether, Fischer-Tropsch fuel, synthetic natural gas, and hydrogen via gasification and synthesis seems to be hopeful (Zhang, 2010). Furthermore, from the macroeconomic point of view, decentralized energy generation from biomass appears to be beneficial over centralized energy generation. This process reduces the costs of investment in additional networks such as transportation and the costs of losses on the transmission network (ISAT, 1999).

Comprehensive knowledge of the performance of various types of biofuels is essential to make choices between promising biofuels, whether to use as neat fuel or blended with existing fossil fuels. It is essential to guarantee the adaptability of engine configurations to a specific biofuel. This in turn has to be widely available at the fuel filling stations to make conversion worthwhile. The development of standards for clean liquid and gaseous biofuels as well as for fossil fuels blended with biofuels is a prerequisite (Public Consultation Draft, 2007). In addition, it may be necessary to develop advanced vehicle and engine technologies compatible with biofuel/biofuel blends (e.g., E85, BTL). Developing alternative biofuels can open up the potential to deliver specifically designed engine technologies. For example, Volkswagen has developed a combined combustion system coupled to fuel derived from the Fischer–Tropsch process (Royal Society, 2008).

Considering all biofeedstock opportunities, the bioeconomy is expected to save 1-2.5 billion tons of CO_2 emission (0.27–0.7 billion tons carbon) by 2030. To reach this target, if syngas can be used as a source of industrial carbon, aquatic plant biomass might prove to be a valuable feedstock (Kircher, 2012). Some of the possible barriers to biomass gasification in heat and power markets are lack of industrial-scale demonstration of biomass technology, availability of fewer number of suitable gas turbines and limited expertise in high-pressure biomass gasification, challenges in ash sintering and fouling in gasification of agro-biomass, high cost of the present gas-cleaning systems, limited knowledge in the recovery of valuable metals and recycling of nutrients from gasification ash, limited feedstock flexibility of small-scale gasifiers, low efficiency of small gas engines, and high investment costs in small-scale combined heat and power production (Girio et al., 2013).

To avoid heavy costs incurred in transporting the raw materials to the refineries, pre-processing of feedstocks can be carried out in the place of their availability. This provides an opportunity for local people to gain employment in these biorefineries. Providing sufficient renewable carbon is the biggest challenge in unfolding the bioeconomy (Bio-economy Innovation, 2010). In this process, governments should play a significant role in promoting biofuel technologies by providing incentives and tax benefits to all the stakeholders involved (such as feedstock providers, automobile industries, marketing sector, etc.). Around 70% of the global biofuel production occurred in the United States of America (45.4%) and Brazil (22.5%). The only other countries worth mentioning are Germany (4.8%), Argentina (3.8%), and France (3%), with the rest of the world lagging behind considerably with regard to overall biofuel production (Energy Academy and Centre for Economic Reform and Transformation, 2013).

Transition from a fossil-based economy to bioeconomy is justified by the need for an integrated response to several global trends such as high dependence on fossil-based resources. This can be addressed by strengthening energy security, which calls for a more diversified supply option range; increasing demand of biological resources for bio-based products; increasing sustainability concerns such as reduction of GHG emissions, moving toward a zero waste society, environmental sustainability of primary production systems, and increasing land-use competition. To tackle all of these interconnected global drivers and constraints, an integrated management of renewable biological resources in agriculture, food, bio-based, and energy industries appears to be the most desirable (Nita et al., 2013).

8 Conclusions

Biogas systems using phytoremediated phytomass of aquatic macrophytes have the potential to be an effective strategy in combating some of the important environmental problems relating climate change, phytoremediation, eutrophication, acidification, and water pollution. This study establishes that biogas production from phytoremediated biomass of aquatic macrophytes is a commercially potential and sustainable process. Aquatic macrophytic biomass is a globally available resource. Such a novel treatment system not only directly improves water quality but also benefits indirectly in the reduction of air pollution by replacing fossil fuels. This study is important in improving phytoremediation systems and can be referred for sustainable disposal of plants used in the bioremediation. The availability of aquatic macrophytes for biogas production in water bodies needs to be explored. Proper combination of green technologies is more essential in an era of increased energy crisis, pollution, and global warming.

A critical analysis of the literature reveals that thermal gasification is the most reliable technology to displace the usage of fossil fuels and to reduce CO_2 emission, mainly when the plant biomass is used as feedstock. These biofuels should mimic the properties of petrol and diesel such as in octane equivalent, energy density, cetane number, and hydrophobicity. It is essential to design and develop new engine technologies that can directly substitute fossil fuels with advanced biofuel delivering, in addition to significantly lowering GHG emissions. For vehicle manufacturers to make the investments needed, a long-term market for transport fuels containing a high blend of biofuels must be established. Economic and regulatory incentives are needed to accelerate the technology developments needed to deliver biofuel supply chains that can provide more substantial reduction of emissions.

However, the use of bioenergy resulting from phytoremediation is constrained by the lack of knowledge regarding the emissions that may be generated and the issues associated with pollution transfer, especially heavy metals. The key goal for bioeconomy of biofuels is to fulfil social, economic, and environmental criteria in a sustainable manner with the integration of desperate fields.

Acknowledgment

The author would like to thank Pondicherry University for providing Internet access to collect the information essential for writing this chapter.

References

Abbas, Z., Arooj, F., Ali, S., Zaheer, I.E., Rizwan, M., Riaz, M.A., 2019. Phytoremediation of landfill leachate waste contaminants through floating bed technique using water hyacinth and water lettuce. Int. J. Phytoremediation 21 (13), 1356–1367.

Abdullah, S.H.Y.S., Hanapi, N.H.M., Azid, A., Umar, R., Juahir, H., Khatoon, H., Endut, A., 2017. A review of biomass-derived heterogeneous catalyst for a sustainable biodiesel production. Renew. Sust. Energ. Rev. 70, 1040–1051.

Adeleye, B.A., Adetunji, A., Bamgboye, I., 2013. Towards deriving renewable energy from aquatic macrophytes polluting water bodies in Niger Delta region of Nigeria. Res. J. Appl. Sci. Eng. Technol. 5 (2), 387–391.

Ahila, K.G., Ravindran, B., Muthunarayanan, V., Nguyen, D.D., Nguyen, X.C., Chang, S.W., Nguyen, V.K., Thamaraiselvi, C., 2020. Phytoremediation potential of freshwater macrophytes for treating dye-containing wastewater. Sustainability 13 (1), 329.

Akhtar, A.B.T., Yasar, A., Ali, R., Irfan, R., 2017. Phytoremediation using aquatic macrophytes. In: Phytoremediation. Springer, Cham, pp. 259–276.

Alam, R., Khan, S.U., Basheer, F., Farooqi, I.H., 2021. February. Nutrients and organics removal from slaughterhouse wastewater using phytoremediation: a comparative study on different aquatic plant species. In: IOP Conference Series: Materials Science and Engineering. vol. 1058(1). IOP Publishing, p. 012068.

Alauddin, Z.A.B., Lahijani, P., Mohammadi, M., Mohamed, A.R., 2010. Gasification of lignocellulosic biomass in fluidized beds for renewable energy development: a review. Renew. Sust. Energy Rev. 14 (9), 2852–2862.

Anand, S., Bharti, S.K., Dviwedi, N., Barman, S.C., Kumar, N., 2017. Macrophytes for the reclamation of degraded waterbodies with potential for bioenergy production. In: Phytoremediation Potential of Bioenergy Plants. Springer, Singapore, pp. 333–351.

Anand, S., Bharti, S.K., Kumar, S., Barman, S.C., Kumar, N., 2019. Phytoremediation of heavy metals and pesticides present in water using aquatic macrophytes. In: Phyto and Rhizo Remediation, pp. 89–119.

Anwar-ul-Haq, M., Ibrahim, M., Yousaf, B., Al-Huqail, A.A., Ali, H.M., 2022. Role of bacterial consortium and synthetic surfactants in promoting the phytoremediation of crude oil-contaminated soil using *Brachiaria mutica*. Energy Res. 10, 874492.

Appenroth, K.J., Ziegler, P., Sree, K.S., 2021. Accumulation of starch in Duckweeds (Lemnaceae), potential energy plants. Physiol. Mol. Biol. Plants, 1–13.

Aqdas, A., Hashmi, I., 2022. Role of water hyacinth (*Eichhornia crassipes*) in integrated constructed wetlands: a review on its phytoremediation potential. Int. J. Environ. Sci. Technol., 1–8.

Aswad, Z.S., Ali, A.H., Al-Mhana, N.M., 2020. Energy production and wastewater treatment using *Juncus, S. triqueter, P. australis, T. latifolia*, and *C. alternifolius* plants in sediment microbial fuel cell. Desalin. Water Treat. 205, 153–160.

Attah-Olottah, R., Adeniyi, K.A., 2018. Application of phytoremediation in the management of oil spillage: a review. Global J. Earth Environ. Sci. 3 (3), 16–22.

Auchterlonie, J., Eden, C.L., Sheridan, C., 2021. The phytoremediation potential of water hyacinth: a case study from Hartbeespoort Dam, South Africa. South Afr. J. Chem. Eng. 37, 31–36.

Awan, B., Sabeen, M., Shaheen, S., Mahmood, Q., Ebadi, A., Toughani, M., 2020. Phytoremediation of zinc contaminated water by marigold (*Tagetes Minuta* L). Central Asian J. Environ. Sci. Technol. Innov. 1 (3), 150–158.

Awogbemi, O., Von Kallon, D.V., Aigbodion, V.S., 2021. Trends in the development and utilization of agricultural wastes as heterogeneous catalyst for biodiesel production. J. Energy Inst. 98, 244–258.

Azadbakht, M., Safieddin Ardebili, S.M., Rahmani, M., 2021. Potential for the production of biofuels from agricultural waste, livestock, and slaughterhouse waste in Golestan province, Iran. Biomass Conver. Biorefinery, 1–11.

Babu, S.O.F., Hossain, M.B., Rahman, M.S., Rahman, M., Ahmed, A.S., Hasan, M.M., Rakib, A., Emran, T.B., Xiao, J., Simal-Gandara, J., 2021. Phytoremediation of toxic metals: a sustainable green solution for clean environment. Appl. Sci. 11 (21), 10348.

Basit, A., Shah, S.T., Ullah, I., Muntha, S.T., Mohamed, H.I., 2021. Microbe-assisted phytoremediation of environmental pollutants and energy recycling in sustainable agriculture. Arch. Microbiol. 203 (10), 5859–5885.

Baştabak, B., Gödekmerdan, E., Koçar, G., 2021. A holistic approach to soil contamination and sustainable phytoremediation with energy crops in the Aegean Region of Turkey. Chemosphere 276, 130192.

Basumatary, S., Nath, B., Kalita, P., 2018. Application of agro-waste derived materials as heterogeneous base catalysts for biodiesel synthesis. J. Renew. Sustain. Energy 10 (4), 043105.

Bayoumi, T.A., Saleh, H.M., 2018. Characterization of biological waste stabilized by cement during immersion in aqueous media to develop disposal strategies for phytomediated radioactive waste. Prog. Nucl. Energy 107, 83–89.

Berglund, M., Borjesson, P., 2006. Assessment of energy performance in the life-cycle of biogas production. Biomass Bioenergy 30, 254–266.

Bharathiraja, B., Jayamuthunagai, J., Praveenkumar, R., Iyyappan, J., 2018. Phytoremediation techniques for the removal of dye in wastewater. In: Bioremediation: Applications for Environmental Protection and Management. Springer, Singapore, pp. 243–252.

Bhat, M., Shukla, R.N., Yunus, M., 2020. Urban pond ecosystems: preservation and management through phytoremediation. In: Fresh water pollution dynamics and remediation. Springer, Singapore, pp. 263–291.

Bhattacharya, A., Kumar, P., 2010. Water hyacinth as a potential biofuel crop. Electron. J. Environ. Agric. Food Chem. 9 (1), 112–122.

Bhowmik, D., Chetri, S., Enerijiofi, K.E., Naha, A., Kanungo, T.D., Shah, M.P., Nath, S., 2023. Multitudinous approaches, challenges and opportunities of bioelectrochemical systems in conversion of waste to energy from wastewater treatment plants. Cleaner Circ. Bioecon., 100040.

Bio-economy Innovation, 2010. Bioeconomy Council Report: Research and Technology Development for Food Security, Sustainable Resource Use and Competitiveness. Research and Technology Council, Berlin.

Bora, M.S., Sarma, K.P., 2020. Phytoremediation of heavy metals/metalloids by native herbaceous macrophytes of wetlands: current research and perspectives. In: Emerging Issues in the Water Environment during Anthropocene, pp. 261–284.

Bosmans, A., Wasan, S., Helsen, L., Jones, P.T., Geysen, D., 2013, October. Waste-to-clean syngas: avoiding tar problems. In: Proceedings of the 2nd International Academic Symposium on Enhanced Landfill Mining. Haletra, Houthalen-Helchteren.

Bote, M.A., Naik, V.R., Jagadeeshgouda, K.B., 2020. Review on water hyacinth weed as a potential bio fuel crop to meet collective energy needs. Mater. Sci. Energy Technol. 3, 397–406.

Bridgwater, A.V., 2012. Review of fast pyrolysis of biomass and product upgrading. Biomass Bioenergy 38, 68–94.

Cao, Z., Wang, S., Wang, T., Chang, Z., Shen, Z., Chen, Y., 2015. Using contaminated plants involved in phytoremediation for anaerobic digestion. Int. J. Phytoremediation 17 (3), 201–207. https://doi.org/10.1080/15226514.2013.876967.

Caputo, A.C., Palumbo, M., Pelagagge, P.M., Scacchia, F., 2005. Economics of biomass energy utilization in combustion and gasification plants: effects of logistic variables. Biomass Bioenergy 28, 35–51.

Carreño-Sayago, U.F., Rodríguez-Parra, C., 2019. *Eichhornia crassipes* (Mart.) Solms: an integrated phytoremediation and bioenergy system. Rev. Chapingo Serie Ciencias Forestales Del Ambiente 25 (3), 399–411.

Chakraborty, S., Mishra, A., Verma, E., Tiwari, B., Mishra, A.K., Singh, S.S., 2019. Physiological mechanisms of aluminum (Al) toxicity tolerance in nitrogen-fixing aquatic macrophyte *Azolla microphylla* Kaulf: phytoremediation, metabolic rearrangements, and antioxidative enzyme responses. Environ. Sci. Pollut. Res. 26 (9), 9041–9054.

Chander, P.D., Fai, C.M., Kin, C.M., 2018. Removal of pesticides using aquatic plants in water resources: a review. In: IOP Conference Series: Earth and Environmental Science. vol. 164(1). IOP Publishing, p. 012027.

Chaudhuri, D., Bandyopadhyay, K., Majumder, A., Misra, A.K., 2017. Heavy metal removal by aquatic plants and its disposal by using as a concrete ingredient. In: Geoenvironmental Practices and Sustainability. Springer, Singapore, pp. 285–291.

Choudhary, M.C., 2020. Phytoremediation: tools & technique. Recent Trends Mol. Biol. Biotechnol., 55 (Chapter 3).

Christian, E., Beniah, I., 2019. Phytoremediation of polluted waterbodies with aquatic plants: Recent progress on heavy metal and organic pollutants. https://doi.org/10.20944/preprints201909.0020.v1. Preprints, 201909.0020.

Chusov, A., Maslikov, V., Badenko, V., Zhazhkov, V., Molodtsov, D., Pavlushkina, Y., 2021. Biogas potential assessment of the composite mixture from Duckweed biomass. Sustainability 14 (1), 351.

Costa, M.B., Tavares, F.V., Martinez, C.B., Colares, I.G., Martins, C.D.M.G., 2018. Accumulation and effects of copper on aquatic macrophytes *Potamogeton pectinatus* L.: potential application to environmental monitoring and phytoremediation. Ecotoxicol. Environ. Saf. 155, 117–124.

da Silva Santos, J., da Silva Pontes, M., Grillo, R., Fiorucci, A.R., de Arruda, G.J., Santiago, E.F., 2020. Physiological mechanisms and phytoremediation potential of the macrophyte *Salvinia biloba* towards a commercial formulation and an analytical standard of glyphosate. Chemosphere 259, 127417.

DalCorso, G., Fasani, E., Manara, A., Visioli, G., Furini, A., 2019. Heavy metal pollutions: state of the art and innovation in phytoremediation. Int. J. Mol. Sci. 20 (14), 3412.

Das, B.K., Najmul Hoque, S.M., 2014. Assessment of the potential of biomass gasification for electricity generation in Bangladesh. J. Renewable Energy 2014. https://doi.org/10.1155/2014/429518. Article ID 429518, 10 pp.

Das, S., Das, S., Ghangrekar, M.M., 2022. Efficacious bioremediation of heavy metals and radionuclides from wastewater employing aquatic macro-and microphytes. J. Basic Microbiol. 62 (3–4), 260–278.

de Alkimin, G.D., Paisio, C., Agostini, E., Nunes, B., 2020. Phytoremediation processes of domestic and textile effluents: evaluation of the efficacy and toxicological effects in *Lemna minor* and *Daphnia magna*. Environ. Sci. Pollut. Res. 27 (4), 4423–4441.

De Beukelaar, M.F., Zeinstra, G.G., Mes, J.J., Fischer, A.R., 2019. Duckweed as human food. The influence of meal context and information on duckweed acceptability of Dutch consumers. Food Qual. Prefer. 71, 76–86.

de Queiroz, R.D.C.S., Maranduba, H.L., Hafner, M.B., Rodrigues, L.B., de Almeida Neto, J.A., 2020. Life cycle thinking applied to phytoremediation of dairy wastewater using aquatic macrophytes for treatment and biomass production. J. Clean. Prod. 267, 122006.

de Souza, C.B., Silva, G.R., 2019. Phytoremediation of Effluents Contaminated With Heavy Metals by Floating Aquatic Macrophytes Species. IntechOpen.

Deshmukh, H.V., 2012. Economic feasibility and pollution abetment study of biogas production process utilizing admixture of *Ipomoea carnea* and distillery waste. J. Environ. Res. Dev. 7 (2), 633–642.

Devi, P., Kumar, P., 2020. Concept and application of phytoremediation in the fight of heavy metal toxicity. J. Pharm. Sci. Res. 12 (6), 795–804.

Dipu, S., Kumar, A.A., Thanga, S.G., V., 2011. Potential application of macrophytes used in phytoremediation. World Appl. Sci. J. 13 (3), 482–486.

Duran, V., Brdecka, M., Seigerroth, J., Jang, B., 2023. Utilization of reeds to sequester Ni and/or Cu from wastewater and to produce valuable products. https://doi.org/10.21203/rs.3.rs-2350984/v1.

Ebrahimbabaie, P., Meeinkuirt, W., Pichtel, J., 2020. Phytoremediation of engineered nanoparticles using aquatic plants: mechanisms and practical feasibility. J. Environ. Sci. 93, 151–163.

EC, 2004. Towards a European Knowledge-Based Bioeconomy. Workshop Conclusions on the Use of Plant Biotechnology for the Production of Industrial Biobased Products. EUR 21459. European Commission, Directorate-General for Research, Brussels, Belgium. Available online at: http://ec.europa.eu/research/agriculture/library_en.htm.

Eid, E.M., Shaltout, K.H., Moghanm, F.S., Youssef, M.S., El-Mohsnawy, E., Haroun, S.A., 2019. Bioaccumulation and translocation of nine heavy metals by *Eichhornia crassipes* in Nile Delta, Egypt: perspectives for phytoremediation. Int. J. Phytoremediation 21 (8), 821–830.

Ekperusi, A.O., Sikoki, F.D., Nwachukwu, E.O., 2019. Application of common Duckweed (*Lemna minor*) in phytoremediation of chemicals in the environment: state and future perspective. Chemosphere 223, 285–309.

Ekperusi, A.O., Nwachukwu, E.O., Sikoki, F.D., 2020. Assessing and modelling the efficacy of *Lemna paucicostata* for the phytoremediation of petroleum hydrocarbons in crude oil-contaminated wetlands. Sci. Rep. 10 (1), 1–9.

Elbehri, A., Segerstedt, A., Liu, P., 2013. Biofuels and Sustainability Challenge: A Global Assessment of Sustainability Issues, Trends and Policies for Biofuels and Related Feedstocks. Food and Agriculture Organization of the United Nations, Rome, p. 42.

Energy Academy and Centre for Economic Reform and Transformation, 2013. BP Statistical Review of World Energy. Heriot-Watt University. www.bp.com/…/bp/…/statistical-review/statistical_review_of_world_energy.

Enyoh, C.E., Isiuku, B.O., 2021. 2,4,6-Trichlorophenol (TCP) removal from aqueous solution using *Canna indica* L.: kinetic, isotherm and thermodynamic studies. Chem. Ecol. 37 (1), 64–82.

Farraji, H., Zaman, N.Q., Tajuddin, R., Faraji, H., 2016. Advantages and disadvantages of phytoremediation: a concise review. Int. J. Environ. Tech. Sci. 2, 69–75.

Farrajia, H., Qamaruzaman, N., Sa'ata, S.M., Dashtia, A.F., 2017. Phytoremediation of suspended solids and turbidity of palm oil mill effluent (POME) by *Ipomea aquatica*. Eng. Heritage J. 1 (1), 36–40.

Favas, P.J., Pratas, J., Paul, M.S., Prasad, M.N.V., 2019. Remediation of uranium-contaminated sites by phytoremediation and natural attenuation. In: Phytomanagement of Polluted Sites. Elsevier, pp. 277–300.

Fernando, A.L., Barbosa, B., Costa, J., Papazoglou, E.G., 2016. Giant reed (*Arundo donax* L.): A multipurpose crop bridging phytoremediation with sustainable bioeconomy. In: Bioremediation and Bioeconomy. Elsevier, pp. 77–95.

Fjellerup, J., Ahrenfeldt, J., Henriksen, U., Gobel, B., 2005. Formation, Decomposition and Cracking of Biomass Tars in Gasification. Technical University of Denmark (DTU) Biomass Gasification Group, Denmark.

Fletcher, J., Willby, N., Oliver, D.M., Quilliam, R.S., 2020. Phytoremediation using aquatic plants. In: Phytoremediation. Springer, Cham, pp. 205–260.

Galal, T.M., Gharib, F.A., Ghazi, S.M., Mansour, K.H., 2017. Phytostabilization of heavy metals by the emergent macrophyte *Vossia cuspidata* (Roxb.) Griff.: a phytoremediation approach. Int. J. Phytoremediation 19 (11), 992–999.

Gautam, S., Tiwari, A., Kumar, R., Singh, S., 2021. Bio-methanol production from lignocellulosic biomass: a review. Renew. Sust. Energ. Rev. 148, 111394. https://doi.org/10.1016/j.rser.2021.111394.

Gautam, P., Upadhyay, S.N., Dubey, S.K., 2020. Bio-methanol as a renewable fuel from waste biomass: current trends and future perspective. Fuel 273, 117783.

Ghanem, H., Chalak, L., Baydoun, S., 2019. Phytoremediation of Lebanese polluted waters: a review of current initiatives. In: MATEC Web of Conferences. vol. 281. EDP Sciences, p. 03007.

Girio, F., Kurkela, E., Kiel, J., Lankhorst, R.K., 2013. Longer Term R&D Needs and Priorities on Bioenergy—Bioenergy Beyond 2020. European Energy Research Alliance, European Industrial Bioenergy Initiative Workshop Report.

Goswami, S., Das, S., 2017. Screening of cadmium and copper phytoremediation ability of *Tagetes erecta*, using biochemical parameters and scanning electron microscopy—energy-dispersive X-ray microanalysis. Environ. Toxicol. Chem. 36 (9), 2533–2542.

Govere, E.M., 2021. Environmental phytoremediation and analytical technologies for heavy metal removal and assessment. In: Plant Biotechnology. Springer, Cham, pp. 203–213.

Grzegórska, A., Rybarczyk, P., Rogala, A., Zabrocki, D., 2020. Phytoremediation—from environment cleaning to energy generation—current status and future perspectives. Energies 13 (11), 2905.

Habibul, N., Chen, J.J., Hu, Y.Y., Hu, Y., Yin, H., Sheng, G.P., Yu, H.Q., 2019. Uptake, accumulation and metabolization of 1-butyl-3-methylimidazolium bromide by ryegrass from water: prospects for phytoremediation. Water Res. 156, 82–91.

Hamza, M., Ayoub, M., Shamsuddin, R.B., Mukhtar, A., Saqib, S., Zahid, I., Ibrahim, M., 2021. A review on the waste biomass derived catalysts for biodiesel production. Environ. Technol. Innov. 21, 101200.

Hauptvogl, M., Kotrla, M., Prčík, M., Pauková, Ž., Kováčik, M., Lošák, T., 2019. Phytoremediation potential of fast-growing energy plants: challenges and perspectives—a review. Pol. J. Environ. Stud. 29 (1), 505–516.

Haziq, J.M., Amalina, I.F., Syukor, A.A., Sulaiman, S., Siddique, M.N.I., Woon, S.X.R., 2020. Phytoremediation: treating euthrophic lake at KotaSAS Lakeside, Kuantan by aquatic macrophytes. In: IOP Conference Series: Materials Science and Engineering. vol. 736(2). IOP Publishing, p. 022017.

He, Y., Chi, J., 2016. Phytoremediation of sediments polluted with phenanthrene and pyrene by four submerged aquatic plants. J. Soils Sediments 16 (1), 309–317.

Hishida, M., Shinoda, K., Akiba, T., Amari, T., Yamamoto, T., Matsumoto, K., 2011. Biomass syngas production technology by gasification for liquid fuel and other chemicals. Mitsubishi Heavy Ind. Tech. Rev. 48 (3), 37–41.

Hossain, N., Zaini, J.H., Mahlia, T.M.I., 2017. A review of bioethanol production from plant-based waste biomass by yeast fermentation. Int. J. Technol. 8 (1).

Hu, H., Li, X., Wu, S., Yang, C., 2020. Sustainable livestock wastewater treatment via phytoremediation: current status and future perspectives. Bioresour. Technol. 315, 123809.

Huang, D., Xiao, R., Du, L., Zhang, G., Yin, L., Deng, R., Wang, G., 2021. Phytoremediation of poly- and perfluoroalkyl substances: a review on aquatic plants, influencing factors, and phytotoxicity. J. Hazard. Mater. 418, 126314.

ISAT, 1999. Biogas Digest, Vol. III: Biogas—Costs and Benefits and Biogas. Information and Advisory Service on Appropriate Technology, Frankfurt, pp. 2–61.

Jani, Y., Mutafela, R., Ferrans, L., Ling, G., Burlakovs, J., Hogland, W., 2019. Phytoremediation as a promising method for the treatment of contaminated sediments. Iran. J. Energy Environ. 10 (1), 58–64.

Jasrotia, S., Kansal, A., Mehra, A., 2017. Performance of aquatic plant species for phytoremediation of arsenic-contaminated water. Appl. Water Sci. 7 (2), 889–896.

Jha, S., Nanda, S., Acharya, B., Dalai, A.K., 2022. A review of thermochemical conversion of waste biomass to biofuels. Energies 15 (17), 6352.

Jiang, B., Xing, Y., Zhang, B., Cai, R., Zhang, D., Sun, G., 2018. Effective phytoremediation of low-level heavy metals by native macrophytes in a vanadium mining area, China. Environ. Sci. Pollut. Res. 25 (31), 31272–31282.

Jisha, C.K., Bauddh, K., Shukla, S.K., 2017. Phytoremediation and bioenergy production efficiency of medicinal and aromatic plants. In: Phytoremediation Potential of Bioenergy Plants. Springer, Singapore, pp. 287–304.

Kacprzak, A., Matyka, M., Krzystek, L., Ledakowicz, S., 2012. Evaluation of biogas collection from reed canary grass, depending on nitrogen fertilisation levels. Chem. Process. Eng. 33 (4), 697–701.

Kaewpanha, M., Guan, G., Hao, X., Wang, Z., Kasai, Y., Kusakabe, K., Abudula, A., 2014. Steam co-gasification of brown seaweed and land-based biomass. Fuel Process. Technol. 120, 106–112.

Kafle, A., Timilsina, A., Gautam, A., Adhikari, K., Bhattarai, A., Aryal, N., 2022. Phytoremediation: mechanisms, plant selection and enhancement by natural and synthetic agents. Environ. Adv., 100203.

Kathi, S., 2016. Bioenergy from phytoremediated phytomass of aquatic plants via gasification. In: Bioremediation and Bioeconomy. Elsevier, pp. 111–128.

Kathi, S., 2022. Phytoremediation of heavy metals and petroleum hydrocarbons using Cynodon dactylon (L.) Pers. In: Cost Effective Technologies for Solid Waste and Wastewater Treatment. Elsevier, pp. 135–145.

Kathi, S., Khan, A.B., 2011. Phytoremediation approaches to PAH contaminated soil. Indian J. Sci. Technol. 4 (1), 56–63.

Kathi, S., Padmavathy, A., 2019. E-waste: global scenario, constituents, and biological strategies for remediation. In: Electronic Waste Pollution: Environmental Occurrence and Treatment Technologies, pp. 75–96.

Kaur, M., Kumar, M., Sachdeva, S., Puri, S.K., 2018. Aquatic weeds as the next generation feedstock for sustainable bioenergy production. Bioresour. Technol. 251, 390–402.

Kaur, M., Kumar, M., Singh, D., Sachdeva, S., Puri, S.K., 2019. A sustainable biorefinery approach for efficient conversion of aquatic weeds into bioethanol and biomethane. Energy Convers. Manag. 187, 133–147.

II. Biomass energy and biofuel from contaminated substrates

Kaushal, J., Mahajan, P., 2022. Kinetic evaluation for removal of an anionic diazo Direct Red 28 by using phytoremediation potential of *Salvinia molesta* Mitchell. Bull. Environ. Contam. Toxicol. 108 (3), 437–442.

Khan, A.G., 2020. In situ phytoremediation of uranium contaminated soils. Phytoremediation, 123–151.

Khan, A.G., 2021a. Nano-phytoremediation of heavy metals contaminated wastewater ecosystems and wetlands by constructed wetlands planted with waterlogging-tolerant mycorrhizal fungi and vetiver grass. Environ. Sci. Proc. 6 (1), 25.

Khan, A.G., 2021b. Potential of coupling heavy metal (HM) phytoremediation by bioenergy plants and their associated HM-adapted rhizosphere microbiota (arbuscular mycorrhizal fungi and plant growth promoting microbes) for bioenergy production. J. Energy Power Technol. 3 (4), 1.

Khan, M.A., Marwat, K.B., Gul, B., Wahid, F., Khan, H., Hashim, S., 2014. *Pistia stratiotes* L. (Araceae): phytochemistry, use in medicines, phytoremediation, biogas and management options. Pak. J. Bot. 46 (3), 851–860.

Khankhane, P.J., Varshney, J.G., 2009. Possible use of giant reed, *Arundo donax* for phytoremediation of runoff water in a catchment area. In: National Consultation on Weed Utilization, Paper Presented in National Consultation on Weed Utilization, 20–21 October, 2009, DWSR Jabalpur, p. 28.

Kinidi, L., Salleh, S., 2017. Phytoremediation of nitrogen as green chemistry for wastewater treatment system. Int. J. Chem. Eng.

Kircher, M., 2012. The transition to a bio-economy: emerging from the oil age. Biofuels Bioprod. Biorefin. 6, 369–375. https://doi.org/10.1002/bbb.

Kishore, V.V.N. (Ed.), 2008. Renewable Energy Engineering and Technology a Knowledge Compendium. The Energy and Resources Institute, New Delhi.

Kocabas, B.B., Attar, A., Yuka, S.A., Yapaoz, M.A., 2023. Biogenic synthesis, molecular docking, biomedical and environmental applications of multifunctional CuO nanoparticles mediated *Phragmites australis*. Bioorg. Chem. 133, 106414.

Koley, A., Bray, D., Banerjee, S., Sarhar, S., Thahur, R.G., Hazra, A.K., Mandal, N.C., Chaudhury, S., Ross, A.B., Camargo-Valero, M.A., Balachandran, S., 2023. Water Hyacinth (*Eichhornia crassipes*) a sustainable strategy for heavy metals removal from contaminated waterbodies. In: Bioremediation of Toxic Metal (loid)s. CRC Press, pp. 95–114.

Kumar, S., Deswal, S., 2020. Phytoremediation capabilities of *Salvinia molesta*, water hyacinth, water lettuce, and Duckweed to reduce phosphorus in rice mill wastewater. Int. J. Phytoremediation 22 (11), 1097–1109.

Kumar, V., Singh, J., Pathak, V.V., Ahmad, S., Kothari, R., 2017. Experimental and kinetics study for phytoremediation of sugar mill effluent using water lettuce (*Pistia stratiotes* L.) and its end use for biogas production. 3 Biotech 7 (5), 1–10.

Kumar, D., Anand, S., Tiwari, J., Kisku, G.C., Kumar, N., 2019. Removal of inorganic and organic contaminants from terrestrial and aquatic ecosystems through phytoremediation and biosorption. In: Environmental Biotechnology: for Sustainable Future. Springer, Singapore, pp. 45–71.

Kumar, A.N., Sarkar, O., Chandrasekhar, K., Raj, T., Narisetty, V., Mohan, S.V., Kim, S.H., 2022. Upgrading the value of anaerobic fermentation via renewable chemicals production: a sustainable integration for circular bioeconomy. Sci. Total Environ. 806, 150312.

Kurniawan, A., Khasanah, K., Jayatri, F.N.M., 2022. Study on the application of phytoremediation of phosphate content to eutrophication in Cengklik reservoir, Boyolali Regency. In: IOP Conference Series: Earth and Environmental Science. vol. 986(1). IOP Publishing, p. 012075.

Lakaniemi, A.M., Koskinen, P.E.P., Nevatalo, L.M., Kaksonen, A.H., Puhakka, J.A., 2011. Biogenic hydrogen and methane production from reed canary grass. Biomass Bioenergy 35 (2), 773–780.

Lakra, K.C., Lal, B., Banerjee, T.K., 2017. Decontamination of coal mine effluent generated at the Rajrappa coal mine using phytoremediation technology. Int. J. Phytoremediation 19 (6), 530–536.

Lakra, K.C., Lal, B., Banerjee, T.K., 2019. Application of phytoremediation technology in decontamination of a fish culture pond fed with coal mine effluent using three aquatic macrophytes. Int. J. Phytoremediation 21 (9), 840–848.

Lata, S., 2021. Sustainable and eco-friendly approach for controlling industrial wastewater quality imparting succour in water-energy nexus system. Energy Nexus 3, 100020.

Li, C., Wang, M., Luo, X., Liang, L., Han, X., Lin, X., 2019. Accumulation and effects of uranium on aquatic macrophyte *Nymphaea tetragona* Georgi: potential application to phytoremediation and environmental monitoring. J. Environ. Radioact. 198, 43–49.

Li, F., He, X., Srishti, A., Song, S., Tan, H.T.W., Sweeney, D.J., Ghosh, S., Wang, C.H., 2021. Water hyacinth for energy and environmental applications: a review. Bioresour. Technol. 327, 124809.

Liu, Z., Tran, K.Q., 2021. A review on disposal and utilization of phytoremediation plants containing heavy metals. Ecotoxicol. Environ. Saf. 226, 112821.

Liu, Y., Xu, H., Yu, C., Zhou, G., 2021. Multifaceted roles of Duckweed in aquatic phytoremediation and bioproducts synthesis. GCB Bioenergy 13 (1), 70–82.

Loría, K.C., Emiliani, J., Bergara, C.D., Herrero, M.S., Salvatierra, L.M., Pérez, L.M., 2019. Effect of daily exposure to Pb-contaminated water on *Salvinia biloba* physiology and phytoremediation performance. Aquat. Toxicol. 210, 158–166.

Lu, D., Huang, Q., Deng, C., Zheng, Y., 2018. Phytoremediation of copper pollution by eight aquatic plants. Pol. J. Environ. Stud. 27 (1).

Mahajan, P., Kaushal, J., 2018. Role of phytoremediation in reducing cadmium toxicity in soil and water. J. Toxicol. 2018.

Mahajan, P., Kaushal, J., Upmanyu, A., Bhatti, J., 2019. Assessment of phytoremediation potential of *Chara vulgaris* to treat toxic pollutants of textile effluent. J. Toxicol. 2019.

Mahendran, R.P., 2014. Phytoremediation – insights into plants as remedies. Malaya J. Biosci. 1 (1), 41–45.

Mahmood, H., Moniruzzaman, M., Iqbal, T., Khan, M.J., 2019. Recent advances in the pretreatment of lignocellulosic biomass for biofuels and value-added products. Curr. Opin. Green Sustain. Chem. 20, 18–24.

Mahmoud, A.E.D., Kathi, S., 2022. Assessment of biochar application in decontamination of water and wastewater. In: Cost Effective Technologies for Solid Waste and Wastewater Treatment. Elsevier, pp. 69–74.

Majumdar, A., Upadhyay, M.K., Giri, B., Karwadiya, J., Bose, S., Jaiswal, M.K., 2023. Iron oxide doped rice biochar reduces soil-plant arsenic stress, improves nutrient values: an amendment towards sustainable development goals. Chemosphere 312, 137117.

Malaviya, P., Singh, A., Anderson, T.A., 2020. Aquatic phytoremediation strategies for chromium removal. Rev. Environ. Sci. Biotechnol. 19 (4), 897–944.

Mallarino Miranda, L.P., Alvear Alayon, M.R., Moreno Rubio, N., Ortega Villamizar, D., Tejeda Benitez, L.P., 2021, August. Evaluation of the phytoremediation potential of *Lemna minor* and *Eichhornia crassipes* in waters contamination with nickel (II). In: ISEE Conference Abstracts, vol. 2021(1).

Manjunath, S., 2016. Phytoremediation of textile industry effluent using aquatic macrophytes. Int. J. Environ. Sci. 5 (2), 65–74.

Manninen, J., Nieminen-Sundell, R., Belloni, K. (Eds.), 2014. People in the Bioeconomy 2044. VTT Technical Research Centre of Finland, pp. 8–9. www.vtt.fi/publications/index.jsp.

Manorama Thampatti, K.C., Beena, V.I., Meera, A.V., Ajayan, A.S., 2020. Phytoremediation of metals by aquatic macrophytes. In: Phytoremediation. Springer, Cham, pp. 153–204.

Mansour, A.T., Alprol, A.E., Abualnaja, K.M., El-Beltagi, H.S., Ramadan, K.M., Ashour, M., 2022. Dried brown seaweed's phytoremediation potential for methylene blue dye removal from aquatic environments. Polymers 14 (7), 1375.

Mehariya, S., Kumar, P., Marino, T., Casella, P., Iovine, A., Verma, P., Musmarra, D., Molino, A., 2021. Aquatic weeds: a potential pollutant removing agent from wastewater and polluted soil and valuable biofuel feedstock. In: Bioremediation Using Weeds. Springer, Singapore, pp. 59–77.

Miranda, A.F., Muradov, N., Gujar, A., Stevenson, T., Nugegoda, D., Ball, A.S., Mouradov, A., 2014. Application of aquatic plants for the treatment of selenium-rich mining wastewater and production of renewable fuels and petrochemicals. J. Sust. Bioenergy Syst. 4, 97–112. https://doi.org/10.4236/jsbs.2014.41010.

Mohan, S.V., Nikhil, G.N., Chiranjeevi, P., Reddy, C.N., Rohit, M.V., Kumar, A.N., Sarkar, O., 2016. Waste biorefinery models towards sustainable circular bioeconomy: critical review and future perspectives. Bioresour. Technol. 215, 2–12.

Mohd Nizam, N.U., Mohd Hanafiah, M., Mohd Noor, I., Abd Karim, H.I., 2020. Efficiency of five selected aquatic plants in phytoremediation of aquaculture wastewater. Appl. Sci. 10 (8), 2712.

Mohebi, Z., Nazari, M., 2021. Phytoremediation of wastewater using aquatic plants: a review. J. Appl. Res. Water Wastewater 8 (1), 50–58.

Mosoarca, G., Vancea, C., Popa, S., Boran, S., 2018. Adsorption, bioaccumulation and kinetics parameters of the phytoremediation of cobalt from wastewater using *Elodea canadensis*. Bull. Environ. Contam. Toxicol. 100 (5), 733–739.

Mshandete, A.M., 2009. The anaerobic digestion of cattail weeds to produce methane using American cockroach gut microorganisms. ARPN J. Agric. Biol. Sci. 4 (1), 45–57.

Mustafa, H.M., Hayder, G., 2021a. Cultivation of *S. molesta* plants for phytoremediation of secondary treated domestic wastewater. Ain Shams Eng. J. 12 (3), 2585–2592.

Mustafa, H.M., Hayder, G., 2021b. Recent studies on applications of aquatic weed plants in phytoremediation of wastewater: a review article. Ain Shams Eng. J. 12 (1), 355–365.

Mustafa, H.M., Hayder, G., Solihin, M.I., Saeed, R.A., 2021, April. Applications of constructed wetlands and hydroponic systems in phytoremediation of wastewater. In: IOP Conference Series: Earth and Environmental Science. vol. 708(1). IOP Publishing, p. 012087.

Mustafa, H.M., Hayder, G., Mustapa, S.I., 2022. Circular economy framework for energy recovery in phytoremediation of domestic wastewater. Energies 15 (9), 3075.

Muthusaravanan, S., Sivarajasekar, N., Vivek, J.S., Paramasivan, T., Naushad, M., Prakashmaran, J., Gayathri, V., Al-Duaij, O.K., 2018. Phytoremediation of heavy metals: mechanisms, methods and enhancements. Environ. Chem. Lett. 16 (4), 1339–1359.

Nagarajan, P., Sruthy, K.S., Lal, V.P., Devan, V.P., Krishna, A., Lakshman, A., Vineetha, K.M., Madhavan, A., Nair, B.-G., Pal, S., 2017, December. Biological treatment of domestic wastewater by selected aquatic plants. In: 2017 International Conference on Technological Advancements in Power and Energy (TAP Energy). IEEE, pp. 1–4.

Nahar, K., Hoque, S., 2021. Phytoremediation to improve eutrophic ecosystem by the floating aquatic macrophyte, water lettuce (*Pistia stratiotes* L.) at lab scale. Egypt. J. Aquat. Res. 47 (2), 231–237.

Nahar, K., Sunny, S.A., 2020. Duckweed-based clean energy production dynamics (ethanol and biogas) and phytoremediation potential in Bangladesh. Model. Earth Syst. Environ. 6 (1), 1–11.

Nakbanpote, W., Meesungnoen, O., Prasad, M.N.V., 2016. Potential of ornamental plants for phytoremediation of heavy metals and income generation. In: Bioremediation and Bioeconomy. Elsevier, pp. 179–217.

Napaldet, J.T., Buot Jr., I.E., 2020. Absorption of lead and mercury in dominant aquatic macrophytes of Balili River and its implication to phytoremediation of water bodies. Trop. Life Sci. Res. 31 (2), 19.

Nash, D.A.H., Abdullah, S.R.S., Hasan, H.A., Idris, M., Othman, A.R., Al-Baldawi, I.A., Ismail, N.I., 2020. Utilisation of an aquatic plant (*Scirpus grossus*) for phytoremediation of real sago mill effluent. Environ. Technol. Innov. 19, 101033.

Nattrass, M., McGrew, N.R., Morrison, J.I., Baldwin, B.S., 2019. Phytoremediation of selenium-impacted water by aquatic macrophytes. J. Am. Soc. Mining Reclamat. 8 (1).

Newete, S.W., Byrne, M.J., 2016. The capacity of aquatic macrophytes for phytoremediation and their disposal with specific reference to water hyacinth. Environ. Sci. Pollut. Res. 23 (11), 10630–10643.

Ng, Y.S., Chan, D.J.C., 2017. Wastewater phytoremediation by *Salvinia molesta*. J. Water Process Eng. 15, 107–115.

Nguyen, T.Q., Sesin, V., Kisiala, A., Emory, R.N., 2021. Phytohormonal roles in plant responses to heavy metal stress: implications for using macrophytes in phytoremediation of aquatic ecosystems. Environ. Toxicol. Chem. 40 (1), 7–22.

Nita, V., Benini, L., Ciupagea, C., Kavalov, B., Pelletier, N., 2013. Bio-Economy and Sustainability: A Potential Contribution to the Bio-Economy Observatory. European Commission, Joint Research Centre, Institute for Environment and Sustainability.

Nivetha, C., Subraja, S., Sowmya, R., Induja, N.M., 2016. Water lettuce for removal of nitrogen and phosphate from sewage. J. Mech. Civil Eng. (IOSR-JMCE) 13 (2), 104–107.

Odoh, C.K., Zabbey, N., Sam, K., Eze, C.N., 2019. Status, progress and challenges of phytoremediation—an African scenario. J. Environ. Manag. 237, 365–378.

Okolie, J.A., Epelle, E.I., Tabat, M.E., Orivri, U., Amenaghawon, A.N., Okoye, P.U., Gunes, B., 2022. Waste biomass valorization for the production of biofuels and value-added products: a comprehensive review of thermochemical, biological and integrated processes. Process Saf. Environ. Prot. 159, 323–344.

Paisio, C.E., Fernandez, M., González, P.S., Talano, M.A., Medina, M.I., Agostini, E., 2018. Simultaneous phytoremediation of chromium and phenol by *Lemna minuta* Kunth: a promising biotechnological tool. Int. J. Environ. Sci. Technol. 15 (1), 37–48.

Pamintuan, K.R.S., Gonzales, A.J.S., Estefanio, B.M.M., Bartolo, B.L.S., 2018. Simultaneous phytoremediation of Ni^{2+} and bioelectricity generation in a plant-microbial fuel cell assembly using water hyacinth (*Eichhornia crassipes*). In: IOP Conference Series: Earth and Environmental Science. vol. 191(1). IOP Publishing, p. 012093.

Parakh, P.D., Nanda, S., Kozinski, J.A., 2020. Eco-friendly transformation of waste biomass to biofuels. Curr. Biochem. Eng. 6 (2), 120–134.

Patnaik, P., Abbasi, T., Abbasi, S.A., 2022. Salvinia (*Salvinia molesta*) and water Hyacinth (*Eichhornia crassipes*): two pernicious aquatic weeds with high potential in phytoremediation. Adv. Sustain. Dev., 243–260.

Pattanaik, L., Pattnaik, F., Saxena, D.K., Naik, S.N., 2019. Biofuels from agricultural wastes. In: Second and Third Generation of Feedstocks. Elsevier, pp. 103–142.

Petrov, D.S., Korotaeva, A.E., Pashkevich, M.A., Chukaeva, M.A., 2023. Assessment of heavy metal accumulation potential of aquatic plants for bioindication and bioremediation of aquatic environment. Environ. Monit. Assess. 195 (1), 122.

Pettit, T., Irga, P., Torpy, F., 2020. The evolution of botanical biofilters: developing practical phytoremediation of air pollution for the built environment. In: 1st International Conference on Climate Resilient Built Environment iCRBE. World Energy and Environment Technology Ltd-WEENTECH.

Pinho, H.J., Mateus, D.M., 2023. Bioenergy routes for valorizing constructed wetland vegetation: an overview. Ecol. Eng. 187, 106867.

Pirrera, G., Pluchino, A., 2017. Phytoremediation for ecological restauration and industrial ecology. In: International Symposium on Soil and Water Bioengineering in a Changing Climate Glasgow, 7th and 8th September, vol. 2017, p. 13.

PoÁ, F., Fellet, G., Fagnano, M., Fiorentino, N., Marchiol, L., 2019. Linking phytotechnologies to bioeconomy; varietal screening of high biomass and energy crops for phytoremediation of Cr and Cu contaminated soils. Ital. J. Agron. 14 (1), 43–49.

Pogrzeba, M., Krzyżak, J., Rusinowski, S., Werle, S., Hebner, A., Milandru, A., 2018. Case study on phytoremediation driven energy crop production using *Sida hermaphrodita*. Int. J. Phytoremediation 20 (12), 1194–1204.

Prasad, M.N.V., 2013. Heavy Metal Stress in Plants: From Biomolecules to Ecosystems, second ed. Springer Science & Business Media, p. 365.

Prasad, M.N.V., 2019. Prospects for manipulation of molecular mechanisms and transgenic approaches in aquatic macrophytes for remediation of toxic metals and metalloids in wastewaters. In: Transgenic Plant Technology for Remediation of Toxic Metals and Metalloids. Academic Press, pp. 395–428.

Prasad, R.K., Chatterjee, S., Mazumder, P.B., Gupta, S.K., Sharma, S., Vairale, M.G., Datta, S., Dwivedi, S.K., Gupta, D.-K., 2019. Bioethanol production from waste lignocelluloses: a review on microbial degradation potential. Chemosphere 231, 588–606.

Praveen, A., Pandey, V.C., 2020. Pteridophytes in phytoremediation. Environ. Geochem. Health 42 (8), 2399–2411.

Public Consultation Draft, 2007. Biofuels Technology Platform Report, European Biofuels Technology Platform www.biofuelstp.eu/…/070926_PublicConsultationDraftBFTPReport.pdf.

Quarshie, S.D.G., Xiao, X., Zhang, L., 2021. Enhanced phytoremediation of soil heavy metal pollution and commercial utilization of harvested plant biomass: a review. Water Air Soil Pollut. 232 (11), 1–28.

Rahman, R.A., Wintoko, J., Prasetya, A., 2022. Comparison of different phytoremediation strategies for acid mine drainage (AMD). In: IOP Conference Series: Earth and Environmental Science. vol. 963(1). IOP Publishing, p. 012040.

Rai, P.K., 2021. Heavy metals and arsenic phytoremediation potential of invasive alien wetland plants *Phragmites karka* and *Arundo donax*: Water-Energy-Food (WEF) Nexus linked sustainability implications. Bioresource Technol. Rep. 15, 100741.

Rai, P.K., Kumar, V., Lee, S., Raza, N., Kim, K.H., Ok, Y.S., Tsang, D.C., 2018. Nanoparticle-plant interaction: Implications in energy, environment, and agriculture. Environ. Int. 119, 1–19.

Rezania, S., Taib, S.M., Din, M.F.M., Dahalan, F.A., Kamyab, H., 2016. Comprehensive review on phytotechnology: heavy metals removal by diverse aquatic plants species from wastewater. J. Hazard. Mater. 318, 587–599.

Rezania, S., Park, J., Rupani, P.F., Darajeh, N., Xu, X., Shahrokhishahraki, R., 2019. Phytoremediation potential and control of *Phragmites australis* as a green phytomass: an overview. Environ. Sci. Pollut. Res. 26 (8), 7428–7441.

Rezooqi, A.M., Mouhamad, R.S., Jasim, K.A., 2021, March. The potential of *Azolla filiculoides* for in vitro phytoremediation of wastewater. In: Journal of Physics: Conference Series. vol. 1853(1). IOP Publishing, p. 012014.

Romero-Hernández, J.A., Amaya-Chávez, A., Balderas-Hernández, P., Roa-Morales, G., González-Rivas, N., Balderas-Plata, M.Á., 2017. Tolerance and hyperaccumulation of a mixture of heavy metals (Cu, Pb, Hg, and Zn) by four aquatic macrophytes. Int. J. Phytoremediation 19 (3), 239–245.

Roos, C.J., 2010. Clean Heat and Power Using Biomass Gasification for Industrial and Agricultural Projects. US Department of Energy, Clean Energy Application Centre, Northwest.

Royal Society, 2008. Sustainable Biofuels: Prospects and Challenges. The Royal Society, London. https://royalsociety.org/~/media/Royal_Society_Content/.../7980.pdf. Policy document (2008).

Rozman, U., Kokalj, A.J., Dolar, A., Drobne, D., Kalčíková, G., 2022. Long-term interactions between microplastics and floating macrophyte *Lemna minor*: the potential for phytoremediation of microplastics in the aquatic environment. Sci. Total Environ. 831, 154866.

Saba, B., Khan, M., Christy, A.D., Kjellerup, B.V., 2019. Microbial phyto-power systems—a sustainable integration of phytoremediation and microbial fuel cells. Bioelectrochemistry 127, 1–11.

Sadaka, S., 2009. Gasification, Producer Gas and Syngas. University of Arkansas, Little Rock. www.uaex.edu/publications/PDF/FSA-1051.pdf.

Saha, P., Shinde, O., Sarkar, S., 2017. Phytoremediation of industrial mines wastewater using water hyacinth. Int. J. Phytoremediation 19 (1), 87–96.

Sakakibara, M., 2016, June. Phytoremediation of toxic elements-polluted water and soils by aquatic macrophyte *Eleocharis acicularis*. In: AIP Conference Proceedings. vol. 1744(1). AIP Publishing LLC, p. 020038.

Saleh, H.M., Bayoumi, T.A., Mahmoud, H.H., Aglan, R.F., 2017. Uptake of cesium and cobalt radionuclides from simulated radioactive wastewater by *Ludwigia stolonifera* aquatic plant. Nucl. Eng. Des. 315, 194–199.

Saleh, H.M., Aglan, R.F., Mahmoud, H.H., 2020a. Qualification of corroborated real phytoremediated radioactive wastes under leaching and other weathering parameters. Prog. Nucl. Energy 119, 103178.

Saleh, H.M., Moussa, H.R., Mahmoud, H.H., El-Saied, F.A., Dawoud, M., Wahed, R.S.A., 2020b. Potential of the submerged plant *Myriophyllum spicatum* for treatment of aquatic environments contaminated with stable or radioactive cobalt and cesium. Prog. Nucl. Energy 118, 103147.

Sameena, P.P., Janeeshma, E., Sarath, N.G., Puthur, J.T., 2022. Phytoremediation and phycoremediation: a sustainable solution for wastewater treatment. In: Recent Trends in Wastewater Treatment. Springer, Cham, pp. 171–191.

Sarkheil, M., Safari, O., 2020. Phytoremediation of nutrients from water by aquatic floating Duckweed (*Lemna minor*) in rearing of African cichlid (*Labidochromis lividus*) fingerlings. Environ. Technol. Innov. 18, 100747.

Sayago, U.F.C., 2019. Design of a sustainable development process between phytoremediation and production of bioethanol with *Eichhornia crassipes*. Environ. Monit. Assess. 191 (4), 1–8.

Schmitt, N., Apfelbacher, A., Jäger, N., Daschner, R., Stenzel, F., Hornung, A., 2019. Thermo-chemical conversion of biomass and upgrading to biofuel: the thermo-catalytic reforming process—a review. Biofuels Bioprod. Biorefin. 13 (3), 822–837.

Seguil, Y.P., Garay, L.V., Cortez, C.M., Tueros, J.C., Hinostroza, S.C., Guerra, V.C., 2022. Systematic review of the efficiency of aquatic plants in the wastewater treatment. In: IOP Conference Series: Earth and Environmental Science. vol. 1009(1). IOP Publishing, p. 012004.

Shah, K., Pathak, L., 2019. Transgenic energy plants for phytoremediation of toxic metals and metalloids. In: Transgenic Plant Technology for Remediation of Toxic Metals and Metalloids. Academic Press, pp. 319–340.

Shah, A.B., Singh, R.P., Rai, U.N., 2020. Trends in phytomanagement of aquatic ecosystems and evaluation of factors affecting removal of inorganic pollutants from water bodies. In: Fresh Water Pollution Dynamics and Remediation. Springer, Singapore, pp. 247–262.

Sharma, K., Garg, V.K., 2023. Vermicomposting technology for organic waste management. In: Current Developments in Biotechnology and Bioengineering. Elsevier, pp. 29–56.

Sharma, K., Kumar, A., Sharma, R., Singh, N., 2022. Exploring integrated methodology for phytoremediation and biofuel production potential of *Eichhornia crassipes*. Indian J. Biochem. Biophys. 59 (3). http://nopr.niscpr.res.in/handle/123456789/59379.

Shen, L., Gao, Y., Xiao, J., 2008. Simulation of hydrogen production from biomass gasification in interconnected fluidized beds. Biomass Bioenergy 32, 120–127.

Singh, S., Karwadiya, J., Srivastava, S., Patra, P.K., Venugopalan, V.P., 2022. Potential of indigenous plant species for phytoremediation of arsenic contaminated water and soil. Ecol. Eng. 175, 106476.

Singha, K.T., Sebastian, A., Prasad, M.N.V., 2019. Iron plaque formation in the roots of *Pistia stratiotes* L.: importance in phytoremediation of cadmium. Int. J. Phytoremediation 21 (2), 120–128.

Singhal, V., Rai, J.P.N., 2003. Biogas production from water hyacinth and channel grass used for phytoremediation of industrial effluents. Bioresour. Technol. 86, 221–225.

Sood, A., Uniyal, P.L., Prasanna, R., Ahluwalia, A.S., 2012. Phytoremediation potential of aquatic macrophyte, Azolla. Ambio 41, 122–137. https://doi.org/10.1007/s13280-011-0159-z.

Sricoth, T., Meeinkuirt, W., Pichtel, J., Taeprayoon, P., Saengwilai, P., 2018. Synergistic phytoremediation of wastewater by two aquatic plants (*Typha angustifolia* and *Eichhornia crassipes*) and potential as biomass fuel. Environ. Sci. Pollut. Res. 25 (6), 5344–5358.

Sruthi, P., Sinisha, A.K., Ajayan, K.V., 2022. Wastewater remediation and biomass production: a hybrid technology to remove pollutants. In: Bioenergy Crops. CRC Press, pp. 152–163.

Stephen, J.L., Periyasamy, B., 2018. Innovative developments in biofuels production from organic waste materials: a review. Fuel 214, 623–633.

Tabassum-Abbasi, P.P., Abbasi, T., Abbasi, S.A., 2021. Salvinia (*Salvinia Molesta*) and water hyacinth (*Eichhornia crassipes*): two pernicious aquatic weeds with high potential in phytoremediation. J. Adv. Sustain. Dev., 243–260.

Tabinda, A.B., Irfan, R., Yasar, A., Iqbal, A., Mahmood, A., 2018. Phytoremediation potential of *Pistia stratiotes* and *Eichhornia crassipes* to remove chromium and copper. Environ. Technol.

Tabinda, A.B., Arif, R.A., Yasar, A., Baqir, M., Rasheed, R., Mahmood, A., Iqbal, A., 2019. Treatment of textile effluents with *Pistia stratiotes*, *Eichhornia crassipes* and *Oedogonium* sp. Int. J. Phytoremediation 21 (10), 939–943.

Tang, Z.E., Lim, S., Pang, Y.L., Ong, H.C., Lee, K.T., 2018. Synthesis of biomass as heterogeneous catalyst for application in biodiesel production: state of the art and fundamental review. Renew. Sust. Energ. Rev. 92, 235–253.

Tian, Y., Zhang, H., 2016. Producing biogas from agricultural residues generated during phytoremediation process: possibility, threshold, and challenges. Int. J. Green Energy 13 (15), 1556–1563.

Ting, W.H.T., Tan, I.A.W., Salleh, S.F., Wahab, N.A., 2018. Application of water hyacinth (*Eichhornia crassipes*) for phytoremediation of ammoniacal nitrogen: a review. J. Water Process Eng. 22, 239–249.

Tiwari, J., Kumar, S., Korstad, J., Bauddh, K., 2019. Ecorestoration of polluted aquatic ecosystems through rhizofiltration. In: Phytomanagement of Polluted Sites. Elsevier, pp. 179–201.

Tokmak, M., Bertiz, D., Özbey, D., Ilgaz, E.K.Ş.İ., Mehmetali, A.K., GÜNEŞ, A., 2019. Design proposal model for improving rivers with phytoremediation method. Uluslararası Peyzaj Mimarlığı Araştırmaları Dergisi (IJLAR) 3 (1), 31–38. E-ISSN: 2602-4322.

Topal, E.I.A., Topal, M., Erdal, Ö.B.E.K., 2017. Evaluation of recovery of aquatic plants used in wastewater treatment and discharged as waste. Int. J. Multidiscipl. Stud. Innov. Technol. 1 (1), 21–23.

Ubando, A.T., Felix, C.B., Chen, W.H., 2020. Biorefineries in circular bioeconomy: a comprehensive review. Bioresour. Technol. 299, 122585.

Udawat, P., Singh, J., 2020. Phytoremediation: a way towards sustainable agriculture. Int. J. Environ. Agric. Biotechnol. 5, 1167–1173.

Ulrikh, D.V., Timofeeva, S.S., Timofeev, S.S., 2018. Choice of sorbents for the pretreatment in the complex technology of phytoremediation. In: IOP Conference Series: Materials Science and Engineering. vol. 451(1). IOP Publishing, p. 012235.

Upadhyay, A.K., Singh, D.P., Singh, N.K., Pandey, V.C., Rai, U.N., 2019. Sustainable phytoremediation strategies for river water rejuvenation. In: Phytomanagement of Polluted Sites. Elsevier, pp. 301–311.

Usman, K., Al-Ghouti, M.A., Abu-Dieyeh, M.H., 2018. Phytoremediation: halophytes as promising heavy metal hyperaccumulators. Heavy Met. 27, 7378.

van der Drift, A., Boerrigter, H., 2005. Synthesis gas from biomass for fuels and chemicals. In: SYNBIOS Conference, Stockholm.

Varjani, S., Shah, A.V., Vyas, S., Srivastava, V.K., 2021. Processes and prospects on valorizing solid waste for the production of valuable products employing bio-routes: a systematic review. Chemosphere 282, 130954.

Verma, V.K., Singh, Y.P., Rai, J.P.N., 2007. Biogas production from plant biomass used for phytoremediation of industrial wastes. Bioresour. Technol. 98, 1664–1669.

Verma, R.K., Sankhla, M.S., Jadhav, E.B., Parihar, K., Awasthi, K.K., 2021. Phytoremediation of heavy metals extracted soil and aquatic environments: current advances as well as emerging trends. Biointerface Res. Appl. Chem. 12, 5486–5509.

Vieira, L.A., Alves, R.D., Menezes-Silva, P.E., Mendonça, M.A., Silva, M.L., Silva, M.C., Sousa, L.F., Loram-Lourenço, L., da Silva, A.A., Costa, A.C., Silva, F.G., 2021. Water contamination with atrazine: is nitric oxide able to improve *Pistia stratiotes* phytoremediation capacity? Environ. Pollut. 272, 115971.

Vladimirovna Chuparina, E., Igorevna Poletaeva, V., Vladimirovich Pastukhov, M., 2023. Metals Ti, Cr, Mn, Fe, Ni, Cu, Zn and Pb in aquatic plants of man-made water reservoir, Eastern Siberia, Russia: tracking of environment pollution. Pollution 9 (1), 23–38.

Wang, H., Zhang, H., Cai, G., 2011. An application of phytoremediation to river pollution remediation. Procedia Environ. Sci. 10, 1904–1907.

Wang, R., Wang, Y., Sun, S., Cai, C., Zhang, J., 2020. Discussing on "source-sink" landscape theory and phytoremediation for non-point source pollution control in China. Environ. Sci. Pollut. Res. 27 (36), 44797–44806.

Wei, I.T.A., Jamali, N.S., Ting, W.H.T., 2019. Phytoremediation of palm oil mill effluent (POME) using *Eichhornia crassipes*. J. Appl. Sci. Process Eng. 6 (1), 340–354.

Werle, S., Bisorca, D., Katelbach-Woźniak, A., Pogrzeba, M., Krzyżak, J., Ratman-Kłosińska, I., Burnete, D., 2017. Phytoremediation as an effective method to remove heavy metals from contaminated area—TG/FT-IR analysis results of the gasification of heavy metal contaminated energy crops. J. Energy Inst. 90 (3), 408–417.

Witters, N., Mendelsohn, R.O., Van Slycken, S., Weyens, N., Schreurs, E., Meers, E., Tack, F., Carleera, R., Vangronsveld, J., 2012. Phytoremediation, a sustainable remediation technology? Conclusions from a case study. I. Energy production and carbon dioxide abatement. Biomass Bioenergy 39, 454–469.

Xin, J., Ma, S., Li, Y., Zhao, C., Tian, R., 2020. *Pontederia cordata*, an ornamental aquatic macrophyte with great potential in phytoremediation of heavy-metal-contaminated wetlands. Ecotoxicol. Environ. Saf. 203, 111024.

Yadav, K.K., Gupta, N., Kumar, A., Reece, L.M., Singh, N., Rezania, S., Khan, S.A., 2018. Mechanistic understanding and holistic approach of phytoremediation: a review on application and future prospects. Ecol. Eng. 120, 274–298.

Yadav, P., Yadav, S., Singh, D., Giri, B.S., Mishra, P.K., 2022a. Barriers in biogas production from the organic fraction of municipal solid waste: a circular bioeconomy perspective. Bioresour. Technol. 362, 127671.

Yadav, R., Singh, S., Kumar, A., Singh, A.N., 2022b. Phytoremediation: a wonderful cost-effective tool. In: Cost Effective Technologies for Solid Waste and Wastewater Treatment. Elsevier, pp. 179–208.

Yadav, R., Singh, S., Kaur, A., Tokas, D., Kathi, S., Singh, A.N., 2023. Harnessing energy from animal waste: a win–win approach for India. In: Manure Technology and Sustainable Development. Singapore, Springer Nature Singapore, pp. 283–304.

Yadvika, S., Sreekrishnan, T.R., Kohli, S., Rana, V., 2004. Enhancement of biogas production from solid substrates using different techniques—a review. Bioresour. Technol. 95, 1–10.

Yan, Y., Pengmao, Y., Xu, X., Zhang, L., Wang, G., Jin, Q., Chen, L., 2020. Migration of antibiotic ciprofloxacin during phytoremediation of contaminated water and identification of transformation products. Aquat. Toxicol. 219, 105374.

Zhang, W., 2010. Automotive fuels from biomass via gasification. Fuel Process. Technol. 91, 866–876.

Zhao, L., Jiang, J., Chen, C., Zhan, S., Yang, J., Yang, S., 2017. Efficiency and mechanism of the phytoremediation of decabromodiphenyl ether-contaminated sediments by aquatic macrophyte *Scirpus validus*. Environ. Sci. Pollut. Res. 24 (14), 12949–12962.

Jatropha curcas L. cultivation on contaminated and marginal land—Exploring the potential for circular economy, environmental protection, and sustainability

Aliyu Ahmad Warra[a] and Majeti Narasimha Vara Prasad[b]

[a]Centre for Entrepreneurial Development, Federal University, Gusau, Nigeria [b]Department of Plant Sciences, School of Life Sciences, University of Hyderabad (An Institution of Eminence), Hyderabad, Telangana, India

1 Introduction

Jatropha curcas (Euphorbiaceae), a multi-purpose, drought-resistant perennial, is gaining considerable importance in recent past beyond biofuels. It is a tropical plant that can be grown in low to high rainfall areas either in the farms as a commercial crop or on the boundaries as a hedge to protect fields from grazing animals and to prevent erosion. Names used to describe the plant vary in region or country. It is most commonly known as "Physic nut." Its global distribution is shown in Figs. 1 and 2. In Mali it is known as "Pourghere." In the Ivory Coast it is known as "bagani." In Senegal it is known as "tabanani." In Tanzania it is known as "makaen/mmbono." In Nigeria it is known as "binidazugu/cinidazugu" and "lapa lapa" in Hausa and Yoruba languages, respectively. The seed contains 40%–50% viscous oil known as "curcas oil." *J. curcas* L. is a sub-tropical shrub that is unique because of the fruits, which have a high content of non-edible oil suitable for the preparation of cosmetics. Cosmetic potential, Soxhlet extraction, physicochemical analysis and cold process saponification of Nigerian *Jatropha curcas* L. seed oil, and the moisture-dependent physical properties of Jatropha seed were reported.

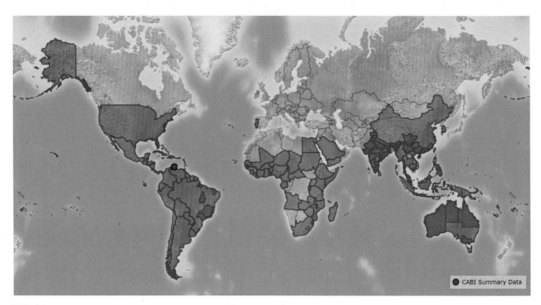

FIG. 1   Global distribution of *Jatropha curcas* (summary data). Invasive species compendium, Wallingford, UK: CAB International. https://www.cabi.org/isc. CABI 2022.

FIG. 2   Global distribution of *Jatropha curcas* introduced and native. In: Invasive species compendium, Wallingford, UK: CAB International. https://www.cabi.org/isc. CABI 2022.

Biotechnological research on *J. curcas* progressed significantly in the last two decades aiming at biodiesel production, environmental remediation, and material supply-chain boosting circular economy (Al-Khayri et al., 2022; Ara et al., 2021; Chandran et al., 2023; de Souza et al., 2023; Kumar and Choudhary, 2023; Kumar et al., 2023; Matsubara et al., 2023; Pandey, 2023; Purwanti et al., 2023; Sano et al., 2012; Shibata et al., 2017; Tamam et al., 2023; Vaikuntapu and Kumar, 2023). Sustainable aviation fuels from *J. curcas* showed higher GHG reduction potential (at least 86%) compared to other conversion processes such as hydrothermal liquefaction (at least 77%) and alcohol-to-jet conversion (at least 60%) (De Jong et al., 2017).

Some recent researches predominantly focused on *Jatropha* for cultivation and establishment of nurseries instead of investing in the entire *Jatropha* value chain, which turned out to be a challenge. Growing a productive *Jatropha* crop was much more complex than initially anticipated. There is also a lack of information on the basic agronomic properties of *J. curcas* cultivation on the marginal lands in the semi-arid. Evaluation of agronomic performance of identified elite strains of *J. curcas* in marginal lands would be of paramount importance for addressing gap areas in their agronomic properties and subsequently for harnessing their optimum economic potentials (Figs. 1 and 2).

1.1 Cultivation on contaminated and marginal land

J. curcas prefers well-drained sandy or gravely soils with good aeration. It doesn't withstand heavy clay soils and all soils with risk of even ephemeral water logging. Soil depth should be at least 45 cm. Investigations are in progress to assess the yield potential of the crop in different agroecological zones of Senegal. According to the preliminary results obtained on the field and the data mentioned in the literature, a minimum annual rainfall of 500 mm seems to be necessary to obtain a profitable seed yield in rainfed agriculture conditions (Warra and Prasad, 2016). Influence of land size on adoption of *J. curcas* in Yatta District, Kenya, was reported. It was found that farmers who had adopted *J. curcas* cultivation were significantly low at 15.4%. It was also found that the total land size owned was not a major factor influencing land size allocated to *Jatropha* cultivation since total land size owned accounted for only 28% variance of land size allocated to *Jatropha* cultivation and 78% of respondents were not willing to convert their pasture lands to *Jatropha* farms. Yatta District that covers 246,700 ha lies in the ASAL regions of the country therefore suitable for *J. curcas* cultivation. *Jatropha* requires extensive land for high yields to be realized and therefore marginal lands and wastelands are deemed appropriate to avoid competition with food. The study revealed that there was a strong negative correlation between the size of land a farmer owned and the size of land allocated for *J. curcas* cultivation implying the bigger the size of land a farmer has, the less likely for them to engage in *Jatropha* cultivation. Farmers with lesser land size were more likely to take up *J. curcas* cultivation because they mostly seek for a more economically rewarding enterprise to boost their low incomes as opposed to their counterparts with bigger land sizes. Size of land owned was not a key factor contributing to increase in adoption of *J. curcas*. The marginal land with the potential for planting energy plant Jatropha inclusive was identified for each 1 km^2 pixel across Asia. The results indicated that the areas with marginal land suitable for *Cassava, Pistacia chinensis,* and *J. curcas* L. were established to be 1.12 million, 2.41 million, and 0.237 million km^2, respectively. Shrubland, sparse forest, and grassland are the major classifications of exploitable land. In China, the total area of marginal land exploitable for the development of energy plant *J. curcas* L. inclusive on a large scale was about 43.75 million ha. As China has fairly limited cultivated land resources, it is widely acknowledged that the development of energy plants should not affect the food security and environment; therefore, the bioenergy development may mainly rely on the marginal land (Warra and Prasad, 2016). To avoid the food versus fuel debate, the use of "marginal" land for biofuel feedstock production (*Jatropha*) has emerged as a dominant narrative. But both the availability and suitability of "marginal" land for the commercial-level jatropha production is not well understood/examined, especially in Africa. In Ethiopia, report examines the process of land identification for jatropha investments, and the agronomic performance of large-scale jatropha plantation on so-called marginal land. When the marginal lands are exploited, the impact on the soil seems to be positive, depending on the used practices and the type of soil but its contribution to soil restoration might be obtained at the expense of biodiversity loss (Warra and Prasad, 2016). However, a report examined whether it would be possible to sustainably produce *J. curcas* L. seeds on the marginal land situated close to the Senegal River, a 6-ha pilot plantation was cultivated under drip irrigation between September 2007 and November 2011, close to the village of Bokhol (Lat. 16° 31′N, Long. 15°23′W). A series of tests were conducted on this plot to identify the best cultivation methods for the area (date, density and method of planting, appropriate type of pruning, fertilizers to be applied, irrigation method, etc.). The average yields obtained at this site, after 4 years of cultivation (less than 500 kg/ha of dry seed) using the best-known production techniques, are significantly lower than anticipated, compared to the available figures for the irrigated cultivation of *Jatropha* in other parts of the world. The main causes of this failure are the plant's limited useful vegetation period of 6 months per year, instead of twelve, and the scale of attacks by a soilborne vascular disease, which destroyed over 60% of the plantation within 4 years. It was also reported that *Jatropha* could be easily integrated in the micro-catchments for water harvesting already used by most farmers (Warra and Prasad, 2016).

1.2 Propagation and plantation establishment

Direct seeding can give good results when it is carried out with good-quality seeds at the beginning of the rainy season in a place where the rains last longer than 4 months. The sowing of three seeds per hole is recommended with an early refilling of the missing holes. In case of direct seeding, regular weeding operations are compulsory to avoid the complete disappearance of *J. curcas* plantlets due to the heavy concurrence of the weeds during the growing season. Inter-cropping of *J. curcas* with annual crops (peanut, pearl millet, and okra) helps achieving this goal. An efficient solution to succeed the establishment of a *J. curcas* plantation at a low cost is to pre-cultivate seedling during 2 months (till the plantlets reach the fifth true leave stage) in seed beds under nursery conditions and to extract and transplant bare root seedlings in the field. The best results for nursery bed preparation were obtained by mixing in equal volumes of sand and local soil found under trees (Warra and Prasad, 2016) (Fig. 3A–G).

FIG. 3   (A) High-yielding stem cuttings. (B) Sapling with fresh flush of leaves. (C) Plantation in constrained land. (D–G) *Jatropha curcas* plantation on constrained land.

1.3 Jatropha performance in semi-arid zone

Generally, establishing *Jatropha* on natural areas has caused a significant loss of carbon stored in the biomass and an initial loss of soil carbon (observed in 3–4 years old plantations; soil carbon stock might, however, increase again in the longer term). If relatively low carbon stock areas are transformed into jatropha plantations, the carbon stock generally increases. Nevertheless, the potential for carbon sequestration (the uptake and storage of carbon-containing substances, in particular, carbon dioxide) is low due to limited biomass build-up caused by continuous pruning and short rotation length. Conversion of Miombo forest in East Africa led to a carbon debt (the amount of biomass and soil carbon released due to land conversion) of more than 30 years in an attempt to show how greenhouse gas (GHG) emissions from land use change associated with the introduction of large-scale *J. curcas* cultivation on Miombo Woodland, using data from extant forestry and ecology studies (Warra and Prasad, 2016). A study indicates that moisture stress is the key factor in the failure of many large-scale jatropha plantations in Ethiopia. Jatropha performance in semi-arid zone has been satisfactory (Abrar et al., 2020). *J. curcas* yield is shown in Fig. 4. India's biofuel and biomass economies highlighting *J. curcas* cultivation on wasteland and its energy flow was analyzed by Baka and Bailis (2014) (Fig. 5).

FIG. 4   (A–E) *J. curcas* yielding high-quality fruits; (F and G) *J. curcas* seeds. High-quality seed oil supports cosmetic and soap industry in addition to production of biodiesel.

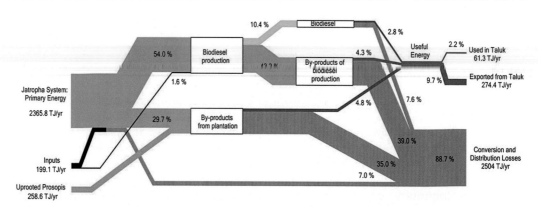

FIG. 5   *Jatropha* energy flow diagram. India's biofuel and biomass economies (Baka and Bailis, 2014) *After Baka, J., Bailis, R., 2014. Wasteland energy-scapes: a comparative energy flow analysis of India's biofuel and biomass economies. Ecol. Econ. 108, 8–17, with permission from Elsevier*

2 Value chain for cultivation and economic growth

Jatropha biodiesel attained global hype to local opportunity (Alherbawi et al., 2021). *Jatropha* value chains can potentially enable more than 40% savings in GHG emissions relative to a fossil reference. Climate change mitigation can only be achieved, however, if

Jatropha is cultivated on land with initially low carbon stocks and if no trees are cut. Circular economy proposal for *Jatropha* model animal farm and other economic products was suggested by Peralta et al. (2022) (Figs. 6 and 7). The carbon sequestration potential of *Jatropha* is limited due to its low growth rate and continuous pruning. It is only possible to achieve a relatively low carbon stock, which is comparable to fallow land. Consequently, cultivating *Jatropha* solely for carbon payments is inadvisable. The cultivation of *Jatropha* as a living fence is popular in Africa and also for land use transformation (Warra and Prasad, 2016).

Jatropha attracted huge interest—it was touted as a wonder crop that could generate biodiesel oil on "marginal lands" in semi-arid areas. Its promise appeared especially great in East Africa, where many projects were launched to grow *Jatropha* in plantations or within contract growing schemes. Today, however, *Jatropha*'s value in East Africa appears to lie primarily in its multi-purpose use by small-scale farmers (Warra and Prasad, 2016). Assessment of the opportunities and challenges of scaling-up *J. curcas* for environmental management and enhanced community livelihoods in Mtito Andei in Kenya was reported (Warra and Prasad, 2016). Rural development obtained through the setup of decentralized *Jatropha* production and marketing chains should be socially, economically, and environmentally more sustainable than the benefits generated by large-scale estates even if job creation in those centralized systems comply with national and international labor standards (Warra and Prasad, 2016). Indonesia focuses *Jatropha* research on four fields including plant breeding, development of cultivation techniques, processing technology, and

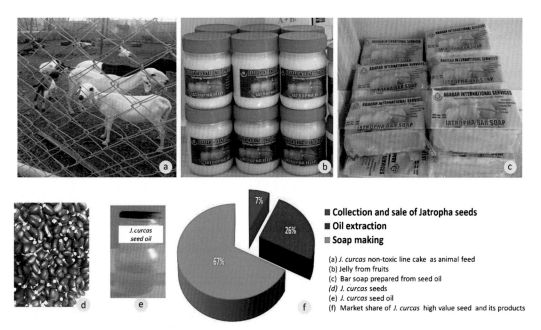

FIG. 6 Circular economy proposal for Jatropha model animal farm (Peralta et al., 2022).

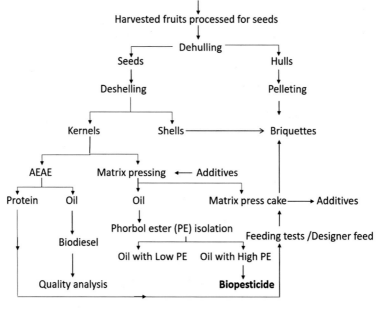

J. curcas cultivation on contaminated and marginal land - Biorefinery and bioeconomy

AEAE = Aqueous enzyme-assisted extraction　　　　　PE = Phorbol ester

FIG. 7　*J. curcas* biorefinery and bioeconomy. Processing of harvested fruits and seeds for biodiesel and biopesticides.

by-product utilization. Researches done by Indonesian researchers in plant breeding and cultivation have resulted in the improvement of jatropha productivity by 60% (from 5 to 8 tons/ha/year) through the improved population method.

2.1 Potential for remediation

Environmentally, *J. curcas* has great potential for soil enrichment, and can replace synthetic fertilizers as the leaves and branches are used as manure for coconut trees. It contributes to carbon sequestration thereby aiding in the mitigation of climate change. It also has the potential of retaining marginal and degraded soil by re-anchoring the soil with substantial roof and reducing the possibility of soil erosion. The study also established that *J. curcas* has the potential of remediating heavy metal- and hydrocarbon-contaminated soils (Warra and Prasad, 2016).

In Ethiopia, approximately 90% of the population lives in areas marked by land degradation and reduced agricultural productivity. A case study in Ethiopia showed that planting jatropha can be an effective prevention and mitigation measure against soil erosion (Warra and Prasad, 2016). The buildup of biomass and collection of seeds to secure local energy needs also allow for GHG savings. Thus, jatropha has the potential to alleviate soil degradation, increase carbon stocks, and contribute to energy security (Warra and Prasad, 2016).

Jatropha plantation is useful as hedges and this makes it a good choice for rehabilitating gullies and for planting along erosion-prone slopes. To stop a gully from expanding, it is necessary to slow the flow of water down the gully during rainstorms. The usual way to do this is to build stone walls across the floor of the gully at intervals. These check dams trap sediment behind them, which gradually builds up to form a series of steps. The pockets of soil behind the dams are level and relatively fertile, and because they are on a natural watercourse, they remain reasonably moist even during the dry season. Many farmers find them a good place to grow crops. The same features that make *Jatropha* useful as hedges make it a good choice for rehabilitating gullies and for planting along erosion-prone slopes.

2.2 Role of products from *J. curcas* L. in rural poverty alleviation

Recent investigations carried out all over the world have demonstrated that *J. curcas* could contribute drastically to the improvement of the living conditions of rural populations in the least developed countries of our planet. *J. curcas* is a multiple-function hardy shrub that can be used for medicinal purpose, to prevent and/or control erosion, to reclaim land, to produce pesticides, to contain or exclude farm animals when grown as living fence, and be planted as a commercial crop. The seed contains a high rate of non-edible oil that can be used for soapmaking in the cosmetic industry (Warra and Prasad, 2016) (Figs. 8 and 9).

2.3 Oil extraction and cosmetic potential

Extraction of oil from *J. curcas* has aroused interest worldwide, the extracted oil is potentially regarded as a source for the local production in developing countries for use in cosmetic preparations. Oils were analyzed through the formation of their corresponding methyl esters and their yield and composition were found to be close to that of the oil of *J. curcas* reported a high oil content (64.4%), dominant triacylglycerol lipid species (88.2%), and an appreciable percentage (47.3%) of linoleic acid after extraction from samples of *J. curcas* seeds obtained from markets in five different towns in Nigeria. Table 1 shows some physicochemical properties of the Malaysian *J. curcas* seed oil compared to the Nigerian *J. curcas* seed oil (Warra and Prasad, 2016). The parameters measured and the values obtained are in favor of utilization of the *J. curcas* seed oil for cosmetic production.

2.4 Prospects for cosmetic products

Even though the current boom for *Jatropha* production is based mainly on the incentive of producing biofuel, however, the possible range of products that can be derived from jatropha is much broader. Hence, based on the expert estimates from 39 countries, the possible use of jatropha-based products (i.e., the utilization of seed oil) is in the cosmetic and soap industries

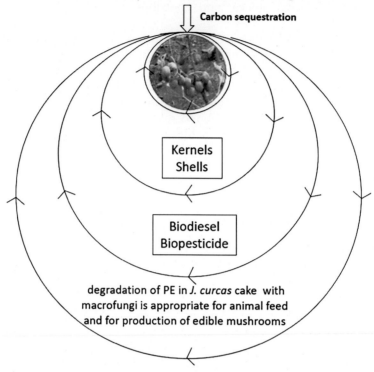

J. curcas cultivation on contaminated and marginal land

Carbon sequestration

Kernels
Shells

Biodiesel
Biopesticide

degradation of PE in *J. curcas* cake with
macrofungi is appropriate for animal feed
and for production of edible mushrooms

FIG. 8 Scope for strengthening circular bioeconomy and biorefinery including degradation of toxic phorbol esters in *J. curcas* cake using macrofungi for the production of animal feed and edible mushrooms.

J. curcas cake (JCC), unfit for animal feed due to toxic phorbol esters (PE)

Biological detoxification using macrofungi offers promise as animal feed

Aurantioporus pulcherrimus, Ganoderma lucidum, Agaricus sp., Agaricus fuscofibrillosus, Agaricus mediofuscus, Ascopolyporus sp., Panaeolus antillarum, Lentinus strigellus, Amylosporus sp. **and Pleurotus pulmonarius**

The above selected macrofungi degraded PE in solid culture over a 30-day period at 28 °C.

P. pulmonarius displayed the highest rate of PE degradation, reaching 97% efficiency. Supplemented with an additional lignocellulosic source, with up to a 99.5% degradation of the toxic compound achieved.

P. pulmonarius was efficient in degradation of PE in JCC, with the residue appropriate for animal feed and for production of edible mushrooms, and as a source of enzymes of biotechnological interest.

FIG. 9 *J. curcas*—model for biorefinery applications (Gomes et al., 2022).

TABLE 1 Physicochemical characteristic of Nigerian *J. curcas* L. seed oil in comparison to Malaysian product (Warra and Prasad, 2016).

Parameter	Increase↑/decrease↓	Actual value for Nigerian product
Iodine value (mg/g)	−12.72%↓	105.20 ± 0.70
Acid value (mg KOH/g)	+40%↑	3.50 ± 0.10
Saponification value (mg/g)	−2.37%↓	198.85 ± 1.40

(Warra and Prasad, 2016). Kakute Ltd., one of the Tanzanian organizations promoting jatropha for oil production, erosion control, and soapmaking, conducted an evaluation in 2003 of the profitability of jatropha-related activities. They found soapmaking to be more profitable than oil extraction, which, in turn, was more profitable than seed collection or production. Soap produced from jatropha is sold as a medical soap, effective in treating skin ailments. It is noted that jatropha soap is sold in dispensaries at a higher price than other soaps on the market. The case of the GTZ project in Mali will be used to demonstrate the economic benefits of physic nut cultivation. The *Jatropha* system is based on existing hedges that were used to fence in fields and to control erosion. The project promotes this system by creating a market for physic nut products. The oil is a raw material for soap production that generates income to local women producers (Warra and Prasad, 2016). In Senegal, a project was carried out by ATI (now Enterprise works), an American NGO, in the region of Ties, planted *Jatropha* hedges and extracted *Jatropha* oil with ram press. The oil was used to run diesel engines (for flour mills) and to make soap. Jatropha soap is perceived in Malawi as a "medicated soap." The soap from unrefined oil is considered, like neem-based soap, to be an effective, gentle anti-scabies wash. In former times, Portugal imported *Jatropha* seeds from the Cape Verde islands to produce soap. In India, Nepal, and Zimbabwe, the prize of tallow or the prize of *Jatropha* and other plant oils is at least 2.5 times the selling price of diesel. Obviously, selling *Jatropha* oil for soap making is far more profitable in these countries than using it as a diesel or kerosene substitute (Warra and Prasad, 2016).

2.5 Recommendations for further research

Participatory data acquisition is needed to build up a consolidated and reliable database that includes documented metadata along the entire land use system. The focus should be primarily on yield data, carbon stock data, and non-CO_2 GHG emissions, which typically show large variations and are highly context specific. It would also be useful to consolidate knowledge related to land use dynamics across both croplands and their surrounding areas. In addition, most studies about the potential for climate change mitigation focus on established *Jatropha* value chains and therefore updated models are required to understand the impact of future *Jatropha* systems. To develop *Jatropha* as a commercially profitable commodity, research collaboration with institutions in countries concerned with upstream to downstream *Jatropha* development is necessary. Genetic diversity assessment for optimizing the yield, seed source variation for better growth performance and post-harvest management, application of plant growth and health-promoting rhizobacteria, progeny evaluation, and influence of soil ameliorants for growth promotion are need to be investigated.

J. curcas is a suitable candidate species for the phytoremediation of highly contaminated mining soils. Salient findings and applications are enumerated beneath (Warra and Prasad, 2016) and the references there in Sánchez-Borrego et al. (2021):

- Suitable for lindane rhizoremediation that has advantages for the environment and human health
- Ethiopia's *Jatropha* Potential on Marginal Land
- *J. curcas*'s role in reducing climate change
- *J. curcas*, a crop with multiple uses for preserving soil and water
- Suitable for planting on contaminated and marginal land
- An oil plant with many advantages
- Known to sequester soil carbon in plantation on diverse ecological and edaphic conditions
- Has a big impact on the management of the environment and sustainable livelihoods in Kibwezi, Kenya.
- A crop suitable for phytoremediate coal fly ash.
- *Jatropha* and *Pongamia* rainfed plantations on Indian wastelands provide better livelihoods and environmental protection.
- *J. curcas* is amenable for plantation on industrial waste-amended and heavy metal-contaminated soil.

J. curcas for bioremediation and bioeconomy, augmenting circular economy is detailed in Table 2. The notable examples are listed below (Warra and Prasad, 2016 and the references there in) Importance of:

- Proteomic analysis of oil bodies in mature *J. curcas* seeds with different lipid
- Bioconversion of *J. curcas* seed cake to hydrogen by a strain of *Enterobacter aerogenes*
- Production of esters from *Jatropha* oil using different short-chain alcohols
- Biodiesel production and de-oiled seed cake nutritional values of a Mexican edible *J. curcas*
- Jatrophidin I, a cyclic peptide from Brazilian *J. curcas* L.: Isolation, characterization, conformational studies and biological activity
- Wasteland energy-scapes:
- Chemical compounds of a native *J. curcas* seed oil from Mexico and their antifungal effect on *Fusarium oxysporum* sp.
- *J. curcas* oil hydroconversion over hydrodesulfurization catalysts for biofuel production
- Extraction of oil from *J. curcas* seeds by subcritical fluid extraction
- Production of biodiesel from Vietnamese *J. curcas* oil by a co-solvent
- *J. curcas* leaves as a mineral source for low sodium diets
- Performance of *J. curcas*. in semi-arid zone
- *J. curcas* productive chain: from sowing to biodiesel and by products.
- *J. curcas* as a multi-purpose crop for sustainable energy
- Life cycle assessment of *Jatropha* biodiesel as transportation fuel in rural India.
- *Jatropha*: from global hype to local opportunity.
- Life cycle assessment of hydrotreated vegetable oil from rape, oil palm and Jatropha.
- Biodiesel from *Jatropha*

TABLE 2 Importance of *J. curcas* for bioremediation and bioeconomy, augmenting circular economy.

Salient research finding	Author(s)	Year
J. curcas biodiesel production's erratic economic and environmental impacts in Nepal	Babahmad, et al.	2018
Synthesis, characterization, and reaction kinetics of a waste Ox bone-based heterogeneous catalyst for the production of biodiesel from *J. curcas* oil.	Kamel et al.	2018
An update on *J. curcas* L. biological development: new knowledge and difficulties.	Mazumdar et al.	2018
Biomass waste from *J. curcas* L. and its application.	Primandari et al.	2018
J. curcas use in phytoremediation of heavily polluted mining soils and generation of catalytic carbons from harvested biomass.	Álvarez-Mateos et al.	2019
An ideal control strategy using a mathematical model for integrated pesticide pest management in *J. curcas*.	Chowdhury et al.	2019
The genome sequence of the non-edible biofuel plant *J. curcas* L. offers a resource for enhancing seed-related properties.	Ha et al.	2019
Utilizing oil from Karanja (*Pongamia pinnata*) and Jatropha (*J. curcas*) to make biodiesel.	Jain	2019
Optimal and mechanism-based Jatropha (*J. curcas* Linnaeus) seed usage for biodiesel production.	Kavitha et al.	2019
J. curcas L. seed dynamics of drying modeled mathematically.	Keneni et al.	2019
Effect of oxidation on the lubricating qualities of biodiesel made from *J. curcas*.	Liu et al.	2019
J. curcas L. seed transesterification in situ with homogeneous and heterogeneous basic catalysts.	Martínez et al.	2019
Ecophysiological reactions of different soil types to simulated acid rain in *J. curcas* L. seedlings.	Shu et al.	2019
Metal accumulation by *Jatropha curcas* L. grown on heavy metal-contaminated soil.	García Martín et al.	2020
Assessing the role of physiological, ionic, and growth factors in *J. curcas* ability to withstand salinity and drought stress.	Abrar et al.	2020
Heterogeneous catalysts play a crucial role in the manufacture of biodiesel from *J. curcas* oil.	Aderibigbe et al.	2020
J. curcas oil is pyrolytically deoxygenated to produce renewable fuel using a variety of multi-wall carbon nanotube-based catalysts.	Asikin-Mijan, et al.	2020

Continued

II. Biomass energy and biofuel from contaminated substrates

TABLE 2   Importance of *J. curcas* for bioremediation and bioeconomy, augmenting circular economy—cont'd

Salient research finding	Author(s)	Year
Stochastic economic and environmental footprints of biodiesel production from *J. curcas* Linnaeus in the different federal states of Nepal.	Baral et al.	2020
A brief study of the potential of *J. curcas* L. as a biodiesel feedstock in Malaysia.	Che Hamzah et al.	2020
J. curcas seed quality categorization using radiographic imaging and machine learning.	de Medeiros et al.	2020
J. curcas L. Clonal Propagation and Reproductive Biology.	Dhillon et al.	2020
Adult plants of *J. curcas* L. cultivated on soil contaminated with heavy metals accumulate metal.	García-Martín et al.	2020
Hyperspectral imaging is used to identify copper in the stems and roots of *J. curcas* L.	García-Martín et al.	2020
J. curcas leaf extract used in the green manufacture of copper nanoparticles.	Ghosh et al.	2020
Variation of *J. curcas* seed oil content and fatty acid composition with fruit maturity stage.	Jonas et al.	2020
J. curcas seed oil content and fatty acid composition vary depending on the stage of fruit maturity.	Kaushik et al.	2020
Non-oil materials oxidizing solid fuel from the biomass of *J. curcas* L.	Kethobile et al.	2020
Investigations into the viability of biodiesel made from *J. curcas* oil as a sustainable dielectric fluid in the EDM process.	Khan et al.	2020
A strategy for using *J. curcas* by-products as a source of energy in the agricultural economy. Part A of Energy Sources: Recovery, Use, and Environmental Impacts.	Piloto-Rodríguez et al.	2020
Combinations of *Moringa oleifera* and *J. curcas* methyl ester fuel in a single-cylinder diesel engine are evaluated in terms of cost, performance, and emissions.	Rajak et al.	2020
J. curcas' rhizobacterial population is connected to the PAH-degrading bacterial consortium that biodegrades pyrene.	Singha and Pandey	2021
Possibilities and prospects for *J. curcas* (L.) genetic improvement.	Sujatha	2021
A review on green remediation techniques for hydrocarbons and heavy metals contaminated soil.	Abdelhafeez et al.	2022
Jatropha curcas L. and *Pongamia pinnata* L. exhibited differential growth and bioaccumulation pattern irrigated with wastewater.	Abid et al.	2021

TABLE 2 Importance of *J. curcas* for bioremediation and bioeconomy, augmenting circular economy—cont'd

Salient research finding	Author(s)	Year
J. curcas entire fruit utilization at its most sustainable: a focus on the connections between food, water, and energy.	Alherbawi et al.	2021
A critical review on the development of wax inhibiting agent in facilitating remediation process of contaminated groundwater.	Alpandi et al.	2021
Hexachlorocyclohexane toxicity in water bodies of Pakistan: challenges and possible reclamation technologies.	Ara et al.	2021
Using *Heteropanax fragrans* (Kesseru) as a basic heterogeneous catalyst allows for efficient biodiesel synthesis from *J. curcas* oil.	Basumatary et al.	2021
J. curcas L. seed quality is characterized using multi-spectral and X-ray imaging.	Bianchini et al.	2021
Unique method of analyzing the health of the seeds of *J. curcas* using multi-spectral and resonance imaging methods.	da Silva et al.	2021
Analyzing the physicochemical properties of dual biodiesel blends made from Mahua (*Madhuca indica*) and Jatropha (*J. curcas*) and diesel.	Dugala et al.	2021
A rigorous analysis of the variables influencing *J. curcas'* capacity for producing sustainable biodiesel.	Ewunie et al.	2021b
Analysis of the yield and physicochemical characteristics of the Ethiopian cultivar *J. curcas* in relation to its suitability and potential for biodiesel production.	Ewunie et al.	2021a,b
J. curcas L. oil can be used to produce cleaner, more sustainable biodiesel with the help of the heterogeneous catalyst $NaFeTiO_4/Fe_2O_3$-$FeTiO_3$.	Gutiérrez-López et al.	2021
A thorough evaluation of hetero-homogeneous catalysts for the synthesis of fatty acid methyl esters from non-edible *J. curcas* oil.	Khan et al.	2021
Evaluation of the environmental sustainability of Pakistani biodiesel production from *J. curcas* L. seed oil.	Khanam et al.	2021
For the manufacture of biodiesel from *J. curcas* oil, a mesoporous polysulfonic acid-formaldehyde polymeric catalyst.	Laskar et al.	2021
A review on environmental and socioeconomic perspectives of three promising biofuel plants *Jatropha curcas*, *Pongamia pinnata* and *Mesua ferrea*.	Mitra et al.	2021
An examination of the social and environmental aspects of the three promising biofuel plants *Mesua ferrea*, *Pongamia pinnata*, and *J. curcas*.	Mitra et al.	2021

Continued

TABLE 2   Importance of *J. curcas* for bioremediation and bioeconomy, augmenting circular economy—cont'd

Salient research finding	Author(s)	Year
Soil fertility enrichment potential of *Jatropha curcas* for sustainable agricultural production: a case study of Birnin Kebbi, Nigeria.	Mohammed et al.	2021
Evaluation of the best location for the production of biodiesel in Iran using *J. curcas*.	Najafi et al.	2021
Using plants to remediate or manage metal-polluted soils: an overview on the current state of phytotechnologies.	Nascimento et al.	2021
Phytoremediation of petroleum hydrocarbon-contaminated soils with two plant species: *Jatropha curcas* and *Vetiveria zizanioides* at Ghana Manganese Company Ltd.	Nero	2021
J. curcas L. cultivation as a potential fuel for second-generation biodiesel.	Neupane et al.	2021
Oil from *J. curcas* that has been chemically altered for use in biolubricants.	Nor et al.	2021
Changes in physiology, metabolism, and stomata of *J. curcas* in response to salt stress.	Pompelli et al.	2021
Phytoremediation of heavy metals contaminated soil in artisanal gold mining at Selogiri, Wonogiri District, Central Java, Indonesia.	Pranoto and Budianta	2020
Phytoremediation of heavy metals contaminated soil in Artisanal Gold Mining at Selogiri, Wonogiri District, Central Java, Indonesia.	Pranoto and Budianta	2020
Effect of ultrasonic waves on degradation of phenol and *para*-nitrophenol by iron nanoparticles synthesized from Jatropha leaf extract.	Rawat et al.	2021
Using a batch reactor to hydrotreat mixes of gas oil and *J. curcas* L. to create hybrid fuels.	Sánchez-Anaya et al.	2021
A thorough analysis of the *J. curcas* feedstock used to produce second-generation biodiesel, including its physical and chemical qualities, manufacturing process, performance, and emissions characteristics.	Singh et al.	2021
Rhizobacterial community of *Jatropha curcas* associated with pyrene biodegradation by consortium of PAH-degrading bacteria.	Singha and Pandey	2020
Arsenic remediation through sustainable phytoremediation approaches.	Srivastava et al.	2021
Biodiesel made from *J. curcas* is a profitable product for Pakistan's energy industry.	Yaqoob et al.	2021

II. Biomass energy and biofuel from contaminated substrates

TABLE 2   Importance of *J. curcas* for bioremediation and bioeconomy, augmenting circular economy—cont'd

Salient research finding	Author(s)	Year
Mechanical extraction of *J. curcas* L. oil for the creation of biofuels.	Yate et al.	2020
In order to produce biodiesel from *J. curcas* oil, calcium oxide is supported on coal fly ash (CaO/CFA).	Yusuff et al.	2021
Iron toxicity in plants: Impacts and remediation.	Zahra et al.	2021
An effective nano-biocatalyst for improved biodiesel generation from *J. curcas* oil is lipase-PDA-TiO$_2$ NPs.	Zulfiqar et al.	2021
Phytobial remediation by bacteria and fungi.	Gajić et al.	2022
Waste Ox bone based heterogeneous catalyst synthesis, characterization, utilization and reaction kinetics of biodiesel generation from *J. curcas* oil.	Jayakumar et al.	2022
Circular economy proposal for the non-toxic physic nut crop (*J. curcas*) in Mexico.	Peralta et al.	2022

References

Abdelhafeez, I., El-Tohamy, S., Abdel-Raheem, S., El-Dars, F., 2022. A review on green remediation techniques for hydrocarbons and heavy metals contaminated soil. Curr. Chem. Lett. 11 (1), 43–62.

Abid, R.A.F.I.A., Mahmood, S., Zahra, S., Ghaffar, S., Malik, M., Noreen, S.I.B.G.H.A., 2021. *Jatropha curcas* L. and *Pongamia pinnata* L. exhibited differential growth and bioaccumulation pattern irrigated with wastewater. Sains Malays. 50 (3), 559–570.

Abrar, M.M., Saqib, M., Abbas, G., Atiq-ur-Rahman, M., Mustafa, A., Shah, S.A.A., Mehmood, K., Maitlo, A.A., Sun, N., Xu, M., 2020. Evaluating the contribution of growth, physiological, and ionic components towards salinity and drought stress tolerance in *Jatropha curcas*. Plants 9 (11), 1574.

Aderibigbe, F.A., Mustapha, S.I., Adewoye, T.L., Mohammed, I.A., Gbadegesin, A.B., Niyi, F.E., Olowu, O.I., Soretire, A.G., Saka, H.B., 2020. Qualitative role of heterogeneous catalysts in biodiesel production from *Jatropha curcas* oil. Biofuel Res. J. 7 (2), 1159–1169.

Alherbawi, M., AlNouss, A., McKay, G., Al-Ansari, T., 2021. Optimum sustainable utilisation of the whole fruit of *Jatropha curcas*: an energy, water and food nexus approach. Renew. Sust. Energ. Rev. 137, 110605.

Al-Khayri, J.M., Sudheer, W.N., Preetha, T.R., Nagella, P., Rezk, A.A., Shehata, W.F., 2022. Biotechnological research progress in *Jatropha*, a biodiesel-yielding plant. Plants 11 (10), 1292.

Alpandi, A.H., Husin, H., Sidek, A., 2021. A critical review on the development of wax inhibiting agent in facilitating remediation process of contaminated groundwater. Environ. Sci. Pollut. Res., 1–11.

Álvarez-Mateos, P., Alés-Álvarez, F.J., García-Martín, J.F., 2019. Phytoremediation of highly contaminated mining soils by *Jatropha curcas* L. and production of catalytic carbons from the generated biomass. J. Environ. Manag. 231, 886–895.

Ara, T., Nisa, W.U., Anjum, M., Riaz, L., Saleem, A.R., Hayat, M.T., 2021. Hexachlorocyclohexane toxicity in water bodies of Pakistan: challenges and possible reclamation technologies. Water Sci. Technol. 83 (10), 2345–2362.

Asikin-Mijan, N., Rosman, N.A., AbdulKareem-Alsultan, G., Mastuli, M.S., Lee, H.V., Nabihah-Fauzi, N., Lokman, I.M., Alharthi, F.A., Alghamdi, A.A., Aisyahi, A.A., Taufiq-Yap, Y.H., 2020. Production of renewable diesel from *Jatropha curcas* oil via pyrolytic-deoxygenation over various multi-wall carbon nanotube-based catalysts. Process. Saf. Environ. Prot. 142, 336–349.

Babahmad, R.A., Aghraz, A., Boutafda, A., Papazoglou, E.G., Tarantilis, P.A., Kanakis, C., Hafidi, M., Ouhdouch, Y., Outzourhit, A., Ouhammou, A., 2018. Chemical composition of essential oil of *Jatropha curcas* L. leaves and its antioxidant and antimicrobial activities. Ind. Crop. Prod. 121, 405–410.

Baka, J., Bailis, R., 2014. Wasteland energy-scapes: a comparative energy flow analysis of India's biofuel and biomass economies. Ecol. Econ. 108, 8–17.

Baral, N.R., Neupane, P., Ale, B.B., Quiroz-Arita, C., Manandhar, S., Bradley, T.H., 2020. Stochastic economic and environmental footprints of biodiesel production from *Jatropha curcas* Linnaeus in the different federal states of Nepal. Renew. Sust. Energ. Rev. 120, 109619.

Basumatary, S., Nath, B., Das, B., Kalita, P., Basumatary, B., 2021. Utilization of renewable and sustainable basic heterogeneous catalyst from *Heteropanax fragrans* (Kesseru) for effective synthesis of biodiesel from *Jatropha curcas* oil. Fuel 286, 119357.

Bianchini, V.D.J.M., Mascarin, G.M., Silva, L.C.A.S., Arthur, V., Carstensen, J.M., Boelt, B., Barboza da Silva, C., 2021. Multispectral and X-ray images for characterization of *Jatropha curcas* L. seed quality. Plant Methods 17 (1), 1–13.

Chandran, D., Raviadaran, R., Lau, H.L.N., Numan, A., Elumalai, P.V., Samuel, O.D., 2023. Corrosion characteristic of stainless steel and galvanized steel in water emulsified diesel, diesel and palm biodiesel. Eng. Fail. Anal. 147, 107129.

Che Hamzah, N.H., Khairuddin, N., Siddique, B.M., Hassan, M.A., 2020. Potential of *Jatropha curcas* L. as biodiesel feedstock in Malaysia: a concise review. Processes 8 (7), 786.

Chowdhury, J., Al Basir, F., Takeuchi, Y., Ghosh, M., Roy, P.K., 2019. A mathematical model for pest management in *Jatropha curcas* with integrated pesticides—an optimal control approach. Ecol. Complex. 37, 24–31.

da Silva, C.B., Bianchini, V.D.J.M., de Medeiros, A.D., de Moraes, M.H.D., Marassi, A.G., Tannus, A., 2021. A novel approach for *Jatropha curcas* seed health analysis based on multispectral and resonance imaging techniques. Ind. Crop. Prod. 161, 113186.

De Jong, S., Antonissen, K., Hoefnagels, R., Lonza, L., Wang, M., Faaij, A., et al., 2017. Life-cycle analysis of greenhouse gas emissions from renewable jet fuel production. Biotechnol. Biofuels 10, 64.

de Medeiros, A.D., Pinheiro, D.T., Xavier, W.A., da Silva, L.J., dos Santos Dias, D.C.F., 2020. Quality classification of *Jatropha curcas* seeds using radiographic images and machine learning. Ind. Crop. Prod. 146, 112162.

de Souza, M.F., da Silva, H.N., Rodrigues, J.F.B., Macêdo, M.D.M., de Sousa, W.J.B., Barbosa, R.C., Fook, M.V.L., 2023. Chitosan/gelatin scaffolds loaded with *Jatropha mollissima* extract as potential skin tissue engineering materials. Polymers 15 (3), 603.

Dhillon, R.S., Hooda, M.S., Handa, A.K., Ahlawat, K.S., Kumar, Y., Kumar, S., Singh, N., 2020. Clonal Propagation and Reproductive Biology in *Jatropha curcas* L

Dugala, N.S., Goindi, G.S., Sharma, A., 2021. Evaluation of physicochemical characteristics of Mahua (*Madhuca indica*) and Jatropha (*Jatropha curcas*) dual biodiesel blends with diesel. J. King Saud Univ. Eng. Sci. 33 (6), 424–436.

Ewunie, G.A., Lekang, O.I., Morken, J., Yigezu, Z.D., 2021a. Characterizing the potential and suitability of Ethiopian variety *Jatropha curcas* for biodiesel production: variation in yield and physicochemical properties of oil across different growing areas. Energy Rep. 7, 439–452.

Ewunie, G.A., Morken, J., Lekang, O.I., Yigezu, Z.D., 2021b. Factors affecting the potential of *Jatropha curcas* for sustainable biodiesel production: a critical review. Renew. Sust. Energ. Rev. 137, 110500.

Gajić, G., Mitrović, M., Pavlović, P., 2022. Phytobial remediation by bacteria and fungi. In: Assisted Phytoremediation. Elsevier, pp. 285–344.

García-Martín, J.F., Badaró, A.T., Barbin, D.F., Álvarez-Mateos, P., 2020. Identification of copper in stems and roots of *Jatropha curcas* L. by hyperspectral imaging. Processes 8 (7), 823.

Ghosh, M.K., Sahu, S., Gupta, I., Ghorai, T.K., 2020. Green synthesis of copper nanoparticles from an extract of *Jatropha curcas* leaves: characterization, optical properties, CT-DNA binding and photocatalytic activity. RSC Adv. 10 (37), 22027–22035.

Gomes, T.G., Hadi, S.I.I.A., de Aquino Ribeiro, J.A., Segatto, R., Mendes, T.D., Helm, C.V., Júnior, A.F.C., Miller, R.N.G., Mendonça, S., de Siqueira, F.G., 2022. Phorbol ester biodegradation in *Jatropha curcas* cake and potential as a substrate for enzyme and *Pleurotus pulmonarius* edible mushroom production. *Biocatalysis and agricultural*. Biotechnology, 102498.

Gutiérrez-López, A.N., Mena-Cervantes, V.Y., García-Solares, S.M., Vazquez-Arenas, J., Hernández-Altamirano, R., 2021. NaFeTiO4/Fe2O3–FeTiO3 as heterogeneous catalyst towards a cleaner and sustainable biodiesel production from *Jatropha curcas* L. oil. J. Clean. Prod. 304, 127106.

Ha, J., Shim, S., Lee, T., Kang, Y.J., Hwang, W.J., Jeong, H., Laosatit, K., Lee, J., Kim, S.K., Satyawan, D., Lestari, P., 2019. Genome sequence of *Jatropha curcas* L., a non-edible biodiesel plant, provides a resource to improve seed-related traits. Plant Biotechnol. J. 17 (2), 517–530.

Jain, S., 2019. The production of biodiesel using Karanja (*Pongamia pinnata*) and Jatropha (*Jatropha curcas*) Oil. In: Biomass, Biopolymer-Based Materials, and Bioenergy. Woodhead Publishing, pp. 397–408.

Jayakumar, M., Gebeyehu, K.B., Selvakumar, K.V., Parvathy, S., Kim, W., Karmegam, N., 2022. Waste Ox bone based heterogeneous catalyst synthesis, characterization, utilization and reaction kinetics of biodiesel generation from *Jatropha curcas* oil. Chemosphere 288, 132534.

Jonas, M., Ketlogetswe, C., Gandure, J., 2020. Variation of *Jatropha curcas* seed oil content and fatty acid composition with fruit maturity stage. Heliyon 6 (1), e03285.

Kamel, D.A., Farag, H.A., Amin, N.K., Zatout, A.A., Ali, R.M., 2018. Smart utilization of jatropha (*Jatropha curcas* Linnaeus) seeds for biodiesel production: optimization and mechanism. Ind. Crop. Prod. 111, 407–413.

Kaushik, N., Roy, S., Biswas, G.C., 2020. Screening of Indian Germplasm of *Jatropha curcas* for Selection of High Oil Yielding Plants.

Kavitha, K.R., Beemkumar, N., Rajasekar, R., 2019. Experimental investigation of diesel engine performance fuelled with the blends of *Jatropha curcas*, ethanol, and diesel. Environ. Sci. Pollut. Res. 26 (9), 8633–8639.

Keneni, Y.G., Hvoslef-Eide, A.T., Marchetti, J.M., 2019. Mathematical modelling of the drying kinetics of *Jatropha curcas* L. seeds. Ind. Crop. Prod. 132, 12–20.

Kethobile, E., Ketlogetswe, C., Gandure, J., 2020. Torrefaction of non-oil *Jatropha curcas* L. (Jatropha) biomass for solid fuel. Heliyon 6 (12), e05657.

Khan, M.Y., Rao, P.S., Pabla, B.S., 2020. Investigations on the feasibility of *Jatropha curcas* oil based biodiesel for sustainable dielectric fluid in EDM process. Mater. Today Proc. 26, 335–340.

Khan, K., Ul-Haq, N., Rahman, W.U., Ali, M., Rashid, U., Ul-Haq, A., Jamil, F., Ahmed, A., Ahmed, F., Moser, B.R., Alsalme, A., 2021. Comprehensive comparison of hetero-homogeneous catalysts for fatty acid methyl ester production from non-edible *Jatropha curcas* oil. Catalysts 11 (12), 1420.

Khanam, T., Khalid, F., Manzoor, W., Rasheedi, A., Hadi, R., Ullah, F., Rehman, F., Akhtar, A., Babu, N.K., Hussain, M., 2021. Environmental sustainability assessment of biodiesel production from *Jatropha curcas* L. seeds oil in Pakistan. PLoS ONE 16 (11), e0258409.

Kumar, V., Choudhary, A.K., 2023. A hybrid response surface methodology and multi-criteria decision making model to investigate the performance and emission characteristics of a diesel engine fueled with phenolic antioxidant additive and biodiesel blends. J. Energy Resour. Technol., 1–22.

Kumar, S., Singhal, M.K., Sharma, M.P., 2023. Analysis of oil mixing for improvement of biodiesel quality with the application of mixture design method. Renew. Energy 202, 809–821.

Laskar, I.B., Changmai, B., Gupta, R., Shi, D., Jenkinson, K.J., Wheatley, A.E., Rokhum, L., 2021. A mesoporous polysulfonic acid-formaldehyde polymeric catalyst for biodiesel production from *Jatropha curcas* oil. Renew. Energy 173, 415–421.

Liu, Z., Li, F., Shen, J., 2019. Effect of oxidation of *Jatropha curcas*-derived biodiesel on its lubricating properties. Energy Sustain. Dev. 52, 33–39.

Martínez, A., Mijangos, G.E., Romero-Ibarra, I.C., Hernández-Altamirano, R., Mena-Cervantes, V.Y., 2019. In-situ transesterification of *Jatropha curcas* L. seeds using homogeneous and heterogeneous basic catalysts. Fuel 235, 277–287.

Matsubara, T., Thanh, T.C., Hue, B.T.B., Takenaka, N., Maeda, Y., 2023. Bio-diesel fuel material supply in Vietnam and its current and future potentiality with a focus on rubber seed oil. J. Oleo Sci. 72 (2), 219–232.

Mazumdar, P., Singh, P., Babu, S., Siva, R., Harikrishna, J.A., 2018. An update on biological advancement of *Jatropha curcas* L.: new insight and challenges. Renew. Sust. Energ. Rev. 91, 903–917.

Mitra, S., Ghose, A., Gujre, N., Senthilkumar, S., Borah, P., Paul, A., Rangan, L., 2021. A review on environmental and socioeconomic perspectives of three promising biofuel plants *Jatropha curcas*, *Pongamia pinnata* and *Mesua ferrea*. Biomass Bioenergy 151, 106173.

Mohammed, A.T., Jaafar, M.N.M., Othman, N., Veza, I., Mohammed, B., Oshadumi, F.A., Sanda, H.Y., 2021. Soil fertility enrichment potential of *Jatropha curcas* for sustainable agricultural production: a case study of Birnin Kebbi, Nigeria. Ann. Rom. Soc. Cell Biol., 21061–21073.

Najafi, F., Sedaghat, A., Mostafaeipour, A., Issakhov, A., 2021. Location assessment for producing biodiesel fuel from *Jatropha curcas* in Iran. Energy 236, 121446.

Nascimento, C.W.A.D., Biondi, C.M., Silva, F.B.V.D., Lima, L.H.V., 2021. Using plants to remediate or manage metal-polluted soils: an overview on the current state of phytotechnologies. Acta Sci. Agron. 43.

Nero, B.F., 2021. Phytoremediation of petroleum hydrocarbon-contaminated soils with two plant species: *Jatropha curcas* and *Vetiveria zizanioides* at Ghana Manganese Company Ltd. Int. J. Phytoremediation 23 (2), 171–180.

Neupane, D., Bhattarai, D., Ahmed, Z., Das, B., Pandey, S., Solomon, J.K., Qin, R., Adhikari, P., 2021. Growing Jatropha (*Jatropha curcas* L.) as a potential second-generation biodiesel feedstock. Inventions 6 (4), 60.

Nor, N.M., Salih, N., Salimon, J., 2021. Chemically modified *Jatropha curcas* oil for biolubricant applications. Hem. Ind. 75 (2), 117–128.

Pandey, K.K., 2023. Effect of synthetic antioxidant-doped biodiesel in the low heat rejection engine. Biofuels 14 (3), 243–258.

Peralta, H., Avila-Ortega, D.I., García-Flores, J.C., 2022. Jatropha farm: a circular economy proposal for the non-toxic physic nut crop in Mexico. Environ. Sci. Proc. 15 (1), 10.

Piloto-Rodríguez, R., Tobío, I., Ortiz-Alvarez, M., Díaz, Y., Konradi, S., Pohl, S., 2020. An approach to the use of *Jatropha curcas* by-products as energy source in agroindustry. Energy Sources A: Recovery Util. Environ. Eff., 1–21.

Pompelli, M.F., Ferreira, P.P., Chaves, A.R., Figueiredo, R.C., Martins, A.O., Jarma-Orozco, A., Bhatt, A., Batista-Silva, W., Endres, L., Araújo, W.L., 2021. Physiological, metabolic, and stomatal adjustments in response to salt stress in *Jatropha curcas*. Plant Physiol. Biochem. 168, 116–127.

Pranoto, B.S.M., Budianta, W., 2020. Phytoremediation of heavy metals contaminated soil in artisanal gold mining at Selogiri, Wonogiri District, Central Java, Indonesia. J. Appl. Geol. 5 (2), 64–72.

Primandari, S.R.P., Islam, A.K.M.A., Yaakob, Z., Chakrabarty, S., 2018. *Jatropha curcas* L. biomass waste and its utilization. In: Advances in Biofuels and Bioenergy.

Purwanti, E., Pantiwati, Y., Nurrohman, E., Lorosae, A.D., 2023. Ethnobotany of medicinal plants used by the local community at the foot of the mount uyelewun, East Nusa Tenggara. Bioedukasi 21 (1), 8–20.

Rajak, U., Chaurasiya, P.K., Nashine, P., Verma, M., Kota, T.R., Verma, T.N., 2020. Financial assessment, performance and emission analysis of *Moringa oleifera* and *Jatropha curcas* methyl ester fuel blends in a single-cylinder diesel engine. Energy Convers. Manag. 224, 113362.

Rawat, S., Singh, J., Koduru, J.R., 2021. Effect of ultrasonic waves on degradation of phenol and Para-nitrophenol by iron nanoparticles synthesized from Jatropha leaf extract. Environ. Technol. Innov. 24, 101857.

Sánchez-Anaya, O., Mederos-Nieto, F.S., Elizalde, I., Sánchez-Minero, J.F., Trejo-Zárraga, F., 2021. Producing hybrid fuels by hydrotreating *Jatropha curcas* L. and gasoil mixtures in a batch reactor. J. Taiwan Inst. Chem. Eng. 128, 140–147.

Sánchez-Borrego, F.J., Alvarez-Mateos, P., García-Martín, J.F., 2021. Biodiesel and other value added products from bio-oil obtained from agrifood waste. Processes 9 (5), 797.

Sano, R., Ara, T., Akimoto, N., Sakurai, N., Suzuki, H., Fukuzawa, Y., Kawamitsu, Y., Ueno, M., Shibata, D., 2012. Dynamic metabolic changes during fruit maturation in *Jatropha curcas* L. Plant Biotechnol. 29 (2), 175–178.

Shibata, D., Sano, R., Ara, T., 2017. *Jatropha* metabolomics. In: The Jatropha Genome, pp. 83–96.

Shu, X., Zhang, K., Zhang, Q., Wang, W., 2019. Ecophysiological responses of *Jatropha curcas* L. seedlings to simulated acid rain under different soil types. Ecotoxicol. Environ. Saf. 185, 109705.

Singh, D., Sharma, D., Soni, S.L., Inda, C.S., Sharma, S., Sharma, P.K., Jhalani, A., 2021. A comprehensive review of physicochemical properties, production process, performance and emissions characteristics of 2nd generation biodiesel feedstock: *Jatropha curcas*. Fuel 285, 119110.

Singha, L.P., Pandey, P., 2020. Rhizobacterial community of *Jatropha curcas* associated with pyrene biodegradation by consortium of PAH-degrading bacteria. Appl. Soil Ecol. 155, 103685.

Singha, L.P., Pandey, P., 2021. Rhizosphere assisted bioengineering approaches for the mitigation of petroleum hydrocarbons contamination in soil. Crit. Rev. Biotechnol. 41 (5), 749–766.

Srivastava, S., Shukla, A., Rajput, V.D., Kumar, K., Minkina, T., Mandzhieva, S., Shmaraeva, A., Suprasanna, P., 2021. Arsenic remediation through sustainable phytoremediation approaches. Fortschr. Mineral. 11 (9), 936.

Sujatha, M., 2021. Genetic Improvement of *Jatropha curcas* (L.) Possibilities and Prospects. https://agris.fao.org/agris-search/search.do?recordID=IN2022000780.

Tamam, M.Q.M., Yahya, W.J., Ithnin, A.M., Abdullah, N.R., Kadir, H.A., Rahman, M.M., Rahman, H.A., Mansor, M.R.A., Noge, H., 2023. Performance and emission studies of a common rail turbocharged diesel electric generator fueled with emulsifier free water/diesel emulsion. Energy 268, 126704.

Vaikuntapu, P.R., Kumar, V.D., 2023. Applications and challenges of harnessing genome editing in oilseed crops. J. Plant Biochem. Biotechnol., 1–22.

Warra, A.A., Prasad, M.N.V., 2016. Jatropha curcas L. cultivation on constrained land: exploring the potential for economic growth and environmental protection. In: Bioremediation and Bioeconomy. Elsevier, pp. 129–147.

Yaqoob, H., Teoh, Y.H., Sher, F., Ashraf, M.U., Amjad, S., Jamil, M.A., Jamil, M.M., Mujtaba, M.A., 2021. *Jatropha curcas* biodiesel: a lucrative recipe for Pakistan's energy sector. Processes 9 (7), 1129.

Yate, A.V., Narváez, P.C., Orjuela, A., Hernández, A., Acevedo, H., 2020. A systematic evaluation of the mechanical extraction of *Jatropha curcas* L. oil for biofuels production. Food Bioprod. Process. 122, 72–81.

Yusuff, A.S., Kumar, M., Obe, B.O., Mudashiru, L.O., 2021. Calcium oxide supported on coal fly ash (CaO/CFA) as an efficient catalyst for biodiesel production from *Jatropha curcas* oil. Top. Catal., 1–13.

Zahra, N., Hafeez, M.B., Shaukat, K., Wahid, A., Hasanuzzaman, M., 2021. Fe toxicity in plants: impacts and remediation. Physiol. Plant. 173 (1), 201–222.

Zulfiqar, A., Mumtaz, M.W., Mukhtar, H., Najeeb, J., Irfan, A., Akram, S., Touqeer, T., Nabi, G., 2021. Lipase-PDA-TiO2 NPs: an emphatic nano-biocatalyst for optimized biodiesel production from *Jatropha curcas* oil. Renew. Energy 169, 1026–1037.

Ornamentals and tree crops for contaminated substrates

Potential of ornamental plants for phytoremediation and income generation

Woranan Nakbanpote[a], Nichanun Kutrasaeng[b], Ponlakit Jitto[c], and Majeti Narasimha Vara Prasad[d]

[a]Department of Biology, Faculty of Science, Mahasarakham University, Khamriang, Kantarawichi, Mahasarakham, Thailand [b]Science Center, Faculty of Science and Technology, Sakon Nakhon Rajabhat University, Sakon Nakhon, Thailand [c]Faculty of Environment and Resource Studies, Mahasarakham University, Kham Riang, Kantharawichai, Maha Sarakham, Thailand [d]Department of Plant Sciences, School of Life Sciences, University of Hyderabad (An Institution of Eminence), Hyderabad, Telangana, India

1 Introduction

Soils and waters contaminated/polluted with toxic metals pose a major environmental problem that needs an effective and affordable technological solution. Phytoremediation is the process through which contaminated substrates are ameliorated by growing plants that have the ability to remove the contaminants. Phytoremediation of heavy metals includes processes such as phytostabilization, phytoextraction, phytovolatilization, and rhizofiltration. A key factor for successful phytoremediation is the identification of a plant that is tolerant and suitable for specific area and conditions. Plant design for successful phytoremediation in a chosen contaminated area should not have an adverse effect on the local biodiversity. Other important aspects for the success of phytoremediation are economic benefits and by-product production in order to convince local people and governments of the benefits of phytoremediation. Therefore, ornamental plants are considered for phytoremediation and phytomanagement. Ornamental plants for the rehabilitation of stressful areas such as heavy metals, salt, and drought are studied and reported dramatically (Fig. 1).

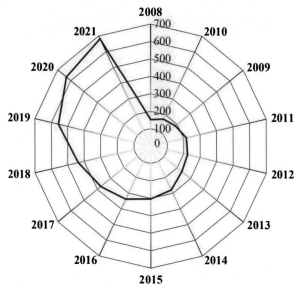

FIG. 1   Ornamental plants for remediation of stressed substrates, for example, heavy metals, salt, and drought. *Source: Zhang et al. (2022).*

Utilizing ornamental plants to clean up the contaminated environment would also change the landscape for ecotourism. The scientific names of ornamental plants frequently selected for the restoration and decoration of the landscape are shown in Fig. 2. Cut flowers such as marigold (*Tagetes* spp.) and lotus (*Nelumbo* spp.) are considered ornamental plants and are used to restore contaminated areas and generate income. Applying a growth regulator, soil amendments, manure, fertilizers, and pesticides to accelerate remediation and economic benefits should be balanced to the benefit of local people and the environment.

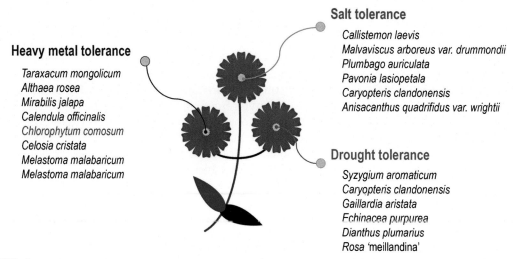

FIG. 2   Ornamental plants in phytoremediation (selected examples).

2 Contamination of heavy metals and phytoremediation

Essential heavy metals include iron (Fe), zinc (Zn), copper (Cu), manganese (Mn), and cobalt (Co). In contrast, cadmium (Cd), lead (Pb), and mercury (Hg) do not have any biological function. However, even as essential micronutrients, they can cause toxicity when concentrations are too high. Heavy metals are elements, so they are not biodegradable. Heavy metals have accumulated in the soil over decades because of the low mobility and binding to soil Fe and Mn oxides and organic fractions, and bioavailable heavy metals depend on soil conditions and plant factors (Greger, 1999). In soils, they occur as free metal ions, exchangeable metal ions, soluble metal complexes (sequestered to ligands), organically bound, precipitated or insoluble compounds such as oxides, carbonates, and hydroxides, or they may form silicate (indigenous soil content) (Kabata-Pendias and Pendias, 2011). Anthropogenic activity leads to the accumulation of many harmful substances such as herbicides, radioactive elements, and heavy metals. Huge quantities of solid wastes (tailings) are generated during the beneficiation of precious ores, which are disposed of in the nearby areas, creating vast barren land as it could hardly support any vegetation. Agricultural soils and reservoirs in many parts of the world are slightly to moderately contaminated by heavy metal toxicity such as Cd, Cu, Zn, Co, Pb, nickel (Ni), chromium (Cr), and arsenic (As). This may be due to the long-term use of commercial mineral fertilizers, agrochemical products, compost, sewage sludge, and waste disposal (Marmiroli and Maestri, 2008).

The anthropogenic discharge of Cu, Zn, and Cd into the environment is ubiquitous, giving its use as a biocide and in various metallurgical industries. Cu and Zn are also present in some sewage sludges and organic and inorganic fertilizers. Excess accumulation of Cu and Zn in plant tissues can cause many physiological and biochemical changes and a reduction in growth (Balafrej et al., 2020; Mir et al., 2021). Excessive amount of Cd can cause various phytotoxic symptoms such as chlorosis, growth inhibition, damage to root tips, and reduced absorption of water and nutrients. In the case of human toxicity, Cd uptake occurs through ingestion and inhalation, and prolonged exposure may lead to various types of cancer (Kubier et al., 2019). Cr is commonly used as leather tanning agents, textile pigments and preservatives, antifouling paints, wood treatment/preservation, steel processing, aluminum alloys, electroplating, and microbial growth inhibition such as cooling towers of power plants. The high solubility of hexavalent chromium (Cr(VI)) is considered a hazardous ion, which contaminates groundwater and can be transferred throughout the food chain. The toxic effects of Cr correlated with the generation of reactive oxygen species (ROS), which cause oxidative stress in plants (Sharma et al., 2020). Certain amount of Hg may be added to agricultural land with sludge, fertilizer, lime, and manures, especially, the use of organic mercury as a seed-coat dressing to prevent fungal diseases in seeds. The possible causal mechanisms of mercury toxicity are changes in the permeability of the cell membrane, reactions of sulfydryl (—SH) groups with cations, affinity for reacting with phosphate groups and active groups of adenosine diphosphate (ADP) or adenosine triphosphate (ATP), and replacement of essential ions, mainly major cations (Patra and Sharma, 2000). Hg has profound cellular, cardiovascular, hematological, pulmonary, renal, immunological, neurological, endocrine, reproductive, and embryonic toxicological effects (Rice et al., 2014). Arsenic (As) has been used in various fields such as medicine, electronics, agriculture (pesticides, herbicides, insecticides, fertilizers, etc.), livestock (cattle and sheep dips), and wood preservatives. Arsenic has both metallic and non-metallic properties. The As and its compounds are ranked as a Group 1 human carcinogen (Abbas et al., 2018). The anthropogenic sources of Pb are mining activities, metal smelting, coal combustion, automotive exhaust fumes, domestic utilization of Pb-based paints, land application of municipal wastewater and sludge, agricultural use of chemical fertilizers and insecticides, as well as waste disposal in landfills. Toxic effects of Pb on plants include inhibition of photosynthesis, enzymatic activities, deficient mineral nutrition uptake and water imbalance, and inhibition of root elongation, which considerable reduces both the vegetative and reproductive growth of plants. Pb induces visible symptoms such as chlorosis, necrosis, and growth inhibition of roots and shoots and changes in the branching pattern of roots. Pb is considered as carcinogenic (Group 2B) to humans (Kumar et al., 2020). Radiocesium (^{137}Cs) is an anthropogenic radioactive isotope released from nuclear weapons tests, reprocessing nuclear-spent fuel facilities, and nuclear accidents. ^{137}Cs is the most radiologically remarkable radionuclide with a physical half-life of 30.17 years. Its chemical properties are similar to those of potassium (K^+), high solvability in seawater, and can be transported by marine ecosystem to biota. Therefore, the potential hazards of ^{137}Cs in the marine environment are constantly being studied and monitored (Abbasi et al., 2021; Saniewski et al., 2022). *Calotropis gigantea* R.Br. (crown flower) had a potential ornamental plant for phytoremediation of ^{137}Cs (Singh et al., 2009). Other plants and weeds that have role in the remediation of radionuclides ^{137}Cs and strontium (^{90}Sr) were presented in Sharma et al. (2015).

The remediation of soil and water contaminated with heavy metals is a significant concern. Various technologies are in place to clean up or reduce metal exposure from contact with metal-contaminated soil and water. Technologies for remediation of metal-contaminated soil include excavation, immobilization, vetrification, soil washing/flushing, precipitation, membrane filtration, adsorption, ion exchange, permeable reactive barriers, biological treatment, and phytoremediation. Phytoremediation is a cost-effective technology that uses plants to degrade, assimilate, metabolize, or detoxify metal and organic chemical contamination. Phytoremediation is one of the in situ procedures, environmental-friendly reducing soil erosion, enriching soil organic matter leading to enhance soil fertility, and sustainable technology for site restoration. It also helps in removing the carbon dioxide (CO_2) from the air during the photosynthesis process. Furthermore, the use of plants to remediate contaminated soils and water could preserve the structure and biological functions of the environments. Phytoremediation utilizes physical, chemical, and biological processes to remove, degrade, transform, or stabilize contaminants within soil and groundwater (Prasad and Prasad, 2012).

Proper selection of plant species for phytoremediation plays an important role in the development of remediation methods (decontamination or stabilization), especially on metal-polluted soil. Hyperaccumulative plants for heavy metal have defense reactions and detoxification mechanisms. After exposing heavy metals, some hyperaccumulative plants can change structural and ultrastructural and localize the heavy metals to sub-cellular compartments with a few or no sensitive metabolic activities (Barceló and Poschenrieder, 2004; Rattanapolsan et al., 2013; Mongkhonsin et al., 2011). The accumulation of metals in the cell wall may involve flavonoid compounds in plant biomass structures (Mongkhonsin et al., 2016). One of the consequences of metal accumulation is an increase in ROS contents, which are destructive if protective antioxidant mechanisms do not operate efficiently. Many antioxidant substances such as ascorbate and the enzymes ascorbate peroxidase (APX), superoxide dismutase (SOD), and peroxidase (POD) are increased under heavy metal stress (Sytar et al., 2013; Zhang et al., 2021). The total phenolic content (TPC), total flavonoid content (TFC), and free radical scavenging activity values also correspond to the metal dose-response curves (Mongkhonsin et al., 2016). Improvements in carotenoid and polyamine content are also related to plant tolerance (Sytar et al., 2013). In addition, glutathione and phytochelatins play a role in resistance to stress in heavy metals (Yadav, 2010; Panitlertumpai et al., 2013).

A further important criterion for the potential plants used for phytoremediation is the provision of economic benefits and environmental remediation. In developing countries, including Thailand, it is difficult to convince local people to grow plants that accumulate metals in their fields to remove pollutants unless a financial incentive is offered. Phytomanagement studies should focus on combining phytoremediation with enhancing soil fertility, chemical chelate, plant cultivation techniques, microorganisms, and minimizing pesticide management, as well as the investigation of new species for phytoremediation, which may also increase economic benefits.

3 Ornamental plants for heavy metals phytoremediation

Ornamentals are cultivated for decorative purposes in gardens and landscapes to cut flowers and embellish the landscape. Most ornamental plants are grown to display aesthetic characteristics, including flowers, leaves, fragrance, overall foliage texture, fruits, stem, and bark. Ornamental plants are separated into two groups of terrestrial plants and aquatic plants. Terrestrial plants are important for flowers and foliage. Aquatic plants are divided into three groups of emergent, floating, and submerged plants. The principal factor controlling the distribution of aquatic plants is the depth and duration of flooding. Other factors may also affect their distribution, abundance, and growth form, including nutrients, disturbance from waves, grazing, salinity, and contaminants. Ornamental plants commonly used in landscaping not only make the environment colorful but also remediate the contaminated environment. They will visually decorate the environment of metal-impacted areas. As many of them are not edible plants, the risk of the food chain is reduced. Some ornamental plants also have utilitarian purposes such as perfume, oil, pigments, and phytochemicals (Gupta et al., 2021; Khan et al., 2021).

3.1 Terrestrial-ornamental plants for phytoremediation

The accumulation of heavy metals in plants depends on the metal(s) and their bioavailable in soils, species, and growth stages of plants, nutritions, seasons, geological and environmental factors (Prasad and Freitas, 2003). Table 1 shows selected species of ornamental plants for heavy metal(s) accumulation studied in pot and field system. Many reports show that ornamental plants accumulated metal(s) in their roots. They can thus remedy contaminated soils to some extent as the manner of phytostabilization and at the same time beautify the environment and sell for incomes. Growth stages of plants also have effect on heavy metal accumulation in each plant part. In case of *Helianthus annus* (sunflower), adult plants showed a lower sensitivity to metal toxic than young seedlings. The order of Cd content in

TABLE 1 Terrestrial-ornamental plants for phytoremediation and phytostabilization.

Botanical name	Metals	Treatment	Results	References
Helianthus annuus	Cr	*System*: pot *Treatment*: *Glomus intraradices* *Stress*: soil and sand (1:1) contaminated with $CrCl_3$ for Cr^{3+}, $Na_2Cr_2O_7$ for $Cr_2O_7^-$	Cr(III) and Cr(VI) depressed plant growth, decreased stomata conductance and net photosysnthesis, which Cr(VI) was more toxic than Cr(III). Arbuscularmicrorhiza (AM) fungus alleviated Cr toxicity and decreased Cr accumulation	Davies Jr. et al. (2002)
Helianthus annus, Salvia splendens, Tagetes eracta	Cd	*System*: pot *Stress*: soil spiked with $CdSO_4$ (1–10 mg-Cd L^{-3})	The uptake of Cd by marigold, sunflower and scarlet sage increased with the increase of Cd doses, while the yield of these plants decreased	Bosiacki (2008)
Nugget marigold, (triploid hydbrid between *T. erecta* L. and *T. patula*)	As	*System*: pot and field *Treatment*: P-fertilizer, 30–50 days *Stress*: As contaminated soil	P-fertilizer increased As uptake in flowering stage of nugget marigold. Arsenite (As(III)) and Arsenate (As (V)) were found in the roots, stems and leaves	Chintakovid et al. (2008)
Tagetes erecta, Chrysanthemum indicum, Gladiolus grandiflorus	Cd	*System*: pot *Stress*: soil spiked with $CdCl_2$ (5–100 mg-Cd kg^{-1})	Cd uptake increased with contents in soils and the maximum accumulation occurred in leaves. In view of higher biomass, Cd removal was the maximum with chrysanthemum > gladiolus > marigold, while gladiolus with highest tolerance and Cd-content holds potential to clean up the contaminated soil	Lal et al. (2008)
Calendula officinalis, Althaea rosea	Cd	*System*: pot *Treatment*: EGTA and SDS *Stress*: soil spiked with $CdCl_2$ (10–100 mg-Cd kg^{-1})	EGTA and SDS could increase plant biomass, and promote Cd accumulation. The two ornamental plants can be regarded as a potential Cd-hyperacumulator through applying the chemical agents	Liu et al. (2008a)
Impatiens balsamina, Calendula officinalis, A. rosea	Cd, Pb	*System*: pot and hydroponic *Stress*: soil spiked with $CdCl_2$ (10–100 mg-Cd kg^{-1}), Hoagland solution spiked with $CdCl_2$ (1–10 mg-Cd L^{-1}) plus $Pb(NO_3)_2$ (50, 100 mg-Pb L^{-1})	*C. offinalis* showed great tolerance to Cd and stronger ability to Cd accumulation. Cd in the roots was greater than in the shoots, apotential for phytostabilization. The relation between the Pb and Cd in solution showed highly significant negative correlation	Liu et al. (2008b)
Ricinus communis, T. erecta	Ni, Pb	*System*: pot *Treatment*: Farm yard manure *Stress*: Ni and Pb contaminated soil	*R. communis* accumulated more Ni than *T. erecta*. Root of both crops contaminated higher Ni than aerial parts. Farm yard manure application enhanced Ni accumulation and reduced time requirement	Malarkodi et al. (2008)

TABLE 1 Terrestrial-ornamental plants for phytoremediation and phytostabilization—cont'd

Botanical name	Metals	Treatment	Results	References
T. erecta	Cd, Pb	*System*: pot *Stress*: soil spiked with Cd ($1–10\,mg\,L^{-1}$) and Pb ($100–1000\,mg\,L^{-1}$)	Marigold accumulated Cd as leaves > stalks > inflorescences. Pb were found in stalks > leaves > inflorescences. Cd and Pb contents in plants depended on the increase concentrations	Bosiacki (2009)
Thlaspi caerulescens, Salix dasyclados Vimm	Cd, Pb, Zn, As	*System*: monoculture and co-cropping in pot *Treatment*: N, P, K fertilized topsoil *Stress*: anthropogenically contaminated soil	Combined cropping with T. caerulescens enabled willows (S. dasyclados) to survive, and increased bioaccumulation of Zn in T. caerulescens shoots. Both separate and combined cropping had no effect on the phytoextraction of As and Pb	Fuksová et al. (2009)
Lonicera japonica	Cd	*System*: pot *Stress*: soil spiked with CdCl$_2$ ($5–200\,mg\text{-}Cd\,kg^{-1}$)	Synergistic interaction was found in accumulation and translocation between Cd and Fe, and a significantly negative correlation between Cd and Cu or Zn concentrations in plant tissues	Liu et al. (2009a)
A. rosea	Cd	System: pot Treatment: EGTA or EDTA and/or SDS Stress: Soil spiked with CdCl$_2$ ($30, 100\,mg\text{-}Cd\,kg^{-1}$)	EGTA and SDS increased plant biomass and promoted Cd accumulation, while EDTA was toxic to plant and ineffective in this regard.	Liu et al. (2009b)
C. officinalis	Cd	*System*: pot *Treatment*: EDTA and EGTA, and/or SDS Stress: Soil spiked with CdCl$_2$ ($30, 100\,mg\text{-}Cd\,kg^{-1}$)	EDTA toxic led to retarded growth. EGTA and SDS, and EGTA alone increased the total Cd accumulation	Liu et al. (2010)
Calendula alata	Cs, Pb	System: hydroponic Stress: nutrient solution spiked with CsCl (0.6 to $5\,mg\text{-}Cs\,L^{-1}$) or Pb(C$_2$H$_3$O$_2$)$_2$ ($0.6–5\,mg\text{-}Pb\,L^{-1}$)	*Calendula* had an extremely fast growth rate and could remediate Cs and Pb. The present of Pb influenced uptake of Cs, and the effect is significantly inhibitory	Borghei et al. (2011)
T. erecta	Cu	*System*: pot *Treatment*: Glomus intraradices *Stress*: CuO (500 to $2500\,mg\text{-}Cu\,kg^{-1}$) spiked in perlite and peat moss (2:1)	Marigold colonized with arbuscular mycorrhiza accumulated more Cu in the roots as well as the whole plant. TF and BCF suggest the system marigold-mycorrhiza can potentially phytostabilize Cu in contaminated soils	Castillo et al. (2011)

Continued

TABLE 1   Terrestrial-ornamental plants for phytoremediation and phytostabilization—cont'd

Botanical name	Metals	Treatment	Results	References
T. erecta	Cd	*System*: pot *Treatment*: *Glomus intraradices, Gomphonema constrictum* and *G. mosseae* *Stress*: soil and sand, spiked with $CdCl_2$ (0–50 mg-Cd kg^{-1})	AM fungi improved the capability of reactive oxygen species (ROS) scavenging, and reduced Cd concentration in plants to alleviate marigold from Cd stress	Liu et al. (2011a)
Cosmos sulphureus, T. erecta, H. annuus	Cd	*System*: pot *Stress*: Cd contaminated soil (50–400 mg kg^{-1})	Marigold showed higher potential, Cd accumulation in shoots was >100 mg kg^{-1}, TF > 1 and BCF was >1 (Cd treatment of 50 mg kg^{-1}). Due to higher biomass and numerous roots, the total uptake of Cd by Guinea grass could be maximized	Rungruang et al. (2011)
Tagetes patula	Cd, Cu, Pb and benzo[a] pyrene (B[a]P)	*System*: pot *Stress*: soil, single B[a]P treatment (2–50 mg kg^{-1}), and co-contamination of B[a]P and $CdCl_2$ (20, 50 mg-Cd kg^{-1}), $CuSO_4$ (100, 500 mg-Cu kg^{-1}) or $Pb(CH_3COO)_2$ (1000, 3000 mg-Pb kg^{-1})	Marigold could be useful for phytoremediation of B[a]P and B[a]P-Cd contaminated sites. Low concentration of B[a]P ($\leq$10 mg kg^{-1}) could facilitate plant growth, but Cd, Cu and Pb inhibited effects on plant growth and B[a]P uptake and accumulation	Sun et al. (2011)
H. annuus	Hg	*System*: pot *Treatment*: cytokinin and ammonium thiosulfate *Stress*: contaminated soil from petrochemical plant	Addition of cytokinine and ammonium thiosulfate increased Hg uptake and translocation	Cassina et al. (2012)
H. annuus, T. patula, Celocia cristata	Ca, Cr, Mn, Fe, Cu, Zn, Pb	*System*: pot *Stress*: soil contaminated with tannery effluent	Sunflower, marigold and cock's comb were found to be good accumulator of different elements. Concentration of different elements in different parts followed a trend of root > leaves > stem > flower	Chatterjee et al. (2012)
H. annuus	Cd, Zn	*System*: pot *Treatment*: amendments of swine manure salicylic acid (SA) and potassium chloride (KCl) *Stress*: Cd and Zn contaminated soil	The amendments of swine manure, SA, and KCl increased height, flower diameter, and biomass. KCl increased Zn and Cd accumulations and suitable for environment and economy	Hao et al. (2012)

TABLE 1 Terrestrial-ornamental plants for phytoremediation and phytostabilization—cont'd

Botanical name	Metals	Treatment	Results	References
H. annuus	Zn, Cd, Cu	*System*: pot *Stress*: metals-contaminated soil	Metals showed a decrease in shoot biomass and chlorophyll concentration. Increase ascorbate peroxidase (APX) activity in adult indicated an elevated use of ascorbate after exposure to metal stress	Nehnevajova et al. (2012)
Nerium oleander	Pb	*System*: hydroponic *Stress*: nutrient solution spiked with $Pb(NO_3)_2$ (20–100 μM Pb)	BCF and TF were <1, which may be useful with regard to phytostabilizing Pb. Pb inhibited plant growth and increased malondialdehyde (MDA) content in leaves	Trigueros et al. (2012)
T. patula, Impatiens walleriana	Cd	*System*: pot *Treatment*: EDTA *Stress*: Soil spiked with $Cd(NO_3)_2$ (20–80 mg-Cd kg^{-1})	Impatiens and marigold accumulated Cd at a rate of 200–1200 mg kg^{-1} in shoot, with BCFs and TFs of 8.5–15 and 1.7–2.6, respectively. EDTA application increased the translocation of Cd from roots to shoots	Wei et al. (2012)
Quamolit pennata, Antirrhinum majus, Celosia cristata var. *pyramidalis*	Pb	*System*: pot *Stress*: soil spiked with $Pb(NO_3)_2$ (1000–5000 mg-Pb kg^{-1})	BCF and TF of these plants were <1. Only *C. cristata* var. *pyramidalis* could be identified as a Pb-accumulator	Cui et al. (2013)
Chrysantemum maximum	Cd, Cu, Ni, Pb	*System*: soilless *Treatment*: *Glomus mosseae*, mycorrhizal fungus *Stress*: perlite exposed to Cd, Cu, and Ni (0.5–10 mg L^{-1}) and Pb (6–100 mg L^{-1})	Plants wasbehaving as Pb-excluders. Cd, Cu, and Ni were accumulated in roots. Low accumulation in flowers was observed for Cd and Cu, concentration-dependent. Ni and Pb were not translocated to flowers. Mycorrhizal plants accumulated less Pb and Cu in both shoots and roots than non-mycorrhizal plants	González-Chávez and Carrillo-González (2013)
Amaranthus hypochondriacus	Cd	*System*: pot *Treatment*: NPK fertilizer, repeated harvest *Stress*: soil spiked with $3CdSO_4$ (5 mg-Cd kg^{-1})	NPK increased dry biomass, resulting in a large increment of Cd accumulation. Repeated harvests had a significant effect on the plant biomass and thus on overall Cd removal. Plant growth time was found to significantly affect the amount of Cd extracted by the plants	Li et al. (2013)

Continued

TABLE 1   Terrestrial-ornamental plants for phytoremediation and phytostabilization—cont'd

Botanical name	Metals	Treatment	Results	References
H. annuus	Zn, Cd	*System*: pot *Treatment*: PGPB *Stress*: Soil spiked with $CdCl_2$ (10–30 mg-$Cd\,kg^{-1}$) and $ZnCl_2$ (100–1000 mg-$Zn\,kg^{-1}$)	The PGPB strains *Ralstonia eutropha* and *Chrysiobacterium humi* reduced losses of weight in metal exposed plants, and induced changes in metal bioaccumulation. *C. humi* enhanced the short-term stabilization, lowering losses in plant biomass, and decreasing aboveground tissue contamination	Marques et al. (2013)
Mirabilis jalapa, Impatiens balsamin, T. erecta	Cr	*System*: pot *Stress*: Cr in tannery sludge	*M. jalapa* showed the strongest ability to tolerate and enrich Cr for phytoextraction, as Cr accumulation in roots > stems > leaves > inflorescence	Miao and Yan (2013)
T. patula	Fe	*System*: pot *Stress*: Fe ore tailing	Marigold qualified well as a potential tool for phytostabilization, and survives excess of heavy metals present in the Fe ore tailing	Chaturvedi et al. (2014)
Erica andevalensis, Erica australis	Al, As, Fe, Mn	*System*: field *Stress*: sulfide-mining waste	*Erica* plants can be considered as a Mn-accumulator and acid-, Al-, As-, Fe- and Mn-tolerant. Both Erica species can be considered suitable for phytostabilization of metal(loid) polluted sites of mine tailing	Pérez-López et al. (2014)
H. annuus	Cd	*System*: pot *Treatment*: chemical chelator, organic chelator, and/or electrical fields *Stress*: Cd contaminated soil (10.2 $mg\,kg^{-1}$)	EDTA as chemical chelator, cow manure extract and poultry manure extract as organic chelator, and electrical fields had no significant impacts on shoot and roots dry weights. Treatment with EDTA and electrical field (10 and 30 V) increased the Cd accumulated in shoot	Tahmasbian and Safari Sinegani (2014)
M. jalapa	Cd	*System*: pot *Treatment*: EGTA or EDTA, +/− SDS *Stress*: soil spiked with $CdCl_2$ (25 mg-$Cd\,kg^{-1}$)	EGTA has better effectiveness and can bring lesser metal leaching risk than EDTA. Cd translocation ability under the EGTA was higher than the EDTA treatment. Single SDS treatment could not enhance the shoot Cd extraction	Wang and Liu (2014)
Sedum alfredii, Brassica napus (Oilseed rape)	Cd	*System*: monoculture and intercropping in pot *Treatment*: *Acinetobacter calcoaceticus* Sasm3 *Stress*: $CdCl_2$ added to 10-year-old vegetable soil to keep 2 mM Cd.	The Sasm3 inoculation increased the biomass (shoot, root, and seeds) of rape and Cd concentration in rape. The intercropping treatment decreased Cd accumulation in rape	Chen et al. (2015)

TABLE 1   Terrestrial-ornamental plants for phytoremediation and phytostabilization—cont'd

Botanical name	Metals	Treatment	Results	References
Sedum alfredii *Medicago sativa*	Pb, Cd, Zn, Cu	*System*: greenhouse, co-cropping in pot *Treatment*: *Phyllobacterium myrsinacearum* strain RC6b *Stress*: Pb/Zn mine tailing	With the inoculation of RC6b, biomass yields of plant shoots were significantly increased as a result of increased microbial activity and carbon utilization. The inoculation increased the efficiency of phytoextraction of Pb, Cd and Zn by the shoots of the co-planting system, but the phytoextraction of Cu was not improved	Liu et al. (2015)
Tagetes crecta	Cr(III)	*System*: hydroponic *Stress*: Hoagland and Arnon nutrient solution containing 0–0.24 mmol L^{-1} Cr(III)	Increasing Cr(III) concentrations resulted in a higher bioaccumulation of total Cr in the tissues. Cr bioaccumulation was greater in roots. Reduction of Cr in xilem vases resulted to a plastic effect in leaves	Coelho et al. (2017)
T. erecta	Cr	*System*: pot *Treatment*: rhizobacteria *Bacillus cereus*-CK505 and *Enterobacter cloacae*-CK555 *Stress*: Mixed 1:1 (*w*/w) ratio of red soil and sandy-clay loam soil, spiked with electroplating effluent to obtain Cr 2–6 mg kg^{-1}	*T. erecta* tolerated 6 mg kg^{-1} of Cr without affecting the plant growth. However, Cr-inducted physiological changes were observed as the decrease of Chlorophyll content and the increase of biomass. The rhizobacteria support the plant to accumulate high levels of Cr	Chitraprabha and Sathyavathi (2018)
H. annuus	Cd, Cu, Pb, Cr, Zn, and Ni	*System*: pot *Treatment*: *Funneliformis mosseae* and *F. caledonium*, arbuscular mycorrhiza (AM) *Stress*: soil from waste electrical and electronic equipment (WEEE)-recycling site	Inoculations of *F. mosseae* and *F. caledonium* decreased soil pH and increased of soil available P and DTPA-extractable Zn. Therefore, AM fungi helped in reducing HM stress and promoting phytoextraction of the heavy metals by *H. annuus*	Zhang et al. (2018)
Canna red dazzler, Canna flaccida, and *Canna calalillies*	As	*System*: pot *Stress*: garden soil spiked with Na_3AsO_4 to obtain 10–50 mg kg^{-1} of soil	As uptake and antioxidant activity were in roots higher than in shoot. *Canna* cultivars could be suitable for rhizofiltration or phytostabilization rather than phytoextraction, and antioxidative mechanism involved to reduce the As stress	Praveen et al. (2019)
T. patula	Cd	*System*: pot *Treatment*: amended with N fertilizer and wheat straw *Stress*: Cd contaminated soil, Acidic Ferralsols and Calcaric Cambisols	Adding N fertilizer and straw decreased soil pH and increased dissolute organic carbon concentration. Then soil Cd availability was raised and resulted in an increase in Cd uptake by *T. patula*	Ye et al. (2019)

Continued

TABLE 1   Terrestrial-ornamental plants for phytoremediation and phytostabilization—cont'd

Botanical name	Metals	Treatment	Results	References
Pelargonium hortorum	Pb	*System*: pot, greenhouse *Treatment*: Soil amendment with EDTA and di-isopropanol amine (DIPA) at dosage of 0–$10\,mmol\,kg^{-1}$ *Stress*: Soil spiked with Pb (0–$1500\,mg\,kg^{-1}$)	Root/shoot length of *Philanthus hortorum* was decreased with increase in Pb concentration and chelating agents. EDTA+Pb showed maximum toxicity, and DIPA showed less toxicity than EDTA	Arshad et al. (2020)
T. erecta	Cd, Pb, Zn	*System*: pot *Stress*: Lateritic soil spiked with $Cd(NO_3)_2$, $Pb(NO_3)_2$, and $Zn(NO_3)_2$ to achieve metals concentrations (0–$160\,mg\,kg^{-1}$ soil)	The order of accumulation of heavy metals in *T. erecta* shoots was in the order of Cd > Zn > Pb, and the roots was Zn > Cd > Pb. The plant had an effective phytoextractor for Zn and Cd, and an excluder for Pb. BCF > 1 for Cd and Zn, and BCF < 1 for Pb. TF > 1 for Cd	Madanan et al. (2021)
C. officinalis var. *officinalis* (hybrid)	Fly ash (Ca, Mg, Al, As, Co, Zn, Fe, Mn, and Ni)	*System*: pot Stress: various ratios of coal fly ash (FA) and garden soil (0:100, 20: 80, 60:40, 80:20, 100:0)	At a low ratio of 40% FA and 60% garden soul, the plant growth was increased in comparison to plants grown in 100% garden soil. However, at high FA doses (60%, 80%, and 100%), the plant growth tended to decrease	Varshney et al. (2021)
Canna orchioides	Cd	*System*: pot, glasshouse *Stress*: Soil spiked with $CdCl_2$ to reach Cd of 100 and $200\,mg\,kg^{-1}$ dry soil	BCF was the highest in the roots, followed by the leaves, rhizomes, and stems. The protective mechanisms in the leaves differ from in the roots. Cd stress increased the activity of superoxide dismutase (SOD), peroxidase (POD), and ascorbate peroxidase (APX) in the leaves, whereas only SOD was raised in the roots. The increases of antioxidant enzymatic activity guaranteed tolerance to Cd exposure	Zhang et al. (2021)
Hibiscus rosa-sinensis, *Rosa* sp.	Fe, Mn, Zn, Cu	*System*: pot *Stress*: Sludge amendments in soil (0%-control, 10%–100% sewage sludge)	The highest heavy metals removal was in the order of Fe > Mn > Zn > Cu with the treatment of 100% sludge. Roots contained the highest concentration of heavy metals in the plant parts. Rose can remove more heavy metals than hibiscus plants	Ishak and Hum (2022)
Cosmos bipinnatus, *Catharanthus*	As, Pb, Hg	*System*: pot *Treatment*: zero valent iron (ZVI) nanoparticles	*C. bipinnatus* showed the maximum accumulation of As, Pb, and Hg. The augmentation of ZVI nanoparticles at	Majumdar et al. (2022)

TABLE 1 Terrestrial-ornamental plants for phytoremediation and phytostabilization—cont'd

Botanical name	Metals	Treatment	Results	References
roseus, Gomphrena globosa, I. balsamina		(Fe NPs) Stress: soil spiked with As (70 mg kg^{-1}), Pb (600 mg kg^{-1}) and Hg (15 mg kg^{-1})	a dosage of 20 mg kg^{-1} enhanced the accumulation of metal(loid)s without inhibiting the growth of plants	
Salvia virgata	Pb	System: pot experiment Treatment: natural tea saponin (TS) and citric acid (CA) Stress: soil spiked with Pb (0 to 100 mg kg^{-1})	The addition of TS and CA increased available Pb(II) in the soil. An appropriate application of TA and CA enhanced the uptake of Pb(II) by S. virgata	Cay (2023)
T. erecta, Verbena hybrida	Zn. Ni, Cd, Cr, Pb, As, Hg	System: greenhouse, pot experiments. Stress: soil amended by municipal sewage sludge (SS)	T. erecta and V. hybrida were the potential metal-hyperaccumulative plants. The plants grew well in 3% SS in soil. The BCF values of T. erecta for Zn and Cd, and V. hybrida for Zn were >1	Tepecik et al. (2023)

roots > leaves > stems in young plants; stems > leaves > roots > flowers in adult plants (Nehnevajova et al., 2012). Imbalanced trace element concentrations in *Tagetes erecta* (marigold) influenced detoxification processes of heavy metals (Liu et al., 2011a, b). The parameters indicating the effectiveness of metal remediation in Tables 1 and 2 are the following equations, where the substance means soil or water.

$$Bioconcentration\ factor\ (BCF) = \frac{metal\ concentration\ in\ root\ (mg\ kg^{-1})}{metal\ concentration\ in\ substance\ (mg\ kg^{-1})}$$

$$Translocation\ factor\ (TF) = \frac{metal\ concentration\ in\ shoot\ (mg\ kg^{-1})}{metal\ concentration\ in\ root\ (mg\ kg^{-1})}$$

$$Biological\ accumulation\ coefficient = \frac{metal\ concentration\ in\ shoot\ (mg\ kg^{-1})}{metal\ concentration\ in\ substance\ (mg\ kg^{-1})}$$

$$Phytoextraction\ coefficient\ (PEC) = \frac{metal\ concentration\ in\ plant\ (mg\ kg^{-1})}{metal\ concentration\ in\ substance\ (mg\ kg^{-1})}$$

$$Total\ metal\ bioaccumulation\ (BA) = metal\ concentration\ in\ plant\ (mg\ kg^{-1}) \times dry\ biomass\ (kg)$$

$$Relative\ dry\ biomass\ yield\ (RBY, \%) = \frac{dry\ biomass\ in\ the\ presence\ of\ metal}{dry\ biomass\ in\ the\ absence\ of\ metal} \times 100$$

TABLE 2　Aquatic-ornamental plants for phytoremediation.

Botanical name	Metal	Treatment	Results	References
Nuphar variegatum	Cd	*System*: field, peatlands in the Algonquin Park-Haliburton region of Ontario	Yellow pond lily accumulated more Cd in its leaves from peatlands with low pH, low alkalinity, and low DOC, and the petioles of *Nuphar* accumulate more Cd form the peatlands with low pH. The organic content of the sediment had no effect on *Nuphar* Cd levels	Thompson et al. (1997)
Nymphaea odorata	Pb	*System*: field sampling	About 28% of the Pb in shoots was accumulated directly from water, whereas there was no evidence of root uptake of Pb from sediments	Outridge (2000)
Nymphaea violacea	Cd	*System*: pot *Stress*: heavy clay soil exposed to $Cd(NO_3)_2$ (50 mg-Cd L^{-1}) and/or $CaCl_2$ (500 mg-Ca L^{-1})	Maximum Cd and Ca accumulation were in the mature leaf lamina in daylight. The accumulation was inhibited by the herbicide 3-(3′,4′-dichlorophenyl)-1,1-dimethylure. Deposition and storage of heavy metals by epidermal glands represented a stage in the sequestration and detoxification of the metals	Lavid et al. (2001)
Nymphaea spontanea	Cr(VI)	*System*: hydroponics *Stress*: $K_2Cr_2O_7$ (1–10 mg L^{-1}), binary metals Cr(VI) 2.5 mg L^{-1} and $Cu(NO_3)_2$ (Cu(II) 0.5 mg L^{-1})	Cr(VI) accumulation in water lilies follows the order: roots > leaves > petioles. Roots play an important role. The maturity of plant exerts a great effect on the removal and accumulation of Cr(VI). Removal of Cr(VI) was more efficient when the metal was present singly than in the presence of Cu(II). Significant toxicity effect was evident as shown in the reduction of chlorophyll, protein and sugar contents in plants exposed to Cr(VI)	Choo et al. (2006)
Iris lactea and *Iris tectorum*	Cd	*System*: hydroponic *Stress*: Hoagland nutrient solution spiked with $CdCl_2$ (1–160 mg-Cd L^{-1})	Cd accumulated in the two irises were in shoots > roots. TEM showed Cd localizing in cell wall, cytoplasm, and inner surface of xylem in root tip	Han et al. (2007)

TABLE 2   Aquatic-ornamental plants for phytoremediation—cont'd

Botanical name	Metal	Treatment	Results	References
Iris lacteal and *I. tectorum*	Pb	*System*: hydroponic *Stress*: Hoagland nutrient solution spiked with $Pb(NO_3)_2$ (2–10 mg-Pb L^{-1})	Pb accumulated in the two irises were in roots > shoots. *I. lacteal* was more tolerant to Pb than *I. tectorum*. TEM showed Pb deposits were found along the plasma membrane of some root tip cells	Han et al. (2008)
Iris pseudacorus	Pb	*System*: pot *Treatment*: $FeSO_4$ (100–500 mg Fe kg^{-1}) *Stress*: clay from paddy field and vermiculite, spiked with $Pb(NO_3)_2$ (100–1000 mg-Pb kg^{-1})	Intermediate iron dose supply (100 mg kg^{-1}) generally enhanced Pb absorption and accumulation. Plant growth was inhibited by iron toxicity at high iron dose (500 mg kg^{-1})	Zhong et al. (2010)
I. pseudacorus	Cr, Zn	*System*: hydroponic *Stress*: peat-perlite (50/50), $ZnCl_2$ (0.07–1.5 mM Zn) and $CrCl_3$ (0.04–0.8 mM Cr(III))	Iris showed a great capacity to tolerate and accumulate both Zn and Cr. The species were a good candidate for Cr rhizofiltration and Zn phytoextraction	Caldelas et al. (2012)
Nelumbo spp.	Cr, Mn, Cu, Zn, As, Cd, Hg, and Pb	*System*: field experiment *Stress*: rare earth mining area	The contents of the heavy metals Cr, Mn, Cu, Zn, As, Cd, Hg, and Pb in lotus seeds were no higher than reference area and meet national food safety standards of China	Dingjian (2012)
Nelumbo nucifera Gaertn	Pb, Cd	*System*: big pot *Stress*: Soil polluted with Pb (100, 500 mg kg^{-1}), Cd (0.2, 1.0 mg kg^{-1}) and Pb+Cd (100 mg kg^{-1} Pb + 0.2 mg kg^{-1} Cd, and 500 mg kg^{-1} Pb + 1.0 mg kg^{-1} Cd)	Pb and Cd accumulations in roots were positive correlation with the levels of Pb and Cd. The metals were more in leaf, swollen stem, petiole and creeping stem. Pb and/or Cd exceeded the national food hygienic standards. Yield of swollen stem dropped down along with the concentration increases	Xiong et al. (2012)
Zantedeschia aethiopica	Fe	*System*: hydroponic *Stress*: nutrient solution spiked with $FeSO_4$ (100 or 200 mg-Fe L^{-1})	White calla lilies plants were moderately tolerant to excess iron toxicity, depending on the pH and moisture system. The plant would allow for the phytoremediation and rehabilitation of iron contaminated wetlands	Casierra-Posada et al. (2014)

Continued

III. Ornamentals and tree crops for contaminated substrates

TABLE 2   Aquatic-ornamental plants for phytoremediation—cont'd

Botanical name	Metal	Treatment	Results	References
Filamentous algae, *Jussiaea* sp., *Pistia* sp., *Eichhornea* sp., *Hydrilla* sp.	U	*System*: field sampling *Stress*: the treated tailings effluent, a small pond inside tailings disposal area and residual water of this area	The U accumulated in filamentous algae and sediment rooted plants were correlated with U concentrations. Filamentous algae, *Jussiaea* and *Pistia* owing to their high bioproductivity, biomass, uranium accumulation and concentration ratio can be useful for prospecting phytoremediation of stream carrying treated or untreated uranium mill tailings effluent	Jha et al. (2016)
Canna indica, *Hydrocotyle umbellate*, *Typha angustifolia*	Cr	*System*: surface flow contracted wetland *Stress*: andosol soil media, Cr(VI) from $K_2Cr_2O_7$ (10–50 mg L^{-1}), hydraulic retention time of 9 days	*C. indica*, *H. umbellate*, and *T. angustifolia* had efficiency to reduce 50 mg L^{-1} of Cr to 99.8%, 99.7%, and 86.4%, respectively, Cr accumulated in the root than the shoot (TF < 1)	Taufikurahman et al. (2018)
Pistia stratiotes	Ni	*System*: plant growth cabinet *Stress*: 10% Hoagland solution spiked with $NiCl_2·6H_2O$ to different concentrations of 1, 5, 10, and 20 mg L^{-1}	*P. stratiotes* plant accumulates Ni in high quantities (20 mg L^{-1}) and has a high rhizofiltration capacity. However, relative growth rate, chlorophyll *a* and carotenoid showed a negative correlation with the increase of the Ni concentrations	Leblebici et al. (2019)
C. indica, *Orya sativa*	Mg, Al, Ca, Fe	*System*: Batch-fed horizontal sub-surface flow constructed wetlands *Stress*: piggery effluent containing nutrient (N, P) and metals (Mg, Al, Ca and Fe), hydraulic retention time of 3 days	*O. sativa* removes more nutrients and metals than *C. indica*. The BCF for *O. sativa* was higher than for *C. indica*. TF for *C. indica* was higher for *O. sativa*. However, *C. indica*, a non-food crop, can be grown continuously throughout the year for piggery effluent treatment	Olawale et al. (2021)
Phragmites australis, *I. pseudacorus*	Cd	*System*: plastic containers in greenhouse with floating mat *Stress*: Stormwater amended with N (4 mg L^{-1}), P (1.8 mg L^{-1}) and various doses of Cd (1, 2, and 4 mg L^{-1})	*P. australis* and *I. pseudacorus* could be appropriated to remediate contaminated sites. Both species had TF < 1 and BCF > 1 for Cd accumulation, which indicated phytostabilization potential	Mohsin et al. (2023)

TABLE 2 Aquatic-ornamental plants for phytoremediation—cont'd

Botanical name	Metal	Treatment	Results	References
Eichhornia crassipes	Cr, Li	*System*: pot *Stress*: Hoagland solution amended with Cr ($4\,mg\,L^{-1}$), Li ($4\,mg\,L^{-1}$), and Cr plus Li (2, 4, 6, and $8\,mg\,L^{-1}$). Cr and Li were prepared from $K_2Cr_2O_7$ and LiCl	*E. crassipes* can effectively remove Cr and Li at low concentrations. Cr did not show any harmful effects, while Li showed some negative effects. BCF indicated that *E. crassipes* effectively accumulated the metals in the roots as compared to the stems and leaves	Hayyat et al. (2023)
P. stratiotes, Eicchornia crassipes	Fe, Mn	*System*: pot, static condition *Stress*: Combination of various wastewater from the coal mining area, car and motorcycle wash, kitchen, bathroom, and laundry has a high concentration of heavy metals (Fe and Mn) and other pollutants (BOD, COD, phosphate, DO, ammonia, nitrite, nitrate, pH)	*E. crassipes* and *P. stratiotes* showed high potential for phytoremediation of Fe, Mn, BOD, COD, phosphate, ammonia, nitrate and nitrite. The contact time affected the removal efficiency. BCF showed that the Fe and Mn accumulation in root higher that shoot. TF < 1 indicated the low mobility of the metals in the plant	Wibowo et al. (2023)
Eichhornia crasssipes	Cd, As, Hg	*System*: field experiment	*E. crasssipes* uptake Cd from sewage water in highest amount as compare to As and Hg. *E. crasssipes* plants could be helpful to remove and minimize the heavy metal pollution of water resource	Nazir et al. (2020)

$$Translocation\ index\ (TI,\%) = \frac{metal\ concentration\ in\ shoot}{metal\ concentration\ in\ whole\ plant} \times 100$$

$$Green\ leaf\ ratio\ (GLR,\%) = \frac{green\ leaf\ number}{total\ leaf\ number} \times 100$$

$$Efficiency\ (\%) = \frac{(initial\ concentraion - remaining\ concentration)}{initial\ concentration} \times 100$$

The extraction of heavy metals by plants is usually limited by the availability of heavy metals in soils. Methods of increasing the availability of heavy metals in soils and their translocation to the plant shoots are vital to facilitate phytoextraction. Ethylene-diamine-tetra-acetic acid (EDTA) as chemical chelator could be applied to increase metals bioavailability (Tahmasbian and Safari Sinegani, 2014; Wei et al., 2012). However, the improvement of plant extraction by the application of EDTA is not an appropriate economic and environmental practice. Sodium dodecyl sulfate (SDS) as an important an ionic surfactant may be used together with EDTA and ethylene glycolbis (2-aminoethyl) tetraacetic acid—EGTA) (Wang and Liu, 2014; Liu et al., 2008a, 2009b, 2010). In addition, the use of di-isopropanol amine (DIPA) is less harmful to plants than EDTA (Arshad et al., 2020). Citric acid can increase the solubility of Fe, Mn, and Cu; therefore, it is possible to use citric acid to improve plant nutrition and metal mobility (Tapia et al., 2013). The addition of citric acid, salicylic acid, and saponin enhances the availability of metals and the accumulation of metals by plants (Hao et al., 2012; Cay, 2023). Thioligand (e.g., thiosulfate) is a soil additive used to increase the availability of metals for plant uptake. Furthermore, halogen salts such as potassium chloride (KCl) are commonly used in agriculture for their fertilizing properties (Cassina et al., 2012; Hao et al., 2012). However, caution must be taken with the metal leaching into deeper soil horizons. Tahmasbian and Safari Sinegani (2014) studied ammonium combination of electrokinetic remediation and phytoremediation to decontaminate metal-polluted soil.

Various amendments are required for plant growth. Some agronomic measures such as organic manures, chemical fertilizers, and plant hormones would have an effect on element uptake and plant growth. Phytohormones, especially citokinines (CKs), can also play an important role in the processes related to metal stress. The exogenous application of CKs produces an alleviation of heavy metal toxicity in plants used for phytoremediation, especially in terms of biomass production. Moreover, the proven effects of CKs on stomatal opening and the consequent increase in transpiration efficiency can lead to an increase in the uptake of pollutants by plants (Cassina et al., 2012). Cow manure extract, poultry manure extract, and farmyard manures from cow, swine, and poultry may be applied as amendments to increase bioavailability and phytoremediation (Malarkodi et al., 2008; Hao et al., 2012; Tahmasbian and Safari Sinegani, 2014). Nitrogen fertilization and the addition of organic carbon such as wheat straw reduce the pH, increase the availability of metal, and promote the absorption of metals by plants (Ye et al., 2019).

The mutual symbiosis between arbuscular mycorrhizal (AM) fungi and roots of terrestrial plants has been reported to be a direct physical link between soil and plant roots, increasing soil nutrient exploitation and transfer of minerals to the roots (Smith and Read, 2008). The role of the symbiosis of the AM fungi, for example, *Glomus intraradices* (Davies Jr. et al., 2002; Castillo et al., 2011), *Glomus constrictrum* and *Glomus mosseae* (Liu et al., 2011a; González-Chávez and Carrillo-González, 2013), *Funneliformis mosseae*, and *Funneliformis caledonium* (Zhang et al., 2018) with a plant in phytoaccumulation is associated with the protecton of the plant from heavy metal-induced toxicity, while improving phytocumulative yield indexes. AM fungi can increase available phosphorus (P), improve the capability of ROS scavenging to alleviate plants from heavy metal stress. The plant-mycorrhizal systems are not only being able to tolerate and accumulate high levels of heavy metals in harvestable parts but also have a rapid growth rate and the potential to produce high biomass in the field. Therefore, the symbiotic system provides an attractive system to advance the phytostabilization and/or the phytoextraction of many terrestrial ornamental plants (Riaz et al., 2021).

Plant growth-promoting bacteria (PGPB) are capable of colonizing the plant root, promoting plant growth by various mechanisms—namely through the fixation of atmospheric nitrogen, utilization of 1-aminocyclopropane-1-carboxylic acid (ACC) as a sole nitrogen (N) source, phosphate (P) solubilization, production of siderophores and antipathogenic substances, supply and regulate phytohormones to alter the homeostasis of the plant hormones (auxins, ethylene, cytokinin, gibberellin, abscisic acid, jasmonic acid, and salicylic acid) and also through the transformation of nutrient elements (Tsukanova et al., 2017; Alves et al., 2022). In particular, PGPB can assist in reducing the toxicity of heavy metals to plants in polluted environments. The elimination and detoxification of metals by microbes through processes involving adsorption, chelation, transformation and precipitation, to name a few, are also critically considered (Alves et al., 2022). The PGPB strains *Ralstonia eutropha* and *Chrysiobacterium humi* reduced losses of weight in metal-exposed sunflower plants and induced changes in metal bioaccumulation and bioconcentration (Marques et al., 2013). Inoculation of *Acinetobacter calcoaceticus* Sasm3 (Chen et al., 2015) and *Phyllobacterium myrsinacearum* RC6b (Liu et al., 2015) significantly increased plant biomass under metal stress, but there was no improvement in phytoextraction. While rhizobacteria from *T. erecta* (e.g., *Bacillus cereus* CK505 and *Enterobacter cloacae* CK555) contribute to metal accumulation in plants (Chitraprabha and Sathyavathi, 2018). Under the effects of Zn and/or Cd stress, *Pseudomonas* sp. PDMZnCd2003, which are a Zn/Cd-tolerant PGPB, indicated that pyocyanin was produced in all treatments, while pyoverdine was induced by stress from the metals (Meesungnoen et al., 2021). Heavy metals also affect the dynamics of endophytic PGPB populations and their persistence in plants. Based on the in vitro system, treatment with Zn plus Cd allowed *Cupriavidus plantarum* MDR5 to persist in *Murdannia spectabilis* (Kurz) Faden (Rattanapolsan et al., 2021).

The goal is to evaluate the phytoremediation potential of plants with the assistance of different agricultural management. The bioavailable fraction of heavy metals in the soil can be reduced effectively in a short period of time and with the affordable costs of regular agronomic practices. In terms of best practices, it is recommended that the phytoextraction efficiency of plants be maximized through repeated crops. Plants are harvested at the squaring stage (shortly after the flower begins to appear), and nitrogen-phosphorus-potassium (N-P-K) compounds or manure is applied to increase metal accumulation and plant biomass.

Marigolds (*Tagetes* spp.) are a potential ornamental plant that can be applied for both remediation and income. Marigolds can be used in combination of other plant species being used for the same purpose for protection purpose as the pungent odor of these plants can keep insects away, and the vibrant color of these flowers will help in modifying the aesthetics of the concerned area, which is badly needed for making the area more inhabitable. In addition, marigold secretes the chemical alpha-terthienyl from root tissue, which is an effective nematicide for root-knot nematodes and lesion nematodes control. Because of its allelopathic effect, the marigold was used as a companion plant or a cover crop in cropping systems to protect crops against nematodes (Evenhuis et al., 2004). *T. erecta* (American marigold), *Tagetes patula* (French marigold), and nugget marigold, a triploid hybrid between *T. erecta* and *T. patula*, could also provide economic benefits to the remediators with large-scale plantation on wastelands through marketing flower. Many research studies reported that *T. erecta* can be applied to soil contaminated with Cd (Lal et al., 2008; Liu et al., 2011a, 2011b; Rungruang et al., 2011), Cu (Castillo et al., 2011), Cr (Miao and Yan, 2013; Coelho et al., 2017; Chitraprabha and Sathyavathi, 2018), Ni and Pb (Malarkodi et al., 2008), Cd and Pb (Bosiacki, 2009), Cd, Pb, and

Zn (Madanan et al., 2021), and soil contaminated with sewage sludge (Zn, Ni, Cd, Cr, Pb, As, Hg) (Tepecik et al., 2023). *T. patula* have potential for phytoremediation of soil contaminated with tannery effluent (Ca, Cr, Mn, Fe, Cu, Zn, Pb) (Chatterjee et al., 2012), Cd (Wei et al., 2012; Ye et al., 2019), Fe ore tailing (Chaturvedi et al., 2014), and co-contaminated between metals (Cd, Cu, Pb) and benzopyrene (Sun et al., 2011). Nugget marigold has potential for As-contaminated area (Chintakovid et al., 2008). Many research studies showed that heavy metals were mainly accumulated in leaves > stems > influorescence of marigolds (Chintakovid et al., 2008; Lal et al., 2008; Bosiacki, 2009; Chatterjee et al., 2012; Miao and Yan, 2013). A symplastic pathway rather than an apoplastic bypass contributed greatly to root uptake, xylem loading, and translocation of heavy metals to the shoots. A hydroponic experiment of ^{108}Cd showed that the root to aboveground Cd translocation via phloem was as an important and common physiological process as xylem determination of the Cd accumulation in stems and leaves of marigold seedlings (Qin et al., 2013). *T. patula* was able to tolerate with Cd-induced toxicity by activation of its antioxidative defense system. Cd, which is known to stimulate the formation of free radicals, induced the antioxydative enzymatic acitvites of APX, SOD (CuSOD and CuZnSOD), and different isozymes of glutathione reductase in leaves (Liu et al., 2011b).

In conclusion, the advantages of marigold for phytoremediation are as follows: (i) it is not an edible crop, which will prevent accumulated heavy metals from entering the food chain; (ii) it grows with high shoot biomass when exposed to heavy metals; (iii) it has wide adaptablility to different soils and climate conditions; and (iv) it secretes phytochemicals from root tissues and therefore can serve as a nematicide for the control nematodes. Thus, using marigold may offer a new possibility for rendering agricultural soils or heavy metal contaminated areas, that is, of Cd contamination in Mae Sod District, Tak province, As contamination in Ron Phibun District, Nakhon Si Thammarat Province, and Pb contamination in Klity Creek in the ThungYaiNaresuan Wildlife Sanctuary, Thong PhaPhum District, Kanchanaburi Province, in Thailand (Prasad and Nakbanpote, 2015).

Phytomanangement suggests that fast-growing trees with short rotation coppice system such as eucalyptus and willow should be grown to create green belts around the contaminated land (Pulford and Watson, 2003). Flowers and foliage leaves sold at flower markets in Bangkok and Chiangmai, Thailand, are shown in Fig. 3A–C and Fig. 4A–K. Fern foliage is in great demand for floral bouquets. *Pityrogramma calomelanos* and *Pteris vittata*, indigenous hyperaccumulating fern species are well acknowledged in literature for their arsenic (As) phytoremediation potential (Visoottiviseth et al., 2002; Cao et al., 2003). *Selaginella myosurus*, a native plant in the coal mine soil in Okaba, Nigeria, was suitable for Cd phytostabilization (Ameh and Aina, 2020). Although *Coleus blumei*, a garden plant, behaves as a non-accumulator for Al, this plant can play an important role in the treatment of polluted water with available Al at pH below 5.0 (Panizza de León et al., 2011). *Chlorophytum comosum* (spider plant) and *Amaranthus hypochondriacus* (amaranth) and *Amaranthus caudatus* had application value in the treatment of Cd-contaminated soils (Wang et al., 2012; Li et al., 2013; Bosiacki et al., 2013). *Gynura pseudochina*, a perennial herbal plant, has potential to accumulate Cr (Mongkhonsin et al., 2011), Cd, and Zn (Panitlertumpai et al., 2013; Mongkhonsin et al., 2016). *Panicum maximum* (guinea grass) and *Cyperus rotundus* (nut grass) have properties to grow for covering the terrestrial area contaminated with Cd (Rungruang et al., 2011; Sao

FIG. 3   Variety of flowers at the flower market in Chiang Mai province (A). Examples of (B) Wreaths for auspicious events and funerals, and (C) Krathongs for Loy Krathong Festival. Fields of flowers grown in the north of Thailand. (D) *Aster* sp. White (Cutter flower), (E) *Argyranthemum frutescens* (Marguerite), (F) *Echinacea* spp. (Coneflowers), (G) *Celosia argentea* L. cv. *Plumosa* (Plumed Celosia), (H) *Rudbeckia hirta* (Black-Eyed Susan), and (I) *Helianthus annuus* (Sunflower).

et al., 2007). Lemon-scented geraniums (*Pelargonium* sp. "Frensham", or scented geranium) accumulated large amounts of Cd, Pb, Ni, and Cu from soil in greenhouse experiments (Dan et al., 2000). Pellegrineschi et al. (1994) improved the ornamental quality of scented *Pelargonium* spp. This plant has pleasant odor that adds scent to the toxic metal contaminated soil. Inter-planting of *Allium sativum* with *P. vittata* and *Lolium perenne* promoted the absorption of Cd and Pb and slightly increased the proportion of *Proteobacteria*, *Acidobacteria*, and *Actinobacteria* in the soil (Hussain et al., 2021). Additionally, there are also many herbaceous plants that can tolerate metal-contaminated areas. Some species have beautiful flowers, leaves, and/or shapes that may be planted for restoration and income generation (Mongkhonsin et al., 2019; Khan et al., 2021).

3.2 Aquatic-ornamental plants for phytoremediation

Freshwater as well as seawater resources are being contaminated by various toxic elements through anthropogenic activities and from natural sources. Remediation of contaminated aquatic environment is important as it is for terrestrial environment. Among the toxic substances found in water resources, heavy metals deserve special attention. They are highly

FIG. 4   Foliage leaves for sale in flower markets (A). Popular foliage leaves for decorating bouquets, vases and decorations for various occasions; (B) *Philodendron giganteum* (Giant Philodendron), (C) *Rhapis excelsa* (Thunb.) Henry ex rehder (lady palm), (D) *Polyscias fuilfoylei* (L.) Harms, (E) *Philodendron bipinnatifidum* Schott ex Endl., (F) *Nephrolepis cordifolia* (Sword fern), (G) *Gardenia jasminoides* (Gardenia jasmine), (H) *Melaleuca bracteata* (Revolution Gold), (I) *Cryptomeria japonica* (Japanese cedar), (J) *Dracaena surculose* (Gold-dust Dracaena), (k) *Dracaena fragrans* (cornstalk dracaena).

toxic at low doses, strongly persistent in the environment and living tissues, and easily transferred to the food chain. In addition, their monitoring and removal are costly. Point and non-point sources of pollution are major causes of water-quality degradation. Concentration-based controls of discharge at point sources are insufficient for attaining water-quality targets. Non-point source pollution is difficult to control, because it comes from many different sources and locations. The major sources of non-point pollution are agriculture and urban/rural activity. Agriculture is one of the largest consumers of chemical fertilizers, pesticides, and manure-based composts.

The most important role of plants in wetlands is that they increase the residence time of water, which means that they reduce the velocity and thereby increase the sedimentation of particles and associated pollutants. Aqua-plant treatment systems have been increasingly recognized for controlling eutrophication, main pathways of nutrient removal are estimated that direct assimilation by plant accounted for nitrogen (N) and phosphorus (P) reduction.

III. Ornamentals and tree crops for contaminated substrates

Plants provide oxygen to the water. Nutrients are removed simultaneously by plants and the diverse microbiological communities that depend on the roots of aquatic plants. Constructed wetlands are widely used to remove heavy metals from wastewater (Kataki et al., 2021). Sedimentation, sorption, co-precipitation, cation exchange, phytoaccumulation, and microbial activity are the main processes for the removal of heavy metals (Dhir, 2013). Iron toxicity is a serious problem in crop production in waterlogged soils. Iron, which is essential for plant growth causes toxicity in plants at high concentration, in aerobic soils is found as insoluble Fe^{3+}, but a large amount of this element could be reduced to the more soluble Fe^{2+} in waterlogged soils characterized by anaerobic conditions and a low pH. In this case, wetland plants develop aerenchyma cells to transfer O_2 from aerial parts to the roots, inducing the precipitation of iron oxides or hydroxides on the root surfaces. In addition, iron oxide deposit on the roots (iron plaque) of many wetland plants has high capacity of functional groups on iron(hydr) oxides to sequester metals by adsorption and/or coprecipitation (Zhong et al., 2010; Kataki et al., 2021).

Phytoremediation of heavy metals by means of constructed wetlands constitutes a low-cost, environmentally friendly alternative to conventional cleanup techniques. Furthermore, as metals accumulate mainly in roots, part of the biomass harvested from such wetland has many potential uses in non-food industries. Some of these side products that could yield substantial economic benefits for affected communities are biogas and compost, fibers, and ornamental plants. Plants that have potentials for phytoremediation in aquatic systems are categorized into three groups of aquatic ornamental plants, adapted terrestrial plants, and aquatic macrophytes.

Aquatic-ornamental plants: Aquatic-ornamental plants are known for the high biomass production, tolerance to polluted environments. Some plants show better performance in removing the total N and P, chemical oxygen demand (COD), biological oxygen demand (BOD), and heavy metals (Cr, Pb, CD, Fe, Cu, Mn, etc.) from sewage. N and P have been identified as the two main nutrients needed to control for mitigating the serious situation of eutrophication. *Zantedeschia aethiopica* (white calla lilies) significantly influenced the removal rate of N (Belmont and Metcalfe, 2003). *Nelumbo nucifera* had efficiency to treat total N and total P from zoo wastewater (Jiang and Xinyuan, 1998). *N. nucifera* was reported in bioaccumulation of (−)-enantiomers of chiral polychlorinated biphenyls (PCBs) in leaves > stems > roots (Dai et al., 2014).

Aquatic-ornamental plants would have the dual benefits of water reclamation and flower production with financial returns. The aquatic plant species with ornamental flowers evaluated for heavy metal(s) removal and their high market values are shown in Table 2. *Iris* species could widely use as phytoaccumulators to remediate Cd (Han et al., 2007; Mohsin et al., 2023), Pb (Han et al., 2008; Zhong et al., 2010), Cr and Zn (Caldelas et al., 2012). The high biomass production and metal extraction capacity make *Iris pseudacorus* a good candidate for Cr rhizofiltration and Zn phytoextraction, as reflected by the level of exportation of each metal to leaves. The reduced exportation of Cr to leaves can be advantageous for flower production, yield of emerged parts, and human safety (Caldelas et al., 2012). *Z. aethiopica* also have potential to remediate iron (Casierra-Posada et al., 2014).

Lotus (*Nelumbonaceae*) and water lily (*Nymphaeaceae*) are commonly found in general pond or wetland. *N. nucifera* can accumulate Pb and Cd in root and leaves (Xiong et al., 2012). Water lily (*Nymphaea* spp.) is planted in the mud and its laminae, carried on long petioles, float on the water's surface. These plants are not only tolerant of high levels of toxic heavy metals, but they can also accumulate large amounts of the metals. Water lilies have been reported to accumulate heavy metals such as Cd, Pb, and Cr (Table 2). Epidermal glands on the submerged surfaces of water lily leaf laminae, petioles, and the rhizome accumulate heavy metals. Typical co-crystallization of Cd and Ca accumulating in epidermal glands results in Mn depletion, sometimes causes depletion of other elements such as Fe, Pb, and Hg accumulating in laminar epidermal glands. Lavid et al. (2001) showed deposition and storage of Cd by water lily. Translocation of Cd was not observed in the plant. Cd is apparently immobilized in the epidermal glands in the early stages of heavy-metal accumulation and subsequently in other epidermal cells. This immobilization would explain the high tolerance to heavy metal of water lily. Photosynthetic energy or products may be involved in heavy-metal uptake by the water lily epidermal glands, as metal uptake was lower in the dark, and green tissues in the leaves (laminae and petioles) accumulated a higher level of metals than the non-green rhizome tissues.

Wastewater is one of the most serious environmental problems in Thailand. The major source of water pollution in the country is domestic wastewater. According to the national policy concerning domestic wastewater management, the national target of untreated discharge to receiving water and the environment should be less than 50% of generated domestic wastewater. Therefore, wastewater management at local level and improvement of management capacity need to be supported (Simachaya, 2009). Floriculture activities in constructed wetlands could provide the economic benefits necessary in encouraging small communities to maintain wastewater treatment system. It is feasible to treat domestic wastewater in small rural communities using constructed wetland. These systems are especially valuable for on-site wastewater treatment in developing countries because they involve simple technology, and the costs of construction and operation are low.

Nong (Lake) Han Kumphawapi in northeast Thailand is one of the largest natural freshwater lakes in the country. Kumphawapi is a shallow lake (<4 m water depth), but it has considerable seasonal fluctuations in water level (Chawchai et al., 2013, 2015). The lake's extensive floating herbaceous swamp vegetation has been described by Penny (1999), who noted a domination of grasses (*Poaceae* including *Phragmites* sp.) and sedges (*Cyperaceae*), as well as *Eichhornia crassipes, N. nucifera, Nymphaea lotus, Nymphoides indica, Utricularia aurea, Limnocharis flava, Monochoria hastata,* and *Salvinia cucullata* (floating moss). From more shallow, marginal areas, *Sagittaria sagittifolia, Alocasia macrorhiza, Ludwigia octovalis, Ludwigia adscendens* (creeping water primrose), *Ipomoea aquatica, Alternanthera philoxeroides, Alternanthera sessilis, Typha angustifolia, Persicaria attenuate,* and *Saccharum* spp. are found (Kealhofer and Penny, 1998). Several fern taxa occur as epiphytic elements on the floating or partially rooted herbaceous substrate (Kealhofer and Penny, 1998). The lake is used for water supply, flood control, waste disposal, fisheries, aquaculture, and tourism. The Red Lotus Lake is the famous wetland (under Ramsar convention) and ecotourism place in Nong Han district, Udon Thani province, Thailand (Fig. 5A). Red water lily (*N. lotus* Linn. var. *pubescens*) (Fig. 5B) is a dominant aquatic plant in Nong Han wetland, and it overspreads in large areas that makes this lake a beautiful place for ecotourism. A lot of tourists visit Nong Han and provide income for local people. This point is important to motivate them for phytoremediation. In addition, some edible plants can be harvested for cooking and

FIG. 5 Red lotus lake in Nong Han district, Udon Thani province. (A) Long-tailed boat eco-tourism to see the beauty of red lotus in full bloom at 6:00 a.m. to 11:00 a.m., (B) *Nymphaea pubescens* Willd, (C) drink, savory and sweet dishes prepared from rhizomes, stalks and lotus roots. Some lakeside edible plants, (D) Tinghara nut from *Trapa bicorni*, (E) *Diplazium esculentum*, (F) *Nelumbo nucifera*, (G) *Ipomoea aquatica* Forsk, (H–I) *Neptunia oleracea*.

marketing purpose, as shown in Fig. 5D–I. Therefore, heavy metal contamination and other accumulated toxins in edible aquatic plants should be regularly investigated and controlled.

Adapted terrestrial plants: Terrestrial plants can develop much longer, fibrous root systems covered with root hairs that create an extremely high surface area. Hydroponically cultivated roots of several terrestrial plants were discovered to be effective in absorbing, concentrating, or precipitating toxic metals from polluted effluents. This process was termed rhizofiltration. The stem cuttings of the terrestrial, ornamental plant, *Talinum triangulare* (waterleaf) in hydroponic medium can be definitely classified as a Cu hyperaccumulator, which has potential to support rhizofiltration (Rajkumar et al., 2009). *Talinum portulacifolium* (Homotypic Synonyms: *Talinum cuneifolium*) is a succulent shrub of about 60 cm height with cuneate to obovate leaves, flowers in terminal panicles, and purple-colored corolla. The plant produces flowers and fruits throughout the year. It is widely distributed in India, Arabia, and Africa. Cuttings are a ready means of propagation of these plants (Kumar and Prasad, 2010). *T. portulacifolium* was reported to accumulate high amounts of Cu in leaves. However, the plant has limited metal uptake from soil with high metal concentrations (Tiagi and Aery, 1986). *T. portulacifolium* was able to uptake Cd and Zn in the stems and leaves, but high concentrations of the metals inhibited the regeneration of stem cuttings and stunting of roots (Muthukumar et al., 2018; Muthukumar and Dinesh-Babu, 2020).

Canna or Canna lilies are ornamental plants that can be planted to restore metal contaminants in terrestrial soils (Table 1). Some species can adapt to grow in wetlands and can be used to remove many pollutants in a wetland environment. *Canna flaccida* (golden canna) is a native herbaceous perennial wetland plant that grows in marshes, wetlands, and savannahs and around the edges of lakes and ponds (Gilman et al., 2015), and it has potential in phytoremediation of As (Praveen et al., 2019). *Canna indica* significantly contributed to remove N, P, COD, and BOD_5 from gray water effluent of wastewater treatment (Jamwal et al., 2021). *C. indica* has a significant impact on the treatment of heavy metals, industrial chemicals, pesticides, and pharmaceuticals in constructed wetland (Karungamye, 2022). *C. indica* showed a good potential of phytoremediation for wastewater containing nutrient (N, P) and metals Cr, Mg, Al, Ca, and Fe (Taufikurahman et al., 2018; Olawale et al., 2021). The rhizome of some *Canna* species such as *C. indica* and *Canna edulis* L. are edible and cultivated for starch production (Saranraj et al., 2019). Therefore, the *Canna* plants used to treat heavy metals must be controlled to not be consumed.

Aquatic macrophytes: Aquatic macrophytes can readily achieve biosorption and bioaccumulation of the soluble and bioavailable contaminants from water. A macrophyte is an aquatic plant that grows in or near water and is either emergent, or submergent, or floating. Floating aquatic hyperaccumulating plants absorb or accumulate contaminants by their roots, while submerged plants accumulate metals as a whole plant (Rahman and Hasegawa, 2011). Benthic rooted macrophytes (both submerged and emergent) play an important role in metal bioavailability from sediments through rhizosphere exchanges and other carrier chelates. Macrophytes readily take up metals in their reduced form from sediments, which exist in anaerobic situations due to lack of oxygen and oxidize them in the plant tissues making them immobile and thus get bioconcentrated in their tissues (Okurut et al., 1999). The use of plant roots to absorb, concentrate, and/or precipitate pollutants or harmful metals from a water stream may be defined as phytofiltration. Potential aquatic macrophytes for metal phytoremediation are *E. crassipes* (water hyacinth), *Lemna* sp. (duckweed) and *Spirodela polyrhiza* (greater duckweed), *Azolla* sp. (water fern), *Salvinia molesta* (butterfly fern), *Pistia stratiotes* (water lettuce), *Nasturtium officinale* (watercess), *Elodea canadensis* (waterweed), *Eleocharis acicularis* (needle spikerush), *Hydrilla verticillata* (Esthwaite waterweed), *Hydrocotyle umbellate* (pennywort), *A. philoxeroides* (Alligator weed), and *Typha* spp. (bulrush or cattail) can remove various heavy metals from water. However, the effective removal of metals by these macrophytes depends on the size and root systems of the plants (Prasad, 2007; Newete and Byrne, 2016).

P. stratiotes (Water lettuce or Water cabbage) (Araceae) is a monotypic genus with wide ecological amplitude. It floats on the surface of the water, and its long, numerous and feathery roots hanging submersed beneath floating leaves (Fig. 6A). It is a common aquatic weed in eutrophic Indian water bodies. *P. stratiotes* can be used to treat metals in wastewater (e.g., U, Ni, Fe, Mn, Zn, Cd, and As) (Jha et al., 2016; Leblebici et al., 2019; Wibowo et al., 2023), treat urban sewage (Zimmels et al., 2006), and used in floating wetland in oxidation pond (Savio et al., 2023). *P. stratiotes* have great potential for the phytoremediation of co-contaminated metals, and deposition of the metals occurs in the roots rather than the leaves by the mechanism of apoplasts (Li et al., 2022; Coelho et al., 2023). It can maintain physiological processes under Fe excess, and visual symptoms of Fe toxicity of bronzing leaf edges could be used for bioindicating purposes (Coelho et al., 2023). *P. stratioles* could be

FIG. 6 *Salvinia molesta* (Giant Salvinia) roots (A), *Pistia stratiotes* (Water Lettuce) roots (B), and Giant Molesta's fast-growing problem affecting the wetland ecosystem and the beauty of the Red Lotus Lake (C). The competition in the area of aquatic macrophytes, (D) *S. molesta* and *Azolla microphylla* (Mosquito fern), (E) *P. stratiotes* and *Lemna minor* (duckweed).

bioindication of As-contaminated aquatic environments as chlorosis and necrosis in the leaves (Farnese et al., 2014). However, Lee et al. (1991) reported that arsenic translocation in *P. stratiotes* was slow, and most of the arsenic was strongly adsorbed onto root surfaces from solution. Iron plaque on *Pistia* roots and its significance in many plants, particularly marshy plants and aquatic macrophytes, have shown the deposits of ion oxides or hydroxides on their root surface. Large surface of the iron plaque provides a reactive substrate to seques-ter metals (Lu et al., 2011; Singha et al., 2019). Further, the transfer of metals to herbivores via metal-rich iron plaque on macrophyte roots needs critical investigation. This might be a sig-nificant method for assessing the food chain transfer and concomitant toxic functions. *P. stratiotes* mats degrade water quality by blocking the air-water interface, reducing oxygen levels in the water, and thus threatening aquatic life.

Typha domingensis (cattail) species can be used to remediate both heavy metals and organic compounds. *T. domingensis* (cattail) has reported the ability to uptake heavy metals (e.g., Pb, Fe, Al, Zn, Ni, Cd, Cu, and Cr) (Hegazy et al., 2011; Mojiri et al., 2013; Soudani et al., 2022). Furthermore, the cattail maintains removal efficiency in co-contaminated metal environ-ments (Mufarrege et al., 2021). The accumulation of metal(s) in plant organs attained the highest values in roots, rhizomes, and old leaves, with the order tissue of roots > stems > leaves > flowers. The cattail was capable of accumulating the heavy metal ions preferentially from wastewater than from sediments (Hegazy et al., 2011; Soudani et al., 2022).

The transfer factors of nutrients and heavy metals indicated that *T. domingensis* is appropriate for nitrogen phytoextraction from water and sediment, whereas it is suitable for metals phytostabilization (Eid et al., 2012). *T. angustifolia* (narrow-leaved cattail) have potential to remove Cr and synthetic reactive dye from wastewater (Nilratnisakorn et al., 2009; Taufikurahman et al., 2018). Rhizofiltration was found to be the best mechanism to explant *Typha* phytoremediation capability.

E. crassipes (water hyacinth) has the ability to absorb heavy metals and nutrients and is suitable for wastewater treatment. The use of water hyacinth can reduce eutrophication by removing ammonia, phosphorous, and nitrate from the municipal wastewater treatment plant effluence prior to discharge (Kutty et al., 2009). *E. crassipes* can bioconcentrate emerging organic pollutants such as endocrine disruptors and neonicotinoids by rhizofiltration, and it can prevent the dispersion of the pollutants into environment (Laet et al., 2019). It is beneficial for phenol and cyanide elimination, both pollutants are absorbed through the root by plasmalemma and accumulated into the root cells and stem (Singh and Balomajumder, 2021). *E. crassipes* is an alternative species for metal remediation in aquatic system and/or wastewater polluted with a low level of Zn, Cr, Cu, Cd, Pb, Ag, and Ni (Odjegba and Fasidi, 2007), Cr and Li (Hayyat et al., 2023), Cd, As, and Hg from sewage water (Nazir et al., 2020). It also showed removal efficiency in pH, BOD, COD, total dissolved solid (TDS), Cr, and Pb presenting in river nearby dye industry (Panneerselvam and Shunmuga, 2021). *E. crassipes* effectively accumulated the metals in the roots as compared with the stems and leaves. However, *E. crassipes* (water hyacinth) may cause severe water management problems because of its huge vegetative reproduction and high growth rate. Therefore, the use of water hyacinth in phytoremediation technology should be considered carefully.

E. canadensis (waterweed) grows rapidly in favorable conditions and can choke shallow ponds, canals, and the margins of some slow-flowing rivers. The accumulation of Cd, Cr, Cu, Mn, Ni, Pb, Zn, and Fe collected from streams and ponds in polluted sites by *E. canadensis* was varied according to site, metal, and season (Thiebaut et al., 2010). *Pedosphaeraceae* and *Parasegetibacter*, a group of specialized sediment bacteria in the rhizosphere of *E. canadensis*, posed significant potentials for promoting plant and Cd phytoremediation (Yuan et al., 2022).

Lemna sp. (duckweed) and *S. polyrhiza* (greater duckweed) have received the greatest attention for toxicity tests as they are relevant to many aquatic environments, including lakes, streams, and effluents. *Lemna minor* contributed to the removal of Cd, Cu, and Ni and was more efficient in extracting Pb from industrial effluents (Bokhari et al., 2015). *S. polyrhiza* survived in high concentration of As(V) solution and may be applied for As phytofiltration in contaminated water or paddy soils (Zhang et al., 2011). *Wolffia globosa* (Asian water meal), a local food from water resources, has been reported as a strong Cd accumulator due to passive adsorption of Cd by the apoplast. It could be used for phytofiltration of the co-contaminated Cd and As in fresh aquatic environment (Xie et al., 2013). Therefore, *W. globosa* from water contaminated by heavy metals should not be consumed. *Hydrocotyle umbellata* (pennywort) is an aquatic plant commonly found in many tropical countries. The plants grow very rapidly and serve as an ornamental and decorative purpose. In addition, *H. umbellate* had phytofiltration efficiency to treat BOD, TDS, electrical conductivity (EC), accumulated heavy metals (e.g., Cd, Cu, Pb, Cr, and Zn) in both roots and shoots of the plant (Taufikurahman et al., 2018; Bokhari et al., 2022).

Salvinia is a water fern with the ability to remove contaminants such as heavy metals and inorganic nutrients. *S. molesta* (butterfly fern) with a third modified root-like frond that hangs in the water is shown in Fig. 6B. *S. molesta* had ability to absorb Pb and remove Cr from Batik industrial effluent (Rachmadiarti et al., 2022). It was able to survive healthy in an industrial effluent containing Cr, Pb, Cu, and Cd and accumulated the metals (Ranjitha et al., 2016). *Salvinia cucullate* (foaling moss) accumulated Cd and Pb in the root cells and was partially transported to the leaves; however, a high Cd and Pb caused chlorosis on the leaves (Phetsombat et al., 2006). Kumari et al. (2016) showed that *S. molesta*, *Salvinia natans*, and *Salvinia auriculata* had the innate capability to accumulate Cd, Cu, Cr, Hg, Pb, Ni, and Zn, and increasing the metal concentration leads to an increase in the bioaccumulation of metal in tissues. Seriously, *S. molesta* is a weed of National Significance. It is regarded as one of the worst weeds because of its invasiveness, potential for spread, and economic and environmental impacts. It floats on the slow-moving water and can grow rapidly to cover the entire water surface with a thick mat of vegetation. This shades out any submerged plant life and impedes oxygen exchange, making the water unsuitable for fish and other animals (Kumari et al., 2016). *S. molesta* is an invasive alien aquatic plant in Thailand. Although the Act prohibiting its introduction and cultivation has been in force since 1978, there remains a problem of controlling its distribution and eradication. The problem of overgrowth of *S. molesta* in the Nong Han Kumphawapi affecting the wetland ecosystem and the beauty of the Red Lotus Lake is shown in Fig. 6C. Fig. 6D–E shows examples of competitive growth of several macrophytes in water bodies.

Consequently, phytoremediation of heavy metals and other contaminants in aquatic system should be constructed by growing ornamental plants for phytostabilization, aesthetically appealing and cut-flower for income. However, aquatic macrophytes should be applied to the same wetland. Aquatic macrophytes and floating islands commonly found in wetland ecosystem of aquatic ornamental plants are shown in Fig. 7. Peatlands low in alkalinity ($<100\,\mu eq/L$ of Ca) have been classified as being the most suspectable to the process of acidification. Therefore, the aquatic plants in these sensitive peatlands may begin to accumulate potentially toxic metals, such as Cd, with concurrent decreases in levels of essential elements. If so, then these effects may be transferred through the food chain. The effects of peatland parameters [pH, alkalinity, dissolved organic carbon (DOC), and sediment organic matter] on the accumulation of Cd in macrophytes is an important step in determining the variables that may lead to higher Cd in peatland vegetation and subsequently wildlife (Thompson et al., 1997). Mine tailings rich in sulfides, for example, pyrite, can form an acid mine drainage (AMD) if it reacts with atmospheric oxygen and water, which may also promote the release of metals and As. To prevent AMD formation, mine tailings rich in sulfides may be saturated with water to reduce the penetration of atmospheric oxygen. An organic layer with plants on top of the mine tailings would consume oxygen, as would plant roots through respiration. Thus, phytostabilization on water-covered mine tailings may further reduce the oxygen penetration into the mine tailings and prevent the release of elevated levels of elements into the surroundings. Since, some wetland plant species have been found with the latter property, for example, *Typha latifolia*, *Glyceria fluitans*, and *Phragmites australis*, wetland communities may easily establish on submerged mine tailings (Sheoran and Sheoran, 2006; Williams, 2002; Wood and Mcatamney, 1994; Woulds and Ngwenya, 2004; Ye et al., 2001). Water lily (*Nymphaea* spp.) is aesthetically appealing to provide a

FIG. 7 Aquatic macrophytes and floating islands commonly found in wetland ecosystem of aquatic ornamental plants; (A) floating island and *Salvinia cucullata* (floating moss), (B) floating island and *Phragmites* sp. (reed), (C) mass of grasses, reed, and cattail in floating island with *Ipomoea aquatica* (water morning glory) and *Ludwigia adscendens* (creeping water primrose), (D) *Utricularia aurea* (common bladderwort).

landscape-pleasing environment. and they are potentially hyperaccumulators of nutrients and metals as they have extensive roots and provide a large surface area for the biofilms formation and thus enhance the microbial activities. Nitrogen concentrations in water lily also generally increase in response to increased P loads. The high affinity of water lily for P, combined with their subsequent influence on N uptake, suggests that these components can play an important role in wetland nutrient cycling (Newman et al., 2004). *Nymphae capensis* could be used as a biomonitor of the discharge of acid waters from Reserve Creek, Australia, containing high concentrations of metals to identify priority sites (hotspots) in three acid sulfate soil impacted enrichments. Water lily lamina concentrations of a suite of metal(loid)s (Al and Fe) were significantly higher than plants collected from an unpolluted "reference" drainage channel, thus validating the concept of using this species as a biomonitor (Stroud and Collins, 2014).

4 Ornamental plant for phytoremediation, sustainability, and bioeconomy

Floriculture is the best option for such alternative agricultural practices in contaminated areas. Ornamental plants not only reduce the concentration of metals and clean up

Marigold farm

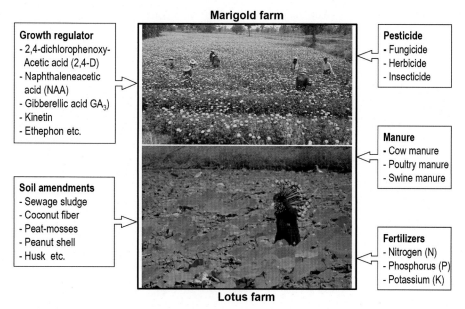

Growth regulator
- 2,4-dichlorophenoxy-
  Acetic acid (2,4-D)
- Naphthaleneacetic
  acid (NAA)
- Gibberellic acid GA$_3$)
- Kinetin
- Ethephon etc.

Pesticide
- Fungicide
- Herbicide
- Insecticide

Manure
- Cow manure
- Poultry manure
- Swine manure

Soil amendments
- Sewage sludge
- Coconut fiber
- Peat-mosses
- Peanut shell
- Husk etc.

Fertilizers
- Nitrogen (N)
- Phosphorus (P)
- Potassium (K)

Lotus farm

FIG. 8   Marigold and lotus cultivation in contaminated environment.

degraded soils, but they also generate income from sales that cut flowers and foliage in flower markets and landscapes for tourism. In addition, growing non-edible crops reduce the risk of metals entering the human food chain. Marigolds (*Tagetes* spp.) and lotus (*Nelumbonaceae*) can be modeled ornamental plants that are successful in both the phytoremediation of heavy metals and the economic benefits sold such as offering sacrifices in Thailand. Furthermore, the application of growth regulators, soil amendments, pesticides, manure, and fertilizers for commercial ornamental plants should be optimized for both income and remediation (Fig. 8).

(a) Marigold (*Tagetes* spp.)

Marigold is used for adoration and found generally in festivals in Thailand. This is a short-lived plant that takes 45–55 days to be cut and sold, and it can keep the produce for 1–4 months depending on the maintenance and planting area conditions. Planting costs are not high because it does not use many chemicals, and fermentation of organic fertilizers can help to reduce costs. Some agriculturists manage entire processes of marigold flower production (seed maturing, planting, flower cutting, and direct selling to markets) and selling their own marigold seeds (by breeding) (Fig. 9A and B). In addition to generating a good income for the owner of the area, it also creates jobs and careers for community members. Marigolds can be harvested throughout the year, especially from April to July and have been planted a lot in the northern and northeastern regions of Thailand. In addition, marigolds use less water and can be grown to generate income during the dry season, replacing rice and vegetables, which require much water. The high demand of the flowers is on Buddhist Sabbath days and

FIG. 9 Processes for commercial planting of marigold and lotus farm. Ornamentals used in remediation generate employment for harvesting (A), (D); sorting (B), (C); and marketing in a flower market (C), (F).

Thai tradition festivals. Some celebrations including weddings, funerals, and canvassing have mostly used marigold flowers for decoration. Approximately 1 billion marigold flowers are sold in flower markets that day. Plenty of marigolds selling in a store in the biggest fresh flower market in Bangkok (Pak Khlong Talat in Thai) are presented in Fig. 9C.

Commercial cultivation of marigolds for soil remediation should be selected based on market requirements and metal tolerance. There are various species of marigolds including *T. erecta* (African marigold), *T. patula* (French marigold), *Tagetes minuta* (wild marigold), *Tagetes lucida* (Maxican mint marigold), *Tagetes tenuifolia* (signet marigold), and Nugget marigold (triploid hydbrid between *T. erecta* and *T. patula*). Many large marigold fields can also be eco-tourism sites such as marigold fields in Pobpra district, Tak province, near the Mae Sot province (this area rich in Zn and other heavy metals, especially Cd contaminations). Chintakovid et al. (2008) studied the potential of nugget marigold for arsenic (As) phytoremediation in Ron Phibun district, Nakhon Si Thammarat Province and found that the marigold accumulated high As concentration and growing well in the field As-contaminated area. French marigold was qualified as a potential tool for phytostabilization and survives excess of heavy metals present in the Fe ore tailing (Chaturvedi et al., 2014). African marigold was the potential metal-hyperaccumulative plant for growing in soil contaminated with Cr from electroplating (Chitraprabha and Sathyavathi, 2018) and soil amended by municipal sewage sludge containing metals (Zn, Ni, Cd, Cr, Pb, As, and Hg) (Tepecik et al., 2023).

The planting of marigolds for commercial purpose is specially attended starting from seedling until harvesting. For complete seedlings, the sowing method in plug tray is recommended. Sowing media and growing support materials such as peat-mosses, coconut fiber, husk, and peanut shell are good for growing media. Pesticides such as fungicide (e.g., *Pythium* spp. prevention) are added to seedling for preventing plant pathogens to damage seedling. Soil amendments are applied during soil preparation. Fertilizers containing N and P are essential for healthy seedlings to grow well. During the tillage process, fertilizer (N-P-K) is added to the soil, manure, and organic matter to promote plant growth. The amount of fertilizer per plot/plot depends on the characteristics and properties of the soil because of differences in each area. Pesticides are an important factor in controlling plant diseases. Plant growth regulators are essential for all stages of growth. In particular, growing marigolds in metal-contaminated soil should increase the capacity of metals to accumulate in plants and increase plant growth rates with manure (Malarkodi et al., 2008), arbuscular mycorrhiza fungi (Castillo et al., 2011; Liu et al., 2011a), rhizobacteria (Chitraprabha and Sathyavathi, 2018), fertilizer, and soil amendment (Ye et al., 2019), etc.

(b) Lotus (*Nelumbonaceae*)

The lotus flower is in high market demand as it is used to decorate many traditional events in Thailand, including religious ceremonies and decorations. More than a million lotus flowers are sold per day at flower markets in Thailand. Currently, many rice fields are converted to lotus fields. There are four varieties of lotus that are commonly planted and in demand in the flower market; Roseum Plenum (*N. nucifera* Gaertn. cv. Roseum Plenum), Magnolia Lotus (*N. nucifera* Gaertn. cv. Album Plenum), Hindu Lotus (*N. nucifera* Gaertn.), and Sacred Lotus (*N. nucifera* Gaertn.). Fig. 9D–F shows a field of lotus, picking lotus buds, sorting, and selling in the flower market. Lotus planting is done by planting it deep into the muddy soil and flooding the pond with a height of about 30–50 cm in order to prevent the flower stalks being too long and producing large flowers. One-month-old lotuses begin to flower and emerge above the water, and the 4- to 5-day-old buds are harvested for sale. If the lotus pond is properly managed, the lotus plants can blossom for 4–5 months. In Thailand, lotus production is higher in the sunny summer and less in the rainy and winter seasons, with prices varying depending on the festival and produce. Old lotus plants, which produce fewer flowers, need to be destroyed by blending with mud to become fertilizer before planting again. The addition of bioextract fermented from citrus fruit peels and banana stalks is a bioaugmentation of beneficial microorganisms and a biostimulation of nutrients to promote indigenous microorganisms. Lotus requires a large amount of essential nutrients (e.g., nitrogen and phosphorus) from either fertilizer or compost for growth. Nutrients mass balance is required in lotus ponds to effectively treat nonpoint sources (e.g., agricultural wastewater) of pollution discharged from agricultural lands (Seo et al., 2010).

Lotus plant (*N. nucifera*) is a valued aquatic plant grown and consumed throughout Asia. All parts of the lotus, stamens, pollen, flowers, stems, rhizomes, and leaves, are used as food and for various therapeutic purposes in India and China. In Thailand, the lotus is an edible plant and can be utilized and sold in all parts. Young stem, rhizomes, flowers, and seeds are fresh vegetables for cooking both savory dishes, sweet dishes, and making beverages

(Fig. 5C). Lotus stamen, seeds, rhizomes can be dried and/or extracted to make teas, herbal medicines, and cosmetics. The large lotus leaves can be used to wrap steamed foods to add aroma and to be molded into biodegradable containers.

Lotus plant has metal dominate hypertolerance as it adapts accumulation strategy to accumulate heavy metals. Leong et al. (2012) reported that the highest concentration of phenolic content, Pb, and Cr was in the seeds of lotus plants grown in ex-tin mining pond. The concentration of Pb found exceeded the permissible limits of Codex Alimenttarius. Although there was a report that the contents of heavy metals in lotus seeds from mining areas are no higher than in reference area (Dingjian, 2012) and heavy metals in lotus roots accumulated in primary independent form or partially negatively correlated with soil (Xiong et al., 2012, 2013), assessment of metal concentration should be a mandatory priority to the utilization of contaminated area for edible plant cultivation.

Lotus (*Nelumbonaceae*) and Water lily (*Nymphaeaceae*) are important to Thai people's lives. Therefore, in order to conserve, develop species, and promote research and utilization, a lotus museum was established at the Rajamangala University of Technology Thanyaburi (Fig. 10A), in which, some collection and plantation of lotus and water lily species are shown in Fig. 10B–G. Water lily (*Nymphaeaceae*) is popular for breeding and selling as an ornamental plant. Many species of water lily and materials for its cultivation and propagation are sold in the flower and tree markets (Fig. 10H–I). In addition to earning income from water lily breeding as an ornamental plant, the stalks of the water lily flower can be harvested and sold

FIG. 10   Lotus Museum at the Rajamangala University of Technology Thanyaburi under the Royal Project of Her Royal Highness Princess Maha Chakri Sirindhorn, (A) collection and plantation, (B)–(G) some of the lotus (*Nelumbonaceae*) and water lily (*Nymphaeaceae*) species, (H) a flower shop that sells many species of lotus and water lily and materials for its cultivation and propagation, and (I) many colors of water lily.

as ingredients in many Thai dishes. However, there were reports of heavy metal accumulation in *Nymphaea* spp. Outridge (2000) reported the accumulation of Pb in the shoot of *Nymphaea odorata*. A high accumulation of Cd and Ca occurred in the mature leaf lamina of *Nymphaea violacea* (Lavid et al., 2001), and a main role of the root was to accumulate Cr(VI) and translocate to the leaves of *Nymphaea spontanea* (Choo et al., 2006). Thus, the use of water lilies in food and medicine should be concerned with the source and the possibility of contamination by heavy metals.

The landscape for ecotourism is large ornamental fields that tourists enjoy visiting to learn how to plant and take pictures with blooming flowers as a backdrop. In addition to marigold fields in Tak Province and red lotus lake in Udon Thani Province, which are unseen, there are many other best flower fields in Thailand such as Singha Park in Chiang Rai Province and Jim Thompson Farm in Nakhon Ratchasima Province. Mae Rim District, Chiang Mai Province, is a large winter ornamental plantation area that has been successful in managing agritourism. The topography of Mae Rim District is hilly and high mountains sloping into valleys with slopes between 20% and 35% hilly and 35%–50% steep. The average annual temperature, humidity, and rainfall are 23.3°C, 82.1%, and 1344.1 mL, respectively. Most soils have low fertility and have deteriorated from erosion. Therefore, the promotion of winter flower planting is to restore the area and generate income for farmers. Some flower beds planted in Mae Rim District are shown in Fig. 3D–I. Fig. 11A shows flower

FIG. 11 Ornamental plants in eco-development, (A) Homestays around the flower fields in Mae Rim district, Chiangmai province, Thailand, (B and C) flower and foliage plants market, (D) lighting to stimulate flowering, (E) essential oil extraction and perfumery, and (F and G) edible flowers from a greenhouse on abandoned farmland during the restoration.

III. Ornamentals and tree crops for contaminated substrates

plots along the hill slopes and homestays for tourists surrounding. Some parks or gardens are near wastewater treatment plants or ponds. These places are the potential to organize and apply sewage sludge for soil productivity for planting. Sewage sludge is useful for soil amendment because it was highly digested organic matters, N, and P. Nevertheless, the use of sewage sludge could contaminate with heavy metals, pathogenic microorganism, methane production, bad smell, and high nitrate could be toxic. However, phytoremediation could be resolved these problems, and planting ornamental plants also enhances aesthetics. Moreover, the floral garden festival could be making income and also reduce the area for disposing sewage sludge. In addition, waste lands are considered to utilize and some to increase fertility by agriculture. Some areas in urban can plant ornamental plants for aesthetics and reduce air and soil pollution, and used water from houses could be applied. This point could be motivating local people in and around contaminated areas or urban to cooperate for cleaning up, aesthetics, and also income.

The popularity of ornamentals has led to careers in breeding and selling flower plants and foliage plants in the flower tree markets (Fig. 11B and C), including the development of related knowledge such as accelerating the flowering of trees by turning on the lights (Fig. 11D), and extracting of perfume from flowers (Fig. 11E). Edible flowers are gaining attention as they can be sold at high prices for garnishing in bakeries and Fine Dining (Fig. 11F–G). However, growing edible flowers must be done in areas that are not contaminated with heavy metals and toxins. Edible flowers can be propagated in greenhouses on farmland abandoned during restoration. In addition, making garlands and arranging Thai flower stands are also a good income-generating occupation (Fig. 12A–C). The main ingredients in Thai flower

FIG. 12   A career in making garlands (A) and arranging Thai-style flower stands (B) in a flower market in Bangkok. An example of an elegant Thai double-flowered panicle (C). The main ingredient in Thai flower arrangements: (D) banana leaves and trees, (E) *Pandanus amaryllifolius* (Bai Toey), (F) *Gomphrena globose* (bachelor's button), (G) *Calotropis gigantes* (crown flower), and (H) *Jasminum officinale* (jasmine).

arrangements are banana leaves and trees, *Pandanus amaryllifolius* (Bai Toey), *Gomphrena globose* (bachelor's button), *C. gigantea* (crown flower), and *Jasminum officinale* (jasmine), as shown in Fig. 12D–H. Therefore, the cultivation of these plants should be studied in the phytoremediation properties and the generating income to promote sustainable development.

5 Management of phytoremediating neornamental biomass

Once ornamental plants are used for contaminated soil remediation, they may be harvested and then incinerated together to recover the residual heavy metals. Alternatively, cut flowers that contain a limit of contaminants can be sold to households for their ornamental values. The other parts of plants containing high heavy metal(s) content should not be dispersedly discarded after death. Management and disposal of the mass amounts of plants with high metal content will be an important concern. After harvesting, metal-enriched biomass needs to be safely disposed of, a technical issue that remains partly unresolved. Suitable techniques are carbonization and incineration, hydrolysis and fermentation, anaerobic digestion, and biogas production (Ghosh and Singh, 2005; Tan et al., 2023).

Carbonization and incineration: Sun drying to reduce the water content is required. In addition, the ash content of air-dried plants is too high to get a good fuel as the end product. In the case of metal hyperaccumulating plants, incineration would be a suitable management.

Hydrolysis and fermentation: Liquid fuel, such as ethanol, may be produced from phytoremediating plants by hydrolysis together with fermentation, which would make the plants a good substrate. Therefore, some pre-treatment is necessary to make the sugar more readily available for chemical hydrolysis, which requires high temperatures, strong acids, and pressurized reactors. Even it is economically feasible to produce fuel from phytoremediating plants, metal(s) content in by-product sludge and its recontamination possibility should be tested.

Anaerobic digestion and production of biogas: Anaerobic digestion is a biological process by which organic matter is degraded in the absence of oxygen, and biogas is produced as a by-product. Biogas production from phytoremediation biomass would be a viable, interesting, and environmentally sound idea for phytoremediating plant management. However, there are some limitations in biogas production from phytomass. The high lignin content can reduce the actual production, and the low bulk density could result in large voids with poor compaction and low feed rates. The mixture of phytomass and manures may produce more biogas. There is a good possibility to use phytoremediation plants in biogas production. However, metal content and speciation in sludge from the anaerobic digestion should be investigated to prevent its redistribution in the environment.

Although there are some processes for the disposal of high metals, plants, economically and environmentally, have to be concerned to elucidate and convince people with phytoremediation. In addition, some aquatic macrophytes, for example, *Cyperrus* spp. are a biomaterial for making mats (Fig. 13A and B). Phytostabilization is the major remediation processes of lotus and water lily; therefore, herbal tea could be prepared from the petal of the lotus blossom (Fig. 13C), and the compost generated from the waste biomass (Fig. 13D) could serve as compost and growing media.

FIG. 13 Management of phytoremediating ornamental biomass: (A, B) mats from *Cyperus* spp., (C) tea from lotus petal, and (D) composite biomaterials.

6 Conclusions

Phytoremediation is a technology that uses plants to degrade, assimilate, metabolize, or detoxify metal and organic chemical contamination. The important criterion for the potential plants used for phytoremediation, that is, provision of economic benefits, harvesting management, and by-product utilization can motivate local people for cultivation the plants for phytoremediation. Growing ornamental plants by their habitat and phytoremediation potential not only makes the environment colorful but also remediates the contaminants in terrestrial and aquatic area environments. Many ornamental plants are not edible plants; the risk of entering metals into the food chain is reduced. Incomes from cut-flower and/or tourism may convince local people to remediate their contaminated lands and reservoir. Management and disposal of phytoremediating plants with high contaminants are an important concern. However, ornamental plants will add a new dimension to the field of phytoremediation and phytomanagement of contaminated aquatic and terrestrial environments. Ornamental plants have the additional benefit of improving the aesthetics of the environment in addition to remediating the environment and generating additional income, including additional employment opportunities.

References

Abbas, G., Murtaza, B., Bibi, I., Shahid, M., Niazi, N.K., Khan, M.I., Amjad, M., Hussain, M., Natasha, 2018. Arsenic uptake, toxicity, detoxification, and speciation in plants: physiological, biochemical, and molecular aspects. Int. J. Environ. Res. Public Health 15, 59.

Abbasi, A., Zakaly, H.M.H., Badawi, A., 2021. The anthropogenic radiotoxic element of ^{137}Cs accumulate to biota in the Mediterranean Sea. Mar. Pollut. Bull. 164, 112043.

Alves, A.R.A., Yin, Q., Oliveira, R.S., Silva, E.F., Novo, L., 2022. Plant growth-promoting bacteria in phytoremediation of metal-polluted soils: current knowledge and future directions. Sci. Total Environ. 838, 156435.

Ameh, E.G., Aina, D.O., 2020. Search for autochthonous plants as accumulators and translocators in a toxic metal-polluted coal mine soil in Okaba, Nigeria. Sci. Afr. 10, e00630.

Arshad, M., Naqvi, N., Gul, I., Yaqoob, K., Bilal, M., Kallerhoff, J., 2020. Lead phytoextraction by *Pelargonium hortorum*: comparative assessment of EDTA and DIPA for mobility and toxicity. Sci. Total Environ. 748, 141496.

Balafrej, H., Bogusz, D., Triqui, Z.A., Guedira, A., Bendaou, N., Smouni, A., Fahr, M., 2020. Zinc hyperaccumulation in plants: a review. Plan. Theory 9, 562.

Barceló, J., Poschenrieder, C., 2004. Structural and ultrastructural changes in heavy metal exposed plants. In: Prasad, M.N.V. (Ed.), Heavy Metal Stress in Plant, second ed. Springer-Verlag, Berlin Heidelberg, India, pp. 223–247.

Belmont, M.A., Metcalfe, C.D., 2003. Feasibility of using ornamental plants (*Zantedeschia aethiopica*) in subsurface flow treatment wetlands to remove nitrogen, chemical oxygen demand and nonylphenol ethoxylate surfactants—a laboratory-scale study. Ecol. Eng. 21, 233–247.

Bokhari, S.H., Ahmad, I., Mahmood-ul-Hassan, M., Mohammad, A., 2015. Phytoremediation potential of *Lemna minor* L. for heavy metals. Int. J. Phytoremediation 18, 25–32.

Bokhari, S.H., Nawaz, G., Azizullah, A., 2022. Heavy metals phytofiltration potential of Hydrocotyle umbellate from Nullah Lai wastewater and its environmental risk. Int. J. Phytoremediation 24, 1465–1474.

Borghei, M., Arjmandi, R., Moogouei, R., 2011. Potential of *Calendula alata* for phytoremediation of stable cesium and lead from solutions. Environ. Monit. Assess. 181, 63–68.

Bosiacki, M., 2008. Accumulation of cadmium in selected species of ornamental plants. Acta Sci. Pol. Hortoru. 7, 21–31.

Bosiacki, M., 2009. Phytoextraction of cadmium and lead by selected cultivars of *Tegetes electa* L. Part II. Content of Cd and Pb in plants. Acta Sci. Pol. Hortoru. 8, 15–26.

Bosiacki, M., Kleiber, T., Kaczmarek, J., 2013. Evaluation of suitability of *Amaranthus caudatus* L. and *Ricinus communis* L. in phytoextration of cadmium and lead from contaminated substrates. Arch. Environ. Prot. 39, 47–59.

Caldelas, C., Araus, J.L., Febrero, A., Bort, J., 2012. Accumulation and toxic effects of chromium and zinc in *Iris pseudacorus* L. Acta Physiol. Plant. 34, 1217–1228.

Cao, X., Ma, L.Q., Shiralipour, A., 2003. Effects of compost and phosphate amendments on arsenic mobility in soils and arsenic uptake by the hyperaccumulator, *Pteris vittata* L. Environ. Pollut. 126, 157–167.

Casierra-Posada, F., Blanke, M.M., Guerrero-Guío, J., 2014. Iron tolerance in Calla lilies (*Zantedeschia aethiopica*). Gesunde Pflanz. 66, 63–68.

Cassina, L., Tassi, E., Pedron, F., Petruzzelli, G., Ambrosini, P., Barbafieri, M., 2012. Using a plant hormone and a thioligand to improve phytoremediation of Hg-contaminated soil from a petrochemical plant. J. Hazard. Mater. 231–232, 36–42.

Castillo, O.S., Dasgupta-Schubert, N., Alvarado, C.J., Zaragoza, E.M., Villegas, H.J., 2011. The effect of the symbiosis between *Tagetes erecta* L. (marigold) and *Glomus intraradices* in the uptake of Copper(II) and its implications for phytoremediation. New Biotechnol. 29, 156–164.

Cay, S., 2023. Assessment of tea saponin and citric acid–assisted phytoextraction of Pb-contaminated soil by *Salvia virgata* Jacq. Environ. Sci. Pollut. Res. 30, 49771–49778.

Chatterjee, S., Singh, L., Chattopadhyay, B., Datta, S., Mukhopadhyay, S.K., 2012. A study on the waste metal remediation using floriculture at East Calcutta Wetlands, a Ramsar site in India. Environ. Monit. Assess. 184, 5139–5150.

Chaturvedi, N., Ahmed, M.J., Dhal, N.K., 2014. Effects of iron ore tailings on growth and physiological activities of *Tagetes patula* L. J. Soils Sediments 14, 721–730.

Chawchai, S., Chabangborn, A., Kylander, M., Löwemark, L., Mörth, C.-M., Blaauw, M., Klubseang, W., Reimer, P.J., Fritz, S.C., Wohlfarth, B., 2013. Lake Kumphawapi—an archive of Holocene palaeoenvironmental and palaeoclimatic changes in Northeast Thailand. Quat. Sci. Rev. 68, 59–75.

Chawchai, S., Yamoah, K.A., Smittenberg, R.H., Kurkela, J., Väliranta, M., Chabangborn, A., Blaauw, M., Fritz, S.C., Reimer, P.J., Wohlfarth, B., 2015. Lake Kumphawapi revisited—the complex climatic and environmental record of a tropical wetland in NE Thailand. The Holocene 26, 614–626.

Chen, B., Ma, X., Liu, G., Xu, X., Pan, F., Zhang, J., Tian, S., Feng, Y., Yang, X., 2015. An endophytic bacterium *Acinetobacter calcoaceticus* Sasm3-enhanced phytoremediation of nitrate-cadmium compound polluted soil by intercropping *Sedum alfredii* with oilseed rape. Environ. Sci. Pollut. Res. 22, 17625–17635.

Chintakovid, W., Visoottiviseth, P., Khokiattiwong, S., Lauengsuchonkul, S., 2008. Potential of the hybrid marigolds for arsenic phytoremediation and income generation of remediators in Ron Phibun District, Thailand. Chemosphere 70, 1532–1537.

Chitraprabha, K., Sathyavathi, S., 2018. Phytoextraction of chromium from electroplating effluent by *Tagetes erecta* (L.). Sustain. Environ. Res. 28, 128–134.

Choo, T.P., Lee, C.K., Low, K.S., Hishamuddin, O., 2006. Accumulation of chromium (VI) from aqueous solutions using water lilies (*Nymphaea spontanea*). Chemosphere 62, 961–967.

Coelho, L.C., Bastos, A.R.R., Pinho, P.J., Souza, G.A., Carvalho, J.G., Coelho, V.A.T., Oliverira, L.C.A., Domingues, R.R., Faquin, V., 2017. Marigold (*Tagetes erecta*): the potential value in the phytoremediation of chromium. Pedosphere 27, 559–568.

Coelho, D.G., de Silva, V.M., Filho, A.A.P.G., Oliveira, L.A., de Arujo, H.H., Farnese, F.S., 2023. Bioaccumulation and physiological traits qualify *Pista stratiotes* as a suitable species for phytoremediation and bioindication of iron-contaminated water. J. Hazard. Mater. 446, 130701.

Cui, S., Zhang, T., Zhao, S., Li, P., Zhou, Q., Zhang, Q., Han, Q., 2013. Evaluation of three ornamental plants for phytoremediation of Pb-contaminated soil. Int. J. Phytoremediation 15, 299–306.

Dai, S., Wong, C.S., Qiu, J., Wang, M., Chai, T., Fan, L., Yang, S., 2014. Enantioselective accumulation of chiral polychlorinated biphenyls in lotus plant (*Nelumbo nucifera* spp.). J. Hazard. Mater. 280, 612–618.

Dan, T.V., Raj, K.S., Saxena, P.K., 2000. Metal tolerance of scented geranium (*Pelargonium* sp. 'Frensham'): effects of cadmium and nickel on chlorophyll fluorescence kinetics. Int. J. Phytoremediation 2, 91–104.

Davies Jr., F.T., Puryear, J.D., Newton, R.J., Egilla, J.N., Saraiva Grossi, J.A., 2002. Mycorrhizal fungi increase chromium uptake by sunflower plants: influence on tissue mineral concentration, growth, and gas exchange. J. Plant Nutr. 25, 2389–2407.

Dhir, B., 2013. Phytoremediation: role of aquatic plants in environmental clean up, Springer India, New Delhi. J. Saudi Chem. Soc. 16, 175–176.

Dingjian, C., 2012. Contents of heavy metals in lotus seed from REEs mining area. J. Saudi Chem. Soc. 16, 175–176.

Eid, E.M., Shaltout, K.H., El-Sheikh, M.A., Asaeda, T., 2012. Seasonal courses of nutrients and heavy metals in water, sediment and above-and below-ground *Typha domingensis* biomass in lake Burullus (Egypt): perspectives for phytoremediation. Flora 207, 783–794.

Evenhuis, A., Korthals, G.W., Molendijk, L.P.G., 2004. *Tagetes patula* as an effective catch crop for longterm control of *Pratylenchus penetrans*. Nematology 6, 877–881.

Farnese, F.S., Oliveira, J.A., Lima, F.S., Leao, G.A., Gusman, G.S., Silva, L.C., 2014. Evaluation of the potential of *Pistia stratiotes* L. (water lettuce) for bioindication and phytoremediation of aquatic environments contaminated with arsenic. Braz. J. Biol. 74, 103–112.

Fuksová, Z., Sazákova, J., Tlustoš, P., 2009. Effects of co-cropping on bioaccumulation of trace elements in *Thlaspi caerulescens* and *Salix dasyclados*. Plant Soil Environ. 11, 461–467.

Ghosh, M., Singh, S.P., 2005. A review on phytoremediation of heavy metals and utilization of its by-products. Appl. Ecol. Environ. Res. 3, 1–18.

Gilman, E.F., Carl, J.D.T., Lyn, A.G., 2015. Golden Canna: *Canna flaccida*. EDIS 6 (4), 1–4.

González-Chávez, M.D.C.A., Carrillo-González, R., 2013. Tolerance of *Chrysantemum maximum* to heavy metals: the potential for its use in the revegetation of tailings heaps. J. Environ. Sci. (China) 25, 367–375.

Greger, M., 1999. Meter availability, uptake, transport and accumulation in plants. In: Prasad, M.N.V. (Ed.), Heavy Metal Stress in Plant, second ed. Springer-Verlag, Berlin Heidelberg, India, pp. 1–27.

Gupta, A.K., Tomar, J.M.S., Kaushal, R., Kadam, D.M., Rathore, A.C., Mehta, H., Ojasvi, P.R., 2021. Aromatic plants based environmental sustainability with speial reference to degraded land management. J. Appl. Res. Med. Aromat. Plants 22, 100298.

Han, Y.L., Yuan, H.Y., Huang, S.Z., Guo, Z., Xia, B., Gu, J., 2007. Cadmium tolerance and accumulation by two species of Iris. Ecotoxicology 16, 557–563.

Han, Y.L., Huang, S.Z., Gu, J.G., Qiu, S., Chen, J.M., 2008. Tolerance and accumulation of lead by species of *Iris* L. Ecotoxicology 17, 853–859.

Hao, X.Z., Zhou, D.M., Li, D.D., Jiang, P., 2012. Growth, cadmium and zinc accumulation of ornamental sunflower (*Helianthus annuus* L.) in contaminated soil with different amendments. Pedosphere 22, 631–639.

Hayyat, M.U., Nawaz, R., Irfan, A., Al-Hussain, S.A., Aziz, M., Siddiq, Z., Ahmad, S., Zaki, M.E.A., 2023. Evaluating the phytoremediation potential of *Eichhornia crassipes* for the removal of Cr and Li from synthetic polluted water. Int. J. Environ. Health Res. 20, 3512.

Hegazy, A.K., Abdel-Ghani, N.T., Abdel-Ghani, G.A., 2011. Phytoremediation of industrial wastewater potentiality by *Typha domingensis*. Int. J. Environ. Sci. Technol. 8, 639–648.

Hussain, J., Wei, X., Xue-Gang, L., Shah, S.R.U., Aslam, M., Ahmed, I., Abdullah, S., Babar, A., Jakhar, A.M., Azam, T., 2021. Garlic (*Allium sativum*) based interplanting alters the heavy metals absorption and bacterial diversity in neighbouring plants. Sci. Rep. 11, 5833.

Ishak, I.A., Hum, N.N.M.F., 2022. Phytoremediation using ornamental plants in removing heavy metals from wastewater sludge. IOP Conf. Ser.: Earth Environ. Sci. 1019, 012009.

Jamwal, P., Raj, A.V., Raveendran, L., Shirin, S., Connelly, S., Yeluripati, J., Richards, S., Rao, L., Helliwell, R., Tamburini, M., 2021. Evaluating the performance of horizontal sub-surface flow constructed wetlands: a case study from southern India. Ecol. Eng. 162, 106170.

Jha, V.N., Tripathi, R.M., Sethy, N.K., Sahoo, S.K., 2016. Uptake of uranium by aquatic plants growing in fresh water ecosystem around uranium mill tailings pond at Jaduguda, India. Sci. Total Environ. 539, 175–184.

Jiang, Z., Xinyuan, Z., 1998. Treatment and utilization of wastewater in the Beijing Zoo by an aquatic macrophyte system. Ecol. Eng. 11, 101–110.

Kabata-Pendias, A., Pendias, H., 2011. Trace Elements in Soils and Plants, fourth ed. CRC Press, Boca Raton, FL.

Karungamye, P.N., 2022. Potential of *Canna indica* in constructed wetlands for wastewater treatment: a review. Conservation 2, 499–513.

Kataki, S., Chatterjee, S., Vairale, M.G., Dwivedi, S.K., Gupta, D.K., 2021. Constructed wetland, an eco-technology for wastewater treatment: a review on types of wastewater treated and components of the technology (macrophyte, biolfilm and substrate). J. Environ. Manag. 283, 111986.

Kealhofer, L., Penny, D., 1998. A combined pollen and phytolith record for fourteen thousand years of vegetation change in Northeastern Thailand. Rev. Palaeobot. Palynol. 103, 83–93.

Khan, A.H.A., Kiyani, A., Mirza, C.R., Butt, T.A., Barros, R., Ali, B., Iqbal, M., Yousaf, S., 2021. Ornamental plants for the phytoremediation of heavy metals: present knowledge and future perspectives. Environ. Res. 195, 110780.

Kubier, A., Wikin, R.T., Pichler, T., 2019. Cadmium in soils and groundwater: a review. Appl. Geochem. 108, 1–16.

Kumar, A., Prasad, M.N.V., 2010. Propagation of *Talinum cuneifolium* L. (Portulacaceae), an ornamental plant and leafy vegetable, by stem cuttings. Floric. Ornam. Biotechnol. 4 (SI1), 68–71.

Kumar, A., Kumar, A., Cabral-Pinto, M.M.S., Chaturvedi, A.K., Shabnum, A.A., Subrahmanyam, G., Mondal, R., Kumar, D., Gupta, D.K., Malyan, S.K., Kumar, S.S., Khan, S.A., Yadav, K.K., 2020. Lead toxicity: health hazards, influence on food chain, and sustainable remediation approaches. Int. J. Environ. Res. Public Health 17, 2179.

Kumari, S., Kumar, B., Sheel, R., 2016. Bioremediation of heavy metals by serious aquatic weed, Salvinia. Int. J. Curr. Microbiol. Appl. Sci. 5, 355–368.

Kutty, S.R.M., Ngatenah, S.N.I., Isa, N.H., Malakahmad, A., 2009. Nutrient removal from municipal wastewater treatment plant effluent using *Eichhornia crassipes*. World Acad. Eng. Technol. 3, 12–22.

Laet, C.D., Matringe, T., Petit, E., Grison, C., 2019. *Eichhornia crassipes*: a powerful bio-indicator for water pollution by emerging pollutants. Sci. Rep. 9, 7326.

Lal, K., Minhas, P.S., Shipra, Chaturvedi, R.K., Yadav, R.K., 2008. Extraction of cadmium and tolerance of three annual cut flowers on Cd-contaminated soils. Bioresour. Technol. 99, 1006–1011.

Lavid, N., Barkay, Z., Tel-Or, E., 2001. Accumulation of heavy metals in epidermal grands of waterlily (*Nymphaeaceae*). Planta 212, 313–322.

Leblebici, Z., Dalmiş, E., Andeden, E.E., 2019. Determination of the potential of *Pistia stratiotes* L. in removing nickel from the environment by utilizing its rhizofiltration capacity. Braz. Arch. Biol. Technol. 62, e19180487.

Lee, C.K., Low, K.S., Hew, N.S., 1991. Accumulation of arsenic by aquatic plants. Sci. Total Environ. 103, 215–227.

Leong, E.S., Tan, S., Chang, Y.P., 2012. Antioxidant properties and heavy metal content of lotus plant (*Nelumbo nucifera* Gaertn) grown in ex-tin mining pond near Kampar, Malaysia. Food Sci. Technol. Res. 18, 461–465.

Li, N., Li, Z., Fu, Q., Zhuang, P., Guo, B., Li, H., 2013. Agricultural technologies for enhancing the phytoremediation of cadmium-contaminated soil by *Amaranthus hypochondriacus* L. Water Air Soil Pollut. 224, 1673.

Li, Y., Xin, J., Ge, W., Tian, R., 2022. Tolerance mechanism and phytoremediation potential of *Pistia stratiotes* to zinc and cadmium co-contamination. Int. J. Phytoremediation 24, 1259–1266.

Liu, J.N., Zhou, Q.X., Sun, T., Ma, L.Q., Wang, S., 2008a. Identification and chemical enhancement of two ornamental plants for phytoremediation. Bull. Environ. Contam. Toxicol. 80, 260–265.

Liu, J., Zhou, Q., Sun, T., Ma, L.Q., Wang, S., 2008b. Growth responses of three ornamental plants to Cd and Cd-Pb stress and their metal accumulation characteristics. J. Hazard. Mater. 151, 261–267.

Liu, Z., He, X., Chen, W., Yuan, F., Yan, K., Tao, D., 2009a. Accumulation and tolerance characteristics of cadmium in a potential hyperaccumulator-*Lonicera japonica* Thunb. J. Hazard. Mater. 169, 170–175.

Liu, J.N., Zhou, Q.X., Wang, S., Sun, T., 2009b. Cadmium tolerance and accumulation of *Althaea rosea* Cav. and its potential as a hyperaccumulator under chemical enhancement. Environ. Monit. Assess. 149, 419–427.

Liu, J., Zhou, Q., Wang, S., 2010. Evaluation of chemical enhancement on phytoremediation effect of CD-contaminated soils with *Calendula officinalis* L. Int. J. Phytoremediation 12, 503–515.

Liu, L.Z., Gong, Z.Q., Zhang, Y.L., Li, P.J., 2011a. Growth, cadmium accumulation and physiology of marigold (*Tagetes erecta* L.) as affected by arbuscular mycorrhizal fungi. Pedosphere 21, 319–327.

Liu, Y.T., Chen, Z.S., Hong, C.Y., 2011b. Cadmium-induced physiological response and antioxidant enzyme changes in the novel cadmium accumulator, *Tagetes patula*. J. Hazard. Mater. 189, 724–731.

Liu, Z., Ge, H., Li, C., Zhao, Z., Song, F., Hu, S., 2015. Enhanced phytoextraction of heavy metals from contaminated soil by plant Co-cropping associated with PGPR. Water Air Soil Pollut. 226, 29.

Lu, Q., He, Z.L., Graetz, D.A., Stoffella, P.J., Yang, X., 2011. Uptake and distribution of metals by water lettuce (*Pistia stratiotes* L.). Environ. Sci. Pollut. Res. 18, 978–986.

Madanan, M.T., Shah, I.K., Varghese, G.K., Kaushal, R.K., 2021. Application of Aztec Marigold (*Tagetes erecta* L.) for phytoremediation of heavy metal polluted lateritic soil. Environ. Chem. Ecotoxicol. 3, 17–22.

Majumdar, A., Upadhyay, M.K., Ojha, M., Afsal, F., Giri, B., Srivastava, S., Bose, S., 2022. Enhanced phytoremediation of metal (loid)s via spiked ZVI nanoparticles: an urban clean-up strategy with ornamental plants. Chemosphere 288, 132588.

Malarkodi, M., Krishnasamy, R., Chitdeshwari, T., 2008. Phytoextraction of nickel contaminated soil using castor phytoextractor. J. Plant Nutr. 31, 219–229.

Marmiroli, N., Maestri, E., 2008. Health implications of trace elements in the environment and the food chain. In: Prasad, M.N.V. (Ed.), Trace Elements as Contaminants and Nutrients. John Willy & Sons, New Jersey, pp. 23–53.

Marques, A.P.G.C., Moreira, H., Franco, A.R., Rangel, A.O.S.S., Castro, P.M.L., 2013. Inoculating *Helianthus annuus* (sunflower) grown in zinc and cadmium contaminated soils with plant growth promoting bacteria—effects on phytoremediation strategies. Chemosphere 92, 74–83.

Meesungnoen, O., Chantiratikul, P., Thumanu, K., Nuengchamnong, N., Hokura, A., Nakbanpote, W., 2021. Elucidation of crude siderophore extracts from supernatants of *Pseudononas* sp. ZnCd2003 cultivated in nutrient broth supplemented with Zn, Cd, and Zn plus Cd. Arch. Microbiol. 203, 2863–2874.

Miao, Q., Yan, J., 2013. Comparison of three ornamental plants for phytoextraction potential of chromium removal from tannery sludge. J. Mater. Cycles Waste Manage. 15, 98–105.

Mir, A.R., Pichtel, J., Hayat, S., 2021. Copper: uptake, toxicity and tolerance in plants and management of Cu-contaminated soil. Biometals 34, 737–759.

Mohsin, M., Nawrot, N., Wojciechowska, E., Kuittinen, S., Szczepańska, K., Dembska, G., Pappinen, A., 2023. Cadmium accumulation by *Phragmites australis* and *Iris pseudacorus* from stormwater in floating treatment wetlands microcosms: insights into plant tolerance and utility for phytoremediation. J. Environ. Manag. 331, 117339.

Mojiri, A., Aziz, H.A., Zahed, M.A., Aziz, S.Q., Selamat, M.R.B., 2013. Phytoremediaiton of heavy metals from urban waste leachate by southern cattail (*Typha domingensis*). Int. J. Sci. Res. Environ. Sci. 1, 63–70.

Mongkhonsin, B., Nakbanpote, W., Nakai, I., Hokura, A., Jearanaikoon, N., 2011. Distribution and speciation of chromium accumulated in *Gynura pseudochina* (L.) DC. Environ. Exp. Bot. 47, 56–64.

Mongkhonsin, B., Nakbanpote, W., Hokura, A., Nuengchamnong, N., Maneechai, S., 2016. Phenolic compounds responding to zinc and/or cadmium treatments in *Gynura pseudochina* (L.) DC. Extracts and biomass. Plant Physiol. Biochem. 109, 549–560.

Mongkhonsin, B., Nakbanpote, W., Meesungnoen, O., Prasad, M.N.V., 2019. Adaptive and tolerance mechanisms in herbaceous plants exposed to cadmium. In: Hasanuzzaman, M., Prasad, M.N.V., Fujita, M. (Eds.), Cadmium Toxicity and Tolerance in Plants. Academic Press, London, pp. 73–109.

Mufarrege, M.D.L., Luca, G.A., Hadad, H.R., Maine, M.A., 2021. Exposure of *Typha domingensis* to high concentrations of multi-metal and nutrient solutions: study of tolerance and removal efficiency. Ecol. Eng. 159, 106118.

Muthukumar, T., Dinesh-Babu, S., 2020. Cadmium affects the regeneration of the leafy vegetable *Talinum portulacifolium* stem cuttings in nutrient solution. An. Biol. 42, 147–159.

Muthukumar, T., Jaison, S., Babu, S.D., 2018. Zinc influences regeneration of *Talinum portulacifolium* stem cuttings in nutrient solution. Not. Sci. Biol. 10, 530–539.

Nazir, M.I., Idrees, I., Danish, P., Ahmad, S., Ali, Q., Malik, A., 2020. Potential of water hyacinth (*Eichhornia crassipes* L.) for phytoremediation of heavy metals from waste water. Biol. Clin. Sci. Res. J. 2020, e006.

Nehnevajova, E., Lyubenova, L., Herzig, R., Schröder, P., Schwitzguébel, J.P., Schmülling, T., 2012. Metal accumulation and response of antioxidant enzymes in seedlings and adult sunflower mutants with improved metal removal traits on a metal-contaminated soil. Environ. Exp. Bot. 76, 39–48.

Newete, S.W., Byrne, M.J., 2016. The capacity of aquatic macrophytes for phytoremediation and their disposal with specific reference to water hyacinth. Environ. Sci. Pollut. Res. 23, 10630–10643.

Newman, S., McCormick, P.V., Miao, S.L., Laing, J.A., Kennedy, W.C., O'Dell, M.B., 2004. The effect of phosphorus enrichment on the nutrient status of a northern Everglades slough. Wetl. Ecol. Manag. 12, 63–79.

Nilratnisakorn, S., Thiravetyan, P., Nakbanpote, W., 2009. A constructed wetland model for synthetic reactive dye wastewater treatment by narrow-leaved cattails (*Typha angustifolia* Linn.). Water Sci. Technol. 60, 1565–1574.

Odjegba, V.J., Fasidi, I.O., 2007. Phytoremediation of heavy metals by *Eichhornia crassipes*. Environmentalist 27, 349–355.

Okurut, T.O., Rijs, G.B.J., Van Bruggen, J.J.A., 1999. Design and performance of experimental constructed wetlands in Unganda, planted with *Cyperus papyrus* and *Phragmites mauritianus*. Water Sci. Technol. 40, 265–271.

Olawale, O., Raphael, D.O., Akinbile, C.O., Ishuwa, K., 2021. Comparison of heavy metal and nutrients removal in *Canna indica* and *Oryza sativa* L. based constructed wetlands for piggery effluent treatment in North-Central Nigeria. Int. J. Phytoremediation 23, 1382–1390.

Outridge, P.M., 2000. Lead biogeochemistry in the littoral zones of South-Central Ontario Lakes, Canada, after the elimination of gasoline lead additives. Water Air Soil Pollut. 118, 179–201.

Panitlertumpai, N., Nakbanpote, W., Sangdee, A., Thumanu, K., Nakai, I., Hokura, A., 2013. Zinc and/or cadmium accumulation in *Gynura pseudochina* (L.) DC. Studied *in vitro* and the effect on crude protein. J. Mol. Struct. 1036, 279–291.

Panizza de León, A., González, R.C., González, M.B., Mier, M.V., Durán-Domínguez-de-Bazúa, C., 2011. Exploration of the ability of *Coleus blumei* to accumulate aluminum. Int. J. Phytoremediation 13, 421–433.

Panneerselvam, B., Shunmuga, P.K., 2021. Phytoremediation potential of water hyacinth in heavy metal removal in chromium and lead contaminated water. Int. J. Environ. Anal. Chem. https://doi.org/10.1080/03067319.2021.1901896.

Patra, M., Sharma, A., 2000. Mercury toxicity in plants. Bot. Rev. 66, 379–422.

Pellegrineschi, A., Damon, J.P., Valtorta, N., Paillard, N., Tepfer, D., 1994. Improvement of ornamental characters and fragrance production in lemon scented geranium through genetic transformation by *Agrobacterium rhizogenes*. Nat. Biotechnol. 12, 64–68.

Penny, D., 1999. Palaeoenvironmental analysis of the Sakon Nakhon Basin, Northeast Thailand: palynological perspectives on climate change and human occupation. Bull. Indo-Pacific Prehistory Assoc. 18, 139–149.

Pérez-López, R., Márquez-García, B., Abreu, M.M., Nieto, J.M., Córdoba, F., 2014. *Erica andevalensis* and *Erica australis* growing in the same extreme environments: phytostabilization potential of mining areas. Geoderma 230-231, 194–203.

Phetsombat, S., Kruatrachue, M., Pokethitiyook, P., Upatham, S., 2006. Toxicity and bioaccumulation of cadmium and lead in *Salvinia cucullata*. J. Environ. Biol. 27, 645–652.

Prasad, M.N.V., 2007. Phytoremediation in India. In: Willey, N. (Ed.), Phytoremediation, Methods and Reviews. Humana Press, New Jersey.

Prasad, M.N.V., Freitas, H.M.O., 2003. Metal hyperaccumulation in plants—biodiversity prospecting for phytoremediation technology. Electron. J. Biotechnol. 6, 276–312.

III. Ornamentals and tree crops for contaminated substrates

Prasad, M.N.V., Nakbanpote, W., 2015. Integrated management of mine waste using biogeotechnologies focusing Thai mines. In: Thangavel, P., Sridevi, G. (Eds.), Environmental Sustainability: Role of Green Technologies. Springer, Berlin, pp. 229–249.

Prasad, M.N.V., Prasad, R., 2012. Nature's cure for cleanup of contaminated environment—a review of bioremediation strategies. Rev. Environ. Health 28, 181–189.

Praveen, A., Pandey, V.C., Mehrotra, S., Singh, N., 2019. Arsenic accumulation in Canna: effect on antioxidative defense system. Appl. Geochem. 108, 104360.

Pulford, I.D., Watson, C., 2003. Phytoremediation of trace metal-contaminated land by trees—a review. Environ. Int. 29, 529–540.

Qin, Q., Li, X., Wu, H., Zhang, Y., Feng, Q., Tai, P., 2013. Characterization of cadmium (^{108}Cd) distribution and accumulation in *Tagetes erecta* L. seedlings: effect of split-root and of remove-xylem/phloem. Chemosphere 93, 2284–2288.

Rachmadiarti, F., Trimulyono, G., Utomo, W.H., 2022. Analyzing the efficacy of *Salvinia molesta* Mitchell as phytoremediation agent for lead (Pb). Nat. Environ. Pollut. Technol. 21, 733–738.

Rahman, M.A., Hasegawa, H., 2011. Aquatic arsenic: phytoremediation using floating macrophytes. Chemosphere 83, 633–646.

Rajkumar, K., Sivakumar, S., Senthilkumar, P., Prabha, D., Subbhuraam, C.V., Song, Y.C., 2009. Effects of selected heavy metals (Pb, Cu, Ni, and Cd) in the aquatic medium on the restoration potential and accumulation in the stem cuttings of the terrestrial plant, *Talinum triangulare* Linn. Ecotoxicology 18, 952–960.

Ranjitha, J., Amrit, R., Rajneesh, K., Vijayalakshmi-, S., Michael, D., 2016. Removal of heavy metals from industrial effluent using *Salvinia molesta*. Int. J. ChemTech Res. 9, 608–613.

Rattanapolsan, L., Nakbanpote, W., Saensouke, P., 2013. Metals accumulation and leaf anatomy of *Murdannia spectabilis* growing in Zn/Cd contaminated soil. EnvionmentAsia 6, 71–82.

Rattanapolsan, L., Nakbanpote, W., Sangdee, A., 2021. Zinc- and cadmium-tolerant endophytic bacteria from *Murdannia spectabilis* (Kurz) Faden. Studied for plant growth-promoting properties, in vitro inoculation, and antagonism. Arch. Microbiol. 203, 1131–1148.

Riaz, M., Kamran, M., Fang, Y., Wang, Q., Cao, H., Yang, G., Deng, L., Wang, Y., Zhou, Y., Anastopoulos, I., Wang, X., 2021. Arbuscular mycorrhizal fungi-induced mitigation of heavy metal phytotoxicity in metal contaminated soils: a critical review. J. Hazard. Mater. 402, 123919.

Rice, K.M., Walker, E.M., Wu, M., Gillette, C., Blough, E.R., 2014. Environmental mercury and its toxic effects. J. Prev. Med. Public Health 47, 74–83.

Rungruang, N., Babola, S., Parkpian, P., 2011. Screening of potential hyperaccumulator for cadmium from contaminated soil. Desalin. Water Treat. 32, 19–26.

Saniewski, M., Zalewska, T., Kraśniewski, W., 2022. Benthic macroinvertebrates as reference indicators for monitoring of anthropogenic isotope 137Cs contamination in the marine environment. Environ. Sci. Pollut. Res. 29, 13822–13834.

Sao, V., Nakbanpote, W., Thiravetyan, P., 2007. Cadmium accumulation by *Axonopus compressus* (Sw.) P. Beauv and *Cyperus rotundas* Linn growing in cadmium solution and cadmium-zinc contaminated soil. Songklanakarin J. Sci. Technol. 29, 881–892.

Saranraj, P., Behera, S.S., Ray, R.C., 2019. Traditional foods from tropical root and tuber crops: innovations and challenges. In: Galanakis, C.M. (Ed.), Innovations in Traditional Foods. Woodhead Publishing, Cambridge, pp. 159–191.

Savio, N., Pandey, D., Srivastava, R.K., 2023. Potentialities of plant-based hybrid wetland systems for the treatment of household waste water using *Canna indica*, *Agave americana*, *Pistia stratiotes* and *Tagetes erecta*. Mater. Today: Proc. 77, 217–222.

Seo, D.C., DeLaune, R.D., Han, M.J., Lee, Y.C., Bang, S.B., Oh, E.J., Chae, J.H., Kim, K.S., Park, J.H., Cho, J.S., 2010. Nutrient uptake and release in ponds under long-term and short-term lotus (*Nelumbo nucifera*) cultivation: influence of compost application. Ecol. Eng. 36, 1373–1382.

Sharma, A., Kapoor, D., Wang, J., Shahzad, B., Kumar, V., Bali, A.S., Jasrotia, S., Zheng, B., Yuan, H., Yan, D., 2020. Chromium bioaccumulation and its impacts on plant: an overview. Plants 9, 100.

Sharma, S., Singh, B., Manchanda, V.K., 2015. Phytoremediation: role of terrestrial plants and aquatic macrophytes in the remediation of radionuclides and heavy metal contaminated soil and water. Environ. Sci. Pollut. Res. 22, 946–962.

Sheoran, A.S., Sheoran, V., 2006. Heavy metal removal mechanism of acid mine drainage in wetlands: a critical review. Miner. Eng. 19, 105–116.

Simachaya, W., 2009. Wastewater tariffs in Thailand. Ocean Coast. Manag. 52, 378–382.

Singh, N., Balomajumder, C., 2021. Phytoremediation potential of water hyacinth (*Eichhornia crassipes*) for phenol and cyanide elimination from synthetic/simulated wastewater. Appl Water Sci 11, 144.

Singh, S.S., Thorat, V., Kaushik, C.P., Raj, K., Eapen, S., D'Souza, S.F., 2009. Potential of *Chromolaena odorata* for phytoremediation of 137Cs from solution and low level nuclear waste. J. Hazard. Mater. 162, 743–745.

Singha, K.T., Sebastian, A., Prasad, M.N.V., 2019. Iron plaque formation in the roots of Pista stratiotes L.: importance in phytoremediation of cadmium. Int. J. Phytoremediation 21, 120–128.

Smith, S.E., Read, D.J., 2008. Mycorrhizal Symbiosis, third ed. Academic Press, San Diego and London.

Soudani, A., Gholami, A., Roozbahani, M.M., Sabzalipour, S., Mojiri, A., 2022. Heavy metal phytoremediation of aqueous solution by *Typha domingensis*. Aquat. Ecol. 56, 513–523.

Stroud, J.L., Collins, R.N., 2014. Improved detection of coastal acid sulfate soil hotspots through biomonitoring of metal(loid) accumulation in water lilies (*Nymphaea capensis*). Sci. Total Environ. 487, 500–505.

Sun, Y., Zhou, Q., Xu, Y., Wang, L., Liang, X., 2011. Phytoremediation for co-contaminated soils of benzo[a]pyrene (B[a]P) and heavy metals using ornamental plant *Tagetes patula*. J. Hazard. Mater. 2-3, 2075–2082.

Sytar, O., Kumar, A., Latowski, D., Kuczynska, P., Strzalka, K., Prasad, M.N.V., 2013. Heavy metal-induced oxidative damage, defense reactions, and detoxification mechanisms in plants. Acta Physiol. Plant. 35, 985–999.

Tahmasbian, I., Safari Sinegani, A.A., 2014. Chelate-assisted phytoextraction of cadmium from a mine soil by negatively charged sunflower. Int. J. Environ. Sci. Technol. 11, 695–702.

Tan, H.W., Pang, Y.L., Lim, S., Chong, W.C., 2023. A state-of-the-art of phytoremediation approach for sustainable management of heavy metals recovery. Environ. Technol. Innov. 30, 103043.

Tapia, Y., Eymar, E., Gárate, A., Masaguer, A., 2013. Effect of citric acid on metals mobility in pruning wastes and biosolids compost and metals uptake in *Atriplex halimus* and *Rosmarinus officinalis*. Environ. Monit. Assess. 185, 4221–4229.

Taufikurahman, T., Praisa, M.A.S., Amalia, S.G., Hutahaean, G.E.M., 2018. Phytoremediation of chromium (Cr) using *Typha angustifolia* L., *Canna indica* L. and *Hydrocotyle umbellata* L. in surface flow system of constructed wetland. IOP Conf. Ser.: Earth Environ. Sci. 308, 012020.

Tepecik, M., Ongun, A.R., Kayikcioglu, H.H., Delibacak, S., Birişçi, T., Aktaş, E., Kalayci Önaç, A., Balik, G., 2023. Effect of sewage sludge on marigold (*Tagetes erecta* L.) and garden verbena (*Verbena hybrida*) plant and soil nutrient elements and heavy metal. KSU J. Agric. Nat. 26, 161–171.

Thiébaut, G., Gross, Y., Gierlinski, P., Boiché, A., 2010. Accumulation of metals in *Elodea canadensis* and *Elodea nuttallii*: implications for plant-macroinvertibrate interactions. Sci. Total Environ. 408, 5499–5505.

Thompson, E.S., Pick, F.R., Bendell-Young, L.I., 1997. The accumulation of cadmium by the yellow pond lily, *Nuphar variegatum*, in Ontario peatlands. Arch. Environ. Contam. Toxicol. 32, 161–165.

Tiagi, Y.D., Aery, N.C., 1986. Biogeochemical studies at the Khetri copper deposits of Rajasthan, India. J. Geochem. Explor. 26, 267–274.

Trigueros, D., Mingorance, M.D., Rossini Oliva, S., 2012. Evaluation of the ability of *Nerium oleander* L. to remediate Pb-contaminated soils. J. Geochem. Explor. 114, 126–133.

Tsukanova, K.A., Chebotar, V.K., Meyer, J.J.M., Bibikova, T.N., 2017. Effect of plant growth-promoting rhizobacteria on plant hormone homeostasis. S. Afr. J. Bot. 113, 91–102.

Varshney, A., Mohan, S., Dahiya, P., 2021. Assessment of leaf morphological characteristics, phenolics content and metal(loid)s concentrations in *Calendula officinalis* L. grown on fly ash amended soil. Ind. Crop. Prod. 174, 114233.

Visoottiviseth, P., Francesconi, K., Sridokchan, W., 2002. The potential of Thai indigenous plant species for the phytoremediation of arsenic contaminated land. Environ. Pollut. 118, 453–461.

Wang, S., Liu, J., 2014. The effectiveness and risk comparison of EDTA with EGTA in enhancing cd phytoextraction by *Mirabilis jalapa* L. Environ. Monit. Assess. 186, 751–759.

Wang, Y., Yan, A., Dai, J., Wang, N., Wu, D., 2012. Accumulation and tolerance characteristics of cadmium in *Chlorophytum comosum*: a popular ornamental plant and potential Cd hyperaccumulator. Environ. Monit. Assess. 184, 929–937.

Wei, J.-L., Lai, H.-Y., Chen, Z.-S., 2012. Chelator effects on bioconcentration and translocation of cadmium by hyperaccumulators, *Tagetes patula* and *Impatiens walleriana*. Ecotoxicol. Environ. Saf. 84, 173–178.

Wibowo, Y.G., Nugraha, A.T., Rohman, A., 2023. Phytoremediation of several wastewater sources using *Pistia stratiotes* and *Eichhornia crassipes* in Indonesia. Environ. Nanotechnol. Monit. Manag. 20, 100781.

Williams, J.B., 2002. Phytoremediation in wetland ecosystems: progress, problems, and potential. Crit. Rev. Plant Sci. 21, 607–635.

Wood, B., Mcatamney, C., 1994. The use of macrophytes in bioremediation. Biotechnol. Adv. 12, 653–662.

Woulds, C., Ngwenya, B.T., 2004. Geochemical processes governing the performance of a constructed wetland treating acid mine drainage, Central Scotland. Appl. Geochem. 19, 1773–1783.

Xie, W.Y., Huang, Q., Li, G., Rensing, C., Zhu, Y.G., 2013. Cadmium accumulation in the rootless macrophyte *Wolffia globose* and its potential for phytoremediation. Int. J. Phytoremediation 15, 385–397.

Xiong, C., Xu, X., Lu, Y., Ouyang, B., Zhang, Y., Ye, Z., Li, H., 2012. Effects of heavy metals Pb and Cd on the physiological response and accumulation in lotus root. Acta Hortic. Sin. 39, 2385–2394.

Xiong, C., Zhang, Y., Xu, X., Lu, Y., Ouyang, B., Ye, Z., Li, X., 2013. Lotus roots accumulate heavy metals independently from soil in main production regions of China. Sci. Hortic. 164, 295–302.

Yadav, S.K., 2010. Heavy metals toxicity in plants: an overview on the role of glutathione and phytochelatins in heavy metal stress tolerance of plants. S. Afr. J. Bot. 76, 167–179.

Ye, Z.H., Whiting, S.N., Qian, J.H., Lytle, C.M., Lin, Z.Q., Terry, N., 2001. Wetlands and aquatic processes, trace elements removal from coal ash leachate by a 10-year-old constructed wetland. J. Environ. Qual. 30, 1710–1719.

Ye, X., Hu, H., Li, H., Xiong, Q., Gao, H., 2019. Combined nitrogen fertilizer and wheat straw increases the cadmium phytoextraction efficiency of *Tagetes patula*. Ecotoxicol. Environ. Saf. 170, 210–217.

Yuan, Q., Wang, P., Wang, X., Hu, B., Tao, L., 2022. Phytoremediation of cadmium-contaminated sediment using Hydrilla verticillate and Elodea canadensis harbor two same keystone rhizobacteria *Pedosphaeraceae* and *Parasegetibacter*. Chemosphere 286, 131648.

Zhang, X., Hu, Y., Liu, Y., Chen, B., 2011. Arsenic uptake, accumulation and phytofiltration by duckweed (*Spirodela polyrhiza* L.). J. Environ. Sci. 23, 601–606.

Zhang, Y., Bai, J., Wang, J., Yin, R., Wang, J., Lin, X., 2018. Arbuscular mycorrhizal fungi alleviate the heavy metal toxicity on sunflower (*Helianthus annuus* L.) plants cultivated on a heavily contaminated field soil at a WEEE-recycling site. Sci. Total Environ. 628-629, 282–290.

Zhang, W., Pan, X., Zhao, Q., Zhao, T., 2021. Plant growth, antioxidative enzyme, and cadmium tolerance responses to cadmium stress in *Canna orchioides*. Hortic. Plant J. 7, 256–266.

Zhong, S., Shi, J., Xu, J., 2010. Influence of iron plaque on accumulation of lead by yellow flag (*Iris pseudacorus* L.) grown in artificial Pb-contaminated soil. J. Soils Sed. 10, 964–970.

Zimmolo, Y., Kirshner, F., Malkovalaja, A., 2006. Application of *Eichhornia crassipes* and *Pistia stratiotes* for treatment of urban sewage in Israel. J. Environ. Manag. 81, 420–428.

Native trees on abandoned mine land: From environmental remediation to bioeconomy

P.J.C. Favas[a,b], J. Pratas[c], R. Chaturvedi[d], M.S. Paul[d], and Majeti Narasimha Vara Prasad[e]

[a]University of Trás-os-Montes e Alto Douro, UTAD, School of Life Sciences and the Environment, Vila Real, Portugal [b]University of Coimbra, MARE—Marine and Environmental Sciences Centre/ARNET—Aquatic Research Network, Department of Life Sciences, Calçada Martim de Freitas, Coimbra, Portugal [c]University of Coimbra, Faculty of Sciences and Technology, Department of Earth Sciences, Coimbra, Portugal [d]Department of Botany, St. John's College, Dr. B.R. Ambedkar University, Agra, India [e]Department of Plant Sciences, School of Life Sciences, University of Hyderabad (An Institution of Eminence), Hyderabad, Telangana, India

1 Introduction

Many plant species can colonize highly contaminated areas with heavy metals and metalloids, such as mine tailings or soils degraded and contaminated by mining/industrial activities. These species are resistant to imposed stress conditions (metal contamination and nutrient deficiency) and can achieve the goals of stabilization, pollution reduction, and aesthetic enhancement. Many of these plants have been recognized as candidates for phytoremediation or eco-remediation programs. In fact, considerable research has been carried out globally on the restoration of abandoned mines and the remediation of mine waste (Table 1).

Phytoremediation is the use of plants (trees, shrubs, grasses, and aquatic plants) and their associated microorganisms to remove, degrade, or isolate toxic substances from the environment (e.g., Favas et al., 2016a; Chowdhury et al., 2016; Masarovičová and Králová, 2018; Issaka and Ashraf, 2021; Nissim and Labrecque, 2021). The term *phytoremediation* is derived

TABLE 1   Resume of 2023 literature (selected) on the restoration of mine waste and abandoned mine land using trees (not exhaustive).

Salient observation	Author(s)
The dynamics of the vegetation in the southeast Peruvian Amazon's gold-mining-affected areas	Alarcón-Aguirre et al. (2023)
Geotrails' ability to promote sustainable development in the Campi Flegrei Active Volcanic Area	Alberico et al. (2023)
Creating a multi-criteria decision matrix for plant species selection that combines phytoremediation with bioenergy production	Amabogha et al. (2023)
Queensland mine rehabilitation options for native ecosystems: stakeholder survey report. Brisbane: The Queensland Government's Office of the Mine Rehabilitation Commissioner	Baskerville et al. (2023)
In a karst location in the Cévennes region of southern France, the effects of previous mining operations on water quality	Bondu et al. (2023)
Automated urban tree survey using Google Street View pictures, remote sensing data, and plant species identification software	Capecchi et al. (2023)
A land use program at the nexus of polluted areas and renewable energy is the Colorado Brightfields Project	Carman (2023)
Response of plant element characteristics to soil arsenic stress and its implications for vegetation restoration in a region that has been mined	Chen et al. (2023)
Effect of lead-zinc mining on socio-economic and health situations in lead-zinc mining districts of Enyigba and Ishiagu in Southeast Nigeria	Chukwu and Obiora (2023)
Impact of artisanal gold mining on woody vegetation in Chewore South Safari Area, northern Zimbabwe	Dzoro et al. (2023)
Native plant species: a tool for mining land restoration	Gairola et al. (2023)
Using random forest classification and regression to identify the sources of the arsenic in private well water	Giri et al. (2023)
In hard-rock acid mine drainage and mine waste, rare earth element recovery is studied using a case study from Idaho Springs, Colorado	Goodman et al. (2023)
Synthetic aperture radar technology for monitoring underground mine deformation: Rajgamar coal mine in Korba, Chhattisgarh, India as a case study	Govil and Guha (2023)
Recent case studies on phytoremediation of metalloid and heavy metal polluted soils in Latin America	Hernández Guiance et al. (2023)
Comparison of soil bacterial communities in a reclaimed waste dump under *Pinus tabulaeformis* and *Populus euramericana* canopies	Hou et al. (2023)
Can reflectance spectroscopy identify the speciation of cobalt and nickel on a mining waste dump surface in addition to content quantification?	Khosravi et al. (2023)
Soil amendment and forest closeness are key factors in the effectiveness of restoration efforts in former mine sites in the Amazon	König et al. (2023)
Utilizing *Solidago chilensis* Meyen, *Haplopappus foliosus* DC, and *Lycium chilense* Miers ex Bertero to remove heavy metals from mine tailings in central chile	Lazo et al. (2023)

TABLE 1 Resume of 2023 literature (selected) on the restoration of mine waste and abandoned mine land using trees (not exhaustive)—cont'd

Salient observation	Author(s)
Positive benefits of vegetation regeneration on the post-mining soil characteristics	Li et al. (2023)
Infrastructure on mining sites is blue-green	Lord et al. (2023)
The problems of abandoned mines in South Africa: the negative effects of mining	Mhlongo (2023)
The potential for *Gliricidia sepium* (Fabaceae) to bioaccumulate heavy metals in mining tailings	Mussali-Galante et al. (2023)
With the aid of microorganisms, woody plants have an advantage in the phytoremediation of manganese ore	Nong et al. (2023)
Cleanup of land inspired by biology	Pandey (2023)
Climate change implications of soil organic matter loss in the mining landscape	Punia and Bharti (2023)
Ecological restoration of degraded land using bioenergy crops	Ranđelović and Pandey (2023)
What type of trees make the best soil? Assessment of physical and chemical properties in 22 tree species growing in reclaimed mine soil 50 years later	Spasić et al. (2023)
After 70 years of natural recovery, the plant community, soil, and microclimate attributes of a defunct limestone quarry were studied	Stephan and Hubbart (2023)
Macromycetes (Fungi: Basidiomycota, Ascomycota), carabid beetle (Coleoptera: Carabidae), and spider (Araneae) assemblages are affected by abandoned kaolin quarries	Walter et al. (2023)
Ectomycorrhizal fungi's contributions to a recovered Populus yunnanensis forest in a tailings pond from a closed metal mine in southwest China	Xiao et al. (2023)
Assessment, modeling, and application for soil environment criteria of the bioaccessibility of arsenic, lead, and cadmium in polluted mining/smelting soils derivation	Xie et al. (2023)
Research review on soil and plant restoration technology development in open-pit coal mines	Xu et al. (2023)
Vegetation restoration in rare earth tailings in Northeast Guangdong	Yang et al. (2023)
Landfills and mining sites. Bridging nature and humans: a transdisciplinary approach in restoration of ecosystems	Zerbe (2023)
Using a combination of UAV hyperspectral pictures and simulated annealing deep neural network to map the soil-accessible copper content in the mining tailings pond	Zhang et al. (2023)
Microbial composition in the Komi Republic's reclaimed and abandoned mining sites (North Russia)	Zverev et al. (2023)

from the Greek word *phyton* (plant) and the Latin word *remedium* (to remedy or correct). Although mine-contaminated soils and tailings can be remedied using a variety of methods (e.g., removal, isolation, incineration, solidification/stabilization, vitrification, thermal treatment, solvent extraction, and chemical oxidation), they are prohibitively expensive and, in some cases, involve the transport of contaminated materials to treatment sites, increasing the risk of secondary contamination. As a result, in situ methods that are less environmentally disruptive and more cost-effective are currently preferred. Phytoremediation techniques emerge as a viable option in this context.

However, many factors, such as low biomass production at metal concentrations, sensitivity to multi-metals, poor growth, and a shallow root system, limited the plants' phytoremediation efficiency. As a result, improving plant phytoremediation efficiency at higher metal concentrations is a promising approach that can be achieved by utilizing biological amendments [e.g., PGPB (plant growth-promoting bacteria)], chelating agents [e.g., EDTA (ethylenediaminetetraacetic acid)], and organic manure (Kumar et al., 2022a, b).

The most common is the use of chelating agents, such as EDTA, to increase metal bioavailability (e.g., Chaturvedi et al., 2019; You et al., 2022). They may, however, persist in the environment and cause groundwater contamination. Organic alteration, such as organic manure, compost, urban solid wastes, and biosolids, has been used in large quantities as a source of nutrients or as a conditioner to improve the physical properties and fertility of soils (Adekiya et al., 2020). Organic matter modification can alter soil pH and thus have an indirect effect on metal bioavailability (Kelvin et al., 2020).

Because plants and soil microbes (PGPB) complement each other so well, phytoremediation based on the combined use of plants and PGPB is currently a research hotspot (Kong and Glick, 2017; Mesa-Marín et al., 2020; Kumar et al., 2022a). When PGPB is used as a bioinoculant, it increases plant biomass and root growth by recycling nutrients, stabilizing soil structure, and modulating heavy metal bioavailability and toxicity (Ahemad, 2019; Chaturvedi et al., 2021; Kumar et al., 2022b).

Perennials and nitrogen-fixing native trees have a significant role to play in abandoned mine land. The role of plants, especially perennials (trees), microbes, and edaphic interactions have key role in mine waste restoration to bio-economy (Fig. 1).

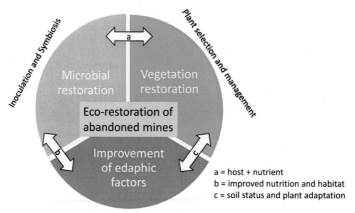

FIG. 1   Mine waste restoration: plant, microbe, and edaphic interactions.

2 Trees: From environmental remediation to industrial feedstock

Phytoremediation of contaminated soil can be accomplished using a variety of techniques, depending on the chemical nature and properties of the contaminant (whether it is inert, volatile, or degradable in the plant or soil) and the plant characteristics. In the context of this chapter and in light of our research on Portuguese mines (Favas et al., 2016a, b, 2018, 2019), it is appropriate to focus on the use of trees for phytostabilization, phytoextraction, and phytocoverage. These are the most effective methods for remediating metal- and metalloid-contaminated soils.

The goal of phytostabilization is to reduce the mobility of contaminants in the soil. This technique covers contaminated soil with vegetation that is resistant to high concentrations of toxic elements, limiting soil erosion and contaminant leaching into groundwater. Surface adsorption/accumulation in roots, as well as precipitation in the rhizosphere by induced changes in pH or oxidation of the root environment, can reduce the mobility of contaminants, organic or inorganic (e.g., Varun et al., 2017; Stylianou et al., 2020; Nyenda et al., 2020). As a result, some contaminants are incorporated into the lignin of the cell wall of the roots' cells or into humus, while others are precipitated as insoluble forms by direct action of root exudates and subsequently trapped in the soil matrix (Fig. 2). The primary goal is to prevent contaminants from being mobilized and to limit their diffusion in the soil.

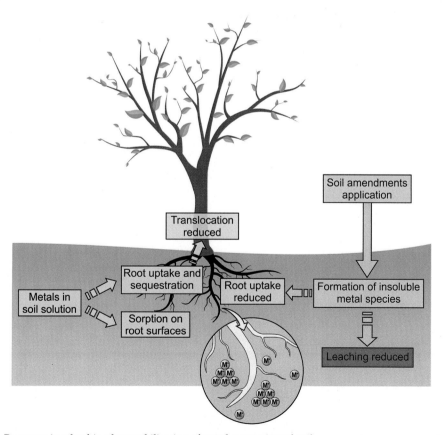

FIG. 2   Processes involved in phytostabilization of metal-contaminated soil.

III. Ornamentals and tree crops for contaminated substrates

Plant species that can exude large amounts of chelating substances can also promote phytostabilization. These substances immobilize contaminants by preventing their absorption while also reducing their mobility in soil. Thus, plants with phytostabilization potential can be extremely beneficial for revegetating mine tailings and contaminated soils. Furthermore, several amendments (e.g., phosphate fertilizers, Fe oxide materials, organic materials, clay minerals) can be applied to the soil to promote the formation of insoluble contaminant forms (e.g., Madejón et al., 2018; Berslin et al., 2021; Kumari and Maiti, 2022).

In contrast to phytostabilization, the goal of phytoextraction is to remove contaminants from the soil through root absorption, followed by translocation and accumulation in the aerial parts (Fig. 3). It is primarily used with metals (Cd, Ni, Cu, Zn, Pb), but it can also be used with other elements (Se, As) and organic compounds. Phytoextraction has received much attention owing to its high efficiency and potential economic value (in metal recovery and energy production) (e.g., Varun et al., 2017; Chowdhury et al., 2017; Huff et al., 2020). Plants

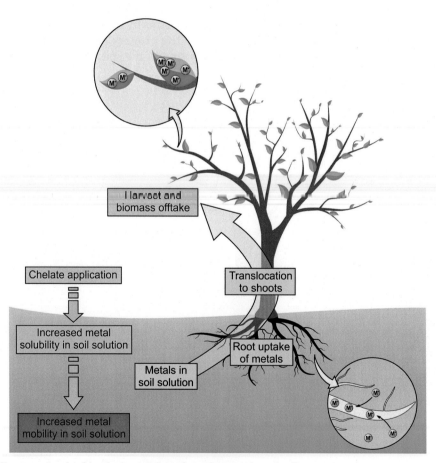

FIG. 3 Processes involved in phytoextraction of metal-contaminated soil.

used in phytoextraction should ideally have the following characteristics, among others: (a) tolerance to high metal concentrations; (b) ability to accumulate high concentrations in aerial tissues; (c) rapid growth; (d) high biomass production; (e) extensive root system; and (f) ease of cultivation and harvesting.

Phytoextraction preferentially uses hyperaccumulator plants, which have the ability to store high concentrations of specific metals in their aerial parts (0.01%–1% dry weight, DW, depending on the metal) (e.g., Dilks et al., 2016; Antoniadisa et al., 2017; Naila et al., 2019). Although hyperaccumulators are excellent phytoextractors, they typically produce little biomass. As a result, it is widely accepted that plants with a significant biomass production capacity can compensate for their relatively lower metal accumulation capacity to the point where the amount of metal removed can be higher.

The ability to hyperaccumulate metals and biomass production determines phytoextraction efficiency. If these factors have an effect on phytoextraction, they can be optimized to improve the phytoremediation process. One possibility is to introduce chemical agents into the soil to increase metal bioavailability and root uptake. This type of assisted phytoextraction, also known as induced phytoextraction, has shown great promise and has received a lot of attention.

Only when the accumulated contaminant is removed through harvesting can phytoextraction be considered effective (Favas et al., 2016a; Antoniadisa et al., 2017; Masarovičová and Kráľová, 2018). Traditional farming methods can be used for harvesting if the majority of the captured heavy metals are translocated to shoots. It is critical to harvest the plants prior to leaf fall, death, and decomposition to prevent contaminants from dispersing or returning to the soil.

The bioconcentration factor (BCF) (or biological absorption coefficient) and translocation factor (TF) can be used to calculate phytoextraction potential (e.g., Antoniadisa et al., 2017). The BCF is calculated as the ratio of the total concentration of elements in harvested plant tissue (C_{plant}) to its concentration in the soil in which the plant was growing (C_{soil}):

$$BCF = \frac{C_{plant}}{C_{soil}}$$

The TF is calculated as the ratio of the total concentration of elements in the plant's aerial parts (C_{shoot}) to the concentration in the root (C_{root}):

$$TF = \frac{C_{shoot}}{C_{root}}$$

The phytoextraction capacity of the plant can be assessed using both the BCF and the TF. A high root-to-shoot translocation (TF) of metals is a fundamental characteristic for a plant to be classified as effective in phytoextraction. The rate of metal accumulation and biomass production can be used to estimate the commercial efficiency of phytoextraction. The metal removal value (g/kg of metal per hectare and per year) is calculated by multiplying the rate of accumulation, metal (g)/plant tissue (kg), by the growth rate, plant tissue (kg)/hectare/year (e.g., Favas et al., 2016a). For the species to be commercially useful, the rate of removal or extraction must reach several hundred, or at least 1 kg/ha/year, and even then, the remediation process may take 15–20 years.

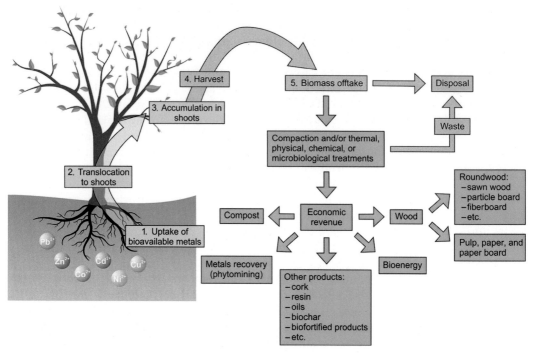

FIG. 4   Biomass-processing opportunities involved in phytoextraction of metal-contaminated soil.

Following harvesting, biomass disposal or processing is required (Fig. 4). Thermal, physical, chemical, or microbiological processes can be used to reduce the volume/weight of biomass in landfill deposition. In the case of incineration, the energy produced is an economic opportunity, and the ash can be further processed for the extraction and recovery of commercially valuable metals (phytomining) (e.g., Favas and Pratas, 2019; Favas et al., 2020; Akinbile et al., 2021).

Metals with commercial value, such as Ni, Zn, Cu, or Co, may encourage phytoremediation/phytomining processes. Several studies using *Alyssum* species have shown that simple agricultural practices such as fertilization and weed control can greatly improve the economic results of phytomining.

According to Li et al. (2003), *Alyssum murale* and *Alyssum corsicum* can achieve a biomass yield of 20t/ha with the addition of fertilizers and herbicides, among other agricultural management practices, and the resulting Ni phytoextraction production can be up to 400kg Ni/ha. Li et al. (2003) estimated that with this intensively managed phytoextraction, the net value of an annual phytomining crop would be US $1749 net ha^{-1}. Following a 5-year field study, Bani et al. (2015) reported maximum values of *Alyssum murale* biomass production of 9.0t/ha and Ni phytoextraction yield of 105kg Ni/ha. These results, obtained with fertilizers and weed control, enabled the authors to estimate the net value of an annual phytomining crop at US $1055ha^{-1}.

Thus, future phytomining research could concentrate on optimizing plant cultivation processes through fertilization, the effects of pH variation, weed control, the addition of chelating agents, planting practices and harvest methods, hybridization or cloning research to increase metal hyperaccumulation potential, and biomass production, among other things (e.g., Li et al., 2003; Bani et al., 2015; Favas and Pratas, 2019; Akinbile et al., 2021).

According to Evangelou et al. (2015), biomass can be used directly as heat in bioenergy production or processed into gases or liquids (e.g., ethanol, biodiesel). Aside from bioenergy, other potential economic products include wood, biochar, and biofortified products. However, these economic revenue opportunities should be carefully examined, taking into account and evaluating all potential environmental and health risks.

Another interesting option is to use tree biomass (e.g., agricultural and forestry wastes) as heavy metal adsorbents for the treatment of industrial and domestic wastewaters (e.g., Jain et al., 2020; Singh et al., 2020; Vieira et al., 2021; Cemin et al., 2021). Several biomaterials derived from tree species, such as *Quercus*, *Eucalyptus*, *Sterculia*, *Juglans*, and *Pinus*, have been tested as adsorbents.

These phytoremediation techniques have several advantages, but they also have some drawbacks that should be considered before implementing this technology. If low cost is a benefit, the time required to observe the results can be lengthy. The pollutant concentration and the presence of other toxins should be within the plant's tolerance limits. It is difficult to find plants that are efficient at remediating multiple contaminants at the same time. When using this technology, these limitations, as well as the possibility of these plants entering the food chain, should be considered.

Some soils are so contaminated that removing metals with plants would take an inordinate amount of time. Traditional remediation technologies (physical, chemical, and thermal methods), on the other hand, can be prohibitively expensive due to high economic costs. Choosing drought-resistant, fast-growing crops or fodder that can grow in metal-contaminated, nutrient-deficient soils is a common practice. This technique covers contaminated soil with vegetation that is resistant to high concentrations of toxic elements, limiting soil erosion and contaminant leaching into groundwater (vegetation covers, vegetative caps, or phytocovers). Thus, herbs (usually grasses), eventually shrubs or trees, are planted on landfills or tailings to reduce rainwater infiltration and limit the spread of pollutants. Roots improve soil aeration by encouraging biodegradation, evaporation, and transpiration (e.g., Seshadri et al., 2016; Favas et al., 2018; Khapre et al., 2019).

The disadvantage of this method is that tailings are generally unsuitable for developing plant roots. To facilitate the revegetation of contaminated soils, adding amendments such as fertilizers and organic residues is a common practice. Incorporating organic residues (e.g., composts, manure, sewage sludge, and biosolids) is also a viable method of recycling waste products.

Vegetative covers are frequently combined with traditional engineering technologies such as hazardous waste enclosure, isolation, or encapsulation (Fig. 5). Hazardous wastes are isolated and contained in this case by a multilayer cover that includes a vegetative cover, topsoil, protective cover layer, geocomposite, geotextile, and high-density polyethylene barrier.

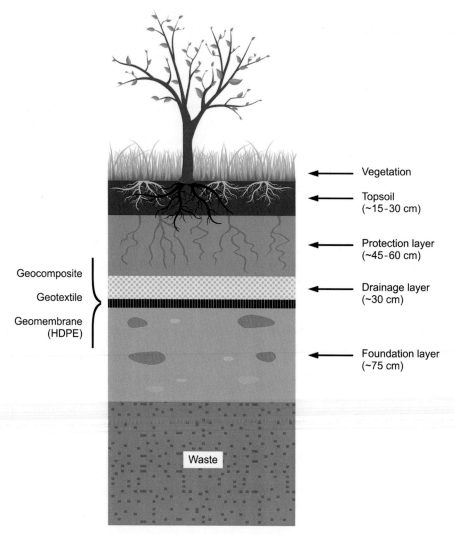

FIG. 5 Typical engineered phytocover integrated with the encapsulation technique.

Portugal is a significant producer of forestry products, accounting for approximately 35% of the total land area forested (ICNF, 2015). The most representative forest species in the Portuguese landscape are *Quercus* spp., *Pinus pinaster*, and *Eucalyptus* spp. Portuguese forests contribute significantly to the country's employment and GDP by providing a variety of services and industrial products. These services include both market services (such as timber exploitation, cork, resin, and pine nuts) and non-market services (recreation, landscape, carbon sequestration, watershed protection, protection from soil erosion, and biodiversity). It also emphasizes the growing importance of biomass as a renewable energy source.

3 Native trees in Portugal's abandoned mines: from phytoremediation to bioeconomy

3.1 Study areas

Many studies have been conducted in contaminated mining and industrial areas, as well as in natural metalliferous soils, in recent decades to list and screen indigenous species, as well as to assess their potential for phytoremediation of contaminated soils and ecological restoration. Several studies have also been conducted in Portugal to survey the indigenous plant species of various contaminated areas and assess their potential for environmental remediation.

Important findings from several studies evaluating the phytotechnological potential (phytoremediation, bioindication, and biogeochemical prospecting) of trees grown on soils enriched with metals and metalloids in distinct abandoned mining areas of tin/tungsten (Sn/W), copper (Cu), and lead (Pb) in Portugal are presented in this chapter.

In terms of Sn/W mines, the results came from the Sarzedas mine (central Portugal), Fragas do Cavalo mine (central Portugal), and Vale das Gatas mine (northern Portugal). As examples of Cu mines, the results from the São Domingos mine (southeast Portugal) and Borralhal mine (central Portugal) are presented. Abandoned Pb mines include the Barbadalhos and Sanheiro mines in central Portugal.

The Sarzedas research area includes two abandoned mines, Gatas and Santa, which comprise the Sarzedas mine. The area is close to the village of Sarzedas (Castelo Branco county, central Portugal). The Sarzedas mineralizations are emplaced in quartzous veins striking N60°E (Santa mine) and N20°W (Gatas mine), filling late Hercynian fractures, which cut the "Schist-Greywacke Complex." Acid rock veins cut the complex frequently and are identical to ante-Ordovician veins found in the area. These veins have hydrothermal alteration and are mineralized by disseminated sulfides. The main mineralization is vein-type, with wolframite (ferberite), stibnite, pyrite, arsenopyrite, and, more sparingly, chalcopyrite, sphalerite, and galena present. Gold is found in its native form. Soils are underdeveloped, and the majority are mountain soils, particularly cambisoils and lithosoils.

The Fragas do Cavalo mine is close to the village of Oleiros (Castelo Branco district, central Portugal). Mineralization takes place in quartz veins embedded in metasedimentary rocks of the Schist-Greywacke Complex, which is almost entirely composed of shales with a few thick layers of siltstones. Wolframite, arsenopyrite, pyrite, and chalcopyrite dominate the mineralization.

The Vale das Gatas mine is located in the Vila Real district of northern Portugal. This sector's geological units include metasedimentary rocks (Schist-Greywacke Complex), Hercynian granites, and veinous rocks (mineralized and non-mineralized). Wolframite, cassiterite, scheelite, several sulfides (pyrite, chalcopyrite, sphalerite, galena, arsenopyrite, pyrrhotite, stannite, covellite, marcasite), silver (Ag), Pb, and bismuth (Bi) sulfosalts, and native Bi were all found in the veins. Mineralization is primarily supported by quartz, fluorite, and muscovite. The typical soil units in this region are Leptosols, Cambisols, and Anthrosols. The dominant soil unit in the study area is the Leptosol type, which consists of soils distinguished by the presence of bedrock 20 cm below the surface. These soils are classified into two types: distric (acid) soils associated with metasedimentary rocks and umbric (acid and organic-rich) soils associated with granitic rocks.

The São Domingos mine, located in southeast Portugal, is one of the historical mining centers, with the extraction of Au, Ag, and primarily Cu dating back to pre-Roman times. The São Domingos mine is part of the Iberian pyrite belt, with outcropping area sequences formed by a distinct vertical mass of cupriferous pyrite associated with zinc and lead sulfide. From a genetic, morphological, and mineralogical standpoint, it is similar to many other volcanogenic massive sulfide mineralizations found in this belt. The primary mineralization consists of massive sulfides (over 85% sulfides), primarily pyrite, sphalerite, chalcopyrite, galena, and sulfosalts (e.g., Vieira et al., 2020; Košek et al., 2020).

The Borralhal mine is an abandoned copper mine in the county of Oleiros (Castelo Branco district, central Portugal). The mineralizations are located in quartzose veins that strike N70°E and cut through the Schist-Greywacke Complex, and they are composed primarily of chalcopyrite, pyrite, bournonite, and other sulfosalts with quartz-carbonate gangue.

The abandoned Barbadalhos mine (also known as the Zorro mine) is located in central Portugal, near the city of Coimbra. The mineralized quartz veins are located near the contact between the Central Iberian Zone and the Ossa Morena Zone. The mineralogy is primarily composed of argentiferous galena and sphalerite. Small amounts of chalcopyrite and arsenopyrite are also present. The mineral that supports mineralization is primarily quartz, but calcite, dolomite, and ankerite are also present. This area's soils are underdeveloped and predominantly acidic, with variable thicknesses ranging from 10 to 80 cm and an overall clay texture with clay loam in higher areas.

The Sanheiro mine (also known as the Sanguinheiro mine) is in Penacova County, near Coimbra (central Portugal). The mineralized quartz veins strike N39°E and cut through the Schist-Greywacke Complex in the Central Iberian Zone, near the Ossa Morena Zone. Galena and sphalerite dominate the mineralization, with minor amounts of pyrite, chalcopyrite, and marcasite in a gangue of quartz, barite, and siderite.

3.2 Sampling and chemical analysis of soils and plants

Several line transects were made in mineralized and non-mineralized zones, as well as in tailings, in the studied areas. Soils and plants were collected at 20 m intervals along line transects (0, 20, 40 m, etc.) in 2 m radius circles.

Four random partial soil samples weighing 0.5 kg each were collected from 0 to 20 cm depth at each location and mixed to create one composite sample. Soil samples were oven-dried at a constant temperature, manually homogenized, and quartered in the laboratory. Each quartered sample yielded two equivalent fractions. One was used to determine pH, while the other was used for chemical analysis. To remove plant matter, the samples for chemical analysis were sieved through a 2 mm mesh sieve and then screened through a 250 μm screen.

Samples were also collected from all plant species found growing within a 2 m radius of each sampling point. The plant sample focused on the aerial parts, considering plant maturity and the proportionality of different types of tissues or the separation of different types of tissues (leaves and stems) in some species. The samples were collected separately into clean cellulose bags and delivered to the laboratory on the same day. Despite the fact that all of the species present in the sample points were sampled, this chapter only presents the results of the tree species.

The leaves/needles and stem samples were collected from trees of similar size and age. Leaves, needles, and stems were collected from outer branches of the middle canopy at south, west, east, and north directions and then were homogeneously mixed. When the vegetal material was sampled, different organs (stems and leaves/needles) were separated, and the tissues were separated by age (in the case of *P. pinaster*). Stems and needles (0-, 1-, 2-, 3-, and 4-year-old) were collected.

The vegetal material was thoroughly washed in the laboratory, first in running water, then in distilled water, and then dried in a glasshouse. When the material was dry, it was milled into a homogeneous powder. The chemical analyses were carried out in the laboratories of the University of Coimbra's Department of Earth Sciences (Portugal). The pH of the soil was determined in a water extract (1:2.5 v/v). For elemental analysis, soil and plant samples were acid-digested. Colorimetry was used for W, atomic absorption spectrophotometry (AAS, Perkin-Elmer, 2380) was used for Co, Cr, Cu, Fe, Mn, Ni, Pb, and Zn, and a hydride generation system was used for As, Sb, and Sn. Data quality control was performed by inserting triplicate samples into each batch. Certified reference materials (Virginia tobacco leaves CTA-VTL-2 and NIST 2709-San Joaquin Soil) were also used. The agreement between the certified reference values and the values determined by the analytical methods was in the range of 86%–105%.

3.3 Trace metals and metalloids in soil

Fig. 6 summarizes the pH and trace element data in soil from the tin/tungsten (Sn/W) mines of Sarzedas, Fragas do Cavalo, and Vale das Gatas. Silver, As, Pb, Sb, and W were the most abundant elements found in the soils of the Sarzedas mine area. The presence of sulfides in the mineralization can explain the low pH values observed near the mineralized area. As a result, soil pH was found to be negatively related to mineralization. High levels of sulfides, particularly pyrite and arsenopyrite, which are easily weathered, favored toxic element dissolution, dispersion, and bioavailability.

The anomalies of As, Pb, and W are noticeable in the soils of the Fragas do Cavalo mine (Fig. 6). These anomalies are unmistakably linked to the area's mineralization. However, these anomalies are limited to the vicinity of the mineralization, with small secondary dispersion aureoles.

The Vale das Gatas mine recorded extremely high maximum values for Pb (6299 mg/kg), As (5770 mg/kg), and W (636 mg/kg) (Fig. 6). The Cu-Mn-W-As-Pb-Zn association is inversely related to pH and reflects the presence of mineralized veins in the area.

Fig. 7 summarizes soil pH and trace element data from the copper (Cu) mines of São Domingos and Borralhal. Soil samples from the São Domingos mine area contained high levels of Ag, As, Cu, Pb, and Zn. The maximum As concentration in these soils reached 1291 mg/kg (Fig. 7). As a result of previous activities at the site (copper smelter), Cu concentrations in these soils reached up to 1829 mg/kg (Fig. 7). The concentration of Pb in the soil was also very high, with an average value of 2694 mg/kg (Fig. 7). The average Zn concentration in soil was 218 mg/kg, but it could rise to 714 mg/kg, which is extremely toxic to plants. Silver concentrations in soils ranged from 2.5 to 16.6 mg/kg, while Co, Cr, and Ni concentrations ranged from 20.1 to 54.3 mg/kg, 5.1 to 84.6 mg/kg, and 27.2 to 52.9 mg/kg, respectively.

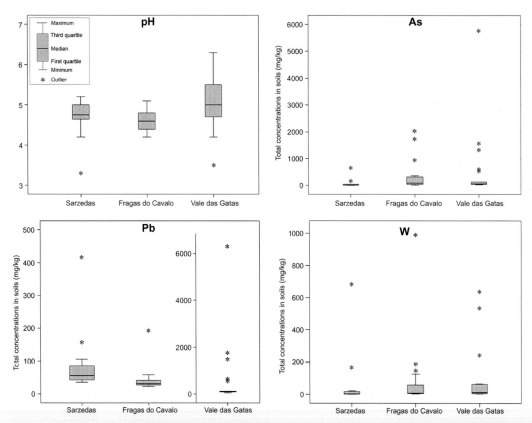

FIG. 6   Arsenic, Pb, and W concentrations (mg/kg) and pH of soil samples of the Sarzedas mine (*n* = 24), Fragas do Cavalo mine (*n* = 22), and Vale das Gatas mine (*n* = 69).

The soils in the Borralhal mine area contained high levels of Ag, As, Cu, Pb, and Sb. These anomalies are unmistakably linked to the area's mineralization (chalcopyrite, pyrite, bournonite, and other sulfosalts). The concentrations of As, Cu, and Pb in these soils reached 555, 1773, and 451 mg/kg, respectively (Fig. 7).

Fig. 8 summarizes the pH and trace element data in soils from the Pb mines of Barbadalhos and Sanheiro. The soils of the Barbadalhos mine contained high levels of Ag, As, Zn, and, most notably, Pb. The Pb concentration in these soils reached 9331 mg/kg, with the average value being 928 mg/kg (Fig. 7), owing to galena mining at the site. The mean Pb concentration (2380 mg/kg) in mineralized zone soils was nearly nine times the threshold for industrial soils suggested by the Canadian Environmental Quality Guidelines.

The situation in the Sanheiro mine is similar to that in the Barbadalhos mine. High levels of Ag, Zn, and, most notably, Pb were also detected in the Sanheiro mine (Fig. 8). The Pb concentration in these soils reached 11,314 mg/kg, with an average of 536 mg/kg.

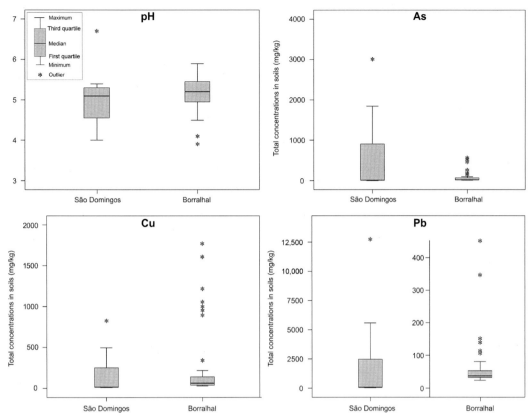

FIG. 7   Arsenic, Cu, and Pb concentrations (mg/kg) and pH of soil samples of the São Domingos mine ($n = 21$) and Borralhal mine ($n = 43$).

3.4 Trace metals and metalloids in plants

Arsenic was preferentially accumulated in the needles of *P. pinaster* in the tree species of the Sarzedas mine area. In fact, As accumulation was high in needles, and it increased in older tissues (Fig. 9). As a result, these tissues are well suited for detecting the anomaly. Because young leaves have higher metabolic activity, this translocation is a common mechanism in plants to avoid toxicity.

In plants, the accumulated concentrations of Pb were found to be within the normal range (Fig. 10). Lead concentrations were highest in *Quercus rotundifolia* tissues.

Among the tree species capable of accumulating W, *P. pinaster* (old needles) stood out (Fig. 11).

Considering other native flora species, shrubs, and herbaceous species, it was discovered that (data not shown in this chapter): arsenic is accumulated mostly in aerial tissues of *Digitalis purpurea*; Pb is accumulated by *Halimium ocymoides*, *D. purpurea*, and *Helichrysum stoechas*; and W is accumulated by *D. purpurea*, *Cistus ladanifer*, *Calluna vulgaris*, and

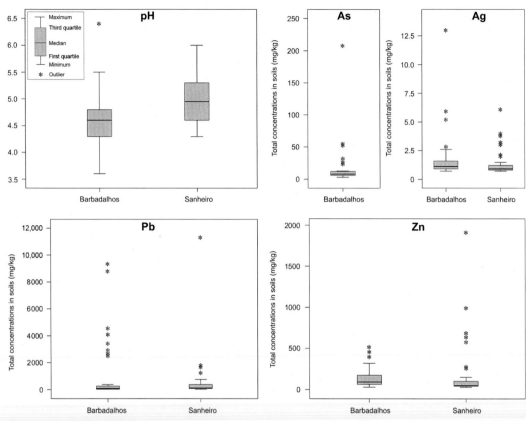

FIG. 8　Arsenic, Ag, Pb, and Zn concentrations (mg/kg) and pH of soil samples of the Barbadalhos mine ($n = 45$) and Sanheiro mine ($n = 50$).

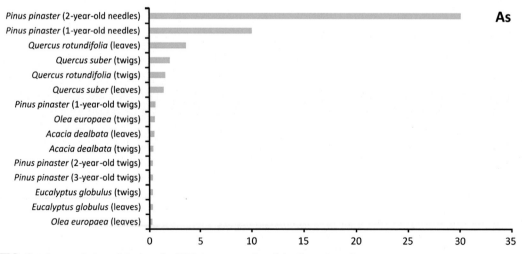

FIG. 9　Accumulation of As (mg/kg DW) in tree species of the Sarzedas mine area.

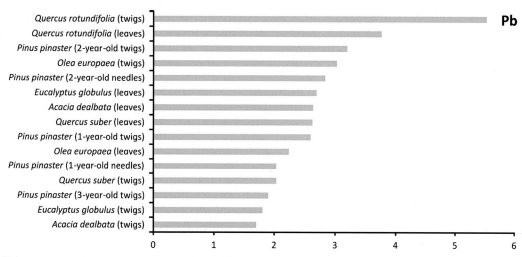

FIG. 10   Accumulation of Pb (mg/kg DW) in tree species of the Sarzedas mine area.

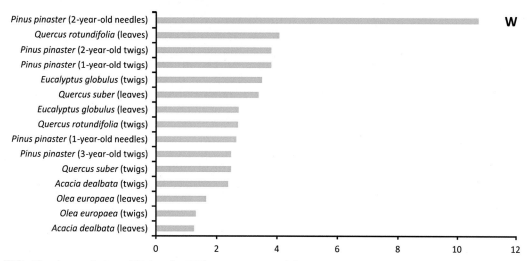

FIG. 11   Accumulation of W (mg/kg DW) in tree species of the Sarzedas mine area.

H. stoechas. Although the concentrations in the soil were low, *D. purpurea* accumulated a significant amount of Sb, indicating its tolerance to this element.

Based on the findings in the Sarzedas mine area, it was determined that the tree species and tissues best suited for metal(loid) bioindicating and with potential for mine restoration are, in descending order of importance: (1) As: old needles of *P. pinaster* and leaves of *Q. rotundifolia*; (2) W: old stems and needles of *P. pinaster*, *Q. rotundifolia* stems and leaves, and *Eucalyptus globulus* stems; (3) Sb: *P. pinaster* stems. When it comes to shrubs and herbaceous plants, the following species/tissues are best suited: (1) As: aerial tissues of *D. purpurea*, *C. vulgaris*, *Chamaespartium tridentatum*, leaves of *C. ladanifer*, and *Erica umbellata*; (2) W: aerial tissues of *D. purpurea*, *C. tridentatum*, stem and leaves of *C. ladanifer*, and *E. umbellata*; (3) Sb: *D. purpurea*, *E. umbellata*; stems of *C. ladanifer*, *C. vulgaris*, and *C. tridentatum*.

Metal(loid) concentrations in tree species are relatively low in the Fragas do Cavalo mine area when compared to other Sn/W studied mines. *Q. rotundifolia* had the highest As concentration (Fig. 12), with values higher than normal levels in plants. Lead concentrations are within the normal range for plants in all tree species (Fig. 13). Tungsten concentrations in *Q. rotundifolia* leaves and twigs and *P. pinaster* twig shoots ranged from 0.27 to 0.78 mg/kg DW (Fig. 14), which is above the normal range for plants.

The content variations in plant samples were, in general, strongly related to the content variations in soils in the Vale das Gatas mine area. It has also been established that metal concentrations in plant tissues are higher in contaminated sites or tailings due to higher metal concentrations in the soil.

The *P. pinaster* trees that grew on the tailings and contaminated soils of the Vale das Gatas mine accumulated the studied elements in greater quantities than plants from the local geochemical background. These values were also higher than expected for this species. In *P. pinaster* samples from tailings and contaminated soil sites, older needles (2 and 3 years old) accumulate higher concentrations of As, Pb, and W, as well as Fe and Zn (Figs. 15–17), whereas Ni and Cu were preferentially accumulated in young needles and stems (1 year old). As a result of the findings, it was possible to conclude that metal(loid)

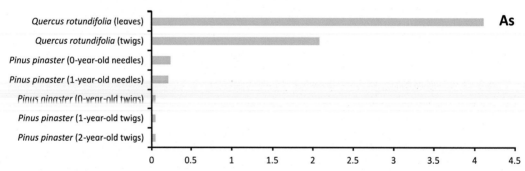

FIG. 12   Accumulation of As (mg/kg DW) in tree species of the Fragas do Cavalo mine area.

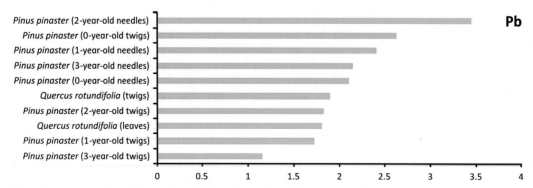

FIG. 13   Accumulation of Pb (mg/kg DW) in tree species of the Fragas do Cavalo mine area.

III. Ornamentals and tree crops for contaminated substrates

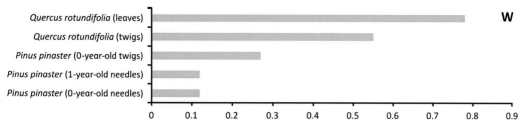

FIG. 14 Accumulation of W (mg/kg DW) in tree species of the Fragas do Cavalo mine area.

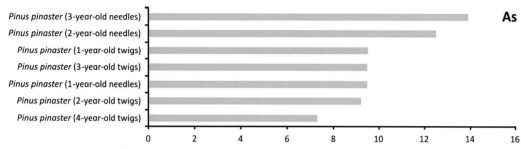

FIG. 15 Accumulation of As (mg/kg DW) in *Pinus pinaster* tissues of the Vale das Gatas mine area.

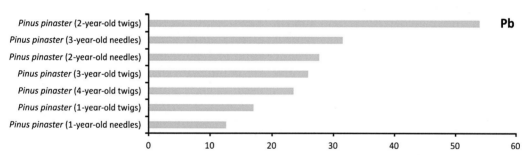

FIG. 16 Accumulation of Pb (mg/kg DW) in *Pinus pinaster* tissues of the Vale das Gatas mine area.

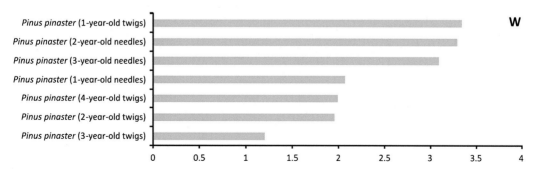

FIG. 17 Accumulation of W (mg/kg DW) in *Pinus pinaster* tissues of the Vale das Gatas mine area.

III. Ornamentals and tree crops for contaminated substrates

concentrations in plants depend as much on the plant organ as on its age, and that in biogeo-chemical studies, foliar and woody material should not be mixed in the same sample.

The accumulation behavior of As, Pb, and W, as well as Fe, Mn, Cu, Zn, and Ni, varied greatly with organ age in the species. As a result, the older needles had higher levels of As (Fig. 15), Pb (Fig. 16), W (Fig. 17), Fe, and Zn. The 2-year-old stems could be used to detect higher levels of Fe, Zn, and Pb, whereas the 1-year-old needles and stems accumulated higher levels of Cu and Ni. The 2-year-old stems could also be used to detect higher levels of Pb (Fig. 16), Fe, and Zn.

Considering the shrub and herbaceous species (data not shown in this chapter), it was found that: The leaves of *Agrostis castellana* and *Holcus lanatus* reflect the Cu, Pb, and Ni pedogeochemical anomalies; the aerial parts of *Pteridium aquilinum* and *Juncus effusus* seem to be indicative of Zn anomalies in the soil; *H. lanatus* and *A. castellana* were the main accumulators of As, Cu, Fe, and Pb and were good accumulators of Zn. *P. aquilinum* accumulated As, Pb, and Zn well. *J. effusus* was discovered to be a Zn accumulator.

Cu concentrations in plant tissues ranged from 4.60 to 23.8 mg/kg DW in tree samples from the São Domingos mine area (Fig. 18). These Cu values are within the normal range for plants.

The concentrations of As in tree tissues ranged from 0.3 to 1.4 mg/kg DW (Fig. 19). The highest As concentrations were found in herbaceous species such as *Juncus conglomeratus*, *Thymus mastichina*, *J. effusus*, and *Scirpus holoschoenus*.

Lead concentrations in tree tissues ranged from 2.9 to 9.1 mg/kg DW (Fig. 20), which is normal for plants. However, Pb concentrations ranged from 3.5 to 84.9 mg/kg DW for some herbaceous species. *J. conglomeratus* and *S. holoschoenus*, two semi-aquatic species collected in the mining area, had high Pb accumulation in plant tissues. Lead concentrations greater than 20 mg/kg DW were found in the leaves of two *Cistus* species, which are common Mediterranean shrubs known for their drought tolerance and low nutrient availability.

Zinc concentrations in tree tissues are also within the plant's normal range (Fig. 21). The herbaceous species *Cistus monspeliensis* and *Daphne gnidium* had the highest Zn concentrations, as did other metals.

In the contaminated area, a few trees, *Eucalyptus*, *Quercus*, and *Pinus* species, were discovered with minor accumulations of different metal(loid)s in the above-ground tissues.

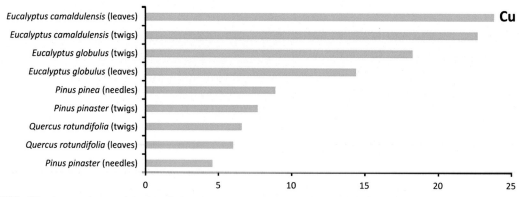

FIG. 18 Accumulation of Cu (mg/kg DW) in tree species of the São Domingos mine area.

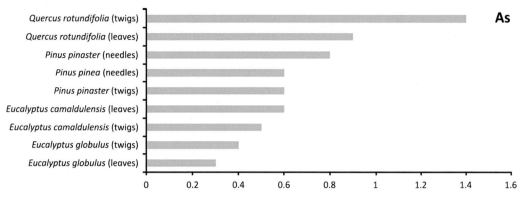

FIG. 19 Accumulation of As (mg/kg DW) in tree species of the São Domingos mine area.

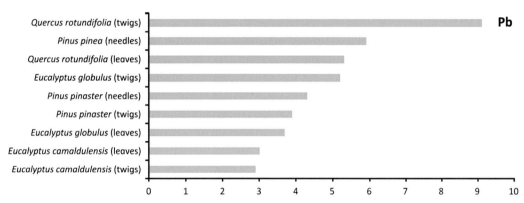

FIG. 20 Accumulation of Pb (mg/kg DW) in tree species of the São Domingos mine area.

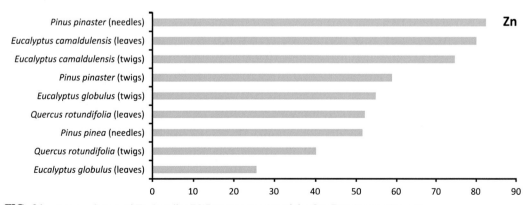

FIG. 21 Accumulation of Zn (mg/kg DW) in tree species of the São Domingos mine area.

III. Ornamentals and tree crops for contaminated substrates

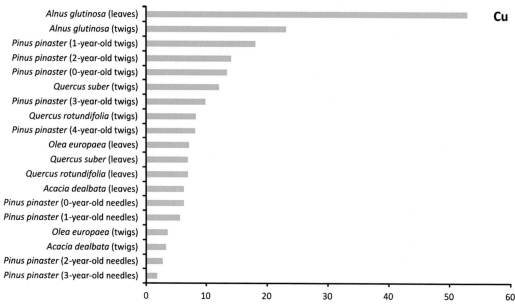

FIG. 22   Accumulation of Cu (mg/kg DW) in tree species of the Borralhal mine area.

However, because of their high biomass, they can be very effective for metal(loid)s phytoextraction and phytostabilization, particularly when established in less contaminated soils on the study area's outskirts.

Copper was preferentially accumulated in the tissues of *Alnus glutinosa* (both leaves and twigs) and *P. pinaster* twigs in the Borralhal mine area tree species (Fig. 22). Copper concentrations in *A. glutinosa* leaves reached 52.8 mg/kg. The concentrations of As in tree tissues ranged from 0.2 to 18 mg/kg DW (Fig. 23). In fact, high As accumulation was found in 2-year-old *P. pinaster* needles. Lead concentrations in tree tissues ranged from 0.9 to 12 mg/kg DW, with 3-year-old needles and 2-year-old twigs containing the highest levels (Fig. 24).

Seven tree species from the native flora of the Barbadalhos mine area were studied, as well as 18 types of shrubs, 17 herbs, four grasses, and three ferns.

Most plants were found to be Pb-tolerant. Lead concentrations in plants ranged from 1.11 to 548 mg/kg DW in the mineralized zone. This is far above the toxic range of 100–400 mg Pb/kg for most plants.

Significant accumulation of Pb was observed in the following shrub and herbaceous species (data not shown in this chapter): *Cistus salvifolius* (548 mg/kg), *Lonicera periclymenum* (318 mg/kg), *Anarrhinum bellidifolium*, *Phytolacca americana*, *D. purpurea*, and *Mentha suaveolens* (255–217 mg/kg). Pteridophytes with 117–251 mg/kg Pb in aerial parts included *Polystichum setiferum*, *P. aquilinum*, and *Asplenium onopteris*. Lead content in plants from the non-mineralized zone ranged from 0.94 to 11.6 mg/kg.

Although the maximum Pb content (Fig. 25) observed in trees such as *Acacia dealbata* (84 mg/kg in leaves; 56.5 mg/kg in twigs), *Olea europaea* (62 mg/kg in twigs; 58 mg/kg in leaves), and *Quercus suber* (57.5 mg/kg in twigs) from mineralized zone is not as impressive

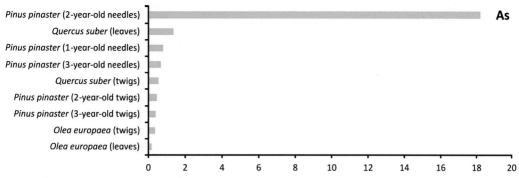

FIG. 23   Accumulation of As (mg/kg DW) in tree species of the Borralhal mine area.

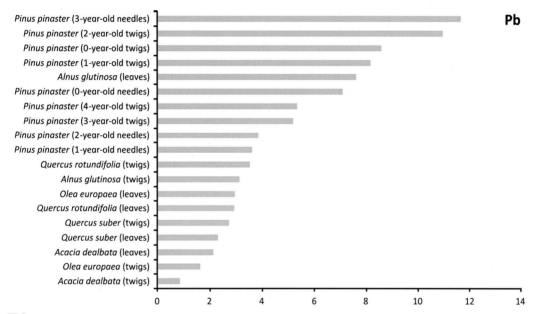

FIG. 24   Accumulation of Pb (mg/kg DW) in tree species of the Borralhal mine area.

as that of the smaller plants mentioned above, these trees can be very effective due to their higher biomass. When combined with the species' hardiness, biomass, and abundance, the moderate accumulation indicates enormous potential for phytoextraction of Pb in the area.

However, in the future, any use of *Acacia* species in revegetation actions should be avoided. These are aggressive invasive plants that have a negative impact on native biodiversity and community structure in Portuguese ecosystems. As a result, these species have been subjected to containment and even eradication efforts.

Zinc concentrations in trees reached 140 mg/kg in *Q. suber* leaves (Fig. 26). Concentrations in *D. purpurea*, on the other hand, reached 1020 mg/kg and ranged from 262 to 887 mg/kg in

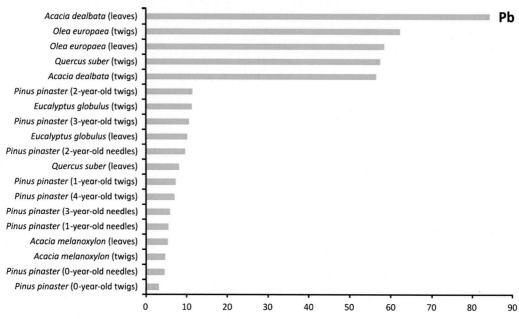

FIG. 25 Accumulation of Pb (mg/kg DW) in tree species of the Barbadalhos mine area.

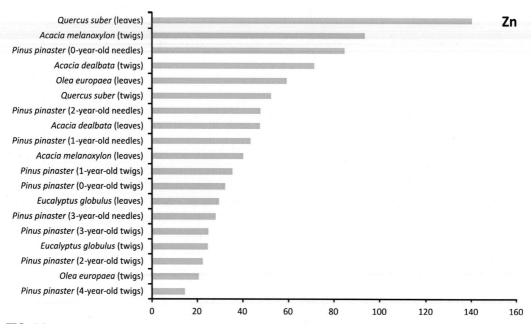

FIG. 26 Accumulation of Zn (mg/kg DW) in tree species of the Barbadalhos mine area.

III. Ornamentals and tree crops for contaminated substrates

L. periclymenum, P. americana, Solanum nigrum, P. setiferum, M. suaveolens, Viola riviniana, and *A. bellidifolium,* in decreasing order.

The metal budget of a plant under the influence of a contaminated rhizosphere may also differ depending on the organ studied. No such difference was observed in *P. pinaster* or *O. europaea* for any metal. Most metals were similarly concentrated in the leaves of the other five trees, with the exception of Fe, which was more concentrated in the leaves of *E. globulus* and *A. melanoxylon.* Lead concentrations in *Q. suber* stem from the mineralized zone were approximately five times higher.

Lead concentrations in trees in the Sanheiro mine area ranged from 2.9 to 37 mg/kg DW (Fig. 27). Lead concentrations are high in the species *E. globulus, Q. robur,* and *Q. suber,* exceeding normal levels in plants. Zinc concentrations in tree tissues are within the normal range for plants, with *Q. robur* and *P. pinaster* having higher levels (Fig. 28).

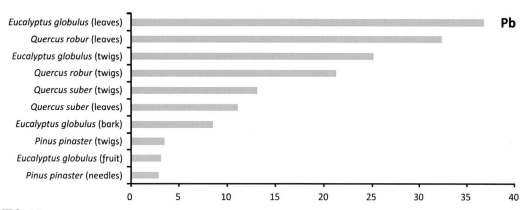

FIG. 27   Accumulation of Pb (mg/kg DW) in tree species of the Sanheiro mine area.

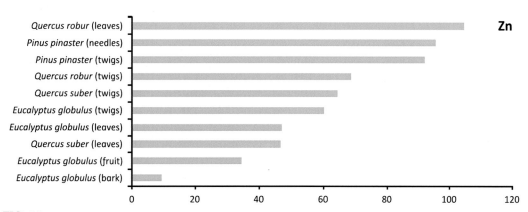

FIG. 28   Accumulation of Zn (mg/kg DW) in tree species of the Sanheiro mine area.

III. Ornamentals and tree crops for contaminated substrates

4 Final considerations and future prospects

Tree species growing in tailings and mine-contaminated soils are resistant to imposed stress conditions (metal contamination and nutrient deficiency) and can achieve the goals of stabilization, pollution reduction, and visual enhancement.

A few trees, *Pinus*, *Eucalyptus*, *Quercus*, *Acacia*, *Alnus*, and *Olea* species, found in tailings and contaminated soils in abandoned mines in Portugal, show metal accumulation in above-ground tissues. As a result, these trees, in conjunction with the spontaneous flora of the herbaceous and arbustive strata, form communities that are resistant to toxic trace elements, and play an important role in the remediation of degraded mine soils. The *Pinus*, *Eucalyptus*, and *Quercus* species should be highlighted. These species are widely distributed throughout the territory and are extremely important economically. Furthermore, these trees could grow and propagate in low-nutrient substrata, which would be a huge benefit in the revegetation of mine tailings.

Metalliferous or metal-contaminated soils have physicochemical properties that inhibit soil formation and plant growth. Other negative factors, in addition to elevated metal(loid) concentrations, include a lack of topsoil, erosion, drought, compaction, wide temperature fluctuations, a lack of soil-forming fine materials, and a lack of essential nutrients. Mine-degraded soils typically have low concentrations of important nutrients such as N, P, and K. Toxic metals can also reduce the number, diversity, and activity of soil organisms, impeding the decomposition of organic matter and the mineralization of nitrogen. The chemical form of the potentially toxic metal, the presence of other chemicals that may aggravate or mitigate metal toxicity, the prevailing pH, and the nutrient status of contaminated soil all influence how plants respond to it.

It was also observed that, despite their lower accumulation when compared to herbaceous and shrub species, trees can be very effective for metals phytoextraction and phytostabilization due to their higher biomass and bioproductivity, particularly when established in less contaminated soils on the studied areas' periphery. Nonetheless, the importance of small and herbaceous plants and shrubs cannot be overstated, as their rapid growth rate, shorter life span, and rapid successive generations can accelerate the process of soil reclamation by providing repeated short-duration rapid phases or bursts of remediation. The timely removal of the biomass of plants and shrubs from the site can make life easier for these candidate trees because the former have already done their part, and their disposal prevents the remediated metals from returning to the soil.

Metal toxicity issues rarely arise in the case of native flora, because native plants adapt to locally elevated metal levels over time. Because native plants are generally slow-growing with shallow root systems and low biomass, they may be better phytoremediators for contaminated lands than known metal hyperaccumulators. Plants that are resistant to toxic metals and have a low nutrient status but have a high rate of growth and biomass are ideal for remediating degraded soils and habitats such as those near mines. The native flora demonstrated resistance to high concentrations of heavy metals in the soil. Some species also showed variable metal accumulation patterns at different soil concentrations. This variation was also observed in different parts of the same plant, implying that plant-soil interactions should be fully considered when selecting plant species for developing and utilizing methods such as phytoremediation.

The existing natural plant cover at abandoned mine sites can be greatly increased by large-scale planting and maintenance of native species with higher metal accumulation potential for a number of years. Even seed dispersal from plants grown on-site is encouraged. Organic amendment is required to aid in the establishment and colonization of these "pioneer plants." They can eventually change the man-made habitat to make it more suitable for future plant communities. Allowing native species to remediate soils is an appealing proposition because native wild species do not require frequent irrigation, fertilization, or pesticide treatments, while simultaneously establishing a plant community comparable to that found nearby.

As a result, mine restoration may benefit from a broader perspective that includes various groups of plant species that can play distinct functional roles in the remediation process. The use of leguminous plants, for example, can enrich the nutrient content, and the combination of perennials and annuals can provide significant inputs in terms of organic matter and nutrient recycling, contributing to soil development in different ways. More information about plant communities growing on metal-contaminated soils is needed for this approach to accurately determine their potential for the remediation of polluted soils at abandoned mines. Suitable conditions/amendments can be created to develop them as good competitors with enhanced growth and proliferation to their counterparts growing on the same metal-contaminated nutrient-depleted soils after screening out ideal phytoremedial candidates from the native flora and assessing their individual requirements.

The studied mining areas' spontaneous tree species demonstrate their ability to withstand high concentrations of heavy metal(loid)s in the soil. Metal(loid) accumulation patterns in the plants tested, on the other hand, varied. Because metal concentrations in above-ground parts were kept low, metal tolerance in most cases may be based primarily on their metal-excluding ability. Metal(loid) concentrations above the toxic level in some species, however, suggest that internal detoxification and metal-tolerance mechanisms may exist. Furthermore, the plants could grow and propagate in low-nutrient substrata, which would be a significant advantage in the revegetation of mine tailings. It was also observed that, despite lower accumulation, trees in the studied areas can be extremely effective due to their higher biomass.

As a result of the findings, the aforementioned tree species may be useful for mine restoration and mitigating mining impacts. This is accomplished by partially removing bioavailable toxic elements and/or retaining them to reduce toxic element output to the ecosystem. The biomass can be used to generate economic value, especially when metal concentrations in above-ground parts are kept low. We can use biomass as a renewable resource for energy production (bioenergy), timber extraction for wooden furniture and other manufactured wood products, or sawn wood and wood-processed products (e.g., particle board, fiberboard, and other panels) for carpentry and construction as examples.

References

Adekiya, A.O., Ejue, W.S., Olayanju, A., 2020. Different organic manure sources and NPK fertilizer on soil chemical properties, growth, yield and quality of okra. Sci. Rep. 10, 16083.

Ahemad, M., 2019. Remediation of metalliferous soils through the heavy metal resistant plant growth promoting bacteria: paradigms and prospects. Arab. J. Chem. 12 (7), 1365–1377. https://doi.org/10.1016/j.arabjc.2014.11.020.

Akinbile, B.J., Makhubela, B.C.E., Ambushe, A.A., 2021. Phytomining of valuable metals: status and prospective—a review. Int. J. Environ. Anal. Chem. https://doi.org/10.1080/03067319.2021.1917557.

Alarcón-Aguirre, G., Quispe, E.S., Zavaleta, T.V., Tejada, L.V.P., Enciso, D.R., Achata, L.R., Garate-Quispe, J., 2023. Vegetation dynamics in lands degraded by gold mining in the southeastern Peruvian Amazon. Trees Forests People, 100369.

Alberico, I., Alessio, G., Fagnano, M., Petrosino, P., 2023. The effectiveness of geotrails to support sustainable development in the campi flegrei active volcanic area. Geoheritage 15 (1), 15.

Amabogha, O.N., Garelick, H., Jones, H., Purchase, D., 2023. Combining phytoremediation with bioenergy production: developing a multi-criteria decision matrix for plant species selection. Environ. Sci. Pollut. Res., 1–14.

Antoniadisa, V., Levizoua, E., Shaheenb, S.M., Okc, Y.S., Sebastiand, A., Baume, C., Prasad, M.N.V., Wenzelf, W.W., Rinklebeg, J., 2017. Trace elements in the soil-plant interface: phytoavailability, translocation, and phytoremediation—a review. Earth Sci. Rev. 171, 621–645.

Bani, A., Echevarria, G., Sulçe, S., Morel, J.L., 2015. Improving the agronomy of *Alyssum murale* for extensive phytomining: a five-year field study. Int. J. Phytoremediation 17, 117–127.

Baskerville, L., Spain, C.S., Nuske, S., Gagen, E.J., 2023. Options for Native Ecosystem Mine Site Rehabilitation in Queensland: Stakeholder Survey Report. Office of the Queensland Mine Rehabilitation Commissioner, Queensland Government, Brisbane.

Berslin, D., Reshmi, A., Sivaprakash, B., Rajamohan, N., Kumar, P.S., 2021. Remediation of emerging metal pollutants using environment friendly biochar—review on applications and mechanism. Chemosphere, 133384.

Bondu, R., Casiot, C., Pistre, S., Batiot-Guilhe, C., 2023. Impact of past mining activities on water quality in a karst area in the Cévennes region, Southern France. Sci. Total Environ., 162274.

Capecchi, I., Borghini, T., Bernetti, I., 2023. Automated urban tree survey using remote sensing data, Google street view images, and plant species recognition apps. Eur. J. Remote Sens. 56 (1), 2162441.

Carman, S., 2023. The Colorado brightfields project: a land use initiative at the intersection of renewable energy and contaminated sites. Res. Rec., 101.

Cemin, A., Ferrarini, F., Poletto, M., Bonetto, L.R., Bortoluz, J., Lemée, L., Guégan, R., Esteves, V.I., Giovanela, M., 2021. Characterization and use of a lignin sample extracted from *Eucalyptus grandis* sawdust for the removal of methylene blue dye. Int. J. Biol. Macromol. 170, 375–389.

Chaturvedi, R., Favas, P., Pratas, J., Varun, M., Paul, M.S., 2019. EDTA-assisted metal uptake in *Raphanus sativus* L. and *Brassica oleracea* L.: assessment of toxicity and food safety. Bull. Environ. Contam. Toxicol. 103, 490–495. https://doi.org/10.1007/s00128-019-02651-9.

Chaturvedi, R., Favas, P.J.C., Pratas, J., Varun, M., Paul, M.S., 2021. Harnessing *Pisum sativum-Glomus mosseae* symbiosis for phytoremediation of soil contaminated with lead, cadmium, and arsenic. Int. J. Phytoremediation 23 (3), 279–290. https://doi.org/10.1080/15226514.2020.1812507.

Chen, R., Han, L., Zhao, Y., Liu, Z., Fan, Y., Li, R., Xia, L., 2023. Response of plant element traits to soil arsenic stress and its implications for vegetation restoration in a post-mining area. Ecol. Indic. 146, 109931.

Chowdhury, R., Lyubun, Y., Favas, P.J.C., Sarkar, S.K., 2016. Phytoremediation potential of selected mangrove plants for trace metal contamination in Indian Sundarban Wetland. In: Ansari, A.A., Gill, S.S., Gill, R., Lanza, G.R., Newman, L. (Eds.), Phytoremediation: Management of Environmental Contaminants. vol. 4. Springer, Amsterdam, ISBN: 978-3-319-41810-0, pp. 283–310, https://doi.org/10.1007/978-3-319-41811-7_15.

Chowdhury, R., Favas, P.J.C., Jonathan, M.P., Venkatachalam, P., Raja, P., Sarkar, S.K., 2017. Bioremoval of trace metals from rhizosediment by mangrove plants in Indian Sundarban Wetland. Mar. Pollut. Bull. 124 (2), 1078–1088. https://doi.org/10.1016/j.marpolbul.2017.01.047.

Chukwu, A., Obiora, S.C., 2023. Effect of lead-zinc mining on socio-economic and health conditions of Enyigba and Ishiagu lead-zinc mining districts of Southeastern Nigeria. Mining Metall. Explorat., 1–11.

Dilks, R.T., Monette, F., Glaus, M., 2016. The major parameters on biomass pyrolysis for hyperaccumulative plants: a review. Chemosphere 146, 385–395.

Dzoro, M., Mashapa, C., Gandiwa, E., 2023. Impact of artisanal gold mining on woody vegetation in Chewore South Safari Area, northern Zimbabwe. TreesForests People, 100375.

Evangelou, M.W.H., Papazoglou, E.G., Robinson, B.H., Schulin, R., 2015. Phytomanagement: phytoremediation and the production of biomass for economic revenue on contaminated land. In: Ansari, A.A., Gill, S.S., Gill, R., Lanza, G.R., Newman, L. (Eds.), Phytoremediation: Management of Environmental Contaminants. vol. 1. Springer, Cham, Heidelberg, New York, Dordrecht, London, pp. 115–132.

Favas, P.J.C., Martino, L.E., Prasad, M.N.V., 2018. Abandoned mine land reclamation—challenges and opportunities (holistic approach). In: Prasad, M.N.V., Favas, P.J.C., Maiti, S.K. (Eds.), Bio-Geotechnologies for Mine Site Rehabilitation. Elsevier, ISBN: 978-0-12-812986-9, pp. 3–31, https://doi.org/10.1016/B978-0-12-812986-9.00001-4.

Favas, P.J.C., Morais, I., Campos, J., Pratas, J., 2020. Fitoextração de níquel por uma população nativa de *Alyssum serpyllifolium* subsp. *lusitanicum* em solos ultramáficos (Portugal): perspetivas para fitomineração. Comun. Geol. 107 (Especial I), 1015–1017. ISSN: 0873-948X; e-ISSN: 1647-581X.

Favas, P.J.C., Pratas, J., 2019. In-situ phytoextraction of nickel by *Odontarrhena serpyllifolia* on ultramafic soils of Portugal. In: International Conference on Geosciences GEOLINKS 2019, Conference Proceedings, Book 3. vol. 1. Saima Consult Ltd, Athens, Greece, ISBN: 978-619-7495-04-1, pp. 109–114, https://doi.org/10.32008/GEOLINKS2019/B3/V1/12.

Favas, P.J.C., Pratas, J., Chaturvedi, R., Paul, M.S., Prasad, M.N.V., 2016a. Tree crops on abandoned mines for environmental remediation and industrial feedstock. In: Prasad, M.N.V. (Ed.), Bioremediation and Bioeconomy., ISBN: 978-0-12-802830-8, pp. 219–249, https://doi.org/10.1016/B978-0-12-802830-8.00010-1. Elsevier. Amsterdam.

Favas, P.J.C., Pratas, J., Paul, M.S., Prasad, M.N.V., 2019. Remediation of uranium-contaminated sites by phytoremediation and natural attenuation. In: Pandey, V.C., Bauddh, K. (Eds.), Phytomanagement of Polluted Sites: Market Opportunities in Sustainable Phytoremediation. Elsevier, ISBN: 978-0-12-813912-7, pp. 277–300, https://doi.org/10.1016/B978-0-12-813912-7.00010-7.

Favas, P.J.C., Pratas, J., Paul, M.S., Sarkar, S.K., Prasad, M.N.V., 2016b. Phytofiltration of metal(loid)-contaminated water: the potential of native aquatic plants. In: Ansari, A.A., Gill, S.S., Gill, R., Lanza, G.R., Newman, L. (Eds.), Phytoremediation: Management of Environmental Contaminants. vol. 3. Springer, Amsterdam, ISBN: 978-3-319-40146-1, pp. 305–343, https://doi.org/10.1007/978-3-319-40148-5_11.

Gairola, S.U., Bahuguna, R., Bhatt, S.S., 2023. Native plant species: a tool for restoration of mined lands. J. Soil Sci. Plant Nutr., 1–11.

Giri, S., Kang, Y., MacDonald, K., Tippett, M., Qiu, Z., Lathrop, R.G., Obropta, C.C., 2023. Revealing the sources of arsenic in private well water using random forest classification and regression. Sci. Total Environ. 857, 159360.

Goodman, A.J., Bednar, A.J., Ranville, J.F., 2023. Rare earth element recovery in hard-rock acid mine drainage and mine waste: a case study in Idaho Springs, Colorado. Appl. Geochem. 150, 105584.

Govil, H., Guha, S., 2023. Underground mine deformation monitoring using synthetic aperture radar technique: a case study of Rajgamar coal mine of Korba Chhattisgarh, India. J. Appl. Geophys. 209, 104899.

Hernández Guiance, S.N., Coria, I.D., Faggi, A., Basílico, G., 2023. Phytoremediation of soils polluted by heavy metals and metalloids: recent case studies in Latin America. In: Phytoremediation: Management of Environmental Contaminants. vol. 7. Springer International Publishing, Cham, pp. 317–332.

Hou, H., Liu, H., Xiong, J., Wang, C., Zhang, S., Ding, Z., 2023. Comparison of soil bacterial communities under Canopies of *Pinus tabulaeformis* and *Populus euramericana* in a Reclaimed Waste Dump. Plants 12 (4), 974.

Huff, D.K., Morris, L.A., Sutter, L., Costanza, J., Pennell, K.D., 2020. Accumulation of six PFAS compounds by woody and herbaceous plants: potential for phytoextraction. Int. J. Phytoremediation 22 (14), 1538–1550.

ICNF, 2015. 6.º Inventário Florestal Nacional. Instituto da Conservação da Natureza e das Florestas, Lisboa.

Issaka, S., Ashraf, M.A., 2021. Phytorestoration of mine spoiled: "evaluation of natural phytoremediation process occurring at ex tin mining catchment". In: Phytorestoration of Abandoned Mining and Oil Drilling Sites. Elsevier, pp. 219–248.

Jain, P.P., Nahar Ali, Z., Sisodiya, S.J., Kunnel, S.G., 2020. Tree barks for bioremediation of heavy metals from polluted waters. In: Methods for Bioremediation of Water and Wastewater Pollution. Springer, Cham, pp. 277–288.

Kelvin, R.P., Víctor, M.C., Julio, C.A.O., Warren, R.R., 2020. Bioavailability and solubility of heavy metals and trace elements during composting of cow manure and tree litter. Appl. Envir. Soil Sci. https://doi.org/10.1155/2020/5680169.

Khapre, A., Kumar, S., Rajasekaran, C., 2019. Phytocapping: an alternate cover option for municipal solid waste landfills. Environ. Technol. 40 (17), 2242–2249. https://doi.org/10.1080/09593330.2017.1414314.

Khosravi, V., Gholizadeh, A., Agyeman, P.C., Ardejani, F.D., Yousefi, S., Saberioon, M., 2023. Further to quantification of content, can reflectance spectroscopy determine the speciation of cobalt and nickel on a mine waste dump surface? Sci. Total Environ. 872, 161996.

Kong, Z., Glick, B.R., 2017. The role of plant growth-promoting bacteria in metal phytoremediation. Adv. Microb. Physiol. 71, 97–132. https://doi.org/10.1016/bs.ampbs.2017.04.001.

König, L.A., Medina-Vega, J.A., Longo, R.M., Zuidema, P.A., Jakovac, C.C., 2023. Restoration success in former Amazonian mines is driven by soil amendment and forest proximity. Philos. Trans. R. Soc. B 378 (1867), 20210086.

Košek, F., Culka, A., Fornasini, L., Vandenabeele, P., Rousaki, A., Mirao, J., Bersani, D., Candeias, A., Jehlička, J., 2020. Application of a handheld Raman spectrometer for the screening of colored secondary sulfates in abandoned mining areas—the case of the São Domingos Mine (Iberian Pyrite Belt). J. Raman Spectrosc. 51, 1186–1199. https://doi.org/10.1002/jrs.5873.

Kumar, H., Ishtiyaq, S., Favas, P.J.C., Varun, M., Paul, M.S., 2022a. Effect of metal-resistant PGPB on the metal uptake, antioxidative defense, physiology, and growth of *Atriplex lentiformis* (Torr.) S. Wats. in soil contaminated with cadmium and nickel. J. Plant Growth Regulat. https://doi.org/10.1007/s00344-022-10853-5.

Kumar, H., Ishtiyaq, S., Varun, M., Favas, P.J.C., Ogunkunle, C.O., Paul, M.S., 2022b. Bioremediation: plants and microbes for restoration of heavy metal contaminated soils. In: Puthur, J.T., Dhankher, O.P. (Eds.), Bioenergy Crops: A Sustainable Means of Phytoremediation. CRC Press, Taylor & Francis Group, LLC, ISBN: 978-0-367-48913-7, pp. 37–70, https://doi.org/10.1201/9781003043522.

Kumari, S., Maiti, S.K., 2022. Nitrogen recovery in reclaimed mine soil under different amendment practices in tandem with legume and non-legume revegetation: a review. Soil Use Manag. 38 (2), 1113–1145.

Lazo, P., Lazo, A., Hansen, H.K., Ortiz-Soto, R., Hansen, M.E., Arévalo, F., Gutiérrez, C., 2023. Removal of heavy metals from mine tailings in Central Chile using *Solidago chilensis* Meyen, *Haplopappus foliosus* DC, and *Lycium chilense* Miers ex Bertero. Int. J. Environ. Res. Public Health 20 (3), 2749.

Li, C., Ji, Y., Ma, N., Zhang, J., Zhang, H., Ji, C., Zhu, J., Shao, J., Li, Y., 2023. Positive effects of vegetation restoration on the soil properties of post-mining land. Plant Soil, 1–11.

Li, Y.-M., Chaney, R., Brewer, E., Roseberg, R., Angle, J.S., Baker, A., Reeves, R., Nelkin, J., 2003. Development of a technology for commercial phytoextraction of nickel: economic and technical considerations. Plant Soil 249, 107–115.

Lord, R., Moffat, A., Sinnett, D., Phillips, P., Brignall, D., Manning, D., 2023. Blue Green Infrastructure on Mineral Sites.

Madejón, P., Domínguez, M.T., Gil-Martínez, M., Navarro-Fernández, C.M., Montiel-Rozas, M.M., Madejón, E., Murillo, J.M., Cabrera, F., Marañón, T., 2018. Evaluation of amendment addition and tree planting as measures to remediate contaminated soils: the Guadiamar case study (SW Spain). Catena 166, 34–43.

Masarovičová, E., Králová, K., 2018. Woody species in phytoremediation applications for contaminated soils. In: Phytoremediation. Springer, Cham, pp. 319–373.

Mesa-Marín, J., Pérez-Romero, J.A., Redondo-Gómez, S., Pajuelo, E., Rodríguez-Llorente, I.D., Mateos-Naranjo, E., 2020. Impact of plant growth promoting bacteria on *Salicornia ramosissima* ecophysiology and heavy metal phytoremediation capacity in estuarine soils. Front. Microbiol. https://doi.org/10.3389/fmicb.2020.553018.

Mhlongo, S.E., 2023. The negative impacts of mining interpreted from the problems of abandoned mines in South Africa. In: Implications of Industry 5.0 on Environmental Sustainability. IGI Global, pp. 221–248.

Mussali-Galante, P., Santoyo-Martínez, M., Castrejón-Godínez, M.L., Breton-Deval, L., Rodríguez-Solis, A., Valencia Cuevas, L., Tovar-Sánchez, E., 2023. The bioaccumulation potential of heavy metals by *Gliricidia sepium* (Fabaceae) in mine tailings. Environ. Sci. Pollut. Res., 1–18.

Naila, A., Meerdink, G., Jayasena, V., Sulaiman, A.Z., Ajit, A.B., Berta, G., 2019. A review on global metal accumulators—mechanism, enhancement, commercial application, and research trend. Environ. Sci. Pollut. Res. 26, 26449–26471. https://doi.org/10.1007/s11356-019-05992-4.

Nissim, W.G., Labrecque, M., 2021. Reclamation of urban brownfields through phytoremediation: implications for building sustainable and resilient towns. Urban For. Urban Green. 65, 127364.

Nong, H., Liu, J., Chen, J., Zhao, Y., Wu, L., Tang, Y., Liu, W., Yang, G., Xu, Z., 2023. Woody plants have the advantages in the phytoremediation process of manganese ore with the help of microorganisms. Sci. Total Environ. 863, 160995.

Nyenda, T., Gwenzi, W., Gwata, C., Jacobs, S.M., 2020. Leguminous tree species create islands of fertility and influence the understory vegetation on nickel-mine tailings of different ages. Ecol. Eng. 155, 105902.

Pandey, V.C. (Ed.), 2023. Bio-Inspired Land Remediation. Springer Nature.

Punia, A., Bharti, R., 2023. Loss of soil organic matter in the mining landscape and its implication to climate change. Arab. J. Geosci. 16 (1), 86.

Ranđelović, D., Pandey, V.C., 2023. Bioenergy crop-based ecological restoration of degraded land. In: Bio-Inspired Land Remediation. Springer International Publishing, Cham, pp. 1–29.

Seshadri, B., Bolan, N.S., Thangarajan, R., Jena, U., Das, K.C., Wang, H., Naidu, R., 2016. Biomass energy from revegetation of landfill sites. In: Prasad, M.N.V. (Ed.), Bioremediation and Bioeconomy. Elsevier, ISBN: 9780128028308, pp. 99–109, https://doi.org/10.1016/B978-0-12-802830-8.00005-8 (Chapter 5).

Singh, S., Kumar, V., Datta, S., Dhanjal, D.S., Sharma, K., Samuel, J., Singh, J., 2020. Current advancement and future prospect of biosorbents for bioremediation. Sci. Total Environ. 709, 135895.

Spasić, M., Vacek, O., Vejvodová, K., Tejnecký, V., Vokurková, P., Křižová, P., Polák, F., Borůvka, L., Drábek, O., 2023. Which Trees Form the Best Soil? Reclaimed Mine Soil Properties Under 22 Tree Species-50 Years Later: Assessment of Physical and Chemical Properties. Reclaimed Mine Soil Properties Under 22.

Stephan, K., Hubbart, J.A., 2023. Plant community, soil and microclimate attributes after 70 years of natural recovery of an abandoned limestone quarry. Land 12 (1), 117.

Stylianou, M., Gavriel, I., Vogiatzakis, I.N., Zorpas, A., Agapiou, A., 2020. Native plants for the remediation of abandoned sulphide mines in Cyprus: a preliminary assessment. J. Environ. Manag. 274, 110531.

Varun, M., Ogunkunle, C.O., D'Souza, R., Favas, P., Paul, M.S., 2017. Identification of *Sesbania sesban* (L.) Merr. as an efficient and well adapted phytoremediation tool for Cd polluted soils. Bull. Environ. Contam. Toxicol. 98, 867–873. https://doi.org/10.1007/s00128-017-2094-6.

Vieira, A., Matos, J.X., Lopes, L., Martins, R., 2020. Evaluation of the mining potential of the São Domingos mine wastes, Iberian Pyrite Belt, Portugal. Comun. Geol. 107 (Especial III), 91–100. ISSN: 0873-948X; e-ISSN: 1647-581X.

Vieira, Y., dos Santos, J.M., Georgin, J., Oliveira, M.L., Pinto, D., Dotto, G.L., 2021. An overview of forest residues as promising low-cost adsorbents. Gondwana Res.

Walter, J., Hradská, I., Kout, J., Bureš, J., Konvička, M., 2023. The impact of abandoned kaolin quarries on macromycetes (Fungi: Basidiomycota, Ascomycota), carabid beetle (Coleoptera: Carabidae), and spider (Araneae) assemblages. Biodivers. Conserv., 1–13.

Xiao, Y., Liu, C., Hu, N., Wang, B., Zheng, K., Zhao, Z., Li, T., 2023. Contributions of ectomycorrhizal fungi in a reclaimed poplar forest (*Populus yunnanensis*) in an abandoned metal mine tailings pond, southwest China. J. Hazard. Mater. 448, 130962.

Xie, K., Xie, N., Liao, Z., Luo, X., Peng, W., Yuan, Y., 2023. Bioaccessibility of arsenic, lead, and cadmium in contaminated mining/smelting soils: assessment, modeling, and application for soil environment criteria derivation. J. Hazard. Mater. 443, 130321.

Xu, D., Li, X., Chen, J., Li, J., 2023. Research progress of soil and vegetation restoration technology in open-pit coal mine: a review. Agriculture 13 (2), 226.

Yang, Q., Li, J., Yang, H., Liu, X., Mou, L., 2023. Vegetation restoration in rare earth tailings in Northeast Guangdong. In: Advances in Energy, Environment and Chemical Engineering. vol. 1. CRC Press, pp. 825–836.

You, Y., Dou, J., Xue, Y., Jin, N., Yang, K., 2022. Chelating agents in assisting phytoremediation of uranium-contaminated soils: a review. Sustainability 14, 6379. https://doi.org/10.3390/su14106379.

Zerbe, S., 2023. Mining sites and landfills. In: Restoration of Ecosystems—Bridging Nature and Humans: A Transdisciplinary Approach. Springer Berlin Heidelberg, Berlin, Heidelberg, pp. 441–462.

Zhang, Y., Wei, L., Lu, Q., Zhong, Y., Yuan, Z., Wang, Z., Li, Z., Yang, Y., 2023. Mapping soil available copper content in the mine tailings pond with combined simulated annealing deep neural network and UAV hyperspectral images. Environ. Pollut., 120962.

Zverev, A.O., Gladkov, G.V., Kimeklis, A.K., Kichko, A.A., Andronov, E.E., Abakumov, E.V., 2023. Microbial composition on abandoned and reclaimed mining sites in the Komi Republic (North Russia). Microorganisms 11 (3), 720.

Rice plants for cleanup of trace elements—Bioeconomic perspectives

Abin Sebastian[a] and Majeti Narasimha Vara Prasad[b]

[a]Department of Botany, St George's College, Aruvithura, Kerala, India [b]Department of Plant Sciences, School of Life Sciences, University of Hyderabad (An Institution of Eminence), Hyderabad, Telangana, India

1 Introduction

The rice plant is a high biomass-producing cereal. Even though this crop stands after maize and sugar cane with regard to worldwide cultivation, the importance of rice plants arises from the fact that rice grain contributes to more than one-fifth of the daily calorie intake of the human population (Bhullar and Gruissem, 2013). The cultivation of rice can be done in diverse climatic conditions and is cost-effective (Felkner et al., 2009). Approximately, 162.3 million hectares of land around the world is dedicated to rice cultivation which produces more than 738.1 million tons of rice grain. Asian countries such as China and India produce more than 50% of rice around the world. India, Thailand, and Vietnam are the chief countries involved in the rice trade. These countries export more than 25 million tons of rice annually (Suwannaporn and Lineman, 2008). India exports mainly Basmati grain, whereas Jasmin rice is the main export commodity from Thailand and Vietnam. Major rice-importing countries belong to Africa and the Persian Gulf. Crop productivity is limited in rice-exporting countries because of the lack of advancement in farming technology. It is described that avoidance of post-harvest loss with proper infrastructure development in India alone can feed 100 million people over a year (Basavaraja et al., 2007). But the advent of system intensification of rice has resulted in higher yield from rice paddies (Chowdhury et al., 2014).

Management and irrigation practices make rice cultivation unique. Rice cultivation may be either wet land or dry land. Field management practices in wetland cultivation include submergence, puddling, and addition of nutrients in the form of manure and fertilizers (Kögel-Knabner et al., 2010). These practices bring temporal changes in the solubility of mineral nutrients, organic matter content, and dissolved oxygen. The prevalence of anoxia in rice paddies raises environmental issues such as emission of greenhouse gases, especially methane (Huang et al., 2004). Another matter of concern during rice cultivation is trace element (TE) contamination. It is reported that rice grains harvested from Asian countries are contaminated with TEs such as Cd above the international standard (Simmons et al., 2005; Bian et al., 2014; Zhou et al., 2014). Hence, toxic TE contamination in paddy soil becomes a major topic of health concern. It was the mismanagement of mine wastewater release and usage of TE-contaminated fertilizer that caused toxic TE contamination of paddy soils (Sebastian and Prasad, 2014a). The pedogenic process is also one of the natural processes, which leads to the contamination of paddy soils with TEs. Paddy soil derived from mafic and ultramafic soil was found to be contaminated with Cd, Cr, and Ni (Ali et al., 2020; Bosetti et al., 2011). Soil formed from sphalerite released Cd and Zn in paddy soil after the microbial oxidation processes (Ali et al., 2020; Bosetti et al., 2011). However, capacity of rice plants to accumulate toxic TEs dependent on the type of the metal as well as physiochemical conditions of the soil. Metal accumulation capacity, wide habitat range, and wide germplasm adapted to varying climatic conditions makes rice plant ideal for phytoremediation (Rugh, 2004). Low labor cost compare with other cereal crops, well-developed mechanization for cultivation practices, ambient biomass productivity, and plentiful by-products from rice industry help to formulate commercial bioremediation strategies using rice paddies.

2 Rice plant as a phytoremediation crop

Phytoremediation crops are characterized by substantial metal tolerance with high biomass yield (Rascio and Navari-Izzo, 2011). Apart from these; extensive root systems that provide more capacity to absorb large amounts of water from the soil is also important in phytoremediation perspective. Even though many of the herbs and shrubs pose metal accumulation characteristics, they require enormous time to grow and require specific environmental conditions. Most of the plants can accumulate about 100 ppm zinc and 1 ppm Cd in shoots, whereas metal hyperaccumulator plants accumulate these metals up to 30,000 and 1500 ppm, respectively (Rajkumar et al., 2012). On the other hand, metal accumulation characteristics could be classified into natural hyperaccumulation or assisted hyperaccumulation where in the latter case accumulation is promoted with help of external agents such as soil amendments or microbes (Wang et al., 2011). Rice plants pose many of the beneficial aspects of the above-mentioned phytoremediation characteristics that makes this plant as a potential phytoremediation crop (Fig. 1).

Rice plants accumulate TEs (Table 1). Rice plants are suitable for commercial bioremediation because of short life cycle, adaptability to grow well in diverse environmental conditions, minimum growth requirements and cost effectiveness. Typically, rice plants complete life cycle in 3–4 months (El-Habet et al., 2014). This makes four harvests during a year and removal of large quantity of biomass contaminated with metals from the field. For example, rice variety that accumulates $10.0 \mu g\,g^{-1}$ tissue and produce minimum biomass of 15.0 t/ha is able to remove 0.15 kg metal from per hectare during harvest and approximately half kilogram of metal per hectare in a year, respectively. Typically, 40–60 kg of rice seed which worth about 30 US dollar is sufficient for sawing a hectare of land. Apart from this, availability of ambient nutrient reserve in the seeds help to germinate rice seeds more

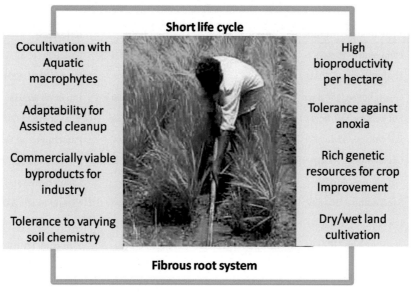

FIG. 1   Features of rice plant as phytoremediation crop.

efficiently compare with other phytoremediation crop such as *Brassica juncea* where food re-serve limit germination events. Attractive rice paddies also meet the esthetic demands such as incorporation greenery in metal-contaminated sites which are usually barren lands. Another noticeable feature of rice plants is the ability to grow in both water logged and dry land soils. All these characteristics are in the favor of remediation of toxic TEs contaminated sites using rice plants. Extent to which rice plants accumulate toxic TEs vary with rice varieties and met-abolic status of the roots (Ramakrishnaiaha and Somashekar, 2010). TEs pollution from an-thropogenic activities is often restricted to 25 cm soil depth (Owen and Otton, 1995). Interestingly, the root system of rice plant is located at this depth and indicates maximum accumulation potential of toxic TEs.

Soils in wetlands trap toxic TEs release from mine tailings (Violante et al., 2010). Ability of rice plants to grow well in wetland conditions makes this plant a phytoremediation agent of mine leachate. It is clear that rice plants grow well in presence of Cd at the rate 20 μg g^{-1} of soil (Table 1). This concentration is often meeting the requirement of phytoremediation of metal contaminated agricultural soils using rice plants where major input of Cd is rock phosphate fertilizer. It must be noted that absorption ability of rice roots is determined by presence of oxygen and physiochemical properties of the soil. Prevalence of anoxia reduces liable metals in the solution (Meia et al., 2009). Wetland grown rice plants overcome anoxia by radical ox-ygen loss that mobilizes metals from the soil (Fu-Zhong et al., 2011). Thus, adaptability to an-oxia along with ability to grow well in both dry and wet land conditions makes rice plants suitable for phytoremediation.

Rice plants belong to Poaceae. Members of this plant family are characterized by fibrous adventitious root system that helps to pump water efficiently from the soil. This character is important to meet the primary requirement of a metal accumulator plant. Fibrous

TABLE 1   Selected studies on toxic trace element accumulation with rice plants.

Metal	Type of study/duration	Metal added	Accumulation	Reference
Cadmium	Field study/90 days	$1.2 \, mg \, Cd \, kg^{-1}$	$0.29 \, mg \, kg^{-1}$ rice	Uraguchi et al. (2009)
	Hydrophonics/10 days	$0.18 \, \mu M \, Cd \, L^{-1}$	$40 \, \mu g \, g^{-1}$ shoot	Pereira et al. (2011)
	Green house/180 days	$6.0 \, mg \, Cd \, dm^{-3}$	$180 \, mg \, kg^{-1}$ plant	Liu et al. (2005)
	Pot culture /90 days	$100 \, mg \, Cd \, kg^{-1}$	$2.86 \, mg \, kg^{-1}$ rice	Vijayarengan (2011)
	Pot culture/45 days	$60 \, mg \, kg^{-1}$	$23.4 \, \mu g \, g^{-1}$ root	Panichpat and Srinives (2009)
Lead	Green house/700 days	$16 \, mg \, kg^{-1}$	$400 \, mg \, g^{-1}$ root	Ma et al. (2012)
	Pot culture/40 days	$2 \, g \, kg^{-1}$	$14 \, mg \, g^{-1}$ root	Chatterjee et al. (2004)
	Sand culture/104 days	$1 \, mM \, kg^{-1}$	$70 \, \mu g \, g^{-1}$ leaf	Yizong et al. (2009)
	Pot culture/60 days	$300 \, mg \, kg^{-1}$	$35 \, mg \, g^{-1}$ leaf	Zong et al. (2010)
Chromium	Hydroponics/10 days	$1 \, mg \, L^{-1}$	$5.7 \, mg \, kg^{-1}$ leaf	Zhu et al. (2010)
	Pot culture/grain filling	$400 \, mg \, kg^{-1}$	$137 \, mg \, kg^{-1}$ root	Schuhmacher et al. (1994)
	Field study/harvest	$6.22 \, \mu g \, g^{-1}$	$0.65 \, \mu g \, g^{-1}$ rice	Bhattacharyya et al. (2005)
	Field study/harvest	$48 \, mg \, kg^{-1}$	$5.6 \, mg \, kg^{-1}$ straw	Alam and Rahman (2003)
Arsenic	Field study/harvest	$3.30 \, mg \, kg^{-1}$	$4.95 \, mg \, kg^{-1}$ shoot	Liu et al. (2007)
	Feild study/harvest	$7.64 \, mg \, kg^{-1}$	$30.42 \, mg \, kg^{-1}$ root	Bhattacharya et al. (2010a)
	Field study/harvest	$14.09 \, mg \, kg^{-1}$	$28.63 \, mg \, kg^{-1}$ root	Bhattacharya et al. (2010b)
	Field study/harvest	$5.85 \, mg \, kg^{-1}$	$1.65 \, mg \, kg^{-1}$ straw	Meng et al. (2012)
Mercury	Field study/120 days	$29 \, mg \, kg^{-1}$	$3.2 \, mg \, kg^{-1}$ root	Liu et al. (2010)
	Pot culture/harvest	$4 \, mg \, kg^{-1}$	$170 \, \mu g \, kg^{-1}$ grain	Wang et al. (2014)
	Pot culture/150 days	$46 \, \mu g \, kg^{-1}$	$14 \, \mu g \, g^{-1}$ root	Li and Cui (2014)

adventitious roots of rice plants often help to cope up with nutrient deficiency too. Extensive fibrous root systems firmly cling with soil particles and prevent paddy soil erosion. Thus, the morphological features of roots of rice plants help to avoid outflow of TEs from the contaminated soil through soil erosion. Plants with higher root biomass tend to accumulate more TEs (Wang et al., 2009). Thus, cultivation strategies for the enhancement of root biomass trigger higher rate of mobilization of toxic TEs from soil to rice plants. Available information on rice roots from germplasm collection also enables selection of ideal rice variety for the remediation of particular metal-contaminated sites. Cadmium-tolerant rice cultivar tends to accumulate more Cd in the root compared with Cd-sensitive cultivar (Pereira et al., 2011). All these indicate that the characteristics of roots could be the prime reason that makes rice plant a promising plant for soil remediation. Recent studies pointed that nodal region play an important role in determining TEs accumulation in rice plants. Studies also showed that mobilization of As in rice increase with the application of phytochelatin synthesis inhibitors such as sulfoximine resulting in the transfer of As from node I region to flag leaf and grain

(Yamazaki et al., 2018). Elemental mapping with synchrotron X-ray fluorescence and high-resolution SIMS elemental mapping showed that nodes, internodes, and leaf sheaths of rice plants accumulate As-sulfur complex in the vacuoles of phloem companion cells (Moore et al., 2014). These studies also pointed that transfer of TEs in rice plants mainly controlled at node I where there is intervascular transport of ions.

Shoot system of rice also pose promising metal accumulation characteristics. But transgenic rice is reported to accumulate Pb up to $1000\,mg\,kg^{-1}$ of straw and indicate potential of rice leaves to accumulate toxic TEs (Yamazaki et al., 2018). Metals such as Cr found to accumulate more in shoots compare with root. Studies conducted on typical rice growing plate indicated that Mn accumulated more in shoot than in root (Sasaki et al., 2011). Bioaccumulation factor for TEs in rice plants was in the decreasing order of $Zn > Mn > Cd > Cu > Cr > Pb$ (Satpathy et al., 2014). Studies with plant hormones indicated that the supplement of ABA and gibberllic acid (GA3) restricts translocation of toxic TEs such as Ni and Cd (Rubio et al., 1994). Application of GA3 also resulted the mobilization of carbohydrate reserve in the rice seeds along with mobilization of TEs. Plant hormones also affect accumulation of heavy metals via controlling the synthesis of phytochelatins (Bücker-Neto et al., 2017). Exogenous application of IAA found to alleviate As toxicity in rice plants. However, treatment of IAA decreased the accumulation of As which in turn helped the plants to maintain growth (He et al., 2022). The presence of strigolactones also found to relocate metal ions such as Cd in rice plants (Xu et al., 2019). This molecule is found to influence nitrous oxide synthesis in rice plants under Cd stress. Also, the activity of anti-oxidant enzymes such as superoxide dismutase and peroxidase were recovered in the course of Cd treatment with strigolactones. The pools of reduced glutathione as well as ascorbic acid also found to increase in rice plants after treatment with strigolactone. Thus, it is clear that plant hormones and signaling molecules help to alleviate toxic effects of heavy metals in rice plants used for phytoremediation.

Translocation factor is among the critical factor that makes plants as phytoremediation crop. Increase of Cd in the shoot was more in Cd-susceptible variety compared with Cd-tolerant variety (Pereira et al., 2011). Higher degree of translocation of Cr and Zn was also observed in rice plants (Payus and Talip, 2014). It has been reported that translocation coefficient is in the decreasing order of $Cd > As > Pb$ in rice plants that were grown in the Changjiang Delta Region (Yuan et al., 2012). Arsenic also shows tendency to concentrate more in rice grain rather than other plant part. Studies on differential metal accumulation of rice plant during irrigation with water of Ramgarh Lake, India, indicate that metal accumulation pattern in the rice root follows decreasing order of $As > Pb > Mn > Cd > Cr > Zn > Cu > Hg$, whereas in the leaf the decreasing order is $As > Cd > Pb > Mn > Cr > Zn > Cu > Hg$ (Singh et al., 2011). Apart from these, root to leaf translocation of Cd found to retard by fertilization with ammonium-based fertilizers (Sebastian and Prasad, 2014b). Most often metal accumulation factor in rice was described in the decreasing order root > shoot > leaf > grain. Abovementioned studies figure out that rice plant parts differentially accumulate metals based on their specific chemical nature.

Chemical speciation of heavy metals in plant tissues enabled metal tolerance in plants which in turn increases the bioaccumulation of TEs. The XRF mapping had shown that Cd accumulation in rice is associated with speciation of Cd with thiolate. The accumulation of Cd in rice was also found to be associated with the presence of Se. It is reported that addition of Se counteracts toxic effect of Cd by increasing viability of root cells (Li et al., 2021). Rice plants found as an ideal plant for the removal of As from submerged conditions. Reductive dissolution of Fe oxides under the flooded environments will release As from the soil which in turn enhance As accumulation in rice. Studies with zirconium-oxide diffusive gradients in thin films (DGT) showed that DGT-As encompassing the root zone decreased steeply from 331.0 µg/L in the seedling stage to 136.0 µg/L in the heading and flowering stage and further to 118.0 µg/L at harvest. Metal accumulation in rice also varies with rice varieties. Germplasm of rice plant is rich with approximately 108,256 varieties (Yadav et al., 2013). Based on origin, Indian rice varieties cover largest number of accessions approximately 16,013 and are followed by Lao PDR having 15,280 accessions at the International Rice Gene bank Collection at IRRI. Most of the studies on heavy metal accumulation were carried out in japonica varieties because of the issues of metal contamination in regions where these cultivars are being cultivated. Diverse Japonica rice germplasm had shown natural variation of about 13- to 23-fold difference in grain Cd concentration (Arao and Ae, 2003; Ueno et al., 2009). It is the higher root activity, high shoot-to-root ratio and water consumption that made rice plant to accumulate more heavy metals (Ishikawa et al., 2010). Quantitative traits of Cd accumulation in rice plants are found to locate at chromosome 2, 7, 3, 4, 6, 8, 5, 11, and 10 (Ueno et al., 2009; Ishikawa et al., 2010; Xue et al., 2009). These traits are found to be similar in low land as well as upland conditions and hence open way to improve rice plants for metal accumulation. It has been reported that genotype variation affect As accumulation in rice irrespective of nature of the cultivation practice, that is dry or wetland cultivation (Wu et al., 2011). All these findings indicate that natural metal accumulation capacities of rice plants are stable irrespective of cultivation practices. These studies also pointed out stability of rice plants in metal accumulation characteristics irrespective of varying edaphic factors such as soils, metals, pH, organic carbon, cation exchange capacity (CEC), and ecological niches. It can be summarized that minimum growth requirement, fast growth rate, and genetic plasticity makes rice plants ideal for phytoremediation.

3 Assisted clean up using rice plants

Nutritional requirements of rice plants are well studied and this allows feeding of rice plants for enhanced metal accumulation. Fertilization often leads higher biomass productivity in rice paddies (Zhang et al., 2012). Nitrogen fertilization is crucial during rice cultivation for enhancement of yield. Ammonium as N source is favored during wet land cultivation compared with nitrate because of problems with denitrification and leaching of nitrate (Smolders et al., 2010). Dissolved ammonium bound tightly with soil and hence prevents leaching of N. Organic manure as well as synthetic fertilizers such as urea is a practical solution for improvement of rice biomass. Humphreys et al. (1987) reported that enhanced agronomic efficiency (56 kg grain per kg applied N) of surface applied N fertilizer was greatest where N application was prior to permanent flooding in a combine-sown rice crop (Humphreys et al., 1987). But agronomic efficiency was 8.2 kg grain per kg N when N applied at the time of sowing. Mechanical disturbance of soil during ground preparation and combine sowing increased N mineralization of high clay soils (Craswell and Waring,

1972). N mineralization leads increased nitrate levels in soils and cause N loss during permanent flooding because of denitrification. It must be noted that nitrogen fertilization of rice plants often practice during tillering stage to increase the number of panicles. But for the phytoremediation point of view, it can be before tillering stage which increase plant biomass essential for metal sequestration. The increase in biomass often leads tissue dilution of the metals and hence lesser the toxicity symptoms (Wright et al., 1998). It has been reported that nitrogen fertilization leads to enhanced accumulation of Cd in rice plants while limiting Cd translocation (He et al., 2022). On the other hand, increase of nicotianamine during nitrogen fertilization leads to more accumulation of Zn, Fe, and Cu in the aerial plant part of cereals (Barunawati et al., 2013). Typical rice fertilization strategy is based on nitrogen, phosphorous, and potassium, which are essential to produce high rice plant biomass. Application of ammonium-based nitrogen fertilizers leads to the accumulation of more amount of toxic TEs in rice plants because of soil acidification (Casova et al., 2009; Eriksson, 1990). Hence, the application of ammonium fertilizer helps in the release of plant available Cd in soil plus more accumulation of metal. Monoammonium phosphates are also potential fertilizers that help to enhance Cd accumulation in rice because of soil acidification capability which enhance metal accumulation (McLaughlin et al., 1995). Since more than 70% of arable soils are acidic, chance of removal of metal from the field with rice plants is very high due to radical oxygen loss that prevent metal toxicity by formation of metal plaque on the surface of roots (Chlopecka and Adriano, 1997). Application of S and Ca leads to the formation of insoluble TEs complexes in soil solution. So in order to avoid metal precipitation, application of S and Ca must be restricted. Chlorinated water could be another approach that enhances metal solubility in the field to extract more metals. The well-developed nutrient management practices such as site-specific nutrient management (SSNM) and system of rice intensification that enhance rice biomass productivity also points higher possibility of improvement of metal accumulation in rice plants by meeting nutrient requirements (Johnston et al., 2009).

Silicon confers TEs tolerance in rice plants (Shi et al., 2005). Amount of bioavailable Si in the soil range from 50.0 to 250.0 mg kg^{-1} soil. The availability of Si to plants is influenced by soil pH, water content, temperature, and mineral ions. The addition of Si had effects such as decrease in uptake of metal ions, reduced metal uptake, root-to-shoot translocation, chelation, and stimulation of anti-oxidant systems in plants, complexation, and co-precipitation of toxic metals with Si in different plant parts, compartmentation, and structural alterations in plants and regulation of the expression of metal transport genes, respectively. Activity of anti-oxidant enzymes such as peroxidase and catalase increased in the course of Si treatment under TEs stress (Bhat et al., 2019). Activity of these enzymes found to prevent membrane damage during exposure to TEs. Therefore, silicon-based soil additives are ideal to increase phytoremediation capability of rice plants. Salts of Si with Ca, K, and Na are used for the addition of Si in the soil. But foliar spray of Si is also one of the promising ways for the application of Si which help to avoid immobilization of Si in the soil. Application of nanocomposites of Si was also effective to mitigate toxic effects of Cd and As in plants.

Organic manure application-accelerated plant growth is a promising strategy of assisted clean up because these amendments are ecofriendly as well as economically feasible. Organic manure is practicing prior to planting of rice plants. This approach usually helps to enhance solubility of metals in the soil along with increase of metal chelator pool. Application of organic matter improve soil properties like CEC, aeration, water holding capacity, and the amount of plant nutrients all which influence Cd accumulation in plants (Adeniyan et al., 2011). Incorporation of organic matter in to soil for enhanced growth of rice plant require attention because of the dynamics of N content under flooded condition. Mineralization of organic N retarded under flooded condition because of anoxia (Ehrenfelda and Yu, 2012). But the draining of the soil ensures aerobic condition that leads NH_4^+ conversion to NO_3^- via nitrification processes, and thus organic matter breakdown is accelerated. Hence, draining of the field ensures maximum mineralization of organic material added. This points intermittent draining of the field support the growth of rice plants because of more N availability which result higher biomass productivity. Germination and growth of rice seedlings are accelerated in organic matter rich field. Nutrient uptake studies in rice fields indicate that the crop acquires more than half the nitrogen (N) and micronutrients from mineralized organic matters such as crop residues (Myint et al., 2010). Livestock wastes in paddy soils help to replenish organic matter, nutrients, and metal complexing agents such as humic acids (Antil and Singh, 2007). The higher amount of nutrients with held in bioorganic waste make plant to combat with toxic TEs (Park et al., 2011). Studies with application of vermicompost indicate that rice plants accumulate more Cd during vermicompost application (Sebastian and Prasad, 2013). It was the enhancement of root growth during vermicompost amendment application which in turn favored Cd accumulation. Cd tolerance achieved with the addition of farmyard manures in this study also points potential of these manures to support establishment of rice seedlings in the presence of toxic TEs. In short, organic manure helps rice plants to combat with toxicity of TEs during early growth period of rice plants.

Microbe-assisted metal cleanup with rice plants also have tremendous potential. It has been reported that microbial methylation contribute to accumulation of methylated As in rice grains (Zheng et al., 2013). Microbial oxidation of As(III) to As(V) and methylation of As are the As detoxification mechanism operate in the soil that makes As unavailable for plant uptake. Increase of Fe-reducing bacteria in the rhizosphere of rice plant is supposed to increase As uptake by rice. But it is must be noted that decrease of nitrate-dependent ferrous ion-reducing bacteria in the soil mobilize As in the soil and hence enhance As uptake (Burton et al., 2008). Iron plaque formation in rice roots assisted with aerobes in the soil act as high affinity adsorbing surface for number of TEs. Iron plaque not only act as adsorbing surface but also support growth of rice plants by protecting roots from metal toxicity. This ensures enhanced biomass production including that of root and results more removal of metals from the soil during a harvest which include roots. This also points the importance of removal of rice roots which accumulate metals during a phytoremediation program because of anaerobic decomposition of rice roots that release the bound metal again in to soil. *Azotobacter* is a noticeable bacterium that improves rice plant growth (Wani et al., 2013). Recommended application of this bacterium is 0.5 kg/ha for promoting plant growth. These bacteria enhance plant growth through increase of N content in the soil. Other than bacteria, arbuscular mychorrhizal fungi are reported to form symbiosis with rice plants. Colonization of rice roots with this group of fungi is supposed to down regulate Si transporters that mediate As uptake which in turn reduced As accumulation in the rice plants (Yeasmin et al., 2007). Mycorrhizal colonization also reported to reduce uptake of Cd by creating a barrier before Cd entry in to roots (Khan et al., 2000). In short, adaptability to varying nutrient management practices and ability to form symbiosis with microbe promise assisted clean up using rice plants.

4 Rice industry by-products for bioremediation

Environmental remediation of TEs in paddy soils are performed with the help of chemicals such as lime stone, steel slag, and acid mine drainage sludge (Kim et al., 2017). It is reported that the application of dolomitic lime and oyster shell will decrease the accumulation of heavy metals in rice because of increase in soil pH (Huang et al., 2018). But many of these materials cause chemical changes in soil. Rice cultivation and related industries offer various by-products, which can be utilized for phytoremediation. It is noteworthy that by-products of rice industry can be utilized for commercial purposes even after phytoremediation (Fig. 2). Straw, husk, bran, germ, and broken rice are the chief by-products of rice industry (Esa et al., 2013). One ton of rice paddy produces approximately 220 kg straw and husk. Rice straw acts as a potential absorbent of TEs such as Fe, Mn, Zn, Pb, and Cd (Nawar et al., 2013). Metal absorption capacity of straw was found to depend on type of the metal, pH, metal concentration, and contact time. More percentage of absorption with increase in absorbent as well as contact time indicates that rice straw act as a potential metal biosorbent. Commercial ion exchange resins made up of petroleum products are not biodegradable. Hence, these resins have more negative environmental impact. Apart from this, the efficiency of these resins is low because of less availability of functional groups. Biosorbents with high ion exchange capacity prepared from rice straw efficiently removed toxic TEs from waste water (Rungrodnimitchai, 2014). These lignocellulosic biosorbents are biodegradable and hence disposal is convenient. It is also noticeable that sodium hydroxide treatment helps in high-efficient removal of lignin in rice straw and hence easier the preparation of cellulose phosphate from straw, which is an efficient biosorbent. Grafting of dimethylaminoethyl methacrylate (DMAEM) on rice straw with the help of potassium permanganate/nitric acid redox system also reported to act as efficient metal absorbent (Mostafa et al., 2012).

Rice husk is the main residue during polishing of rice grain. Rice husk had shown higher rate of adsorption of Pb and Zn (Elham et al., 2010). It is also a potential adsorbent of As(V) and the adsorption rate is found to increase with chemical treatment (Lata and Samadder, 2014). Higher rate of metal adsorption capacity makes husk as a potential amendment which immobilizes toxic TEs in the soil. Cellulose phosphate prepared with rice husk

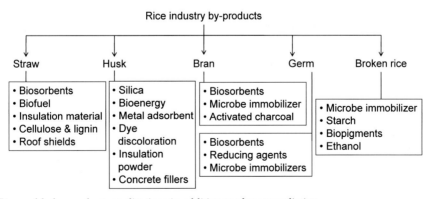

FIG. 2 Rice paddy by-products applications in addition to phytoremediation.

had shown 100% removal of divalent Cd from the solution (Athinarayanan et al., 2014). The ability to produce higher surface area with help of mechanical methods makes husk more attractive material for biosorption. Rice husk is rich in cellulose and hemicellulose which account for more than 50% bulk of the material. These sugar fibers make husk insoluble in water and hence husk can be efficiently used for wastewater treatment. Apart from cellulose part, silica in the rice husk also helps in adsorption (Awizar et al., 2013). Husk can be processed as a biosorbent by washing it with tap water or chemical treatment. Treatment with epichlorohydrin, tartaric acid, orthophosphoric acid and polyaniline are reported to increase metal adsorption capacity of husk (Lata and Samadder, 2014). Briefly, metal absorption capacity as well as easier manipulation of ion exchange properties via chemical treatments makes rice husk an efficient metal binding agent.

Rice husk can also be utilized for production of silica, activated carbon, tetramethoxysilane, insulation materials, concrete fillers, oils absorbents, and insulation powder for steel mills (Thiravetyan, 2012). Rice husk contain amorphous silica which is being used in low permeability concrete for use in bridges, marine environments, and nuclear power plants. Rice husk ash has more than 90% amorphous silica that could be used as a substitute for silica fume used in industry. Silica is prepared from the rice husk ash by application of high temperature. Digestion of the rice husk ash with caustic leads extraction of silica as sodium silicate in the solution. Sodium is precipitated by passage of carbon dioxide and the filtrate containing silica is dried and packaged. Through this method 1 ton of silica can be produced from 1.6 ton of husk ash and the estimated selling cost of silica is 1 US dollar per kilogram of silica. Apart from the industrial applications, rice husk is also a source of energy. Husk produced from rice industry around the world has an energy content of about 14 GJ/ton which is equivalent to energy produced from 1 billion barrels of oil per year. The usage of husk as an energy source would have economic advantage of approximately 7 billion US dollar per year.

Rice bran can be used for metal removal process because of higher capacity of sorption (Montanher et al., 2005). These residues are also characterized by the abundance of water-soluble vitamins, carbohydrates, potassium, nitrogen, and phosphorus. The presence of water-soluble components makes higher affinity contact of water with rice bran and hence bran has high affinity to bind dissolved metals. Being nutritious, rice bran also act as immobilizer of microbes. *Rhizopus arrhizus* immobilized in to rice bran had shown more Ni uptake compare with pure rice bran used for biosorption (Gurel et al., 2010). Application of activated charcoal is not economically feasible because of low availability of the material and higher production cost. But removal of metals using activated charcoal is technically easier. Rice bran can be utilized to produce activated charcoal which is economically feasible. It is reported that rice bran carbon efficiently bind with hexavalent Cr (Ranjan and Hasan, 2010). Abundance of functional groups makes rice bran as an important adsorbent where chemical conversion of elements is frequent. This allows reduction of metallic ions such as conversion of Cr^{6+} to Cr^{3+}. The ability to support chemical changes is the reason for high-efficient removal of Cd and Pb with rice bran compared with other green materials being used for metal removal such as pine saw dust. Rice germ is often being removed with rice bran. It is rich in polyunsaturated fatty acids and the presence of these fatty acids enhances reducing power of rice germ (Nagendraprasad et al., 2011). Thus, interaction of metals with rice germ often ends up in reduction of metals. Germ oil can be extracted and use for energy purposes too. Rice germ is also rich in lectins which help microbial immobilization in rice bran (Zhao-Wen et al., 1984). It is also important to note that rice bran together with rice germ hold major portion of metal accumulated in rice grain. Hence, the removal of rice bran and the resulting white rice is important with regard to human health perspective.

Broken rice is another noticeable by-product from rice industry, which is reserve of proteases and starch (Gohel and Duan, 2012). This shapeless rice can be utilized for microbial immobilization. Many of the species of Aspergillus are well known to grow well in broken rice (Paranthaman et al., 2009). Biopigment production is another candidate process where broken rice is being utilized. Culturing with fungus *Monascus ruber* MTCC232 found to produce pigments of red, orange, and yellow colors (Dikshit and Tallapragada, 2011). Non-cook process-based ethanol production potential is also high with broken rice. In this method, starch hydrolyzing enzymes are used for conversion of starch to simple sugars which later converted into ethanol by fermentation.

Rice industry also offers the production of biodegradable polymers that can be used to produce mulch films, agriculture net, plant grow bags, and so on. Mulch film is used to maintain good soil structure, prevent contamination, moisture retention, and weed control. Danimer Scientific has produced custom-made biodegradable agricultural mulch film in the name of Nodax™. However, the production cost is high because of the difficulties in the handling of agrowaste which utilized to produce polyhydroxyesters (PHA). Broken rice is also found to be a good source for the production of PHA (Amelia et al., 2019). Broken rice had PHAs levels of up to 5.18 g/L. So broken rice could be a very good alternative to produce biodegradable polymers.

5 Sustainable bioremediation prospects of rice paddies

Rice paddies being agro wetland ecosystem act as an ideal place for growth of various aquatic macrophytes and algae that enhance biomass productivity (Linke et al., 2014). Besides this, remediation of paddy fields also can be achieved with the help of these macrophytes or algae (Fig. 3). Aquatic macrophytes are well known for rhizofiltration, whereas algal biomass is reported to work as biosorbing agents (Rai, 2009). More biomass produced from paddy fields can be utilized for the production of biosorbents or compost. Both these products have tremendous scope in the remediation of metal-contaminated sites.

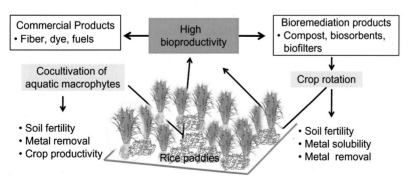

FIG. 3 Benefits of incorporation of sustainable agriculture practices for rice cultivation to boost bioeconomy.

FIG. 4 *Azolla* (nitrogen fixer) culture preparation (A) Construction of $1 \times 2\,m$ pit using clay bricks. (B) Silpaulin lining of the constructed pit. (C) Preparation of sand and manure. (D) Spread of sand and manure in silpaulin lined pit.

Co-cultivation of rice with aquatic macrophyte is useful in phytoremediation. For example, Azolla grows well in rice paddies (Figs. 4–6) (Table 2) (Bocchi and Malgloglio, 2010). This aquatic plant helps to enhance biomass yield from the rice paddies as well as nitrogen content in the field. This makes the genus Azolla as an important component in rice cultivation for boosting the yield and biomass. *Azolla caroliniana* is reported to have capacity for metal hyperaccumulation. This fern is found to accumulate significant level of metals especially Cr(III) (Bennicelli et al., 2004). The usage of azolla is reported to suit at the level of contamination of 1–20 ppm metals in the solution. Growth of aquatic macrophytes such as *Eichhornia crassipes, Lemna minor,* and *Pistia stratiotes* in the low land fields act as efficient rhizofiltration agent (Karkhanis et al., 2005). *Elatine hexandra, Althenia filiformis,* and *Monita rivularis* in the flooded rice fields significantly remove Cd from the rice field. These macrophytes are also potential remediating agent of salinity, which is a serious agricultural problem. Najas is a genus of aquatic plant known by the common name southern water nymph. These plants are very often found in rice paddies. Najas is reported to accumulate significant amount of Pb ($3.5\,mg\,g^{-1}$ dw) in the plant body (Singh et al., 2010). The higher level of Pb accumulation in this plant is achieved with the activity of anti-oxidant enzymes such as catalase and glutathione peroxidase. The higher metal concentrating properties of this plant point out that introduction of this plant in metal-contaminated rice paddies help to increase the efficacy of phytoremediation. *Ceratophyllum demersum* L. is another aquatic plant which helps to improve phytoremediation in rice paddies (Bunluesin et al., 2004). It is reported that this plant

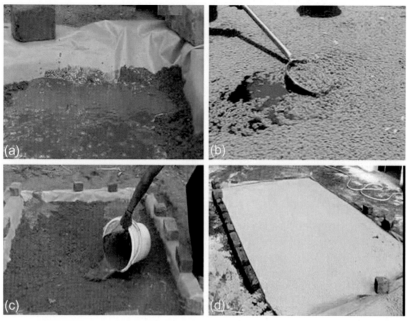

FIG. 5   (A) Cow dung slurry mixture with sand and manure in silpaulin lined pit. (B) *Azolla* source culture for mass multiplication. (C) Introduction of *Azolla* culture to the constructed pit as mentioned earlier. (D) Pity with fully grown *Azolla* ready for spread in rice paddies.

FIG. 6   (A–D) *Azolla* introduction to rice paddies at tillering phase for nitrogen supplementation.

III. Ornamentals and tree crops for contaminated substrates

TABLE 2 *Azolla* culture—Expenditure.

Expenditure per tank/pit of size 1×2m	Per year in INR
88 kg cow dung (Rs 1 for kg cow dung)	88
250 g magnesium sulphate	10
88 g muriate of potash	10
1733 g single super phosphate	18
2 kg mineral mixture	60
4 sq m silpaulin sheet (durable for 3 years)	200
Bricks 30 nos (durable for 3 years)	300
Labor 12 per year @ Rs 300 per day	3600
Total expenditure per year	**4286**

1.5 kg per tank per day for 350 days = 525 kg
Compost 100 kg/tank cost = Rs 300
Expenditure
Azolla production cost per kg = Rs 8.16

accumulate Cd about 1.293 mg g^{-1} dw. Metal tolerance in this plant is also achieved with the help of anti-oxidant enzymes such as peroxidases and glutathione reductases. Algal biomass in the flooded plain is also found to be effective in metal removal (Reniger, 1977). Successive cropping and harvesting of algal biomass reduces levels of toxic TEs in the paddy soil even though the process requires considerable effort to collect the algal biomass (Vandenhove et al., 2001). Incorporation of blue green algae in the field not only removes metal from the field but also add to nitrogen economy.

Aquatic plants obtained from the field can be used for manufacture of biofilters where biomass is packed in to filters. These filters are efficient in the removal of metal ions especially from water bodies. Azolla-binding filters are reported to bind metals efficiently (Pb—10%, Cr—3%, U—3.5%, Cr^{3+}—9%, Ni—3%) (Sachdeva and Sharma, 2012). Metal uptake capacity of algal biomass also can be improved by chemical treatments with CaCl$_2$. It is well known that dead algal biomass have more efficiency to remove metals. Thus, harvested algal biomass from rice paddies has great scope in commercial production of biosorbents. Residual algal-bacterial biomass in a symbiotic relationship can be induced and this approach found to enhance metal biosorption as well as biodegradation of organic pollutants in photo bioreactor (Munoz et al., 2006). Immobilized or granulated algal biomass-filled column is another promising technique that help sorption of metals. Algae turf scrubbers are also promising technique for metal ion removal using algal biomass (Mulbry et al., 2008). But economic gain from these mechanical systems is low because of expenditure for meeting specific growth requirements. Constructed rice paddies with algae can be good alternative for algae turf scrubbers where economic gain is more because of by-products from rice plants and algae.

Rain-fed low land rice cultivation area where one crop of rice practiced is ideal for crop rotation with legume (Seng et al., 2008). Crop rotation with legume not only support biomass productivity but also enhances the fertility of the paddy soil (Graham and Vance, 2003). Noticeable aspects of legumes are short maturation period (55.0–90.0 days), drought tolerance, and ability to fix nitrogen which improves nitrogen content in the soil. Rotation of legume with rice plants help to intensify the land use along with increase of crop productivity per area per year. Cultivation of Sesbania which have short maturation period (approximately 60.0 days) helps green manure production. Cultivation of mung bean or cow pea not only adds to nutrient improvement of the field but also provide economic gain to the farmer. Soil acidification during legume-mediated nitrogen fixation is also beneficial for phytoremediation of paddy fields because soil acidity makes more soluble metal for plant uptake. For example, growth of oilseed plants in rice paddies found to enhance Cd accumulation in rice plants (Yu et al., 2014). This indicates the promising role of legume in mobilization of metals where subsequent removal of metals is achieved by the growth of rice plants. Legume growth also contributes to produce phytoremediation products. It is reported that the application of rapeseed cake together with phytoextraction using *Sedum plumbizincicola* reduces Cd uptake in rice plants (Shen et al., 2010). Legume cotyledon is enriched with phytic acid. The compound is rich in phosphate and shows high affinity binding with metal ions. This indicates the potential use of legume cotyledons for usage in biofilter preparation that help to biosorb metal ions.

Thermochemical decomposition of rice husk as well as straw in an oxygen-limited environment help to produce biochar. Biochar mainly composed of carbon and ashes that are well known to adsorb metal ions. Hydrothermal carbonization procedure at temperature of 180.0°C and pressure of 70.0 bar reported to result biochar from rice husk. The zeta potential value of biochar prepared via above method was -30 to -10 mV. This indicates higher suspension stability, improvement of surface area and porosity in the rice husk-derived biochar, which is ideal for the adsorption of metals (Hossain et al., 2020). However, increase of temperature above 400.0°C will impart alkaline nature to biochar which will lead to the precipitation of metal ions in the form of hydroxides and phosphates (Wu et al., 2012). The incorporation of biochar not only increases metal binding capacity but also helps to improve holding capacity of greenhouse gases and nutrients in the soil, respectively. So the conversion of rice paddy biomass in to biochar is a promising strategy which helps environmental remediation of TE-contaminated sites.

Compost preparation is another promising approach which helps to utilize biomass produced from rice paddies. Composts improve soil properties including aeration and porosity (Pagliai et al., 2004). Increase of both these properties help to enhance root growth. Vermicompost can be produced from rice straw, rice husk, and rice bran. Additions of vermicompost to soil lead to Cd accumulation in rice seedlings due to the enhancement of root growth (Sebastian and Prasad, 2013). This indicates that the addition of compost in soil improves phytoextraction process. Commercially important intercrops that produce fiber, dye, biofuels, and so on also can be practiced in rice paddies. It can be summarized that rice paddy offers great opportunities for the enhancement of biomass, and the biomass obtained can be converted in to commercially important phytoremediation agents. Thus, it is clear that rice paddies also contribute to bioeconomy through biomass and commercially important by-products of co-cultivated or crop-rotated plants that help in sustainable agriculture.

6 Outlook

Rice plants pose all the essential characters of metal accumulation. Root growth is the most important factor, which controls metal tolerance in plants. The available information from rice germplasm with regard to root characteristics and the information about higher root biomass producing rice gene such as *OsEXPA8* can be utilized for the development of metal hyperaccumulator rice plants. There exists tremendous scope of assisted metal accumulation using rice plants. Chemical fertilizers that are practiced during rice cultivation must be screened for enhanced metal accumulation. It is also important to utilize the by-products from rice industry for bioremediation process. Co-cultivation as well as crop rotation opportunities in rice paddies must be studied for a successful sustainable bioremediation which provide economic gain.

References

Adeniyan, O.N., Ojo, A.O., Akinbode, O.A., Adediran, J.A., 2011. Comparative study of different organic manures and NPK fertilizer for improvement of soil chemical properties and dry matter yield of maize in two different soils. J. Soil Sci. Environ. Manage. 2 (1), 9–13.

Alam, M.Z., Rahman, M., 2003. Accumulation of arsenic in rice plant from arsenic contaminated irrigation water and effect on nutrient content. In: Accumulation of Arsenic in Rice Plant. Presented in the BUET-UNU Symposium, 5–6 February, Dhaka, Bangladesh, pp. 131–135.

Ali, W., Mao, K., Zhang, H., Junaid, M., Xu, N., Rasool, A., Feng, X., Yang, Z., 2020. Comprehensive review of the basic chemical behaviours, sources, processes, and endpoints of trace element contamination in paddy soil-rice systems in rice-growing countries. J. Hazard. Mater. 397, 122720.

Amelia, T.S.M., Govindasamy, S., Tamothran, A.M., Vigneswari, S., Bhubalan, K., 2019. Applications of PHA in agriculture. In: Biotechnological Applications of Polyhydroxyalkanoates. Springer, pp. 347–361.

Antil, R.S., Singh, M., 2007. Effects of organic maures and fertilizers on organic matter and nutrients status of the soil. Arch. Agron. Soil Sci. 53, 519–528.

Arao, T., Ae, N., 2003. Genotypic variations in cadmium levels of rice grain. Soil Sci. Plant Nutr. 287, 223–233.

Athinarayanan, J., Periasamy, V.S., Alshatwi, A.A., 2014. Biogenic silica-metal phosphate (metal = Ca, Fe or Zn) nanocomposites: fabrication from rice husk and their biomedical applications. J. Mater. Sci. Mater. Med. 25 (7), 1637–1644.

Awizar, D.A., Othman, N.K., Jalar, A., Daud, A.R., Rahman, I.A., Al-hardan, N.H., 2013. Nanosilicate extraction from rice husk ash as green corrosion inhibitor. Int. J. Electrochem. Sci. 8, 1759–1769.

Barunawati, N., Giehl, R.F.H., Bauer, B., Wirén, N.V., 2013. The influence of inorganic nitrogen fertilizer forms on micronutrient retranslocation and accumulation in grains of winter wheat. Front. Plant Sci. 4, 1–11.

Basavaraja, H., Mahajanashetti, S.B., Udagatti, N.C., 2007. Economic analysis of post-harvest losses in food grains in India: a case study of Karnataka. Agric. Econ. Res. Rev. 20, 117–126.

Bennicelli, R., Stepniewska, Z., Banach, A., Szajnocha, K., Ostrowski, J., 2004. The ability of *Azolla caroliniana* to re-move heavy metals (Hg(II), Cr(III), Cr(VI)) from municipal waste water. Chemosphere 55, 141–146.

Bhat, J.A., Shivaraj, S.M., Singh, P., Navadagi, D.B., Tripathi, D.K., Dash, P.K., Solanke, A.U., Sonah, H., Deshmukh, R., 2019. Role of silicon in mitigation of heavy metal stresses in crop plants. Plants 8 (3), 71.

Bhattacharya, P., Samal, A.C., Majumdar, J., Sant, S.C., 2010a. Uptake of arsenic in rice plant varieties cultivated with arsenic rich groundwater. Environ. Asia 3 (2), 34–37.

Bhattacharya, P., Samal, A.C., Majumdar, J., Santra, S.C., 2010b. Accumulation of arsenic and its distribution in rice plant (*Oryza sativa* L.) in Gangetic West Bengal, India. Paddy Water Environ. 8, 63–67.

Bhattacharyya, P., Chakraborty, A., Chakrabarti, K., Tripathy, S., Powell, M.A., 2005. Chromium uptake by rice and accumulation in soil amended with municipal solid waste compost. Chemosphere 60, 1481–1486.

Bhullar, N.K., Gruissem, W., 2013. Nutritional enhancement of rice for human health: the contribution of biotechnology. Biotechnol. Adv. 31, 50–57.

Bian, R., Joseph, S., Cui, L., Pan, G., Li, L., Liu, X., Zhang, A., Rutlidge, H., Wong, S., Chia, C., Marjo, C., Gong, B., Munroe, P., Donne, S., 2014. A three-year experiment confirms continuous immobilization of cadmium and lead in contaminated paddy field with biochar amendment. J. Hazard. Mater. 272, 121–128.

Bocchi, S., Malgioglio, A., 2010. Azolla-Anabaena as a biofertilizer for rice paddy fields in the Po Valley, a temperate rice area in Northern Italy. Int. J. Agron. 2010, 1–5.

Bosetti, F., Zucchi, M.I., Pinheiro, J.B., 2011. Molecular and morphological diversity in Japanese rice germplasm. Plant Genet. Resour. 9, 229–232.

Bücker-Neto, L., Paiva, A.L.S., Machado, R.D., Arenhart, R.A., Margis-Pinheiro, M., 2017. Interactions between plant hormones and heavy metals responses. Genet. Mol. Biol. 40, 373–386.

Bunluesin, S., Kruatrachue, M., Pokethitiyook, P., et al., 2004. Plant screening and comparison of *Ceratophyllum demersum* and *Hydrilla verticillata* for cadmium accumulation. Bull. Environ. Contam. Toxicol. 73, 591–598.

Burton, E.D., Bush, R.T., Leigh, A., Sullivan, L.A., Johnston, S.G., Hocking, R.K., 2008. Mobility of arsenic and selected metals during re-flooding of iron- and organic-rich acid-sulfate soil. Chem. Geol. 258, 64–73.

Casova, K., Cerny, J., Szakova, J., Balik, J., Tlustos, P., 2009. Cadmium balance in soils under different fertilization managements including sewage sludge application. Plant Soil Environ. 55 (8), 353–361.

Chatterjee, C., Dube, B.K., Sinha, P., Srivastava, P., 2004. Detrimental effects of lead phytotoxicity on growth, yield, and metabolism of rice. Commun. Soil Sci. Plant Anal. 35, 255–265.

Chlopecka, A., Adriano, D.C., 1997. Influence of zeolite, apatite and Fe-oxide on Cd and Pb uptake by crops. Sci. Total Environ. 207, 195–206.

Chowdhury, M.D., Riton, K.V., Sattar, A., Brahmachari, K., 2014. Studies on the water use efficiency and nutrient uptake by rice under system of intensification. Bioscan 9 (1), 85–88.

Craswell, E.T., Waring, S.A., 1972. Effect of grinding on the decomposition of soil organic matter. I. The mineralization of organic nitrogen in relation to soil type. Soil Biol. Biochem. 4, 427–433.

Dikshit, R., Tallapragada, P., 2011. *Monascus purpureus*: a potential source for natural pigment production. J. Microbiol. Biotechnol. Res. 1 (4), 164–174.

Ehrenfelda, J.G., Yu, S., 2012. Patterns of nitrogen mineralization in wetlands of the New Jersey pinelands along a shallow water table gradient. Am. Midl. Nat. 167 (2), 322–335.

El-Habet, H.B., Naeem, E.S., Abel-Meeged, T.M., Sedeek, S., 2014. Evaluation of genotypic variation in lead and cadmium accumulation of rice (*Oryza sativa*) in different water conditions in Egypt. Int. J. Plant Soil Sci. 3, 911–933.

Elham, A., Hossein, T., Mahnoosh, H., 2010. Removal of Zn(II) and Pb(II) ions using rice husk in food industrial wastewater. J. Appl. Sci. Environ. Manag. 14 (4), 159–162.

Eriksson, J.E., 1990. Effects of nitrogen-containing fertilizers on solubility and plant uptake of cadmium. Water Air Soil Pollut. 49, 355–368.

Esa, N.M., Ling, T.B., Peng, L.S., 2013. By-products of rice processing: an overview of health benefits and applications. J. Rice Res. 1, 1–11.

Felkner, J., Tazhibayeva, K., Townsend, R., 2009. Impact of climate change on rice production in Thailand. Am. Econ. Rev. 99, 205–210.

Fu-Zhong, W., Wan-Qin, Y., Zhang, J., Li-Qiang, Z., 2011. Growth responses and metal accumulation in an ornamental plant (*Osmanthus fragrans* var. thunbergii) submitted to different Cd levels. ISRN Ecol. 2011, 1–7.

Gohel, V., Duan, G., 2012. No-cook process for ethanol production using Indian broken rice and pearl millet. Int. J. Microbiol. 2012, 1–9.

Graham, P.H., Vance, C.P., 2003. Legumes: importance and constraints to greater use. Plant Physiol. 131, 872–877.

Gurel, L., Senturk, I., Bahadir, T., Buyukgungor, H., 2010. Treatment of nickel plating industrial wastewater by fungus immobilized onto rice bran. J. Microb. Biochem. Technol. 2, 34–37.

He, Y., Zhang, T., Sun, Y., et al., 2022. Exogenous IAA alleviates arsenic toxicity to rice and reduces arsenic accumulation in rice grains. J. Plant Growth Regul. 41, 734–741.

Hossain, N., Nizamuddin, S., Griffin, G., et al., 2020. Synthesis and characterization of rice husk biochar via hydrothermal carbonization for wastewater treatment and biofuel production. Sci. Rep. 10, 18851.

Huang, Y., Zhang, W., Zheng, X., Li, J., Yu, Y., 2004. Modeling methane emission from rice paddies with various agricultural practices. J. Geophys. Res. Atmos. 109, 1–12.

Huang, T.H., Lai, Y.J., Hseu, Z.Y., 2018. Efficacy of cheap amendments for stabilizing trace elements in contaminated paddy fields. Chemosphere 198, 130–138.

Humphreys, E., Chalk, P.M., Muirhead, W.A., Melhuish, F.M., White, R.J.G., 1987. Effects of time of urea application on combine-sown Calr ose rice in South-East Australia. III. Fertiliser nitrogen recovery, efficiency of fertilization and soil nitrogen supply. Aust. J. Agric. Res. 38, 129–138.

Ishikawa, S., Abe, T., Kuramata, M., Yamaguchi, M.O.T., Yamamoto, T., Yano, M., 2010. Major quantitative trait locus for increasing cadmium specific concentration in rice grain is located on the short arm of chromosome 7. J. Exp. Bot. 61 (3), 923–934.

Johnston, A.M., Khurana, H.S., Majumdar, K., Satyanarayana, T., 2009. Site-specific nutrient management—concept, current research and future challenges in Indian agriculture. J. Indian Soc. Soil Sci. 57, 1–10.

Karkhanis, M., Jadia, C.D., Fulekar, M.H., 2005. Rhizofilteration of metals from coal ash leachate. Asian J. Water Environ. Pollut. 3 (1), 91–94.

Khan, A.G., Kuek, C., Chaudhry, T.M., Khoo, C.S., Hayes, W.J., 2000. Role of plants, mycorrhizae and phytochelators in heavy metal contaminated land remediation. Chemosphere 41, 197–207.

Kim, S.C., Hong, Y.K., Oh, S.J., Oh, S.M., Lee, S.P., Kim, D.H., Yang, J.E., 2017. Effect of chemical amendments on remediation of potentially toxic trace elements (PTEs) and soil quality improvement in paddy fields. Environ. Geochem. Health 39 (2), 345–352.

Kögel-Knabner, I., Amelung, W., Cao, Z., Fiedler, S., Frenzel, P., Jahn, R., Kalbitz, K., Kölbl, A., Schloter, M., 2010. Biogeochemistry of paddy soils. Geoderma 157 (1–2), 1–14.

Lata, S., Samadder, S.R., 2014. Removal of heavy metals using rice husk: a review. Int. J. Environ. Res. Dev. 4, 165–170.

Li, W., Cui, X., 2014. Focus on rice: towards better understanding of the life cycle of crop plants. Mol. Plant 7 (6), 931–933.

Li, Z., Liang, Y., Hu, H., Shaheen, S.M., Zhong, H., Tack, F.M., Wu, M., Li, Y.F., Gao, Y., Rinklebe, J., Zhao, J., 2021. Speciation, transportation, and pathways of cadmium in soil-rice systems: a review on the environmental implications and remediation approaches for food safety. Environ. Int. 156, 106749.

Linke, M.G., Godoy, R.S., Rolon, A.S., Maltchik, L., 2014. Can organic rice crops help conserve aquatic plants in southern Brazil wetlands? Appl. Veg. Sci. 17, 346–355.

Liu, J., Zhu, Q., Zhang, Z., Xu, J., Yang, J., Wong, M.H., 2005. Variations in cadmium accumulation among rice cultivars and types and the selection of cultivars for reducing cadmium in the diet. J. Sci. Food Agric. 85, 147–153.

Liu, W.X., Shen, L.F., Liu, J.W., Wang, Y.W., Li, S.R., 2007. Uptake of toxic heavy metals by rice (Oryza sativa L.) cultivated in the agricultural soil near Zhengzhou City, People's Republic of China. Bull. Environ. Contam. Toxicol. 79, 209–221.

Liu, C., Wu, C., Rafiq, M.T., Aziz, R., Hou, D., Ding, Z., Lin, Z., Lou, L., Feng, Y., Li, T., Yang, X., 2013. Accumulation of mercury in rice grain and cabbage grown on representative Chinese soils. J. Zhejiang Univ. Sci. 14, 1144–1151.

Ma, X., Liu, J., Wan, M., 2012. Differences between rice cultivars in iron plaque formation on roots and plant lead tolerance. Adv. J. Food Sci. Technol. 5 (2), 160–163.

McLaughlin, M.J., Maier, N.A., Freeman, K., Tiller, K.G., Williams, C.M.J., Smart, M.K., 1995. Effect of potassic and phosphatic fertilizer type, fertilizer Cd concentration and zinc rate on cadmium uptake by potatoes. Fertil. Res. 40, 63–70.

Meia, X.Q., Yea, Z.H., Wong, M.H., 2009. The relationship of root porosity and radial oxygen loss on arsenic tolerance and uptake in rice grains and straw. Environ. Pollut. 157, 2550–2557.

Meng, B.O., Feng, X., Qiu, G., Li, P., Liang, P., Li, P., Shang, L., 2012. Inorganic mercury accumulation in rice (Oryza sativa L.). Environ. Toxicol. Chem. 31, 2093–2098.

Montanher, S.F., Oliveira, E.A., Rollemberg, M.C., 2005. Removal of metal ions from aqueous solutions by sorption onto rice bran. J. Hazard. Mater. 117, 207–211.

Moore, K.L., Chen, Y., van de Meene, A.M.L., Hughes, L., Liu, W., Geraki, T., Mosselmans, F., McGrath, S.P., Grovenor, C., Zhao, F.J., 2014. Combined nanoSIMS and synchrotron X-ray fluorescence reveal distinct cellular and subcellular distribution patterns of trace elements in rice tissues. New Phytol. 201, 104–115.

Mostafa, K.M., Samarkandy, A.R., El-Sanabary, A.W., 2012. Harnessing of chemically modified rice straw plant waste as unique adsorbent for reducing organic and inorganic pollutants. Int. J. Org. Chem. 2, 143–151.

Mulbry, W., Kondrad, S., Pizarro, C., Kebede-Westhead, E., 2008. Treatment of dairy manure effluent using freshwater algae: algal productivity and recovery of manure nutrients using pilot-scale algal turf scrubbers. Bioresour. Technol. 99, 8137–8142.

Munoz, R., Alvarez, M.T., Munoz, A., Terrazas, E., Guieysse, B., Mattiasson, B., 2006. Sequential removal of heavy metals ions and organic pollutants using an algal-bacterial consortium. Chemosphere 63, 903–911.

Myint, A.K., Yamakawa, T., Kajihara, Y., Zenmyo, T., 2010. Application of different organic and mineral fertilizers on the growth, yield and nutrient accumulation of rice in a Japanese ordinary paddy field. Sci. World J. 5 (2), 47–54.

Nagendraprasad, M.N., Sanjay, K.R., Khatokar, M.S., Vismaya, M.N., Swamy, S.N., 2011. Health benefits of rice bran—a review. J. Nutr. Food Sci. 1, 1–7.

Nawar, N., Ebrahim, M., Sami, E., 2013. Removal of heavy metals Fe^{3+}, Mn^{2+}, Zn^{2+}, Pb^{2+} and Cd^{2+} from wastewater by using rice straw as low cost adsorbent. Acad. J. Interdiscip. Stud. 2, 85–95.

Owen, D.E., Otton, J.K., 1995. Mountain wetlands: efficient uranium filters—potential impacts. Ecol. Eng. 5, 77–93.

Pagliai, M., Vignozzi, N., Pellegrini, S., 2004. Soil structure and the effect of management practices. Soil Tillage Res. 79, 131–143.

Panichpat, T., Srinives, P., 2009. Partitioning of lead accumulation in rice plants. Thai J. Agric. Sci. 42 (1), 35–40.

Paranthaman, R., Alagusundaram, K., Indhumathi, J., 2009. Production of protease from rice mill wastes by *Aspergillus niger* in solid state fermentation. World J. Agric. Sci. 5 (3), 308–312.

Park, J.H., Lamb, D., Paneerselvam, P., Choppala, G., Bolan, N.S., Chung, J.W., 2011. Role of organic amendments on enhanced bioremediation of heavy metal(loid) contaminated soils. J. Hazard. Mater. 183 (3), 549–574.

Payus, C., Talip, A.F.A., 2014. Assessment of heavy metals accumulation in paddy rice (*Oryza sativa* L.). Afr. J. Agric. Res. 9 (41), 3082–3090.

Pereira, B.F.F., Rozane, D.E., Araújo, R.S., Barth, G., Queiroz, R.J.B., Nogueira, T.A.R., Moraes, M.F., Cabral, C.P., Boaretto, A.E., Malavolta, E., 2011. Cadmium availability and accumulation by lettuce and rice. Braz. J. Soil Sci. 35, 645–646.

Rai, P.K., 2009. Heavy metal phytoremediation from aquatic ecosystems with special reference to macrophytes. Crit. Rev. Environ. Sci. Technol. 39, 697–753.

Rajkumar, M., Sandhya, S., Prasad, M.N.V., Freitas, H., 2012. Perspectives of plant-associated microbes in heavy metal phytoremediation. Biotechnol. Adv. 30 (6), 1562–1574.

Ramakrishnaiaha, H., Somashekar, R.K., 2010. Heavy metal contamination in roadside soil and their mobility in relations to pH and organic carbon. Soil Sediment Contam. Int. J. 11, 643–654.

Ranjan, D., Hasan, S.H., 2010. Rice bran carbon: an alternative to commercial activated carbon for the removal of hexavalent chromium from aqueous solution. Bioresources 5, 1661–1664.

Rascio, N., Navari-Izzo, F., 2011. Heavy metal hyper accumulating plants: how and why do they do it? And what makes them so interesting? Plant Sci. 180, 169–181.

Reniger, P., 1977. Concentration of cadmium in aquatic plants and algal mass in flooded rice culture. Environ. Pollut. 14, 297–302.

Rubio, M.I., Escrig, C., Martinez-Cortina, F., Lopez-Benet, J., Sanz, A., 1994. Cadmium and nickel accumulation in rice plants. Effects on mineral nutrition and possible interactions of abscisic and gibberellic acids. Plant Growth Regul. 14, 151–157.

Rugh, C.L., 2004. Genetically engineered phytoremediation: one man's trash is another man's transgene. Trends Biotechnol. 22, 496–498.

Rungrodnimitchai, R., 2014. Rapid preparation of biosorbents with high ion exchange capacity from rice straw and bagasse for removal of heavy metals. Sci. World J. 2014, 1–9.

Sachdeva, S., Sharma, A., 2012. Azolla: role in phytoremediation of heavy metals. In: Proceeding of the National Conference "Science in Media 2012" Organized by YMCA University of Science and Technology, Faridabad, Haryana (India). IJMRS's International Journal of Engineering Sciences, ISSN (Online): 2277-9698.

Sasaki, A., Yamaji, N., Ma, J.F., 2011. OsYSL6 is involved in the detoxification of excess manganese in rice. Plant Physiol. 157, 1832–1840.

Satpathy, D., Reddy, M.V., Dhal, S.P., 2014. Risk assessment of heavy metal contamination in paddy soil, plants, and grains (*Oryza sativa* L.) at the East Coast of India. Biomed. Res. Int. 2014, 545473. 1–11.

Schuhmacher, M., Domingo, J.L., Llobet, J.M., Corbella, J., 1994. Cadmium, chromium, copper, and zinc in rice and rice field soil from Southern Catalonia, Spain. Bull. Environ. Contam. Toxicol. 53, 54–60.

Sebastian, A., Prasad, M.N.V., 2013. Cadmium accumulation retard activity of functional components of photo assimilation and growth of rice cultivars amended with vermicompost. Int. J. Phytoremediation 15, 965–978.

Sebastian, A., Prasad, M.N.V., 2014a. Cadmium minimization in rice. A review. Agron. Sustain. Dev. 34, 155–173.

Sebastian, A., Prasad, M.N.V., 2014b. Photosynthesis-mediated decrease in cadmium translocation protects shoot growth of *Oryza sativa* seedlings up on ammonium phosphate-sulfur fertilization. Environ. Sci. Pollut. Res. 21, 986–997.

III. Ornamentals and tree crops for contaminated substrates

Seng, V., Eastick, R., Fukai, S., Ouk, M., Men, S., Chan, S.Y., Nget, S., 2008. Crop diversification in lowland rice cropping systems in Cambodia: effect of soil type on legume production global issues. Paddock action. In: Proceedings of 14th Agronomy Conference 2008, 21–25 September 2008, Adelaide, South Australia.

Shen, L.B., Wu, L.H., Tan, W.N., Han, X.R., Luo, Y.M., Ouyang, Y.N., Jin, Q.Y., Jiang, Y.G., 2010. Effects of *Sedum plumbizincicola—Oryza sativa* rotation and phosphate amendment on Cd and Zn uptake by *O. sativa*. Ying Yong Sheng Tai Xue Bao 21 (11), 2952–2958.

Shi, X., Zhang, C., Wang, H., Zhang, F., 2005. Effect of Si on the distribution of Cd in rice seedlings. Plant Soil 272, 53–60.

Simmons, R.W., Pongsakul, P., Saiyasitpanich, D., Klinphoklap, S., 2005. Elevated levels of Cd and zinc in paddy soils and elevated levels of Cd in rice grain downstream of a zinc mineralized area in Thailand: implications for public health. Environ. Geochem. Health 27, 501–511.

Singh, R., Tripathi, R.D., Dwivedi, S., Kumar, A., Trivedi, P.K., Chakrabarty, D., 2010. Lead bioaccumulation potential of an aquatic macrophyte Najas indica are related to antioxidant system. Bioresour. Technol. 101 (9), 3025–3032.

Singh, J., Upadhyay, S.K., Pathak, R.K., Gupta, V., 2011. Accumulation of heavy metals in soil and paddy crop (*Oryza sativa*), irrigated with water of Ramgarh Lake, Gorakhpur, UP, India. Toxicol. Environ. Chem. 93 (3), 462–473.

Smolders, A.J.P., Lucassen, E.C.H.E.T., Bobbink, R., Roelofs, J.G.M., Lamers, L.P.M., 2010. How nitrate leaching from agricultural lands provokes phosphate eutrophication in groundwater fed wetlands: the sulphur bridge. Biogeochemistry 98, 1–7.

Suwannaporn, P., Lineman, A., 2008. Rice-eating quality among consumers in different rice grain preference countries. J. Sens. Stud. 23, 1–13.

Thiravetyan, P., 2012. Application of rice husk and bagasse ash as adsorbents in water treatment. In: Bhatnagar, A. (Ed.), Application of Adsorbents for Water Pollution Control. Bentham Science Publishers, pp. 432–454.

Ueno, D., Koyama, E., Kono, I.O.T., Yano, M., Ma, J.F., 2009. Identification of a novel major quantitative trait locus controlling distribution of Cd between roots and shoots in rice. Plant Cell Physiol. 50 (12), 2223–2233.

Uraguchi, S., Mori, S., Kuramata, M., Kawasaki, A., Arao, T., Ishikawa, S., 2009. Root-to-shoot Cd translocation via the xylem is the major process determining shoot and grain cadmium accumulation in rice. J. Exp. Bot. 60 (9), 2677–2688.

Vandenhove, H., Van Hees, M., Van Winkel, S., 2001. Feasibility of phytoextraction to clean up low-level uranium-contaminated soil. Int. J. Phytoremediation 3, 301–320.

Vijayarengan, P., 2011. Mineral nutrient variations in rice (*Oryza sativa*) after treatment with exogenous cadmium. Int. J. Environ. Biol. 2 (3), 147–152.

Violante, A., Cozzolino, V., Perelomov, L., Caporale, A.C., Pigna, M., 2010. Mobility and bioavailability of heavy metals and metalloids in soil environments. J. Soil Sci. Plant Nutr. 10 (3), 268–292.

Wang, H., Huang, J., Ye, Q., Wu, D., Chen, Z., 2009. Modified accumulation of selected heavy metals in Bt transgenic rice. J. Environ. Sci. (China) 21 (11), 1607–1612.

Wang, M.Y., Chen, A.K., Wong, M.H., Qiu, R.L., Cheng, H., Ye, Z.H., 2011. Cadmium accumulation in and tolerance of rice (*Oryza sativa* L.) varieties with different rates of radial oxygen loss. Environ. Pollut. 159, 1730–1736.

Wang, X., Ye, Z., Li, B., Huang, L., Meng, M., Shi, J., Jiang, G., 2014. Growing rice aerobically markedly decreases mercury accumulation by reducing both Hg bioavailability and the production of MeHg. Environ. Sci. Technol. 48, 1878–1885.

Wani, S., Chand, S., Ali, T., 2013. Potential use of *Azotobacter chroococcum* in crop production: an overview. Curr. Agric. Res. J. 1, 35–38.

Wright, D.P., Scholes, J.D., Read, D.J., 1998. Effects of VA mycorrhizal colonization on photosynthesis and biomass production of *Trifolium repens* L. Plant Cell Environ. 21 (2), 209–216.

Wu, Z., Ren, H., McGrath, S.P., Wu, P., Zhao, F.J., 2011. Investigating the contribution of the phosphate transport pathway to arsenic accumulation in rice. Plant Physiol. 157 (1), 498–508.

Wu, W., Yang, M., Feng, Q., McGrouther, K., Wang, H., Lu, H., Chen, Y., 2012. Chemical characterization of rice straw-derived biochar for soil amendment. Biomass Bioenergy 47, 268–276.

Xu, B., Yu, Y., Zhong, Y., Guo, Y., Ding, J., Chen, Y., Wang, G., 2019. Influence of Br24 and Gr24 on the accumulation and uptake of Cd and As by rice seedlings grown in nutrient solution. Pol. J. Environ. Stud. 28 (5), 3951–3958.

Xue, D., Chen, M., Zhang, G., 2009. Mapping of QTLs associated with cadmium tolerance and accumulation during seedling stage in rice (*Oryza sativa* L.). Euphytica 165, 587–596.

III. Ornamentals and tree crops for contaminated substrates

Yadav, S., Singh, A., Singh, M.R., Goel, N., Vinod, K.K., Mohapatra, T., Singh, A.K., 2013. Assessment of genetic diversity in Indian rice germplasm (*Oryza sativa* L.): use of random versus trait-linked microsatellite markers. J. Genet. 92, 545–557.

Yamazaki, S., Ueda, Y., Mukai, A., Ochiai, K., Matoh, T., 2018. Rice phytochelatin synthases OsPCS1 and OsPCS2 make different contributions to cadmium and arsenic tolerance. Plant Direct 2 (1), e00034.

Yeasmin, T., Zaman, P., Rahman, A., Absar, N., Khanum, N.S., 2007. Arbuscular mycorrhizal fungus inoculums production in rice plants. Afr. J. Agric. Res. 4, 463–467.

Yizong, H., Ying, H., Yunxia, L., 2009. Heavy metal accumulation in iron plaque and growth of rice plants upon exposure to single and combined contamination by copper, cadmium and lead. Acta Ecol. Sin. 29, 320–326.

Yu, L., Zhu, J., Huang, Q., Su, D., Jiang, R., Li, H., 2014. Application of a rotation system to oilseed rape and rice fields in cd-contaminated agricultural land to ensure food safety. Ecotoxicol. Environ. Saf. 108, 287–293.

Yuan, X., Li, T., Li, J., 2012. The influence of paddy soil components and As, Pb and Cd speciation on their uptake by rice in the Changjiang Delta Region. In: Asia Pacific Conference on Environmental Science and Technology Advances in Biomedical Engineering 6, pp. 147–152.

Zhang, Y., Zhang, G., Xiao, N., Wang, L., Fu, Y., Sun, Z., Fang, R., Chen, X., 2012. The rice 'nutrition response and root growth' (NRR) gene regulates heading date. Mol. Plant 6 (2), 585–588.

Zhao-Wen, S., Sun, C., Zhu, Z., Xi-Hua, T., Rui-Juan, S., 1984. Purification and properties of rice germ lectin. Can. J. Biochem. Cell Biol. 62 (10), 1027–1032.

Zheng, R., Sun, G.X., Zhu, Y.G., 2013. Effects of microbial processes on the fate of arsenic in paddy soil. Chin. Sci. Bull. 58, 186–193.

Zhou, H., Zhou, X., Zeng, M., Liao, B.H., Liu, L., Yang, W.T., Wu, Y.M., Qiu, Q.Y., Wang, Y.J., 2014. Effects of combined amendments on heavy metal accumulation in rice (*Oryza sativa* L.) planted on contaminated paddy soil. Ecotoxicol. Environ. Saf. 101, 226–232.

Zhu, X., Lin, L., Zhang, Q., Liu, Q., Ma, X., Ye, L., Sha, J., 2010. Effects of zinc and chromium stresses on heavy metal accumulation of rice roots at different growth stages of rice plants. In: iCBBE Proceedings. icbbe.org/2011/Proceeding2010.aspx.

Zong, H.Y., Ying, H., Yun Xia, L., 2010. Combined effects of chromium and arsenic on rice seedlings (*Oryza sativa* L.) growth in a solution culture supplied with or without P fertilizer. Sci. China Life Sci. 53 (12), 1459–1466.

Duckweeds: Bioremediation of surface wastewater and biorefinery

K. Sowjanya Sree[a] and Klaus J. Appenroth[b]

[a]Department of Environmental Science, Central University of Kerala, Periye, India [b]Matthias Schleiden Institute-Plant Physiology, Friedrich Schiller University, Jena, Germany

1 Introduction

In the recent days, an increasing contamination of surface fresh waters has been observed. There are a variety of living and nonliving components that contaminate the surface waters. These nonliving contaminants can belong to the following categories—common salts (as nitrates and phosphates), heavy metals (as mainly inorganic salts), xenobiotic (organic compounds), and complex mixture of unwanted compounds. The source of these contaminants can be from nature, e.g., from weathering of rocks. However, in most cases, these contaminations have anthropogenic sources. An effective method to prevent such anthropogenic contamination is to have stringent regulations and implementation. However, since their implementation is often inadequate, cleaning of surface waters contaminated by wastewater becomes inevitable. For the remediation of surface wastewater, a broad spectrum of effective methods that are based on physical or chemical principles exists. However, these methods are expensive, not always available, and can be an environmental threat themselves, for instance, by the production of contaminated sludge. In the last decades, several methods of bioremediation, in specific, phytoremediation were developed. These are eco-friendly, cost-efficient, and generally applicable. Especially in developing countries, this is an attractive alternative to chemical or physical wastewater treatment methods.

The source of high amounts of common salts of nitrates and phosphates in surface waters is mostly agriculture. In order to obtain maximum yield during a harvest season, often enormous amounts of mineral fertilizers are used in agricultural fields. Only a part of these fertilizers is used by the crop plants and a majority of them are lost in the runoff waters or they leach out. The degree of leaching can depend on the chemical composition of the fertilizer and the soil structure. One of the sinks of such runoffs and leachates is surface water, from ponds, lakes, rivers, and finally to oceans. Again there are methods to decrease at least the amount of fertilizers leaching from the agricultural fields, e.g., simply by optimizing the amount of fertilizer application. Another interesting method is to apply organic waste as a fertilizer to the crop plants, which is one step toward circular economy. Although these methods cannot be applied to all agricultural areas and crops, it is certainly of significant importance. These applications would dramatically reduce the amount of fertilizer leachates and runoffs into their sinks. Alternatively, the wastewater containing such fertilizer runoff or leachate contaminants can be reused to irrigate agricultural fields and is perfectly suitable for growing crop plants without the application of additional fertilizer. Nowadays, however, we have to deal with a high rate of leachate and runoff of these salts into the surface waters which results in eutrophication. Eutrophication can have multifold negative effects on the physico-chemical and biological characteristics of the surface water. A significant challenge of eutrophication is the contamination of surface waters by common salts like sodium chloride which is a part of several types of wastewaters, e.g., the wastewater generated from salt mining. The high concentration of sodium chloride in the water bodies represents a serious threat to fresh water organisms and must be considered separately from eutrophication in general.

Heavy metals are often the waste products in mining, tannery, plating, and other industrial productions. When these waste products are driven into the surface waters without proper treatment, the water bodies become polluted with heavy metals. Heavy metals are available as a broad group of salts with different chemical properties (Appenroth, 2010). These chemical moieties are capable of bioaccumulation and biomagnification across the trophic levels. In these polluted aquatic ecosystems, heavy metals are absorbed by plants and are in turn supplied into the food chain and food web, thereby, creating a threat of heavy metal toxicity to higher-order organisms including humans. On the other hand, this bioaccumulation capacity of plants makes them a good vehicle for bioremediation. Over the past several decades, phytoremediation of different polluted sites has been demonstrated to be an effective method.

Xenobiotics are a very heterogeneous group of organic compounds ranging from herbicides, pesticides, and antibiotics to personal healthcare and pharmaceutical products. Usually, they are present in the environment in low concentrations but nevertheless might be very toxic. For each group of compounds, a special method of remediation is required. Proper treatment of the xenobiotic-containing wastewater released from sources such as hospitals or pharmaceutical companies is most effective. Apart from their property of bioaccumulation and biomagnification, some of the xenobiotics are recalcitrant and are resistant to degradation causing environmental problems.

Complex mixture of unwanted compounds is often released by industries. Their entry into the surface waters presents a high requirement for purification as the different components have different physical and chemical properties. Examples of such complex mixtures are the waste from dairy industry, or textile dyeing. When purification of wastewater at the source is not effectively performed, these complex mixtures of unwanted substances end up highly diluted in large amount of surface waters, and often, cleaning of these large volumes of surface water contaminated with such wastes is a serious challenge to deal with, a typical example being large contaminated rivers.

2 Bioremediation of surface wastewater

There are several physico-chemical and biological methods that have been established for the effective treatment of wastewater. These include conventional methods like coagulation or flocculation, precipitation, and filtration, and other established methods like oxidation, electrochemical treatment, ion-exchange methods, and so on. The technique to be employed for wastewater treatment largely depends on the characteristics of that wastewater. The popularity of the usage of a given method by the industries also depends on the economics of the system (Crini and Lichtfouse, 2019). In search of more cost-effective alternatives, for several decades, research has been focused on the use of the potential of different living organisms in terms of uptake, bioaccumulation, and removal of waste products and pollutants from the environment. One of the groups of organisms employed for this purpose is the higher plants.

Several aquatic plant species have been investigated in the past concerning their capacity to remove pollutants from surface water. To name a few, water hyacinth (*Eichhornia crassipes*), water lettuce (*Pistia stratiotes*), Brazilian waterweed (*Egeria densa*), giant salvinia (*Salvinia molesta*), water chest nut (*Trapa natans*), water spinach (*Ipomea aquatica*) (Mustafa and Hayder, 2021). Plants of the family Lemnaceae (duckweeds) were especially well investigated concerning this aspect (Acosta et al., 2021).

Lemnaceae Martinov represents a family of aquatic floating flowering plants comprising of 36 species categorized in 5 genera, *Spirodela*, *Landoltia*, *Lemna*, *Wolffiella*, and *Wolffia* (Bog et al., 2019). These plants range from a centimeter to less than a millimeter in size (Bog et al., 2020). Being flowering plants, they do form flowers and seeds but this is only rarely observed (Acosta et al., 2021). Most prominent mode of propagation is vegetative by budding of daughter fronds from mother fronds (Fig. 1), forming genetically uniform populations, i.e., clones (Sree et al., 2015c). They represent the fastest growing angiosperms with clones of *Wolffia microscopica* having a doubling time of 29 h (Sree et al., 2015d) and can be further improved by optimization of the growth conditions (unpublished data). As a consequence, large amount of biomass can be produced at a fast pace which is an important prerequisite for use in bioremediation. Moreover, these plants have a much reduced morphological

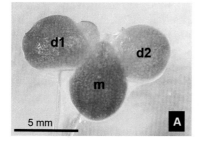

FIG. 1 Vegetatively propagating duckweed species. (A) *Lemna gibba*. Note the daughter fronds (d1 and d2) arising from two lateral pouches of the mother frond (m). (B) *Wolffia arrhiza*. Note that only one vegetative pouch is present from which one daughter frond (d) is budding from the mother frond (m) at a given time.

III. Ornamentals and tree crops for contaminated substrates

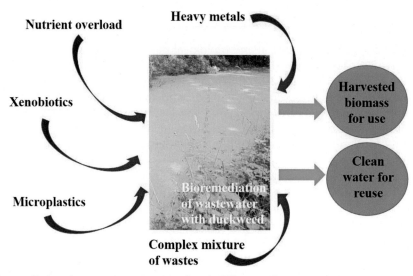

FIG. 2 Bioremediation of wastewater contaminated with different pollutants using duckweeds can yield two important resources, biomass for use in different practical applications and clean water which can be recirculated into the ecosystem for reuse.

structure of a leaf-like floating frond with species in the genera *Spirodela*, *Landoltia*, and *Lemna* having one to many adventitious roots and those of the genera *Wolffiella* and *Wolffia*, not bearing roots at all. The whole submerged part of the plant body, which is in contact with water, is able to take up substances from the aquatic medium (Landolt, 1986; Acosta et al., 2021).

As long as 300 years ago, it was reported that wastewater can be cleaned by higher plants thriving in the aquatic ecosystem (McCutcheon and Rock, 2001; Ansari et al., 2020). From a recent ethnobotanical historical study, it was found that in Yemeni culture the water from a cistern was considered to be pure for drinking purpose as well as for religious rituals only when duckweeds were floating atop. This was reported almost 200 years back by Botta (1841), a French naturalist. This practice indicates that the Yemenis considered duckweeds as the water-purifying agents in such of the cisterns (Edelman et al., 2022). With the scientific advancements of today, we know that duckweeds can be used for the remediation of water polluted with several different types of waste substances as has been detailed below.

In the following first section, we describe the use of different duckweed species in phytoremediation of surface wastewater. Fig. 2 illustrates the availability of two important resources as a result of this bioremediation process, which are duckweed biomass for use in different practical applications and the availability of clean water for reuse. The duckweed species commonly employed in the bioremediation of wastewater are presented in Table 1.

2 Bioremediation of surface wastewater   315

TABLE 1 Most commonly used duckweed species in wastewater remediation and their geographical distribution. Information concerning species distribution were taken from Bog et al. (2020).

Duckweed species	Species distribution
Spirodela polyrhiza (L.) Schleid	All over the world, except Arctic and Antarctic
Landoltia punctata (G.Mey.) Les and D.J. Crawford	Throughout South America, South-east Asia, Australia and Africa but today also widely distributed in Europe
Lemna aequinoctialis Welw	Very common, even dominating species in several Asian countries
Lemna turionifera Landolt	In temperate regions of North America and continental parts of Eurasia
Lemna minor L.	Native to the cooler, rather oceanic regions of North America, Europe, Africa and western Asia. Does not occur in East Asia or is at least extremely rare
Lemna gibba L.	On all continents except Australia and Antarctica
Wolffia globosa (Roxb.) Hartog and Plas	In warm temperate, subtropical and tropical regions of eastern and southern Asia and Africa (especially south of equator) with mild winters. Very common also in India, Nepal, Sri Lanka, China and Japan

2.1 Eutrophic wastewater

The main source of generation of eutrophic wastewater is the use of mineral fertilizers in agricultural fields. The major contributors are salts of nitrate, ammonium, and phosphate (Zhou et al., 2023). The water-soluble nitrogen-containing compounds are easily taken up by aquatic plants. Interestingly, duckweed prefer the uptake of ammonium over nitrate. This feature is especially important because ammonia, which is formed from ammonium at higher pH conditions in the polluted waters, is toxic to several aquatic organisms. So, removal of ammonium from the wastewater is an essential step in the process of phytoremediation.

Zhou et al. (2022) investigated six different duckweed species (*Spirodela polyrhiza, Landoltia punctata, Lemna aequinoctialis, Lemna turionifera, Lemna minor,* and *Wolffia globosa*) for studying the dynamics of nitrate and ammonium uptake in these plants. All the six investigated duckweed species preferentially took up ammonium from the medium. The uptake of nitrate was observed to take place only after depletion of ammonium in the aqueous medium. In order to understand the background biochemistry of this preferential uptake by duckweeds, the authors investigated nitrate and nitrite reductases, glutamine synthetases, and glutamate synthetases in duckweeds that were exposed to ammonium and nitrate. These analyses showed that several species of duckweeds are well suitable for remediation of wastewaters contaminated with nitrogen-containing compounds, especially ammonium. Similar observations were also made by several other investigators using different species of the genus *Lemna*. *Lemna minor* reduced the ammonium N content by 97% in the medium (Petersen et al., 2022a). Over 90% of ammonium was removed by several species of *Lemna* (Benshalom et al., 2014). Sun et al. (2021) reported that *L. aequinoctialis* removed 84% of ammonium in 66 days from a synthetic wastewater medium. It was observed that the species *L. minor* readily grows on medium containing ammonium up to a concentration of 84 mg/L, and at concentrations higher than this limit, growth inhibition was observed (Wang et al., 2014, cf. Ziegler et al., 2016).

Nevertheless, duckweeds can also efficiently uptake and accumulate nitrates and phosphates from wastewater. *Lemna aequinoctialis* efficiently removed 98% of phosphates from a synthetic wastewater medium in about 10 weeks (Sun et al., 2021). Petersen et al. (2022a) showed that *L. minor* reduced the nitrate concentration by 13% (but ammonium by 97%) and phosphate by 53% in the growth medium. In another study, over 70% of nitrates and more than 33% of phosphates were removed by several species of *Lemna*. Moreover, during this treatment process, a neutral pH was maintained, the oxygen demand was reduced, and even the biological contamination by mosquito larvae and coliform bacteria were decreased (Benshalom et al., 2014).

It is assumed that duckweeds can accumulate up to 9.1 t/hectare/year of total nitrogen and 0.8 t/hectare/year of total phosphorous (Zhou et al., 2023). The same research group (Zhou et al., 2018) reported that four different duckweed species are able to remove more than 93% of total N and P from local municipal wastewater and that their final concentrations after phytoremediation process was lower than the required mark according to water quality standards. The greater duckweed, *S. polyrhiza*, can also efficiently remove 84% of total N and 89% of total P from a wastewater lagoon in 8 weeks (Sun et al., 2021). Also, several other nutrients are accumulated to a substantial amount by duckweeds (Appenroth et al., 2018) which is an important feature for using in the purification of eutrophic wastewater.

The high growth rates of several duckweed species stands advantageous to the use of duckweeds for bioremediation, wherein the high amounts of uptake of ammonium, nitrate, and phosphates are combined with the availability of fresh biomass for this purpose produced at a fast pace. This usually leads to the formation of thick mats of duckweeds on the surface of the eutrophic water bodies. A secondary effect is the inhibition of growth of algae, or the occurrence of algal blooms is prevented in these nutrient rich waters (El-Sheekh et al., 2021). Duckweeds, hence, present as promising plants for controlling eutrophication in the wastewaters.

However, under these nutrient-rich eutrophic conditions, the mass propagation of duckweed species forming thick mats of biomass thereby preventing light penetration leads to the senescence of the plants in the lower section of the duckweed mat. As a result, nutrients will again be released into the waterbody from these necromass, obscuring the effect of the bioremediation process. Therefore, a careful management of the duckweed-based wastewater treatment system is essential (Ceschin et al., 2019; see Box 1 for an interesting read).

There is another important facet of the potential of duckweeds to uptake and accumulate the above-discussed mineral substances. Ammonium can be produced via chemical conversion of nitrogen from air. However, phosphorus has to be mined from natural stocks that are available only in a few countries (Poirier et al., 2022). The different sources of phosphate in nature are depleting and it has become very important that phosphate is recaptured from several processes. This can be performed by duckweeds, for instance, by *S. polyrhiza*, several species of *Lemna* and *Wolffia* (Zhou et al., 2022). Duckweeds not only take up phosphates but also store it in large amounts as phytates in the whole plant tissue and can be recovered (Landolt and Kandeler, 1987).

BOX 1

Management of duckweed overgrowths

Duckweeds are known to be the fastest growing angiosperms known to us on Earth. In conducive environments, these plants are able to use the rich nutrient supplies and grow extensively at a very fast pace. This results in a thick surface coverage of the water body by the duckweeds, often called duckweed mats. These mats can lead to ill effects in the aquatic ecosystem like poor penetration of sunlight to lower depths of the water body, inadequate exchange of gases between the hydrosphere and atmosphere, including a few others, finally affecting the life and existence of several submerged aquatic organisms. Traditionally spoken, these duckweed overgrowths are seen as a menace and several physical and chemical methods are applied to get rid of the floating duckweeds, which is usually a tough job—with damaging effects on the ecological system. And, it is also noticed that an increase of the density and coverage of the duckweed on the water body can keep recurring with seasonal changes and with the changes in the physico-chemical properties of the water body. Hence, instead

of the use of conventional tools and techniques to manage the duckweed overgrowths, we suggest the following. The duckweeds overgrowing on the surface of a water body could be harvested up to a point that only 50% of the water surface is covered by a thin single layer of duckweeds. From here onwards, depending on the growth rate or doubling time (which is in most cases around 2 to 3 days) of the duckweed present, almost every third day half of the duckweed biomass could be harvested. Duckweeds take up nutrients for their growth and harvesting the biomass finally results in removing the excessive nutrients from the surface water. This is the most natural way of preventing further formation of duckweed mats. The harvested duckweed can be sample tested for its composition and can accordingly be utilized and be integrated into a circular economy mode (See the upcoming content of the chapter for pertinent examples). This can be adopted by individuals or by communities and villages depending on the size of the water body as well as its ownership.

2.2 Heavy metals

Most heavy metals are toxic to duckweeds at different levels and concentrations reducing their growth and decreasing their chlorophyll and carotenoid content. *Lemna minor* was used to investigate the effect of nine different heavy metals on these plants and the following sequence of toxicity was uncovered: $Ag^+ > Cd^{2+} > Hg^{2+} > Tl^+ > Cu^{2+} > Ni^{2+} > Zn^{2+} > Co^{2+} > Cr^{6+}$ (Naumann et al., 2007). This holds true also for some of the metalloids like As and Se (Appenroth, 2010). It has been observed that even the nonessential heavy metals are taken up by plants, most probably by ion channels that have low specificity. For example, the uptake of chromate is seriously inhibited by the presence of sulphates, demonstrating that chromate is taken up by the sulphate channels (Appenroth et al., 2008).

Despite the toxicity, duckweeds have been tested and used for the phytoremediation of heavy metal-contaminated surface waters. Bioremediation of wastewater using different species of duckweeds especially *S. polyrhiza*, *L. punctata*, *L. minor*, *Lemna gibba*, *Lemna japonica*, and *Wolffia arrhiza* was summarized for numerous heavy metals as well as for As and Se in several studies as reviewed by Ziegler et al. (2016, 2017). In the following, we will discuss some of the newer results of detailed investigations on heavy metal phytoremediation.

Many macrophytes have been tested and used for remediation of heavy metal-polluted waters and species of duckweed family are especially successful in this application (Ziegler et al., 2016). Pang et al. (2023) reviewed the phytoremediation capacity of floating aquatic plants for their efficiency in removal of heavy metals. The efficiency of duckweeds (*L. minor*) was compared together with other aquatic macrophytes like water lettuce, water hyacinth, and water moss. It was found that duckweeds have a high removal efficiency for As (90.9%), Cd (97.8%), Cr (90.2%), Cu (98.5%), Ni (98.1%), Pb (99.9%), and Zn (98%) from industrial wastewater at an interval of 25 days (Tufaner, 2020).

Cadmium (Cd) is a nonessential heavy metal. Attempts were made to remediate the Cd-polluted wastewaters using the duckweed species, *L. punctata* (Wang et al., 2021). The tolerance and detoxification mechanisms of hydroponically cultivated plants was investigated. Cd was taken up to a significant amount by these rooted duckweeds. It was observed that Cd was localized to a greater degree in the soluble fraction (23%–55%) and in the cell wall fraction (21%–54%) but was only up to half (14%–23%) in the cell organelles of these plants. Cd was mainly found to be bound to different proteins, dominantly to albumins and globulins. The two main strategies of Cd tolerance and detoxification as identified in *L. punctata* were cell wall immobilization and vacuolar dissociation.

Cobalt can be found in minerals like cobaltite and others but the significance of natural sources in terms of contamination of surface water with cobalt is very low, owing to low mobility of Co (Mahey et al., 2020). More significant points of concern are the anthropogenic sources like waste from cement industries and e-waste. Although the heavy metal cobalt is an essential element involved in the functioning of some of the enzymes especially those employed in biological nitrogen fixation (Hu et al., 2021) in several organisms, the question whether it has an essential role in angiosperms is still open for investigation. Upon exposure to higher concentrations of cobalt, some of the higher plant species like *L. minor* have shown serious toxic effects, as investigated by Sree et al. (2015b). These effects include inhibition of the vegetative growth, decrease of chlorophyll content, and decrease of the photosynthetic efficiency because of the inhibition of biosynthesis of chlorophyll. The extent of ion uptake and accumulation in the fronds of *L. minor* proves that this plant is well fit for phytoremediation of surface wastewater (Sree et al., 2015b). Similar observations were also made by Mahey et al. (2020).

Chromium is mainly released into the environment as chromate (Cr^{6+}) by waste generated from metallurgy, leather tanning, electroplating, and other anthropogenic activities. As in many other cases of wastewater treatment, a low-cost alternative would be to prevent the release of these toxic heavy metals to the nature. As this is very often not well performed, remediation of large amounts of surface waters becomes necessary. Dead and live biomass have shown immense capabilities in remediating chromium-polluted environment. It has been found that the reduction of chromate to Cr^{3+} decreases the mobility of this heavy metal, thereby lowering the degree of toxic effects (Singh et al., 2021). Plants take up chromate via several of the transporters designed for phosphate or sulphate. Therefore, the analysis of the composition of the wastewater is very important. High concentrations of phosphate and sulphate ions in the chromate-contaminated wastewater will compete with the binding of chromate ions to the transporters and inhibit the uptake of this heavy metal. Upon uptake of chromium by plants, several mechanisms of detoxification exist (Kaszycki et al., 2005) and the genes coding for the proteins and enzymes involved in the detoxification process strongly respond to the presence of chromium, making the process of phytoremediation especially effective (Appenroth et al., 2008; Srivastava et al., 2021). One of the toxic effects of chromate on duckweed is the inhibition of photosynthesis, as demonstrated by direct florescence measurements of PSII (Appenroth et al., 2001).

The potential sources of mercury are coal and gold mining (Sitarska et al., 2023). Mercury is present in contaminated surface water mainly as Hg^{2+} but can also exist as elemental Hg. While Hg^{2+} is usually localized, elemental Hg is very volatile and can be transported over large distances in the air. Moreover, it can be converted to methyl mercury and can be incorporated into the aquatic food chain and food web. Uptake of Hg by duckweed like *L. minor* is very effective and 90% of the Hg is taken up by *L. minor* regardless of the external Hg concentrations in the aquatic medium. However, when the concentration of Hg is higher than 0.3 mg/L (1.5 μM), the growth of plants is inhibited, thereby hindering the uptake of the heavy metal also.

Other than heavy metals, some of the metalloids like arsenic (As) also have toxic effects and pose environmental threats. Duckweeds have been used in phytoremediation of As-contaminated wastewater. Although rarely used for such applications, *W. globosa* was employed for phytoremediation of As-containing wastewaters. Ansari et al. (2020) showed that these plants can accumulate as much as 400 mg of As/kg/dry weight. Consequently, this duckweed species is effective in phytoremediation of As-polluted waters.

Of important note is that, at several environmentally relevant concentrations of heavy metals present in the aquatic ecosystems, growth of duckweed is inhibited but not yet the photosynthesis. As a consequence, photosynthesis products are not used for plant growth but stored as starch. This starch is thus accumulated in the whole plant body and not only in specific organs as seeds or caryopses as in other crop plants. This results in the production starch rich biomass (Sree et al., 2015b). However, this approach should be solely limited to phytoremediation of waters already polluted. Heavy metals should not be applied to water bodies for production of starch-rich biomass. This is in view of the fact that artificial addition of heavy metals would lead to further contamination of the environment. Hence, for production of starch-rich biomass, other sustainable strategies for accumulation of starch should be used (Appenroth et al., 2021).

2.3 Microplastics

The concentration of microplastics in the environment increased dramatically over the last decade, and its presence in the marine and the aquatic ecosystems is especially of high concern. The release of this material into the environment has been considered as the second most alarming environmental problem after global warming (Ceschin et al., 2023; Azeem et al., 2021). Microplastics can enter the environment from the discharges of water treatment plants

(Masiá et al., 2020) or from degradation of macroplastics, among several other sources. The macroplastic sources, accounting to about 80% of the total plastic available, are different synthetic materials like polystyrene, polyurethane, polyvinylchloride, polypropylene, polyethylene, and polyethylene terephthalate (Almroth and Eggert, 2019). For biomonitoring tests concerning microplastics in the aquatic fresh water ecosystems, *L. minor* has been used. Usually, the phytotoxic effects of microplastics are small and can be observed only after long-term application (Rozman et al., 2022). Careful investigation of several parameters like frond and root size, plant growth, and chlorophyll content revealed the toxic effects of microplastics like poly (styrene-*co*-methyl methycrylate) microplastics. These substances mainly adsorb on the outer plant surface (Rozman et al., 2022; Rozman and Kalčíková, 2022; Ceschin et al., 2023). Several European initiatives, for instance, the European Chemical Agency in 2019 and the European parliament in 2019, have banned the addition of plastic microbeads to several goods and products. Its implementation would reduce the microplastic pollution in the European Union by 85%–95% (Masiá et al., 2020). Several researchers could show that duckweeds like *L. minor* are able to remediate wastewater contaminated with microplastics (Masiá et al., 2020; Rozman et al., 2022; Ceschin et al., 2023).

2.4 Xenobiotics

Xenobiotics like antibiotics represent a serious threat to contamination of surface fresh water. *Lemna aequinoctialis* was tested and used to remove model antibiotics like Ciprofloxacin (fluoroquinolone group) and Sulfamethoxazole (sulfonamide group) from wastewater. Both of these antibiotics are often present in pharmaceutical wastewater. The presence of these antibiotics in the wastewater inhibited the growth of duckweeds, but nevertheless, there was effective uptake by the duckweed fronds resulting in cleaning of wastewater. Beside hydrolysis and photolysis, uptake of the antibiotics by duckweed fronds was found to be an effective technique for removal of these xenobiotics (Habaki et al., 2023). Wastewater contaminated by antibiotics, tetracycline, and chloramphenicol, with concentrations between 5 and 20 mg/L was shown to be effectively phytoremediated by co-culture of two of the aquatic macrophytes, *L. gibba* and *Azolla filiculoides*. The removal efficiency in some cases reached up to 100%. The antibiotic tetracycline was removed from wastewater with a greater efficiency than chloramphenicol. The tetracycline removal reached 100% with *A. filiculoides* and 84% with *L. gibba*. And, the chloramphenicol removal reached 70% with *A. filiculoides* and 64% with *L. gibba* (Maldonado et al., 2022).

The antibiotic Flumequine can be taken up by *L. minor* and will be degraded (Ziegler et al., 2017; Cascone et al., 2004). *Lemna trisulca* was able to take up Anatoxin-A (cyanobaterial neurotoxin) and the plants subsequently degraded the toxin (Kaminski et al., 2014).

Thakuria et al. (2023) reviewed the removal of pesticides from surface wastewater. The following duckweeds were used in the process: *S. polyrhiza, L. minor, L. gibba, L. aequinoctialis,* and *L. valdiviana*. It could be shown that the degradation of several toxic compounds like organophosphorus is caused directly by the duckweed plants (Gao et al., 2000).

Polycyclic aromatic hydrocarbons are difficult to remove from wastewater. These include phenanthrenes and pyrenes. Using *L. minor*, Zazouli et al. (2023) showed that these substances inhibit the growth of these plants. But at concentrations between 10 and 20 mg/L, uptake and bioaccumulation of these xenobiotics was effective. Moreover, both of these compounds are metabolized in the plant cells and several intermediates have been identified.

Benzotriazoles were removed from wastewater using *W. arrhiza* (Polinska et al., 2022). The following benzotriazoles (BTRs) were investigated: 1H-benzotriazole (1H-BTR), 4-methyl-1H-benzotriazole (4M-BTR), 5-methyl-1H-benzotriazole (5M-BTR), and 5-chlorobenzotriazole (5Cl-BTR). The conditions for remediation were optimized concerning plant biomass, light exposure, and pH. In most cases 92%–100% were removed. Uptake by the plants was the main mechanism of decreasing concentration resulting in oxidative stress in *W. arrhiza*.

2.5 Complex mixture of waste substances

In laboratory experiments, usually the toxicity of single compounds is tested. The situation is quite different concerning wastewaters from various industries, agriculture, hospitals, or mines. Here, the wastewater usually consists of a number of toxic compounds. In these cases, it is very difficult to employ the physical or chemical methods of remediation of wastewater as different techniques and strategies need to be utilized for removal of each of the components of the waste mixture, which is not a potential solution for waste clean-up. However, phytoremediation techniques can be very well applied in this process. Interaction and uptake of heavy metals, xenobiotics, nanoparticles, etc. is easily possible using different plant species because the uptake of these substances is usually unspecific. As already stated earlier, transporters for sulphate can also uptake chromate, cleaning the wastewater in this manner (Kaszycki et al., 2005). Ziegler et al. (2017) reported quite a number of practical examples for remediation of wastewaters where such complex mixtures are dominant: mine gallery water, activated sludge, treated domestic wastewater, university wastewater effluent, oxygen-treated steel plant effluent, cooking wastewater, and raw dairy wastewater. In the following, we report some recent examples of complex wastewater treatment by duckweed.

Wastewater from textile industry containing cadmium, copper, lead, chromium, and nickel was successfully remediated using *L. minor*. Acidification by acetic acid increased the heavy metal remediation efficiency by a factor of 3–4 (Farid et al., 2022). Textile wastewater containing heavy metals like Ni and Cd and several organic compounds was remediated using *S. polyrhiza*. Parihar and Malaviya (2023) showed that a consortium of several bacterial strains in association with *S. polyrhiza* played an important role in this bioremediation procedure. Intensive investigations on using duckweed for bioremediation of dairy wastewater were carried out in the last few years by the group of M.A.K. Jansen in Cork, Ireland. They studied the potential of *L. minor* in this process (Walsh et al., 2020). The ionic composition of diluted dairy wastewater was optimized; however, this did not improve the growth rate of the duckweed species. In some of the key experiments, the authors discovered that the ratio of Ca to Mg is the critical point, concluding that their molar ratio should be approximately 1 to 1.6. Stejskal et al. (2022) used *L. minor* and *L. gibba* for cleaning dairy wastewater and used the resulting biomass to feed fish, especially Eurasian Perch and Rainbow Trout. It turned out that it is absolutely necessary to pre-treat the dairy wastewater by anaerobic digestion before duckweed were cultivated on it (O'Mahoney et al., 2022). *Lemna minor* was also used as an extractive species for the uptake of ammonium

(NH$_4^+$-N). Duckweeds are able to uptake ammonium very rapidly and preferentially (see section above—on eutrophic wastewater of this chapter). But these plants are also able to uptake substantial amounts of nitrate—N and phosphate—P. Using *L. minor*, the following uptake rates were observed: for NH$_4^+$—N it was 158 mg/m^2/day, for NO$_3^-$—N it was 206 mg/m^2/day and for PO$_4^{3-}$—P it was 32 mg/m^2/day (Paolacci et al., 2021). These authors also showed that there is an optimum in the surface coverage degree; in case of *L. minor* it is at 80% surface density (Paolacci et al., 2021) and that the intensity of light has even a dramatic influence on the effect of phytoremediation (Walsh et al., 2021). Calculations using different models demonstrated that duckweeds cultivated on dairy wastewater can substitute alfalfa for animal feeding (Femeena et al., 2023).

2.6 Microbial associations in bioremediation potential of duckweeds

In nature, there is always an interaction between the biotic components which results either in positive or negative interactions. The present section deals with the positive interaction of duckweeds with microorganisms, i.e., bacteria and fungi for improving the process and efficiency of bioremediation of wastewater. Usually under laboratory conditions, sterile axenic duckweeds were investigated. Use of such gnotobiotic systems is good for investigating the one-to-one effect; however, the structure of the system is far away from what is naturally occurring. The duckweed-microbe interaction modifies the properties of duckweeds in a dramatic way. A large group of bacteria stimulate the growth of duckweed and therefore are called the Plant Growth Promoting Bacteria (PGPB) (Appenroth et al., 2016; Ishizawa et al., 2020). PGPB were isolated from unsterile duckweeds, e.g., from *L. minor*, and were used for growth promotion studies (Makino et al., 2022). Association with the bacterial strain MRB1–4 increased the growth of *L. minor* by 2.1- to 3.8-fold. These duckweed-associated bacteria not only stimulate plant growth but also support duckweed-led bioremediation (Pramanic et al., 2023). Association of *L. gibba* with nitrifying bacteria (*Acinetobacter* sp.) enhanced the removal of ammonium and total N from wastewater (Shen et al., 2019). However, it has been observed that the use of duckweed populations together with artificially associated bacteria poses some problems during bioremediation of wastewater. The artificial association between duckweed and bacteria is often not very stable under non-axenic conditions or upon introduction into nature. This is because of the fact that the native bacteria compete with the introduced ones for their association on the plant surface that is in contact with the aqueous medium. Therefore, Ishizawa et al. (2020) suggested a two-step process for the use of duckweed-microbe association for bioremediation. In the first step called the colonization step, few duckweed fronds were incubated together with the bacterial culture and thereafter in a subsequent step called the mass cultivation step the biomass necessary for remediation was produced. This two-step process was demonstrated using the bacterium, *Acinetobacter calcoaceticus* P 23. The duckweed growth promotion effect in this association has been attributed to plant growth promotion factors which were produced by bacteria and additionally by the nitrogen fixation capacity of the bacteria. Moreover, there is an additional effect of *L. minor* on bacteria elevating the nitrogen fixing activity as has been shown for *Azotobacter vinelandii* (Shuvro et al., 2023). Most often, the duckweed species, *L. minor* and *S. polyrhiza*, were used to investigate duckweed-microbe interactions (Dash et al., 2021). The following bacterial families, *Methylobacteriaceae*,

Pseudomonadaceae, Brucellaceae, Rhodobacteraceae, Acetobacteraceae, showed relatively high abundance in association with duckweed surfaces (Singh et al., 2023). While it is a standard procedure to identify the families of bacteria, Acosta et al. (2020, 2023) reported a new genomic-based method for the taxonomic identification of species or even strains.

In case of bioremediation of some of the xenobiotics, the interaction between duckweed and bacteria is essential. Radulović et al. (2020) reported the isolation of six phenol-resistant bacterial strains associated with *L. minor* and its use in removing phenol from wastewater. After 14 days of treatment, 90% of the initial concentration of phenol present in the wastewater was removed. This system was effective even at relatively high phenol concentrations in the wastewater. Bacteria associated with *S. polyrhiza* were able to even degrade 4-tert-butylphenol, a substance which is very resistant to several other methods of remediation. Interestingly, the associated bacterium, *Sphingobium fuliginis,* was able to use this phenolic compound as the sole carbon source (Ogata et al., 2013). Muerdter and LeFevre (2019) reported the degradation of the insecticides Imidacloprid and Thiacloprid using unsterile cultures of *Lemna* sp. It was observed that there was no degradation of these insecticides by the surface sterilized plant cultures alone. This demonstrated the role of duckweed associated microorganisms in the breakdown of these insecticides.

3 Biorefinery for duckweed biomass

3.1 Need and importance of alternative resources

Increasing global population and the increasing modernization of the different anthropological processes has been constantly rising the demand for the fuel available to meet all the requirements. Generating fuel has evolved to a great extent over the history. Starting from the ancient Stone Age practice of creating fire by rubbing the rocks together, mankind has progressed remarkably through continuous innovations in generating fuel from various naturally available resources. Although fossil fuels had been a huge fuel generation resource over ages, their nonrenewable nature and negative impact on the current global climate change has led the researchers to revolutionize the fuel generation industry for inventing sustainable alternatives for energy production. Technological advancements in the field have led to the development, optimization, and management of natural resources like solar power, hydro power, wind, ocean and tidal forces, geothermal energy, and so on for the purpose. One of the natural resources in this category is biomass from different organisms. Although used for ages as fuel, e.g., as fire wood or cow dung cakes for cooking, it has been realized, over the last few decades, that, like any other resource utilization, optimization of the process chain for cost efficient and sustainable usage of biomass as a fuel resource is essential.

3.2 Biomass as alternative energy resource

The efficient use of biomass, a renewable resource, is achieved by the conversion of biomass to different biofuels instead of direct burning of biomass for generation of heat which was the conventional usage of biomass for energy. Biomass of various origins can be converted to biofuel. The first generation of biofuels included the production of biofuels from the biomass that originated from food crop plants such as different grains, e.g., corn, sugarcane, and sugar beets. Over the years, the efficiency of production and utilization of these first generation of biofuels has increased with the effective use of modern technologies. However, the use of these plants for bioenergy production competes with its use for human nutrition and animal feeding. The second generation of biofuels are produced from biomass like wood resources which are available in large amounts. However, because of the high lignin content in these tissues, the required biotechnological processes for conversion of biomass to biofuel are energy demanding and are consequently less effective. The third generation of biofuels use the biomass generated from aquatic plants. To this third generation of bioenergy plants, belong algae. Algae grow very fast but because of their small size, harvesting is very difficult (Chen et al., 2022). Therefore, several higher aquatic macrophytes are being investigated for the use of their biomass for conversion to biofuel. This category of plants includes water hyacinth (*E. crassipes*), water lettuce (*P. stratiotes*), Brazilian waterweed (*E. densa*) or duckweed (Verma and Suthar, 2015).

3.3 Duckweed biomass for bioenergy production

Duckweeds are a sustainable bioenergy resource. As already stated in the earlier section, these floating aquatic plants include some of the fastest growing angiosperms. The growth of these plants does not impose pressure on the environment; on the contrary, duckweeds take up the excessive nutrients present in wastewater (see the section above), do not produce any foul odor in the growth system and increase their biomass to a great extent at a fast pace which can be harvested easily from the surface of the aquatic system. Significantly, the production of duckweed biomass does not compete with arable lands for its growth. Moreover, the wastelands and unused barren lands can be used for setting up the pond systems for duckweed growth and biomass production. Duckweed biomass is also advantageous in terms of its conversion to biofuel owing to its low lignin content.

The use of duckweed biomass for biofuel conversion especially depends on its biochemical composition. The major biochemical components of duckweed biomass that are of interest for conversion to biofuel are starch, lipids, proteins, and fibers. Starch can be easily converted to produce bio alcohol. Lipids are mainly suitable to produce biodiesel, and fiber-containing biomass can be best used as solid fuel. The characteristics of the duckweed biomass produced depends largely on the conditions in which the duckweeds were grown or cultivated. The biomass can be characteristically starch-rich or can be rich in proteins. When duckweeds grow under optimal growth conditions with the required nutrients, light intensity and ideal temperature, the biomass is rich in proteins (Appenroth et al., 2017; Table 2), and when these plants are exposed to any kind of stress conditions like lack of nutrients or salinity, the plants accumulate high amounts of starch and the biomass subsequently produced is starch-rich (Sree et al., 2015a; Appenroth et al., 2021; Table 3). Although duckweeds can contain high amounts of protein when grown under optimal growth conditions, protein-rich biomass is not recommended for use in production of bio-energy, instead it should be used for the purpose of food and feed. The content of lipids and fiber is rather low which makes the duckweed biomass not very suitable for

TABLE 2 Composition of *Wolffiella hyalina* grown under optimal cultivation conditions.

Component	Mean ± SD
Dry weight (DW) (%)	7.66 ± 0.71
Protein (% DW)	35.2 ± 0.9
Total fat (% DW)	6.7 ± 1.29
Starch (% DW)	6.0 ± 0.7

Calculated from Appenroth, K.-J., Sree, K.S., Böhm, V., Hammann, S., Vetter, W., Leiterer, M., Jahreis, G., 2017 Nutritional value of duckweeds (Lemnaceae) as human food. Food Chem. 217, 266–273.

TABLE 3 Starch accumulation in duckweeds under varying growth conditions.

S. No.	Growth condition	Duckweed species (Clone no.)	Starch content (% DW)	Duration (days)	Reference
1	Salt stress (150 mM NaCl)	*Lemna minor* (9441)	52.3 ± 0.9	7 to 14	Sree et al. (2015a)
2		*Spirodela intermedia* (7450)	48.2 ± 0.2	7	Sree et al. (2015a)
3	Phosphate limitation	*L. minor* (9441)	46.3 ± 0.8	14	Sree and Appenroth (2022)
4		*S. intermedia* (7797)	40.0 ± 1.1	14	Sree and Appenroth (2022)
5	Nitrate limitation	*S. intermedia* (7797)	36.8 ± 0.6	14	Sree and Appenroth (2022)
6		*Landoltia punctata* (9637)	37.2 ± 0.5	14	Sree and Appenroth (2022)
7	Sulphate limitation	*Wolffiella hyalina* (9525)	29.1 ± 0.9	14	Sree and Appenroth (2022)
8		*S. intermedia* (7797)	18.9 ± 0.4	14	Sree and Appenroth (2022)
9	Heavy metal stress (100 μM Co^{2+})	*L. minor* (9441)	49.8 ± 1.7	4	Sree et al. (2015b)

biodiesel production but to be used as solid fuel. Therefore, the duckweed biomass should be mainly used for production of bio-alcohols through bio-conversion of starch.

The content of starch in duckweeds depends on the one side on the species and clones of duckweed being used but on the other side it depends to a very high degree on the cultivation conditions, especially the nutrient composition of the aqueous medium. The most important and well investigated players for induction of starch in duckweeds are nutrient deficiencies in the growth medium. Of special mention are low concentrations of nitrogen, in the form of nitrate or ammonium, and phosphate which can induce the accumulation of high amount of starch, usually between 40% and 50% of dry weight in the whole frond body (Sree and

Appenroth, 2022). Besides limitation of nutrients in the cultivation medium, several other molecules, growth-retardant chemicals, can also be used as inducers of starch accumulation in duckweeds. These include phytohormones especially abscisic acid, heavy metals, and salt (Appenroth et al., 2021). Combining these factors should result in even higher starch content, perhaps exceeding 50% on dry weight basis. The exposure of duckweeds to heavy metals is not recommended for large-scale applications because of the resulting threat of pollution to the environment. However, the biomass obtained from remediation of a heavy metal-contaminated wastewater using duckweeds can be tested for the starch content and can be suitably processed further, so as to make an economic gain from the bioremediation process. Application of phytohormones on duckweeds is limited to experiments in a small scale in the laboratory. So, for large-scale practical applications and for efficient production of starch-rich biomass at large amounts, exposure of duckweeds to nutrient stress and salt stress are the most suitable technologies. As already described in the previous section, duckweeds can be well used for cleaning different kinds of wastewater. Wastewater is normally not an optimal nutrient solution for growth of aquatic plants, having either an excess of some nutrients, e.g., Ammonium (Zhou et al., 2022) or lack of other components, e.g., nitrates and phosphates. Therefore, it can be suggested that duckweed biomass produced from bioremediation of such nonoptimal wastewater can accumulate starch and therefore are suitable for use as raw material for bioethanol production.

As one can imagine, exposure of duckweeds to stressors as mentioned above has an impact on the growth of the plant itself, thereby impacting the final output from the bioethanol production process. Hence, for increasing the efficiency of the system, a two-step process has been suggested (Sree and Appenroth, 2022). The first step includes the production of biomass under optimal plant growth conditions wherein there would be an increase in the plant biomass at a fast pace. The subsequent step would include the exposure of this well-grown biomass to a stress condition for 2 weeks or less which will accumulate starch in the plants, thereby producing high amounts of duckweed biomass that is rich in starch. Under such growth-restricting conditions, starch accumulation in duckweeds takes place either in the whole plant tissue or in specialized resting organs like turions, depending on the species of duckweed in use. Some of the species that characteristically produce turions are *S. polyrhiza* and *L. turionifera*. Turions being resting fronds, accumulate huge amounts of starch (up to 70% dry weight), higher than that compared to the whole frond accumulation scenario as described above. However, the yield and end-product efficiency of turion usage for generation of starch rich biomass is not cost effective. The turion production is induced under similar conditions as starch accumulation in the fronds, e.g., phosphate limitation and a specialized organ majorly accumulates starch which is budded off from the vegetative pouch of the frond. These turions have less aerenchyma tissue. And, their density increases because of the high starch deposits. Together, this leads to the sinking of the turions after being detached from the floating fronds. So, harvesting these starch-rich turions becomes a labor-intensive process. Moreover, the majority of the biomass that is afloat cannot be used in combination with turion biomass. Hence, instead of using turions as the starch-rich raw material, we suggest the application of the two-step process as described above for the production of starch-rich biomass wherein the biomass has up to 50% dry weight starch accumulated in its whole plant body. So, the total biomass afloat on the water surface can be harvested and used as a starting material. This process is feasible and less labor-intensive, making the production chain cost-efficient.

In the biorefinery of starch-rich duckweed biomass for production of bioalcohols, the first step consists of decomposition of the high molecular weight carbohydrates mainly to glucose. This can be done by enzymatic hydrolysis as described by Pagliuso et al. (2020). These authors described the use of an enzyme cocktail, Cellic Ctec2 (Novozymes), for this process which was very effective in producing glucose. The reason for this high efficiency was assumed to be the very low lignin content in duckweeds. Alternatively, Rana et al. (2021) investigated the effect of acid hydrolysis using 0.1% sulfuric acid. The result was that when compared with enzymatic degradation (See also Su et al., 2014), a higher yield was found after acid hydrolysis.

The next step in this biorefinery process is the transformation of the low molecular weight carbohydrates to alcohols, mainly ethanol. This is usually done by using the yeast *Saccharomyces cerevisiae* (Cui and Cheng, 2015). However, it is also possible to produce higher alcohols like butanol which is even better suitable for energy production than ethanol. For that purpose, bacterial strains like those belonging to *Clostridium acetobutylicum* or even the genetically modified *Escherichia coli* have been employed (Cui and Cheng, 2015; Su et al., 2014).

In the past few years, the process of producing energy from duckweeds was made more efficient by using a combinatorial strategy wherein a combination of different methods was employed. Kaur et al. (2019) combined several methods of energy production using duckweed biomass such as, hydrothermal treatment, anaerobic digestion, and ethanol fermentation. Although the production of methane in combination with ethanol fermentation is not very effective using duckweed biomass, the production of this biogas provides an additional way to produce energy. In another procedure, mainly bioethanol was produced but the waste obtained from this process was used for biogas production (Patel and Bhatt, 2021). Calicioglu et al. (2019) optimized the sequential bioprocesses like ethanol fermentation, acidogenic digestion and methanogenic digestion. Duckweeds were pre-treated either by drying or by liquid hot water pre-treatment or by enzymatic saccharification. The combination of these sequential bioprocesses enhanced essentially the efficiency of bioethanol production from duckweeds. More recently, Djandja et al. (2021) reported pyrolysis of duckweed biomass at high temperature and by hydrothermal liquefaction and the process parameters were also optimized (Chen et al., 2020a,b).

Biogas which majorly contains methane as an energy product, is another possible alternative for duckweed biomass biorefinery. However, there are only a few reports in this direction. There are two main reasons. The one is low methane production rate from duckweed biomass and the second is decrease in the efficiency of the process because of the acidification of organic mixture during anaerobic digestion. There are two ways to circumvent these disadvantages. Mixing duckweed biomass with other waste residues e.g., livestock manure and co-digestion prevents the decrease of the pH and increases methane production (Chen et al., 2022). An alternative way is to use the residual biomass of duckweeds or by-products generated during the production of ethanol for the production of biogas. This alternative procedure was carried out by Chusov et al. (2022) using *L. minor*. However, surprising and contrasting results were presented by Rana et al. (2021). These authors compared the efficiency of production of ethanol to that of biogas from duckweed biomass. The bio-ethanol production followed acidic conversion of starch to glucose, followed by a fermentation step. The production of bio methane was carried out only by anaerobic digestion. They came to the conclusion that duckweed has more potential for biogas production than that of ethanol.

On the whole, there has been an intensive progress in the field of duckweed-based bioenergy production. However, it is far away from large-scale applications. Process optimization at large scale to improve the efficiency of the process and to make the biorefinery more cost-efficient will be the key components of future research in this field.

4 Circular economy of duckweeds

In view of the increasing world population and the increasing demand for energy, the period of exploiting the resources through linear economy has to come to an end and should be replaced by circular economy. Circular economy follows a regenerative design by using the waste products from one step as a valuable starting material to a subsequent step. Using duckweeds for cleaning wastewater and using the resulting duckweed biomass to produce bio-energy is an excellent example of circular economy. Further, the clean water can be recirculated into the bio-systems (Fig. 3). In the following, we reviewed some of the most recent examples of circular economy using duckweeds as key players.

Jansen and co-workers in the south of Ireland had demonstrated this in an outdoor water treatment plant using *L. minor* and *L. gibba*. In a traditional fish farm, fresh water is taken up from upstream of a river to fill up the fish ponds and the wastewater generated in the process of fish cultivation is regularly released into the same river at a downstream location. This typical linear system of production was interrupted by using the wastewater from the fish cultivation as a nutrient source for duckweed cultivation. The nutrient extractive capability of

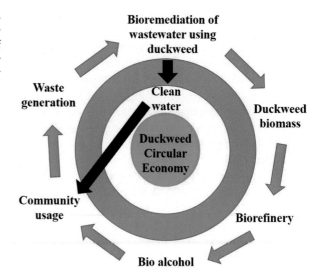

FIG. 3   Circular economy of duckweeds, illustrating the loop of flow of resources, aided by duckweed led processes. Note the formation of smaller loops as illustrated by reuse of clean water, as a result of duckweed based bioremediation of wastewater.

duckweed was set in equilibrium with the waste production rate of fish such as rainbow trout and Perch. The thus produced biomass is reportedly planned to be used in Irish food industry, for instance, by extracting the plant proteins. The water area used for this circular process is approximately $11,000 m^2$ (Stejskal et al., 2022). The group reported that the plant was constructed in 2019 and the water area was completely covered by duckweed by 2021 (Paolacci et al., 2021). The duckweeds were harvested bi-weekly from this system and it was shown that the bioremediation capacity of the duckweed culture was high enough to keep the concentration of nutrients, especially ammonia which is toxic for fish cultures, below the required limit. Duckweeds were cultivated in separate ponds other than the fish ponds but both the waterbodies were connected by water channels. The necessary energy supply for the system was provided by wind turbines. It was reported that it was possible to produce $0.49 kg/m^2/year$ of fish using this outdoor system (O'Mahoney et al., 2022).

In the second example from the same group from Cork, Ireland, they used dairy wastewater for feeding the duckweed ponds as a nutrient source. In the dairy industry, based on the milk product being produced, between 0.2 L of wastewater and 11 L of wastewater is produced per every liter of milk that is processed. This wastewater has a very complex composition consisting of ammonia, nitrate, phosphate, calcium, magnesium, potassium, iron, sodium, and chlorides together with sugars, fat, and proteins. The authors reported that it was not possible to cultivate duckweed directly on this complex mixture of wastewater. One reason was that the bacteria took over and prevented the growth of duckweeds. The cultivation of duckweed on this waste water was possible only after the pre-treatment of the wastewater by microbiological cleaning. As a final result, the minerals lost into the wastewater were restored into the duckweed biomass. This biomass can be used for different purposes like fertilization of agricultural land or use in biorefinery. This purification of dairy wastewater was not carried out outdoor but was in an indoor system, in a high-tech multitier production system with artificial illumination and under controlled conditions. The advantage of this procedure is that it can be used around the year and round the clock (Walsh et al., 2021; Jansen et al., 2023).

A similar indoor growth equipment was constructed by Petersen et al. (2022a) using LED illumination and by use of automatic regulatory points for the nutrient input after measuring the composition of the effluent. This equipment produced almost 1 Kg of fresh weight of duckweed biomass, *L. minor*, per day. The nutrient medium as well as the method of illumination was optimized in small-scale experiments (Petersen et al., 2021, 2022b). Petersen et al. (2021) found that the optimal growth rates and protein content of two duckweed species *L. minor* and *W. hyalina* were obtained with 75% nitrate N and 25% ammonium N in the nutrient medium. The total N concentration was 15 mM in the medium. This equipment was not primarily constructed for dealing with wastewater treatment but was to produce duckweed protein rich biomass for animal feeding (Demann et al., 2023). However, it can also be used for phytoremediation.

Another example of circular economy was described by Azwar et al. (2022) in connection with shell fish aquaculture. The wastewater of the shell fish aquaculture was used as a nutrient substrate for cultivating several aquatic plants including duckweed. Instead of considering the resulting biomass of duckweed as organic waste and disposing it in landfills, the authors used it for refinery for different methods of pyrolysis and thermochemical conversion.

5 Conclusion

Significant progress has been reported in the recent decade concerning the cultivation of different duckweed species on different kinds of wastewater. This wastewater could be contaminated by high concentrations of common minerals (eutrophication), sodium chloride, heavy metals, microplastics, xenobiotics, or complex mixtures from industrial production (e.g., from diary industry). As a consequence of the suboptimal cultivation of duckweeds in wastewater, often plants with high starch content were obtained. However, most of the results were reported from small-scale, laboratory experiments. A bottleneck for further progress is the cultivation on large, industrial scale producing tons of biomass instead of gram or kilogram, for biorefinery applications. This will be the challenge for the next decade and should become the main focus of applied research.

The result of cultivation of duckweed on wastewater is not only clean water, which can be reused but is also the production of starch-rich biomass. This biomass can be used for saccharification and fermentation to produce bio-alcohols, which is probably the most effective way of using duckweeds as energy plants. At the moment, effective methods are not yet clearly developed, e.g., acid hydrolysis of starch versus enzymatic degradation. This, however, will become clear in the next few years. Following this path, it can be expected that wastewater cleaning and production of bio alcohol using duckweed will become an outstanding example for circular economy.

References

Acosta, K., Appenroth, K.J., Borisjuk, L., Edelman, M., Heinig, U., Jansen, M.A.K., Oyama, T., Pasaribu, B., Schubert, I., Sorrels, S., Sree, K.S., Xu, S., Michael, T., Lam, E., 2021. Return of the Lemnaceae: duckweed as a model plant system in the genomics and postgenomics era. Plant Cell 33, 3207–3234. https://doi.org/10.1093/plcell/koab189.

Acosta, K., Sorrels, S., Chrisler, W., Huang, W., Gilbert, S., Brinkman, T., Michael, T.P., Lebeis, S.L., Lam, E., 2023. Optimization of molecular methods for detecting duckweed-associated bacteria. Plants 12, 872. https://doi.org/10.3390/plants12040872.

Acosta, K., Xu, J., Gilbert, S., Denison, E., Brinkman, T., Lebeis, S., et al., 2020. Duckweed hosts a taxonomically similar bacterial assemblage as the terrestrial leaf microbiome. PLoS One 15, e0228560. https://doi.org/10.1371/journal.pone.0228560.

Almroth, C.B., Eggert, H., 2019. Marine plastic pollution: sources, impacts, and policy issues. Rev. Environ. Econ. Policy 13, 317–326. https://doi.org/10.1093/reep/rez012.

Ansari, A.A., Naeem, M., Gill, S.S., AlZuaibr, F.M., 2020. Phytoremediation of contaminated waters: an eco-friendly technology based on aquatic macrophytes application. Egypt. J. Aquat. Res. 46, 371–376. https://doi.org/10.1016/j.ejar.2020.03.002.

Appenroth, K.-J., 2010. What are "heavy metals" in plant science? Acta Physiol. Plant. 32, 615–619. https://doi.org/10.1007/s11738-009-0455-4.

Appenroth, K.-J., Luther, A., Jetschke, G., Gabrys, H., 2008. Modification of chromate toxicity by sulphate in duckweeds (Lemnaceae). Aquat. Toxicol. 89, 167–171. https://doi.org/10.1016/j.aquatox.2008.06.012.

Appenroth, K.-J., Sree, K.S., Bog, M., Ecker, J., Seeliger, C., Böhm, V., Lorkowski, S., Sommer, K., Vetter, W., Tolzin-Banasch, K., Kirmse, R., Leiterer, M., Dawczynski, C., Liebisch, G., Jahreis, G., 2018. Nutritional value of the duckweed species of the genus *Wolffia* (Lemnaceae) as human food. Front. Chem. 6, 483. https://doi.org/10.3389/fchem.2018.00483.

Appenroth, K.-J., Sree, K.S., Böhm, V., Hammann, S., Vetter, W., Leiterer, M., Jahreis, G., 2017. Nutritional value of duckweeds (Lemnaceae) as human food. Food Chem. 217, 266–273.

Appenroth, K.-J., Stöckel, J., Srivastava, A., Strasser, R.J., 2001. Multiple effects of chromate on the photosynthetic apparatus of *Spirodela polyrhiza* as probed by OJIP chlorophyll a fluorescence measurements. Environ. Pollut. 115, 49–64. https://doi.org/10.1016/S0269-7491(01)00091-4.

Appenroth, K.-J., Ziegler, P., Sree, K.S., 2016. Duckweed as a model organism for investigating plant-microbe interactions in an aquatic environment and its applications. Endocytobiosis Cell Res. 27, 94–106. https://doi.org/10.1016/j.chemosphere.2019.124682.

Appenroth, K.-J., Ziegler, P., Sree, K.S., 2021. Accumulation of starch in duckweeds (Lemnaceae), potential energy plants. Physiol. Mol. Biol. Plants 27, 2621–2633. https://doi.org/10.1007/s12298-021-01100-4.

Azeem, I., Adeel, M., Ahmad, M.A., Shakoor, N., Jiangcuo, G.D., Azeem, K., Ishfaq, M., Shakoor, A., Ayaz, M., Xu, M., et al., 2021. Uptake and accumulation of nano/microplastics in plants: a critical review. Nano 11, 2935. https://doi.org/10.3390/nano11112935.

Azwar, E., Mahari, W.A.W., Rastegari, H., Tabatabaei, M., Peng, W., Tsang, Y.F., Park, Y.-K., Chen, W.H., Lam, S.S., 2022. Progress in thermochemical conversion of aquatic weeds in shellfish aquaculture for biofuel generation: technical and economic perspectives. Bioresour. Technol. 344 (Part A), 126202. https://doi.org/10.1016/j.biortech.2021.126202.

Benshalom, M., Shandalov, S., Brenner, A., Oron, G., 2014. The effect of aeration and effluent recycling on domestic wastewater treatment in a pilot-plant system of duckweed ponds. Water Sci. Technol. 69, 350–357. https://doi.org/10.2166/wst.2013.720.

Bog, M., Appenroth, K.-J., Sree, K.S., 2019. Duckweed (Lemnaceae): its molecular taxonomy. Front. Sustain. Food Syst. 3, 117. https://doi.org/10.3389/fsufs.2019.00117.

Bog, M., Appenroth, K.-J., Sree, K.S., 2020. Key to the determination of taxa of Lemnaceae: an update. Nord. J. Bot. 38, e02658. https://doi.org/10.1111/njb.02658.

Botta, P.E., 1841. Relations d'un Voyage Dans l'Yémen, Entrepris en 1837 Pour le Museum D'Histoire Naturelle de Paris. B. Duprat, Paris, France, p. 148. Available online: https://gallica.bnf.fr/ark:/12148/bpt6k58013253/f21.item.texteImage.zoom. (Accessed 8 May 2023).

Calicioglu, O., Richard, T.L., Brennan, R.A., 2019. Anaerobic bioprocessing of wastewater-derived duckweed: maximizing product yields in a biorefinery value cascade. Bioresour. Technol. 289, 121716. https://doi.org/10.1016/j.biortech.2019.121716.

Cascone, A., Forni, C., Migliore, L., 2004. Flumequine uptake and the aquatic duckweed, *Lemna minor* L. Water Air Soil Pollut. 156, 241–249. https://doi.org/10.1023/B:WATE.0000036816.15999.53.

Ceschin, S., Mariani, F., Di Lernia, D., Venditti, I., Pelella, E., Iannelli, M.A., 2023. Effects of microplastic contamination on the aquatic plant *Lemna minuta* (least duckweed). Plants 12, 207. https://doi.org/10.3390/plants12010207.

Ceschin, S., Sgambato, V., Ellwood, N.T.W., Zuccarello, V., 2019. Phytoremediation performance of *Lemna* communities in a constructed wetland system for wastewater treatment. Environ. Exp. Bot. 162, 67–71. https://doi.org/10.1016/j.envexpbot.2019.02.007.

Chen, G., Yu, Y., Li, W., Yan, B., Zhao, K., Dong, X., Cheng, Z., Lin, F., Li, L., Zhao, H., Fang, Y., 2020a. Effects of reaction conditions on products and elements distribution via hydrothermal liquefaction of duckweed for wastewater treatment. Bioresour. Technol. 317, 124033. https://doi.org/10.1016/j.biortech.2020.124033.

Chen, D., Zhang, H., Wang, Q., Shao, M., Li, X., Chen, D., Zeng, R., Song, Y., 2020b. Intraspecific variations in cadmium tolerance and phytoaccumulation in giant duckweed (*Spirodela polyrhiza*). J. Hazard. Mater. 395, 122672. https://doi.org/10.1016/j.jhazmat.2020.122672.

Chen, G., Zhao, K., Li, W., Yan, B., Yu, Y., Li, J., Zhang, Y., Xia, S., Cheng, Z., Lin, F., Li, L., Zhao, H., Fang, Y., 2022. A review on bioenergy production from duckweed. Biomass Bioenergy 161, 106468. https://doi.org/10.1016/j.biombioe.2022.106468.

Chusov, A., Maslikov, V., Badenko, V., Zhazhkov, V., Molodtsov, D., Pavlushkina, Y., 2022. Biogas potential assessment of the composite mixture from duckweed biomass. Sustainability 14, 351. https://doi.org/10.3390/su14010351.

Crini, G., Lichtfouse, E., 2019. Advantages and disadvantages of techniques used for wastewater treatment. Environ. Chem. Lett. 17, 145–155. https://doi.org/10.1007/s10311-018-0785-9.

Cui, W., Cheng, J.J., 2015. Growing duckweed for biofuel production: a review. Plant Biol. 17 (Suppl 1), 16–23. https://doi.org/10.1111/plb.12216.

Dash, T., Appenroth, K.-J., Sree, K.S., 2021. Microbial symbionts of aquatic plants. In: Shrivastava, N., Mahajan, S., Varma, A. (Eds.), Symbiotic Soil Microorganisms. Biology and Applications. Springer Nature, Cham, Switzerland, pp. 229–240.

Demann, J., Petersen, F., Dusel, G., Bog, M., Devlamynck, R., Ulbrich, A., Olfs, H.-W., Westendarp, H., 2023. Nutritional value of duckweed as protein feed for broiler chickens—digestibility of crude protein, amino acids and phosphorus. Animals 13, 130. https://doi.org/10.3390/ani13010130.

Djandja, O.S., Yin, L.X., Wang, Z.C., Guo, Y., Zhang, X.X., Duan, P.G., 2021. Progress in thermochemical conversion of duckweed and upgrading of the bio-oil: a critical review. Sci. Total Environ. 769, 144660. https://doi.org/10.1016/j.scitotenv.2020.144660.

Edelman, M., Appenroth, K.-J., Sree, K.S., Oyama, T., 2022. Ethnobotanical history: duckweeds in different civilizations. Plants 11, 2124. https://doi.org/10.3390/plants11162124.

El-Sheekh, M., Abdel-Daim, M.M., Okba, M., Gharib, S., Soliman, A., El-Kassas, H., 2021. Green technology for bioremediation of the eutrophication phenomenon in aquatic ecosystems: a review. Afr. J. Aquat. Sci. 46 (3), 274–292. https://doi.org/10.2989/16085914.2020.1860892.

Farid, M., Sajjad, A., Asam, Z.U.Z., Zubair, M., Rizwan, M., Abbas, M., Farid, S., Ali, S., Alharby, H.F., Alzahrani, Y.M., Alabdallah, N.M., 2022. Phytoremediation of contaminated industrial wastewater by duckweed (*Lemna minor* L.): growth and physiological response under acetic acid application. Chemosphere 304. https://doi.org/10.1016/j.chemosphere.2022.135262.

Femeena, P.V., Costello, C., Brennan, R.A., 2023. Spatial optimization of nutrient recovery from dairy farms to support economically viable load reductions in the Chesapeake Bay watershed. Agric. Syst. 207, 103640. https://doi.org/10.1016/j.agsy.2023.103640.

Gao, J.P., Garrison, A.W., Hoehamer, C., Mazur, C.S., Wolfe, N.L., 2000. Uptake and phytotransformation of organophosphorus pesticides by axenically cultivated aquatic plants. J. Agric. Food Chem. 48, 6114–6120. https://doi.org/10.1021/jf9904968.

Habaki, H., Thyagarajan, N., Li, Z.H., Wang, S.Y., Zhang, J., Egashira, R., 2023. Removal of antibiotics from pharmaceutical wastewater using *Lemna aoukikusa* (duckweed). Sep. Sci. Technol. 58, 1491–1501. https://doi.org/10.1080/01496395.2023.2195544.

Hu, X., Wei, X., Ling, J., Chen, J., 2021. Cobalt: an essential micronutrient for plant growth? Front. Plant Sci. 12, 768523. https://doi.org/10.3389/fpls.2021.768523.

Ishizawa, H., Ogata, Y., Hachiya, Y., Tokura, K.-I., Kuroda, M., Inoue, D., Toyama, T., Tanaka, Y., Mori, K., Morikawa, M., Ike, M., 2020. Enhanced biomass production and nutrient removal capacity of duckweed via two-step cultivation process with a plant growth-promoting bacterium, *Acinetobacter calcoaceticus* P23. Chemosphere 238, 124682. https://doi.org/10.1016/j.chemosphere.2019.124682.

Jansen, M.A.K., Paolacci, S., Stejskal, V., Walsh, É., Kühnhold, H., Coughlan, N.E., 2023. Duckweed research and applications on the emerald isle. Duckweed Forum 11, 2, 7.

Kaminski, A., Bober, B., Chrapusta, E., Bialczyk, J., 2014. Phytoremediation of anatoxin-a by aquatic macrophyte *Lemna trisulca* L. Chemosphere 112, 305–310. https://doi.org/10.1016/j.chemosphere.2014.04.064.

Kaszycki, P., Gabrys, H., Appenroth, K.J., Jaglarz, A., Sedziwy, S., Walczak, T., Koloczek, H., 2005. Exogenously applied sulphate as a tool to investigate transport and reduction of chromate in the duckweed *Spirodela polyrhiza*. Plant Cell Environ. 28, 260–268. https://doi.org/10.1111/j.1365-3040.2004.01276.x.

Kaur, M., Kumar, M., Singh, D., Sachdeva, S., Puri, S.K., 2019. A sustainable biorefinery for efficient conversion of aquatic weeds into bioethanol and biomethane. Energy Convers. Manag. 187, 133–147. https://doi.org/10.1016/j.enconman.2019.03.018.

Landolt, E., 1986. The family of *Lemnaceae*—A monographic study. In: Biosystematic Investigations in the Family of Duckweeds (*Lemnaceae*). vol. 1. Veröffentlichung des Geobotanischen Instistuts der Eidgenössisch Technischen Hochscule, Stiftung Rübel Zürich.

Landolt, E., Kandeler, R., 1987. The family of Lemnaceae—A monographic study. In: Biosystematic Investigations in the Family of Duckweeds (*Lemnaceae*). vol. 2. Veröffentlichung des Geobotanischen Instistuts der Eidgenössisch Technischen Hochscule, Stiftung Rübel Zürich.

Mahey, S., Kumar, R., Sharma, M., et al., 2020. A critical review on toxicity of cobalt and its bioremediation strategies. SN Appl. Sci. 2, 1279. https://doi.org/10.1007/s42452-020-3020-9.

Makino, A., Nakai, R., Yoneda, Y., Toyama, T., Tanaka, Y., Meng, X.Y., Mori, K., Ike, M., Morikawa, M., Kamagata, Y., Tamaki, H., 2022. Isolation of aquatic plant growth-promoting bacteria for the floating plant duckweed (*Lemna minor*). Chemosphere 238, 124682. https://doi.org/10.1016/j.chemosphere.2019.124682.

Maldonado, I., Quispe, A.P.V., Chacca, D.M., Vilca, F.Z., 2022. Optimization of the elimination of antibiotics by *Lemna gibba* and *Azolla filiculoides* using response surface methodology (RSM). Front. Environ. Sci. 10, 940971. https://doi.org/10.3389/fenvs.2022.940971.

Masiá, P., Sol, D., Ardura, A., Laca, A., Borrell, Y.J., Dopico, E., Laca, A., Machado-Schiaffino, G., Díaz, M., Garcia-Vazquez, E., 2020. Bioremediation as a promising strategy for microplastics removal in wastewater treatment plants. Mar. Pollut. Bull. 156, 111252. https://doi.org/10.1016/j.marpolbul.2020.111252.

McCutcheon, S.C., Rock, S.A., 2001. Phytoremediation: state of the science conference and other developments. Int. J. Phytoremed. 3, 1–11. https://doi.org/10.1080/15226510108500047.

Muerdter, C.P., LeFevre, G.H., 2019. Synergistic *Lemna* duckweed and microbial transformation of imidacloprid and thiacloprid neonicotinoids. Environ. Sci. Technol. Lett. 6, 761–767. https://doi.org/10.1021/acs.estlett.9b00638.

Mustafa, H.M., Hayder, G., 2021. Recent studies on applications of aquatic weed plants in phytoremediation of wastewater; a review article. Ain Shams Eng. J. 12, 355–365. http://creativecommons.org/licenses/by-nc-nd/4.0/.

Naumann, B., Eberius, M., Appenroth, K.-J., 2007. Growth rate based dose-response relationships and EC-values of ten heavy metals using the duckweed growth inhibition test (ISO 20079) with *Lemna minor* L. clone St. J. Plant Physiol. 164, 1656–1664. https://doi.org/10.1016/j.jplph.2006.10.011.

O'Mahoney, R., Coughlan, N.E., Walsh, É., Jansen, M.A.K., 2022. Cultivation of *Lemna minor* on industry-derived, anaerobically digested, dairy processing wastewater. Plants 11, 3027. https://doi.org/10.3390/plants11223027.

Ogata, Y., Toyama, T., Wang, N.Y.X., Sei, K., Ike, M., 2013. Occurrence of 4-*tert*-butylphenol (4-tert-BP) biodegradation in aquatic sample caused by the presence of *Spirodela polyrhiza* and isolation of a 4-tert-BP utilizing bacterium. Biodegradation 24, 191–202. https://doi.org/10.1007/s10532-012-9570-9.

Pagliuso, D., Gradis, A., Lam, E., Buckeridge, M.S., 2020. High saccharification, low lignin, and high sustainability potential make duckweeds adequate as bioenergy feedstocks. Bioenergy Res. 14, 1082–1092. https://doi.org/10.1007/s12155-020-10211-x.

Pang, Y.L., Quek, Y.Y., Lim, S., Shuit, S.H., 2023. Review on phytoremediation potential of floating aquatic plants for heavy metals: a promising approach. Sustainability 15, 1290. https://doi.org/10.3390/su15021290.

Paolacci, S., Stejskal, V., Jansen, M.A.K., 2021. Estimation of the potential of *Lemna minor* for effluent remediation in integrated multi-trophic aquaculture using newly developed synthetic aquaculture wastewater. Aquac. Int. 29, 2101–2118. https://doi.org/10.1007/s10530-021-02530-7.

Parihar, A., Malaviya, P., 2023. Textile wastewater phytoremediation using *Spirodela polyrhiza* (L.) Schleid. Assisted by novel bacterial consortium in a two-step remediation system. Environ. Res. 221, 115307. https://doi.org/10.1016/j.envres.2023.115307.

Patel, V.R., Bhatt, N., 2021. Aquatic weed *Spirodela polyrhiza*, a potential source for energy generation and other commodity chemicals production. Renew. Energy 173, 455–465. https://doi.org/10.1016/j.renene.2021.03.054.

Petersen, F., Demann, J., Restemeyer, D., Olfs, H.-W., Westendarp, H., Appenroth, K.-J., Ulbrich, A., 2022a. Influence of light intensity and spectrum on duckweed growth and proteins in a small-scale, re-circulating indoor vertical farm. Plants 11, 1010. https://doi.org/10.3390/plants11081010.

Petersen, F., Demann, J., Restemeyer, D., Ulbrich, A., Olfs, H.-W., Westendarp, H., Appenroth, K.-J., 2021. Influence of the nitrate-N to ammonium-N ratio on relative growth rate and crude protein content in the duckweeds *Lemna minor* and *Wolffiella hyalina*. Plants 10, 1741. https://doi.org/10.3390/plants10081741.

Petersen, F., Demann, J., von Salzen, J., Olfs, H.-W., Westendarp, H., Wolf, P., Appenroth, K.-J., Ulbrich, A., 2022b. Re-circulating indoor vertical farm: technicalities of an automated duckweed biomass production system and protein feed product quality evaluation. J. Clean. Prod. 380, 134894. https://doi.org/10.1016/j.jclepro.2022.134894.

Poirier, Y., Jaskolowski, A., Clúa, J., 2022. Phosphate acquisition and metabolism in plants. Curr. Biol. 32, R589–R683 (WOS:000822390600028).

Polinska, W., Piotrowska-Niczyporuk, A., Karpinska, J., Struk-Sokolowska, J., Kotowska, U., 2022. Mechanisms, toxicity and optimal conditions—research on the removal of benzotriazoles from water using *Wolffia arrhiza*. Sci. Total Environ. 847, 157571. https://doi.org/10.1016/j.scitotenv.2022.157571.

Pramanic, A., Sharma, S., Dhanorkar, M., et al., 2023. Endophytic microbiota of floating aquatic plants: recent developments and environmental prospects. World J. Microbiol. Biotechnol. 39, 96. https://doi.org/10.1007/s11274-023-03543-1.

Radulović, O., Stanković, S., Uzelac, B., Tadić, V., Trifunović-Momčilov, M., Lozo, J., Marković, M., 2020. Phenol removal capacity of the common duckweed (*Lemna minor* L.) and six phenol-resistant bacterial strains from its rhizosphere: in vitro evaluation at high phenol concentrations. Plants 9, 599. https://doi.org/10.3390/plants9050599.

Rana, Q.U.A., Khan, M.A.N., Shiekh, Z., et al., 2021. Production of bioethanol and biogas from *Spirodela polyrhiza* in a biorefinery concept and output energy analysis of the process. Biomass Convers. Bioref. https://doi.org/10.1007/s13399-021-02066-9.

Rozman, U., Kalčíková, G., 2022. The response of duckweed *Lemna minor* to microplastics and its potential use as a bioindicator of microplastic pollution. Plants 11, 2953. https://doi.org/10.3390/plants11212953.

III. Ornamentals and tree crops for contaminated substrates

Rozman, U., Kokalj, A.J., Dolar, A., Drobne, D., Kalčíková, G., 2022. Long-term interactions between microplastics and floating macrophyte *Lemna minor*: the potential for phytoremediation of microplastics in the aquatic environment. Sci. Total Environ. 831, 154866. https://doi.org/10.1016/j.scitotenv.2022.154866.

Shen, M., Yin, Z., Xia, D., Zhao, Q., Kang, Y., 2019. Combination of heterotrophic nitrifying bacterium and duckweed (*Lemna gibba* L.) enhances ammonium nitrogen removal efficiency in aquaculture water via mutual growth promotion. J. Gen. Appl. Microbiol. 65, 151–160. https://doi.org/10.2323/jgam.2018.08.002.

Shuvro, S.K., Jog, R., Morikawa, M., 2023. Diazotrophic bacterium *Azotobacter vinelandii* as a mutualistic growth promoter of an aquatic plant: *Lemna minor*. Plant Growth Regul. 100, 171–180.

Singh, P., Itankar, N., Patil, Y., 2021. Biomanagement of hexavalent chromium: current trends and promising perspectives. J. Environ. Manag. 279, 111547. https://doi.org/10.1016/j.jenvman.2020.111547.

Singh, P., Jani, K., Sharma, S., et al., 2023. Microbial population dynamics in Lemnaceae (duckweed)-based wastewater treatment system. Curr. Microbiol. 80, 56. https://doi.org/10.1007/s00284-022-03149-0.

Sitarska, M., Traczewska, T., Filarowska, W., Hołtra, A., Zamorska-Wojdyła, D., Hanus-Lorenz, B., 2023. Phytoremediation of mercury from water by monocultures and mixed cultures pleustophytes. J. Water Process Eng. 52, 103529. https://doi.org/10.1016/j.jwpe.2023.103529.

Sree, K.S., Adelmann, K., Garcia, C., Lam, E., Appenroth, K.-J., 2015a. Natural variance in salt tolerance and induction of starch accumulation in duckweeds. Planta 241, 1395–1404. https://doi.org/10.1007/s00425-015-2264-x.

Sree, K.S., Appenroth, K.J., 2022. Starch accumulation in duckweeds (Lemnaceae) induced by nutrient deficiency. Emirates J. Food Agric. 34, 3. https://doi.org/10.9755/ejfa.2022.v34.i3.2846.

Sree, K.S., Keresztes, Á., Mueller-Roeber, B., Brandt, R., Eberius, M., Fischer, W., Appenroth, K.-J., 2015b. Phytotoxicity of cobalt ions on the duckweed *Lemna minor*—morphology, ion uptake, and starch accumulation. Chemosphere 131, 149–156. https://doi.org/10.1016/j.chemosphere.2015.03.008.

Sree, K.S., Maheshwari, S.C., Boka, K., Khurana, J., Keresztes, Á., Appenroth, K.-J., 2015c. The duckweed *Wolffia microscopica*: a unique aquatic monocot. Flora 210, 31–39. https://doi.org/10.1016/j.flora.2014.10.006.

Sree, K.S., Sudakaran, S., Appenroth, K.J., 2015d. How fast can angiosperms grow? Species and clonal diversity of growth rates in the genus *Wolffia* (Lemnaceae). Acta Physiol. Plant. 37, 204. https://doi.org/10.1007/s11738-015-1951-3.

Srivastava, D., Tiwari, M., Dutta, P., Singh, P., Chawda, K., Kumari, M., Chakrabarty, D., 2021. Chromium stress in plants: toxicity, tolerance and phytoremediation. Sustainability 13, 4629. https://doi.org/10.3390/su13094629.

Stejskal, V., Paolacci, S., Toner, D., Jansen, M.A.K., 2022. A novel multitrophic concept for the cultivation of fish and duckweed: a technical note. J. Clean. Prod. 366, 132881. https://doi.org/10.1016/j.jclepro.2022.132881.

Su, H., Zhao, Y., Jiang, J., Lu, Q., Li, Q., Luo, Y., Zhao, H., Wang, M., 2014. Use of duckweed (*Landoltia punctata*) as a fermentation substrate for the production of higher alcohols as biofuels. Energy Fuel 28, 3206–3216. https://doi.org/10.1021/ef500335h.

Sun, L., Zhao, H., Liu, J., Li, B., Chang, Y., Yao, D.A., 2021. New green model for the bioremediation and resource utilization of livestock wastewater. Int. J. Environ. Res. Public Health 18, 8634. https://doi.org/10.3390/ijerph18168634.

Thakuria, A., Singh, K.K., Dutta, A., Corton, E., Stom, D., Barbora, L., Goswami, P., 2023. Phytoremediation of toxic chemicals in aquatic environment with special emphasis on duckweed mediated approaches. Int. J. Phytoremed. https://doi.org/10.1080/15226514.2023.2188423.

Tufaner, F., 2020. Post-treatment of effluents from UASB reactor treating industrial wastewater sediment by constructed wetland. Environ. Technol. 41, 912–920. https://doi.org/10.1080/09593330.2018.1514073.

Verma, R., Suthar, S., 2015. Utility of duckweeds as source of biomass energy: a review. Bioenergy Res. 8, 1589–1597. https://doi.org/10.1007/s12155-015-9639-5.

Walsh, É., Kuehnhold, H., O'Brien, S., et al., 2021. Light intensity alters the phytoremediation potential of *Lemna minor*. Environ. Sci. Pollut. Res. 28, 16394–16407. https://doi.org/10.1007/s11356-020-11792-y.

Walsh, É., Paolacci, S., Burnell, G., Jansen, M.A.K., 2020. The importance of the calcium-to-magnesium ratio for phytoremediation of dairy industry wastewater using the aquatic plant *Lemna minor* L. Int. J. Phytoremed. 22, 694–702. https://doi.org/10.1080/15226514.2019.1707478.

Wang, W., Yang, C., Tang, X., Gu, X., Zhu, Q., Pan, K., Hu, Q., Ma, D., 2014. Effects of high ammonium levels on biomass accumulation of common duckweed *Lemna minor* L. Environ. Sci. Pollut. Res. 21, 14202–14210. https://doi.org/10.1007/s11356-014-3353-2.

Wang, X., Zhang, B., Wu, D., Hu, L., Huang, T., Gao, G., Huang, S., Wu, S., 2021. Chemical forms governing cd tolerance and detoxification in duckweed (*Landoltia punctata*). Ecotoxicol. Environ. Saf. 207, 111553. https://doi.org/10.1016/j.ecoenv.2020.111553.

Zazouli, M.A., Asghari, S., Tarrahi, R., Lisar, S.Y.S., Babanezhad, E., Dashtban, N., 2023. The potential of common duckweed (*Lemna minor*) in phytoremediation of phenanthrene and pyrene. Environ. Eng. Res. 28, 210592. https://doi.org/10.4491/eer.2021.592.

Zhou, Y., Chen, G., Peterson, A., Zha, X., Cheng, J., Li, S., Cui, D., Zhu, H., Kishchenko, O., Borisjuk, N., 2018. Biodiversity of duckweeds in eastern China and their potential for bioremediation of municipal and industrial wastewater. J. Geosci. Environ. Protect. 6, 108–116. https://doi.org/10.4236/gep.2018.63010.

Zhou, Y., Kishchenko, O., Stepanenko, A., Chen, G., Wang, W., Zhou, J., Pan, C., Borisjuk, N., 2022. The dynamics of NO_3^- and NH_4^+ uptake in duckweed are coordinated with the expression of major nitrogen assimilation genes. Plants 11, 11. https://doi.org/10.3390/plants11010011.

Zhou, Y., Stepanenko, A., Kishchenko, O., Chen, G., Wang, W., Zhou, J., Pan, C., Borisjuk, N., 2023. Duckweed for phytoremediation of polluted water. Plants 12, 589. https://doi.org/10.3390/plants12030589.

Ziegler, P., Sree, K.S., Appenroth, K.-J., 2016. Duckweeds for water remediation and toxicity testing. Toxicol. Environ. Chem. 98, 1127–1154. https://doi.org/10.1080/02772248.2015.1094701.

Ziegler, P., Sree, K.S., Appenroth, K.-J., 2017. The uses of duckweed in relation to water remediation. Desalin. Water Treat. 63, 327–342. https://doi.org/10.5004/dwt.2017.0479.

Cultivation of sweet sorghum on heavy metal-contaminated soils by phytoremediation approach for production of bioethanol

R. Pasumarthi[a], P. Srinivasa Rao[b], and S. Gopalakrishnan[a]

[a]International Crops Research Institute for the Semi-Arid Tropics (ICRISAT), Patancheru, Telangana, India [b]Western Colorado Research Centre, Fruita, CO, United States

1 Introduction

The term heavy metals (HMs) has a wide range of meanings, and there has been no consistent definition by any authoritative body such as the International Union of Pure and Applied Chemistry (IUPAC) over the past 60 years (Duffus, 2002). However, over the past two decades, this term has been used by numerous publications and legislations for indicating a group name for metals or semi-metals that cause toxicity to humans, plants, and animals and ecotoxicity. Though the imprecise term is defined by several researchers at various levels including density, atomic number, atomic weight, chemical properties, and toxicity, there is no connectivity between these properties. For instance, Koller and Saleh (2018) stated that HMs are elements having an atomic number greater than 20 and atomic density above $5 \, g \, cm^{-3}$ and exhibit properties of metals. HMs have been in use by human beings for several hundred years. Recalcitrance in environment, toxicity, and bioaccumulative nature of HMs have proven them to be environmental pollutants (Ali et al., 2019). Lead has been used in water pipes and pigments for glazing ceramics. Mercury was reported to be used as infant's teething pain killer and was also used as medicine for Syphilis by Romans. Cadmium is majorly used in rechargeable batteries, which were discarded along with household wastes (Järup, 2003). As this chapter deals with bioremediation aspects, HMs causing human and ecotoxic effects were considered further. Three kinds of HMs are of concern, including toxic metals

(Hg, Cr, Pb, Zn, Cu, Ni, Cd, As, Co, Sn, etc.), precious metals (Pd, Pt, Ag, Au, Ru, etc.), and radionuclides (U, Th, Ra, Am, etc.) (Wang and Chen, 2006).

The stability and non-degradability of metals, mean higher exposure of HMs to humans and animals result in their pollutant nature, and numerous reports are available for the related health and ecological issues (Caussy et al., 2003; Li et al., 1995). HMs such as Fe, Co, and Zn are essential nutrients for human beings; however, their consumption at higher dosage leads to toxicity (Rama Jyothi, 2021). Toxic effects of HMs may be chronic or acute, which depends on the route of transfer and reactive forms. For instance, for Cd, all forms are toxic; for Pb, organic forms are highly toxic; for As, inorganic arsenate [As (+5)] or [As (+3)] is highly toxic; for Hg and Hg (II), organomercurials, mainly methylmercury (Mudgal et al., 2010) are toxic. Toxic effects of these major HMs are periodically reviewed by many researchers (Burbacher et al., 1990; Duruibe et al., 2007; Wongsasuluk et al., 2014; Zatta, 2001). Representatives of HM-related health issues and their tolerable limits of action are summarized in Table 1. The presence of HMs in air, water, and soil at higher levels is of concern due to their carcinogenic nature. Arsenic, Ca, Ni, and hexavalent chromium are categorized as group 1 carcinogens by International Agency for Research on Cancer (IARC, 2012).

The HMs are sourced from both natural and anthropogenic activities (Chopra et al., 2009; Li et al., 2009). Parent rocks are the natural contributors, and their HM content is usually found to be low depending on the parent rock composition. Various anthropogenic activities transfer the HMs through air, water, and soil, the major transmitters of any kind of pollutants. Such routes and sources are (1) air, which has mining, smelting, and refining of fossil fuels; smoke from production units of metallic goods; and vehicular exhaust; (2) water, having domestic and industrial sewage and effluents, thermal power plants, and atmospheric fallout;

TABLE 1 Effect of HMs on humans.

HMs	Effect on human health	JECFA tolerable limits (x/kg bw/day)
As	Bronchitis, dermatitis, poisoning	2.1 µg
Cd	Renal dysfunction, bone defects, blood pressure, bronchitis, cancer	25 µg
Pb	Mental retardation in children, developmental delay, fatal infant encephalopathy, congenital paralysis, sensor neural deafness, epilepsy, acute/chronic damage of CNS, liver, kidney, and GI tract	0.025 µg
Hg	Tremors, gingivitis, psychological changes, acrodynia, spontaneous abortion, protoplasm poisoning, CNS damage	4 µg
Zn	CNS damage, corrosive effects on skin	0.3–1 mg
Cr	CNS damage, fatigue, irritability	–
Cu	Anemia, liver and kidney damage, stomach and intestinal irritation	0.5 mg

JECFA, Joint FAO/WHO Expert Committee on Food Additives.
Source: JECFA, 1982. Evaluation of Certain Food Additives and Contaminants (Twenty-Sixth Report of the Joint FAO/WHO Expert Committee on Food Additives). WHO Technical Report Series, No. 683. JECFA, 2011. Evaluation of Certain Contaminants in Food (Seventy-Second Report of the Joint FAO/WHO Expert Committee on Food Additives). WHO Technical Report Series, No. 959. JECFA, 2013. Evaluation of Certain Food Additives and Contaminants (Seventy-Seventh Report of the Joint FAO/WHO Expert Committee on Food Additives). WHO Technical Report Series, No. 983.

and (3) soil having agricultural and animal wastes, municipal and industrial sewage, coal ashes, fertilizers, discarded metal goods, and atmospheric fallout. Anthropogenic sources are the maximum contributors for metals rather than natural sources (McGowen et al., 2001; Nriagu and Pacyna, 1988). A recent study by Millward and Turner (2001) also states that anthropogenic factors are the major metal contributors as they alter the natural biogeo-chemical cycles.

About 600,000 HMs-contaminated brownfield sites are present in America that need to be cleaned up (Mahar et al., 2016a, b). In addition, 100,000 ha of cropland, 55,000 ha of pasture, and 50,000 ha of forest have been lost by HM contamination and demands for reclamation (McGrath et al., 2001; Ragnarsdottir and Hawkins, 2005). About 20 million ha of land globally was reported to have soil with concentration of metals (As, Cd, Cr, Hg, Pb, Co, Cu, Ni, Zn, and Se) higher than geo-baseline or regulatory levels (Liu et al., 2018). In Europe, around 2 million sites were contaminated with HMs, cyanide, mineral oil, and chlorinated hydrocarbons by 2005 (EEA, 2005). A review by Liedekerke et al. (2013) stated that about 2.5 million sites of contaminated soil were present in Europe, to which different sectors have contributed, namely 38% by municipal and industrial waste, 34% by industrial/commercial sector, and 60% by mineral oil and HMs. Sharma et al. (2023) estimated that there are around 3.5 million potentially contaminated sites and 0.5 million highly contaminated sites in the European Union. In developing countries, particularly in India, China, Pakistan, and Bangladesh, HM pollution occurs by the release of untreated industrial effluents into the surface drains, which further spreads to agricultural lands. Untreated effluent is sometimes used for irrigation due to water scarcity, where it acts as an HM source for agricultural croplands (Ragnarsdottir and Hawkins, 2005). This scenario occurs majorly in *peri* urban and urban cultivation, and the HMs bioaccumulate into crops and therefore food chains (Singh and Kumar, 2006; Mehmood et al., 2019). Urbanization and construction of industrial infrastructure are one of the major reasons for decrease in agricultural lands. The increasing demand for lands for meeting the needs of increasing population has forced farmers to use contaminated sites for crop cultivation. This chapter deals with remediation of HM-contaminated sites, specifically by phytoremediation, and the role of phytoremediation in improving the economy.

2 Remediation measures for HMs

Wide range of conventional remediation techniques involving physical (soil replacement and thermal desorption) and chemical (leaching, adsorption, electro-dialysis, precipitation, ion exchange, and fixation) methods, along with use of a broad range of chemical additives, are available for HM removal from soil (Bricka et al., 1993; Yao et al., 2012). Due to the difficulties involving scale-up processing, adaptability, site conditions, low efficiency, loss of soil structure and fertility, metal specificity, and cost-effectiveness, both physical and chemical methods have not been promoted on a large scale. For instance, precipitation is not specific and does not remove HM at very low concentrations. Although ion exchange is efficient, it is an expensive process. Moreover, the contaminated sludge generated through conventional physical and chemical methods required safe disposal (Kapahi and Sachdeva, 2019). Figs. 1 and 2 depict various physical and chemical remediation measures with their pros and cons.

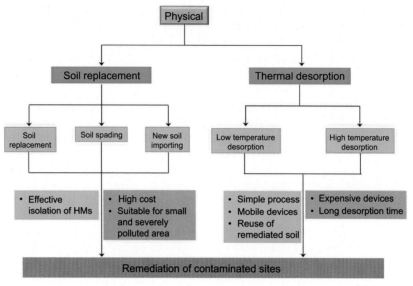

FIG. 1 Physical remediation measures for HM removal. Strategies/technologies involved for HM remediation are in blue shapes. Positive impacts and risk factors associated with the strategies have been indicated in green and orange shapes respectively.

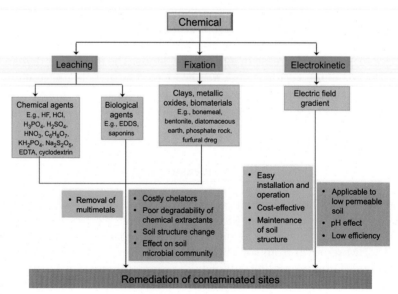

FIG. 2 Chemical remediation measures for HM removal. Strategies/chemicals involved for HM remediation are in blue shapes. Positive impacts and risk factors associated with the strategies have been indicated in green and orange shapes, respectively.

The phytotoxic effects of HMs include reduction in plant growth and protein content, loss of mineral homeostasis, and hence loss in yield and crop quality (Chibuike and Obiora, 2014). HMs Ni, Pb, Cu, Cd are reviewed and reported by Sethy and Ghosh (2013) that they negatively influence different metabolic process associated with seed germination. Yet, some plant species are able to tolerate the negative effects of HM in addition to having the ability to extract in their tissues. The plants that cope with elevated levels of HMs survive the situation either by avoidance mechanism or tolerance mechanism. The avoidance mechanism reduces the HM movement into plant by minimizing mobility of HMs, root exudates (organic acids, amino acids, soluble sugars, and proteins), regionalization of cell walls, membrane, and vacuoles. Tolerance mechanism is to reduce the toxicity of HMs through chelation, osmotic adjustment, and antioxidant system (Yu et al., 2019). This paved the way for the development of a technology called phytoremediation, also known as botanical bioremediation or green remediation, which involves the use of plant species for the extraction, removal, sequestration, detoxification, immobilization of toxic substances through various mechanisms. Phytoremediation being ecofriendly also overcomes the negative effects associated with the physical and chemical remediation methods mentioned earlier. Phytoremediation can be applied for different matrixes such as soil, water, and sediment. All these aspects have been summarized by many reviews from past decades to the current scenario (Chaney et al., 1997; Gomes, 2012; Moffat, 1995; Salt et al., 1977; Sharma and Pandey, 2014; Shehata et al., 2019).

Besides the chemical states of HMs, there are different physical forms: (a) dissolved (in soil solution), (b) exchangeable (organic and inorganic components), (c) as structural components of the lattices of soil minerals, and (d) as insoluble precipitates with other soil components. The former two forms are available to the plants. The latter two forms remain recalcitrant (Aydinalp and Marinova, 2003). Depending on the physical and chemical states of HMs, phytoremediation in soil occurs through one of the following modes: phytoextraction, phytostabilization, and phytovolatilization (Sharma and Pandey, 2014), which is depicted in Fig. 3.

Phytoextraction: Also known as phytoaccumulation, in phytoextraction, the HMs in soil are translocated to aerial plant parts with the aid of roots followed by sequestration and compartmentation of HMs in plant tissues (Ali et al., 2012) through the processes of absorption, concentration, and precipitation. Most of the HMs including Ni (Kukier and Chaney, 2004), Cu (Delorme et al., 2001), and Zn (Sun et al., 2010) have been extracted by this mode.

Phytostabilization: In phytostabilization, tolerant plant species are employed to change toxic forms of HMs into non-toxic or less toxic forms and reduce the bioavailability. The HMs are absorbed or accumulated in roots or precipitated into the root zone, either naturally or by chemical amendments. As this process is confined to the rhizosphere, migration of HMs can be inhibited, which helps in conserving ground- and surface water and reduces likelihood of HMs entering the food chain (Mench et al., 2010; Shackira and Puthur, 2019). Many HMs including As, Pb, Cd, Cr, Cu, and Zn can be stabilized by this mode (Brennan and Shelley, 1999). Despite being said that phytostabilization has advantage over phytoextraction as it does not require disposal of hazardous biomass (Wuana and Okieimen, 2011), it can be seen as a drawback as well because of contaminant remaining in soil and requirement for extensive fertilizers and soil amendments application (Sharma and Pandey, 2014; Zgorelec et al., 2020).

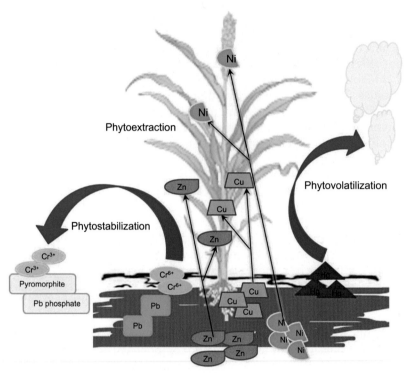

FIG. 3   HM removal mechanisms in soil by phytoremediation.

Phytovolatilization: Phytovolatilization involves the transformation of HMs into volatile forms through leaves, which are further transpired into atmosphere. Hg is the major HM, phytoremediated by phytovolatilization, Rugh et al., 1996. But the negative impact associated with Hg is that recycling by precipitation and redeposition of atmospheric Hg into lakes and oceans leads to production of methylmercury. Phytovolatilization is not meant for completely removing the pollutant, and it transfers pollutants to the atmosphere. Phytovolatilization required risk assessment before field-level application (Yan et al., 2020).

An analysis of strengths, weaknesses, opportunities, and threats done on phytoremediation by Gomes (2012) revealed that phytoremediation has its own pros and cons as do physical and chemical remediation measures. Major weaknesses observed were (i) plant selection with requirement of multi-traits such as fast growth, high biomass, deep roots, and easy harvesting; (ii) slow process; and (iii) limited practical experience; (iv) proper disposal of polluted biomass post-phytoremediation; (v) low bioavailability of tightly bound metal ions present in soil; (vi) influence of pest, diseases, climate, and weather on accumulation capacity of plants (Mahar et al., 2016a, b). Though phytoremediation has these weaknesses, it has many supporting strengths such as (i) high public acceptance; (ii) maintenance of soil biological components; (iii) environmental benefits such as control of soil erosion, carbon sequestration, and creation of wildlife habitat; (iv) generation of recyclable metal-rich plant tissues; (v) socioeconomic benefits via local labor employment and buildup of value-added industry; (vi) cost-effectiveness; and (vii) sustainability.

3 Phytoremediation of HMs: Hyperaccumulators

Plant selection is a crucial first step in phytoremediation. It is already known that the elements present in the soil will be reflected to some extent in the plants; this extent is represented by concentration, a platform for differentiating and selecting a plant for phytoremediation purpose. HM hyperaccumulators belong to a category of plants that can accumulate high concentrations of metals, which is hundreds to thousand-fold higher than accumulation by normal plants. Jaffré et al. (1976) coined the term "hyperaccumulators" to describe *Sebertia acuminata*'s greater metal absorption potential (25% on dry matter (DM)). They suggested that this is the discriminatory concentration threshold for differentiating between normal plants and hyperaccumulating plants. Brooks et al. (1977) used it to define plants with dry tissues having more than 1000 $\mu g \, Ni \, g^{-1}$. Jaffré (1980) has refined the nomenclature by using the terms *hypermanganesophores* and *hypernickelophores* to describe metal extraction specificity of the plants. However, the public exposure toward the use of hyperaccumulating plants for HM removal has increased in later years (Chaney, 1983; Anonymous, 1990). The necessary features that distinguish hyperaccumulators from non-hyperaccumulating plant species include higher rate of (i) HM uptake, (ii) root-to-shoot translocation, and (iii) detoxification by chelation and vacuolar sequestration of HM (Rascio and Navari-Izzo, 2011; Fasani, 2012).

Baker and Brooks (1989) reviewed the distribution of terrestrial plants with hyperaccumulating potential and found 145 hyperaccumulators for Ni with a distribution in 6 suborders, 17 orders, and 22 families including herbs, shrubs, and trees. This indicates that hyperaccumulators are not closely related, but they possess the common feature of growth on metalliferous soils without phytotoxic effects (Rascio and Navari-Izzo, 2011). According to one recent report on the database of HM accumulators (Reeves et al., 2018), plants grown in their natural habitat with dry weight foliar tissue concentrations of $>100 \, \mu g \, g^{-1}$ of Cd, Ti, and Se, $>300 \, \mu g \, g^{-1}$ of Co, Cu, and Cr, $>1000 \, \mu g \, g^{-1}$ of Ni, As, Pb, and rare earth elements (REEs), $>3000 \, \mu g \, g^{-1}$ of Zinc, and $> 10,000 \, \mu g \, g^{-1}$ Mn are said to be under the acceptable limits (Baker and Brooks, 1989; van der Ent et al., 2012). This evaluation also identified more than 500 plant taxa that can be categorized as hyperaccumulators for one or more elements including Asteraceae, Brassicaceae, Caryophyllaceae, Cunouniaceae, Cyperaceae, Euphorbiaceae, Fabaceae, Flacourtiaceae, Lamiaceae, Poaceae, and Violaceae. The most-researched studies of hyperaccumulation model systems include *Thlaspi* sp., (Delorme et al., 2001; Idris et al., 2004), *Brassica* sp., (Quartacci et al., 2014; Wu et al., 2013), and *Alyssum* sp. (Barzanti et al., 2011; Bayramoglu et al., 2012). However, features found in these plant families such as slow growth, shallow root system, small biomass, and unknown agronomic potential of hyperaccumulators have made it necessary to find alternative plant species for phytoremediation.

4 Phytoremediation of HMs: Energy crops

Various research groups differ on plant selection for phytoremediation. Many research groups have suggested that the use of hyperaccumulators for HM remediation is of prime importance rather than biomass (Chaney et al., 1997; van der Ent et al., 2012). However, the success of phytoremediation depends not only on the complete removal of toxic substances but also on the generation of valuable biomass including timber, bioenergy, feedstock for pyrolysis, and biofortified products or ecologically important species in order to demonstrate cost-effectiveness (Conesa et al., 2012; van der Lelie et al., 2001). In addition to hyper accumulation and biomass production, the selection criteria for phytoremediation plants include: adapting to site-specific conditions, the ability to thrive in the soil type, rapid growth rate, and ease of plating (Muthusaravanan et al., 2018). Bioaccumulation factor (f) is the ratio of plant's metal concentration to the total metal concentration in soil. The bioaccumulation factor (BCF) and quantity of biomass produced per hectare are indicators of plant accumulation behavior. Maize has higher BCF and is an effective phytoextractor (Wuana and Okieimen, 2010). Report of Meers et al. (2010) strongly supports the use of bioenergy crops for HM phytoremediation. Maize was tested under field conditions in Flanders, Belgium, in soil contaminated with HMs, Pb, Cd, Zn, and As by historic smelter activities. They stated that cultivation of energy maize in this region could result in the production of 30,000–42,000 kWh including electrical and thermal renewable energy per hectare. This could replace a coal-fed power plant with the reduction of up to 21 tons ha^{-1} year^{-1} CO$_2$ along with the HM removal.

Energy crops fall into two categories: annuals: sweet sorghum and fiber sorghum, kenaf, and rapeseed; and perennials, a category further sub-divided into (a) agricultural: wheat, sugar beet, cardoon, reeds, miscanthus, switchgrass, and canary reed grass; and (b) forest: willows, poplars, eucalyptus, and black locust (Simpson et al., 2009). High biomass crops with HM tolerance such as Indian mustard, oat, maize, barley, sunflower, ryegrass, fast-growing willow, and poplars have been studied (Komárek et al., 2007; Meers et al., 2005; Shen et al., 2002; Vervaeke et al., 2003). Table 2 summarizes the HM removal efficiency of several

TABLE 2   Comparison of HM removal efficiency by agricultural crops.

Crop	Binomial name	Biomass (tons ha^{-1})	Heavy metal uptake (kg ha^{-1})				References
			Pb	Cd	Zn	Cu	
Sorghum	*Sorghum bicolor*	25.8 (dw)	0.35	0.052	1.44	0.24	Zhuang et al. (2009)
		22.1 (dw)	0.38	0.006	1.22	0.64	Marchiol et al. (2007)
Sunflower	*Helianthus annuus*	–	0.091	0.002	0.41	0.12	Marchiol et al. (2007)
		24 (dw)	0.016	–	2.14	–	Madejón et al. (2003)
Indian mustard	*Brassica juncea*	7.3 (dw)	–	0.007	0.89	0.15	Keller et al. (2003)
Tobacco	*Nicotiana tabacum*	12.6 (dw)	–	0.042	1.83	0.47	
Alfalfa	*Medicago sativa*	45.9 (ww)	0.115	0.013	0.438	0.124	Ciura et al. (2005)
Maize	*Zea mays*	92.7 (ww)	0.042	0.0093	0.45	0.096	
Barley	*Hordeum vulgare*	5.84 (dw)	0.035	0.014	1.07	0.057	Soriano and Fereres (2003)

dw, dry weight; ww, wet weight.
Source: Zhuang, P., Shu, W., Li, Z., Liao, B., Li, J., Shao, J., 2009. Removal of metals by sorghum plants from contaminated land. J. Environ. Sci. 21, 1432–1437.

agricultural crops (Zhuang et al., 2009). In conclusion, knowledge of energy crop cultivation under contaminated conditions will provide new avenues for bioeconomy and also for reclamation of contaminated soils.

5 Phytoremediation of HMs by sugar crops: Sweet sorghum

Among the energy crops, sugar crops (sugarcane, sugar beet, and sweet sorghum) with HM remediation capacity have a great impact on the bioenergy-bioethanol (Yadav et al., 2011). Reports of Rayment et al. (2002), Yadav et al. (2010), and Jain et al. (2010) on sugarcane reveal their HM accumulation potential for Cd, Zn, and Hg. Sugarcane has the ability not only to generate high biomass but also to accumulate Cu, Cd, Se, and Mn into its parts above the ground. A pot experiment conducted using different concentrations of Cd by Xia et al. (2009) and the Cd accumulation by sugar cane is observed to be as follows: $0.55-63.32 \, mg \, kg^{-1}$ in leaves, $0.37-486.26 \, mg \, kg^{-1}$ in stems, and $2.28-2776.626 \, mg \, kg^{-1}$ in roots. Sugarcane grown near the municipal landfill site and medical waste treatment system in Brazil is also found to accumulate the HM such as Cd, Cr, Cu, Hg, Mn, Pb, and Zn (Segura-Muñoz et al., 2006). A study on HM content of vegetables in Pakistan revealed that sugar beet has the potential to accumulate Cd, Pb, As, and Hg but at the safest and legally permissible level (Abbas et al., 2010). Pandey et al. (2016) conducted a study on cultivating sugarcane in soils irrigated with paper mill industrial effluent, and it was found that $10.54-60.22 \, mg \, kg^{-1}$ of Cr, $3.56-22.38 \, mg \, kg^{-1}$ of Cu, $0.001-0.15 \, mg \, kg^{-1}$ of Cd, $0.01-1.11 \, mg \, kg^{-1}$ of Pb, $4.55-48.9 \, mg \, kg^{-1}$ of Zn, and $1.87-4.42 \, mg \, kg^{-1}$ of Ni were present in the edible parts of sugar cane. They stated that Cr and Ni levels are higher than the permissible limits for food as per WHO (World Health Organization), FAO (Food and Agriculture Organization), and IS (Indian Standards). Sharma et al. (2023) have reviewed different plants possessing phytoremediation abilities, namely *Zea mays* (Maize), *Cannabis sativa* (Hemp), *Ricinus communis* (Castor), *Brassica* spp., and sweet sorghum and stated that the end use of sweet sorghum biomass is bioethanol production.

Sorghum (*Sorghum bicolor* L. Moench), a C4 annual grass valued for food, feed, fiber, and feedstock, is known by many names such as jowar (India), kaoliang (China), great millet and guinea corn (west Africa), kafir corn (south Africa), mtama (north Africa), dura (Sudan), and milo or milo-maize (Unites States) (Purseglove, 1972). It is also known as "sugarcane of the desert" and "camel among crop" for its hardiness in drought. It can be grown in tropical, subtropical, temperate, and semi-arid regions and also in poor-quality soils (Sanderson et al., 1992). It has many salient features such as rapid growth, high sugar content (10%–15%), higher biomass, wider adaptability to harsh agroclimatic conditions, and metal-absorbing property (Zhuang et al., 2009; Rao et al., 2009). Sweet sorghum is a type of *S. bicolor*, generally cultivated for syrup and also forage and feed. Sweet sorghum was reported to exhibit high tolerance to drought, water logging, and salinity compared with sugarcane and corn (Almodares and Hadi, 2009; Rao et al., 2019). It has sweet juicy stalks and higher quantity of sugars (both glucose and fructose) than grain sorghum; hence, it is called sweet sorghum. Therefore, it can serve as the best candidate for playing dual role in phytoremediation and bioeconomy via bioethanol "sweet fuel" production and also related coproduct generation (Rao et al., 2012). The sweet sorghum varieties developed by International Crops Research

Institute for the Semi-Arid Tropics (ICRISAT) and the related research work done on biofuel will be discussed later in this chapter.

The first hint of the HM absorption of sorghum was shown by An (2004) during an ecotoxic assessment, when he found that sorghum can accumulate more Cd than cucumber, wheat, and sweet corn. Subsequent proof was given by Kaplan et al. (2005) during the analysis on sulfur-containing waste as a soil amendment under pot trials. It has been observed that Ni, Cr, and Co accumulation occurs in roots, whereas Cd accumulation occurs in the straw of sorghum. This indicates the efficiency of sorghum fibrous roots in HM absorption. A similar trend was observed on a short-term study on hydroponically grown *S. bicolor* for Pb, Cd, and Zn removal (Hernández-Allica et al., 2008; Soudek et al., 2012). Five hybrid sweet sorghum species were tested by Yuan et al. (2019) for their phytoremediation ability on multi-metal (Cd, Zn, Pb) contaminated soil. BCF and translocation factor (TF) of the selected species are higher toward Cd, Zn, and Pb and generated more biomass compared with local sweet sorghum. In a 3-month microcosm study with artificially polluted soil containing Cd and Zn, sorghum was found to have $122\,mg\,Cd\,kg^{-1}$ dry weight (DW) in shoot. This is higher than the threshold value of $100\,mg\,kg^{-1}$ DW set for hyperaccumulators. From the various soil physico-chemical and biological properties evaluated, it is noted that phytoremediation with sorghum is able to recover the soil function, though the experimental soil has more phytotoxicity than the control treatments (Epelde et al., 2009). In a greenhouse pot experiment, different germplasms of sweet sorghum (KFJT-3, FJT-CK, KFJT-1) were tested for germination and phytoremediation efficiency under Cd toxicity conditions. In this membership function value (MFCV) of cadmium is taken as a comprehensive index for Cd tolerance in sweet sorghum. The MFCV values of KFJT-3, KFJT-CK, and KFJT-1 are 0.473, 0.456, and 0.413, respectively. In addition, KFJT-3 yielded high biomass, bioaccumulation facto (3.80), and TF (1.02). Therefore KFJT-3 was proved to be suitable candidate for CD phytoremediation in comparison to KFJT-CK and KFJT-1 (Dong et al., 2021).

The first in situ phytoremediation pilot plant was established during 2005 in Torviscosa, Italy, where Marchiol et al. (2007) designed a study to evaluate the phytoremediation effect of high biomass crops including *S. bicolor* and *Helianthus annuus*. The experimental site possesses multi-metal contaminants including As, Cd, Cu, Zn, and Fe, and it is under the national priority list of polluted sites in Italy. Limited metal extraction is observed in both the plants; however, the experimental design did not involve any practices such as soil amendments for enhancing metal bioavailability. It is understood that the factors considered for successful agriculture should also be considered for successful phytoremediation. Phytoremediation is in fact an agricultural practice. Agricultural practices such as soil fertilization (for increased biomass production), crop rotation (to curtail pest and disease incidence), and weed control are reported to enhance phytoremediation. The application of nitrate has enhanced Cd accumulation by sweet sorghum (Wu et al., 2023). A pot experiment was conducted using $Ca(NO_3)_2$ as source of nitrate (0.5, 4, and 8) and two Cd application levels (5 and $50\,\mu M$). It was found that Cd accumulation by sweet sorghum was enhanced by nitrate application. It was stated that nitrogen and sulfur metabolism is intervened, and high Cd accumulation was realized along with ATP sulfurylase (ATPS) activities and non-protein thiol (NPT) in leaves of sweet sorghum. Excess nitrates positively influence ATPS, which in turn increases NTP and thereby high Cd accumulation. Nanomaterial also plays a role in facilitating the uptake of HMs. Nano zero-valent iron (Nzvi) in combination of arbuscular mycorrhizal fungi (AMF) was tested for its effect on phytoremediation of Cd, Pb, and Zn in acidic soils. Results indicated that ZVI had minimal phytotoxicity, whereas AM inoculation was successful at immobilizing HMs, and they could work together to contribute to the phytoremediation of HM-contaminated soils using sweet sorghum (Cheng et al., 2021).

A study by Zhuang et al. (2009) is the first study conducted on HM remediation by sweet sorghum in a field at Lechang city, China, which has been contaminated with Pb, Cd, Zn, and Cu via atmospheric emissions and surface irrigation with mining wastewaters. The study was designed to evaluate the sweet sorghum varieties Keller, Rio, and Mary (bred for ethanol production in the United States) alone and in combination with soil amendments such as $(NH_4)_2SO_4$, NH_4NO_3, and ethylenediaminetetraacetic acid (EDTA). It was observed that there was no difference between the cultivars and biomass yield, whereas HM extraction efficiency was in the order of Keller > Mary > Rio. Among the soil amendments, EDTA promotes Pb accumulation, whereas $(NH_4)_2SO_4$ and NH_4NO_3 promote Zn and Cd accumulation. Therefore, this assisted phytoextraction with facilitated agronomic practices serves as a sustainable remediation measure. The effect of chelators (ethylene diamine tetra acetic acid -EDTA, ethylene diamine di succinate-EDDS, humic acids-Has) on phytoremediation of multi-metal-contaminated soil by *S. bicolor* and three other herbaceous plants was studied by Lee and Sung (2014). Usage of EDTA was most effective soil amendment but had negative impact on soil microbial properties. The soil dehydrogenase activity and microbial population in EDTA introduced soils were less compared with soil amended with EDDS and humic acids (Lee and Sung, 2014). In peri-urban areas, treated sewage water is a resource for agriculture, but if handled improperly, it can lead to soil pollution with HMs and other contaminants. Sweet sorghum has been reported for its capability to accumulate HMs (Cd, As, Pb) (Shafiei Darabi et al., 2016).

6 Sweet Sorghum: A feedstock for "sweet fuel" bioethanol

Every year, fossil fuel sources are getting depleted as the natural regenerative carbon cycle of fossil fuels cannot compete with their current utilization rate, and they are anticipated to run out within the next 40–50 years. In India, the petroleum products consumption is increasing by 5%–6% annually. In addition, the consequences of fossil fuels such as global warming, acid rain, and urban smog have necessitated the shift to renewable energy sources such as biofuels, which are less harmful to the environment and also sustainable. Alternative energy sources must be secure, convenient, affordable, efficient, and renewable (Chum and Overend, 2002), and ethanol is one of the promising alternatives among the biofuels. Though it has 68% lower equivalent energy than petroleum fuel, it is gaining importance due to its complete combustion being an oxygenated (35%) biofuel and release of less toxic by-products than other alcoholic and fossil fuels. It is inexpensive to blend with gasoline due to its high concentration of octane (Rodionova et al., 2017). Another contributing factor is its production from a broad range of feedstocks such as sugar (sugarcane, sugar beet, sweet sorghum), starch (corn, cassava), and lignocellulosic (agri-by-products: corn stover and fiber, wheat and barley straws, sugarcane bagasse, seed cake; woody biomass: hardwood and softwood; energy crops: switchgrass, poplar, banagrass, miscanthus, etc.) (Minteer, 2006; Vohra et al., 2014;

Bajpai, 2021). Biorefineries are used to process or break down the feedstock material to their bulking agents/precursors. For instance, lignocellulose can break down to cellulose, hemicellulose, and lignin, and fermentable sugars are produced from cellulose and hemicellulose (Bušić et al., 2018). All these factors will make ethanol the "fuel of the future," to use Henry Ford's phrase coined during his preparation of the Model T Ford. He designed the model to run on either gasoline or ethanol, with the vision of building a vehicle that was affordable for the working family and powered by a fuel that would boost the rural farm economy (Kovarik, 1998).

During the year 2016, the world ethanol production was about 100.2 billion liters, with increasing trend in production, it is anticipated that ethanol production and consumption would increase to 134.5 billion liters by 2024 (Bušić et al., 2018). About 68% produce was used as fuel, 21% by industrial sector, and 11% beverages (Friedl, 2018). E5, E85 bioethanol (85% volume by fraction) is suitable for flexible fuel vehicles, and 5%, 10% blending is suitable for engines without any modifications. Each country has set its own regulations for blending ethanol and even has set target requirements for the future (Cheng and Timilsin, 2011). In India, in 2003, the Government of India (GOI) mandated the use of 5% ethanol blend in gasoline through its ambitious Ethanol Blending Program (EBP). Since 2003, the trade balance for ethanol has been generally negative. The balance has tapered down, however, from its peak of $140 million in 2005 to $11 million in 2012, indicating a gradual rise in export of ethanol and other spirits. To promote biofuels as an alternative energy source, the GOI in 2018 announced a comprehensive National Policy on Biofuels formulated by the Ministry of New and Renewable Energy (MNRE), calling for blending at least 20% of biofuels with diesel (biodiesel) and petrol (bioethanol) by 2030. This target has been advanced by GOI to 2025–26 as the target of 10% blending by MNRE has been achieved much ahead of timeline. The policies are designed to facilitate and bring about optimal development and utilization of indigenous biomass feedstock for biofuel production (Basavaraj et al., 2012). Sugar-based, starch-based, and lignocellulose are the three types of feedstocks are used for ethanol production. Starch-based feedstocks contribute to higher ethanol production than sugar-based feedstocks (60% vs. 40%). However, the use of starch-based feedstock is limited due to higher energy requirement for its saccharification process (Vohra et al., 2014; Mussatto et al., 2010; Martin, 2019; Hoang and Nghiem, 2021). Among the feedstocks, sugarcane is the major stock in tropical areas such as Brazil, Colombia, and India, whereas corn is the major stock in the United States, the European Union, and China (Cheng and Timilsin, 2011). Increased ethanol demand, decreased feedstock production, lack of clear technology, and the question of food/feed versus fuel have created the need for alternative feedstocks (Vohra et al., 2014).

Sweet sorghum has come to play a key role for this interlinked issue, and ethanol production from sweet sorghum is not a new process as it has three decades of history by various technologies (Christakopoulos et al., 1993; Kargi and Curme, 1985; Kargi et al., 1985; Lezinou et al., 1994; Mamma et al., 1995). However, it has received renewed attention for its beneficial characteristics, such as ethanol production at lower cost over sugarcane and sugar beet along with several by-products that enhance farmers' economy (Tables 3 and 4). Sweet sorghum is known for its adoptability to poor soil and drought conditions. It is known to produce high biomass with less demand for agronomic inputs. Sweet sorghum, although not considered as heavy accumulator of Cd, is a high biomass generator. It was reported that about 78% of total biomass accounts for juice, which contains fermentable sugars such as fructose (6%–12%),

TABLE 3 Favorable traits of sweet sorghum.

As crop	As ethanol source	As bagasse	As raw material for industrial products
Short duration (3–4 months)	Eco-friendly processing	High biological value	Paper and pulp making
C4 dryland crop	Less sulfur	Rich in micronutrients	Butanol, lactic acid, acetic acid production
Good tolerance to biotic and abiotic constraints	High octane ring	Ruminant/poultry feed	Alcoholic and non-alcoholic beverage production
Meets fodder and food needs	Automobile friendly (up to 25% of ethanol-petrol mixture without engine modification)	Power generation	Coproduct generation: dry ice, fuel oil, and methane
Non-invasive species		Biocompost	
Low soil N_2O and CO_2 emission		Good for silage making	
Seed propagated			

Source: Rao, P.S., Reddy, B.V.S., Reddy, R.Ch., Blümmel, M., Kumar, A.A., Rao, P.P., Basavaraj, G., 2012. Utilizing co-products of the sweet sorghum based biofuel industry as livestock feed in decentralized systems. In: Makkar, H.P.S. (Ed.), Biofuel Co-Products as Livestock Feed: Opportunities and Challenges. FAO, Rome, Italy, pp. 229–242.

TABLE 4 Comparison between sugarcane, sugar beet, and sweet sorghum on agronomic traits and ethanol production parameters.

Traits	Sugarcane	Sugar beet	Sweet sorghum
Crop duration	About 7 months	About 5–6 months	About 4 months
Growing season	Only one season	Only one season	Temperate: 1 season; Tropical: 2/3 seasons
Soil requirement	Grows well in drain soil	Grows well in sandy loam; also tolerates alkalinity	All types of drained soil
Water management ($m^3 h^{-1}$)	36,000	18,000	4000–8000
Crop management	Requires good management	Greater fertilizer requirement; requires moderate management	Little fertilizer required; less pest and disease complex; easy management
Yield per ha (tons)	70–80	30–40	54–69
Sugar content on weight basis (%)	10–12	15–18	7–12
Sugar yield ($tons ha^{-1}$)	7–8	5–6	6–8
Ethanol production from juice ($L ha^{-1}$)	3000–5000	5000–6000	3000
Harvesting	Mechanically harvested	Very simple; normally manual	Very simple; both manual and mechanical harvest

Source: Almodares, A., Hadi, M.R., 2009. Production of bioethanol from sweet sorghum: a review. Afr. J. Agric. Res. 4 (9), 772–780. Rao, P.S., Rao, S.S., Seetharama, N., Umakath, A.V., Reddy, P.S., Reddy, B.V.S., Gowda, C.L.L., 2009. Sweet Sorghum for Biofuel and Strategies for its Improvement. International Crops Research Institute for the Semi-Arid Tropics, Andhra Pradesh, India.

glucose (9%–33%), and sucrose (53%–85%) (Xiao et al., 2021). Three cultivars of sweet sorghum (Dale, M81 E, and Theis) were evaluated as feedstocks for ethanol production by Ekefre et al. (2017). The juice yield when tested on 85 and 113 days was reported to be in the range of 3724–6724 L ha^{-1} by Dale, 4333–10,915 L ha^{-1} by M81E, and 3735–9289 L ha^{-1} by Theis. Five species of hybrid sweet sorghum were tested for phytoremediation of contaminated soil sites located at Chenzhou, Hunan province, southern China. A local commercial sweet sorghum variety was used as control for the experiment. The hybrid sweet sorghum species were found to be superior to the local variety and produced about 10–24 times higher biomass quantity (Yuan et al., 2019). In addition, the technology for alcoholic fermentation from sugar- and sucrose-containing feedstock is a well-known and mastered process using microorganisms such as *Saccharomyces cerevisiae* (yeast) and *Zymomonas mobilis* (bacteria) (Bai et al., 2008). Some varieties of sweet sorghum have a significant sucrose content (500 gal syrup per hectare), a major feature required for ethanol. In sweet sorghum, the sugar is stored in the main stalk, which can be recovered by pressing the stalks through rollers (similar to sugarcane processing). This yields about 20 gal of ethanol per ton of stalks (Kojima and Johnson, 2005). Approximately, 50–85 tons ha^{-1} of sweet sorghum stalks yields 39.7–42.5 tons ha^{-1} of juice, which after fermentation produces 3450–4132 L ha^{-1} ethanol (Serna-Saldívar et al., 2012). Other studies have shown similar ethanol production levels: 3296 L ha^{-1} (Kim and Day, 2011) and 4750–5220 L ha^{-1} (Wu et al., 2010). The energy gain from production of a feedstock should be more than the energy consumed for its production. According to USDA (United States Department of Agriculture), sweet sorghum produces about 12–16 Btu (BTU) for every BTU used toward its production (Umakanth et al., 2019). In addition, sweet sorghum can rule out the food/feed versus fuel question due to the farmer's benefit from sorghum grains, after the stalk is harvested for juice (Kojima and Johnson, 2005). Besides this, the pressed stalk, called sweet sorghum bagasse, has several avenues in improving rural economy via ruminant/poultry feed and as raw material for biofertilizer production, paper making, and co-product generation including power (Rao et al., 2012). Also, the bagasse of sweet sorghum could be degraded in 50–60 days and made value-added compost, which could be used as plant growth promoter in cereals (Gopalakrishnan et al., 2020a). Physical pretreatment: chipping, milling, grinding, chemical pre-treatment: H_3PO_4, H_2SO_4, NaOH, etc., and recent advancements in production such as solid state fermentation (SSF) have improved the sweet sorghum bioethanol production efficiency (Xiao et al., 2021).

7 Microbe-assisted phytoremediation

Usage of microorganisms to remediate HMs through their biogeochemical process is called microbial remediation or bioremediation. The microbes use various toxic compounds throughout their metabolic activities, and they have acquired different mechanisms to tolerate broad spectrum of metals (Ayangbenro and Babalola, 2017). As phytoremediation is a time-consuming process, it can be enhanced by a plant microbe-mediated approach, popularly referred to as bio-phytoremediation. Bacterial cells (approximately 1.0–1.5 mm^3) have an extremely higher surface-area-to-volume ratio, which influences HM absorption, than inorganic soil components by a metabolism-independent, passive/metabolism-dependent active process (Ledin et al., 1996). The major microbial groups—bacteria, fungi, yeast, and algae play a better role in bioremediation by their biosorption properties. Some of the tested species include *Bacillus subtilis, Rhizopus arrhizus, S. cerevisiae,* and *Scytonema hofmanni* (Vijayaraghavan and Yun, 2008), and microorganisms such as *Agrobacterium, Pseudomonas, Klebsiella, Microbacterium, Mesorhizobium, Rhodococus* exhibited different HM tolerance levels (González Henao and Ghneim-Herrera, 2021). Though the microbial biosorbents are cheaper and more effective alternatives for HM removal, effectiveness can be attained mainly in aqueous solutions (Joshi and Juwarkar, 2009; Kapoor et al., 1999; Wang and Chen, 2009).

Plant growth-promoting (PGP) microbes, a group of microbes found in rhizosphere or in association with the roots or other plant parts as endophytes, can play a key role in this scenario (Glick et al., 1999; Gopalakrishnan et al., 2011a, b, 2012, 2013a, b, 2015a, b, c). The higher bacterial biomass in the rhizosphere occurs because of the nutrient release (especially small molecules such as amino acids, sugars, and organic acids) from the roots. In turn, the PGP bacteria will support for plants by various direct (nitrogen fixation, phosphate solubilization, iron chelation, and phytohormone production) or indirect (suppression of plant pathogenic organisms, induction of resistance in host plants against plant pathogens and abiotic stresses) mechanisms (Penrose and Glick, 2001; Gopalakrishnan et al., 2016a, b, 2020a, b, c, 2021, 2022; Pratyusha and Gopalakrishnan 2023; Sathya et al. 2016).

A wide range of PGP microbes is able to alleviate HM stress in soil and increase plant HM uptake by enhancing the bioavailability through chelating agents, acidification, redox changes, and phosphate solubilization and by increasing plant growth as well. Some of them are *Rhizobium, Pseudomonas, Agrobacterium, Burkholderia, Azospirillum, Bacillus, Azotobacter, Serratia, Alcaligenes,* and *Arthrobacter* (Carlot et al., 2002; Glick, 2003). A detailed review of PGP bacteria and their role in phytoremediation has been provided by many researchers (Ma et al., 2011; Rajkumar et al., 2012, Mishra et al., 2017) and also demonstrated experimentally on various agricultural crops such as *Z. mays, Vigna mungo,* and *H. annuus* (Ganesan, 2008; Jiang et al., 2008; Rajkumar et al., 2008) and hyperaccumulating crops such as *Alyssum murale* and *Salix caprea* (Abou-Shanab et al., 2008; Kuffner et al., 2008). The following mechanisms were suggested for HM removal by microbes.

Siderophores: The low-molecular-mass (400–1000 Da) compounds with high-association constants for complexing Fe and also other metals Al, Cd, Cu, Ga, In, Pb, and Zn (Rajkumar et al., 2010). Production of siderophores by PGP *Pseudomonas* spp., its metal complexes formation with high solubility, and the resulting higher HM uptake by plant have been reported (Wu et al., 2006a, b). *Psudomonas aeruginosa* ZGKD3 was reported to produce siderophores induced by Cd, Cu, Zn, Ni, and Pb (Shi et al., 2017).

Organic acids: Low-molecular-weight organic acids (humic, malic, citric, acetic acids) of PGP microbes bind to metal ions in soil solution and increase the metal bioavailability to plants. However, the stability of the ligand:metal complexes is dependent on several factors such as the nature of organic acids (number of carboxylic groups and their position), the binding form of the HMs, and pH of soil solution (Saravanan et al., 2007; Ebrahimian and Bybordi, 2014).

Biosurfactants: The amphiphilic molecules with non-polar tail and polar/ionic head, produced by microbes, are able to form complexes with HMs at the soil interface, desorbing metals from soil matrix and thus increasing metal solubility/bioavailability in the soil solution. Biosurfactants are known for their low toxicity, environmental compatibility, high emulsifying activity, high biodegradability, and functional at extreme pH and temperatures (Mishra et al., 2021). Several studies have demonstrated the role of microbial biosurfactants in facilitating the release of adsorbed HMs and in enhancing the phytoextraction potential of plants. Still, documentation under field conditions is lacking (Basak and Das, 2014; Franzetti et al., 2014).

Polymeric substances and glycoprotein: Extracellular polymeric substances, mucopolysaccharides, and proteins produced by plant-associated microbes can make complexes with HMs and decrease their mobility in soils. Glomalin, an insoluble glycoprotein produced by arbuscular mycorrhizal fungi (AMF), proved its ability to form complexes with HMs Cu, Pb, and Cd (Bano and Ashfaq, 2013; Foster et al., 2000; Liu et al., 2001, Gupta and Diwan, 2016). *Agrobacterium* spp., *Xnathomonas campestris*, *Leuconostoc*, *Pseudomonas* spp., *Zygomonas mobilis*, are examples of microorganisms that produce exopolysaccharides (EPS), which assist in HM tolerance. Alginate, gellan, hyaluronan, xanthan, galactopol, and fucopol are commercially available EPS (Ojuederie and Babalola, 2017).

Metal reduction and oxidization: Mobility of HMs can be enhanced through oxidation and reduction reactions of plant-associated Fe- and S-oxidizing bacteria. Use of Fe-reducing bacteria and the Fe/S-oxidizing bacteria together showed significantly increased mobility of Cu, Cd, Hg, and Zn, which might be due to coupled and synergistic metabolism of oxidizing and reducing microbes (Beolchini et al., 2009; Shi et al., 2011; Wani et al., 2007).

Biosorption: The plant-associated microbes may contribute to plant metal uptake through biosorption (microbial adsorption of soluble/insoluble organic/inorganic metals). Among the microbes, mycorrhizal fungi are key partners. The large surface area, cell wall (chitin, extracellular slime, etc.), and intracellular compounds (metallothioneins, P-rich amorphic material) of fungi endow them with a strong capacity for HM absorption from soil (He and Chen, 2014; Volesky and Holan, 1995). *Pseudomonas, Bacillus, Micrococcus, Escherichia, Streptococcus, Enterobacter, Staphylococcus* (bacteria) and *Aspergillus niger, Penicillium notatum,* and *Aspergillus fluvus* (fungi) were tested for biosorption of Cu, Cr, Cd, and Ni polluted soil. *Pseudomonas* and *P. notatum* exhibited higher biosorption of Cr and Cr and *P. notatum* highest biosorption of Cd (Oyewole et al., 2019).

The other interesting mechanism by which PGP microbes can alleviate HM stress on plant growth is through the production of enzyme 1-aminocyclopropane-1-carboxylate (ACC) deaminase, which controls the production of ethylene, a stress hormone (Glick, 2014). Similarly production of indole-3-acetic acid by rhizobacteria can enhance HM uptake by plant (Zaidi et al., 2006), IAA was reported to enhance phytoextraction potential of sunflower and reduce the toxic effects of Pb and Zn on the plants (Fässler et al., 2010). The role of endophytic bacteria interacting with their host plant is of significance in the process of phytoremediation. Under HM stress, the endophytes with the ability of stress tolerance can alleviate the stress and allocate the metals to the plant shoot (Ma et al., 2011; Weyens et al., 2009, Franco-Franklin et al., 2021).

Few reports are available on the assistance of PGP microbes in HM remediation by sweet sorghum. The report of Duponnois et al. (2006) documents that the inoculation of Cd-tolerant *Pseudomonas* strains, mainly *Pseudomonas monteilii*, showed significantly improved Cd uptake by sorghum plants under glasshouse conditions. Measurement of catabolic potentials on 16 substrates showed that pseudomonad strains presented a higher use of ketoglutaric and hydroxybutyric acids, as opposed to fumaric acid in control soil samples. It is suggested that fluorescent pseudomonads could act on the effect of small organic acids on phytoextraction of HMs from soil. Subsequently, Abou-Shanab et al. (2008) examined the ability of four bacterial isolates (*B. subtilis*, *B. pumilus*, *Pseudomonas pseudoalcaligenes*, and *Brevibacterium halotolerans*) on HM removal capacity of sorghum. Sorghum roots accumulated higher concentration of Cr followed by metals Pb, Zn, and Cu. A comparative analysis was done on phytoremediation efficiency of sweet sorghum, *Phytolacca acinosa* and *Solanum nigrum*, with the inoculation of PGP endophyte *Bacillus* sp. SLS18 on Mn- and Cd-amended soils. Sweet sorghum was found to have higher metal absorption (Mn vs. Cd; 65% vs. 40%) than *P. acinosa* (Mn vs. Cd; 55% vs. 31%) and *S. nigrum* (Mn vs. Cd; 18% vs. 25%). The effect of this remediation process on biomass was also observed in the order of sweet sorghum > *P. acinosa* > *S. nigrum*. The effect of *Brevibacillus* sp. SR-9 on uptake of Cd by sweet sorghum was studied by Li et al. (2022). The inoculation of SR-9 increased Cd content, in above-ground plant parts by 38.53% and in roots by 11.3%. The PGP bacterial strain SR-9 was found to enhance sweet sorghum plant growth about 21.1% of above ground parts and 17.4% of roots.

PGP bacteria including *Kluyveria intermedia*, *Klebsiella oxytoca*, and *Citrobacter murliniae* were reported to assist *S. bicolor* in phytoremediation of As, Cd, Pb, in abandoned gold ore processing plant (Boechat et al., 2020). The experiment was conducted in contaminated soils and non-contaminated soils. The soils were inoculated with individual PGP and consortiums of their different combinations. PGP *K. oxytoca* and consortiums (*K. intermedia* + *K. oxytoca* and *K. intermedia* + *C. murliniae*) were found to be effective in plant growth promotion and phytoextraction. AMF were tested for phytoremediation of HM-contaminated soils in combination with soil amendment using Hypoxyapatite (HAP), Manure, and Biochar (0.1% and 1%). It was found that the amendment type and dosage determine the effectiveness of remediation. HAP at 1% dosage was found to be more effective. The augmentation of soil with exogenous AMF in combination with soil amendments was proved to be effective strategy (Wang et al., 2022).

8 Work at ICRISAT

ICRISAT has developed several improved hybrid parental lines of sweet sorghum with high stalk sugar content that were tested in pilot studies for sweet sorghum-based ethanol production in India, the Philippines, Mali, and Mozambique. Concerted research efforts under National Agricultural Research System (NARS) have led to the development and release of cultivars such as SSV 84, CSV 19SS, CSH 22SS, and CSV 24SS for all India cultivation with productivity ranging from 40 to 50 tons ha^{-1} (Vinutha et al., 2014). Trial data over 3 years (2005–07) and six seasons indicated that there is no reduction in grain yield while improving the sugar yield. Sugar yield and associated traits have greater genotype × environment interaction; therefore, it is prudent to breed for season-specific hybrids.

ICRISAT launched a global **BioPower** initiative in 2007 to find ways to empower the dryland poor to benefit from emerging opportunities in renewable energies. This involves the collaborative partnership of NARS, particularly India, the Philippines, Mali, and private sector partners in Brazil, the United States, Germany, and Mexico. ICRISAT focuses on hybrids parent development to produce cultivars withstanding biotic and abiotic stresses, thereby strengthening sweet sorghum value chains and their impact. The ICRISAT has made the first attempt in India to evaluate and identify useful high biomass producing sweet sorghum germplasm from world collections. The sweet sorghum program at ICRISAT mainly focuses on developing primarily hybrid parents adapted to rainy and post-rainy seasons due to the highly significant interaction of genotype by environment ($G \times E$). However, about 100 sweet sorghum varieties and restorer lines and 50 improved hybrids were identified. ICSV 93046, ICSV 25274, ICSV 25280, and ICSSH 58 were identified for release owing to their superior performance in All India Coordinated Sorghum Improvement Project (AICSIP) multilocation trials during 2008–12 (Rao et al., 2013). Sweet sorghum improvement aims for simultaneous improvement of stalk sugar traits such as total soluble sugars or (brix %), green stalk yield, juice quantity, girth of the stalk, and grain yield. Conventional breeding approaches are practiced for an increase in sucrose yield; R lines showed a brix percentage of 12%–24% in the rainy season and 9%–19% in the post-rainy season. In total, 600 A/B pairs were screened at ICRISAT, and the brix percentage ranged from 10% to 15% in the rainy season and from 8% to 13% in the post-rainy season (Rao et al., 2009). Sweet sorghum bagasse is highly palatable, and intake by livestock is more than normal sorghum stover (Blummel et al., 2009).

Some insect- and pest-resistant materials have been developed at ICRISAT, such as ICSR 93034 and ICSV 700. ICSV 93046 (ICSV 700 × ICSV 708) is a promising sweet sorghum variety tolerant to shoot fly, stem borer, and leaf diseases; it also displays stay-green stems and leaves even after physiological maturity and has good grain (3.4–4.1 tons ha^{-1}) and biomass yield. Another hybrid, ICSSH 72, shows excellent fodder quality in the rainy season and is resistant to leaf diseases. SPV 422 also exhibits resistance to leaf diseases and other hybrids developed at ICRISAT, India; for example, ICSSH 21 (ICSA 38 × NTJ 2) and ICSSH 58 (ICSA 731 × ICSV 93046) are under advance testing stages. ICSSH 30 variety shows superior grain yields in both rainy and post-rainy seasons, whereas ICSSH 39 and 28 are best for sugar yield. ICSSH 24 variety is supposed to be best suited for the rainy season (Vinutha et al., 2014). Some of the varieties and hybrids developed from ICRISAT are given in Table 5.

TABLE 5 Sweet sorghum varieties and hybrids developed from ICRISAT.

Varieties				Hybrids		
ICSV 93046	**ICSSV 10032**	**ICSSV 10059**	**ICSSV 12021**	**ICSSH 1**	**ICSSH 28**	**ICSSH 55**
ICSSV 10001	ICSSV 10033	ICSSV 10060	ICSSV 12022	ICSSH 2	ICSSH 29	ICSSH 56
ICSSV 10005	ICSSV 10034	ICSSV 10061	ICSSV 12023	ICSSH 3	ICSSH 30	ICSSH 57
ICSSV 10006	ICSSV 10035	ICSSV 10062	ICSSV 12024	ICSSH 4	ICSSH 31	ICSSH 58
ICSSV 10007	ICSSV 10036	ICSSV 10063	ICSSV 12025	ICSSH 5	ICSSH 32	ICSSH 59
ICSSV 10008	ICSSV 10037	ICSSV 10064	ICSSV 12026	ICSSH 6	ICSSH 33	ICSSH 60
ICSSV 10009	ICSSV 10038	ICSSV 10065	ICSSV 12027	ICSSH 7	ICSSH 34	ICSSH 61
ICSSV 10010	ICSSV 10039	ICSSV 10066	ICSSV 14001	ICSSH 8	ICSSH 35	ICSSH 62
ICSSV 10011	ICSSV 10040	ICSSV 10067	ICSSV 14002	ICSSH 9	ICSSH 36	ICSSH 63
ICSSV 10012	ICSSV 10041	ICSSV 10068	ICSSV 14003	ICSSH 10	ICSSH 37	ICSSH 64
ICSSV 10013	ICSSV 10042	ICSSV 10069	ICSSV 14004	ICSSH 11	ICSSH 38	ICSSH 65

TABLE 5 Sweet sorghum varieties and hybrids developed from ICRISAT—cont'd

Varieties				Hybrids		
ICSV 93046	**ICSSV 10032**	**ICSSV 10059**	**ICSSV 12021**	**ICSSH 1**	**ICSSH 28**	**ICSSH 55**
ICSSV 10014	ICSSV 10043	ICSSV 12005	ICSSV 14005	ICSSH 12	ICSSH 39	ICSSH 66
ICSSV 10015	ICSSV 10044	ICSSV 12006	ICSSV 12020	ICSSH 13	ICSSH 40	ICSSH 67
ICSSV 10016	ICSSV 10045	ICSSV 12007	ICSSV 12021	ICSSH 14	ICSSH 41	ICSSH 68
ICSSV 10017	ICSSV 10046	ICSSV 12008	ICSSV 12022	ICSSH 15	ICSSH 42	ICSSH 69
ICSSV 10018	ICSSV 10047	ICSSV 12009	ICSSV 12023	ICSSH 16	ICSSH 43	ICSSH 70
ICSSV 10019	ICSSV 10048	ICSSV 12010	ICSSV 12024	ICSSH 17	ICSSH 44	ICSSH 71
ICSSV 10021	ICSSV 10049	ICSSV 12011	ICSSV 12025	ICSSH 18	ICSSH 45	ICSSH 72
ICSSV 10022	ICSSV 10050	ICSSV 12012	ICSSV 12026	ICSSH 19	ICSSH 46	ICSSH 73
ICSSV 10024	ICSSV 10051	ICSSV 12013	ICSSV 12027	ICSSH 20	ICSSH 47	ICSSH 74
ICSSV 10025	ICSSV 10052	ICSSV 12014	ICSSV 14001	ICSSH 21	ICSSH 48	ICSSH 75
ICSSV 10026	ICSSV 10053	ICSSV 12015	ICSSV 14002	ICSSH 22	ICSSH 49	ICSSH 76
ICSSV 10027	ICSSV 10054	ICSSV 12016	ICSSV 14003	ICSSH 23	ICSSH 50	ICSSH 77
ICSSV 10028	ICSSV 10055	ICSSV 12017	ICSSV 14004	ICSSH 24	ICSSH 51	ICSSH 78
ICSSV 10029	ICSSV 10056	ICSSV 12018	ICSSV 14005	ICSSH 25	ICSSH 52	
ICSSV 10030	ICSSV 10057	ICSSV 12019		ICSSH 26	ICSSH 53	
ICSSV 10031	ICSSV 10058	ICSSV 12020		ICSSH 27	ICSSH 54	

9 Work at Indian NARS

Concerted research efforts at AICSIP centers have resulted in the identification of several promising sweet sorghum varieties such as SSV 96, GSSV 148, SR 350-3, SSV 74, HES 13, HES 4, SSV 119, and SSV 12611 for total soluble solids (TSS%) and juice yield during 1991–92 trials, GSSV 148 for cane sugar during 1993–94 trials, NSS 104 and HES 4 for green cane yield, juice yield, juice extraction, and total sugar content during 1999–2000 trials, and RSSV 48 for better alcohol yield during 2001–02. An evaluation of 11 promising sweet sorghum varieties bred at different AICSIP centers indicated superiority of the varieties NSSV 255 and RSSV 56 for green cane yield, juice yield, juice extractability, commercial cane sugar (CCS) yield (q ha^{-1}), and percent non-reducing sugars over the rest of the varieties. The varieties RSSV 79, PKV 809, NSSV 256, and NSSV 6 excelled the check with superior performance for green cane yield, juice yield, juice extractability, CCS yield, and total sugars (Reddy et al., 2007). Phule Vasundhara (RSSH -50), a hybrid (185A × RSSV260), was developed by Rahuri center and released in 2015 for cultivation in Maharashtra state.

The Rusni distillery, established in 2007 near Sangareddy in the Medak district of Telangana, India, was the first sweet sorghum distillery amenable to use multiple feedstocks for transport-grade ethanol production. It generated 99.4% of fuel ethanol with a total capacity of 40 kiloliters per day (KLPD). It also produced 96% extra neutral alcohol (ENA) and 99.8% pharma alcohol from agro-based raw materials such as sweet sorghum juice, molded grains, broken rice, cassava, and rotten fruits. ICRISAT has incubated sweet sorghum ethanol production in partnership with Rusni Distilleries through its Agri-Business Incubator. A pilot-scale sweet sorghum distillery of 30 KLPD capacity was established in 2009 at Nanded, Maharashtra. It used commercially grown sweet sorghum cultivars such as CSH 22SS, ICSV 93046, sugargrace, JK Recova, and RSSV 9 in the 25 km radius of the distillery to produce transport-grade ethanol and ENA during 2009–10. However, it could not continue operations due to the low mandated ethanol price. The Cabinet Committee of Economic Affairs (CCEA) of GOI on November 22, 2012, recommended 5% mandatory blending of ethanol with gasoline (Aradhey and Lagos, 2013). The government's current target of 5% blending of ethanol in gasoline has been partially successful in years of surplus sugar production and unfulfilled when sugar production declines. The interim price of US $0.44 L^{-1} would no longer hold as the price would now be decided by market forces. It is expected that this decision will have a positive effect on forthcoming distilleries in India and as 85%, 35%, 25% of India's crude oil, natural gas, coal requirements are met through imports. Ethanol production has increased from 38 Cr liters (2013–14) to 173 Cr liters (2019–20) as a result of Government initiatives—NITI Ayog report, 2021 (Basavaraj et al., 2013; Sarawal, et al., 2021).

10 Work in other countries

Other countries involved in sweet sorghum research and development are the United States, Brazil, Colombia, Haiti, Argentina, Italy, Germany, Hungary, France, China, the Philippines, and Indonesia. Non-food crops and materials such as cassava and sweet sorghum are the priority choice for biofuel ethanol production in China. Sorghum Research Institute (SRI) of Liaoning Academy of Agricultural Sciences (LAAS) is the lead organization involved in sweet sorghum research in China since the 1980s. So far 17 promising sweet sorghum hybrids were released nationally. A few industries such as ZTE energy company limited (ZTE, Inner Mongolia), Fuxin Green BioEnergy Corporation (FGBE), Xinjiang Santai Distillery, Liaoning Guofu Bioenergy Development Company Limited, Binzhou Guanghua Biology Energy Company Ltd., Jiangxi Qishengyuan Agri-Biology Science and Technology Company Ltd., Jilin Fuel Alcohol Company Limited, and Heilongjiang Huachuan Siyi Bio-fuel Ethanol Company Ltd. conducted either large-scale sweet sorghum processing trials or are at the commercialization stage (Reddy et al., 2011).

In the Philippines, sweet sorghum has been proven to be a technically and economically viable alternative feedstock for bioethanol production. The plantation, agronomic performance, and actual bioethanol production of sweet sorghum have been evaluated on different plantation sites nationwide. A hectare of sweet sorghum plantation can potentially provide farmers with an annual net income of US $1860.47 at a stalk-selling price of US $22 and grain price of US $0.30. The San Carlos Bioenergy Inc. (SCBI) became the first commercial distillery to process sweet sorghum bioethanol in Southeast Asia under the Department of Agriculture (DA) and produced 14,000 L of fuel-grade ethanol in 2012 (Demafelis et al., 2013). The

Ecofuels 300 KLPD distillery at San Mariano, Isabela, is planning to use sugarcane and sweet sorghum as feedstocks for ethanol production commercially. The sweet sorghum growers are enthusiastic as the ratoon (new shoot) yields are about 20%–25% higher than that of plant crop. The Bapamin enterprises based in Batac have been successfully marketing vinegar and hand sanitizer made from sweet sorghum since 2009 (Reddy et al., 2011). According to European Union's renewable energy directive, it is mandate to blend 5.7% ethanol with petrol.

In Ukraine, focus has been given for sweet sorghum as alternative energy source and germplasm, which is inclusive of introduced and acclimated genotypes, were collected at M.M. Gryshko National Botanical Garden (NBG) of the National Academy of Sciences. Rakhmetova et al. (2020) have studied the potential for ethanol production by sweet sorghum in north and central Ukraine. They have concluded that about $4041 \, \text{L} \, \text{ha}^{-1}$ ethanol from juice and $6389.62 \, \text{L} \, \text{ha}^{-1}$ ethanol from combination of juice and grain are possible by following the optimal cultivation methods (NPK 16,16,16 @135 kg/ha).

In United States, New Blue company enounced construction of a mega production plant that requires 205,000 MT of agricultural residues (Hoang and Nghiem, 2021). The renewable fuel standards (RFS) and policy in united stated are one of the contributing factors for a ten-fold increase in ethanol production and consumption from 2002 to 2019 (Newes et al., 2022). According to EIA (U.S. Energy Information Administration, 2022), in 2021, about 17.5 billion gallons of biofuel was produced out of which 15.01 billion gallons was fuel ethanol. EIA reports that 13.94 billion gallons was consumed, 0.06 billion gallons of imports and 1.25 billion gallons of exports were recorded. The Government of Brazil has identified 1.8 m ha for sweet sorghum plantation to augment fuel-grade ethanol production. In some African countries such as Mozambique, Kenya, South Africa, and Ethiopia, sweet sorghum adaptation trials are being conducted in pilot scale to assess feasibility of sweet sorghum for biofuel production.

11 Future outlook

The phytoremediation potential of sweet sorghum has not been studied extensively enough under field conditions. A review on prospects of sweet sorghum as source of high biomass stated that high feedstock cost, which about 60%–70% of total production cost, is one of the bottlenecks, and therefore, the sweet sorghum biorefineries are expected to face sustainability issues (Xie et al., 2018). Therefore, it may be difficult for the industry to take off under the current scenario of fluctuating ethanol price, feedstock price, and ethanol recovery rate. A well-developed technology for phytoremediation and ethanol production involving sweet sorghum along with well-defined policy support is crucial to meet future blending requirements and also to improve rural while adhering to greenfield biorefinery approach. Adopting soil amendment methods and good agronomic practices in combination of suitable species or hybrids would yield high biomass and effective phytoremediation. Pre-treatment methods (physical and chemical) and inclusion of methods such as solid-state fermentation in production process would aid the industries to curtail production costs.

References

Abbas, M., Parveen, Z., Iqbal, M., Riazuddin, I.S., Ahmed, M., Bhutto, R., 2010. Monitoring of toxic metals (cadmium, lead, arsenic and mercury) in vegetables of Sindh, Pakistan. Kathmandu Univ. J. Sci. Eng. Technol. 6, 60–65.

Abou-Shanab, R.A., Ghanem, K., Ghanem, N., Al-Kolaibe, A., 2008. The role of bacteria on heavy-metal extraction and uptake by plants growing on multi-metal-contaminated soils. World J. Microbiol. Biotechnol. 24, 253–262.

Ali, H., Naseer, M., Sajad, M.A., 2012. Phytoremediation of heavy metals by *Trifolium alexandrinum*. Int. J. Environ. Sci. 2, 1459–1469.

Ali, H., Khan, E., Ilahi, I., 2019. Environmental chemistry and ecotoxicology of hazardous heavy metals: environmental persistence, toxicity, and bioaccumulation. J. Chem. 2019, 6730305.

Almodares, A., Hadi, M.R., 2009. Production of bioethanol from sweet sorghum: a review. Afr. J. Agric. Res. 4 (9), 772–780.

An, Y.J., 2004. Soil ecotoxicity assessment using cadmium sensitive plants. Environ. Pollut. 127 (1), 21–26.

Anonymous, 1990. NEA dumps on science art. Science 250, 1515.

Aradhey, A., Lagos, J., 2013. India—Biofuel Annual. GAIN Report. Last accessed at http://gain.fas.usda.gov/Recent %20GAIN%20Publications/Biofuels%20Annual_New%20Delhi_India_8-13-2013.pdf.

Ayangbenro, A.S., Babalola, O.O., 2017. A new strategy for heavy metal polluted environments: a review of microbial biosorbents. Int. J. Environ. Res. Public Health 14 (1), 94. https://doi.org/10.3390/ijerph14010094.

Aydinalp, C., Marinova, S., 2003. Distribution and forms of heavy metals in some agricultural soils. Pol. J. Environ. Stud. 12 (5), 629–633.

Bai, F.W., Anderson, W.A., Moo-Young, M., 2008. Ethanol fermentation technologies from sugar and starch feedstocks. Biotechnol. Adv. 26 (1), 89–105.

Bajpai, P., 2021. Global production of bioethanol. In: Developments in Bioethanol. Green Energy and Technology. Springer, Singapore, https://doi.org/10.1007/978-981-15-8779-5_10.

Baker, A.J.M., Brooks, R.R., 1989. Terrestrial higher plants which hyperaccumulate metallic elements—a review of their distribution, ecology and phytochemistry. Biorecovery 1, 81–126.

Bano, S.A., Ashfaq, D., 2013. Role of mycorrhiza to reduce heavy metal stress. Nat. Sci. 5 (12A), 16–20.

Barzanti, R., Colzi, I., Arnetoli, M., Gallo, A., Pignattelli, S., Gabbrielli, R., Gonnelli, C., 2011. Cadmium phytoextraction potential of different *Alyssum* species. J. Hazard. Mater. 196, 66–72.

Basak, G., Das, N., 2014. Characterization of sophorolipid biosurfactant produced by *Cryptococcus* sp. VITGBN2 and its application on Zn(II) removal from electroplating wastewater. J. Environ. Biol. 35, 1087–1094.

Basavaraj, G., Rao, P.P., Basu, K., Reddy, C.R., Kumar, A.A., Rao, P.S., Reddy, B.V.S., 2012. A Review of the National Biofuel Policy in India: A Critique of the Need to Promote Alternative Feedstocks. Working Paper Series No. 34. International Crops Research Institute for the Semi-Arid Tropics, Andhra Pradesh, India.

Basavaraj, G., Rao, P.P., Basu, K., Reddy, C.R., Kumar, A.A., Rao, P.S., Reddy, B.V.S., 2013. Assessing viability of bio-ethanol production from sweet sorghum in India. Energy Policy 56, 501–508.

Bayramoglu, G., Arica, M.Y., Adiguzel, N., 2012. Removal of Ni(II) and Cu(II) ions using native and acid treated Ni-hyperaccumulator plant *Alyssum discolor* from Turkish serpentine soil. Chemosphere 89 (3), 302–309.

Beolchini, F., Anno, D.A., Propris, L.D., Ubaldini, S., Cerrone, F., Danovaro, R., 2009. Auto and heterotrophic acidophilic bacteria enhance the bioremediation efficiency of sediments contaminated by heavy metals. Chemosphere 74, 1321–1326.

Blummel, M., Rao, S.S., Palaniswami, S., Shah, L., Reddy, B.V., 2009. Evaluation of sweet sorghum (*Sorghum bicolor* L. Moench) used for bio-ethonol production in the context of optimizing whole plant utilization. Anim. Nutr. Feed. Technol. 9 (1), 1–10.

Boechat, C.L., Carlos, F.S., Nascimento, C.W.A., Quadros, P.D., Sá, E.L.S., Camargo, F.A.O., 2020. Bioaugmentation-assisted phytoremediation of As, Cd, and Pb using *Sorghum bicolor* in a contaminated soil of an abandoned gold ore processing plant. Rev. Bras. Cienc. Solo 44, e0200081.

Brennan, M.A., Shelley, M.L., 1999. A model of the uptake, translocation, and accumulation of lead (Pb) by maize for the purpose of phytoextraction. Ecol. Eng. 12, 271–297.

Bricka, R.M., Williford, C.W., Jones, L.W., 1993. Technology Assessment of Currently Available and Developmental Techniques for Heavy Metals—Contaminated Soils Treatment. Technical Report, IRRP-93-4. U.S. Army Engineer Waterways Experiment Station, USA, pp. 1–3.

Brooks, R.R., Lee, J., Reeves, R.D., Jaffre, T., 1977. Detection of nickeliferous rocks by analysis of herbarium specimens of indicator plants. J. Geochem. Explor. 7, 49–77.

Burbacher, T.M., Rodier, P.M., Weiss, B., 1990. Methylmercury developmental neurotoxicity: a comparison of effects in humans and animals. Neurotoxicol. Teratol. 12 (3), 191–202.

Bušić, A., Marđetko, N., Kundas, S., Morzak, G., Belskaya, H., Ivančić Šantek, M., Komes, D., Novak, S., Šantek, B., 2018. Bioethanol production from renewable raw materials and its separation and purification: a review. Food Technol. Biotechnol. 56 (3), 289–311. https://doi.org/10.17113/ftb.56.03.18.5546.

Carlot, M., Giacomini, A., Casella, S., 2002. Aspects of plant-microbe interactions in heavy metal polluted soil. Acta Biotechnol. 22, 13–20.

Caussy, D., Gochfeld, M., Gurzau, E., Neagu, C., 2003. Lessons from case studies of metals: investigating exposure, bioavailability, and risk. Ecotoxicol. Environ. Saf. 54, 45–51.

Chaney, R.L., 1983. Plant uptake of inorganic waste constituents. In: Parr, J.F., Marsh, P.D., Kla, J.M. (Eds.), Land Treatment of Hazardous Wastes. Noyes Data Corporation, Park Ridge, NJ, pp. 50–76.

Chaney, R.L., Malik, M., Li, Y.M., Brown, S.L., Brewer, E.P., Angle, J.S., Baker, A.J.M., 1997. Phytoremediation of soil metals. Curr. Opin. Biotechnol. 8, 279–284.

Cheng, G.R., Timilsin, A., 2011. Status and barriers of advanced biofuel technologies: a review. Renew. Energy 36, 3541–3549.

Cheng, P., Zhang, S., Wang, Q., Feng, X., Zhang, S., Sun, Y., Wang, F., 2021. Contribution of nano-zero-valent iron and arbuscular mycorrhizal fungi to phytoremediation of heavy metal-contaminated soil. Nanomaterials 11, 1264. https://doi.org/10.3390/nano11051264.

Chibuike, G.U., Obiora, S.C., 2014. Heavy metal polluted soils: effect on plants and bioremediation methods. Appl. Environ. Soil Sci. 12, 752708.

Chopra, A.K., Pathak, C., Prasad, G., 2009. Scenario of heavy metal contamination in agricultural soil and its management. J. Appl. Nat. Sci. 1 (1), 99–108.

Christakopoulos, P., Li, L.W., Kekos, F., Macris, B.J., 1993. Direct conversion of sorghum carbohydrates to ethanol by a mixed microbial culture. Bioresour. Technol. 45, 89–92.

Chum, L.M., Overend, R.P., 2002. Biomass and renewable fuel. Fuel Bioprocess Technol. 17, 187–195.

Ciura, J., Poniedzialek, M., Sekara, A., Jedrszczyk, E., 2005. The possibility of using crops as metal phytoremediation. Pol. J. Environ. Stud. 14 (1), 17–22.

Conesa, H.M., Evangelou, M.W.H., Robinson, B.H., Schulin, R., 2012. A critical view of current state of phytotechnologies to remediate soils: still a promising tool? Sci. World J. 2012, 173829. https://doi.org/10.1100/2012/173829.

Delorme, T.A., Gagliardi, J.V., Angle, J.S., Chaney, R.L., 2001. Influence of the zinc hyperaccumulator *Thlaspi caerulescens* J. & C. Presl. and the nonmetal accumulator *Trifolium pratense* L. on soil microbial population. Can. J. Microbiol. 47, 773–776.

Demafelis, R.B., El Jirie, N.B., Hourani, K.A., Tongko, B., 2013. Potential of bioethanol production from sweet sorghum in the Philippines: an income analyses for farmers and distilleries. Sugar Tech 15 (3), 225–231.

Dong, X., Chang, Y., Zheng, R., et al., 2021. Phytoremediation of cadmium contaminated soil: impacts on morphological traits, proline content and stomata parameters of sweet Sorghum seedlings. Bull. Environ. Contam. Toxicol. 106, 528–535. https://doi.org/10.1007/s00128-021-03125-7.

Duffus, J.H., 2002. Heavy metals—a meaningless term? Pure Appl. Chem. 74 (5), 793–807.

Duponnois, R., Kisa, M., Assigbetse, K., Prin, Y., Thioulouse, J., Issartel, M., Moulin, P., Lepage, M., 2006. Fluorescent pseudomonads occurring in *Macrotermes subhyalinus* mound structures decrease Cd toxicity and improve its accumulation in sorghum plants. Sci. Total Environ. 370, 391–400.

Duruibe, J.O., Ogwuegbu, M.O.C., Egwurugwu, J.N., 2007. Heavy metal pollution and human biotoxic effects. Int. J. Phys. Sci. 2 (5), 112–118.

Ebrahimian, E., Bybordi, A., 2014. Effect of organic acids on heavy-metal uptake and growth of canola grown in contaminated soil. Commun. Soil Sci. Plant Anal. 45 (13), 1715–1725. https://doi.org/10.1080/00103624.2013.875206.

EEA, 2005. The European Environment—State and Outlook 2005. European Environment Agency, Copenhagen.

Ekefre, D.E., Mahapatra, A.K., Latimore Jr., M., Bellmer, D.D., Jena, U., Whitehead, G.J., Williams, L.A., 2017. Evaluation of three cultivars of sweet sorghum as feedstocks for ethanol production in the Southeast United States. Heliyon 3, e00490.

Epelde, L., Mijangos, I., Becerril, J.M., Garbisu, C., 2009. Soil microbial community as bioindicator of the recovery of soil functioning derived from metal phytoextraction with sorghum. Soil Biol. Biochem. 41, 1788–1794.

Fasani, E., 2012. Plants that hyperaccumulate heavy metals. In: Furini, A. (Ed.), Plants and Heavy Metals. Springer Briefs in Molecular Science. Springer, Dordrecht.

Fässler, E., Evangelou, M.W., Robinson, B.H., Schulin, R., 2010. Effects of indole-3-acetic acid (IAA) on sunflower growth and heavy metal uptake in combination with ethylene diamine disuccinic acid (EDDS). Chemosphere 80 (8), 901–907.

Foster, L.J.R., Moy, Y.P., Rogers, P.L., 2000. Metal binding capabilities of Rhizobium etli and its extracellular polymeric substances. Biotechnol. Lett. 22, 1757–1760.

Franco-Franklin, V., Moreno-Riascos, S., Ghneim-Herrera, T., 2021. Are endophytic bacteria an option for increasing heavy metal tolerance of plants? A meta-analysis of the effect size. Front. Environ. Sci. 8, 603668.

Franzetti, A., Gandolfi, I., Fracchia, L., Van Hamme, J., Gkorezis, P., Roger, M., Ibrahim, B.M., 2014. Biosurfactant use in heavy metal removal from industrial effluents and contaminated sites. In: Kosaric, N., Sukan, F.V. (Eds.), Biosurfactants: Production and Utilization—Processes, Technologies and Economics. CRC Press, Boca Raton, FL, pp. 361–369.

Friedl, A., 2018. Bioethanol from sugar and starch. In: Energy from Organic Materials (Biomass). Springer, pp. 905–924, https://doi.org/10.1007/978-1-4939-7813-7_432.

Ganesan, V., 2008. Rhizoremediation of cadmium soil using a cadmium-resistant plant growth-promoting rhizopseudomonad. Curr. Microbiol. 56, 403–407.

Glick, B.R., 2003. Phytoremediation: synergistic use of plants and bacteria to clean up the environment. Biotechnol. Adv. 21, 383–393.

Glick, B.R., 2014. Bacteria with ACC deaminase can promote plant growth and help to feed the world. Microbiol. Res. 169 (1), 30–39. https://doi.org/10.1016/j.micres.2013.09.009.

Glick, B.R., Patten, C.L., Holguin, G., Penrose, D.M., 1999. Biochemical and Genetic Mechanisms Used by Plant Growth Promoting Bacteria. Imperial College Press, London.

Gomes, H.I., 2012. Phytoremediation for bioenergy: challenges and opportunities. Environ. Technol. Rev. 1 (1), 59–66.

González Henao, S., Ghneim-Herrera, T., 2021. Heavy metals in soils and the remediation potential of bacteria associated with the plant microbiome. Front. Environ. Sci. 9, 604216. https://doi.org/10.3389/fenvs.2021.604216.

Gopalakrishnan, S., Pande, S., Sharma, M., Humayun, P., Kiran, B.K., Sandeep, D., Vidya, M.S., Deepthi, K., Rupela, O., 2011a. Evaluation of actinomycete isolates obtained from herbal vermicompost for biological control of Fusarium wilt of chickpea. Crop Prot. 30, 1070–1078.

Gopalakrishnan, S., Kiran, B.K., Humayun, P., Vidya, M.S., Deepthi, K., Rupela, O., 2011b. Biocontrol of charcoal-rot of sorghum by actinomycetes isolated from herbal vermicompost. J. Biotechnol. 10, 18142–18152.

Gopalakrishnan, S., Humayun, P., Srinivas, S., Vijayabharathi, R., Ratnakumari, B., Rupela, O., 2012. Plant growth-promoting traits of biocontrol potential Streptomyces isolated from herbal vermicompost. Biocontrol Sci. Tech. 22, 1199–1210.

Gopalakrishnan, S., Srinivas, V., Vidya, M.S., Rathore, A., 2013a. Plant growth-promoting activities of Streptomyces spp. in sorghum and rice. Springerplus 574, 1–8.

Gopalakrishnan, S., Srinivas, V., Shravya, A., Prakash, B., Ratnakumari, B., Vijayabharathi, R., Rupela, O., 2013b. Evaluation of Streptomyces spp. for their plant growth-promoting traits in rice. Can. J. Microbiol. 59, 534–539.

Gopalakrishnan, S., Srinivas, V., Alekhya, G., Prakash, B., Kudapa, H., Rathore, A., Varshney, R.K., 2015a. The extent of grain yield and plant growth enhancement by plant growth-promoting broad-spectrum Streptomyces sp. in chickpea. Springerplus 4 (31), 1–10.

Gopalakrishnan, S., Srinivas, V., Alekhya, G., Prakash, B., Kudapa, H., Varshney, R.K., 2015b. Evaluation of broad-spectrum Streptomyces sp. for plant growth promotion traits in chickpea (Cicer arietinum L.). Philippine Agric. Sci. 98, 270–278.

Gopalakrishnan, S., Srinivas, V., Alekhya, G., Prakash, B., Kudapa, H., Varshney, R.K., 2015c. Evaluation of Streptomyces sp. obtained from herbal vermicompost for broad spectrum of plant growth-promoting activities in chickpea. Org. Agric. 5, 123–133.

Gopalakrishnan, S., Vadlamudi, S., Samineni, S., Sameer Kumar, C.V., 2016a. Plant growth-promotion and biofortification of chickpea and pigeonpea through inoculation of biocontrol potential bacteria, isolated from organic soils. Springerplus 5, 1882.

Gopalakrishnan, S., Srinivas, V., Sameer Kumar, C.V., 2016b. Plant growth-promotion traits of Streptomyces sp. in pigeonpea. Legume Perspect. 11, 43–44.

Gopalakrishnan, S., Srinivas, V., Ashok Kumar, A., Umakanth, A.V., Addepally, U., Srinivasa Rao, P., 2020a. Composting of sweet sorghum bagasse and its impact on plant growth-promotion. Sugar Tech 22 (1), 143–156.

Gopalakrishnan, S., Sharma, R., Srinivas, V., Naresh, N., Mishra, S.P., Ankati, S., Pratyusha, S., Govindaraj, M., Gonzalez, S.V., Nervik, S., Simic, N., 2020b. Identification and characterization of a Streptomyces albus strain and its secondary metabolite organophosphate against charcoal rot of Sorghum. Plan. Theory 9, 1727.

Gopalakrishnan, S., Thakur, V., Saxena, R.K., Vadlamudi, S., Purohit, S., Kumar, V., Rathore, A., Chitikineni, A., Varshney, R.K., 2020c. Complete genome sequence of sixteen plant growth promoting *Streptomyces* strains. Sci. Rep. 10, 10294.

Gopalakrishnan, S., Srinivas, V., Naresh, N., Mishra, S.P., Ankati, S., Pratyusha, S., Madhuprakash, J., Govindaraj, M., Sharma, R., 2021. Deciphering the antagonistic effect of *Streptomyces* spp. and host-plant resistance induction against charcoal rot of sorghum. Planta 253, 57.

Gopalakrishnan, S., Srinivas, V., Chand, U., Pratyusha, S., Samineni, S., 2022. *Streptomyces* consortia-mediated plant growth-promotion and yield performance in chickpea. 3 Biotech 12, 318.

Gupta, P., Diwan, B., 2016. Bacterial exopolysaccharide mediated heavy metal removal: a review on biosynthesis, mechanism and remediation strategies. Biotechnol. Rep. 13, 58–71. https://doi.org/10.1016/j.btre.2016.12.006.

He, J., Chen, J.P., 2014. A comprehensive review on biosorption of heavy metals by algal biomass: materials, performances, chemistry, and modeling simulation tools. Bioresour. Technol. 160, 67–78.

Hernández-Allica, J., Becerril, J.M., Garbisu, C., 2008. Assessment of the phytoextraction potential of high biomass crop plants. Environ. Pollut. 152, 32–40.

Hoang, T.-D., Nghiem, N., 2021. Recent developments and current status of commercial production of fuel ethanol. Fermentation 7, 314. https://doi.org/10.3390/fermentation7040314.

IARC, 2012. IARC Monographs on the Evaluation of Carcinogenic Risks to Humans. Arsenic, Metals, Fibers and Dusts. A Review of Human Carcinogens. (100C).

Idris, R., Trifonova, R., Puschenreiter, M., Wenze, W.W., Sessitsch, A., 2004. Bacterial communities associated with flowering plants of the Ni hyperaccumulator *Thlaspi goesingense*. Appl. Environ. Microbiol. 70, 2667–2677.

Jaffré, T., 1980. Étude Écologique du Peuplement Végétal des Sols Dérivés de Roches Ultrabasiques en Nouvelle Calédonie. ORSTOM, Paris, p. 273.

Jaffré, T., Brooks, R.R., Lee, J., Reeves, R.D., 1976. *Sebertia acuminata*: a hyperaccumulator of nickel from New Caledonia. Science 193, 579–580.

Jain, R., Srivastava, S., Solomon, S., Shrivastava, A.K., Chandra, A., 2010. Impact of excess zinc on growth parameters, cell division, nutrient accumulation, photosynthetic pigments and oxidative stress of sugar cane. Acta Physiol. Plant. 32, 979–986.

Järup, L., 2003. Hazards of heavy metal contamination. Br. Med. Bull. 68, 167–182. https://doi.org/10.1093/bmb/ldg032.

Jiang, C.Y., Sheng, X.F., Qian, M., Wang, Q.Y., 2008. Isolation and characterization of a heavy metal resistant *Burkholderia* sp. from heavy metal-contaminated paddy field soil and its potential in promoting plant growth and heavy metal accumulation in metal polluted soil. Chemosphere 72, 157–164.

Joshi, P., Juwarkar, A., 2009. In vivo studies to elucidate the role of extracellular polymeric substances from *Azotobacter* in immobilization of heavy metals. Environ. Sci. Technol. 43, 5884–5889.

Kapahi, M., Sachdeva, S., 2019. Bioremediation options for heavy metal pollution. J. Health Pollut. 9 (24), 191203.

Kaplan, M., Orman, S., Kadar, I., Koncz, J., 2005. Heavy metal accumulation in calcareous soil and sorghum plants after addition of sulphur-containing waste as a soil amendment in Turkey. Agric. Ecosyst. Environ. 111, 41–46.

Kapoor, A., Viraraghavan, T., Cullimore, R.D., 1999. Removal of heavy metals using the fungus *Aspergillus niger*. Bioresour. Technol. 70, 95–104.

Kargi, F., Curme, J.A., 1985. Solid state fermentation of sweet sorghum to ethanol in a rotary-drum fermentor. Biotechnol. Bioeng. 27, 1122–1125.

Kargi, F., Curme, J.A., Sheehan, J.J., 1985. Solid state fermentation of sweet sorghum to ethanol. Biotechnol. Bioeng. 27, 34–40.

Keller, C., Hammer, D., Kayser, A., Richner, W., Brodbeck, M., Sennhauser, M., 2003. Root development and heavy metal phytoextraction efficiency: comparison of different plant species in the field. Plant Soil 249 (1), 67–81.

Kim, M., Day, D.F., 2011. Composition of sugar cane, energy cane, and sweet sorghum suitable for ethanol production at Louisiana sugar mills. J. Ind. Microbiol. Biotechnol. 38 (7), 803–807.

Kojima, M., Johnson, T., 2005. Potential for Biofuels for Transport in Developing Countries. Energy Sector Management Assistance Programme Report. The International Bank for Reconstruction and Development/The World Bank, Washington, DC.

Koller, M., Saleh, H.M., 2018. Introductory chapter: introducing heavy metals. In: Heavy Metals. InTech, https://doi.org/10.5772/intechopen.74783.

Komárek, M., Tlustoš, P., Szákova, J., Richner, W., Brodbeck, M., Sennhauser, M., 2007. The use of maize and poplar in chelant-enhanced phytoextraction of lead from contaminated agricultural soils. Chemosphere 67 (4), 640–651.

Kovarik, B., 1998. Henry ford, Charles Kettering and the fuel of the future. Automot. Hist. Rev. 32, 7–27 (Spring).

Kuffner, M., Puschenreiter, M., Wieshammer, G., Gorfer, M., Sessitsch, A., 2008. Rhizosphere bacteria affect growth and metal uptake of heavy metal accumulating willows. Plant Soil 304, 35–44.

Kukier, U., Chaney, R.L., 2004. In situ remediation of nickel phytotoxicity for different plant species. J. Plant Nutr. 27, 465–495.

Ledin, M., Krantz-Rulcker, C., Allard, B., 1996. Zn, Cd, and Hg accumulation by microorganisms, organic and inorganic soil components in multicompartment system. Soil Biol. Biochem. 28, 791–799.

Lee, J., Sung, K., 2014. Effects of chelates on soil microbial properties, plant growth and heavy metal accumulation in plants. Ecol. Eng. 73, 386–394. https://doi.org/10.1016/j.ecoleng.2014.09.053.

Lezinou, V., Christakopoulos, P., Kekos, D., Macris, B.J., 1994. Simultaneous saccharification and fermentation of sweet sorghum polysaccharides to ethanol in a fed-batch process. Biotechnol. Lett. 16, 983–988.

Li, X., Coles, B.J., Ramsey, M.H., Thornton, I., 1995. Chemical partitioning of the new National Institute of Standards and Technology standard reference materials (SRM 2709-2711) by sequential extraction using inductively coupled plasma atomic emission spectrometry. Analyst 120, 1415–1419.

Li, J.L., He, M., Han, W., Gu, Y.F., 2009. Analysis and assessment on heavy metal sources in the coastal soils developed from alluvial deposits using multivariate statistical methods. J. Hazard. Mater. 164, 976–981.

Li, X.Q., Liu, Y.Q., Li, Y.J., Han, H., Zhang, H., Ji, M.F., Chen, Z.J., 2022. Enhancing mechanisms of the plant growth-promoting bacterial strain *Brevibacillus* sp. SR-9 on cadmium enrichment in sweet Sorghum by metagenomic and transcriptomic analysis. Int. J. Environ. Res. Public Health 19, 16309.

Liedekerke, M.V., Yigini, Y., Montanarella, L., 2013. Contaminated sites in Europe: review of the current situation based on data collected through a European network. J. Environ. Hum. Health 2013, 158764.

Liu, Y., Lam, M.C., Fang, H.H.P., 2001. Adsorption of heavy metals by EPS of activated sludge. Water Sci. Technol. 43, 59–66.

Liu, L., Li, W., Song, W., Guo, M., 2018. Remediation techniques for heavy metal-contaminated soils: principles and applicability. Sci. Total Environ. 633, 206–219. https://doi.org/10.1016/j.scitotenv.2018.03.161.

Ma, Y., Prasad, M.N.V., Rajkumar, M., Freitas, H., 2011. Plant growth promoting rhizobacteria and endophytes accelerate phytoremediation of metalliferous soils. Biotechnol. Adv. 29, 248–258.

Madejón, P., Murillo, J.M., Marañón, T., Cabrera, F., Soriano, M.A., 2003. Trace element and nutrient accumulation in sunflower plants two years after the Aznalcóllar spill. Sci. Total Environ. 307 (1–3), 239–257.

Mahar, A., Wang, P., Ali, A., Awasthi, M.K., Lahori, A.H., Wang, Q., Li, R., Zhang, Z., 2016a. Challenges and opportunities in the phytoremediation of heavy metals contaminated soils: a review. Ecotoxicol. Environ. Saf. 126, 111–121. https://doi.org/10.1016/j.ecoenv.2015.12.023.

Mahar, A., Wang, P., Ali, A., Awasthi, M.K., Lahori, A.H., Wang, Q., Li, R., Zhang, Z., 2016b. Challenges and opportunities in the phytoremediation of heavy metals contaminated soils: a review. Ecotoxicol. Environ. Saf. 126, 111–121.

Mamma, D., Christakopoulos, P., Koullas, D., Macris, K.B.J., Koukios, E., 1995. An alternative approach to the bioconversion of sweet sorghum carbohydrates to ethanol. Biomass Bioenergy 8 (2), 99–103.

Marchiol, L., Fellet, G., Perosa, D., Zerbi, G., 2007. Removal of trace metals by *Sorghum bicolor* and *Helianthus annuus* in a site polluted by industrial wastes: a field experience. Plant Physiol. Biochem. 45, 379–387.

Martin, K., 2019. Energy from Organic Materials (Biomass) (A Volume in the Encyclopedia of Sustainability Science and Technology, Second Edition). Bioethanol from Sugar and Starch, Springer, pp. 905–924 (Chapter 432).

McGowen, S.L., Basta, N.T., Brown, G.O., 2001. Use of diammonium phosphate to reduce heavy metal solubility and transport in smelter-contaminated soil. J. Environ. Qual. 30 (2), 493–500. https://doi.org/10.2134/jeq2001.302493x.

McGrath, S.P., Zhao, F.J., Lombi, E., 2001. Plant and rhizosphere process involved in phytoremediation of metal-contaminated soils. Plant Soil 232, 207–214.

Meers, E., Ruttens, A., Hopgood, M., Lesage, E., Tack, F.M.G., 2005. Potential of *Brassica rapa*, *Cannabis sativa*, *Helianthus annuus* and *Zea mays* for phytoextraction of heavy metals from calcareous dredged sediment derived soils. Chemosphere 61 (4), 561–572.

Meers, E., Slycken, S.V., Adriaensen, K., Ruttens, A., Vangronsveld, J., Laing, G.D., Witters, N., Thewys, T., Tack, F.-M.G., 2010. The use of bio-energy crops (*Zea mays*) for phytoattenuation of heavy metals on moderately contaminated soils: a field experiment. Chemosphere 78, 35–41.

Mehmood, K., Ahmad, H., Abbas, R., Murtaza, G., Ullah, S., 2019. Heavy metals in urban and peri-urban soils of a heavily-populated and industrialized city: assessment of ecological risks and human health repercussions. Hum. Ecol. Risk. Assess. https://doi.org/10.1080/10807039.2019.1601004.

Mench, M., Lepp, N., Bert, V., Schwitzguébel, J.P., Gawronski, S.W., Schröder, P., et al., 2010. Successes and limitations of phytotechnologies at field scale: outcomes, assessment and outlook from COST action 859. J. Soils Sediments 10, 1039–1070. https://doi.org/10.1007/s11368-010-0190-x.

Millward, G.E., Turner, A., 2001. Metal pollution. In: Steele, J.H., Turekian, K.K., Thorpe, S.A. (Eds.), Encyclopedia of Ocean Sciences. Academic Press, San Diego, CA, pp. 1730–1737.

Minteer, S., 2006. Alcoholic Fuels. Taylor & Francis, New York.

Mishra, J., Rachna, S., Arora Naveen, K., 2017. Alleviation of heavy metal stress in plants and remediation of soil by rhizosphere microorganisms. Front. Microbiol. 8, 1706.

Mishra, S., Lin, Z., Pang, S., Zang, Y., Bhatt, P., Chen, S., 2021. Biosurfactant is a powerful tool for the bioremediation of heavy metals from contaminated soils. J. Hazard. Mater. 418, 126253.

Moffat, A., 1995. Plants proving their worth in toxic metal cleanup. Science 269, 302–303.

Mudgal, V., Madaan, N., Mudgal, A., Singh, R.B., Mishra, S., 2010. Effect of toxic metals on human health. Open Nutraceut. J. 3, 94–99.

Mussatto, S.I., Dragone, G., Guimarães, P.M.R., Paulo, J., Silva, A., Carneiro, L.M., Roberto, I.C., Vicente, A., Domingues, L., Teixeira, J.A., 2010. Technological trends, global market, and challenges of bio-ethanol production. Biotechnol. Adv. 28, 817–830.

Muthusaravanan, S., Sivarajasekar, N., Vivek, J.S., Paramasivan, T., Naushad, M., Prakashmaran, J., Gayathri, V., Al-Duajj, O.K., 2018. Phytoremediation of heavy metals: mechanisms, methods and enhancements. Environ. Chem. Lett. 16, 1339–1359. https://doi.org/10.1007/s10311-018-0762-3.

Newes, E., Clarck, C.M., Vimmersted, L., Peterson, S., Burkholder, D., Korontey, D., Inman, D., 2022. Ethanol production in the United States: the roles of policy, price, and demand. Energy Policy 161, 112713.

Nriagu, J.O., Pacyna, J.M., 1988. Quantitative assessment of worldwide contamination of air, water and soils by trace metals. Nature 333, 134–139.

Ojuederie, O., Babalola, O., 2017. Microbial and plant-assisted bioremediation of heavy metal polluted environments: a review. Int. J. Environ. Res. Public Health 14 (12), 1504. https://doi.org/10.3390/ijerph14121504.

Oyewole, O.A., Zobeashia, S.S.L.T., Oladoja, E.O., Raji, O.R.O., Odiniya, E.E., Musa, A.M., 2019. Biosorption of heavy metal polluted soil using bacteria and fungi isolated from soil. SN Appl. Sci. 1, 857. https://doi.org/10.1007/s42452-019-0879-4.

Pandey, B., Surindra, S., Vineet, S., 2016. Accumulation and health risk of heavy metals in sugarcane irrigated with industrial effluent in some rural areas of Uttarakhand, India. Process Saf. Environ. Prot. 102, 655–666.

Penrose, D.M., Glick, B.R., 2001. Levels of 1-aminocyclopropane-1-carboxylic acid (ACC) in exudates and extracts of canola seeds treated with plant growth-promoting bacteria. Can. J. Microbiol. 47 (4), 368–372.

Pratyusha, S., Gopalakrishnan, S., 2023. *Streptomyces*-mediated synthesis of silver nanoparticles for enhanced growth, yield and grain nutrients in chickpea. Biocatal. Agric. Biotechnol. 47, 102567.

Purseglove, J.W., 1972. Tropical Crops: Monocotyledons. Longman Group Limited, London, p. 334.

Quartacci, M.F., Micaelli, F., Sgherri, C., 2014. *Brassica carinata* planting pattern influences phytoextraction of metals from a multiple contaminated soil. Agrochimica 58 (1).

Ragnarsdottir, K.V., Hawkins, D., 2005. Trace metals in soils and their relationship with scrapie occurrence. Geochim. Cosmochim. Acta 69, A194–A196.

Rajkumar, M., Ma, Y., Freitas, H., 2008. Characterization of metal-resistant plant-growth promoting *Bacillus weihenstephanensis* isolated from serpentine soil in Portugal. J. Basic Microbiol. 48, 1–9.

Rajkumar, M., Ae, N., Prasad, M.N.V., Freitas, H., 2010. Potential of siderophore-producing bacteria for improving heavy metal phytoextraction. Trends Biotechnol. 28, 142–149.

Rajkumar, M., Sandhya, S., Prasad, M.N.V., Freitas, H., 2012. Perspectives of plant-associated microbes in heavy metal phytoremediation. Biotechnol. Adv. 30, 1562–1574.

Rakhmetova, S.O., Vergun, O.M., Blume, R.Y., Bondarchuk, O.P., Shymanska, O.V., Tsygankov, S.P., et al., 2020. Ethanol production potential of sweet sorghum in North and Central Ukraine. Open Agric. J. 14, 321–338. https://doi.org/10.2174/1874331502014010321.

Rama Jyothi, N., 2021. Heavy metal sources and their effects on human health. In: Heavy Metals—Their Environmental Impacts and Mitigation. Intech Open, https://doi.org/10.5772/intechopen.95370.

Rao, P.S., Rao, S.S., Seetharama, N., Umakath, A.V., Reddy, P.S., Reddy, B.V.S., Gowda, C.L.L., 2009. Sweet Sorghum for Biofuel and Strategies for its Improvement. International Crops Research Institute for the Semi-Arid Tropics, Andhra Pradesh, India.

Rao, P.S., Reddy, B.V.S., Reddy, R.C., Blümmel, M., Kumar, A.A., Rao, P.P., Basavaraj, G., 2012. Utilizing co-products of the sweet sorghum based biofuel industry as livestock feed in decentralized systems. In: Makkar, H.P.S. (Ed.), Biofuel Co-Products as Livestock Feed: Opportunities and Challenges. FAO, Rome, Italy, pp. 229–242.

Rao, P.S., Kumar, C.G., Reddy, B.V., 2013. Sweet sorghum: from theory to practice. In: Rao, P.S., Kumar, C.G. (Eds.), Characterization of Improved Sweet Sorghum Cultivars. Springer, India, pp. 1–15.

Rao, P.S., Vinutha, K.S., Anil Kumar, G.S., Chiranjeevi, T., Uma, A., Lal, P., Prakasham, R.S., et al., 2019. Sorghum: a multipurpose bioenergy crop. In: Sorghum: A State of the Art and Future Perspetives 58. American Society of Agronomy and Crop Science Society of America, Inc, pp. 399–424.

Rascio, N., Navari-Izzo, F., 2011. Heavy metal hyperaccumulating plants: how and why do they do it? And what makes them so interesting? Plant Sci. 180 (2), 169–181.

Rayment, G.E., Jeffrey, A.J., Barry, G.A., 2002. Heavy metals in Australian sugar cane. Commun. Soil Sci. Plant Anal. 33, 3203–3212.

Reddy, B.V., Ashok Kumar, A., Ramesh, S., 2007. Sweet sorghum: a water saving bio-energy crop. In: International Conference on Linkages Between Energy and Water Management for Agriculture in Developing Countries, January 29–30, 2007. IWMI, ICRISAT, Hyderabad, India.

Reddy, B.V.S., Layaoen, H., Dar, W.D., Rao, P.S., Eusebio, J.E., 2011. Sweet Sorghum in the Philippines: Status and Future. International Crops Research Institute for the Semi-Arid Tropics, Andhra Pradesh, India.

Reeves, R.D., Baker, A.J.M., Jaffré, T., Erskine, P.D., Echevarria, G., van der Ent, A., 2018. A global database for plants that hyperaccumulate metal and metalloid trace elements. New Phytol. 218, 407–411. https://doi.org/10.1111/nph.14907.

Rodionova, M.V., Poudyal, R.S., Tiwari, I., Voloshin, R.A., Zharmukhamedov, S.K., Nam, H.G., Zayadan, B.K., Bruce, B.D., Hou, H.J.M., Allakhverdiev, S.I., 2017. Biofuel production: challenges and opportunities. Int. J. Hydrog. Energy. https://doi.org/10.1016/j.ijhydene.2016.11.125.

Rugh, C.L., Wilde, H.D., Stack, N.M., Thompson, D.M., Summers, A.O., Meagher, R.B., 1996. Mercuric ion reduction and resistance in transgenic *Arabidopsis thaliana* plants expressing a modified bacterial *merA* gene. Proc. Natl. Acad. Sci. 93, 3182–3187.

Salt, D.E., Blaylock, M., Kumar, N.P.B.A., Dushenkov, V., Ensley, B.D., Chet, I., Raskin, I., 1977. Phytoremediation: a novel strategy for the removal of toxic metals from the environment using plants. Nat. Biotechnol. 13, 468–475.

Sanderson, M.A., Jones, R.M., Ward, J., Wolfe, R., 1992. Silage sorghum performance trial at Stephen-ville. In: Forage Research in Texas. Report PR-5018. Texas Agricultural Experimental Station, Stephenville, USA.

Saravanan, V.S., Madhaiyan, M., Thangaraju, M., 2007. Solubilization of zinc compounds by the diazotrophic, plant growth promoting bacterium *Gluconacetobacter diazotrophicus*. Chemosphere 66, 1794–1798.

Sarawal, R., Kumar, S., Mehta, A., Varadan, A., Singh, K.S., Ramakumar, S.S.V., Mathai, R., 2021. Road map for ethanol blending in India 2020-25. NITI Ayog. ISBN: 978-81-949510-9-4.

Sathya, A., Vijayabharathi, R., Srinivas, V., Gopalakrishnan, S., 2016. Plant growth-promoting actinobacteria on chickpea seed mineral density: an upcoming complementary tool for sustainable biofortification strategy. 3 Biotech 6, 138.

Segura-Muñoz, S.I., da Silva, O.A., Nikaido, M., Trevilato, T.M., Bocio, A., Takayanagui, A.M., Domingo, J.L., 2006. Metal levels in sugar cane (*Saccharum* spp.) samples from an area under the influence of a municipal landfill and a medical waste treatment system in Brazil. Environ. Int. 32 (1), 52–57.

Serna-Saldívar, S.O., Chuck Hernandez, C., Perez Carrillo, E., Heredia-Olea, E., 2012. Sorghum as a multifunctional crop for the production of fuel ethanol: current status and future trends. In: Lima, M.A.P. (Ed.), Bioethanol. InTech, China, pp. 51–74.

Sethy, S.K., Ghosh, S., 2013. Effect of heavy metals on germination of seeds. J. Nat. Sci. Biol. Med. 4 (2), 272–275.

Shackira, A.M., Puthur, J.T., 2019. Phytostabilization of heavy metals: understanding of principles and practices. In: Srivastava, S., Srivastava, A., Suprasanna, P. (Eds.), Plant-Metal Interactions. Springer, Cham, https://doi.org/10.1007/978-3-030-20732-8_13.

Shafiei Darabi, S.A., Almodares, A., Ebrahimi, M., 2016. Phytoremediation efficiency of *Sorghum bicolor* (L.) Moench in removing cadmium, lead and arsenic. Open J. Environ. Biol. 1 (1), 001–006.

Sharma, P., Pandey, S., 2014. Status of phytoremediation in world scenario. Int. J. Environ. Bioremediation Biodegrad. 2 (4), 178–191.

Sharma, P., Rai, S., Gautam, K., Sharma, S., 2023. Phytoremediation strategies of plants: challenges and opportunities. In: Plants and Their Interactions to Environmental Pollution. Elsevier, https://doi.org/10.1016/B978-0-323-99978-6.00012-1.

Shehata, S.M., Badawy, R.K., Aboulsoud, Y.I.E., 2019. Phytoremediation of some heavy metals in contaminated soil. Bull. Natl. Res. Cent. 43, 189. https://doi.org/10.1186/s42269-019-0214-7.

Shen, Z.G., Li, X.D., Wang, C.C., Chen, H.M., Chua, H., 2002. Lead phytoextraction from contaminated soil with high-biomass plant species. J. Environ. Qual. 31 (6), 1893–1900.

Shi, J.Y., Lin, H.R., Yuan, X.F., Chen, X.C., Shen, C.F., Chen, Y.X., 2011. Enhancement of copper availability and microbial community changes in rice rhizospheres affected by sulfur. Molecules 16, 1409–1417.

Shi, P., Xing, Z., Zhang, X., Chai, T.-Y., 2017. Effect of heavy-metals on synthesis of siderophores by *Pseudomonas aeruginosa* ZGKD3. IOP Conf. Ser.: Earth Environ. Sci. 52, 102103.

Simpson, J.A., Picchi, G., Gordon, A.M., Thevathasan, N.V., Stanturf, J., Nicholas, I., 2009. Short Rotation Crops for Bioenergy Systems. Environmental Benefits Associated with Short-Rotation Woody Crops. IEA, Paris. Technical Review No. 3, IEA Bioenergy.

Singh, S., Kumar, M., 2006. Heavy metal load of soil, water and vegetables in peri-urban Delhi. Environ. Monit. Assess. 120 (1–3), 79–91.

Soriano, M.A., Fereres, E., 2003. Use of crops for in situ phytoremediation of polluted soils following a toxic flood from a mine spill. Plant Soil 256 (2), 253–264.

Soudek, P., Petrová, S., Vaněk, T., 2012. Phytostabilization or accumulation of heavy metals by using of energy crop *Sorghum* sp. Int. Proc. Chem. Biol. Environ. Eng. 46, 25–29.

Sun, L.N., Zhang, Y.F., He, L.Y., Chen, Z.J., Wang, Q.Y., Qian, M., Sheng, X.F., 2010. Genetic diversity and characterization of heavy metal-resistant-endophytic bacteria from two copper-tolerant plant species on copper mine wasteland. Bioresour. Technol. 101, 501–509.

Umakanth, A.V., Kumar, A.A., Vermerris, W., 2019. Sweet sorghum for biofuel industry. In: Breeding Sorghum for Diverse End Uses. Woodhead Publishing, pp. 255–270, https://doi.org/10.1016/b978-0-08-101879-8.00016-4.

U.S. Energy Information Administration, June 2022. Monthly Energy Review (preliminary data) https://www.eia.gov/energyexplained/biofuels/.

van der Ent, A., Baker, A.J.M., Reeves, R.D., Pollard, A.J., Schat, H., 2012. Hyperaccumulators of metal and metalloid trace elements: facts and fiction. Plant Soil 362 (1–2), 319–334. https://doi.org/10.1007/s11104-012-1287-3.

van der Lelie, D., Schwitzguébel, J.P., Glass, D.J., Gronsveld, J.V., Baker, A., 2001. Assessing phytoremediation's progress. Environ. Sci. Technol. 35 (21), 447–452.

Vervaeke, P., Luyssaert, S., Mertens, J., Meers, E., Tack, F.M.G., Lust, N., 2003. Phytoremediation prospects of willow stands on contaminated sediment: a field trial. Environ. Pollut. 126 (2), 275–282.

Vijayaraghavan, K., Yun, Y.S., 2008. Bacterial biosorbents and biosorption. Biotechnol. Adv. 26, 266–291.

Vinutha, K.S., Rayaprolu, L., Yadagiri, K., Umakanth, A.V., Patil, J.V., Rao, P.S., 2014. Sweet sorghum research and development in India: status and prospects. Sugar Tech 16 (2), 133–143.

Vohra, M., Manwar, J., Manmode, R., Padgilwar, S., Patil, S., 2014. Bioethanol production: feedstock and current technologies. J. Environ. Chem. Eng. 2, 573–584.

Volesky, B., Holan, Z.R., 1995. Biosorption of heavy metals. Biotechnol. Prog. 11, 235–250.

Wang, J.L., Chen, C., 2006. Biosorption of heavy metals by *Saccharomyces cerevisiae*: a review. Biotechnol. Adv. 24, 427–451.

Wang, J., Chen, C., 2009. Biosorbents for heavy metals removal and their future. Biotechnol. Adv. 27, 195–226.

Wang, F.Y., Cheng, P., Zhang, S.Q., Zhang, S.W., Sun, Y.H., 2022. Contribution of arbuscular mycorrhizal fungi and soil amendments to remediation of a heavy metal contaminated soil using sweet sorghum. Pedosphere 32 (6), 844–855.

Wani, P.A., Khan, M.S., Zaidi, A., 2007. Chromium reduction, plant growth-promoting potentials, and metal solubilization by *Bacillus* sp. isolated from alluvial soil. Curr. Microbiol. 54, 237–243.

Weyens, N., van der Lelie, D., Taghavi, S., Vangronsveld, J., 2009. Phytoremediation: plant endophyte partnerships take the challenge. Curr. Opin. Biotechnol. 20, 248–254.

Wongsasuluk, P., Chotpantarat, S., Siriwong, W., Robson, M., 2014. Heavy metal contamination and human health risk assessment in drinking water from shallow groundwater wells in an agricultural area in Ubon Ratchathani province, Thailand. Environ. Geochem. Health 36 (1), 169–182.

Wu, C.H., Wood, T.K., Mulchandani, A., Chen, W., 2006a. Engineering plant-microbe symbiosis for rhizoremediation of heavy metals. Appl. Environ. Microbiol. 72, 1129–1134.

Wu, S.C., Cheung, K.C., Luo, Y.M., Wong, M.H., 2006b. Effects of inoculation of plant growth promoting rhizobacteria on metal uptake by *Brassica juncea*. Environ. Pollut. 140, 124–135.

Wu, X., Staggenborg, S., Propheter, J.L., Rooney, W.L., Yu, J., Wang, D., 2010. Features of sweet sorghum juice and their performance in ethanol fermentation. Ind. Crop. Prod. 31 (1), 164–170.

Wu, Z., McGrouther, K., Chen, D., Wu, W., Wang, H., 2013. Subcellular distribution of metals within *Brassica chinensis* L. in response to elevated lead and chromium stress. J. Agric. Food Chem. 61 (20), 4715–4722.

Wu, J., Zhao, N., Zhang, P., Zhu, L., Lu, Y., Lei, X., Bai, Z., 2023. Nitrate enhances cadmium accumulation through modulating sulfur metabolism in sweet sorghum. Chemosphere 313, 137413. https://doi.org/10.1016/j.chemosphere.2022. PMID: 36455657.

Wuana, R.A., Okieimen, F.E., 2010. Phytoremediation potential of maize (*Zea mays* L.). A review. Afr. J. Gen. Agric. 6 (4), 275–287.

Wuana, R.A., Okieimen, F.E., 2011. Heavy metals in contaminated soils: a review of sources, chemistry, risks and best available strategies for remediation. ISRN Ecol., 402647. https://doi.org/10.5402/2011/402647.

Xia, H., Yan, Z., Chi, X., Cheng, W., 2009. Evaluation of the phytoremediation potential of *Saccharum officinarum* for Cd-contaminated soil. In: International Conference on Energy and Environment Technology., https://doi.org/10.1109/iceet.2009.541.

Xiao, M.-Z., Sun, Q., Hong, S., Chen, W.-J., Pong, B., Du, Z.-Y., Yang, W.-B., Sun, Z., Yuan, T.-Q., 2021. Sweet sorghum for phytoremediation and bioethanol production. J. Leather Sci. Eng. 3 (33), 2–23.

Xie, Q., Tang, S., Wang, Z., Chen, C., Xie, P., Xie, Q., 2018. The prospect of sweet sorghum as the source for high biomass crop. J. Agric. Sci. Bot. 2 (3), 5–11.

Yadav, D.V., Jain, R., Rai, R.K., 2010. Impact of heavy metals on sugar cane. In: Sherameti, I., Varma, A. (Eds.), Soil Heavy Metals. Springer, Berlin Heidelberg, pp. 339–367.

Yadav, D.V., Jain, R., Rai, R.K., 2011. Detoxification of heavy metals from soils through sugar crops. In: Sherameti, I., Varma, A. (Eds.), Detoxification of Heavy Metals. Springer, Berlin Heidelberg, pp. 389–405.

Yan, A., Wang, Y., Tan, S.N., Mohd Yusof, M.L., Ghosh, S., Chen, Z., 2020. Phytoremediation: a promising approach for revegetation of heavy metal-polluted land. Front. Plant Sci. 11, 359. https://doi.org/10.3389/fpls.2020.00359.

Yao, Z., Li, J., Xie, H., Yu, C., 2012. Review on remediation technologies of soil contaminated by heavy metals. Procedia Environ. Sci. 16, 722–729.

Yu, G., Ma, J., Jiang, P., Li, J., Gao, J., Qiao, S., Zhao, Z., 2019. IOP Conf. Ser.: Earth Environ. Sci. 310, 052004.

Yuan, X., Xiong, T., Yao, S., Liu, C., Yin, Y., Li, H., Li, N., 2019. A real field phytoremediation of multi-metals contaminated soils by selected hybrid sweet sorghum with high biomass and high accumulation ability. Chemosphere 237, 124536.

Zaidi, S., Usmani, S., Singh, B.R., Musarrat, J., 2006. Significance of *Bacillus subtilis* strain SJ-101 as a bioinoculant for concurrent plant growth promotion and nickel accumulation in *Brassica juncea*. Chemosphere 64, 991–997.

Zatta, P., 2001. Metals and the brain: from neurochemistry to neurodegeneration. Brain Res. Bull. 55 (2), 123–124.

Zgorelec, Z., Bilandzija, N., Knez, K., et al., 2020. Cadmium and mercury phytostabilization from soil using *Miscanthus giganteus*. Sci. Rep. 10, 6685. https://doi.org/10.1038/s41598-020-63488-5.

Zhuang, P., Shu, W., Li, Z., Liao, B., Li, J., Shao, J., 2009. Removal of metals by sorghum plants from contaminated land. J. Environ. Sci. 21, 1432–1437.

14

Bioremediation and bioeconomy potential of sunflower (*Helianthus annuus* L.)

Majeti Narasimha Vara Prasad

Department of Plant Sciences, School of Life Sciences, University of Hyderabad
(An Institution of Eminence), Hyderabad, Telangana, India

1 Introduction

An English patent with the number 408 from 1716 is the first evidence of utilizing sunflower seeds as a source of oil in Europe. Arthur Bunyan was awarded in 1716 for his invention of "How from a certain English seed might be expressed a good sweet oil of great use to all persons concerned in the manufacture of woolen, painters, leather dressers, etc." This oil is to be made from the seed of the flowers commonly called and known by the name of sunflowers of all sorts, both double and single. The patent clearly has two meanings. First, the oil was created for industrial purposes rather than for human use, and second, there were many different plant types.

The vast majority of the 120 million tonnes or more of fats and oils produced globally in 2004 were utilized in diets for humans. There were 25 million tonnes available for oleochemistry. The subsequent utilization of the fatty acids in the oil depends on their makeup. Coconut and palm kernel oils require special consideration due to their large proportion of fatty acids with short or medium chain lengths (mainly C12 and C14). These, for instance, are very well suited for conversion into surfactants. Long-chain fatty acids (C18) are the predominant component of palm, soybean, rapeseed, and sunflower oils, which are utilized as raw materials for lubricants and polymer applications. Sunflower seed oil high oleic acid content has several innovative uses.

2 Research focus on Sunflower industrial applications

Research focuses on the plant's industrial applications (such as the manufacturing of rubber). This chapter will cover agronomic and genetic advancements, as well as food and industrial applications of the usage of sunflower oil. Recently, the price of gasoline rose to as high as 78 US dollars per barrel, with the Golden Sachs Bank officially predicting that it may soon hit 100 dollars per barrel. The primary drivers of this trend are thought to be the rapidly declining natural sources of fossil fuels and the rising global demand for energy, with special emphasis on many emerging Asian nations. According to British Petroleum specialists, humans have already used up 42% of the world's known fossil oil reserves.

Diversifying energy sources is highly desired on a global scale in order to better prepare for any future oil crises. Some tropical nations (Brazil, China, Indonesia, India, Vietnam, etc.) have recently planted several million hectares with oil palm, the crop that produces the most oil, to make biodiesel. To contribute to the diversification of the energy supply, industrialized and emerging temperate countries are also intending to enhance the production of suitable oilseed crops (soybean, sunflower, rape, etc.).

Clean energy technologies are now cost-competitive with their "dirtier" equivalents for the first time in modern history. As nuclear and coal-based energy continue to be plagued by environmental and safety problems and as oil and natural gas prices remain stubbornly high and frustratingly variable on a global scale, renewable energy costs continue their almost inexorable downward trend.

Suddenly, the so-called "alternative" energy technologies are looking pretty mainstream. Clean energy markets are expanding, which shows how well accepted they are. In 2005, the global markets for wind and solar energy reached $11.8 and $11.2 billion, respectively, growing 47% and 55% from the previous year. In 2005, the worldwide market for biofuels reached $15.7 billion, an increase of more than 5% from the year before.

Because sunflowers are drought tolerant and better adapted to Mediterranean environments, they are not grown in Southern Europe where rapeseed yield is abundant and commercially viable. High-oleic sunflower oil produces a premium biodiesel that is well suited to regions with harsh winters.

3 Sunflower for Phytotechnology

There has been a general feeling among scientists that phytoremoval of environmental contaminants and pollutants is only a temporary solution. However, the large body of scientific information that is available as on today erased this disbelief, and new areas are emerging in the field of phytotechnologies (McCutcheon and Schnoor, 2003). Sunflower (*Helianthus annuus* L.) is one of the most promising environmental crops that is being used in diverse situations for environmental cleanup.

Sunflower in early days is a popular ornamental. However, in recent years its importance as an environmental crop is being increasingly recognized. Dehulled seeds are used in poultry feed (Table 1). Agronomic experiments conducted on a farm research site in India using recycled organic manure from integrated farming system (cows, goats, poultry, etc.) have substantially increased the growth and yield (data not shown). In agronomic trials in a typical Mediterranean climate where winter precipitations average about 500 mm, brackish

TABLE 1 Sunflower seed oil constituents and composition.

Whole seed		Fatty acid content in oil	
Constituent	Composition	FA	Range (%)
Hull	21–27	Myristic	5–7
Oil	48–53	Palmitic	3–5
Protein	14–19	Stearic	0.3–0.8
Soluble sugars	7–9	Arachidic	0.6–0.8
Fiber	16–27	Oleic	22–50
Ash	2–3	Linoleic	40–70

water-irrigated sunflower crops' performance and productivity are satisfactory, contributing to sustainable agriculture and also finding an alternative solution to drought.

At a contaminated wastewater site in Ashtabula, Ohio, 4-week-old sunflowers were able to remove more than 95% of uranium in 24h (Dushenkov et al., 1997a, 1997b, 1999). Except for sunflower (*Helianthus annuus*) and tobacco (*Nicotiana tabacum*), other non-Brassica plants had phytoextraction coefficients less than 1. Rhizofiltration has been employed using sunflower at a US Department of Energy (DOE) pilot project with uranium wastes at Ashtabula, Ohio, and on water from a pond near the Chernobyl nuclear plant in Ukraine. Sunflowers accumulated Cs and Sr, with Cs remaining in the roots and Sr moving into the shoots (Dushenkov and Kapulnik, 2002) (Fig. 1). Soils from an abandoned pasture, a forest, and a floodplain near

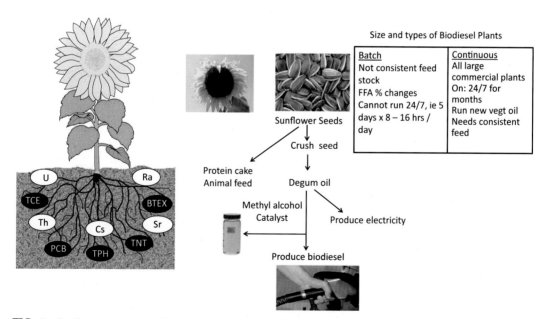

FIG. 1 Sunflower is a potential environmental crop for the cleanup of inorganic and organic contaminants and pollutants. Such an approach is low cost, a green technology, and solar driven. It has government and public support.

III. Ornamentals and tree crops for contaminated substrates

Cincinnati, OH, were cleaned using an association of plants comprising sunflower, timothy grass, and red clover and accelerated the mineralization of 2,4,5-trichlorophenoxyacetic acid (2,4,5-T) (Fig. 1).

Uranium (U) contamination of groundwater poses a serious environmental problem in uranium mining areas and in the vicinity of nuclear processing facilities. Preliminary laboratory experiments and treatability studies indicate that the roots of terrestrial plants can be efficiently used to remove U from aqueous streams (rhizofiltration). Almost all of the U removed from the water in the laboratory using sunflower plants was concentrated in the roots. Rhizofiltration technology has been tested in the field with U-contaminated water at concentrations of $21–874\,\mu g/L$ at a former U processing facility in Ashtabula, OH. The pilot-scale rhizofiltration system provided final treatment to the site source water and reduced the U concentration to $<20\,\mu g/L$ before discharge to the environment. System performance was subsequently evaluated under different flow rates, permitting the development of effectiveness estimates for the approach (Dushenkov et al., 1997a,b) (Figs. 2 and 3).

Terrestrial plants are thought to be more suitable for rhizofiltration because they produce longer, more substantial, and often fibrous root systems with large surface areas for metal sorption. Sunflower (*Helianthus annuus* L.) removed Pb (Dushenkov et al., 1995), U (Dushenkov et al., 1997a), ^{137}Cs, and [90]Sr (Dushenkov et al., 1997b) from hydroponic solutions. Rhizofiltration was found to be appropriate for the cleanup of Sr from surface water using hydroponic and field experiments. A pond near the Chernobyl nuclear reactor was phytoremediated with sunflowers, and their roots accumulated large quantities of radionuclides with a bioaccumulation coefficient greater than 600 for both shoots and roots (Negri and Hinchman, 2000). The role of synthetic chelates and mycorrhizal fungi in the phytoremoval of contaminants has been evaluated in a number of investigations (Table 2).

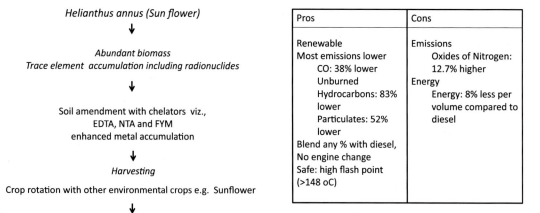

FIG. 2   *Helianthus annus* sunflower as an environmental crop for cultivation on co-contaminated soils (inorganic and organic).

III. Ornamentals and tree crops for contaminated substrates

Input	Output
Seeds	Protein (seed) meal
Methanol (CH_2OH)	Biodiesel
Catalyst (Methoxide)	Glycerine
Small amount acid	Electricity
Electricity from biodiesel (run plant)	Waste water/resin or mg power

Biodiesel Reaction

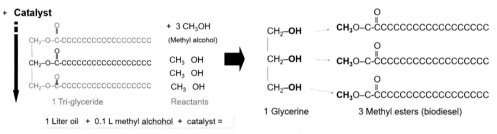

FIG. 3 Sunflower biodiesel production; input and output as per SunBio.

TABLE 2 Salient research findings using sunflower for bioremediation and bioproduct production.

	Author(s)	Year
Sunflower (*Helianthus annuus* L.) bioremediation of potentially hazardous components of sewage sludge in greenhouse and outdoor environments	Bayat et al.	2022
Potential uses of *Helianthus annuus* (sunflower) residue-based adsorbents for dye removal in waste waters	Anastopoulos et al.	2022
Phytoremediation Potentials of Fluted Pumpkin (*Telfairia Occidentalis*), Sorghum (*Sorghum Bicolor*), and Sunflower on Spent Engine Oil Polluted Texturally Contrasting Soils	Chidinma	2022
Indian Mustard and Sunflower Enhanced Phytoremediation of U in Army Test Range Soil	Guo et al.	2022
Bioremediation potential of *Helianthus annuus*	Bello et al.	2022
The use of chromium-resistant bacteria and cerium dioxide nanoparticles decreased the toxicity of chromium in sunflower plants	Ma et al.	2022
Effectiveness of several models for studying how sunflower plants react to Pb contaminations when treated with SiO_2 nanoparticles (NPs) and *Pseudomonas fluorescens*	Mousavi et al.	2022
Production of mushrooms in South America's Southern Cone: Bioeconomy, Sustainable Development, and Current Boom	Postemsky et al.	2022

Continued

III. Ornamentals and tree crops for contaminated substrates

TABLE 2　Salient research findings using sunflower for bioremediation and bioproduct production—cont'd

	Author(s)	Year
Through controlling the microenvironment in the root zone, organic acids support Cd phytoextraction in association with different oilseed sunflower cultivars	Qiao et al.	2022
Influence of a magnetoplasma installation on sunflower growth when oil contamination is present	Sergeeva and Purygin	2022
Identification of lead-resistant rhizobacteria of *Carthamus tinctorius* and their effects on lead absorption of sunflower	Shahraki et al.	2022
Phytoremediation of lead in sunflower (*Helianthus annuus* L.) by growth-promoting rhizobacterium and neem cake	Shaikh et al.	2022
Sunflower physiology and the role of PGPR in producing water with high total dissolved solids (TDS) and its lasting impacts on soil fertility	Urooj et al.	2022
Sunflower growth performance is improved when tannery solid waste biochar with microbes is applied to the soil	Younas et al.	2022
Poultry droppings and sunflower seed husk are used to break down petroleum hydrocarbons in soil that has been contaminated with crude oil	Boopathy et al.	2021
Possibility of combining the phytoremediation of heavy metals (HMs) by bioenergy plants with the HM-adapted rhizosphere microbiota (arbuscular mycorrhizal fungus and plant growth-promoting microorganisms) associated with those plants	Khan	2021
Microcosm experiment to determine whether sunflowers may grow in soil contaminated with PCBs and HM	Ancona et al.	2021
In order to maintain sustainability, *Aspergillus foetidus* controlled the biochemical properties of sunflower and soybean under heat stress conditions	Hamayun et al.	2021
Exogenous plant growth regulator improves the effectiveness of U and Cd remediation while reducing the negative effects of U and Cd stress on sunflower (*Helianthus annuus* L.)	Chen et al.	2021
Economic and environmental revitalization through ecological and ecosystem engineering	Barbosa and Fernando	2021
Adsorbents generated from sunflower biomass for the removal of toxic/heavy metals from waste water	Anastopoulos et al.	2021
Sunflower (*Helianthus annuus*) L. phytoremediation of wasted oil and palm kernel sludge-contaminated soil	Odebode et al.	2021
Fish growth impacts of biodegradation of crude oil-polluted water utilizing bacteria isolated from sunflower husk: Taguchi parametric optimization	Popoola and Yusuff	2021
Pressmud and *Pseudomonas gessardii* BLP141 worked together to improve sunflower growth, physiology, yield, and Pb toxicity	Raza Altaf et al.	2021

TABLE 2 Salient research findings using sunflower for bioremediation and bioproduct production—cont'd

	Author(s)	Year
Oil and bioenergy production along with arsenic and mercury uptake and accumulation in oilseed sunflower accessions chosen to reduce co-contaminated soil	Sahito et al.	2021
Decreased plant absorption of PAHs from biocharized sunflower husk-amended soil	Sushkova et al.	2021
On sunflower husks, charcoal, and goethite in the single/mixed pesticide solutions, carboxin and diuron adsorption mechanisms are present	Szewczuk-Karpisz et al.	2021
Low Cd breeding varieties in sunflower	Yuanzhi et al.	2021
Sunflower (*Helianthus annuus* L.) phytoremediation of uranium and cadmium-contaminated soils enhanced with biodegradable chelating agents	Chen et al.	2020
Husk energy supply systems for sunflower oil mills	Havrysh et al.	2020
Helianthus annuus plants growing in hexavalent chromium-laced soil benefit from the physiological and antioxidative effects of Fe0 nanoparticles	Mohammadi et al.	2020
Effect of various water sources and water availability regimes on the accumulation of heavy metals in two species of sunflower	Surucu et al.	2020
The efficiency of chemical and biological amendments in sunflower plants cultivated in Ni-polluted soils increased Ni phytoextraction	Jarrah et al.	2019
Vegetable depuration module is undergoing pilot testing as part of a bioremediation system for heavy metal-contaminated water and soil	Scotti et al.	2019
Utilizing Mexican sunflower (*Tithonia diversifolia*) and wild groundnut (*Calopogonium mucunoides*) to accelerate the microbial decomposition of soil that has been contaminated by crude oil	Agbor and Antai	2018
Sunflower and *Pseudomonas* species have the potential to bioremediate lead and zinc-contaminated soil	Alori et al.	2018
Sunflower (*Helianthus annuus*) plant-based phytoremediation of Pb and Cd-contaminated soils	Alaboudi et al.	2018
Sunflower (*Helianthus annuus* L.)-based biosurfactant-aided phytoremediation of multicontaminated industrial soil	Liduino et al.	2018
By combining the functions of the sunflower plant and the yeast consortia, benzo[*a*]pyrene in soil can be phyto-mycoremediated	Mandal and Das	2018
Pseudomonas spp. as growth-promoting agents for jack beans and sunflowers in sulfentrazone-treated soil	Melo et al.	2018
Sunflower biomass and ionome: phytomanagement and rehabilitation of Cu-contaminated soils at a former wood preservation facility	Mench et al.	2018
Soil-to-plant transfer factors of natural uranium and radium	Rodriguez et al.	2006
Comparative effect of Al, Se, and Mo toxicity on NO_3 assimilation	Ruiz et al.	2007

Continued

TABLE 2   Salient research findings using sunflower for bioremediation and bioproduct production—cont'd

	Author(s)	Year
Phytoextraction of excess soil phosphorus	Sharma et al.	2007
Trace metal accumulation, movement, and remediation in soils receiving animal manure	Sistani and Novak	2005
Accumulation of radioiodine from hydroponic system	Soudek et al.	2006a
^{137}Cs and ^{90}Sr uptake	Soudek et al.	2006b
The influence of EDDS on the uptake of heavy metals in hydroponic system	Tandy et al.	2006
Removal of polycyclic aromatic (PAH) hydrocarbons from contaminated soils	Gong et al.	2005a, 2006
Dissolution and removal of PAHs	Gong et al.	2005b
Leaching and uptake of heavy metals in EDTA-assisted phytoextraction process	Chen et al.	2004
Response of antioxidants grown on different amendments of tannery sludge: metal accumulation potential	Singh et al.	2004
^{137}Cs uptake	Soudek et al.	2004
EDTA and citric acid role on phytoremediation of Cd, Cr, and N	Turgut et al.	2004
Accumulation of copper	Lin et al.	2003
EDTA-assisted heavy-metal uptake in an association with poplar at a long-term sewage-sludge farm	Liphadzi et al.	2003
Trace element and nutrient accumulation 2 years after the Aznalcollar mine spill, Spain	Madejon et al.	2003
Uptake and translocation of plutonium in hydroponics	Lee et al.	2002a
Uptake of plutonium from soils—a comparative study with *Brassica juncea*	Lee et al.	2002b
Mycorrhizal fungi enhanced accumulation and tolerance of chromium	Davies et al.	2001
Accumulation of heavy metals in an association with sorghum as affected by the Guadiamar spill	Murillo et al.	1999
Removal of Cd^{2+}, Cr^{6+}, Cu^{2+}, Mn^{2+}, Ni^{2+}, and Pb^{2+} in miniature rhizofiltration batch experiments	Dushenkov et al.	1997a
Rhizofiltration of U, Sr, and Cs using 4-week-old plants	Dushenkov et al.	1997b

Application of sunflower in phytotechnologies for the cleanup of heavy metals, radionuclides, and organic contaminants and pollutants (in reverse chronology) (Table 3).[a]

[a]The list may not be exhaustive.

TABLE 3 The potential of sunflower as a energy and rubber-producing crop.

	Author(s)	Year
A new energy crop on *Onopordum* spp.: a research on biofuel properties	Acaroglu et al.	2022
Bioethanol and biohydrogen production from agricultural waste	Aggarwal et al.	2022
Bioethanol from oil producing plants. Biorefinery of Oil Producing Plants for Value-Added Products	Cheng et al.	2022
Renewable diesel as fossil fuel substitution in Malaysia	Chia et al.	2022
Biodiesel production from nonedible feedstocks catalyzed by nanocatalysts: a review	Chimezie et al.	2022
On-farm energy production: biofuels. In Regional Perspectives on Farm Energy	Ciolkosz and Steiman	2022
Genomic regions associate with major axes of variation driven by gas exchange and leaf construction traits in cultivated sunflower (*Helianthus annuus* L.)	Earley et al.	2022
Amino acids from oil producing plants. Biorefinery of Oil Producing Plants for Value-Added Products	Hussin et al.	2022
Effect of exogenous fibrolytic enzymes supplementation or functional feed additives on in vitro ruminal fermentation of chemically pre-treated sunflower heads	Jabri et al.	2022
ICAR-Flexi Rubber Check Dam: an innovative technology for efficient water conservation. Advances in water management technology for enhancing agricultural productivity	Jena	2022
Applying novel eco-exergoenvironmental toxicity index to select the best irrigation system of sunflower production	Khanali et al.	2022
Agricultural land and crop production in Myanmar	Lambrecht et al.	2022
Bio-diesel production of sunflower through sulfur management in a semi-arid subtropical environment	Mahmood et al.	2022
Bioethanol and biodiesel blended fuels—Feasibility analysis of biofuel feedstocks in Bangladesh	Mahmud et al.	2022
Synergistic effect of sunflower oil and soybean oil as alternative processing oil in the development of greener tyre tread compound	Mohamed et al.	2022
Utilization of dairy scum waste as a feedstock for biodiesel production via different heating sources for catalytic transesterification	Mohd Johari et al.	2022
Bio-refinery of oilseeds: oil extraction, secondary metabolites separation towards protein meal valorisation—a review	Nehmeh et al.	2022
Fatty acid composition of oil crops: genetics and genetic engineering	Porokhovinova et al.	2022
Advances in production of biodiesel from vegetable oils and animal fats. Biodiesel Production: Feedstocks, Catalysts, and Technologies	Rashid and Hazmi	2022
Feedstocks for green diesel. In Green Diesel: An Alternative to Biodiesel and Petrodiesel, Springer, Singapore	Sharma et al.	2022

Continued

III. Ornamentals and tree crops for contaminated substrates

TABLE 3 The potential of sunflower as a energy and rubber-producing crop—cont'd

	Author(s)	Year
Heterogeneous catalyzed synthesis of biodiesel from crude sunflower oil	Sivamani et al.	2022
Overview: catalysts, feedstocks in biodiesel production. Research Topics in Bioactivity, Environment and Energy	Taft and Canchaya	2022
When a weed is not a weed: succession management using early seral natives for intermountain rangeland restoration. Rangelands	Tilley et al.	2022
A review on application of nanocatalysts for the production of biodiesel using different feedstocks. Materials Today: Proceedings	Topare et al.	2023
Conditions for efficient alkaline storage of cover crops for biomethane production	Van Vlierberghe et al.	2022
Special issue on recent advances in rubber research and applications	Zakaria and Wong	2022
Pectin films with recovered sunflower waxes produced by electrospraying	Chalapud et al.	2022

4 Sunflower yield and chemical composition

Critical Factors: What is the percentage of free fatty acids? (FFA)

- New veg oil: 0.4%–0.7% FFA
- Over 2.5% of FFA use "pre-treatment"
  - Add sulfuric acid/methanol
  - Esterify free fatty acids to methyl ester
- Type of feed stock?
  - New veg oil
  - Animal fats (2%–6% FFA)
  - Used cooking oil (15% FFA)

Output per 1 hectare

(Depends on seed type and location (geography))

1.	Harvest sunflower seeds	2500 kg (University Unidne, Italy)[a]
2.	Crush seeds	
	• Protein/oil cake	1400 kg
	• Raw oil	940 kg (50% oil × 75% efficient—H_2O)
	• Moisture	150 kg
3.	Degummed oil	923 kg 1.5% removed
4.	Biodiesel ➜ power generation	65 kg to run plant
5.	Glycerin	92 kg
6.	Biodiesel (net)	840 kg 98% conversion

[a] *The list may not be exhaustive.*

1.	Output per 1000 hectares	
2.	Harvest sunflower seeds	2500 MT
3.	Crush seeds	
	– Protein/oil cake	1400 kg
	– Raw oil	940 kg (50% oil × 75% efficient—H_2O)
	– Moisture	150 kg
4.	Degummed oil	923 MT
5.	Biodiesel/oil generator	70 MT
6.	Glycerin	92 MT
7.	Biodiesel (net)	840 MT

5 Conclusion

Sunflower is a good candidate for a domestic rubber crop. Sunflower leaves contain <1% rubber. Latex yield in the leaves and stems of some sunflower accessions is variable. Sunflower latex/rubber yield must be improved for commercial viability. A lot is known about sunflower breeding. However, can this be achieved agronomically? Plant breeding or through genetic engineering? Sunflower molecular weight must be improved for commercial viability via a transgenic approach. Molar Mass (g/mol) order of prominent rubber crops is *Hevea* > *Parthenium* > *Helianthus*.

References

Acaroglu, M., Baser, E., Aydogan, H., Canli, E., 2022. A new energy crop *Onopordum* spp.: a research on biofuel properties. Energy 261, 125305.

Agbor, R.B., Antai, S.P., 2018. Enhancement of microbial degradation of crude oil polluted soil using Mexican sunflower (*Tithonia diversifolia*) and wild groundnut (*Calapogonium mucunoides*). Int. J. Ecol. Ecosolution 5 (1), 1–7.

Aggarwal, N.K., Kumar, N., Mittal, M., 2022. Bioethanol and biohydrogen production from agricultural waste. In: Bioethanol Production. Springer, Cham, pp. 119–136.

Alaboudi, K.A., Ahmed, B., Brodie, G., 2018. Phytoremediation of Pb and Cd contaminated soils by using sunflower (*Helianthus annuus*) plant. Ann. Agric. Sci. 63 (1), 123–127.

Alori, E.T., Joseph, A., Adebiyi, O.T.V., Ajibola, P.A., Onyekankeya, C., 2018. Bioremediation potentials of sunflower and *Pseudomonas* species in soil contaminated with lead and zinc. Afr. J. Biotechnol. 17 (44), 1324–1330.

Anastopoulos, I., Ighalo, J.O., Igwegbe, C.A., Giannakoudakis, D.A., Triantafyllidis, K.S., Pashalidis, I., Kalderis, D., 2021. Sunflower-biomass derived adsorbents for toxic/heavy metals removal from (waste) water. J. Mol. Liq. 342, 117540.

Anastopoulos, I., Giannopoulos, G., Islam, A., Ighalo, J.O., Iwuchukwu, F.U., Pashalidis, I., Kalderis, D., Giannakoudakis, D.A., Nair, V., Lima, E.C., 2022. Potential environmental applications of *Helianthus annuus* (sunflower) residue-based adsorbents for dye removal in (waste) waters. In: Biomass-Derived Materials for Environmental Applications. Elsevier, pp. 307–318.

Ancona, V., Gatto, A., Aimola, G., Rascio, I., Tumolo, M., Losacco, D., Locaputo, V., Grenni, P., Garbini, G.L., Rolando, L., Visca, A., 2021. Microcosm Experiment for Assessing Sunflower Capability to Grow on a PCB and HM Contaminated Soil.

Barbosa, B., Fernando, A.L., 2021. Ecological and ecosystem engineering for economic-environmental revitalization. In: Handbook of Ecological and Ecosystem Engineering, pp. 25–46.

Bayat, M., Faramarzi, A., Ajalli, J., Abdi, M., Nourafcan, H., 2022. Bioremediation of potentially toxic elements of sewage sludge using sunflower (*Heliantus annus* L.) in greenhouse and field conditions. Environ. Geochem. Health 44 (4), 1217–1227.

Bello, M.O., Bello, O.M., Ogbesejana, A.B., 2022. Bioremediation potential of *Helianthus annuus*. In: Bioremediation and Phytoremediation Technologies in Sustainable Soil Management. Apple Academic Press, pp. 47–74.

Boopathy, S., Appavoo, M.S., Radhakrishnan, I., 2021. Sunflower seed husk combined with poultry droppings to degrade petroleum hydrocarbons in crude oil-contaminated soil. Environ. Eng. Res. 26 (5), 200361.

Chalapud, M.C., Baümler, E.R., Carelli, A.A., de la Paz Salgado-Cruz, M., Morales-Sánchez, E., Rentería-Ortega, M., Calderón-Domínguez, G., 2022. Pectin films with recovered sunflower waxes produced by electrospraying. Membranes 12, 560.

Chen, Y., Li, X., Shen, Z., 2004. Leaching and uptake of heavy metals by ten different species of plants during an EDTA-assisted phytoextraction process. Chemosphere 57, 187–196.

Chen, L., Yang, J.Y., Wang, D., 2020. Phytoremediation of uranium and cadmium contaminated soils by sunflower (*Helianthus annuus* L.) enhanced with biodegradable chelating agents. J. Clean. Prod. 263, 121491.

Chen, L., Hu, W.F., Long, C., Wang, D., 2021. Exogenous plant growth regulator alleviate the adverse effects of U and Cd stress in sunflower (*Helianthus annuus* L.) and improve the efficacy of U and Cd remediation. Chemosphere 262, 127809.

Cheng, Y.S., Rattanaporn, K., Sriariyanun, M., 2022. Bioethanol from oil producing plants. In: Biorefinery of Oil Producing Plants for Value-Added Products. 1, pp. 287–306.

Chia, S.R., Nomanbhay, S., Ong, M.Y., Chew, K.W., Show, P.L., 2022. Renewable diesel as fossil fuel substitution in Malaysia: a review. Fuel 314, 123137.

Chidinma, C.E., 2022. Phytoremediation potentials of Sorghum (Sorghum Bicolor), sunflower (Helianthus Amarus) and fluted pumpkin (Telifaria Occidentallis) on spent engine oil polluted texturally contrasting soils. Medicon Agric. Environ. Sci. 3, 27–33.

Chimezie, E.C., Zhang, X., Djandja, O.S., Nonso, U.C., Duan, P.G., 2022. Biodiesel production from nonedible feedstocks catalyzed by nanocatalysts: a review. Biomass Bioenergy 163, 106509.

Ciolkosz, D., Steiman, M., 2022. On-farm energy production: biofuels. In: Regional Perspectives on Farm Energy. Springer, Cham, pp. 139–148.

Davies Jr., F.T., Puryear, J.D., Newton, R.J., Egilla, J.N., Grossi, J.A.S., 2001. Mycorrhizal fungi enhance accumulation and tolerance of chromium in sunflower (*Helianthus annuus* L.). J. Plant Physiol. 158, 777–786.

Dushenkov, S., Kapulnik, Y., 2002. Phytofilttration of metals. In: Raskin, I., Ensley, B.D. (Eds.), Phytoremediation of Toxic Metals: Using Plants to Clean up the Environment. John Wiley and Sons, Inc., New York, pp. 89–106.

Dushenkov, S., Mikheev, A., Prokhnevsky, A., Ruchko, M., Sorochinsky, B., 1999. Phytoremediation of radiocesium-contaminated soil in the vicinity of Chernobyl, Ukraine. Environ. Sci. Technol. 33 (3), 469–475.

Dushenkov, V., Nanda Kumar, P.B.A., Motto, H., Raskin, I., 1995. Rhizofiltration: the use of plants to remove heavy metals from aqueous streams. Environ. Sci. Technol. 29, 1239–1245.

Dushenkov, S., Kapulnik, Y., Blaylock, M., Sorochisky, B., Raskin, I., Ensley, B., 1997a. Phytoremediation: a novel approach to an old problem. In: Wise, D.L. (Ed.), Global Environmental Biotechnology. Elsevier Science B.V., Amsterdam, pp. 563–572.

Dushenkov, S., Vasudev, D., Kapulnik, Y., Gleba, D., Fleisher, D., Ting, K.C., Ensley, B., 1997b. Removal of uranium from water using terrestrial plants. Environ. Sci. Technol. 31, 3468–3474.

Earley, A.M., Temme, A.A., Cotter, C.R., Burke, J.M., 2022. Genomic regions associate with major axes of variation driven by gas exchange and leaf construction traits in cultivated sunflower (*Helianthus annuus* L.). Plant J. 111 (5), 1425–1438.

Gong, Z., Wilke, B.-M., Alef, K., Li, P., 2005a. Influence of soil moisture on sunflower oil extraction of polycyclic aromatic hydrocarbons from a manufactured gas plant soil. Sci. Total Environ. 343, 51–59.

Gong, Z., Alef, W.K., Wilke, B.M., Li, P., 2005b. Dissolution and removal of PAHs from a contaminated soil using sunflower oil. Chemosphere 58, 291–298.

Gong, Z., Wilke, B.-M., Alef, K., Li, P., Zhou, Q., 2006. Removal of polycyclic aromatic hydrocarbons from manufactured gas plant-contaminated soils using sunflower oil: laboratory column experiments. Chemosphere 62, 780–787.

Guo, F., Proctor, G., Larson, S.L., Ballard, J.H., Zan, S., Yang, R., Wang, X., Han, F.X., 2022. Earthworm enhanced phytoremediation of U in army test range soil with Indian mustard and sunflower. ACS Earth Space Chem. 6 (3), 746–754.

Hamayun, M., Hussain, A., Iqbal, A., Khan, S.A., Ahmad, A., Gul, S., Kim, H.Y., Lee, I.J., 2021. *Aspergillus foetidus* regulated the biochemical characteristics of soybean and sunflower under heat stress condition: role in sustainability. Sustainability 13 (13), 7159.

Havrysh, V., Kalinichenko, A., Mentel, G., Mentel, U., Vasbieva, D.G., 2020. Husk energy supply systems for sunflower oil mills. Energies 13 (2), 361.

Hussin, H., Hanafi, N.S., Lee, A.C., Salleh, M.M., Sam, S.C., Abd-Aziz, S., 2022. Amino acids from oil producing plants. In: Biorefinery of Oil Producing Plants for Value-Added Products. vol. 2, pp. 653–671.

Jabri, J., Ammar, H., Abid, K., Beckers, Y., Yaich, H., Malek, A., Rekhis, J., Morsy, A.S., Soltan, Y.A., Soufan, W., Almadani, M.I., 2022. Effect of exogenous fibrolytic enzymes supplementation or functional feed additives on in vitro ruminal fermentation of chemically pre-treated sunflower heads. Agriculture 12 (5), 696.

Jarrah, M., Ghasemi-Fasaei, R., Ronaghi, A., Zarei, M., Mayel, S., 2019. Enhanced Ni phytoextraction by effectiveness of chemical and biological amendments in sunflower plant grown in Ni-polluted soils. Chem. Ecol. 35 (8), 732–745.

Jena, S.K., 2022. ICAR-flexi rubber check dam: an innovative technology for efficient water conservation. In: Advances in Water Management Technology for Enhancing Agricultural Productivity, p. 42.

Khan, A.G., 2021. Potential of coupling heavy metal (HM) phytoremediation by bioenergy plants and their associated HM-adapted rhizosphere microbiota (arbuscular mycorrhizal fungi and plant growth promoting microbes) for bioenergy production. J. Energy Power Technol. 3 (4), 1.

Khanali, M., Ghasemi-Mobtaker, H., Varmazyar, H., Mohammadkashi, N., Chau, K.W., Nabavi-Pelesaraei, A., 2022. Applying novel eco-exergoenvironmental toxicity index to select the best irrigation system of sunflower production. Energy 250, 123822.

Lambrecht, I., Belton, B., Fang, P., Minten, B., Naing, P.T., 2022. Agricultural Land and Crop Production in Myanmar. vol. 24 International Food Policy Research Institute.

Lee, J.H., Hossner, L.R., Attrep Jr., M., Kung, K.S., 2002a. Uptake and translocation of plutonium in two plant species using hydroponics. Environ. Pollut. 117, 61–68.

Lee, J.H., Hossner, L.R., Attrep Jr., M., Kung, K.S., 2002b. Comparative uptake of plutonium from soils by *Brassica juncea* and *Helianthus annuus*. Environ. Pollut. 120, 173–182.

Liduino, V.S., Servulo, E.F., Oliveira, F.J., 2018. Biosurfactant-assisted phytoremediation of multi-contaminated industrial soil using sunflower (*Helianthus annuus* L.). J. Environ. Sci. Health A 53 (7), 609–616.

Lin, J., Jiang, W., Liu, D., 2003. Accumulation of copper by roots, hypocotyls, cotyledons and leaves of sunflower (*Helianthus annuus* L.). Bioresour. Technol. 86 (2), 151–155.

Liphadzi, M.S., Kirkham, M.B., Mankin, K.R., Paulsen, G.M., 2003. EDTA-assisted heavy-metal uptake by poplar and sunflower grown at a long-term sewage-sludge farm. Plant Soil 257, 171–182.

Ma, J., Alshaya, H., Okla, M.K., Alwasel, Y.A., Chen, F., Adrees, M., Hussain, A., Hameed, S., Shahid, M.J., 2022. Application of cerium dioxide nanoparticles and chromium-resistant bacteria reduced chromium toxicity in sunflower plants. Front. Plant Sci. 13, 876119.

Madejon, P., Murillo, J.M., Maranon, T., Cabrera, F., Soriano, M.A., 2003. Trace element and nutrient accumulation in sunflower plants two years after the Aznalcollar mine spill. Sci. Total Environ. 307, 239–257.

Mahmood, A., Awan, M.I., Sadaf, S., Mukhtar, A., Wang, X., Fiaz, S., Khan, S.A., Ali, H., Muhammad, F., Hayat, Z., Gul, F., 2022. Bio-diesel production of sunflower through sulphur management in a semi-arid subtropical environment. Environ. Sci. Pollut. Res. 29 (9), 13268–13278.

Mahmud, S., Haider, A.R., Shahriar, S.T., Salehin, S., Hasan, A.M., Johansson, M.T., 2022. Bioethanol and biodiesel blended fuels—feasibility analysis of biofuel feedstocks in Bangladesh. Energy Rep. 8, 1741–1756.

Mandal, S.K., Das, N., 2018. Phyto-mycoremediation of benzo[a]pyrene in soil by combining the role of yeast consortium and sunflower plant. J. Environ. Biol. 39 (2), 261–268.

McCutcheon, S.C., Schnoor, J.L. (Eds.), 2003. Phytoremediation—Transformation and Control of Contaminants. Wiley Interscience, p. 985.

Melo, C.A.D., de Souza, W.M., Medeiros, W.N., Massenssini, A.M., Ferreira, L.R., Costa, M.D., 2018. Pseudomonas spp. as growth promoting agents of sunflower and jack bean in soil with sulfentrazone. Científica 46 (1), 17–29.

Mench, M.J., Dellise, M., Bes, C.M., Marchand, L., Kolbas, A., Le Coustumer, P., Oustrière, N., 2018. Phytomanagement and remediation of Cu-contaminated soils by high yielding crops at a former wood preservation site: sunflower biomass and ionome. Front. Ecol. Evol. 6, 123.

Mohamed, N.R., Othman, N., Shuib, R.K., 2022. Synergistic effect of sunflower oil and soybean oil as alternative processing oil in the development of greener tyre tread compound. J. Rubber Res. 25, 239–249.

III. Ornamentals and tree crops for contaminated substrates

Mohammadi, H., Amani-Ghadim, A.R., Matin, A.A., Ghorbanpour, M., 2020. Fe0 nanoparticles improve physiological and antioxidative attributes of sunflower (*Helianthus annuus*) plants grown in soil spiked with hexavalent chromium. 3 Biotech 10 (1), 1–11.

Mohd Johari, S.A., Ayoub, M., Inayat, A., Ullah, S., Uroos, M., Naqvi, S.R., Farukkh, S., 2022. Utilization of dairy scum waste as a feedstock for biodiesel production via different heating sources for catalytic transesterification. ChemBioEng. Rev. 9 (6), 605–632.

Mousavi, S.M., Motesharezadeh, B., Hosseini, H.M., Zolfaghari, A.A., Sedaghat, A., Alikhani, H., 2022. Efficiency of different models for investigation of the responses of sunflower plant to Pb contaminations under SiO$_2$ nanoparticles (NPs) and *Pseudomonas fluorescens* treatments. Arab. J. Geosci. 15 (14), 1–12.

Murillo, J.M., Marañón, T., Cabrera, F., López, R., 1999. Accumulation of heavy metals in sunflower and sorghum plants affected by the Guadiamar spill. Sci. Total Environ. 242, 281–292.

Negri, M.C., Hinchman, R.R., 2000. The use of plants for the treatment of radionuclides. In: Raskin, I., Ensley, B.D. (Eds.), Phytoremediation of Toxic Metals: Using Plants to Clean up the Environment. John Wiley and Sons, Inc., New York, pp. 107–132.

Nehmeh, M., Rodriguez-Donis, I., Cavaco-Soares, A., Evon, P., Gerbaud, V., Thiebaud-Roux, S., 2022. Bio-refinery of oilseeds: oil extraction, secondary metabolites separation towards protein meal valorisation—a review. Processes 10 (5), 841.

Odebode, A.J., Njoku, K.L., Adesuyi, A.A., Akinola, M.O., 2021. Phytoremediation of spent oil and palm kernel sludge contaminated soil using sunflower (*Helianthus annuus*) L. J. Appl. Sci. Environ. Manag. 25 (5), 877–885.

Popoola, L.T., Yusuff, A.S., 2021. Biodegradation effects of crude oil-polluted water using bacteria isolates from sunflower husk on fish growth: parametric optimization using taguchi approach. Carpathian J. Earth Environ. Sci. 16 (1), 211–222.

Porokhovinova, E.A., Matveeva, T.V., Khafizova, G.V., Bemova, V.D., Doubovskaya, A.G., Kishlyan, N.V., Podolnaya, L.P., Gavrilova, V.A., 2022. Fatty acid composition of oil crops: genetics and genetic engineering. Genet. Resour. Crop. Evol., 1–17.

Postemsky, P., Bidegain, M., González Matute, R., Figlas, D., Caprile, D., Salazar-Vidal, V., Saparrat, M., 2022. Mushroom Production in the Southern Cone of South America: Bioeconomy, Sustainable Development and Its Current Bloom.

Qiao, D., Han, Y., Zhao, Y., 2022. Organic acids in conjunction with various oilseed sunflower cultivars promote Cd phytoextraction through regulating micro-environment in root zone. Ind. Crop. Prod. 183, 114932.

Rashid, U., Hazmi, B., 2022. Advances in production of biodiesel from vegetable oils and animal fats. In: Biodiesel Production: Feedstocks, Catalysts, and Technologies, pp. 1–21.

Raza Altaf, A., Teng, H., Saleem, M., Raza Ahmad, H., Adil, M., Shahzad, K., 2021. Associative interplay of *Pseudomonas gessardii* BLP141 and pressmud ameliorated growth, physiology, yield, and Pb-toxicity in sunflower. Bioremediat. J. 25 (2), 178–188.

Rodriguez, P.B., Tome, F.V., Fernandez, M.P., Lozano, J.C., 2006. Linearity assumption in soil-to-plant transfer factors of natural uranium and radium in *Helianthus annuus* L. Sci. Total Environ. 361, 1–7.

Ruiz, J.M., Rivero, R.M., Romero, L., 2007. Comparative effect of Al, Se, and Mo toxicity on NO3- assimilation in sunflower (*Helianthus annuus* L.) plants. J. Environ. Manag. 83 (2), 207–212.

Sahito, Z.A., Zehra, A., Tang, L., Ali, Z., Bano, N., Ullah, M.A., He, Z., Yang, X., 2021. Arsenic and mercury uptake and accumulation in oilseed sunflower accessions selected to mitigate co-contaminated soil coupled with oil and bioenergy production. J. Clean. Prod. 291, 125226.

Scotti, A., Silvani, V.A., Cerioni, J., Visciglia, M., Benavidez, M., Godeas, A., 2019. Pilot testing of a bioremediation system for water and soils contaminated with heavy metals: vegetable depuration module. Int. J. Phytoremediation 21 (9), 899–907.

Sergeeva, D.V., Purygin, P.P., 2022. Influence of magnetoplasma installation on growth of sunflower in the presence of oil contamination. In: AIP Conference Proceedings. vol. 2390, No. 1. AIP Publishing LLC, p. 030078.

Shahraki, A., Mohammadi-Sichani, M., Ranjbar, M., 2022. Identification of lead-resistant rhizobacteria of *Carthamus tinctorius* and their effects on lead absorption of sunflower. J. Appl. Microbiol. 132 (4), 3073–3080.

Shaikh, N.G., Rahman, A., Sadiq, M., Ehteshamul-Haque, S., Shams, Z.I., 2022. Phytoremediation of Lead in Sunflower (*Helianthus annuus* L.) by Growth-Promoting Rhizobacterium and Neem Cake.

Sharma, S., Singh, S., Sarma, S.J., Brar, S.K., 2022. Feedstocks for green diesel. In: Green Diesel: An Alternative to Biodiesel and Petrodiesel. Springer, Singapore, pp. 41–53.

Sharma, N.C., Starnes, D.L., Sahi, S.V., 2007. Phytoextraction of excess soil phosphorus. Environ. Pollut. 146 (1), 120–127.

Singh, S., Saxena, R., Pandey, K., Bhatt, K., Sinha, S., 2004. Response of antioxidants in sunflower (*Helianthus annuus* L.) grown on different amendments of tannery sludge: its metal cumulation potential. Chemosphere 57, 1663–1673.

Sistani, K.R., Novak, J.M., 2005. Trace metal accumulation, movement, and remediation in soils receiving animal manure. In: Prasad, M.N.V., Sajwan, K.S., Naidu, R. (Eds.), Trace Elements in the Environment: Biogeochemistry, Biotechnology and Bioremediation. CRC Press, Boca Raton, pp. 689–706.

Sivamani, S., Al Aamri, M.A.S., Jaboob, A.M.A.A., Kashoob, A.M.M., Al-Hakeem, L.K.A., Almashany, M.S.M.S., Safrar, M.A.M., 2022. Heterogeneous catalyzed synthesis of biodiesel from crude sunflower oil. J. Niger. Soc. Phys. Sci. 4, 16–19.

Soudek, P., Tykva, R., Vanek, T., 2004. Laboratory analyses of ^{137}Cs uptake by sunflower, reed and poplar. Chemosphere 55, 1081–1087.

Soudek, P., Tykva, R., Vankova, R., Vanek, T., 2006a. Accumulation of radioiodine from aqueous solution by hydroponically cultivated sunflower (*Helianthus annuus* L.). Environ. Exp. Bot. 57, 220–225.

Soudek, P., Valenova, S., Vavrikova, Z., Vanek, T., 2006b. ^{137}Cs and ^{90}Sr uptake by sunflower cultivated under hydroponic conditions. J. Environ. Radioact. 88, 236–250.

Surucu, A., Marif, A.A., Majid, S.N., Farooq, S., Tahir, N.A.R., 2020. Effect of different water sources and water availability regimes on heavy metal accumulation in two sunflower species. Carpathian J. Earth Environ. Sci. 15, 289–300.

Sushkova, S., Minkina, T., Dudnikova, T., Barbashev, A., Popov, Y., Rajput, V., Bauer, T., Nazarenko, O., Kizilkaya, R., 2021. Reduced plant uptake of PAHs from soil amended with sunflower husk biochar. Eur. J. Soil Sci. 10 (4), 269–277.

Szewczuk-Karpisz, K., Tomczyk, A., Celińska, M., Sokołowska, Z., Kuśmierz, M., 2021. Carboxin and diuron adsorption mechanism on sunflower husks biochar and goethite in the single/mixed pesticide solutions. Materials 14 (10), 2584.

Taft, C.A., Canchaya, J.G.S., 2022. Overview: catalysts, feedstocks in biodiesel production. In: Research Topics in Bioactivity, Environment and Energy, pp. 337–357.

Tandy, S., Schulin, R., Nowack, B., 2006. The influence of EDDS on the uptake of heavy metals in hydroponically grown sunflowers. Chemosphere 62, 1454–1463.

Tilley, D., Hulet, A., Bushman, S., Goebel, C., Karl, J., Love, S., Wolf, M., 2022. When a weed is not a weed: succession management using early seral natives for intermountain rangeland restoration. Rangelands 44 (4), 270–280.

Topare, N.S., Gujarathi, V.S., Bhattacharya, A.A., Bhoyar, V.M., Shastri, T.J., Manewal, S.P., Gomkar, C.S., Khedkar, S.-V., Khan, A., Asiri, A.M., 2023. A review on application of nano-catalysts for production of biodiesel using different feedstocks. Mater. Today Proc. 72, 324–335.

Turgut, C., Pepe, M.K., Cutright, T.J., 2004. The effect of EDTA and citric acid on phytoremediation of Cd, Cr, and Ni from soil using *Helianthus annuus*. Environ. Pollut. 131, 147–154.

Urooj, N., Bano, A., Riaz, A., 2022. Role of PGPR on the physiology of sunflower irrigated with produced water containing high total dissolved solids (TDS) and its residual effects on soil fertility. Int. J. Phytoremediation 24 (6), 567–579.

Van Vlierberghe, C., Escudie, R., Bernet, N., Santa-Catalina, G., Frederic, S., Carrere, H., 2022. Conditions for efficient alkaline storage of cover crops for biomethane production. Bioresour. Technol. 348, 126722.

Younas, H., Nazir, A., Bareen, F.E., 2022. Application of microbe-impregnated tannery solid waste biochar in soil enhances growth performance of sunflower. Environ. Sci. Pollut. Res., 1–19.

Yuanzhi, F., Volodymyr, T., Halyna, Z., 2021. Low CD breeding varieties in sunflower. In: The X International Science Conference "Trends and Prospects Development of Science and Practice in Modern Environment", November 22–24, Geneva, Switzerland, p. 15. 403p.

Zakaria, M.R., Wong, M.Y., 2022. Special issue on recent advances in rubber research and applications. J. Rubber Res. 25 (3), 171–172.

Brownfield development for smart bioeconomy

Chrysopogon zizanioides (vetiver grass) for abandoned mine restoration and phytoremediation: Cogeneration of economical products

Woranan Nakbanpote[a], Ponlakit Jitto[b], Uraiwan Taya[c], and Majeti Narasimha Vara Prasad[d]

[a]Department of Biology, Faculty of Science, Mahasarakham University, Khamriang, Kantarawichi, Mahasarakham, Thailand [b]Faculty of Environment and Resource Studies, Mahasarakham University, Kham Riang, Kantharawichai, Maha Sarakham, Thailand [c]Land Development Regional Office 5, Department of Land Development, Ministry of Agriculture and Cooperatives, Khon Kaen, Thailand [d]Department of Plant Sciences, School of Life Sciences, University of Hyderabad (An Institution of Eminence), Hyderabad, Telangana, India

1 Introduction

Chrysopogon is a genus of tropical and sub-tropical plants in the grass family. *Chrysopogon zizanioides*, also called vetiver and khus, is a perennial grass of the family Poaceae. This grass is reportedly indigenous to at least 70 tropical or sub-tropical countries. This perennial grass can live for up to fifty years, making it an effective, low-cost, long-term species for phytoremediation (National Resource Council, 1995). The benefits of vetiver grass in terms of soil restoration, environmental rehabilitation, and soil and water conservation are slope stabilization, soil erosion control, environmental improvement and protection, and disaster prevention. The advantages of vetiver are concluded as shown in Fig. 1. The shoots of vetiver grass planted on steep slopes help trap sediments up to 60%–80% and reduce the velocity of rainwater runoff by 50%–70%. The roots that are nearly three meters deep can improve and

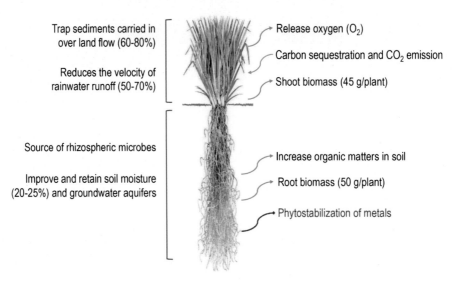

Trap sediments carried in over land flow (60-80%)

Reduces the velocity of rainwater runoff (50-70%)

Source of rhizospheric microbes

Improve and retain soil moisture (20-25%) and groundwater aquifers

Release oxygen (O_2)

Carbon sequestration and CO_2 emission

Shoot biomass (45 g/plant)

Increase organic matters in soil

Root biomass (50 g/plant)

Phytostabilization of metals

FIG. 1   The benefits of vetiver grass in terms of soil restoration, environmental rehabilitation, and soil and water conservation.

maintain 20%–25% moisture in the soil and underground aquifers. Lifespan roots produce the rhizosphere effect and are a source of rhizosphere microorganisms. Vetiver is a C4 plant with a high efficiency in converting solar radiation to biomass. Carbon and nitrogen are all accumulated in the root, stem, and leaves. Vetiver has one of the best carbon sequestrations in the world. Carbon sequestration to become shoot and root biomass conducts to increase organic matter in the soil (Lakshmi and Sekhar, 2020). The anatomical study revealed that vetiver ecotypes have distinct characteristics of hydrophytes, i.e., having large cellular pores that should relate to the storage of water and air, as well as circulating water and air in the vetiver leaves. Owing to its high biomass, vetiver has great potential for phytoremediation even though vetiver is not a hyperaccumulator. Moreover, the unique characteristics of long deep roots can penetrate to clean up the deep soil layer. The majority of accumulated amounts of heavy metals in roots provided this species was considered more suitable for phytostabilizing mine tailings, and the danger of transferring toxic metals to the food chain was minimal. Due to the advantages of vetiver grass, it is used in stabilizing post-open-pit mining areas and overburdened and roadside slopes. The post-Padang zinc mine in Tak province, Thailand, is a model mine that has successfully planted vetiver grass and local plants to rehabilitate overburden and open pit to become the forest. Vetiver is widely used in Thailand for soil and water conservation, rehabilitation of degraded agricultural areas, especially areas with hard and compact soil, and planting along ponds and canals to reduce soil erosion. In addition, the conservation of vetiver ecotypes, the propagation, and the planting and harvesting to flourish in the area have to be continuously processed to maximize the benefits. Finally, the promotion of income generation from vetiver biomass, such as construction materials, handicrafts, essential oils, biofuels, and media for mushrooms is essential for the sustainability of vetiver cultivation.

2 Vetiver ecotype and propagation in Thailand

The study and research on vetiver have begun in Thailand since His Majesty King Bhumibol Adulyadej the Great, King Rama IX, first started the vetiver grass project at Haui Sai, Phetchaburi Province, in 1991. Over the last 30 years, the study and use of vetiver have progressed significantly, especially its use in soil and water conservation through the supervision and support of the Office of the Royal Development Projects Board (ORDPB) and the Land Development Department (LDD). Vetiver is generally known to be originated in India. In addition, vetiver is found widely distributed naturally in all regions of Thailand. There are two species of vetiver in Thailand, namely, *Chrysopogon nemoralis* (Balansa) Holttum and *C. zizanioides* (L.) Roberty. *C. nemoralis* is called upland vetiver or "Faek Don" in Thai (Faek means vetiver, Don means upland). This local vetiver is located in a dry area with moderate to strong sunlight and well-drained soil. *C. zizanioides* is called "Faek Hom" in Thai (Hom means fragrant) because of its fragrant root. *C. zizanioides* is able to adapt and grow well in various environmental conditions (Roongtanakiat, 2006). Major different characteristics between *C. zizanioides* and *C. nemoralis* are concluded in Table 1. Conservative cultivation of vetiver has been carried out in many areas of Thailand, including the Padaeng Zinc Mine, Phra That Pha Daeng District, Tak Province (Fig. 2A). Srifah et al. (2000) showed that Single-Strand Conformational Polymorphism (SSCP) and Random Amplified Polymorphic DNA (RAPD) analyses of the 35 ecotypes of Thai vetiver could be used to classify *C. zizanioides* and *C. nemoralis*. The two species are ecologically distinct and adaptable to different habitats. Thai vetiver ecotypes were named according to the provinces where they were discovered. The

TABLE 1 The difference between *Chrysopogan zizanioides* and *Chrysopogon nemoralis* (Roongtanakiat, 2006; Cuong et al., 2015).

Chrysopogan zizanioides	*Chrysopogon nemoralis*
Clump into bushes with long erect leaves	Clump with long leaves spreading down
Height 150–200 cm	Height 100–150 cm
Leaves 45–100 cm long, 0.6–1.2 cm wide	Leaves 35–80 cm long, 0.4–0.8 cm wide
Dark green and curved on the upper leaf surface and flat near the apex. Paler green color on the lower leaf surface	Pale green on the upper leaf surface. Pale green or paler on the lower leaf surface
Clearly seen septum at the base and middle part of the blade, but not clearly seen midrib	Not clearly seen septum, hard midrib and forms the prominent ridge
Smooth leaf texture and a lot of wax coating	Coarsely rough leaf texture and slightly waxy coating
Large and straight stems	Small stems slightly curved with short internodes
Inflorescence height 150–250 cm	Inflorescence height 100–150 cm
Roots 100–300 cm deep, fragrant root	Roots 80–100 cm deep, no fragrance

FIG. 2   Management of vetiver cultivation at the Padaeng zinc mine. (A) Planting the varieties of upland vetiver or "Faek Don" in Thai for plant conservation, (B) propagation of vetiver grass in polybags, (C) vetiver plantation on bare overburdened dumps.

LDD has performed a comparative study of 28 vetiver ecotypes, and 6 ecotypes of *C. nemoralis* and 4 ecotypes of *C. zizanioides* show outstanding results for soil and water conservation (LDD, 2007). Table 2 shows the 10 ecotypes that have been found suitable for growing in different soil types. *C. zizanioides* is a long-rooted, environmentally adaptable species with wide application for the purposes of erosion control, landslide stabilization, and particularly riverbank stabilization (Cuong et al., 2015). *C. nemoralis* Ratchaburi ecotype showed potential in carbon sequestration at the highest rate of 11 tons/ha/year, while the lowest was the Loei ecotype at 4 tons/ha/year (Thammathavorn and Khanema, 2010). However, the microbial activity such as rhizobacteria and mycorrhiza in the soil associated with the root zones is one of the reasons that vetiver hedgerows produce amounts of biomass and carbon added to the sequestered soil carbon pool (Bhromsiri and Bhromsiri, 2010; Gu et al., 2022).

TABLE 2   Suitability of vetiver ecotypes in various soil types (Roongtanakiat, 2009).

Soil type	*Chrysopogan zizanioides*	*Chrysopogon nemoralis*
Sandy soil	Kamphaeng Phet 2, Songhla 3	Nakhon Sawan, Kamphaeng Phet 1, Roi Et, Ratchaburi
Clay loam soil	Surat Thani, Songkhla 3	Loei, Nakhon Sawan, Ratchaburi, Kamphaeng Phet 1, Prachuap khiri Khan
Leterite soil	Kamphaeng Phet 2, Songkhla 3, Surat Thani, Sri Lanka	Prachuap Khiri Khan, Loei

The main goals in vetiver propagation are quality planting material, low cost, hardiness, and being easy to transport. Although vetiver can be propagated by both sexual and asexual reproduction, vetiver cultivation rarely produces seeds. The ultimate gold of propagation is to grow individual vetiver planting materials in the field, either through the process of multiplication first or directly growing the propagated plants in the field. Vetiver parts used in propagation are tiller or slip, culm, ratoon, and clump (Chomchalow, 2000). Several methods of using tillers to propagate the vetiver plant are employed in Thailand. Planting vetiver in polyethylene bags (commonly called polybags) (Fig. 2B) is easy to maintain; however, it requires proper tools for watering and caring. They are suitable for direct transplanting on land or specific areas. However, propagation in polybags is costly for polybags, nurseries, water, labor, and transportation. Polybags that may be left unattended during planting in the field can create environmental problems later. For multiplication purposes, vetiver can be planted directly on the cultivated plots. Thai government-owned vetiver multiplication centers, the LDD regional stations, are responsible for the cultivation and distribution of the vetiver seedlings and tillers. The tillers used in planting are obtained from the selected clump, and then trim the top to 20 cm and the roots to 5 cm. After that, the shoots are separated and bound together in bunches waiting for distribution, as shown in Fig. 8A. The roots are soaked in water for 4 days, after which they start to grow. It is a labor-saving practice with high survival rates as the roots are not disturbed, as in the case of using polybags. It is also environmentally friendly since there are no waste materials from the used polybags. Additionally, planting vetiver in biodegradable pots, which may be made from vetiver leaves, is another environmentally friendly and easy option as it can be planted in the entire pot. Alternatively, some meristematic tissues of vetiver can be used as plantlets for the tissue culture system. Plantlets obtained from the tissue culture technique have an advantage over tillers as they are small in size and fit well in small structures of the nursery blocks (Chomchalow, 2000). Guidelines for choosing a vetiver propagation have to concern healthy and vigorous growing, plant materials, efficient nursery management (water, nutrient, light, etc.) based on economic consideration, and a period of acclimatization, i.e., exposure to sunlight prior to field planting. The last thing to consider is practical transportation; the plant material and the container should be lightweight and not to be discarded, e.g., biodegradable, nursery blocks, strip planting, and dibbling tubes. The best time to transplant vetiver tillers in the area is the beginning of the rainy season in order to increase the survival rate and increase the success of the vetiver plantation.

3 Phytostabilization of mine waste and metal-contaminated area

The heavy metals contamination of the environment by soil erosion in mining industries attracts world-wide concern, especially in open-pit mining, also known as opencast mining. Opencast mining destroys landscapes, forests, and wildlife habitats at the mining site when trees, plants, and topsoil are cleared from the mining area. Phytoremediation is an affordable and sustainable technology that is most useful when contaminants are within the root zone of the plants (top 1–2 m of soil). Phytostabilization and phytoextraction are two techniques that make up in situ phytoremediation for metals. Revegetation methods are thought to be the most practical and economical method for the rehabilitation of the mine wastelands for the long-term stability of the land surface. A good vegetation cover is beneficial in the restoration of contaminated land and results in enhanced amenity values as well as the prevention of surface soil erosion (Truong, 1999; Baker et al., 1994). However, major limiting factors for plant establishment on mine tailing are the toxicity of heavy metals and nutrient deficiencies (Pichtel and Salt, 1998; Bradshaw, 1987).

Vetiver grass (*C. zizanioides*), is a plant with considerable promise for soil and water conservation as well as phytoremediation applications because of its unique morphological and physiological characteristics (Xia, 2004; Greenfield, 1989; Truong and Loch, 2004). Vetiver is a tall (1–2 m), tufted, perennial, scented grass with a straight stem, long narrow leaves, and a lacework root system that is abundant, complex, and extensive. A long (3–4 m), massive, and complex root system of vetiver grass can penetrate the deeper layers of the soil. Vetiver is a versatile plant that grows fast and can survive in harsh environments. Vetiver was reported in highly tolerant to adverse climatic conditions such as frost, heat, wave, temperature (−15 to 55°C), drought, flood, and inundation], edaphic conditions, and highly tolerant to elevated levels of heavy metals, herbicides, pesticides, and organic wastes (Danh et al., 2009). Some applications using vetiver grass to treat heavy metal contamination in soil and water are presented in Table 3.

Heavy metals tolerance properties of vetiver involve different mechanisms to detoxify reactive oxygen species (ROS) and also lipid peroxidation occurring in different parts of the plant. An increase in metal accumulation in the plant tissues is possibly induced by various antioxidant enzymes viz., superoxide dismutase (SOD), peroxidase (POD), catalase (CAT), and ascorbate peroxidase (APX) (Pang et al., 2003; Banerjee et al., 2016, 2019; Kriti et al., 2021; Fang et al., 2021). Overproduction of proline and abscisic acid (ABA) is one of the mechanisms of vetiver stress response and growth in the presence of metals. Vetiver showed qualitative and quantitative differences in phytochelatins (PC_n) production for detoxification by forming metal-phytochelatin complexes (Pang et al., 2003; Andra et al., 2010). The increases of glutathione peroxidase (GPX), which reduces lipid hydroperoxides, glutathione *S*-transferase (GST), and glutathione reductase (GR), which involves detoxification by glutathione (GSH), are mechanisms of vetiver to the stress of metals (Pang et al., 2003; Banerjee et al., 2016, 2019; Kriti et al., 2021; Kiiskila et al., 2021). In addition, the application of vetiver for phytoremediation depends upon various factors such as physical and chemical properties of growth media as well as agronomic practice. The application of chemical fertilizers, compost or manure, and even sewage sludge, can mitigate the effects of heavy metals and promote the growth of vetiver (Pang et al., 2003; Rotkittikhun et al., 2007; Gautam and Agrawal, 2017; Kriti et al., 2021). Domestic refuse alone and the combination of domestic refuse and artificial fertilizer significantly improved the survival rates and growth of vetiver on the Pb/Zn tailings (Yang et al., 2003). Organic matter application significantly improved soil physical characteristics and nutrient availability of Pb/Zn contaminated soils, while a chelating agent could modify metal bioavailability, plant uptake, and translocation (Walker et al., 2003). Vetiver grass grew better in firing range soil, which was contaminated with Zn, Cu, and Fe when fertilized with Osmocote® fertilizer in comparison to plants fertilized with 10-10-10 (NPK) fertilizer, and application of ethylene diamine tetra-acetic acid (EDTA) significantly increased the amount of Pb that was phytoextracted (Wilde et al., 2005). In the case of iron ore mining rehabilitation, the combination of soil amendment materials, especially diethylenetriminepentaacetic acid (DTPA) and compost, was more effective than sole chelating agents

TABLE 3 Utilization of vetiver (*Chrysopogon zizanioides*) to remediate heavy metals contamination in water and soil.

Botanical name	Heavy metal	Treatment	Results	References
C. zizanioides	Pb, Zn	System: Pot, Greenhouse Treatment: $(NH_4)_2SO_4$ Stress: Pb/Zn tailing mixed with soil compost at proportions of 20%–50% (w/w)	The metals inhibit leaf growth and photosynthesis. *C. zizanioides* had different mechanisms to detoxify ROS that existed in different parts of plants, according to the increased activities of SOD, POD, and CAT. Accumulation of proline and abscisic acid also were stimulated. Physiological responses to heavy metal treatments differed greatly between roots and shoots. Nitrogen fertilizer alleviated the effects of the metals	Pang et al. (2003)
C. zizanioides, four ecotypes (from Thailand: Surat Thani, Songkhla, and Kamphaeng Phet, and one from Sri Lanka), *Thysanolaena maxima*	Pb	*System*: Pot, Glasshouse *Treatment*: Amendment with pig manure (20%, 40%) and inorganic fertilizer (75, 150 mg kg^{-1}) *Stress*: Pb-contaminated soil (up 1% total Pb) with a low amount of organic matter and N-P-K	*T. maxima* and *C. zizanioides* (Surat Thani and Songkhla) could tolerate high Pb concentrations in soil (10.7 mg kg^{-1}). The amendment of pig manure and the fertilizer improved the plant growth and biomass and did not have any influence on the Pb uptake. Both *T. maxima* and vetiver accumulated Pb higher in roots, which indicated to be suitable for phytostabilization in Bo Ngam lead mine	Rotkittikhun et al. (2007)
C. zizanioides, Three vetiver ecotypes, Kamphaeng Phet-2, Sri Lanka and Surat Thani	Fe, Mn, Zn, Cu, Pb	*System*: hydroponic system *Stress*: wastewater taken from a milk factory (W1), a battery manufacturing plant (W2), an electric lamp plant (W3), and an ink manufacturing facility (W4)	The vetiver grown in W1 had the best growth, while the vetiver grown in W4, highly contaminated with Mn, Fe, and Cu, had the worst growth. The three vetiver ecotypes absorbed Fe > Mn > Zn > Cu > Pb, and they concentrated these metals more in roots than in shoots. Sri Lanka had the best growth and the highest heavy metal removal efficiencies	Roongtanakiat et al. (2007)

Continued

TABLE 3 Utilization of vetiver (*Chrysopogon zizanioides*) to remediate heavy metals contamination in water and soil—cont'd

Botanical name	Heavy metal	Treatment	Results	References
C. zizanioides	Pb	*System:* Hydroponic *Stress:* Lead nitrate (0–1200 mg Pb L^{-1}) in nutrient solution.	Vetiver accumulated Pb in roots higher than shoots. The most probable mechanism for Pb detoxification in vetiver is by synthesizing phytochelatins (PCn) and forming Pb-PCn complexes	Andra et al. (2010)
C. zizanioides	Fe, Mn, Zn, Cu, Pb, Ni, Cr, Al	*System:* Pot *Stress:* Heavy metal-rich soil from iron ore mine	A high metal inhibited chlorophyll content of leaves but stimulated the accumulation of proline and lipid peroxidation. There were different mechanisms to detoxify ROS in the shoots. The root parts of the plant accumulated higher metal concentrations than the shoot. Fe was highest, followed by Al, Cu, Zn, Cr, Mn, and Ni. Vetiver grass is a good phytostabilizer, and can be used for the remediation and restoration of iron-ore mine spoil dumps	Banerjee et al. (2016)
C. zizanioides	Fe, Mn, Mg, Zn, Cu, Ni, Pb, Cd, and Cr	*System:* Field, Botanical Garden *Stress:* Red mud 0%–15%, amendment with sewage sludge (soil: sludge 2:1 w/w)	Red mud (10%) amendment with sludge increased the organic matter and nutrients and supported the plant growth. Vetiver had efficient translocation of Mn and Cu from root to shoot, and it has a potential phytostabilization for Fe, Zn, Mg, Cd, Pb, Ni, and Cr	Gautam and Agrawal (2017)
C. zizanioides, Surat Thani, and Ratchaburi ecotypes	[134]Cs	*System:* Hydroponic culture *Stress:* [134]Cs culture solution (2.5, 5.0 and 7.5 MBq/L)	The Surat Thani ecotype was stronger tolerated to [134]Cs than the Ratchaburi ecotype. Both ecotypes were suitable for phytostabilization, because they accumulated more [134]Cs in the roots than in shoots	Roongtanakiat and Akharawutchayanon (2017)

Plant species	Metals	System / Treatment / Stress	Findings	Reference
C. zizanioides, Typha latifolia	Hg, As, Pb, Cu, Zn	*System:* Pot *Treatment:* EDTA, $Al_2(SO_4)_3$ *Stress:* Contaminated soil from small-scale mining	The amendment with EDTA and $Al_2(SO_4)_3$ aided the accumulation and translocation of the metals in the cattail root, but it had no effect on the metal accumulation in vetiver	Anning and Akoto (2018)
C. zizanioides	Cr(VI), Fe, Mn, Ni, Cd, Cu, Zn	*System:* Pot *Treatment: Bacillus cereus* T1B3 isolated from chromite tailing *Stress:* chromite ore tailings	*B. cereus* T1B3 was resistant to very high metals. The strain had plant growth-promoting activity by producing ACC, IAA, and siderophores and solubilizing the phosphate. Inoculating vetiver with T1B3 strain protected vetiver from the metals stress and improved phytoremediation efficiency	Nayak et al. (2018)
C. zizanioides, Four diverse genotypes: S2 (diploid variety), S4 (tetraploid derivative of S2), TH (originated from Thailand), BL (a broad leaf)	Fe, Zn, Mn, Cr, Cu	*System:* Field *Stress:* Iron ore spoil dumpsites generated from the Joda East Iron mine located in Odisha, India	Accumulation of metals in plant tissues led to oxidative damage induced by ROS. As a consequence, antioxidative enzymes (SOD, CAT, GPOD, GR, GPX) were increased to scavenge ROS. The activity of intracellular GSSG concentration increases for the maintenance of glutathione in its reduced form (GSH) and the production of phytochelatins. The major amounts of metals were retained in the roots. The genotype BL followed by S4 was more efficient in remediating toxic metals	Banerjee et al. (2019)
C. zizanioides	Fe, Pb, Ni, Zn, Mn, Cr, Al, Cu and SO_4^{2-}	*System:* Floating treatment wetland *Stress:* Acid mine drainage (AMD)	Metals were mainly localized on the root surface as Fe plaques, whereas Mn and Zn showed greater translocation from root to shoot. There was near complete removal of SO_4^{2-} (91%) and metals (90%–100%), with the exception of Pb and Cu. Vetiver could effectively remediate AMD-impacted waters over an extended period of time	Kiiskila et al. (2019)

Continued

TABLE 3 Utilization of vetiver (*Chrysopogon zizanioides*) to remediate heavy metals contamination in water and soil—cont'd

Botanical name	Heavy metal	Treatment	Results	References
C. zizanioides	Fe, Ni, Zn, Pb, and As	*System*: greenhouse column, *Treatment*: drinking water treatment residuals (WTRs)-soil media *Stress*: AMD-impacted soil collected from an abandoned coal mine	Aluminum (Al)-based and calcium (Ca)-based WTRs were used to treat AMD-impacted soil. Vetiver grass was grown on the soil-WTR mixed media. Soil pH increased from 2.69 to 7.2, and soil erosion indicators such as turbidity (99%) and TSS (95%) in leachates were significantly reduced	RoyChowdhury et al. (2019)
C. zizanioides, Typha domingensis	As, Co, Cu, Mn, Ni, Pb, Zn	*System*: Constructed wetland with horizontal sub-surface flow *Stress*: Gold mine tailing storage facility seepage	After treatment for 75 days, the above-ground biomass of *T. domingensis* (12.30–14.18 g per plant) was higher than the biomass of *C. zizanioides* (6.65 g per plant). Although the TFs of *C. zizanioides* for the metals were higher compared to *T. domingensis*, *C. zizanioides* revealed very low growth rates	Compaore et al. (2020)
C. zizanioides	Zn, Pb, Cd, Cu, Cr	*System*: Pot *Treatment*: Coal gangue *Stress*: Cu mine tailing	Coal gangue converted the exchangeable and carbonate fractions of Zn, Pb, Cd, and Cu into unavailable forms of Fe-Mn oxide and inhibited Zn, Pb, Cd, and Cu translocation. In contrast, the accumulation of Cr in the plant was increased	Chu et al. (2020)
C. zizanioides	Pb, Cd	*System*: Floating hydroponic system *Stress*: Paperboard mill wastewater	The vetiver system with aeration was able to eliminate the nutrients and heavy metals from the paper board mill effluent. SEM-EDAX indicated that Pb and Cd uptake and storage in the vacuoles were the essential characteristics of vetiver grass, enabling it to grow at a high metal concentration	Davamani et al. (2021)

Plant species	Metal(s)	System/Treatment/Stress	Findings	Reference
C. zizanioides, Cymbopogon citratus	Ni, Cd	System: Pot Treatment: Amendment with garden soil and compost Stress: Ni-Cd battery electrolyte waste (EW) (1%, 2%, and 4%)	Vetiver was more tolerant towards EW toxicity than lemongrass. The bioaccumulation factor of metals in both grasses increased with EW contamination. pH affected the Ni and Cd uptake, whereas phosphate, organic matter, and bioavailable metals influenced negatively. Increase metal concentration induced parameters, i.e., lipid peroxidation and antioxidant enzymes (GST, GR, SOD, and POD) in vetiver. The roots exhibited greater phytoremediation potential than their shoots, and the phytoremediation of Ni was higher than the Cd	Kriti et al. (2021)
C. zizanioides	Cr(VI)	System: Pot Stress: Synthetic Cr(VI) wastewater (5, 10, and 30 mg L^{-1}) Low-medium-high density, with 5, 10, and 15 slips/pot. The pHs of 30 mg L^{-1} of Cr(VI) were adjusted to 3.5, 5, 7.5, and 10.5	Vetiver could be used for phytostabilization of heavy metal in aqueous solutions. At a low concentration, Cr was located in the subterranean component. At a high concentration, Cr accumulated in leaves after the roots became saturated. The rate of removal depended on the initial concentration, plant density, and pH. Increased plant density resulted in increased Cr(VI) uptake	Masinire et al. (2021)
C. zizanioides	Mn, Cu, Zn, Cd	System: Pot Treatment: Herbaspirillum sp. p5-19, with and without soil Stress: Cu tailing	Herbaspirillum sp. p5-19 inoculation alleviated the oxidative stress from the metals because the decrease of malondialdehyde (MDA) and the increase of antioxidant enzymes of APX, SOD, POD, and CAT were found in vetiver. It also improved the phytoremediation ability of vetiver	Fang et al. (2021)

Continued

TABLE 3 Utilization of vetiver (*Chrysopogon zizanioides*) to remediate heavy metals contamination in water and soil—cont'd

Botanical name	Heavy metal	Treatment	Results	References
C. zizanioides	Cd	*System:* Glass bowl, 10000 Lux, 25°C ± 3°C, and 70% humidity *Treatment:* Humic acid (HA) (150, 200, 250 mg L^{-1}) *Stress:* Cd (_5, 25, 30 mg L^{-1})	A low HA concentration (150 mg L^{-1}) improved the biological absorption efficiency because HA influenced Cd distribution in vetiver cells and significantly increased the proportion of Cd in the root cytoplasm. Cd accumulation was in the root higher than in the shoots. The Cd distribution followed the trend: cell wall > organelle > soluble substance	Wu et al. (2022)
C. zizanioides	Cd, Zn, Pb, Mn, Cr	*System:* Pot *Treatment:* multi-metal-tolerant *Bacillus cereus* *Stress:* Meta-contaminated soil, which contains Cd (31 mg kg^{-1}), Zn (7696 mg kg^{-1}), Pb (326 mg kg^{-1}), Mn (2519 mg kg^{-1}), and Cr (302 mg kg^{-1})	*B. cereus* has properties of fabrication of hydrogen cyanide, siderophore, Indole Acetic Acid, N fixation, P solubilization, and metal solubility. The bacterial bioaugmentation enhances the growth and phytoremediation potency of *C. zizanioides* on the metal-contaminated soil	Lu et al. (2023)

and sole compost in enhancing vetiver growth, nutrient, and heavy metals uptake (Roongtanakiat et al., 2008). EDTA could enhance vetiver growing on zinc mine soil collected from Tak province to uptake Zn, Mn, and Cu (but not Fe), while DTPA increased the phytoavailable metals but not their uptake (Roongtanakiat, 2009). The translocation ratio of Pb from vetiver roots to shoots was significantly increased with the application of EDTA (Chen et al., 2004). In addition, phytoremediation properties of vetiver were different in ecotypes and metals. Roongtanakiat and Chairoj (2001) planted three vetiver grass ecotypes of *C. zizanioides* ecotypes of Kamphaeng Phet and Surat Thani, *C. nemoralis* ecotype of Ratchaburi in soil supplemented with different amounts of Mn, Zn, Cu, Cd, and Pb. The root of Ratchaburi ecotype could also absorb significantly higher amounts of Zn, Cd, and Pb than those of Surat Thani and Kamphaeng Phet ecotypes. For Mn and Cu, Ratchaburi and Surat Thani ecotypes could uptake more than those of Kamphaeng Phet. In addition, Ratchaburi ecotype growing in Zn/Cd/Pb contaminated soil, which was collected near an operational zinc mine in Mae Sot district, Tak province, accumulated more heavy metals in their roots than shoots (Roongtanakiat and Sanoh, 2011). Chantachon et al. (2004) also showed *C. zizanioides* could uptake more Pb from the soil than *C. nemoralis*, and *C. zizanioides* showed greater biomass than *C. nemoralis*. Therefore, the use of vetiver grass for phytoremediation of tailings and metal-contaminated soils requires the selection of local species, soil chemical fertilizer, and amendment materials such as organic matter and chelating agents often required to establish successful vegetation and metal uptake.

In terms of phytostabilizing metal-contaminated sites, a lower metal concentration in the shoot is preferred in order to prevent metal from entering the ecosystem through the food chain. Phytostabilization which stabilizes the pollutants in soils, reduces their mobility into the soil and, therefore, the risk of leaching into ground water. Therefore, the extensive root system of vetiver grass has good potential to stabilize tailings particles and avoid erosion (Aziz and Islam, 2022; (Danh et al., 2009). The majority of the accumulated amounts of heavy metals in roots provided by this species were considered more suitable for stabilizing mine tailings, and the danger of transferring toxic metals to grassing animals was minimal (Yang et al., 2003). However, inoculation of plant growth-promoting bacteria, which have heavy metals tolerant properties, could improve the phytoextraction ability of vetiver to accumulate the metals such as *Bacillus cereus* T1B3 (Nayak et al., 2018) and *Herbaspirillum* sp. p5-19 (Fang et al., 2021). Vetiver is also very effective for the rehabilitation of old quarries and mines. It is able to stabilize the erodible surface first so that other species can colonize the areas later (Truong, 1999). In South Africa, it was used effectively to stabilize waste and slime dams from platinum and gold mines (Truong, 1999). The Vetiver system should provide a powerful phytoremediation tool for the attenuation of the mercury pollution problem in Yolo and Lake counties by trapping and containing both the air and water-born insoluble mercury (Hg) at sources and by reducing the soluble fraction in acid mine drainage (Truong, 2000). In Australia, *C. zizanioides* has been successfully used to stabilize mining overburden, coal and gold mine tailings (Truong and Baker, 1996). Vetiver grass was used to stabilize landfill and industrial waste sites contaminated with heavy metals such as As, Cd, Cr, Ni, Cu, Pb, and Hg (Truong and Baker, 1998). In China, vetiver grass was planted on a large scale for pollution control and mine tail stabilization (Chen, 2000). In India, vetiver showed an excellent candidate for remediation and restoration of the coal fly ash containing heavy metals (Zn, Pb, Cu, Ni, and As) and acidic/alkaline (pH) (Ghosh et al., 2015; Verma et al. (2014).

Phytoremediation of iron mine spoil dump soil of Joda mine was undertaken with vetiver system technology to stabilize and maintain ecological balance (Banerjee et al., 2018). *C. zizanioides* genotypes S4 (a tetraploid derivative of S2) and BL (a broad leaf) were more efficient in remediating toxic metals from the iron ore spoil dumpsites (Banerjee et al., 2019). Although vetiver grass is recognized for its efficiency in the recovery of degraded areas, Brazilian environmental legislation imposes restrictions on the use of exotic species. Thus, geotextile mats with mixtures of grass seeds, legumes, and other local herbaceous species of fast growth and good ground cover were used in most areas of the Uranium Concentrate Unit of Caetité in the southwestern Bahia state, Serra Geral region, Northeast Brazil. The vetiver grass was only used experimentally in the recovery area as a way to guarantee greater slope safety in the clarify water basin (Alves et al., 2018).

Due to environmental problems caused by mining operations, abandoned mines, and mine waste have generated significant environmental burdens in Thailand (Nakbanpote et al., 2018), mining activities in Thailand are currently more regulated, especially on environmental issues. According to the Ministry of Science, Technology and Environment's announcement concerning the Enhancement and Conservation of the National Environmental Quality Act B.E. 2535 (1992), mining activities of all sizes must prepare a report on environmental impact assessment (EIA). The EIA report must be submitted to the Office of Natural Resources and Environmental Policy and Planning (ONEP) for review and approval prior to further proceedings of the mining project. For the mined-out areas, the holder of a Prathanabat (a mining concession) must (i) either backfill the pits, mine sump, or shafts that are no longer in use, (ii) reclaim the area to the original landforms, (iii) degrade the slope area to provide safety and stability, and (iv) rehabilitate by re-vegetation, regardless of whether the Prathanabat has expired, unless the conditions in the Prathanabat stipulates otherwise. Mine reclamation and rehabilitation must be conducted in successive stages as proposed in the mining plan. Before mine closure, all mine structures and facilities must be completely removed, regardless of whether the Prathanabat has expired, except where such mining estates are private-owned properties. In addition, the rehabilitation of mined-out areas must comply with the conditions proposed in the EIA report that is approved and reviewed by ONEP (DPIM, 2009). The important mines that have been successful in rehabilitation and have a vetiver planting campaign are Limestone mining in Saraburi province, Akara gold mine in Phetchabun province, Mae Moh coal mine in Lampang province, and Padand zinc mine in Tak province. Department of Primary Industries and Mines, Ministry of Industry, establishes the Green Mining Award to promote sustainable mining activities. The Padaeng zinc mine is a model for planting vetiver grass for the rehabilitation process.

4 Rehabilitation of post zinc mine by vetiver grass in Thailand

Padaeng Industry Public Company Limited (PDI) was established on 10 April 1981 by the Thai public (Ministry of Finance) and private investors, and a private company from Belgium (Vieille Montagne). The Company engaged in the mining and refining business with the objective of producing zinc metal and zinc alloys to serve its customers. The company's mine is located in the Mae Sot district, and its refinery is in the Muang district of Tak province (Fig. 3A); the roaster plant is in Rayong province, and the head office is in Bangkok. The company was granted its first mining lease in 1982. The mining operation was started in 1984 after mining preparation and finishing of the refinery plant construction. The mining leases and related activity areas cover about 332 ha. The original surface of the zinc ore deposit was on the top of the hill, and the hill slope was without soil covering. Some parts of the leases and permits have not been used. They are left as natural forests and are a buffer zone to the surrounding areas. Therefore, there is only a mining-used area for rehabilitation, which is about 167 ha. Rehabilitation has been done as soon as possible when finished using each part of mining and related activities. PDI has seriously been taking action by coordinating with the Office of the Royal Development Project Board and Huay Sai Royal Development Study centre in Phetchaburi province to bring vetiver for growing on the bare soil surface to prevent erosion and protect soil moisture. It will be suitable for the pioneer plantation before planting other trees, then the bare land becomes the real forest (Fig. 4).

FIG. 3 Location of the Padaeng Zinc mine, Mae Sot, Tak Province, Thailand (A). (B) the rehabilitation areas where vetiver grass and local tree species have been planted to return as a forest area and the open-pit mining area, in 2013, (C) and (D) community participation in annual vetiver planting for soil and water conservation at the zinc mine.

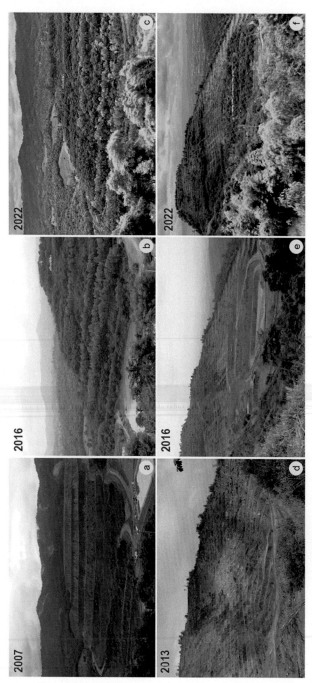

FIG. 4 Success in the restoration and rehabilitation of the overburden collection area (A)–(C) and the last open-pit mining area (D)–(F) of the Padang zine mine after closing.

Rehabilitation was complete in places where finished mine and related activities, namely, mine pit edge, waste collection areas, and others. The first step of land preparation was the land preparation step, which is done by purchasing fertilized soil from outside to cap bare soil at least 30 cm thick before planting. The company provided a budget of 50 million Thai Baht for fertilized soil due to unsuitable soil within the project area. In the second step, bare soil was covered by vetiver growing in principal and added with Ruzi grass. It was aimed at soil rehabilitation, adding more organic matter into the soil, preventing soil erosion, reducing the velocity of running water on the surface, and protecting moisture in the soil as long as possible. Finally, trees of local species, such as Teak, Iron wood, Siamese sal, local Cork tree, and Orchid Tree, were planted in principle. Additional fast-growing trees, such as Eucalyptus, Royal Poinciana, and Leucaena are also planted. Legumes have been important crop plants with the benefit of including improved soil quality (Vandermeer, 1989). For long-term remediation, herbaceous legumes were used as pioneer species to solve the problem of nitrogen deficiencies in mining wastelands because of the N_2-fixing ability of their microbial colonization (Archer et al., 1988; Bradshaw, 1987). Approximately 97,600 trees were planted together. The plantation was done shortly before, and at the start of the rainy season, then there was rainwater to make them survive and grow enough before the coming of summer.

The company started growing vetiver to conserve soil and water on 15 August 2003. Then, annual vetiver growing has been done. An experiment was conducted at Padaeng mine in order to compare the development of two *C. nemoralis* ecotypes, Nakhon Sawan and Prachuap Khiri Khan, and two *C. zizanioides* ecotypes, Kamphaeng Phet 2 and Surat Thani, grown in the zinc mining area. It was found that both *C. zizanioides* ecotypes gave a better growth performance than *C. nemoralis*, while Kamphaeng Phet 2 gave the highest plant height and shoot dry weight (Roongtanakiat, 2009). Therefore, the species of Kamphaengphet No. 2 was the most appropriate for the climate and the soil of the Padaeng mine. Mining rehabilitation was undertaken in the overburdened area along areas where open-pit mining activity continues, as shown in Fig. 3B. A total of 19.17 million vetiver sprouts were grown. An area of 103 ha has been rehabilitated till the end of 2014 from a total of 166 ha (or 62%). Therefore, the Padaeng Zinc mine is one of the biggest mines in Thailand, where the vetiver has been growing as big mountain covering with vetiver (Fig. 3C). The company also invited any government agencies and community, especially groups of teachers and students from local schools, colleges, and universities, to join in the activity (Fig. 3C and D). Successful rehabilitation of the zinc mine for handing over the restoration areas to the Forest Resources Management Office No. 4, Tak Province, both in the overburden area and the last open-pit mining area, are shown in Fig. 4A–C and D–F, respectively.

For phytoremediation of zinc mine tailings, the application of fertilizer, soil amendments, and chelating agents have been researched to establish successful vegetation and metal uptake. The translocation of heavy metal in vetiver was investigated.

4.1 Fertilizer and soil amendments

Primary nutrients were needed in large quantities for plant growth. LDD (1998) reported that concentrations of N, P, and K in vetiver shoot were 2.5%, 0.17%, and 1.5%, respectively. The vetiver grown in zinc mine soil, which has lower fertility than iron ore tailings, had lower concentrations of primary nutrients in the shoots of 2.12%–2.55%, 0.44%–0.50%, and 1.26%–1.40%, respectively. (Roongtanakiat, 2009). The application of organic fertilizer can increase vetiver yield and reduce the toxicity of heavy metals (Yong et al., 1992). The influence of organic and inorganic fertilizers on the growth of vetiver grown in Zinc mine soils was compared in a pot experiment. In the case of vetiver grown in zinc mine soil, the compost elevated vetiver biomass while the inorganic fertilizer decreased vetiver growth which gave biomass significantly different to those in control and compost treatments. For vetiver cultivation on deteriorated land with low fertility, the LDD (1998) recommended filling the bottom of the plant holes with manure or compost. Once the tillers start to sprout, the 15-15-15 inorganic fertilizer should be added to accelerate growth at the rate of 25 kg/rai (0.4 acre) along the contour (ORDPB, 2000). Besides increasing organic matter and nutrient content in the soil, the application of organic amendments, e.g., compost to mine tailings, is known to increase water holding capacity, cation exchange capacity and to improve the structure of mine tailings by forming stable aggregates (Ye et al., 2000; Stevenson and Cole, 1999). A field experiment performed at the Padaeng zinc mine revealed that the application of compost could significantly increase growth and shoot dry weight of vetiver; however, there was no significant difference between 47 tons/rai and 8 tons/rai applications. Hence, the application of 4 tons/rai of compost was suggested for vetiver plantation in this area, as recommended by LDD (1998).

4.2 Chelating agent application

Since plant uptake requires metals in an environmentally mobile form, the negative charge of various soil particles tends to attract and bind heavy metals, which are cations, and prevent them from becoming soluble and diffusing to the root surface. Using chelating agents such as EDTA, DTPA, nitrilotriacetic acid (NTA), and cyclohexanediaminetetraacetic acid (CDTA) have been developed to overcome these problems (Cooper et al., 1999). However, the effects of chelating agents on growth performance and heavy metal uptake can differ among chelating agents, heavy metals, and soils. In the zinc mine soil, the combination of DTPA and compost application actually reduced the growth of vetiver in both height and biomass. The EDTA could enhance concentration and uptakes of Zn, Mn, and Cu but not Fe, while DTPA increases the mentioned heavy metal concentrations but not uptakes. These studies also revealed that sole compost application to zinc mine soil did not affect heavy metal uptakes by vetiver (Roongtanakiat, 2009).

4.3 Translocation of heavy metal in vetiver

The distribution of heavy metals in vetiver plants can be divided into three groups. (i) very little of the As, Cd, Cr, and Hg absorbed were translocated to the shoots (1%–5%); (ii) A moderate proportion of Cu, Pb, Ni, and Se were translocated (16%–33%); (iii) Zn was almost evenly distributed between shoot and root (40%) (Truong, 1999). However, numerous investigators concluded that vetiver root accumulated higher heavy metal concentrations than shoot, and they concluded that vetiver is a non-hyperaccumulator plant

(Truong, 1999; Roongtanakiat, 2006). When vetiver plants were more mature, they could not concentrate higher heavy metal in the shoot. The shoot heavy metal concentration decreased, possibly due to the dilution effect of increasing biomass, whilst the root heavy metal concentrations increased (Roongtanakiat and Chairoj, 2001). The vetiver grown on iron tailings and zinc mine soils could translocate higher quantities of heavy metal from root to shoot with translocation factors (TFs) of 0.55–0.86 and 0.50–0.89, respectively. Even soil amendments could enhance some metal translocations; the compost and chelating agents did not affect the Zn translocation of vetiver grown in both mine soils (Alloway, 1995).

5 Utilization of vetiver grass in roadwork and in wastewater treatment systems

Vetiver grass significantly controls soil erosion and stabilizes highway slope damage caused by erosion or scouring at the curved section (Seutloali and Beckedahl, 2015). *C. zizanioides* is suitable as planting material for roadworks. Some vetiver systems on side ditch lining and highway shoulder are associated with gabion walls, surface drainage systems, and concrete square grid slope protection. Care is needed for 1–2 years after planting to become fully effective in rehabilitation (Sanguankaeo et al., 2003). In order for a vetiver system to be effective, sustainable, and cost-effective in maintenance, proper weed control, fertilization, and optimum planting techniques are required. Planting *Arachis pintoi*, a creeping forage peanut, for mat formation in between rows of vetiver grasses not only prevents and controls weeds but also increases nitrogen via nitrogen fixation (Sanguankaeo et al., 2006).

In the case of wastewater treatment, vetiver grasses have the ability to uptake heavy metals from wastewater, as shown in Table 3. Most of the metals were accumulated in the roots rather than the shoots. This indicated that vetiver could be used for phytostabilization of heavy metals in aqueous solutions. Iron plaque formation on the root part can increase metal stabilization in the root and decrease metal translocation from roots to shoots (Kiiskila et al., 2017). However, the concentration of heavy metals in wastewater plays an important role in vetiver growth and metal accumulation (Masinire et al., 2021; Wu et al., 2022). The efficiency of vetiver grass to remove heavy metals from wastewater depended on ecotypes (Roongtanakiat et al., 2007; Roongtanakiat and Akharawutchayanon, 2017). In addition, the removal efficiency of vetiver depends on root length and plant density due to increased surface area for metal absorption (Suelee et al., 2017). Vetiver (*C. zizanioides*) can be used to treat wastewater from mines like gold mines, either as a monoculture or as a mixed culture with another plant such as cattail (*Typha domingensis*). Fig. 5 shows a horizontally constructed wetland with a layout and piping, which this system is suitable for treating gold mine tailing storage facility seepage (Compaore et al., 2020). Vetiver grass is a good candidate for the remediation of wastewater by constructed wetland technology (Table 4). Vetiver had the potential to eliminate chemical oxygen demand (COD) and biological oxygen demand (BOD) from various wastewater sources., In a floating treatment system (Fig. 6A), a large surface area of the vetiver roots is a factor for biofilm development, including the root-exuded organic carbon. The diffusion of oxygen by the roots due to photosynthesis has strong repercussions on

FIG. 5   View of the experimental plot and the inlet pipe, first experimental setup with monotype species of *Typha domingensis* and *Chrysopogon zizanioides* in a constructed wetland treating gold mine tailing storage facility seepage. *From: Compaore, W.F., Dumoulin, A., Rousseau, D.P., 2020. Metal uptake by spontaneously grown Typha domingensis and introduced Chrysopogon zizanioides in a constructed wetland treating gold mine tailing storage facility seepage. Ecol. Eng. 158, 106037.*

nitrogen transformations, phytotoxin oxidation, and aerobic degradation. In addition, the vetiver root system also uptake nutrients from wastewater (Dorafshan et al., 2023). Therefore, the grass reduced total dissolved solids (TDS), COD, calcium, chlorides, total nitrogen, and sulfates in different wastewater of the sewage farm (Dhanya and Jaya, 2013). In addition, the hydroponically grown vetiver treatment system combined with aeration showed a promising future for developing a sustainable, cost-effective methodology to treat organic and inorganic contaminated wastewater, such as paper board mill effluent (Davamani et al., 2021). Phytoremediation of metal by vetiver can occur by high accumulation of the metal in the roots, then an excess of metal uptake can be transported via xylem and phloem vessels before accumulating in leaf vacuoles (Fig. 6B) (Davamani et al., 2021). Vetiver had an efficiency in removing phenol and developed resistance to phenol-induced ROS through the levels of antioxidant enzymes such as SOD and peroxidase (Singh et al., 2008). Vetiver remediation also increased the diversity and significantly changed the structure of the microbial community; therefore, the application of vetiver in the constructed wetlands provided the elimination of xenobiotics such as Atrazine (Marcacci et al., 2006), Ciprofloxacin (Panja et al., 2019), dinitroanisole (DNAN), nitroguanidine (NQ), and 1 and hexahydro-1,3,5-trinitro-1,3,5-triazine (RDX) (Panja et al., 2018). Furthermore, inoculation of the endophytic bacterial strain *Achromobacter xylosoxidans* F3B, which was able to utilize an aromatic compound as a carbon source, improved phytoremediation of aromatic pollutants (toluene) without largely interfering with the diversity of native endophytes (Ho et al., 2013). Based on the ability in wastewater treatment, vetiver (*C. zizanoides*) filtration constructed wetland in cooperating with oxidation pond has been applied for community wastewater treatment at Laem Phak Bia Sub-District, Ban Laem District, Phetchaburi Province, Thailand (Fig. 7).

TABLE 4 Utilization of vetiver (*Chrysopogon zizanioides*) in the construced wetland (CW) wastewater treatment system.

Pollutant	Treatment	Results	References
Atrazine	*System:* Floating pot (batch) *Stress:* 8 µM of ^{14}C-atrazine (417 Bq mL^{-1}) supplemented with nutrient solution	^{14}C-atrazine was found to accumulate radioactivity at the tip of leaves under moderate transpiring conditions (75% humidity). The metabolic features indicated that conjugation to glutathione was a major metabolic pathway to detoxify atrazine in vetiver	Marcacci et al. (2006)
Institutional kitchen wastewater	*System:* Up flow, 5 drums of 130 L capacity filled with soil, connecting in series *Stress:* pH 4–5, BOD 700–800 mg L^{-1}, COD 1300–1900 mg L^{-1}	Vetiver has significant potential to reclaim wastewater. The vetiver system was able to remove 80%–85% of BOD, 85%–90% of COD, and decrease total coliform	Mathew et al. (2016)
Munition Industry Wastewater	*System:* Floating pot (batch), 1 L HDPE bottles containing 800 mL of wastewater., and growth condition of 25 ± 3°C, 14 h photoperiod *Stress:* Organic explosive compounds of industrial munition facilities; dinitroanisole (DNAN), nitroguanidine (NQ), and 1,2,4-triazol-3-one (NTO), to replace conventional munitions such as trinitrotoluene (TNT) and hexahydro-1,3,5-trinitro-1,3,5-triazine (RDX)	Vetiver was able to remove DNAN, NQ, and RDX by 96%, 79% and 100%, respectively. The transformation products of DNAN, RDX, and HMX were found in vetiver. Therefore, disposal options for spent vetiver grass biomass have to be concerned	Panja et al. (2018)
Ciprofloxacin (CIP) synthetic wastewater	*System:* Batch experiment of floating hydroponic system *Stress:* 1 L HDPE bottles (4% w/v) containing 800 mL of Hoagland's nutrient solution spiked with CIP (0.05, 0.1, 1, and 10 mg L^{-1})	CIP caused oxidative stress, chlorosis, and a moderate decline in plant growth. CIP induced anti-oxidant enzymes and impacted metabolic pathways. However, vetiver grass removed >80% of CIP within 30 days by the plant and root-associated microorganisms in the root zone	Panja et al. (2019)
Paperboard mill wastewater	*System:* Vetiver slips floating in a hydroponic system, with and without aeration *Stress:* Raw effluent; pH 7.83, BOD 156 mg L^{-1}, COD 512 mg L^{-1}, TDS 1200 mg L^{-1}, TSS 1000 mg L^{-1}, Pb 2.01 mg L^{-1}, Cd 1.90 mg L^{-1}, TP 9.25 mg L^{-1}, TN 39.0 mg L^{-1}	The vetiver system with aeration had the efficiency in removing 57% of COD, 81% of BOD, 70% of TSS, 75% of TDS, 33% of Cd, 54% of PB, 68% of TP, and TN	Davamani et al. (2021)

Continued

TABLE 4 Utilization of vetiver (*Chrysopogon zizanioides*) in the constructed wetland (CW) wastewater treatment system—cont'd

Pollutant	Treatment	Results	References
Sewage wastewater	*System*: The Batch system of a pilot scale experiment 150 L PVC Drum, which was three layers of coarse aggregate (stone) from bottom, sand, and soil with vetiver grass growth. *Stress*: Gray water from the household: pH 6.7, BOD 260 mg L^{-1}, COD 561 mg L^{-1}, TSS 315 mg L^{-1}	The vetiver roots were spread both horizontally and vertically. The roots were the main aspects of the purification process. The vetiver system helped to purify water, similar to naturally occurring	Hemalatha et al. (2021)
Surface water contaminated with organic matter	*System* : Vertical flow constructed wetlands (VFCWs) *Stress*: Surface water samples; pH 6.2–7., BOD 155 mg L^{-1}, COD 303 mg L^{-1}, with various treatments of HLRs (500, 1000, and 1500 mL/m^2/min)	The vetiver-based-VFCWs model had the efficiency to remove 90% of BOD and >80% of COD. caused wastewater to come into contact with the filter media and biofilm and promote microorganisms and root system absorption resulting in better removal efficiency	Nguyen et al. (2023)

BOD, biological oxygen demand; COD, chemical oxygen demand; TDS, total dissolved solids; TSS, total suspended solids; TP, total phosphorus; TN, total nitrogen; OLR, organic loading rate; HLR, hydraulic loading rate; HRT, hydraulic retention time.

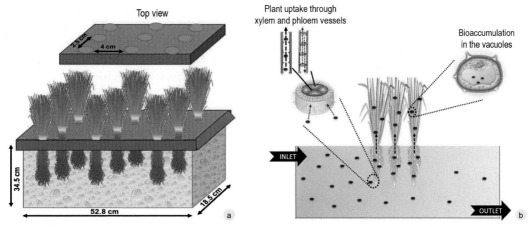

FIG. 6 Experimental setup of the hydroponic approach of paperboard mill wastewater by using vetiver (*Chrysopogon zizanioides*) (A), and (B) a schematic mechanistic of metal uptake from contaminated water, transport, and bioaccumulation in vacuoles. *Form: Davamani, V., Parameshwari, C.I., Arulmani, S., John, J.E., Poornima, R., 2021. Hydroponic phytoremediation of paperboard mill wastewater by using vetiver* (Chrysopogon zizanioides). *J. Environ. Chem. Eng. 9(4), 105528.*

FIG. 7 *Chrysopogon zizanioides* filtration constructed wetland in cooperating with oxidation pond for community wastewater treatment at Laem Phak Bia Sub-District, Ban Laem District, Phetchaburi Province, Thailand.

6 Utilization of vetiver grass for soil and water conservation

Vetiver grass has special characteristics of long deep roots that extend into the ground like a living natural wall. The LDD, a government agency, has been assigned to propagate vetiver plants and distribute them to target farmers sufficiently for soil and water conservation, restoring and improving soil fertility, and reducing soil erosion. There are many different ways of using vetiver in agriculture.

Planting vetiver around the reservoir (Fig. 8B): Planting vetiver grass close together in a row around the edge of the pond can trap sediment from flowing into the pond and prevents the pond from becoming shallow. The number of rows of vetivers depends on the slope of the pond edge. Vetiver is normally grown in three rows. The first row is at flood level, the second is planted on top, and the third is lower, holding the distance between the rows about 20–30 cm. The vetiver should be at least 45–60 days old.

Planting and maintaining vetiver to protect river banks and control water channels (Fig. 8E): Vetiver grass has a deep root system that can help hold the river banks from being washed away by erosion. It keeps the area moist, slowing down the speed of the current and trapping the sediment that flows with the water. The vetiver plants are grown in long rows along with a distance of about 5 cm to control soil erosion around water channels and banks along subcreeks.

FIG. 8   Propagation and distribution of vetiver grass seedlings by the Land Development Department of Thailand for soil and water conservation (A). Examples of utilization of vetiver grass; (B) prevent sediment accumulation in the water source, (C) create the boundaries of the planting plots, (D) maintain soil moisture and use trimmed leaves to control weeds and increase organic matter on the soil surface, and (E) construct the edges of the thoroughfare and the canal embankment in the farm.

Planting vetiver as a watershed weir: Vetiver slows down the flow of water, filters sediment, retains moisture, and strengthens the weir upstream for longer use. The appropriate method is to plant vetiver integrated with a weir. The middle part of the weir is filled with soil and planted rows of vetiver grass on the weir ridge with a distance of 10 cm between the plants and a distance of 20–30 cm between the rows.

Planting and maintenance of vetiver grass for a wet fireproof line: The green vetiver contains moisture that keeps the surrounding area moist, reducing the severity of wildfires and preventing them from spreading to other areas. The method is to plant vetiver close together in several long rows with a distance of 10 cm between the plants and a distance of about 50–100 cm between the rows to form a strip 8–10 m wide. Vetiver plants are needed 10,000 plants/row/km, approximately.

Planting vetiver across the water channel in an inverted V shape: Vetiver species should be *C. zizanioides* that can grow well in wet areas and are at least 45–60 days old. Vetiver grass is planted in two rows close to the creek in an inverted V shape with a distance of 5 cm. The rows of vetiver grass acted as weirs upstream to slow down the flow of water, increases soil moisture, and reduce pollution.

Planting vetiver around the bole of a tree. Planting vetiver around the bole of the tree to have a radius of about 1.5–2 m (around the edge of the canopy). When the tree grows, the planting circle needs to be extended so that the vetiver grass line is on the edge of the canopy. The moisture in the soil is beneficial to the plants for a longer period of time. Vetiver can be planted in fruit orchards and vegetable plots (Fig. 8C). The cutting of vetiver leaves to cover the soil can help reduce the evaporation of water from the soil surface (Fig. 8D).

Planting vetiver for erosion prevention on steep slopes: Planting on steep-slope areas is done by planting vetiver along the level of the area. Each clump of vetiver grass is joined together to form a dense fence, helping to slow down the speed of runoff and increase water seeps into the soil. Rows of vetiver serve to trap sediment. The roots are firmly attached to the soil, reducing erosion rates and increasing the nutrient supply of plants that grow in the region. Planting vetiver grasses together with cover plants will improve soil erosion prevention. However, deeper roots can cause soil to split and water to enter the soil, which may cause soil erosion. Therefore, the cultivation of vetiver grass to prevent erosion of slopes should be carefully studied in the following areas: (i) soil structure, (ii) specie and ecology of vetiver grass and plants, and (iii) appropriate methods for each site. Geotextiles can be applied on bare slopes to control rain splashing and runoff and encourage a microclimate for vegetation growth (ORDPB, 2000). Natural geotextiles are available in many parts of the world, less costly to produce and apply, and are environmentally friendly as they are made of biodegradable materials (Seutloali and Beckedahl, 2015). Therefore, in some sloping areas, the application of natural geotextile should be combined with vetiver grass planting and seed spreading, which would offer a permanent erosion control.

7 Economic product from vetiver

Vetiver grass is a very economical, eco-friendly, and practical tool for phytoremediation and bioremediation of contaminated substrates. For effective management of vetiver grass cultivation, vetiver at the age of 3–4 months is required to leave pruning to stimulate and maintain hedgerows and fertility or a prolific and efficient root system. Disposal of those plant tissues after environmental restoration applications can be a serious consideration due to the associated costs and handling concerns. Vetiver posed several alternative income-generating options. All parts of vetiver, no matter what roots, trunk, and leave could be beneficial e.g., forage, mushroom substrate, mulch, cover crop, toxic substance absorption, wastewater treatment, pest and weed control, aromatic oils, herbs, handicrafts, paper, and building material (Fig. 9A). In addition, vetiver grass can be used as a biofuel (both as a direct fuel and its potential conversion to cellulosic ethanol) and its overall potential relating to climate change and the sequestering of atmospheric carbon. The promotion of vetiver grass as an income-generating crop and its development from merely being an agricultural waste to economical partial substitutes or raw materials would reduce environmental deterioration and deforestation (Gnansounou et al., 2017).

Construction materials: Vetiver could be transformed from simply being an agricultural material for soil conservation into an industrial, low-cost construction material by exploration of the use of vetiver grass ash as cement replacement material, furthering its utilization as vetiver fiber-clay composite for construction and using vetiver pulps as fiber reinforcement. Hengsakeekul and Nimityongskul (2004) utilized vetiver grass and clay to construct a paddy storage silo. The construction of a cylindrical silo demonstration having a diameter and height of 3 m and a capacity of 10 tons at the Royal Chitralada project was done as a pilot project. The quality of paddy stored in a vetiver-clay silo was unchanged for a period of six months. The Association of Siamese Architects decided to build its new center in order to support and provide service to the users. The creation of design from environmentally manufactured materials stimulated the demand of the market for the use of alternative materials. The actual

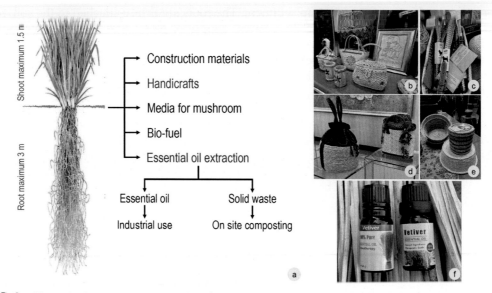

FIG. 9   Alternative income-generating options from vetiver grass (A), Examples of handicrafts made from vetiver grass (B)–(E), and essential oil for aromatherapy (F).

structure was assembled from the vetiver board (VB) and rubber inserted with wood to en-hance the structure's durability and strength. The VB covered the inner structure, giving the exterior of different visual effect. In addition to the walls, ceiling, and floors of the rooms, VB has also been used in making furniture such as cupboards, tables, and chairs. (Anon, 2009).

Handicrafts: Handicraft made from vetiver benefits those who make the handicrafts and those who produce the raw material. Special programs in Thailand and Venezuela have used specially treated vetiver leaves for the material of the high-end handicraft market. Examples of vetiver handicrafts exhibited in Thailand are shown in Fig. 9B–E. The Vetiver Network In-ternational, in cooperation with the Royal Development Projects Board, has supported the training of Indians and Chinese in vetiver handicrafts by Thai experts. Zehra Tyabji and Rashmi Ranade, from Women Weavers, received training and are now available to run train-ing workshops in India. In India, traditionally, vetiver roots have been used for handcrafts, but the proper treatment and use of vetiver leaves results in a much more finely crafted prod-uct that can be sold in top-end markets where there is a growing demand for "natural" prod-ucts (Grimshaw and Helfer, 1995).

Media for mushroom: Vetiver showed the opportunity for substitution of sawdust with cul-tivar on the media in mycelia growth of *Ganoderma lucidum* in a plastic bag. The ratio of saw-dust and vetiver leaves at 100:0 and 20:80, supplemented with $CaCO_3$ 0.5%, $MgSO_4.7H_2O$ 0.2%, and fine rice barn 6%, gave the biological efficiency (%*B.E. = Fresh weight of mushroom × 100/Dry weight of substrates*) with 42.72% and 40.71%, respectively (Sornprasert and Aroonsrimorakot, 2014).

Biofuels: Vetiver is well placed as a potential furnace feedstock and has the added advan-tage in that it could be grown on marginal lands to feed local steam-based generating plants creating energy for nearby communities. When grown as a field crop at about 100,000 plants per ha ($0.3 × 0.3$ m spacing), vetiver can produce up to 40–100 tons of dry biomass on soils of reasonable depth and fertility. Vetiver has an energy value of 7000 BTU/lbs compared to pe-troleum -18000; coal 12-13000, dry wood 8500; and sugar cane bagasse 4000. These non-vetiver biomass sources are used as feedstock to generate electricity. The advantage of vetiver feedstock over other fuels is that it is clean, it does not emit noxious chemicals when burnt, it is entirely renewable, and the burning process does not add net carbon amounts to the atmo-sphere (Grimshaw and Helfer, 1995).

Aromatic oil: Spongy root mass of certain cultivars of *C. zizanioides* contains a trace amount of essential or volatile oil, known as vetiver oil or "khus oil." Vetiver oil is a viscous light-brown or golden-brown oil. It has a rich green-woody earthy, and nut-like fragrance. Fig. 9F shows an example of essential oil extracted from vetiver grass sold online to the general public. The grass has been cultivated for its roots which contain the essential oil used exten-sively in perfumery, cosmetics, and biomedical utilization (Chomchalow, 2001; Dowthwaite and Rajani, 2000). In addition, vetiver root aromatic oil has been used as a repellent for the control of termites (Zhu et al., 2001; Bahri et al., 2021). The aromatic oil extracts are used for antiseptic, aphrodisiac, cicatrisant, nervine, sedative, tonic, sedative, and vulnerary. Vet-iver oil production is closely related to plant metabolism and gene regulation (George et al., 2017). The major compounds of vetiver oils are khusimone, khusimol, khusol, khusilal, zizanoic acid, β-vetivone, α-vetiverol, isovalencenol, nootkatone, camphor, junenol, juniper, (*E*)-isovalencenol, (*Z*)-9,10-dehydro-2-norzizaene, (*E*)-opposita-4(15),7(11)-dien-12-al (Dethier et al., 1997; Pripdeevech et al., 2006; Massardo et al., 2006; Santos et al., 2019). Oil yield and composition in vetiver could be influenced by several factors such as genotypes (Lal et al., 2020), drought and salt stress (George et al., 2017), geographical origin, cultivar, cultivation methods, soil fertilization, the presence of microorganisms, photoperiods and other environmental parameters (Pripdeevech et al., 2006; Dethier et al., 1997; Adams et al., 2003). Low temperature (such as in winter) causes a decrease in plant metabolic activities and hence a decrease in oil production (Massardo et al., 2006). Aromatic grasses are unpalatable, industrial/commercial crops, perennial in nature with multiple harvests, and tolerate to stress conditions (pH variability, heavy metal toxicity, and drought) (Gupta et al., 2013). Most importantly, essential oil is extracted through hydro distillation and free from the risk of toxic heavy metals (Zheljazkov et al., 2006; Khajanchi et al., 2013; Gautam and Agrawal, 2021). In addition, Rotkittikhun et al. (2010) reported that Pb could increase the oil production of vetiver; the predominant compound was khusimol (10.7%–18.1%), followed by (E)-isovalencenol (10.3%–15.6%). Vetiver growing in 10% of Bauxite residue in sludge-amended soil improved the quality and quantity of essential oil of vetiver; the key compounds of vetiver oil, i.e., isovalencene and vetivone contents were increased (Gautam and Agrawal, 2021).

8 Conclusions

With vetiver phytoremediation, a massive and deep root system of vetiver can prevent the leaching and runoff of heavy metals to nearby areas and ground water by phytostabilization. In addition, this grass can be an excellent pioneer plant to conserve water and improve soil quality. For non-hyperaccumulator like vetiver, improving biomass and propagation is necessary for the high efficiency of phytoremediation. The application of organic fertilizer can increase vetiver yield and may reduce the toxicity of heavy metals. Once the vetiver is fully grown, the aerial growth should be harvested periodically to accelerate new growth. An important advantage of harvested vetiver is that it is not considered hazardous waste, unlike hyperaccumulator residual. It can be used safely for bioenergy production, compost, or even as material for handicrafts. Vetiver grass is not a weed; it is not invasive. The cultivar of vetiver is easily asexually reproduced using simple agricultural practices by root sub-division; each slip normally consists of 2–3 tillers. For long-term remediation, metal-tolerant species are commonly used for the revegetation of mine tailings, and herbaceous legumes can be used as pioneer species to solve the problem of nitrogen deficiencies. Vetiver is also very useful in wastewater treatment and soil and water conservation in agriculture. However, vetiver plantation requires leaf pruning to stimulate growth and maintain hedgerows. Therefore, cogeneration of economic products should be promoted in order to generate income from vetiver planting in order to create sustainability. International cooperation, vetiver information, uses, raw materials, and handicrafts are included on the website of The Vetiver Network International Patron - H.R.H Princess Maha Chakri Sirindhorn of Thailand (https://www.vetiver.org/).

Acknowledgments

The authors would like to thank Dr. Chaiwat PHADERMROD, Former head of the security department Occupational Health and Environment, Padaeng Industry Public Co. Ltd. (Mae Sot Office, Tak Province, Thailand) for permission to conduct this phytostabilization study before the mine closed. In addition, Nakbanpote, W. would like to acknowledge the Learning Center for Conserving Forest Resources under the Royal Initiative, Forest Resources Management Office No.4 (Tak), Tak Province, Thailand, for taking to visit the progress of forest rehabilitation.

References

Adams, R.P., Pandey, R.N., Dafforn, M.R., James, S.A., 2003. Vetiver DNA-fingerprinted cultivars: effects of environment on growth, oil yields and composition. J. Essent. Oil Res. 15, 363–371.

Alloway, B.J., 1995. Heavy Metals in Soils. Blackie Academic & Professional, London, UK.

Alves, L.d.J., Nunes, F.C., Prasad, M.N.V., Mangabeira, P.A.O., Gross, E., Loureiro, D.M., Medrado, H.H.S., Bomfim, P.S.F., 2018. Uranium mine waste phytostabilization with native plants—a case study from brazil. In: Prasad, M.N.V., Favas, P.J.C., Maiti, S.K. (Eds.), Bio-Geotechnologies for Mine Site Rehabilitation. Elsevier, Amsterdam, pp. 299–322.

Andra, S.S., Datta, R., Sarkar, D., Makris, K.C., Mullens, C.P., Sahi, S.V., Bach, S.B.H., 2010. Synthesis of phytochelatins in vetiver grass upon lead exposure in the presence of phosphorus. Plant Soil 326, 171–185.

Anning, A.K., Akoto, R., 2018. Assisted phytoremediation of heavy metal contaminated soil from a mined site with *Typha latifolia* and *Chrysopogon zizanioides*. Ecotoxicol. Environ. Saf. 148, 97–104.

Anon, 2009. Vetiver grass board decorates ASA Centre in Bangkok. Vetiverim 49, 4.

Archer, I.M., Marshman, N.A., Salomons, W., 1988. Development of a revegetation programme from copper and sulphide-bearing mine waste in the humid tropic. In: Solomons, W., Forstner, U. (Eds.), Environmental Management of Solid Waste. Overseas Typographers, Makati, pp. 166–184.

Aziz, S., Islam, M.S., 2022. Mechanical effect of vetiver grass root for stabilization of natural and terraced hill slope. Geotech. Geol. Eng. 40, 3267–3286.

Bahri, S., Raharjo, T.T., Ambarwati, Y., Nurhasanah, Y., 2021. Isolation and identification of Terpenoid compound from vetiver grass-root (*Vetiveria zizanioides* Stapf) as a repellent against termite (*Cyrptotermes* sp.) through bioactivity assay. J. Phys. Conf. Ser. 1751, 1–15.

Baker, A.J.M., McGrath, S.P., Sidoli, C.M.D., Reeves, R.D., 1994. The possibility of *in situ* heavy metal decontamination of polluted soils using crops of metal-accumulating plants. Resour. Conserv. Recycl. 11, 41–49.

Banerjee, R., Goswami, P., Pathak, K., Mukherjee, A., 2016. Vetiver grass: an environment clean-up tool for heavy metal contaminated iron ore mine-soil. Ecol. Eng. 90, 25–34.

Banerjee, R., Goswami, P., Mukherjee, A., 2018. Stabilization of iron ore mine spoil dump sites with vetiver system. In: Prasad, M.N.V., Favas, P.J.C., Maiti, S.K. (Eds.), Bio-Geotechnologies for Mine Site Rehabilitation. Elsevier, Amsterdam, pp. 393–413.

Banerjee, R., Goswami, P., Lavania, S., Mukherjee, A., Lavania, U.C., 2019. Vetiver grass is a potential candidate for phytoremediation of iron ore mine spoil dumps. Ecol. Eng. 132, 120–136.

Bhromsiri, C., Bhromsiri, A., 2010. The effects of plant growth promoting rhizobacteria and *Arbuscular mycorrhizal* fungi on the growth, development and nutrient uptake of different vetiver ecotypes. Thai J. Agric. Sci. 43, 239–249.

Bradshaw, D., 1987. Reclamation of land and ecology of ecosystem. In: William, R.J., Gilpin, M.E., Aber, J.D. (Eds.), Restoration Ecology. Cambridge University Press, Cambridge, pp. 53–74.

Chantachon, S., Kruatrachue, M., Pokethitiyook, P., Upatham, S., Tantanasarit, S., Soonthornsarathool, V., 2004. Phytoextraction and accumulation of lead from contaminated soil by vetiver grass: laboratory and simulated field study. Water Air Soil Pollut. 154, 37–55.

Chen, H., 2000. Chemical method and phytoremediation of soil contaminated with heavy metals. Chemosphere 41, 229–234.

Chen, Y., Shen, Z., Li, X., 2004. The use of vetiver grass (*Vetiveria zizanioides*) in the phytoremediation of soils contaminated with heavy metals. Appl. Geochem. 19, 1553–1565.

Chomchalow, N., 2000. Technique of Vetiver Propagation With Special Reference to Thailand, Pacific Rim Vetiver Network, Technical Pollution No. 2000/1. Pacific Rim. Office of the Royal Development Project Board, Bangkok, Thailand.

Chomchalow, N., 2001. The Utilization of Vetiver as Medicinal and Aromatic Plants With Special Reference to Thailand, Pacific Rim Vetiver Network, Techical Pollution No. 2001/1. Pacific Rim. Office of the Royal Development Project Board, Bangkok, Thailand.

Chu, Z., Wang, X., Wang, Y., Zha, F., Dong, Z., Fan, T., Xu, X., 2020. Influence of coal gangue aided phytostabilization on metal availability and mobility in copper mine tailings. Environ. Earth Sci. 79, 1–14.

Compaore, W.F., Dumoulin, A., Rousseau, D.P., 2020. Metal uptake by spontaneously grown *Typha domingensis* and introduced *Chrysopogon zizanioides* in a constructed wetland treating gold mine tailing storage facility seepage. Ecol. Eng. 158, 106037.

Cooper, E.M., Sims, J.T., Cunningham, S.D., Huang, J.W., Berti, W.R., 1999. Chelate-assisted phytoextraction of lead from contaminated soils. J. Environ. Qual. 28, 1709–1719.

Cuong, D.C., Truong, P., Minh, V.V., Minh, P.T., Truong, P.N., 2015. Confusion between *Chrysopogon nemoralis* and *Chrysopogon zizanioides* at Bo Bo mountain in Quang Nam province Vietnam. In: 6th International Conference on Vetiver (ICV6) Da Nang, Vietnam. May 5–8th 2015.

Danh, L.T., Truong, P., Mammucari, R., Tran, T., Foster, N., 2009. Vetiver grass, *Vetiveria zizanioides*: a choice plant for phytoremediation of heavy metals and organic wastes. Int. J. Phytoremediation 11, 664–691.

Davamani, V., Parameshwari, C.I., Arulmani, S., John, J.E., Poornima, R., 2021. Hydroponic phytoremediation of paperboard mill wastewater by using vetiver (*Chrysopogon zizanioides*). J. Environ. Chem. Eng. 9, 105528.

Dethier, M., Sakubu, S., Ciza, A., Cordier, Y., Menut, C., Lamaty, G., 1997. Aromatic plants of tropical central Africa. XXVIII. Influence of cultural treatment and harvest time on vetiver oil quality in Burundi. J. Essent. Oil Res. 9, 447–451.

Dhanya, G., Jaya, D.S., 2013. Pollutant removal in wastewater by vetiver grass in constructed wetland system. Int. J. Eng. Res. Technol. 2, 1361–1368.

Dorafshan, M.M., Abedi-Koupai, J., Eslamian, S., Amiri, M.J., 2023. Vetiver grass (*Chrysopogon zizanoides* L.): a hyperaccumulator crop for bioremediation of unconventional water. Sustainability 15, 3529.

Dowthwaite, S.V., Rajani, S., 2000. Vetiver: perfumer's liquid gold. In: Proc. ICV-2 held in Cha-am, Phethchaburi, Thailand, 18–22 January, pp. 478–481.

DPIM, 2009. Mining in Thailand an Investment Guide. Department of Primary Industries and Mines Ministry of Industry. https://www.dpim.go.th/en/media/003_mine.pdf.

Fang, Q., Huang, T., Wang, N., Ding, Z., Sun, Q., 2021. Effects of *Herbaspirillum* sp. p5-19 assisted with alien soil improvement on the phytoremediation of copper tailings by *Vetiveria zizanioides* L. Environ. Sci. Pollut. Res. 28 (45), 64757–64768.

Gautam, M., Agrawal, M., 2017. Phytoremediation of metals using vetiver (*Chrysopogon zizanioides* (L.) Roberty) grown under different levels of red mud in sludge amended soil. J. Geochem. Explor. 182, 218–227.

Gautam, M., Agrawal, M., 2021. Application potential of *Chrysopogon zizanioides* (L.) Roberty for the remediation of red mud-treated soil: an analysis via determining alterations in essential oil content and composition. Int. J. Phytoremediation 23, 1356–1364.

George, S., Manoharan, D., Li, J., Britton, M., Parida, A., 2017. Drought and salt stress in *Chrysopogon zizanioides* leads to common and specific transcriptomic responses and may affect essential oil composition and benzylisoquinoline alkaloids metabolism. Curr. Plant Biol. 11, 12–22.

Ghosh, M., Paul, J., Jana, A., De, A., Mukherjee, A., 2015. Use of the grass, *Vetiveria zizanioides* (L.) Nash for detoxification and phytoremediation of soils contaminated with fly ash from thermal power plants. Ecol. Eng. 74, 258–265.

Gnansounou, E., Alves, C.M., Raman, J.K., 2017. Multiple applications of vetiver grass—a review. Int. J. Environ. Sci. 2, 125–141.

Greenfield, J.C., 1989. Vetiver Grass: The Idea Plant for Vegetative Soil and Moisture Conservation. ASTAG-The World Bank, Washington, DC.

Grimshaw, R.G., Helfer, L., 1995. Vetiver grass for soil and water conservation, land rehabilitation, and embankment stabilization: a collection of papers and newsletter complied by the vetiver network. In: World Bank Technical Paper Number 273. The International Bank for Reconstruction and Development, The World, Bank, Washington. 281 pp.

Gu, T., Mao, Y., Chen, C., Wang, Y., Lu, Q., Wang, H., Cheng, W., 2022. Diversity of arbuscular mycorrhiza fungi in rhizosphere soil and roots in *Vetiveria zizanioides* plantation chronosequence in coal gangue heaps. Symbiosis 86, 111–122.

Gupta, A.K., Verma, S.K., Khan, K., Verma, R.K., 2013. Phytoremediation using aromatic plants: a sustainable approach for remediation of heavy metals polluted sites. Environ. Sci. Technol. 47, 10115–10116.

Hemalatha, G., Uma, S.G., Muthulakshmi, S., 2021. Sewage water treatment using vetiver grass. Mater. Today: Proc. 46, 3795–3798.

Hengsakeekul, T., Nimityongskul, P., 2004. Construction of paddy storage silo using vetiver grass and clay. AU J. Technol. 7, 120–128.

Ho, Y.-N., Hsieh, J.-L., Huang, C.-C., 2013. Construction of a plant-microbe phytoremediation system: combination of vetiver grass with a functional endophytic bacterium, *Achromobacter xylosoxidans* F3B, for aromatic pollutants removal. Bioresour. Technol. 14, 43–47.

Kiiskila, J.D., Sarkar, D., Feuerstein, K.A., Datta, R., 2017. A preliminary study to design a floating treatment wetland for remediating acid mine drainage-impacted water using vetiver grass (*Chrysopogon zizanioides*). Environ. Sci. Pollut. Res. 24 (36), 27985–27993.

Kiiskila, J.D., Sarkar, D., Panja, S., Sahi, S.V., Datta, R., 2019. Remediation of acid mine drainage-impacted water by vetiver grass (*Chrysopogon zizanioides*): a multiscale long-term study. Ecol. Eng. 129, 97–108.

Kiiskila, J.D., Sarkar, D., Datta, R., 2021. Differential protein abundance of vetiver grass in response to acid mine drainage. Physiol. Plant. 173 (3), 829–842.

Khajanchi, L., Yadava, R.K., Kaurb, R., Bundelaa, D.S., Khana, M.I., Chaudharya, M., Meenaa, R.L., Dara, S.R., Singha, G., 2013. Productivity essential oil yield, and heavy metal accumulation in lemon grass (*Cymbopogon flexuosus*) under varied wastewater-groundwater irrigation regimes. Ind. Crop. Prod. 45, 270–278.

Kriti, Basant, N., Singh, J., Kumari, B., Sinam, G., Gautam, A., Singh, G., Swapnil, Mishra, K., Mallick, S., 2021. Nickel and cadmium phytoextraction efficiencies of vetiver and lemongrass grown on Ni-Cd battery waste contaminated soil: a comparative study of linear and nonlinear models. J. Environ. Manag. 295, 113144.

Lakshmi, C.S., Sekhar, C.C., 2020. Role of *Vetiveria zizanioides* in soil protection and carbon sequestration. J. Pharm. Innov. 9, 492–494.

Lal, R.K., Maurya, R., Chanotiya, C.S., Gupta, P., Mishra, A., Srivastava, S., Yadav, A., Sarkar, S., Pant, Y., Pandey, S.S., Shukla, S., 2020. On carbon sequestration efficient clones/genotypes selection for high essential oil yield over environments in Khus (*Chrysopogon zizanioides* (L.) Roberty). Ind. Crop. Prod. 145, 112139.

LDD, 1998. Vetiver Grass Overview. Land Development Department/Ministry of Agriculture and Cooperatives, Bangkok, Thailand.

LDD, 2007. Vetiver Grass Species in Soil and Water Conservation. Land Development Department/Ministry of Agriculture and Cooperatives, Bangkok, Thailand (in Thai).

Lu, H., Xia, C., Chinnathambi, A., Nasif, O., Narayanan, M., Shanmugam, S., Chi, N.T.L., Pugazhendhi, A., On-Uma, R., Jutamas, K., Anupong, W., 2023. Optimistic influence of multi-metal tolerant Bacillus species on phytoremediation potential of *Chrysopogon zizanioides* on metal contaminated soil. Chemosphere 311, 136889.

Marcacci, S., Raveton, M., Ravanel, P., Schwitzguebel, J.-P., 2006. Conjugation of atrazine in vetiver (*Chrysopogon zizanioides* Nash) grown in hydroponics. Environ. Exp. Bot. 56, 205–215.

Masinire, F., Adenuga, D.O., Tichapondwa, S.M., Chirwa, E.M., 2021. Phytoremediation of Cr(VI) in wastewater using the vetiver grass (*Chrysopogon zizanioides*). Miner. Eng. 172, 107141.

Massardo, D., Senatore, F., Alifano, P., Giudice, L.D., Pontieri, P., 2006. Vetiver oil production correlates with early root growth. Biochem. Syst. Ecol. 34, 376–382.

Mathew, M., Sebastian, M., Cherian, S.M., 2016. Effectiveness of vetiver system for the treatment of wastewater from an institutional kitchen. Proc. Technol. 24, 203–209.

Nakbanpote, W., Prasad, M.N.V., Mongkhonsin, B., Panitlertumpai, N., Munjit, R., Rattanapolsan, L., 2018. Strategies for rehabilitation of mine waste/leachate in Thailand. In: Prasad, M.N.V., Favas, P.J.C., Maiti, S.K. (Eds.), Bio-Geotechnologies for Mine Site Rehabilitation. Elsevier, Amsterdam, pp. 617–643.

National Resource Council, 1995. Vetiver Grass: A Thin Green Line Against Erosion. National Academy Press, Washington, DC.

Nayak, A.K., Panda, S.S., Basu, A., Dhal, N.K., 2018. Enhancement of toxic Cr(VI), Fe, and other heavy metals phytoremediation by the synergistic combination of native *Bacillus cereus* strain and *Vetiveria zizanioides* L. Int. J. Phytoremediation 20, 682–691.

Nguyen, M.K., Hung, N.T.Q., Nguyen, C.M., Lin, C., Nguyen, T.A., Nguyen, H.-L., 2023. Application of vetiver grass (*Vetiveria Zizanioides* L.) for organic matter removal from contaminated surface water. Bioresour. Technol. Rep. 22, 101431.

ORDPB, 2000. Manual of the International Training Course on the Vetiver System. 19–30 November 2000, Bangkok, Thailand.

IV. Brownfield development for smart bioeconomy

Pang, J., Chan, G.S.Y., Zhang, J., Liang, J., Wong, M.H., 2003. Physiological aspects of vetiver grass for rehabilitation in abandoned metalliferous mine wastes. Chemosphere 52, 1559–1570.

Panja, S., Sarkar, D., Datta, R., 2018. Vetiver grass (*Chrysopogon zizanioides*) is capable of removing insensitive high explosives from munition industry wastewater. Chemosphere 209, 920–927.

Panja, S., Sarkar, D., Li, K., Datta, R., 2019. Uptake and transformation of ciprofloxacin by vetiver grass (*Chrysopogon zizanioides*). Int. Biodeterior. Biodegrad. 142, 200–210.

Pichtel, J., Salt, C.A., 1998. Vegetative growth and trace metal accumulation on metalliferous wastes. J. Environ. Qual. 27, 618–642.

Pripdeevech, P., Wongpornchai, S., Promsiri, A., 2006. Highly volatile constituents of *Vetiveria zizanioides* roots grown under different cultivation conditions. Molecules 11, 817–826.

Roongtanakiat, N., Chairoj, P., 2001. Vetiver grass for the remediation of soil contaminated with heavy metals. Kasetsart J. Nat. Sci. 35, 433–440.

Roongtanakiat, N., 2006. Vetiver in Thailand: general aspects and basic studies. KU Sci. J. 24, 13–19.

Roongtanakiat, N., Tangruangkiat, S., Meesat, R., 2007. Utilization of vetiver grass (*Vetiveria zizanioides*) for removal of heavy metals from industrial wastewaters. ScienceAsia 33, 397–403.

Roongtanakiat, N., Osotsapar, Y., Yindiram, C., 2008. Effects of soil amendment on growth and heavy metals content in vetiver grown on iron ore tailings. Kasetsart J. Nat. Sci. 42, 397–406.

Roongtanakiat, N., 2009. Vetiver Phytoremediation for Heavy Metal Decontamination, Pacific Rim Vetiver Network, Technical Pollution No. 2009/1. Pacific Rim. Office of the Royal Development Project Board, Bangkok, Thailand.

Rotkittikhun, P., Kruatrachue, M., Pokethitiyook, P., Baker, A.J.M., 2010. Tolerance and accumulation of lead in *Vetiveria zizanioides* and its effect on oil production. J. Environ. Biol. 31, 329–334.

Roongtanakiat, N., Sanoh, S., 2011. Phytoextraction of zinc, cadmium and lead from contaminated soil by vetiver grass. Kasetsart J. Nat. Sci. 45, 603–612.

Roongtanakiat, N., Akharawutchayanon, T., 2017. Evaluation of vetiver grass for radiocesium absorption ability. Agric. Nat. Resour. 51, 173–180.

Rotkittikhun, P., Chaiyarat, R., Kruatrachue, M., Pokethitiyook, P., Baker, A.J.M., 2007. Growth and lead accumulation by the grasses *Vetiveria zizanioides* and *Thysanolaena maxima* in lead-contaminated soil amended with pig manure and fertilizer: a glasshouse study. Chemosphere 66, 45–53.

RoyChowdhury, A., Sarkar, D., Datta, R., 2019. A combined chemical and phytoremediation method for reclamation of acid mine drainage-impacted soils. Environ. Sci. Pollut. Res. 26, 14414–14425.

Sanguankaeo, S., Chaisintarakul, S., Veerapunth, E., 2003. The application of the vetiver system in erosion control and stabilization for highways construction and maintenance in Thailand. In: Proceedings of the 3rd International Conference on Vetiver and Exhibition Guangzhou.

Sanguankaeo, S., Sungam, S., Sawasdimongkol, L., Veerapunth, E., 2006. Improving the Efficiency of the Vetiver System in the Highway Slope Stabilization for Sustainability and Saving of Maintenance Cost. ICV-4, CARACUS, 24/10/2006 https://www.vetiver.org/ICV4-ppt/BA11-PP.pdf.

Santos, K.A., Klein, E.J., da Silva, C., da Silva, E.A., Cardozo-Filho, L., 2019. Extraction of vetiver (*Chrysopogon zizanioides*) root oil by supercritical CO_2, pressurized-liquid, and ultrasound-assisted methods and modeling of supercritical extraction kinetics. J. Supercrit. Fluids 150, 30–39.

Seutloali, K.E., Beckedahl, H.R., 2015. A review of road-related soil erosion: an assessment of causes, evaluation techniques and available control measures. Earth Sci. Res. J. 19, 73–80.

Singh, S., Melo, J.S., Eapen, S., D'Souza, S.F., 2008. Potential of vetiver (*Vetiveria zizanioides* (L.) Nash) for phytoremediation of phenol. Ecotoxicol. Environ. Saf. 71, 671–676.

Sornprasert, R., Aroonsrimorakot, S., 2014. Utilization of *Vetiveria zizaniodes* (L.) Nash leaves in Ganoderma lucidum cultivated. APCBEE Procedia 8, 47–52.

Srifah, P., Pomthong, B., Hongtrakul, V., Sangduen, N., 2000. DNA polymorphisms generated by single-strand conformational polymorphism and random amplified polymorphic DNA technique are useful as tools for Thai vetiver genome analysis. In: Vetiver and the Environment. The Chaipattana Foundation (Thailand), Bangkok (Thailand), pp. 412–416.

Stevenson, F.L., Cole, M.A., 1999. Cycles of Soil: Carbon, Nitrogen, Phosphorus, Sulfur, Micronutrients. John Wiley & Sons, Inc., New York, USA.

Suelee, A.L., Sharifah, N.M.S.H., Faradiella, M.K., Ferdaus, M.Y., Zelina, Z.I., 2017. Phytoremediation potential of vetiver grass (*Vetiveria zizanioides*) for treatment of metal-contaminated water. Water Air Soil Pollut. 228, 158.

Thammathavorn, S., Khanema, P., 2010. Sequestration Potential of Vetiver. Report to the Office of the Royal Development Projects Board by the Biology Branch, Science Office, Suranaree Technical University, Nakhon Ratchasima, Thailand (in Thai).

Truong, P.N., Baker, D., 1996. Vetiver grass for the stabilization and rehabitation of acid sulfate soils. In: Proc. Second National Conference on Acid Sulfate Soils. Coffs Harbour, Australia, pp. 196–198.

Truong, P.N., Baker, D., 1998. Vetiver Grass for Environment Protection, Pacific Rim Vetiver Network, Technical Pollution No. 1998/1. Pacific Rim. Office of the Royal Development Project Board, Bangkok, Thailand.

Truong, P., 1999. Vetiver grass technology for land stabilisation, erosion and sediment control in the Asia Pacific region. In: First Asia Pacific Conference on Ground and Water Bioengineering for Erosion Control and Slope Stabilisation, Manila, Philippines.

Truong, P., 2000. Vetiver grass technology for environmental protection. In: Proc. Second International Vetiver Conferences: Vetiver and the Environment, Cha Am, Thailand, January 2000.

Truong, P.N.V., Loch, R., 2004. Vetiver system for erosion and sediment control. In: ISCO 2004—13th International Soil Conservation Organisation Conference—Brisbane, July 2004 Conserving Soil and Water for Society: Sharing Solutions.

Vandermeer, J., 1989. The Ecology of Intercropping. Cambridge University Press, Cambridge, p. 237.

Verma, S.K., Singh, K., Gupta, A.K., Pandey, V.C., Trivedi, P., Verma, R.K., Patra, D.D., 2014. Aromatic grasses for phytomanagement of coal fly ash hazards. Ecol. Eng. 73, 425–428.

Walker, D.J., Clemente, R., Roig, A., Bernal, M.P., 2003. The effects of soil amendments on heavy metal bioavailability in two contaminated Mediterranean soils. Environ. Pollut. 122, 303–312.

Wilde, E.W., Brigmon, R.L., Dunn, D.L., Heitkamp, M.A., Dagnan, D.C., 2005. Phytoextraction of lead from firing range soil by vetiver grass. Chemosphere 61, 1451–1457.

Wu, Y., Liu, Z., Yang, G.X., Yang, P., Peng, Y.P., Chen, C., Xue, F., Liu, T., Lium, H.L., Liu, S.Q., 2022. Combined effect of humic acid and vetiver grass on remediation of cadmium-polluted water. Ecotoxicol. Environ. Saf. 244, 114026.

Xia, H.P., 2004. Ecological rehabilitation and phytoremediation with four grasses in oil shale mined land. Chemosphere 54, 345–353.

Yang, B., Shu, W.S., Ye, Z.H., Lan, C.Y., Wong, H.M., 2003. Growth and metal accumulation in vetiver and two Sesbania species on lead/zinc mine tailings. Chemosphere 52, 1593–1600.

Ye, Z.H., Wong, J.W.C., Wong, M.H., 2000. Vegetation response to lime and manure compost amendments on acid lead/zinc mine tailings: a green house study. Restor. Ecol. 8, 289–295.

Yong, R.N., Mohamed, A.M.O., Warkentin, B.P., 1992. Principles of Contaminant Transport in Soils. Elsevier, Amsterdam.

Zheljazkov, V.D., Craker, L.E., Baoshan, X., 2006. Effects of Cd, Pb and Cu on growth and essential oil contents in dill pepper mint and basil. Environ. Exp. Bot. 58, 9–16.

Zhu, B.C.R., Henderson, G., Chen, F., Fei, H.X., Laine, R.A., 2001. Evaluation of vetiver oil and seven insect active essential oils against the Formosan subterranean termite. J. Chem. Ecol. 27, 1617–1672.

16

Mulberry (*Morus* spp.) for phytostabilization of coal mine overburden: Co-generation of economic products

Majeti Narasimha Vara Prasad

Department of Plant Sciences, School of Life Sciences, University of Hyderabad
(An Institution of Eminence), Hyderabad, Telangana, India

1 Introduction

India has rich sources of minerals and coal (Fig. 1). The Jharkhand state is rich in mines (Fig. 2). The mine waste is a source of toxic metal leachates, resulting in acid mine drainage, ultimately leading to environmental degradation and human health risks (Prasad and Jeeva, 2009). Mine waste is generally unfit for vegetation due to a lack of essential nutrients, drought, extreme pH, and a lack of plant health- and growth-promoting microbial consortia (Prasad et al., 2016).

A possible solution for such mine waste and coal mine overburden is phytostabilization. Establishment of vegetation cover with desirable and economically important species would decrease the spread of contamination through wind and water erosion and the leaching of heavy metals and acid mine drainage into the groundwater. Vegetation, possibly aided by microbiota or soil amendments, promotes soil development, nutrient recycling, and the development of microbial communities. It is a non-invasive, cost-efficient method to reclaim mine waste areas and reduce the risks of heavy metal pollution and acid mine drainage.

Recultivation of brown fields has been successful both in temperate and tropical environments. Mycorrhizae and plant growth-promoting bacteria have been extensively researched to aid plants in overcoming these hostile situations. Mycorrhiza and growth-promoting bacteria can also decrease the amount of soil amendment needed, lowering costs.

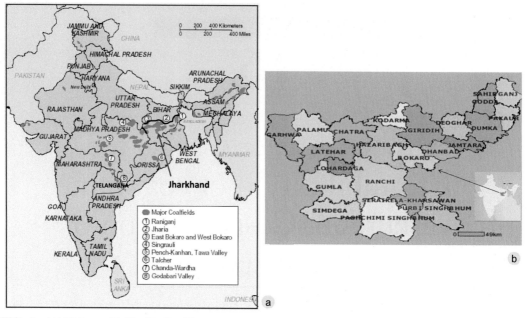

FIG. 1 (A) Major coalfield of India. (B) Mine-rich Jharkhand state of India, one of the top states in India for sericulture generating a wide variety of products.

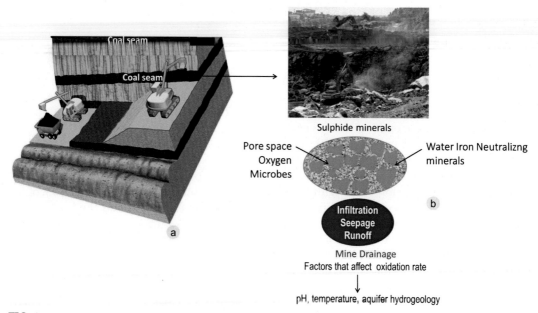

FIG. 2 (A) Cross section of coal-bearing rocks showing coal seam. (B) Extraction of coal generates a large amount of waste rocks, including acid mine drainage (AMD). Factors affecting AMD are also briefly depicted.

2 Environmental issues associated with mine overburdens

Mining, especially open cast mining, is an intensive process. Unless planned properly, it can result in enormous environmental degradation of the mined area. The environmental implications of mine waste are the following:

(a) Release of toxic metals
(b) Damage to heritage
(c) Air, water, and soil pollution
(d) Sulfide minerals and acid drainage

The physical and chemical characteristics of the land also undergo a drastic change due to the ecosystem degradation.

Metal mobility is influenced by many processes driven by the plant, bacteria, mycorrhiza, and soil status. Phytostabilization has great potential as a cost-efficient and non-invasive method for the reclamation of mine waste areas, but more research is needed on the effectiveness of revegetation on risk reduction and to determine which processes mostly contribute to positive effects.

Through surface runoff, eolian (wind) dispersion, and leaching, heavy metals can spread for tens of kilometers in the environment (Prasad et al., 2016).

Surface runoff and leaching may pollute the groundwater, while the dust spread by the wind settles on vegetation, including agricultural crops, where it enters the food chain when consumed. Both acidity and heavy metals can have a severe impact on ecological and human health (Prasad et al., 2016).

Traditional methods to treat mine waste are soil washing, capping, and storing the waste in exhausted open-pit mines (Prasad et al., 2016). A possible non-invasive remediation measure to improve mine waste and other polluted areas is phytoremediation. Phytoremediation is the use of plants and associated microorganisms to remove, immobilizes, or degrade harmful environmental contaminants (Rahman et al., 2022; Yasin et al., 2021).

3 Risks associated with coal mining overburden pollution

Acid mine drainage (AMD) is the largest source of environmental problems caused by the mining industry. Acid mine drainage is the result of tailings and overburden being exposed to air and water. The oxidation of pyrite and other sulfide minerals in the presence of oxidizing bacteria results in the production of acids. This oxidation process occurs slowly in all soils. Tailings and overburden, with their small grain size and thus greater surface exposure, are more prone to generating AMD. The oxidation process is accelerated beyond the natural buffering capacity of the host rock and water resources (Prasad et al., 2016). Mine tailings range from highly acidic (pH2) to alkaline (pH9), depending on the carbonate content and acid-generating potential (Fig. 2).

4 Coal mine overburdens

An overburden may be defined as the area that lies above a zone of scientific or economic interest. During mining, the term is usually associated with the rock, soil, and biota found lying above the coal seams (Fig. 3).

Before beginning any mining activity, if appropriate measures are taken, the overburdens may be protected from being contaminated by toxic compounds and thus used for reclamation purposes. For this purpose, the top soil and subsoils removed before mining are carefully removed and conserved. The overburdens are removed from the location using shovels and draglines and deposited in another location, but care is taken to contain them so that they do not erode into close-lying water sources.

Interburdens are the areas found between two coal seams and are usually contaminated by the explosives used during blasting and acid runoff. They generally consist of shale and sandstone and are numbered from zero to one going from bottom to top. Once the mining activity is complete, ideally, the overburdens should be used for re-afforestation programs, but the loss of the top soil, higher concentration of inorganic elements, and the presence of organic pollutants such as those present in the bombs used for blasting make re-afforestation a difficult endeavor. Another peculiarity of certain overburdens is that they tend to catch fire during the summer months. A probable cause could be the production of combustible gases such as methane.

FIG. 3 (a) West Bokaro coalfield mine waste dump. (B) Revegetation of the mine waste dump. (C) Phytostabilization by Vetiver. (D) Phytostabilization by native vegetation.

5 Environmental concerns

One ton of coal leaves about 25 tons of overburden. Pyrite is contained in overburden; when pyrite is exposed to water and air, it forms dangerous substances like sulfuric acid and iron hydroxide (Ashfaq et al., 2009, 2010). When this mixes with water, it forms acid mine drainage, which is highly destructive to the environment (Fig. 3).

The overburden of waste and uneconomic mineralized rock is required to be removed to mine the useful mineral resource in a surface mining operation. In this process, a dump is formed by casting the waste material and dumping it in a nearby area. The dump so formed is known as the mine waste dump. Waste dump may be classified as internal and external dump. External dump is created outside the pit, whereas internal dump is created inside back of the mining area (Si et al., 2021; Singh and Behal, 2002).

Dump configuration is based on the height of the dump, volume, slope angle, slope and degree of confinement, foundation conditions, dump material properties, dumping method, dumping rate, seismicity and dynamic stability, topography, dump drainage condition, bulk gradation, plasticity of fines, index properties and classification, hydraulic conductivity, consolidation, strength, mineralogy and soil chemistry, in situ density, and compaction.

Tailing is the product that is discarded after mining and processing to remove the economic products. Processing may range from simple mechanical sorting to crushing and grinding followed by physical or chemical processing. It contains all other constituents of the ore except majority of the extracted metal. It may also contain heavy metals and other substances at concentration levels that can be toxic to biota in the environment.

Phytostabilization is described as the establishment of vegetation cover with desirable and economically important species on the overburdens produced by mining activities. They often consist of sandstone and shale and separate two coal seams from each other as interburden. In a coal mine, the regular arrangement of the coal seams is seen to be in levels numbered from zero to one going from bottom to top. Each coal seam alternates with an interburden, and hence all mining activities give rise to a substantial amount of overburden when blasting, drilling, and fracturing activities are carried out. The overburden produced may be stored and used to fill the quarry back once all mining activity has been completed. Ideally, the overburdens should be used for re-afforestation programs, but the loss of the top soil, higher concentration of inorganic elements, and the presence of organic pollutants such as those present in the bombs used for blasting make re-afforestation a difficult endeavor (Figs. 4 and 5) (Juwarkar and Jambhulkar, 2008).

6 Mulberry cultivation on coal mine over burden

The study site lies along NH-23 on the road connecting Ranchi to Hazaribagh, a little off the district capital of Ramgarh in the town of West Bokaro. The location coordinates are 23°48′0″N and 85°45′0″E. As discussed earlier, mining may be carried out via surface mining or underground methods. Mines like the one seen above in the West Bokaro coalfields are the kinds in which most coal is present near the surface, making surface mining a viable option. Before mining is begun, the area is cleared of all vegetation and the land is dug up using heavy machinery like draglines, trucks, etc.

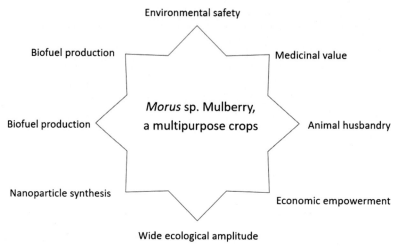

FIG. 4 Mulberry as a multipurpose crop.

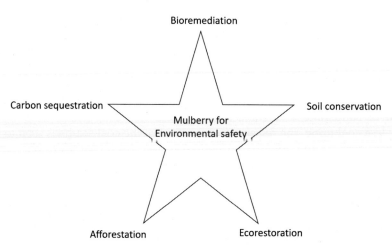

FIG. 5 Mulberry is an ideal plant for phytostabilization of coal mine waste dump.

Cultivation of mulberry on the overburdens has been seen as an economically viable activity (Table 1). *Morus alba*, the plant on which the larvae of the silkworm feed, takes up heavy metals and sequesters them within their idioblasts. The sequestration process causes no harm to the plant, which thrives under appropriate cultivation conditions. The leaves are then plucked off and fed to the larvae, leading to the production of silk (Fig. 6). Thus, it makes this reclamation strategy an ideal cottage industry practice as well (Guo et al., 2021).

Using the approach of natural attenuation, where natural processes are used to reduce the effects of toxic contaminants present in a given system, in this case the plants, locally growing plant species, viz., *Shorea*, *Prosopis*, etc., functioning as hyperaccumulators, are grown along with other exotic species, viz., Mulberry, *Cassia*, etc., which also have the ability to

TABLE 1 *Morus alba* in abatement of mine overburden pollution.

	Author(s)	Year
Politics of the natural vegetation to balance the hazardous level of elements in marble polluted ecosystem through phytoremediation and physiological responses.	Ahmad et al.	2021
Physiological, anatomical, and transcriptional responses of mulberry (*Morus alba* L.) to Cd stress in contaminated soil.	Guo et al.	2021
Phytoremediation potential of forage mulberry (*Morus atropurpurea* Roxb.) for cadmium contaminated paddy soils.	Jiang et al.	2019, 2022
Phytoremediation and the way forward: challenges and opportunities.	Shinomol and Bhanu	2021
Accumulation and translocation of food chain in soil-mulberry (*Morus alba* L.)-silkworm (*Bombyx mori*) under single and combined stress of lead and cadmium.	Si et al.	2021
Comparative study of three *Pteris vittata*-crop intercropping modes in arsenic accumulation and phytoremediation efficiency.	Wan et al.	2021
Glutathione and calcium biomineralization of mulberry (*Morus alba* L.) involved in the heavy metal detoxification of lead-contaminated soil.	Wang and Ji	2021
Phytoremediation for urban landscaping and air pollution control—a case study in Trivandrum city, Kerala, India.	Watson and Bai	2021
Phytoremediation Potential of *E. camaldulensis* and *M. alba* for Copper, Cadmium, and Lead Absorption in Urban Areas of Faisalabad City, Pakistan.	Yasin et al.	2021
Tolerance capacities of *Broussonetia papyrifera* to heavy metal(loid)s and its phytoremediation potential of the contaminated soil.	Zeng et al.	2022
Phytoremediation of Cd-contaminated farmland soil via various *Sedum alfredii*-oilseed rape cropping systems: Efficiency comparison and cost-benefit analysis.	Zhang et al.	2021

hyperaccumulate the heavy metals present in the system, in this case the overburdens (waste generated during mining activities), thus leading to their reclamation and restoration.

Special emphasis is laid on plants that can also lead to some sort of economic activity, viz., *Morus alba*, which finds application in the cottage industry. Sericulture is therefore of both environmental and economic advantage by benefiting reclamation activities as well as providing the locals with economic security (Jha et al., 2022; Jiang et al., 2017; Juwarkar et al., 2009).

Though re-filling the quarries with the overburdens is an ideal practice, it has been seen that the activity doesn't lead to anything except maybe clearing space. With the original vegetation cover having been lost, the overburdens are often loose and dry. In the case of wind and rains, they often erode into the nearby water sources, a must for refineries, potentially contaminating kilometers of area lying downstream. This is why it is extremely important to conserve the top soil before carrying out mining activity in any area. Most importantly, the top soil hosts soil microorganisms that are absolutely essential for plants to grow and a vegetative cover to be established.

Plants are imperative toward carrying out the reclamation activity on overburdens due to the reasons discussed earlier. As an efficient and cost-effective method, especially in areas that don't require much maintenance, the approach has been to investigate the existing plant

FIG. 6   Mulberry is an environmental crop.

species and analyze the ones best suited toward growing in such contaminated sites. Since the native species are already adapted to the local area and climate, efforts are being made to identify plants that can act as hyperaccumulators, along with established exotic species, of the contaminants present in the overburdens and help in establishing a primary vegetative cover.

Insects are fast growing with less feed compared to higher animals and produce highly invaluable biomass. The use of insect larvae for the production of feed and value additions is gaining considerable importance in the developed world. The growing demand for arable land is increasing. Since insects are enriched with nutrients, vitamins, and minerals, rearing insects on phytomass harvested/produced from mine overburden is a usable feed for insect larvae. Insects can feed on phytomass produced from mine overburden. *Morus* is a classic example of feed for silkworm larvae which can be further utilized for industral and domestic products (Figs. 6–8) in addition to the silk production and silk-producing cottage industry, which support rural employment and revenue generation. In India, Jharkhand state is rich in coal mines, and incidentally silkworm rearing and silk production top the list in India.

FIG. 7 Bioeconomics of silkworm larvae cultured with mulberry leaves. Silkworm larvae thus reared are a resource for a wide variety of products.

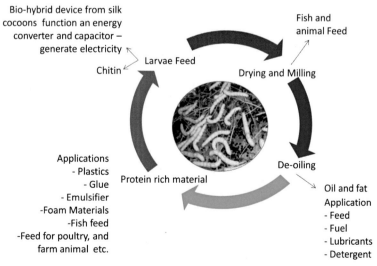

FIG. 8 Bioeconomics of silkworm larvae cultured with mulberry leaves. Silkworm larvae thus reared are a resource for a wide variety of products.

IV. Brownfield development for smart bioeconomy

Mining activities are a must and need to be carried out for economic growth and social security; their associated drawbacks such as air, soil, and water pollution, loss of biodiversity, shrinkage of forests, heavy metal contamination, etc., cannot be overlooked anymore and need immediate attention.

7 Mulberry cultivation on coal mine overburden

Cultivation of mulberry on the overburdens has been seen as an economically viable activity. *Morus alba*, the plant on which the larvae of the silk worm feed, takes up heavy metals and sequesters them within their idioblasts (Katayama et al., 2013; Zhou et al., 2015). The sequestration process causes no harm to the plant, which thrives under appropriate cultivation conditions. The leaves are then plucked off and fed to the larvae, leading to the production of silk (Figs. 9–11). Thus, it makes this reclamation strategy an ideal cottage industry practice as well (Prince et al., 2001, 2002).

Using the approach of natural attenuation where natural processes are used to reduce the effects of toxic contaminants present in a given system, in this case the plants, locally growing plant species, viz., *Shorea*, *Prosopis*, etc., functioning as hyperaccumulators, are grown along

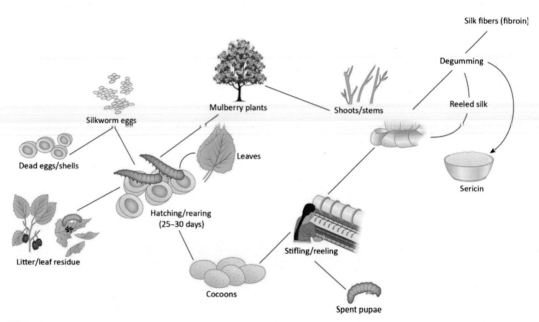

FIG. 9 Schematic representation of the steps involved in rearing and processing of silk cocoons and the by-products and co-products generated during the process. Litter and leaf residues are generated continually throughout the rearing of the silkworms, and pupae are generated after the silk has been unraveled from the cocoons (reeled). Degumming of the silk fibers provides sericin as the by-product. *With permission from Reddy, R., Jiang, Q., Aramwit, P., Reddy, N., 2021. Litter to leaf: the unexplored potential of silk byproducts. Trends Biotechnol. 39(7), 706 718.*

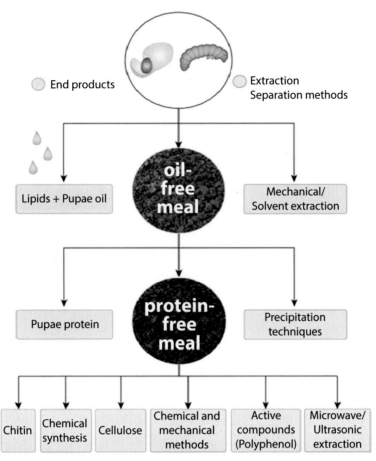

FIG. 10   Process of sequential extraction of components from spent silkworm pupae. Although it is difficult to completely extract the individual components at each step, partial remnants of the components do not substantially affect subsequent processing and use. *With permission from Reddy, R., Jiang, Q., Aramwit, P., Reddy, N., 2021. Litter to leaf: the unexplored potential of silk byproducts. Trends Biotechnol. 39(7), 706–718.*

with other exotic species, viz., Mulberry, *Cassia*, etc., which also have the ability to hyperaccumulate the heavy metals present in the system, in this case the overburdens (waste generated during mining activities), thus leading to their reclamation and restoration.

Special emphasis is laid on plants that can also lead to some sort of economic activity, viz., *Morus alba*, which finds application in the cottage industry (Huang et al., 2018; Huihui et al., 2020; Mallick and Sengupta, 2022). There are toxic effects of heavy metals Pb and Cd on mulberry (*Morus alba*). Sericulture is therefore of both environmental and economic advantage by benefiting reclamation activities as well as providing the locals with economic security.

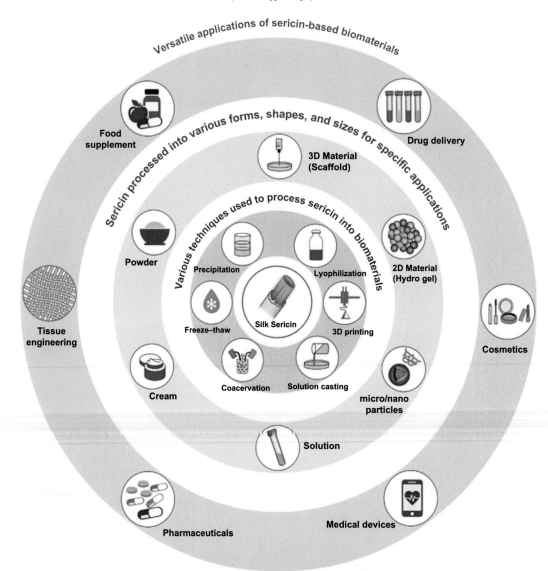

FIG. 11   A comprehensive representation of the methods used to convert sericin into various biomaterials and subsequent applications of the biomaterials developed from sericin With permission Shankar et al 2023.

Acknowledgments

Thanks are due to TATA Steel, West Bokaro Coalfields, for permission to fieldwork with necessary logistics. Special thanks to Col. (Retd.) Bhavani and Mr. Roshan Kumar for their guidance and help in field work. Thanks are due to Mr. A. Sampath, research scholar, Kakatiya University, Warangal, and Dr. Satish Suthari for their help in field work and providing some silkworm photographs.

Glossary

Biodiesel Diesel obtained from plants or animals as opposed to deriving from petroleum resources.

Biorefineries Facilities where one or more renewable resources can be used to produce multiple value-added products. Here, the residue generated from one process becomes the raw material for a subsequent process.

Bioremediation The process of treating contaminated soil, water, or air using natural or biological sources such as biopolymers or microorganisms.

Byproducts Materials that are inevitably generated but are not intended for use. For example, the stems of the mulberry plants compared with leaves that are necessary for feeding the silkworms.

Co-products Products that are generated during the processing of the main product. For example, pupae that are generated after the cocoons are reeled for their fibers. Degumming the process of removing the gum or glue protein sericin from the silk threads. Litter excreta produced by the insects during their rearing time.

Rearing The process of caring for the silkworms and generating the cocoons.

Reeling The process of unraveling the fibers from the cocoons and winding the fibers onto a designated frame.

Sericin The "glue" protein that forms the outer coating of the fibroin protein and helps to adhere the fibers together.

References

Ahmad, Z., Khan, S.M., Page, S., 2021. Politics of the natural vegetation to balance the hazardous level of elements in marble polluted ecosystem through phytoremediation and physiological responses. J. Hazard. Mater. 414, 125451.

Ashfaq, M., Ali, S., Hanif, M.A., 2009. Bioaccumulation of cobalt in silkworm (*Bombyx mori* L.) in relation to mulberry, soil and wastewater metal concentrations. Process Biochem. 44, 1179–1184.

Ashfaq, M., Afzal, W., Hanif, M.A., 2010. Effect of ZnII deposition in soil on mulberry-silk worm food chain. Afr. J. Biotechnol. 9 (11), 1665–1672.

Guo, Z., Zeng, P., Xiao, X., Peng, C., 2021. Physiological, anatomical, and transcriptional responses of mulberry (*Morus alba* L.) to cd stress in contaminated soil. Environ. Pollut. 284, 117387.

Huang, R., Jiang, Y., Jia, C., Jiang, S., Yan, X., 2018. Subcellular distribution and chemical forms of cadmium in *Morus alba* L. Int. J. Phytoremediation 20, 460–465.

Huihui, Z., Xin, L., Zisong, X., Yue, W., Zhiyuan, T., Meijun, A., Yuehui, Z., Wenxu, Z., Nan, X., Guangyu, S., 2020. Toxic effects of heavy metals Pb and Cd on mulberry (*Morus alba* L.) seedling leaves: photosynthetic function and reactive oxygen species (ROS) metabolism responses. Ecotoxicol. Environ. Saf. 195, 110–469.

Jha, H., Mandal, V., Sen, K.K., Tandey, R., Chouhan, K.B.S., Mehta, R., 2022. Environmental biotechnology: public perception. In: Emerging Trends in Environmental Biotechnology. CRC Press, pp. 265–273.

Jiang, Y., Huang, R., Yan, X., Jia, C., Jiang, S., Long, T., 2017. Mulberry for environmental protection. Pak. J. Bot. 49, 781–788.

Jiang, Y.B., Jiang, S., Li, Z.B., Yan, X.P., Qinand, Z.X., Huang, R.Z., 2019. Field scale remediation of Cd and Pb contaminated paddy soil using three mulberry (*Morus alba* L.) cultivars. Ecol. Eng. 129, 38–44.

Jiang, Y., Jiang, S., Huang, R., Wang, M., Cao, H., Li, Z., 2022. Phytoremediation potential of forage mulberry (*Morus atropurpurea* Roxb.) for cadmium contaminated paddy soils. Int. J. Phytoremediation 24 (5), 518–524.

Juwarkar, A., Jambhulkar, H.P., 2008. Phytoremediation of coal mine spoil dump through integrated biotechnological approach. Bioresour. Technol. 99, 4732–4741.

Juwarkar, A.A., Yadav, S.K., Thawale, P.R., Kumar, P., Singh, S.K., Chakrabarti, T., 2009. Developmental strategies for sustainable ecosystem on mine spoil dumps: a case of study. Environ. Monit. Assess. 157 (1–4), 471–481.

Katayama, H., Banba, N., Sugimura, Y., Tatsumi, M., Kusakari, S., Oyama, H., Nakahira, A., 2013. Subcellular compartmentation of strontium and zinc in mulberry idioblasts in relation to phytoremediation potential. Environ. Exp. Bot. 0098-8472. 85s, 30–35.

Mallick, P., Sengupta, M., 2022. Prospect and commercial production of economically important plant mulberry (*Morus* Sp.) towards the upliftment of rural economy. In: Commercial Scale Tissue Culture for Horticulture and Plantation Crops. Springer, Singapore, pp. 219–243.

Prasad, M.N.V., Jeeva, S., 2009. Coal mining and its leachate are potential threats to *Nepenthes khasiana* (Nepenthaceae) that preys on insects - an endemic plant in north eastern India. BioDiCon 2 (3), 29–33.

Prasad, M.N.V., Nakbanpote, W., Phadermrod, C., Rose, D., Suthari, S., 2016. Mulberry and vetiver for phytostabilization of mine overburden: cogeneration of economic products. In: Bioremediation and Bioeconomy. Elsevier, pp. 295–328.

Prince, S., Senthilkumar, P., Subburam, V., 2001. Mulberry-silkworm food chain—a templet to assess heavy metal mobility in terrestrial ecosystems. Environ. Monit. Assess. 69, 231–238.

Prince, W.S., Kumar, P.S., Doberschutz, K.D., Subburam, V., 2002. Cadmium toxicity in mulberry plants with special reference to the nutritional quality of leaves. J. Plant Nutr. 25, 689–700.

Rahman, S.U., Yasin, G., Nawaz, M.F., Cheng, H., Azhar, M.F., Riaz, L., Javed, A., Li, Y., 2022. Evaluation of heavy metal phytoremediation potential of six tree species of Faisalabad city of Pakistan during summer and winter seasons. J. Environ. Manag. 320, 115801.

Shankar, S., Murthy, A.N., Rachitha, P., Raghavendra, V.B., Sunayana, N., Chinnathambi, A., Alharbi, S.A., Basavegowda, N., Brindhadevi, K., Pugazhendhi, A., 2023. Biosynthesis of silk sericin conjugated magnesium oxide nanoparticles for its antioxidant, antiaging, and anti-biofilm activities. Environ. Res. 223, 115421.

Shinomol, G.K., Bhanu, R.K., 2021. Phytoremediation and the way forward: challenges and opportunities. In: Applied Water Science: Remediation Technologies. vol. 2. Wiley, pp. 405–436.

Si, L., Zhang, J., Hussain, A., Qiao, Y., Zhou, J., Wang, X., 2021. Accumulation and translocation of food chain in soil-mulberry (*Morus alba* L.)-silkworm (*Bombyx mori*) under single and combined stress of lead and cadmium. Ecotoxicol. Environ. Saf. 208, 111582.

Singh, V.K., Behal, K.K., 2002. Effect of coal ash amended soil on growth of mulberry plant (*Morus alba*). J. Ecophysiol. Occup. Health 2 (3&4), 243–254.

Wan, T., Dong, X., Yu, L., Huang, H., Li, D., Han, H., Jia, Y., Zhang, Y., Liu, Z., Zhang, Q., Tu, S., 2021. Comparative study of three Pteris vittata-crop intercropping modes in arsenic accumulation and phytoremediation efficiency. Environ. Technol. Innov. 24, 101923.

Wang, L., Ji, G., 2021. Glutathione and calcium biomineralization of mulberry (*Morus alba* L.) involved in the heavy metal detoxification of lead-contaminated soil. J. Soil Sci. Plant Nutr. 21, 1182–1190.

Yasin, G., Ur Rahman, S., Yousaf, M.T.B., Azhar, M.F., Zahid, D.M., Imtiaz, M., Hussain, B., 2021. Phytoremediation potential of *E. camaldulensis* and *M. alba* for copper, cadmium, and lead absorption in urban areas of Faisalabad City, Pakistan. Int. J. Environ. Res. 15, 597–612.

Zeng, P., Guo, Z., Xiao, X., Zhou, H., Gu, J., Liao, B., 2022. Tolerance capacities of *Broussonetia papyrifera* to heavy metal (loid) s and its phytoremediation potential of the contaminated soil. Int. J. Phytoremediation 24 (6), 580–589.

Zhang, J., Cao, X., Yao, Z., Lin, Q., Yan, B., Cui, X., He, Z., Yang, X., Wang, C.H., Chen, G., 2021. Phytoremediation of cd-contaminated farmland soil via various *Sedum alfredii*-oilseed rape cropping systems: efficiency comparison and cost-benefit analysis. J. Hazard. Mater. 419, 126489.

Zhou, L., Zhao, Y., Wang, S., Han, S., Liu, J., 2015. Lead in the soil–mulberry (*Morus alba* L.)–silkworm (*Bombyx mori*) food chain: translocation and detoxification. Chemosphere 128, 171–177.

17

Utilization of contaminated lands for cultivation of dye producing plants for dyestuff

K. Schmidt-Przewoźna

Institute of Natural Fibres and Medicinal Plants—PIB, Poznan, Poland

1 Introduction

Natural plants that serve as a source of color are of great economic importance. In recent years, we have observed a revival of interest in natural raw materials and extraction techniques that had often been completely forgotten. The sudden revival of interest in natural dyes concerns not only textile dyeing but also dyes used in cosmetics, such as hair dye, food coloring, and other products. We can find dyes in fruits, flowers, roots, bark, leaves, seeds of trees, plants, insects, and lichens. Different parts of the same plants may give different colors and shades. Moreover, there is a lot of potential for the application of the compounds found in dyeing trees in various fields of science and economy. Natural dyes obtained from plants are seen as ecological solutions and eco-friendly dyes with natural antibacterial and antimicrobial properties. Plant extracts contain tannins, flavonoids, saponins, essential oils, mucilage, vitamins, and many other valuable nutritional substances. They have various properties, for example, medicinal, soothing, caring, and disinfecting. From this point of view, they are well known for their curative properties. Many of them are extensively used in Ayurvedic and homeopathic medicine in Asia and Africa (Cardon, 2010).

Many plants with dyeing properties do not require special conditions for cultivation. They can grow on sand and wetlands as well as contaminated lands. The environment adapts to climate and soil changes. Natural dyes derived from plant materials have a wide range of applications. They can be used to dye textiles, food products, cosmetics, composites, leather, and wood. Of course, in the food and cosmetics industries, numerous requirements must be met for the cultivation, acquisition, and extraction of the dye. In contrast, dyes obtained for dyeing fabrics, wood, and composites can be obtained from contaminated lands. Given the huge potential of plant materials, there is a niche demand for products dyed with plant materials. These dyes are intended for use in the development of organic clothing collections with health properties. They can be and are used in the cosmetic and food industries. The research field concerns the issues of the rational use of natural resources as well as environmental management and protection. This research requires the participation and cooperation of scientists from several scientific fields: the natural sciences, technology, and social science.

2 Historical background

The search for color is present in all cultures in the world and dates back to the dawn of human civilization. Archeological findings indicate that dyes from natural sources have been used to color textiles for the last 6000 years. Since the beginning of their existence, people have tried to mark their ethnic and individual character by means of, for example, colors. Color determined their affiliation to the group—the identity that gave them a sense of security and safety; it also expressed their emotional states. Color became a means of expression and an essential element of the evolutionary change from the world of nature to the world of culture. Observing colorful nature, primitive people also wished to use nature's colors for themselves. They found it in roots, berries, branches, the bark of trees, leaves, insects, shellfish, and minerals. The art of using vegetable dyes, and particularly the skill of using auxiliary chemicals, was a secret art for ages. The art of dyeing developed on different continents independently and as a result of migrations of people and the import of dyes. The cradle of dyeing knowledge and art is central and south Asia and the great civilizations of antiquity (Schmidt-Przewozna and Skrobak, 2013). In ancient Egypt, dyeing was at an advanced level. The earliest methods of dyeing using mordant and mineral dyes were well known. This knowledge was adopted from the Egyptians by the Israelites and Phoenicians and later by the Greeks and Romans (Schmidt-Przewozna, 2009).

The discovery of America and its exploration revealed that independent dyeing techniques had been developed in what is today Peru and Mexico. Thanks to the existing sea routes to East India, the following dyes became known there: cochineal, indigo, dyeing trees, red sandalwood *Pterocarpus santalinus*, cutch *Acacia catechu*, brazilwood *Caesalpinia brasiliensis* (red), and logwood *Haemotoxylon campechianum* (blue and violet). The ground root turmeric, *Curcuma longa*, was imported from China and India. This dyestuff allowed for the obtaining of yellow and orange colors. Annatto *Bixa orellana* and indigo *Indigofera tinctoria* were also known and used in small amounts. Color can be obtained from virtually every plant; however, its durability and saturation differ. The keys to dyeing are knowledge and Wisniak, which have been passed on from generation to generation over the course of human culture (Schmidt-Przewozna, 2001, 2009; Cardon, 2007; Wisniak, 2004; Melo, 2009).

(A) *Bixa orellana* (lip stick plant). (B–D) Harvesting, transportation and extraction of indigo dyer traditionally.

Until 1856, all dyes were obtained from natural raw materials. The first synthetic dye, mauveine, was invented by William Henry Perkin. Some sources say that it was magenta, which was invented by the Polish chemist Jakub Natanson in 1855. The outstanding success of Perkin's mauve gave great impetus to the discovery of new dyes in a period receptive to new scientific advances. The development of chemistry slowly led to a departure from complex dyeing technologies based on natural raw materials. It is widely believed that synthetic dyes are easier to use, that the colors are more intense, and that their resistance to washing and light is better.

In taking part in this discussion, one should pay attention to historic artifacts. Many of them, for hundreds and even thousands of years, kept their beautiful and clean colors. Looking at these historic fabrics, we may notice the difference in the colors' durability. Textiles found in the graves of the pharaohs in Egypt are more than 4000 years old. Some colors were partially degraded, but many others are of fascinating intensity (Bechtold and Mussak, 2009; Jose et al., 2019; Schmidt-Przewozna, 2002). Natural dyeing techniques require extensive knowledge related not only to the dyeing of fabrics but also to the fabrics' preparation for dyeing. The knowledge of plants and the coloring capabilities that can be achieved through the use of various mortars were important. The dyeing process lasted longer and was carried out in stages. In modern times, characterized by rapid consumption and seasonally changing fashion, color does not have to last for years. In the twentieth century, historical techniques of dyeing with natural plant extracts, developed over generations, slowly disappeared. However, at the turn of the 21st century, the positive impact of natural-fiber fabrics dyed with plant extracts on the human body began to be noticed. These interests are connected to the general processes of increased use of ecological natural resources, which are tested for their health benefits as well as their wide application in fashion and design. Accordingly, in the fibers of modern clothing, among others, through the use of natural dyeing, the individual compounds contained therein act on the human skin.

Such research is conducted in India, in other Asian countries, and in Europe, as well as in Poland, at the Institute of Natural Fibers and Medicinal Plants in Poznań. There are regions in India, East Asia, and Central and South America where natural dyeing has been preserved on a very limited scale. In the 1990s, there was a debate in India (one of the main centers of cultivation of dyeing plants) as to whether using natural dyes was useful. As a result, Indian research institutes launched a series of studies on the improvement of the fastness of dyes. Such studies are carried out at the India Institute of Technology in New Delhi. As a result of these works, collections of clothes dyes naturally produced by local societies of manufacturing engineers have been introduced to the market. One of the companies that works on launching fabrics dyed with natural dyestuffs is Aura Herbal Textiles Ltd.

Plant extracts are essences from plants: their flowers, leaves, bark, fruit, or roots. They contain the following: tannins, flavonoids, saponins, essential oils, vitamins, and other nourishing substances. Depending on the content of active substances and their precious ingredients, they can have the following properties: medicinal, soothing, moistening, anti-inflammatory, regenerative, antiviral, antifungal, antioxidant, and protection against UV radiation (UVR) (Schmidt-Przewozna and Zimniewska, 2005). Many of them are extensively used in Ayurvedic and homeopathic medicine in Asia and Africa. Oak galls, *Terminalia chebula*, pomegranate peel, *Punica granatum*, and oak, willow, and buckthorn bark are rich sources of tannins. When used in pre-mordanting, they affect the hue and the durability of colors. Some of them have also been used to intensify the color of other natural dyes and improve their fastness (Table 1).

3 Potential and future prospects of dyeing plants

All around us, new factories, highways, and concrete housing estates are being created. People do not think about how they change the environment until they begin to suffer from civilizational diseases such as allergies, atherosclerosis, or cancer. Pollution is predominant in industrial and urban areas. In areas where there are a lot of cars, the air is filled with gaseous pollutants and suspended particulate matter. Car exhaust gases include heavy metals, toxic nitrogen oxides, and—most recently recognized as one of the most dangerous—suspended microscopic particulates. These particles penetrate the lung alveoli, and the body is unable to defend against them. In the surroundings of communication routes, just a few dozen meters from the lanes, there is not only contaminated air but also water and soil contamination. It turns out that the only solution in such a changed environment is plants. Plants, as organisms leading a sedentary lifestyle, had to develop defense mechanisms in the process of evolution to allow survival even in extreme conditions. They not only produce oxygen and water vapor but are also capable of taking up and accumulating contaminants from soil, water, and air in their leaves, stems, and roots. They retain them thanks to the wax and hairs on the leaf surface, or when the poison penetrates beneath the surface of the skin of the leaf, they immobilize and

TABLE 1   Dye-yielding plants from contaminated sites.

Research observations	Author(s)	Year
Characterization and evaluation of *Nasturtium officinale* R. Br. dye as an alternative to synthetic dye	Ganie et al.	(n.d.)
Optimization of the dyeing process of natural dye extracted from *Polyalthia longifolia* leaves on silk and cotton fabrics	Khatun and Mostafa	(2022)
3D design of clothing in medical applications	Cichocka et al.	(2021)
Green dyeing of woolen yarns with weld and madder natural dyes in the presence of biomordant	Hosseinnezhad et al.	(2021)
Phytomanagement of a trace element-contaminated site to produce a natural dye: first screening of an emerging biomass valorization chain	Perlein et al.	(2021)
Effect of natural extracts and mordants on the dyeability of cotton fabrics for UV protection	Yamuna and Sudha	(2021)
Phytochemical screening and extraction of natural dye from the leaves of *Polyalthia longifolia* and its application on silk fabric	Khatun and Mostafa	(2019)
Coreopsis tinctoria Nutt. is a source of many colors	Schmidt-Przewozna et al.	(2019)
Natural dyes and pigments: extraction and applications	Mansour	(2018)
Natural and safe dyeing of curing clothing is intended for patients with dermatoses	Schmidt-Przewoźna et al.	(2018)
Utilization of temple waste flower, *Tagetus erecta*, for dyeing cotton, wool, and silk on an industrial scale	Vankar et al.	(2009)

neutralize them. In this way, plants also catch the dangerous microdust emitted by car engines. The cleaning activity of plants also involves cooperation with the soil microflora. Microorganisms working on the breakdown of complex chemical compounds appear in the vicinity of the roots. Scholars claim that the number of microorganisms there is 10 times higher than in plantless areas. The treatment of the environment with the use of plants is called phytoremediation. This method has only two advantages: it is effective and relatively inexpensive. In areas degraded by industry, perennial plantations of so-called energy plants are created, such as miscanthus, willow, and poplar. Herbaceous plants are also useful (Fig. 1).

In order to breathe clean air, it is worth having an oak, ash, maple, or common pear tree near the house. Thanks to their leaf hairs or wax coating, these trees will remove the most suspended particulate matter. Common weeds also play an important role in areas where municipal waste is stored. They form a vegetation cover over the polluted area. We can find the following perennials in landfills: tansy *Tanacetum vulgare*, broadleaf dock *Rumex obtusifolius*, stinging nettle *Urtica dioica*, greater celandine *Chelidonium majus*, annual bedstraw *Galium aparine*, and common yarrow *Achillea millefolium* (Fig. 2).

On former landfill sites, especially on the top of slag heaps and their gentle slopes, perennial plants have made their home. Many of them have dyeing properties. Researchers from the University of Ecology and Management in Warsaw have defined sets of the most expansive plants

FIG. 1   Common yarrow *Achillea millefolium* on contaminated lands. *Photo by K. Schmidt-Przewoźna.*

FIG. 2   Goldenrod *Solidago gigantea. Photo by K. Schmidt-Przewoźna.*

FIG. 3 Elderberry *Sambucus nigra* in areas degraded by municipal landfills in Poland. *Photo by K. Schmidt-Przewoźna.*

in areas degraded by municipal landfills in Poland: late goldenrod *Solidago gigantea*, lady's thumb *Polygonum persicaria*, knotgrass *Polygonum aviculare*, tansy *T. vulgare*, bitter dock *R. obtusifolius*, greater burdock *Arctium lappa*, stinging nettle *U. dioica*, dandelion *Taraxacum officinale*, common yarrow *A. millefolium*, and greater celandine *C. majus* (Dygus et al., 2012). These common herbs are also dyeing plants. After a few years, post-mining sites were populated with shrubs, such as willows *Salix* sp. And trees. Among the trees, the most common ones were silver birch *Betula pendula*, aspen *Populus tremula*, Scots pine *Pinus sylvestris*, and the English oak *Quercus robur*. The fastest-growing species used in the recultivation of degraded areas are the "Royal Purple" smoke tree, *Cotinus coggygria*, and staghorn sumac, *Rhus typhina* (Fig. 3).

4 Sources of natural dyeing plants

4.1 Dyeing trees

Many trees, thanks to the compounds contained in them, are extensively used in traditional medicine and homeopathy. The ingredients derived from trees, such as bark, seeds, pith, and leaves, constitute a valuable resource for traditional and unconventional medicine thanks to the compounds they contain. Moreover, as they contain biologically active substances, they have a wide spectrum of pharmacological effects. They have also been used in the furniture, cosmetics, dyeing, and tanning industries. Dyeing trees are not only a rich source of dyes but are also used in pre-mordating fabrics due to their high content of tannins. Oak galls, myrabalan, oak bark, pomegranate peel, and other plants have been used in the process of mordating fabrics in different cultures around the world. This process is important in achieving lasting, more intense colors and their resistance to external factors such as water and light (Przewozna et al., 2011).

4.1.1 Sandalwood

Common names: red sanders, red sandalwood, sanders wood (Fig. 4).

Red sandalwood *P. santalinus* is a medium-sized tree of about 10–15 m in height. It grows on dry, rocky soils as well as the dry slopes of deciduous forests. It has been a valuable resource for the centime memorial (Dean, 1999).

It can be found in India, Indonesia, Ceylon, Australia, and the islands of the Pacific. Documented sources confirm that it was used 4000 years ago in India, Egypt, Greece, and Rome. Many temples and buildings have been built using this material. In Egypt, sandalwood was also used to cremate dead bodies.

The plant contains alkaloids, phenolics, saponins, glycosides, flavonoids, and tannins. Valuable essential oils, dyes, and wood are obtained from red sandalwood (Cardon, 2007).

Valuable components are present in the leaves, fruit, and bark of the tree.

Medical and pharmaceutical uses

Essential oils support the treatment of urethral and urinary tract infections. Moreover, sandalwood oil helps to remove bloating and colic as it acts as a relaxant on the intestine. It also acts as an expectorant.

Red sandalwood tree oil has both excellent medicinal and cosmetic properties. It helps in the treatment of urethral and urinary tract infections. Moreover, sandalwood oil helps to remove bloating and colic as it acts as a relaxant on the intestine. It also acts as an expectorant. It has astringent, antiseptic, antibacterial, disinfectant, and anti-inflammatory properties.

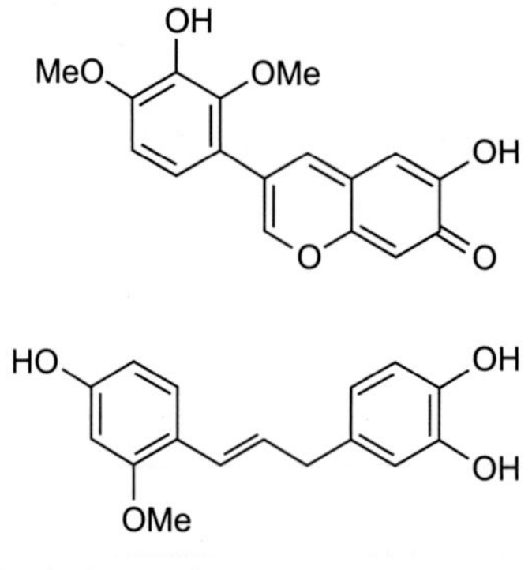

FIG. 4   Santalin (dye from red sanders) chemical structure. http://onlinelibrary.wiley.com/doi/10.1002/anie. 201302317/abstract (Accessed 13 April 2015) 13:52 a.m.

It has a positive effect on dry skin and is efficient in the treatment of acne and lichens. It is widely used in aromatherapy for its calming effect, stimulating imagination, and relieving anxiety.

The antibacterial activity of the leaves and stem bark of *Pterocarpu santalinus* was investigated by Manjunatha from the National Institute of Technology, Karnataka, India (Manjunatha, 2006). The stem bark extract showed maximum activity against *Enterobacter aerogenes*, *Alcaligenes faecalis*, *Escherichia coli*, *Pseudomonas aeruginosa*, *Proteus vulgaris*, *Bacillus cereus*, *Bacillus subtilis*, and *Staphylococcus aureus*. The leaf extract showed maximum activity against *E. coli*, *A. faecalis*, *E. aerogenes*, and *Pseudomonas aeruginosa* (Manjunatha, 2006; Bojase et al., 2002; Vanita et al., 2013).

Other uses

The oil is used in the production of cosmetics—shampoos, shower gels, body lotions, and aromatic oils.

The dyeing extracts are used to dye pharmaceuticals, food (coloring sodas reddish-orange and intensifying the color of seafood, meat, bread, and alcoholic drinks), paper, and textiles, as well as in the tanning and cosmetic industries. It is a component of perfumes as well as an aphrodisiac, valued by many peoples in Asia and Africa.

Sandalwood is also a valuable building and sculpting material.

Red sandalwood: Natural dye

The red wood yields a natural dye, santalin, which dyes objects red, brown, and russet/ginger. The dye has been widely used to color wool in different parts of the world. It was imported to Europe in mass amounts from India, as a substitute for other red dyes. In the eighteenth century, Jan Hellot, a French chemist and tanner, recommended the following recipe to obtain a red color:

Tanning process: 12 pounds of alder bark, 10 pounds of sumac, and ½ pounds of ground gallnuts.
Dyeing process: 4 pounds of red sandalwood for each piece of cloth (Cardon, 2007).

4.1.2 *Indian mulberry Morinda citrifolia*

Common names: great morinda, Indian mulberry, mengkudu-malai, noni, and beach mulberry, cheese fruit (Fig. 5).

Indian mulberry *Morinda citrifolia* is a tree that grows in warm climates in Asia, Australia, and Africa. It grows in shady forests as well as on open, rocky, and sandy shores. It grows well on sand, saline soils, and degraded areas. On the volcanic island of Krakatoa, degraded by lava flows, the tree grows without any problems. Indian mulberry does not require good soil. The tree reaches a height of approximately 5–9 m.

Its leaves, fruit, bark, and roots are used for different purposes.

Medical and pharmaceutical uses

The leaves, fruit, and roots of *M. citrifolia* have been used in folk medicine for hundreds of years.

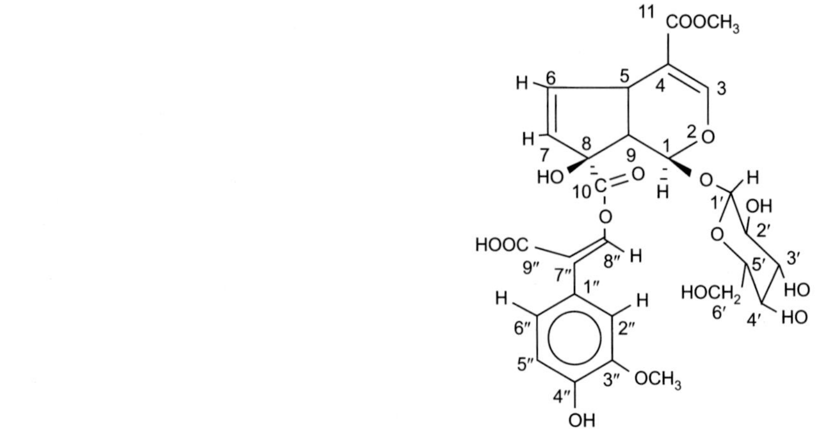

FIG. 5   *Morinda citrifolia* dye (alizarin) chemical structure. http://www.prodifact.com/1morinda-citrifolia.html (Accessed 13 April 2015), 13:56 a.m.

Noni fruit juice contains many vitamins, minerals, and natural antioxidants, as well as proteins and enzymes that work synergistically and support the work of the whole body in many ways (Morton, 1992). The juice supports the treatment of many diseases, including diabetes, arthritis, gout, and cancer. It contains compounds with analgesic, antibacterial, and antiviral properties that support the body's resistance to infections (Saleem et al., 2005). The characteristic polysaccharides discovered in it enhance the immune system. It is recommended for people with chronic diseases and convalescents. Thanks to the wealth of natural antioxidants contained in it—polyphenols, phenolic acids, and coumarin, which support the fight against free radicals and delay the aging process of the body—noni fruit juice constitutes a great component of slimming and detox diets. It helps remove toxins from the body. The fruit of the Indian mulberry is consumed to remedy lumbago, asthma, and dysentery in India and China (West et al., 2008). Noni fruit juice is considered an elixir of vitality. It regulates metabolism, helps in removing toxins, improves digestion, boosts energy levels, and restores the natural homeostasis of the organism. It also accelerates the regeneration of damaged skin cells, promoting faster healing of wounds, bruises, and burns. The populations of Africa, Asia, and Australia often use fruits in their diets and medicine (Mohd Zin et al., 2007).

In Hawaii, fruit juice is used to treat lice and make shampoo.

In Malaysia, mengkudu leaf tea is drunk to alleviate the symptoms of coughing and to treat nausea and colic. Most of the folk remedies for diabetes involve chewing the leaves or a combination of the plant and leaves.

Other uses

Noni fruits are not tasty and have an unpleasant odor. However, they are consumed in some places in the world. On the islands of Samoa and Fiji, they are food for the poor. Morinda extract is also used to stain food and alcohol.

The young leaves can also be eaten as a vegetable, and they contain protein (4%–6%). The seeds can be eaten after being roasted.

Morinda is also used as a dietary supplement in tablets.

Indian mulberry: natural dye

Drury (1873) wrote that a dye was derived from the "heart of the wood in older trees."

Chemists at Delhi University, India, isolated morindone, damnacanthal, and nordamnacanthal from the shavings of the heartwood, the latter not previously found in plants (Murti et al., 1959).

Later, Balakrishna et al. (1960) discovered a new glycoside of morindone, which they named morindonin, in the root bark. The main dyeing substance in *Morinda* is alizarin. Other important substances are rubiadin, lucidin, damnacanthal, normancanthal, soranjidiol, antharagallol, pseudopurpurin, purpurin, and morindone (Dominique P. 676, 777). Morindone is a mordant dye that gives a yellowish-red color when used with an aluminum mordant, brown with a chromium mordant, and dull purple to black with an iron mordant. The dye from the roots of the tree is very popular in India and the Pacific islands. In Borneo, they call it *Iban mengkudu* and use it to dye fabrics called *ikat*, and in Java, batik painting. The red color is obtained in the complex process of dyeing, where an important step is the pre-mordating of the cotton used for *ikat*. In Rumah Garie's longhouses in Borneo, the cotton yarn is mordated for 2 weeks. Ginger, palm salt, and palm oil are used in the process. Red-maroon and russet colors are obtained during dyeing. To get a deeper red color, sappan wood and alum are often added (Fig. 6).

FIG. 6 Indian mulberry *Morinda citrifolia. Photo by K. Schmidt-Przewoźna.*

The bark of the Indian mulberry plant produces a reddish-purple and brown-colored dye, which is used in making *batik*. The tree is extensively grown for the purpose of obtaining dye in Java.

4.1.3 *Kamala Mallotus philippensis*

Common Names: kamala tree, dyer's rottlera, monkey face tree, orange kamala, red kamala, scarlet croton, kamala, raini, rohan, rohini, and jia ma la.

Kamala *Mallotus philippensis* is a tree that grows up to 25 m in height. It can be found in the Himalayas, India, Sri Lanka, China, Taiwan, Australia, and Africa. The tree's leaves, fruits, and boughs are used. The fruit is covered with red granules, from which powder is obtained. The powder constitutes only 1.5%–4% of their weight, which makes this valuable resource expensive.

Medical and pharmaceutical uses

Kamala medicinal uses include treatment for afflictions of the skin, inflammations, eye diseases, bronchitis, abdominal diseases, and spleen enlargement, but there is no scientific evidence to prove its efficiency (Sharma and Varma, 2011).

The roots are used for dissolving coagulated blood. The leaves and powdered fruit are used to treat skin diseases and hard-to-heal wounds. Kamala extracts help in the treatment of anorexia and increase the feeling of hunger. *M. philippensis* has antibacterial, antifungal, antitumor, and laxative properties (Thakur et al., 2015; Rao and Seshadri, 1947).

Other uses

The fruit of *M. philippensis* is covered with a red powder called *kamala*, and it is used locally to make textile dye, syrup, and as an old remedy for tapeworm, because it has a laxative effect. The wood is used for rafters, tool handles, and matchboxes. The oil of the seeds is used as a substitute for tung oil (*Vernicia* Lour., Euphorbiaceae) in the formulation of rapid-drying paints, varnishes, hair conditioners, and ointments.

The dye composition of kamala makes it useful in food and drinks. The leaves are used as fodder, and the wood is used to make paper.

Kamala: Natural dye

The powder obtained from ripe fruits contains rottlerin (Fig. 7), isorotten, citric acid, tannin, and wax. Kamala is a mortar dye, and, therefore, to obtain a durable color, cotton fabrics should be mordated before dyeing. It is used primarily for dyeing silk, but it can also be used to dye wool, cotton, and linen. In Japan, on the island of Mijako, *M. philippensis* leaves are used in the traditional indigo dyeing's finishing processes. As a result, the color blue acquires a beautiful copper shade. The colors that we obtain from kamala using different dyes (Fig. 8) and methods are yellow, cream, olive, and red.

Experimental findings

See Table 2.

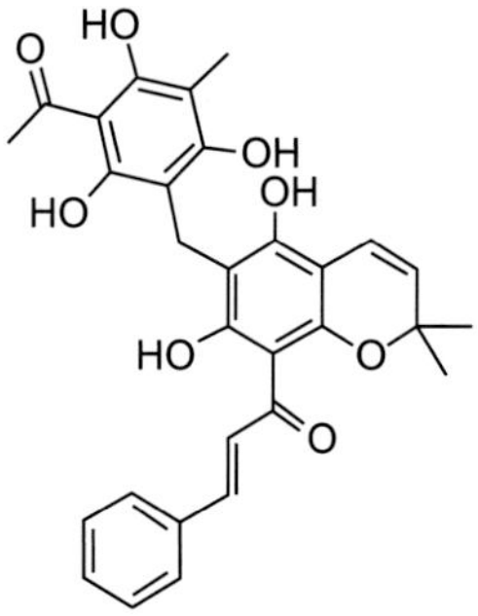

FIG. 7 Rottlerin chemical structure. http://en.wikipedia.org/wiki/Rottlerin (Accessed 13 April 2015), 13:51 a.m.

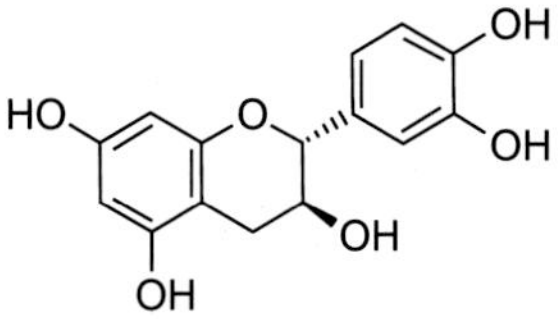

FIG. 8 Catechin chemical compound. http://en.wikipedia.org/wiki/Catechin (Accessed 13 April 2015), 13:50 a.m.

TABLE 2 Kamala dyeing results on linen and cotton samples.

No.	Fabric	Mordant	Spectrophotometric result					Fastness	
			L*	a*	b*	C*	h°	Wash	Light
1	Linen—5g	No mordant	78.7	4.5	33.8	34.1	82.4	5	4
2	Cotton—5g	No mordant	82.1	3.2	31.8	32.0	84.2	3	3–4
3	Linen—5g	Alum—2g	84.8	3.8	31.5	31.7	83.1	5	4
4	Cotton—5g	Alum—2g	84.4	2.0	29.2	29.2	86.0	4	4
5	Linen—5g	Soda—2g	87.1	2.3	15.8	16.0	81.9	3	3
6	Cotton—5g	Soda—2g	89.4	1.1	12.9	13.0	85.2	3	3
7	Linen—5g	Citric acid—2g	87.3	−0.7	24.2	24.3	91.6	3	3
8	Cotton—5g	Citric acid—2g	88.2	−1.6	21.2	21.3	94.3	3–4	3
9	Linen—5g	Copper—2g	82.7	−2.5	19.0	19.2	97.5	4	4
10	Cotton—5g	Copper—2g	85.4	−2.1	18.2	18.3	96.6	4	4
11	Linen—5g	Iron—2g	66.6	4.6	18.0	18.6	75.6	4	4
12	Cotton—5g	Iron—2g	70.5	4.5	20.5	21.0	77.7	4	4

Fabrics: L, linen; C, cotton; Mordant: Alum, $Al_2K_2(SO_4)_4 \cdot 24H_2O$; Copper, $CuSO_4 \cdot 5H_2O$; Iron, $FeSO_4 \cdot 7H_2O$; Citric acid, $C_6H_8O_7 \cdot H_2O$; Washing soda, Na_2CO_3.

4.1.4 Logwood Haematoxylum campechianum L.

Common names: bois ampeche, bloodwood tree, ampeche, ampeche wood, logwood, palo de ampeche, palo de tinta, palo de tinte, palo negro, tinto, ek (Figs. 9 and 10).

Logwood, *H. campechianum* L., is a small, spiny tree native to the tropical regions of America and Mexico, extending south through the tropical dry forests of Oaxaca, Guatemala, Costa Rica, and Colombia (Cardon, 2007).

The tree grows up to 12 m in height. Its wood is hard and dark brown.

The generic name of logwood *Haematoxylum* (often spelled *Haematoxylon*) means *bloodwood*, referring to the dark red heartwood that is a source of dye. The name *campechianum* refers to the city of Campeche on the Yucatan Peninsula. It is possible to find this tree, which contains valuable heartwood, in the vicinity of this town.

The parts of the tree that are used are the wood, heartwood, and leaves. It was brought to Europe after Christopher Columbus's discovery of America. The English brought huge amounts of it in large pieces from their colonies. During the Industrial Revolution in the UK, the tree's dye was needed for dyeing cotton and wool.

As it was not a durable dye, its use was banned at the turn of the seventeenth and eighteenth centuries in many European countries.

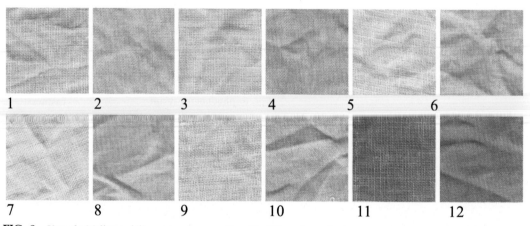

FIG. 9   Kamala *Mallotus philippensis* colors results with different mordants.

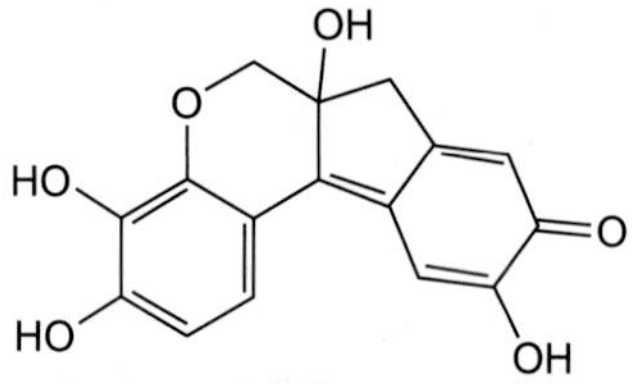

FIG. 10   Hematoxylin chemical structure. http://pl.wikipedia.org/wiki/Hematoksylina (Accessed 13 April 2015), 13:50 a.m.

Medical and pharmaceutical uses

Logwood has also been used traditionally in folk medicine. The tree's bough and its other parts were used to treat many diseases. In a 1981 publication describing medical plants from Mexico, one of the 184 plants is *H. campechianum* L. It was used in the pre-Columbian era and was one of the most important medical plants.

It is used as a mild astringent and tonic. It is administered in the form of a decoction and a liquid extract. It is also useful against diarrhea, dysentery, and dyspepsia, as well as for menstrual disorders.

Other uses

The heartwood may be used to dye not only linen and cotton but also leather, fur, silk, feathers, paper, bones, and in the manufacturing of inks (Schmidt-Przewozna and Brandys, 2014).

Logwood, which is called palo de tinto in Mayan cultures, was used a 1000 years ago to make wooden constructions in temples and palaces. Even today, the people living in that area build wooden houses from this hardwood. The Mayans used the natural shapes of the branches and tree trunks in their constructions.

Logwood: Natural dye

The dye has been known since pre-Columbian times. In the Mayan language, the name of the tree, "ek," also means the color black.

The dye obtained from logwood comes from its heartwood. The main dyeing substance in logwood is hematoxylin, which was isolated for the first time by Chevreul in 1810. Other dyes in this plant are small amounts of brazilein, hematein, and large amounts of tannins. Hematoxylin is extracted by boiling chips or shavings of logwood in water. Logwood powder has a red color, and in the dyeing process, it can be a source of violet, gray-violet, lavender, brown, black, light brown, purplish-red, and indigo blue colors.

The whole range of colors that can be obtained from logwood has to do with the use of dyeing mortars. The pH of the water is also important. Violet and lavender are obtained with the use of alum and copper sulfate; blacks and navy blues are obtained with the use of ferrous sulfate; and the colors red and purple are obtained without the use of mortar (Schmidt-Przewozna and Brandys, 2014).

Experimental findings

See Table 3 and Fig. 11.

4.1.5 Cutch tree A. catechu

Common Names: cutch, cashoo, khoyer, terra Japonica, acacia à cachou, caciu, kath, kachu, cacho, Japan earth, black cutch, cachou, sa-che, seesit, sha (Fig. 12).

The source of color from in cutch tree is the heartwood. The cutch tree is small and usually typical of arid regions with sandy soils (e.g., in southern Pakistan, India, Burma, Thailand, and China). The heart of cutch wood, when extracted, evaporated, cooled, and filtrated, is a source of red-brown color (Cardon, 2007).

TABLE 3 Logwood dyeing results on linen and cotton samples.

No.	Fabric	Mordant	Spectrophotometric result					Fastness	
			L^*	a^*	b^*	C^*	h°	Wash	Light
1	Linen—5 g	No mordant	57.8	3.7	8.5	9.3	66.5	4	3
2	Linen/G—5 g	No mordant	43.5	4.5	8.8	9.9	63.1	4–5	3
3	Cotton—5 g	No mordant	33.6	3.9	3.8	5.5	43.8	4	2
4	Linen—5 g	Alum—0.35 g	51.2	8.7	−137	16.3	302.4	4	3
5	Linen/G—5 g	Alum—0.35 g	36.4	9.0	−14.5	17.1	301.9	4–5	3–4
6	Cotton—5 g	Alum—0.35 g	54.2	6.9	−12.3	14.1	299.3	4	2
7	Linen—5 g	Soda—0.35 g	79.3	2.0	8.9	9.1	77.5	4	3
8	Linen/G—5 g	Soda—0.35 g	66.5	3.6	17.6	17.9	78.4	4	3–4
9	Cotton—5 g	Soda—0.35 g	72.1	3.3	10.8	11.3	73.0	4	2
10	Linen—5 g	Citric acid—0.35 g	60.9	7.0	5.0	8.6	35.1	4	3
11	Linen/G—5 g	Citric acid—0.35 g	47.3	7.3	4.5	8.6	31.5	4–5	3–4
12	Cotton—5 g	Citric acid—0.35 g	53.2	9.5	8.1	12.5	40.4	4	2
13	Linen—5 g	Copper—0.35 g	43.3	−3.9	−3.9	5.5	225.1	4	3
14	Linen/G—5 g	Copper—0.35 g	33.2	−1.8	−2.4	3.0	232.5	4–5	3–4
15	Cotton—5 g	Copper—0.35 g	41.3	−4.5	−4.9	6.6	227.6	4	3
16	Linen—5 g	Iron—0.35 g	53.6	1.2	−1.8	2.2	304.0	4	4
17	Linen/G—5 g	Iron—0.35 g	29.7	2.2	−4.2	4.7	297.8	4–5	4–5
18	Cotton—5 g	Iron—0.35 g	41.6	1.6	−3.3	3.7	296.4	4	3

Fabrics: L, linen; C, cotton; Linen/G, linen pre-mordating in oak gall; Mordant: Alum, $Al_2K_2(SO_4)_4 \cdot 24H_2O$; copper, $CuSO_4 \cdot 5H_2O$; Iron, $FeSO_4 \cdot 7H_2O$; Citric acid, $C_6H_8O_7 \cdot H_2O$; Washing soda, Na_2CO_3.

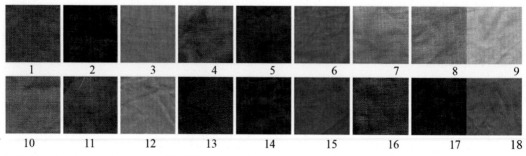

FIG. 11 Logwood color results with different mordants.

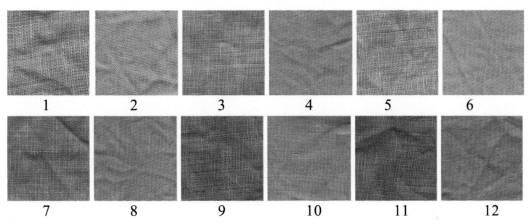

FIG. 12 Cuth color results with different mordants.

Medical and pharmaceutical uses

The bark and heartwood of the cutch tree are used in traditional medicine. Catechu wood extract has been commonly used in medicine, health products, and the food and chemical industries. It has the following properties: antioxidative, antiaging, reducing blood lipids, reducing blood sugar, anticancer, antiradiation, antiultraviolet, and removing free radicals.

Other uses

Cutch heart extract is used in tanning leather, dyeing, and as a preservative for fishing nets. Blavod black vodka is very popular in the UK. Its unique black color is derived from the black catechu tree. The most important use for the cutch tree is its high usage in the furniture industry.

Cutch tree: Natural dye

The dried aqueous extract of the wood and bark of the cutch tree is a well-known natural dye. This is obtained by leaching wood. Extract is the best for cotton and silk dyeing and pretanning leathers. The extract of the cutch tree also has astringent properties that facilitate penetration and brightening of the skin.

Experimental findings

See Table 4 and Fig. 12.

4.1.6 Annatto B. orellana

Common names: lipstick tree, achiote, bija, roucou, onoto, atsuete, and colorau (Fig. 13).

The annatto plant grows in tropical areas. Annatto is a profusely fruiting shrub or small tree that grows 5–10 m in height. Approximately 50 seeds grow inside prickly, reddish-orange, heart-shaped pods. One small annatto tree can produce up to 270 kg of seeds.

Annatto is found in Ecuador, Bolivia, Brazil, and Mexico. Annatto's Latin designation (*B. orellana* L.) was chosen for the Spanish conquistador Francisco de Orellana during his exploration of the Amazon River (Cardon, 2007).

TABLE 4   Cutch dyeing results on linen and cotton samples.

No.	Fabric	Mordant	Spectrophotometric result					Fastness	
			L*	a*	b*	C*	h°	Wash	Light
1	Linen	No mordant	52.9	17.7	16.7	24.3	43.4	5	5
2	Cotton	No mordant	53.4	16.4	18.0	24.3	47.8	4	4
3	Linen	Alum—2 g	64.0	13.5	21.2	25.1	57.4	5	5–6
4	Cotton	Alum—2 g	68.1	13.7	21.1	25.1	57.0	4	5
5	Linen	Soda—2 g	76.2	11.5	9.9	15.2	40.6	5	5
6	Cotton	Soda—2 g	69.7	11.2	9.6	14.7	40.6	4	4–5
7	Linen	Citric acid—2 g	61.8	12.8	17.4	21.5	53.7	4–5	4
8	Cotton	Citric acid—2 g	63.7	11.1	14.8	18.5	53.0	3	3
9	Linen	Copper—2 g	57.9	10.2	15.2	18.3	56.1	4–5	5
10	Cotton	Copper—2 g	63.1	8.8	14.0	16.5	57.9	4	4–5
11	Linen	Iron—0.5 g	54.2	4.7	7.4	8.8	57.8	5	5
12	Cotton	Iron—0.5 g	51.2	4.5	7.05	8.3	57.5	4–5	5

Fabrics: L, linen; C, cotton. Mordant: Alum, $Al_2K_2(SO_4)_4 \cdot 24H_2O$; Copper, $CuSO_4 \cdot 5H_2O$; Iron, $FeSO_4 \cdot 7H_2O$; Citric acid, $C_6H_8O_7 \cdot H_2O$; Washing soda, Na_2CO_3.

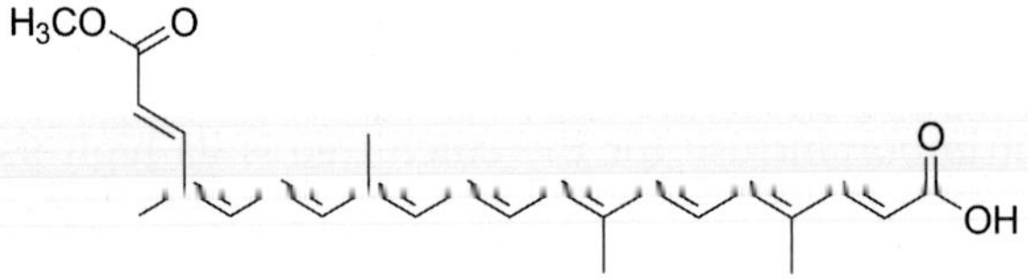

FIG. 13   Bixin chemical structure. The color of annatto comes from various carotenoid pigments. http://pl.wikipedia.org/wiki/Biksyna (Accessed 13 April 2015), 13:49 a.m.

Medical and pharmaceutical uses

Annatto seeds contain a characteristically pleasant-smelling oil. The composition of the seed oil is similar to that of soybean oil. Fats in plants can be used for the production of butter (Lauro and Jack, 2000; Morton, 1960). A decoction of annatto root is taken orally to control asthma. A macerated seed decoction is taken orally for relief of fever, and the pulp surrounding the seed is made into an astringent drink used to treat dysentery and kidney infections. Infusions of root in water and rum are used to treat venereal diseases. The dye is used as an antidote for prussic acid poisoning. Extracts from different parts of the tree kill bacteria, fight free radicals, kill parasites, increase urination, stimulate digestion, lower blood pressure, act as mild laxatives, protect the liver, reduce fever, arterial hypertension, high cholesterol, cystitis, obesity, renal insufficiency, and eliminate uric acid.

Annatto is a non-carcinogenic and non-toxic native plant used as a hemostatic, antioxidant, astringent, antibacterial, antidysenteric, diuretic, aphrodisiac, and effective febrifuge, digestive, and gentle purge. It is also prescribed for epilepsy, erysipelas, sundry skin diseases, and throat infections.

Other uses

Annatto has the common name "lipstick tree." In the past, it was used for coloring women's lips and for body painting. From seeds, we can obtain dye and oil extract for cosmetics, essential oils for shampoos, soaps, etc. From the color cosmetics list, the code for annatto seeds is the same as extract from annatto. Code is E160b; CI number is 75120.

Annatto is very popular as a food colorant. It is used for cheese, sausages, meats, and candies, as well as beverages, cosmetics, pharmaceutical products, and traditional spices for food.

Annatto: Natural dye

Pigment from the seeds of this tree contains bixin (Fig. 14) and norbixine. Because of this, annatto is an excellent natural colorant used in the food industry, as a natural dye for textiles, and also for coloring other things like feathers, sheepskin, ivory, bones, bamboo mats, and rattan. The color may be fixed with tamarind leaves or with bark containing tannin. These pigments are a source of tones between yellow and red. Specialist literature reports that green shades can be obtained from the leaves.

The powder from the seeds is soaked in hot water overnight, boiled for 1 h, and filtered.

In addition to bixin and norbixin, annatto contains bixaghanene, bixein, bixol, crocetin, ellagic acid, ishwarane, isobixin, phenylalanine, salicylic acid, threonine, tomentosic acid, and tryptophan (Silva et al., 2008).

Experimental findings

See Table 5 and Fig. 14.

4.2 Tannin sources: Tannin compounds

Natural sources rich in tannin include oak gall, *Quercus infectoria* Oliv., Chebulic Myrobalan, *T. chebula*, oak bark, *Q. robur*, and pomegranate, *P. granatum*. Descriptions and details of this plant follow.

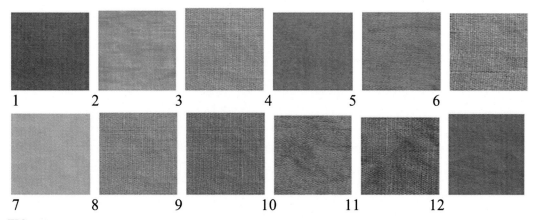

FIG. 14   Annatto colors results with different mordants.

TABLE 5 Annatto dyeing results on linen and silk samples.

No.	Fabric	Mordant	Spectrophotometric result					Fastness	
			L^*	a^*	b^*	C^*	$h°$	Wash	Light
1	Silk—4.5 g	No mordant	59.0	33.3	43.6	54.9	52.6	4	4
2	Linen—4.5 g	No mordant	68.1	28.2	32.1	42.8	48.7	4	3
3	Linen—4.5 g	Alum—2 g	70.1	24.5	28.1	44.7	50.6	5	5
4	Silk/G—4.5 g	Alum—2 g	77.2	26.4	32.7	46.3	57.3	5	5
5	Linen/G—4.5 g	Soda—2 g	75.5	16.1	38.4	41.6	67.0	5	5
6	Linen—4.5 g	Soda—2 g	70.5	22.5	22.4	31.8	44.9	5	4
7	S/B—4.5 g	Citric acid—2 g	54.3	31.8	33.8	44.2	52.0	4	4
8	Linen/G—4.5 g	$CuSO_4$—2 g	28.2	11.8	7.4	13.8	33.5	5	5
9	Linen—4.5 g	$CuSO_4$—2 g	24.0	10.2	6.6	12.2	32.6	4	5
10	L/B—4.5 g	Iron—2 g	51.8	30.9	41.7	56.2	48.9	5	5
11	Linen/G—4.5 g	Iron—2 g	53.2	32.1	44.3	58.1	50.2	5	5
12	S/B—4.5 g	Iron—2 g	58.1	35.0	49.0	60.2	54.5	5	5

Fabrics: Linen-Linen/G, linen pre-mordating in galas; Linen/O, linen pre-mordating in bark of oak; Silk-Silk/G, silk pre-mordating in galas; Silk/O, silk pre-mordating in bark of oak. Mordant: Alum, $Al_2K_2(SO_4)_4 \cdot 24H_2O$; Copper, $CuSO_4 \cdot 5H_2O$; Iron, $FeSO_4 \cdot 7H_2O$; Citric acid, $C_6H_8O_7 \cdot H_2O$; Washing soda, Na_2CO_3.

4.2.1 Oak gall Quercus infectoria

Q. infectoria Oliv is a small tree and shrub mainly found in Greece, Asia Minor, Syria, and Iran. Galls are formed on the leaves, roots, and bark of trees. We can usually find them on oaks and Chinese sumac *Rhus* species (Fig. 15).

Oak galls contain tannic and gallic acids. Gallnuts are used to obtain tannin, which allows for better dyeing of fabrics and is used in tanning. In some galls, the compounds (tannins) constitute half or even more of the substance's dry matter. The tannin compound is gallotannin, which is an ester of glucose, ellagic acid, and gallic acid. Gallnuts, which are

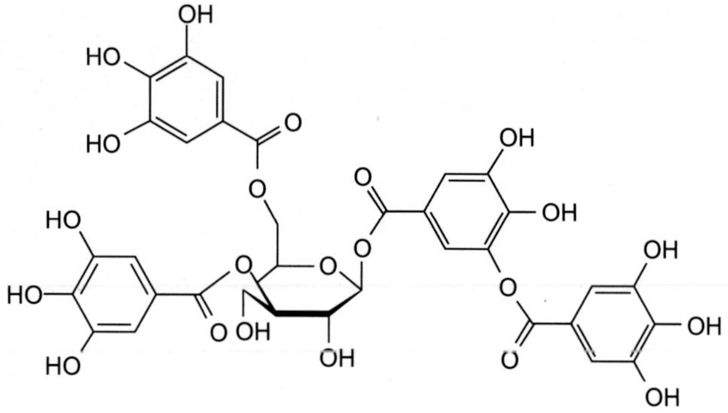

FIG. 15 Tannin acid chemical structure. http://pl.wikipedia.org/wiki/Kwas_taninowy (Accessed 13 April 2015), 13:48 a.m.

widely used in medicine, usually come from oaks. They have an astringent, sedative, antipyretic, and antidiabetic effect, and they are widely used in the medical and pharmaceutical industries. They are also widely used as an additive to food and animal feed, as well as in paint and ink production and metallurgy. Galls give black, navy blue, and brown colors (Kannan et al., 2009; Suchalata and Devi, 2005) (Fig. 16).

Experimental findings

See Table 6 and Fig. 17.

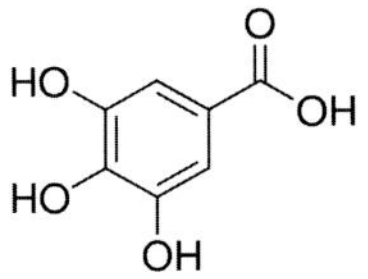

FIG. 16 Gallic acid chemical structure. https://en.wikipedia.org/wiki/Gallic_acid.

TABLE 6 Gall oak dyeing results on linen and cotton samples.

| No. | Fabric | Mordant | Spectrophotometric result | | | | | Fastness | |
			L^*	a^*	b^*	C^*	$h°$	Wash	Light
1	Linen—5 g	No mordant	70.2	3.9	19.4	19.8	78.7	4	5
2	Linen/G—5 g	No mordant	69.2	2.4	23.5	23.6	84.1	5	6
3	Cotton—5 g	No mordant	74.6	3.3	18.9	19.2	80.0	4	5
4	Linen—5 g	Alum—0.35 g	68.2	4.9	21.7	22.3	77.2	4–5	5
5	Linen/G—5 g	Alum—0.35 g	66.4	5.7	23.6	24.3	76.5	5	6
6	Cotton—5 g	Alum—0.35 g	72.5	5.1	20.4	21.0	76.0	4	5
7	Linen—5 g	Soda—0.35 g	62.4	4.2	21.4	21.8	78.8	4–5	5
8	Linen/G—4.5 g	Soda—0.35 g	58.3	4.6	22.8	23.2	78.5	4–5	6
9	Cotton—5 g	Soda—0.35 g	69.1	3.4	20.2	20.5	80.3	4	5
10	Linen— g	Citric acid—0.2 g	69.5	4.9	19.8	20.4	76.1	4–5	5
11	Linen/G—5 g	Citric acid—0.2 g	70.5	4.5	21.2	21.6	78.1	5	6
12	Cotton—5 g	Citric acid—0.2 g	73.4	3.9	19.2	19.6	78.5	4	5
13	Linen—5 g	CuSO$_4$—0.35 g	50.2	5.6	24.3	24.9	77.1	5	5
14	Linen/G—4.5 g	CuSO$_4$—0.35 g	51.2	5.1	23.5	24.1	77.7	5	5–6
15	Cotton—5 g	CuSO$_4$—0.35 g	56.9	5.2	25.1	25.7	78.3	4	5
16	Linen—5 g	Fe—0.35 g	39.5	3.7	3.0	4.7	38.8	4	5
17	Linen/G—5 g	Fe—0.35 g	35.7	3.2	2.8	4.3	41.3	4–5	5–6
18	Cotton—5 g	Fe—0.35 g	45.7	3.2	4.3	5.4	53.5	4	5

Fabrics: Linen—Linen/G, linen pre-mordating in galas; Cotton. Mordant: Alum, $Al_2K_2(SO_4)_4 \cdot 24H_2O$; Copper, $CuSO_4 \cdot 5H_2O$; Iron, $FeSO_4 \cdot 7H_2O$; Citric acid, $C_6H_8O_7 \cdot H_2O$; Washing soda, Na_2CO_3.

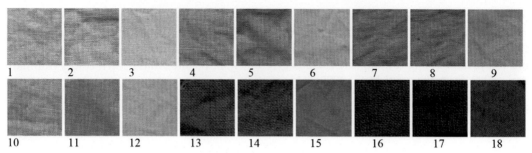

FIG. 17   Oak galls colors results with different mordants.

4.2.2 Chebulic Myrobalan Terminalia chebula

Myrobalan is a tree that grows up to approximately 30 m. It can be found in the natural environment of Asia, in the Himalayas, northern India, Sri Lanka, Burma, Thailand, Indochina, and southern China. It is an undemanding tree because it can grow on sand and clay soil. Thanks to the compounds contained in it, such as chebulinic acid, ellagic acid, gallic acid, and flavonoids (routine and quercetin), it is often used in homeopathic medicine in Asia and Africa. *T. chebula* has antibacterial, antiviral, and antifungal properties (Kannan et al., 2009; Suchalata and Devi, 2005). Thanks to the wealth of its ingredients, it is used in making therapeutic infusions for rinsing the mouth, throat, and eyes. Drugs containing Myrobalan lower cholesterol levels, regulate metabolism, and digestive disorders are used in the treatment of dysentery and, auxiliarly, of HIV infections. Infusions prepared from its fruit have proven helpful for people with chronic fever, and they are also used as a mild sedative. *T. chebula* contains 58%–60% of tannin compounds, which makes it a valuable dye and mordant. It is the source of the yellow color.

4.2.3 Oak bark, Quercus robur

The most important parts of the oak bark are tannins, which have an astringent effect. Oak bark is used externally in the treatment of inflammatory conditions of the skin, mucous membranes, and mild burns. Oak bark decoction has an astringent effect on damaged skin and mucous membranes. Apart from tannins, oak bark also contains free phenolic acids (ellagic and gallic acids), flavonoids (quercetin), resin compounds, and mineral salts. Oak bark is the source of brown colors that, due to the presence of tannin, are used in the pre-mordating of fabrics (Hwang et al., 2000; Nagesh et al., 2012) (Figs. 18 and 19).

4.2.4 Pomegranate Punica granatum

Pomegranate is a shrub or small tree of up to 5 m in height. We can find valuable resources in the peel of the fruit and the bark of the branches and roots. Pomegranate can be found in most warm countries. The rind of the pomegranate contains a considerable amount of tannin, about 19% with pelletierine (Adeel et al., 2009; Tiwari et al., 2010). The main coloring agent is granatoine. Pomegranates also have antioxidant, antiviral, and antitumor properties. The rind of the fruit is a valuable source of vitamins A and E and folic acid. In folk medicine,

FIG. 18 *Quercus robur. Photo by K. Schmidt-Przewoźna.*

unripe fruit and peel infusions are administered to stop diarrhea and treat dysentery. In Africa, it was used to treat leprosy.

Its fruits are used to make refreshing juices, sorbets, and even wine. Pomegranate was used as a dye in antiquity, in Mediterranean civilizations, and in the cultures of the East. In ancient civilizations, it was seen as a symbol of love. Its dried peels are used to dye fabrics yellow, brown, and olive. Often, in order to revive the color, in India and South Asia, turmeric root extract is added. In conjunction with indigo, green or dark green colors have been obtained. Pomegranate extract is used in tanning, particularly saffian.

FIG. 19   *Quercus robur* bark. *Photo by K. Schmidt-Przewoźna.*

4.3 Wild and cultivated plants

4.3.1 Criteria for classification and basic types of natural dyes

There are a lot of different types of dyes used in the textile industry. The main division is based on the origin of the substance. Therefore, the following types can be distinguished:

• Natural dyes: plant, mineral, and animal dyes.

There are several divisions of these dyes based on their chemical structure and the ways they are used in fabrics.

4.3.2 Natural dyes are classified into two groups: Substantive and adjective

Substantive dyestuffs dye the fibers directly. They were probably the first dyes used by early humans. Substantive dyes are usually rich in tannins from barks, leaves, and fruits.

Vat dyes, Indigo *I. tinctoria* and Woad *Isatis tinctoria,* also belong to substantive dyes. The coloring matter of vat dyes is insoluble in water. The color is formed in the presence of oxygen in the air. In Indygo and Woad, the dye matter is a substance called indican. Adjective dyestuffs need mordant to fix permanently to the textile fibers. Metallic salts of aluminum, chrome, iron, and tin enable affinity between the fiber and natural dyeing extracts.

The use of mordant often results in much stronger color on fibers. Examples of such dyes include Madder, Logwood, Ccochineal, Fustic, and many others. Some dyes do not have a solubilizing group. In such cases, a temporary solubilizing group is introduced into the dye at the time of application. A classic example of such a dye is indigo. Indigo is insoluble in water. After application, the reduced form of the indigo is converted back to its original form by air oxidation of the material. This process is known as "vatting," and indigo is classified as a vat dye. Most of the vat dyes are based on two chemical structures: indigo and anthraquinone.

4.3.3 Indigoid dyes

Indigo *I. tinctoria, Marsdenia Tinctoria*, Woad *I. tinctoria*, and *Tyrian purple*.

4.3.4 Anthraquinone dyes

These dyes can be both of plant and insect origin. Within the group of dyeing plants, the following ones can be mentioned: Madder (Hosseinnezhad et al., 2021) *Rubia tinctorum*, Indian Madder (Hosseinnezhad et al., 2021) *Rubia cordifolia*, and Lady's Bedstraw *Galium verum*. On the other hand, within the group of insects, the following ones can be mentioned: Lac dye *Kerria lacca Kerr*, Kermes *Kermes vermilio* Planchon, Cochineal *Dactylopius coccus* Costa, and Polish cochineal *Poryrophora polonica*.

4.3.5 Flavonoid dyes

These are dyes that give plants various colors, from yellow in citrus fruits to dark blue in berries. They have valuable properties; they protect from the harmful effects of ultraviolet radiation, fungi, and insects. Within this group, the following plants can be distinguished: Weld, *Reseda luteola*, Dyer's broom, *Genista tinctoria*, Venetian sumac, *C. coggygria*, Bastard hemp, *Datisca cannabina*, Dyers mulberry, *Maclura tinctoria*, Buckthorn, *Rhamnus cathartica*, Kamala, *M. phillipensis*, and Brazil wood, *Caesalpinia echinata*.

4.3.6 Naphthoquinone and benzoquinone dyes

These dyes can be found in the following raw materials: Henna, *Lawsone inermis leaves*, Black Walnuts shells, dyer's alkanet *Alkanna tinctoria*, and Sanderswood, *P. santalinus*.

4.3.7 Carotenoid dyes

These dyes are red, orange, or yellow pigments present in many plants and animals. The attractive colors of many red and yellow fruits and vegetables are also connected with their carotenoid content. These dyes can be found in: Annatto, *B. orellana* seeds, Saffron, *Crocus sativus*, and Tagetes, *Tagetes erecta*.

4.3.8 Tannin-based dyes

Tannins are found mainly in plant cell vacuoles and are concentrated in epidermal tissues (e.g. Seigler, 1998). They are present in wood, bark, roots, leaves, fruits, and galas. Tannins protect plants from fungi as well as insects, acting as antiseptics. They show similar effects toward human skin: they have drying properties, accelerate wound healing, inhibit bleeding, and prevent the growth of microorganisms. The two most important components of tannins are tannic acid and gallic acid. Tannic acid is widely used in therapeutics, and it is derived from galls on the leaves of oaks, which are the result of parasitic insect activity. Turkish tannin comes from the galls of oak *Q. infectoria*, infested with *Andricus infectoria*, while Chinese tannins are derived from the sumac *Rhus semialata* galls, which are the result of the activity of aphids, *Aphis sinensis*. There are several natural sources rich in tannins, such as oak galls, the bark of Cutch *A. catechu, Acacia nilotica*, Chebulic Myrobalam *T. chebula*, Buckthorn *Frangula alnus*, Oak *Quercus sessilis*, Silver birch *B. pendula*, Sumach *Rhus semiala*, and others.

4.3.9 Application classes

Direct dyes are dyes that combine with the fiber and the fabric without the presence of additional agents. They are easy to use. A fabric can be simply put into the ready decoction from the plant and boiled until we obtain the desired color. Some examples of plants that act as direct dyes are turmeric, pomegranate fruit peel, annatto seeds, and husks of Terminalia catalpa.

Mordant dyes are a large group of plant dyes. The extracts of plants will not automatically dye fabrics. They require more sophisticated methods of coloration and the use of mordants, also known as stains. These mordants cause the permanent fixation of the dye with the fiber. The color depends on the type of mordant and its concentration. By using different mordants, the same dye can have various colors. These compounds not only fix the dye but also influence its hue and color. Basic mordants are, e.g. alum, copper, iron, soda, and others.

Vat dyes are water-insoluble. The color does not develop in the dyebath but by oxidation when exposed to oxygen. The process of coloration is also called vatting. The basic dyes in this group are indigo and woad. These dyes are applied to the dyeing and printing of cellulosic fibers (e.g., flax, hemp, and cotton) and result in permanent and bright colors. The linen yarn, when placed in the dyer vat through the entire process, remains white only after taking it out of the vat does it change color, initially to green and later to blue, as a result of the contact with oxygen. It may seem like a magical process, as the green cannot be kept for long; it will always turn indigo blue in the end.

Reactive dyes are the group of dyes that have the ability to form a permanent fixation with the fiber. Dyes obtained from lichens are natural dyes representing this group. Lichens, small organisms growing on the bark of trees, stones, and soil, can dye beautiful shades of purple because they contain orcein.

An essential element of the dyeing process was using fixers (post-mordants) such as alum, sodium lye, kvass, whey, kvass made of flour mixed with hot water, oak bark, oak gall, pieces of rusty iron, and urine.

4.3.10 The classification of dyes based on color

Experiments have been conducted with wild plants, trying to achieve three primary colors—red, yellow, and blue. The next step was a combination of all shades and tints of this color to get a wide palette of hues for cellulose and animal fibers. During our research, a group of herbal plants showing good dyeing properties was selected. Those plants have been classified into three color groups: blue, yellow, and red (Tables 7–9 and Fig. 20).

5 Experimental findings

Experiments have been conducted with wild plants, trying to achieve three primary colors—red, yellow, and blue. The next step was a combination between all shades and tints of this color to get a wide palette of hues for cellulose and animal fibers. During our research a group of herbal plants showing good dyeing properties has been selected. Those plants have been classified into three color groups: blue, yellow, and red (Tables 7–9 and Fig. 20).

TABLE 7 Wild plants with yellow color properties.

Family	Scientific name	Common name	Parts containing dyestuff
Papaveraceae	*Fumaria officinalis* L.	Drug fumitory	Plant
Equisetaceae	*Equisetum avense* L.	Common horsetail	Plant
Polygonaceae	*Polygonum persicaria* L.	Heart's-ease	Plant
Polygonaceae	*Polygonum aviculare* L.	Common knotweed	Plant
Polygonaceae	*Rumex acetosella* L. *Rumex acetosa* L.	Sorrel	Plant
Iridaceae	*Iris pseudoscorus* L.	Yellow flag	Flowers
Compositae	*Bidens tripartitus* L.	Trifid bur-marigold	Plant
Compositae	*Achillea millefolium* L.	Yarrow	Plant
Compositae	*Arctium lappa* L.	Łopian większy	Plant
Compositae	*Taraxacum officinale* L.	Dandelion	Flowers
Compositae	*Calendula officinalis* L.	Pot marigold	Flowers
Lycopodiaceae	*Lycopodium complanatum* L.	Ground cedar	Plant
Papilionaceae	*Genista tinctoria*	Dyer's greenweed	Plant tops
Resedaceae	*Reseda luteola* L.	Weld	Plant tops
Compositae	*Solidagp serotina* Ait.	November goldenrod	Plant tops
Compositae	*Tanaceum vulgare* L.	Tansy	Plant tops
Compositae	*Anthemis tinctoria* L.	Chamomile	Plant tops
Compositae	*Anthemis arvensis* L.	Chamomile	Plant tops
Compositae	*Matricaria chamomilla* L.	Wild chamomile	Plant tops
Geraniamceae	*Geranium robertianum* L.	Herb-Robert	
	Serratula tinctoria L.	Saw-wort	
Rdestowate	*Polygonium persarica* L.	Spotted ladysthumb	Plant
	Urtica dioica L.	Nettle	Yellow, green
Ericaceae	*Calluna vulgaris* L.	Heather	Yellow, green

5.1 Method of dyeing

5.1.1 *Extraction of dye*

The plants are crushed into small pieces and soaked in hot water overnight, boiled for 1 h, and filtrated.

5.1.2 *Mordants*

In our methods, we used oak galls, sodium carbonate anhydrous, copper sulfate, citric acid, iron (ferrous sulfate), and alum (potassium aluminum sulfate).

TABLE 8   Wild plants with red color properties.

Family	Scientific name	Common name	Parts containing dyestuff
Rubiceae	*Galium aparine* L.	Annual bedstraw	Roots
Rubiceae	*Galium mollugo* L.	White bedstraw, wild madder	Roots
Labiatae	*Origanum vulgare* L.	Wild marjoran	Plant
Rubiceae	*Asperula tinctoria* L.	Dyer's woodruff	Roots
Guttiferae	*Hypericum perforatum* L.	St. John's wort	Plant
Rubiceae	*Galium verum* L.	Yellow bedstraw	Roots
Rubiceae	*Galium palustre* L.	Marsh bedstraw	Roots
Boraginaceae	*Lithospermum officinale* L.	Common gromwell	Plant
Boraginaceae	*Lithospermum arvense*	Corn gromwell	Plant

TABLE 9   Wild plants with blue color properties.

Family	Scientific name	Common name	Parts containing dyestuff
Cruciferae	*Isatis tinctoria* L.	Woad	Leaves
Polygonaceae	*Polygonum aviculare* L.	Common knotweed	Plant
Compositae	*Cirsium arvense* L.	Canada thistle	Flowers
Polygonaceae	*Polygonum tinctorium* L.	Rdest barwierski	Plant
Papilionaceae	*Indygofera tinctoria* L.	Indigo	Leaves
Ericaceae	*Vaccinum myrtillus* L.	Bilberry	Fruits, leaves

5.1.3 *Development of color on linen, cotton, and silk fabrics*

Equipment

Laboratory dyeing machine: the Easykrome (Ugolini).

5.2 Color spectrophotometric measurements and color fastness

The main research goal was the application of vegetable dyestuffs to natural fabrics and the development of a fashion collection dyed with those dyestuffs. After the preliminary selection of plants showing dyeing properties, a few species were selected for cultivation and laboratory trials. A dyeing method was developed (temperature, kind, and quantity of mordant, pH) for plant dyestuff testing to obtain a suitable and diverse palette of colors for one dye. When the dyeing plant database was established, the following

FIG. 20 Golden chamomile *Anthemis tinctoria* in Wielkopolska region in Poland—reclaimed land. *Photo by K. Schmidt-Przewoźna.*

species were selected: annatto, wild madder, juniper, and turmeric. The color of samples prepared during trials has been expressed by color parameters obtained in spectrophotometric measurements, according to the CIELab system of color spacing. Changes in color parameters of samples (*L*, *a*, *b*) were tested using three textile raw materials dyed with five natural dyestuffs with different metallic mordants added. The resulting color palette covers many shades of yellow, orange, red, green, and brown. Testing of spectrum characteristics in the visible light wavelength range was performed using a Macbeth 2020 +spectrophotometer. The measurements were used to determine the chromaticity coordinates CIE: *X*, *Y*, and *Z*, which were used to calculate the coordinates of color in the CIELab system. The system uses three coordinates—*L*, *a*, and *b*—that can be calculated from the trichromatic components *X*, *Y*, *Z* of color, thus facilitating an understandable description of color. The *L*, *a*, and *b* changes in color parameters of all tested samples were tested against the effects of different mordants and raw materials. Different mordants yield differences in color parameter values. Similarly, different raw materials dyed with the same dye and with the same mordant added produce differences in color (Δ*E*—a difference of color). The obtained *L*, *a*, and *b* color parameters are presented in Tables 2–6 and illustrated graphically. The samples dyed with mordants display good resistance of the color to washing, sweat, friction, and light. Here at least is one example where a return to an ancient technology enhanced with modern science can result in a more efficient commercial process and one that is more environmentally friendly. For the results of color measurement, see Table 2 for kamala, Table 3 for logwood, Table 4 for cutch, Table 5 for annatto, and Table 6 for gall oak.

5.2.1 Light fastness

Color fastness to sunlight

To observe the effect of sunlight on color fastness, linen and silk samples were tested on the Laboratory Machine Xenotest 150. The test was carried out according to the standard PN-ISO 105-B02:1997.

Thirty naturally dyed samples were exposed to sunlight for 200 h, then graded for color fastness.

Measurement of color fastness to wash

The changes of color on linen and silk samples were assessed on a gray scale (1–5).

Testing Washing Fastness with the Laboratory Dyer Ugolini according to the standard PN-ISO 105-C06:1996:

Preparation of the washing bath: 4 g of washing agent per 1 L of water
Preparation of the samples of naturally dyed and reference fabrics
For Tests A and B: Reference fabrics for linen were linen and wool:
Linen: reference fabrics—linen and wool
Silk: reference fabrics—silk and cotton
Test conditions A1M: temperature 40°C, time 45 min
For natural silk crepe and silk shantung, temperature of 30°C and a duration of 45 min were applied.

5.2.2 UV protection factor of naturally dyed linen and silk (Schmidt-Przewozna and Kowalinski, 2006)

The Laboratory of Physiological Influence of Textiles on Human Body has done research to compare the results for ultraviolet protection factor (UPF) on linen and silk samples dyed with natural dyes. The transmission, absorption, and reflection of UVR are in turn dependent on the fiber, fabric construction (thickness and porosity), and finishing. Many dyes used in the finishing process absorb UVR. Darker colors of the same fabric type (black, navy, and dark red) will usually absorb UVR more than light pastel shades and consequently have a higher UPF rating. See Table 10 for the UPF classification system.

In this study, all shades of red and purple applied to silk and linen were analyzed. The determination of the UVR transmission of a dry textile was done in accordance with

TABLE 10   UPF classification system (according to the australian standard).

UPF range	UVR protection category	UPF ratings
15–24	Very good but insufficient protection	15, 20
25–39	Very good protection	25, 30, 35
40–50, 50 +	Excellent protection	40, 45, 50, 50 +

the Australian/New Zealand Standard and British Standard for sun protection clothing with the use of the Cary 50 Solascreen apparatus.

The study also determined the UPF of linen, hemp, and silk fabrics dyed with natural dyestuffs. It also examined the influence of fabrics' structure, color, and methods of dyeing on UV protection. Results are shown in Table 11 and detailed below.

TABLE 11 The results of UPF on linen and silk samples dyed by natural dyestuffs.

No.	Samples	Natural dyestuff	Mordant	Color	UVA	UVB	UPF
	Fabric A—linen		No mordant	White	13.828	19.205	5
	Fabric B—linen		No mordant	White	3.966	2.878	20
	Silk A		No mordant	White	21.911	27.249	0
1.	Fabric A—linen	Wild madder	No mordant	Light violet	7.948	7.821	10
2.	Fabric A—linen	Wild madder	Copper sulfate	Dark violet	3.270	3.334	20
3.	Linen A	Gallas	Ferrous sulfate	Dark beige	1.180	1.230	50 +
4.	Fabric A—linen	Logwood	Washing soda	Dark violet	2.951	2.771	30
5.	Fabric A—linen	Wild madder	Citric acid	Coral	6.043	6.297	10
6.	Fabric B—linen	Indian mulberry	Copper sulfate	Light violet	1.326	1.299	45
7.	Fabric B—linen	Logwood	Washing soda	Dark violet	1.823	1.714	35
8.	Fabric B—linen	Indian mulberry	Citric acid	Dark pink coral	1.966	2.133	35
9.	Fabric B—linen	India madder	No mordant	Pink bisque	0.922	0.992	50 +
10.	Fabric B—linen	India madder	Washing soda	Light salmon	1.058	1.089	50 +
11.	Fabric A—linen	Wild madder	Pre-mordant, alum	Red	2.221	2.275	25
13.	Silk B	Bilberry	Alum + iron	Violet	3.645	3.885	20
14.	Silk B	Sandalwood	Citric acid	Madder red	1.857	1.781	35
15.	Linen	Turmeric	Pre-mordant	Yellow	4.456	3.036	25
16.	Linen	Turmeric	No mordant	Light yellow	3.543	3.305	20
17.	Linen	Turmeric	Pre-mordant + copper	Olive yellow	2.907	2.011	30
18.	Linen	Turmeric	Pre-mordant + citric acid	Sun yellow	4.022	3.364	30
19.	Linen	Dyer's Coreopsis	No mordant	Old gold	1.646	1.979	30
20.	Linen	Turmeric	Pre-mordant + soda	Sahara yellow	3.302	2.53	30
21.	Linen	Turmeric	Pre-mordant + iron	Olive brown	2.772	2.532	30
22.	Linen	Dyer's Coreopsis	No mordant	Old gold	1.646	1.979	30
23.	Linen	Dyer's Coreopsis	Soda	Old gold	1.854	2.137	40

Continued

TABLE 11   The results of UPF on linen and silk samples dyed by natural dyestuffs—cont'd

No.	Samples	Natural dyestuff	Mordant	Color	UVA	UVB	UPF
24.	Linen	Dyer's Coreopsis	Citric acid	Gold	2.237	2.939	35
25.	Linen	Dyer's Coreopsis	Copper	Old gold	0.967	1.255	50
26.	Linen	Dyer's Coreopsis	Iron	Dark brown	0.881	1.117	50
27.	Linen	Common knotweed	No mordant	Brown	1.816	2.178	35
28.	Linen	Indigo	No mordant	Dark blue	0.840	0.794	50 +
29.	Linen	Henna	No mordant	Rust	1.739	2.438	35
30.	Linen A	Fengurek	Copper	Yellow	1.83	1.94	40
31.	Linen A	Dyer's chamomile	Alum	Light yellow	1.383	1.249	50 +
32.	Linen A	Black Myrobalan	Iron	Beige	1.156	1.276	50 +

Fabrics

Fabric A: 43002 thin linen Silk A—silk knitwear 100%
Fabric B: 30187 thick linen Silk B—silk shantung 100%

Results

1. The value of UPF linen and silk fabrics depends on product structure, density of thread, thickness, type of dyestuffs used, color, and kind of fabrics.
2. The result of the comparison of UPF on linen and silk fabrics:
3. Excellent UVR protection was obtained on samples:
4. Very good protection was obtained on samples:
   - India madder (linen B)—no mordant, 50 +
   - India madder (linen B)—soda, 50 +
   - Dyer's Coreopsis (linen)—soda, 50 +
   - Dyer's Coreopsis (linen)—copper, 50 +
   - Dyer's Coreopsis (linen)—iron, 50 +
   - Gall oak (linen A)—iron, 50 +
   - Black Myrobalan (linen Fabric A) iron, 50 +
   - Laka (linen B)—copper sulfate, 45
   - Indian mulberry (linen B) copper, 45
   - Dyer's Coreopsis (linen)—soda, 40
   - Cochineal (silk B)—citric acid, 35
   - Cochineal (silk B)—alum + ferrous sulfate, 35
   - Henna (linen)—no mordant, 35
   - Common knotweed (linen)—no mordant, 35
   - Dyer's Coreopsis (linen)—citric acid, 35
   - Dyer's Coreopsis (linen)—no mordant, 30

- Laka (linen Fabric A)—copper sulfate, 20
- Laka (linen Fabric A)—washing soda, 30
- Laka (linen Fabric B)—washing soda, 35
- Wild Madder (linen A)—pre-mordant, alum, 25

6 Conclusion

Plants play a huge role in the remediation of sites contaminated by external factors. They restore the site by cleaning, loosening, and extracting heavy metals from the soil. They fertilize the areas affected by drought and regulate soil water management (Figs. 21–26).

Dyeing trees has valuable properties that can be used for multifarious purposes. Dyes derived from them have a number of compounds with health benefits. Primitive peoples used them to paint their bodies, dye fabrics, and perform magic and ritual ceremonies. Alternative

FIG. 21 The collection from hemp fabrics—dyed weld. *Photo by Cezary Hładki.*

FIG. 22   The collection from hemp and linen fabrics—dyed weld. *Photo by Cezary Hładki.*

medicine is based on a deep knowledge of the compounds found in plants. Furthermore, tannin compounds, flavonoids, and anthocyanins extracted from plants have great medical potential. This special group of plants shows antibacterial, antiviral, antifungal, and antioxidant properties. Anthocyanins have antioxidant and anti-inflammatory effects. They play a significant role in the strengthening of blood vessels, and they lower cholesterol levels in the blood. They play an important role in the prevention of cancer and atherosclerosis, as well as in the treatment of irregularities in the functioning of the eye. Flavonoids have antioxidant and anti-inflammatory properties, and they reduce the harmful effects of UVR.

A wide range of colors can be obtained from dyeing plants. The use of a mortar has a significant impact on the color as it modulates it. In view of increasing pollution and emerging civilizational diseases, including allergies, scientific research has been conducted on the construction of fabrics that have a positive, even healing effect on human skin. Such research has been pursued at the Institute of Natural Fibers and Medicinal Plants in Poznań under Project BIOAKOD (bioactive curing of clothing based on natural fibers). Fabrics made from natural fibers, such as linen and organic cotton, were dyed using plant dyes: Dyer's broom, *G. tintoria*,

FIG. 23 The collection from hemp fabrics—dyed logwood. *Photo by Cezary Hładki.*

Coreopsis *Coreopsis tintoria*, logwood, *H. campechianum*, madder, *R. tinctorium* L., oak gall, *Q. infectoria* Oliv., and Chebulic Myrobalan, *Terminalia cebula*. Microbiological and medical tests have confirmed the positive effect of naturally dyed fabrics on human skin.

Moreover, some natural dyes increase skin protection against UVR. Linen fabrics dyed with Dyer's chamomile, Black Myrobalan, Dyer's Coreopsis, and India madder received a +50 UV factor, which is very good protection. The finishing of textiles plays a very important role in eco-production. Natural fibers show good sun protection due to components of natural pigments like lignin, waxes, and pectins that act as UVR absorbents. The UPF barrier effect can also be achieved with the use of special UV blockers, which are generally used in medicinal products and cosmetics. Herbal ingredients contained in dyeing trees and herbs have many important medicinal properties. The results of the analyses showed that many linen/hemp fabrics have a high degree of UV protection of +50. As it turned out, not only the fabric itself, but also the dyeing processes significantly influenced the protection effect. A particularly good result was obtained by dyeing the logwood with Hematoxylum campechianum to light purple and dark aubergine colors. When comparing linen and hemp fabrics, the latter fabrics definitely show greater protection.

The second major project carried out by the Laboratory of Natural Dyeing, which resulted in industrial implementations, was the project funded by the Ministry of Agriculture and

FIG. 24   The collection from linen and silk fabrics—dyed logwood. *Photo by Cezary Hładki*

Rural Development *"Reconstruction and sustainable development of production and processing of natural fiber raw materials for agriculture and the economy"*. As part of this project, the task titled, 'Technology for the production of innovative, naturally dyed, functional textile products' was carried out from 2017 to 2020. The main objective of the task was to apply dyeing technology to the development and manufacture of functional garments. (Schmidt-Przewozna 2020).

There are approximately two thousand plant species recognized as medicinal plants that can find applications in the pharmaceutical and cosmetic industries. They also exhibit moisturizing, softening, and soothing properties. The plants protect the skin from the sun's rays, whiten the skin and reduce discoloration, stimulate blood circulation in the capillaries, and have toning and astringent properties. They contain various bioactive compounds: glycosides, tannins, essential oils, mucilages, and other substances used in medicine and cosmetology. Glycosides include flavonoids, saponins, phenolic compounds, and anthraquinone compounds; they are natural antioxidants. They improve the blood supply to the skin and rejuvenate it. Tannins also play an important role thanks to their antiviral, antibacterial, anti-inflammatory, and strong astringent properties. The tannins found in plants have a beneficial effect on the skin and the human body. Tannins also play a significant role in dyeing processes. A selection of plant-based natural dyes was initiated, and many trials were

FIG. 25 The collection from linen—dyed new extract Madder R. *Photo by Paulina Buczynska.*

FIG. 26 The collection from linen and silk—dyed logwood and Madder R. *Photo by Paulina Buczynska.*

conducted to select dye plants with health-promoting properties and, importantly, that produce reproducible, lightfast, and laundry-resistant colors during the dyeing process. After testing more than 50 different natural raw materials, five plants were selected for further experiments. The selection depended on the potential health-promoting components present in the plants, the repeatability of the color, and its resistance to washing and light.

Alcoholic extracts of Madder (*R. tinctorum*), Weld (*R. luteola*), Logwood (*H. campechianum*), and Marigold (*Tagetes patula*) were used for dyeing. The research work resulted in a range of yellows and olives, a range of reds and salmon, and colors of lavender, aubergine, and violet. Samples of linen and hemp fabrics and functional garments were examined spectrophotometrically, and the colors were determined according to the Pantone numerical system. The choice of colors also depended on the resistance of the fabrics to washing, perspiration, and light. As part of the project, an innovative extract, Madder R, was produced at the Łukasiewicz Research Network, Institute of New Chemical Syntheses in Puławy. An innovative, ecological dyeing method was also developed with CO_2.

The antibiotic activity of dyeing plant extracts was assessed and studied. The studies aimed at determining the antimicrobial activity of the extracts (Table 12).

The research conducted showed that all plant extracts showed antibacterial activity. Among them, indigo and madder R, with 4000 JA/g of extract, had the greatest antibacterial activity. The remaining extracts tested showed slightly lower antibacterial activity, ranging from 670 to 1330 JA/g of extract.

Studies of extracts have fully demonstrated that the strengths of natural plant dyes are the bioactive compounds they contain: glycosides, tannins, essential oils, mucilages, and others. The plant extracts used in the textile dyeing process, depending on the content of the active substances and their valuable components, can have the following properties:

- Compounds contained in plants are great for skin hydration, smoothing, and wound healing and have anti-inflammatory, regenerative, antiviral, antifungal, and antioxidant properties.
- Fabrics dyed with natural dyes have shown a higher absorption of UVR, providing excellent protection for the skin.

TABLE 12 Antibacterial activity of the tested extracts from dye plants.

Sample number	Name of extract	MIC (mg/mL)	MBC (mg/mL)	JA (g)
1.	Logwood	0.5	0.75	2000
2.	Indigo	0.25	7.5	4000
3.	Madder	1.50	5.0	670
4.	Madder N	0.75	2.5	1330
5.	Madder R	0.25	0.5	4000
6.	Dyer'broom	2.0	5.0	500
7.	Weld	1.0	1.5	1000
8.	Marigold	2.5	4.5	400

MIC, minimum inhibitory concentration; MBC, minimum bactericidal concentration; JA, activity unit.

- The spectrum of colors and shades is achieved through the use of dye mordants, which can bring out several shades and even colors from a single plant, also acting as stabilizers.
- They cause a positive sensation of wearing natural, airy clothing with a sublime range of colors.
- Many natural dyes show high color fastness, resistance to washing, and fading when exposed to the sun.
- Most of the plant compounds used in dyeing processes are biodegradable, non-toxic, and non-allergenic, and the waste produced can, in some cases, be used as biofertilizers. Thus, it can be said that the entire dyeing process as well as the resulting effect are humans and environmentally friendly.
- Natural dyes of plant origin are a source of renewable matter, obtainable through the agricultural process, unlike synthetic dyes, which are produced from non-renewable sources (oil, coal).
- These seem to be factors that sufficiently underline the legitimacy of the trend toward the use of natural dyes, which is increasing year by year, not only in niche forms but also in the prospects for the application of these technologies at the scale of the textile industry.

References

Adeel, S., Ali, S., Bhatti, I.A., Zsila, F., 2009. Dyeing of cotton fabric using pomegranate (*Punica granatum*) aqueous extract. Asian J. Chem. 21 (5), 3493–3499.
Balakrishna, S., Seshadri, T.R., Venkataramani, B., 1960. Special chemical components of commercial woods and related plant materials. IX. Morindonin, a new glycoside of morindone. J. Sci. Ind. Res. 19, 433–436.
Bechtold, T., Mussak, R. (Eds.), 2009. Handbook of Natural Colorants, vol. 8. John Wiley & Sons.
Bojase, G., Wanjala, C.C.W., Gashe, B.A., Majinda, R.R.T., 2002. Antimicrobial flavonoids from, *Bolusanthus speciosus*. Planta Med. 68, 615–620.
Cardon, D., 2007. Natural Dyes: Sources, Tradition, Technology and Science. Archetype Publications, London, UK. pp. 66, 115, 146, 143–148, 163–164,193, 693–694.
Cardon, D., 2010. Natural dyes, our global heritage of colors. In: Textile Society of America Symposium Proceedings.
Cichocka, A., Frydrych, I., Zimniewska, M., Muzyczek, M., Mikołajczak, P., Schmidt-Przewoźna, K., Romanowska, B., Pawlaczyk, M., Krucińska, I., Komisarczyk, A., Kowalska, S., 2021. 3D design of clothing in medical applications. Autex Res. J. 21 (4), 408–412.
Dean, J., 1999. Wild Colour. Octopus Publishing Group, London, UK, p. 114.
Drury, H., 1873. The Useful Plants of India, second ed. William H. Allen & Co., London.
Dygus, K., Wasiak, J.S.G., Madej, M., 2012. Roslinosc odpadow komunalnych i przemysłowych. Wydawnictwo Naukowe Gabriel Borowski, Warszawa, pp. 100–104.
Ganie, S.A., Ganaie, R.A., Ara, S., Agarwal, S., Dar, Z.A. and ul Mehmood, M.A., Characterization and evaluation of *Nasturtium officinale*R. Br. dye as an alternative to synthetic dye. Int. J. Adv. Res. Sci. Tech. 12(4) 84–93.
Hosseinnezhad, M., Gharanjig, K., Jafari, R., Imani, H., 2021. Green dyeing of woolen yarns with weld and madder natural dyes in the presences of biomordant. Progr. Color Color. Coat. (PCCC) 14 (1), 35–45.
Hwang, J.K., Kong, T.W., Baek, N.I., Pyun, Y.R., 2000. Alpha-glycosidase inhibitory activity of hexagalloylglucose from the galls of *Quercus infectoria*. Planta Med. 66, 273–274.
Jose, S., Pandit, P., Pandey, R., 2019. Chickpea husk—a potential agro waste for coloration and functional finishing of textiles. Ind. Crop. Prod. 142, 111833.
Kannan, P., Ramadevi, S.R., Hopper, W., 2009. Antibacterial activity of *Terminalia chebula* fruit extract. Afr. J. Microbiol. Res. 3 (4), 180–184.
Khatun, M.H., Mostafa, M.G., 2019. Phytochemical screening and extraction of natural dye from leaves of *Polyalthia longifolia* and its application on silk fabric. J. Life Earth Sci. 14, 87–92.

Khatun, M.H., Mostafa, M.G., 2022. Optimization of dyeing process of natural dye extracted from *Polyalthia longifolia* leaves on silk and cotton fabrics. J. Nat. Fibers, 1–16.

Lauro, G.J., Jack, F.F., 2000. Natural Food Colorants: Science and Technology. IFT Basic Symposium Series, Marcel Dekker, New York.

Manjunatha, B.K., 2006. Antibacterial activity of *Pterocarpus santalinus*. Indian J. Pharm. Sci., 115–116.

Mansour, R., 2018. Natural dyes and pigments: extraction and applications. In: Handbook of Renewable Materials for Coloration and Finishing. 9, pp. 75–102.

Melo, M.J.2009 History of Natural Dyes in the Ancient Mediterranean World.

Mohd Zin, Z., Abdul Hamid, A., Osman, A., Saari, N., Misran, A., 2007. Isolation and identification of antioxidative compound from fruit of Mengkudu (*Morinda citrifolia* L.). Int. J. Food Prop. 10 (2), 363–373.

Morton, J.F., 1960. Can Annatto (*Bixa orellana* L), an Old Source of Food Color, Meet New Needs for Safe Dye. USA, Florida State Horticultural Society.

Morton, J.F., 1992. The ocean-going noni, or Indian mulberry (*Morinda citrifolia*, Rubiaceae) and some of its "colorful" relatives. Econ. Bot. 46 (3), 241–256.

Murti, V.V.S., Neelakantan, S., Seshadri, T.R., Venkataramani, B., 1959. Special chemical components of commercial woods and related plant materials. VIII. Heartwood of *Morinda tinctoria*. J. Sci. Ind. Res. 18B, 367–370.

Nagesh, L., Sivasamy, S., Muralikrishna, K.S., Bhat, K.G., 2012. Antibacterial potential of gall extract of *Quercus infectoria* against *enterococcus faecalis*—an *in vitro*. Pharmacogn. J. 4 (30), 47–50.

Perlein, A., Bert, V., Fernandes de Souza, M., Gaucher, R., Papin, A., Geuens, J., Wens, A., Meers, E., 2021. Phytomanagement of a trace element-contaminated site to produce a natural dye: first screening of an emerging biomass valorization chain. Appl. Sci. 11 (22), 10613.

Przewozna, K., Kowaliski, J., Tomassini, M., 2011. Potential and Future Prospects of Dyeing Plants: Madder *Rubia tinctorium* L., Annatto *Bixa orellana* L., Dyer's Broom *Genista tinctoria* L., Woad *Isatis tinctoria* L. NOVA Science, New York, pp. 49–57.

Rao, V.S., Seshadri, T.R., 1947. Kamala dye as an anthelmintic. Proc. Indian Acad. Sci. A 26 (3), 178–181. Redaction of Natutal Science University.

Saleem, M., Kim, H.J., Ali, M.S., Lee, Y.S., 2005. An update on bioactive plant lignans. Nat. Prod. Rep. 22 (6), 696–716.

Schmidt-Przewozna, K., 2001. The sources and properties of natural dyestuff applied in east and west tradition. J. Dyes Hist. Archaeol. 20. Archetype Publication, UK.

Schmidt-Przewozna, K., 2002. The return to natural dyeing methods—properties and threats. In: Procdings of the Textile Institute 82nd World Conference, Cairo, Egypt, pp. 220–226.

Schmidt-Przewozna, K., 2009. Barwienie Metodami Naturalnymi. ECO-Press, Białystok, ISBN: 978-83-929329-0-1, pp. 1–20.

Schmidt-Przewozna, K., 2020. Dyeing With Natural Methods, Dyeing Plants and Their Potential.

Schmidt-Przewozna, K., Brandys, A., 2014. Dyeing Trees as a Source of Dyestuffs, Cosmetic and Herbal Ingredients. Proc.Narossa.cd.

Schmidt-Przewozna, K., Kowalinski, J., 2006. Light fastness properties & UV protection factor on naturally dyed linen and hemp. In: Proceedings Saskatoon Conference.

Schmidt-Przewozna, K., Skrobak, E., 2013. A Thread of Colour Yarn. Regional Invest-Druk, Warszawa, pp. 1–10.

Schmidt-Przewozna, K., Zimniewska, M., 2005. The Effect of Natural Dyes Used for Linen on UV Blocking Renewable Resources and Plant Biotechnology. NOVA Science, New York, pp. 110–117.

Schmidt-Przewoźna, K., Urbaniak, M., Zimniewska, M., Brandys, A., Banach, J., Gromek, M., Mikołajczak, P., Pawlaczyk, M., Krucińska, I., Frydrych, I., Komisarczyk, A., 2018. Natural and safe dyeing of curing clothing intended for patients with dermatoses. Postępy Fitoterapii 1, 32–39.

Schmidt-Przewozna, K., Morales Villavicencio, A., Kicinska-Jakubowska, A., Zajaczek, K., Brandys, A., 2019. *Coreopsis tinctoria* Nutt. as a source of many colours. Herba Polonica 65 (4).

Seigler, D.S., 1998. Plant Secondary Metabolism. Kluwer Academic Publishers, Dordrecht, The Netherlands. 759 pages.

Sharma, J., Varma, R., 2011. A review on endangered plant of *Mallotus philippensis* (Lam.). Pharmacologyonline 3, 1256–1265.

Silva, G.F., Gamarra, F.M.C., Oliveira, A.L., Cabral, F.A., 2008. Extraction of bixin from annatto seeds using supercritical carbon dioxide. Braz. J. Chem. Eng. 25 (2), 419–426.

Suchalata, S., Devi, C.S., 2005. Antioxidant activity of ethanolic extract of *Terminalia chebula* fruit against isoproterenol—induced oxidative stress in rats. Indian J. Biochem. Biophys. 42, 242–249.

Thakur, S., Thakur, S., Chaube, S., Singh, S., 2015. An etheral extract of Kamala (*Mallotus philippinensis (Moll. Arg)* Lam.) seed induce adverse effects on reproductive parameters of female rats. In: Reproductive Toxicology. vol. 20. Elsevier, USA, pp. 149–156.

Tiwari, H.C., Singh, P., Mishra, P.K., Srivastava, P., 2010. Evaluation of various techniques for extraction of natural colorants from pomrgranate rind—ultrasound and enzyme assisted extraction. Indian J. Fibre Text. Res. 25, 272–276.

Vanita, P., Nirali, A., Khyati, P., 2013. Effect of phytochemical constituents of *Ricinus communis, Pterocarpus santalinus, Terminalia belerica* on antibacterial, antifungal and cytotixic activity. Int. J. Toxicol. Pharmacol. Res. 5 (2), 47–54.

Vankar, P.S., Shanker, R., Wijayapala, S., 2009. Utilization of temple waste flower-tagetus erecta for dyeing of cotton, wool, and silk on industrial scale. J. Text. Apparel Technol. Manage. 6 (1), 1–15.

West, B.J., Claude, J.J., Westendorf, J., 2008. A new vegetable oil from noni (*Morinda citrifolia*) seeds. Int. J. Food Sci. Technol. 43 (11), 1988–1992.

Wisniak, J., 2004. Dyes from antiquy to synthesis. Indian J. History Sci. 39 (1), 76–100.

Yamuna, V., Sudha, S., 2021. Effect of natural extracts and mordants on dyeability of cotton fabrics for UV protection. Int. J. Mech. Eng. 6 (3), 383–386.

Bioeconomy and circular economy approaches for the greening of urban wastelands: Focus on biodiversity to achieve sustainability

Guillaume Lemoine[a], Majeti Narasimha Vara Prasad[b], and Patricia Dubois[c]

[a]Biodiversity and Ecological Engineering Referent, Etablissement Public Foncier Hauts-de-France, Euralille, France [b]Department of Plant Sciences, School of Life Sciences, University of Hyderabad (An Institution of Eminence), Hyderabad, Telangana, India [c]Responsable du Service Stratégie et Partenariat, Établissement Public Foncier Hauts-de-France, Euralille, France

1 Introduction

Brownfield restoration as a smart economic growth option for promoting ecotourism, biodiversity, and leisure: two case studies in Nord, Pas-de-Calais were presented in the previous edition of this book (Lemoine, 2016). Brownfield issues have a significant impact on both human behavior and the environment, according to urban environmental research. Sustainable urban policies may be given higher priority by governments and financial partners for the redevelopment of brownfields. Brownfield redevelopment is essential to the sustainability of metropolitan areas. To make it simple to incorporate sustainable development into a brownfield rehabilitation project, multiple research directions are needed (Alvernia and Soesilo, 2019; Ansari et al., 2016).

Ecosystem restoration and rehabilitation in cities can be relatively cheap or expensive although even expensive efforts can be beneficial if they successfully enhance ecosystem services like recreation or climate change mitigation. Our cities will become more sustainable in the future if urban green space is successfully reshaped and rethought with the participation of residents and other stakeholders.

475

Due to its importance for wildlife conservation, human well-being, and climate change adaptation, urban green space has attracted a lot of attention in recent years. Worldwide biodiversity loss and ecosystem degradation necessitate the development of fresh ideas for ecological restoration and rehabilitation with the goal of enhancing ecosystem services and biodiversity preservation in urban areas (Bonthoux et al., 2014; Farfán-Beltrán et al., 2022; Figueroa et al., 2022). Although remnants of natural and seminatural ecosystems can be found in urban areas, most urban ecosystems have highly altered environmental conditions and species compositions, which has led to the emergence of novel and hybrid ecosystems. The absence of (semi)natural reference systems for defining restoration aims and gauging restoration performance in urban contexts is a result of this ecological uniqueness. This makes it more difficult to perform ecological restoration in urban areas. We introduce a new conceptual framework that takes these difficulties into account and offers direction and support for urban ecological restoration and rehabilitation by creating restoration targets for various degrees of ecological innovation (i.e., historic, hybrid, and novel ecosystems). We advise utilizing well-established, species-rich, and functional urban ecosystems as a model for the restoration and rehabilitation of novel urban ecosystems. Numerous cities are likely to have such urban reference systems. By highlighting how valuable they are in relation to damaged ecosystems, restoration efforts can be encouraged and directed. It may also be wise to concentrate the restoration of severely damaged urban ecosystems on certain ecosystem functions, services, and/or biodiversity values given that urban restoration approaches must take into account local history and site characteristics as well as inhabitants' demands (Bonthoux et al., 2014). Details are elaborated in Fig. 1 (Bostenaru Dan and Bostenaru-Dan, 2021; Klaus and Kiehl, 2021).

Fig. 1A shows a conceptual model of transformation of historical natural and seminatural ecosystems into hybrid or irreversibly altered novel ecosystems (Kollmann et al., 2019); (B) the ecological restoration of historically degraded and hybrid ecosystems in urban areas (black arrows); and (C) the ecological rehabilitation of novel urban ecosystems. Surfaces that have been transformed from gray to (partially) green are indicated by dashed arrows. Note that the arrow at G may result from either the establishment of a green roof using native species from a gray roof or the restoration of an existing green roof.

The French Government in December 1990, on the request of the regional authority, created the Establishment Public Foncier Hauts-de-France (EPF), an urban regeneration operator (public body) in the territories (districts) of Nord, Pas-de-Calais, and Somme, whose primary mission is to acquire land and demolish and depollute former built-up areas to enable housing, especially social housing to be built by promoting greater urban density through recycling industrial and urban wasteland. The aim is to make a "city within the city" (or "dense city"), and EPF thus helps to fight urban sprawl and avoid using new agricultural land. In the departments of Nord and Pas-de-Calais, in accordance with its multiannual roadmap (PPI [Plan Pluriannuel d'Intervention]) for 2015 through 2019, the work of EPF has several purposes. In addition to urban regeneration, which is its "core business," EPF is also an organization working for biodiversity, social balance, and economic initiatives. EPF is therefore one of the land management solutions at the disposal of local authorities (town councils and joint authority cooperation organizations—PCIs [Établissements Publics de Coopération Inter communale] to enable them to fulfill some of their projects. To these ends, Articles L-321-1 *et seq.* of the French Urban Development Code state that "EPFs

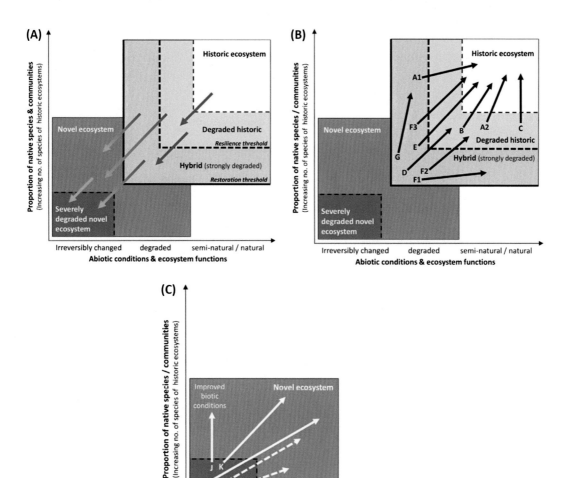

FIG. 1 Conceptual model visualizing (A) the degradation of historic natural and seminatural ecosystems to hybrid or irreversibly changed novel ecosystems; (B) ecological restoration of degraded historic and hybrid ecosystems in urban areas (*black arrows*) and (C) ecological rehabilitation of novel urban ecosystems. Dashed arrows represent the conversion of *gray* into (partially) *green* surfaces. Note that, arrow at G could be a result of either the restoration of an existing *green* roof or the development of a *green* roof with native species from a *gray* roof.

Letters and numbers refer to ecosystem types.

(A) Ancient forests in cities.

(B) Ancient wet-lands, ponds, creeks, rivers, lakeshores, and floodplains in cities.

(C) Ancient grass-land (including dune grasslands and prairies) in old parks, vacant lots, cemeteries and so on.

(D) Urbanized ponds, wetlands, creeks, rivers, lakeshores, and floodplains in cities.

(E) Lawns, roadside vegetation, grass on vacant lots, and other perennial herbaceous vegetation.

(F) Urban brownfields, wastelands, and abandoned military areas (various soil conditions from slightly to highly altered).

(G) Conventional green roof or gray roof suitable for greening with native pants.

(H) Built ponds, basins, artificial wetlands, and creeks including cased streams.

(J) Planted flower and vegetable beds in gardens, intensively managed parks, cemeteries, and similar horticultural systems.

(K) Small patches of herbaceous vegetation such as traffic islands, flower decorations, or shrubberies.

(L) Gray roof suitable for conventional greening.

(M) Sealed ground and buildings, e.g. streets, tram tracks, arking lots, facades, strongly degraded brownfields with high amounts of sealed and polluted soils, and other gray infrastructure.

Part (A) adapted from Kollmann, J., Kirmer, A., Tischew, S., Hölzel, N., Kiehl, K., Kollmann, J., 2019. Neuartige Ökosysteme und invasive Neobiota. Renaturierungsökologie, 435–447, strongly modified.

implement land management strategies to promote sustainable development and the fight against urban sprawl. As part of this activity, contributed to policies to protect against technological and natural risks." In its 32 years of existence, it has thus treated nearly 5500 ha of wasteland, and each year it purchases approximately 50 ha of derelict land in municipalities and joint authorities for urban wasteland development (Tables 1–3).

TABLE 1   Synopsis of recent research on sustainable brownfield redevelopment and planning.

	Author(s) (Year)
A conceptual framework for urban ecological restoration and rehabilitation	Klaus and Kiehl (2021)
A brief assessment of ecological restoration in a Mexico City urban ecological preserve based on trophic guilds and arthropod communities	Farfán-Beltrán et al. (2022)
Agroecological food production, ecological restoration, peasants' well-being, and agri-food biocultural heritage in Xochimilco	Figueroa et al. (2022)
Spatiotemporal evolution and optimization of landscape patterns based on the ecological restoration of territorial space	Hu et al. (2022)
Research on residual plaque landscape construction rapid urbanization and the ecological restoration model	Leichang et al. (2022)
Plans for ecological restoration and urban country park planning using GIS	Li et al. (2022a)
Establishing and improving an ecological security system in Shaanxi province, China, in order to restore the ecological integrity of the land	Li et al. (2022b)
A case study of Jingmen, China, illustrates an integrated method for creating ecological security patterns and locating ecological restoration and protection areas	Li et al. (2022c)
Ecological restoration over a long period of time in a typical shallow urban lake	Liu et al. (2022)
A case of central urban areas in Chongqing, China: ecological restoration strategies for mountainous cities based on ecological security patterns and circuit theory	Lv et al. (2022)
Ecosystem services' role in the revitalization of tiny mono-industrial communities is part of urban ecological restoration in a post-industrial context	Măgureanu and Sîrbu (2022)
Chinese case study of the central Yunnan urban agglomeration identifies key regions for territorial ecological restoration based on ecological security patterns	Ran et al. (2022)
Finding important green space for ecological restoration in China's Fujian province using ecological security patterns	Wang et al. (2022)
Thoughts on the construction method of stock space based on the perspective of ecological restoration and urban repair	Xiaojian and Yuewen (2022)
Construction of ecological security pattern and identification of ecological restoration zones in the city of Changchun, China	Xu et al. (2022)
Opinions on the method of stock space construction considering ecological restoration and urban repair construction of ecological security pattern and identification of ecological restoration zones in the city of Changchun, China	Yue et al. (2022)

TABLE 1 Synopsis of recent research on sustainable brownfield redevelopment and planning—cont'd

	Author(s) (Year)
The yellow river basin's ecological vulnerability has changed across time and space as a result of ecological restoration efforts.	Zhang et al. (2022a)
Initial ecological restoration evaluation of a metropolitan river in China's subtropics	Zhang et al. (2022b)
Heavy metal accumulation and source apportionment in urban river under ecological restoration: relationships with land use and risk assessment based on Monte Carlo simulation	Gao et al. (2023)
A case study of Fuzhou city, China, identifies priority ecological restoration areas based on human disturbance and ecological security patterns	Ke et al. (2023)
Planning and development of sustainable brownfield sites using bibliometric and visual analyses	Zheng and Masrabaye (2023)

TABLE 2 The top 5 journals' names.

Ranks	Journals	1990–2000	2001–2010	2011–2021	Total of documents	Impact factor (2019)	h-index
01	Chemosphere	X	95	425	520	7.086	126
02	Environmental Science and Pollution Research	X	X	434	434	3.056	113
03	Science of The Total Environment	17	86	264	367	7.96	180
04	Journal of Hazardous Materials	X	71	251	322	9.52	120
05	Environmental Pollution	X	78	144	222	8.071	132

Source: Zheng, B., Masrabaye, F., 2023. Sustainable brownfield redevelopment and planning: bibliometric and visual analysis. Heliyon 9(2), e13280.

Towns and cities are living territories shaped by economic, commercial, social, and land-related developments. In various places, industrial and/or urban wasteland appears (former substandard housing or real estate, commercial, railway, military wasteland, etc.). Urban wastelands can therefore be found everywhere in France and Europe, in particular in former industrial regions (10,000 ha identified in 1990 in the French region of Nord, Pas-de-Calais, for example). These areas disappear when urban rehabilitation projects are conducted upon them, but are also created at a steady rate as a consequence of economic hardship in many regions, company relocations and the departure of inhabitants to the outskirts of towns or to more attractive sectors and regions. Even small and medium-sized towns in rural France are affected by wasteland, since each region evolves in its own way, with its own urban dynamics and transformations. In some cases, once buildings have been demolished and cleared, the derelict land (or the non-built-up parts of land development projects) can remain derelict for many years pending repurposing.

TABLE 3 The top 15 worldwide affiliations of publications related to research progress of sustainable brownfield redevelopment.

Affiliation	TP	TP (%)	Country
Chinese Academy of Sciences	403	5.59	China
Ministry of Education China	170	2.36	China
University of Chinese Academy of Sciences	142	1.97	China
Zhejiang University	78	1.08	China
The United States Environmental Protection Agency	75	1.04	USA
CNRS	72	1	France
Ministry of Agriculture of the People's Republic of China	70	0.97	China
Chinese Research Academy of Environmental Sciences	63	0.87	China
Consejo Superior de Investigaciones Científicas	61	0.84	Spain
Research Center for Eco-Environmental Sciences Chinese Academy of Sciences	61	0.84	China
University of Lorraine	58	0.80	France
University of South Australia	57	0.79	Australia
Sun Yat-Sen University	53	0.73	China
Tsinghua University	53	0.73	China
Nanjing Agricultural University	52	0.72	China

Source: Zheng, B., Masrabaye, F., 2023. Sustainable brownfield redevelopment and planning; bibliometric and visual analysis. Heliyon 9(2), e13280.

The soils of wasteland and derelict urban ground are often characterized by the "industrial" or urban history of the location. These artificially developed areas are highly mineralized (paving blocks, asphalt, ballast) or often appear as "technosoils" (zones backfilled with rubble or crushed concrete), thus displaying severe limitations in terms of soil structure, ecology and economy. Finally, they also have a specific hot and dry microclimate.

These areas can subsequently accommodate natural phenomena, in particular, that of plant growth, which may take on several forms, depending on the type of soil on the site, the age, and size of the site, and their situation in the urban ecological matrix (proximity to more or less natural "source" environments).

2 Current situation: significant plant colonization

The vegetation that grows there can be of different types. Alongside ruderal species, which can develop on soils that are more or less natural and more or less eutrophic after organic waste and matter have been deposited—these species include nettles, thistles, European

Black Nightshade, etc.—the artificially developed areas in towns frequently see thermophilous flora: The wasteland and derelict areas of large cities, cosmopolitan locations *par excellence*, also host a quantity of exotic or even invasive species. Flower species in towns, whatever the location, are starting to blend into a majority of exogenous species, which are also becoming cosmopolitan (Lemoine, 2016).

3 Problem: a situation deemed unsatisfactory

In many cases, the established urban vegetation can cause problems for inhabitants, local authorities and property developers. First, the presence of ruderal species may displease and adversely affect the living conditions of inhabitants. It is expensive to regularly maintain these spaces (using often environmentally unfriendly methods such as plant shredding). The woodland succession based on exotic species or more regional wood types (mainly willow, birch, and sycamore) can also be a source of concern for public officials and inhabitants, who may fear the emergence of illegal activity (drug trafficking, prostitution, and squatters) on these spaces, part of which is hidden by vegetation (afforestation with trees). While we may welcome the spontaneous protected species, since the dynamics of vegetation and shrub growing on urban wasteland are outside of human control, corresponding to a certain form of "nature," the appearance of protected species (flora and fauna) cannot be ruled out and may lead to the situation becoming more complicated. Indeed, among drought-resistant and thermophilous species, some appear on the list of protected plants, such as *Linaria supina*, Flat Pea (*Lathyrus sylvestris*), Bee Orchid (*Ophrys apifera*), Liquorice Milkvetch (*Astragalus glycyphyllos*), and others, for example in the departments of Nord and Pas-de-Calais, or on the national list of protected fauna (Common Wall Lizard, Natterjack Toad, etc.). Their presence is not conducive to the emergence of urban projects that may be at the origin of land recycling operations. The presence of protected species will entail regulatory constraints for developers and their contracting authorities, such as organizing wildlife diagnostics, the application of the ERC sequence (inquiry timeframes, additional costs relating to compensation and worksite delay), etc. In the departments of Nord and Pas-de-Calais, EPF—which contributes to urban regeneration by recycling urban and industrial wasteland—frequently faces this type of problem.

4 Wasteland: a habitat with fuzzy definition and outlines

Wastelands and abandoned urban lands do not represent the most popular environments for botanists and naturalists. The growing place of cities and intensive agricultural practices realized in the landscape matrix, however, requires us to reconsider the role that these "empty" spaces can play for the conservation of fauna and flora, even that which is more or less common. Various inventories even tend to show that in our artificialized regions, the biodiversity of wastelands is far from commonplace, even if they do not host extraordinary species, except probably in some special environments under heavy edaphological constraints (ballast, technosoils).

We thus discover from the review of the consulted bibliography that the biodiversity of wastelands is not negligible. In order to have maximum biodiversity on wastelands, they are required to be heterogeneous (in situation, on the surface, in kinds of "soils," and especially in age). In light of the consideration for activity zones in the Netherlands and in quarries (spontaneously or voluntarily), the concepts of biodiversity "in motion" and "temporary nature" are probably part of the ideas to be developed to preserve this natural heritage in a context where any wasteland is possibly concerned or threatened by urban reallocation.

As uncultivated, abandoned lands, wastelands are—in essence—indefinite spaces. These are land in suspension, "pending" projects, or abandoned as enclosed, polluted, or unusable inherently, as are the edges along the infrastructure. These environments can convey an image that is "not right," of spontaneous spaces, spaces of "nature" (that is to say, that are not controlled by humans) or freely changing spaces. Wastelands are often inaccessible places and sometimes form safe havens for excluded human populations. In these spaces, we find everything from plants to species of various origins (native, exotic, sometimes invasive or cultivated), wild domestic (and sometimes savage), objects positioned and/or abandoned, or even squats and precarious shelters. They show strong heterogeneity (in size, age, origin, substrate, history, etc.) and multiple habitats (in the phytocenotic sense) ranging from mineral, compact, and waterproofed spaces, with "spontaneous" afforestation.

Wastelands and brownfields can have three roles/uses: an economic use, as a reserve of land to accommodate future buildings (including housing) in an urban renewal context; a role providing inhabitants with a breathing space, as well as a place to live and even relax; and, if they are easily accessible and appropriable, in some cases, they also play the role of safe havens for marginalized populations by offering places of reception and freedom, which sometimes become mysterious, disturbing or even dangerous places that are uninviting; wastelands also provide a third environmental role: In addition to their relevance for biodiversity, like other green spaces, they allow for local climatic regulation of local urban areas (excess heat cooling, increase in the humidity of the air, etc.) (Paris, 2012). Wastelands are relay areas for flora and fauna, and ecosystems by definition, ephemeral, that human interventions denature. We cannot preserve the flora of a wasteland by freezing it in time. It is therefore on the ecosystemic scale of the urban unit that it is appropriate to ask ourselves about how they are managed, or indeed not managed, before converting them (Zheng and Masrabaye, 2023) (Figs. 2 and 3).

5 A progressive awareness of their relevance for urban flora

The study of the vernacular flora of cities is relatively recent. It involves inventories, describing "botanical curiosities," appeared in 1561 with Conrad Gessner, who was in Germany looking for cultivated plants and species brought from the Americas. Other old inventories concerned very specific places such as monuments, walls, ruins and rubble. The flora thus observed took the name of ruderal (*rudus*: ruins, rubble). Botanists in the late nineteenth century and during the twentieth century also followed the vegetative dynamics of bombed sites with interest, such as plant reclaims observed on volcano lava flows.

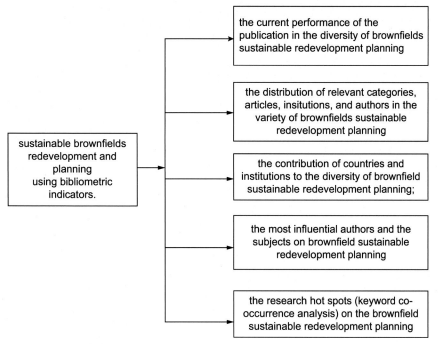

FIG. 2 Evaluating of progress on the sustainable brownfield's redevelopment and planning. *Source: Zheng, B., Masrabaye, F., 2023. Sustainable brownfield redevelopment and planning: bibliometric and visual analysis. Heliyon 9(2), e13280.*

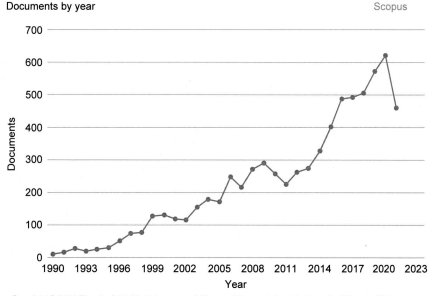

FIG. 3 Research progress of sustainable brownfield redevelopment planning and design by year (1990–2021). *Source: Zheng, B., Masrabaye, F., 2023. Sustainable brownfield redevelopment and planning: bibliometric and visual analysis. Heliyon 9(2), e13280.*

In France, Vallot began an inventory of the flora of the sidewalks of Paris at the end of the nineteenth century, and Jovet performed the first urban botanical inventory in Paris in 1926. The first urban studies were conducted in the 1970s and 1980s by the Anglo-Saxons, then by teams of naturalists and biologists from Central Europe, especially about birds by Luniak in Poland, and on plants by the Sukopp team in Germany. The first atlas of urban flora was made for 1983s London by Borton in 1986. For French towns, the MNHN (National Natural History Museum [Musée National d'Histoire Naturelle]) produced a good little guide: "Wild Plants on my Street" ("Sauvages de ma rue"). Often richer than those of the surrounding countryside, urban vegetation is characterized by its high variability, both spatial and temporal, and is composed of flora with a great ecological amplitude and largely exotic part. Several recent works have also shown the homogenizing effect of the urbanization of fauna and flora, as most cities in the world have an identical structure. Thus, at the same latitudes, we find groups of species that tend to be close enough, whereas regional biodiversity can be very different. The introduction of species—which can be significant in urban areas, particularly with many exotic horticultural plants in gardens and with animals that are new to society—participate greatly in this "trivialization" because they are often the same species that are sold around the world through garden centers and pet shops. The city of Wolfsburg hosted more than 800 species of flowering plants and ferns, including 400 in urban areas. Among them, 160 are exclusive to the city. Elias Landolt listed 1200 plant species in the territory of the city of Zurich, 50% of them in urban areas. The freight station north of Basel hosted 450 different species of flowering plants, including xero-thermophilic ones, some of which were at the time considered rare north of the Alps or had practically disappeared from the territory of Switzerland. Muratet et al. (2007) meanwhile identified a set of 277 species in the wastelands of the city of Düsseldorf, representing 51% of the species present in this territory. The city of Birmingham, for its part, hosts 378 species found on 50 abandoned sites (Figs. 4 and 5).

6 Spaces that are ever more studied and relatively abundant flora

Recent work in urban ecology has shown that wastelands are essential in maintaining urban biodiversity as they host species-rich communities. The floristic relevance of wastelands largely depends on the urban structure in which they are inserted and their size. In the department of Seine-Saint-Denis, out of 17 wastelands totaling 26.8 ha, the following were monitored as part of the "Wasteland Research Program": 379 plant species, 42 bird species, and 17 butterfly species were recorded. Compared with the data collected by the Observatory Department of Urban Biodiversity, the 17 wastelands studied host biodiversity that represents one third of the total plant biodiversity observed throughout the department. These results clearly show the importance of wastelands as reservoirs of biodiversity in such a very urban department. The "meadows" of the wastelands are the systems that house end welcome the most species. Then come the thickets and bare ground spaces. Woods are the least diversified habitats. Spaces abundantly used by humans (waste dumps, latrines) are colonized by nitrophilous flora (nettle, elderberry), whereas spaces that are waterproofed/sterilized by asphalt welcome saxicole flora such as *Sedum* sp. and *Saxifraga tridactylites*. Soils and environments disturbed by human occupation are also colonized by specific species. Nearly 50% of plant species listed are however, exotic! (Fig. 6)

Documents per year by source

Compare the document counts for up to 10 sources. Compare sources and view CiteScore, SJR, and SNIP data

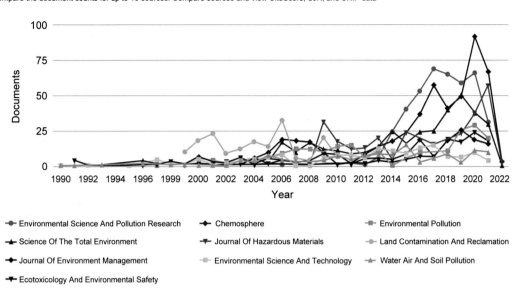

- Environmental Science And Pollution Research
- Chemosphere
- Environmental Pollution
- Science Of The Total Environment
- Journal Of Hazardous Materials
- Land Contamination And Reclamation
- Journal Of Environment Management
- Environmental Science And Technology
- Water Air And Soil Pollution
- Ecotoxicology And Environmental Safety

FIG. 4 The top 10 journals' names. *Source: Zheng, B., Masrabaye, F., 2023. Sustainable brownfield redevelopment and planning: bibliometric and visual analysis. Heliyon 9(2), e13280.*

Documents by country or territory

Compare the document counts for up to 15 countries/territories.

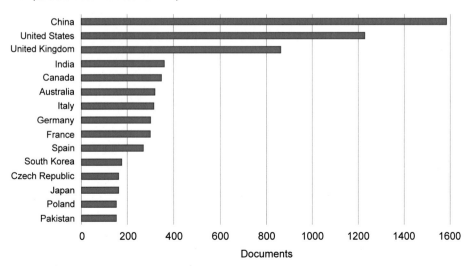

FIG. 5 The international academic cooperation of the 15 most productive countries. *Source: Zheng, B., Masrabaye, F., 2023. Sustainable brownfield redevelopment and planning: bibliometric and visual analysis. Heliyon 9(2), e13280.*

Documents by affiliation   Scopus

Compare the document counts for up to 15 affiliations.

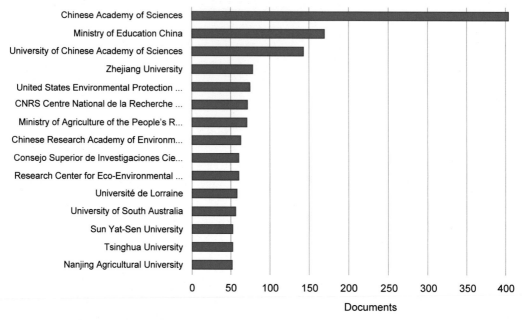

FIG. 6 The top 15 worldwide affiliations of publications related to research progress of sustainable brownfield re-development planning and design. *Source: Zheng, B., Masrabaye, F., 2023, Sustainable brownfield redevelopment and planning: bibliometric and visual analysis. Heliyon 9(2), e13280.*

In the French department of Hauts-de-Seine (Paris area), the flora of 98 wastelands was also studied by Muratet et al. (2007). A total of 365 vascular plants were identified, counting for 58% of taxa present in the department. On various sizes ranging from $12\,m^2$ to almost 2 ha, wastelands host an average of 39 species (from 5 to 92 species). The most observed species were (Common Mugwort (*Artemisia vulgaris*), Creeping Thistle (*Cirsium arvense*), Ribwort Plantain (*Plantago lanceolota*), then Hawkweed Oxtongue (*Picris hieracioides*), Common Nettle (*Urtica dioica*), Tansy (*Taraxacum campylodes*), Hedge Bindweed (*Calystegia sepium*), Broad-leaved Dock (*Rumex obstufolius*), and Butterfly-bush (*Buddleja davidii*). Among those, 109 species were found only once, and although none of them are protected except the Narrow-leaved Bitter-cress (*Cardamine impatiens*). Some are considered rare in this territory; these are Grass-poly (*Lythrum hyssopifolia*), Cypress Spurge (*Euphorbia cyparissias*) and White Mullein (*Verbascum lychnitis*). The exotic species rates vary between 0% and 46% depending on the location (Muratet et al., 2007). This study shows that the greatest diversity of species is found in relatively recent wastelands (4–13 years old), which welcome species from young environments, and in wastelands surrounded by parks and green spaces within a radius of 200 m

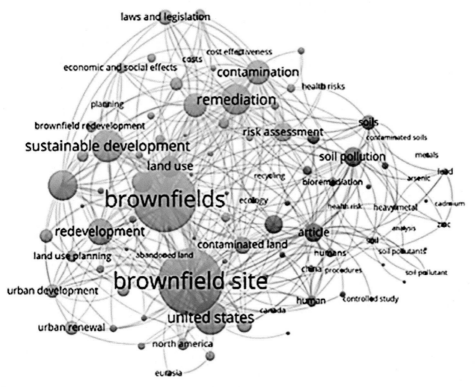

FIG. 7 Key areas of research (brownfield). *Source: Zheng, B., Masrabaye, F., 2023. Sustainable brownfield redevelopment and planning: bibliometric and visual analysis. Heliyon 9(2), e13280.*

(219 yards). Inventories made in 179 wastelands in the cities of Tours and Blois (in the Loire valley) counted 542 species, or 26% of the known flora in the French region of Centre. 382 species were counted in the agglomeration of Blois and 470 species in that of Tours. The most regularly observed species were Hawkweed Oxtongue (*P. hieracioides*), Wild Carrot or Bishop's Lace (*Daucus carota*) and Common Vetch (*Vicia sativa*). The inventories identified 36 protected species in the region, the most common being the Mouse-ear (*Cerastium dubium*). A species with national protected status, the Spring Pheasant's Eye (*Adonis vernalis*) was also observed on a single wasteland in Tours. 105 species are exotic (19.3% of the total) and among them 21 are considered invasive. The most common were Common Field-speedwell (*Veronica persica*), Annual Fleabane (*Erigeron annuus*), Compact Dock (*Rumex thyrsiflorus*), and Chinese Mugwort (*Artemisia verlotiorum*) (Fig. 7).

7 In the region of Hauts-de-France

In the north of France (the departments of Nord and Pas-de-Calais), similar findings are observed and show the richness of the abandoned spaces of our territory. Toussaint identified 301 plants on the 24 sites sampled along the major transport infrastructure of the Lille-Kortrijk-Tournai Eurometropolis. On these sites, few or no heritage species were found. The edges of the Eurometropolis infrastructures welcome an "ordinary" biodiversity, which does not prevent these sites, in an urban and agricultural intensive context, from playing a significant role in the ecological functioning of the territory. The sites that present the most biodiversity (13 sites) host about 55 species per site. Lemoine (2016) found 190 species in 6 ha of infrastructure borders in the same situation during "walking inventories" experiments. These are broad verges overwhelmingly in peri-urban situations.

On the side of the old former industrial spaces, the brownfield of La Lainière (Roubaix-Wattrelos, near Lille) hosts 167 species of plants over about 30 ha. The Saint-Sauveur wasteland in Lille *intramuros* situated on a former freight station of 23 ha hosts 159 species of which 23 are considered heritage. In addition to Prostrate Toadflax (*Linaria supina*) and the Bee Ophrys (*Ophrys apifera*), which are protected, we can encounter Savoy Hawkweed (*Hieracium sabaudum*), Brome of Madrid (*Bromus madritensis*), Hare Tail (*Lagurus ovatus*), Basil Thyme (*Acinos arvensis*), Proliferous Pink (*Petrorhagia prolifera*), Buck's-horn Plantain (*Plantago coronopus*), Roof Brome (*Bromus tectorum*), Silver Cinquefoil (*Potentilla argentea*), Dense-flowered Mullein (*Verbascum densiflora*), Mediterranean Hair-grass (*Rostraria cristata*), and Fern-grass (*Catapodium rigidium*). On the whole, the abundance of xerophytic-Mediterranean species is noticeable. The Butel and Season wasteland in Isques (department of Pas-de-Calais) hosts 151 plant species over 1.5 ha (Lemoine perso obs). Four of them are protected, Liquorice Milkvetch (*Astragalus glycyphyllos*) Marsh Helleborine (*Epipactis palustris*), Mahaleb cherry *Prunus mahaleb* and Large wintergreen (*Pyrola rotundifolia* var. arenaria). One hundred and eighty-six plant taxa were recorded on the Malinxoff wasteland (2–3 ha) in Saint Omer including three heritage species: Bee Ophrys (*Ophrys apifera*), Narrow-leaved Everlasting-Pea (*Lathyrus sylvestris*), and Grass Vetchling (*Lathyrus nissolia*) (Fig. 8).

FIG. 8   (A) Without intervention, urban wastelands are rapidly colonized by vegetation, La Lainière site in Roubaix and Wattrelos. (B) State of the site (and soil) of the former Saint-Liévin spinning mill (Wattrelos-59) after deconstruction of the factories.

Other significant "nature" spaces in the city—squares and public gardens, which have a great social role- also host an important part of urban biodiversity. They are maintained somewhat intensively or in a somewhat ecological way (differentiated management). However, their diversity is less than that of wastelands, but they welcome a procession of different species. Urban parks and vacant lots therefore appear to be complementary. Studies carried out in Seine-Saint-Denis show, for example, that Cultivated Vetch (*Vicia sativa*), Dioecious White Bryony (*Bryona dioica*), Parsnip and High Oats, found in 50% of wastelands, are absent from spaces maintained by humans. Conversely, parklands are home to *Oxalis corniculata*, Creeping Wood Sorrel, and Scarlet Pimpernel, commonly known as Blue-scarlet Pimpernel, and Red Chickweed (*Anagallis arvensis*); there are also Gallant Soldier (*Galinsoga parviflora*) and Spotted Ladysthumb (*Persicaria maculosa*); these species, found in 50% of parks, are absent from the 17 wastelands monitored by Muratet et al. (2007).

8 Flora potentially favorable to wild pollinators

The inventory of pollinating insects on the wild flowers of Seine-Saint-Denis made it possible to identify a wide range of species composed of Hymenoptera (40%), Diptera (40%) and, to a lesser extent, Coleoptera (10%), Hemiptera (5%), and Lepidoptera (4%). A photographic follow-up of the plants (Spipoll Protocol) showed that the honeybee represents only 6% of the Hymenoptera identified, while bumblebees and solitary bees (20 different genera) corresponded, respectively, to 28% and 54% of the photographed species. Another monitoring of pollinators on six wastelands, with sampling on two wastelands in each municipality (Saint-Denis, Villateneuse and Stains) in June 2010 made it possible to observe 160 species of pollinating insects on about 50 species of flowering plants. Hymenoptera (50%) prevail over Diptera (35%), Heteroptera, Coleoptera, and Lepidoptera (1% of observations). The honeybee only represents 3% of the total observations. In terms of Hymenoptera, the honeybee accounts for 6%, while bumblebees and solitary bees represent 22% and 60%. The plants most visited by pollinators are *Picris* (Oxtongue group), Wild Carrot, and Thistle (Figs. 9 and 10).

As part of the Wasteland Project, various follow-ups have, as we saw above, been performed. The identification of various pollinating insects and plants visited by them have made it possible to identify the nature of the interrelations between insects and plants and the 10 most attractive plants, as well as the 10 most common insects. The plants of the urban wastelands of Seine-Saint-Denis most visited by pollinating insects are Wild Carrot (*Daucus carota*), Little or Lesser Burdock (*Arctium minus*), Creeping Thistle (*Cirsium arvense*), Scentless False Mayweed (*Tripleurospermum inodorum*), Scentless Chamomile (*Matricaria inodorum*), Narrow-leaved or Cape Ragwort (*Senecio inaequidens*), Hawkweed Oxtongue (*Picris hieraciodes*), Butterfly-bush (*Buddleja davidii*), Hoary Mustard (*Hirschfeldia incana*), Wild Teasel (*Dipsacus fullonum*), and Broad-leaved Everlasting-Pea (*Lathyrus latifolius*). Forty percent of these plants are exotic: Cape Ragwort, Hoary Mustard, Butterfly-bush, and Broad-leaved Pea.

FIG. 9   (A) Lean meadow with *Rhinanthus crest-of-coq* established on the site of Saint-Liévin. (B) Messicole-based vegetation on the saint-liévin site. (C) Plot of Sainfoin grown on the Saint-Liévin site. (D) Plot of *Anthyllides vulneraires* on the site of Saint-Liévin.

FIG. 10   (A) Grassland mixture on the arc international wasteland in Arques. (B) Establishment of a flower meadow on the wasteland of arc international in Arques.

9 Dynamic spaces for insect communities

In wastelands and brownfield areas, especially xero-thermophilic ones, in Freiburg-im-Breisgau, biologists identified nearly a quarter of the bee species known in Germany. On a single species of Vergerette (*Erigeron* specie), they counted 64 different species of bees, syrphids, and butterflies. Parks and gardens connected to wastelands can also offer habitats (hunting and feeding areas) to vertebrate and invertebrate species from wastelands and thus form as many relay areas of an urban green network (Muratet et al., 2007).

The relevance that the alternation of demolitions and reconstructions can have in the dynamics of urban "habitats" and different plant successions is addressed by Bonthoux et al. (2014). It is clear to some authors that urban densification, which encourages a virtuous policy of recycling wastelands instead of urban sprawl, has a negative impact on urban biodiversity (Bonthoux et al., 2014). Analysis of the data provided as part of the participatory monitoring "Spipoll," shows that urbanization also homogenizes communities of pollinating insects in France. The artificialization of soils (urbanization) and pollution harm flower-dependent insects, which are facing a reduction in their food resources and breeding grounds.

Some authors (Bonthoux et al., 2014) showed that urban wastelands slow down this phenomenon and can provide habitats for rare or declining animal or plant species. Fischer et al. (2013) suggested and tested various techniques for restoring wild lands to create habitats favorable to the conservation of wild meadow plants, which are becoming increasingly rare in an intensive agricultural context, via reintroduction projects. It also involved imagining spaces that require weak management efforts. In another publication, Fischer et al. (2016) presented the value of good management of urban grasslands for maintaining wild bee communities. Bonthoux et al. (2014) describe the temporary role of young urban wastelands as habitats for many phytophagous insects. Meanwhile, found a constant number of Orthoptera and Leafhoppers in the various kinds of vegetation on the wasteland monitored. These communities, however, vary in composition according to the different stages of vegetation. The authors stress the importance of maintaining a mosaic of wasteland of different ages for the conservation of these species' communities.

The place of wastelands is all the more important when they are part of a mosaic of eco-landscapes of different shapes, surface areas, ages and structures (substrates, micro-climates, etc.). The conservation of urban biodiversity must therefore be designed in a dynamic approach and with the maintenance of pioneering spaces and be simultaneously more "enfallowed." Bonthoux et al. (2014) and Kattwinkel et al. (2011) recommend that the concept of biodiversity in motion, where series of many small connected spaces of different ages are more efficient than a few large, homogeneous areas lying as fallow. The goal of conserving urban biodiversity should not be limited to the preservation of emblematic species, but must, on the contrary, be oriented toward the preservation of stable and robust communities of common species.

10 Reflection on temporary biodiversity in activity zones

Kattwinkel et al. (2011), for their part, analyzed the biodiversity of an activity zone of 550 ha that is constantly changing near Bremen (Germany) as part of a project on "Biodiversity and

Temporary Construction." The richness of flora and insects (Orthoptera and Leafhoppers) was thus modeled and compared in the various plots, after an inventory of 133 sampling sites and a description of the ecological requirements for 38 plants and 43 insect species. In advance, the authors also recall that recycling is relevant to avoid urban sprawl, but this paradigm, however, is contrary to the set of objectives that aim to preserve urban biodiversity. The results of their study show that urban biodiversity is maximized when there is a dynamic and permanent renewal in the use of land plots between constructed spaces and deconstructed spaces. The highest level of biodiversity is observed when the space is occupied by 50%–60% of wastelands of about 15 years of age. The greatest floristic diversity is indeed reached at the age of 15 years, whereas concerning insects on wastelands, it is between 10 and 15 years old. It is also better to have a large number of wasteland plots of smaller size and different ages than large wastelands in smaller numbers.

It is also appropriate to understand the conservation of urban biodiversity at the scale of a territory and not at the scale of a single building development operation. The authors develop their analysis by proposing, in addition to space management, a concept of temporary and reversible constructions that easily allows for a return to pioneer stages after departure from the buildings. They also propose that the lifespan of buildings should be short (recyclable, reusable, easily removable, etc.), proposing this as an answer that could be developed by the companies that own the zone for better adaptation to an uncertain and ever-changing economy. The conservation of (unmanaged) wasteland nearby, which is not necessarily easily accepted by users, could be, for the authors, considered as one of the compensatory measures for easier construction of their reversible buildings.

11 Initiatives of the Etablissement Public Foncier Hauts-de-France: treating temporarily available wasteland through green space operations

As discussed, after EPF's works, these "anthropoosoils" (elevated rubble) made up of various soil conditions, nurture the establishment of spontaneous wildlife and vegetation with the arrival of afforestation, pervasive exotic species, allergenic species (birch, ragweed), ruderal species and even protected species that may adversely affect future housing projects or give a site or district a negative image. Faced with this unsatisfactory situation, over the past 7 years, EPF Hauts-de-France has carried out an almost systematic grassing-over of building demolition sites. It shows the areas with seeds adapted to the challenging requirements of the soil, and for which the mixture of seeds has been selected for its aesthetic look and for the low cost of upkeep of the vegetation. These moves favor "chosen biodiversity," which will not bring about any legal drawbacks in the long term. This moves allow urban areas that are awaiting housing developments to have a temporary lease of life that, in turn, improves the living environment.

To test the method, EPF carried out two pilot operations on a large scale. Two sites of more than 7 ha each were thus seeded in Spring 2015 in Arques (Pas-de-Calais), and in Summer 2015, in Wattrelos (Nord). In Arques, dry grassland was installed to benefit from the oligotrophy of the soil and the visible "pebble fields" (rubble from demolition). Thermophilious species were selected from among the range of regional flora, and were of certified regional origin (Ecosem firm, in Wallonia - Belgium). This led to the sowing of diversified flowering species, drought-resistant and thermophilious blends, wild grasses, and Bladder Campion (*Silene vulgaris*). Since the end of 2016, have been sheep grazing on part of the site. On the second site (Wattrelos), the plant blends implanted are based on more common and diversified Poaceae (*Festuca* sp., *Poa* sp., *Dactylis glomerata, Agrostis capillaris, Phleum pratense*) and Fabaceae (agricultural mixes in this case), in order to cover the ground quickly. The choice of plants on all the sites treated aimed to be conducive to invertebrates, and mainly to wild pollinators, both for feeding larvae and adults such as Lepidoptera, Orthoptera, Hymenoptera (Leafcutter Bee, Mason Bee, Bumblebee, etc.). Flowering/colored species with slow growth were selected to achieve the lowest management costs and obtain better social acceptance. For Fabaceae, the smallest species grown is the white clover, but in order to vary and limit the risks of difficulties or failure to grow given the wide diversity of anthroposoils on the site, other species were tested such as Common Bird's-foot Trefoil (*Lotus corniculatus*) and Red Clover (*Trifolium pratense*), as well as Common Kidneyvetch (*Anthyllis vulneraria*) and Common Sainfoin (*Onobrychis viciifolia*). The latter species were selected for their appeal to wild and domestic Hymenoptera (bees: *Anthophila*) and particularly long-tongue species (such as certain bumblebees and *Megachilidae*), which have become particularly threatened species in Europe and France following the gradual phasing out—since the end of the Second World War—of rotations conducted with legume peas that contributed nitrogen to fields.

On smaller surface areas (on the various sites of EPF), other species and mixes are also being tested to find the "perfect" mix offering the best compromise between planting cost and upkeep, growth, aesthetic, and ecological benefit.

12 Evaluation

All the herbaceous mixtures introduced are being assessed both for their suitability to the soil and their upkeep requirements, as well as the benefits for wild fauna (mainly invertebrates). Here, the aim is to find low-growing and dense coverage mixes that do not entail upkeep costs.

On many sites, occasional monitoring exercises have led to the observation of many species of *Bombus, Osmia, Anthidium,* and *Megachile* from the Apoidea family. As regards Lepidoptera, "planted" urban wasteland has become home to many *Argus, Colias,* swallowtails, common coppers, satyrid butterflies, etc. The *Anthidium punctatum, Anthidium strigatum, Osmia tridentata,* mallow skipper, six-belted clearwing, sickle-bearing bush-cricket, and many other species were also encountered on the EPF wastelands in the Lille area and Arques, whereas they had been previously unheard of in these parts of the region.

These seeding operations, which are easy to implement, happen on the majority of sites on which EPF works (with the exception of sites that are quickly redeveloped). These actions are inexpensive to carry out (€1000–€3000 per hectare) and are fully paid for by EPF. The overall cost of the operation as it has been run over the past 3 years can be estimated at between

€100,000 and €150,000. These actions have today been applied to several dozen hectares in total, drawing inspiration from small-scale initiatives that EPF carried out on certain wastelands 10–15 years ago, where, to restore environments (former quarry) and stabilize the slopes of slag heaps, herbaceous mixtures were introduced comprising Common Bird's-foot Trefoil (*L. corniculatus*), Common Kidneyvetch (*A. vulneraria*), and Purple Crown Vetch (*Securigera varia*), inadvertently triggering the arrival of new populations of butterflies such as, respectively, the Small Blue (*Cupido minumus*), Wood White (*Leptidea sinapis*), and Burnet Moth (*Zygaena ephialtes*). Given the high reactivity of populations of invertebrates (Hymenoptera, Lepidoptera and Orthoptera) in these biocide-free environments, EPF wished to incorporate this innovative approach into the framework of the national action plan "France, Terre de pollinisateurs" (France, Land of Pollinators) promoted by the French Ministry of Ecology, and in the cross-border project Interreg Flandre—Wallonie—Hauts-de-France SAPOLL (SAve our POLLinators) for which EPF is a technical partner. A vertebrate monitoring exercise also observed the presence over 3 years of a breeding pair of gray partridges, as well as European hares (*Lepus europaeus*) and European rabbits (*Oryctolagus cuniculus*) on the Wattrelos site.

13 Outlook: perceiving urban wasteland as spaces for dynamic biodiversity

Urban wastelands can be considered as a dynamic component of the urban ecosystem. Just like built-up areas, public spaces (squares) and green spaces, wastelands form a permanent part of the city. They differ, however, from other "urban objects" in so far as they are temporary by nature, as they are waiting to be reallocated to another purpose and reused. They are not always in the same place and "move around" as urban transformations unfold. The place of wastelands in urban spaces can thus be compared to the place of clearings in a primary forest, which regularly form due to natural disruption (a windfall caused by storms or senescence of centuries-old trees) and progressively disappear with the dynamics of replacement growth.

Planted wasteland, for the plentiful communities of invertebrates, could thus disappear with a low ecological impact if the region has other parts of wasteland nearby that might host part of the species that have developed there, and that can take advantage of this temporarily available land that is "provided to them." Wasteland, while its vocation is to promote temporary diversity (3 to 6 years) thereby contributes to the preservation of nature in towns and cities, to the improvement of living conditions, and forms links in the urban green chain in the form of "stepping stones" before they disappear as they are reallocated. These temporary reservoirs of biodiversity (or permanent ones if the planned project ultimately does not happen) should be complementary to green spaces (parks and gardens) if they are well maintained, and to other derelict land on which no work and management is planned. Wasteland left to its own natural devices host plants and animal varieties that are complementary to those sown by EPF. Comparative inventories are currently underway. For best success, the city must connect all these plots with other green spaces. Every urban green space throughout the city should be designed to form a network of interconnected spaces to better support biodiversity (green network).

FIG. 11   (A) Dry grassland with swollen campion and vulnerary warbler on the Arc International site Légende photos. (B) Viper's Bugloss Mason Bee (Arques).

14 Greened-over wastelands: a contribution to sustainable cities

EPF's sites that are awaiting an urban project can therefore contribute to the notion of a sustainable or even desirable city by providing a range of functions and contributing to its resilience, such as reinforcing and preserving biodiversity (ordinary biodiversity, not protected by law), and toward mitigating climate change. The introduction of plants onto urban wasteland (highly mineralized spaces), in addition to preventing any dust from blowing around, can additionally help fight drought in towns and prevent heat islands from forming, while also providing a higher-quality living environment to the neighbors of these former factories and blocks of buildings that are now derelict. Surveys about attitudes to wasteland have concluded that residents show a preference for grassy and colorful spaces (with a lot of dicotyledons). These operations also help to show inhabitants, visitors, and developers the benefits of dynamic spaces and a lively neighborhood that the region hopes to make appealing for possible investors. The introduction of a plant covering using Poaceae and Fabaceae also goes toward treating any residual organic pollution (phytoremediation) (Fig. 11).

15 Conclusion

The approach developed and the actions carried out by EPF are inspired by a vision of ecosystems as well as projects and concepts of "temporary nature." EPF's action responds to multiple objectives: acceptance of populations, aesthetic, economic (territorial attractiveness and reduced management costs), climatic, and naturalistic. The "Promotion/Support of Biodiversity" approach achieved, contrary to European approaches presented and described, is an approach aimed at supporting and strengthening a "chosen" biodiversity, with the prior determination of target species, including wild pollinators. The choice for this type of biodiversity is based, on one hand, on the edaphoclimatic potentialities of the urban anthroposoils in place, characterized by the strong constraints imposed by the substrate characteristic of reclassified wastelands (rubble, crushed concrete) that helps foster invertebrate and even vertebrate communities typical of open spaces, like xerothermic lawns and mesotrophic grasslands. These are backgrounds often considered rare in the regional territory, and often absent in an urban context. On the other hand, the actions put in place aim to favor target species previously determined for their often unfavorable conservation status (Hymenoptera, Lepidoptera, and Orthoptera). The flowering meadows achieved (that are richly colored—see photos), spaces identified as major challenges by many authors (Bonthoux et al., 2014) also correspond to the vegetation of wastelands that are best accepted or tolerated by residents and neighbors. Indeed, the overview of a study currently being published on the perception of wasteland by inhabitants shows that abandoned spaces with meadow-type vegetation structures are better accepted and even better used than those with a lot of bare soil or that are covered with shrubs, Grasslands created *ex nihilo*, by adapting the composition of seeded mixtures, could also be used for (temporary) conservation of some grassland plants also in an unfavorable conservation situation, as proposed by Fischer et al. (2013).

There are many benefits of generalized and temporary greening. Introducing plants in derelict urban wastelands fulfills a range of complementary objectives (holistic approach), such as the improvement of living conditions, reinforcement of biodiversity, combating dust flight and urban heat islands, the fight against invasive exotic species, the possible reduction of residual organic pollution and the promotion of regional appeal, and show that consideration for biodiversity can "target" species that are not legally protected (Lepidoptera, Orthoptera, Hymenoptera, etc.); they also show that actions that might promote "temporary biodiversity" and species that are somewhat common on temporarily available land deserve to be conducted and encouraged. These actions also contribute to the resilience of ecosystems and populations of invertebrates in towns and contribute to the urban green network ("stepping stones" approach), as well as helping wild pollinators, which can also be done in urban spaces.

These operations are much appreciated by the neighbors and inhabitants of these towns and villages for their contribution to improving the living conditions and image of the district. In addition to the benefits to biodiversity, these actions also offer the advantage of avoiding costs. The absence of invasive and even allergenic exotic species can also be cited as a benefit of these almost systematic plant introduction operations. This approach has also been studied from the perspective of an emerging request, that of developing new, so-called transitory, uses on sites that are not quickly reallocated. The development of transitory uses aims to restore social, landscape, economic or ecological value to sites that are often located in town centers (Costa et al., 2018).

The initiatives conducted have today inspired other organizations: EPFs from other regions (EPF of the department of Savoie, for example); semipublic companies such as SEM VR, which develops the same approaches on the "l'Union" site (Tourcoing); a business park managing body; the town of Dunkirk; etc. The EPF's moves are also completely in keeping with the concept of the "sustainable city" favoring the quest to clean up the soil, fight against zones of urban warmth and reduce airborne dust. This innovative measure is one of the best experiments of ecological engineering identified by the French Ministry of Ecologic and Solidarity Transition and by the French Biodiversity Agency (AFB [Agence Française pour la Biodiversité]). It received an award for ecological engineering in the category of "developing public or private areas" in November 2018.

References

Alvernia, P., Soesilo, T.E.B., 2019. Phytoremediation as a sustainable way for land rehabilitation of heavy metal contamination. J. Phys. Conf. Ser. 1381, 012062.

Ansari, A.A., Gill, S.S., Gill, R., Lanza, G.R., Newman, L., 2016. Phytoremediation: Management of Environmental Contaminants. vol. 1 Springer. ISBN-13: 978-3319346342.

Bonthoux, S., Brun, M., Di Pietro, F., Greulich, S., 2014. How can wastelands promote biodiversity in cities? A review. Landsc. Urban Plan. 132 (2014), 79–88.

Bostenaru Dan, M., Bostenaru-Dan, M.M., 2021. Greening the brownfields of thermal power plants in rural areas, an example from Romania, set in the context of developments in the industrialized country of Germany. Sustainability 13, 3800. https://doi.org/10.3390/su13073800.

Costa, S.O., Kellecioglu, I., Weber, R., 2018. Developing Brownfields via Public-Private-People Partnerships Lessons Learned From Baltic Urban Lab. Nordregio Report.

Farfán-Beltrán, M.E., Chávez-Pesqueira, M., Hernández-Cumplido, J., Cano-Santana, Z., 2022. A quick evaluation of ecological restoration based on arthropod communities and trophic guilds in an urban ecological preserve in Mexico City. Rev. Chil. Hist. Nat. 95.

Figueroa, F., Puente-Uribe, M.B., Arteaga-Ledesma, D., Espinosa-García, A.C., Tapia-Palacios, M.A., Silva-Magaña, M.A., Mazari-Hiriart, M., Arroyo-Lambaer, D., Revollo-Fernández, D., Sumano, C., Rivas, M.I., 2022. Integrating agroecological food production, ecological restoration, peasants' wellbeing, and agri-food biocultural heritage in Xochimilco, Mexico City. Sustainability 14 (15), 9641.

Fischer, L.-K., Eichfeld, J., Korwarik, I., Buchholz, S., 2016. Disentangling urban habitat and matrix effects on wild bee species. PeerJ 4, e2729. https://doi.org/10.7717/peerj.2729. p. 19.

Fischer, L.-K., Von Der Lippe, M., Rillig, M.C., Korwarik, I., 2013. Creating novel urban grasslands by reintroducing native species in wasteland vegetation. Biol. Conserv. 59 (2013), 119–126.

Gao, X., Han, G., Liu, J., Zhang, S., Zeng, J., 2023. Heavy metal accumulation and source apportionment in Urban River under ecological restoration: relationships with land use and risk assessment based on Monte Carlo simulation. ACS Earth Space Chem. 7 (3), 642–652.

Hu, X., Xu, W., Li, F., 2022. Spatiotemporal evolution and optimization of landscape patterns based on the ecological restoration of territorial space. Land 11 (12), 2114.

Kattwinkel, M., Biedermann, R., Kleyer, M., 2011. Temporary conservation for urban biodiversity. Biol. Conserv. 144, 2335–2343.

Ke, S., Pan, H., Jin, B., 2023. Identification of priority areas for ecological restoration based on human disturbance and ecological security patterns: a case study of Fuzhou City, China. Sustainability 15 (3), 2842.

Klaus, V.H., Kiehl, K., 2021. A conceptual framework for urban ecological restoration and rehabilitation. Basic Appl. Ecol. 52, 82–94.

Kollmann, J., Kirmer, A., Tischew, S., Hölzel, N., Kiehl, K., Kollmann, J., 2019. Neuartige Ökosysteme und invasive Neobiota. Renaturierungsökologie, 435–447.

Leichang, H., Li, Y., Shanhua, B., Yilu, G., Yu, Z., Xiaoyu, J., Xu, F., 2022. Study on the construction of residual plaque landscape ecological restoration model in the process of rapid urbanization. Nat. Environ. Pollut. Technol. 21 (2), 481–486.

Lemoine, G., 2016. Brownfield restoration as a smart economic growth option for promoting ecotourism, biodiversity, and leisure: two case studies in Nord-Pas De Calais. In: Bio and Bio. Elsevier, pp. 361–388.

Li, H., Pang, S., Zhu, C., Li, Y., Guo, Y., Lei, T., Wang, K., 2022b. Strategies of GIS-Based Urban Country Park Planning and Ecological Restoration. Discrete Dynamics in Nature and Society.

Li, H., Zhang, T., Cao, X.S., Zhang, Q.Q., 2022a. Establishing and optimizing the ecological security pattern in Shaanxi Province (China) for ecological restoration of land space. Forests 13 (5), 766.

Li, Q., Zhou, Y., Yi, S., 2022c. An integrated approach to constructing ecological security patterns and identifying ecological restoration and protection areas: a case study of Jingmen, China. Ecol. Indicators 137, 108723.

Liu, Z., Bai, G., Liu, Y., Zou, Y., Ding, Z., Wang, R., Chen, D., Kong, L., Wang, C., Liu, L., Liu, B., 2022. Long-term study of ecological restoration in a typical shallow urban lake. Sci. Total Environ. 846, 157505.

Lv, L., Zhang, S., Zhu, J., Wang, Z., Wang, Z., Li, G., Yang, C., 2022. Ecological restoration strategies for mountainous cities based on ecological security patterns and circuit theory: a case of central urban areas in Chongqing, China. Int. J. Environ. Res. Public Health 19 (24), 16505.

Măgureanu, I.N., Sîrbu, C., 2022. Urban ecological restoration in a post-industrial context: the contribution of ecosystem services in the process of revitalizing small mono-industrial cities. In: Present Environment and Sustainable Development, pp. 81–82.

Muratet, A., Machon, N., Jiguet, F., Moret, J., Porcher, E., 2007. The role of urban structures in the distribution of wasteland flora in the greater Paris area, France. Ecosystems 10, 661–671.

Ran, Y., Lei, D., Li, J., Gao, L., Mo, J., Liu, X., 2022. Identification of crucial areas of territorial ecological restoration based on ecological security pattern: a case study of the Central Yunnan urban agglomeration, China. Ecol. Indic. 143, 109318.

Wang, Z., Liu, Y., Xie, X., Wang, X., Lin, H., Xie, H., Liu, X., 2022. Identifying key areas of green space for ecological restoration based on ecological security patterns in Fujian Province, China. Land 11 (9), 1496.

Xiaojian, W., Yuewen, Y., 2022. Thoughts on the construction method of stock space based on the perspective of ecological restoration and urban repair. J. Landscape Res. 14 (3), 17–23.

Xu, J., Xu, D., Qu, C., 2022. Construction of ecological security pattern and identification of ecological restoration zones in the City of Changchun, China. Int. J. Environ. Res. Public Health 20 (1), 289.

Yue, A., Mao, C., Zhao, S., 2022. Smart governance of urban ecological environment driven by digital twin technology: A case study on the ecological restoration and management in S island of Chongqing. In: IOP Conference Series: Earth and Environmental Science. vol. 1101. IOP Publishing, p. 072003. No. 7.

Zhang, X., Liu, K., Wang, S., Wu, T., Li, X., Wang, J., Wang, D., Zhu, H., Tan, C., Ji, Y., 2022a. Spatiotemporal evolution of ecological vulnerability in the Yellow River Basin under ecological restoration initiatives. Ecol. Indic. 135, 108586.

Zhang, J., Ma, J., Zhang, Z., He, B., Zhang, Y., Su, L., Wang, B., Shao, J., Tai, Y., Zhang, X., Huang, H., 2022b. Initial ecological restoration assessment of an urban river in the subtropical region in China. Sci. Total Environ. 838, 156156.

Zheng, B., Masrabaye, F., 2023. Sustainable brownfield redevelopment and planning: bibliometric and visual analysis. Heliyon 9 (2), e13280.

Biological recultivation of fly ash dumps strengthening bioeconomy and circular economy in the ural region of Russia

Natalia V. Lukina[a], Elena I. Filimonova[a], Margarita A. Glazyrina[a], Maria G. Maleva[a], Majeti Narasimha Vara Prasad[b], and Tamara S. Chibrik[†]

[a]Ural Federal University Named After First President of Russia B.N. Yeltsin, Yekaterinburg, Russian Federation [b]Department of Plant Sciences, School of Life Sciences, University of Hyderabad (An Institution of Eminence), Hyderabad, Telangana, India

1 Introduction

Fly ash generation, management, treatment, and disposal are all considered to be major solid waste issues worldwide. Fly ash is a byproduct of coal combustion found in thermal power stations using coal or lignite (Coal and Lignite Burning Thermal Power Stations = C/LBTPS). It is anticipated that in the coming 20 years, there will be 50% increase in the world's energy demand. The most prevalent fossil fuel on earth is coal which is relatively inexpensive. China, India, Russia, the United States, and some other countries have the largest deposits of high-grade coal.

[†]Deceased.

To meet the growing energy demand, the dependency on coal for power generation and disposal of fly ash will continue to increase. Due to the physical characteristics and sheer volumes generated, serious problems are being faced with fly ash in several instances:

(1) Due to intensive disposal, fly ash particles, both as dry ash and as pond ash, occupy many hectares of land near power plants.
(2) Fly ash is extremely difficult to handle in its dry state due to its fineness, and flying fine ash particles erode structural surfaces and have an impact on horticulture.
(3) It also disrupts the ecology through soil, air, and water pollution.
(4) Prolonged exposure to fly ash results in a number of dangerous illnesses, including silicosis, lung fibrosis, bronchitis, and pneumonitis.
(5) Toxic trace elements, such as Sb, As, Be, Cd, Pb, Hg, Se, and V, are drawn to the surface of fly ash particles due to the iron and aluminum oxides present there. These elements are observed to be heavily concentrated there.

Cleanup of Industrial deserts or lunarscapes are large areas overloaded with technogenic waste is cost prohibitive. One emerging approach to this problem is biological recultivation (Bolshakov and Chibrik, 2007). It has been satisfactorily implemented in various countries by selecting and growing perennial grasses, trees, and bushy plant species Economic and ecological aspects of land development options are important in the field of bioeconomy (Nesgovorova et al., 2019; Dubois et al., 2020; Sonwani et al., 2022).

On a site-specific basis, there are two types of principal restoration options: Economic and ecological aspects of land development options are important in the field of bioeconomy (Fantinato et al., 2023). The principal restoration options on a site-specific basis are of two types:

- *ameliorative*: improvement of the physical and chemical nature of the site;
- *adaptive*: ecological restoration through establishing ecosystem structure and function and thus biodiversity.

Ecological restoration usually depends on careful selection of suitable substrates for plant growth (Miguel et al., 2023). Species that would provide wildlife habitat (and forage for domestic animals) and improve esthetics are generally preferred. However, native species that are available as propagules often do not satisfy the above criteria. In this situation, a rapid solution to problems can be addressed by selecting species that enable colonization, facilitating succession, and restoration of the native ecosystem. *Agrostis capillaris* L., *Festuca rubra* L., and *Calamagrostis epigejos* (L.) Roth are examples. These grasses have a proven reclamation function on a variety of industrial waste-contaminated soils (Březina et al., 2006; Mitrović et al., 2008; Dulya et al., 2013; Sierka and Kopczyńska, 2014; Radochinskaya et al., 2019). The revegetated contaminated area must meet two basic objectives: forage and habitat for livestock and wildlife.

There is not a single best method in all circumstances for any reclamation operation. The procedures and techniques described in this chapter and in chapter by Lemoine (2016) (Brownfield restoration as a smart economic growth option for promoting ecotourism and leisure: Two case studies in Nord-Pas De Calais) are successful examples.

1.1 Fly ash in a circular economy—Challenges and barriers

There is a difference between stabilizing ash to make it "non-hazardous," processing ash for ease of use to help extract salts, minerals, and metals, and its disposal in a landfill (Jayapal et al., 2023; Lee et al., 2023; Ren et al., 2023). There are also differences between different beneficial species. The latter is more in line with the dynamics guiding the circular economy, and some technology opportunities to support this are already beginning to emerge.

However, there are problems with the effectiveness and economics of these approaches. In particular, these may be complicated by size, variations in ash compositions brought on by feedstock, differences in processes, and the nascent premium market for green or recycled raw materials.

This chapter on waste management makes clear that fly ash valorization is a hot topic, especially in light of the ways in which the circular economy is slowly but surely entering all aspects of society. Several approaches are (or will be) excellent candidates for lowering the amount of hazardous waste dumped in landfills and raising the value and/or reutilization of WtE fly ash. Although this is the case, landfilling is still widely approved by the public and the government in many countries, especially if local opposition to waste processing plants continues to grow in significance. This acceptance is reflected in the fact that many places have cheap landfilling costs.

There is a clear desire to add new technologies to the combustion-based energy-from-waste sector so that it can participate and thrive as the circular economy concepts are implemented (Ashraf et al., 2023; Makhathini et al., 2023; Sapsford et al., 2023). The availability of a range of fly ash valorization technologies on the market in the future years will be helpful for WtE plant owners and operators as well as society at large. The competition and complementarity between the various solutions will allow the WtE sector to contribute to a more sustainable Circular Economy to the greatest extent possible.

The Ural is one of the oldest mining regions of Russia (Fig. 1). It is also one of the largest territories of mineral resources in Russia. This has led to intensive development of metallurgy, construction, chemical industry, and mining operations, including gold mining (Khokhryakov, 2003; Koptsik, 2014). These lands have lost their economic value due to mining and mineral exploration. The mining industry areas have had a harmful impact on the environment because of soil damage and changes in hydrological conditions (Brooks et al., 2005; Belskii and Belskaya, 2009; Belskaya and Vorobeichik, 2013). The environmental impacts of mining-related activities in and around the Ural region including Karabash are extremely severe (Williamson et al., 2004a, b, 2008). The area has been affected by gaseous and particulate emissions from a copper smelter, acid drainage from abandoned mine workings, and leachates and dust from waste dumps (Chukanov et al., 1993; Udachin et al., 2003; Spiro et al., 2004, 2012; Williamson et al., 2008).

Global economy growth is increasing demand for electricity. In many countries, TPS operates on various types of low-quality coal with high ash content (Yao et al., 2015). Waste after coal burning in the form of pulp is thrown through pipelines into natural or artificially created pits, occupying hundreds of hectares in area. Drying up, they turn into dusty spaces, polluting air, soil, and water bodies (Cheung et al., 2000; Haynes, 2009). Unfavorable physicochemical properties of ash, its phytotoxicity, lack of mineral nutrition elements in a form accessible to plants, the absence or extremely low content of free-living and symbiotic

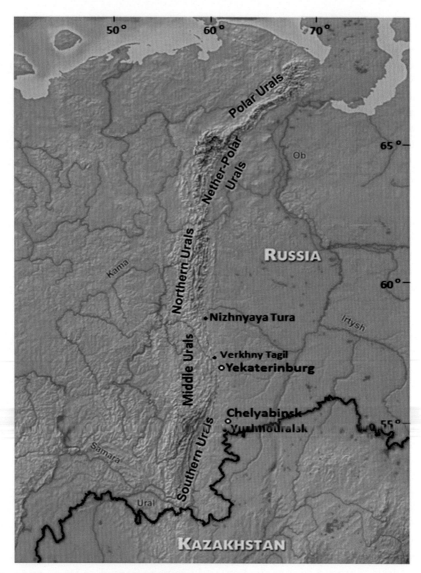

FIG. 1   The Ural is one of the oldest mining regions in Russia.

nitrogen-fixing microorganisms make ash an extremely unfavorable substrate for plants to grow on it (Prasad, 2006; Maiti, 2013). In this regard, a problem arose with the development of cheap and most effective methods of ash dumps biological recultivation.

The most effective way to eliminate the pollution effects of ash dumps is currently by planting them with greenery, which is, creating sanitary and hygienic vegetation covered with partial economic use (hayfields). For this, biological recultivation is carried out at fly ash dumps: a covering with a layer of fertile soil, peat, silt of treatment facilities, etc. with further perennial grasses seeding. The choice of material for coating is primarily determined by the goals of biological recultivation, as well as its availability, and the selection of a range of grasses largely depends

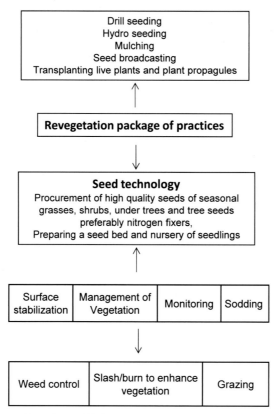

FIG. 2   The generalized procedures for establishing and revegetating mine industry waste-contaminated sites.

on the climatic conditions of the area and substrate characteristics (Fig. 2). In this chapter, some results of biological recultivation of various fly ash dumps located in the Urals (Russia) in various zonal and climatic conditions (taiga zone and forest-steppe zone) are presented.

Key components of biological recultivation of an area include:

- structure and functioning of technogenic ecosystems;
- monitoring of the ecological situation of the mine industry-ravaged lands;
- development of biological recultivation;
- floristic composition, which indicates conditions of the environment;
- structure of the plant populations (finding the dominant species);
- dynamics and structure of populations (mycorrhiza show the community preparedness);
- productivity, chemical compounds, and the quality of phytoproducts;
- transformation of fly ash dumps to phytocoenose that are sterile substrata.

2 The purpose and methods of research

The aim of this work was to assess the success of biological recultivation and transformation of plant communities on fly ash dumps of Verkhnetagilskaya and Nizhneturinskaya thermal power stations (NTTPS) located in the Sverdlovsk region (eastern slope of the Middle Urals, taiga zone, sub-zone of the southern taiga), as well as on Yuzhnouralskaya thermal power station (YUTPS), located in the Chelyabinsk region (eastern slope of the Southern Urals, forest-steppe zone).

The Ural Mountains are a natural border between Europe and Asia. They are conditionally divided into five regions: the Polar Urals, the Subpolar Urals, the Northern Urals, the Middle Urals, and the Southern Urals (Fig. 1). The length of the Ural Mountains from north to south is more than 2800 km, the width, together with the foothill plains, varies from 110 to 120 km in the northern part (Polar Urals) to 400 km in the south (Southern Urals) (Shakirov, 2011).

The diversity of the Urals climate and the adjacent plains is determined by the large extent of the Ural Mountains from north to south in the meridional direction, as well as by significant absolute and relative heights of the surface. In this regard, the northern and southern parts of the Urals are in unequal radiation and circulation conditions, causing a fairly clear division of the territory into climatic zones.

The Verkhnetagilskaya thermal power station (VTTPS) fly ash dump is located in the southwestern part of the Sverdlovsk region on the eastern slope of the Middle Urals (57°20′43″ N 59°56′20″ E; altitude: 276 m). The climate of the district is temperate continental. Compared to the Turin region, the climate here is warmer and much drier. The average annual rainfall is 628 mm. During the warm period from April to October, 436 mm falls. The average annual air temperature is +1.0°C, the average July temperature is +17.2°C, and the average January temperature is −16.0°C. Soils of medium podzolic degree prevail in the region. Coarse gravelly underdeveloped soils are widely distributed. The area is located in the taiga zone, in the southern taiga sub-zone. The area is dominated by pine forests.

The NTTPS fly ash dump is located near the town of Nizhnyaya Tura, 220 km north of Yekaterinburg (58°37′22″ N, 59°52′17″ E; altitude: 194 m). The climate of the Turin district is temperate continental; it is characterized as cool in terms of heat supply, and as humid in terms of moisture. The average annual air temperature is +0.7°C, the average July temperature is +17.7°C, the average January temperature is −16.7°C. The average annual rainfall is 673 mm. During the warm period from April to October, 489 mm falls. The soil cover of the Turin district is represented by podzolic soils, among which there are small islands of marsh soils (Shakirov, 2011). The area is located in the taiga zone, at the transition of the middle taiga to the southern taiga. Coniferous forests (*Picea obovata* Ledeb., *Abies sibirica* Ledeb., *Pinus sibirica* Du Tour, *Pinus sylvestris* L., etc.) with small-leaved (*Betula pendula* Roth and *Betula pubescens* Ehrh.) species prevail here.

The YUTPS fly ash dumps are located in the Chelyabinsk region on the eastern slope of the Southern Urals (southern forest-steppe) 90 km south of the city of Chelyabinsk (54°26′53″ N, 61°13′56″ E; altitude: 211 m). The climate in the area is temperate continental. The area is not sufficiently humid in terms of moisture degree. Winters here are cold and long. There are often droughts in summer. The average annual air temperature is +1.2°C, the average July temperature is +18.6°C, and the average January temperature is −16.6°C. The average annual rainfall is 441 mm. During the warm period from April to October, 312 mm falls (Shakirov, 2011). The soil cover is represented by zonal podzolized, leached, and typical chernozems. In the vegetation cover, birch and aspen-birch groves and island pine forests are combined with meadows, real grass-forb, and petrophyte steppes, as well as floodplain and steppe meadows (Kulikov, 2005).

Fly ash dumps are formed from the thermal power stations working on high-ash content coals. Structurally these fly ash dumps include the presence of slag and are characterized by different-sized particles. The thickness of the fly ash layer varies from 2 to 20 m. Ashes are light or dark gray in color with black inclusions of the not burned coal particles. The fine-grained ash factions contain coarse particles of physical sand (1–0.05 mm) and dust (0.05–0.001 mm).

Granulometric ash substratum structure (accounting in % of the air to dry probe) is shown in Table 1, and the chemical structure of the fly ash substratum is shown in Table 2. The analytical chemistry of water extract of ash substratum is shown in Table 3. The total chemical composition of the ash dump substratum corresponds to aluminosilicate formations (SiO_2, 40.5%–60.3%; Al_2O_3, 12.9%–32.4%). Fly ash contains practically no nitrogen or organic substance. The particles of the not burned coal (potential humus) are connected with the silicate of the ashes and are subjected to slow physicochemical and biochemical transformations that play a crucial role in the restoration of substratum fertility. The content of mobile phosphorus changes from 1.25 up to 32.5 P_2O_5 mg/100 g of ashes; exchangeable potassium changes from 1.6 up to 25.0 mg/100 g of fly ash.

Coal ashes have a small absorption capacity, similar to the structure of soils, because of the low content of highly dispersed organic substances and silt particles.

The absorption ability of the external layers of fly ash dumps can be increased by covering their surface with peat and soil, as well as by introducing mineral fertilizers and neutralized wastewater. The pH of ashes changes from low acid (pH = 5.9) up to alkaline (pH = 8.5), but in the ranges suitable for the growth and development of plants. The ash dump's substratum contains a large macro- and microelements spectrum.

In a temperature mode, the coal ashes belong to a low thermally conductive substratum with sharp oscillation frequencies of temperature from surface to depth.

The water-physical properties of the substratum of ashes are rather peculiar. The coal ashes have a good porosity and good air and water penetrability, from deep to middle at the level of sandy loam and sandy soils. The unfavorable conditions arise during the germination of seeds and in the first period of plant life. At this time, the roots are located in a stratum of 0–10 cm, which is subjected to rapid drying.

Thus, on water-physical properties and chemical structure, the coal ashes can be related to substratum suitable for the existence of plants. However, for the creation of a long, productive cultural phytocoenosis, the agrotechnical measures for improving the properties of coal ashes as a substratum for the cultivation of plants are necessary.

3 Phytocoenosis formation on fly ash dumps

3.1 Verkhnetagilskaya thermal power station (VTTPS) fly ash dump

Biological recultivation on the VTTPS fly ash dump started in 1968–70 (that is 3 years after the end of ash feeding) and continued in subsequent years. Biological recultivation quickly restored conservation on fly ash dump and controlled wind erosion of the substrate. The end result was the creation of vegetation with economic importance. Clay soils, ranging in thickness of 10–15 cm width, were laid to regulate erosion. Most of the bands were sown

TABLE 1 Granulometric ash substratum structure (account in % on is air—dry probe).

The name of ash dumps	Hygroscopical moisture	Sand (diameter, mm)		Amount of particles Dust (diameter, mm)			Silt	The sum of fractions		Substratum mechanical structure
		Average 1–0.25	Small-sized 0.25–0.05	Large 0.05–0.01	Average 0.01–0.005	Small-sized 0.005–0.001	<0.001	Physical sand (<0.01)	Physical clay (>0.01)	
Verkhnetagilskaya thermal power station fly ash dump (forest zone)	–	0.53	4.23	56.08	5.45	9.33	4.91	60.84	19.69	Soup
Nizhneturinskaya thermal power station fly ash dump (forest zone)	2.04	6.15	26.84	23.59	1.78	4.67	6.54	56.58	12.99	Soup
Yuzhnouralskaya thermal power station fly ash dumps (forest-steppe zone)	0.60	27.70	41.26	6.74	3.60	2.30	3.68	75.84	9.58	Sand

TABLE 2 Ash substratum chemical structure.

The name of ash dumps	Loss at calcinations, %	The total contents of basic elements (% in calcinated probe)										The contents of mobile Elements			
		SiO_2	Al_2O_3	Fe_2O_3	CaO	MgO	MnO	P_2O_5	SO_3	K_2O	Na_2O	Nitrogen, %	P_2O_5 mg/100g ashes	K_2O	pH (KCl)
Verkhnetagilskaya thermal power station fly ash dump (forest zone)	2.4	48.4	23.4	14.2	4.9	2.9	–	–	3.8	–	–	Trace	23.5	7.0	8.5
Nizhneturinskaya thermal power station fly ash dump (forest zone)	40.5	32.4	5.5	7.8	–	40.5	–	–	0.4	–	–	0–0.0013	25.0–32.5	8.0–10.1	7.6–8.1
Yuzhnouralskaya thermal power station fly ash dumps (forest-steppe zone)	58.3	31.4	7.2	2.0	0.3	58.3	–	–	0.7	–	–	0.08	2.7	1.6	8.0

TABLE 3   The analysis of the water extract of ash substratum.

The name of ash dump	Dense residual, %	The contents (% from the absolute dry probe)					
		CO_3^{-2}	HCO_3^{-1}	Cl^{-1}	SO_4^{-2}	Ca^{+2}	Mg^{+2}
Forest zone	0.200	Not present	0.023	0.004	0.118	0.034	Not present

by perennial grasses [*Agropyron pectinatum* (Bieb.) Beauv., *Bromus inermis* Leyss., *F. rubra, Medicago media* Pers., *Onobrychis arenaria* (Kit.) DC., etc.]. Part of the territory was left under the self-overgrowing. As a result of this, a diverse range of ecotopes were formed. The study of the formation of vegetation on the ash dump was carried out according to conventional techniques for 30 years, starting with the 10 years old (1980) to the age of 40 plant communities (Fig. 3).

The fly ashes were released by the hydraulic method via pipelines. The recultivation using the strips of ground was conducted on part of ash dump after the termination of the release of ashes. The ground was seeded by perennial grass.

Twenty years after the recultivation work on the ash dumps, a rather diverse ecotope spectrum was observed. This caused the formation of unique biotope and vegetative communities. The recultivation actions influenced this process.

It is possible to describe the initial ecotopes with the following scheme:

I. Non-recultivated area
Ia. Initial ecotope: dry ash dump, "pure" ashes
Ib. Moderate moistering, "pure" ashes, and favorable conditions for drift seeds;
Ic. Cesidual depressions periodically flooded by water (thawed waters, filtration from ash dumps, etc.);

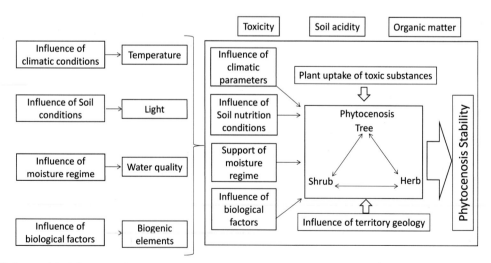

FIG. 3   Model of the external environmental factors impact on the mechanisms of phytocenosis stability.

II. The area with the strips of a soil deposited during the primary recultivation:
IIa. Ashes with deposition of a ground and crop of perennial grasses;
IIb. Ashes with the placement of ground and without any seeding;
IIc. Space between bands of substratum
III. Second phase of recultivation:
▪ bushes sawing, full covering by the stratum of peat.

The scheme of phytocoenosis formation on fly ash dump depending on the ecotope is shown in Table 4.

Concerning phytocenosis formation, the following is understood: development of a vegetative grouping takes place from a stage of a settlement of separation to its grouping with a certain density irrespective of the phytocoenosis dynamic status.

In the formation of communities on industrially disturbed lands with the rather mentioned process of self-overgrowing, the following stages are selected: ecotopic groups (projective cover 0.1%); simple groups (0.1%–5%); complicated groups (6%–50%); and phytocoenosis (projective cover more than 50%).

The scheme of phytocoenosis formation on ash dumps, depending on ecotope was constructed on the basis of actual dates of geobotanical descriptions, which were done on selected ecotopes. Through the generalization of given and similar material from other ash dumps, the creation of a generalized model of phytocenosis formation on ash dumps became possible. This model was used for the phytocoenosises located in different zones and climatic conditions. Phytocoenosis is considered to be a main component of industrial ecosystems that were formed from conditions of thermal power station ash dumps (Fig. 4).

Ten years after biological reclamation in dry areas, the "pure" ash *Chenopodium album* L. (Ia) dominates. Areas with sufficient moisture ash substrate (Ib) are formed by groups of *Puccinellia distans* (Jacq.) Parl. and *Puccinellia hauptiana* (V.I. Krecz.) Kitag. In recultivated areas (IIa), cultural phytocenosis is formed with the domination of seeded grasses, such as *A. pectinatum, B. inermis, F. rubra, M. media*, and others. At overgrown reclaimed territory on strips of soil and grass, forb plant communities (IIb) formed with a predominance of *Elytrigia repens* (L.) Desv. ex Nevski, *Poa pratensis* L., *Deschampsia cespitosa* (L.) P. Beauv., and *Artemisia vulgaris* L. In the space of "pure" fly ash, there is depleted species composition, sparse plant communities (IIc) with a high abundance of *Melilotus officinalis* (L.) Lam. and *Melilotus albus* Medik. and significant participation of *D. cespitosa* and *F. rubra*. The appearance of regrowth of trees and shrubs was noted.

On "pure" ashes (Ia), the overgrowth of *C. epigejos* was generated in 20 years, with minor participation of other species. Salix and Betula were sparse and there was a very dense overgrowth of Salix (6 species) with impurities of *B. pendula, B. pubescens*, and *Populus tremula* L. At favorable humidity (Ib) that makes a substratum stable, the formation of communities is accelerated. Within 10 years, a meadow community was formed on many hectares with the dominance of *D. cespitosa*; co-dominants were *Trifolium repens* L., *Poa trivialis* L., and shrub undergrowth (*Salix* spp.). The development of moss cover (20%–40%) was observed everywhere. Nearby the lowerings, filled with water (Ic) has been colonized by *Equisetum palustre* L.

On a primary recultivated territory, on strips with deposited organic debris and perennial grass (IIa), in the first year, partial cutting was carried out. As a result, the diverseherbaceous-cereal and herbaceous vegetative communities with sparse *P. sylvestris* and *B. pendula* were generated in 20 years. These trees strengthened their role in establishing phytocoenose.

TABLE 4 The scheme of phytocoenosis formation on the ashdump in forest zone depending on ecotope.

Ecotope	Age, years				
	5	10	15	•••••	25
Ia	Ecotopical different-grass vegetative grouping (EG)*		Simple diverseherbacious-cereal or diverseherbacious-cereal vegetative grouping (payload)**	→↑	Over with sparse of willows and birches (P)**** Dense were over of willows with impurity of birches and aspens (P)
Ib		EG: - water plant moss; - diverseherbacious-cereal Payload: - diverseherbacious-cereal	Complicated vegetative grouping (CG)*** - cereal; - diverse-grass-cereal; - bean-cereal	→→→→	Meadow with the growth of birches Fluffy and willows (P) Dense bushes atussock grass meadow (P) White clover-rough meadow grass (P)
Ic			Single islands Vegetative groupings of a hydro-hydrophit type	→↑	Various variants of coastal vegetation
IIa	Culture phytocoenosis + woods – apofites	CG: - highgrass-cereal; - diverseherbacious-cereal from a large share (long) of cultural kinds	Diverseherbacious-cereal phytocoenosis with growing woods	→↑	Wood phytocoenosis with the rather produced circles: wood—sparse of pine, birches, aspens; grassy—different grass cereal (P)

IIb	Payload: - diverseherbacious-cereal with wood	CG: - diverseherbaceous-cereal also grow up—volume of trees and bushes	→→	Wood phytocoenosis: were over wood with poorly produced grass-bush with a circle	Wood phytocoenosis with clearly expressed wood, bush, grassy, and moss by circles (P)
IIc	EG: - diverseherbceous	Payload: - diverseherbceous; - diverseherbacious-cereal; - young growth wood	CG: - diverseherbaceous with wood young growth - diverseherbaceous cereal - power young growth and wood young growth	→→	Were over of deciduous breeds with poorly produced grassy-bush with a circle, strong moss (P)

The note: * EG—ecotopical vegetative grouping—0.1%; ** payload—simple vegetative grouping—0.1–5%; *** CG—complicated vegetative grouping—6–50%; **** P—phytocoenosis—more than 50%. Percent means the projective cover of ashdump surface by plants.

FIG. 4   (A) Agrophytocenosis of *Bromus inermis* Leyss. on the Verkhnetagilskaya power station ash dump 20 years (Verhniy Tagil). (B) Self over growing of non-recultivated sites at ash dump of Verkhnetagilskaya power station 40 years (VerhniyTagil, formation of forest phytocenosis: *Betula pendula* Roth, *Picea obovata* Ledeb). (C) Selfovergrowing of recultivated sites at ash dump of Verkhnetagilskaya power station 50 years (VerhniyTagil, formation of forest phytocenosis: *Picea obovata* Ledeb., *Betula pendula* Roth). (D) *Chimaphila umbellata* (L.) W.P.C. Barton (fly ash dump of Verkhnetagilskaya power station). (E) *Epipactis helleborine* (L.) Crantz (fly ash dump of Verkhnetagilskaya power station). (F) *Botrychium multifidum* (S.G. Gmel.) Rupr. (fly ash dump of Verkhnetagilskaya power station). (G) Ash dump of Nizneturinskaya power plant 60 years (Nizhnaya Tura, dominated by *Calamagrostis epigejos* (L.) Roth and willow stand). (H) The ash dump of Yuzhnouralskaya power station 50 years (Yuzhnouralsk, dominated: *Agropyron pectinatum* (Bieb.) Beauv., *Elaeagnus angustifolia* L.).

On strips with organic debris without crops of grass (IIb), the formation of woody phytocoenosis was accelerated. These reasons exclude cutting, retarding of the sodding of a surface and formation of grassy communities of a meadow type.

It is necessary to take into account that the delivered organic debris contained certain propagules of wood phytocoenosis species. As a result, a mixed forest with a prevalence of *P. sylvestris*, and less often with *B. pendula* was formed.

On the "pure" ashes between the sod strips, the forest communities composed of *B. pendula*, sparse *P. sylvestris*, and *P. tremula* formed with a delay of 5–10 years (IIc). A layer of grassy bushes was poorly produced. It is possible that the formation of this phytocoenosis ash dump was connected with the receipt of seeds from an adjacent, earlier overgrowth. It is natural that improving ash properties by proper management controlled wind erosion. The opposite process, drift of ashes by wind on bands of a ground, also takes place.

The bushes sowing, continuous plotting of a stratum of peat, a crop of long-term grass was conducted on the second time recultivated area. With the use of a complex organic and mineral fertilizer, productive moving and grazing fields were created.

By 2013, on the VTTPS fly ash dump (40 years after biological recultivation), at the site of the "pure" ash *C. epigejos* thickets of vegetation had been formed (Ia), with codominants of *Cirsium setosum* (Willd.) Bess., and *D. cespitosa*, with tree groups of *B. pendula* and *B. pubescens*. Meadows were formed (Ib) dominated by *D. cespitosa*, with codominants *C. epigejos*, *P. pratensis*, *Hieracium umbellatum* L., and *Chamaenerion angustifolium* (L.) Scop. Tree species were classified as undergrowth (up to 1.5 m) *P. tremula*, *B. pendula*, *P. sylvestris*, and

Salix myrsinifolia Salisb. On the fly ash along the periphery, a small-leaved forest was formed, characterized by a relatively high canopy cover of trees and a complex vertical structure. The upper tree layer was dominated by small-leaved species, such as *P. tremula*, *B. pendula*, and *B. pubescens*, *Salix caprea* L., coniferous species – *P. sylvestris*, and *P. obovata* were included in the lower group. The shrub layer was composed of *Chamaecytisus ruthenicus* (Fisch. ex Woł.) Klásk., *Rosa acicularis* Lindl., *S. myrsinifolia*, and *Salix pentandra* L. The undergrowth is composed of *Sorbus aucuparia* L., *Viburnum opulus* L., and *Prunus padus* L., the height of which varied from 0.7–0.8 to 3.5 m (projective cover is 15%–20%, sometimes up to 30%). The total projective cover of herbaceous species was 30%–35%. The greatest values of herbaceous plants had *T. repens*, and *Trifolium pratense* L., *F. rubra*, *P. pratensis*, *C. epigejos*, and *Vicia cracca* L. There were also typical forest species of the boreal zone: *Pyrola chlorantha* Sw., *Chimaphila umbellata* (L.) W.P.C. Barton, as well as a young population of Orchidaceae Juss. (*Platanthera bifolia* (L.) Rich.).

In the primary recultivated territories on strips coated with organic debris (IIa), forb grass and forb plant communities were formed. Total projective cover on soil strips reached 90%–100%, and on the "pure" ash was 60%–80%. The species on the fly ash and the soil strips included *Pimpinella saxifraga* L., *Euphorbia virgata* Waldst. & Kit., *Achillea millefolium* L., *Picris hieracioides* L., *F. rubra* in high abundance. *P. pratensis*, *Centaurea scabiosa* L., *Lathyrus pratensis* L., *V. cracca*, *Stellaria graminea* L., and *Silene latifolia* Poir. prevailed on the strips of "pure" ash.

At a substantial part of the fly ash dump, as a result of ash and soil overgrowth (IIb), forest communities close to the zonal type with substantial interests, and sometimes with the dominance of *P. sylvestris*, *B. pendula*, *B. pubescens*, *P. tremula*, were formed. *P. obovata*, *P. sibirica* along with *Larix sibirica* Ledeb. and *A. sibirica* were found in the undergrowth.

The shrub layer was formed by *S. aucuparia* and *P. padus*. With the increasing age and degree of development of the fly ash dump, forest communities enhanced their impact on the environment. The transformation of herbaceous vegetation in these communities towards an increase in the diversity of forest species was noted. This was accompanied by a decrease in the number and loss of some weed-ruderal species from the emerging plant communities. Bush cover species have appeared, such as *Vaccinium vitis-idaea* L., *Orthilia secunda* (L.) House, *Pyrola rotundifolia* L., *Pyrola media* Sw., and *Moneses uniflora* (L.) A. Gray. The herbaceous layer was dominated by *Fragaria vesca* L., *Aegopodium podagraria* L., *Rubus saxatilis* L. In addition, rare for the Urals species of the Orchidaceae family (*P. bifolia*, *Listera ovata* (L.) R.Br., and *Malaxis monophyllos* (L.) Sw., *Epipactis helleborine* (L.) Crantz) were found (Rankou, 2011).

By 2021, meadow phytocoenoses with a predominance of *C. epigejos*, *D. cespitosa*, *P. pratensis*, etc. remained on the "pure" ash (Ia) at the fly ash dump of the VTTPS (50 years after biological recultivation). The total cover of herbaceous vegetation was 65–85%. Single trees grow in these phytocoenoses: *B. pendula*, *P. sylvestris*, shrubs of the genus *Salix* spp. and *Rosa* sp.

In areas of "pure" ash (Ib), in the place of a meadow phytocoenosis dominated by *D. cespitosa*, young forest phytocenoses are formed with a high proportion of *B. pendula*, *P. tremula*, less often *B. pubescens* and *P. sylvestris*. Crown density varied from 0.3 up to 0.5. The height of the trees ranged from 3.5 to 7.5 m. The shrub layer was dominated by *S. myrsinifolia*, *S. caprea*, *Salix cinerea* L. Grass species (*D. cespitosa*, *C. epigejos*, *Poa palustris* L., and *P. pratensis*) predominated in the herbaceous layer; the abundance of *Ch. angustifolium*, *O. secunda*, *P. rotundifolia*. *Botrychium multifidum* (S.G. Gmel.) Rupr. and *P. bifolia* occurred in groups (Glazyrina et al., 2017). The total coverage of herbaceous vegetation was 45%–65%.

In the marginal areas of the fly ash dump on "pure" ash (Ib), in place of a small-leaved forest, a mixed forest is being formed. The stand height reached 12–15 m; the tree crown density was 0.5–0.6. The upper layer was dominated by *B. pendula*, *P. tremula*, and *P. sylvestris*. In the undergrowth, there were seedlings of the main crops, as well as *P. obovata*. and singly *A. sibirica*. In the undergrowth (coverage 10%–30%) were found *S. myrsinifolia*, *S. aucuparia*, *Ch. ruthenicus*, *P. padus*, and *V. opulus*. In the herb-shrub layer (the projective cover varied from 7% to 80%), the following species dominated: *T. repens*, *C. epigejos*, *P. pratensis*, *Agrostis gigantea* Roth, *V. cracca*. Six species of the tribe Pyroleae Dumort. (family Ericaceae Juss.) grew in this area. *O. secunda* and *P. rotundifolia* were found in high abundance. Small groups of *Ch. umbellata*, *Pyrola minor* L., *P. chlorantha*, and *M. uniflora* were found throughout the site. Three species of the Orchidaceae family grew in this area: *P. bifolia* (sp gr), which restores abundance in the Urals, as well as *L. ovata* and *M. monophyllos* (Maleva et al., 2021a).

A forest phytocoenosis with a crown density of 0.6–0.8 was formed on the recultivated site with strip soil application (IIb). The stand height reached 14–18 m. The tree layer was dominated by *B. pendula*, *P. tremula*, and *P. sylvestris*; single individuals of *B. pubescens*, *Alnus incana* (L.) Moench, and *Populus* sp. were also found. In the undergrowth, in addition to the main crops, *P. obovata*, *P. sibirica*, *A. sibirica*, *L. sibirica* grew. In the grass-shrub layer (the projective cover varied from 10% to 65%), the following species prevailed: *O. secunda*, *L. pratensis*, *Poa angustifolia* L., *Vicia sepium* L. Such species of the Orchidaceae family as *P. bifolia*, and *L. ovata*, grew scattered and in groups, in addition, *M. monophyllos*, *E. helleborine*, and *Goodyera repens* (L.) R.Br. found in single individuals.

Thus, studies have shown that disturbed areas can be «refugiums» temporary for the conservation of the gene pool of many rare and valuable species (Das et al., 2022; Bazzicalupo et al., 2023; Maleva et al., 2021b).

The Fig. 5 shows the dynamics of the share participation of tree species in the formation of forest phytocoenoses in the non-recultivated and recultivated sites of the fly ash dump. For

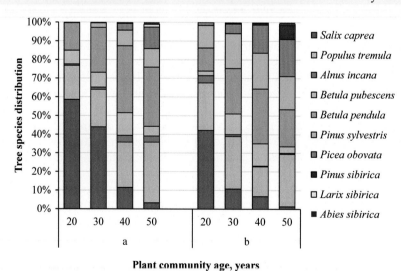

FIG. 5 Dynamics of the percentage distribution of tree species in at the Verkhnetagilskaya thermal power station fly ash dump forest phytocoenoses: A—non-recultivated; B—recultivated sites.

30 years (from 20 to 50 years of age) in forest phytocoenoses, the proportion of coniferous trees has increased significantly: from 13 to 48% in the recultivated site, and from 1% to 14% in the non-recultivated site. During the same period, natural elimination (death) of trees and bushes of the genus Salix (*S. caprea, S. cinerea, Salix viminalis* L., *S. pentandra, Salix gmelinii* Pall.) occurred on the sites. Studies have shown that early successional coniferous species *P. sylvestris* and late successional coniferous species (*P. obovate, L. sibirica, A. sibirica*) settle in the non-recultivated site for 10–15 years, while *P. sibirica* for 20–25 years later than in the recultivated one.

The dynamics of floristic reachness (number of species in community as a whole) was traced in two forest phytocoenoses formed in the recultivated site with strip soil application and in the non-recultivated one (Fig. 6).

The total number of species recorded in a recultivated site increased rapidly in the first years of succession. This is primarily due to the fact that the strip application of soil creates more favorable conditions for the growth and settlement of plants, in addition, the imported soil contains a seed bank. By the age of 10, the number of species stabilizes, fluctuating in the range of 70–90 species. In the future, a change in the species composition occurs: the number and proportion of forest species increase while the number and proportion of weed-ruderal species decrease. On the non-recultivated site of the fly ash dump, the total number of species does not stabilize but continues to grow. By 37–45 years of overgrowth, the total species diversity was no more than 70 species. As well as on the recultivated site, on the "clean" ash there was a change in the species composition, but this process took a longer time (Lukina et al., 2021).

In the secondary recultivated area (III), after uprooting of shrubs (species of the genus Salix) and a continuous application of a layer of peat (10–15 cm), perennial grasses were sown in the early 1990s. Productive haylands were created on this territory, which was maintained for 15 years (introduction of a complex of organic and mineral fertilizers and mowing).

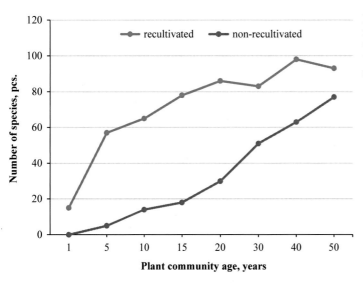

FIG. 6 Dynamics of floristic reachness during overgrowing of reculti-vated and non-recultivated sites in Verkhnetagilskaya thermal power station fly ash dump.

FIG. 7 Productivity 10 years of herbaceous communities at different sites of Verkhnetagilskaya thermal power station fly ash dump: 1—cultural phytocoenosis (dominant species *Bromus inermis* Leyss.), 2—meadow on fly ash (dominant species—*Puccinellia distans* (Jacq.) Parl. and *Puccinellia hauptiana* (V.I. Krecz.) Kitag.), 3—cereal herbaceous phytocoenosis (a—strips of soils, b—strips of "pure" ash).

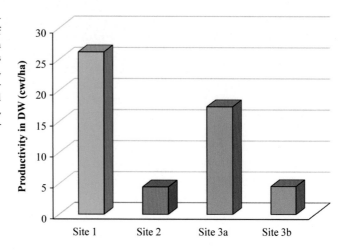

By the age of 10 years (in 2004), these plant communities were dominated by *B. inermis* and *F. rubra* with a total projective cover of 90%–100%. The productivity of cultural phytocoenosis was 26.18 kg/ha (Fig. 7).

After 25 years (in 2019), after the cessation of haymaking and fertilization, the projective grass cover in the cultural phytocoenosis decreased to 60%–70%. *B. inermis* remained the dominant plant community; *F. rubra* and *E. virgata* were codominants. The abundance of meadow and meadow-weed species increased in the grass stand: *Ch. angustifolium*, *A. millefolium*, *Rumex acetosella* L., *P. trivialis*, (sp gr–cop$_1$), *S. graminea*, *Silene noctiflora* L., and etc.

The productivity of cultural phytocoenosis decreased to 15.06 kg/ha (2019). The share of *B. inermis* in the total phytomass decreased from 69.0% (2004) to 51.3% (2019). In 1.63 times there was a decrease in the density of shoots (density of hayfields) *B. inermis* from 408 pcs. per m^2 (2004) up to 235 pcs. per m^2 (2019). Over 15 years, the proportion of *B. inermis* vegetative shoots in the herbage structure increased from 83.0% to 98.0%, while the proportion of generative shoots decreased from 17.0% to 2.0% (Glazyrina et al., 2022).

Despite the fact that with the increase in the age of the plant community created at the VTTPS fly ash dump, the sowing qualities of *B. inermis* seeds deteriorate significantly (seed germination energy decreased from 75.3% to 15.5%, seed germination—from 83.6% to 57.5%), this species has been preserved in the cultural phytocoenosis, renewing vegetatively.

B. inermis is a perennial grass, up to 120 cm high. This rhizomatous grass is the most valuable in terms of fodder and reclamation. Possesses exceptional shoot-forming ability. *B. inermis* is the most stable and promising for growing on fly ash dumps with peat in the Middle Urals. The biological feature of this species is the development of a powerful root system with a sufficient content of mineral nutrition elements in peat and intensive shoot formation, which contributed to its active vegetative renewal and the creation of long-term (more than 20 years) productive artificial phytocoenoses.

Thus, reclamation activities at the VTTPS fly ash dump (application of peat and sowing of perennial grasses) contributed to the creation of artificial phytocoenoses with a high potential productive longevity.

The cessation of economic activity (haymaking and fertilization) led to a decrease in the productivity of cultural phytocoenoses. In the economic use of hayfields, it is important to assess the quality of the products obtained. It is known that fly ash from dumps contains an increased amount of microelements, including those toxic to plants. Previously, it was shown that covering the fly ash dump with a layer of peat prevents the entry of heavy metals into the aerial parts of plants (Chibrik and Spirina, 2013).

When the economic assessment of plant communities is concerned, one of the most important indicators is their productivity. At 10-year-old plant grouping, the average weight of the air-dried shoots was 4.56 cwt/ha (ranged from 0.22 cwt/ha to 7.32 cwt/ha) on only ash; however, with organic matter, the air-dry weight of shoots was 17.4 t/ha (ranged from 4.04 to 24.6 cwt/ha). Thus, the average productivity of the communities on ashes with organic matter was four times greater compared to productivity on ash alone.

3.2 Nizhneturinskaya thermal power station

The fly ash dump of the NTTPS is located near the city of Nizhnyaya Tura, Sverdlovsk Region (Middle Urals, taiga zone, southern taiga sub-zone). NTTPS operated on coal from the Volchansky (Bogoslovsky) deposit, and on coal from the Ekibasstuz deposit (Kazakhstan); in 2015 it was switched to use gas. The fly ash field with an area of 60 ha was located two km northwest of NTTPS in a natural pit, the depth of the fly ash layer is 12–15 m. After coal combustion, the fly ash was dominated by compounds of silicon (SiO_2—40.5%), iron (Fe_2O_3—5.5%), and aluminum (Al_2O_3—32.4%). The substrate had a dense-platy structure, due to the high content of calcium in the afly ash (up to 12%), which cements the ash particles. According to agrochemical analysis, nitrogen is almost completely absent in the fly ash substrate, at the same time it contains a lot of phosphorus (up to 30 mg/100 g of fly ash) and potassium (11.4 mg/100 g of fly ash). The operation of the fly ash dump was completed in 1957. In 1959–61, biological reclamation was carried out at the fly ash dump.

At Site 1, studies were carried out on the formation of vegetation on "pure" ashes in the process of self-overgrowing.

Site 2, with an area of about 15 ha, was covered with a 2 cm layer of soil and left for self-overgrowth. Subsequently, most of this territory was built up with a garage complex and treatment facilities.

On Site 3, an area of 25 ha was applied irrigation with wastewater. Disinfected water was supplied to the area from the treatment plant through pipes. The water contained a small amount of nitrogen (7 mg/L), iron (0.22 mg/L), a lot of chlorine (15.6 mg/L), sulfur compounds (36.1 mg/L), and a dense residue (up to 291 mg/L). Perennial grasses were sown on 10 ha of the irrigated area: *B. inermis*, *A. pectinatum*, *Elymus trachycaulus* (Link) Gould & Shinners, *Festuca pratensis* Huds., *Phleum pratense* L., *Elymus fibrosus* (Schrenk) Tzvelev, *Medicago sativa* L., *M. albus*, *O. arenaria*. Watering was carried out for 10 years in the spring-summer period before hay mowing, watering was resumed in autumn. The rest of the irrigated area was self-overgrowing.

The studies were carried out from 1959 to 2019. The study of the processes of self-overgrowing of "pure" ashes (Site 1) showed that the first plants began to settle only 5 years after the drying of the substrate. Settlement and growth of plants were hampered by the poverty of the substrate with mineral nutrition elements, lack of moisture, alkaline reaction of the environment, and wind erosion. After 5 years, only a few plant species were found along the dams: *P. distans, P. hauptiana, Ch. album, Polygonum aviculare* L., *Tussilago farfara* L., *T. repens, E. repens*. After 10 years, the species composition of plants increased to 13 species: *Ch. angustifolium, Lappula squarrosa* (Retz.) Dumort., *Atriplex tatarica* L., *Rumex crispus* L., *P. pratensis, A. millefolium, M. albus*, etc. The weight of above-ground phytomass in this area was 4.8 cwt/ha.

In Site 2, formed by applying a 2 cm layer of soil, as a result of self-overgrowth, ruderal species settled in the first two years: *Ch. album, Lepidium ruderale* L., *R. acetosella, Silene nutans* L.

Three years later, the active introduction of species of the family Poaceae Barnhart (*B. inermis, F. pratensis, Ph. pratense*) and the family Fabaceae Lindl. (*T. pratense, T. repens, M. sativa, M. albus, M. officinalis*) began, which migrated from the experimental sites. By the age of 10, a meadow phytocoenosis with the dominance of *M. officinalis* and *M. albus* had formed in this area; *M. sativa, T. pratense, T. repens, A. gigantea, E. repens*, etc., were present as co-dominants; additional undergrowth of trees (*P. sylvestris, B. pendula, P. tremula*) and shrubs (*S. caprea, Rosa majalis* Herrm., *Rubus idaeus* L.) appeared. The weight of the aboveground phytomass of herbaceous plants in this area was 51.6 cwt/ha (Fig. 8).

In Site 3a, with sewage irrigation and sowing of perennial grasses, plants completely covered the territory surface in 2–3 years, protecting it from dusting. The species of the Poaceae family turned out to be more stable in the conditions of the fly ash dump, and the representatives of the Fabaceae family were less stable. By the age of 10, a meadow phytocoenosis with a high abundance of sown species was formed in Site 3: *B. inermis, Ph. pratense, E. fibrosus, Dactylis glomerata* L., *A. pectinatum*, etc., and local flora species: *Beckmannia eruciformis* (L.) Host, *D. cespitosa, T. repens*, etc. The weight of aboveground phytomass in this area was 60.8 cwt/ha (Fig. 8).

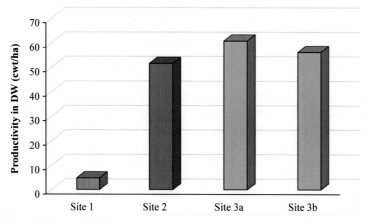

FIG. 8 Productivity of 10-year-old herbaceous communities in different sites of the NTTPS fly ash dump: 1—fly ash, 2—fly ash + 2 cm of soil, 3—fly ash + wastewater irrigation (a—with sowing perennial grasses, b—without grass sowing).

In Site 3b with wastewater irrigation, but without grass sowing, a shrubby meadow phytocoenosis was also formed from species of local flora and species migrating from neighboring ash dump sites: *E. repens, Ph. pratense, F. pratensis, F. rubra, D. glomerata, B. inermis, M. officinalis, M. albus, M. sativa, T. repens, T. farfara, D. cespitosa*, etc.

By the age of ten, tree species grew here: *S. caprea, S. viminalis, P. tremula, B. pendula*, some individuals reached a height of 3.5 m. The weight of the aboveground phytomass of herbaceous species in this area was 56.2 cwt/ha (Fig. 8). The entire area with sewage irrigation was used as hayland.

An assessment of the impact of reclamation measures on the agrochemical composition of substrates carried out after 10 years, showed that watering with wastewater, as well as applying a layer of soil to the surface of the ash dump, improved some properties of the root layer. In the reclaimed areas, there was a decrease in the alkalinity of the environment (pH), especially in the upper horizons, and the accumulation of mineral nutrients (N, P_2O_5, K_2O) in the root layer of the substrate, especially in the area with the application of the soil layer. There was no significant accumulation of mineral nutrients in the area with wastewater irrigation, since regular mowing of grasses was carried out here (Chibrik et al., 2022).

Further studies conducted in 2019 (60 years after the biological reclamation) showed that in Site 1 of the self-overgrowing of "pure" ash, a forb-reed phytocoenosis was formed with the dominance of *C. epigejos*, and a high abundance of species such as *M. albus, M. officinalis, P. saxifraga, P. pratensis, F. rubra, B. inermis*. The projective vegetation cover was 80%–100%. The weight of aboveground phytomass was 22.6 cwt/ha. The share of *C. epigejos* in the total phytomass was 27.0%. Such species as *P. pratensis* (23.9%), *B. inermis* (9.7%), and *M. albus* (8.6%) had a high share in the herbage composition.

In Site 2, a mixed forest was formed (closeness 0.3) with the dominance of *P. sylvestris, B. pendula*, and *P. tremula*. The anthropogenic environment (proximity to the urban area) affected the composition of the herbaceous layer: in it, along with forest species (*Pulmonaria obscura* Dumort., *P. rotundifolia, Geranium sylvaticum* L., *F. vesca*), weed-ruderal and meadow-weed species (*A. vulgaris, Plantago major* L., *Sonchus arvensis* L., etc.) had a high abundance.

In Site 3, after the cessation of irrigation, the substrate began to dry out and the sown species began to fall out. Sixty years after the biological recultivation, a meadow phytocoenosis was formed here with the dominance of *C. epigejos*; such species as *A. millefolium, B. inermis*, and *P. saxifraga* were found sporadically. The projective vegetation cover varied from 70% to 80%. The weight of aboveground phytomass was 15.2 cwt/ha. The share of *C. epigejos* in the composition of phytomass in different parts of the phytocenosis varied from 95.0% to 98.0%.

On the experimental sites of the NTTPS ash dump, the study of the dynamics of floristic reachness (number of species in community as a whole) restoration was carried out. The study of the species composition of plant communities formed on fly ash substrates gives an idea of the ecological and floristic potential of these territories. The study of the species composition of plant communities formed on fly ash substrates gives an idea of the ecological and floristic potential of these territories (Pandey, 2015). An analysis of the dynamics of biodiversity at the ash dump showed that over 60 years, in all the studied plant communities, there was an increase in species richness. In the area of non-reclaimed ash, the number of species increased from 7 to 41 species over the entire period under study (Fig. 9).

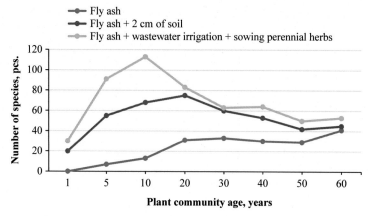

FIG. 9 Dynamics of floristic reachness on recultivated and non-recultivated fly ash dump of Nizhneturinskaya thermal power station.

Already in the 2nd year, the total number of species increased from 25 to 61 species in the area with soil application. Species *Sinapis arvensis* L., *L. ruderale*, *Ch. album*, and others had a high abundance. The growth trend of biodiversity was noted in the next 10–20 years (up to 75 species). By the age of 50–60 years, with the increasing influence of the tree layer, biodiversity decreased to 42–45 species).

In the area with sewage irrigation during the first 5 years, there was a rapid increase in the total number of species from 25 to 113. Irrigation and regular mowing have led to the growth of species of the Poaceae family with well-developed root systems that have formed a dense vegetation cover. In the subsequent 10-year period after the cessation of irrigation, the species richness of the plant community decreased to 63 species. By the age of 50–60 years, the species composition on the site stabilized at the level of 50–53 species.

3.3 Yuzhnouralskaya thermal power station

The analysis of the results on the formation of phytocoenoses in the forest-steppe zone was carried out using the example of two sites of fly ash dumps of YUTPS (recultivated and non-recultivated) located in the South Urals in the Chelyabinsk region (Fig. 1).

The area of the recultivated fly ash dump was 68 ha. In 1965–66, this site has undergone biological recultivation. The entire surface of the fly ash dump was covered with a layer of chernozem soil 10–15 cm thick and planted with the following species: *A. pectinatum*, *B. inermis*, *M. media*, and *O. arenaria*.

The non-recultivated fly ash dump was located five km south of YUTPS. The area of the first and second sections of the fly ash dump was 213 ha, the operation of which was completed by 1975.

The fly ash of Chelyabinsk coal is loose, structureless; it was the smallest fraction of dust with a large admixture of crushed slag. Agrochemical analysis of the ash showed that it practically does not contain nitrogen, P_2O_5—2.5 mg/100 g of ash, K_2O—10.0 mg/100 g of ash, pH 7.8.

In the forest-steppe zone, the leading limiting factor affecting the formation of vegetation in ash dumps is the lack of moisture in the upper ash horizons. In addition, studies have shown that the surface of the fly ash heats up more than the surface of the soil by 3–4°C. Therefore, at an air temperature of 25°C, the temperature on the ash surface reached 32°C, while on the soil surface, it was 28°C.

Monitoring studies carried out at the RS of fly ash dump from 1966 to 2016 showed that after the biological reclamation, productive and economically valuable plant communities were immediately formed with a predominance of planted species: *O. arenaria*, *M. media*, *B. inermis*, *A. pectinatum*.

In the first 5 years of ashes covered with soil, the species of the Fabaceae family performed well. In the second year after sowing, the height of some individuals of *M. media*. reached 72.0 cm, the average height was 30.8 cm. The length of the roots of *M. media* reached 57.5 cm, averaging 34.2 cm. The height of *O. arenaria* individuals reached 69.4 cm, averaging 43.3 cm. The length of the roots of *O. arenaria* reached 46.9 cm.

In the first years after sowing on soil-covered fly ash, species of the Poaceae family developed more slowly than on zonal soils. The average height of *A. pectinatum* individuals was 29.16 cm, and *B. inermis* was 27.9 cm.

Subsequently, over 10–15 years, the planted species (especially species of the Fabaceae family) were gradually replaced by wild species. The lack of care for crops, as well as sheep grazing, accelerated the degradation of cultural phytocoenoses. *A. pectinatum* turned out to be the most resistant species. It successfully settled in the fly ash dump, mainly due to seed propagation, partially penetrating into communities on "pure ash." The weight of air-dry aboveground phytomass at the recultivated fly ash dump was 8.9 cwt/ha.

Studies conducted 25 years after the biological reclamation showed that a forb-grass phytocoenosis with the dominance of *A. pectinatum*, *E. repens*, *P. pratensis*, *B. inermis*, *E. virgata*, *Artemisia dracunculus* L., and *Potentilla impolita* Wahlenb. was formed on the recultivated fly ash dump. The total projective vegetation cover varied from 50% to 90%. Tree species were represented by rare bushes *Elaeagnus angustifolia* L. up to 1.5 m high. The weight of the air-dry aboveground phytomass of herbaceous plants at the recultivated fly ash dump was 13.5 cwt/ha. The transformation of cultural phytocoenoses was accompanied by their xerophytization and followed the path of formation of herbaceous communities approaching meadow steppes with a predominance of *A. pectinatum* and *P. pratensis* (Lukina, 2010).

The studies conducted 50 years after the formation of cultural phytocoenoses showed that recultivated fly ash dump preserved a forb-grass phytocoenosis with the dominance of *A. pectinatum* (occurrence rate was 81%), *Artemisia austriaca* Jacq. (27%) and *E. virgata* (48%) became co-dominants. The total projective cover of herbaceous species averaged 37%, varying from 10% to 90%. The weight of air-dry aboveground phytomass was 27.7 cwt/ha; 66% of the total weight is represented by *A. pectinatum*.

On the recultivated territory, full-fledged individuals of *A. pectinatum* developed, forming benign seeds (weight of 1000 seeds was 0.34 g; seed germination energy was 50%; germination was 70%), which was confirmed by a high level of renewal.

A. pectinatum is a perennial grass, 25–75 cm high, fibrous roots, that reach a depth of up to 2.5 m, forms a large number of vegetative shoots. The spikes are thick. A valuable fodder plant used in hayfields and pastures, it is characterized by very high fodder qualities. This species is resistant to high anthropogenic pressure (mowing, eating by livestock, etc.) and has high productivity. It is drought-resistant and maintains salinization of the soil. It is a long-lived plant (up to 25 years) (Kulikov, 2005). *A. pectinatum* is the most promising species for the biological reclamation of fly ash dumps in the forest-steppe zone in the Southern Urals.

Monitoring studies on sections I and II of the non-recultivated fly ash dump of YUTPS showed that the formation of vegetation cover on the ash is delayed by 10–15 years. Some of the first species to settle on the fly ash substrate were *Salsola collina* Pall., *Ceratocarpus arenarius* L., *Ch. album*, *Lipandra polysperma* (L.) S. Fuentes, Uotila & Borsch, *Bassia prostrata* (L.) Beck, etc.

After 5 years (in 1980), a forb-wormwood complex plant group with a protective cover from 20 to 50% was formed in this area. The species of local flora most resistant to growing on "pure" fly ash substrate were found with a high abundance: *Bassia laniflora* (S.G. Gmel.) A.I. Scott (cop$_2$), *Artemisia absinthium* L. (cop$_1$), and *P. aviculare* (sp gr).

By the age of 15 (in 1990) at the non-recultivated fly ash dump, a wormwood-reed grass plant group was formed on the ashes with the dominance of *C. epigejos* (cop$_{1-2}$) and *A. absinthium* (cop$_1$), with a high abundance of *Cynoglossum officinale* L., *Echium vulgare* L. (sp). The total projective vegetation cover increased to 40%–50%. The weight of air-dry aboveground phytomass was 2.8 cwt/ha. Tree species were represented by rare bushes of *El. angustifolia* which was 2–3 m tall.

Studies conducted 40 years later (in 2016) showed that a forb-grass phytocoenosis was formed on the fly ashes, with rare *El. angustifolia* (sp), dominant of the community was *C. epigejos* (cop$_3$–soc), co-dominants were *P. pratensis* (sp–cop$_1$), *A. dracunculus*, *A. austriaca*, *Lactuca tatarica* (L.) C.A. Mey., *Lathyrus tuberosus* L., *P. impolita*, *Potentilla argentea* L. (sp–cop$_1$ gr). The total projective vegetation cover varied from 60% to 90%. The weight of air-dry aboveground phytomass was 25.2 cwt/ha.

The study of the dynamics of the weight of air-dry aboveground phytomass of coeval herbaceous plants in different ecotopes of fly ash dumps showed that with increasing age of plant communities, an increase in aboveground phytomass occurs (Fig. 10).

In 5-year-old plant communities in the conditions of a recultivated fly ash dump, the weight of air-dry aboveground phytomass was 12.2 cwt/ha, which was 13 times more than in a non-recultivated one. In 15-year-old plant communities in the conditions of a recultivated fly ash dump, the weight of air-dry aboveground phytomass decreased to 8.9 cwt/ha due to the loss of species of the Fabaceae family from the composition of the herbage (trampling and eating by sheep), which was 3 times more than in the non-recultivated fly ash dump (2.8 cwt/ha). In 25-year-old plant communities on recultivated and non-recultivated fly ash dumps, the weight of air-dry aboveground phytomass was 13.5 and 13.3 cwt/ha, respectively. In 40- and 50-year-old plant communities under conditions of different fly ash dumps, the weight of aboveground phytomass increased to 28.0 and 25.2 cwt/ha, respectively. At the reclaimed fly ash dump, 66.1% of the phytomass was *A. pectinatum*, 9.9% *P. impolita*, 8.1% *A. austriaca*. On the non-recultivated part of fly ash dump, 61.3% was *C. epigejos*, 22.67% *P. pratensis*.

The dynamics of floristic reachness (number of species in community as a whole) was traced during the restoration of vegetation on two sites of fly ash dump: on the recultivated territory due application soil to the surface and on the non-recultivated one (Fig. 11). On the recultivated fly ash dump, the total number of species grows rapidly in the first years of succession (from 28 to 87 species). This is primarily due to the fact that the application of the soil

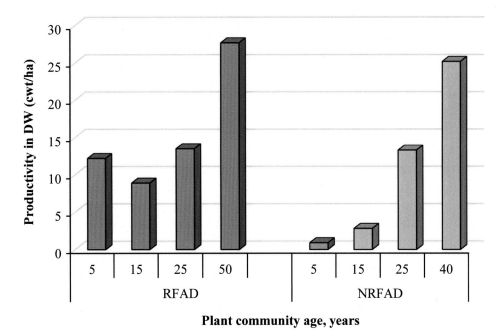

FIG. 10 Productivity dynamics of herbaceous communities on recultivated and non-recultivated fly ash dumps of Yuzhnouralskaya thermal power station.

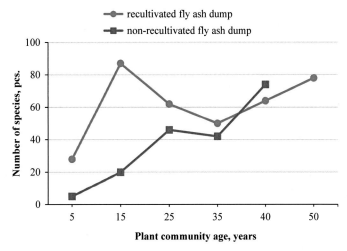

FIG. 11 Dynamics of floristic reachness on the recultivated and non-recultivated fly ash dumps of Yuzhnouralskaya thermal power station.

creates more favorable conditions for the growth and settlement of plants, in addition, the imported soil contains a seed bank. From 15 to 35 years of age, degradation of cultural phytocoenoses was observed due to pasture digression (load) and growth of turf grasses. The number of species decreased to 50. By the age of 50, after the cessation of sheep grazing, the number of species increased again to 78. Throughout the study period, the species composition changed: the number and proportion of meadow-steppe species increased while the number and proportion of weed-ruderal species decreased. In the non-recultivated area of the fly ash dump, the formation of vegetation cover on the ash was delayed by 10–15 years. The total number of species gradually increased from 5 (mostly ruderal) species to 74 species. As well on the recultivated part of fly ash dump, the species composition of the ash changes, but this process is longer.

4 Conclusions

The conducted observations enable an evaluation of the biological recultivation of VTTPS fly ash dump with strips, by covering the sites with ground and perennial grass. The stabilization of the substratum of the ashes is achieved at the expense of bands. This results in the diminution or even the termination of dust storms.

It is necessary to recognize the importance of the improvement of the water-physical and agrochemical properties of ashes, via the washing of ground from bands to space between the strips with "pure" ashes. The crop of perennial grass and its consequent cutting accelerates the gardening of bands with a ground but retards the establishment of trees, bushes, and the formation of a wood circle of the formed forest phytocoenosis. The application of a layer of peat on fly ash has made it possible to create productive cultural phytocenoses, but their quality must be controlled.

Experience gained in creating phytocenosis on the fly ash dump of the VTTPS showed that *B. inermis* is a key species for restoration. One of the most important indicators of the economic assessment of phytocoenosis is community productivity. At 10 years, the shoots of plant groups in fly ash dump conditions had the average dry weight phytomass of 4.56 cwt/ha (min, 0.22 cwt/ha; max, 7.32 cwt/ha) on the ash, and 17.4 cwt/ha on the ash with organic matter (min, 4.04 cwt/ha; max, 24.6 cwt/ha). On average, the productivity of forming communities on the ash-organic matter mixture is four times higher than the productivity on empty ash.

After 40 years of biological recultivation, the productivity of plant communities at soil strips is 1.3 times higher than the productivity at ash strips and 1.5 times higher than the productivity at clean fly ash.

Studies conducted at the NTTPS fly ash dump showed that reclamation activities contributed to the growth of species diversity and changed the direction of vegetation formation. Covering the surface of the fly ash substrate with a 2 cm soil layer and the proximity of the anthropogenic environment accelerated the formation of forest phytocoenoses dominated by *P. sylvestris*, *B. pendula*, and *P. tremula* with a high proportion of meadow and weed-ruderal species in the herbaceous layer. When irrigating with wastewater and sowing perennial grasses on the fly ash dump, stable highly productive meadow plant communities were created over 10 years with the dominance of *B. inermis*, *Ph. pratense*, *E. fibrosus*, and *D. glomerata*, which transformed (after the termination of agrotechnical measures) into meadow phytocoenoses with the dominance of *C. epigejos*.

In 10-year plant communities in the conditions of fly ash dump, the weight of the average air-dry phytomass was 4.8 c/ha on fly ash, 51.6 cwt/ha on fly ash with soil cover, and

56.2–60.8 cwt/ha on ash with sewage irrigation. On average, the productivity of plant communities on 10-year-old recultivated plots was 10–12 times higher than the productivity on "pure" ash.

The study of the transformation of vegetation at the YUTPS fly ash dump, located in the forest-steppe zone, showed that after biological recultivation (application of a 10 cm layer of soil), productive and economically valuable plant communities are immediately formed with a predominance of planted species: *O. arenaria*, *M. media*, *B. inermis*, *A. pectinatum*. Subsequently, over 10–15 years, the sown species (especially the Fabaceae family) were gradually replaced by wild ones. *A. pectinatum* has successfully settled in the ash dump and, apparently, is the most promising species for the biological reclamation of fly ash dumps in the forest-steppe zone in the Southern Urals. In general, the formation of plant communities follows the path of rapprochement with meadow steppes.

On the non-recultivated fly ash dump, the formation of vegetation cover on the "pure" ash was delayed by 10–15 years. The formation of vegetation comes from simple open plant groups with a predominance of *S. collina*, *C. arenarius*, *Ch. album*, *L. polysperma*, *B. prostrata*, etc., to sagebrush-reed grass phytocoenoses dominated by *C. epigejos*, *P. pratensis*, *A. dracunculus*, *A. austriaca*, *L. tatarica*, etc., with a low proportion of fodder plants.

Acknowledgments

The study was funded by the Ministry of Science and Higher Education of the Russian Federation as part of the state task of the Ural Federal University, FEUZ-2023-0019. MNVP is thankful to the Ural Federal University (UrFU), Ekaterinburg, for the invitation as "Visiting Professor" in 2014 and 2015. Authors express profound sorrow for the sad demise of Biological recultivation research group leader "Tamara S Chibrik" while preparing this chapter.

References

Ashraf, S., Anwar, A., Ali, Q., Safdar, A., 2023. Plant response to industrial waste. In: Plants and their Interaction to Environmental Pollution. Elsevier, pp. 265–282.

Bazzicalupo, M., Calevo, J., Smeriglio, A., Cornara, L., 2023. Traditional, therapeutic uses and phytochemistry of terrestrial European orchids and implications for conservation. Plants 12 (2), 257.

Belskaya, E.A., Vorobeichik, E.L., 2013. Responses of leaf-eating insects feeding on aspen to emissions from the Middle Ural copper smelter. Russ. J. Ecol. 44 (2), 108–117.

Belskii, E.A., Belskaya, E.A., 2009. Composition of pied flycatcher (*Ficedula hypoleuca* Pall.) nestling diet in industrially polluted area. Russ. J. Ecol. 40 (5), 368–371.

Bolshakov, V., Chibrik, T., 2007. Biological recultivation. Sci. Russ. 3, 106–112. Russian Academy of Sciences.

Březina, S., Koubek, T., Münzbergová, Z., Herben, T., 2006. Ecological benefits of integration of *Calamagrostis epigejos* ramets under field conditions. Flora-Morphol. Distribut. Funct. Ecol. Plants 201 (6), 461–467.

Brooks, S.J., Udachin, V., Williamson, B.J., 2005. Impact of copper smelting on lakes in the southern Ural Mountains, Russia, inferred from chironomids. J. Paleolimnol. 33 (2), 229–241.

Cheung, K.C., Wong, J.P.K., Zhang, Z.Q., Wong, J.W.C., Wong, M.H., 2000. Revegetation of lagoon ash using the legume species *Acacia auriculiformis* and *Leucaena leucocephala*. Environ. Pollut. 109, 75–82. https://doi.org/10.1016/S0269-7491(99)00235-3.

Chibrik, T.S., Spirina, M.S., 2013. Natural-Technogenic Complexes: Recultivation and Sustainable Functioning: Sat. Materials int. Scientific. Conf. June 10–15, 2013. Okarina Publishing House, Novosibirsk, pp. 212–215 (in Russian).

Chibrik, T.S., Lukina, N.V., Filimonova, E.I., Glazyrina, M.A., Rakov, E.A., 2022. Influence of recultivation methods on formation of ash dump phytocenosis in taiga zone (Middle Urals). AIP Conf. Proc. 2390, 030010. https://doi.org/10.1063/5.0069037.

Chukanov, V.N., Volobuev, P.V., Poddubnyj, V.A., Trapeznikov, A.V., 1993. Ecological problem of Ural. Defektoskopiya 7, 38–46.

Das, N., Samantaray, S., Ghosh, C., Kushwaha, K., Sircar, D., Roy, P., 2022. *Chimaphila umbellata* extract exerts antiproliferative effect on human breast cancer cells via RIP1K/RIP3K-mediated necroptosis. Phytomed. Plus 2 (1), 100159.

Dubois, H., Verkasalo, E., Claessens, H., 2020. Potential of birch (*Betula pendula* Roth and *B. pubescens* Ehrh.) for forestry and forest-based industry sector within the changing climatic and socio-economic context of Western Europe. Forests 11 (3), 336.

Dulya, O.V., Mikryukov, V.S., Vorobeichik, E.L., 2013. Strategies of adaptation to heavy metal pollution in *Deschampsia caespitosa* and *Lychnis flos-cuculi*: analysis based on dose-response relationship. Russ. J. Ecol. 44, 4271–4281.

Fantinato, E., Fiorentin, R., Della Bella, A., Buffa, G., 2023. Growth-survival trade-offs and the restoration of non-forested open ecosystems. Global Ecol. Conserv. 41, e02383.

Glazyrina, M.A., Filimonova, E.I., Lukina, N.V., Fateeva, S.S., 2017. *Botrychium multifidum* (S.G. Gmel.) Rupr in natural and technogeneous habitat in the Middle Ural. Scientific J. Proc. Petrozavodsk State Univ. 6 (167), 53–61.

Glazyrina, M.A., Lukina, N.V., Filimonova, E.I., Chibrik, T.S., 2022. Transformation of artificial phytocenosis created on ash dumps of the Middle Urals. AIP Conf. Proc. 2390, 030024. https://doi.org/10.1063/5.0069045.

Haynes, R.J., 2009. Reclamation and revegetation of fly ash disposal sites—challenges and research needs. J. Environ. Manag. 90 (1), 43–53. https://doi.org/10.1016/j.jenvman.2008.07.003.

Jayapal, A., Chaterjee, T., Sahariah, B.P., 2023. Bioremediation techniques for the treatment of mine tailings: a review. Soil Ecol. Lett. 5 (2), 1–15.

Khokhryakov, A.V., 2003. Environmental problems of mining in the Ural. Erzmetall: J. Explorat. Min. Metall. 56 (10), 595–600.

Koptsik, G.N., 2014. Modern approaches to remediation of heavy metal polluted soils: a review. Eurasian Soil Sci. 47 (7), 707–722. © Pleiades, 2014. Original Russian Text published in Pochvovedenie. No. 7, pp. 851–868.

Kulikov, P.V., 2005. Summary of the Flora of the Chelyabinsk Region (Vascular Plants). Geotour, Yekaterinburg-Miass. 537 pp. (in Russian).

Lee, S.H., Park, H., Kim, J.G., 2023. Current status of and challenges for phytoremediation as a sustainable environmental management plan for abandoned mine areas in Korea. Sustainability 15 (3), 2761.

Lemoine, G., 2016. Brownfield restoration as a smart economic growth option for promoting ecotourism and leisure: two case studies in Nord-Pas De Calais. Bioremediat. Bioecon. 15, 361–388. https://doi.org/10.1016/B978-0-12-802830-8.00015-0

Lukina, N.V., 2010. Formation of the phytocoenosises on the ash dumps of Uzhnouralsk Power Station. Arid. Ecosyst. 4 (44), 62–69.

Lukina, N., Chibrik, T., Filimonova, E., Glazyrina, M., Rakov, E., Veselkin, D., 2021. Strip clay application accelerates for 15–20 years the vegetation formation in ash dump (Middle Urals, Russia). BIO Web Conf. 31, 00016. https://doi.org/10.1051/bioconf/20213100016.

Maiti, S.K., 2013. Ecorestoration of the Coalmine Degraded Lands. Springer, New Delhi, https://doi.org/10.1007/978-81-322-0851-8. 375 pp.

Makhathini, T.P., Bwapwa, J.K., Mtsweni, S., 2023. Various options for mining and metallurgical waste in the circular economy: a review. Sustainability 15 (3), 2518.

Maleva, M., Borisova, G., Chukina, N., Sinenko, O., Filimonova, E., Lukina, N., Glazyrina, M., 2021a. Adaptive morphophysiological features of *Neottia ovata* (Orchidaceae) contributing to its natural colonization on fly ash deposits. Horticultura 7 (5), 109. https://doi.org/10.3390/horticulturae7050109.

Maleva, M.G., Borisova, G.G., Chukina, N.V., Novikov, P.E., Filimonova, E.I., Lukina, N.V., Glazyrina, M.A., 2021b. Foliar content of phenolic compounds in Platanthera bifolia from natural and transformed ecosystems at different stages of orchid development. J. Sib. Fed. Univ. Biol. 14 (3), 274–286. https://doi.org/10.17516/1997-1389-0349.

Miguel, P., Stumpf, L., Pinto, L.F.S., Pauletto, E.A., Rodrigues, M.F., Barboza, L.S., Leidemer, J.D., Duarte, T.B., Pinto, M.A.B., de Garcia Fernandez, M.B., Islabão, L.O., 2023. Physical restoration of a minesoil after 10.6 years of revegetation. Soil Tillage Res. 227, 105599.

Mitrović, M., Pavlović, P., Lakušić, D., Djurdjević, L., Stevanović, B., Kostić, O., Gajić, G., 2008. The potential of *Festuca rubra* and *Calamagrostis epigejos* for the revegetation of fly ash deposits. Sci. Total Environ. 407 (1), 338–347.

Nesgovorova, N.P., Saveliev, V.G., Ivantsova, G.V., Purtova, A.A., Ufimtseva, M.G., 2019. Frameworks for formation and stability of Kurgan Regional arboretum phytocenosis. KnE Life Sci. 4 (14), 1065–1077. https://doi.org/10.18502/kls.v4i14.570.

Pandey, V.C., 2015. Assisted phytoremediation of fly ash dumps through naturally colonized plants. Ecol. Eng. 82, 1–5. https://doi.org/10.1016/j.ecoleng.2015.04.002.

Prasad, M.N.V., 2006. In: Prasad, M.N.V., et al. (Eds.), Trace Elements in the Environment: Biogeochemistry, Biotechnology, and Bioremediation. CRC Taylor and Francis, Boca Raton, FL, USA, pp. 405–424, https://doi.org/10.1201/9781420032048.

Radochinskaya, L.P., Kladiev, A.K., Rybashlykova, L.P., 2019. Production potential of restored pastures of the Northwestern Caspian. Arid. Ecosyst. 9, 51–58. https://doi.org/10.1134/S2079096119010086.

Rankou, H., 2011. *Epipactis helleborine*. The IUCN Red List of Threatened Species 2011. e.T175992A7164692. Available at: https://www.iucnredlist.org/. (Accessed 22 September 2022).

Ren, Z., Cheng, R., Chen, P., Xue, Y., Xu, H., Yin, Y., Huang, G., Zhang, W., Zhang, L., 2023. Plant-associated microbe system in treatment of heavy metals-contaminated soil: mechanisms and applications. Water Air Soil Pollut. 234 (1), 39.

Sapsford, D.J., Stewart, D.I., Sinnett, D.E., Burke, I.T., Cleall, P.J., Harbottle, M.J., Mayes, W., Owen, N., Sardo, A.M., Weightman, A., 2023. Circular economy landfills for temporary storage and treatment of mineral-rich wastes. In: Proceedings of the Institution of Civil Engineers-Waste and Resource Management. Thomas Telford Ltd, pp. 1–30.

Shakirov, A.V., 2011. Physical-Geographical Zoning of the Urals. UB RAC, Yekaterinburg, Russia. 617 pp. (in Russian with English Summary).

Sierka, E., Kopczyńska, S., 2014. Participation of *Calamagrostis epigejos* (L.) Roth in plant communities of the River Bytomka valley in terms of its biomass use in the power industry. Environ. Socio-econ. Stud. 2 (2). https://doi.org/10.1515/environ-2015-0036. s.38–44.

Sonwani, S., Hussain, S., Saxena, P., 2022. Air pollution and climate change impact on forest ecosystems in Asian region—a review. Ecosyst. Health Sustain. 8 (1), 2090448.

Spiro, B., Weiss, D., Purvis, O., Mikhailova, I., Williamson, B., Coles, B., Udachin, V., 2004. Lead isotopes in lichen transplants around a Cu smelter in Russia determined by MC-ICP-MS reveal transient records of multiple sources. Environ. Sci. Technol. 38, 6522–6528.

Spiro, B., Udachin, V., Williamson, B., Purvis, O., Tessalina, S., Weiss, D., 2012. Lacustrine sediments and lichen transplants: two contrasting and complimentary environmental archives of natural and anthropogenic lead in the South Urals, Russia. Aquat. Sci. 75 (2), 185–198. https://doi.org/10.1007/s00027-012-0266-3.

Udachin, V., Williamson, B.J., Purvis, O.W., Dubbin, W., Brooks, S., Coste, B., Herrington, R.J., Mikhailova, I., 2003. Assessment of environmental impacts of active smelter operations and abandonet mines in Karabash, Ural mountains of Russia. Sustain. Dev. 11, 1–10.

Williamson, B.J., Mikhailova, I., Purvis, O.W., Spiro, B., Udachin, V., 2004a. SEM-EDX analysis in the source apportionment of particulate matter on Hypogymnia Physodes lichen transplants around the Cu smelter and former mining town of Karabash, South Urals, Russia. Sci. Total Environ. 322 (1–3), 139–154.

Williamson, B.J., Udachin, V., Purvis, O.W., Spiro, B., Cressey, G., Jones, G.C., 2004b. Characterisation of airborne particulate pollution in the Cu smelter and former mining town of Karabash, South Ural Mountains of Russia. Environ. Monit. Assess. 98 (1–3), 235–259.

Williamson, B.J., Purvis, O.W., Mikhailova, I.N., Spiro, B., Udachin, V., 2008. The lichen transplant methodology in the source apportionment of metal deposition around a copper smelter in the former mining town of Karabash, Russia. Environ. Monit. Assess. 141 (1–3), 227–236.

Yao, Z.T., Ji, X.S., Sarker, P.K., Tang, J.H., Ge, L.Q., Xia, M.S., Xi, Y.Q., 2015. A comprehensive review on the applications of coal fly ash. Earth Sci. Rev. 141, 105–121. https://doi.org/10.1016/j.earscirev.2014.11.016.

Algal bioproducts, biofuels, and biorefinery for business opportunities

20

Phycoremediation and business prospects

V. Sivasubramanian[a] and Majeti Narasimha Vara Prasad[b]

[a]Phycospectrum Environmental Research Centre (PERC), Chennai, Tamil Nadu, India
[b]Department of Plant Sciences, School of Life Sciences, University of Hyderabad
(An Institution of Eminence), Hyderabad, Telangana, India

1 Introduction

Much progress has been made in last 20 years in developing biotechnology for algal mass culture. Commercial production of microalgae has been practiced since 1950s commencing with the production of *Chlorella vulgaris* in Japan and Taiwan and later followed with the production of *Spirulina*. These algae are being produced for use as human health food. Microalgae are considered one of the most promising feedstock for biofuels. Although microalgae are not yet produced at large scale for bulk applications, recent advances—particularly in the methods of systems biology, genetic engineering, and biorefining—present opportunities to develop this process in a sustainable and economical way within the next 10–15 years.

2 Phycoremediation

Phycoremediation is defined as the use of algae to remove pollutants from the environment or to render them harmless. In the last couple of years, large number of publications highlighted the use of algae in removing pollutants from wastewater and promoting biorefinery and circular economy (Almeida et al., 2022; Prajapati et al., 2013a,b; Andrade et al., 2022; Angelaalincy et al., 2023; Apandi et al., 2022; Aranguren Díaz et al., 2022; Arrojo et al., 2022; Arutselvan et al., 2022; Bauddh and Korstad, 2022; Chaudhary et al., 2022; Danouche et al., 2022; Devi et al., 2022; Gani et al., 2022; Goswami et al., 2022; Hamed et al., 2022; Hu et al., 2022; Kiran and Mohan, 2022; Koul et al., 2022; Kumar et al. 2022; Kumari et al., 2022; Kurade et al., 2022; Lakhani et al., 2022; Michelon et al., 2022a,b; Pathy et al., 2022;

Pradhan et al., 2022; Rambabu et al., 2022; Razaviarani et al., 2022; Rezasoltani and Champagne, 2023; Ricky and Shanthakumar, 2022; Ricky et al., 2022; Sarkar et al., 2022; Selvaraj et al., 2022; Shackira et al., 2022; Shakoor et al., 2022; Sharma et al., 2023; Shayesteh et al., 2022; Silva et al., 2022; Sisman-Aydin and Simsek, 2022; Tan et al., 2022; Ummalyma and Singh, 2022; Verma et al., 2022; Waqas et al., 2022; Angelaalincy et al., 2023). Globally phycoremediation research gained considerable momentum (Fig. 1). Currently, phycoremediation is highly diversified with multiple applications (Table 1, Fig. 2). A large number of successful applications of phycoremediation have been implemented for the treatment of a wide variety of industrial effluents (Fig. 3).

Large-scale phycoremediation of industrial effluent has been done successfully in some industries in India (Sivasubramanian, 2016 and the references there in). Using algae-based treatment technology, efficient pH correction, sludge reduction, and reduction of biological oxygen demand (BOD) and chemical oxygen demand (COD) could be achieved while avoiding toxic chemicals by these industries. During effluent treatment process, huge amount of valuable algal biomass is also being generated by these industries. The growing demand for a safer, cost-effective method of cleaning up of contaminated soil and wastewaters makes phycoremediation a popular and reasonable alternative throughout the world. It takes advantage of the alga's natural ability to take up, accumulate, and sometimes degrade constituents that are present in their growth environment. Algal species are relatively easy to grow, adapt, and manipulate within a laboratory setting and appear to be ideal organisms for use in remediation studies (Sivasubramanian, 2016 and the references there in). There are several reasons for choosing algae for bioremediation purposes, namely (a) ease of their cultivation, (b) minimal requirements of special chemicals for growth, and (c) they make their own oxygen using sunlight as a primary source for energy. Further, India being a tropical country with plenty of sunshine is well suited for implementation of phycoremediation as a technology.

Phycoremediation is a remediation technology based primarily on algae (Sivasubramanian, 2016 and the references there in). It has a very good market in India. The first industry we have installed 9 years back is saving lot of money on chemicals and through sale of products such as biofertilizer, compost, and aquaculture feed; this industry is making US $2.5 million. In our

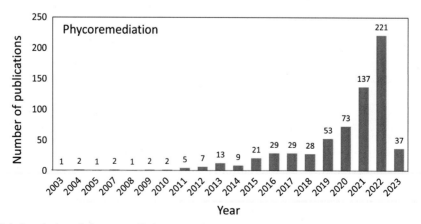

FIG. 1   Global explosion of phycoremediation research.

TABLE 1 Salient phycoremediation research findings during 2022 and 2023.

	Author(s)
Phycoremediation of Coastal Marine Water Contaminated with Dissolved Oil by *Nannochloropsis oculata*	Almeida et al. (2022)
Phycoremediation of Copper by *Chlorella protothecoides* (UTEX 256): Proteomics of Protein Biosynthesis and Stress Response	Andrade et al. (2022)
Phycoremediation of Arsenic and biodiesel production using green microalgae Coelastrella sp. M60—an integrated approach	Angelaalincy et al. (2023)
Scenedesmus sp. Harvesting by Using Natural Coagulant after Phycoremediation of Heavy Metals in Different Concentrations of Wet Market	Apandi et al. (2022)
Phycoremediation as a Strategy for the Recovery of Marsh and Wetland with Potential in Colombia	Aranguren Díaz et al. (2022)
Potential of the microalgae *Chlorella fusca* (Trebouxiophyceae, Chlorophyta) for biomass production and urban wastewater phycoremediation	Arrojo et al. (2022)
Phycoremediation of textile and tannery industrial effluents using microalgae and their consortium for biodiesel production	Arutselvan et al. (2022)
Phycoremediation: use of algae to sequester heavy metals	Bauddh and Korstad (2022)
Newly isolated Acutodesmus obliquus assisted phycoremediation and sequestration of CETP wastewater: an integrated management of remediation. Biomass Conversion and Biorefinery, pp. 1–11	Chaudhary et al. (2022)
Pb (II)-phycoremediation mechanism using *Scenedesmus obliquus*: cells physicochemical properties and metabolomic profiling	Danouche et al. (2022)
Trends in Waste Water Treatment using Phycoremediation for Biofuel Production. In Bioremediation of Toxic Metal (loid)s	Devi et al. (2022)
Outdoor phycoremediation and biomass harvesting optimization of microalgae *Botryococcus* sp. cultivated in food processing wastewater using an enclosed photobioreactor	Gani et al. (2022)
Phycoremediation of nitrogen and phosphate from wastewater using Picochlorum sp.: A tenable approach	Goswami et al. (2022)
Evaluation of the phycoremediation potential of microalgae for captan removal: Comprehensive analysis on toxicity, detoxification and antioxidants modulation	Hamed et al. (2022)
Progress on microalgae biomass production from wastewater phycoremediation: metabolic mechanism, response behavior, improvement strategy and principle	Hu et al. (2022)
Phycoremediation potential of *Tetradesmus* sp. SVMIICT4 in treating dairy wastewater using Flat-Panel photobioreactor	Kiran and Mohan (2022)
Phycoremediation: A sustainable alternative in wastewater treatment (WWT) regime	Koul et al. (2022)
Phycoremediation of cashew nut processing wastewater and production of biodiesel using *Planktochlorella nurekis* and *Chlamydomonas reinhardtii*	Kumar et al. (2023a)

Continued

TABLE 1   Salient phycoremediation research findings during 2022 and 2023—cont'd

	Author(s)
Experimental and optimization studies on phycoremediation of dairy wastewater and biomass production efficiency of *Chlorella vulgaris* isolated from Ganga River, Haridwar, India	Kumari et al. (2022)
Integrated phycoremediation and ultrasonic-irradiation treatment (iPUT) for the enhanced removal of pharmaceutical contaminants in wastewater	Kurade et al. (2022)
A comprehensive study of bioremediation for pharmaceutical wastewater treatment	Lakhani et al. (2022)
Amino acids, fatty acids, and peptides in microalgae biomass harvested from phycoremediation of swine wastewaters	Michelon et al. (2022a)
Virucidal activity of microalgae extracts harvested during phycoremediation of swine wastewater	Michelon et al. (2022b)
Malachite green removal using algal biochar and its composites with kombucha SCOBY: An integrated biosorption and phycoremediation approach	Pathy et al. (2022)
Microalgal Phycoremediation	Pradhan et al. (2022)
Phycoremediation for carbon neutrality and circular economy: Potential, trends, and challenges	Rambabu et al. (2022)
Algal biomass dual roles in phycoremediation of wastewater and production of bioenergy and value-added products	Razaviarani et al. (2022)
An integrated approach for the phycoremediation of Pb (II) and the production of biofertilizer using nitrogen-fixing cyanobacteria	Rezasoltani and Champagne (2023)
Phycoremediation integrated approach for the removal of pharmaceuticals and personal care products from wastewater—a review	Ricky and Shanthakumar (2022)
A pilot-scale study of the integrated phycoremediation-photolytic ozonation based municipal solid waste leachate treatment process	Ricky et al. (2022)
Modeling and optimization of phycoremediation of heavy metals from simulated ash pond water through robust hybrid artificial intelligence approach	Sarkar et al. (2022)
An insight on pollutant removal mechanisms in phycoremediation of textile wastewater	Selvaraj et al. (2022)
Phycoremediation: a means for restoration of water contamination	Shackira et al. (2022)
Autochthonous Arthrospira platensis Gomont Driven Nickel (Ni) Phycoremediation from Cooking Oil Industrial Effluent	Shakoor et al. (2022)
A waste-based circular economy approach for phycoremediation of X-ray developer solution	Sharma et al. (2023)
Long term outdoor microalgal phycoremediation of anaerobically digested abattoir effluent	Shayesteh et al. (2022)
Fluoxetine and Nutrients Removal from Aqueous Solutions by Phycoremediation	Silva et al. (2022)

TABLE 1 Salient phycoremediation research findings during 2022 and 2023—cont'd

	Author(s)
Investigation of the phycoremediation potential of freshwater green algae *Golenkinia radiata* for municipal wastewater	Sisman-Aydin and Simsek (2022)
Advances in POME treatment methods: potentials of phycoremediation, with a focus on South East Asia	Tan et al. (2022)
Biomass production and phycoremediation of microalgae cultivated in polluted river water	Ummalyma and Singh (2022)
Phycoremediation of milk processing wastewater and lipid-rich biomass production using *C. vulgaris* under continuous batch system	Verma et al. (2022)
Phycoremediation of textile effluents with enhanced efficacy of biodiesel production by algae and potential use of remediated effluent for improving growth of wheat	Waqas et al. (2022)
Phycoremediation of Arsenic and biodiesel production using green microalgae *Coelastrella* sp. M60–an integrated approach	Angelaalincy et al. (2023)
Green synthesis of AgNPs, alginate microbeads and *Chlorella minutissima* laden alginate microbeads for tertiary treatment of municipal wastewater	Abdo et al. (2023)
Microbial assisted multifaceted amelioration processes of heavy-metal remediation	Agrawal et al. (2023)
Spirogyra sp. and *Chlorella* sp. in Bioremediation of Mine Drainage	Almeida et al. (2023)
Sustainable Technologies for Liquid Waste Treatment	Al Zohbi (2023)
Novel Electrokinetic-Assisted Membrane Photobioreactor (EK-MPBR) for Wastewater Treatment	Amini et al. (2022)
Algae and bacteria consortia for wastewater decontamination and transformation into biodiesel, bioethanol, biohydrogen, biofertilizers and animal feed	Anand et al. (2023)
Phycoremediation of Arsenic and biodiesel production using green microalgae *Coelastrella* sp. M60	Angelaalincy et al. (2023)
Towards Viable Eco-Friendly Local Treatment of Blackwater in Sparsely Populated Regions	Anusuyadevi et al. (2023)
The Effect of Tofu Wastewater and pH on the Growth Kinetics and Biomass Composition of *Euglena* sp	Asiandu et al. (2023)
The effect of temperature and photoperiod on bioremediation potentials of mono and mixed cultures of chlorophyte and cyanobacteria for municipal wastewater treatment	Badamasi et al. (2023)
Genetic engineering of algae	Badiefar et al. (2023)
Recycling drinking water RO reject for microalgae-mediated resource recovery	Bhandari et al. (2023)
Chlorella pyrenoidosa-mediated removal of pathogenic bacteria from municipal wastewater—multivariate process optimization and application in the real sewage	Bhatt et al. (2023)

Continued

V. Algal bioproducts, biofuels, and biorefinery for business opportunities

TABLE 1   Salient phycoremediation research findings during 2022 and 2023—cont'd

	Author(s)
Enhanced decolourisation and degradation of azo dyes using wild versus mutagenic improved bacterial strain: a review	Bhayana et al. (2023)
Phycoremediation of Toxic Metals for Industrial Effluent Treatment	Bloch and Ghosh (2023)
Circular bioeconomy in palm oil industry: Current practices and future perspectives	Cheah et al. (2023)
Omics Approaches for Microalgal Applications in Wastewater Treatment	Chowdhury et al. (2023)
Evaluation of the potential of *C. vulgaris* for the removal of pollutants from standing water	Cordero et al. (2023a)
Use Of Chlorophyll Microorganisms To Improve Physico-Chemical Properties Of Wastewater	Cordero et al. (2023b)
Addressing the Strategies of Algal Biomass Production with Wastewater Treatment	Dasauni and Nailwal (2023)
Microalgae-based wastewater treatment for micropollutant removal in swine effluent: High-rate algal ponds performance under different zinc concentrations	de Sousa Oliveira et al. (2023)
Bioremediation of organic and inorganic contaminants by microbes	Dey et al. (2023)
Removal of PFAS by Biological Methods	Douna and Yousefi (2023)
Genetic Engineering of Algae	Durairaj et al. (2023)
Using Algae as a Renewable Source in the Production of Biodiesel	Dursun (2023)
Coupling wastewater treatment, biomass, lipids, and biodiesel production of some green microalgae	El-Sheekh et al. (2023)
Bioremediation Using Microalgae and Cyanobacteria and Biomass Valorisation	Encarnação et al. (2023)
Integrated microalgae-based biorefinery for wastewater treatment, industrial CO_2 sequestration and microalgal biomass valorization: A circular bioeconomy approach	Fal et al. (2023)
Exploring the Benefits of Phycocyanin: From *Spirulina* Cultivation to Its Widespread Applications	Fernandes et al. (2023)
Reuse of wastewater from the production of microalgae and its effect on the growth of *Pelargonium x hortorum*	Garcia-Corral et al. (2023)
Microalgal-induced remediation of wastewaters loaded with organic and inorganic pollutants: An overview	Ghaffar et al. (2023)
Multifaceted role of microalgae for municipal wastewater treatment: a futuristic outlook towards wastewater management	Goswami et al. (2023)
Variation in phytoplankton diversity during phycoremediation in a polluted Colombian Caribbean swamp	Gutiérrez-Hoyos et al. (2023)
Research on the Tolerance and Degradation of o-Cresol by Microalgae	Han et al. (2023)

TABLE 1   Salient phycoremediation research findings during 2022 and 2023—cont'd

	Author(s)
Nutrient uptake and biomass productivity performance comparison among freshwater filamentous algae species on mesocosm-scale FANS under ambient summer and winter conditions	Hariz et al. (2023)
Integration of Real Industrial Wastewater Streams to Enhance *C. vulgaris* Growth: Nutrient Sequestration and Biomass Production.	Javed et al. (2023)
Phyco-remediation: Role of Microalgae in Remediation of Emerging Contaminants	Jha et al. (2023)
Biosorption of pollutants from chemically derived wastewater using *Microcoleus* sp. AQUA-Water Infrastructure, Ecosystems and Society	Kabir et al. (2023)
Myco-and phyco-remediation of polychlorinated biphenyls in the environment: a review	Kaleem et al. (2023)
Utilization of nitrogen-rich agricultural waste streams by microalgae for the production of protein and value-added compounds	Khan et al. (2023)
Bioremediation of Arsenic A Sustainable Approach in Managing	Khanikar and Ahmaruzzaman (2023)
Cyanidiales-Based Bioremediation of Heavy Metals	Kharel et al. (2023)
Metal (loid)s Sources, Toxicity and Bioremediation	Kidwai et al. (2023)
An Overview of Algae for Biodiesel Production Using Bibliometric Indicators. International Journal of Energy Research,	Kombe (2023)
Bioremediation of ethanol wash by microalgae and generation of bioenergy feedstock	Kookal et al. (2023)
Exploring the evolution, trends and scope of microalgal biochar through scientometrics	Krishnamoorthy et al. (2023)
Phycoremediation of cashew nut processing wastewater and production of biodiesel using *Planktochlorella nurekis* and *C. reinhardtii*	Kumar et al. (2023a)
Phycoremediation of sewage wastewater and industrial flue gases for biomass generation from microalgae	Kumar et al. (2018)
Management of tannery waste effluents towards the reclamation of clean water using an integrated membrane system	Kumar et al. (2023b)
Algal Biomass Production Coupled to Wastewater Treatment	Kumar et al. (2023c)
Agriculture of microalgae *C. vulgaris* for polyunsaturated fatty acids (PUFAs) production employing palm oil mill effluents (POME) for future food, wastewater, and energy nexus	Kumaran et al. (2023)
Global scenario and technologies for the treatment of textile wastewater	Kurade et al. (2023a)
Integrated phycoremediation and ultrasonic-irradiation treatment (iPUT) for the enhanced removal of pharmaceutical contaminants in wastewater	Kurade et al. (2023b)
Integration of microalgae cultivation and anaerobic co-digestion with dairy wastewater to enhance bioenergy and biochemicals production	Kusmayadi et al. (2023)

Continued

TABLE 1 Salient phycoremediation research findings during 2022 and 2023—cont'd

	Author(s)
Remediation of Environmental Contaminants Through Phytotechnology	Latif et al. (2023)
Water reclamation from palm oil mill effluent (POME): Recent technologies, by-product recovery, and challenges	Mahmod et al. (2023)
A Comprehensive Review on Microalgae-Based Biorefinery as Two-Way Source of Wastewater Treatment and Bioresource Recovery	Malik et al. (2023)
Techno-economic identification of production factors threatening the competitiveness of algae biodiesel	Maroušek et al. (2023)
Phycoremediation: algae-based bioremediation	Mathias et al. (2023)
Recent update on gasification and pyrolysis processes of lignocellulosic and algal biomass for hydrogen production	Mishra et al. (2023)
Potential of microalgae for phytoremediation of various wastewaters for nutrient removal and biodiesel production through nanocatalytic transesterification	Mittal and Ghosh (2023)
Impact of Microalgae in Domestic Wastewater Treatment: A Lab-Scale Experimental Study	Moondra et al. (2023)
Effects of Bioabsorption of Chromium and Lead on Cyanobacterial Species (Oersted ex Gomont *Spirulina sub salsa* and *Lemmermann*) *Calothrix marchica*	Mukherjee and Ray (2023)
Bioleaching of critical metals using microalgae	Mukherjee et al. (2023)
Review on Microalgae Potential Innovative Biotechnological Applications	Mulluye et al. (2023)
Reformation of dairy effluent—a phycoremediation approach	Nachiappan and Chandrasekaran (2023)
Photobioreactors for microalgae-based wastewater treatment	Nagarajan et al. (2023)
Bioremediation Approaches for Genomic Microalgal Applications in Wastewater Treatment	Nirmala et al. (2023)
Organic pollutant and dye degradation with nanocomposites	Noreen and Ahmad (2023)
Biomass Fuel Production through Cultivation of Microalgae *Coccomyxa dispar* and *Scenedesmus parvus* in Palm Oil Mill Effluent and Simultaneous Phycoremediation	Ooi et al. (2023)
Wetland Removal Mechanisms for Emerging Contaminants	Overton et al. (2023)
Microalgae as Biological Cleanser for Wastewater Treatment	Palekar et al. (2023)
Current Developments in Bioengineering and Biotechnology	Pandey et al. (2023)
Tailoring Microalgae for Efficient Biofuel Production	Patel et al. (2023)
The Bioremediation Potential of *Ulva lactuca* (Chlorophyta) Causing Green Tide in Marchica Lagoon (NE Morocco, Mediterranean Sea)	Rahhou et al. (2023)
Heavy metal pollution of river water and eco-friendly remediation using potent microalgal species	Raj et al. (2023)

TABLE 1 Salient phycoremediation research findings during 2022 and 2023—cont'd

	Author(s)
Understanding the bioaccumulation of pharmaceuticals and personal care products	Rajput et al. (2023)
A review on algae biosorption for the removal of hazardous pollutants from wastewater: Limiting factors, prospects and recommendations	Ramesh et al. (2023)
Nutrient recovery from wastewaters by algal biofilm for fertilizer production Part 1: Case study on the techno-economical aspects at pilot-scale	Reinecke et al. (2023)
An integrated approach for the phycoremediation of Pb (II) and the production of biofertilizer using nitrogen-fixing cyanobacteria	Rezasoltani and Champagne (2023)
Microalgal cultivation on digestate: Process efficiency and economics.	Rossi et al. (2023)
Byproducts of a Microalgal Biorefinery as a Resource for a Circular Bioeconomy	Roy and Mohanty (2023)
Algae-Based Treatment of Domestic and Industrial Wastewater	Sakarya et al. (2023)
Sustainability index analysis of microalgae cultivation from biorefinery palm oil mill effluent	Santoso et al. (2023)
Application of multi-gene genetic programming technique for modeling and optimization of phycoremediation of Cr (VI) from wastewater	Sarkar et al. (2023)
Photobioreactor design and parameters essential for algal cultivation using industrial wastewater	Sathinathan et al. (2023)
Algal biorefinery: an integrated process for industrial effluent treatment and improved lipid production in bioenergy application	Saxena et al. (2023)
Microalgal cultivation on grass juice as a novel process for a green biorefinery	Schoeters et al. (2023)
Algae-Powered Buildings: A Review of an Innovative, Sustainable Approach	Sedighi et al. (2023)
A critical review on phycoremediation of Pollutants from Wastewater—A Novel Algae Based Secondary Treatment with the Opportunities of Production of Value-Added Products	Sengupta et al. (2023)
Comparative analysis of biodegradation and characterization study on agal-assisted wastewater treatment in a bubble column photobioreactor	Sevugamoorthy and Rangarajan (2023)
Phycoremediation Processes in Industrial Wastewater Treatment. CRC Press	Shah (2023)
Ecosystem services and climate action from a circular bioeconomy perspective	Sharma and Malaviya (2023)
A waste-based circular economy approach for phycoremediation of X-ray developer solution	Sharma et al. (2023)
Influence of subsoil microbial community across different depths on the biodegradation of pesticides	Sheikh et al. (2023)
Anaerobic digestion as a tool to manage eutrophication and associated greenhouse gas emission	Singh et al. (2023a)
Dairy wastewater treatment using *Monoraphidium* sp. KMC4 and its potential as hydrothermal liquefaction feedstock	Singh et al. (2023b)

Continued

TABLE 1   Salient phycoremediation research findings during 2022 and 2023—cont'd

	Author(s)
Sustainable Use of CO_2 and Wastewater from Mushroom Farm for *C. vulgaris* Cultivation: Experimental and Kinetic Studies on Algal Growth and Pollutant Removal	Širić et al. (2023)
Current Developments in Biotechnology and Bioengineering: Photobioreactors: Design and Applications	Sirohi et al. (2023)
Influence of supplementary carbon on reducing the hydraulic retention time in microalgae-bacteria (MaB) treatment of municipal wastewater	Soroosh et al. (2023)
Bioremediation of Textile Dyes for Sustainable Environment—A Review	Sridharan and Krishnaswamy (2023)
Immobilized Micro Algae for Removing Wastewater Pollutants and Ecotoxicological View of Adsorbed Nanoparticles—An Overview	Suganya et al. (2023)
Comparative studies on phycoremediation efficiency of different water samples by microalgae	Talib et al. (2023)
A review on microalgal-bacterial Co-culture: The multifaceted role of beneficial bacteria towards enhancement of microalgal metabolite production	Tong et al. (2023)
Sustainable microalgal cultivation in poultry slaughterhouse wastewater for biorefinery products and pollutant removal	Ummalyma et al. (2023)
Microalgae-based biotechnologies for treatment of textile wastewater.	Xiong et al. (2023)
Omics Approach in Bioremediation of Heavy Metals	Yadav et al. (2023)
Evaluation of bioactive compounds from robust consortium microalgae species under actual flue gas exposure	Yahya et al. (2023)
Integrated route of fast hydrothermal liquefaction of microalgae and sludge by recycling the waste aqueous phase for microalgal growth	Yuan et al. (2023)
Microalgae-based bioremediation of pharmaceuticals wastewater	Zahra et al. (2023)
Phycoremediation in anaerobically digested palm oil mill effluent using cyanobacterium, *Spirulina platensis*	Zainal et al. (2012)
Microalgae biopolymers: a review	Zatta et al. (2023)
Efficient nutrient removal of Pyropia-processing wastewater and rapid algal biomass harvesting by *S. obliquus* combined with chitosan	Zheng et al. (2023)
Microalgal Carbon Dioxide (CO_2) Capture and Utilization from the European Union Perspective	Zieliński et al. (2023)

recent project at an acrylon fiber dyeing company, we have intervened in a crucial step of already existing effluent treatment plant (ETP) resulting in color removal and detoxification of gel dyeing effluent, which was very difficult to handle with conventional technology, and on chemicals, this industry is saving US $130,000 per year. Our successful demonstration of petrochemical waste at Pacific Rubiales oil drilling site in Colombia (Latin America) is a proof of marketability in other countries too.

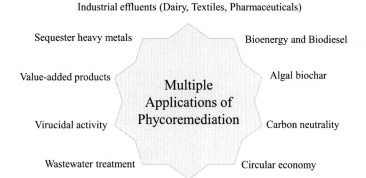

FIG. 2   Multiple applications of phycoremediation.

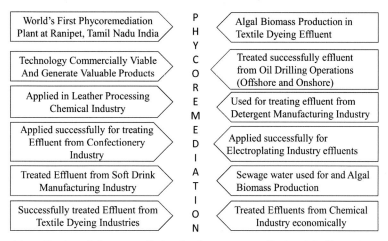

FIG. 3   Successful application of phycoremediation for the treatment of industrial effluents.

Our approach is always to implement algal technology in one of the identified steps of existing ETP and improving the efficiency without employing the chemicals that are otherwise used by regular ETPs saving lots of money for the industry. When we visit an industry, we do not go with a rigid design, and we make a customized design of the process only after the visit and feasibility studies, we usually conduct with typical samples of effluent. Since, when algal technology works fine, the operation cost will be reduced by 90% and more it will be very attractive and very easy to convince any industry. The present paper deals with author's own experience in dealing with a variety of industrial effluents, wastewaters, and sludge to develop effective microalgae-based remediation technology and to generate valuable algal biomass during that process.

3 Major challenges in algal biomass production

Sustainable production of algal biomass, which is also economically viable, is a major challenge. Providing good-quality nutrients in sufficient quantities has been the major hurdle in the production of biomass. The production of large quantities of biomass also requires a large amount of CO_2. A total of 1.8 tons of CO_2 is needed to produce 1 ton of algal biomass. The main nutrients needed for the production of microalgae are nitrogen and phosphorus. For sustainable production, it will be important to make use of residual nutrient sources and to recycle nutrients as much as possible.

After production, the biomass needs to be harvested, the lipids extracted, and the remaining cell components recovered. Harvesting of microalgae is currently expensive because of the high energy requirements and capital costs involved. Major inputs that will determine the cost of production of algal biomass are the nutrients. If we can substitute the costly nutrients with wastewater, effluent and sludge from industries, and sewage, the cost of production could be minimized to a great extent. Integrating phototrophic system with heterotrophic bioreactors can improve not only productivity but also the oil content. Waste from sugar mill can be a better choice for growing algae in bioreactors heterotrophically.

4 World's first phycoremediation plant at Ranipet, India

Large-scale phycoremediation projects at lab/pilot and filed scale have been considered for their dependable service to treat wastewater and industrial effluents (Sivasubramanian, 2016 and the references there in).

India's largest phytoremediation plant is in operation at SNAP Natural and Alginate Products, Ranipet, India, from September 2006 (Fig. 4). The industry generates 30–40 kL of highly acidic effluent every day, which is being pH-corrected and evaporated using an algae-based treatment technology developed by Sivasubramanian and his team from Phycospectrum Environmental Research Centre (PERC), Chennai, India (Sivasubramanian, 2016 and the references there in). The physicochemical characteristics of effluent are provided in Table 2. There is 100% reduction in sludge by phycoremediation.

FIG. 4 Phycoremediation plant in operation at SNAP industries, India, from 2006.

TABLE 2 A comparison of physicochemical parameters of raw effluent with effluent taken from the bottom of the tank after 2 years of phycoremediation (evaporation at 30 kL/day/2 years) (values±SD) (Sivasubramanian, 2016).

Parameters	Raw effluent	Effluent taken from the bottom of the tank after 2 years of phycoremediation
Turbidity NTU	106±7.20	5.9±0.15
TDS (mg/L)	27,600±600	49,220±230
Electrical conductivity (μmho/cm)	36,430±120	69,896±354
pH	1.66±0.15	7.02±0.17
Alkalinity pH (mg/L)	0	0
Alkalinity (mg/L)	Nil	2916±160.8
Total hardness (mg/L)	2100±123	5125±110
Ca (mg/L)	520±15.7	1120±49.3
Mg (mg/L)	192±4.2	558±7.7
Na (mg/L)	6800±143.7	7750±123.5
K (mg/L)	700±15.3	8125±150.7
Fe (mg/L)	17.99±0.34	4.13±0.19
Mn (mg/L)	Nil	Nil
Free ammonia (mg/L)	56±7.2	13.44±0.28
NO_2 (mg/L)	0.43±0.04	Nil
NO_3 (mg/L)	22±1.5	16±1.2
Chloride (mg/L)	3216±14.6	12,189±110
Fluoride (mg/L)	0.62±0.14	0
Sulfate (mg/L)	5221±110.8	1195±89.3
Phosphate (mg/L)	28.62±1.22	169±5.67
SiO_2 (mg/L)	5.48±0.23	79.49±3.45
BOD (mg/L)	44±5.6	960±23.5
COD (mg/L)	148±12.5	3266±24.6

5 The technology and biomass-based commercially valuable products

Phycoremediation plant is used to treat the acidic effluent from this alginate industry. The liquid effluent is highly acidic. Conventionally, sodium hydroxide has been used for the neutralization of the acidic effluent, which results in an increase in total dissolved solids (TDSs) and the generation of solid waste. The study was conducted in three stages. In the first stage,

FIG. 5 Biofertilizer and aquaculture product produced by SNAP industry.

the solar ponds used for evaporating the effluent were converted into high-rate algal ponds with *Chroococcus turgidus*, a blue green alga. Based on the results of pilot plant studies, a full scaling-up of the slope tank was made. With the addition of around 30 kL of acidic effluent every day, the pH of the effluent remained constant around 7.02, and total dissolved salts stabilized at 49 g/L. There was no sludge formation even after 2 years of operation. With just one circulation of the effluent at a pumping rate of 80 kL/h on the slopes, the desired evaporation of 30 kL was achieved. Algal cell density is maintained at $2300 \times 10^4 \, mL^{-1}$ (0.75 g/L on dry weight basis). Excess algal biomass is regularly harvested by the industry to produce two important products, namely biocompost, biofertilizer, and aquaculture feed (Fig. 5). Both these products (06 EMMA and PLANK-10) are produced at 2 tons per month ($2 million per year). PLANK-10 is widely used aquaculture farms all over India to reduce harmful algae, improve plankton levels, reduce BOD and COD, and improve dissolved oxygen in fisheries ponds. Algal mats growing on slope surface of ETP plant are also harvested on a regular basis. Algal biomass also settles down at the bottom of ETP plant, which is used as manure.

Phycoremediation is a sophisticated and multifaceted effluent treatment process. And each case of application will address a different set of operational imperatives. The summary given in Table 3 depicts the business case for this technology in the case of this particular application.

6 Leather processing chemical industry

Leather processing industry in Tamil Nadu, India, was selected for investigation. The industry manufactures dyes, binders, and pigments. The effluent and sludge generated have heavy metals and residual chemicals used in production. Algal technology is employed in treating the effluent and sludge generated by the industry (60–70 kL/day). Table 4 gives

TABLE 3 SNAP phycoremediation plant—operational and financial comparison.

Cost parameter	Conventional effluent treatment	Phycoremediation	Annual cost benefit
Acidity—high levels of dissolved carbon dioxide	Neutralization with caustic soda	Algal treatment to absorb the acidic contents and neutralize the effluent	Rs. 50 lakhs spent for caustic soda annually is saved (100%) The total cost for the utilities (labor/electricity, etc.) used in the operation is almost identical. At around Rs. 2 lakhs p.a.[a]
Sludge formation	Evaporation of effluents deposits sludge. That needs to be buried in a land fill About 290 tons of sludge produced annually from this treatment	Algal remediation produces a nutrient rich, commercially valuable fertilizer that is highly demanded in the market There is no residual sludge	The sludge disposal used to cost an estimated Rs. 3 lakhs annually. This cost is saved[a] Additionally revenues from the sale of algal biomass fertilizer
Structures and space	11,000 m² of masonry tank for evaporating the effluent	3000 m² of tank for containing and evaporating the effluent	About 75% of the effluent treatment facility space is released. This very valuable real estate structural space is now being used for other productive uses

[a] 3000 USD.

TABLE 4 Physicochemical characteristics of effluent from a leather processing industry in Tamil Nadu, India.

Parameter	Raw	Treated	Parameter	Raw	Treated
Turbidity NTU	87.3	21.3	Total Kjeldahl Nitrogen (TKN)	72.8	63.84
Total solids, mg/L	3401	3275	Free ammonia (as NH_3), mg/L	70.56	50.4
Total suspended solids, mg/L	187	503	Nitrite (as NO_2), mg/L	0.85	0.10
Total dissolved solids, mg/L	3214	2772	Nitrate (as NO_3), mg/L	19	3
Electrical conductivity, μmho/cm	4561	3876	BOD, mg/L	210	35
pH	8.08	8.55	COD, mg/L	602	103
Alkalinity total (as $CaCO_3$), mg/L	2048	1512	Iron (as Fe), mg/L	2.43	0.13
Total hardness (as $CaCO_3$), mg/L	500	323	Chloride (as Cl), mg/L	458	208
Calcium (as Ca), mg/L	110	81	Sulfate (as SO_4), mg/L	184	52
Magnesium (as Mg), mg/L	54	29	Phosphate (as PO_4), mg/L	83.66	0.16
Sodium (as Na), mg/L	800	880	Silica (as SiO_2), mg/L	54.12	45.46
Potassium (as K), mg/L	150	70	Tidy's test (as O), mg/L	5.5	16

information on the physicochemical parameters of the effluent. Analysis shows that effluent has almost all the essential nutrients needed for algal growth.

Treated water is recycled by the industry. The sludge produced by the industry also supports very good growth of algae. *C. vulgaris* grows luxuriantly in leather processing chemical industry (reaching 1 g/L dry weight) (Figs. 6 and 7). Regular harvesting of microalgal biomass is done, and the treated water is reused.

FIG. 6   Leather processing industry in Tamil Nadu, India: (A) raw effluent, (B) after treatment *Chlorella vulgaris* (450 cells $\times$ 104 mL^{-1} of) in.

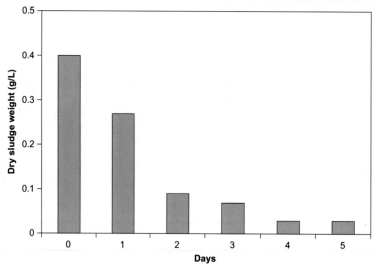

FIG. 7   Growth of *Chlorella vulgaris* and sludge reduction in effluent from a leather processing industry.

7 Effluent from confectionery industry

A confectionery industry in Tamil Nadu, India, was selected for the study. The total effluent generated per day amounts to an average of 50–70 kL. The plant effluent generated is divided into two streams, namely industrial effluent stream from the production process and sewage effluent stream from the human activities. These are mixed prior to sending to the equalization tank. The effluent for phycoremediation treatment is taken after it goes through the dissolved air floatation (DAF) in the conventional treatment method (Fig. 8). pH is conventionally corrected by adding caustic soda at the equalization tank stage, which results in doubling of TDS. After pH correction, the effluent is sent to buffer tank and anaerobic reactors (ARs) to digest rest of the organic compounds. After digestion in AR, effluent goes through a series of clarifiers and sand filter and finally taken to R/O for recycling.

8 Effluent characteristics

The characteristic raw effluent produced by this confectionery industry is characterized by its organic content, which is composed of easily biodegradable compounds such as sugars, sweeteners, casein, vegetable oils, acacia gums, condensed milk, and food coloring and

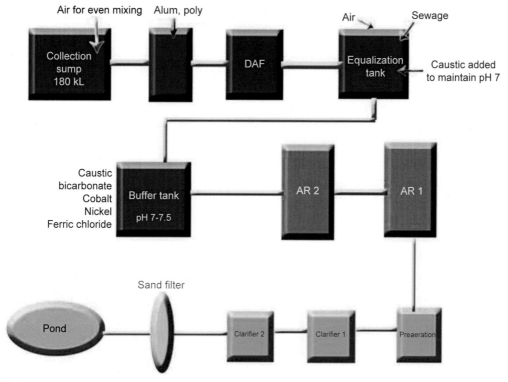

FIG. 8 ETP facility at the confectionery industry (algal remediation was introduced at the fourth stage, equalization tank, to correct pH, reduce sugar levels, reduce BOD and COD).

flavoring agents. This confectionery industry is using major ingredients such as sugar and sweeteners, natural colors, acacia gum, sugar substitutes, gum base, and flavors for all type of confectionery products. They use number a chemicals during the process, and they all become a part of the effluent. This liquid effluent is acidic in pH, dominated by yeast cell population. Table 5 shows the physicochemical parameters of effluent from confectionery industry.

9 Algal treatment

C. vulgaris grows very well in the raw effluent. It utilizes the sugar present in the effluent. The effluent becomes less turbid, and this reduces the load for the AR (Table 6). Table 7 shows growth of *C. vulgaris* in pilot tank. Phycoremediation helps to correct pH, reduce sugar levels,

TABLE 5 Physicochemical characteristics of effluent from a confectionery industry near Chennai, India.

S. No.	Parameters	Raw effluent
1	pH	6.5
2	Turbidity NTU	116.3
3	Total dissolved solids, mg/L	3528
4	Sodium (as Na), mg/L	925
5	Potassium (as K), mg/L	75
6	Electrical conductivity, µmho/cm	5606
7	Alkalinity total (as $CaCO_3$), mg/L	1859
8	Nitrite (as NO_2), mg/L	Nil
9	Nitrate (as NO_3), mg/L	29
10	Phosphate (as PO_4), mg/L	5.78
11	Sulfate (as SO_4), mg/L	135
12	Tidy's test (as O), mg/L	781
13	Oil and grease, mg/L	0.0094

TABLE 6 Uptake of sugar from confectionery effluent by *Chlorella vulgaris*.

Organism name and treatment method	Total chlorophyll	Total carbohydrate	Total protein	Total lipid
Before treatment (raw effluent), mg/L	–	13,918	15,918	18,972
After treatment (supernatant) (treated with *C. vulgaris* VIAT027), mg/L	–	9896	10,961	10,987
After treatment (microalgal pellet), µg/10^6 cells	0.0068	0.4323	0.6001	0.6923

TABLE 7 Growth of *Chlorella vulgaris* (Circulation Flow Rate 900 L/h).

Day	Initial dip (cm)	Addition of raw effluent in cm (whole day)	Evaporation (cm)	Final dip (cm)	Cell count×10⁴ cells/mL *C. vulgaris* VIAT027		Cell count×10⁴ cells/mL yeast cells	
					Initial	Final	Initial	Final
1	20.0	5.0	1	24	230	262	35	5
2	17.5	7.5	0.5	24.5	262	273	56	21
3	12.5	12.5	1	24	273	285	120	15
4	7.5	17.5	1	24	285	305	131	19
5	5.0	20.0	1	24	305	320	152	10
6	1.0	24.0	0.7	24.3	320	338	160	2

and reduce BOD and COD. Effluent is added and removed from the algal treatment tank at 3500 L/h, which is the flow rate requirement to anaerobic digesters. Algal biomass reaches 1.5 g dry weight/L.

9.1 Biochemical analysis of *C. vulgaris* grown in confectionery industry effluent

Fig. 5 shows the results of biochemical analysis of *Chlorella vulgaris* grown in confectionery industry effluent compared with control. There is considerable increase in proteins and total lipids when grown in effluent. A preliminary trial conducted using effluent from a confectionery industry in Trivandrum, India, showed rich growth of algae in the raw effluent (Figs. 9 and 10). *Chlorococcum humicola* grew luxuriantly reaching a cell number of $3160 \times 10^4 \, mL^{-1}$ (4.8 g dry weight/L).

10 Effluent from soft drink manufacturing industry

Soft drink manufacturing units in India use groundwater for the production. The groundwater is filtered and softened using chemicals and sent to R/O and nanofiltration (N/F) for further TDS reduction. The effluent generated by the industry includes R/O reject, reject from N/F, utilities, cleaning, softener regeneration, bottle wash, and cleaning in process (CIP). In some units, the treated effluent is sent to R/O for recycling. The effluent is treated with conventional chemical and physical methods (Fig. 11). Removal of nutrients, especially nitrates and phosphates, is the major problem faced by this industry. The effluent contains high inorganic nutrients (especially nitrates and phosphates) and low pH. Microalgal technology is effectively employed to remove nitrates and phosphates. Table 8 shows the physicochemical characteristics of effluent from a soft drink manufacturing industry. The effluent has all the essential nutrients for algal growth. Fig. 12A and B shows nutrient removal efficiency of

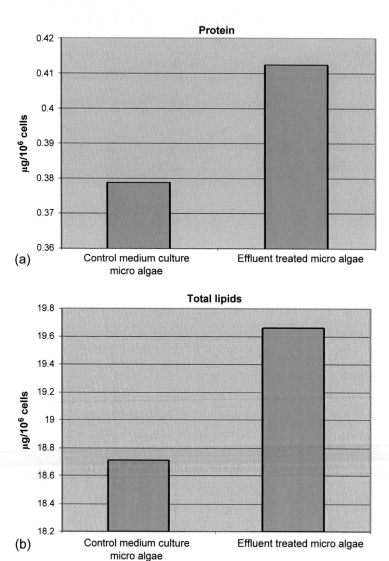

FIG. 9 Biochemical characteristics of *Chlorella vulgaris* grown in effluent from a confectionery industry near Chennai, India.

FIG. 10 Growth of *Chlorococcum humicola* in confectionery effluent from Trivandrum, India (productivity $= 3160 \times 104$ cell/mL; 68 g/L wet weight; 4.8 g/L dry weight).

FIG. 11 ETP plant at a soft drink industry (raw effluent as well as treated effluent is rich in nutrients suitable for algal biomass production).

V. Algal bioproducts, biofuels, and biorefinery for business opportunities

TABLE 8 Physicochemical characteristics of effluent from a soft drink manufacturing industry in Ahmedabad, India—treated with microalgae.

Parameters	Soft drink industry effluent		
	Initial	Final	% Reduction
Physical examination			
Turbidity NTU	17.8	40	–
Total solids, mg/L	3960	3621	8.6
Total dissolved solids (TDS), mg/L	3864	3518	8.9
Total suspended solids (TSS), mg/L	96	103	–
Electrical conductivity, μmho/cm	5496	4872	10.9
Chemical examination			
pH	7.21	8.78	–
Alkalinity pH (as $CaCO_3$), mg/L	0	36	–
Alkalinity total (as $CaCO_3$), mg/L	1296	1325	–
Total hardness (as $CaCO_3$), mg/L	1220	360	70.5
Calcium (as Ca), mg/L	320	81	74.7
Magnesium (as Mg), mg/L	101	36	64.4
Sodium (as Na), mg/L	690	690	0
Potassium (as K), mg/L	20	40	–
Iron (as Fe), mg/L	2.91	3	–
Manganese (as Mn), mg/L	Nil	Nil	–
Free ammonia (as NH_3), mg/L	21.28	32.8	–
Nitrite (as NO_2), mg/L	Nil	Nil	–
Nitrate (as NO_3), mg/L	21	8	61.9
Chloride (as Cl), mg/L	792	760	4.0
Fluoride (as F), mg/L	1.08	1	–
Sulfate (as SO_4), mg/L	288	121	58
Phosphate (as PO_4), mg/L	8.05	3	62
Tidy's test (as O), mg/L	106	13.2	–
Silica (as SiO_2), mg/L	44.18	24	45.7
BOD, mg/L	360	50	86.1
COD, mg/L	998	141	85.9
Total Kjeldhal nitrogen, mg/L	22.4	50	–
Copper (as Cu), mg/L	0.00321	0.0003	–
Zinc (as Zn), mg/L	0.148	0.1	–
Chromium (as Cr), mg/L	0.00236	0.0001	–

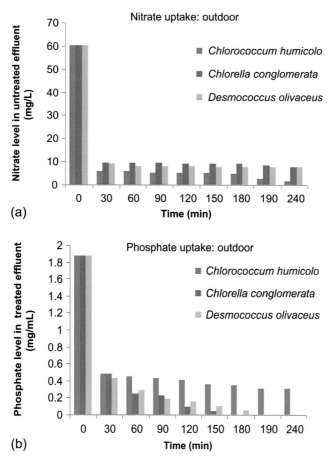

FIG. 12 Nutrient removal from soft drink industry effluent using microalga.

microalgae employed in this industry. Algae employed could remove nutrients (nitrate and phosphate) at a rapid rate well within the requirement of the industry.

11 Effluent from textile dyeing industries

Phycoremediation technology is employed in a few textile dyeing industries by the author and his team. Two textile dyeing industries were selected for the present discussion (one near Chennai and the second one in Ahmedabad; Figs. 13 and 14). In Chennai industry, the effluent is generated from various sources such as dye bath, mercerizer, wash water, desizing water, and printing. The industry in Tamil Nadu generates around 200 kL of effluent per day. The effluent is treated with conventional physical and chemical methods resulting in huge amount of sludge and water effluent. The dyeing industrial effluents are rich in various dyes and high pH and TDS, because of various chemicals being used, such as sodium bicarbonate and sodium chloride. The second industry generates around $84,000\,m^3$ effluents every day.

V. Algal bioproducts, biofuels, and biorefinery for business opportunities

FIG. 13 Phycoremediation of textile dyeing effluent (near Chennai) and algal biomass production.

The effluent generated is highly alkaline and treated with conventional chemical methods. Dye removal, reduction of BOD and COD are the major problems associated with effluent treatment. This effluent contains various dyes and high pH, TDS, BOD, and COD because of various chemicals being used, such as sodium bicarbonate and sodium chloride.

V. Algal bioproducts, biofuels, and biorefinery for business opportunities

FIG. 14   Algal growth in the raw effluent of textile dyeing industry in Ahmedabad, India. *Chlorococcum humicola, Chroococcus turgidus,* and *Desmococcus olivaceous* grow very well in the raw effluent. Color removal, pH correction, reduction of sludge, BOD, and COD achieved without using chemicals.

12 Phycoremediation of textile dyeing effluents

Table 9 gives details of the characteristics of textile dyeing effluent from a factory near Chennai, India, treated with microalgae in the lab as well as field. Effective color removal, pH correction, BOD and COD reduction, and sludge reduction achieved by algal treatment. Effluent provides all the necessary nutrients for the algal growth. Cell number is maintained around $250 \times 10^4\,\mathrm{mL^{-1}}$.

Table 10 provides details of the physicochemical characteristics of textile dyeing effluent from Ahmedabad treated with microalgae. In the Chennai unit, the algal biomass is harvested, dried, and used as fuel. The dried algal pellet has a high calorific value.

13 Algal biomass production in textile dyeing effluent

Both the effluents from the textile dyeing industry support very good growth of algae (*Chlorococcum humicola*). The treated effluent after harvesting algal biomass is sent to R/O for recycling. The R/O reject is fed into phycoremediation plant as source of nutrients. The biomass generated was harvested and analyzed for assessing the quality. Table 11 shows the qualitative analysis of algal biomass generated. When compared with the control (lab-grown biomass), effluent-grown biomass exhibits higher values for most of the biochemicals. Table 12 shows results of the FAME analysis of oil extracted from algal biomass. The analysis shows the existence of a single fatty acid in major composition indicating that it is highly suitable for biodiesel production, and a very little unsaturation is a good sign of hope in the process of biodiesel production from algae, and the alga *C. humicola* has a great potential of being a feedstock for biodiesel production. Recently, algal biodiesel is gaining considerable importance (Sivasubramanian, 2016 and the references there in).

TABLE 9 Phycoremediation efficiency of *Chlorococcum humicola* in textile industrial (near Chennai) combined dyeing effluent under lab and field conditions—a comparison.

Parameters	Raw combined effluent	Effluent treated with alga (lab)	Effluent treated with alga (field)	% Reduction	% Reduction (field)
Physical examination					
Turbidity NTU	170.9	73.2	12.2	57.16	92.82
Total solids, mg/L	7704	5842	7811	24.16	–
Total dissolved solids (TDS), mg/L	7380	5772	7294	21.78	1.16
Total suspended solids (TSS), mg/L	324	70	117	78.39	63.88
Electrical conductivity, μmho/cm	10,379	8390	11,077	19.16	–
Chemical examination					
pH	11.27	10.32	9.48	–	–
Alkalinity pH (as $CaCO_3$), mg/L	140	238	993	–	–
Alkalinity total (as $CaCO_3$), mg/L	800	388	488	51.5	39
Total hardness (as $CaCO_3$), mg/L	875	650	110	25.71	87.42
Calcium (as Ca), mg/L	190	160	24	15.78	87.36
Magnesium (as Mg), mg/L	96	60	12	37.5	87.5
Sodium (as Na), mg/L	710	1463	2500	–	–
Potassium (as K), mg/L	120	175	300	–	–
Iron (as Fe), mg/L	9.52	1.48	1.01	84.45	89.39
Manganese (as Mn), mg/L	Nil	Nil	Nil	Nil	Nil
Free ammonia (as NH_3), mg/L	5.6	58.24	3.44	–	38.57
Nitrite (as NO_2), mg/L	Nil	Nil	0.10	–	–
Nitrate (as NO_3), mg/L	18	63	17	–	5.55
Chloride (as Cl), mg/L	846	1020	1153	–	–
Sulfate (as SO_4), mg/L	262	555	178	–	32.06
Phosphate (as PO_4), mg/L	15.16	4.13	29.66	72.75	–
Tidy's test (as O), mg/L	283.5	85.2	51.0	69.94	82.01
Silica (as SiO_2), mg/L	88.28	22.66	41.69	74.33	52.77
BOD, mg/L	900	240	180	73.33	80

TABLE 9 Phycoremediation efficiency of *Chlorococcum humicola* in textile industrial (near Chennai) combined dyeing effluent under lab and field conditions—a comparison—cont'd

Parameters	Raw combined effluent	Effluent treated with alga (lab)	Effluent treated with alga (field)	% Reduction	% Reduction (field)
COD, mg/L	2663	772	502	71.01	81.14
Total Kjeldhal nitrogen, mg/L	19.04	98.56	15.68	–	17.64
Copper (as Cu), mg/L	0.00876	0.00416	0.00917	52.51	–
Zinc (as Zn), mg/L	0.116	0.095	0.157	18.10	–
Chromium (as Cr), mg/L	0.0055	0.00224	0.00336	59.27	38.90

TABLE 10 Phycoremediation efficiency of *Chlorococcum humicola* in textile dyeing effluent in Ahmedabad: laboratory trials.

Parameters	Dye bath effluent	Dye bath effluent treated with alga	% Reduction
Physical examination			
Turbidity NTU	0.2	7.2	–
Total solids, mg/L	7055	6851	02.89
Total dissolved solids (TDS), mg/L	6658	6384	04.11
Total suspended solids (TSS), mg/L	397	467	–
Electrical conductivity, µmho/cm	9488	9013	05.00
Chemical examination			
pH	10.18	9.30	–
Alkalinity pH (as $CaCO_3$), mg/L	127	139	–
Alkalinity total (as $CaCO_3$), mg/L	764	1393	–
Total hardness (as $CaCO_3$), mg/L	500	400	20.00
Calcium (as Ca), mg/L	110	90	18.18
Magnesium (as Mg), mg/L	54	42	22.22
Sodium (as Na), mg/L	1975	1975	–
Potassium (as K), mg/L	300	250	16.66
Iron (as Fe), mg/L	2.33	2.20	05.57
Manganese (as Mn), mg/L	Nil	Nil	–
Free ammonia (as NH_3), mg/L	42.56	27	36.56
Nitrite (as NO_2), mg/L	Nil	Nil	Nil
Nitrate (as NO_3), mg/L	32	24	25.00
Chloride (as Cl), mg/L	1542	1500	02.72

Continued

TABLE 10   Phycoremediation efficiency of *Chlorococcum humicola* in textile dyeing effluent in Ahmedabad: laboratory trials—cont'd

Parameters	Dye bath effluent	Dye bath effluent treated with alga	% Reduction
Sulfate (as SO_4), mg/L	1019	887	12.95
Phosphate (as PO_4), mg/L	12.25	6.5	46.93
Tidy's test (as O), mg/L	150	125.2	16.53
Silica (as SiO_2), mg/L	138.68	49.72	64.14
BOD, mg/L	490	360	26.53
COD, mg/L	1608	1042	35.19
Total Kjeldhal nitrogen, mg/L	74	27	63.51
Copper (as Cu), mg/L	0.00565	0.0052	07.96
Zinc (as Zn), mg/L	0.147	0.112	23.80
Chromium (as Cr), mg/L	0.0041	0.00373	09.02

TABLE 11   Value-added biochemicals of C. *humicola* grown in lab and field conditions.

Parameters	Grown in CFTRI medium—Lab	Alga grown in dyeing effluents	
		Field tank—A	Field tank—B
Chl_a (µg/10^6 cells)	0.1714	0.5179	0.9890
β-Carotene (µg/10^6 cells)	0.1025	0.4892	0.5683
Protein (mg/10^6 cells)	2.77	8.80	9.89
Carbohydrates (mg/10^6 cells)	2.41	17.80	8.70
Lipids (mg/10^6 cells)	12.00	24.00	45.00
Estimation of peroxidase activity (units/10^6 cells)	0.0012	0.0051	0.0007
Estimation of superoxide dismutase activity (units/10^6 cells)	2.402×10^{-5}	2.245×10^{-5}	3.012×10^{-6}
Estimation of polyphenol oxidase activity (units/10^6 cells)	7.018×10^{-6}	4.4×10^{-7}	5×10^{-6}
Estimation of catalase activity (units/10^6 cells)	0.00002	0.00281	0.00003
Estimation of polyphenols (µg/10^6 cells)	0.0368	0.4467	0.0636
Estimation of glutathione reductase activity (units/10^6 cells)	2.132×10^{-5}	3.021×10^{-5}	2.034×10^{-6}
Estimation of nitrate reductase (units/10^6 cells)	6.135×10^{-5}	9.622×10^{-4}	7.314×10^{-5}
Determination of total antioxidant property (µg/10^6 cells)			
1. Ascorbic acid	0.0108	0.1659	0.0219
2. α-Tocopherol	0.0511	0.7793	0.0362

TABLE 12 FAME analysis of algal oil extracted from *Chlorococcum humicola*.

Oil from alga	Alga grown in lab (%)	Alga grown in combined dyeing effluent (field) (%)	Alga grown in dye bath effluent (field) (%)
Heneicosanoic acid methyl ester (C21:0)	67	73.3	75.8
Arachidic acide methyl ester (C20:0)	14.7	13	10.7
cis-11,14,17-Eicosapentanic acid methyl ester (C20:5n3)	3.15	5.14	6.3
Lignoceric acid methyl ester (C24:0)	10.35	8.5	6.7
Total amount of saturated fatty acid content comes to around	92	94	91

14 Effluent from oil drilling operations (offshore and onshore)

Kakinada (Andhra Pradesh, India) based industry is providing total solutions in waste chain management for off shore and Onshore Oil and Gas Operators. It specializes in environmental services with focus on reduction of waste, recycling, and recovery/reuse of valuable base oil and oil-based mud with the prime motto of converting waste to value. The offshore disposal of wastes is allowed with limitation on discharges. The regulations in India are becoming more and more stringent, and the Ministry of Environment Forest (MoEF) is insisting on cleaner/better disposal practices of drilling wastes, and these are expected to become mandatory in the near future. At the quay side, a truck-mounted vacuum system using high-pressure washers (up to 200 bar) is used to empty the contents of the vessel tanks into skips. The skips are transported to the treatment facility, and the slops transferred to a storage tank for treatment. The composition of slop is normally in the range of 90% water, 8%–9% oil, and 1%–2% solids. Slop is allowed to settle for some time to facilitate natural separation of oil, water, and sediment phases. The next step is to reduce the amount of water from the liquid phase. This process is done in a mixing tank. Proprietary chemicals—emulsion breakers—are added to break the emulsion between base oil and water. These emulsion breakers disturb the surface tension of the emulsified water, with the result that free water can be separated. There are two phases in the mixing tank after adding emulsion breakers, one water phase, and one SOBM (synthetic-oil based mud) fluid phase. The water phase is transferred to a storage tank for further polishing, and the SOBM phase can be reused. The final polishing of the water phase is done by flocculation/filtration. The clean water can be reused as washing water or discharged into "the green belt." Drill cuttings/drilling waste contaminated with oil-based mud/synthetic-oil based mud (OBM/SOBM) during drilling is collected in sealed skips and transported from the base/quay side. The treated drill cuttings/drilling waste, which is devoid of oil to the extent of less than 10 g/kg of drill cuttings, is stored and sent for safe disposal in a treated drill cuttings disposal pit as per regulatory requirements. The trucks carrying the treated drilling waste will unload the waste in a safe disposal facility/lined pit. The facility is designed on the premise that it is exclusively used for disposal of treated drilling waste from oil and gas industry, which is provided with HDPE liner and leachate collection system. The treated drilling waste will have oil below 10 g/kg and no other toxic elements, thus meeting the MoEF guidelines for disposal into lined pit. The oil drilling industrial effluents have higher inorganic chemicals from oil separation process, which utilizes HCl, Triton X, sodium bicarbonate, etc. This effluent is alkaline and high in TDS. Phycoremediation technology is now being employed to treat wastewater and mud with mineral oil (Fig. 15). Excellent sludge reduction and reduction of mineral oil (<2%) have been achieved. The wastewater and sludge support excellent growth of selected microalgae. Table 13 gives a comparison of analytical

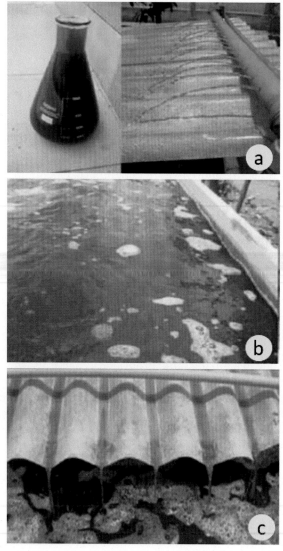

FIG. 15   Showing growth of *Chlorococcum humicola* in detergent effluent treatment using pilot tanks. Cell no 450×10^4 mL^{-1}.

TABLE 13 Phycoremediation of oil drilling industrial effluent: A comparison of lab, pilot plant, and scaled-up plant trials.

Parameters	Raw effluent	Effluent treated with alga (lab)	Effluent from pilot slope tank treated with alga	Effluent from scaled up tank treated with alga	% Reduction (lab)	% Reduction (pilot tank)	% Reduction (scaled up tank)
Physical examination							
Turbidity NTU	56.3	23.2	42.5	7.3	32.55	24.51	87.03
Total solids, mg/L	16,564	15,316	14,544	3175	7.53	12.19	80.83
Total dissolved solids (TDS), mg/L	16,492	15,262	14,410	3168	7.45	12.62	80.79
Total suspended solids (TSS), mg/L	72	54	134	7	25	–	90.27
Electrical conductivity, μmho/cm	23,488	21,770	20,561	4501	7.31	12.46	80.83
Chemical examination							
pH	6.45	7.20	7.48	7.78	–	–	–
Alkalinity pH (as $CaCO_3$), mg/L	0	0	0	0	0	0	0
Alkalinity total (as $CaCO_3$), mg/L	390	198	115	310	49.23	70.51	20.51
Total hardness (as $CaCO_3$), mg/L	375	975	5850	875	–	–	–
Calcium (as Ca), mg/L	90	210	1300	200	–	–	–
Magnesium (as Mg), mg/L	36	108	624	90	–	–	–
Sodium (as Na), mg/L	525	1725	6900	810	–	–	–
Potassium (as K), mg/L	88	175	850	35	–	–	60.22
Iron (as Fe), mg/L	1.38	1.56	22.64	20	–	–	–
Manganese (as Mn), mg/L	Nil	Nil	Nil	Nil	Nil	Nil	Nil
Free ammonia (as NH_3), mg/L	2.24	3.24	10.08	22	–	–	–
Nitrite (as NO_2), mg/L	Nil	0.22	Nil	Nil	–	Nil	–

Continued

V. Algal bioproducts, biofuels, and biorefinery for business opportunities

TABLE 13 Phycoremediation of oil drilling industrial effluent: A comparison of lab, pilot plant, and scaled-up plant trials—cont'd

Parameters	Raw effluent	Effluent treated with alga (lab)	Effluent from pilot slope tank treated with alga	Effluent from scaled up tank treated with alga	% Reduction (lab)	% Reduction (pilot tank)	% Reduction (scaled up tank)
Nitrate (as NO₃), mg/L	41	9	48	38	78.04	–	7.31
Chloride (as Cl), mg/L	351	1224	6393	896	–	–	–
Sulfate (as SO₄), mg/L	215	200	9	235	6.97	95.81	–
Phosphate (as PO₄), mg/L	3.42	2.82	19.01	6.48	17.54	–	–
Tidy's test (as O), mg/L	445.2	43	328	90	90.34	26.32	79.78
Silica (as SiO₂), mg/L	13.53	32.47	49.38	23	–	–	–
BOD, mg/L	4710	180	1400	320	96.17	70.27	93.20
COD, mg/L	1050	502	3423	892	52.19	–	15.04
Total Kjeldhal nitrogen, mg/L	3.36	31.36	10.08	114	–	–	–
Copper (as Cu), mg/L	0.00812	0.00096	0.0080	0.00628	88.17	1.47	22.66
Zinc (as Zn), mg/L	0.147	0.090	0.143	0.114	38.77	2.72	22.44
Chromium (as Cr), mg/L	0.00485	0.00036	0.00485	0.00410	92.57	–	15.46

report on the physicochemical characteristics of oil drilling effluent treated with microalgae in lab, pilot tank, and scaled-up tank. There is a significant reduction in all major parameters, and there was 70% reduction in sludge. The mud supports very good growth of algae. Microalgal remediation technology is now employed to reduce mineral oil in the mud. Phycoremediation has improved the texture, organic content, and water-holding capacity of mud, and it is now used to grow plant.

15 Effluent from detergent manufacturing industry

The detergent company at Ranipet, India, generates nearly 60–70 kL of effluent per day during production. Blue wash and sulfonation plants are two different plants in the industry. Effluent is generated from both the plants. Blue wash plant works in a stepwise process.

Each step leads to generate effluent as first wash, second wash, third wash, and fourth wash. The TDS will be very high in the first wash as compared with the fourth wash. The effluent generated from different washes is accumulated. The first blue wash has a TDS of 150,000 mg/L, and the last wash will have the lowest TDS of 30,000 mg/L. The blue wash had very high TDS due to its high inorganic salt content. The pH of the blue wash is acidic in nature. Blue wash effluent was not taken for the study due to its very minimum output. The effluent generated from the sulfonation plant has maximum output. The effluent of sulfonation plant was taken for the phycoremediation study. The TDS of the effluent of sulfonation plant was in the range of 75,000–150,000. The pH of the sulfonation effluent was also acidic in nature. The high TDS content and acidic pH of the effluent have made it difficult for the industry in disposing it in the environment.

The company was dealing this problem in a conventional method. They treated the effluent with lime to increase its pH. The treated effluent was kept in solar evaporation ponds. Over a period of time, the water of the effluent evaporates into the atmosphere. The sludge was remaining in the ponds after evaporation. The sludge accumulated in the solar evaporating ponds was removed and stored. The sludge generated due to this method amassed to an extent of 500 tons per annum.

The TDS in the effluent was due to high organic and inorganic content. The chemical constituents such as calcium, chloride, magnesium, nitrate, nitrite, potassium, sodium, and sulfate were responsible for the high level of TDS in the ultramarine effluent.

Phycoremediation technology is now employed in the detergent manufacturing industry to correct pH and to reduce sludge. Water is completely evaporated. There is more than 70% reduction in sludge achieved in the treatment process. The industry now avoids chemicals and treats the effluent with algae. Detergent industry effluent seems to support luxuriant growth of selected species of microalgae. Microalgae degrade sulfates and sulfites present in the effluent. Table 14 shows the physicochemical characteristics of effluent from the detergent manufacturing industry treated with algae. Fig. 16 shows growth of *C. humicola* in detergent effluent. The cell density reaches up to $450 \times 10^4\,mL^{-1}$ (1.5 g dry weight/L) showing thereby detergent effluent can be a very good medium for growing algae.

16 Effluent from electroplating industry

The electroplating industrial effluents are both acidic and alkaline. The acidic effluent contains heavy metals such as chromium, nickel, and cadmium and high TDS. Alkaline effluent is generated from coating of paint to wheels; this effluent contains calcium chloride and heavy metals. The sludge is rich in heavy metals especially chromium (Sivasubramanian, 2016 and the references there in). Table 15 shows the physicochemical characteristics of effluent from electroplating industry treated with algae. Fig. 17 shows algal growth in electroplating sludge. Chrome sludge from the electroplating industry supported very good algal growth (1.5 g dry weight/L). Using open raceway pond, chrome sludge was treated with microalga, *Desmococcus olivaceus*. There was a considerable amount of sludge reduction and biomass production in open raceway pond amended with chrome sludge. A remarkable reduction was found in TDS, sodium, potassium, and phosphate.

TABLE 14 Physicochemical characteristics of effluent from a detergent industry treated with microalgae (pilot plant trials).

Parameters	Physicochemical parameters of raw effluent	Physicochemical parameters of treated effluent
Physical examination		
Appearance	Whitish	Whitish
Odor	Offensive smell	Offensive smell
Turbidity NTU	14.1	74.6
Total solids, mg/L	63,650	56,662
Total suspended solids, mg/L	140	2280
Total dissolved solids, mg/L	63,510	54,382
Electrical conductivity, μmho/cm	90,846	77,707
Chemical examination		
pH	3.8	6.7
Alkalinity pH (as $CaCO_3$), mg/L	0	0
Alkalinity total (as $CaCO_3$), mg/L	Nil	Nil
Total hardness (as $CaCO_3$), mg/L	5800	2100
Calcium (as Ca), mg/L	1360	560
Magnesium (as Mg), mg/L	576	168
Sodium (as Na), mg/L	18,500	16,100
Potassium (as K), mg/L	500	600
Iron (as Fe), mg/L	0.82	0.99
Manganese (as Mn), mg/L	Nil	Nil
Free ammonia (as NH_3), mg/L	0.37	16.28
Nitrite (as NO_2), mg/L	2.67	2.58
Nitrate (as NO_3), mg/L	11	12
Chloride (as Cl), mg/L	27,400	24,200
Fluoride (as F), mg/L	0.97	0.84
Sulfate (as SO_4), mg/L	2007	2920
Phosphate (as PO_4), mg/L	3.38	73.84
Copper (as Cu), mg/L	0.00128	0.00634
Chromium (as Cr), mg/L	0.079	0.057
Zinc (as Zn), mg/L	0.304	1.843
Tidy's test (as O), mg/L	651	516
Silica (as SiO_2), mg/L	31.66	71.9
Total Kjeldahl nitrogen (as N), mg/L	11.2	34.72
BOD, mg/L	860	740
COD, mg/L	2894	2512
Oil and grease, mg/L	0.0064	0.0152

FIG. 16 ETP and phycoremediation plant at oil drilling industrial treatment plant—Kakinada, Andhra Pradesh (effluent storage tanks, algal growth on sludge, phycoremediation plant, algal growth in effluent are shown).

TABLE 15 Comparison of parameters of raw and algae-treated electroplating industrial chrome sludge in open pond.

Parameters	Raw sludge	Algal-treated sludge	% of Reduction
Physical examination			
Turbidity NTU	69.0	18.9	41.6
Total solids, mg/L	3022	1955	35.30
Total dissolved solids (TDS), mg/L	66	17	74.24
Total suspended solids (TSS), mg/L	2956	1938	34.43
Electrical conductivity, μmho/cm	4702	2745	41.42
Chemical examination			
pH	7.95	8.91	
Alkalinity pH (as $CaCO_3$), mg/L	8	16	
Alkalinity total (as $CaCO_3$), mg/L	1393	1027	26.27
Total hardness (as $CaCO_3$), mg/L	680	520	23.52
Calcium (as Ca), mg/L	176	198	
Magnesium (as Mg), mg/L	58	48	17.24
Sodium (as Na), mg/L	680	340	50
Potassium (as K), mg/L	50	20	60
Iron (as Fe), mg/L	6.48	4.70	27.46
Manganese (as Mn), mg/L	Nil	Nil	Nil
Free ammonia (as NH_3), mg/L	38.08	23.52	38.23
Nitrite (as NO_2), mg/L	0.84	0.36	41.14
Nitrate (as NO_3), mg/L	45	23	48.88
Chloride (as Cl), mg/L	351	176	49.85
Fluoride (as F), mg/L	1.12	0.99	11.60
Sulfate (as SO_4), mg/L	186	129	30.64
Phosphate (as PO_4), mg/L	43.51	6.02	86.16
Tidy's test (as O), mg/L	67.2	58.4	13.09
Silica (as SiO_2), mg/L	40.12	47.68	15.85
BOD, mg/L	643	497	22.70
COD, mg/L	210	160	23.80
Total Kjeldhal nitrogen, mg/L	51.52	59.36	
Copper (as Cu), mg/L	0.01371	0.01256	08.38
Zinc (as Zn), mg/L	0.285	0.238	16.49
Chromium (as Cr), mg/L	0.024	0.016	33.33

FIG. 17 Phycoremediation of electroplating industrial chrome sludge in experimental raceway pond: (A) Before treatment and (B) After treatment.

17 Sewage and algal biomass production

Sewage is the most favorable medium for algae production (Sivasubramanian, 2016 and the references there in). It provides all the necessary nutrients needed for algal growth. New Zealand's Aquaflow Bionomic Corp. has become the World's first producer of biofuel from sewage-pond-grown algae (Sivasubramanian, 2016 and the references there in). One particular advantage of the human-sewage approach is that algae from sewage tend to have a lot of oil according to Cary Bullock, CEO of Greenfuel Technologies, a company cultivating algae to convert emissions into biofuel.

Bacteria digestion of organics is the known method to reduce BOD, COD, TSS, TDS, etc., in sewage. Anaerobic bacteria proliferate in untreated sewage giving rise to H_2S gas that produces the obnoxious smell. Aerobic bacteria require plenty of dissolved oxygen to perform organic digestion. Untreated sewage is let out into the nearest water bodies where anaerobic bacteria slowly consume the organics and produce H_2S that gives rise to bad smell. Further the waters become infested with water weeds and plants such as hyacinth. These then become the breeding place of mosquitoes and disease-producing organisms. Utilization of domestic sewage for algal biomass production has been studied extensively by employing a variety of microalgal species (Sivasubramanian, 2016 and the references there in).

Sewage samples collected from the city of Chennai, India, were screened and treated with microalgae. *C. humicola, Chlorella conglomerata, C. turgidus,* and *Desmococcus olivaceus* were inoculated into sewage samples. Growth and biochemical composition of microalgae and physicochemical parameters of sewage samples were monitored. Biodiesel potentials of algal biomass were also determined. *Chlorella* sp. grew very well followed by *Desmococcus* sp. Table 16 shows the physicochemical characteristics of sewage treated with algae. Phycoremediation could reduce important parameters such as BOD, COD, and other major nutrients and minerals to a great extent. Fig. 18 shows the results of FAME analysis of algal biomass grown in sewage compared with control. FAME % increases by 50% when alga was grown in sewage.

TABLE 16　Physicochemical characteristics of sewage treated with microlgae.

Parameters	Raw	*Chlorococcum* sp.	*Chroococcus* sp.	*Chlorella* sp.	*Desmococcus* sp.
Physical examination					
Appearance	Blackish	Greenish	Greenish	Greenish	Greenish
Odor	Offensive odor	Algal smell	Algal smell	Algal smell	Algal smell
Turbidity NTU	14.5	22.9	38.1	53.7	35.1
Total solids, mg/L	3224	2174 (32.57)	3240	2914 (9.61)	2990 (7.26)
Total dissolved solids (TDS), mg/L	2784	1736 (37.64)	2572 (7.61)	2548 (8.48)	2630 (5.53)
Total suspended solids (TSS), mg/L	440	438 (0.45)	668	366 (16.81)	360 (18.18)
Electrical conductivity, μmho/cm	3974	2486 (37.44)	3673 (7.57)	3643 (8.33)	3982
Chemical examination					
pH	7.05	9.1	9.54	9.37	9.44
Alkalinity pH (as $CaCO_3$), mg/L	0	20	20	40	40
Alkalinity total (as $CaCO_3$), mg/L	255	159 (37.65)	195 (23.53)	160 (37.25)	147 (42.35)
Total hardness (as $CaCO_3$), mg/L	520	184 (64.61)	164 (68.46)	156 (70)	112 (78.46)
Calcium (as Ca), mg/L	120	42 (65)	37 (69.17)	34 (71.67)	26 (78.33)
Magnesium (as Mg), mg/L	53	19 (64.15)	17 (67.92)	17 (67.92)	11 (79.24)

TABLE 16 Physicochemical characteristics of sewage treated with microlgae—cont'd

Parameters	Raw	*Chlorococcum* sp.	*Chroococcus* sp.	*Chlorella* sp.	*Desmococcus* sp.
Sodium (as Na), mg/L	580	420 (27.59)	550 (5.17)	540 (6.9)	550 (5.17)
Potassium (as K), mg/L	50	40 (20)	50 (0)	40 (20)	40 (20)
Iron (as Fe), mg/L	0.78	0.65 (16.67)	0.64 (17.95)	0.70 (10.26)	0.76 (2.56)
Manganese (as Mn), mg/L	0	0	0	0	0
Free ammonia (as NH_3), mg/L	1.69	1.05 (37.87)	1.24 (26.63)	0.83 (50.89)	1.29 (23.67)
Nitrite (as NO_2), mg/L	1.41	0.43 (69.5)	0.39 (72.34)	0.57 (59.57)	0.66 (53.19)
Nitrate (as NO_3), mg/L	14	11 (21.43)	12 (14.28)	12 (14.28)	13 (7.14)
Chloride (as Cl), mg/L	1061	680 (35.9)	1002 (5.56)	1020 (3.86)	1050 (1.04)
Sulfate (as SO_4), mg/L	90	72 (20)	86 (4.44)	88 (2.22)	89 (1.11)
Phosphate (as PO_4), mg/L	0.63	0.50 (20.63)	0.58 (7.94)	0.34 (46.03)	0.60 (4.76)
Tidy's test (as O), mg/L	236	59 (75)	38 (83.9)	36 (84.74)	37 (84.32)
Silica (as SiO_2), mg/L	28	27 (3.57)	25 (10.71)	23 (17.86)	28
BOD, mg/L	2612	190 (92.72)	260 (90.04)	140 (94.64)	120 (95.4)
COD, mg/L	5200	657 (87.36)	1680 (67.7)	555 (89.33)	384 (92.61)
Total Kjeldhal nitrogen, mg/L	37	13 (64.86)	9 (75.67)	8 (78.38)	7 (81.08)
	0.00092	0.00044 (52.17)	0.00018 (80.43)	0.00023 (75)	0.00011 (88.04)
Zinc (as Zn), mg/L	0.9	0.4 (55.55)	0.6 (33.33)	0.09 (90)	0.45 (50)
Chromium (as Cr), mg/L	0.00567	0.00448 (20.99)	0 (100)	0.00224 (60.49)	0.0034 (40.03)
Oil and grease, mg/L	0.0148	0.0044 (70.27)	0.0018 (87.84)	0.00122 (91.76)	0.00244 (83.51)

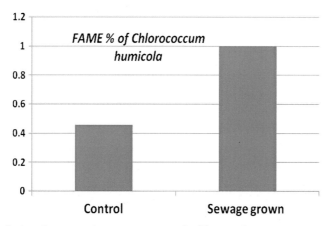

FIG. 18 FAME % of microalga grown in sewage compared with control.

18 Effluent from a chemical industry that produces organic acids

The main products of the chemical industry are petrochemicals and a range of food acids. Phthalic anhydride, maleic anhydride, succinic anhydride derivatives, and various downstream products, including food ingredients, etc., generate different streams of effluents, which are treated in various conventional ways to finally bring down the COD to the required level. The effluent produced is highly acidic (pH 0.6–1.0) and with very high TDS. For an initial screening, acidic effluent (malic acid—1%; citraconic acid—2.3%; fumaric acid—1.5%; phthalic acid—1.7%; benzoic acid—0.8%; pH—less than 2) was added to the cultures daily (amount of addition was determined based on pH change). Feasibility study conducted in the laboratory gave encouraging results. Selected microalgae were tried in the pilot tanks. *Desmococcus olivaceous* and *C. humicola* grew very well and selected for scaling-up (Fig. 19). Microalgae could correct the pH and completely degrade all the organic acids present in the effluent. There was a huge reduction in BOD, COD, and sludge. Algal biomass could be developed to a maximum of 0.75 g dry weight/L/day. Biochemical analysis done on the algal biomass grown in effluent revealed very high increase in proteins, carbohydrates, and lipids when compared with controls (Figs. 20–23).

19 Rejuvenation of highly acidic rivers in east Jaintia Hills, Meghalaya, India

The present contract involves restoration and rejuvenation of two highly acidic Rivers in East Jaintia Hills: Kyrhuhkhla and lunar River. These Rivers are highly acidic (pH 2.8 to 3.5) due to acid mine discharge from coal mines. The Meghalayan coal is having high calorific

FIG. 19 Growth of microalgae in the acidic effluent from a chemical industry producing organic acids.

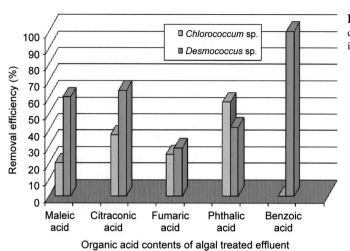

FIG. 20 Organic acid removal efficiency of microalgae from the chemical industry effluent at the lab trial.

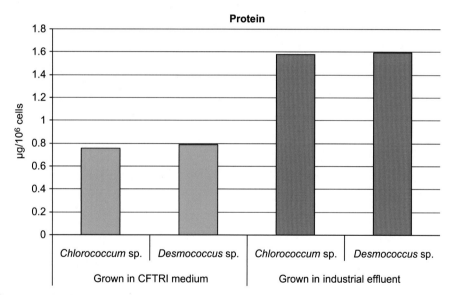

FIG. 21 Protein levels in the microalgae grown in the chemical industry effluent.

value but rich in sulfur. As there is heavy rainfall in Meghalaya and water table is very low, water under current from these mines carries sulfur with it and pollutes the flowing streams and makes them highly acidic. Though there are many acidic streams in Meghalaya, present contract involves restoration of Kyrhuhkhla and lunar as mentioned above. The salient features of the Project are:

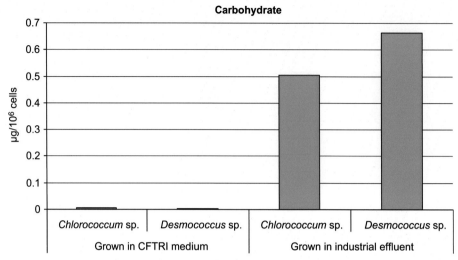

FIG. 22   Carbohydrate levels in the microalgae grown in chemical industry effluent.

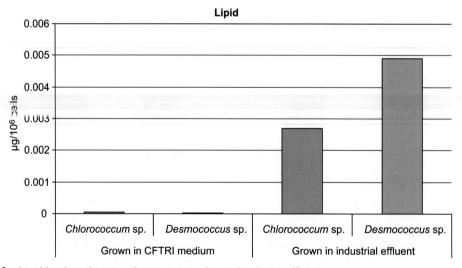

FIG. 23   Lipid levels in the microalgae grown in chemical industry effluent.

(1) Time period is – 24 months wef April 22

(2) Flow: The Average flow of Kyrhuhkhla in Lean season is 60 MLD+ while average flow in lean season for Lunar is 200 Mld+

(3) Length of Rivers to be Rejuvenated

   (a) Kyrhuhkhla—10 km

   (b) Lunar—13.3 km

FIG. 24 (A–C) Rejuvenation of highly acidic rivers in East Jaintia Hills, Meghalaya, North-East India. Please note the 5 million Liters capacity of tanks with Algae along the river banks.

The contract has been awarded by the Meghalaya Government, Department of Mines Resources on recommendation of Oversight Committee of Ministry of Forest and Environment, Government of India. This committee has been constituted by NGT New Delhi. The Committee is headed by Addl Secy MoEF CC, who is also Chairman, CPCB.

This project has created 5 million liters of Algae capacity along the banks of the river and dozing approx. 2.5 million liters per day (both) till pH starts improving and then dozing according to requirement. We have created nine sites with approx 5.5 Lakh liter capacity sites. The work is going on in full swings (Fig. 24A–C).

20 Conclusions

From the discussion so far regarding phycoremediation, it is evident that most of the industrial effluents and sludge support very good growth of microalgae (Table 17). This is very significant information to be accommodated when the economics of mass cultivation of algae is taken up for discussion. The economic viability is based on the twin benefits of phycoremediation. On the one hand, it handles the removal and degradation of most of the toxic chemicals, and on the other, it provides the same chemicals as nutrients for the growing algae. Thus, we not only save on the nutrients but also see that the effluent is treated and that most of the toxic chemicals degraded and converted into valuable algal biomass. It is equally pertinent here to note that the algal strains capable of delivering this twin advantage have to be developed under careful lab and field studies and be kept in scrupulously maintained culture banks. Any presence of the residue of the degraded toxic chemicals from the effluent reflecting in the resulting algal biomass is negated in all our experiments through standard studies. However, this should become a standard quality control procedure during large-scale productions.

Availability of the feedstock consistently throughout the year and the ensuing economic viability are the major hurdles in the biomass-based technologies. Algal biomass production generally suffers due to expensive nutrient inputs, water scarcity, and non-availability of land. Most of the industries dealing with mass cultivation of microalgae depend on chemical fertilizers for cheaper nutrient inputs. Integrating algal biomass production with phycoremediation seems to address most of these problems associated with mass cultivation.

TABLE 17 Advantages of integration of algal biomass production with phycoremediation.

Industry in India	Type of effluent/ sludge/ Waste	Benefits of phycoremediation				Algal biomass production potential (kg dry biomass/kL/ day)
		pH correction	Reduction of BOD, COD	Sludge reduction	Reduction in operation cost	
SNAP Natural and alginate products, Ranipet	Acidic effluent with high TDS	H	H	H	H	0.75
Leather processing chemical industry, Ranipet	Effluent and sludge	L	H	H	H	1.0
Confectionery industry 1—Chennai	Acidic effluent	H	H	H	H	1.5
Confectionery industry 2—Trivandrum	Acidic effluent	H	H	M	H	4.8
Soft drink industry, Ahmedabad	Neutral effluent	L	L	L	H	0.5
Textile dyeing industry, Chennai	Effluent and sludge	L	H	H	H	0.75
Oil drilling industry, Kakinada	Effluent and sludge	L	H	H	H	0.5
Detergent industry, Ranipet	Acidic effluent with high TDS	H	H	H	H	1.5
Electroplating industry, Chennai	Effluent and sludge	L	H	H	H	1.5
Chemical industry, Ranipet	Acidic effluent with high TDS	L	H	H	H	0.75

L: low; M: medium; H: high.

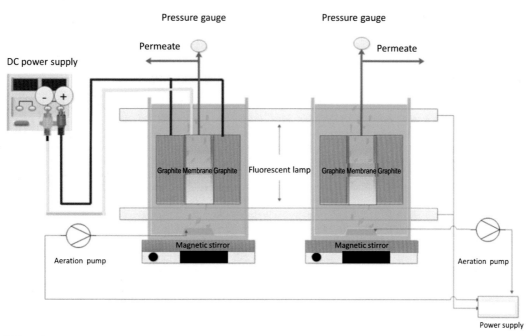

FIG. 25 Novel electrokinetic-assisted membrane photobioreactor (EK-MPBR) for wastewater treatment. *Source MDPI: Amini, M.; Mohamedelhassan, E.; Liao, B. The biological performance of a novel electrokinetic-assisted membrane photobioreactor (EK-MPBR) for wastewater treatment. Membranes 2022, 12, 587. https://doi.org/10.3390/membranes12060587. Copyright: © 2022 by the authors. Licensee MDPI, Basel, Switzerland. This article is an open-access article distributed under the terms and conditions of the Creative Commons Attribution (CC BY) license (https://creativecommons.org/licenses/by/4.0/).*

Apart from cleaning the environment from the onslaught of ever-increasing pollution, valuable, and cheaper biomass is generated from wastewater, effluents, and sludge if appropriate algal species are identified and grown in industrial wastes. Using Chlorella removal of nutrients (N and P) from municipal wastewater using novel Electrokinetic-Assisted Membrane Photobioreactor (EK-MPBR) yielded promising results (Fig. 25).

References

Abdo, M.M., Abdel-Hamid, M.I., El-Sherbiny, I.M., El-Sherbeny, G., Abdel-Aal, E.I., 2023. Green synthesis of AgNPs, alginate microbeads and *Chlorella minutissima* laden alginate microbeads for tertiary treatment of municipal wastewater. Bioresource Technol. Rep. 21, 101300.

Agrawal, K., Ruhil, T., Gupta, V.K., Verma, P., 2023. Microbial assisted multifaceted amelioration processes of heavy-metal remediation: a clean perspective toward sustainable and greener future. Crit. Rev. Biotechnol., 1–19.

Al Zohbi, G., 2023. Review on sustainable technologies for liquid waste treatment. In: Waste Recovery and Management: An Approach Toward Sustainable Development Goals, p. 47.

Almeida, Â., Cotas, J., Pereira, L., Carvalho, P., 2023. Potential role of *Spirogyra* sp. and Chlorella sp. in bioremediation of mine drainage: a review. Phycology 3 (1), 186–201.

Almeida, J.C., Marques, I.M., de Souza, J.R.B., Moreira, Í.T.A., de Oliveira, O.M.C., 2022. Phycoremediation of coastal marine water contaminated with dissolved oil by nannochloropsis oculata. Water Air Soil Pollut. 233 (11), 1–13.

Amini, M., Mohamedelhassan, E., Liao, B., 2022. The biological performance of a novel electrokinetic-assisted membrane photobioreactor (EK-MPBR) for wastewater treatment. Membranes 12 (6), 587.

Anand, U., Dey, S., Parial, D., Federici, S., Ducoli, S., Bolan, N.S., Dey, A., Bontempi, E., 2023. Algae and bacteria consortia for wastewater decontamination and transformation into biodiesel, bioethanol, biohydrogen, biofertilizers and animal feed: a review. Environ. Chem. Lett., 1–25.

Andrade, L.M., Tito, C.A., Mascarenhas, C., Lima, F.A., Dias, M., Andrade, C.J., Mendes, M.A., Nascimento, C.A.O., 2022. Phycoremediation of copper by *Chlorella prototothecoides* (UTEX 256): proteomics of protein biosynthesis and stress response. Biomass 2 (3), 116–129.

Angelaalincy, M., Nishtha, P., Ajithkumar, V., Ashokkumar, B., Moorthy, I.M.G., Brindhadevi, K., Chi, N.T.L., Pugazhendhi, A., Varalakshmi, P., 2023. Phycoremediation of Arsenic and biodiesel production using green microalgae Coelastrella sp. M60—an integrated approach. Fuel 333, 126427.

Anusuyadevi, P.R., Kumar, D.J.P., Jyothi, A.O., Patwardhan, N.S., Janani, V., Mol, A., 2023. Towards viable eco-friendly local treatment of blackwater in sparsely populated regions. Water 15 (3), 542.

Apandi, N.M., Gani, P., Sunar, N.M., Mohamed, R.M.S.R., AlGheethi, A., Apandi, A.M., Nagarajah, R., Shaari, N.A.R., Cheong, K., Rahman, R.A., 2022. Scenedesmus sp. harvesting by using natural coagulant after phycoremediation of heavy metals in different concentrations of wet market wastewater for potential fish feeds. Sustainability 14 (9), 5090.

Aranguren Díaz, Y., Monterroza Martínez, E., Carillo García, L., Serrano, M.C., Machado Sierra, E., 2022. Phycoremediation as a strategy for the recovery of marsh and wetland with potential in Colombia. Resources 11 (2), 15.

Arrojo, M.Á., Regaldo, L., Calvo Orquín, J., Figueroa, F.L., Abdala Díaz, R.T., 2022. Potential of the microalgae Chlorella fusca (Trebouxiophyceae, Chlorophyta) for biomass production and urban wastewater phycoremediation. AMB Express 12 (1), 1–14.

Arutselvan, C., Narchonai, G., Pugazhendhi, A., Kumar Seenivasan, H., Lewis Oscar, F., Thajuddin, N., 2022. Phycoremediation of textile and tannery industrial effluents using microalgae and their consortium for biodiesel production. J. Clean. Prod. 367, 133100.

Asiandu, A.P., Nugroho, A.P., Naser, A.S., Sadewo, B.R., Koernlawan, M.D., Budiman, A., Siregar, U.J., Suwanti, S., Suyono, E.A., 2023. The effect of tofu wastewater and pH on the growth kinetics and biomass composition of Euglena sp. Curr. Appl. Sci. Technol., 10–55003.

Badamasi, M., Abdullahi, B., Suleiman, K., Habib, A., 2023. The effect of temperature and photoperiod on bioremediation potentials of mono and mixed cultures of chlorophyte and cyanobacteria for municipal wastewater treatment. Ethiopian J. Environ. Studies Manag. 16 (1), 88–105.

Badiefar, L., Rodríguez-Couto, S., Riazalhosseini, B., 2023. Genetic engineering of algae. In: Emerging Technologies in Applied and Environmental Microbiology. Academic Press, pp. 149–179.

Bauddh, K., Korstad, J., 2022. Phycoremediation: use of algae to sequester heavy metals. Hydrobiology 1 (3), 288–303.

Bhandari, M., Kharkwal, S., Prajapati, S.K., 2023. Recycling drinking water RO reject for microalgae-mediated resource recovery. Resour. Conserv. Recycl. 188, 106699.

Bhatt, A., Arora, P., Prajapati, S.K., 2023. *Chlorella pyrenoidosa*-mediated removal of pathogenic bacteria from municipal wastewater—multivariate process optimization and application in the real sewage. J. Environ. Chem. Eng. 11 (2), 109494.

Bhayana, T., Saxena, A., Gupta, S., Dubey, A.K., 2023. Enhanced decolourisation and degradation of azo dyes using wild versus mutagenic improved bacterial strain: a review. Vegetos 36 (1), 28–37.

Bloch, K., Ghosh, S., 2023. Phycoremediation of toxic metals for industrial effluent treatment. In: Phycoremediation Processes in Industrial Wastewater Treatment. CRC Press, pp. 57–75.

Chaudhary, S., Ahmad, S., Sachdev, S., Pathak, V.V., Pathak, A.K., 2022. Newly isolated Acutodesmus obliquus assisted phycoremediation and sequestration of CETP wastewater: an integrated management of remediation. Biomass Convers. Biorefin., 1–11.

Cheah, W.Y., Pahri, S.D.R., Leng, S.T.K., Er, A.C., Show, P.L., 2023. Circular bioeconomy in palm oil industry: current practices and future perspectives. Environ. Technol. Innov., 103050.

Chowdhury, B.R., Das, S., Bardhan, S., Lahiri, D., 2023. Omics approaches for microalgal applications in wastewater treatment. In: Genomics Approach to Bioremediation: Principles, Tools, and Emerging Technologies, pp. 143–156.

Cordero, A.P., Vergara, D.E.M., Mendoza, Y.A., 2023a. Evaluation of the potential of Chlorella vulgaris for the removal of pollutants from standing water. J. Positive Psychol. Wellbeing, 626–631.

Cordero, A.P., Vergara, D.E.M., Mendoza, Y.A., 2023b. Use of Chlorophyll microorganisms to improve physico-chemical properties of wastewater. J. Positive School Psychol., 940–946.

Danouche, M., El Ghachtouli, N., Aasfar, A., Bennis, I., El Arroussi, H., 2022. Pb (II)-phycoremediation mechanism using Scenedesmus obliquus: cells physicochemical properties and metabolomic profiling. Heliyon 8 (2), e08967.

Dasauni, K., Nailwal, T.K., 2023. Addressing the strategies of algal biomass production with wastewater treatment. In: Phycoremediation Processes in Industrial Wastewater Treatment. CRC Press, pp. 1–20.

de Sousa Oliveira, A.P., Assemany, P., Covell, L., Tavares, G.P., Calijuri, M.L., 2023. Microalgae-based wastewater treatment for micropollutant removal in swine effluent: high-rate algal ponds performance under different zinc concentrations. Algal Res. 69, 102930.

Devi, A., Singh, A., Mahajan, M., Sahab, S., Srivastava, V., Singh, P., Singh, R.P., 2022. Trends in waste water treatment using phycoremediation for biofuel production. In: Bioremediation of Toxic Metal (Loid)s. CRC Press, pp. 153–167.

Dey, P., Krishna Murthy, T.P., Divyashri, G., Raghavendra, A., Singh, A., Girish, A., Shaik, A., Poola, A.A., Gopinath, D.K., Prabhu, P., Konety, R., 2023. Bioremediation of organic and inorganic contaminants by microbes. Microbial Degrad. Detoxif. Pollut. 2, 1.

Douna, B.K., Yousefi, H., 2023. Removal of PFAS by biological methods. Asian Pacific J. Environ. Cancer 6 (1), 53–68.

Durairaj, S., Srikanth, P., Durairaj, S., 2023. Genetic engineering of algae. In: Phycoremediation Processes in Industrial Wastewater Treatment. CRC Press, pp. 127–145.

Dursun, N., 2023. Using algae as a renewable source in the production of biodiesel. In: Basic Research Advancement for Algal Biofuels Production. Springer, Singapore, pp. 105–145.

El-Sheekh, M.M., Galal, H.R., Mousa, A.S.H., Farghl, A.A., 2023. Coupling wastewater treatment, biomass, lipids, and biodiesel production of some green microalgae. Environ. Sci. Pollut. Res., 1–13.

Encarnação, T., Ramos, P., Mohammed, D., McDonald, J., Lizzul, M., Nicolau, N., da Graça Campos, M., Sobral, A.J., 2023. Bioremediation using microalgae and cyanobacteria and biomass valorisation. In: Marine Organisms: A Solution to Environmental Pollution? Uses in Bioremediation and in Biorefinery. Springer International Publishing, Cham, pp. 5–28.

Fal, S., Smouni, A., El Arroussi, H., 2023. Integrated microalgae-based biorefinery for wastewater treatment, industrial CO_2 sequestration and microalgal biomass valorization: a circular bioeconomy approach. Environ. Adv., 100365.

Fernandes, R., Campos, J., Serra, M., Fidalgo, J., Almeida, H., Casas, A., Toubarro, D., Barros, A.I., 2023. Exploring the benefits of phycocyanin: from spirulina cultivation to its widespread applications. Pharmaceuticals 16 (4), 592.

Gani, P., Apandi, N.M., Mohamed Sunar, N., Matias-Peralta, H.M., Kean Hua, A., Mohd Dzulkifli, S.N., Parjo, U.K., 2022. Outdoor phycoremediation and biomass harvesting optimization of microalgae Botryococcus sp. cultivated in food processing wastewater using an enclosed photobioreactor. Int. J. Phytoremed., 1–13.

Garcia-Corral, I., Morillas-Espana, A., Ciardi, M., Massa, D., Jimenez-Becker, S., 2023. Reuse of wastewater from the production of microalgae and its effect on the growth of pelargonium x hortorum. J. Appl. Phycol. 35 (1), 173–181.

Ghaffar, I., Hussain, A., Hasan, A., Deepanraj, B., 2023. Microalgal-induced remediation of wastewaters loaded with organic and inorganic pollutants: an overview. Chemosphere, 137921.

Goswami, R.K., Agrawal, K., Verma, P., 2022. Phycoremediation of nitrogen and phosphate from wastewater using Picochlorum sp.: a tenable approach. J. Basic Microbiol. 62 (3–4), 279–295.

Goswami, R.K., Agrawal, K., Verma, P., 2023. Multifaceted role of microalgae for municipal wastewater treatment: a futuristic outlook toward wastewater management. Clean—Soil, Air, Water 51 (3), 2100286.

Gutiérrez-Hoyos, N., Sánchez, C., Gutiérrez, J.E., 2023. Variation in phytoplankton diversity during phycoremediation in a polluted Colombian Caribbean swamp. Environ. Monit. Assess. 195 (2), 327.

Hamed, S.M., Okla, M.K., Al-Saadi, L.S., Hozzein, W.N., Mohamed, H.S., Selim, S., AbdElgawad, H., 2022. Evaluation of the phycoremediation potential of microalgae for captan removal: comprehensive analysis on toxicity, detoxification and antioxidants modulation. J. Hazard. Mater. 427, 128177.

Han, G., Ma, L., Zhang, C., Wang, B., Sheng, X., Wang, Z., Wang, X., Wang, L., 2023. Research on the tolerance and degradation of o-cresol by microalgae. Water 15 (8), 1522.

578

20. Phycoremediation and business prospects

Hariz, H.B., Lawton, R.J., Craggs, R.J., 2023. Nutrient uptake and biomass productivity performance comparison among freshwater filamentous algae species on mesocosm-scale FANS under ambient summer and winter conditions. Ecol. Eng. 189, 106910.

Hu, R., Cao, Y., Chen, X., Zhan, J., Luo, G., Ngo, H.H., Zhang, S., 2022. Progress on microalgae biomass production from wastewater phycoremediation: metabolic mechanism, response behavior, improvement strategy and principle. Chem. Eng. J., 137187.

Javed, F., Rashid, N., Fazal, T., Hafeez, A., Rehman, F., 2023. Integration of real industrial wastewater streams to enhance *Chlorella vulgaris* growth: nutrient sequestration and biomass production. Water Air Soil Pollut. 234 (3), 164.

Jha, S., Shukla, R., Singh, R., Shukla, M., Srivastava, P., Pandey, A., Dikshit, A., 2023. Phyco-remediation: role of microalgae in remediation of emerging contaminants. In: Emerging Contaminants and Plants: Interactions, Adaptations and Remediation Technologies. Springer International Publishing, Cham, pp. 163–192.

Kabir, M., Chowdhury, S.A., Banik, B.K., Hoque, M.A., Marzan, A.A., 2023. Biosorption of pollutants from chemically derived wastewater using *Microcoleus* sp. In: AQUA-Water Infrastructure, Ecosystems and Society.

Kaleem, M., Mumtaz, A.S., Hashmi, M.Z., Saeed, A., Inam, F., Waqar, R., Jabeen, A., 2023. Myco-and phycoremediation of polychlorinated biphenyls in the environment: a review. Environ. Sci. Pollut. Res. 30 (6), 13994–14007.

Khan, S., Das, P., Thaher, M.I., AbdulQuadir, M., Mahata, C., Al Jabri, H., 2023. Utilization of nitrogen-rich agricultural waste streams by microalgae for the production of protein and value-added compounds. Curr. Opin. Green Sustain. Chem., 100797.

Khanikar, L., Ahmaruzzaman, M., 2023. Bioremediation of arsenic A sustainable approach in managing arsenic contamination. In: Bioremediation of Toxic Metal (Loid)s. CRC Press, pp. 185–203.

Kharel, H.L., Shrestha, I., Tan, M., Nikookar, M., Saraei, N., Selvaratnam, T., 2023. Cyanidiales-based bioremediation of heavy metals. Biotech 12 (2), 29.

Kidwai, M.K., Malik, A., Garg, V.K., 2023. Metal (loid)s sources, toxicity and bioremediation. In: Bioremediation of Toxic Metal (Loid)s. CRC Press, pp. 3–32.

Kiran, B.R., Mohan, S.V., 2022. Phycoremediation potential of Tetradesmus sp. SVMIICT4 in treating dairy wastewater using flat-panel photobioreactor. Bioresour. Technol. 345, 126446.

Kombe, G.G., 2023. An overview of algae for biodiesel production using bibliometric indicators. Int. J. Energy Res. 2023.

Kookal, S.K., Nawkarkar, P., Gaur, N.A., Kumar, S., 2023. Bioremediation of ethanol wash by microalgae and generation of bioenergy feedstock. J. Appl. Phycol. 35 (1), 183–194.

Koul, B., Sharma, K., Shah, M.P., 2022. Phycoremediation: a sustainable alternative in wastewater treatment (WWT) regime. Environ. Technol. Innov. 25, 102040.

Krishnamoorthy, N., Pathy, A., Kapoor, A., Paramasivan, B., 2023. Exploring the evolution, trends and scope of microalgal biochar through scientometrics. Algal Res. 69, 102944.

Kumar, C.P., Sylas, V.P., Mechery, J., Ambily, V., Kabeer, R., Sunila, C.T., 2023a. Phycoremediation of cashew nut processing wastewater and production of biodiesel using *Planktochlorella nurekis* and *Chlamydomonas reinhardtii*. Algal Res., 102924.

Kumar, P.K., Krishna, S.V., Verma, K., Pooja, K., Bhagawan, D., Himabindu, V., 2018. Phycoremediation of sewage wastewater and industrial flue gases for biomass generation from microalgae. South African journal of chemical engineering 25, 133–146.

Kumar, R., Basu, A., Bishayee, B., Chatterjee, R.P., Behera, M., Ang, W.L., Pal, P., Shah, M., Tripathy, S.K., Ambika, S., Janani, V.A., 2023b. Management of tannery waste effluents towards the reclamation of clean water using an integrated membrane system: a state-of-the-art review. Environ. Res., 115881.

Kumar, S., Singh, A., Kishor, R., 2023c. Algal biomass production coupled to wastewater treatment. In: Phycoremediation Processes in Industrial Wastewater Treatment. CRC Press, pp. 77–108.

Kumaran, M., Palanisamy, K.M., Bhuyar, P., Maniam, G.P., Rahim, M.H.A., Govindan, N., 2023. Agriculture of microalgae *Chlorella vulgaris* for polyunsaturated fatty acids (PUFAs) production employing palm oil mill effluents (POME) for future food, wastewater, and energy nexus. Energy Nexus 9, 100169.

Kumari, S., Kumar, V., Kothari, R., Kumar, P., 2022. Experimental and optimization studies on phycoremediation of dairy wastewater and biomass production efficiency of Chlorella vulgaris isolated from Ganga River, Haridwar, India. Environ. Sci. Pollut. Res., 1–12.

Kurade, M.B., Jadhav, U.U., Phugare, S.S., Kalyani, D.C., Govindwar, S.P., 2023a. Global scenario and technologies for the treatment of textile wastewater. In: Current Developments in Bioengineering and Biotechnology. Elsevier, pp. 1–43.

V. Algal bioproducts, biofuels, and biorefinery for business opportunities

Kurade, M.B., Mustafa, G., Zahid, M.T., Awasthi, M.K., Chakankar, M., Pollmann, K., Khan, M.A., Park, Y.K., Chang, S.W., Chung, W., Jeon, B.H., 2022. Integrated phycoremediation and ultrasonic-irradiation treatment (iPUT) for the enhanced removal of pharmaceutical contaminants in wastewater. Chem. Eng. J., 140884.

Kurade, M.B., Mustafa, G., Zahid, M.T., Awasthi, M.K., Chakankar, M., Pollmann, K., Khan, M.A., Park, Y.K., Chang, S.W., Chung, W., Jeon, B.H., 2023b. Integrated phycoremediation and ultrasonic-irradiation treatment (iPUT) for the enhanced removal of pharmaceutical contaminants in wastewater. Chem. Eng. J. 455, 140884.

Kusmayadi, A., Huang, C.Y., Leong, Y.K., Lu, P.H., Yen, H.W., Lee, D.J., Chang, J.S., 2023. Integration of microalgae cultivation and anaerobic co-digestion with dairy wastewater to enhance bioenergy and biochemicals production. Bioresour. Technol. 376, 128858.

Lakhani, S., Acharya, D., Sakariya, R., Sharma, D., Patel, P., Shah, M., Prajapati, M., 2022. A comprehensive study of bioremediation for pharmaceutical wastewater treatment. Chem. Eng. Sci. 4, 100073.

Latif, A., Abbas, A., Iqbal, J., Azeem, M., Asghar, W., Ullah, R., Bilal, M., Arsalan, M., Khan, M., Latif, R., Ehsan, M., 2023. Remediation of environmental contaminants through phytotechnology. Water Air Soil Pollut. 234 (3), 139.

Mahmod, S.S., Takriff, M.S., Al-Rajabi, M.M., Abdul, P.M., Gunny, A.A.N., Silvamany, H., Jahim, J.M., 2023. Water reclamation from palm oil mill effluent (POME): recent technologies, by-product recovery, and challenges. J. Water Process Eng. 52, 103488.

Malik, S., Kishore, S., Bora, J., Chaudhary, V., Kumari, A., Kumari, P., Kumar, L., Bhardwaj, A., 2023. A comprehensive review on microalgae-based biorefinery as two-way source of wastewater treatment and bioresource recovery. Clean—Soil, Air, Water 51 (3), 2200044.

Maroušek, J., Gavurová, B., Strunecký, O., Maroušková, A., Sekar, M., Marek, V., 2023. Techno-economic identification of production factors threatening the competitiveness of algae biodiesel. Fuel 344, 128056.

Mathias, R.C., Kushalan, S., Hegde, H., Jathanna, N.N., Hegde, S., 2023. Phycoremediation: algae-based bioremediation. In: Algae Materials. Academic Press, pp. 451–469.

Michelon, W., da Silva, M.L.B., Matthiensen, A., de Andrade, C.J., de Andrade, L.M., Soares, H.M., 2022a. Amino acids, fatty acids, and peptides in microalgae biomass harvested from phycoremediation of swine wastewaters. Biomass Convers. Biorefin. 12 (3), 869–880.

Michelon, W., Zuchi, I.D.P., Reis, J.G., Matthiensen, A., Viancelli, A., da Cruz, A.C.C., Silva, I.T., Fongaro, G., Soares, H.M., 2022b. Virucidal activity of microalgae extracts harvested during phycoremediation of swine wastewater. Environ. Sci. Pollut. Res. 29 (19), 28565–28571.

Mishra, K., Siwal, S.S., Saini, A.K., Thakur, V.K., 2023. Recent update on gasification and pyrolysis processes of lignocellulosic and algal biomass for hydrogen production. Fuel 332, 126169.

Mittal, V., Ghosh, U.K., 2023. Potential of microalgae for phytoremediation of various wastewaters for nutrient removal and biodiesel production through nanocatalytic transesterification. Asia Pac. J. Chem. Eng. 18 (1), e2847.

Moondra, N., Jariwala, N., Christian, R., 2023. Impact of microalgae in domestic wastewater treatment: a lab-scale experimental study. Pollution 9 (1), 211–221.

Mukherjee, D., Ray, S., 2023. Effects of bioabsorption of chromium and Lead on cyanobacterial species (Oersted ex Gomont *Spirulina subsalsa* and *Lemmermann*) Calothrix marchica. Indian J. Ecol. 50 (1), 114–118.

Mukherjee, S., Paul, S., Bhattacharjee, S., Nath, S., Sharma, U., Paul, S., 2023. Bioleaching of critical metals using microalgae. AIMS Environ. Sci. 10 (2), 226–244.

Mulluye, K., Bogale, Y., Bayle, D., Atnafu, Y., 2023. Review on microalgae potential innovative biotechnological applications. Biosci. Biotechnol. Res. Asia 20 (1), 35–43.

Nachiappan, K., Chandrasekaran, R., 2023. Reformation of dairy effluent—a phycoremediation approach. Environ. Monit. Assess. 195 (3), 405.

Nagarajan, D., Chen, C.Y., Lee, D.J., Chang, J.S., 2023. Photobioreactors for microalgae-based wastewater treatment. In: Current Developments in Biotechnology and Bioengineering. Elsevier, pp. 121–152.

Nirmala, N., Dawn, S.S., Arun, J., 2023. Bioremediation approaches for genomic microalgal applications in wastewater treatment. In: Genomics Approach to Bioremediation: Principles, Tools, and Emerging Technologies, pp. 199–208.

Noreen, R., Ahmad, A., 2023. Organic pollutant and dye degradation with nanocomposites. In: Sodium Alginate-Based Nanomaterials for Wastewater Treatment. Elsevier, pp. 97–136.

Ooi, W.C., Dominic, D., Kassim, M.A., Baidurah, S., 2023. Biomass fuel production through cultivation of microalgae *Coccomyxa dispar* and *Scenedesmus parvus* in palm oil mill effluent and simultaneous phycoremediation. Agriculture 13 (2), 336.

Overton, O.C., Olson, L.H., Majumder, S.D., Shwiyyat, H., Foltz, M.E., Nairn, R.W., 2023. Wetland removal mechanisms for emerging contaminants. Land 12 (2), 472.

Palekar, S., Menon, S., Girish, N., Wagh, N.S., Lakkakula, J., 2023. Microalgae as biological cleanser for wastewater treatment. In: Phycoremediation Processes in Industrial Wastewater Treatment. CRC Press, pp. 43–55.

Pandey, A., Govindwar, S.P., Kurade, M.B., Jeon, B.H. (Eds.), 2023. Current Developments in Bioengineering and Biotechnology: Advances in Eco-Friendly and Sustainable Technologies for the Treatment of Textile Wastewater. Elsevier.

Patel, H.K., Dobariya, J., Kalaria, R.K., 2023. Tailoring microalgae for efficient biofuel production. In: Phycoremediation Processes in Industrial Wastewater Treatment. CRC Press, pp. 159–180.

Pathy, A., Krishnamoorthy, N., Chang, S.X., Paramasivan, B., 2022. Malachite green removal using algal biochar and its composites with kombucha SCOBY: an integrated biosorption and phycoremediation approach. Surf. Interfaces 30, 101880.

Pradhan, B., Bhuyan, P.P., Nayak, R., Patra, S., Behera, C., Ki, J.S., Ragusa, A., Lukatkin, A.S., Jena, M., 2022. Microalgal phycoremediation: a glimpse into a sustainable environment. Toxics 10 (9), 525.

Prajapati, S.K., Kaushik, P., Malik, A., Vijay, V.K., 2013a. Phycoremediation coupled production of algal biomass, harvesting and anaerobic digestion: possibilities and challenges. Biotechnol. Adv. 31 (8), 1408–1425.

Prajapati, S.K., Kaushik, P., Malik, A., Vijay, V.K., 2013b. Phycoremediation and biogas potential of native algal isolates from soil and wastewater. Bioresour. Technol. 135, 232–238.

Rahhou, A., Layachi, M., Akodad, M., El Ouamari, N., Rezzoum, N.E., Skalli, A., Oudra, B., El Bakali, M., Kolar, M., Imperl, J., Petrova, P., 2023. The bioremediation potential of *Ulva lactuca* (Chlorophyta) causing green tide in Marchica lagoon (NE Morocco, Mediterranean Sea): biomass, heavy metals, and health risk assessment. Water 15 (7), 1310.

Raj, A.R.A., Mylsamy, P., Sivasankar, V., Kumar, B.S., Omine, K., Sunitha, T.G., 2023. Heavy metal pollution of river water and eco-friendly remediation using potent microalgal species. In: Water Science and Engineering.

Rajput, A.P., Kulkarni, M., Pingale, P.L., Tekade, M., Shakya, A.K., Tekade, R.K., 2023. Understanding the bioaccumulation of pharmaceuticals and personal care products. In: Essentials of Pharmatoxicology in Drug Research. vol. 1. Academic Press, pp. 393–434.

Rambabu, K., Avornyo, A., Gomathi, T., Thanigaivelan, A., Show, P.L., Banat, F., 2022. Phycoremediation for carbon neutrality and circular economy: potential, trends, and challenges. Bioresour. Technol., 128257.

Ramesh, B., Saravanan, A., Kumar, P.S., Yaashikaa, P.R., Thamarai, P., Shaji, A., Rangasamy, G., 2023. A review on algae biosorption for the removal of hazardous pollutants from wastewater: limiting factors, prospects and recommendations. Environ. Pollut., 121572.

Razaviarani, V., Arab, G., Lerdwanawattana, N., Gadia, Y., 2022. Algal biomass dual roles in phycoremediation of wastewater and production of bioenergy and value-added products. Int. J. Environ. Sci. Technol., 1–18.

Reinecke, D., Bischoff, L.S., Klassen, V., Bliferne-Klassen, O., Grimm, P., Kruse, O., Klose, H., Schurr, U., 2023. Nutrient recovery from wastewaters by algal biofilm for fertilizer production part 1: case study on the techno-economical aspects at pilot-scale. Sep. Purif. Technol. 305, 122471.

Rezasoltani, S., Champagne, P., 2023. An integrated approach for the phycoremediation of pb (II) and the production of biofertilizer using nitrogen-fixing cyanobacteria. J. Hazard. Mater. 445, 130448.

Ricky, R., Shanthakumar, S., 2022. Phycoremediation integrated approach for the removal of pharmaceuticals and personal care products from wastewater—a review. J. Environ. Manag. 302, 113998.

Ricky, R., Shanthakumar, S., Gothandam, K.M., 2022. A pilot-scale study of the integrated phycoremediation-photolytic ozonation based municipal solid waste leachate treatment process. J. Environ. Manag. 323, 116237.

Rossi, S., Mantovani, M., Marazzi, F., Bellucci, M., Casagli, F., Mezzanotte, V., Ficara, E., 2023. Microalgal cultivation on digestate: process efficiency and economics. Chem. Eng. J. 460, 141753.

Roy, M., Mohanty, K., 2023. Byproducts of a microalgal biorefinery as a resource for a circular bioeconomy. In: Biotic Resources. CRC Press, pp. 109–138.

Sakarya, F.K., Ertekin, E., Haznedaroglu, B.Z., 2023. Algae-based treatment of domestic and industrial wastewater. In: A Sustainable Green Future. Springer, Cham, pp. 409–428.

Santoso, A.D., Hariyanti, J., Pinardi, D., Kusretuwardani, K., Widyastuti, N., Djarot, I.N., Handayani, T., Sitomurni, A.I., Apriyanto, H., 2023. Sustainability index analysis of microalgae cultivation from biorefinery palm oil mill effluent. Global J. Environ. Sci. Manage. 9 (3), 559–576.

Sarkar, B., Dutta, S., Lahiri, S.K., 2022. Modeling and optimization of phycoremediation of heavy metals from simulated ash pond water through robust hybrid artificial intelligence approach. J. Chemom. 36 (7), e3427.

Sarkar, B., Sen, S., Dutta, S., Lahiri, S.K., 2023. Application of multi-gene genetic programming technique for modeling and optimization of phycoremediation of Cr (VI) from wastewater. Beni-Suef Univ. J. Basic Appl. Sci. 12 (1), 1–17.

Sathinathan, P., Parab, H.M., Yusoff, R., Ibrahim, S., Vello, V., Ngoh, G.C., 2023. Photobioreactor design and parameters essential for algal cultivation using industrial wastewater: a review. Renew. Sust. Energ. Rev. 173, 113096.

Saxena, N., Vasistha, S., Rai, M.P., 2023. Algal biorefinery: an integrated process for industrial effluent treatment and improved lipid production in bioenergy application. Vegetos 36 (1), 259–267.

Schoeters, F., Thoré, E.S., De Cuyper, A., Noyens, I., Goossens, S., Lybaert, S., Meers, E., Van Miert, S., de Souza, M.F., 2023. Microalgal cultivation on grass juice as a novel process for a green biorefinery. Algal Res. 69, 102941.

Sedighi, M., Pourmoghaddam Qhazvini, P., Amidpour, M., 2023. Algae-powered buildings: A review of an innovative, sustainable approach in the built environment. Sustainability 15 (4), 3729.

Selvaraj, D., Dhayabaran, N.K., Mahizhnan, A., 2022. An insight on pollutant removal mechanisms in phycoremediation of textile wastewater. Environ. Sci. Pollut. Res., 1–21.

Sengupta, S.L., Chaudhuri, R.G., Dutta, S., 2023. A critical review on phycoremediation of pollutants from wastewater—a novel algae based secondary treatment with the opportunities of production of value-added products. Environ. Sci. Pollut. Res. https://doi.org/10.21203/rs.3.rs-2349815/v1.

Sevugamoorthy, D., Rangarajan, S., 2023. Comparative analysis of biodegradation and characterization study on agal-assisted wastewater treatment in a bubble column photobioreactor. Environ. Challenges 10, 100659.

Shackira, A.M., Sarath, N.G., Puthur, J.T., 2022. Phycoremediation: a means for restoration of water contamination. Environ. Sustain., 1–14.

Shah, M.P. (Ed.), 2023. Phycoremediation Processes in Industrial Wastewater Treatment. CRC Press.

Shakoor, I., Nazir, A., Chaudhry, S., Capareda, S.C., 2022. Autochthonous Arthrospira platensis Gomont driven nickel (Ni) phycoremediation from cooking oil industrial effluent. Molecules 27 (16), 5353.

Sharma, R., Malaviya, P., 2023. Ecosystem services and climate action from a circular bioeconomy perspective. Renew. Sust. Energ. Rev. 175, 113164.

Sharma, S., Kant, A., Sevda, S., Aminabhavi, T.M., Garlapati, V.K., 2023. A waste-based circular economy approach for phycoremediation of X-ray developer solution. Environ. Pollut. 316, 120530.

Shayesteh, H., Vadiveloo, A., Bahri, P.A., Moheimani, N.R., 2022. Long term outdoor microalgal phycoremediation of anaerobically digested abattoir effluent. J. Environ. Manag. 323, 116322.

Sheikh, S., Miranda, J., Rai, V., Sheikh, S., Shameena, K.A., 2023. Influence of subsoil microbial community across different depths on the biodegradation of pesticides. In: Current Developments in Biotechnology and Bioengineering. Elsevier, pp. 379–400.

Silva, A.D., Fernandes, D.F., Figueiredo, S.A., Freitas, O.M., Delerue-Matos, C., 2022. Fluoxetine and nutrients removal from aqueous solutions by phycoremediation. Int. J. Environ. Res. Public Health 19 (10), 6081.

Singh, A., Rana, M.S., Tiwari, H., Kumar, M., Saxena, S., Anand, V., Prajapati, S.K., 2023a. Anaerobic digestion as a tool to manage eutrophication and associated greenhouse gas emission. Sci. Total Environ. 861, 160722.

Singh, P., Mohan, S.V., Mohanty, K., 2023b. Dairy wastewater treatment using Monoraphidium sp. KMC4 and its potential as hydrothermal liquefaction feedstock. Bioresour. Technol. 376, 128877.

Širić, I., Abou Fayssal, S., Adelodun, B., Mioč, B., Andabaka, Ž., Bachheti, A., Goala, M., Kumar, P., Al-Huqail, A.A., Taher, M.A., Eid, E.M., 2023. Sustainable use of CO_2 and wastewater from mushroom farm for Chlorella vulgaris cultivation: experimental and kinetic studies on algal growth and pollutant removal. Horticulturae 9 (3), 308.

Sirohi, R., Pandey, A., Sim, S.J., Chang, J.S., Lee, D.J. (Eds.), 2023. Current Developments in Biotechnology and Bioengineering: Photobioreactors: Design and Applications. Elsevier.

Sisman-Aydin, G., Simsek, K., 2022. Investigation of the phycoremediation potential of freshwater green algae Golenkinia radiata for municipal wastewater. Sustainability 14 (23), 15705.

Sivasubramanian, V., 2016. Phycoremediation and business prospects. In: Bioremediation and Bioeconomy. Elsevier, pp. 421–456.

Soroosh, H., Otterpohl, R., Hanelt, D., 2023. Influence of supplementary carbon on reducing the hydraulic retention time in microalgae-bacteria (MaB) treatment of municipal wastewater. J. Water Process Eng. 51, 103447.

Sridharan, R., Krishnaswamy, V.G., 2023. Bioremediation of textile dyes for sustainable environment—a review. Modern approaches in waste bioremediation. Environ. Microbiol., 447–460.

Suganya, D., Rajan, M.R., Mahendhiran, M., Durairaj, S., 2023. Immobilized micro algae for removing wastewater pollutants and ecotoxicological view of adsorbed nanoparticles—an overview. In: Phycoremediation Processes in Industrial Wastewater Treatment, pp. 147–158.

V. Algal bioproducts, biofuels, and biorefinery for business opportunities

Talib, S.L.A., Yasin, N.H.M., Takriff, M.S., Japar, A.S., 2023. Comparative studies on phycoremediation efficiency of different water samples by microalgae. J. Water Process Eng. 52, 103584.

Tan, K.A., Wan Maznah, W.O., Morad, N., Lalung, J., Ismail, N., Talebi, A., Oyekanmi, A.A., 2022. Advances in POME treatment methods: potentials of phycoremediation, with a focus on South East Asia. Int. J. Environ. Sci. Technol. 19 (8), 8113–8130.

Tong, C.Y., Honda, K., Derek, C.J.C., 2023. A review on microalgal-bacterial co-culture: the multifaceted role of beneficial bacteria towards enhancement of microalgal metabolite production. Environ. Res., 115872.

Ummalyma, S.B., Singh, A., 2022. Biomass production and phycoremediation of microalgae cultivated in polluted river water. Bioresour. Technol. 351, 126948.

Ummalyma, S.B., Chiang, A., Herojit, N., Arumugam, M., 2023. Sustainable microalgal cultivation in poultry slaughterhouse wastewater for biorefinery products and pollutant removal. Bioresour. Technol. 374, 128790.

Verma, R., Suthar, S., Chand, N., Mutiyar, P.K., 2022. Phycoremediation of milk processing wastewater and lipid-rich biomass production using Chlorella vulgaris under continuous batch system. Sci. Total Environ. 833, 155110.

Waqas, M.R., Nadeem, S.M., Khan, M.Y., Ahmad, Z., Ali, L., Asghar, H.N., Khalid, A., 2022. Phycoremediation of textile effluents with enhanced efficacy of biodiesel production by algae and potential use of remediated effluent for improving growth of wheat. Environ. Sci. Pollut. Res., 1–9.

Xiong, J.Q., Kabra, A.N., Salama, E.S., 2023. Microalgae-based biotechnologies for treatment of textile wastewater. In: Current Developments in Bioengineering and Biotechnology. Elsevier, pp. 457–471.

Yadav, N., Thakur, I.S., Srivastava, S., 2023. Omics approach in bioremediation of heavy metals (HMs) in industrial wastewater. In: Genomics Approach to Bioremediation: Principles, Tools, and Emerging Technologies, pp. 343–361.

Yahya, L., Harun, R., Zainal, A., 2023. Evaluation of bioactive compounds from robust consortium microalgae species under actual flue gas exposure. Biochem. Eng. J. 190, 108739.

Yuan, C., Zhao, S., Ni, J., He, Y., Cao, B., Hu, Y., Wang, S., Qian, L., Abomohra, A., 2023. Integrated route of fast hydrothermal liquefaction of microalgae and sludge by recycling the waste aqueous phase for microalgal growth. Fuel 334, 126488.

Zahra, S.A., Ahmad, I., Abdullah, N., Iwamoto, K., Yuzir, A., 2023. Microalgae-based bioremediation of pharmaceuticals wastewater. In: The Treatment of Pharmaceutical Wastewater. Elsevier, pp. 277–309.

Zainal, A., Yaakob, Z., Takriff, M.S., Rajkumar, R., Ghani, J.A., 2012. Phycoremediation in anaerobically digested palm oil mill effluent using cyanobacterium, *Spirulina platensis*. J. Biobased Mater. Bioenergy 6 (6), 704–709.

Zatta, P.H.S., Jacob-Furlan, B., Mirabile, R.C., Gonçalves, R.S.R., da Silva, P.A.S., Taher, D.M., Lira, G.S., Martins, L.S., Ordonez, J.C., 2023. Microalgae biopolymers: a review. Rev. Engenh. Térm. 21 (4), 44–52.

Zheng, S., Chen, S., Wu, A., Wang, H., Sun, S., He, M., Wang, C., Chen, H., Wang, Q., 2023. Efficient nutrient removal of Pyropia-processing wastewater and rapid algal biomass harvesting by Scenedesmus obliquus combined with chitosan. J. Water Process Eng. 51, 103365.

Zieliński, M., Dębowski, M., Kazimierowicz, J., Świca, I., 2023. Microalgal carbon dioxide (CO_2) capture and utilization from the European Union perspective. Energies 16 (3), 1446.

21

Algae-based bioremediation bioproducts and biofuels for biobusiness

Raman Kumar[a,b],*, *Yograj Neha*[a],*, *G.A. Ravishankar*[c], *and Vidyashankar Srivatsan*[a]

[a]Applied Phycology and Food Technology Laboratory, Biotechnology Division, CSIR-Institute of Himalayan Bioresource Technology, Palampur, Himachal Pradesh, India [b]Academy of Scientific and Innovative Research (AcSIR), CSIR- Human Resource Development Centre, (CSIR-HRDC) Campus, Ghaziabad, Uttar Pradesh, India [c]Dayananda Sagar Institutions, Bengaluru, Karnataka, India

1 Introduction

Microalgae have been advocated as a green cellular factory for the production of high-value nutraceuticals (anti-oxidants, polyunsaturated fatty acids, and carotenoids) and commodity products (lipids, hydrocarbons, and starch) (Forján et al., 2015). Their ability to grow in non-potable waters, non-arable land, and higher photosynthetic efficiencies makes them an attractive industrial source of aforesaid products (Hosseinkhani et al., 2022). Despite their high value, commercial realization of microalgae-mediated products is yet to be achieved owing to the water-intensive cultivation processes and energy-intensive downstream processes (Parmar et al., 2023). Integration of bioremediation and simultaneous microalgae cultivation offers a sustainable route for microalgae biomass production (Dagnaisser et al., 2022). Microalgae have been shown to grow abundantly in nutrient-rich wastewaters containing high concentrations of nitrates, ammonia, and phosphates obtained from industrial manufacturing processes such as steel slag, cement, food processing, distillery, aquaculture,

*Equal contribution.

583

and leather tannery (You et al., 2021; Rambabu et al., 2022; Venugopal, 2022). In addition, microalgae have been successfully utilized for capturing industrial gases rich in CO_2, NO_X, SO_X, etc. (Yen et al., 2015). Several reports on the successful utilization of microalgae for wastewater remediation coupled with biomass production and downstream resource recovery are available from recent years (Carraro et al., 2022; Nagarajan et al., 2022; Satya et al., 2023; Yan et al., 2022).

Conventional waste treatment processes involving physicochemical processes such as activated sludge process, chemical precipitation (coagulation), microfiltration, carbon adsorption, ozonation, and reverse osmosis have been successfully deployed in several industries for removal of organic wastes, inorganic phosphates, and toxins, respectively (Doble and Kumar, 2005). However, some of the major disadvantages associated with these conventional treatment technologies are high maintenance and operational costs, high energy requirements, low throughput, and generation of high quantities of sludge (Chan et al., 2022). Microalgae-based wastewater treatment varies from these conventional technologies with the ability to correct pH, reduce total dissolved solids (TDS), biochemical oxygen demand (BOD), and chemical oxygen demand (COD), and remove nutrients simultaneously unlike the conventional methods where each of these unit processes and treatment steps is carried out individually (Park et al., 2011a, b; Sivasubramanian et al., 2012). Microalgae have been consistently shown to remove more than 70% of nutrients from wastewater and in some cases greater than 90% from the initial levels. In addition to the bioremediation, microalgae biomass generated could be exploited for several downstream applications including production of animal feed, pigments (natural colorants), biofertilizers, adsorbents, and biofuels, making the whole process economically viable (Mehariya et al., 2021). Some critical differences between conventional WWT and microalgae-mediated WWT are presented in Table 1.

2 Microalgae for wastewater treatment and xenobiotics breakdown

2.1 Composition of industrial effluents

Industrial wastewaters are characterized by the presence of pollutants such as heavy metals (Cd, Ni, Pb, Hg, As, Cu, Cr), high organic matter content, synthetic dyes and chemicals, suspended particles, and infectious microorganisms. It is estimated that global wastewater production is approximately 359.4 billion m^3 annually, and almost 80% of the wastewater is untreated and released into the environment (Jones et al., 2021; Amin et al., 2022). The most abundant nutrients present in wastewater are nitrogen (N) and phosphorus (P). The N&P concentrations are highest in sewage/municipal wastewaters, specifically effluents from agriculture and livestock effluents such as that of swine (1180 mg L^{-1}) compared to industrial effluents (up to 480 mg L^{-1}) (Cheng et al., 2020a). The N/P ratio was relatively higher in agriculture, livestock, and municipal wastewaters (Díaz et al., 2022). Industrial effluents are generally characterized by high BOD, COD, total organic carbon, and varying pH. The chemical composition of effluents varies from different industries with agricultural and food processing wastewaters containing high BOD, suspended solids, and ammonia nitrogen. Effluents from the confectionary industry contain simple sugars while those of meat processing industry contain body fluids, antibiotics, hormones, and other organic wastes.

TABLE 1   Differences between microalgae-mediated and conventional wastewater treatments.

Algae-mediated wastewater treatment	Conventional waste treatment
Rely on microalgae and macroalgae	Rely on physical and chemical treatments
Utilize natural oxygen generated from photosynthesis for oxidation of pollutants	Utilize artificial oxidation process such as stirred tank reactors and activated sludge process
Wide range of waste substrate specificity (municipal/domestic wastewaters, industrial effluents, agricultural/food processing wastes, etc.), including waste gases	Substrate specific methods of treatment for different effluent types
Generally in situ process	Either in situ or ex situ process
Compatible with conventional processes	Highly specific and non-compatible
Single-step processes for pH correction, TDS reduction, BOD/COD removal, color or odor removal	Multi-stage process for each parameter
Negligible generation of sludge and treatment rejects	High sludge generation
Waste remediation can be coupled with biomass valorization	Sludge used as fertilizers or as landfills
Process can be linked with algae-based bioenergy production (biodiesel, bioethanol, and biogas)	Highly energy consuming
Lower capital and operational costs	High capital and maintenance costs
Slower process with long retention times	Faster process

Adapted from Vidyashankar, S., Ravishankar, G.A., 2016. Algae-based bioremediation: bioproducts and biofuels for biobusiness. Bioremediat. Bioecon. 457–493.

In the case of paper and pulp industries, waters are contaminated with high levels of suspended solids and are abundant in salts of inorganic acids and alkali. In steel industries, the waters are contaminated with gasification products such as benzene, naphthalene, anthracene, cyanide, ammonia, phenols, cresols together with a range of more complex organic compounds known collectively as polycyclic aromatic hydrocarbons (PAHs) (Doble and Kumar, 2005). The composition of different industrial effluents is presented in Table 2. In addition to PAHs, industrial effluents such as that of agro-chemical and pharmaceutical industries contain complex organic molecules that are unnatural to aquatic ecosystem. These include pesticides, organochlorides, dye effluents, plasticizers (phthalates, bisphenol), and dioxins, which are commonly termed as xenobiotics. These xenobiotics are generally persistent, i.e., recalcitrant to conventional WWT processes and generally termed as persistent organic pollutants (POPs) (Rieger et al., 2002). Release of nutrient-rich effluents and untreated industrial effluents into waterbodies causes eutrophication, bioaccumulation of heavy metals and organic compounds that are detrimental to aquatic ecosystem and human health (Nagarajan et al., 2022).

2.2 Microalgae for effluent treatment

Microalgae offer a successful option for wastewater remediation owing to its special nutrient absorption properties and high photosynthetic efficiencies. The maximum attainable

TABLE 2 Composition of different industrial effluents.

Industrial effluent	Effluent composition								
	pH	Conductivity (ms cm⁻¹)	BOD (mg L⁻¹)	COD (mg L⁻¹)	TSS (mg L⁻¹)	TOC (mg L⁻¹)	Ammonia nitrogen (mg L⁻¹)	Total nitrates (mg L⁻¹)	Total nitrogen (mg L⁻¹)
Pharmaceutical industry	5.0–9.5	8.06	—	2580–6840	89	2100	64.2	0.24	228
Petrochemical industry	1.0–13.41	—	4710	130–20,150	1891.5	3321.3	1.5–41.7	1.5	200
Paper and pulp industry	2.7–8.1	2156	—	9200–25,634	89	—	200	0.5–2.8	194
Food processing industry	3.65–5.92	866	900–30,000	135–357,725	16–10,793.33	—	1.80–2925	0.106–26.29	23.08
Tannery industry	7.6–8.1	1760–2220	1270	157–3350	1374	—	8.4–86	2.9–4.4	17.5–166.7
Textile industry	8.75–11.53	2.5	660	325–2430	49–1200	52	2.72–7.8	7.97–11	40
Brewery wastewater	3–12	—	1200–3600	2000–6000	2900–3000	—	29.4		25–80

BOD, biological oxygen demand; COD, chemical oxygen demand; TSS, total suspended solids; TOC, total organic carbon.

Source: Doble, M., Kumar, A., 2005. Biotreatment of Industrial Effluents. Elsevier; Nagarajan, D., Lee D.-J., Varjani, S., Lam, S.S., Allakhverdiev, S.I., Chang, J.-S., 2022. Microalgae-based wastewater treatment—microalgae-bacteria consortia, multi-omics approaches and algal stress response. Sci. Total Environ. 157110; Amin, M., Tahir, F., Ashfaq, H., Akbar, I., Razzaque, N., Haider, M.N., Xu, J., Zhu, H., Wang, N., Shahid, A., 2022. Decontamination of industrial wastewater using microalgae integrated with biotransformation of the biomass to green products. Energy Nexus 6, 100089; Ferreira, A., Ribeiro, B., Ferreira, A.F., Tavares, M.L.A., Vladic, J., Vidović, S., Cvetkovic, D., Melkonyan, L., Avetisova, G., Goginyan, V., 2019. Scenedesmus obliquus microalga-based biorefinery—from brewery effluent to bioactive compounds, biofuels and biofertilizers—aiming at a circular bioeconomy. Biofuels Bioprod. Biorefin. 13, 1169-86.

photosynthetic efficiency for microalgae is 8.3% while that of a typical C3 plant does not exceed 2.4% (Chisti, 2013; Nagarajan et al., 2020). Microalgae often show very high nutrient removal capacity (>80%), and this ability could be attributed to special adaptive mechanisms such as carbon concentration, lipid remodeling, and phosphorus allocation (luxury uptake mechanism) in fast-growing microalgae as described by Çakirsoy et al. (2022). Nutrient removal capacity of few microalgae strains is presented in Table 3. Wastewater-mediated cultivation of microalgae resulted in high biomass productivities in several microalgae species such as *Chlorella* sp., *Scenedesmus* sp., *Tetraselmis* sp., *Tetradesmus* sp., and *Synechocystis* sp. (Amin et al., 2022; Sharma et al., 2022). It has been determined that to produce 1 kg of microalgae biomass, approximately 367 g carbon, 88 g nitrogen, 2 g of phosphorus, and close to 0.001–$0.11 \, m^3$ of water depending upon the type of species and cultivation are required. Further, it is estimated that approximately 1.5–2.2 kg gaseous CO_2 could be fixed for the production of production of 1 kg biomass conditions (Mayers et al., 2016; Nagarajan et al., 2020). Comparison of industrial effluent composition (Table 2) with aforesaid microalgae nutrient requirements suggests the possibility of utilizing wastewaters as primary source of nutrients such as N and P during mass cultivation of microalgae, improving the economic viability of the whole process.

From the literature, it is evident that microalgae have the ability to grow on a wide range of wastewater substrates containing varied nitrogen sources (ammonia N, nitrates, and organic nitrogen sources), organic phosphorus or inorganic phosphates, organic and inorganic carbon sources. The most commonly found microalgae species in domestic/municipal wastewater are cyanobacteria and chlorophycean members; however, cyanobacteria dominate the planktonic community with an abundance of >90% (Furtado et al., 2009).

Among the cyanobacterial community, strains belonging to *Phormidium* sp., *Plankothrix* sp., *Limnotrix* sp., and *Synechocystis* sp., are predominant (Martins et al., 2010; Gupta et al., 2013). In aerobic treatment ponds of wastewaters derived from agro-food industries, *Oscillatoria* sp. and *Lyngbya* sp. were pre-dominant (Vasconcelos and Pereira, 2001). The pre-dominant chlorophycean species used for wastewater remediation are cyanobacteria and chlorophycean strains belonging to genus *Chlorella, Scenedesmus, Monoraphidium* sp., *Desmodesmus, Chlamydomonas,* and *Tetraselmis.* These species have relatively higher specific growth rate and nutrient removal capacity compared with other groups. *Chlorella* and *Scendemsus* species are generally known as "hyperaccumulators" and "hyper adsorbents" of contaminants such as heavy metals and POPs (Maryjoseph and Ketheesan, 2020). These two species show a very high nutrient removal efficiency, >80% for both N and P, and almost 90% reduction in COD from secondary treated wastewater (Cho et al., 2011). Particularly, strains belonging to the genus *Chlorella* have gained global attention owing to their ability to grow in autotrophic, mixotrophic, and heterotrophic conditions utilizing a wide range of substrates (Goh et al., 2023). In addition to chlorophycean members, several euglenophytes (*Euglena gracilis, Euglena viridis*) have been identified in wastewaters with higher organic loading. These euglenophytes derive nutrients from the environments using mixotropic mode of nutrition (Chanakya et al., 2012). Apart from domestic sewage waters, algal forms were identified to remove nutrients from agricultural wastewater and livestock waste slurries that are high in N and P. For example, *Botryococcus braunii* was reported to grow at higher growth rates in piggery wastewater comprising $800 \, mg \, L^{-1}$ nitrates and showed ~80% removal of nitrate-N (An et al., 2003).

TABLE 3　Nutrient removal properties of different microalgae.

Type of wastewater	Microalgae strain	Cultivation parameters	Nutrient removal capacity	References
Municipal wastewater	*Galdieria sulphuraria*	Outdoor cultivation Raceway ponds aerated with 2% CO_2	PO_4^{3-}—>60% BOD—>60% NH_4:N—>70%	Li et al. (2019)
	Consortium of algae—*Phormidium* and *Chlorella pyrenoidosa*	Outdoor film bioreactor attached to HRT pond	COD—53% NO_3-N—81% TP—75% NH_4-N—81%	Naaz et al. (2019)
	Scendesmus obliquus	Panel bioreactors attached to HRT pond	NH_4-N—76.5% PO_4^{3-}—70.6%	Ling et al. (2019)
	Chlorella sorokiniana	Flat panel bioreactors	DIC—46-56% PO_4^{3-} up to 60% NH_4-N—100%	de Souza Leite et al. (2019)
	Consortium of *Scenedesmus and Chlorella* sp.	Flat panel bioreactors	Up to 60% removal of total N&P	Gonzalez-Camejo et al. (2019)
	Scenedesmus dimorphus, S. quadricauda, Desmodesmus armatus, Chlorella sp.	Raceway ponds	PO_4^{3-}—90% NH_4-N—99%	Lage and Gentili (2023)
Animal manure wastewater (Piggery waste)	*Chlorella vulgaris* FACHB 30	Photo-bioreactors	COD—99% NH_4-N—100% TP—95%	Zhang et al. (2019b)
	Synechocystis sp.	NR	COD—68.6% TN—75.8% TP—71.4%	Huo et al. (2020)
	Tribonema sp.	NR	COD—56.6% TN—89.9% TP—72.7%	Cheng et al. (2020b)
Animal manure wastewater (Aquaculture)	Chlorella sp. GD	Bubble column photo-bioreactors	COD—80% TN—90% NH_4-N—77% NO_3^--N—83% TP—99%	Kuo et al. (2016)
Textile industry effluent	*Scenedesmus abundans* NCIM No. 2897	Photo-bioreactors	BOD—13.9% COD—86.8% NO_3^--N—68.8% PO_4^{3-}—70% Total chlorides—44.4%	Brar et al. (2019)

TABLE 3 Nutrient removal properties of different microalgae—cont'd

Type of wastewater	Microalgae strain	Cultivation parameters	Nutrient removal capacity	References
	Chlorella pyrenoidosa NCIM 2738	Photo-bioreactors	BOD—24.05% COD—85% NO_3^--N—74.43% PO_4^{3-}—28.01% Total chlorides—61%	Brar et al. (2019)
	Consortium of microalgae—*Chlorella* sp., *Scenedesmus* sp., *Botyococcus braunii*	Fed batch bioreactor	TP—95% TN—70% Color—68–72%	Kumar et al. (2018)
Food industry—dairy effluent	*Scenedesmus quadricauda*	Photo-bioreactor	TOC—76.77% TN—92.15% TP—100% SO_4^{2-}—100%	Daneshvar et al. (2019b)
	Scenedesmus sp. ASK22	NR	COD: 90.5% TN—100% TP—91.24%	Pandey et al. (2019)
	Tetraselmis suecica	Photo-bioreactor	TN—44.92% PO_4^{3-}—42.18% TOC—40.16%	Daneshvar et al. (2018b)
	Ascochloris sp. ADW007	Photo-bioreactor	COD—95.1% TN—79.7% TP—98.1%	Kumar et al. (2019)
	Chlorella vulgaris, *Chlorella sorokiniana*, *Chlorella saccharophila*, *Auxenochlorella protothecoides*, and *Chlamydomonas reinhardtii*	NR	Below discharge limits	Gramegna et al. (2020)
Brewery & Distillery Industry effluents	*Scenedesmus obliquus*	Bubble column photo-bioreactor	TN—88% TP—30% COD—71%	Ferreira et al. (2019)
	Chlorella sp. L166, *Chlorella* sp. UTEX1602, *Scenedesmus* sp. 336 and *Spirulina* sp. FACHB-439	Laboratory scale—Flasks	NH_4-N—89.99% TN—75.96% TP—95.71% COD—73.66%	Song et al. (2020)
Paper & pulp industry effluent	Mixed culture—*Scenedesmus* sp.	Outdoor circular ponds	BOD—82% COD—75% NO_3^-—65% TP—71.29%	Usha et al. (2016)
	Scenedesmus acuminatus SAG 38.81	Photobioreactors	COD—36% NH_4-N—99% PO_4^{3-}—96.9%	Tao et al. (2017)

Continued

TABLE 3   Nutrient removal properties of different microalgae—cont'd

Type of wastewater	Microalgae strain	Cultivation parameters	Nutrient removal capacity	References
	Chlorella vulgaris	Laboratory scale	TN—76.56% TP—92.72% COD—75.48% TOC—70.67%	Daneshvar et al. (2018a)
Slaughterhouse/ Tannery processing effluents	*Phormidium* sp.	Photobioreactors	COD—90% TN—57% TP—52%	Maroneze et al. (2014)
	Mixed microalgae culture	Photobioreactors	TOC—89.6% TN—70.2% TP—96.2%	Taşkan (2016)
	Chlorella vulgaris, Scenedesmus dimorphus, Chlorococcum sp., and *Chlamydomonas* sp.	Laboratory scale—Flasks	NH_4-N—55% NO_3^--N—85.6% PO_4^{3-}—60.5% COD—43.4%	Nagabalaji et al. (2019)
	Scenedesmus sp.	Laboratory scale—Flasks	COD—80.33% NH_4-N—85.63% TP—96.78%	da Fontoura et al. (2017)
Petrochemical industry effluents	*Tribonema* sp.	Vertical tubular photo-bioreactors	COD—98.4% TP—93.71% TN—96.8%	Huo et al. (2019)
	Chlorella vulgaris	Laboratory scale—Flasks	COD—10.89% BOD—30.36% TP—44.84% TN—92.59%	Madadi et al. (2021)
Pharmaceutical industry effluents	*Scenedesmus abundans*	Photobioreactor	COD—97.24% TP—93.71% TN—97.12%	Nayak and Ghosh (2019)
	Mixed culture—*Tetradesmus obliquus* and *Chlorella sorokiniana*	Bubble column photobioreactor	TP—95% TN—82-97% Degradation of active pharmaceutical ingredients—Up to 40%	Escapa et al. (2017a)
	Tetraselmis indica BDU 123	Laboratory scale—Flasks	COD—66.30% TOC—78.14% TP—70.03% TN—67.17%	Nayak and Ghosh (2020)

COD, chemical oxygen demand; BOD, biological oxygen demand; TOC, total organic carbon; DIC, dissolved inorganic carbon; TP, total phosphorus; TN, total nitrogen; NH_4-N, ammoniacal nitrogen, PO_4^{3-}, phosphates; SO_4^{2-}, sulfates; NO_3^-, nitrates; NR, not reported.

3 Microalgae bacterial consortia for bioremediation

Degradation of organic pollutants and nutrient removal from wastewaters is a complex process involving symbiotic activity between microalgae, fungi, and bacteria. In most cases, microalgae and fungi or bacteria exist naturally as a consortium and form microbial mats constituted by cyanobacteria, diatoms, anoxygenic phototrophic bacteria, sulfate-reducing bacteria, etc. These microbial members are arranged in layers depending on the sequence of metabolic reactions determined by gradients of light and redox potentials (Subashchandrabose et al., 2011). The interactions between microalgae and bacteria can be either cooperative mainly for nutrient exchange or competitive (Amin et al., 2022). The most commonly exchanged nutrients are O_2 and CO_2 between algae and bacteria where O_2 released by microalgae after photosynthesis is utilized by bacteria for degradation of organic carbon and pollutants releasing CO_2 that are in turn utilized by microalgae for photosynthesis and growth (Saravanan et al., 2021). Further, the bacteria in the consortia cater to the additional nutritional requirements of microalgae such as vitamins (biotin, thiamine, cobalamine) and bacterial siderophores sequester iron during iron-deficient conditions (Croft et al., 2006; Butler, 1998), while the extracellular polymeric substances (EPS) and exudates such as sugars and alcohol (mannitol, arabinose, acetate, propionate, lactate, and ethanol) secreted by microalgal or cyanobacteria serve as bacterial growth substrates (Abed et al., 2007). In addition to nutrient exchange, the algae-bacteria consortia interact through other mechanisms such as signal transduction and horizontal gene transfers (Jiang et al., 2021). In case of POP treatments, the algae-bacterial consortia prove beneficial where typically microalgae decolorize and break down complex large molecules to smaller units, and bacteria utilize these small molecules as energy sources. Further, the consortia enhance nitrogen and phosphorus assimilation, which in turn can be recovered from biomass (Lin et al., 2019). The mechanisms of interaction observed between microalgae and bacterial consortia could be exploited for enhanced productivity and bioresource production in microalgae. For example, co-cultivation of axenic cultures of *Chlorella* sp., specifically *C. vulgaris*, and *C. varaibilis* with bacteria such as *Rhizobium* sp., *Hyphomonas* sp., or *Sphingomonas* sp. improved microalgae growth, biomass productivity, and chlorophyll content (reference). The enhanced productivity in microalgae was attributed to secretion of indole acetic acid (IAA), decreased O_2, release of metabolizable N_2, micronutrients such as iron (Jiang et al., 2021). Further, several volatile organic compounds (VOCs) have been identified to be involved in communication between microalgae and bacterial association such as 1-decanone, 1,2,3-butanetriol, and quinolone, 2-pentadecanone, 1,2-propanediol. Highly diversified VOCs in large numbers were detected during co-cultivation of *C. vulgaris* with bacteria such as *Rhizobium* sp., *Agrobacterium* sp., resulting in three times enhanced biomass productivity in microalgae compared with axenic cultures of *C. vulgaris* (Chegukrishnamurthi et al., 2022). Various microalgae and bacterial consortia have been reported to utilize or degrade complex organic compounds such as phenols, non-chlorinated aliphatic, aromatic hydrocarbons, and petroleum hydrocarbons as sources of carbon (Table 4).

Apart from the enhanced ability to remove nutrients, co-cultivation of bacteria with microalgae improves the operational efficiency of microalgae resource recovery. The major benefits of a microalgae-bacterial consortia bioremediation are better nutrient recovery,

TABLE 4 Microalgae-bacteria/fungi consortia in wastewater remediation

Microalgae/Microalga consortia	Bacteria/fungi—consortia	Nature of wastewater/POPs	Removal efficiency N (%)	Removal efficiency P (%)	Removal efficiency C (%)	Initial POP concentration	Removal efficiency (POPs)	Reference
Synechocystis sp. PCC6803	*Pseudomonas* related strain GM41	Phenanthrene	NR	NR	NR	$0.15\,mmol\,L^{-1}$	$0.8\,\mu g\,day^{-1}$	Abed (2010)
Pseudoanabaena PP16	*Pseudomonas* sp. P1	Phenol	NR	NR	NR	$1\,mmol\,L^{-1}$	95%	Kirkwood et al. (2006)
Consortia of • *Phormidium* sp., • *Oscillatoria* sp., • *Chroococcus* sp.	*Burkholderia cepacia*	Diesel	NR	NR	NR	0.6% v/v	Hydrocarbon removal up to 99%	Chavan and Mukherji (2010)
Chlorella sorokiniana 211/8k	*Ralstonia basilensis*	Sodium salicylate	NR	NR	NR	$5\,mmol\,L^{-1}$	$1\,mmol\,L^{-1}\,day^{-1}$	Guieysse et al. (2002)
Chlorella sorokiniana	*Comamonas* sp.	Acetonitrile	NR	NR	NR	$1\,g\,L^{-1}$	$0.44\,g\,L^{-1}\,day^{-1}$	Munoz et al. (2005)
Chlorella sorokiniana IAM C-212	*Microbacterium* sp. CSSB-3	Propionate	NR	NR	NR	$125\,mg\,L^{-1}$	100% removal	Imase et al. (2008)
Chlorella sorokiniana	*Pseudomonas migulae*	Phenanthrene	NR	NR	NR	$200\text{--}500\,mg\,L^{-1}$	$24.2\,g\,m^{-3}\,h^{-1}$	Munoz et al. (2003) and Muñoz et al. (2003)
Scenedesmus obliquus GH2	Consortia of • *Sphingomonas* sp., GY2B • *B. cepacia* GS3C • *Pseudomonas* GP3A • *Pandoraea pnomenusa* GP3B	Straight-chain alkanes, alkyl cycloalkanes, alkyl benzenes, naphthalene, fuorene and Phenanthrene; Crude oil	NR	NR	NR	0.3% *v/v*	100% removal	Tang et al. (2010)

Microalgae	Bacteria/community	Wastewater/substrate						Reference
Consortia of • Chlorella sp., • Scenedesmus obliquus, • Stichococcus sp. • Phormidium sp.		Phenols Oils	NR	NR	NR	0.48 mg L^{-1} 40 mg L^{-1}	85% removal 96% removal	Safonova et al. (2004)
Auxenochlorella protothecoides	Wastewater-borne microbial community	Effluent from oxidation tank at a winery plant	MABC: 100±0 A: 99±1 B: 16±17	MABC: 100±0 A: 100±0 B: 1±1	MABC: 38±6 A: 44±11 B: 36±14	NA	NA	Higgins et al. (2018)
Chlorella sp.	Acinetobacter sp. isolated from wastewater	Sterilized centrate from WWTP	MABC:79.12 A: 73.10 B: 73.10	MABC: 96.26 A: 83.43 B: 48.75	MABC: 79.11 B: 63.91	NA	NA	Liu et al. (2017)
Chlorella vulgaris	Rhizobium sp.	Sterile synthetic wastewater based on BG11 medium supplemented with glucose, peptone, meat extract, urea and KH$_2$PO$_4$	MABC: 54.7 A: 46.8 B: 50.7	MABC: ~96 A: <58 B: <69	MABC: 48.7 A: 18.3 B: 42.2	NA	NA	Ferro et al. (2019)
Chlorella vulgaris	Exiguobacterium sp. and Bacillus licheniformis	Piggery wastewater after filtration, decolouration, and autoclave treatment	MABC: ~80 A: 70	MABC: ~82 A: 60	MABC: ~85 A: 34	NA	NA	Wang et al. (2020b)
Mixed algae (dominated by Leptolyngbya sp.)	Microbes from aerobic granular sludge	Synthetic domestic wastewater	MABC: 44.3	MABC: 65.4	MABC: 84.6	NA	NA	Cai et al. (2019)
Scenedesmus sp. 336 and Chlorella sp. 1602	Microbes from activated sludge	Artificial municipal sewage	MABC: 100% NO$_3$-N removal on the first day)	MABC: 100%	MABC: 94.42	NA	NA	Chen et al. (2019)

MABC: microalgae-bacterial consortium; A: axenic algae; B: axenic bacteria; NO$_3^-$: nitrates.

improved biomass productivity, and facilitation of biomass recovery during downstream processing (Jiang et al., 2021). Presence of bacteria improved the harvesting efficiency of microalgae biomass through formation of flocs and increasing the rate of gravity sedimentation. Microalgae bacteria consortia have shown enhanced degradation of organic compounds, mainly POPs and xenobiotics such as phenolic compounds, pesticides, azo compounds, and antibiotics (Amin et al., 2022).

4 Mechanism of xenobiotics degradation by microalgae

Effluents derived from petrochemical, textile, pesticides, and pharmaceutical industries possess a variety of xenobiotics such as phenols, benzene compounds, polychlorinated biphenyl (PCB), and antibiotics, which are resistant to conventional bioremediation processes. Microalgae have developed unique mechanisms to degrade these xenobiotics, which are bioadsorption, bioaccumulation, biodegradation/biotransformation, and photolysis (Jain et al., 2022). In general, chlorophycean microalgae such as *Chlorella* sp., *Scenedesmus* sp., and *Selenastrum capricornutum* have been widely reported in breakdown of these POPs. Microalgae cell wall contains many charged functional groups that interact with pollutants through chelation, electrostatic interaction, surface precipitation, and ion exchange (Amin et al., 2022). Bioaccumulation is typically involved during heavy metal sequestration and certain hydrophobic compounds. Bioaccumulation phenomenon involves three types of uptakes of contaminants, namely concentration gradient-driven passive diffusion, facilitated diffusion supported by specific transmembrane transporter proteins, and active uptake of contaminants. Microalgae have shown to accumulate phenols such as triclosan, radiolabeled compounds (Wu et al., 2022) (Jain et al., 2022). Biodegradation/biotransformation mechanisms are frequently observed with remediation of phenols, azo compounds, and pesticides where the xenobiotics are degraded through an active enzymatic degradation process. This method of bioremediation is more promising as compared with bioadsorption or bioaccumulation where contaminants are removed from aqueous media but still persistent inside the microalgae cell. In biodegradation phenomenon, the contaminants such as phenolic compounds act as carbon source and electron donors for microalgae cells and other non-growth substrates (Xiong et al., 2021). Oxidative hydroxylation, glycolysation, deamination, hydrogenation, carboxylation, demethylation, ring cleavage, and oxidation are few metabolic processes that are involved in degradation of organic compounds (Wu et al., 2022). Biotransformation of azo dyes to aromatic amine and subsequently to simple amine compounds has been observed in textile effluents when treated with microalgae (Touliabah et al., 2022). Photolysis is light-dependent degradation of contaminants mainly phenols. Microalgae enhance removal and degradation of phenolic compounds through indirect photolysis where microalgae act as photosensitizers where the various reactive oxygen species such as hydroxyl radicals ($\cdot$OH), singlet oxygen ($^1O_2$), superoxide ($O_2^-\cdot$), and hydrogen peroxide (H_2O_2) generated during photosynthetic process degrade the phenols (Wei et al., 2021). The degradation of xenobiotics and POPs by different microalgae species and their possible mechanism are presented in Table 5.

TABLE 5   Degradation of xenobiotics by microalgae.

Organic pollutants	Microalgae starin	Percent removal efficiency	Possible mechanism of action	References
Pharmaceutical active compounds				
Sulfadimethoxine	*Chlorella* sp. L38	88%	Biodegradation	Li et al. (2022)
	Phaeodactylum tricornutum MASCC-0025	100%		
Cefradine	*Chlorella* sp. L166	97.27%	Hydrolysis, biodegradation	Wang et al. (2022a)
	Scenedesmus quadricauda	98.50%		
Methylparaben	*Chlorella vulgaris* AGF002	33.16%	Biodegradation, photolysis	Vale et al. (2022)
Tetracycline	*Microcystis aeruginosa*	99%	Biodegradation, photolysis, hydrolysis, bioaccumulation, and bio sorption	Pan et al. (2021)
	Chlorella pyrenoidosa	99%		
Chlortetracycline	*Spirulina platensis*	99.83%	Bio sorption, biodegradation, bioaccumulation, and photolysis	Zhou et al. (2021)
Ciprofloxacin	*Chlamydomonas* sp. Tai-03	100%	Biodegradation, photolysis, bio sorption, and hydrolysis	Xie et al. (2020)
Sulfadiazine	*Chlamydomonas* sp. Tai-03	54.51%		
	C. vulgaris	29%		Chen et al. (2020)
2,4-Dichlorophenol	*Picocystis* sp.	73%	Biodegradation	Ouada et al. (2019)
	C. pyrenoidosa	100%	Biodegradation	Li et al. (2018)
Naproxen	*Scenedesmus quadricauda*	58.8%	Biodegradation, bioaccumulation, and photolysis	Ding et al. (2017a)
Ibuprofen	*Navicula* sp.	60.0%	Biodegradation, bioaccumulation	Ding et al. (2017b)
Levofloxacin	*C. vulgaris*	11.77%	Bio sorption, bioaccumulation, and biodegradation	Xiong et al. (2017)
	Chlamydomonas mexicana	7.23%		
	Ourococcus multisporus	5.73%		
	Micractinium resseri	2.03%		

Continued

TABLE 5 Degradation of xenobiotics by microalgae—cont'd

Organic pollutants	Microalgae starin	Percent removal efficiency	Possible mechanism of action	References
Paracetamol	*Synechocystis* sp	26.5%	Adsorption	Escapa et al. (2017b)
	Tetradesmus obliquus	40%		
	Chlorella sorokiniana	21%		
Salicylic acid	*C. sorokiniana*	25%		
	Tetradesmus obliquus	95%		
Phenolic compounds				
Phenol	*Scenedesmus abundans*	87.8%	Adsorption	Fawzy and Alharthi (2021)
	Arthrospira platensis	10%	Photo degradation	Nur et al. (2021)
	Chlorella vulgaris	98%	Biodegradation	Baldiris-Navarro et al. (2018)
Triclosan	*Nannochloris* sp.	100%	Photo degradation	Bai and Acharya (2019)
p-Cresol	*C. vulgaris*	55.8	Biodegradation	Xiao et al. (2019)
Bisphenol A	*C. pyrenoidosa*	15%	Bioaccumulation	Guo et al. (2017)
	Desmodesmus sp.	57%	Biodegradation	Wang et al. (2017)
p-Nitrophenols	*Chlorella* sp.	50.24%	Adsorption	Zheng et al. (2017)
Hydrocarbons				
Diethyl phthalate	*Cylindrotheca closterium*	97.7%	Biodegradation	Chi et al. (2019)
	Dunaliella salina	91.2%		
Di-*n*-butylphthalate	*Cylindrotheca closterium*	91.4%		
	Dunaliella salina	34.5%		
Pyrene	*Chlorella* spp.	78.71%	NR	Aldaby and Mawad (2019)

NR—not reported.

5 Heavy metal removal by microalgae

Heavy metals are major persistent pollutants of aquatic ecosystem and cause a serious threat to human health and other life forms if not treated properly. Heavy metals exist in different forms (speciation) such as free ions, vapors, salts, minerals, or bound to organic compounds, and these forms constantly change owing to various biotic and abiotic interactions. The toxicity of the heavy metals depends on the chemical form of speciation. Factors that affect the equilibrium between these different forms are type of microorganisms, medium pH, alkalinity, and temperature (Haferburg and Kothe, 2007). Among the various forms, the free ionic species are most toxic to aquatic biota. Microalgae have developed a variety of mechanisms to counter the heavy metals such as extracellular detoxification, reduced metal uptake, efflux, and metal sequestration by peptides such as phytochelatins (Tripathi et al., 2006). Further, microalgae have developed enzymatic and non-enzymatic anti-oxidant defense systems to counter the heavy metal stress (Danouche et al., 2020). The heavy metal removal efficiency of microalgae is listed in Table 6.

Microalgae have developed a number of strategies for metal detoxification and removal. As discussed earlier, microalgae cell wall possesses numerous charged functional groups that can bind to heavy metals and cause bio-precipitation. In certain cases, microalgae secrete chelating compounds and polysaccharides that bind to heavy metals. Further, heavy metals are absorbed into cells through active transport through different transporter proteins (Vidyashankar and Ravishankar, 2016). The absorbed heavy metals are bound to phytochelatins or metallothieneins, a special class of polypeptides that sequester heavy metals to vacuoles (Perales-Vela et al., 2006) The sequestered heavy metal ions are then neutralized by binding to polyphosphate bodies, polyphenols, and organic acids (Nowicka, 2022). Although microalgae-mediated heavy metal sequestration and detoxification strategies are efficient, scaling up of the technology is a challenge as the microalgae biomass containing heavy metal has limited applications. Till date, there are no pilot-scale studies on heavy metal sequestration and further toxicity assessment of the resultant biomass.

6 Value-added bioproducts from wastewater-cultivated microalgae

Microalgae are known to produce a wide range of bioproducts such as pigments, lipids, proteins, and polysaccharides that have high commercial value (Kumar et al., 2022a, b). However, the major challenges associated with commercialization of microalgae-based bioproducts are high water requirements during cultivation and energy-intensive downstream process (Singla et al., 2021). Integration of wastewater treatment with microalgae cultivation would enhance the process efficiency. Although wastewater-derived biomass has limited applications compared with traditional cultivation practices, yet they could be used for production of biofuels, agro-chemicals (biofertilizers/biostimulants), and biopolymers (Amin et al., 2022). However, the application is entirely dependent on the nature of wastewater and its composition.

6.1 Lipids and biodiesel

The most common application is lipid production from wastewater-generated microalgae biomass. Microalgae adapt to environmental conditions, specifically unfavorable growth conditions, by modulating their lipid metabolism. Under nutrient limitation conditions, microalgae accumulate lipids, specifically triacylglycerols (TAGs) (Klok et al., 2014) that are precursors for biodiesel production (Brennan and Owende, 2010). Further, the heating value of microalgae generated from wastewater treatment processes was found to be

TABLE 6 Heavy metal removal efficiency of microalgae.

Microalgae	Mode of cultivation	Initial concentration of heavy metal (mg L^{-1})	Heavy metal removal efficiency	References
Arsenic				
Spirulina sp.	Laboratory scale (flasks)	7276	60.2%	Doshi et al. (2009)
Ulothrix cylindricum	Laboratory scale (flasks)	10	98%	Tuzen et al. (2009)
Maugeotia genuflexa	Laboratory scale (flasks)	10	96%	Sarı et al. (2011)
Mixture of green (Chlorophyta) algae and blue-green (Cyanobacteria) algae	Laboratory scale (flasks)	50	70%	Sulaymon et al. (2013a, b)
Chlamydomonas reinhardtii	Laboratory scale (flasks)	12	38.6%	Saavedra et al. (2018)
Chlorella vulgaris		12	32.6%	
Scenedesmus almeriensis		12	41.6%	
Cadmium(II)				
Scenedesmus-24	Self-designed photobiorecator	200	60.5%	Jena et al. (2015)
Chlorella minutissima	Laboratory scale (flasks)	NR	NR	Yang et al. (2015)
• Lipid-extracted *Chlamydomonas* sp. • Lipid-extracted *Chlorella* sp. • Lipid-extracted *Coelastrum* sp.	Laboratory scale (flasks)	NR	NR	Zheng et al (2016)
Parachlorella sp	Laboratory scale (flasks)	100		Dirbaz and Roosta (2018)
Immobilized *Chlorella* sp	Laboratory scale (flasks)	10	92.5%	Shen et al. (2018)
Chromium (VI)				
Spirulina platensis	Airlift reactor	250	59.5%	Gokhale et al. (2008)
Pigment-extracted *Spirulina platensis*		250	69.9%	
C. vulgaris	Stirred photobioreactor	250	56%	
Methylated *Spirulina platensis*	Outdoor ponds	18–25	>80%	Finocchio et al. (2010)
Immobilized *Chlorella minutissima*		100	99.7%	

TABLE 6 Heavy metal removal efficiency of microalgae—cont'd

Microalgae	Mode of cultivation	Initial concentration of heavy metal (mg L^{-1})	Heavy metal removal efficiency	References
	Photo bioreactors			Singh et al. (2012)
Rhizoclonium hookeri	Laboratory scale (flasks)	1000	6.7%	Kayalvizhi et al. (2015)
Immobilized *Spirulina platensis*	Laboratory scale (flasks)	250	19.6%	Kwak et al. (2015)
Scenedesmus quadricauda	10 L Carboys	100	98.3% for Cr(III)	Khoubestani et al. (2015)
			47.6% for Cr(VI)	
C. vulgaris	Laboratory scale (flasks)	147	43%	Sibi (2016)
S. quadricauda biochar	Laboratory scale (flasks)	10	100%	Daneshvar et al. (2019a)
• *Spirulina platensis* • Lipid-extracted *Spirulina platensis*	Fiber reinforced plastic (540 L)	500	–	Nithya et al. (2019)
Lead (Pb (II))				
Immobilized *Chlamydomonas reinhardti*	Laboratory scale (flasks)	500	–	Bayramoğlu et al. (2006)
Chaetoceros sp.	Batch reactor	20	60%	Molazadeh et al. (2015)
Chlorella sp.		20	78%	
Phormidium sp.	Laboratory scale (flasks)	10	92.2%	Das et al. (2016)
Mercury (Hg^{2+})				
Immobilized *Chlamydomonas reinhardti*	Laboratory scale (flasks)	500	NR	Bayramoğlu et al. (2006)
Chlorella sp. DT	Laboratory scale (flasks)	8	12.5%	Huang et al. (2006)
Transgenic *Chlorella* sp.		8	27.5%	
Spirogyra sp.	Laboratory scale (flasks)	1	76%	Rezaee et al. (2006)
C. vulgaris	Tubular photobioreactor	48	72.9%	Solisio et al. (2019)
Scenedesmus obtusus XJ-15	Laboratory scale (flasks)	20	NR	Huang et al. (2019)

NR—Not Reported.

$18-22\ 10^6\ MJ\ ton^{-1}$, making them an excellent bioenergy source (Park et al., 2013; Mehrabadi et al., 2015). Microalgae accumulate lipids and hydrocarbons up to 75% w/w of the total biomass in certain species such as *B. braunii*; however, in general, the lipid content ranges between 17% and 40% w/w (Udayan et al., 2022). Some hyper lipid accumulating microalgae are presented in Table 7. The most commonly followed route of bioenergy production is extraction of lipids followed by transesterification and conversion to biodiesel (Karpagam et al., 2021). The lipid productivity of microalgae biomass is affected by growth parameters such as medium composition, for example, N and P regime, CO_2 supplementation, light intensity, and salinity levels (Sarma et al., 2021).

There are several examples of lipid production from microalgae utilizing industrial effluents. Microalgae such as *Coelastrella* sp., *C. vulgaris*, *Desmodesmus* sp. accumulated lipid up to 30% w/w of total biomass when cultivated in municipal wastewater (Ferro et al., 2018). Cultivation of microalgae such as *Acutodesmus obliquus*, *C. vulgaris*, and *Desmodesmus* sp. in municipal wastewater under continuous light exposure improved lipid content with higher saturated and mono-unsaturated fatty acids (oleic acid), however, with a reduced biomass productivity (Purba et al., 2022). Bioremediation of effluents from dairy processing, soybean processing, distillery, and aquaculture industries coupled with lipid production using microalgae such as *Chlorella* sp., *Scenedesmus* sp., and *Desmodesmus* sp. has been demonstrated (Table 8).

In general, the lipid productivity of microalgae biomass increases under abiotic stress conditions, mainly N and P deprivation in growth medium and higher salinity (NaCl) levels (Vidyashankar et al., 2013). However, contrary to the requirements, wastewater streams contain high amounts of N and P limiting lipid production in microalgae biomass. Pre-dominant microalgae strains that occur in wastewaters are *Chlorella*, *Scenedesmus*, and cyanobacteria. They are relatively low lipid accumulators under nutrient replete conditions. However, supplementation of external CO_2 and two-stage cultivation have been most popular methods for enhancing the lipid yield in microalgae biomass (Liyanaarachchi et al., 2021). Two-stage cultivation involves two growth phases, where the first phase focuses on enhanced biomass production utilizing the ability of microalgae to grow under mixotrophic/heterotrophic conditions while the second phase involves induction of lipid accumulation in cells through abiotic stresses such as nutrient deprivation, increased salinity, high light intensities, and CO_2 enrichment or excessive carbon supplementation (Nagappan et al., 2019). Two-stage cultivation generally increases lipid productivity in microalgae by multi-folds. For example, *C. vulgaris* UTEX-265 cultivated in anaerobically digested municipal wastewater under phototrophic mode in the first phase followed by mixotrophic cultivation in second phase with glucose supplementation resulted in lipid productivity of $108\ mg\ L^{-1}\ day^{-1}$ with more than 80% removal of N and P (Farooq et al., 2013). Similar examples of concomitant bioremediation and lipid production through two-phase cultivation strategy have been demonstrated with other microalgae species such as *Chlorella sorokiniana* (Hena et al., 2015), *Scenedesmus obliquus* SAG 276.10 (Álvarez-Díaz et al., 2015), and *Auxenchlorella protothecoides* UMN 280 (Zhou et al., 2012).

As discussed earlier, biodiesel production through the transesterification process is the most widely employed route of bioenergy production owing to higher conversion efficiency (up 90%) of the process (Karpagam et al., 2021). The biodiesel quality of microalgae lipids

TABLE 7 Hyper lipid accumulating microalgae.

Microalgae	Class	Oil/lipid content (% dry weight)	Habitat
Scenedesmus dimorphus	Chlorophyceae	6–7/16–40	Freshwater
C. vulgaris	Chlorophyceae	14–40/56	Freshwater
Chlorella protothecoides	Chlorophyceae	23/55	Freshwater
Chlorella sorokiana	Chlorophyceae	22	Freshwater
C. vulgaris	Chlorophyceae	25±5	Freshwater
Neochloris oleoabundans	Chlorophyceae	35–65	Freshwater
Crypthecodinium cohni	Dinoflagellate	20	Marine
Tetraselmis suecia	Chlorophyceae	36.4	Marine
Dunaliella primolecta	Chlorophyceae	23	Marine
Isochrysis spp.	Haptophyceae	25–33	Marine
Anabaena cylindrical	Cyanophyceae	4–7	Freshwater
Chlamydomonas rheinhardii	Chlorophyceae	21	Stagnant water
Pavlova sp.	Haptophyceae	25–30	Marine
Skeletonema costatum	Bacillariophyceae	17.4	Marine
Chlamydomonas polypyrenoideum	Chlorophyceae	42	Stagnant water
Dunaliella salina	Chlorophyceae	14–20	Brackish water
Porphyridium cruentum	Porphyridiophyceae	9–14	Marine
Scenedesmus obliquus	Chlorophyceae	11–22/35–55	Freshwater
Spirulina maxima	Cyanophyceae	6–7	Freshwater
Chaetoceros muelleri	Bacillariophyceae	21.8	Marine
Nitzschia spp.	Bacillariophyceae	45–47	Marine
Nannochloropsis spp.	Eumastigophyceae	31–68	Marine
Schizochytrium spp.	Thraustochytriidae	50–77	Marine
Monodus subterraneus	Eumastigophyceae	30.4	Marine
Dinoflagellate	Dinophycaea	82.9	Marine
Spirogyra sp.	Zygnematophyceae	7.3	Freshwater
C. protothecoides	Chlorophyceae	52	Freshwater
Neochloris oleabundans	Chlorophyceae	15.9–56.0	Freshwater

Source: Amin, M., Tahir, F., Ashfaq, H., Akbar, I., Razzaque, N., Haider, M.N., Xu, J., Zhu, H., Wang, N., Shahid, A., 2022. Decontamination of industrial wastewater using microalgae integrated with biotransformation of the biomass to green products. Energy Nexus 6, 100089; Vidyashankar, S., Ravishankar, G.A., 2016. Algae-based bioremediation: bioproducts and biofuels for biobusiness. Bioremediat. Bioecon. 457–493.

TABLE 8 Lipid production from wastewater-cultivated microalgae biomass.

Wastewater source	Microalgae used	Cultivation parameters	Biomass yield	Nutrient removal efficiency	Lipid yield (%)	Fatty acid composition	References
Municipal wastewater	Coelastrella sp.	Cultivation type—Lab scale multi-cultivator PBR; Working volume—80 mL; Light intensity—100 μmol m^{-2} s^{-1}; Light cycle—24:0	1.46 g L^{-1}	TN->90%, TP->99%	30.8%	NR	Ferro et al. (2018)
	Chlorella vulgaris		1.15 g L^{-1}	TN->90%, TP->99%	34.2%	NR	
	Desmodesmus sp. 2-6		1.08 g L^{-1}	TN->90%, TP->99%	36.7%		
	Desmodesmus sp. RUC-2		1.18 g L^{-1}	TN->90%, TP->99%	29.8%		
Municipal wastewater-landfill leachate	Scenedesmus obliquus	Cultivation type—Lab scale non-aerated bottles; Working volume—4 L; Light intensity—53 μmol m^{-2} s^{-1}; Light cycle—12:12	1.2 g L^{-1}	PO$_4^{3-}$—60%, COD—86%, NH$_4$N—95%, TN—90%, NO$_3$—50%	16%	NR	Hernández-García et al. (2019)
	Desmodesmus sp.		1.3 g L^{-1}	PO$_4^{3-}$—57%, COD—86%, NH$_4$N—96%, TN—93%, NO$_3^-$—60%	20% Carbs 41%	Increased PUFAs	
Municipal wastewater	S. obliquus	Cultivation type—Lab scale; Working volume—100 mL; Light intensity—120 μEm^{-2} s^{-1}; Light cycle—16:8	1.2 g L^{-1}	TN—75%, TP—75%	11.6%	Increased C-18:1 and decreased ALA	Nzayisenga et al. (2020)
	Desmodesmus sp.		1.4 g L^{-1}	TN—75%, TP—75%	6.2%	No major changes	
	A. obliquus CN01	Cultivation type—Lab scale; Working volume—NR; Light intensity—80 μmol m^{-2} s^{-1}	0.62 g L^{-1}	TN—96%, TP—3.8 mg g^{-1} day^{-1}	26.3%	Increased LC-SFA	(Purba et al., 2022)
	D. maximus CN06		0.70 g L^{-1}	TN—91%, TP—1.37 mg g^{-1} day^{-1}	57.82%	Decreased MUFA and PUFA	

Wastewater	Microalgae	Cultivation conditions	Biomass	Nutrient removal	Lipid content	Remarks	References
	C. vulgaris NIES 1269	Light cycle—24:0	0.5 g L^{-1}	TN—90% TP—0.82 mg g^{-1} day^{-1}	46.38%	Higher C-18:1	
Untreated municipal wastewater	Algal consortium, PA6: Phormidium and Chlorella pyrenoidosa	Cultivation type—Pilot scale Working volume—100L Light intensity—NR Light cycle—Outdoor	3.48 ± 0.44 g m^{-2} day^{-1}	COD—53% TN—81% NH$_4$-N—81% TP—75%	35.2%	NR	Naaz et al. (2019)
Synthetic saline wastewater	Chlorella pyrenoidosa	Cultivation type—Lab scale PBR Working volume—1 L Light intensity—NR Light cycle—12:12	53.62 mg L^{-1} day^{-1}	NH$_4$-N—99.6% DIN—99.2% DIP—98.2%	65.2%	Increased salinity increases SFA	Yang et al. (2022)
Dairy wastewater	Acutodesmus dimorphus	Cultivation type—Lab scale Working volume—1 L Light intensity—60 µmol m^{-2} s^{-1} Light cycle—12:12	0.84 g L^{-1}	COD—91.90% NO$_2^-$—100% NH$_4$-N—100% TP—100% TS—31.24% TIC—30.56% TOC—44.54%	25.05%	NR	Chokshi et al. (2016)
Dairy wastewater	Chlorella protothecoides	Cultivation type—Lab scale Working volume—500 mL Light intensity—150 µmol m^{-2} s^{-1} Light cycle—NA	4.54 g L^{-1}	COD—92.6% TN—99.7% TP—79.9%	27%	Higher LCFA content Pre-dominant fatty acids—C-18:1, C-16:3, C-16:2, and C16:1	Patel et al. (2020)

Continued

TABLE 8 Lipid production from wastewater-cultivated microalgae biomass—cont'd

Wastewater source	Microalgae used	Cultivation parameters	Biomass yield	Nutrient removal efficiency	Lipid yield (%)	Fatty acid composition	References
Soyabean wastewater	*S. obliquus* NIES-2280	**Cultivation type**—Lab scale heterotrophic cultivation **Working volume**—NR **Light intensity**—Dark **Light cycle**—NR	416 mg L^{-1} day^{-1}	COD—72% TN—95% TP—54%	99.3 mg L^{-1} day^{-1}	Up to 50% C-18:1	Shen et al. (2020)
Alcohal wastewater: anaerobically digested starch wastewater	*C.pyrenoidosa* (FACHB-9)	**Cultivation type**—outdoor cultivation **Working volume**—160 L **Light intensity**—NR **Light cycle**—12:12	2.42 g L^{-} day^{-1}	COD-49.37% NH$_4$-N—87.71% TP—87.14%	19.68%	NR	Tan et al. (2018)
Soyabean wastewater	*Chlorella* sp. L38	**Cultivation type**—Lab scale **Working volume**—200 mL **Light intensity**—6K lux **Light cycle**—12:12	NA	COD—70.5% TN—84.7% TP—97.3%	3.89 mg L^{-1} day^{-1}	37.7% SFA and 62.3% PUFA C-18:2 (LA) is major fatty acid	Qiu et al. (2019)
	Chlorella sp. L166		NA	COD- 61.1% TN- 67.3% TP- 90.4%	7.22 mg L^{-1} day^{-1}	41.9% SFA	
Aquaculture wastewater	*Chlorella sorokiniana* SDEC-18	**Cultivation type**—Lab scale **Working volume**—1 L **Light intensity**—81 μmol photon m^{-2} s^{-1} **Light cycle**—NA	2.91 mg L^{-1} day^{-1}	NA	43%	Higher MUFA Enhanced PUFA accumulation	Zhang et al. (2019a)

Wastewater	Species/strain	Cultivation conditions	Biomass	Nutrient removal	Lipid productivity/content	Fatty acid composition	References
Anaerobic digestion effluent from cattle manure	*Scenedesmus* sp. L-1	**Cultivation type**—Lab scale **Working volume**—250 mL **Light intensity**—5000 lux **Light cycle**—12:12	4.65 g L^{-1}	COD—90% TN—90% NH$_4$-N—90% TP—88%	81.9 mg L^{-1} day^{-1}	60.52% MUFA 39.48 SFA	Luo et al. (2019)
Acid and heat treated palm oil mill effluent	*C. sorokiniana* CY-1	**Cultivation type**—Lab scale PBR **Working colume**—1 L **Light intensity**—8000 lux **Light cycle**—NR	2.12 g L^{-1}	COD—47.09% TN—62.07% TP—30.77%	11.21%	NR	Cheah et al. (2018)
Raw and anaerobically digested food processing industry wastewate	*S. obliquus* NCIM 5586	**Cultivation type**—Lab scale **Working volume**—500 mL **Light intensity**—81 μmol photon m^{-2} s^{-1} **Light cycle**—10:14	211 mg L^{-1} day^{-1}	COD—89% TN—84% TP—70%	27.5 mg L^{-1} day^{-1}	C-16:0 and ALA contribute >55% of total fatty acids	Gupta and Pawar (2018)
Molasses wastewater with 48%–56% total sugar	*Scenedesmus* sp. Z-4	**Cultivation type**—Lab scale **Working volume**—250 mL **Light intensity**—3K lux **Light cycle**—12:12	3.5 g L^{-1}	COD—87.2% TN—90.5% TP—88.6%	28.9%	NR	Ma et al. (2017)

PBR, photobioreactor; COD, chemical oxygen demand; BOD, biological oxygen demand; TN, total nitrogen; TP, total phosphorus; NH$_4$-N, ammoniacal nitrogen; TS, total sulfates; DIC, dissolved inorganic carbon; TIC, total inorganic carbon; TOC, total organic carbon; DIP, dissolved inorganic phosphate; DIN, dissolved inorganic nitrogen; NO$_2^-$, nitrites; NO$_3^-$, nitrates; PO$_4^{3-}$, phosphates; SFA, saturated fatty acids; PUFA, polyunsaturated fatty acids; MUFA, mono-unsaturated fatty acids; ALA, alpha-linolenic acid; LA, linoleic acid; LCFA, long chain fatty acids; NR, not reported.

depends on fatty acid composition. Lipids with aliphatic carbon chain with 16–18 atoms specifically with higher percentage of saturated and mono-unsaturated fatty acids such as palmitic and oleic acids are most preferred for biodiesel production (Koutra et al., 2018). Palmitic acid (C-16:0), oleic acid (C-18:1), linoleic acid (C-18:1), and alpha linolenic acid (C-18:1) were the pre-dominant fatty acids observed among most of the wastewater-cultivated microalgae. The ratio of saturated to unsaturated fatty acids depended on the growth environment especially nutrient composition of wastewaters. In general, it was observed that higher concentration of polyunsaturated fatty acids resulted in poor oxidative stability of the biodiesel. According to the European standards (EN14214), the percentage composition of linolenic acid must not exceed 12% of total fatty acids (Knothe, 2006). For obtaining optimum biomass and lipid productivity, dilution of effluents was necessary (Koutra et al., 2018). Effluents are generally colored with both high suspended and dissolved solids. Utilization of these effluents directly results in lower penetration of sunlight and gas exchange resulting in poor growth of microalgae and consequently lower lipid productivity. It has been observed that dilution of different effluents such as agro-waste, municipal and sewage waste, up to 10-fold generally resulted in enhanced lipid accumulation with a favorable fatty acid composition (Zuliani et al., 2016). Apart from lipid-mediated biodiesel production, microalgae offer other routes of bioenergy production (Fig. 1).

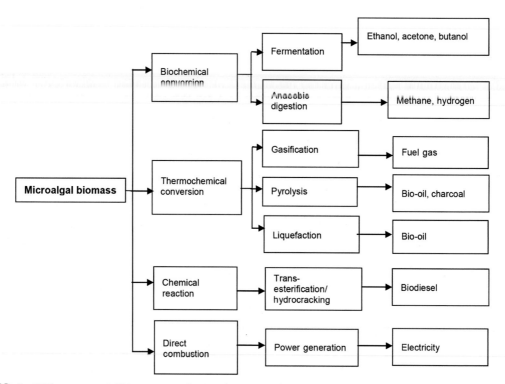

FIG. 1   Different routes of bioenergy production from microalgae biomass.

6.2 Bioethanol

Bioethanol from microalgae is another attractive route for bioenergy production and is considered as third-generation biofuels. The major advantage with microalgae biomass as feedstock for bioethanol production is the absence of lignin in its cell wall allowing milder treatment conditions for obtaining maximum ethanol productivity (Lakatos et al., 2019). In addition, microalgae accumulate significant quantities of carbohydrates that can be converted to ethanol through conventional microbial fermentation. The total carbohydrate content of microalgae biomass ranges between 14% and 50% dry weight mainly constituted by starch, glycogen, and cellulose (Phwan et al., 2018). Microalgae such as *Chlorella sorokiniana*, *Tetraselmis* sp., *Chlorococcum* sp., *Dunaliella* sp., *Porphyridium cruentum*, and *Anaebena cylindrica* are few high-carbohydrate-accumulating species (Phwan et al., 2018; Dragone et al., 2010). The bioethanol production from microalgae can be achieved by three processes, namely (i) microbial fermentation of pre-treated microalgae biomass, (ii) dark fermentation of reserved carbohydrates, and (iii) direct "photo-fermentation" from carbon dioxide to bioethanol using light energy (Lakatos et al., 2019). Dark fermentation involves cultivation of microalgae strains in the absence of light where the reserve carbohydrates (polysaccharides) are converted to simple sugars through amylase activity, which proceeds to pyruvate synthesis through glycolysis pathway. The intermediate pyruvate can be converted to acetate, ethanol, formate, glycerol, lactate, H_2, and CO_2 via various fermentative pathways depending upon the culture conditions. *Chlamydomonas reinhardtii*, *Chlorella fusca*, and *Chlorogonium elongatum* are some microalgae species that can ferment intracellular starch (Lakatos et al., 2019). In case of photo-fermentation, cyanobacteria such as *Synechocystis* sp. PCC 6803, *Synechococcus* sp. PCC 7942, *Synechococcus* sp. PCC 7002 were pre-dominantly utilized for bioethanol production under phototrophic conditions. This was achieved by heterologous expression of two gene cassettes namely pyruvate decarboxylase and alcohol dehydrogenase II from *Zymomonas mobilis*, which converted pyruvate produced from photosynthesis to acetaldehyde and ethanol (Dexter and Pengcheng, 2009). However, the major challenges in these two routes of bioethanol production are inhibition of microalgal growth by ethanol and lower ethanol titer (de Farias Silva and Bertucco, 2016). The microbial fermentation of hydrolyzed microalgae biomass is the most common route of bioethanol production. This involves physical, chemical, or enzymatic treatment of microalgae cells to release the carbohydrate followed by saccharification. Chemical pre-treatment includes acid or alkali hydrolysis while physical treatment involves pressure homogenization, bead milling, ultrasonication, and steam treatment (Phwan et al., 2018). Some common microbes that have been used for fermentation process are *Saccharomyces cerevisiae*, *Z. mobilis*, *Escherichia coli*, and *Saccharomyces bayanus*. Chlorophycean algae such as *C. vulgaris*, *Scenedesmus obliquus*, *Chlorococcum* sp., and *C. reinhardtii* are some commonly evaluated microalgae species owing to their high carbohydrate content. The average yield of ethanol ranged from 0.2 g to 0.4 g ethanol g^{-1} biomass with a productivity up to 22.6 g L^{-1} observed in *C. vulgaris* (Phwan et al., 2018; Lakatos et al., 2019).

6.3 Syngas and bio-oil

Thermochemical conversion of dry microalgae biomass at high temperatures (800–1000°C) and pressure (~30 MPa) in the presence of O_2, CO_2, and steam partially oxidizes the biomass to generate gaseous product, syngas, comprising mainly H_2 and CO and to some extent CO_2 and methane. This process is known as thermochemical gasification. Syngas is combustible and used as precursor for production of methanol and ammonia (Razzak et al., 2022). Bio-oil is derived from a thermochemical process called pyrolysis where the dry biomass is heated at very high temperatures ($\sim$500–550°C) in the absence of O_2. The gaseous products generated during thermal treatment are cooled instantly to obtain bio-oil. The typical energy values of microalgae derived bio-oil are $29\,MJ\,kg^{-1}$ biomass much higher compared with conventional lingo-cellulosic biomass (Yang et al., 2019). The additional products obtained from pyrolysis are biochar (the solid residue) and gaseous products (methane, H_2). The bio-oils are generally upgraded to a variety of fuel products such as gasoline, jet fuels by the process of deoxygenation cracking and fractionation similar to petroleum crude. The upgradation involves removal of nitrogen and oxygen in the bio-crude (Yang et al., 2019; Razzak et al., 2022).

6.4 Biofertilizer applications

Other potential applications of microalgae biomass generated after wastewater treatment are biofertilizer and biostimulant. Microalgae biomass contains significantly higher N and P levels compared with other organic fertilizers. The average N content ranges between 4.9% and 7.1% N while the P content ranges from 1.5% to 2.1% (Khan et al., 2019). Further, the ability of microalgae to accumulate N and P through luxury uptake mechanism makes microalgae biomass a concentrated source of nutrients for plant growth promotion. The use of wastewater-derived microalgae biomass as biofertilizers and source of biostimulants is trending with several case studies from recent years (Álvarez-González et al., 2022; Arashiro et al., 2022; Dagnaisser et al., 2022). Microalgae belonging to chlorophycea group such as *Chlorella* sp., *Scenedesmus* sp., *Acutodesmus* sp., *Desmodesmus* sp., *Dunaliella* sp., and cyanobacteria such as *Spirulina* sp., *Nostoc* sp., *Anabaena* sp., are predominantly utilized for biofertilizer applications (Parmar et al., 2023). Some typical benefits of applying microalgae as biofertilizers and biostimulants are improved germination in seeds, enhanced vigor of seedlings, increased number of root nodules, increased root and shoot length, increased root and shoot weight, enhancement in photosynthetic pigment content in leaves, increased flowering and fruiting, and enhanced fruit/grain yield (Kapoore et al., 2021). A list of few case studies demonstrating the biofertilizer and biostimulant properties of microalgae biomass generated from wastewater treatment is presented in Table 9.

The ability of microalgae to elicit positive effects on plant growth and metabolism could be attributed to the presence of various metabolites such as polysaccharides, phenolic compounds, phyto-hormones such as auxins, cytokinins, gibberlic acids, vitamins, and antioxidants that modulate plant signaling and physiology (Lu and Jian, 2015; Ronga et al., 2019; Kapoore et al., 2021). Addition of microalgae biomass enhances the water retention capacity of soil, soil organic carbon content and promotes soil microflora in the rhizosphere region (Abinandan et al., 2019). In addition, they fix atmospheric nitrogen, mobilize phosphorus, and enhance micronutrient availability (Fe) to plants and rhizospheric microorganisms (Alvarez et al., 2021). It has been observed that application of microalgae biomass and live microalgae (specifically cyanobacteria consortia) reduced the demand for NPK fertilizers up to 50% of the total requirements during cultivation of rice (Prasanna et al., 2015). Microalgae extracts have been shown to possess biostimulatory properties, specifically

TABLE 9 Biofertilizer applications of wastewater-cultivated microalgae biomass.

Waste source	Microalgae used	Nutrient removal efficiency	Method of application	Crop (s) evaluated	Observation	References
Dairy effluent	*Chlorella pyrenoidosa*	BOD—88% COD—85% NH_4-N—99% PO_4^{3-} —97%	Encapsulated algal biomass as biofertilizers	*Oryza sativa*	• Enhanced the length of root and shoot in rice seedlings	Yadavalli and Heggers (2013)
Aquaculture wastewater	*Spirulina platensis*	NO_3^-—50% NH_4-N—95%	• Soil drenching [admixtures of soil and dry biomass] • Seed soaking	• *Eruca sativa* • *Ameranthus gangeticus* • *Brassica rapa* ssp. *Chinensis*	• Enhanced germination rate • Increased seedling vigor • Enhanced plant height and root length • Increased biomass yield • Increased chlorophyll content	Wuang et al. (2016)
Brewery effluents	*Scenedesmus obliquus*	COD—71% TN—88% TP—30% NH_4-N—81%	• Algal cell suspension • Deoiled biomass fertilizers	• *Triticum aestivum* • *Hordeum vulgare*	Increased the germination and sprouting	Ferreira et al. (2019)
Domestic wastewater	*Chlorella* sp.	Selenium (sodium selenite)—44% TSS—86% COD—70% TC—67% TP—77% NH_4-N—93%	• Foliar spray • Soil drenching • Seed treatment	*Phaseolus vulgaris*	• Enhanced germination rate • Selenium enrichment in seeds and leaves [bio-fortification]	Li et al. (2021)
Domestic wastewater	Consortia of *Chlorella* sp., and *Scenedesmus* sp.	NO_3^-—96% NH_4-N—98% PO_4^{3-} —95% COD—83% TOC—86% TN—94%	Deoiled algal biomass as biofertilizer	*Solanum lycopersicum*	• Enhanced shoot and root weight • Enhanced macro [N, P, K] and micronutrients [Ca, Mg, Fe] in biomass • Increased tomato yields	Silambarasan et al. (2021)

Continued

TABLE 9 Biofertilizer applications of wastewater-cultivated microalgae biomass—cont'd

Waste source	Microalgae used	Nutrient removal efficiency	Method of application	Crop (s) evaluated	Observation	References
Piggery wastewater	• *Tetradesmus obliquus* • *Chlorella protothecoides* • *Chlorella vulgaris* • *Synechocystis* sp. • *Neochloris oleoabundans* • *Nostoc* sp.	COD—62%–79% NH$_4$-N—79%—92% PO$_4^{3-}$—90%–98%	Seed treatments with algae biomass	• *S. lycopersicum* • *Cucumis sativus* • *H. vulgare* • *Glycine max* • *Nasturium officinale* • *T. aestivum*	• Increased germination index • Increased root length in wheat and cucumber • Biopesticidal activity against *Fusarium oxysporum*	Ferreira et al. (2021)
Paddy-soaked rice mill wastewater	*C. pyrenoidosa*	NH$_4$-N—69.4% PO$_4^{3-}$—64.7%	• Seed treatment with microalgae gel suspensions • Soil drenching with live algae biomass	*Abelmoschus angulosus*	• Increased germination percentage • Increase number of leaves • Increased plant height	Umamaheswari and Shanthakumar (2021)
Sewage wastewater	*Chlorella minutissima*	TDS—96% TP—70% K—45% NH$_4$-N—90% BOD—>90% COD—80% NO$_3$—89%	Soil drenching (admixtures of dry microalga biomass and soil)	• *Spinacia oleraceae* • *Zea mays*	• Enhanced available nitrogen and organic carbon in soil • Enhanced biomass and root yield • Increased soil enzymes activity [urease, nitrate reductase & dehydrogenase]	Sharma et al. (2021)
Domestic wastewater	*Scenedesmus* sp.	COD—69% TIN—91% TP—81% NH$_4$-N—95%	Soil drenching (admixtures of microalga biomass paste and substrate)	*Ocimum basilicum* L.	• Enhanced leaf weight • Increased magnesium content in leaves • Overall performance similar to inorganic fertilizers	Álvarez-González et al. (2022)

Wastewater	Microalgae	Removal efficiency	Application	Plant species	Effects	References
Sewage wastewater	C. vulgaris	NO_3^-—93% COD—95% BOD—92%	• Algal treated sewage water as biofertilizers	Solenum lycopersicum	• Increased number of fruit per plants	Pooja et al. (2022)
Urban wastewater	C. vulgaris UAL-1	NH_4-N—93.8% NO_3^-—73.1% PO_4^{3-}—80.5% COD—85.2%	• Microalgal biomass extract as biostimulants	• Lepidium sativum L. • G. max L. • C. sativus L.	• Improved germination index in watercress seeds • Improved the formation of adventitious roots • Enhanced the weight of cotyledons	Amaya-Santos et al. (2022)
Municipal wastewater	Tetradesmus obliquus	NH_4-N-98.5% PO_4^{3-}—89% TN—68% TP—51%	• Soil Drenching (microalgal biomass extract as biostimulants in root)	• L. sativum L. • Vigna radiata L. Wilczek	• Show auxin like activity in Mung bean • Anti-bacterial activity against various pathogens	Grivalský et al. (2022)

TP, total phosphorus; TN, total nitrogen; COD, chemical oxygen demand; BOD, biological oxygen demand; TSS, total suspended solids; TDS, total dissolved solids; TC, total carbon; TIN, total inorganic nitrogen; PO_4^{3-}, phosphate; NO_3^-, nitrate; NH_4-N, ammoniacal nitrogen.

improving the biotic and abiotic stress tolerance to plants against drought, high salinity, and high temperature (Parmar et al., 2023). Microalgal biostimulants improve the enzymatic anti-oxidant defense systems such as superoxide dismutase, catalase, peroxidase, and glutathione that reduces the oxidative stress in plants. In addition, microalgal extracts have been shown to possess pesticidal activity, specifically nematicidal properties. They induce systemic immunity in plants against biotic stress and infections (Lee and Ryu, 2021).

6.5 Biochar

Another major bioproduct with agricultural applications derived from microalgae biomass is biochar. Biochar is a by-product derived from thermochemical conversion processes such as hydrothermal carbonization, pyrolysis, and torrefaction of deoiled algae biomass, which could be used for agriculture applications (Mona et al., 2021). Application of microalgae-derived biochar enhanced the soil moisture through enhanced water retention, improved soil texture, and soil cation exchange capacities. This increased the soil pH and nutrient bioavailability to plants (Wang et al., 2020a). The elemental composition of biochar obtained from microalgae such as *Chlorella* sp., *Scenedesmus* sp., and *Nannochloropsis* sp., contains 33%–55% w/w carbon, 25%–57% w/w oxygen, 2%–10% w/w hydrogen, 2%–3% sulfur, and 4%–12% w/w N along with minerals such as Ca, Fe, Mg, and K (Chang et al., 2015; Borges et al., 2014; Arun et al., 2020). Field trials of bioachar obtained from wastewater-cultivated *Scendesmus abundans* improved the growth, leaf weight, shoot weight, and leaf chlorophyll content in tomato plants (Arun et al., 2020). Some major mechanisms by which biochar enhances the plant growth are by improving soil enzyme activity, promoting soil microflora and soil microbial carbon content, and enhancing the nutrient availability in the rhizosphere region (Palanisamy et al., 2017; Ullah et al., 2020).

6.6 High-value products: Nutraceuticals and animal feed

Microalgae are known as industrially important sources of nutraceuticals such as carotenoids, polyunsaturated fatty acids, phycobiliproteins, proteins, and polysaccharides (Kumar et al., 2022a, b). Microalgae carotenoids such as beta-carotene from *Dunaliella* sp., astaxanthin from *Haematococcus pluvialis*, lutein from *C. vulgaris*, fucoxanthin from *Phaeodactylum tricornutum* have been attributed with strong anti-oxidant and anti-inflammatory properties (Patel et al., 2022; Fu et al., 2023). In addition, they demonstrate strong anti-cancerous and hepato-protective properties. Recently, fucoxanthin has been reported to ameliorate symptoms of non-alcoholic fatty liver disease (Fu et al., 2023). Microalgae-derived carotenoids are GRAS-approved and have been used extensively as dietary supplements (Wang et al., 2022b). In addition, aforesaid carotenoids have been used for enhancing the quality of poultry and aquaculture feeds. Supplementation of beta-carotene and astaxanthin improved the egg yolk carotenoids content in poultry and skin coloration in fishes and crustaceans (Fernandes et al., 2020). Further, these carotenoids improved the disease resistance and survival rates of larvae in aquaculture (Yu et al., 2018). Similarly, phycobiliproteins, namely phycocyanin from cyanobacterial species and phycoerythrin from *Porphyridum* sp., have been reported to possess strong anti-oxidant and anti-inflammatory properties and have been used as natural colorant and food additive (El-Araby et al., 2022). Another most important nutraceutical derived from microalgae are long-chain PUFAs such as gamma linolenic acid from *Spirulina platensis*, arachidonic acid from *Porphyridium* sp., eicosapentaenic acid from *Nannochloropsis* sp., and *P. tricornutum* and docosahexaenoic acid from *Schizochytrium* sp., which find application in production of infant foods, omega-3 supplements, and a variety of functional foods for improved cardiovascular health, cognition, and neural development (Liu et al., 2022). Microalgae are also used to combat protein and micronutrient deficiencies. *S. platensis* and *Chlorella* species are used as food ingredients for improving protein and micronutrient status through direct feeding and incorporation in functional foods (Kumar et al., 2022a, b, 2023). There are several recent reports that have compiled the status of nutraceutical, pharmaceutical, and animal feed production from microalgae including the marketed products, health claims, and key industry players (Kumar et al., 2022a, b; Puri et al., 2022).

Commercial production of these nutraceuticals typically involves cultivation of microalgae in potable waters using standardized growth medium composition in open raceway ponds or photobioreactors (PBRs) (Singla et al., 2021). Use of wastewater for production of nutraceuticals is not generally practiced owing to the risks of bioaccumulation of heavy metals, presence of xenobiotic residues, and potential presence of antibiotic resistance genes obtained through horizontal gene transfers during wastewater remediation (Kumar et al., 2022a, b). However, biorefinery approaches involving multi-product recovery from wastewater generated biomass have paved the way for production and isolation of specific nutraceuticals. Some case studies involving bioremediation and nutraceuticals production are presented in Table 10.

Astaxanthin production from *H. pluvialis* cultivated using a variety of effluents such as swine wastewaters (Ledda et al., 2016), minkery wasters (Liu, 2018), oil mill effluents (Fernando et al., 2021), bioethanol process waste (Haque et al., 2017), and dairy wastewater was demonstrated with excellent nutrient removal efficiencies. Aquaculture industry produces large quantities of effluents containing ammoniacal nitrogen, phosphorus, and organic carbon. Bioremediation of aquaculture wastewaters coupled with biomass and high-value nutraceuticals has been reported (Dourou et al., 2020). Microalgae such as *Chlorella* sp., *Nannochloropsis gaditana*, *Scenedsmus* sp., *Tetraselmis chuii*, and *Chaetocerols calcitrans* have been cultivated in aquaculture wastewaters of catfish, shrimps, tilapia, and Nile tilapia for simultaneous removal of nutrients and production of PUFA-rich lipids, cell mass that was in turn utilized as live feed for production of these aquatic species (Ansari et al., 2019; Malibari et al., 2018; Tossavainen et al., 2019). Other successful demonstration of bioremediation-mediated high-value product recovery is production of DHA from *Schizochytrium* sp., utilizing glycerol-rich residues derived from biofuel production (Chilakamarry et al., 2021). A typical biodiesel production produces close to 10% glycerol as process waste during transesterification process. *Schizochytrium limnacium* SR 21 was successfully cultivated with crude glycerol achieving a biomass productivity of $49.2\,g\,L^{-1}$ and a lipid yield of $10.15\,g\,L^{-1}$ rich in DHA. The strain was able to remediate 40% organic carbon content from the crude glycerine (Bouras et al., 2020). Similar observations were made for another *Schizochytrium* strain with crude glycerol as carbon source for DHA production. The strain achieved a biomass yield of $66.69\,g\,L^{-1}$ and DHA yield of $17.25\,g\,L^{-1}$ (Kujawska et al., 2021). It was observed that *Schizochytrium* could utilize soap containing crude glycerol for production of DHA (Chi et al., 2007).

TABLE 10 Nutraceutical production from wastewater-cultivated microalgae.

Wastewater source	Microalgae used	Cultivation parameters	Growth rate	Nutrient removal efficiency	Yield (%)	References
Polysaccharides						
Soyabean wastewater	*Chlorella* sp. L38	**Cultivation type**—Lab scale **Working volume**—200 mL **Light intensity**—lux **Light cycle**—12:12	NR	COD—70.5% TN—84.7% TP—97.3%	Polysaccharide—1.38 mg L^{-1} day^{-1}	Qiu et al. (2019)
	Chlorella sp. L166		NR	COD—61.1% TN—67.3% TP—90.4%	2.86 mg L^{-1} day^{-1}	
Municipal wastewater	*Chroococcus turgidus*	**Cultivation type**—Lab scale **Working volume**—400 mL **Light intensity**—100 µmol m^{-2} S^{-1} **Light cycle**—24:0	NR	NH$_4$-N—74.6%–83% NO$_3^-$16%–71.2% NO$_2^-$—22.2%–63.6% PO$_4^{3-}$—89%–95.3% BOD—50%–76.2% COD—70.3%–78.6%	43.93%	Sisman-Aydin and Simsek (2022)
Astaxanthin						
Minkery wastewater	*Haematococcus pluvialis* CPCC 93	**Cultivation type**—Bench scale PBR **Working volume**—1 L **Light intensity**—6700 K lux **Light cycle**—NR	0.90 g L^{-1}	TN—100% TP—100%	67.95 mg L^{-1}	Liu (2018)
Bioethanol waste effluent	*H. pluvialis* CPCC 93	**Cultivation type**—Air lift PBR **Working volume**—2.2 L **Light intensity**—15–30 K lux **Light cycle**—12:12	4.37 g L^{-1}	TOC—67% TN—91.67% TP—100%	1.109 mg g^{-1}	Haque et al. (2017)

Substrate	Species	Cultivation conditions	Removal efficiency	Biomass	Product yield	Reference
Swine wastewater	*H. pluvialis* 34-1D	**Cultivation type**—Lab scale **Working volume**—300 mL **Light intensity**—150 μE m^{-2} s^{-1} **Light cycle**—NA	TN—100% TP—98.08%	1.31 g L^{-1}	1.27%	Ledda et al. (2016)
Palm oil mill effluent	*H. pluvialis*	**Cultivation type**—Lab scale **Working volume**—300 mL **Light intensity**—6K lux **Light cycle**—24:0	COD—50.9% TN—49.3% TP—69.4%	0.52 g L^{-1}	22.43 mg L^{-1}	Fernando et al. (2021)
Synthetic dairy wastewater	*H. pluvialis*	**Cultivation type**—Lab scale PBR **Working volume**—2 L **Light intensity**—200 lux **Light cycle**—12:12	COD—95.5% TN—79.6% TP—56.5%	0.55 g L^{-1}	2.1%	Nishshanka et al. (2022)
Glucose as carbon source	*Chlorella zofingiensis*	**Cultivation type**—Lab scale PBR **Working volume**—2 L **Light intensity**—200 lux **Light cycle**—12:12	NR	10 g L^{-1}	10.3 mg L^{-1}	Ip and Chen (2005)

Fucoxanthin

Substrate	Species	Cultivation conditions	Removal efficiency	Biomass	Product yield	Reference
Industrial wastewater	*Chlorella sorokiniana*	NR	NR	4.97 mg g^{-1}	104.8 μg g^{-1}	Safafar et al. (2015)
	Nannochloropsis limnetica			2.56 mg g^{-1}	183.2 μg g^{-1}	
	Nannochloropsis salina			5.2 mg g^{-1}	13.05 μg g^{-1}	

Continued

TABLE 10 Nutraceutical production from wastewater-cultivated microalgae—cont'd

Wastewater source	Microalgae used	Cultivation parameters	Growth rate	Nutrient removal efficiency	Yield (%)	References
Lutein						
Glycerol as carbon source	*Chlamydomonas acidophilia*	NR	$2.7\,g\,m^{-2}\,day^{-1}$	NR	$6\,mg\,g^{-1}$	Cuaresma et al. (2011)
Carotenoids						
Urban wastewater	*Coelastrum* cf. *pseudomicroporum Korshikov*	**Cultivation type**—Lab scale **Working volume**—300 mL **Light intensity**—600 μmol photon m^{-2} s^{-1} **Light cycle**—NR	3.65×10^6 cell mL^{-1}	NA	$0.48\,mg\,L^{-1}$	Úbeda et al. (2017)
Urban wastewater:saline water (50:50)			$2 \cdot 10^5$ cell mL^{-1}	NA	$1.58\,mg\,L^{-1}$	
Aquaculture wastewater	*C. sorokiniana* SDEC-18	**Cultivation type**—Lab scale **Working volume**—1 L **Light intensity**—81 μmol photon m^{-2} s^{-1} **Light cycle**—NA	$2.91\,mg\,L^{-1}\,day^{-1}$	NA	$1\,mg\,g^{-1}$	Zhang et al. (2019a)
Poultry wastewater	*Dunaliella* FACHB-558	**Cultivation type**—Lab scale **Working volume**—500 mL **Light intensity**—200 μmol m^{-2} S^{-1} **Light cycle**—NA	$678\,mg\,L^{-1}$	TOC—64.1% TN—63.8% TP—87.2%	$4.02\,mg\,L^{-1}$	Han et al. (2019)
Poultry slaughterhouse wastewater	*Neochloris* sp. SK57	**Cultivation type**—Lab scale **Working volume**—500 mL **Light intensity**—NR **Light cycle**—11:13	$1.4\,g\,L^{-1}$	COD—96.8% NO$_2^-$—95% PO$_4^{3-}$—79%	$38\,mg\,g^{-1}$	Ummalyma et al. (2022)

Phycocyanin

Substrate/wastewater	Species	Cultivation conditions	Biomass	Removal efficiency	Product yield	References
Tofu wastewater	*Spirulina*	NR	NR	NR	26.93 mg L^{-1} day^{-1}	Nur et al. (2023)
Anaerobically digested dairy manure wastewater	*Spirulina*	**Cultivation type**—Lab scale **Working volume**—1L **Light intensity**—4K lux **Light cycle**—16:08	NR	NR	0.77 g g^{-1}	Taufikurahman et al. (2020)
Glucose as carbon source	*Galdieria sulphuraria*	NR	NR	NR	3.6 mg g^{-1}	Schmidt et al. (2005)

Phycoerythrin

Substrate/wastewater	Species	Cultivation conditions	Biomass	Removal efficiency	Product yield	References
Swine wastewater	*Porphyridium purpureum*	**Cultivation type**—Lab scale **Working volume**—1L **Light intensity**—4K lux **Light cycle**—16:08	9.44 ± 0.44 g L^{-1}	COD—94.85% TN—92.69% TP—96.08% NH$_4$-N—100%	54.45 mg L^{-1}	Zhang et al. (2022)

Proteins

Substrate/wastewater	Species	Cultivation conditions	Biomass	Removal efficiency	Product yield	References
Poultry slaughterhouse wastewater	*Neochloris* sp. SK57	**Cultivation type**—Lab scale **Working volume**—500 mL **Light intensity**—NR **Light cycle**—11:13	1.4 g L^{-1}	COD—96.8% NO$_2^-$—95% PO$_4^{3-}$—79%	41.7%	Ummalyma et al. (2022)
Aquaculture wastewater	C. sorokiniana	NR	1.93 g L^{-1}	TOC—82.27% TN—86.42% NH$_4$-N—93.25%	21.54%	He et al. (2023)
Tilapia aquaculture wastewater	Chlorella pyrenoidosa	NR	NR	COD—88.53% TN—96.96% TP—85.15% NH$_4$-N—98.01%	0.60 g L^{-1}	Wang et al. (2023)
Dairy wastewater	C. sorokiniana SU-1	**Cultivation type**—Lab scale PBR **Working volume**—200 mL **Light intensity**—140 µmol m^{-2} s^{-1} **Light cycle**—24:0	3.2 ± 0.1 g L^{-1}	COD—86.8 ± 6% TP—94.6 ± 3% TN—80.7 ± 1%	24.5%—26%	Kusmayadi et al. (2023)

Continued

TABLE 10 Nutraceutical production from wastewater-cultivated microalgae—cont'd

Wastewater source	Microalgae used	Cultivation parameters	Growth rate	Nutrient removal efficiency	Yield (%)	References
Municipal wastewater	C. turgidus	**Cultivation type—** Lab scale **Working volume—** 400 mL **Light intensity—** $100\,\mu\text{mol}\,\text{m}^{-2}\,\text{s}^{-1}$ **Light cycle—24:0**	NR	NH_4-N—74.6%–83% NO_3^- 16%–71.2% NO_2^-—22.2%–63.6% PO_4^{3-}—89%–95.3% BOD—50%–76.2% COD—70.3%–78.6%	35.25%	Sisman-Aydin and Simsek (2022)
Simulated wastewater	Chlorella vulgaris FACHB-31	**Cultivation type—** Lab scale **Working volume—**1 L **Light intensity—** 3000 lux **Light cycle—12:12**	NR	PO_4^{3-}-P—98% TP—96% NH_4-N—98% TN—96%	25.01–365.49 mg	Kong et al. (2021)
Swine wastewater	Coelastrella sp. KU505	**Cultivation type—** Lab scale **Working volume—**500 mL **Intensity—** $80 \pm 5\,\mu\text{mol}\,\text{m}^{-2}\,\text{s}^-$ **Light cycle—14:10**	NR	NH_4-N—62.3% TP—77.6%	$0.207\,\text{g}\,\text{g}^{-1}$	Li et al. (2020)
DHA						
Glycerol wastewater	Schizochytrium	NR	$66.69\,\text{g}\,\text{dm}^{-3}$	NR	$17.25\,\text{g}\,\text{dm}^{-3}$	Kujawska et al. (2021)
Tofu whey wastewater and saline wastewater	Schizochytrium	NR	$19.08\,\text{g}\,\text{L}^{-1}$	COD—60% TN—60% TP—60%	$2.66\,\text{g}\,\text{L}^{-1}$	Wang et al. (2022c)
Cheese whey demineralization saline wastewater	Schizochytrium limacinum PA-968	NR	$28.40\,\text{g}\,\text{L}^{-1}$	NR	$0.62\,\text{g}\,\text{L}^{-1}\,\text{day}^{-1}$	Humhal et al. (2017)

NR, not reported; PBR, photobioreactor; BOD, biological oxygen demand; COD, chemical oxygen demand; TN, total nitrogen; TP, total phosphorus; TOC, total organic carbon; PO_4^{3-}, phosphate; NO_3^-, nitrate; NH_4-N, ammoniacal nitrogen; NO_2^-, nitrites.

7 Bioremediation and biorefinery-mediated circular economy

The sustainability of microalgae-based processes can be achieved only through a multi-product utilization through a biorefinery approach (Mehariya et al., 2021). Despite the extensive research on the benefits of microalgae-based bioproducts, their commercialization has not been successful mainly due to the energy-intensive downstream processing creating a net negative energy balance (Subhash et al., 2022). A biorefinery approach of complete valorization of microalgae biomass has been mandated for achieving sustainability and economic feasibility. A typical microalgae biorefinery would be to integrate wastewater treatment coupled with biomass production followed by downstream processing for biodiesel production followed by utilization of the deoiled biomass for production of various value-added products such as nutraceuticals, animal feed, and agrochemicals or diversification to other bioenergy routes (Kholssi et al., 2021). Some commonly suggested valorization strategies are (i) production of animal feed after biofuel extraction, (ii) gasification of deoiled biomass for generation of syngas, (iii) liquefaction of deoiled biomass for production of bio crude/bio char, (iv) anaerobic digestion of biomass to produce methane, (v) fermentation of deoiled microalgae biomass for bioethanol, (vi) application of deoiled biomass as biofertilizers and soil conditioners (Katiyar et al., 2021). Integrating wastewater remediation and flue gas treatment with microalgae cultivation was found to reduce the energy requirements for microalgae-based technologies. A typical microalga-based biorefinery scheme is presented in Fig. 2.

The most successful biorefinery application integrating bioremediation, lipid extraction followed by value addition has been in utilization of algae biomass as biofertilizers. Deoiled *Scenedesmus* sp. biomass obtained from domestic wastewater coupled with flue gas treatment (coal-fired flue gas, 2.5% *v*/v) was utilized as biofertilizer in rice cultivation (Nayak and Ghosh, 2019). Deoiled *Scenedemsus* biomass effectively reduced the requirements of commercial NPK fertilizers promoted plant growth. Similarly, treatment of sewage water with *Chlorella minutissima* followed by utilization of the biomass as biofertilizer resulted in enhanced growth of vegetable crops such as spinach and sweet corn (Sharma et al., 2021). Few studies have been reported on biostimulatory potential of deoiled biomass obtained after flue gas treatment and bioremediation such as *Tetradesmus obliquus* CT02 (Sinha et al., 2021) and *Nostoc* sp., (Silambarasan et al., 2021) where algal extracts improved the growth, yield, and fresh weight of commercially important vegetable crops such as tomato and lettuce.

A closed-loop circular economy model can be proposed involving agriculture or food process wastewaters, microalgae biomass production, and sequential valorization to multiple products. A typical scenario would be integrating hydroponics and microalgae cultivation where the wastewaters generated after a cultivation cycle in hydroponics could be effectively used for cultivation of microalgae and further multi-product recovery (Parmar et al., 2023). Hydroponics residual wastewaters, commonly called system drainage, typically contain high concentrations of N (150–600 mg L^{-1}), P (30–100 mg L^{-1}), and other nutrients such as sulfates, K, Ca, trace minerals, and organic substances such as humic acids and root exudates (Saxena and Bassi, 2013). These residual waters can be effectively utilized for growing microalgae biomass. It has been reported that co-cultivation of microalgae such as *C. vulgaris*, *Scenedesmus quadricauda* in a hydroponic system growing tomato enhanced both the crop yield and

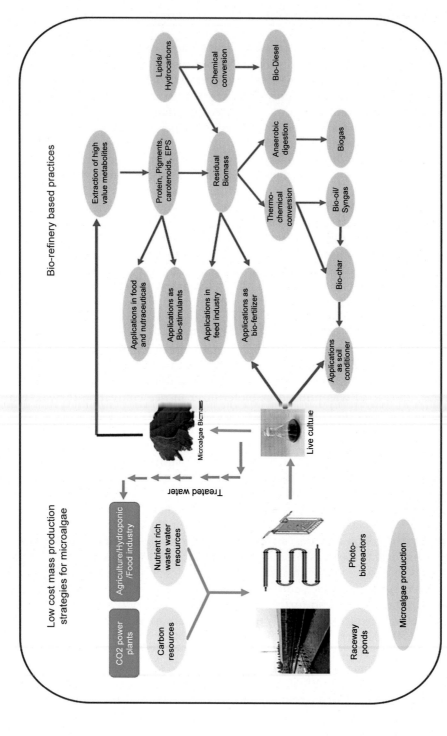

FIG. 2 Integration of bioremediation and biorefinery routes for microalgae-based bioproducts generation. *Source: Parmar, P., Kumar, R., Neha, Y., Srivatsan, V., 2023. Microalgae as next generation plant growth additives: functions, applications, challenges and circular bioeconomy based solutions. Front. Plant Sci. 14. (self-publication) reproduced under Creative Commons Attribution on CC-BY version 4.0 (http://creativecommons.org/licenses/by/4.0/).*

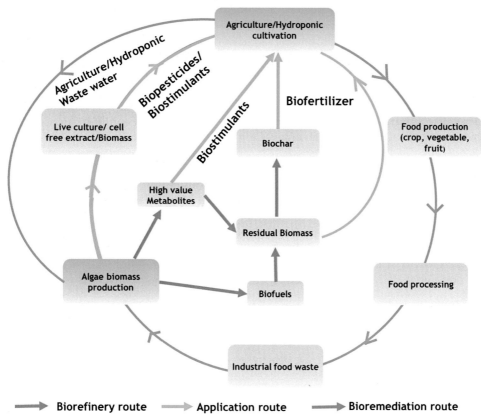

FIG. 3 A typical closed-loop circular economy scheme integrating bioremediation and biorefinery. *Source: Parmar, P., Kumar, R., Neha, Y., Srivatsan, V., 2023. Microalgae as next generation plant growth additives: functions, applications, challenges and circular bioeconomy based solutions. Front. Plant Sci. 14, (self-publication) reproduced under Creative Commons Attribution CC-BY version 4.0 (http://creativecommons.org/licenses/by/4.0/).*

microalgae productivity (Barone et al., 2019). Co-cultivation of microalgae consortia consisting of *Chlorella* sp., *Scenedesmus* sp., *Synechocystis* sp., and *Spirulina* sp., with tomato plants in a hydroponic drain water resulted in enhanced growth of plants and microalgae productivity. The consortia could reduce the nutrient concentrations >80% from the initial levels (Supraja et al., 2020). A closed-loop circular economy model utilizing agriculture waste with sequential bioproduct from microalgae biomass is presented in Fig. 3.

8 Challenges in utilization of wastewater-grown microalgae

Literature amply demonstrates the ability of microalgae to grow prolifically utilizing wastewaters and process effluents. However, industrial scalability of bioremediation coupled biomass production has several challenges. Some critical challenges are (i) scale-up of cultivation, (ii) presence of POPs and xenobiotic residues and breakdown products, (iii) accumulation of heavy metals, and (iv) contamination of cultures with bacteria and fungi (Abdelfattah et al., 2022; Amin et al., 2022). From mass cultivation point of view, use of industrial effluents such as textile waste, dairy waste, livestock, and aquaculture waste is challenging as they are colored, contain high suspended and dissolved solids reducing the light penetration in cultures (Abdelfattah et al., 2022). Light penetration is essential as most of the microalgae strains are phototrophic in nature, and light significantly influences growth and biomass productivity. The average biomass yield of microalgae species such as *Scenedesmus* sp., *Chlorella* sp., *Tetradesmus* sp., and *Selenastrum* sp., which are commonly utilized for bioremediation, ranges between 0.7 and $1.4\,g\,L^{-1}$ (Rani et al., 2021). However, for achieving commercial feasibility, higher biomass productivity is required coupled with metabolite accumulation. Factors that commonly affect biomass production utilizing wastewaters include effluent pH, nutrient composition, and presence of colored compounds (Amin et al., 2022). For achieving good light penetration, the effluents or digestates must be diluted appropriately. It has been observed that high dilution (5% *v/v*) of anaerobic digestates improved the growth of microalgae and metabolite composition, specifically lipids and fatty acids (Koutra et al., 2018). This necessitates the inclusion of fresh or treated water to the effluent stream before cultivation affecting the sustainability of the process. Microalgae bioremediation is generally achieved by open pond cultivation in circular or raceway ponds where the cell densities are maintained very low for effective light penetration and prevention of shading and dark fermentation of cultures at the bottom of the ponds (Vidyashankar and Ravishankar, 2016; Shekh et al., 2022). This poses a crucial challenge in scaling up of the process, especially during the downstream processing. The total solids content of microalgae culture is less than 1%–1.5%, (approximately 0.7–$1.4\,g\,L^{-1}$ culture) and requires concentration to ﬔﬔ ﬔﬔﬔ ﬔ ﬔﬔ ﬔﬔ ﬔﬔﬔﬔ ﬔﬔﬔﬔ ﬔﬔﬔﬔ ﬔﬔﬔﬔﬔﬔﬔ ﬔﬔﬔﬔﬔﬔ ﬔﬔﬔﬔﬔ ﬔﬔﬔ ﬔﬔﬔﬔﬔ, 2013). Harvesting and dewatering are achieved by a variety of processes such as flocculation, electro-coagulation, centrifugation, flotation, and membrane filtration (Vandamme et al., 2013; Barros et al., 2015). An efficient nutrient removal in effluents is achieved by consortia of microalgae where each species possesses unique shape, size, and surface charge requiring customized harvesting solutions. Typically harvesting and dewatering are achieved by two stages where the first stage involves harvesting of cells to a concentration of 2%–3% followed by dewatering to 15% solids content (Milledge and Heaven, 2013). This problem can be countered by cultivation using photobioreactors (PBRs) where high cell densities and biomass productivities can be achieved. The biomass productivity in PBRs ($1.5\,kg$ dry biomass $m^{-3}\,day^{-1}$) is approximately 12 times higher compared with open pond cultivation ($0.12\,kg$ dry biomass $m^{-3}\,day^{-1}$) (Silva et al., 2015). Although highly productive, PBRs are associated with high capital costs and energy-intensive operations such as maintenance of temperatures, continuous mixing of cultures, and injection of CO_2 (Ighalo et al., 2022). This negates the purpose of low-cost bioremediation strategies using microalgae. Calculation of net energy returns (NER), a simple ratio of energy produced over energy consumed for various cultivation strategies, indicated that biomass production using horizontal or tubular PBRs had NER <1, while cultivation in open ponds or flat-panel bioreactor resulted in NER >1 indicating their suitability in mass cultivation (Jorquera et al., 2010).

Apart from the scale-up challenge, effluents obtained from tannery, textile, pharmaceutical, and pesticide industries possess POPs, antibiotics, and heavy metals. Although microalgae are effective in their biotransformation of these xenobiotics, the biomass may contain breakdown products and sequestered heavy metal that are harmful for human health (Bhatt et al., 2022). Till date, there no systematic studies performed on the additive toxicity of microalgae biomass obtained after treatment of effluents rich in POPs and heavy metals. This limits the application of wastewater-cultivated microalgae biomass in nutraceutical, animal feed, or biofertilizer applications.

As discussed earlier, microalgae-based effluent treatment is associated with bacteria and fungi that augment the process and increase the nutrient removal rates (Jiang et al., 2021). However, these bacteria and fungi may be virulent and possess threat when incorporated into food chain. It has been reported that horizontal gene transfers are a common mode of interaction between bacteria and microalgae that offer wide benefits to microalgae during biotransformation of xenobiotics and heavy metal sequestration (Saravanan et al., 2021). Although beneficial for bioremediation, horizontal gene transfers come with the risk of transfer of virulent genes, antibiotic resistance genes into the ecosystem. The major issue in deployment of bioremediation-derived algae biomass is lack of long-term toxicity studies limiting the biomass utilization in bioenergy applications.

9 Conclusion and future prospects

Literature amply demonstrates the possibility of microalgae-mediated bioremediation coupled with bioproduct generation. However, the major challenges still persistent are mass cultivation and energy-intensive downstream processes. Integration of bioremediation and biorefinery of microalgae biomass is a viable solution to improve the economic feasibility of microalgae technologies. However, there are several challenges in utilizing microalgae biomass derived from wastewater treatments such as heavy metal and xenobiotic residues, microbial contamination, and possible presence of antibiotic resistant genes and pathogenic properties from associated microorganisms. This necessitates detailed research in the following areas:

- Deciphering the mechanism of action in xenobiotics breakdown and identification of responsible genes in microalgae.
- Development of high-density cultivation strategies using PBRs for (i) reduction of land and water footprint, (ii) saving energy in harvesting and dewatering of biomass.
- Utilization of wet biomass for bioenergy and bioproducts generation.
- Evaluation of the microbiome composition of wastewater-generated biomass.
- Short-term and long-term toxicity studies on microalgal biomass cultivated using industrial effluents.
- Validation of different circular economy schemes for scale-up.

Acknowledgments

VS acknowledges the financial support from CSIR-Institute of Himalayan Bioresource Technology from MLP-0201 project and Director, CSIR-IHBT for encouragement. RK acknowledges Indian Council of Medical Research for Senior Research Fellowship. The CSIR-IHBT manuscript number for this chapter is 5378.

References

Abdelfattah, A., Ali, S.S., Ramadan, H., El-Aswar, E.I., Eltawab, R., Ho, S.-H., Elsamahy, T., Li, S., El-Sheekh, M.M., Schagerl, M., 2022. Microalgae-based wastewater treatment: mechanisms, challenges, recent advances, and future prospects. Environ. Sci. Ecotechnol., 100205.

Abed, R.M.M., 2010. Interaction between cyanobacteria and aerobic heterotrophic bacteria in the degradation of hydrocarbons. Int. Biodeter. Biodegr. 64, 58–64.

Abed, R.M.M., Zein, B., Al-Thukair, A., de Beer, D., 2007. Phylogenetic diversity and activity of aerobic heterotrophic bacteria from a hypersaline oil-polluted microbial mat. Syst. Appl. Microbiol. 30, 319–330.

Abinandan, S., Subashchandrabose, S.R., Venkateswarlu, K., Megharaj, M., 2019. Soil microalgae and cyanobacteria: the biotechnological potential in the maintenance of soil fertility and health. Crit. Rev. Biotechnol. 39, 981–998.

Aldaby, E.S.E., Mawad, A.M.M., 2019. Pyrene biodegradation capability of two different microalgal strains. Global NEST J. 21, 290–295.

Álvarez-Díaz, P.D., Ruiz, J., Zouhayr Arbib, J., Barragán, M.C.G.-P., Perales, J.A., 2015. Wastewater treatment and biodiesel production by Scenedesmus obliquus in a two-stage cultivation process. Bioresour. Technol. 181, 90–96.

Álvarez-González, A., Uggetti, E., Serrano, L., Gorchs, G., Ferrer, I., Díez-Montero, R., 2022. Can microalgae grown in wastewater reduce the use of inorganic fertilizers? J. Environ. Manage. 323, 116224.

Alvarez, A.L., Weyers, S.L., Goemann, H.M., Peyton, B.M., Gardner, R.D., 2021. Microalgae, soil and plants: a critical review of microalgae as renewable resources for agriculture. Algal Res. 54, 102200.

Amaya-Santos, G., Ruiz-Nieto, Á., Sánchez-Zurano, A., Ciardi, M., Gómez-Serrano, C., Acién, G., Lafarga, T., 2022. Production of Chlorella vulgaris using urban wastewater: assessment of the nutrient recovery capacity of the biomass and its plant biostimulant effects. J. Appl. Phycol. 34, 2971–2979.

Amin, M., Tahir, F., Ashfaq, H., Akbar, I., Razzaque, N., Haider, M.N., Jianren, X., Zhu, H., Wang, N., Shahid, A., 2022. Decontamination of industrial wastewater using microalgae integrated with biotransformation of the biomass to green products. Energy Nexus 6, 100089.

An, J.-Y., Sim, S.-J., Lee, J.S., Kim, B.W., 2003. Hydrocarbon production from secondarily treated piggery wastewater by the green alga Botryococcus braunii. J. Appl. Phycol. 15, 185–191.

Ansari, F.A., Gupta, S.K., Bux, F., 2019. Microalgae: a biorefinary approach to the treatment of aquaculture wastewater. In: Gupta, S., Bux, F. (Eds.), Application of Microalgae in Wastewater Treatment. Volume 2: Biorefinery Approaches of Wastewater Treatment. Springer, Cham, pp. 69–83, https://doi.org/10.1007/978-3-030-13909-4_4.

Arcchino, I.T., Jong, J., Berman, I., Van Hulle, C.W.H., Rousseau, D.P.L., Ouiff, M., 2022. Life cycle assessment of microalgae systems for wastewater treatment and bioproducts recovery: natural pigments, biofertilizer and biogas. Sci. Total Environ. 847, 157615.

Arun, J., Gopinath, K.P., Vigneshwar, S.S., Swetha, A., 2020. Sustainable and eco-friendly approach for phosphorus recovery from wastewater by hydrothermally carbonized microalgae: study on spent bio-char as fertilizer. J. Water Process Eng. 38, 101567.

Bai, X., Acharya, K., 2019. Removal of seven endocrine disrupting chemicals (EDCs) from municipal wastewater effluents by a freshwater green alga. Environ. Pollut. 247, 534–540.

Baldiris-Navarro, I., Sanchez-Aponte, J., González-Delgado, A., Jimenez, A.R., Acevedo-Morantes, M., 2018. Removal and biodegradation of phenol by the freshwater microalgae Chlorella vulgaris. Methods 6, 7.

Barone, V., Puglisi, I., Fragalà, F., Piero, A.R.L., Giuffrida, F., Baglieri, A., 2019. Novel bioprocess for the cultivation of microalgae in hydroponic growing system of tomato plants. J. Appl. Phycol. 31, 465–470.

Barros, A.I., Gonçalves, A.L., Simões, M., Pires, J.C.M., 2015. Harvesting techniques applied to microalgae: a review. Renew. Sustain. Energy Rev. 41, 1489–1500.

Bayramoğlu, G., Tuzun, I., Celik, G., Yilmaz, M., Arica, M.Y., 2006. Biosorption of mercury (II), cadmium (II) and lead (II) ions from aqueous system by microalgae Chlamydomonas reinhardtii immobilized in alginate beads. Int. J. Miner. Process. 81, 35–43.

Bhatt, P., Bhandari, G., Bhatt, K., Simsek, H., 2022. Microalgae-based removal of pollutants from wastewaters: occurrence, toxicity and circular economy. Chemosphere 306, 135576.

Borges, F.C., Xie, Q., Min, M., Muniz, L.A.R., Farenzena, M., Trierweiler, J.O., Chen, P., Ruan, R., 2014. Fast microwave-assisted pyrolysis of microalgae using microwave absorbent and HZSM-5 catalyst. Bioresour. Technol. 166, 518–526.

Bouras, S., Katsoulas, N., Antoniadis, D., Karapanagiotidis, I.T., 2020. Use of biofuel industry wastes as alternative nutrient sources for DHA-yielding Schizochytrium limacinum production. Appl. Sci. 10, 4398.

Brar, A., Kumar, M., Vivekanand, V., Pareek, N., 2019. Phycoremediation of textile effluent-contaminated water bodies employing microalgae: nutrient sequestration and biomass production studies. Int. J. Environ. Sci. Technol. 16, 7757–7768.

Brennan, L., Owende, P., 2010. Biofuels from microalgae—a review of technologies for production, processing, and extractions of biofuels and co-products. Renew. Sustain. Energy Rev. 14, 557–577.

Butler, A., 1998. Acquisition and utilization of transition metal ions by marine organisms. Science 281, 207–209.

Cai, W., Zhao, Z., Li, D., Lei, Z., Zhang, Z., Lee, D.-J., 2019. Algae granulation for nutrients uptake and algae harvesting during wastewater treatment. Chemosphere 214, 55–59.

Çakirsoy, I., Miyamoto, T., Ohtake, N., 2022. Physiology of microalgae and their application to sustainable agriculture: a mini-review. Front. Plant Sci. 13.

Carraro, C.d.F.F., Loures, C.C.A., de Castro, J.A., 2022. Microalgae bioremediation and CO_2 fixation of industrial wastewater. Clean. Eng. Technol. 8, 100466.

Chan, S.S., Khoo, K.S., Chew, K.W., Ling, T.C., Show, P.L., 2022. Recent advances biodegradation and biosorption of organic compounds from wastewater: microalgae-bacteria consortium—a review. Bioresour. Technol. 344, 126159.

Chanakya, H.N., Mahapatra, D.M., Ravi, S., Chauhan, V.S., Abitha, R., 2012. Sustainability of large-scale algal biofuel production in India. J. Indian Inst. Sci. 92, 63–98.

Chang, Y.-M., Tsai, W.-T., Li, M.-H., 2015. Chemical characterization of char derived from slow pyrolysis of microalgal residue. J. Anal. Appl. Pyrolysis 111, 88–93.

Chavan, A., Mukherji, S., 2010. Effect of co-contaminant phenol on performance of a laboratory-scale RBC with algal-bacterial biofilm treating petroleum hydrocarbon-rich wastewater. J. Chem. Technol. Biotechnol. 85, 851–859.

Cheah, W.Y., Show, P.L., Juan, J.C., Chang, J.-S., Ling, T.C., 2018. Microalgae cultivation in palm oil mill effluent (POME) for lipid production and pollutants removal. Energ. Conver. Manage. 174, 430–438.

Chegukrishnamurthi, M., Shekh, A., Ravi, S., Mudliar, S.N., 2022. Volatile organic compounds involved in the communication of microalgae-bacterial association extracted through headspace-solid phase microextraction and confirmed using gas chromatography-mass spectrophotometry. Bioresour. Technol. 348, 126775.

Chen, S., Wang, L., Feng, W., Yuan, M., Li, J., Houtao, X., Zheng, X., Zhang, W., 2020. Sulfonamides-induced oxidative stress in freshwater microalga *Chlorella vulgaris*: evaluation of growth, photosynthesis, antioxidants, ultrastructure, and nucleic acids. Sci. Rep. 10, 8243.

Chen, X., Zhan, H., Qi, Y., Song, C., Chen, G., 2019. The interactions of algae-activated sludge symbiotic system and its effects on wastewater treatment and lipid accumulation. Bioresour. Technol. 292, 122017.

Cheng, H.-H., Narindri, B., Chu, H., Whang, L.-M., 2020a. Recent advancement on biological technologies and strategies for resource recovery from swine wastewater. Bioresour. Technol. 303, 122861.

Cheng, P., Chen, D., Liu, W., Cobb, K., Zhou, N., Liu, Y., Liu, H., Wang, Q., Chen, P., Zhou, C., 2020b. Autoflocculation microalgae species *Tribonema* sp. and *Synechocystis* sp. with T-IPL pretreatment to improve swine wastewater nutrient removal. Sci. Total Environ. 725, 138263.

Chi, J., Li, Y., Gao, J., 2019. Interaction between three marine microalgae and two phthalate acid esters. Ecotoxicol. Environ. Saf. 170, 407–411.

Chi, Z., Pyle, D., Wen, Z., Frear, C., Chen, S., 2007. A laboratory study of producing docosahexaenoic acid from biodiesel-waste glycerol by microalgal fermentation. Process Biochem. 42, 1537–1545.

Chilakamarry, C.R., Mimi Sakinah, A.M., Zularisam, A.W., Pandey, A., 2021. Glycerol waste to value added products and its potential applications. Syst. Microbiol. Biomanuf. 1, 378–396.

Chisti, Y., 2013. Constraints to commercialization of algal fuels. J. Biotechnol. 167, 201–214.

Cho, S., Luong, T.T., Lee, D., You-Kwan, O., Lee, T., 2011. Reuse of effluent water from a municipal wastewater treatment plant in microalgae cultivation for biofuel production. Bioresour. Technol. 102, 8639–8645.

Chokshi, K., Pancha, I., Ghosh, A., Mishra, S., 2016. Microalgal biomass generation by phycoremediation of dairy industry wastewater: an integrated approach towards sustainable biofuel production. Bioresour. Technol. 221, 455–460.

Croft, M.T., Warren, M.J., Smith, A.G., 2006. Algae need their vitamins. Eukaryot. Cell 5, 1175–1183.

Cuaresma, M., Casal, C., Forján, E., Vílchez, C., 2011. Productivity and selective accumulation of carotenoids of the novel extremophile microalga *Chlamydomonas acidophila* grown with different carbon sources in batch systems. J. Ind. Microbiol. Biotechnol. 38, 167–177.

da Fontoura, J., Tolfo, G.S., Rolim, M.F., Gutterres, M., 2017. Influence of light intensity and tannery wastewater concentration on biomass production and nutrient removal by microalgae Scenedesmus sp. Process Saf. Environ. Prot. 111, 355–362.

V. Algal bioproducts, biofuels, and biorefinery for business opportunities

Dagnaisser, L.S., Barbosa, M.G., dos Santos, A., Rita, V.S., Cardoso, J.C., de Carvalho, D.F., de Mendonça, H.V., 2022. Microalgae as bio-fertilizer: a new strategy for advancing modern agriculture, wastewater bioremediation, and atmospheric carbon mitigation. Water Air Soil Pollut. 233, 1–23.

Daneshvar, E., Antikainen, L., Koutra, E., Kornaros, M., Bhatnagar, A., 2018a. Investigation on the feasibility of *Chlorella vulgaris* cultivation in a mixture of pulp and aquaculture effluents: treatment of wastewater and lipid extraction. Bioresour. Technol. 255, 104–110.

Daneshvar, E., Zarrinmehr, M.J., Hashtjin, A.M., Farhadian, O., Bhatnagar, A., 2018b. Versatile applications of freshwater and marine water microalgae in dairy wastewater treatment, lipid extraction and tetracycline biosorption. Bioresour. Technol. 268, 523–530.

Daneshvar, E., Zarrinmehr, M.J., Kousha, M., Hashtjin, A.M., Saratale, G.D., Maiti, A., Vithanage, M., Bhatnagar, A., 2019a. Hexavalent chromium removal from water by microalgal-based materials: adsorption, desorption and recovery studies. Bioresour. Technol. 293, 122064.

Daneshvar, E., Zarrinmehr, M.J., Koutra, E., Kornaros, M., Farhadian, O., Bhatnagar, A., 2019b. Sequential cultivation of microalgae in raw and recycled dairy wastewater: microalgal growth, wastewater treatment and biochemical composition. Bioresour. Technol. 273, 556–564.

Danouche, M., El Ghachtouli, N., El Baouchi, A., El Arroussi, H., 2020. Heavy metals phycoremediation using tolerant green microalgae: enzymatic and non-enzymatic antioxidant systems for the management of oxidative stress. J. Environ. Chem. Eng. 8, 104460.

Das, D., Chakraborty, S., Bhattacharjee, C., Chowdhury, R., 2016. Biosorption of lead ions (Pb^{2+}) from simulated wastewater using residual biomass of microalgae. Desalin. Water Treat. 57, 4576–4586.

de Farias Silva, C.E., Bertucco, A., 2016. Bioethanol from microalgae and cyanobacteria: a review and technological outlook. Process Biochem. 51, 1833–1842.

de Souza Leite, L., Hoffmann, M.T., Daniel, L.A., 2019. Microalgae cultivation for municipal and piggery wastewater treatment in Brazil. J. Water Process Eng. 31, 100821.

Dexter, J., Pengcheng, F., 2009. Metabolic engineering of cyanobacteria for ethanol production. Energ. Environ. Sci. 2, 857–864.

Díaz, V., Leyva-Díaz, J.C., Almécija, M.C., Poyatos, J.M., del Mar Muñío, M., Martín-Pascual, J., 2022. Microalgae bioreactor for nutrient removal and resource recovery from wastewater in the paradigm of circular economy. Bioresour. Technol., 127968.

Ding, T., Lin, K., Yang, B., Yang, M., Li, J., Li, W., Gan, J., 2017a. Biodegradation of naproxen by freshwater algae [illegible]

Ding, T., Yang, M., Zhang, J., Yang, B., Lin, K., Li, J., Gan, J., 2017b. Toxicity, degradation and metabolic fate of ibuprofen on freshwater diatom *Navicula* sp. J. Hazard. Mater. 330, 127–134.

Dirbaz, M., Roosta, A., 2018. Adsorption, kinetic and thermodynamic studies for the biosorption of cadmium onto microalgae *Parachlorella* sp. J. Environ. Chem. Eng. 6, 2302–2309.

Doble, M., Kumar, A., 2005. Biotreatment of Industrial Effluents. Elsevier.

Doshi, H., Ray, A., Kothari, I.L., 2009. Live and dead *Spirulina* sp. to remove arsenic (V) from water. Int. J. Phytoremediation 11, 53–64.

Dourou, M., Dritsas, P., Baeshen, M.N., Elazzazy, A., Al-Farga, A., Aggelis, G., 2020. High-added value products from microalgae and prospects of aquaculture wastewaters as microalgae growth media. FEMS Microbiol. Lett. 367, fnaa081.

Dragone, G., Fernandes, B., Vicente, A., Teixeira, J.A., 2010. Third generation biofuels from microalgae, current research, technology and education. Appl. Microbiol. Biotechnol. 2, 1355–1366. Formatex Research Center.

El-Araby, D.A., Amer, S.A., Attia, G.A., Osman, A., Fahmy, E.M., Altohamy, D.E., Alkafafy, M., Elakkad, H.A., Tolba, S.A., 2022. Dietary *Spirulina platensis* phycocyanin improves growth, tissue histoarchitecture, and immune responses, with modulating immunoexpression of CD3 and CD20 in Nile tilapia, *Oreochromis niloticus*. Aquaculture 546, 737413.

Escapa, C., Coimbra, R.N., Paniagua, S., García, A.I., Otero, M., 2017a. Comparison of the culture and harvesting of *Chlorella vulgaris* and *Tetradesmus obliquus* for the removal of pharmaceuticals from water. J. Appl. Phycol. 29, 1179–1193.

Escapa, C., Coimbra, R.N., Nuevo, C., Vega, S., Paniagua, S., García, A.I., Calvo, L.F., Otero, M., 2017b. Valorization of microalgae biomass by its use for the removal of paracetamol from contaminated water. Water 9, 312.

Farooq, W., Lee, Y.-C., Ryu, B.-G., Kim, B.-H., Kim, H.-S., Choi, Y.-E., Yang, J.-W., 2013. Two-stage cultivation of two *Chlorella* sp. strains by simultaneous treatment of brewery wastewater and maximizing lipid productivity. Bioresour. Technol. 132, 230–238.

Fawzy, M.A., Alharthi, S., 2021. Cellular responses and phenol bioremoval by green alga *Scenedesmus abundans*: equilibrium, kinetic and thermodynamic studies. Environ. Technol. Innov. 22, 101463.

Fernandes, R.T.V., Gonçalves, A.A., de Arruda, A.M.V., 2020. Production, egg quality, and intestinal morphometry of laying hens fed marine microalga. Rev. Bras. Zootec. 49.

Fernando, J.S., Ravishan, M.P., Dinalankara, D.M.S.D., Perera, G.L.N.J., Ariyadasa, T.U., 2021. Cultivation of microalgae in palm oil mill effluent (POME) for astaxanthin production and simultaneous phycoremediation. J. Environ. Chem. Eng. 9, 105375.

Ferreira, A., Melkonyan, L., Carapinha, S., Ribeiro, B., Figueiredo, D., Avetisova, G., Gouveia, L., 2021. Biostimulant and biopesticide potential of microalgae growing in piggery wastewater. Environ. Adv. 4, 100062.

Ferreira, A., Ribeiro, B., Ferreira, A.F., Tavares, M.L.A., Vladic, J., Vidović, S., Cvetkovic, D., Melkonyan, L., Avetisova, G., Goginyan, V., 2019. *Scenedesmus obliquus* microalga-based biorefinery—from brewery effluent to bioactive compounds, biofuels and biofertilizers—aiming at a circular bioeconomy. Biofuels Bioprod. Biorefin. 13, 1169–1186.

Ferro, L., Gentili, F.G., Funk, C., 2018. Isolation and characterization of microalgal strains for biomass production and wastewater reclamation in Northern Sweden. Algal Res. 32, 44–53.

Ferro, L., Gojkovic, Z., Muñoz, R., Funk, C., 2019. Growth performance and nutrient removal of a *Chlorella vulgaris-Rhizobium* sp. co-culture during mixotrophic feed-batch cultivation in synthetic wastewater. Algal Res. 44, 101690.

Finocchio, E., Lodi, A., Solisio, C., Converti, A., 2010. Chromium (VI) removal by methylated biomass of *Spirulina platensis*: the effect of methylation process. Chem. Eng. J. 156, 264–269.

Forján, E., Navarro, F., Cuaresma, M., Vaquero, I., Ruíz-Domínguez, M.C., Gojkovic, Ž., Vázquez, M., Márquez, M., Mogedas, B., Bermejo, E., 2015. Microalgae: fast-growth sustainable green factories. Crit. Rev. Environ. Sci. Technol. 45, 1705–1755.

Fu, Y., Wang, Y., Yi, L., Liu, J., Yang, S., Liu, B., Chen, F., Sun, H., 2023. Lutein production from microalgae: a review. Bioresour. Technol., 128875.

Furtado, A.L.F.F., do Carmo Calijuri, M., Lorenzi, A.S., Honda, R.Y., Genuário, D.B., Fiore, M.F., 2009. Morphological and molecular characterization of cyanobacteria from a Brazilian facultative wastewater stabilization pond and evaluation of microcystin production. Hydrobiologia 627, 195–209.

Goh, P.S., Lau, W.J., Ismail, A.F., Samawati, Z., Liang, Y.Y., Kanakaraju, D., 2023. Microalgae-enabled wastewater treatment: a sustainable strategy for bioremediation of pesticides. Water 15, 70.

Gokhale, S.V., Jyoti, K.K., Lele, S.S., 2008. Kinetic and equilibrium modeling of chromium (VI) biosorption on fresh and spent *Spirulina platensis/Chlorella vulgaris* biomass. Bioresour. Technol. 99, 3600–3608.

Gonzalez-Camejo, J., Viruela, A., Ruano, M.V., Barat, R., Seco, A., Ferrer, J., 2019. Effect of light intensity, light duration and photoperiods in the performance of an outdoor photobioreactor for urban wastewater treatment. Algal Res. 40, 101511.

Gramegna, G., Scortica, A., Scafati, V., Ferella, F., Gurrieri, L., Giovannoni, M., Bassi, R., Sparla, F., Mattei, B., Benedetti, M., 2020. Exploring the potential of microalgae in the recycling of dairy wastes. Bioresour. Technol. Rep. 12, 100604.

Grivalský, T., Ranglová, K., Lakatos, G.E., Manoel, J.A.C., Černá, T., Barceló-Villalobos, M., Estrella, F.S., Ördög, V., Masojídek, J., 2022. Bioactivity assessment, micropollutant and nutrient removal ability of *Tetradesmus obliquus* cultivated outdoors in centrate from urban wastewater. J. Appl. Phycol., 1–16.

Guieysse, B., Borde, X., Muñoz, R., Hatti-Kaul, R., Nugier-Chauvin, C., Patin, H., Mattiasson, B., 2002. Influence of the initial composition of algal-bacterial microcosms on the degradation of salicylate in a fed-batch culture. Biotechnol. Lett. 24, 531–538.

Guo, R., Yingxiang, D., Zheng, F., Wang, J., Wang, Z., Ji, R., Chen, J., 2017. Bioaccumulation and elimination of bisphenol a (BPA) in the alga *Chlorella pyrenoidosa* and the potential for trophic transfer to the rotifer *Brachionus calyciflorus*. Environ. Pollut. 227, 460–467.

Gupta, S., Pawar, S.B., 2018. An integrated approach for microalgae cultivation using raw and anaerobic digested wastewaters from food processing industry. Bioresour. Technol. 269, 571–576.

Gupta, V., Ratha, S.K., Sood, A., Chaudhary, V., Prasanna, R., 2013. New insights into the biodiversity and applications of cyanobacteria (blue-green algae)—prospects and challenges. Algal Res. 2, 79–97.

Haferburg, G., Kothe, E., 2007. Microbes and metals: interactions in the environment. J. Basic Microbiol. 47, 453–467.

Han, T., Haifeng, L., Zhao, Y., Hong, X., Zhang, Y., Li, B., 2019. Two-step strategy for obtaining *Dunaliella* sp. biomass and β-carotene from anaerobically digested poultry litter wastewater. Int. Biodeter. Biodegr. 143, 104714.

Haque, F., Dutta, A., Thimmanagari, M., Chiang, Y.W., 2017. Integrated *Haematococcus pluvialis* biomass production and nutrient removal using bioethanol plant waste effluent. Process Saf. Environ. Prot. 111, 128–137.

He, Y., Lian, J., Lan Wang, L., Tan, F.K., Li, Y., Wang, H., Rebours, C., Han, D., Qiang, H., 2023. Recovery of nutrients from aquaculture wastewater: effects of light quality on the growth, biochemical composition, and nutrient removal of Chlorella sorokiniana. Algal Res. 69, 102965.

Hena, S., Fatihah, N., Tabassum, S., Ismail, N., 2015. Three stage cultivation process of facultative strain of Chlorella sorokiniana for treating dairy farm effluent and lipid enhancement. Water Res. 80, 346–356.

Hernández-García, A., Velásquez-Orta, S.B., Novelo, E., Yáñez-Noguez, I., Monje-Ramírez, I., Orta Ledesma, M.T., 2019. Wastewater-leachate treatment by microalgae: biomass, carbohydrate and lipid production. Ecotoxicol. Environ. Saf. 174, 435–444.

Higgins, B.T., Gennity, I., Fitzgerald, P.S., Ceballos, S.J., Fiehn, O., VanderGheynst, J.S., 2018. Algal-bacterial synergy in treatment of winery wastewater. NPJ Clean Water 1, 6.

Hosseinkhani, N., McCauley, J.I., Ralph, P.J., 2022. Key challenges for the commercial expansion of ingredients from algae into human food products. Algal Res. 64, 102696.

Huang, C.-C., Chen, M.-W., Hsieh, J.-L., Lin, W.-H., Chen, P.-C., Chien, L.-F., 2006. Expression of mercuric reductase from Bacillus megaterium MB1 in eukaryotic microalga Chlorella sp. DT: an approach for mercury phytoremediation. Appl. Microbiol. Biotechnol. 72, 197–205.

Huang, R., Huo, G., Song, S., Li, Y., Xia, L., Gaillard, J.-F., 2019. Immobilization of mercury using high-phosphate culture-modified microalgae. Environ. Pollut. 254, 112966.

Humhal, T., Kastanek, P., Jezkova, Z., Cadkova, A., Kohoutkova, J., Branyik, T., 2017. Use of saline waste water from demineralization of cheese whey for cultivation of Schizochytrium limacinum PA-968 and Japonochytrium marinum AN-4. Bioprocess Biosyst. Eng. 40, 395–402.

Huo, S., Chen, J., Zhu, F., Zou, B., Chen, X., Basheer, S., Cui, F., Qian, J., 2019. Filamentous microalgae Tribonema sp. cultivation in the anaerobic/oxic effluents of petrochemical wastewater for evaluating the efficiency of recycling and treatment. Biochem. Eng. J. 145, 27–32.

Huo, S., Kong, M., Zhu, F., Qian, J., Huang, D., Chen, P., Ruan, R., 2020. Co-culture of Chlorella and wastewater-borne bacteria in vinegar production wastewater: enhancement of nutrients removal and influence of algal biomass generation. Algal Res. 45, 101744.

Ighalo, J.O., Dulta, K., Kurniawan, S.B., Omoarukhe, F.O., Ewuzie, U., Eshiemogie, S.O., Ojo, A.U., Abdullah, S.R.S., 2022. Progress in microalgae application for CO_2 sequestration. Clean. Chem. Eng., 100044.

Imase, M., Watanabe, K., Aoyagi, H., Tanaka, H., 2008. Construction of an artificial symbiotic community using a Chlorella-symbiont association as a model. FEMS Microbiol. Ecol. 63, 273–282.

Ip, P.-F., Chen, F., 2005. Production of astaxanthin by the green microalga Chlorella zofingiensis in the dark. Process Biochem. 40, 733–738.

Jain, M., Khan, S.A., Sharma, K., Jadhao, P.R., Pant, K.K., Ziora, Z.M., Blaskovich, M.A.T., 2022. Current perspective of innovative strategies for bioremediation of organic pollutants from wastewater. Bioresour. Technol. 344, 126305.

Jena, J., Nilotpala Pradhan, V., Aishvarya, R.R., Nayak, B.P., Dash, L.B., Sukla, P.K.P., Mishra, B.K., 2015. Biological sequestration and retention of cadmium as CdS nanoparticles by the microalga Scenedesmus-24. J. Appl. Phycol. 27, 2251–2260.

Jiang, L., Li, Y., Pei, H., 2021. Algal-bacterial consortia for bioproduct generation and wastewater treatment. Renew. Sustain. Energy Rev. 149, 111395.

Jones, E.R., Van Vliet, M.T.H., Qadir, M., Bierkens, M.F.P., 2021. Country-level and gridded estimates of wastewater production, collection, treatment and reuse. Earth Syst. Sci. Data 13, 237–254.

Jorquera, O., Kiperstok, A., Sales, E.A., Embiruçu, M., Ghirardi, M.L., 2010. Comparative energy life-cycle analyses of microalgal biomass production in open ponds and photobioreactors. Bioresour. Technol. 101, 1406–1413.

Kapoore, R.V., Wood, E.E., Llewellyn, C.A., 2021. Algae biostimulants: a critical look at microalgal biostimulants for sustainable agricultural practices. Biotechnol. Adv. 49, 107754.

Karpagam, R., Jawaharraj, K., Gnanam, R., 2021. Review on integrated biofuel production from microalgal biomass through the outset of transesterification route: a cascade approach for sustainable bioenergy. Sci. Total Environ. 766, 144236.

Katiyar, R., Banerjee, S., Arora, A., 2021. Recent advances in the integrated biorefinery concept for the valorization of algal biomass through sustainable routes. Biofuels Bioprod. Biorefin. 15, 879–898.

Kayalvizhi, K., Vijayaraghavan, K., Velan, M., 2015. Biosorption of Cr (VI) using a novel microalga Rhizoclonium hookeri: equilibrium, kinetics and thermodynamic studies. Desalin. Water Treat. 56, 194–203.

Khan, S.A., Sharma, G.K., Malla, F.A., Kumar, A., Gupta, N., 2019. Microalgae based biofertilizers: a biorefinery approach to phycoremediate wastewater and harvest biodiesel and manure. J. Clean. Prod. 211, 1412–1419.

Kholssi, R., Ramos, P.V., Marks, E.A.N., Montero, O., Rad, C., 2021. 2Biotechnological uses of microalgae: a review on the state of the art and challenges for the circular economy. Biocatal. Agric. Biotechnol. 36, 102114.

Kirkwood, A.E., Nalewajko, C., Fulthorpe, R.R., 2006. The effects of cyanobacterial exudates on bacterial growth and biodegradation of organic contaminants. Microb. Ecol. 51, 4–12.

Klok, A.J., Lamers, P.P., Martens, D.E., Draaisma, R.B., Wijffels, R.H., 2014. Edible oils from microalgae: insights in TAG accumulation. Trends Biotechnol. 32, 521–528.

Knothe, G., 2006. Analyzing biodiesel: standards and other methods. J. Am. Oil Chem. Soc. 83, 823–833.

Kong, W., Kong, J., Ma, J., Lyu, H., Feng, S., Wang, Z., Yuan, P., Shen, B., 2021. *Chlorella vulgaris* cultivation in simulated wastewater for the biomass production, nutrients removal and CO_2 fixation simultaneously. J. Environ. Manage. 284, 112070.

Koutra, E., Economou, C.N., Tsafrakidou, P., Kornaros, M., 2018. Bio-based products from microalgae cultivated in digestates. Trends Biotechnol. 36, 819–833.

Kujawska, N., Talbierz, S., Dębowski, M., Kazimierowicz, J., Zieliński, M., 2021. Optimizing docosahexaenoic acid (DHA) production by *Schizochytrium* sp. grown on waste glycerol. Energies 14, 1685.

Kumar, A.K., Sharma, S., Patel, A., Dixit, G., Shah, E., 2019. Comprehensive evaluation of microalgal based dairy effluent treatment process for clean water generation and other value added products. Int. J. Phytoremediation 21, 519–530.

Kumar, G., Huy, M., Bakonyi, P., Bélafi-Bakó, K., Kim, S.-H., 2018. Evaluation of gradual adaptation of mixed microalgae consortia cultivation using textile wastewater via fed batch operation. Biotechnol. Rep. 20, e00289.

Kumar, R., Hedge, A.S., Gopta, S., Srivatsan, V., 2022a. Microalgal product basket: portfolio positioning across food, feed and fuel segments with industrial growth projections. In: Microalgae for Sustainable Products: The Green Synthetic Biology Platform. Royal Society of Chemistry, pp. 191–229.

Kumar, R., Hegde, A.S., Sharma, K., Parmar, P., Srivatsan, V., 2022b. Microalgae as a sustainable source of edible proteins and bioactive peptides-current trends and future prospects. Food Res. Int., 111338.

Kumar, R., Sharma, V., Das, S., Patial, V., Srivatsan, V., 2023. *Arthrospira platensis* (Spirulina) fortified functional foods ameliorate iron and protein malnutrition by improving growth and modulating oxidative stress and gut microbiota in rats. Food Funct. 14, 1160–1178.

Kuo, C.-M., Jian, J.-F., Lin, T.-H., Chang, Y.-B., Wan, X.-H., Lai, J.-T., Chang, J.-S., Lin, C.-S., 2016. Simultaneous microalgal biomass production and CO_2 fixation by cultivating *Chlorella* sp. GD with aquaculture wastewater and boiler flue gas. Bioresour. Technol. 221, 241–250.

Kusmayadi, A., Huang, C.-Y., Leong, Y.K., Po-Han, L., Yen, H.-W., Lee, D.-J., Chang, J.-S., 2023. Integration of microalgae cultivation and anaerobic co-digestion with dairy wastewater to enhance bioenergy and biochemicals production. Bioresour. Technol. 376, 128858.

Kwak, H.W., Kim, M.K., Lee, J.Y., Yun, H., Kim, M.H., Park, Y.H., Lee, K.H., 2015. Preparation of bead-type biosorbent from water-soluble *Spirulina platensis* extracts for chromium (VI) removal. Algal Res. 7, 92–99.

Lage, S., Gentili, F.G., 2023. Chemical composition and species identification of microalgal biomass grown at pilot-scale with municipal wastewater and CO_2 from flue gases. Chemosphere 313, 137344.

Lakatos, G.E., Ranglová, K., Manoel, J.C., Grivalský, T., Kopecký, J., Masojídek, J., 2019. Bioethanol production from microalgae polysaccharides. Folia Microbiol. 64, 627–644.

Ledda, C., Tamiazzo, J., Borin, M., Adani, F., 2016. A simplified process of swine slurry treatment by primary filtration and *Haematococcus pluvialis* culture to produce low cost astaxanthin. Ecol. Eng. 90, 244–250.

Lee, S.-M., Ryu, C.-M., 2021. Algae as new kids in the beneficial plant microbiome. Front. Plant Sci. 12, 599742.

Li, B., Di, W., Li, Y., Shi, Y., Wang, C., Sun, J., Song, C., 2022. Metabolic Mechanism of Sulfadimethoxine Biodegradation by *Chlorella* sp. L38 and *Phaeodactylum tricornutum* MASCC-0025. Front. Microbiol. 13.

Li, J., Lens, P.N.L., Ferrer, I., Du Laing, G., 2021. Evaluation of selenium-enriched microalgae produced on domestic wastewater as biostimulant and biofertilizer for growth of selenium-enriched crops. J. Appl. Phycol. 33, 3027–3039.

Li, X., Yang, C., Zeng, G., Shaohua, W., Lin, Y., Zhou, Q., Lou, W., Cheng, D., Nie, L., Zhong, Y., 2020. Nutrient removal from swine wastewater with growing microalgae at various zinc concentrations. Algal Res. 46, 101804.

Li, Y., Slouka, S.A., Henkanatte-Gedera, S.M., Nirmalakhandan, N., Strathmann, T.J., 2019. Seasonal treatment and economic evaluation of an algal wastewater system for energy and nutrient recovery. Environ. Sci. Water Res. Technol. 5, 1545–1557.

Li, F., Zhao, L., Jinxu, Y., Shi, W., Zhou, S., Yuan, K., Sheng, G.D., 2018. Removal of dichlorophenol by Chlorella pyrenoidosa through self-regulating mechanism in air-tight test environment. Ecotoxicol. Environ. Saf. 164, 109–117.

Lin, C., Cao, P., Xiaolin, X., Ye, B., 2019. Algal-bacterial symbiosis system treating high-load printing and dyeing wastewater in continuous-flow reactors under natural light. Water 11, 469.

Ling, Y., Sun, L.-p., Wang, S.-y., Lin, C.S.K., Sun, Z., Zhou, Z.-g., 2019. Cultivation of oleaginous microalga *Scenedesmus obliquus* coupled with wastewater treatment for enhanced biomass and lipid production. Biochem. Eng. J. 148, 162–169.

Liu, H., Qian, L., Wang, Q., Liu, W., Wei, Q., Ren, H., Ming, C., Min, M., Chen, P., Ruan, R., 2017. Isolation of a bacterial strain, *Acinetobacter* sp. from centrate wastewater and study of its cooperation with algae in nutrients removal. Bioresour. Technol. 235, 59–69.

Liu, Y., Ren, X., Fan, C., Wenzhong, W., Zhang, W., Wang, Y., 2022. Health benefits, food applications, and sustainability of microalgae-derived N-3 PUFA. Foods 11, 1883.

Liu, Y., 2018. Optimization Study of Biomass and Astaxanthin Production by Haematococcus pluvialis Under Minkery Wastewater Cultures, Doctoral dissertation.

Liyanaarachchi, V.C., Premaratne, M., Ariyadasa, T.U., Nimarshana, P.H.V., Malik, A., 2021. Two-stage cultivation of microalgae for production of high-value compounds and biofuels: a review. Algal Res. 57, 102353.

Lu, Y., Jian, X., 2015. Phytohormones in microalgae: a new opportunity for microalgal biotechnology? Trends Plant Sci. 20, 273–282.

Luo, L., Ren, H., Pei, X., Xie, G., Xing, D., Dai, Y., Ren, N., Liu, B., 2019. Simultaneous nutrition removal and high-efficiency biomass and lipid accumulation by microalgae using anaerobic digested effluent from cattle manure combined with municipal wastewater. Biotechnol. Biofuels 12, 1–15.

Ma, C., Wen, H., Xing, D., Pei, X., Zhu, J., Ren, N., Liu, B., 2017. Molasses wastewater treatment and lipid production at low temperature conditions by a microalgal mutant *Scenedesmus* sp. Z-4. Biotechnol. Biofuels 10, 1–13.

Madadi, R., Zahed, M.A., Pourbabaee, A.A., Tabatabaei, M., Naghavi, M.R., 2021. Simultaneous phycoremediation of petrochemical wastewater and lipid production by *Chlorella vulgaris*. SN Appl. Sci. 3, 1–10.

Malibari, R., Sayegh, F., Elazzazy, A.M., Baeshen, M.N., Dourou, M., Aggelis, G., 2018. Reuse of shrimp farm wastewater as growth medium for marine microalgae isolated from Red Sea–Jeddah. J. Clean. Prod. 198, 160–169.

Maroneze, M.M., Barin, J.S., Ragagnin, C., de Menezes, M., Queiroz, I., Zepka, L.Q., Jacob-Lopes, E., 2014. Treatment of cattle-slaughterhouse wastewater and the reuse of sludge for biodiesel production by microalgal heterotrophic bioreactors. Sci. Agric. 71, 521–524.

Martins, J., Peixe, L., Vasconcelos, V., 2010. Cyanobacteria and bacteria co-occurrence in a wastewater treatment plant: absence of allelopathic effects. Water Sci. Technol. 62, 1954–1962.

Maryjoseph, S., Kothooran, B., 2020. Microalgae based wastewater treatment for the removal of emerging contaminants: a review of challenges and opportunities. Case Stud. Chem. Environ. Eng. 2, 100046.

Mayers, J.J., Nilsson, A.E., Svensson, E., Albers, E., 2016. Integrating microalgal production with industrial outputs—reducing process inputs and quantifying the benefits. Ind. Biotechnol. 12, 219–234.

Mehariya, S., Goswami, R.K., Karthikeysan, O.P., Verma, P., 2021. Microalgae for high-value products: a way towards green nutraceutical and pharmaceutical compounds. Chemosphere 280, 130553.

Mehrabadi, A., Craggs, R., Farid, M.M., 2015. Wastewater treatment high rate algal ponds (WWT HRAP) for low-cost biofuel production. Bioresour. Technol. 184, 202–214.

Milledge, J.J., Heaven, S., 2013. A review of the harvesting of micro-algae for biofuel production. Rev. Environ. Sci. Bio/Technol. 12, 165–178.

Molazadeh, P., Khanjani, N., Rahimi, M.R., Nasiri, A., 2015. Adsorption of lead by microalgae *Chaetoceros* sp. and *Chlorella* sp. from aqueous solution. J. Community Health Res. 4, 114–127.

Mona, S., Malyan, S.K., Saini, N., Deepak, B., Pugazhendhi, A., Kumar, S.S., 2021. Towards sustainable agriculture with carbon sequestration, and greenhouse gas mitigation using algal biochar. Chemosphere 275, 129856.

Munoz, R., Guieysse, B., Mattiasson, B., 2003. Phenanthrene biodegradation by an algal-bacterial consortium in two-phase partitioning bioreactors. Appl. Microbiol. Biotechnol. 61, 261–267.

Muñoz, R., Köllner, C., Guieysse, B., Mattiasson, B., 2003. Salicylate biodegradation by various algal-bacterial consortia under photosynthetic oxygenation. Biotechnol. Lett. 25, 1905–1911.

Munoz, R., Rolvering, C., Guieysse, B., Mattiasson, B., 2005. Aerobic phenanthrene biodegradation in a two-phase partitioning bioreactor. Water Sci. Technol. 52, 265–271.

Naaz, F., Bhattacharya, A., Pant, K.K., Malik, A., 2019. Investigations on energy efficiency of biomethane/biocrude production from pilot scale wastewater grown algal biomass. Appl. Energy 254, 113656.

Nagabalaji, V., Sivasankari, G., Srinivasan, S.V., Suthanthararajan, R., Ravindranath, E., 2019. Nutrient removal from synthetic and secondary treated sewage and tannery wastewater through phycoremediation. Environ. Technol. 40, 784–792.

Nagappan, S., Devendran, S., Tsai, P.-C., Dahms, H.-U., Ponnusamy, V.K., 2019. Potential of two-stage cultivation in microalgae biofuel production. Fuel 252, 339–349.

Nagarajan, D., Lee, D.-J., Chen, C.-Y., Chang, J.-S., 2020. Resource recovery from wastewaters using microalgae-based approaches: a circular bioeconomy perspective. Bioresour. Technol. 302, 122817.

Nagarajan, D., Lee, D.-J., Sunita Varjani, S., Lam, S., Allakhverdiev, S.I., Chang, J.-S., 2022. Microalgae-based wastewater treatment—microalgae-bacteria consortia, multi-omics approaches and algal stress response. Sci. Total Environ., 157110.

Nayak, J.K., Ghosh, U.K., 2019. Post treatment of microalgae treated pharmaceutical wastewater in photosynthetic microbial fuel cell (PMFC) and biodiesel production. Biomass Bioenergy 131, 105415.

Nayak, J.K., Ghosh, U.K., 2020. Microalgal remediation of anaerobic pretreated pharmaceutical wastewater for sustainable biodiesel production and electricity generation. J. Water Process Eng. 35, 101192.

Nishshanka, G.K., Hasara, S., Liyanaarachchi, V.C., Premaratne, M., Ariyadasa, T.U., Nimarshana, P.H.V., 2022. Sustainable cultivation of *Haematococcus pluvialis* and *Chromochloris zofingiensis* for the production of astaxanthin and co-products. Can. J. Chem. Eng. 100, 2835–2849.

Nithya, K., Sathish, A., Pradeep, K., Baalaji, S.K., 2019. Algal biomass waste residues of *Spirulina platensis* for chromium adsorption and modeling studies. J. Environ. Chem. Eng. 7, 103273.

Nowicka, B., 2022. Heavy metal-induced stress in eukaryotic algae—mechanisms of heavy metal toxicity and tolerance with particular emphasis on oxidative stress in exposed cells and the role of antioxidant response. Environ. Sci. Pollut. Res. 29, 16860–16911.

Nur, M.M.A., Garcia, G.M., Boelen, P., Buma, A.G.J., 2021. Influence of photodegradation on the removal of color and phenolic compounds from palm oil mill effluent by *Arthrospira platensis*. J. Appl. Phycol. 33, 901–915.

Nur, M.M., Azimatun, S.D., Rahmawati, I.W., Sari, Z.A., Setyoningrum, T.M., Jaya, D., Murni, S.W., Djarot, I.N., 2023. Enhancement of phycocyanin and carbohydrate production from *Spirulina platensis* growing on tofu wastewater by employing mixotrophic cultivation condition. Biocatal. Agric. Biotechnol., 102600.

Nzayisenga, J.C., Farge, X., Groll, S.L., Sellstedt, A., 2020. Effects of light intensity on growth and lipid production in microalgae grown in wastewater. Biotechnol. Biofuels 13, 1–8.

Ouada, S.B., Ali, R.B., Cimetiere, N., Leboulanger, C., Ouada, H.B., Sayadi, S., 2019. Biodegradation of diclofenac by two green microalgae: *Picocystis* sp. and *Graesiella* sp. Ecotoxicol. Environ. Saf. 186, 109769.

Palanisamy, M., Iniyakumar, M., Elaiyaraju, P., Mukund, S., Uthandi, S., Sivasubramanian, V., 2017. Effect of application of algal biochar on soil enzymes. J. Algal Biomass Util. 8, 1–9.

Pan, M., Lyu, T., Zhan, L., Matamoros, V., Angelidaki, I., Cooper, M., Pan, G., 2021. Mitigating antibiotic pollution using cyanobacteria: removal efficiency, pathways and metabolism. Water Res. 190, 116735.

Pandey, A., Srivastava, S., Kumar, S., 2019. Isolation, screening and comprehensive characterization of candidate microalgae for biofuel feedstock production and dairy effluent treatment: a sustainable approach. Bioresour. Technol. 293, 121998.

Park, J.B.K., Craggs, R.J., Shilton, A.N., 2011a. Recycling algae to improve species control and harvest efficiency from a high rate algal pond. Water Res. 45, 6637–6649.

Park, J.B.K., Craggs, R.J., Shilton, A.N., 2011b. Wastewater treatment high rate algal ponds for biofuel production. Bioresour. Technol. 102, 35–42.

Park, J.B.K., Craggs, R.J., Shilton, A.N., 2013. Enhancing biomass energy yield from pilot-scale high rate algal ponds with recycling. Water Res. 47, 4422–4432.

Parmar, P., Kumar, R., Neha, Y., Srivatsan, V., 2023. Microalgae as next generation plant growth additives: functions, applications, challenges and circular bioeconomy based solutions. Front. Plant Sci. 14.

Patel, A.K., Joun, J., Sim, S.J., 2020. A sustainable mixotrophic microalgae cultivation from dairy wastes for carbon credit, bioremediation and lucrative biofuels. Bioresour. Technol. 313, 123681.

Patel, A.K., Tambat, V.S., Chen, C.-W., Chauhan, A.S., Kumar, P., Vadrale, A.P., Huang, C.-Y., Dong, C.-D., Singhania, R.R., 2022. Recent advancements in astaxanthin production from microalgae: a review. Bioresour. Technol., 128030.

Perales-Vela, H.V., Peña-Castro, J.M., Canizares-Villanueva, R.O., 2006. Heavy metal detoxification in eukaryotic microalgae. Chemosphere 64, 1–10.

Phwan, C.K., Ong, H.C., Chen, W.-H., Ling, T.C., Ng, E.P., Show, P.L., 2018. Overview: comparison of pretreatment technologies and fermentation processes of bioethanol from microalgae. Energ. Conver. Manage. 173, 81–94.

Pooja, K., Priyanka, V., Chandra Sekhar Rao, B., Raghavender, V., 2022. Cost-effective treatment of sewage wastewater using microalgae *Chlorella vulgaris* and its application as bio-fertilizer. Energy Nexus 7, 100122.

Prasanna, R., Adak, A., Verma, S., Bidyarani, N., Babu, S., Pal, M., Shivay, Y.S., Nain, L., 2015. Cyanobacterial inoculation in rice grown under flooded and SRI modes of cultivation elicits differential effects on plant growth and nutrient dynamics. Ecol. Eng. 84, 532–541.

Purba, L.D., Amalia, F.S., Othman, A.Y., Mohamad, S.E., Iwamoto, K., Abdullah, N., Shimizu, K., Hermana, J., 2022. Enhanced cultivation and lipid production of isolated microalgae strains using municipal wastewater. Environ. Technol. Innov. 27, 102444.

Puri, M., Gupta, A., McKinnon, R.A., Abraham, R.E., 2022. Marine bioactives: from energy to nutrition. Trends Biotechnol. 40, 271–280.

Qiu, Y., Yilin, Z., Song, C., Xie, M., Qi, Y., Kansha, Y., Kitamura, Y., 2019. Soybean processing wastewater purification via *Chlorella* L166 and L38 with potential value-added ingredients production. Bioresour. Technol. Rep. 7, 100195.

Rambabu, K., Amos Avornyo, T., Gomathi, A.T., Show, P.L., Banat, F., 2022. Phycoremediation for carbon neutrality and circular economy: potential, trends, and challenges. Bioresour. Technol., 128257.

Rani, S., Gunjyal, N., Ojha, C.S.P., Singh, R.P., 2021. Review of challenges for algae-based wastewater treatment: strain selection, wastewater characteristics, abiotic, and biotic factors. J. Hazard. Toxic Radioact. Waste 25, 03120004.

Razzak, S.A., Lucky, R.A., Hossain, M.M., Hugo deLasa., 2022. Valorization of microalgae biomass to biofuel production: a review. Energy Nexus, 100139.

Rezaee, A., Ramavandi, B., Ganati, F., Ansari, M., Solimanian, A., 2006. Biosorption of mercury by biomass of filamentous algae *Spirogyra* species. J. Biol. Sci. 6, 695–700.

Rieger, P.-G., Meier, H.-M., Gerle, M., Vogt, U., Groth, T., Knackmuss, H.-J., 2002. Xenobiotics in the environment: present and future strategies to obviate the problem of biological persistence. J. Biotechnol. 94, 101–123.

Ronga, D., Biazzi, E., Parati, K., Carminati, D., Carminati, E., Tava, A., 2019. Microalgal biostimulants and biofertilisers in crop productions. Agronomy 9, 192.

Saavedra, R., Muñoz, R., Taboada, M.E., Vega, M., Bolado, S., 2018. Comparative uptake study of arsenic, boron, copper, manganese and zinc from water by different green microalgae. Bioresour. Technol. 263, 49–57.

Safafar, H., Van Wagenen, J., Møller, P., Jacobsen, C., 2015. Carotenoids, phenolic compounds and tocopherols contribute to the antioxidative properties of some microalgae species grown on industrial wastewater. Mar. Drugs 13, 7339–7356.

Safonova, E., Kvitko, K.V., Iankevitch, M.I., Surgko, L.F., Afti, I.A., Reisser, W., 2004. Biotreatment of industrial wastewater by selected algal-bacterial consortia. Eng. Life Sci. 4, 347–353.

Saravanan, A., Senthil Kumar, P., Sunita Varjani, S., Jeevanantham, P.R.Y., Thamarai, P., Abirami, B., George, C.S., 2021. A review on algal-bacterial symbiotic system for effective treatment of wastewater. Chemosphere 271, 129540.

Sarı, A., Uluozlü, Ö.D., Tüzen, M., 2011. Equilibrium, thermodynamic and kinetic investigations on biosorption of arsenic from aqueous solution by algae (*Maugeotia genuflexa*) biomass. Chem. Eng. J. 167, 155–161.

Sarma, S., Sharma, S., Rudakiya, D., Upadhyay, J., Rathod, V., Patel, A., Narra, M., 2021. Valorization of microalgae biomass into bioproducts promoting circular bioeconomy: a holistic approach of bioremediation and biorefinery. 3 Biotech 11, 1–29.

Satya, A.D., Maysarah, W.Y., Cheah, S.K., Yazdi, Y.-S.C., Khoo, K.S., Vo, D.-V.N., Bui, X.D., Vithanage, M., Show, P.L., 2023. Progress on microalgae cultivation in wastewater for bioremediation and circular bioeconomy. Environ. Res. 218, 114948.

Saxena, P., Bassi, A., 2013. Removal of nutrients from hydroponic greenhouse effluent by alkali precipitation and algae cultivation method. J. Chem. Technol. Biotechnol. 88, 858–863.

Schmidt, R.A., Wiebe, M.G., Eriksen, N.T., 2005. Heterotrophic high cell-density fed-batch cultures of the phycocyanin-producing red alga *Galdieria sulphuraria*. Biotechnol. Bioeng. 90, 77–84.

Sharma, G.K., Khan, S.A., Shrivastava, M., Bhattacharyya, R., Sharma, A., Gupta, D.K., Kishore, P., Gupta, N., 2021. Circular economy fertilization: phycoremediated algal biomass as biofertilizers for sustainable crop production. J. Environ. Manage. 287, 112295.

Sharma, R., Mishra, A., Pant, D., Malaviya, P., 2022. Recent advances in microalgae-based remediation of industrial and non-industrial wastewaters with simultaneous recovery of value-added products. Bioresour. Technol. 344, 126129.

Shekh, A., Sharma, A., Schenk, P.M., Kumar, G., Mudliar, S., 2022. Microalgae cultivation: photobioreactors, CO_2 utilization, and value-added products of industrial importance. J. Chem. Technol. Biotechnol. 97, 1064–1085.

Shen, X.-F., Gao, L.-J., Zhou, S.-B., Huang, J.-L., Chen-Zhi, W., Qin, Q.-W., Zeng, R.J., 2020. High fatty acid productivity from *Scenedesmus obliquus* in heterotrophic cultivation with glucose and soybean processing wastewater via nitrogen and phosphorus regulation. Sci. Total Environ. 708, 134596.

Shen, Y., Zhu, W., Li, H., Ho, S.-H., Chen, J., Xie, Y., Shi, X., 2018. Enhancing cadmium bioremediation by a complex of water-hyacinth derived pellets immobilized with *Chlorella* sp. Bioresour. Technol. 257, 157–163.

Khoubestani, S., Roghayeh, N.M., Farhadian, O., 2015. Removal of three and hexavalent chromium from aqueous solutions using a microalgae biomass-derived biosorbent. Environ. Prog. Sustain. Energy 34, 949–956.

Sibi, G., 2016. Biosorption of chromium from electroplating and galvanizing industrial effluents under extreme conditions using *Chlorella vulgaris*. Green Energy Environ. 1, 172–177.

Silambarasan, S., Logeswari, P., Sivaramakrishnan, R., Kamaraj, B., Chi, N.T.L., Cornejo, P., 2021. Cultivation of *Nostoc* sp. LS04 in municipal wastewater for biodiesel production and their deoiled biomass cellular extracts as biostimulants for *Lactuca sativa* growth improvement. Chemosphere 280, 130644.

Silva, A.G., Carter, R., Merss, F.L.M., Corrêa, D.O., Vargas, J.V.C., Mariano, A.B., Ordonez, J.C., Scherer, M.D., 2015. Life cycle assessment of biomass production in microalgae compact photobioreactors. GCB Bioenergy 7, 184–194.

Singh, S.K., Ajay Bansal, M.K., Jha, and Apurba Dey., 2012. An integrated approach to remove Cr (VI) using immobilized *Chlorella minutissima* grown in nutrient rich sewage wastewater. Bioresour. Technol. 104, 257–265.

Singla, R., Parmar, P., Anand, S.G., Vidyashankar, S.V., 2021. Application of microalgae for food supplements and animal feed: scientific, sustainability and socioeconomic challenges. In: Shekh, A., Dasgupta, S. (Eds.), Microalgal Biotechnology: Recent Advances, Market Potential, and Sustainability. The Royal Society of Chemistry, pp. 325–359.

Sinha, A., Goswami, G., Kumar, R., Das, D., 2021. A microalgal biorefinery approach for bioactive molecules, biofuel, and biofertilizer using a novel carbon dioxide-tolerant strain *Tetradesmus obliquus* CT02. Biomass Convers. Biorefinery, 1–14.

Sisman-Aydin, G., Simsek, K., 2022. Municipal wastewater effects on the performance of nutrient removal, and lipid, carbohydrate, and protein productivity of blue-green algae *Chroococcus turgidus*. Sustainability 14, 17021.

Sivasubramanian, V., Subramanian, V., Muthukumaran, M., 2012. Phycoremediation of effluent from a soft drink manufacturing industry with a special emphasis on nutrient removal—a laboratory study. J. Algal Biomass Util. 3, 21–29.

Solisio, C., Al Arni, S., Converti, A., 2019. Adsorption of inorganic mercury from aqueous solutions onto dry biomass of *Chlorella vulgaris*: kinetic and isotherm study. Environ. Technol. 40, 664–672.

Song, C., Xiaofang, H., Liu, Z., Li, S., Kitamura, Y., 2020. Combination of brewery wastewater purification and CO_2 fixation with potential value-added ingredients production via different microalgae strains cultivation. J. Clean. Prod. 268, 122332.

Subashchandrabose, S.R., Ramakrishnan, B., Megharaj, M., Venkateswarlu, K., Naidu, R., 2011. Consortia of cyanobacteria/microalgae and bacteria: biotechnological potential. Biotechnol. Adv. 29, 896–907.

Subhash, G.V., Meghna Rajvanshi, G., Kumar, R.K., Sagaram, U.S., Prasad, V., Govindachary, S., Dasgupta, S., 2022. Challenges in microalgal biofuel production: a perspective on techno economic feasibility under biorefinery stratagem. Bioresour. Technol. 343, 126155.

Sulaymon, A.H., Mohammed, A.A., Al-Musawi, T.J., 2013a. Competitive biosorption of lead, cadmium, copper, and arsenic ions using algae. Environ. Sci. Pollut. Res. 20, 3011–3023.

Sulaymon, A.H., Mohammed, A.A., Al-Musawi, T.J., 2013b. Column biosorption of lead, cadmium, copper, and arsenic ions onto algae. J. Bioprocess. Biotechniq. 3, 1–7.

Supraja, K.V., Behera, B., Balasubramanian, P., 2020. Performance evaluation of hydroponic system for co-cultivation of microalgae and tomato plant. J. Clean. Prod. 272, 122823.

Tan, X.-B., Zhao, X.-C., Zhang, Y.-L., Zhou, Y.-Y., Yang, L.-B., Zhang, W.-W., 2018. Enhanced lipid and biomass production using alcohol wastewater as carbon source for *Chlorella pyrenoidosa* cultivation in anaerobically digested starch wastewater in outdoors. Bioresour. Technol. 247, 784–793.

Tang, X., He, L.Y., Tao, X.Q., Dang, Z., Guo, C.L., Lu, G.N., Yi, X.Y., 2010. Construction of an artificial microalgal-bacterial consortium that efficiently degrades crude oil. J. Hazard. Mater. 181, 1158–1162.

Tao, R., Kinnunen, V., Praveenkumar, R., Lakaniemi, A.-M., Rintala, J.A., 2017. Comparison of *Scenedesmus acuminatus* and *Chlorella vulgaris* cultivation in liquid digestates from anaerobic digestion of pulp and paper industry and municipal wastewater treatment sludge. J. Appl. Phycol. 29, 2845–2856.

Taşkan, E., 2016. Performance of mixed algae for treatment of slaughterhouse wastewater and microbial community analysis. Environ. Sci. Pollut. Res. 23, 20474–20482.

Taufikurahman, T., Ilhamsyah, D.P.A., Rosanti, S., Ardiansyah, M.A., 2020. Preliminary design of phycocyanin production from *Spirulina platensis* using anaerobically digested dairy manure wastewater. IOP Conf. Ser. Earth Environ. Sci., 012007. IOP Publishing.

Tossavainen, M., Lahti, K., Edelmann, M., Eskola, R., Lampi, A.-M., Piironen, V., Korvonen, P., Ojala, A., Romantschuk, M., 2019. Integrated utilization of microalgae cultured in aquaculture wastewater: wastewater treatment and production of valuable fatty acids and tocopherols. J. Appl. Phycol. 31, 1753–1763.

Touliabah, H.E.-S., El-Sheekh, M.M., Ismail, M.M., El-Kassas, H., 2022. A review of microalgae- and cyanobacteria-based biodegradation of organic pollutants. Molecules 27, 1141.

Tripathi, B.N., Mehta, S.K., Amar, A., Gaur, J.P., 2006. Oxidative stress in *Scenedesmus* sp. during short-and long-term exposure to Cu^{2+} and Zn^{2+}. Chemosphere 62, 538–544.

Tuzen, M., Sarı, A., Mendil, D., Uluozlu, O.D., Soylak, M., Dogan, M., 2009. Characterization of biosorption process of As (III) on green algae *Ulothrix cylindricum*. J. Hazard. Mater. 165, 566–572.

Úbeda, B., Gálvez, J.Á., Michel, M., Bartual, A., 2017. Microalgae cultivation in urban wastewater: *Coelastrum* cf. *pseudomicroporum* as a novel carotenoid source and a potential microalgae harvesting tool. Bioresour. Technol. 228, 210–217.

Udayan, A., Pandey, A.K., Sirohi, R., Sreekumar, N., Sang, B.-I., Sim, S.J., Kim, S.H., Pandey, A., 2022. Production of microalgae with high lipid content and their potential as sources of nutraceuticals. Phytochem. Rev., 1–28.

Ullah, N., Ditta, A., Khalid, A., Mehmood, S., Rizwan, M.S., Ashraf, M., Mubeen, F., Imtiaz, M., Iqbal, M.M., 2020. Integrated effect of algal biochar and plant growth promoting rhizobacteria on physiology and growth of maize under deficit irrigations. J. Soil Sci. Plant Nutr. 20, 346–356.

Umamaheswari, J., Shanthakumar, S., 2021. Paddy-soaked rice mill wastewater treatment by phycoremediation and feasibility study on use of algal biomass as biofertilizer. J. Chem. Technol. Biotechnol. 96, 394–403.

Ummalyma, S.B., Sirohi, R., Udayan, A., Yadav, P., Raj, A., Sim, S.J., Pandey, A., 2022. Sustainable microalgal biomass production in food industry wastewater for low-cost biorefinery products: a review. Phytochem. Rev., 1–23.

Usha, M.T., Sarat Chandra, T., Sarada, R., Chauhan, V.S., 2016. Removal of nutrients and organic pollution load from pulp and paper mill effluent by microalgae in outdoor open pond. Bioresour. Technol. 214, 856–860.

Vale, F., Sousa, C.A., Sousa, H., Santos, L., Simões, M., 2022. Impact of parabens on microalgae bioremediation of wastewaters: a mechanistic study. Chem. Eng. J. 442, 136374.

Vandamme, D., Foubert, I., Muylaert, K., 2013. Flocculation as a low-cost method for harvesting microalgae for bulk biomass production. Trends Biotechnol. 31, 233–239.

Vasconcelos, V.M., Pereira, E., 2001. Cyanobacteria diversity and toxicity in a wastewater treatment plant (Portugal). Water Res. 35, 1354–1357.

Venugopal, V., 2022. Green processing of seafood waste biomass towards blue economy. Curr. Res. Environ. Sustain. 4, 100164.

Vidyashankar, S., Deviprasad, K., Chauhan, V.S., Ravishankar, G.A., Sarada, R., 2013. Selection and evaluation of CO₂ tolerant indigenous microalga *Scenedesmus dimorphus* for unsaturated fatty acid rich lipid production under different culture conditions. Bioresour. Technol. 144, 28–37.

Vidyashankar, S., Ravishankar, G.A., 2016. Algae-based bioremediation: bioproducts and biofuels for biobusiness. Bioremediat. Bioecon., 457–493.

Wang, C., Zheng, Y., Li, R., Yin, Q., Song, C., 2022a. Removal of cefradine by *Chlorella* sp. L166 and *Scenedesmus quadricauda*: toxicity investigation, degradation mechanism and metabolic pathways. Process. Saf. Environ. Prot. 160, 632–640.

Wang, D., Jiang, P., Zhang, H., Yuan, W., 2020a. Biochar production and applications in agro and forestry systems: a review. Sci. Total Environ. 723, 137775.

Wang, J., Wang, Y., Ziqiang, G., Mou, H., Sun, H., 2023. Stimulating carbon and nitrogen metabolism of *Chlorella pyrenoidosa* to treat aquaculture wastewater and produce high-quality protein in plate photobioreactors. Sci. Total Environ. 878, 163061.

Wang, J., Xinge, H., Chen, J., Wang, T., Huang, X., Chen, G., 2022b. The extraction of β-carotene from microalgae for testing their health benefits. Foods 11, 502.

Wang, R., Diao, P., Chen, Q., Hao, W., Ning, X., Duan, S., 2017. Identification of novel pathways for biodegradation of bisphenol A by the green alga *Desmodesmus* sp. WR1, combined with mechanistic analysis at the transcriptome level. Chem. Eng. J. 321, 424–431.

Wang, S.-K., Tian, Y.-T., Dai, Y.-R., Wang, D., Liu, K.-C., Cui, Y.-H., 2022c. Development of an alternative medium via completely replaces the medium components by mixed wastewater and crude glycerol for efficient production of docosahexaenoic acid by Schizochytrium sp. Chemosphere 291, 132868.

Wang, Y., Wang, S., Sun, L., Sun, Z., Li, D., 2020b. Screening of a *Chlorella*-bacteria consortium and research on piggery wastewater purification. Algal Res. 47, 101840.

Wei, C., Wang, S., Xiaochen, H., Zhang, J., Chen, J., Zhang, Q., Wang, Y., Wei, J., 2021. Exposure to dibutyl phthalate induced the growth inhibition and oxidative injury in Dunaliella salina. IOP Conf. Ser. Earth Environ. Sci., 042038. IOP Publishing.

Wu, P., Zhang, Z., Luo, Y., Bai, Y., Fan, J., 2022. Bioremediation of phenolic pollutants by algae-current status and challenges. Bioresour. Technol., 126930.

Wuang, S.C., Khin, M.C., Chua, P.Q.D., Luo, Y.D., 2016. Use of *Spirulina* biomass produced from treatment of aquaculture wastewater as agricultural fertilizers. Algal Res. 15, 59–64.

Xiao, M., Ma, H., Sun, M., Yin, X., Feng, Q., Song, H., Gai, H., 2019. Characterization of cometabolic degradation of p-cresol with phenol as growth substrate by *Chlorella vulgaris*. Bioresour. Technol. 281, 296–302.

Xie, P., Chen, C., Zhang, C., Guanyong, S., Ren, N., Ho, S.-H., 2020. Revealing the role of adsorption in ciprofloxacin and sulfadiazine elimination routes in microalgae. Water Res. 172, 115475.

Xiong, J.-Q., Kurade, M.B., Jeon, B.-H., 2017. Biodegradation of levofloxacin by an acclimated freshwater microalga, *Chlorella vulgaris*. Chem. Eng. J. 313, 1251–1257.

Xiong, Q., Li-Xin, H., Liu, Y.-S., Zhao, J.-L., He, L.-Y., Ying, G.-G., 2021. Microalgae-based technology for antibiotics removal: from mechanisms to application of innovational hybrid systems. Environ. Int. 155, 106594.

Yadavalli, R., Heggers, G.R.V.N., 2013. Two stage treatment of dairy effluent using immobilized *Chlorella pyrenoidosa*. J. Environ. Health Sci. Eng. 11, 1–6.

Yan, C., Zhengzhe, Q., Wang, J., Cao, L., Han, Q., 2022. Microalgal bioremediation of heavy metal pollution in water: recent advances, challenges, and prospects. Chemosphere 286, 131870.

Yang, C., Li, R., Zhang, B., Qiu, Q., Wang, B., Yang, H., Ding, Y., Wang, C., 2019. Pyrolysis of microalgae: a critical review. Fuel Process. Technol. 186, 53–72.

Yang, J.S., Cao, J., Xing, G.L., Yuan, H.L., 2015. Lipid production combined with biosorption and bioaccumulation of cadmium, copper, manganese and zinc by oleaginous microalgae *Chlorella minutissima* UTEX2341. Bioresour. Technol. 175, 537–544.

Yang, Z.-Y., Gao, F., Liu, J.-Z., Yang, J.-S., Liu, M., Ge, Y.-M., Chen, D.-Z., Chen, J.-M., 2022. Improving sedimentation and lipid production of microalgae in the photobioreactor using saline wastewater. Bioresour. Technol. 347, 126392.

Yen, H.-.W., Ho, S.-.H., Chen, C.-.Y., Chang, J.-.S., 2015. CO_2, NOx and SOx removal from flue gas via microalgae cultivation: a critical review. Biotechnol. J. 10, 829–839.

You, X., Nan, X., Yang, X., Sun, W., 2021. Pollutants affect algae-bacteria interactions: a critical review. Environ. Pollut. 276, 116723.

Yu, W., Wen, G., Lin, H., Yang, Y., Huang, X., Zhou, C., Zhang, Z., Duan, Y., Huang, Z., Li, T., 2018. Effects of dietary *Spirulina platensis* on growth performance, hematological and serum biochemical parameters, hepatic antioxidant status, immune responses and disease resistance of Coral trout *Plectropomus leopardus* (Lacepede, 1802). Fish Shellfish Immunol. 74, 649–655.

Zhang, A.H., Feng, B., Zhang, H., Jiang, J., Zhang, D., Yi, D., Cheng, Z., Huang, J., 2022. Efficient cultivation of *Porphyridium purpureum* integrated with swine wastewater treatment to produce phycoerythrin and polysaccharide. J. Appl. Phycol. 34, 2315–2326.

Zhang, L., Pei, H., Yang, Z., Wang, X., Chen, S., Li, Y., Xie, Z., 2019a. Microalgae nourished by mariculture wastewater aids aquaculture self-reliance with desirable biochemical composition. Bioresour. Technol. 278, 205–213.

Zhang, Z., Yan, K., Zhang, L., Wang, Q., Guo, R., Yan, Z., Chen, J., 2019b. A novel cadmium-containing wastewater treatment method: bio-immobilization by microalgae cell and their mechanism. J. Hazard. Mater. 374, 420–427.

Zheng, H., Guo, W., Li, S., Chen, Y., Qinglian, W., Feng, X., Yin, R., Ho, S.-H., Ren, N., Chang, J.-S., 2017. Adsorption of p-nitrophenols (PNP) on microalgal biochar: analysis of high adsorption capacity and mechanism. Bioresour. Technol. 244, 1456–1464.

Zheng, H., Guo, W., Li, S., Qinglian, W., Yin, R., Feng, X., Juanshan, D., Ren, N., Chang, J.-S., 2016. Biosorption of cadmium by a lipid extraction residue of lipid-rich microalgae. RSC Adv. 6, 20051–20057.

Zhou, T., Cao, L., Zhang, Q., Liu, Y., Xiang, S., Liu, T., Ruan, R., 2021. Effect of chlortetracycline on the growth and intracellular components of *Spirulina platensis* and its biodegradation pathway. J. Hazard. Mater. 413, 125310.

Zhou, W., Min, M., Li, Y., Bing, H., Ma, X., Cheng, Y., Liu, Y., Chen, P., Ruan, R., 2012. A hetero-photoautotrophic two-stage cultivation process to improve wastewater nutrient removal and enhance algal lipid accumulation. Bioresour. Technol. 110, 448–455.

Zuliani, L., Frison, N., Jelic, A., Fatone, F., Bolzonella, D., Ballottari, M., 2016. Microalgae cultivation on anaerobic digestate of municipal wastewater, sewage sludge and agro-waste. Int. J. Mol. Sci. 17, 1692.

Bioprocesses and bioengineering for boosting biobased economy

Building circular bio-based economy through sustainable waste management

K. Amulya[a], Shikha Dahiya[b], and S. Venkata Mohan[c]

[a]National University of Ireland, Galway, Ireland [b]Biochemical Process Engineering, Chemical Engineering, Department of Civil, Environmental and Natural Resources Engineering, Lulea Technical University, Luleå, Sweden [c]Bioengineering and Environmental Sciences Lab, Department of Energy and Environmental Engineering (DEE), CSIR-Indian Institute of Chemical Technology (CSIR-IICT), Hyderabad, Telangana, India

1 Introduction

Over the past few decades, a profound transformation has been observed in the relationship between humans and the natural environment (Donner et al., 2020). The economy on a global scale has also dramatically expanded over the past half century (Korhonen et al., 2018). This economic expansion was possible due to the fossil fuels and energy inputs that aided in the easy extraction of resources, their transformation into products, and the transport of these products across the globe (Girardet, 2020). The profits obtained were reinvested in extracting more resources, establishing a linear economic model of wasteful production and consumption patterns. Assessments conducted by various global bodies, such as the ecological footprint analysis (Global Footprint Network, 2016; Katakojwala et al., 2023), planetary boundaries (Rockström et al., 2009), and Intergovernmental Panel on Climate Change (2014) (IPCC, 2014), suggest that extracting fossil fuels, manufacturing products, and disposing waste will undermine the future prosperity of the planet. Leading to grand challenges including climate change, ecological disorders, biodiversity loss, acidification of oceans and wide-scale habitat destruction (Kelly et al., 2020).

As these side effects of the grand challenges are compounding rapidly, various strategies are being put forth to devise an economic model that encourages growth in the economy without compromising the quality of the environment (Iacovidou et al., 2017). The concept of a circular economy (CE) has gained favorable momentum in this direction as an alternative economic model that favors economic prosperity while encouraging production and consumption patterns that have negligible impact on the environment (Webster, 2015). CE can be defined as an economic model that is restorative and regenerative by design, relies on a systemic perspective, and aims to design out waste by focusing on restoration and decoupling economic growth from the use of nonrenewable resources (EMF, 2015). In this economic model, by classifying waste into biological and technical "nutrients" that circulate in loops or cascading approaches, the amount of waste generated and the extraction of new raw materials can be significantly reduced (EMF, 2012). In this future economic model, a general consensus was reached to introduce biological renewable waste as a core element, favoring sustainable development and also providing a solution to deal with the huge quantities of waste being generated worldwide (Vanhamäki et al., 2020). Thus, an economic model wherein biological renewable resources such as wastes are valorized into energy, value-added products, chemicals, and materials can be termed circular bioeconomy (CBE) (Lokesh et al., 2018).

Globally, wastes generated can be classified into agricultural residues, municipal solid waste (MSW), industrial waste, hazardous waste, demolition and construction waste, medical waste, electronic waste, and gaseous emissions (CO_2) (Usmani et al., 2020). Among these, industrial waste is generated the most, while agricultural waste is produced nearly five times as much as MSW (Usmani et al., 2020). Counties in which agriculture contributes a huge share of the economy primarily produce agricultural organic waste (Usmani et al., 2020). Hazardous waste generally originates from manufacturing industries and hospitals, and these wastes require specialized facilities or disassembly centers for their treatment. Demolition and construction wastes are less frequently upcycled and are usually sent to landfills (Usmani et al., 2020). Among these, wastes with organic carbon content can be used as feedstocks to advance the CBE. Various feedstocks that can be incorporated into a CBE are discussed in Fig. 1. These renewable feedstocks can be processed and converted into new products in the CBE. In addition to reuse and recovery, the management and minimization of waste also take a prominent position in the vision of a CBE.

In line with the earlier mentioned facts, this chapter attempts to elucidate the sustainable technologies that could maximize the value of waste by producing industrially important chemicals, materials, and fuels that not only reduce our dependency on fossil-based fuels and products but also help in mitigating environmental pollution. It also discusses the cascading approach that could be implemented using these feedstocks for the development of waste biorefineries. Since sustainability and economics are important parameters that need consideration, the chapter further discusses the various aspects that need to be taken into account for developing full-scale, sustainable, and cost-effective waste biorefineries.

2 Value addition from waste

2.1 Acidogenic fermentation

Acidogenic fermentation/acidogenesis has gained immense interest in recent times due to its ability to convert organic matter into a spectrum of value-added bioproducts like mixtures

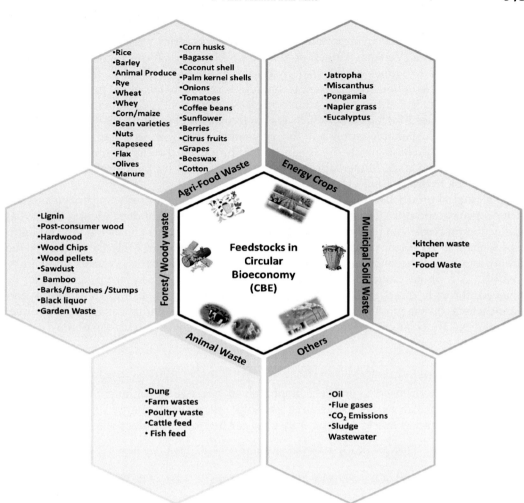

FIG. 1 Various renewable feedstocks that contribute for a circular bioeconomy (CBE).

of volatile fatty acids (VFAs), biohydrogen (bio-H_2), alcohols, and other by-products (Dahiya et al., 2018). The wide range of applications of these acidogenic products are being used as commodity/platform chemicals (VFAs) or biofuels (bio-H_2 and alcohol) (Venkata Mohan et al., 2016). Acidogenic fermentation is now majorly exploited for bio-H_2 and VFA production. Efficiency and sustainability have been the focus of recent developments in acidogenic fermentation and the generation of its products, which have also addressed challenges associated with process optimization (Venkata Mohan et al., 2016). The production and optimization of two major acidogenic fermentation products, bio-H_2 and VFAs, are explained in the following sections.

2.1.1 Bio-hydrogen

Hydrogen is considered an important alternative energy carrier and a bridge to sustainable energy in the future (Venkata Mohan et al., 2009c). Hydrogen has a high energy yield of 122 kJ/g, which is 2.75 times greater than hydrocarbon fuels (Tawfik et al., 2011). Hydrogen production is carried out using different technologies, of which half of the production is carried out by thermocatalytic or gasification processes using fossil fuels. Alternatively, hydrogen can be produced biologically, which is also termed bio-H_2. Biological production is a promising and sustainable approach for generating hydrogen, a clean energy source (Sarkar et al., 2018). Biologically, hydrogen can be produced through two main pathways: photosynthesis and dark fermentation. Hydrogen generated by biological machinery can be termed "biohydrogen" (bio-H_2).

All biological production methods are controlled by hydrogen-producing enzymes, such as hydrogenase and nitrogenase (Skizim et al., 2012). Based on light dependency, the bio-H_2 production processes can be classified into light-independent (dark) fermentation and light-dependent photosynthetic processes (Chandra and Venkata Mohan, 2014). Depending on the carbon source and the microorganisms involved, the photosynthetic process can again be classified as a photosynthetic or fermentation process. Light-dependent processes can be achieved through direct and indirect biophotolysis of water using green algae and cyanobacteria or via photofermentation mediated by photosynthetic bacteria. In direct biophotolysis, the solar energy directly converts water into hydrogen via the photosynthetic reaction (Eq. 1).

$$H_2O + hv \text{ (solar energy)} \rightarrow 2\,H_2 + O_2 \tag{1}$$

The cyanobacteria used in indirect biophotolysis possess the unique characteristics of using CO_2 from the air as a carbon source and solar energy as an energy source to produce biomass (Eqs. 2 and 3) that is subsequently used for hydrogen production.

$$6H_2O + 6CO_2 + hv \text{ (solar energy)} \rightarrow C_6H_{12}O_6 + 6O_2 \tag{2}$$

$$C_6H_{12}O_6 + 6H_2O + hv \text{ (solar energy)} \rightarrow 12H_2 + 6CO_2 \tag{3}$$

Photosynthetic bacteria use simple organic acids (acetic and butyric acids) as electron donors and liberate molecular hydrogen using the nitrogenase enzyme under nitrogen-deficient, anaerobic conditions and in the presence of light energy. Dark fermentative hydrogen production is carried out by obligate and facultative anaerobes under strictly anaerobic conditions. The dark fermentation process is confined to anaerobic metabolism, where anaerobic microorganisms (mostly acidogenic bacteria) generate H_2 metabolically through an acetogenic process along with the generation of VFAs and CO_2 (Eqs. 4–8).

$$C_6H_{12}O_6 + 2H_2O \rightarrow 2CH_3COOH + 2CO_2 + 4H_2 \tag{4}$$

$$C_6H_{12}O_6 \rightarrow CH_3CH_2CH_2COOH + 2\,CO_2 + 2\,H_2 \tag{5}$$

$$C_6H_{12}O_6 + 2H_2 \rightarrow 2CH_3CH_2COOH + 2\,H_2O \tag{6}$$

$$C_6H_{12}O_6 + 2H_2 \rightarrow COOHCH_2CH_2OCOOH + CO_2 \tag{7}$$

$$C_6H_{12}O_6 \rightarrow CH_3CH_2OH + CO_2 \tag{8}$$

Recent advances in the production and scale-up of bio-H_2 have focused on improving efficiency and reducing costs (Dahiya et al., 2021). Acidogenic fermentation for bio-H_2 production can use a variety of substrates, like carbohydrates (derived from sugarcane, corn, wheat, and other starchy crops), lignocellulosic biomass (such as wood, sawdust, and agricultural waste), algal biomass, and organic wastes (such as sewage sludge, food waste, and animal manure) (Saratale et al., 2008; Venkata Mohan et al., 2009a; Subhash and Venkata Mohan, 2014a; Venkata Mohan, 2009a; Katakojwala and Venkata Mohan, 2022; Rathi et al., 2022). Biomass derived from plant crops, agricultural residues, and woody biomass are also being used for generating bio-H_2 by both thermochemical and biological routes (Venkata Mohan et al., 2009b; Subhash and Venkata Mohan, 2014b; Venkata Mohan, 2009b). In order to make the organic fraction available for the bacteria, the cellulosic material requires an initial pretreatment step, which adds to the cost of the entire process (Mohanakrishna and Modestra, 2023). Unlike cellulosic materials, food waste contains readily available carbon, thus making it a highly preferred substrate for bio-H_2 production. The dark fermentation process is widely used for bio-H_2 production as it is relatively less energy-intensive and more environmentally sustainable due to its operational feasibility (Fig. 2) (Mohanakrishna and Venkata Mohan, 2013). Despite the advantages, low substrate conversion efficiency, accumulation of

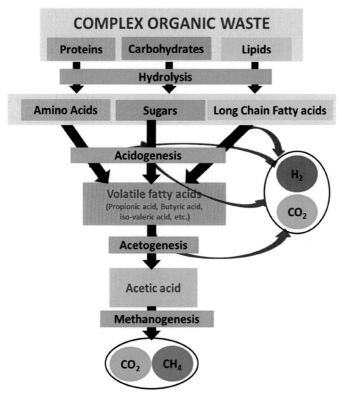

FIG. 2 Schematic representation of the acidogenic fermentation process, generating volatile fatty acids and biohydrogen.

carbon-rich acid intermediates, and drop in the system pH impede upscaling of bio-H_2 production (Pasupuleti et al., 2014). If these factors are optimized, then they can be game changers, and bio-H_2 production at a practical scale in a cost-effective manner with high efficiency might become a reality.

The scientific community has now established the acidogenic fermentation platform for scale-up, but still improvements are required to reach the Thauer limit (Ergal et al., 2020). Among the strategies adapted, a few of them include, selecting feasible strains or genetic engineering (Kamshybayeva et al., 2023), designing microbial consortia (Ergal et al., 2020), optimizing the bioreactor designs (Asrul et al., 2022), process modeling (Yahaya et al., 2022), pretreatment of feedstocks (Kovalev et al., 2022), and integration with other processes (Ubando et al., 2022). However, further research is needed to fully optimize these processes and ensure their economic viability. Also, the major focus in bio-H_2 production is now on production, transportation, and storage facilities (Dahiya et al., 2020). bio-H_2 production can be limited by factors such as low yields, low reaction rates, and the availability of suitable feedstocks (Dahiya et al., 2021). Additionally, there are challenges associated with the purification of bio-H_2 gas from other by-products, such as carbon dioxide and methane (Chai et al., 2021). Lower yields, slow reaction rates, and the availability of potential feedstocks can all have an impact on the amount of bio-H_2 that can be produced (Prabakar et al., 2018). Since bio-H_2 production is performed in smaller and dispersed facilities, it can pose challenges for transportation to end users. High-pressure tanks or cryogenic storage are needed for transporting hydrogen gas, which can be costly and unsafe. Also, there is currently a dearth of infrastructure, like pipelines and refueling facilities, for the transportation of hydrogen. Another challenge is its storage, due to its low density and high reactivity (Thepsithar et al., 2020). It is currently stored at high pressures or at cryogenic temperatures, which require expensive and specialized equipment (Thepsithar et al., 2020). Other storage technologies, like storing hydrogen in carrier molecules like carbon nanotubes, metal hydrides, formic acid, and methanol, are still being developed (He et al., 2016). Another challenge that needs attention is the cost of bio-H_2. However, the price of bio-H_2 is anticipated to drop as production technology advances and economies of scale are achieved. Ongoing research and development are needed to overcome these challenges and ensure the safety and economic viability of biohydrogen production, storage, and transportation. With petroleum-based fuels reaching the verge of depletion and the concern about continued emissions of additional carbon dioxide into the atmosphere intensifying, bio-hydrogen from waste materials will be considered a key technology for future sustainable energy supply. Renewable shares of 36% (2025) and 69% (2050) of the total energy demand will lead to hydrogen shares of 11% in 2025 and 34% in 2050. With increasing reports on bio-H_2 production and several funded programs from many national government agencies all over the world, it is quite evident that hydrogen is being propagated as the fuel of the future (Balat and Kırtay, 2010).

2.1.2 Volatile fatty acids

Considering the various possibilities of resource recovery, the production of VFAs as intermediates of anaerobic digestion (AD) from different organic wastes is deemed to be one of the prominent areas with good commercial interest (Fig. 2) (Singhania et al., 2013). VFA are short-chain fatty acids (acetic, propionic, butyric, etc.) consisting of six or fewer carbon atoms

(C2–C6) with pK_a values between 4.7 and 4.9 and can be distilled at atmospheric pressure. VFA can be produced chemically or synthetically from fossil resources, or as metabolic intermediates during acidogenic/dark fermentation (acidogenesis, acetogenesis, and methanogenesis). These VFAs are a part of bacterial metabolism and can be produced either using mixed cultures or by pure strains under anaerobic conditions. Although commercial production of VFA is mostly accomplished by chemical routes (Huang et al., 2002), its production using biological methods and organic-rich wastes is steadily increasing. Biologically produced VFA, when separated properly, can be used as one of the most important and valuable chemical compounds with diverse applications in the market.

In biological VFA production, substrates ranging from pure sugars, viz., glucose and sucrose, to organic-rich wastes such as sludge generated from wastewater treatment plants, food waste, MSW, and industrial wastewaters have been commonly employed as the main carbon source for VFA production (Dahiya et al., 2015; Amulya et al., 2014b; Reddy et al., 2011; Dahiya et al., 2023). Production of VFAs is the function of individual bacteria in a pure culture or the syntrophic relationship between the groups of organisms present in the open microbiome. In an open-mixed microbiome, acetic acid production generally occurs through the action of acetogens. These acetogens can degrade long-chain fatty acids into acetic acid and H_2. These acidogens are mainly of two types. Obligate H_2-producing acetogenic bacteria, which are capable of producing acetate and H_2 from higher fatty acids. Second are the hydrogen-consuming acetogenic bacteria such as *Clostridium aceticum*, which are also considered homoacetogens. These bacteria utilize H_2 and CO_2 to form acetic acid. In general, the overall equation for the production of acetic acid from glucose is given by Eq. (9).

$$C_6H_{12}O_6 + 2NAD^+ + 2ADP + 2\,Pi \rightarrow 2CH_3COCOO^- + 2NADH + 2ATP + 2H_2O + 2H^+ \quad (9)$$

In broader terms, acetogens are obligately anaerobic bacteria that use the reductive acetyl-CoA or Wood-Ljungdahl pathway as their main mechanism for energy conservation and for synthesis of acetyl-CoA and cell carbon from CO_2. An acetogen is sometimes called a "homoacetogen" (meaning that it produces only acetate as its fermentation product) or a "CO_2-reducing acetogen." The Wood-Ljungdahl pathway is the hallmark of acetogens but is also used by other bacteria, including methanogens and sulfate-reducing bacteria, for both catabolic and anabolic purposes. The key aspects of the acetogenic pathway are several reactions that include the reduction of carbon dioxide to carbon monoxide and the attachment of the carbon monoxide to a methyl group. The first process is catalyzed by an enzyme called carbon monoxide dehydrogenase. The coupling of the methyl group (provided by methylcobalamin) and the CO is catalyzed by acetyl CoA synthetase.

$$2CO_2 + 4H_2 \rightarrow CH_3COOH + 2H_2O \quad\quad\quad (10)$$

Syntrophic bacteria produce acetic acid from higher fatty acids (propionic acid and butyric acid). Two pathways for propionate metabolism are known: the methylmalonyl-CoA pathway and the dismutation pathway, where two propionate molecules are converted to acetate and butyrate, with the butyrate being degraded to acetate and hydrogen. In the methylmalonyl-CoA pathway, propionate is first activated to propionyl-CoA and then carboxylated to methylmalonyl-CoA. Methylmalonyl-CoA is rearranged to form succinyl-CoA, which is converted to succinate. Succinate is oxidized to fumarate, which is then hydrated to malate and oxidized to oxaloacetate. Pyruvate is formed by decarboxylation and is further oxidized in an HS-CoA-dependent decarboxylation to acetyl-CoA and finally to acetate. Butyrate is degraded via the β-oxidation pathway (McInerney et al., 2008). In a series of reactions, acetyl groups are cleaved off, yielding acetate and hydrogen. To metabolize fatty acids, the first activation of a HS-CoA derivative takes place. The HS-CoA derivative is then dehydrogenated to form an enoyl-CoA. After water addition, a second dehydrogenation takes place to form ketoacylacetyl-CoA. After hydrolysis, acetyl-CoA and an acyl-CoA are formed, which enters another cycle of dehydrogenation and the cleaving off of acetyl-CoA, leading to acetic acid generation.

Propionic acid is produced by the propionic acid bacteria, which include *Corynebacteria*, *Propionibacterium* and *Bifidobacterium*. The formation of propionate is a complex and indirect process involving five or six reactions. Overall, three moles of lactate are converted to two moles of propionate+one mole of acetate+one mole of CO_2, and one mole of ATP is squeezed out in the process. Most of the butyric acid-producing bacteria produce acetic acid in addition to butyric acid as their major fermentation products. In general, glucose (hexose) is catabolized via the Embden-Meyerhof-Parnas pathway, and xylose (pentose) is catabolized by the Hexose Monophosphae pathway to pyruvate, which is then oxidized to acetyl-CoA and carbon dioxide with the concomitant reduction of ferredoxin (Fd) to FdH_2. FdH_2 is then oxidized to Fd to produce hydrogen, which is catalyzed by hydrogenase, with excess electrons released to convert NAD^+ to NADH. Acetyl-CoA is the key metabolic intermediate at the node dividing the acetate formation branch from the butyrate formation branch. Phosphotransacetylase and acetate kinase are two enzymes that convert acetyl-CoA to acetic acid, whereas phosphotransbutyrylase and butyrate kinase catalyze the production of butyric acid from butyryl-CoA.

VFA production is still in the optimization stage, where substrate variability, microbial structure, product feedback inhibition, process parameters, and product recovery are still challenges for scalability (Dahiya et al., 2023; Atasoy et al., 2018; Nagarajan et al., 2022). The quality and composition of the substrate used for VFA production can vary widely, which can affect the yield and quality of the VFAs produced (Vázquez-Fernández et al., 2022). To overcome this challenge, it is important to optimize the selection and preparation of the substrate as well as the operating conditions of the production process. Furthermore, parameters like pH, temperature, HRT, and substrate feedback inhibition might impact the metabolism of microorganisms involved in the production of VFA (Lv et al., 2022). This can be overcome by choosing and engineering microbial strains that are more resistant to inhibiting substances and optimizing the environment for microbial growth and metabolism (Bhatia and Yang, 2017). The recovery of products from acidogenic fermentation is a major challenge. Recent research has focused on developing novel separation technologies, including membrane filtration, solvent extraction, and adsorption, to selectively recover VFAs and other products from the fermentation broth (Dahiya et al., 2023; Bhatia and Yang, 2017; Atasoy et al., 2018). VFAs are primarily produced biologically at laboratory or pilot scales, which can restrict the process's scalability and commercial viability. Developing and improving production procedures that can be scaled up to commercial scales while retaining high yields and product quality can address these challenges (Dahiya et al., 2023).

The VFA produced from acidogenic fermentation of waste is the building block for many chemical synthesis reactions and serves as a valuable substrate for a variety of applications, such as the production of biodegradable plastics, the generation of bioenergy, and biological nutrient removal. It is often feasible, and certainly highly desirable, to utilize the VFA-rich fermented waste directly in these applications. VFAs serve as a potential substrate for the production of hydrogen via photofermentation by mixed microbial cultures (Srikanth et al., 2009b). VFA can also serve as substrate for microbial fuel cells (MFCs) (Chen et al., 2012), as only soluble organic compounds can be directly fed into the anode of MFCs to generate electricity, and single-chain fatty acids are the most preferred (Venkata Mohan et al., 2008b). Apart from bio-H_2 and electricity, they can also serve as a carbon source for lipid storage in microalgae (Venkata Mohan and Devi, 2012), the synthesis of single-cell oils by oleaginous yeast (Christophe et al., 2012), and the production of bioplastics (polyhydraoxyalkanates). In addition, VFA also assists the biological removal of nitrogen and phosphorus from wastewater through aerobic nitrification followed by anoxic denitrification. All these applications show the potential of VFA to bring about a revolution in the field of bioenergy. VFA production can be considered an effective strategy to bridge the gap between waste remediation and product recovery.

2.2 Biomethane

Anaerobic microbes are employed in the biomethanation process to convert organic matter into methane (CH_4) gas using a series of microbial consortia, which can be used as a sustainable energy source. Current developments in biomethanation have concentrated on improving biogas quality, process efficiency, and the variety of feedstocks that can be employed (Mahla et al., 2022a). Maintaining ideal conditions for microbial growth and activity is one of the fundamental issues in biomethanation. In order to increase process efficiency, recent research has concentrated on optimizing process variables like temperature, pH, and nutrient supply. Co-digestion of various feedstocks, such as industrial effluents, food waste, and agricultural wastes, can increase the biomethanation process' efficiency and stability (Azarmanesh et al., 2023). Finding the best feedstock combinations and creating efficient co-digestion procedures have been the main topics of recent research (Azarmanesh et al., 2023). The efficiency and stability of the biomethanation process can be improved by engineering microbial populations. In order to maximize methane production and improve the decomposition of complex organic materials, recent research has concentrated on designing synthetic microbial communities utilizing genetic engineering and synthetic biology techniques (Yadav et al., 2022). Also, depending on the feedstock utilized and the effectiveness of the process, the quality of the biogas generated during the biomethanation process can change (Bareha et al., 2022). The development of more effective and affordable biogas purification technologies to eliminate contaminants like carbon dioxide and hydrogen sulfide has been the focus of recent research (Mahla et al., 2022b). Recent advances have shown that biomethanation can be coupled with carbon capture and utilization technologies to further reduce greenhouse gas emissions and enhance the economic viability of the process (Menin et al., 2022).

2.3 Biohythane

A renewable energy source known as biohythane combines the advantages of hydrogen and biogas. It is produced using a two-step process that starts with AD and ends with dark fermentation (Santhosh et al., 2021). Current developments in the production of biohythane have now focused on enhancing the process's sustainability and efficiency while also addressing issues with feedstock availability, process optimization, and product recovery (Rawoof et al., 2021). Biohythane can be effectively produced from a variety of organic waste materials, including sewage sludge, food waste, animal manure, crop residues, and industrial waste (Lay et al., 2020). Like the other biofuels, changes in the composition of the feedstock, pH, temperature, and other variables might affect the stability of the biohythane-generating process, which can affect the productivity and quality of the biogas and hydrogen produced (Shanmugam et al., 2021). The consumption of the biogas and hydrogen produced during the biohythane-generating process may be constrained due to the availability of acceptable end-use applications, such as transportation fuel or electricity generation. It might be difficult and expensive to upgrade and purify biohythane to get rid of contaminants and boost its energy content; thus, several routes for its purification are being adapted (Luo et al., 2017). Further, the infrastructure and distribution networks for biohythane are not as well developed as those for conventional fossil fuels, making it more difficult to transport and store biohythane (Pandey et al., 2019). Due to this, even though biohythane generation has enormous promise as a renewable energy source, more research and development are needed to solve these problems and ensure the process's economic and environmental viability.

2.4 Biopolymers

The flexibility and durability of synthetic plastics have made them one of the most widely used materials in a variety of industrial and day-to-day applications (Keshavarz and Roy, 2010). Ever since their discovery, they have been used indiscriminately, but now people have realized the harmful impact of these plastics on the environment. Environmental issues like global warming and the greenhouse effect are caused when these materials are used for a short time span and then are often dumped in landfills or incinerated. In addition to the environmental issues, the dwindling price of crude oil is another factor of immense uncertainty, especially for the production of petroleum-based plastics. In order to avert the problems associated with synthetic plastics, the demand for developing biodegradable polymeric materials is increasing.

Bio-based plastics are polymeric materials produced from renewable resources. These can be either biodegradable, for example, starch and polyhydroxyalaknotes (PHA)) or nonbiodegradable (bio-PE). The term biodegradable polymers can be defined as materials that can deteriorate in anaerobic, aerobic, or microbial processes, and their degradability is affected by size, thickness, composition, etc. Furthermore, not all biodegradable plastics, such as polycaprolactone, can be classified as bio-based. In addition to classifying them based on the raw material and degradability, they can also be categorized into two other groups. One is the drop-in plastics, which are produced from bio-based raw materials and depict properties similar to petrochemical equivalents, and the other is chemical novel types, which are unique and hence cannot be recycled in conventional methods (Fig. 3).

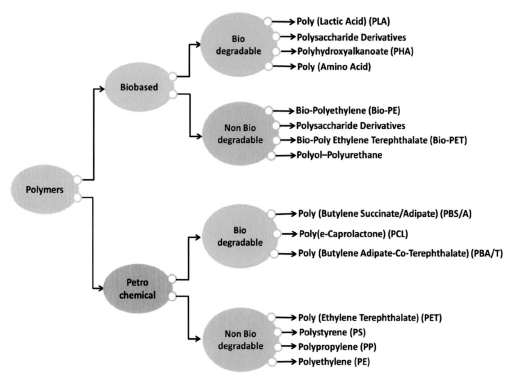

FIG. 3 Classification of biopolymers based on their degradability and raw materials. *Adapted from Amulya, K., Katakojwala, R., Venkata Mohan, S., 2021. Low carbon biodegradable polymer matrices for sustainable future. Compos. C: Open Access.*

Bio-based plastics can be produced in three principal routes. In the first route, natural polymers such as starch and cellulose are directly used, or they are partially modified or catalyzed to obtain other plastics like cellulose acetate, lactide, furandicarboxylic acid, etc. The other route that is widely being researched is the use of organic waste residues for the production of these polymers. The bio-based monomers are first synthesized via microbiological or green chemistry routes and are then polymerized. Examples of such kind of monomers and polymers include succinic acid, adipic acid, lactic acid, poly lactic acid (PLA), polybytyl succinate (PBS), and PE. The use of microorganisms for intracellular synthesis of these polymers is the third route for their production. Examples of such polymers include PHA, exopolysaccharide, bacterial cellulose, etc.

Among the various aforementioned biopolymers, the most feasible biopolymer alternative and also the most commercially successful product is PLA. PLA is a thermoplastic polyester produced either by self-condensation of lactic acid or ring-opening polymerization of lactide (Keshavarz and Roy, 2010). PLA is generally produced in two consecutive steps: formation of lactic acid through fermentation of renewable feedstock in the first step, followed by successive homo-polymerization/condensation of lactic acid (Nagarajan et al., 2016). PLA has good structural stability with high transparency and is used in the food industry for sensitive food product packaging (Zhang et al., 2018). The degradability of PLA is dependent on variations in its composition and quality. Due to this variation, it either readily degrades in a short duration or can last for several years. Since its properties and manufacturing process are similar to those of conventional plastics such as PE, polypropylene, and polyethylene terephthalate, it helps negate an additional capital expenditure for its processing.

Another biodegradable polymer that is completely biologically synthesized is PHAs. These were first discovered in the 1920s by the French microbiologist Maurice Lemoigne as intracellular granules in the Gram-positive bacterium *Bacillus megaterium*. Many years later, in 1974, PHA containing 3HB and 3-hydroxyvalerate was found in activated sludge by Wallen and Rohwedder (1974). After nearly a decade, 11 varieties of PHAs with linear and branched repeating units of 4 to 8 were reported. These are usually produced as cytoplasmic aggregation in various bacteria under specific nutrients and growing conditions, like a surplus amount of carbon and a deficiency of one or more vital nutrients such as oxygen, nitrogen, phosphorus, sulfur, and other trace elements like Mg, Ca, and Fe (Venkata Mohan and Reddy, 2013; Amulya et al., 2014a). These are biocompatible, insoluble, thermally stable, and degrade in nature without leaving any residues (Reddy and Mohan, 2012). They hold applications in the industrial, medical, and agricultural sectors (Adeleye et al., 2020). Nevertheless, several factors, such as specific medium conditions, the usage of pure cultures, downstream processes (extraction), and purification of the biopolymer, limit their commercial production.

Other biodegradable polymers that are synthesized chemically include PBS and polybutylene adipate terephthalate (PBAT) (Dahiya et al., 2020). Interestingly, biological routes for the production of these products are now being widely explored. In biological processes, their monomers, such as succinic acid, adipic acid, 1,4-butanediol, and terephthalic acid, are first synthesized and then chemically converted to PBS or PBAT. PBS and PBT hold applications in the packaging industry, textiles, mulching films in agriculture, and hygiene supplies (Sohn et al., 2020). Other plant-based polymers include cellulose and starch. Cellulose has lower solubility and hence cannot be used for packaging. Hence, it is usually blended with plasticizers, or its surface can be modified to make it suitable for various applications (Chen, 2015). Apart from direct use, nanocellulose fibers can be obtained from cellulose and used as nanofillers due to their high mechanical stability and optical transparency (Aliotta et al., 2019). In addition, cellulose acetate fiber, another derivate of cellulose, is also used for coatings, photographic films, and the manufacture of playing cards and cigarette filters (Aliotta et al., 2019). Starch is accumulated as granules in the plant cells, and the commercially available starch is obtained majorly from potatoes; other sources include corn, rice, wheat, etc. (Bertolini, 2009). Starch-based polymers can effectively be used in a diverse range of applications because they can be easily blended with several oil-based or bio-based polymers to produce exclusive composite materials (Bertolini, 2009). Starch-based biopolymers are widely recommended in the medical sector due to their biocompatibility, low toxicity, and degradability with excellent mechanical properties.

2.5 Bio-electrochemical systems

2.5.1 *Microbial fuel cells for bioelectricity*

The use of MFC for power generation was first documented in 1915 by Potter and his co-workers (Potter, 1910). Many years later, in 1980, studies on mediators that improve the performance of MFCs were conducted (Du et al., 2007; Venkata Mohan et al., 2014d). Nearly two decades later, studies on electrochemically active bacteria that transfer electrons from the cytoplasm to the electrode surface (Bond and Lovley, 2003) started to emerge, resulting in the development of mediator-less MFCs (Pasupuleti et al., 2014). Since then, intense research has been carried out to improve the power densities of MFCs.

2.5.1.1 Bioelectrogenesis in MFC

Normally, bacteria utilize the carbon (substrate) available in the wastewater to generate reducing equivalents [protons (H^+) and electrons (e^-)] in order to carry out their metabolic activities. Redox components/carriers (NAD, FAD, FMN, etc.) carry these e^- and H^+ toward an available terminal electron acceptor, thus generating a proton motive force, which in turn facilitates the generation of energy-rich phosphate bonds that are useful for microbial growth and metabolism (Kim et al., 2002; Suresh Babu et al., 2014). In MFCs, the electrodes introduced into the cell act as intermediary electron acceptors. Oxidation occurs at the anode, generating (H^+) and electrons (e^-), while reduction takes place at the cathode. The electrons remain at the anode, generating negative anodic potential, while H^+ moves toward the cathode via the proton exchange membrane, generating positive cathode potential. The reactions occurring at the anode and cathode can be represented as follows:

$$C_6H_{12}O_6 + 6H_2O \rightarrow 6CO_2 + 24H^+ + 24e^- \text{ (Anode)} \tag{11}$$

$$4e^- + 4H^+ + O_2 \rightarrow 2H_2O \text{ (Cathode)} \tag{12}$$

$$C_6H_{12}O_6 + 6H_2O + 6O_2 \rightarrow 6CO_2 + 12H_2O \text{ (Overall)} \tag{13}$$

The transfer of electrons from the bacterial cytoplasm to the electrode surface is an important factor that governs the performance of MFC. There are two possible mechanisms by which the electrons are transferred onto the electrode surface: (i) direct electron transfer (DET) and (ii) mediated electron transfer (MET). DET occurs via the physical contact of the microbial cell wall and anode without the involvement of any redox species or mediators (Zhou et al., 2012). The membrane-bound electron transport proteins could permit the transfer of electrons from the outer membrane of the bacteria to the anode (Kim et al., 2002). The formation of biofilm through the coic acids that help in the adherence of the bacteria to the electrode surface by many of the Gram-positive bacteria also helps in DET (Venkata Mohan et al., 2014c). C-type cytochromes and multi-heme proteins are identified as one of the possible routes for DET. One of the major disadvantages of this process is that a monolayer of cells participates in the transfer of electrons to the anode (Zhao et al., 2009). Studies have identified that a few electrochemically active bacteria have shown efficient DET mechanisms, e.g., *Geobacter, Rhodoferax,* and *Shewanella* (Schroder, 2007; Liu et al., 2012). Some bacteria, like *Geobacter* and *Shewanella,* carry out DET through conductive pilli (electrically conductive nanowires) formed on the bacterial cell surface (Gupta et al., 2013). MET is carried out through the redox shuttles that mediate the electron flow from bacterial metabolism toward electrode. The oxidized mediators penetrate the membrane of the organism and get reduced by the electrons. These reduced mediators again pass through the bacterial membrane and reach the anode, where they get reoxidized by losing electrons (Lovley and Nevin, 2013). In this manner, the electrons are transferred from inside the bacterial cell to the electrode, thus accelerating electron transfer and enhancing power output. MET occurs either by the addition of artificial mediators or by the secretion of soluble mediators such as primary and secondary metabolites from bacterial metabolism. Examples of these mediators include inorganic (potassiumferricyanide) or organic (benzoquinone) group dyes and metallorganics such as neutral red, methyleneblue, thionine, Meldola's blue, 2-hydroxy-1,4-naphthoquinone, and Fe (III) EDTA (Chang et al., 2006; Malvankar et al., 2012). Since the synthetic mediators are unstable and toxic, their use in MFCs is limited. However, various naturally secreted secondary metabolites by the bacteria act as good mediators. These include phenazines, phenoxazines, quinines, pyocyanin, etc. (Neto et al., 2010; Choi et al., 2003).

MFC can use a diverse range of substrates (anolyte fuel) as electron donors for anodic oxidation to generate reducing equivalents that in turn generate power. Utilization of wastewater, as well as solid waste as anodic fuel, has been well established since it facilitates simultaneous treatment (McKinlay and Zeikus, 2004; Bennetto et al., 1983; Hernandez and Newman, 2001). The performance of MFC depends on several factors such as fuel cell configuration, nature of anolyte, electrode materials, spacing between the electrodes, proton exchange membrane properties, nature of microbes, electron transfer mechanism, redox condition (pH), anolyte microenvironment, etc. (Venkata Mohan et al., 2008a; Venkata Mohan et al., 2013b). MFCs primarily make use of anaerobic bacteria but other types of organisms like photosynthetic bacteria have also been used in MFCs for tapping solar energy thus widening the scope of MFC application (Venkata Mohan et al., 2013a; Venkata Mohan and Chandrasekhar, 2011). These fuel cells can also be designed in different configurations, for example, benthic MFCs, plant-based MFCs, and stacked MFCs. In benthic MFCs, bioelectricity is harnessed from an aquatic eco-system by using natural habitants (Raghavulu et al., 2009), while in plant-based MFCs, indirect utilization of solar radiation takes place for the generation of green electricity by integrating the roots of living plants (rhizosphere) with electrodes. Rhizodeposits secreted from roots are organic in nature and provide favorable microenvironment for proliferation of bacteria (Srikanth et al., 2009a). In stacked MFCs, individual MFCs are stacked in series and parallel connections to increase the power output (Raghavulu et al., 2013).

2.5.2 Bioelectrochemical treatment system

If microbially catalyzed electro chemical cells (MCES) are used exclusively for the treatment of wastewater, then they are termed a bioelectrochemical treatment system (BET). Multiple reactions, namely biochemical, physical, physico-chemical, electrochemical, oxidation, etc., can be triggered in the anodic chamber of BET operation for the treatment of complex pollutants present in wastewaters (Chandra et al., 2012; Chiranjeevi et al., 2013a). Although the anode chamber of BET resembles the conventional anaerobic bioreactor and electrochemical cell, the occurrence of both oxidation and reduction reactions gives it an edge over conventional treatment processes where reduction reactions are predominant (Chiranjeevi et al., 2013b). Designing an appropriate configuration of BET can facilitate its integration with existing wastewater treatment processes, which is a critical aspect for further development of this process (Kim et al., 2012; Venkata Mohan et al., 2009d).

2.5.3 Microbial electro-synthesis

Like BET, the configuration of MFC can be modified to generate bio-H_2. Applying a small amount of external potential to achieve value addition in the form of bio-H_2 is the main principle underlying the functioning of microbial electro-synthesis (MEC). It facilitates the conversion of electron equivalents in organic compounds to H_2 gas by combining microbial metabolism with bio-electrochemical reactions. Low energy consumption compared to conventional water electrolysis, high product (H_2) recovery, and substrate degradation compared to the dark-fermentation process are some of the potential benefits that make MEC an alternate process (Velvizhi et al., 2015).

2.5.4 Bioelectrochemical electrosynthesis

In addition, certain microorganisms in a defined fuel cell system are capable of generating biofuels, hydrogen gas, methane, acetate, ethanol, hydrogen peroxide, butanol, and other valuable inorganic and organic chemicals, and this electrically driven reduction of electron acceptors in the cathode chamber is termed as bioelectrochemical electrosynthesis (Velvizhi et al., 2015; Chandrasekhar and Venkata Mohan, 2012). The essence of these MCES is the feasibility of achieving value addition with simultaneous waste remediation (Fig. 4). Despite the advantages and advances made in MCES research, the output achieved either in terms of power or value-added products in these systems is a major limitation. The performance of these systems is dependent on the efficiency of the electron transfer machinery in the organisms. From the extensive research being carried out in microbial dynamics, it is not too far to realize that microbial limitations will have a profound influence on the performance of these systems rather than limitations in configurations. Therefore, attention is required to

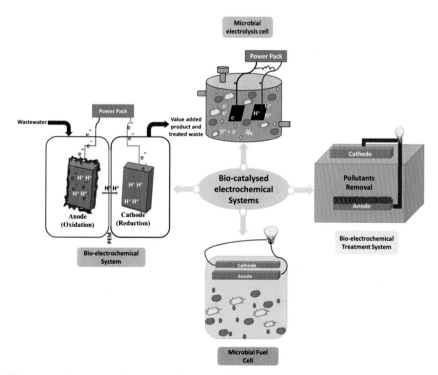

FIG. 4 Different configurations of microbial electrochemical systems (MECs) based on their applications.

enhance the electron transfer capabilities of the organisms. Genetic manipulation of specific proteins involved in electron transfer can help increase electron transfer by organisms. A new era for MFC research might evolve, where efforts will be directed on gaining an understanding of the biocatalyst's capabilities, electron transfer mechanisms, and kinetics (Kelly and He, 2014; Vamshi Krishna et al., 2014; Cheng and Logan, 2007; Rabaey and Rozendal, 2010; Gong et al., 2013).

2.6 Biodiesel

2.6.1 Microalgal biodiesel

In 1900, Rudolph Diesel used peanut oil as fuel at the World Exhibition in Paris, marking that date in history as the first use of vegetable oil as an alternative fuel. However, the major problem with vegetable oil when used directly as fuel is its high viscosity, which interferes with the engine's functioning and leads to engine failure (Knothe, 2005; Singh and Rastogi, 2009). Therefore, vegetable oils were modified to bring their combustion-related properties closer to those of diesel fuel, and in the present day, these fuels are known as biodiesel. Currently, biodiesel is one of the renewable biofuels that is being explored as an alternate fuel and for abatement of greenhouse gas emissions. The first-generation biodiesel was produced using edible sources like soybean, rapeseed, sunflower, and safflower. The second-generation biodiesel was produced by using nonedible oil sources like used frying oil, grease, tallow, lard, karanja, jatropha, and mahua oils (Francis and Becker, 2002; Dorado et al., 2002; Alcantara et al., 2000). First- and second-generation sources are costly, initiate land clearing, and potentially compete with net food production (Chisti, 2008; Marsh, 2009). The third generation of sources are oleaginous microorganisms that have lipid compositions similar to those found in plant or animal sources. These microorganisms are capable of accumulating lipids up to 20%–50% of their dry cell weight and storing them as triacylglycerols (TAG) in oil vacuoles. Their lipid compositions make them potential options for biodiesel generation since they share similarities with vegetable oils. In this category of renewable sources, algae are the forerunners (Venkata Mohan et al., 2011; Venkata Mohan et al., 2014a). Using algae for biodiesel production can solve the problems of reducing large land requirements, more CO_2 sequestration, and rapid growth when compared with crop-based terrestrial sources of biomass.

Algae are photosynthetic microorganisms that live in water, grow hydroponically, and are capable of assimilating carbon heterotrophically and mixotrophically (Gouveia and Oliveira, 2009; Subhash et al., 2014). This property of the microalgae is an advantage, and hence algal biodiesel production can be integrated with wastewater treatment (Subhash and Venkata Mohan, 2014c). Several wastewaters are potential sources of carbon, which can be taken up by microalgae and converted to biodiesel. Under stress conditions, microalgae synthesize TAG in the form of storage lipids. Several species of microalgae, like *Prymnesium paryum*, 22%–38%; *Chlorella vulgaris*, 22%; *Chlamydomonas rheinhardii*, 21%; *Spirogyra* sp., 11%–21%; *Scenedesmus dimorphous*, 16.40%; *Scenedesmus obliquus*, 12%–12%; *Synchoccus* sp., 11%; *Porphyridium cruentum*, 4%–14%; *Dunaliella bioculata*, 8%; *Tetraselmis maculate*, 3%; based on dry biomass (Chandra et al., 2014; Devi et al., 2012), are rich in lipids, which can be converted to biodiesel using extraction strategies.

2.6.2 Yeast biodiesel

Yeasts can be considered a more favorable choice compared to bacterial species for the production of biodiesel due to their lack of endotoxins, which can be more prevalent in bacterial species. Yeasts can accumulate excess carbon as glycogen or lipid in the form of TAG. Yeasts can produce lipids from different carbon sources and even form fatty acids found in the culture medium (Chatterjee and Venkata Mohan, 2022). Several oleaginous microorganisms can accumulate more than 70% lipid on a cell dry weight basis and can accumulate lipids that are similar in composition to those of vegetable oils (Venkata Mohan et al., 2014b). Non-oleaginous yeasts, such as *Saccharomyces cerevisae* and food yeast (*Candida ultilis*) cannot accumulate lipid content greater than 10% of their total dry biomass (Becker, 1994; Becker, 2004). However, in nitrogen-limited medium with an excess carbon source, the contents of mannans and glucans increase, while in the case of oleaginous yeast, the excess carbon source gets converted into lipids.

The process involves three phases of growth: the exponential growth phase, the lipid accumulation phase, and the stationary phase, or late accumulation phase. When nitrogen exhausts from the culture medium, adenosine monophosphate deaminase gets activated and catalyzes the conversion of AMP into inosine 5′-monophosphate and ammonium. The phosphatidic acid (PA) is formed either by glycerol-3-phosphate (G-3-P) or by dihydroxyacetone phosphate, which is dependent on the availability of carbon sources. Three phosphatidate phosphatase isoenzymes are required to convert PA into Diacylglycerol (DAG) through the dephosphorylation reaction. The last steps of de-novo lipogenesis in oleaginous yeasts are varied in different strains and show dependency on acetyl-Co-A, which binds with the DAG backbone by diacylglycerol acyltransferases and finally forms TAG. The overall reaction of TAG synthesis is known as the "Kennedy pathway." The fatty acid profile of oleaginous yeast is dependent on the provided culture conditions. The fatty acids obtained are similar to those produced by plants and algae. Various nutrient limitations, such as nitrogen, phosphorus, zinc, or iron, affect fatty acid production by different oleaginous yeasts. Temperature, pH, the presence of trace elements, aeration rate, and dissolved oxygen are some other factors that affect lipid accumulation.

3 Sustainability of bioprocesses

A system or process is said to be sustainable if it can be maintained or continued over time without negatively affecting the economy, society, or environment. In bioprocesses, which use living organisms to produce a wide range of products like biofuels, chemicals, and pharmaceuticals, sustainability is a crucial factor. Life cycle assessment (LCA) is used to examine a product's or process's environmental impact over the duration of its full life cycle, from raw material extraction to end-of-life disposal (Dahiya et al., 2020). By providing a thorough analysis of their environmental impact, including greenhouse gas emissions, energy consumption, water use, and waste generation, LCA can be used as a tool to evaluate the sustainability of bioprocesses (Sun et al., 2019). These tools can be used to evaluate how bioprocesses affect the economies and communities in which they operate as well as the supply chain's overall sustainability. The purpose of LCA and sustainability tools in bioprocesses is to give a comprehensive evaluation of their sustainability, taking into account all of their effects on the environment, society, and the economy (Dahiya et al., 2020). With the objective of enhancing their sustainability and minimizing their total impact on the environment and society, bioprocesses can be designed, operated, and managed in a way that takes this information into account (Ng et al., 2014). We can make sure that these processes are sustainable over the long term and satisfy the increasing demand for bio-based products in an environmentally responsible manner by including sustainability considerations into bioprocess design and decision-making.

4 Integrated biorefineries

The use of renewable resources is essential in a CBE, and the idea of cascading use shows potential for this purpose. Cascading use involves utilizing resources in a flexible and multi-purpose manner to maximize their usage. A biorefinery is a useful approach for implementing the cascading use concept as it allows for the conversion of renewable resources into valuable products through a hierarchical valorization process. While the concept is not new, it has gained significant attention recently. Biorefineries can produce marketable bio-based products while promoting environmental and economic sustainability (Katakojwala and Mohan, 2021a). Biorefineries can be classified based on their ability to process different feedstocks, conversion process characteristics, and product diversification. As they produce bio-based products, they can play a significant role in the development of a CBE. Integrated biorefineries are a combination of resource recovery techniques that not only enhance flexibility but also reduce costs (Ng et al., 2014; Stegmann et al., 2020). Integrated biorefineries utilize the byproducts generated from one process as a resource for another process, with the aim of producing high-value products like platform chemicals (Usmani et al., 2021). These biorefineries operate using three major routes – physicochemical, thermochemical, and biological processes.

Physicochemical processes involve the use of physical and chemical techniques to convert renewable feedstocks. Thermochemical processes such as pyrolysis and gasification use high temperatures for the conversion of renewable feedstocks to energy, electricity, fuel heat, and value-added products (Kopperi and Venkata Mohan, 2022). In physiochemical processes, techniques such as rendering and transesterification that make use of chemical agents for the conversion of renewable resources to fuels and bioproducts are employed. In the biological process, microorganisms, through their metabolic processes, utilize the organic carbon in renewable resources and produce various products. Pretreatment techniques are essential in biorefineries because they allow for the extraction of valuable components from complex and heterogeneous feedstocks while removing unwanted materials.

The objective of these techniques is to improve the downstream recovery of products, and they can involve changes to the physical properties of the feedstocks. Waste biorefineries can have simple or complex layouts, depending on the desired products. Biorefineries have the potential to produce a wide range of products and energy sources from renewable feedstocks, and integrating high-value, low-volume products with bioenergy and biofuels can help make the overall process more sustainable and cost-effective.

Retrofitting existing biofuel infrastructure with bio-based production processes that generate platform chemicals can also make biofuels more competitive with petrochemical alternatives. They should be designed to produce a minimum of two platform chemicals. This would enable increasing the selling price of the biofuels and making them competitive with their petrochemical counterparts. This mandates the integration of a range of treatment processes that, in turn, depend on the nature of the waste, the pretreatment process employed, and the final product that needs to be produced (Alibardi et al., 2020). A few examples of biorefineries are lignocellulosic biorefineries, algae biorefineries, whole crop biorefineries, etc. (Alibardi et al., 2020).

Besides AD, which is a well-established process currently being used for the treatment of heterogeneous and complex waste, acidogenesis for bio-H_2 production might as well be considered suitable for playing a central role in waste biorefineries (Dahiya et al., 2020). Route one of a multi-product biorefinery that integrates acidogenesis involves the production of bio-H_2 and CO_2, with the effluent rich in VFAs being further processed biologically for the production of biopolymers such as PHA. This approach utilizes acidogenesis to break down complex waste materials into simpler compounds, with the resulting bio-H_2 and CO_2 being used to produce other valuable products. The effluent, which contains VFAs, can then be further processed to produce biopolymers, providing an additional source of value from the waste stream. The different routes mentioned, such as the production of biopolymers from VFAs or the use of CO_2 and H_2 in gas fermentation, demonstrate the potential for the waste biorefinery concept to generate a range of platform chemicals and biofuels. The integration of acidogenesis with photofermentation, microbial electrolysis cells, and MFCs also highlights the potential for the waste biorefinery to generate renewable energy in addition to valuable products. Furthermore, microalgal and yeast cultivation can be integrated for the production of biodiesel, lipids, pigments, and other high-value products, demonstrating the versatility and potential of the waste biorefinery concept. This approach has the potential to enable efficient, economic, and sustainable waste management practices while producing valuable products.

5 Transition toward a bio-based economy

Four main societal challenges that can be seen as drivers for the bio-based economy are

- Food security (production, quality, and fair consumption)
- Climate change (mitigation and adaptation)
- Resource security (energy and scarce materials)
- Ecosystem services (soil, water, and biodiversity)

Addressing these multi-dimensional issues requires a strategic and comprehensive approach involving different policies (European Commission, 2019). Although the concept of developing biorefineries and bio-based products, including biofuels, has been extensively supported in different regions of the world, the lack of comprehensive policies limits the development of a bio-based economy. Thus, the transition toward a bio-based economy can be achieved by developing policies that affect multiple sectors such as agriculture, research, industry, and trade. Gauging the impacts of the policies is also important, given the variety of policy instruments available (taxes, subsidies, price support, etc.) and the way they are applied. The policies implemented should be able to stimulate the efficient use of resources to achieve a sustainable bio-based economy.

Besides focusing on reuse, attention should also be given to lower CO_2 emissions (Ramakrishna, 2021). The low-carbon economy precisely focuses on the CO_2 emitted right from raw materials to the entire product value chain and the end of life (EoL) processes it undergoes. It is meant to address climate change, specifically CO_2 emissions, and aims at developing processes that generate low CO_2 right from manufacturing to the EoL of a product (Venkata Mohan and Katakojwala, 2021). Low-carbon economic development focuses on the "carbon" right from the production stage to circulation, consumption, and waste disposal. In addition, a low-carbon economy takes into account the impact of raw materials and the manufacturing value chain on the climate change. This should be integrated with the CE to achieve a low-carbon CE (Venkata Mohan et al., 2020). The recyclability/reusability promoted by the CE can be complemented by the optimized climate impact promoted by the low-carbon economy. Low-carbon CE can aid in designing concepts that can differentiate between designs that have the scope to be reused, repaired, or remanufactured and product designs aimed specifically to reduce GHG emissions. The adoption of a low-carbon CE can be viewed as a strategy to improve other aspects of society, such as encouraging the new labor force required for reprocessing goods, which could be looped within the circular economic model (Venkata Mohan et al., 2020). CE, BE/BBE, and low-carbon economies are all part of the overarching concept of a green economy (Rathi et al., 2022) (Fig. 5). Policies that promote low-

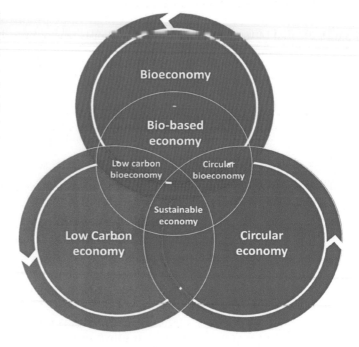

FIG. 5 Venn diagram representing the overarching green economy (GE) encompassing the circular economy (CE), bioeconomy (BE), bio-based economy (BBE), low carbon economy (LCE). *Adapted from Dahiya, S., Katakojwala, R., Ramakrishna, S., Mohan, S. V., 2020. Biobased products and life cycle assessment in the context of circular economy and sustainability. Mater. Circular Econ., 2, 1-28.*

carbon CE, decouple economic growth from resource use, and lower CO_2 emissions need to be introduced (Katakojwala and Mohan, 2021b). If implemented, the economic benefits could be huge and would contribute US$1 trillion/yr to the global economy (Ramakrishna, 2021).

Not only is it the responsibility of the government, but it is also the responsibility of consumers and producers to act in a responsive way and support the realization of a sustainable bio-based economy. Fortunately, there is a growing population that realizes this grave situation, but more awareness and education are required to realize the implementation of a sustainable bio-based economy. Although the bio-based economy proposes to offer various prospects, many efforts are still needed in order to fully exploit its advantages. In order to achieve this, renewable sources of energy should be available at affordable prices, the market value of the products should be stimulated, and research must be more focused on developing innovative products. Regulatory moves will also play a role in ensuring that new technologies come to market (Katakojwala and Mohan, 2021b). Shifting society's dependence away from petroleum to renewable resources and wringing the full potential of waste in an integrated and holistic approach is essential and will help in developing a sustainable and competitive bio-based economy. We can succeed in transitioning toward a bio-based economy if food security, environmental protection, climate protection, and biodiversity protection are coalesced with it.

6 Future prospects

The current industrial system that generates products required by the continuously developing world is not sustainable. Earth is being depleted of its resources at an alarming rate that needs to be halted. Recent data show that 25% of the population from industrialized countries consumes about 75% of the world's natural resources and controls about 88% of the total production, 80% of trade, and 94% of industrial products in the entire world (Gong et al., 2013). Additionally, 90% of the resources used during the production process approach obsolescence and end up as waste, creating environmental problems. All these factors generate premises to evaluate and develop ways of producing products in a sustainable and environmentally friendly manner. These processes should be based on the use of renewable sources like biomass or biowastes, which are abundantly available, are the most promising carbon-neutral source of energy that can mitigate greenhouse emissions, and are perceived to be an essential part of the bio-based economy.

In order to increase the sustainability of integrated systems, it is important to promote the creation and enhancement of methodologies and evaluation using technological resources at every level. The emphasis must be on recognizing the consequences of the goods' whole life cycle with the use of meticulous sustainability metrics and resources. In order to establish accurate indicators and minimize uncertainty, data obtained from regional aspects can be incorporated into the existing databases. Although fossil fuels will still be available for some time to come, their price is unlikely to reduce. Therefore, it will be important to formulate policies for the production of bioproducts that can promote CBE and low-carbon economies. CBE promotes the use of materials in a circular loop and the design-out-waste principle, while the low-carbon economy encourages the sustainability of these systems. Low-carbon economies and CBE thus need to complement each other in order to achieve success. A low-carbon economy will help to meet the climate commitments of the Paris Agreement and even the UN Sustainable Development Goals for 2030.

Acknowledgments

The authors wish to thank the Director, CSIR-IICT, for support and encouragement in carrying out this work.

References

Adeleye, A.T., Odoh, C.K., Enudi, O.C., Banjoko, O.O., Osiboye, O.O., Odediran, E.T., Louis, H., 2020. Sustainable synthesis and applications of polyhydroxyalkanoates (PHAs) from biomass. Process Biochem. 96, 174–193.

Alcantara, R., Amores, J., Canoira, L., Fidalgo, E., Franco, M.J., Navarro, A., 2000. Catalytic production of biodiesel from soybean oil, used frying oil and tallow. Biomass Bioenergy 18, 515–527.

Alibardi, L., Astrup, T.F., Asunis, F., Clarke, W.P., De Gioannis, G., Dessì, P., Lens, P.N., Lavagnolo, M.C., Lombardi, L., Muntoni, A., Pivato, A., 2020. Organic waste biorefineries: looking towards implementation. Waste Manag. 114, 274–286.

Aliotta, L., Gigante, V., Coltelli, M.B., Cinelli, P., Lazzeri, A., 2019. Evaluation of mechanical and interfacial properties of bio-composites based on poly (lactic acid) with natural cellulose fibers. Int. J. Mol. Sci. 20 (4), 960.

Amulya, K., Reddy, M.V., Mohan, S.V., 2014a. Acidogenic spent wash valorization through polyhydroxyalkanoate (PHA) synthesis coupled with fermentative biohydrogen production. Bioresour. Technol. 158, 336–342.

Amulya, K., Venkateswar Reddy, M., Venkata Mohan, S., 2014b. Acidogenic spent wash valorization through polyhydroxyalkanoate (PHA) synthesis coupled with fermentative biohydrogen production. Bioresour. Technol. 158, 336–342.

Asrul, M.A.M., Atan, M.F., Yun, H.A.H., Lai, J.C.H., 2022. A review of advanced optimization strategies for fermentative biohydrogen production processes. Int. J. Hydrog. Energy 47.

Atasoy, M., Owusu-Agyeman, I., Plaza, E., Cetecioglu, Z., 2018. Bio-based volatile fatty acid production and recovery from waste streams: current status and future challenges. Bioresour. Technol. 268, 773–786.

Azarmanesh, R., Qaretapeh, M.Z., Zonoozi, M.H., Chiasinejad, H., Zhang, Y., 2023. Anaerobic co-digestion of sewage sludge with other organic wastes: a comprehensive review focusing on selection criteria, operational conditions, and microbiology. Chem. Eng. J. Adv., 100453.

Balat, H., Kırtay, E., 2010. Hydrogen from biomass—present scenario and future prospects. Int. J. Hydrog. Energy 35, 7416–7426.

Bareha, Y., Faucher, J.P., Michel, M., Houdon, M., Vaneeckhaute, C., 2022. Evaluating the impact of substrate addition for anaerobic co-digestion on biogas production and digestate quality: the case of deinking sludge. J. Environ. Manag. 319, 115657.

Becker, E.W., 1994. Microalgae: Biotechnology and Microbiology. Cambridge University Press, Cambridge, UK.

Becker, E.W., 2004. Microalgae in human and animal nutrition. Biotechnol. Appl. Phycol., 312–351.

Bennetto, H.P., Stirling, J.L., Tanaka, K., Vega, C.A., 1983. Anodic reactions in microbial fuel cells. Biotechnol. Bioeng. 983, 559–568.

Bertolini, A. (Ed.), 2009. Starches: Characterization, Properties, and Applications. CRC Press.

Bhatia, S.K., Yang, Y.H., 2017. Microbial production of volatile fatty acids: current status and future perspectives. Rev. Environ. Sci. Biotechnol. 16, 327–345.

Bond, D.R., Lovley, D.R., 2003. Electricity production by Geobacter sulfur reducens attached to electrodes. Appl. Environ. Microbiol. 69, 1548–1555.

Chai, S., Zhang, G., Li, G., Zhang, Y., 2021. Industrial hydrogen production technology and development status in China: a review. Clean Techn. Environ. Policy 23 (7), 1931–1946.

Chandra, R., Rohit, M.V., Swamy, Y.V., Venkata Mohan, S., 2014. Regulatory function of organic carbon supplementation on biodiesel production during growth and nutrient stress phases of mixotrophic microalgae cultivation. Bioresour. Technol. 165, 279–287.

Chandra, R., Subhash, G.V., Venkata Mohan, S., 2012. Mixotrophic operation of photo-bioelectrocatalytic fuel cell under anoxygenic microenvironment enhances the light dependent bioelectrogenic activity. Bioresour. Technol. 109, 46–56.

Chandra, R., Venkata Mohan, S., 2014. Enhanced bio-hydrogenesis by co-culturing photosynthetic bacteria with acidogenic process: augmented dark-photo fermentative hybrid system to regulate volatile fatty acid inhibition. Int. J. Hydrog. Energy 39, 7604–7615.

Chandrasekhar, K., Venkata Mohan, S., 2012. Bio-electrochemical remediation of real field petroleum sludge as an electron donor with simultaneous power generation facilitates biotransformation of PAH: effect of substrate concentration. Bioresour. Technol. 110, 517–525.

Chang, S., Moon, H., Bretschger, O., Jang, J.K., Park, H.I., Nealson, K.H., Kim, B.H., 2006. Electrochemically active bacteria (EAB) and mediator-less microbial fuel cells. J. Microbiol. Biotechnol. 16, 163–167.

Chatterjee, S., Venkata Mohan, S., 2022. Fungal biorefinery for sustainable resource recovery from waste. Bioresour. Technol. 345.

Chen, J., 2015. Synthetic textile fibers: regenerated cellulose fibers. In: Textiles and Fashion. Woodhead Publishing, pp. 79–95.

Chen, Z., Huang, Y.C., Liang, J.H., Zhao, F., Zhu, Y.G., 2012. A novel sediment microbial fuel cell with a biocathode in the rice rhizosphere. Bioresour. Technol. 108, 55–59.

Cheng, S., Logan, B.E., 2007. Ammonia treatment of carbon cloth anodes to enhance power generation of microbial fuel cells. J. Electrochem. Commun. 9, 492–496.

Chiranjeevi, P., Chandra, R., Venkata Mohan, S., 2013a. Ecologically engineered submerged and emergent macrophyte based system: an integrated ecoelectrogenic design for harnessing power with simultaneous wastewater treatment. Ecol. Eng. 51, 181–190.

Chiranjeevi, P., Chandra, R., Venkata Mohan, S., 2013b. Rhizosphere mediated electrogenesis with the function of anode placement for harnessing bioenergy through CO2 sequestration. Ecol. Eng. 51, 181–190.

Chisti, Y., 2008. Biodiesel from microalgae beats bioethanol. Trends Biotechnol. 26, 126–131.

Choi, Y., Kim, N., Kim, S., Jung, S., 2003. Dynamic behaviors of redox mediators within the hydrophobic layers as an important factor for effective microbial fuel cell operation. Bulletin-Korean Chem. Soc. 24, 437–440.

Christophe, G., Lara Deo, J., Kumar, V., Nouaille, R., Fontanille, P., Larroche, C., 2012. Production of oils from acetic acid by the oleaginous yeast Cryptococcus curvatus. Appl. Biochem. Biotechnol. 167, 1270–1279.

Dahiya, S., Chatterjee, S., Sarkar, O., Mohan, S.V., 2021. Renewable hydrogen production by dark-fermentation: current status, challenges and perspectives. Bioresour. Technol. 321, 124354.

Dahiya, S., Katakojwala, R., Ramakrishna, S., Mohan, S.V., 2020. Biobased products and life cycle assessment in the context of circular economy and sustainability. Mater. Circular Econ. 2, 1–28.

Dahiya, S., Kumar, A.N., Sravan, J.S., Chatterjee, S., Sarkar, O., Mohan, S.V., 2018. Food waste biorefinery: sustainable strategy for circular bioeconomy. Bioresour. Technol. 248, 2–12.

Dahiya, S., Lingam, Y., Mohan, S.V., 2023. Understanding acidogenesis towards green hydrogen and volatile fatty acid production—critical analysis and circular economy perspective. Chem. Eng. J., 141550.

Dahiya, S., Sarkar, O., Swamy, Y.V., Mohan, V., S., 2015. Acidogenic fermentation of food waste for volatile fatty acid production along with co-generation of biohydrogen. Bioresour. Technol. https://doi.org/10.1016/j.biortech.2015.01.007.

Devi, M.P., Subhash, G.V., Venkata Mohan, S., 2012. Heterotrophic cultivation of mixed microalgae for lipid accumulation during sequential growth and starvation phase operation: effect of nutrient supplementation. Renew. Energy 43, 276–283.

Donner, M., Gohier, R., de Vries, H., 2020. A new circular business model typology for creating value from agrowaste. Sci. Total Environ. 716, 137065.

Dorado, M.P., Ballesteros, E., Almeida, J.A., Schellert, C., Lohrlein, H.P., Krause, R., 2002. An alkali-catalyzed transesterificationprocess for high free fatty acid waste oils. Trans. ASAE 45, 525–529.

Du, Z., Li, H., Gu, T., 2007. A state of the art review on microbial fuel cells: a promising technology for wastewater treatment and bioenergy. Biotechnol. Adv. 25, 464–482.

EMF, 2012. Towards the Circular Economy. vol. 1 Ellen MacArthur Foundation Publishing.

EMF, 2015. Towards a Circular Economy: Business Rationale for an Accelerated Transition.

Ergal, İ., Gräf, O., Hasibar, B., Steiner, M., Vukotić, S., Bochmann, G., Rittmann, S.K.M., 2020. Biohydrogen production beyond the Thauer limit by precision design of artificial microbial consortia. Commun. Biol. 3 (1), 443.

European Commission, 2019. The European green deal. Eur. Comm. 53 (9), 24.

Francis, G., Becker, K., 2002. Biodiesel From Jatropha Plantations Ondegraded Land. University of Hohenheim, Stuttgart, Germany, p. 9. http://www.youmanitas.nl/pdf/Bio-diesel.pdf.

Girardet, H., 2020. People and nature in an urban world. One Earth 2 (2), 135–137.

Global Footprint Network, 2016. National Footprint Accounts, 2016 ed. Available at: Www.footprintnetwork.org/en/index.php/GFN/blog/national_footprint_accounts_2016_carbon_makes_up_60_of_worlds_footprint. (Accessed January 2023).

Gong, Y., Ebrahim, A., Feist, A.M., Embree, M., Zhang, T., Lovely, D., Zengler, K., 2013. Sulfide-driven microbial electrosynthesis. Environ. Sci. Technol. 47, 568–573.

Gouveia, L., Oliveira, A.C., 2009. Microalgae as a raw material forbiofuels production. J. Ind. Microbiol. Biotechnol. 36, 269–274.

Gupta, V.G., Tuohy, M., Kubicek, C.P., Saddler, J., Xu, F. (Eds.), 2013. Bioenergy Research: Advances and Applications., ISBN: 978-0-444-59561-4.

He, T., Pachfule, P., Wu, H., Xu, Q., Chen, P., 2016. Hydrogen carriers. Nat. Rev. Mater. 1 (12), 1–17.

Hernandez, M.E., Newman, D.K., 2001. Extra cellular electron transfer cell. Cell. Mol. Life Sci. 58, 1562–1571.

Huang, Y.L., Wu, Z., Zhang, L., Cheung, C.M., Yang, S., 2002. Production of carboxylic acids from hydrolyzed corn meal by immobilized cell fermentation in a fibrous-bed bioreactor. Bioresour. Technol. 82, 51–59.

Iacovidou, E., Millward-Hopkins, J., Busch, J., Purnell, P., Velis, C.A., Hahladakis, J.N., Zwirner, O., Brown, A., 2017. A pathway to circular economy: developing a conceptual framework for complex value assessment of resources recovered from waste. J. Clean. Prod. 168, 1279–1288.

IPCC, 2014. Climate change 2014: synthesis report. In: Pachauri, R.K., Meyer, L.A. (Eds.), Contribution of Working Groups I, II and III to the Fifth Assessment Report of the Intergovernmental Panel on Climate Change. IPCC, Geneva, Switzerland, pp. 117–130. Core Writing Team.

Kamshybayeva, G.K., Kossalbayev, B.D., Sadvakasova, A.K., Kakimova, A.B., Bauenova, M.O., Zayadan, B.K., Allakhverdiev, S.I., 2023. Genetic engineering contribution to developing cyanobacteria-based hydrogen energy to reduce carbon emissions and establish a hydrogen economy. Int. J. Hydrog. Energy.

Katakojwala, R., Advaitha, K., Venkata Mohan, S., 2023. Circular economy induced resilience in socio-ecological systems: an ecolonomic perspective. Mater. Circ. Econ. 5.

Katakojwala, R., Mohan, S.V., 2021a. A critical view on the environmental sustainability of biorefinery systems. Curr. Opin. Green Sustain. Chem. 27, 100392.

Katakojwala, R., Mohan, S.V., 2021b. A critical view on the environmental sustainability of biorefinery systems. Curr. Opin. Green Sustain. Chem. 27, 100392.

Katakojwala, R., Venkata Mohan, S., 2022. Multi-product biorefinery with sugarcane bagasse: process development for nanocellulose, lignin and biohydrogen production and lifecycle analysis. Chem. Eng. 4 (446).

Kelly, P.T., He, Z., 2014. Review nutrients removal and recovery in bioelectrochemical systems: a review. Bioresour. Technol. 153, 351–360.

Kelly, R., Nettlefold, J., Mossop, D., Bettiol, S., Corney, S., Cullen-Knox, C., Fleming, A., Leith, P., Melbourne-Thomas, J., Ogier, E., van Putten, I., 2020. Let's talk about climate change: developing effective conversations between scientists and communities. One Earth 3 (4), 415–419.

Keshavarz, T., Roy, I., 2010. Polyhydroxyalkanoates: bioplastics with a green agenda. Curr. Opin. Microbiol. 13, 321–326.

Kim, D., An, J., Kim, B., Jang, J.K., Kim, B.H., Chang, I.S., 2012. Scaling-up microbial fuel cells: configuration and potential drop phenomenon at series connection of unit cells in shared anolyte. ChemSusChem 5, 1086–1091.

Kim, H.J., Park, H.S., Hyun, M.S., Chang, I.S., Kim, M., Kima, B.H., 2002. A mediator-less microbial fuel cell using a metal reducing bacterium, Shewanella putrefaciens. Enzym. Microb. Technol. 30, 145–152.

Knothe, G., 2005. Dependence of biodiesel fuel properties on the structure of fatty acid alkyl esters. Fuel Process. Technol. 86, 1059–1070.

Kopperi, H., Venkata Mohan, S., 2022. Comparative appraisal of nutrient recovery, bio-crude, and bio-hydrogen production using Coelestrella sp. in a closed-loop biorefinery. Front. Bioeng. Biotechnol. 10.

Korhonen, J., Honkasalo, A., Seppälä, J., 2018. Circular economy: the concept and its limitations. Ecol. Econ. 143, 37–46.

Kovalev, A.A., Kovalev, D.A., Panchenko, V.A., Zhuravleva, E.A., Laikova, A.A., Shekhurdina, S.V., Vivekanand, V., Litti, Y.V., 2022. Approbation of an innovative method of pretreatment of dark fermentation feedstocks. Int. J. Hydrog. Energy 47 (78), 33272–33281.

Lay, C.II., Kumar, G., Mudhoo, A., Lin, C.Y., Leu, H.J., Shobana, S., Nguyen, M.L.T., 2020. Recent trends and prospects in biohythane research: an overview. Int. J. Hydrog. Energy 45 (10), 5864–5873.

Liu, L., Bryan, S.J., Huang, F., Yu, J., Nixon, P.J., Rich, P.R., Mullineaux, C.W., 2012. Control of electron transport routes through redox regulated redistribution of respiratory complexes. Proc. Natl. Acad. Sci. 109, 6431–6436.

Lokesh, K., Ladu, L., Summerton, L., 2018. Bridging the gaps for a 'circular' bioeconomy: selection criteria, bio-based value chain and stakeholder mapping. Sustainability 10 (6), 1695.

Lovley, D.R., Nevin, K.P., 2013. Electro biocommodities: powering microbial production of fuels and commodity chemicals from carbon dioxide with electricity. Curr. Opin. Biotechnol. 24, 385–390.

Luo, S., Jain, A., Aguilera, A., He, Z., 2017. Effective control of biohythane composition through operational strategies in an innovative microbial electrolysis cell. Appl. Energy 206, 879–886.

Lv, N., Cai, G., Pan, X., Li, Y., Wang, R., Li, J., Li, C., Zhu, G., 2022. pH and hydraulic retention time regulation for anaerobic fermentation: focus on volatile fatty acids production/distribution, microbial community succession and interactive correlation. Bioresour. Technol. 347, 126310.

Mahla, P.K., Vithalani, P.C., Bhatt, N.S., 2022a. Biomethanation: advancements for upgrading biomethane using biogas technologies. In: Industrial Microbiology and Biotechnology, pp. 487–504.

Mahla, P.K., Vithalani, P.C., Bhatt, N.S., 2022b. Biomethanation: advancements for upgrading biomethane using biogas technologies. In: Industrial Microbiology and Biotechnology, pp. 487–504.

Malvankar, N.S., Tuominen, M.T., Lovley, D.R., 2012. Lack of cytochrome involvement in long-range electron transport through conductive biofilms and nanowires of *Geobacter sulfurreducens*. Energy Environ. Sci. 5, 8651–8659.

Marsh, G., 2009. Small wonders: biomass from algae. Renew. Energy Focus 9, 74–78.

McInerney, M.J., Struchtemeyer, C.G., Sieber, J., Mouttaki, H., Stams, A.J., Schink, B., Rohlin, L., Gunsalus, R.P., 2008. Physiology, ecology, phylogeny, and genomics of microorganisms capable of syntrophic metabolism. Ann. N. Y. Acad. Sci. 1125, 58–72.

McKinlay, J.B., Zeikus, J.G., 2004. Extracellular iron reduction is mediated in part by neutral red and hydrogenase in Escherichia coli. Appl. Environ. Microbiol. 70, 3467–3474.

Menin, L., Asimakopoulos, K., Sukumara, S., Rasmussen, N.B., Patuzzi, F., Baratieri, M., Gavala, H.N., Skiadas, I.V., 2022. Competitiveness of syngas biomethanation integrated with carbon capture and storage, power-to-gas and biomethane liquefaction services: techno-economic modeling of process scenarios and evaluation of subsidization requirements. Biomass Bioenergy 161, 106475.

Mohanakrishna, G., Modestra, J.A., 2023. Value addition through biohydrogen production and integrated processes from hydrothermal pretreatment of lignocellulosic biomass. Bioresour. Technol. 369, 128386.

Mohanakrishna, G., Venkata Mohan, S., 2013. Multiple process integrations for broad perspective analysis of fermentative H2 production from wastewater treatment: technical and environmental considerations. Appl. Energy 107, 244–254.

Nagarajan, S., Jones, R.J., Oram, L., Massanet-Nicolau, J., Guwy, A., 2022. Intensification of acidogenic fermentation for the production of biohydrogen and volatile fatty acids—a perspective. Fermentation 8 (7), 325.

Nagarajan, V., Mohanty, A.K., Misra, M., 2016. Perspective on polylactic acid (PLA) based sustainable materials for durable applications: focus on toughness and heat resistance. ACS Sustain. Chem. Eng. 4 (6), 2899–2916. https://bioplasticsnews.com.

Neto, S.A., Forti, J.C., Andrade, A.R., 2010. An overview of enzymatic biofuel cells. Electrocatalysis 1, 87–94.

Ng, K.S., Hernandez, E.M., Sadhukhan, J., 2014. Biorefineries and Chemical Processes: Design, Integration and Sustainability Analysis. John Wiley & Sons.

Pandey, A., Mohan, S.V., Chang, J.S., Hallenbeck, P.C., Larroche, C. (Eds.), 2019. Biomass, Biofuels, Biochemicals: Biohydrogen. Elsevier.

Pasupuleti, S.B., Sarkar, O., Venkata, M.S., 2014. Upscaling of biohydrogen production process in semi-pilot scale biofilm reactor: evaluation with food waste at variable organic loads. Int. J. Hydrog. Energy 39 (14), 7587–7596.

Potter, M.C., 1910. On the difference of potential due to the vital activity of microorganisms. Proc. Univ. Durham Philos. Soc. 3, 245–249.

Prabakar, D., Manimudi, V.T., Sampath, S., Mahapatra, D.M., Rajendran, K., Pugazhendhi, A., 2018. Advanced biohydrogen production using pretreated industrial waste: outlook and prospects. Renew. Sust. Energ. Rev. 96, 306–324.

Rabaey, K., Rozendal, R.A., 2010. Microbial electrosynthesis-revisiting the electrical route for microbial production. Nat. Rev. Microbiol. 8, 706–716.

Raghavulu, S.V., Modestra, A.J., Amulya, K., Reddy, C.N., Venkata Mohan, S., 2013. Relative effect of bioaugmentation with electrochemically active and non-active bacteria on bioelectrogenesis in microbial fuel cell. Bioresour. Technol. 146, 696–703.

Raghavulu, S.V., Venkata Mohan, S., Venkateswar Reddy, M., Mohanakrishna, G., Sarma, P.N., 2009. Behavior of single chambered mediatorless microbial fuel cell (MFC) at acidophilic, neutral and alkaline microenvironments during chemical wastewater treatment. Int. J. Hydrog. Energy 34, 7547–7554.

Ramakrishna, S., 2021. Circular economy and sustainability pathways to build a new-modern society. Dry. Technol. 39 (6), 711–712.

Rathi, B.S., Kumar, P.S., Rangasamy, G., Rajendran, S., 2022. A critical review on biohydrogen generation from biomass. Int. J. Hydrog. Energy.

Rawoof, S.A.A., Kumar, P.S., Vo, D.V.N., Devaraj, T., Subramanian, S., 2021. Biohythane as a high potential fuel from anaerobic digestion of organic waste: a review. Renew. Sust. Energ. Rev. 152, 111700.

Reddy, M.V., Chandrasekhar, K., Venkata Mohan, S., 2011. Influence of carbohydrates and proteins concentration on fermentative hydrogen production using canteen based waste under acidophilic microenvironment. J. Biotechnol. 155, 387–395.

Reddy, M.V., Mohan, S.V., 2012. Effect of substrate load and nutrients concentration on the polyhydroxyalkanoates (PHA) production using mixed consortia through wastewater treatment. Bioresour. Technol. 114, 573–582.

Rockström, J., Steffen, W., Noone, K., Persson, Å., Chapin III, F.S., Lambin, E.F., Lenton, T.M., Scheffer, M., Folke, C., Schellnhuber, H.J., et al., 2009. Planetary boundaries: exploring the safe operating space for humanity. Ecol. Soc. 14 (2), 32.

Santhosh, J., Sarkar, O., Venkata Mohan, S., 2021. Green hydrogen-compressed natural gas (bio-H-CNG) production from food waste: organic load influence on hydrogen and methane fusion. Bioresour. Technol. 340, 125643.

Saratale, G.D., Chen, S., Lo, Y., Saratale, J.L.G., Chang, J., 2008. Outlook of biohydrogen production from lignocellulosic feedstock using dark fermentation: a review. J. Sci. Ind. Res. 67, 962–979.

Sarkar, O., Butti, S.K., Mohan, S.V., 2018. Acidogenic biorefinery: food waste valorization to biogas and platform chemicals. Waste Biorefin., 203–218.

Schroder, U., 2007. Anodic electron transfer mechanisms in microbial fuel cells and their energy efficiency. Phys. Chem. Chem. Phys. 9, 2619–2629.

Shanmugam, S., Sekar, M., Sivaramakrishnan, R., Raj, T., Ong, E.S., Rabbani, A.H., Rene, E.R., Mathimani, T., Brindhadevi, K., Pugazhendhi, A., 2021. Pretreatment of second and third generation feedstock for enhanced biohythane production: challenges, recent trends and perspectives. Int. J. Hydrog. Energy 46 (20), 11252–11268.

Singh, I., Rastogi, V., 2009. Performance analysis of a modified 4-stroke engine using biodiesel fuel for irrigation purpose. Int. J. Environ. Sci. 4, 229–242.

Singhania, R.R., Patel, A.K., Christophe, G., Fontanille, P., Larroche, C., 2013. Biological upgrading of volatile fatty acids, key intermediates for the valorization of biowaste through dark anaerobic fermentation. Bioresour. Technol. 145, 166–174.

Skizim, N.J., Ananyev, G.M., Krishnan, A., Dismukes, G.C., 2012. Metabolic pathways for photobiological hydrogen production by nitrogenase and hydrogenase-containing unicellular cyanobacteria cyanothece. J. Biol. Chem. 287, 2777–2786.

Sohn, Y.J., Kim, H.T., Baritugo, K.A., Jo, S.Y., Song, H.M., Park, S.Y., Park, S.K., Pyo, J., Cha, H.G., Kim, H., Na, J.G., 2020. Recent advances in sustainable plastic upcycling and biopolymers. Biotechnol. J. 15 (6), 1900489.

Srikanth, S., Venkata Mohan, S., Devi, M.P., Dinikar, P., Sarma, P.N., 2009a. Acetate and butyrate as substrates for hydrogen production through photo-fermentation: process optimization and combined performance evaluation. Int. J. Hydrog. Energy 34, 7513–7522.

Srikanth, S., Venkata Mohan, S., Devi, M.P., Peri, D., Sarma, P.N., 2009b. Acetate and butyrate as substrates for hydrogen production through photo-fermentation: process optimization and combined performance evaluation. Int. J. Hydrog. Energy 34 (17), 7513–7522.

Stegmann, P., Londo, M., Junginger, M., 2020. The circular bioeconomy: its elements and role in European bioeconomy clusters. Resources Conserv. Recycl.: X 6, 100029.

Subhash, G.V., Rohit, M.V., Devi, M.P., Swamy, Y.V., Venkata Mohan, S., 2014. Temperature induced stress influence on biodiesel productivity during mixotrophic microalgae cultivation with wastewater. Bioresour. Technol. 169, 789–793.

Subhash, G.V., Venkata Mohan, S., 2014a. Deoiled algal cake as feedstock for dark fermentative biohydrogen production: an integrated biorefinery approach. Int. J. Hydrog. Energy 39 (18), 9573–9579.

Subhash, G.V., Venkata Mohan, S., 2014b. Deoiled algal cake as feedstock for dark fermentative biohydrogen production: an integrated biorefinery approach. Int. J. Hydrog. Energy 39 (18), 9573–9579.

Subhash, G.V., Venkata Mohan, S., 2014c. Sustainable biodiesel production through bioconversion of lignocellulosic wastewater by oleaginous fungi. Biomass Convers. Biorefin. https://doi.org/10.1007/s13399-014-0128-4.

Sun, C., Xia, A., Liao, Q., Fu, Q., Huang, Y., Zhu, X., 2019. Life-cycle assessment of biohythane production via two-stage anaerobic fermentation from microalgae and food waste. Renew. Sust. Energ. Rev. 112, 395–410.

Suresh Babu, P., Srikanth, S., Dominguez-Benetton, X., Venkata Mohan, S., Pant, D., 2014. Dual gas diffusion cathode design for microbial fuel cell (MFC): optimizing the suitable mode of operation in terms of bioelectrochemical and bioelectro-kinetic evaluation. J. Chem. Technol. Biotechnol. https://doi.org/10.1002/jctb.4613.

Tawfik, A., Salem, A., El-Qelish, M., Abdullah, A.M., Abou Taleb, E., 2011. Feasibility of biological hydrogen production from kitchen waste via anaerobic baffled reactor (ABR). Int. J. Sustain. Water Environ. Syst. 2, 117–122.

Thepsithar, P., Kiong, M.K.E., Piga, M.B., Zengqi, M.X., Yin, S.J., Ming, L., Li, M.P., Xueni, M.G., Rosario, M.M.K.P., 2020. Alternative Fuels for International Shipping. Maritime Energy & Sustainable Development (MESD) Centre of Excellence, Nanyang Technological University.

Ubando, A.T., Chen, W.H., Hurt, D.A., Rajendran, S., Lin, S.L., 2022. Biohydrogen in a circular bioeconomy: a critical review. Bioresour. Technol., 128168.

Usmani, Z., Sharma, M., Awasthi, A.K., Sivakumar, N., Lukk, T., Pecoraro, L., Thakur, V.K., Roberts, D., Newbold, J., Gupta, V.K., 2021. Bioprocessing of waste biomass for sustainable product development and minimizing environmental impact. Bioresour. Technol. 322, 124548.

Usmani, Z., Sharma, M., Karpichev, Y., Pandey, A., Kuhad, R.C., Bhat, R., Punia, R., Aghbashlo, M., Tabatabaei, M., Gupta, V.K., 2020. Advancement in valorization technologies to improve utilization of bio-based waste in bioeconomy context. Renew. Sust. Energ. Rev. 131, 109965.

Vamshi Krishna, K., Sarkar, O., Venkata Mohan, S., 2014. Bioelectrochemical treatment of paper and pulp wastewater in comparison with anaerobic process: integrating chemical coagulation with simultaneous power production. Bioresour. Technol. 174, 142–151.

Vanhamäki, S., Virtanen, M., Luste, S., Manskinen, K., 2020. Transition towards a circular economy at a regional level: a case study on closing biological loops. Resour. Conserv. Recycl. 156, 104716.

Vázquez-Fernández, A., Suárez-Ojeda, M.E., Carrera, J., 2022. Bioproduction of volatile fatty acids from wastes and wastewaters: influence of operating conditions and organic composition of the substrate. J. Environ. Chem. Eng., 107917.

Velvizhi, G., Mohan, V., S., 2015. Bioelectrogenic role of anoxic microbial anode in the treatment of chemical wastewater: microbial dynamics with bioelectro-characterization. Water Res. 70 (1), 52–63.

Venkata Mohan, S., 2009a. Harnessing of biohydrogen from wastewater treatment using mixed fermentative consortia: process evaluation towards optimization. Int. J. Hydrog. Energy 34, 7460–7474.

Venkata Mohan, S., 2009b. Harnessing of biohydrogen from wastewater treatment using mixed fermentative consortia: process evaluation towards optimization. Int. J. Hydrog. Energy 34, 7460–7474.

Venkata Mohan, S., Amulya, K., Modestra, J.A., 2020. Urban biocycles—closing metabolic loops for resilient and regenerative ecosystem: a perspective. Bioresour. Technol. 306, 123098.

Venkata Mohan, S., Babu, M.L., Mohanakrishna, G., Sarma, P., 2009a. Harnessing of biohydrogen by acidogenic fermentation of Citrus limetta peelings: effect of extraction procedure and pretreatment of biocatalyst. Int. J. Hydrog. Energy 34 (15), 6149–6156.

Venkata Mohan, S., Babu, M.L., Mohanakrishna, G., Sarma, P., 2009b. Harnessing of biohydrogen by acidogenic fermentation of Citrus limetta peelings: effect of extraction procedure and pretreatment of biocatalyst. Int. J. Hydrog. Energy 34 (15), 6149–6156.

Venkata Mohan, S., Chandrasekhar, K., 2011. Solid phase microbial fuel cell (SMFC) for harnessing bioelectricity from composite food waste fermentation: influence of electrode assembly and buffering capacity. Bioresour. Technol. 102 (14), 7077–7085.

Venkata Mohan, S., Devi, M.P., 2012. Fatty acid rich effluents from acidogenic biohydrogen reactor as substrates for lipid accumulation in heterotrophic microalgae with simultaneous treatment. Bioresour. Technol. 123, 627–635.

Venkata Mohan, S., Devi, M.P., Subhash, G.V., Chandra, R., 2014a. In: Pandey, A., Lee, D.J., Chisti, Y. (Eds.), Biofuels From Algae. Elsevier, CJL, School, ISBN: 9780444595584, pp. 155–187. Chapter 8.

Venkata Mohan, S., Katakojwala, R., 2021. The circular chemistry conceptual framework: a way forward to sustainability in industry 4.0. Curr. Opin. Green Sustain. Chem. 28, 100434.

Venkata Mohan, S., Mohana Krishna, G., Sarma, P.N., 2008a. Effect of anodic metabolic function on bioelectricity generation and substrate degradation in single chambered microbial fuel cell. Environ. Sci. Technol. 42, 8088–8094.

Venkata Mohan, S., Mohanakrishna, G., Goud, R.K., Sarma, P.N., 2009c. Acidogenic fermentation of vegetable based market waste to harness biohydrogen with simultaneous stabilization. Bioresour. Technol. 100, 3061–3068.

Venkata Mohan, S., Mohanakrishna, G., Reddy, B.P., Sarvanan, R., Sarma, P.N., 2008b. Bioelectricity generation from chemical wastewater treatment in mediator less (anode) microbial fuel cell (MFC) using selectively enriched hydrogen producing mixed culture under acidophilic microenvironment. Biochem. Eng. J. 39, 121–130.

Venkata Mohan, S., Nikhil, G.N., Chiranjeevi, P., Reddy, C.N., Rohit, M.V., Kumar, A.N., Sarkar, O., 2016. Waste biorefinery models towards sustainable circular bioeconomy: critical review and future perspectives. Bioresour. Technol. 215, 2–12.

Venkata Mohan, S., Prathima Devi, M., Mohanakrishna, G., Amarnath, N., Lenin Babu, M., Sarma, P.N., 2011. Potential of mixed microalgae to harness biodiesel from ecological water-bodies with simultaneous treatment. Bioresour. Technol. 102, 1109–1117.

Venkata Mohan, S, Reddy, M.V., 2013. Optimization of critical factors to enhance polyhydroxyalkanoates (PHA) synthesis by mixed culture using Taguchi design of experimental methodology. Bioresour. Technol. 128, 409–416.

Venkata Mohan, S., Rohit, M.V., Chiranjeevi, P., Chandra, R., Navaneeth, B., 2014b. Heterotrophic microalgae cultivation to synergize biodiesel production with waste remediation: progress and perspectives. Bioresour. Technol. https://doi.org/10.1016/j.biortech.2014.10.056.

Venkata Mohan, S., Srikanth, S., Velvizhi, G., Lenin Babu, M., 2013b. Microbial fuel cells for sustainable bioenergy generation: principles and perspective applications. In: Gupta, V.K., Tuohy, M.G. (Eds.), Biofuel Technologies: Recent Developments. Springer, ISBN: 978-3-642-34518-0. Chapter 11.

Venkata Mohan, S., Veer Raghuvulu, S.V., Dinakar, P., Sarma, P.N., 2009d. Integrated function of microbial fuel cell (MFC) as bio-electrochemical treatment system associated with bioelectricity generation under higher substrate load. Biosens. Bioelectron. 24, 2021–2027.

Venkata Mohan, S., Velvizhi, G., Modestra, A.J., Srikanth, S., 2014c. Microbial fuel cell: critical factors regulating biocatalyzed electrochemical process and recent advancements. Renew. Sust. Energ. Rev. 40, 779–797.

Venkata Mohan, S., Velvizhi, G., Vamshi Krishna, K., Babu, M.L., 2014d. Microbial catalyzed electrochemical systems: a bio-factory with multi-facet applications. Bioresour. Technol. 165, 355–364.

Venkata Mohan, S., Venkateswar Reddy, M., Chandra, R., Venkata Subhash, G., Prathima Devi, M., Srikanth, S., 2013a. Bacteria for bioenergy: a sustainable approach towards renewability. In: Gaspard, S., Ncib, M.C. (Eds.), Biomass for Sustainable Applications: Pollution, Remediation and Energy. RSC Publishers, ISBN: 9781849736008.

Wallen, L.L, Rohwedder, W.K, 1974. Poly-beta-hydroxyalkanoate from activated sludge. Environ. Sci. Technol. 8 (6), 576 579.

Webster, K., 2015. The Circular Economy: A Wealth of Flows. Ellen MacArthur Foundation, Isle of Wight.

Yadav, M., Joshi, C., Paritosh, K., Thakur, J., Pareek, N., Masakapalli, S.K., Vivekanand, V., 2022. Reprint of: organic waste conversion through anaerobic digestion: a critical insight into the metabolic pathways and microbial interactions. Metab. Eng. 71.

Yahaya, E., Lim, S.W., Yeo, W.S., Nandong, J., 2022. A review on process modeling and design of biohydrogen. Int. J. Hydrog. Energy 47.

Zhang, X., Fevre, M., Jones, G.O., Waymouth, R.M., 2018. Catalysis as an enabling science for sustainable polymers. Chem. Rev. 118 (2), 839–885.

Zhao, F., Slade, R.C., Varcoe, J.R., 2009. Techniques for the study and development of microbial fuel cells: an electrochemical perspective. Chem. Soc. Rev. 38, 1926–1939.

Zhou, M., Jin, T., Wu, Z., Chi, M., Gu, T., 2012. Microbial fuel cells for bioenergy and bioproducts. In: Gopalakrishnan, K., Leeuwen, J.V., Brown, R. (Eds.), Sustainable Bioenergy and Bioproducts. Springer-Verlag, Berlin-New York, pp. 131–172.

Case studies

Bioremediation in Brazil: Recent evolutions and remaining challenges to boost up the bioeconomy

Geórgia Labuto[a,b], Lucélia Alcantara Barros[a],
Marcus Leonan Costa Guimaraes[a], Ricardo Santos Silva[a],
and Taciana Guarnieri Soares Guimarães[a]

[a]Laboratory of Integrated Sciences, Universidade Federal de São Paulo, São Paulo, Brazil
[b]Department of Chemistry, Universidade Federal de São Paulo, São Paulo, Brazil

1 Brazil and bioeconomy

There is a worldwide consensus that Brazil has enormous potential for developing the bioeconomy, given that it boasts approximately 20% of the planet's biodiversity and is a significant producer of biomass (Valli et al., 2018). With diverse biomes such as the Amazon, Cerrado, Atlantic Forest, Pampa, Caatinga, and Pantanal, each influenced by different latitudes and geographical aspects of its vast territory, and a coastal area that spans over 7000 km, Brazil has the opportunity to establish itself as one of the key players globally in the economic transition toward circular economy principles and sustainability (Maximo et al., 2022; Valli et al., 2018).

Several Brazilian economic sectors are making remarkable progress toward transitioning to sustainable practices, particularly in energy production from renewable sources that utilize different raw materials. The effects of climate change have propelled the search for decarbonizing processes, leading to the creation of the Environmental Social Governance (ESG) agenda that outlines a set of practices that companies committed to their social and environmental responsibilities must implement (Alkaraan et al., 2022). The sugar and alcohol sector, which accounts for 25% of Brazil's energy matrix by producing ethanol from sugarcane juice through fermentation, has been motivated to improve its practices by the ESG agenda.

The sector has the potential to increase its production by up to 50% while simultaneously reducing its greenhouse gas emissions by 30% through the conversion of its agro-industrial residues, such as bagasse and sugarcane straw, to 2.0 ethanol using an enzymatic cocktail derived from microorganisms found in Brazil (Macedo et al., 2008). In addition, another renewable fuel source is palm oil production, which is currently used in association with diesel, and in 2022, the production and commercialization of aviation fuels produced from palm oil were announced (Mobilidade Sustentável, 2022).

Brazil has been a pioneer in the production, use, and regulation of biofuels. However, other sectors require investment for development. Bio-innovation is urgent and is a fundamental principle for achieving economic development with responsibility and sustainability while respecting the ecosystem and promoting diversity and biodiversity.

This example must inspire and promote advances in other areas, such as bioenergy, chemical products from renewable sources, drug development, the production of biomaterials, and bio inputs. There is an urgent need to coordinate efforts between different sectors of society, including government, academia, businesses, and civil society, to share risks and provide guarantees between sectors that promote innovation, breaking away from traditional business ideas. In this way, Brazil can fully exploit its potential to explore raw materials from renewable sources and its biodiversity within the principles of bioeconomy, circular economy, and ecological economy.

Almost 9 years after the first edition, we are reviewing the Brazilian bioeconomy scenario, especially regarding the use of biomass in bioremediation processes. Bioremediation can be of great value in achieving two of the goals established by the United Nations in the 2030 Agenda for Sustainable Development, which must be met by 2030 (UN, 2015). These objectives are Zero Hunger (Goal 2), due to the potential for recovery of arable areas, and Clean Water and Sanitation (Goal 6), which is committed to improving water quality, reducing pollution, eliminating dumping, and minimizing the release of hazardous chemicals and materials, halving the proportion of untreated wastewater and substantially increasing recycling and safe reuse globally.

2 Overview of research in bioremediation in Brazil

Using the same search terms as the previous edition, namely bioremediation, biosorption, and phytoremediation, we searched the Web of Science and Scopus databases from January 1, 2015, to January 31, 2023. The search using all fields in Web of Science returned 32,012 scientific articles, while the search limited to title, abstract, and keywords in Scopus returned 40,398 articles. These numbers indicate a 23%–30% growth in global scientific production and interest in this field. Furthermore, of the reported articles, 1510 and 1716 were associated with Brazilian researchers in Web of Science and Scopus databases, respectively, representing a growth of 72%–98%, higher than the global scenario. This growth is even more significant, considering that only 8 years have elapsed since the previous edition.

During the 20 years analyzed in the previous chapter (1994–2014), Brazil ranked seventh in scientific production related to bioremediation, biosorption, and phytoremediation (Fig. 1A). However, in the new time interval (Fig. 1B), Brazil has risen significantly in this field and now

A) 1994 to 2014

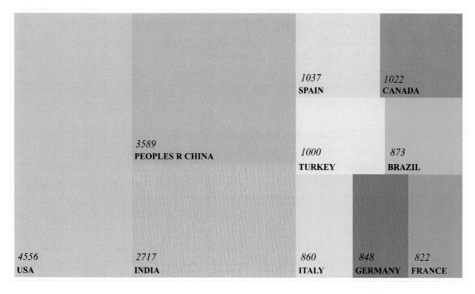

B) 2015 to 2023

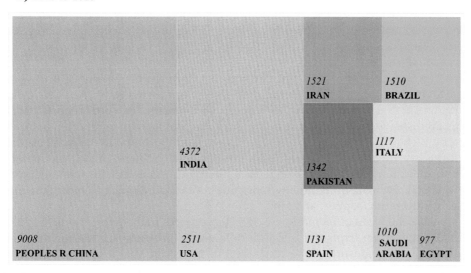

FIG. 1 Country-wise distribution of scientific articles related to bioremediation, biosorption, and phytoremediation according to the Web of Science database. (A) 1994–2014 (first 20 years) and (B) 2015–2023 (last 9 years).

occupies the fifth position in the world with a growth of 72% in the number of articles published over the last 8 years, compared with the previously period. Despite this growth, Brazil's participation in the total number of publications with the target terms remains around 4%. New countries have also entered the top 10, and there has been a change in the position of countries that previously led related research, such as the United States of America (USA) (8%

of the total), which is no longer the first producer of research, having been surpassed by China (27.4%), followed by India (14.8% of total).

We computed 50,304 citations for the 2383 Brazilian articles reposted on the Web of Science from 1994 to 2023 (Fig. 2A). This is a significant increase considering the 31,416 citations reported until 2014 (Fig. 2A) and represents an increment of 60% in the overall number of citations. Brazil is the second-highest country regarding citation growth, with China first with an 85% increase. Comparatively, India and the USA grew at just 47% and 17%, respectively. These numbers denote the interest and recognition of the quality of Brazilian research.

Overall, the trend of biomass application by Brazilian researchers appears to have remained steady in recent years. Bioremediation (1144 articles) continues to be the most widely used method, followed by phytoremediation (296 articles) and biosorption (270 articles). There has been a significant increase in scientific interest, with a 194% and 44% in the number of articles published on bioremediation and phytoremediation, respectively. Around 29% of the 383 articles published in the previous period were for biosorption. However, it should be noted that we are now dealing with only 8 years (2015–23), while previously, we presented results for 20 years (1994–2014). Unlike previously, bioremediation methods were more commonly used for inorganic species (370 articles, 32%) than organic species (244 articles, 21%). The same was observed for phytoremediation, with 171 articles aimed at removing inorganic species (42%) and 23 articles for organic species (22%), as well as for biosorption with 134 articles for inorganic species (49%) and 55 articles for organic species (20%). These numbers can be justified by the metallic species exhibiting recalcitrant and labile behavior, needing removal as they are not biodegradable and can undergo interconversions leading to more or less toxic oxidation states (da Silva et al., 2017). The use of microbial consortia has grown to remove both inorganic and organic contaminants.

To analyze the interconnections of publications on bioremediation, biosorption, and phytoremediation in Brazil, a bibliometric analysis was conducted using the VOSviewer software (version 1.6.15) (Andrade et al., 2019) based on the main keywords reported in the articles. Fig. 2B displays the bibliometric map for the 36 main keywords reported in all articles retrieved by Web of Science. The size of the clusters is proportional to the number of occurrences of the keyword, while different colors correspond to themes and lines interconnect themes by applications. Brazilian research has mainly concentrated on the treatment of water, sewage, and soil, focusing on removing inorganic contaminants such as heavy metals, cadmium, lead, chromium, copper, and zinc.

In comparing the types of biomasses used as biosorbents in the two periods (Fig. 3A), the reduction of applications of plant materials is noteworthy, whose space was mainly occupied by the use of microbial consortia, followed by animal materials. Microbial consortia have been gaining prominence mainly due to their increasing use in wastewater treatment, showing an excellent capacity for removal and degradation of organic and inorganic pollutants such as petroleum, emerging contaminants, nitrogen, and phosphate (Valentim dos Santos et al., 2016; Villela et al., 2019). This fact was evidenced by a search in the Scopus database, which returned 46 articles restricted to the keyword "microbial consortium," of which 14 referred to wastewater treatment (30% of the total). Considerable growth has also been noted in the use of fungi (10%), animal materials (8%), and yeast (5%), mainly related to soil and wastewater bioremediation, with yeast being applicable for the production of biomaterials such as

A)

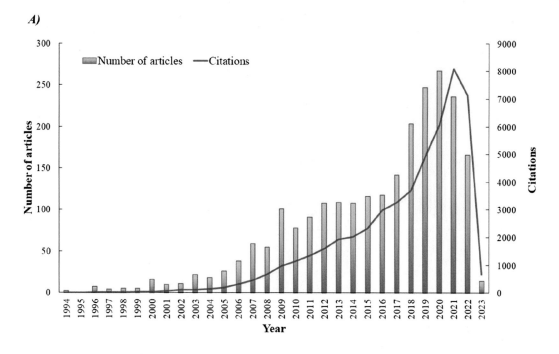

B)

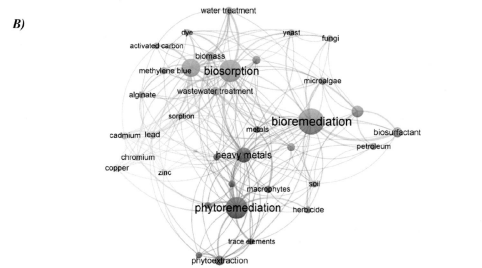

FIG. 2 (A) Number of articles published by Brazilian scientists and their corresponding citations between 1994 and 2023. (B) Bibliometric map generated using the 36 primary keywords from articles retrieved from the Web of Science using the target terms bioremediation, biosorption, and phytoremediation. *Source: Web of Science and Scopus websites accessed on 09/02/2023; VOSviewer (version 1.6.15).*

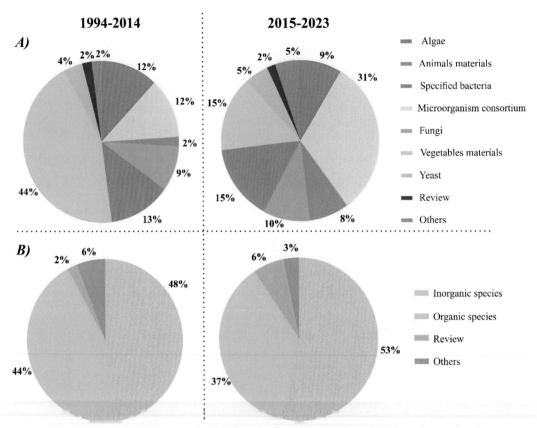

FIG. 3 (A) Comparison of biomass types used as biosorbents in articles published by Brazilian researchers between 1994 and 2014 and between 2015 and 2023. (B) Comparison of contaminant or analyte types of interest in biosorption studies published by Brazilian researchers from 1994 to 2014 and from 2015 to 2023. Target terms searched in the Web of Science: bioremediation, biosorption, and phytoremediation.

biofilms, bioseparators, and biosynthesizers (Abe et al., 2022; Salvadori et al., 2022). Curiously, researchers reporting the use of agricultural residues and lignocellulosic materials have decreased compared with other biomasses, despite being low-cost and abundant options in Brazil.

According to Scopus, 603 articles were published between 2015 and 2023, exploring the application of bioremediation, biosorption, and phytoremediation methods for inorganic species, representing 53% of the total (Fig. 3B) and focusing on metals such as iron, copper, zinc, manganese, and nickel and also for nitrogen and phosphorus. On the other hand, organic species were found in 419 articles, making up 37% of the publications (Fig. 3B). The primary applications for organic species have been focused on removing dyes, herbicides, hydrocarbons, and petroleum.

To verify which main funding agencies contributed to the funding of articles published by Brazilian scientists, the same target terms were used to search the Web of Science from 1994 to 2023, with filters for Brazil and the Funding agency. The primary sources of reported funding, with 1051 and 893 mentions, respectively, were the Brazilian National Council for Scientific and Technological Development (Conselho Nacional de Desenvolvimento Centífico e Tecnológico, CNPq) and the Coordination of Superior Level Staff Improvement (Coordenação de Aperfeiçoamento de Pessoal de Nível Superior, CAPES) maintained by the Brazilian federal government. It is worth noting that both CNPq and CAPES, with the latter contributing more, are institutions that fund scholarships. The other funding agencies that stood out with mentions of 339, 178, and 85, respectively, were state agencies, namely the São Paulo Research Foundation (Fundação de Amparo à Pesquisa do Estado de São Paulo, FAPESP), the research supporting foundation of Minas Gerais state (Fundação de Amparo à Pesquisa do Estado de Minas Gerais, FAPEMIG), and the Carlos Chagas Filho Foundation for Research Support in the State of Rio de Janeiro (Fundação de Amparo à Pesquisa do Estado do Rio de Janeiro, FAPERJ). This situation indicates the preponderance of public sector funding for research.

Searches were conducted on the Lattes Platform, which consolidates the academic profiles of all active Brazilian researchers, and the Directory of Research Groups of CNPq to assess changes in the number of Brazilian researchers and research groups associated with the target terms bioremediation, biosorption, or phytoremediation. Additionally, the number of theses and dissertations defended in Brazil were surveyed using the repository maintained by CAPES (CAPES, n.d.). Table 1 presents the compiled data.

There appears to be an insignificant reduction in the number of research groups working on bioremediation. However, the number of doctors working on this topic has more than doubled in the last 8 years, showing a growth of 209%. For the term phytoremediation, five new research groups were established. The number of research groups that include biosorption and bioeconomy in their area of activity also grew, with the emergence of eight and 53 groups of researchers, respectively. The growth observed for the number of research groups is probably the result of an increase in the number of doctors who declare themselves active in research with bioremediation (109%), phytoremediation (133%), biosorption (129%), or bioeconomy (1577%).

The FAPESP is one of Brazil's most prominent research promotion institutions. The FAPESP Virtual Library (FAPESP, n.d.) was searched to determine the number of research

TABLE 1 Overview of research groups, PhD researchers, and completed theses and dissertations associated with the target terms of bioremediation, phytoremediation, biosorption, and bioeconomy in Brazil for 2014 and 2023.

Keyword	Scenery in 2014				Scenery in 2023			
	Research groups	Number of PhDs	Concluded Dissertation	Thesis	Research groups	Number of PhDs	Concluded Dissertation	Thesis
Bioremediation	144	1158	150	48	142	2430	858	338
Phytoremediation	59	587	29	17	64	1371	360	135
Biosorption	18	357	7	2	26	819	294	138
Bioeconomy	8	53	2	2	61	889	84	48

Data source: Lattes Platform (CNPq) and Carlos Chagas Platform (CNPq), Accessed 31 January 2023.

grants and scholarships financed by the institution related to the target terms of this chapter (FAPESP, n.d.). A total of 107 research projects and 170 scholarships, completed and in progress, were identified for the four keywords used in this chapter. Additional keywords associated with each target term were also surveyed, and the results are listed below:

- Bioremediation: biodegradation, bacteria, biosurfactants, wastewater treatment, fungi, hydrocarbons, environmental microbiology, pesticides, enzymes, toxicity.
- Phytoremediation: soil pollution, lead, heavy metals, plant metabolism, legumes, edetic acid, sugarcane, experimental design, phytotoxicity, and herbicides.
- Biosorption: yeasts, biodegradation, cholecalciferol, microencapsulation, *Saccharomyces cerevisiae*, *Allium cepa*, biomass, meat and derivatives, europium, and brewing industry.
- Bioeconomy: bioenergy, biorefineries, biomass, sustainability, biotechnology, innovation, mathematical models, sustainable agriculture, and computational learning.

Fig. 4 displays the distributions of Brazil's most commonly used types of biomasses and their main application areas, comparing publications between the previously discussed period and the years 2015–23.

3 Advancements in regulatory frameworks associated with the bioeconomy in Brazil

Brazil does not have a dedicated state organization for coordinating the development of the bioeconomy. Instead, governance of this issue is shared by various organizations, ministries, and institutions, both for-profit and non-profit, without an established strategy for cooperation or coordination. This lack of coordination is reflected in the regulatory framework, which comprises a set of laws, decrees, norms, and treaties that are the legal basis for developing the bioeconomy in Brazil. Table 2 presents Brazil's most relevant regulatory frameworks related to biomass exploration, including recent updates and additions highlighted in bold.

Although there is no specific regulatory framework for bioeconomy in Brazil, a legal framework enables the process of encouraging its development. Furthermore, since the previous chapter, there has been an evolution in the Brazilian regulatory framework concerning issues related to the bioeconomy.

Several laws and regulatory decrees have been established, which create conditions for developing products and services that consider Brazil's vast biological diversity. For example, in 2015, Law N° 13123 (Brasil, 2015) was enacted, providing access to Brazilian genetic heritage or associated traditional knowledge by foreign legal entities, provided that they are linked to Brazilian scientific research institutions. The Brazilian Institute of Environment and Renewable Natural Resources (IBAMA) is responsible for issuing special authorization for access to genetic heritage for scientific research. The Genetic Heritage Management Council (Conselho de Gestão do Patrimônio Genético, CGen) controls access to traditional knowledge.

In 2016, a new Decree N° 8772 (Brasil, 2016a) was formulated, establishing new rules and conditions for accessing genetic samples and related information for research and technological development. Another important aspect of this new regulatory framework was the creation of the National System for Management of Genetic Heritage and Associated Traditional

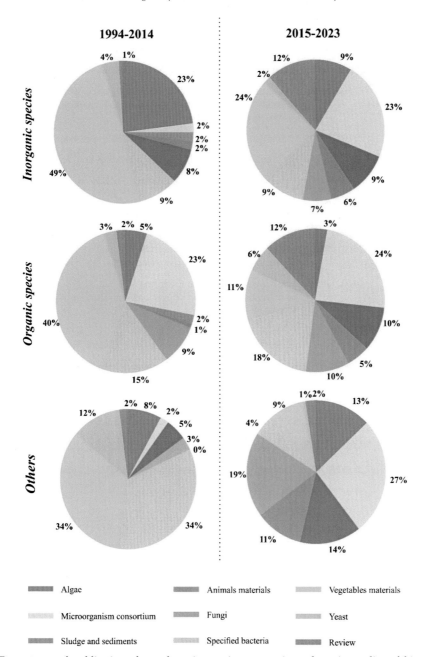

FIG. 4 Percentages of publications focused on inorganic or organic analytes in studies of bioremediation, biosorption, or phytoremediation by Brazilian researchers between 2015 and 2023. *Source: Web of Science and Scopus website on 09/02/2023.*

TABLE 2   Main Regulatory Frameworks Related to Biomass Exploration in Brazil. In bold are detached the new documents.

Regulatory framework	Date	Determination of subject of regulatory framework
Law N° 6938	31/08/1981	Defines the National Environmental Policy, its purposes and mechanisms of formulation and implementation, and settles other measures
Law N° 8078	11/09/1990	Explain consumer protection and settles other measures
Law N° 9279	14/05/1996	Regulates rights and obligations relating to industrial property
Law N° 9456	25/04/1997	Establishes the Law of Cultivated Plant Protection and settles other measures
Law N° 9610	19/02/1998	Changes, updates, and consolidates copyright law and settles other measures
Law N° 9782	26/01/1999	Defines the National Sanitary Surveillance System, creates the National Health Surveillance Agency and settles other measures
Law N° 10196	14/02/2001	Changes and adds to the terms of Law N° 9279 of May 14, 1996, regulating rights and obligations regarding industrial property and presents other terms
Law N° 10603	17/12/2002	Resolves the protection of information not disclosed and submitted for approval to commercialization of products and settles other measures
Law N° 10973	02/12/2004	Predict incentive to innovation and scientific and technological research in the productive environment and makes other measures
Law N° 11097	13/01/2005	Predict the introduction of biodiesel in the Brazilian energy matrix
Law N° 11105	24/03/2005	Predict technical support to the federal government and the implementation of the national biosafety policy regarding genetically modified organisms
Law N° 11196	21/11/2005	Also called Lei do Bem (Law of Good) and grants tax incentives to companies that carry out research and development of technological innovation
Law N° 11516	28/08/2017	Creates the Institute for Biodiversity Conservation—Chico Mendes Institute
Law N° 12490	16/09/2011	Regulates the policy and supervision of activities related to the national fuel supply
Law N° 13123	20/05/2015	Resolves the access to genetic resources, protection and access to associated traditional knowledge, and benefits sharing for the conservation and sustainable use of biodiversity. Revoke the provisional measure N° 2186—16th of August 2001, and settles other measures
Law N° 13243	11/01/2016	Predict incentives for scientific development, research, scientific and technical education and innovation
Law N° 13576	26/12/2017	Deals with the National Biofuels Policy (RenovaBio) and other provisions

TABLE 2   Main Regulatory Frameworks Related to Biomass Exploration in Brazil. In bold are detached the new documents—cont'd

Regulatory framework	Date	Determination of subject of regulatory framework
Complementary law N° 177	12/01/2021	Modify the nature and sources of income of the National Fund for Scientific and Technological Development (Fundo Nacional de Desenvolvimento Científico e Tecnológico, FNDCT). Include programs developed by social organizations among the institutions that can access FNDCT resources, including support for Science, Technology and Innovation, projects and activities earmarked for neutralizing greenhouse gas emissions and promoting the development of the bioeconomy sector
Decree N° 75572	08/04/1975	Promulgates the Paris Convention for the Protection of Industrial Property Review, Stockholm, 1967
Decree N° 635	21/08/1992	Promulgates the Paris Convention for the Protection of Industrial Property, revised in Stockholm on July 14th, 1967
Legislative Decree N° 2	03/02/1994	Approves the text of the convention on biological diversity, signed at the United Nations Conference on Environment and Development held in Rio de Janeiro from June 5 to June 14th, 1992
Decree N° 2366	05/11/1997	Regulates Law N° 9456, of April 25, 1997, establishing the cultivated plant protection on the National Service for Plant Variety Protection (Serviço Nacional de Proteção de Cultivares, SNPC) and other measures
Decree N° 2519	16/03/1998	Enacts the convention on biological diversity signed in Rio de Janeiro on June 5th, 1992
Decree N° 4074	04/01/2002	Regulates Law N° 7802, of July 11, 1989, which provides for research, experimentation, production, packaging and labeling, transport, storage, marketing, commercial advertising, use, import, export, waste disposal and packaging, registration, classification, control, inspection and surveillance of pesticides, their components and the like, and settles other measures
Decree N° 4339	22/08/2002	Establishes principles and guidelines for the implementation of the National Biodiversity Policy
Decree N° 4680	24/04/2003	Regulates the right to information, as provided by Law N° 8078 of September 11, 1990, as foods and food ingredients intended for human or animal consumption containing or produced from GMOs, subject to compliance with other applicable rules
Decree N° 4703	21/05/2003	Creation of the National Biological Diversity Program (Programa Nacional da Diversidade Biológica, Pronabio) to organize commitments to implement the Convention on Biological Diversity
Decree N° 5591	22/11/2005	Regulates provisions of Law N° 11105, of March 24, 2005, which regulates items II, IV, and V of Paragraph 1 of Article 225 of the Brazilian Federal Constitution and settles other measures
Decree N° 5950	31/10/2006	Regulates Article 57-A of Law N° 9985 of July 18, 2000, to establish the limits for planting genetically modified organisms (GMOs) in surrounding protected areas

Continued

TABLE 2 Main Regulatory Frameworks Related to Biomass Exploration in Brazil. In bold are detached the new documents—cont'd

Regulatory framework	Date	Determination of subject of regulatory framework
Decree N° 6041	08/02/2007	Establishes the development policy of biotechnology, creates the National Biotechnology Committee, and settles other measures
Decree N° 5	10/01/2011	Establish measures aimed at promoting the production and use of forest biomass
Decree N° 179	03/08/2012	Makes the second amendment to Decree-Law N° 5/2011, of 10 January, which establishes measures to promote the production and use of forest biomass
Decree N° 166	21/08/2015	Makes the first change to Decree-Law N° 5/2011, of 10 January, which establishes measures to promote the production and use of biomass to ensure the supply of dedicated forestry biomass plants
Decree N° 8772	20/05/2016	Regulates Law N° 13123, of May 20, 2015, which provides for access to genetic heritage, protection and access to associated traditional knowledge and sharing of benefits for the conservation and sustainable use of biodiversity
Decree N° 64	12/06/2017	Approves the system for the new forest biomass plants
Decree N° 9283	07/02/2018	Establishes measures to develop innovation and scientific and technological research in the productive environment with a view to technical training, technological autonomy and the development of national and regional productive systems
Decree N° 9888	27/06/2019	Provides for the definition of mandatory annual targets for reducing greenhouse gas emissions for the sale of fuels dealt with in Law N° 13576 of December 26, 2017, and establishes the National Biofuel Policy Committee—RenovaBio Committee
Decree N° 136	12/08/2020	Approves the Nagoya Protocol text on access to genetic resources and fair and equitable sharing of benefits derived from their use to the convention on biological diversity
Decree N° 11427	02/03/2023	Approves an organizational structure and the positions and functions of trust of the Ministry of Development, Industry, Commerce and Services
Paris convention of union for the protection of intellectual property	20/03/1883	Provides international criteria for the granting and validity of industrial patents
Treaty of cooperation in patents (PCT)	19/06/1970	Establishes the principle of a single application for a patent valid for all countries adhering to the PCT
Treaty of Budapest	28/04/1977	Provides for the International Recognition of the Deposit of Microorganisms for patent proceedings
CTNBio[a] resolution N° 1	05/09/1996	Approving the Internal Regulations of the National Technique Commission of Biosafety (CTNBio)

TABLE 2 Main Regulatory Frameworks Related to Biomass Exploration in Brazil. In bold are detached the new documents—cont'd

Regulatory framework	Date	Determination of subject of regulatory framework
CNS[b] resolution N° 196	10/10/1996	Approves the regulatory guidelines and standards for research involving human subjects
Normative Act N° 127	05/03/1997	Provides for applying the Industrial Property Law regarding patents and certificates of addition to the invention
CNS[b] resolution N° 251	07/08/1997	Provides for applying the Industrial Property Law regarding patents and certificates of addition to the invention
CNS[b] resolution N° 292	08/07/1999	Complementary rule to CNS Resolution 196/96, on the specific area of research on human beings, outside of coordinates or with foreign participation and research involving the sending of biological material abroad
CONAMA[c] resolution N° 314 revoked Replaced by CONAMA[c] resolution N° 463	29/07/2014	Deals with environmental control of remedial products
Normative resolution N° 02	27/11/2006	Provides for classifying risks of genetically modified organisms (GMO) and the biosafety levels to be applied in activities and projects with GMOs and their derivatives in containment
Normative resolution N° 5	12/03/2008	Provides rules for the commercial release of genetically modified organisms and their derivatives
IBAMA[d] normative instruction N° 5 revoked Replaced by IBAMA[d] normative instruction N° 11	17/10/2022	Establishes the procedures and requirements to be adopted for registration, registration renewal, and prior consent to conduct research, importation and experimentation with remediation products
IBAMA[d] normative technical guideline—OTN N°3-DIQUA	21/10/2022	Establishes the classification of products intended for remediation
ANP[e] resolution N° 758	27/11/2018	Regulates the certification of efficient production or importation of biofuels referred to in art. 18 of Law N° 13576, of December 26, 2017, and the accreditation of inspection firms
ANP[e] resolution N° 802	05/12/2019	Establishes the procedures for generating the ballast necessary for the primary issuance of Decarbonization Credits, referred to in art. 14 of Law N° 13576, of December 26, 2017, and amends ANP Resolution N° 758, of November 23, 2018
MME[f] ordinance N° 56	21/12/2022	The Ordinance regulates the issuance, bookkeeping, registration, marketing and withdrawal of Decarbonization Credits based on the National Biofuel Policy (RenovaBio)

Continued

TABLE 2   Main Regulatory Frameworks Related to Biomass Exploration in Brazil. In bold are detached the new documents—cont'd

Regulatory framework	Date	Determination of subject of regulatory framework
MAPA[g] ordinance N° 121	18/06/2019	Establishes actions to strengthen production chains that sustainably use natural resources. Structured in five thematic axes, the Ministry of Agriculture will act in each with 05 objectives
MCTI[h] ordinance N° 3877	09/10/2020	Establishes, within the scope of the Ministry of Science, Technology and Innovation, the MCTI Bioeconomy Productive Chains Program
FNRB[i] decree N° 8772	11/05/2016	Created by Law N° 13123/2015 and is regulated by Regulation N° 8772/2016. The foundation's main objective is to promote the appreciation and sustainable use of genetic heritage and related traditional knowledge

[a] *Comissão Técnica Nacional de Biossegurança (National Technical Commission on Biosafety).*
[b] *Conselho Nacional de Saúde (National Health Council).*
[c] *Conselho Nacional do Meio Ambiente (National Environment Council).*
[d] *Instituto Brasileiro do Meio Ambiente e dos Recursos Renováveis (Brazilian Institute of Environment and Renewable Natural Resources).*
[e] *Agência Nacional do Petróleo, Gás Natural e Biocombustíveis (National Agency for Petroleum, Natural Gas and Biofuels).*
[f] *Ministério de Minas e Energia (Ministry of Mines and Energy).*
[g] *Ministério da Agricultura e Pecuária e abastecimento (Ministry of Agriculture and Livestock and Food Supply).*
[h] *Ministério da Ciência, Tecnologia e Inovação (Ministry of Science, Technology and Innovation).*
[i] *Fundo Nacional para a Repartição de Benefícios (National Fund for Benefit Sharing).*

Knowledge (Sistema Nacional de Gestão do Patrimônio Genético e do Conhecimento Tradicional Associado, SisGen) (SisGen, n.d.). This electronic platform aims to facilitate the law's implementation, streamlining the process and allowing users to register their research activities online without prior authorization from SisGen.

These initiatives can potentially accelerate the development of research processes and the utilization of natural resources. However, some academics argue that Decree 8772/2016 (Brasil, 2016a) could be further streamlined to reduce bureaucracy (Silva and Oliveira, 2018). Nevertheless, foreign access to Brazilian genetic heritage remains a contentious issue for researchers, jurists, and civil society organizations. They are concerned that such access could lead to inefficient transfer of benefits generated by exploitation, business sector influence on legislation, biopiracy, loss of autonomy over Brazilian genetic heritage, and so on (Fittipaldy et al., 2020; Machado et al., 2017). Brazil has recently faced instances of biopiracy, such as patent applications for Brazilian plant species (Fittipaldy et al., 2020; Nogueira et al., 2010; Rezende and Ribeiro, 2009; Rojahn, 2010).

Brazil has established a legal framework in biosafety ough Law N° 11,105 (Brasil, 2005a), enacted on November 24, 2005, and Decree N° 5591 (Brasil, 2005b), issued on the same day. These regulations establish safety guidelines, inspection mechanisms, and activity records for genetically modified organisms (GMO) to promote scientific progress in biotechnology and biosafety while ensuring human, animal, and plant health protection. In addition, the biosafety law also created the National Biosafety Council (Conselho Nacional de Biossegurança, CNBS) and redefined the National Technical Commission on Biosafety (Comissão Técnica Nacional de Biossegurança, CTNBio). These bodies are responsible for establishing safety standards and regulations for inspecting activities involving GMO and formulating and implementing the National Biosafety Policy.

On January 11, 2016, Law N° 13243 (Brasil, 2016b) was enacted to encourage scientific and technological innovation in the productive environment and to alleviate challenges related to research and recall activities. Additionally, Decree N° 9283 (Brasil, 2018) was issued on July 2, 2018, to regulate and foster the development of Brazil's innovation environment. This regulatory framework outlines measures to stimulate innovation and scientific and technological research to promote technological autonomy and the development of national and regional production systems.

A significant development during this period was the creation of a substitute normative instruction by the Ministry of the Environment (Ministério do Meio Ambiente, MMA) in conjunction with the Brazilian Institute of Environment and Renewable Natural Resources. This instruction provides guidelines for registering remedial products, renewals, obtaining prior consent for importation, authorizations for research and experimentation, and other necessary measures (Normative Instruction N° 11, issued on October 17, 2022) (IBAMA, 2022a).

This normative instruction mandates that all environmental remediators must be registered with the Brazilian Institute of Environment and Renewable Natural Resources (IBAMA). Additionally, it clarifies and defines the products used for bioremediation, as presented in Table 3.

The normative also establishes legal requirements for product registration, commercialization, importation, and research authorization. It defines research and experimentation in the area as activities related to the preparation or application of a remedy on a pilot scale and under controlled conditions, aimed at obtaining knowledge related to it for recording purposes or for changing the characteristics or indications for the use of a remedial product already registered. Additionally, the Normative Technical Guideline—OTN N° 3-DIQUA on October 21, 2022 (IBAMA, 2022b) classifies remediators based on three different aspects:

TABLE 3 Definitions of uses for products for environmental bioremediation in brazil established by the Ministry of the Environment (MMA) in conjunction with the Brazilian Institute of the Environment and Renewable Natural Resources, by N° 11 (October 17th, 2022) (IBAMA, 2022a).

Kind of product	Definition
Remediator	Physical, chemical, or biological process product or agent intended for the recovery of contaminated environments and ecosystems and for the treatment of effluents and waste
Biorremediator	A remediator that contains microorganisms as the active ingredient, which are capable of reproducing and biochemically degrading contaminating compounds and substances
Bioestimulator	Remediator that favors the growth of microorganisms naturally present in the environment and capable of accelerating the degradation process of contaminating compounds and substances
Chemical or physicochemical remediator	Remediator that presents as an active ingredient substance or chemical compound capable of degrading, adsorbing, or absorbing compounds and contaminating substances
Phytoremediator	A plant used as a remedial agent in order to remove, immobilize, or reduce the potential of organic and inorganic contaminants present in soil or water

TABLE 4 Classification of products used for environmental bioremediation in Brazil established by the Ministry of the Environment (MMA) in conjunction with the Brazilian Institute of the Environment and Renewable Natural Resources by N° 3 (October 21st, 2022) (IBAMA, 2022b).

Classification	Detailing
According to the nature of the treatment: whether it is based on physical, chemical, or biological phenomena	Biological
	Physical
	Physicochemical
	Chemical
	Thermic
According to the type of technique used: if it removes, separates, stabilizes or contains the contaminant	Containment of contaminants
	Removal or reduction of contaminants
	Solidification of contaminants
According to the mode of action: by degradation or sorption	Abiotic degradation
	Biotic degradation (biodegradation)
	Sorption

nature of the treatment, type of technique used, and mode of action. Table 4 provides a detailed breakdown of the classifications.

The recent Decree N° 11427 of March 2, 2023 (Brasil, 2023), approved the organizational structure and list of trust positions and functions of the Ministry of Development, Industry, Commerce, and Services (Ministério do Desenvolvimento, Indústria, Comércio e Serviços). This federal government agency has areas of activity that include the development policy of intellectual property, technology transfer, commerce, industry and services, and foreign trade policies. Additionally, it is responsible for participation in international negotiations, development of the green economy, decarbonization, and bioeconomy, providing for the reallocation and transformation of trust positions and functions. In Article 33 of the Decree in question, the responsibilities of the Secretariat for Green Economy, Decarbonization, and Bioindustry are established, which include proposing, implementing, and evaluating public policies to encourage businesses that generate social and environmental impact, stimulating access to public and private capital for such businesses, promoting regulatory advancement, and supporting the development of businesses that protect the environment. Additionally, the Secretariat must propose policies for professional and technological training related to a sustainable economy, improve national and international legal frameworks related to biodiversity and climate change, and collaborate with civil society organizations to implement policies within its competence.

4 Registration and trade of bioremediators in Brazil

As previously mentioned, authorization and registration at IBAMA are required for an environmental bioremediation product to be marketed or used for research in Brazil. Currently, there are 79 registered products classified as physicochemical or bioremediators, with only one product being used for research and experimentation (IBAMA) (IBAMA, 2022c, d, e). Table 5 presents the number of products according to their type and indication for use.

All 49 bioremediators registered in Brazil are formulated using bacteria, mostly belonging to the genus *Bacillus*. These bioremediators are used for organic composting waste, treating effluents, and restoring water bodies and in septic tanks, grease traps, and sanitary sewers. The commercialized remediators are classified as physicochemical and consist of different chemical substances used for the same purposes mentioned above or for specific purposes.

An online search was conducted for the 79 products registered with IBAMA to evaluate the availability of these remedial and bioremediation products in the Brazilian market, and only two of the products were not found for sale. In the search for bioremediation products marketed in Brazil, three products did not have registrations with IBAMA, which is necessary according to Brazilian regulations (Frias, 2020; Frias and Santos, 2020). Nevertheless, these products were registered as biological sanitizing products with the National Health Surveillance Agency (Agência Nacional de Vigilância Sanitária, ANVISA) (ANVISA, 2021a, b, c).

A survey was carried out using the target terms Bioremediation, Biosorption, and Phytoremediation, along with the keyword Remediation and inserting these words separately in the title and summary fields to evaluate the growth in the number of patent registrations deposited at the National Institute of Industrial Property (Instituto Nacional da Propriedade Industrial, INPI) (INPI, 2022). The data were collected and separated to obtain the number of patents filed in the last 8 years and the 20 years described. In addition, the results were checked to avoid duplicates. Fig. 5 compares Brazilian patent filings for the searched terms between the two periods.

Although in total numbers, the number of patents deposited has not yet surpassed those reported in the previous period of 20 years, there is a growing trend since, in just 8 years, 88% and 62% have already been reached in deposits containing the target terms bioremediation and remediation, respectively. More than half the number of patents containing these terms were produced in less than half the period in years. For the target term phytoremediation, there is a reduction in the number of new patents filed.

TABLE 5 Bioremediators registred at IBAMA in Brazil.

Kind of product	Number of registers	Indication of use
Physicochemical remediators	30	Effluent, water, and soil treatment
Bioremediator	49	Composting, effluent, water, and soil treatment, petrochemical industry
Remediator to search and experimentation	1	Effluent, water, and soil treatment

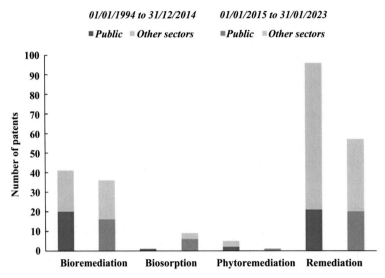

FIG. 5 Brazilian patent deposit records from 01/01/1994 to 31/12/2014 and from 01/01/2015 to 31/01/2023 were obtained from the INPI database by searching for the target terms Bioremediation, Biosorption, and Phytoremediation and keyword Remediation with the respective deposit percentages carried out by Brazilian universities. *Source: INPI.*

For the patents associated with the target term biosorption, there was a growth of 900%, which is noteworthy, considering biosorption as a field to boost the bioeconomy in Brazil if the relevant products and/or processes are inserted in the market.

Advanced research was conducted to identify the main sectors depositing those patents. First, the field name of the depositor/holder was searched using the terms university, college, or institute. Then, a comparative search was conducted for both reported periods, using the keywords established in the title and abstract fields. The results showed a vital contribution from public research institutions, mainly universities, indicated by the darker colors in the graph (Fig. 5).

Brazil has shown great potential for depositing patents related to bioremediation, including using residues such as those from agro-industrial processes (e.g., açaí seeds) and utilizing microorganisms and microalgae. In April 2012, INPI created the Priority Program for Green Patents to accelerate the analysis of patent applications involving environmentally friendly technologies so that Brazil can keep up with global trends in this area. Although the program was only implemented in 2016, it has already led to the granting of patents within a few years, promoting innovation in the bioeconomy and the rapid availability of technologies designed to benefit the environment and society. This initiative may have contributed to the recent increase in patent numbers.

4.1 Patents related to activated carbon in Brazil

Activated carbon is a commodity widely employed in different industrial sectors related to water treatment, air purification, food and beverage processing, pharmaceutical industry, environmental remediation, removing impurities and contaminants in the refining of petrochemicals and gas purification, production of personal care products (e.g., toothpaste, soaps, and face masks), energy storage, and catalysis (bcc Research, 2018; Fortune Business Insights, 2021; Grand View Research, 2018). As well as biomass, activated carbon can be functionalized to obtain specific characteristics for a specific use (Carvalho et al., 2021; José et al., 2019).

Activated carbon is conventionally produced from non-renewable sources such as coal and petroleum, raising concerns about sustainability and environmental impact (Gan, 2021; Ioannidou and Zabaniotou, 2007; Köseoğlu and Akmil-Başar, 2015). However, in theory, any carbon-rich material can be converted to charcoal and activated carbon. That means that biomass from various sources such as agricultural waste, forest residues, fermentative processes, and energy crops can be used to produce activated carbon, reducing dependence on non-renewable resources and promoting the use of waste streams (Danish and Ahmad, 2018; Rashidi et al., 2012; Reza et al., 2020). Utilizing biomass from renewable sources to produce activated carbon offers several advantages and can boost the bioeconomy (Modesto et al., 2020).

There are several advantages associated with using biomass-based activated carbon. Firstly, the biomasses are derived from renewable sources, making them abundant and ensuring a sustainable raw material supply. In addition, plant-based activated carbon has a lower carbon footprint than conventional activated carbon, as plants absorb carbon dioxide from the atmosphere during photosynthesis. Moreover, biomass-based activated carbon can be customized to suit specific applications, as the properties of activated carbon can be fine-tuned by varying the raw material source, processing conditions, and activation method. This results in activated carbon with tailored pore structures, surface chemistries, and surface areas suitable for various applications (Danish and Ahmad, 2018; Joseph et al., 2020; Khajeh et al., 2013; Ouyang et al., 2020).

According to a report by Market Research Future (Market and Markets, 2023), the global activated carbon market was valued at US$ 5.7 billion in 2021 and is projected to reach US$ 8.9 billion by 2026, growing at a granular activated carbon of 9.3% during the forecast period. However, the amount of activated carbon used globally is difficult to estimate due to the vast number of applications and variations in usage rates.

Brazil is highly dependent on imported activated carbon, and there is a deficit in the commercial market. According to data from the United Nations Comtrade database (search for 3802 and 38,010 HS commodity codes for activated carbon), Brazil imported around 38,000 and 7000 tons of these respective codes in 2022, with a total value of US$ 50,351,900 and US$ 26,489.278. The main countries of origin for Brazil's activated carbon imports in 2022 were China, the United States, and India. At the same time, the commodities described for both codes' exportation were around US$ 7000.000 (OEC, 2023; UN Comtrade Database, 2022).

The main reason for this deficit is the lack of investment in the production of activated carbon within the country, combined with the high demand for the product in various industrial sectors (OEC, 2023; UN Comtrade Database, 2022). This deficit affects the Brazilian economy and limits the country's ability to implement sustainable development practices that require activated carbon, such as wastewater treatment and air pollution control. This situation is a counter-census given the potential generator of biomass in Brazil.

Therefore, increasing the production of activated carbon from renewable sources in Brazil would not only contribute to reducing the dependence on imports but also promote the development of the bioeconomy and stimulate the sustainable use of residues, natural and agro-industrial resources, converting Brazil from an importer to an exporter of this commodity.

Brazilian researchers have been dedicated to proposing the production of activated carbon from various sources of biomass derived from agro-industrial processes, biomes, and waste (Adebayo et al., 2014; Castro et al., 2019; Duarte et al., 2021; Gonçalves et al., 2013; Mafra et al., 2022; Silva et al., 2010). Materials such as yeast biomass from fermentation processes in the sugar and alcohol sector, sugarcane bagasse, coffee grounds, and other sources have been successfully utilized to obtain high-surface-area activated carbon. This activated carbon has been applied for catalysis and the removal of contaminants from aqueous mediums such as pesticides, herbicides, textile dyes, drugs, metals, and organic acids (Barbosa et al., 2021; de Freitas et al., 2022; Franco et al., 2022; Georgin et al., 2022; Labuto et al., 2022; Limousy et al., 2017; Nascimento et al., 2021; Queiroz et al., 2020; Ramos et al., 2022; Rodrigues et al., 2021).

In research conducted in the Web of Science and Scopus databases considering articles published by Brazilian researchers with two sets of keywords (Activated Carbon and Biomass, and Activated Charcoal and Biomass) for the period 01/01/1994—31/12/2014, 63 and 10 articles were located, respectively. Between 01/01/2015 and 31/01/2023, 274 and 48 articles were identified, respectively. The growth in article production in the evaluated time intervals was over 300% for both sets of keywords, which may be related to the increase in the number of patents observed in recent years. According to Brazilian researchers, biomass used to produce activated carbon, coconut shells, sugarcane bagasse, seaweed, and coffee waste stands out. The complete set of articles published by Brazilian scientists with the keywords mentioned above had 8675 and 6298 citations in the period, for the Web of Science and Scopus databases, with averages of 471 and 273 citations per article, respectively, which is considerable given the still timid contribution of Brazil in biomass conversion studies to activated carbon. Such data show that there is still much to be explored by Brazil concerning the production of activated carbon for numerous applications, and this could be an important niche to boost the Brazilian bioeconomy.

Concerning patents involving the production of activated carbon from biomass, a search was carried out using the exact expression "activated carbon" in the title and "activated carbon from biomass" (exact terms) in the summary of patents registered with the INPI (INPI, 2022) resulting in 29 patents for the synthesis of coal made exclusively from biomass. Of these, 11 were registered between 01/01/1994 and 31/12/2015, and 18 were registered between 01/01/2015 and 31/01/2023.

The increasing numbers indicate a rising interest in domestically producing activated carbon. Various agro-industrial residues, such as olives, sugarcane bagasse, cocoa, pineapple crown, coffee beans, coconut, and sisal bagasse, have been used as biomass sources. Additionally, dry yeast from industrial fermentation processes can also be used to produce activated carbon, in line with circular economy principles. The process involves the physical activation of yeast biomass residue, a by-product of the sugar and alcohol industry, using CO_2 and water vapor. These by-products are also generated within the same industry, further contributing to the circular economy approach. (Modesto, 2019). However, Brazil still has a huge potential to explore concerning the production of activated carbon for numerous applications, which can serve as an important niche to boost the Brazilian bioeconomy.

5 New milestones in the organization politics and society for bioeconomy in Brazil

Since the chapter was published in the previous edition, several public and private initiatives have emerged in Brazil to organize and promote the bioeconomy. Following are some outstanding initiatives.

5.1 Foundation of the Brazilian Association of Bioinnovation (ABBI)

In the final year of the period covered in the first edition of this book (2014), the ABBI (ABBI, 2023) was founded. ABBI is a civil, non-profit, non-partisan, nationwide organization that sees Brazil as a potential leader of the advanced global bioeconomy. Its objective is to represent companies and institutions from various sectors of the economy that invest in innovative technologies based on biological and renewable resources to create products, processes, or business models that generate collective social and environmental benefits. ABBI's initiatives have been supported by large Brazilian and international companies, some of which are world-renowned.

ABBI defines the bioeconomy as "the entire value chain guided by advanced scientific knowledge and the search for technological innovations in the application of biological and renewable resources in industrial processes to generate circular economic activity and collective social and environmental benefits." According to ABBI, the development of the bioeconomy must consider the interests of civil society and the business and government sectors, with a focus on food safety, environmental preservation, human health, and promoting the industry's competitiveness.

This institution is dedicated to hosting informative events related to Bioeconomy and Circular Economy, which provide interested parties with insights on legal aspects, economic opportunities, and funding for initiatives guided by Bioeconomy. In the third edition of its annual flagship event, the Brazil Bioeconomy Forum 2021, the central theme was "Bioeconomy: from vocation to reality." The theme reflects recognizing Brazil's significant challenges to achieve full Bioeconomic development.

The event featured a thematic lecture entitled "Amazônia 4.0 and the Brazil of the future: putting the potential of Bioeconomy into practice." It also included two panels: (a) Bioeconomy and Decarbonization and (b) Bioeconomy and Investment Attraction in Brazil. Along with the forum, the third Brazil Bioeconomy 2021 Award was presented, recognizing solutions linked to the sustainable development goals (SDGs) established by the UN. Public schools, researchers, startups, and private and public companies participated in this award in three categories: Ideas, Start-ups and Scale-ups, and Anchor Companies.

5.2 Creation of the Joint Parliamentary front for innovation in bioeconomy

On June 26th, 2019, the Joint Parliamentary Front for Innovation in Bioeconomy (FPBioeconomia) (Frente Parlamentar da Bioeconomia, 2019) was founded. It is a non-partisan, non-profit association with an indefinite duration within the Brazilian National Congress. The front includes parliamentarians (deputies and senators), a representative from ABBI serving as executive secretary, and collaborating members from scientists, non-governmental organizations, and the private sector. Notable members include the Brazilian Agricultural Research Corporation (Embrapa) and the University of Brasília (UnB).

FPBioeconomia is guided by a definition of bioeconomy that is included in its statute. According to this definition, bioeconomy encompasses the entire value chain, guided by advanced scientific knowledge and technological innovations that apply biological and renewable resources in industrial processes. The aim is to generate circular economic activity and promote collective social and environmental benefits.

FPBioeconomia has its actions based on ten goals:

- Using biodiversity ethically and sustainably;
- Increase the representation of the bioeconomy in the Brazilian GDP;
- Consolidate the country as a leader and technological reference in the circular economy;
- Modernize federal legislation;
- Create an environment conducive to innovation, productivity, and competence;
- Build critical mass in the legislature;
- Establish an industry 4.0 in Brazil leveraged by the bioeconomy;
- Ensure the viability of investments in biotechnology in the country;
- Build an institutional framework favorable to national interests for global participation;
- Collaborate toward attention to the SDGs by the private sector, civil society, and government, generating innovation, productivity, sustainability, and development.

6 Public actions and financing opportunities

The Brazilian federal government has taken various initiatives and established state bodies to provide public financing for research and developing specific economic activities to boost the bioeconomy.

One of these initiatives was the creation of The Brazilian Biodiversity Information System—SiBBr to facilitate the organization, indexing, integration, storage, and availability of data on Brazilian biodiversity and ecosystems (SiBBr, 2018) through Ordinance No. 6223 of November 29, 2018 (MCTIC, 2018). The system provides data obtained both from institutions participating in the Ministry of Science, Technology, and Innovation (Ministério da Ciência, Tecnologia e Inovação MCTI) and from programs promoted by CNPq. Additionally, Brazil joined the GBIF (Global Biodiversity Information Facility) in 2012 as an associate member (Global Biodiversity Information Facility) (GBIF, n.d.). The common objective is to promote the scientific growth and development of knowledge on Brazilian biodiversity, which is an essential step toward sustainable development, as the production and conscious use of renewable and unique biological resources to produce food, energy, chemicals, and various materials such as plastics are results of the implementation of the bioeconomy. Moreover, it is an important step toward access to our genetic heritage (Pylro et al., 2014; Rodrigues, 2018).

The Brazilian Development Bank (Banco Nacional de Desenvolvimento Econômico e Social, BNDES) has started various activities such as training professionals, producing informational materials, and establishing financing lines. One of their noteworthy training initiatives was a public notice published in 2022 to finance the development of a technical and professional training itinerary in bioeconomy in the Amazon—Public Selection BNDES/FEP Fomento N° 01/2022 (BNDES, 2022a). This initiative aims to strengthen the value chains of the bioeconomy in the Amazon region by implementing around 27 high school technical courses to train 50,000 students across the nine states of the region. The initiative aims to increase young people's professional qualifications and employability while making better use of the local economic potential.

Through the BNDES and the MCTI, the Brazilian federal government has been working on initiatives to train and retain researchers in companies and promote university–company interaction. One such initiative involves collaboration with EMBRAPII, a social organization that partners with accredited research centers and the business sector to encourage innovation in companies. EMBRAPII has developed a network of 28 Innovation Units in Bioeconomy to foster research, development, and innovation activities and add value and sustainability to Brazilian biodiversity.

The network offers several advantages to companies, including financing of non-reimbursable resources up to 50% of the project's value, agility in resource disbursement, flexibility to innovate, low bureaucracy for hiring, cost and risk sharing, access to research and development institutions with technical and scientific experience, sharing of infrastructure and technical capacity among EMBRAPII Units, and differentiated service to start-ups, small companies, cooperative arrangements, and companies based in Brazil's North region.

EMBRAPII has established various lines of action in the bioeconomy field, which include biocontrol and biotechnological processes for the sustainable management of agricultural pests, biomass processing, renewable chemistry, green chemistry, environmental biotechnologies for the recovery of contaminated areas, and the valorization of waste from the industrial sector. Over the past 8 years, EMBRAPII has financed 339 research, development, and innovation (RD&I) projects in the bioeconomy, covering enabling technologies and applications in this field (BNDES, 2022b). The forestry bioeconomy has been the area with the highest volume of projects, with seven initiatives, followed by new materials with four projects. These projects have benefited 341 companies, with an investment of approximately US$ 73 million, resulting in 152 completed projects and 78 requests for intellectual property.

To promote innovation in the field of bioeconomy, EMBRAPII has identified it as one of its strategic areas within the Basic Funding Alliance. This funding modality supports projects with innovative solutions with a lower level of technological maturity, which are closer to basic research and carry a higher risk for participating companies. In this modality, the participating company must invest at least 10% of the project's value in return, with the remaining amount financed without reimbursement.

During COP27, EMBRAPII announced funding resources for companies seeking innovative solutions in the forestry bioeconomy and circular economy. These companies will receive technical support from public higher education institutions in the states of the northern region of Brazil, where the Amazon is located. This initiative aims to respond to the urgent demand for industries to align themselves with the climate agenda and develop green technologies. The objective is to expand the scope of existing projects, which include initiatives such as biofertilizers, biofuels, biocosmetics, and solid waste transformation, while also encompassing initiatives that align with the SDGs of the United Nations (Heimann, 2019).

In addition, to encourage master's and doctoral graduates to pursue careers in the industrial sector, CNPq launched a pilot project in 2013 called the Industrial Academic Doctorate, which offered 20 doctoral scholarships for students who were working on research projects linked to industries. The success of this program led to an increase in the number of scholarships provided in subsequent years. As a result, Resolution N° 7, issued on April 9, 2020 (CNPq, 2020), created the Academic Master's and Doctorate Programs for Innovation (MAI/DAI). The program aims to strengthen science, technology, entrepreneurship, and innovation by involving undergraduate and graduate students in projects of interest to companies, government agencies, and third-sector entities. Its objectives are to contribute to training personnel at undergraduate and graduate levels, encourage innovative projects, promote partnerships between Scientific, Technological, and Innovation Institutions (ICTs) and companies, and contribute to developing or improving products, processes, and business services.

In this program, CNPq, ICTs, partner companies, and scholarship holders have specific requirements and obligations that must be met. CNPq is responsible for providing scholarship funds using public resources. The duration of the scholarships can vary, ranging from 1 year for undergraduate-level students (ITI) to 2 years for master's degree students (GM) and 4 years for doctoral scholarship holders (GD). ICTs are responsible for appointing an Institutional Representative for the MAI/DAI Program (RID), selecting candidates through a public selection process, defining cooperation with the partner company, and submitting an electronic summary of the scholarship holder's thesis. Additionally, ICTs must have a post-graduate program recognized by the National Education Council (CNE/MEC) and a formal partnership with interested companies to participate in the MAI/DAI Program.

The partner company is responsible for assigning a supervisor to oversee the project's progress with the student and providing necessary resources for the project's development. The scholarship holder must be regularly enrolled in a post-graduate program and have no employment relationship, dedicating themselves entirely to the project, with few exceptions.

In 2020, CNPq awarded 2102 scholarships in three categories (Doctorate, Master, and Industrial Technological Initiation) to 59 Brazilian institutions, totaling an investment of approximately US$ 1 million (CNPq, 2021). The 2022 notice requires (CNPq, 2022) that research projects align with at least one of the Ministry of Science, Technology, Innovations, and Communications' Priority Technology Areas, which include Strategic Technologies, Enabling Technologies, Production Technologies, Technologies for Sustainable Development, and Technologies for Quality of Life (Ministério da Ciência, 2020a, b). Projects in basic research, social sciences, and humanities that contribute to developing these priority areas are also accepted. The proposer must serve as the Institutional Representative MAI/DAI—RID before CNPq, coordinate the proposal, and have a formal affiliation with the institution responsible for implementing the project.

As a distinguishing characteristic of Brazilian scientific practice, undergraduate students are involved in research groups coordinated by university professors through a program called Scientific Initiation. In recognition of the positive outcomes of this approach to insert students in scientific research, the scholarship announcement requires projects to include participation by undergraduate students who will receive Technological Initiation scholarships and contribute to the ongoing thesis or dissertation project. Moreover, the Partner Company must supplement the funding with a minimum amount of approximately US$

2000 or 4000 for each master's or doctoral scholarship, respectively, which can be paid in installments and used to cover expenses such as equipment, other permanent materials, and maintenance costs.

The BNDES has also implemented the 2022 Pronaf ABC+ Bioeconomy (BNDES, 2022c), which is a financing program for farmers and rural family producers to invest in the use of renewable energy technologies, environmental technologies, water storage, small hydro-electric projects, forestry, and the adoption of conservationist practices. Moreover, the program aims to improve the soil's productive capacity by promoting practices such as correcting soil acidity and fertility. The program offers differentiated annual interest rates for Brazilian parameters, ranging from 5% to 6%, with 100% of financeable amounts and finances around US$ 38,000 per agricultural year, with grace periods of 5–20 years (for initiatives of exploitation of palm oil and rubber trees for latex production). The main goal of this program is to promote sustainable forestry practices that generate different products, including wood.

The Programa Prioritário em Bioeconomia—PPBIO (Priority Program in Bioeconomy) (PPBIO, 2022) was created in 2018, and it is an initiative by the Brazilian government, specifically by the Manaus Free Trade Zone Superintendence (Superintendência da Zona Franca de Manaus, SUFRAMA), aimed at promoting and strengthening the bioeconomy sector in Brazil. The program focuses on the Amazon region and its rich biodiversity, intending to create sustainable economic development through renewable resources and innovative technologies. The program has several proposals and initiatives, including promoting research and development in the bioeconomy, encouraging investments in the sector, modernizing legislation related to using genetic resources, and increasing the employability of qualified professionals, such as masters and doctors, in the private sector. One of the program's main components is the creation of the Bioeconomy Innovation and Technology Development Center (Centro de Desenvolvimento de Inovação e Tecnologia em Bioeconomia, CDIT Bioeconomia), which will serve as a hub for research, development, and innovation in the bioeconomy sector. The center will be located in Manaus, with partnerships with universities, research institutes, and private companies.

The program also proposes the creation of a network of bioeconomy innovation hubs throughout the Amazon region, intending to promote the development of innovative solutions and products based on the region's biodiversity. Additionally, the program aims to increase the participation of small and medium-sized enterprises in the bioeconomy sector by implementing specialized funding mechanisms and incentives for innovation. Overall, the Priority Program in Bioeconomy represents a significant effort by the Brazilian government to promote sustainable economic development and innovation in the bioeconomy sector, focusing on the Amazon region and its rich biodiversity.

Regarding informative material production, the publication "Potential of Bioeconomy for the Sustainable Development of the Amazon and Possible Actions for BNDES" 2021 (Pamplona et al., 2021) is particularly noteworthy. The goal is to provide interested parties with a contextual understanding of bioeconomy in the Amazon context, the challenges of sustainable development, the potential of Amazonian biodiversity for various economic sectors, bottlenecks to expanding credit, and the BNDES's experience in supporting sustainable production arrangements in the region.

A second noteworthy publication is "The Brazilian Bioeconomy in Figures" (Silva et al., 2018), which aimed to assess the value of bioeconomic activities in Brazil and highlight their contribution by sector. The study found that in 2016, the value of sales attributed to the Brazilian bioeconomy was US$ 285.9 billion in Brazil and US$ 40.2 billion for sales of economic activities located in other countries, totaling US$ 326.1 billion. The publication also identified opportunities for adding value to products and pointed out challenges that need to be addressed to help Brazil reach its full potential as a bioeconomic power.

Embrapa, a public company, has added the bioeconomy as one of the impact axes in Embrapa's 6th Master Plan (2020–30) (EMBRAPA, 2020). As such, one of its guiding principles is to develop technologies and knowledge that contribute to the bioeconomy by using bio-based resources to generate bioproducts, bio inputs, and renewable energy. It has even set a goal to make five new renewable raw materials available for use in the context of the bioeconomy by 2030.

A study carried out by the ABBI together with EMBRAPA, the National Laboratory of Biorenewables of the Center for Research in Energy and Materials Laboratório Nacional de Biorrenováveis do Centro de Pesquisa em Energia e Materiais, LNBR/CNPEM), Center for Technology for the Chemical and Textile Industry (Centro de Tecnologia da Indústria Química e Têxtil, SENAI/CETIQT), and Cenergia Laboratory of the Federal University of Rio de Janeiro (Laboratório Cenergia da Universidade Federal do Rio de Janeiro, Cenergia/UFRJ) evaluated a Brazilian scenario in which bioeconomy and energy transition are combined (ABBI, 2022; Santana, 2023). The survey shows an economic assessment in which, compared with a scenario of maintenance of current public policies, there would be an increase of US$ 284 billion/year by 2050 in Brazilian revenues by intensifying biotechnology, even with implementing a low-carbon technology, complying with the Paris Agreement.

This scenario would be achieved by assuming specific changes, such as increasing the production of sugarcane and eucalyptus, recovering 6 million hectares of pastures, replacing beef with cultured meat, and harnessing agricultural waste to produce increased amounts of advanced biofuels and bionaphtha. According to the study, to achieve this scenario, it is necessary to invest around US$ 45 billion and intensify public policies that encourage the necessary initiatives to achieve it.

In 2015, the Chemical Industry Development and Innovation Plan (PADIQ) was launched as a joint initiative between the BNDES and the public Financial Agency for Studies and Projects (Financiadora de Estudos e Projetos, Finep) (Finep, n.d.). The program aims to support projects that include technological development and investment in the manufacture of chemical products. The program had resources estimated at US$ 430 million for operations contracted from 2016 to 2019 and aimed to coordinate actions to promote innovation and productive investments, integrating the support instruments available for financing projects in the Chemical Industry in the country. The first PADIQ notice was launched in November 2015. It targeted the following segments: additives for animal feed, silicon derivatives, composite materials and carbon fibers, chemical additives for oil exploration and production, chemical inputs for personal hygiene, perfumery and cosmetics (HPPC), and chemicals from renewable sources.

The Ministry of Agriculture and Livestock and Food Supply (Ministério da Agricultura, Pecuária e Abastecimento, MAPA) has created the Bioeconomy Brazil—Sociobiodiversity program, which aims to support small farmers, family farmers, traditional communities, and their enterprises. This program was established by Ordinance N° 121 on June 18, 2019 (MAPA, 2019) and is structured into five thematic areas: Pro-Extractivism, Medicinal, Aromatic, and Condimentary Herbs, Olive Oils and Special Teas from Brazil, Sociobiodiversity Routes, Potentialities of Brazilian Agrobiodiversity, and Renewable Energy for Family Farming. The program's main objective is to promote partnerships between the government, small farmers, traditional communities, and businesses to develop sustainable production systems based on socio-biodiversity and extractivism and the production and use of renewable energy sources. This program will expand the participation of these segments in productive and economic arrangements that involve the concept of bioeconomy.

The Brazilian federal government created RenovaBio in 2019 through Law N° 13576/2017, enacted in December of that year. RenovaBio is the most extensive decarbonization program in the world. A state policy recognizes the strategic role of all biofuels, including ethanol, biodiesel, biomethane, biokerosene, and second-generation biofuels, in the Brazilian energy matrix. This policy acknowledges their contribution to energy security, market predictability, and mitigation of greenhouse gas emissions in the fuel sector. RenovaBio comprises three strategic axes: (1) Decarbonization Targets, (2) Biofuel Production Certification, and (3) Decarbonization Credit (CBIO). The targets of Axis 1 were regulated by Decree N° 9888/2019, while certification of biofuel production (Axis 2) was regulated by ANP (National Petroleum Agency) Resolution N° 758/2018. Furthermore, in November 2019, the MME published MME Ordinance N° 419/2019, which regulates transactions with the CBIO. In December 2019, ANP published Resolution N° 802/2019, which established the procedures for generating the necessary ballast for the primary issuance of the CBIO (Ministério de Minas e Energia, n.d.)

In the same vein as renewable energy, FAPESP has created the Bioenergy Research Program (BIOEN) (BIOEN, 2008). According to FAPESP, BIOEN aims to stimulate and coordinate research and development activities in the academic and industrial sectors, promoting the advancement of knowledge and its application in bioenergy production and its derivatives in Brazil. The program connects university researchers with the industrial sector and provides resources to support research in the state of São Paulo, Brazil's largest sugarcane ethanol producer. The goals of BIOEN include increasing biomass productivity, developing efficient and competitive biomass conversion platforms, improving the energy efficiency and sustainability of bioenergy's end use, and accelerating the transition to a bioeconomy.

7 Private sector participation

The Brazilian National Confederation of Industry (Confederação Nacional da Indústria, CNI) continues to carry out initiatives to promote the mobilization of the industrial sector so that it contributes to solutions for the sustainable development of Brazil, which became in 2013 with the publication of the study Bioeconomy: An Agenda for Brazil and more recently with its four-year Strategic Map plan (2018–22) and the SDGs (CNI, 2018; Harvard Business Review Analytic Services and Confederação Nacional da Indústria, 2013). To this end, it established an agenda that aims to move and encourage industry participation in discussions on access to the Brazilian genetic heritage; act in the Brazilian scenario so that there are resources to finance enterprises that aim at the use of biodiversity sustainably, for initiatives that require venture capital and for research and development related to biodiversity. Furthermore, this agenda intends to form an innovation ecosystem in which companies understand the advantages arising from the minimization of environmental risks arising from the sustainable use of biodiversity.

To achieve these objectives, the CNI has put forward proposals for regulations, innovation, and investments. Regarding regulation, the CNI aims to contribute to the governance structure, simplify the relationship between science and technology institutes and the productive sector, support initiatives to increase the INPI efficiency, and train and align the bodies and users of biodiversity. In terms of innovation, the CNI aims to incorporate researchers into industries, disseminate business opportunities in the bioeconomy, prospect and map new species of biodiversity, bring the industry closer to all levels of education, find new uses for products of biological origin, and develop strategies for intercropped bioconversion.

As for investments, the CNI proposes to encourage the articulation of innovation hubs in the bioeconomy, develop specialized funding mechanisms for innovation, encourage risk-sharing of investments between the government and industry in pre-competitive projects, promote corporate venture capital, attract foreign funds to Brazil, and promote research and development at various stages of the process of obtaining new goods and services based on biodiversity resources.

The CNI has proposed various initiatives to modernize legislation for the bioeconomy, which includes improving the Regulatory Framework for Access to Genetic Resources and Benefit Sharing, as well as revising several laws such as the Biosafety Law (Law N° 11105/2005), Normative Resolutions 02 of 2006 and 05 of 2008 of the National Technical Commission on Biosafety (CNTBio), the Industrial Property Law (Law N° 9279/1996), the Innovation Law (Law N° 10973/2004), and the Good Law (Law N° 11196/2005).

To support the development of the bioeconomy, the National Industrial Training Service (Serviço Nacional de Aprendizagem Industrial, SENAI), a private Brazilian institution of public interest, established the Institute SENAI of Innovation in Biomass (ISI Biomass) in the state of Mato Grosso do Sul. ISI Biomass aims to conduct applied research to develop new products, services, and innovative processes in biomass transformation, to add more value to commodities, raw materials, and residual biomass. In addition, it is intended to attract new currency and make Brazilian industries more competitive through the introduction of new technologies, following the political strategy of the Brazilian government described in the "Plano Brasil Maior" (2011–14) (Ministério da Agricultura e Pecuária, 2017). The institute's program has already supported more than 150 projects.

In contrast to the proposals of the industrial sector and despite the incentives provided for technology transfer from universities to the private sector, as well as incentives for researchers to work in industry, there is still significant resistance to hiring foreign master's and doctoral graduates, the majority of them being employed in the public sector and public universities.

According to a study by the Center for Management and Strategic Studies (Centro de Gestão e Estudos Estratégicos, CGEE) social organization supervised by the Brazilian MCTI, there was an increase in the employability of masters and doctors between 2010 2017.

Nevertheless, the unemployment rate among master's and doctoral graduates in Brazil is currently averaging 25%, with an even higher rate of 35% among those with a master's degree. It starkly contrasts with the worldwide average of around 2% (CGEE, 2019).

The employability of master's degree holders is mainly observed in the education (40.9%) and public administration (33.7%) sectors, while the industry and manufacturing sector only employs 4.2% of them, revealing the private sector's limited role in hiring these skilled professionals. On the other hand, for doctoral degree holders, the disparity between employability in the public and private sectors is even more significant. Seventy-five percent of doctors are employed in the education sector, 12.6% in the public sector, and a meager 1.3% in the industry and manufacturing sectors. Notably, the employment situation of these highly qualified professionals would have been even direr if not for the Restructuring and Expansion of Federal Universities program (REUNI), which took place between 2003 and 2015, intending to increase access to and retention in higher education. Under REUNI, federal universities increased from 45 to 59, and the university campuses from 148 to 274, enrolling almost 66,000 undergraduate students today (Paula and Almeida, 2020). This program resulted in the hiring of many doctoral degree holders to join the faculty of the participating institutions, who were likely unemployed or working in jobs that did not utilize their full potential.

The limited employability of masters and doctors undoubtedly represents a loss for Brazilian society and the economy. This is because most of the investment in their training comes from public funds, yet the private sector is not fully utilizing their expertise. Unfortunately, the private sector is generally unwilling to invest in research, innovation, and technological development within the business and industrial realm. If this trend were to be reversed, it would undoubtedly contribute to the growth of the bioeconomy and other industrial and innovation sectors, enabling Brazil to become more competitive in the international market.

8 Challenges and conquers of bioremediation as a driver of the bioeconomy in Brazil

According to the Organization for Economic Co-operation and Development (OECD), the biotechnology sector is expected to generate around 1 trillion dollars/per year until 2030 and create 22 million jobs. This growth will be driven by several factors, including population growth, which demands new ways of producing food and supplying water and energy. Therefore, innovations that focus on renewable products, greater agricultural productivity, and the production of fuels and energy through biomass should be prioritized.

Additionally, the aging population demands new and improved healthcare treatments, which require the emergence of new biotechnological treatments. Climate change and waste generation also demand innovation in how materials are produced, creating environmentally friendly products, and remediation of their impacts.

Brazil has been a leader in producing, using, and regulating biofuels; however, other sectors require more investment for development. Some initiatives integrating agriculture and industry to foster economic development are underway. Nevertheless, these examples must inspire and promote progress in other areas such as bioremediation, bioenergy, chemical products from renewable sources, drug development, production of biomaterials, and bio inputs. Nonetheless, several limitations hamper the implementation of the bioeconomy in Brazil. One of the most significant challenges is the lack of adequate infrastructure to explore biomasses from all Brazilian biomes, such as roads, ports, and energy distribution systems. Another major challenge is the complex regulatory framework, which makes it hard for companies to comply with regulations, obtain permits, and access financing. Additionally, it is imperative to improve funding for research and establish a culture of innovation within the industrial environment, creating research and innovation sectors led by hired researchers (masters and doctors).

In addition, the development of public policies aimed at extending and integrating different sectors is necessary to enable the flow of knowledge, technology, and inputs and to promote the development of new products and services that can add more value to the bioeconomy. To achieve these aims, efforts must be coordinated between different sectors of society, including the government, academia, business, and civil society, so that Brazil can reach its full potential in exploring raw materials from renewable sources and its biodiversity within the principles of Bioeconomy, circular economy, and ecological economy. Addressing these challenges will require concerted efforts from stakeholders in the public and private sectors.

Only by supering these challenges will it be possible to find innovative and sustainable solutions that ensure biodiversity and environmental protection, promoting the migration toward circular and biologically based economies. Combining innovation and sustainability is direction to solving the main global challenges.

9 Conclusions

A lot has been achieved in the last 8 years. The growth in the number of publications and the training of human resources in the academic environment demonstrate that the scientific community is qualified, interested, and attentive to the Bioeconomy. However, there is still a noticeable need for a State policy, rather than a government one, to promote the transition to a Bioeconomic model committed to sustainability principles. It is also urgent for the private sector to proactively create an innovative environment within the manufacturing industry and agribusiness. It can be achieved by developing research within their sectors and hiring researchers.

In this way, public and private initiatives must be reinforced, deepened, developed, and created to allow for a greater scientific understanding of the available resources regarding biomass and Brazilian genetic heritage. Only then will Brazil be able to take advantage of the country's full resource potential in developing and employing new technologies for industrial processes.

Brazil requires agility, commitment, and convergence of all involved sectors to capitalize on the promising opportunity of the Bioeconomy for the economy and the environment. It will allow for the accelerated growth of new knowledge, followed by its utilization by the private sector in generating processes and products that contribute to job creation and social welfare.

Acknowledgments

The authors are grateful to FAPESP for financial grants (2016/06271-4 and 2022/08358-0) and fellowships (2021/06471-1, 2021/14791-6, and 2022/06823-8) to Coordination of Superior Level Staff Improvement (CAPES) for fellowship (M.L.C.G.).

References

ABBI, 2023. Associação Brasileira de Bioinovação. Associação Brasileira de Bioinovação (ABBI). (Accessed 5 February 2023).

ABBI, 2022. Identificação das oportunidades e o potencial do impacto da bioeconomia para a descarbonização do Brasil. Publicações ABBI (Associação Brasileira de Bioinovação).

Abe, M.M., Branciforti, M.C., Nallin Montagnolli, R., Marin Morales, M.A., Jacobus, A.P., Brienzo, M., 2022. Production and assessment of the biodegradation and ecotoxicity of xylan- and starch-based bioplastics. Chemosphere 287, 132290. https://doi.org/10.1016/j.chemosphere.2021.132290.

Adebayo, M.A., Prola, L.D.T., Lima, E.C., Puchana-Rosero, M.J., Cataluña, R., Saucier, C., Umpierres, C.S., Vaghetti, J.C.P., da Silva, L.G., Ruggiero, R., 2014. Adsorption of Procion Blue MX-R dye from aqueous solutions by lignin chemically modified with aluminium and manganese. J. Hazard. Mater. 268, 43–50. https://doi.org/10.1016/j.jhazmat.2014.01.005.

Alkaraan, F., Albitar, K., Hussainey, K., Venkatesh, V., 2022. Corporate transformation toward Industry 4.0 and financial performance: the influence of environmental, social, and governance (ESG). Technol. Forecast. Soc. Change 175, 121423. https://doi.org/10.1016/j.techfore.2021.121423.

Andrade, D.F., Romanelli, J.P., Pereira-Filho, E.R., 2019. Past and emerging topics related to electronic waste management: top countries, trends, and perspectives. Environ. Sci. Pollut. Res. 26, 17135–17151. https://doi.org/10.1007/s11356-019-05089-y.

ANVISA, 2021a. Saneantes—Produtos Registrados. ANVISA. https://consultas.anvisa.gov.br/#/saneantes/produtos/25351505155201571/. (Accessed 18 February 2023).

ANVISA, 2021b. Saneantes—Produtos Registrados. ANVISA. https://consultas.anvisa.gov.br/#/saneantes/produtos/25351004927201471/. (Accessed 18 February 2023).

ANVISA, 2021c. Saneantes—Produtos Registrados. ANVISA. https://consultas.anvisa.gov.br/#/saneantes/produtos/25351285956201785/. (Accessed 18 February 2023).

Barbosa, J.A., Labuto, G., Carrilho, E.N.V.M., 2021. Magnetic nanomodified activated carbon: characterization and use for organic acids sorption in aqueous medium. Chem. Eng. Commun. 208, 1450–1463. https://doi.org/10.1080/00986445.2020.1791832.

bcc Research, 2018. Activated Carbon: Types and Global Markets.

BIOEN FAPESP, n.d. 2008. Fapesp Bioenergy Program—BIOEN. FAPESP. https://bioenfapesp.org/old/fapesp-bioenergy-program-bioen/index.html (Accessed 2 February 2023).

BNDES, 2022a. Seleção Pública FEP Fomento—N° 01/2022—Iniciativa itinerário da formação técnica e profissional em bioeconomia na Amazônia. Banco Nacional do Desenvolvimento. https://www.bndes.gov.br/wps/wcm/connect/site/a6765e5a-0b01-439f-8497-46d5998eabf5/Edital+retificado+12jul22.pdf?MOD=AJPERES&CVID=o7UEQlU. (Accessed 10 February 2023).

BNDES, 2022b. BNDES e Embrapii estimulam inovação em bioeconomia e outros setores. Banco Nacional do Desenvolvimento. https://www.bndes.gov.br/wps/portal/site/home/imprensa/noticias/conteudo/bndes-e-embrapii-estimulam-inovacao-em-bioeconomia-e-outros-setores. (Accessed 10 February 2023).

BNDES, 2022c. Pronaf ABC+ Bioeconomia. Banco Nacional do Desenvolvimento. https://www.bndes.gov.br/wps/portal/site/home/financiamento/produto/pronaf-bioeconomia. (Accessed 8 February 2023).

Brasil, 2023. Decreto N° 11.427, de 2 de Março de 2023. Brasil.

Brasil, 2018. Decreto N° 9.283, de 7 de Fevereiro de 2018.

Brasil, 2016a. Decreto N° 8.772, de 11 de Maio de 2016.

Brasil, 2016b. Lei N° 13.243, de 11 de Janeiro de 2016.

Brasil, 2015. Lei N° 13.123, de 20 de Maio de 2015.

Brasil, 2005a. Lei N° 11.105, de 24 de Março de 2005.

Brasil, 2005b. Decreto N° 5.591, de 22 de Novembro de 2005.

CAPES, n.d. Catálogo de Teses e Dissertações. CAPES. http://catalogodeteses.capes.gov.br/catalogo-teses/#!/ (Accessed 16 February 2023).

Carvalho, J.T.T., Milani, P.A., Consonni, J.L., Labuto, G., Carrilho, E.N.V.M., 2021. Nanomodified sugarcane bagasse biosorbent: synthesis, characterization, and application for Cu(II) removal from aqueous medium. Environ. Sci. Pollut. Res. 28, 24744–24755. https://doi.org/10.1007/s11356-020-11345-3.

Castro, J.P., Nobre, J.R.C., Bianchi, M.L., Trugilho, P.F., Napoli, A., Chiou, B.-S., Williams, T.G., Wood, D.F., Avena-Bustillos, R.J., Orts, W.J., Tonoli, G.H.D., 2019. Activated carbons prepared by physical activation from different pretreatments of amazon piassava fibers. J. Nat. Fibers 16, 961–976. https://doi.org/10.1080/15440478.2018.1442280.

CGEE, 2019. Brasil: Mestres e Doutores 2019. CGEE (Centro de Gestão e Estudos Estratégicos). https://mestresdoutores2019.cgee.org.br/web/guest/inicio. (Accessed 5 February 2023).

CNI, 2018. Mapa estratégico da indústria 2018-2022. Confederação Nacional da Indústria.

CNPq, 2022. Chamada Pública nº 068/2022—Programa de Mestrado e Doutorado Para Inovação—MAI-DAI. http://memoria2.cnpq.br/web/guest/chamadas-publicas?p_p_id=resultadosportlet_WAR_resultadoscnpqportlet_INSTANCE_0ZaM&filtro=encerradas/. (Accessed 20 February 2023).

CNPq, 2021. Programa MAI/DAI. Conselho Nacional de Desenvolvimento Científico e Tecnológico. https://www.gov.br/cnpq/pt-br/acesso-a-informacao/acoes-e-programas/programas/programa-mai-dai. (Accessed 20 February 2023).

CNPq, 2020. Resolução 7/2020 de 9 abril de 2020—Programa de mestrado e doutorado acadêmico para inovação—MAI/DAI.   http://memoria2.cnpq.br/web/guest/view/-/journal_content/56_INSTANCE_0oED/10157/8820750. (Accessed 20 February 2023).

Danish, M., Ahmad, T., 2018. A review on utilization of wood biomass as a sustainable precursor for activated carbon production and application. Renew. Sust. Energ. Rev. 87, 1–21. https://doi.org/10.1016/j.rser.2018.02.003.

da Silva, W.R., da Silva, F.B.V., Araújo, P.R.M., do Nascimento, C.W.A., 2017. Assessing human health risks and strategies for phytoremediation in soils contaminated with As, Cd, Pb, and Zn by slag disposal. Ecotoxicol. Environ. Saf. 144, 522–530. https://doi.org/10.1016/j.ecoenv.2017.06.068.

de Freitas, D.A., Barbosa, J.A., Labuto, G., Nocelli, R.C.F., Carrilho, E.N.V.M., 2022. Removal of the pesticide thiamethoxam from sugarcane juice by magnetic nanomodified activated carbon. Environ. Sci. Pollut. Res. 29, 79855–79865. https://doi.org/10.1007/s11356-021-18484-1.

Duarte, R.V.d.L., Rodrigues, L.A., de Moraes, N.P., Couceiro, P.R.d.C., 2021. Brasil nut mesocarp (Bertholletia excels) as a biomass source for activated carbon production: correlation between thermal treatment and adsorption performance. Biointerface Res. Appl. Chem. 12, 4584–4596. https://doi.org/10.33263/BRIAC124.45044596.

EMBRAPA, 2020. VII Plano Diretor da EMBRAPA 2020–2030—1ª Edição. Infoteca-e—Repositório de Informação Tecnológica da Embrapa (Empresa Brasileira de Pesquisa Agropecuária Ministério da Agricultura, Pecuária e Abastecimento).

FAPESP, n.d. Biblioteca Virtual da FAPESP. FAPESP. https://bv.fapesp.br/pt/ (Accessed 19 February 2023).

Finep, n.d. Plano de Desenvolvimento e Inovação da Indústria Química—PADIQ. Finep (Financiadora de Estudos e Projetos). http://www.finep.gov.br/padiq (Accessed 10 February 2023).

Fittipaldy, M.C.P.d.M., Faria, F.S.E.d.V.D., Rodriguez, A.F.R., 2020. Biodiversidade e conhecimentos tradicionais no contexto da biopirataria e dos marcos legais. South Am. J. Basic Educ. Tech. Technol. 7, 648–677.

Fortune Business Insights, 2021. Activated Carbon Market Size, Share & COVID-19 Impact Analysis, by Type (Powdered, Granular, and Others), by Application (Water Treatment, Air & Gas Purification, Food & Beverage, Pharmaceutical & Healthcare Treatment, and Others), and Regional Forecast, 2022–2030. Market Research Report https://www.fortunebusinessinsights.com/activated-carbon-market-102175. (Accessed 30 January 2023).

Franco, D.S.P., Georgin, J., Netto, M.S., da Boit Martinello, K., Silva, L.F.O., 2022. Preparation of activated carbons from fruit residues for the removal of naproxen (NPX): analytical interpretation via statistical physical model. J. Mol. Liq. 356, 119021. https://doi.org/10.1016/j.molliq.2022.119021.

Frente Parlamentar da Bioeconomia, 2019. Frente Parlamentar Mista pela Inovação na Bioeconomia. Frente Parlamentar da Bioeconomia. https://fpbioeconomia.com.br/#comunicacao. (Accessed 14 February 2023).

Frias, A.R.R.d., 2020. Biorremediação de corpos hídricos: Tecnologias disponíveis e estudo de caso com método THE WATER CLEANSER® (Dissertação de Mestrado). Centro Universitário Estadual da Zona Oeste, Rio de Janeiro.

Frias, A.R.R.d., Santos, E.d.O.S., 2020. Tecnologias comerciais para biorremediação de corpos hídricos. Rev Episteme Transvers 11, 196–221.

Gan, Y.X., 2021. activated carbon from biomass sustainable sources. J Carbon Res 7, 39. https://doi.org/10.3390/c7020039.

GBIF, n.d. GBIF Brasil. GBIF. https://www.gbif.org/pt/country/BR/summary (Accessed 15 February 2023).

Georgin, J., Pinto, D., Franco, D.S.P., Schadeck Netto, M., Lazarotto, J.S., Allasia, D.G., Tassi, R., Silva, L.F.O., Dotto, G.L., 2022. Improved adsorption of the toxic herbicide Diuron asing activated carbon obtained from residual cassava biomass (*Manihot esculenta*). Molecules 27, 7574. https://doi.org/10.3390/molecules27217574.

Gonçalves, M., Guerreiro, M.C., Oliveira, L.C.A., Solar, C., Nazarro, M., Sapag, K., 2013. Micro mesoporous activated carbon from coffee husk as biomass waste for environmental applications. Waste Biomass Valor. 4, 395–400. https://doi.org/10.1007/s12649-012-9163-1.

Grand View Research, 2018. Activated Carbon Market Size, Share & Trends Analysis Report by Type (Powdered, Granular), by Application (Liquid Phase, Gas Phase) by End Use (Water Treatment, Air Purification), by Region, and Segment Forecasts, 2022–2030. Market Analysis Report.

Harvard Business Review Analytic Services, Confederação Nacional da Indústria, 2013. Bioeconomy an Agenda for Brazil.

Heimann, T., 2019. Bioeconomy and SDGs: does the bioeconomy support the achievement of the SDGs? Earths Future 7, 43–57. https://doi.org/10.1029/2018EF001014.

IBAMA, 2022a. Instrução Normativa N° 11, de 17 de Outubro de 2022. Diário Oficial da União.

IBAMA, 2022b. Orientação Técnica Normativa—OTN N° 3-DIQUA. Diário Oficial da União.

IBAMA, 2022c. Produtos remediadores físico-químicos registrados no Ibama. Gov.br.

IBAMA, 2022d. Lista de Produtos Remediadores Registrados no Ibama—Cicam—CGQua-Diqua. Gov.br. https://ibamagovbr-my.sharepoint.com/:x:/g/personal/33638542858_ibama_gov_br/EV4vRMYGSFZLsBSjbze1I7sBHpcVdPfiBKSfjcyoUNuQzA?rtime=uaH4axcZ20g. (Accessed 8 February 2023).

IBAMA, 2022e. Produtos biorremediadores registrados. Gov.br. https://www.gov.br/ibama/pt-br/assuntos/quimicos-e-biologicos/produtos-remediadores/produtos-biorremediadores-registrados. (Accessed 8 February 2023).

INPI, 2022. pePI—Pesquisa em Propriedade Industrial. Instituto Nacional da Propriedade Industrial (INPI). https://busca.inpi.gov.br/pePI/jsp/patentes/PatentSearchBasico.jsp. (Accessed 11 February 2023).

Ioannidou, O., Zabaniotou, A., 2007. Agricultural residues as precursors for activated carbon production—a review. Renew. Sust. Energ. Rev. 11, 1966–2005. https://doi.org/10.1016/j.rser.2006.03.013.

José, J.C., Debs, K.B., Labuto, G., Carrilho, E.N.V.M., 2019. Synthesis, characterization, and application of yeast-based magnetic bionanocomposite for the removal of Cu(II) from water. Chem. Eng. Commun. 206, 1581–1591. https://doi.org/10.1080/00986445.2019.1615468.

Joseph, B., Kaetzl, K., Hensgen, F., Schäfer, B., Wachendorf, M., 2020. Sustainability assessment of activated carbon from residual biomass used for micropollutant removal at a full-scale wastewater treatment plant. Environ. Res. Lett. 15, 064023. https://doi.org/10.1088/1748-9326/ab8330.

Khajeh, M., Laurent, S., Dastafkan, K., 2013. Nanoadsorbents: classification, preparation, and applications (with emphasis on aqueous media). Chem. Rev. 113, 7728–7768. https://doi.org/10.1021/cr400086v.

Köseoğlu, E., Akmil-Başar, C., 2015. Preparation, structural evaluation and adsorptive properties of activated carbon from agricultural waste biomass. Adv. Powder Technol. 26, 811–818. https://doi.org/10.1016/j.apt.2015.02.006.

Labuto, G., Carvalho, A.P., Mestre, A.S., dos Santos, M.S., Modesto, H.R., Martins, T.D., Lemos, S.G., da Silva, H.D.T., Carrilho, E.N.V.M., Carvalho, W.A., 2022. Individual and competitive adsorption of ibuprofen and caffeine from primary sewage effluent by yeast-based activated carbon and magnetic carbon nanocomposite. Sustain. Chem. Pharm. 28, 100703. https://doi.org/10.1016/j.scp.2022.100703.

Limousy, L., Ghouma, I., Ouederni, A., Jeguirim, M., 2017. Amoxicillin removal from aqueous solution using activated carbon prepared by chemical activation of olive stone. Environ. Sci. Pollut. Res. 24, 9993–10004. https://doi.org/10.1007/s11356-016-7404-8.

Macedo, I.C., Seabra, J.E.A., Silva, J.E.A.R., 2008. Green house gases emissions in the production and use of ethanol from sugarcane in Brazil: the 2005/2006 averages and a prediction for 2020. Biomass Bioenergy 32, 582–595. https://doi.org/10.1016/j.biombioe.2007.12.006.

Machado, C.J.S., Godinho, R.d.S., Vilani, R.M., 2017. Decreto 8.772/2016: Riscos e retrocessos na proteção da biodiversidade e dos conhecimentos tradicionais associados. Revista de Direito Ambiental.

Mafra, E.R.M.L., de Paula Protásio, T., Bezerra Bezerra, J., Pedroza, M.M., Barbosa, D.B., Viana, M.F., de Souza, T.M., Bufalino, L., 2022. Comparative analysis of seed biomass from Amazonian fruits for activated carbon production. Biomass Convers Biorefin. https://doi.org/10.1007/s13399-022-03348-6.

VII. Case studies

MAPA, 2019. Portaria MAPA N° 121, de 18 de Junho de 2019. MAPA (Ministério da Agricultura, Pecuária e Abastecimento)—Diário Oficial da União.

Market and Markets, 2023. Activated Carbon Market by Type, Application (Liquid Phase (Water Treatment, Foods & Beverages, Pharmaceutical & Medical), Gas Phase (Industrial, Automotive), and Region (APAC, North America, Europe, Middle East, South America))—Global Forecast to 2026. Market and Markets. https://www. marketsandmarkets.com/Market-Reports/activated-carbon-362.html. (Accessed 13 February 2023).

Maximo, Y.I., Hassegawa, M., Verkerk, P.J., Missio, A.L., 2022. Forest bioeconomy in Brazil: potential innovative products from the forest sector. Land (Basel) 11, 1297. https://doi.org/10.3390/land11081297.

MCTIC, 2018. Portaria N° 6.223, de 29 de Novembro de 2018. Diário Oficial da União.

Ministério da Agricultura e Pecuária, 2017. Plano Brasil Maior.

Ministério da Ciência, T.I. e C. (MCTIC), 2020a. Portaria MCTIC N° 1.329 de 27 de Março de 2020. Diário Oficial da União.

Ministério da Ciência, T.I. e C. (MCTIC), 2020b. Portaria MCTIC N° 1.122, de 19 de Março de 2020. Diário Oficial da União.

Ministério de Minas e Energia, n.d. RenovaBio. Secretaria de Petróleo, Gás Natural e Biocombustíveis http://antigo. mme.gov.br/web/guest/secretarias/petroleo-gas-natural-e-biocombustiveis/acoes-e-programas/programas/ renovabio (Accessed 4 February 2023).

Mobilidade Sustentável, 2022. Etanol 2.0 pode consolidar protagonismo do Brasil em combustíveis limpos. https:// valor.globo.com/patrocinado/movimento-mobilidade-sustentavel/mobilidade-sustentavel/noticia/2022/01/ 19/etanol-20-pode-consolidar-protagonismo-do-brasil-em-combustiveis-limpo.ghtml. (Accessed 13 February 2023).

Modesto, H.R., 2019. Produção e caracterização de carvão ativado obtido a partir de resíduo de biomassa de levedura utilizando CO_2 como agente ativador (Trabalho de conclusão de curso). Universidade Federal de São Paulo, Diadema.

Modesto, H.R., Lemos, S.G., dos Santos, M.S., Komatsu, J.S., Gonçalves, M., Carvalho, W.A., Carrilho, E.N.V.M., Labuto, G., 2020. Activated carbon production from industrial yeast residue to boost up circular bioeconomy. Environ. Sci. Pollut. Res. 28, 24694–24705. https://doi.org/10.1007/s11356-020-10458-z.

Nascimento, J.R., Bezerra, K.C.H., Martins, T.D., Carrilho, E.N.V.M., Rodrigues, C.d.A., Labuto, G., 2021. Textile effluent treatment employing yeast biomass and a new nanomagnetic biocomposite. Environ. Sci. Pollut. Res. 28, 27318–27332. https://doi.org/10.1007/s11356-021-12594-6.

Nogueira, R.C., de Conqueiro, H.E., Soares, M.R.R., 2010. Patenting bioactive molecules from biodiversity: the Brazilian experience. Expert Opin. Ther. Pat. 20, 145–157. https://doi.org/10.1517/13543770903333221.

OEC, 2023. Activated Carbon. The Observatory of Economic Complexity. https://oec.world/en/profile/hs/ activated-carbon. (Accessed 27 February 2023).

Ouyang, J., Zhou, L., Liu, Z., Heng, J.Y.Y., Chen, W., 2020. Biomass-derived activated carbons for the removal of pharmaceutical mircopollutants from wastewater: a review. Sep. Purif. Technol. 253, 117536. https://doi.org/10.1016/ j.seppur.2020.117536.

Pamplona, L., Salarini, J., Kadri, N., 2021. Potential of bioeconomy for the sustainable development of the Amazon and acting possibilities for the BNDES. Rev. BNDES, 55–86.

Paula, C.H.d., Almeida, F.M.d., 2020. O programa Reuni e o desempenho das Ifes brasileiras. Ensaio: Aval. Polít. Públ. Educ. 28, 1054–1075. https://doi.org/10.1590/s0104-40362020002801869.

PPBIO, 2022. Planejamento estratégico do Programa Prioritário de Bioeconomia—PPBIO (2020-2023)—2ª Versão. Programa Prioritário de Bioeconomia.

Pylro, V.S., Roesch, L.F.W., Ortega, J.M., do Amaral, A.M., Tótola, M.R., Hirsch, P.R., Rosado, A.S., Góes-Neto, A., da Costa Silva, A.L., Rosa, C.A., Morais, D.K., Andreote, F.D., Duarte, G.F., de Melo, I.S., Seldin, L., Lambais, M.R., Hungria, M., Peixoto, R.S., Kruger, R.H., Tsai, S.M., Azevedo, V., 2014. Brazilian Microbiome Project: revealing the unexplored microbial diversity—challenges and prospects. Microb. Ecol. 67, 237–241. https://doi.org/10.1007/ s00248-013-0302-4.

Queiroz, L.S., de Souza, L.K.C., Thomaz, K.T.C., Leite Lima, E.T., da Rocha Filho, G.N., do Nascimento, L.A.S., de Oliveira Pires, L.H., Faial, K.d.C.F., da Costa, C.E.F., 2020. Activated carbon obtained from amazonian biomass tailings (açai seed): modification, characterization, and use for removal of metal ions from water. J. Environ. Manag. 270, 110868. https://doi.org/10.1016/j.jenvman.2020.110868.

Ramos, J.L., Monteiro, J.O.F., dos Santos, M.S., Labuto, G., Carrilho, E.N.V.M., 2022. Sustainable alternative for removing pesticides in water: nanomodified activated carbon produced from yeast residue biomass. Sustain. Chem. Pharm. 29, 100794. https://doi.org/10.1016/j.scp.2022.100794.

Rashidi, N.A., Yusup, S., Ahmad, M.M., Mohamed, N.M., Hameed, B.H., 2012. Activated carbon from the renewable agricultural residues using single step physical activation: a preliminary analysis. APCBEE Proc. 3, 84–92. https://doi.org/10.1016/j.apcbee.2012.06.051.

Reza, M.S., Yun, C.S., Afroze, S., Radenahmad, N., Bakar, M.S.A., Saidur, R., Taweekun, J., Azad, A.K., 2020. Preparation of activated carbon from biomass and its' applications in water and gas purification, a review. Arab. J. Basic Appl. Sci. 27, 208–238. https://doi.org/10.1080/25765299.2020.1766799.

Rezende, E.A., Ribeiro, M.T.F., 2009. O cupuaçu é nosso? Aspectos atuais da biopirataria no contexto brasileiro. RGSA—Rev. Gestão Soc. Ambiental 3, 53–74.

Rodrigues, M., 2018. Bioeconomia é a nova fronteira para o futuro da América Latina. Cienc. Cult. 70, 21–22. https://doi.org/10.21800/2317-66602018000400007.

Rodrigues, R., Santos, M.S., Nunes, R.S., Carvalho, W.A., Labuto, G., 2021. Solvent-free solketal production from glycerol promoted by yeast activated carbons. Fuel 299, 120923. https://doi.org/10.1016/j.fuel.2021.120923.

Rojahn, J., 2010. Fair Shares or Biopiracy? Developing Ethical Criteria for the Fair and Equitable Sharing of Benefits From Crop Genetic Resources (Tese de Doutorado). Universidade Eberhard Karls de Tuningen, Lüdenscheid.

Salvadori, M.R., Ando, R.A., Corrêa, B., 2022. Bio-separator and bio-synthesizer of metallic nanoparticles—a new vision in bioremediation. Mater. Lett. 306, 130878. https://doi.org/10.1016/j.matlet.2021.130878.

Santana, I., 2023. Bioeconomia no Brasil pode gerar faturamento de US$ 284 bi anuais. EMBRAPA. https://www.embrapa.br/busca-de-noticias/-/noticia/77870291/bioeconomia-no-brasil-pode-gerar-faturamento-de-us-284-bi-anuais. (Accessed 19 February 2023).

SiBBr, 2018. Sistema de Informação sobre a Biodiversidade Brasileira. Sistema de Informação sobre a Biodiversidade Brasileira. https://sibbr.gov.br/. (Accessed 15 February 2023).

Silva, M.D., Oliveira, D.R.d., 2018. The new Brazilian legislation on access to the biodiversity (Law 13,123/15 and Decree 8772/16). Braz. J. Microbiol. 49, 1–4. https://doi.org/10.1016/j.bjm.2017.12.001.

Silva, E.A., Vaz, L.G.L., Veit, M.T., Fagundes-Klen, M.R., Cossich, E.S., Tavares, C.R.G., Cardozo-Filho, L., Guirardello, R., 2010. Biosorption of chromium(III) and copper(II) ions onto marine alga *Sargassum* sp. in a fixed-bed column. Adsorpt. Sci. Technol. 28, 449–464. https://doi.org/10.1260/0263-6174.28.5.449.

Silva, M.F.d.O.E., Pereira, F.d.S., Martins, J.V.B., 2018. The Brazilian bioeconomy in figures. BNDES Setorial 47, 277–332.

SisGen, n.d. Sistema Nacional de Gestão do Patrimônio Genético e do Conhecimento Tradicional Associado (SisGen). SisGen. https://sisgen.gov.br/paginas/InstallSolution.aspx (Accessed 27 January 2023).

UN, 2015. Agenda 2030 para o Desenvolvimento Sustentável. UN Brasil. https://brasil.un.org/pt-br/91863-agenda-2030-para-o-desenvolvimento-sustentavel. (Accessed 1 February 2023).

UN Comtrade Database, 2022. Free Access to Detailed Global Trade Data. UN. https://comtradeplus.un.org/. (Accessed 27 February 2023).

Valentim dos Santos, J., Varón-López, M., Fonsêca Sousa Soares, C.R., Lopes Leal, P., Siqueira, J.O., de Souza Moreira, F.M., 2016. Biological attributes of rehabilitated soils contaminated with heavy metals. Environ. Sci. Pollut. Res. 23, 6735–6748. https://doi.org/10.1007/s11356-015-5904-6.

Valli, M., Russo, H.M., Bolzani, V.S., 2018. The potential contribution of the natural products from Brazilian biodiversity to bioeconomy. An. Acad. Bras. Cienc. 90, 763–778. https://doi.org/10.1590/0001-3765201820170653.

Villela, H.D.M., Peixoto, R.S., Soriano, A.U., Carmo, F.L., 2019. Microbial bioremediation of oil contaminated seawater: a survey of patent deposits and the characterization of the top genera applied. Sci. Total Environ. 666, 743–758. https://doi.org/10.1016/j.scitotenv.2019.02.153.

Phytomanagement of polycyclic aromatic hydrocarbon and heavy metal contaminated sites in Assam, North Eastern India, for boosting the bioeconomy

Hemen Sarma[a] and Majeti Narasimha Vara Prasad[b]

[a]Bioremedatio Technology Group, Department of Botany, Bodoland University, Kokrajhar, Assam, India [b]Department of Plant Sciences, School of Life Sciences, University of Hyderabad (An Institution of Eminence), Hyderabad, Telangana, India

1 Introduction

The upper Brahmaputra valley agro climatic zone of Assam State in India covers an area of 16,013 km^2 encompassing five districts, viz., Golaghat, Jorhat, Sivasagar, Dibrugarh, and Tinsukia (Fig. 1). This valley is ideally suited for rice and tea cultivation, and these crops make a significant contribution to India's national economy (Sarma et al., 2016). In the upper Brahmaputra valley, crude oil was discovered in 1867 at Digboi, and the first oil well in Asia was drilled (Sarma et al., 2016). Since then, several oil wells have been drilled in various parts of upper Assam. Most oil drilling sites are located at the fringes of human settlements, paddy fields, and tea gardens resulting in heavy metals and hydrocarbon contamination (Sarma et al., 2019). Heavy metals are toxic to humans even at extremely low concentrations and have a high density of 4–5 g/cm^3 (Dhaliwal et al., 2020). Pollutants in the form of heavy metals and hydrocarbons can lead to soil pollution (Cole, 2018; Rajendran et al., 2022). The principal source of PAHs in soils is incomplete combustion of organic material (Liu et al., 2019) in addition to petroleum hydrocarbon. Polyaromatic hydrocarbons are classified as xenobiotics

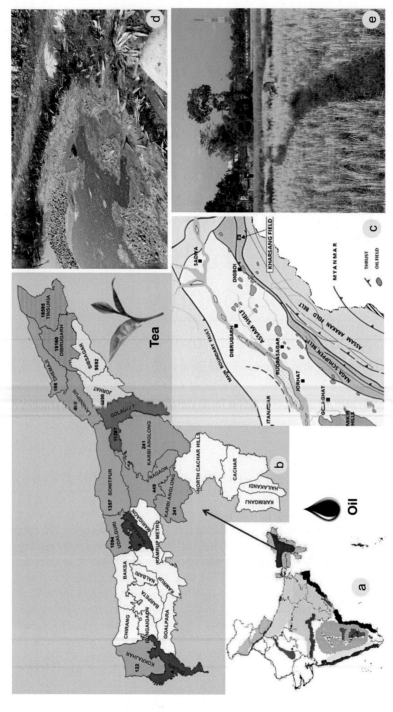

FIG. 1 (A) Northeastern part India showing Assam State (red color), which is rich in tea, rice, and crude oil; (B) Indian sub-continent with 12 distinct ecoregions showing the northeastern part, which is a treasure house of nature's capital (biotic and abiotic) resources; (C) locations of oil fields in upper Assam. Rice paddies and tea plantations are the most commonly cultivated for bioeconomy in the region (D) crude oil seepage through Namti drilling site (Sivasagar), (E) rice field contamination near Borhola group (Jorhat).

and emerging priority pollutants (Logeshwaran et al., 2018). Hydrocarbon-contaminated soil from oil exploration activities creates a threat to crop production and has a detrimental effect on the bio-economic system by making land unsuitable for agriculture and other economic purposes. This problem is of particular concern in India, the second most populated country in the world, where agriculture is the primary economic base to feed the masses (Sarma et al., 2019). Traditional methods for soil remediation are often expensive due to the mass involvement of labor, energy, and infrastructure (Thakare et al., 2021).

The physical or chemical removal of hydrocarbons from contaminated areas may result in the transfer of remediated compounds to uncontaminated areas (Abdel-Razek et al., 2020). Several approaches for treating PAH-contaminated soil have been developed. Soil washing, chemical oxidation, electrokinetics and phytoremediation are thought to be more efficient methods of removing PAHs from polluted soil (Gitipour et al., 2018). Clays are rapidly gaining favor in the petroleum refining sectors for a range of techniques such as adsorption and separation. It is possible to remove a range of chemicals from crude oil and petroleum fractions using clay adsorption, including sulfur, heavy metals, colors, and the separation of various hydrocarbon groups (Abdelwahab Emam, 2018).

As a result, in situ phytotechnology has been developed as a cheaper alternative to conventional methods in areas where remediation would otherwise not be put into practice (Sarma et al., 2021). One of the most effective approaches for treating petroleum-contaminated soil (PCS) and water is phytoremediation, often known as plant-assisted remediation. The key benefits of this method are that it requires no equipment, is potentially cost-effective and is ecologically sound (Abdullah et al., 2020). Microbes are cultured in polluted soil, sediment, surface water, and groundwater in this new green technology to speed the breakdown and removal of inorganic and organic contaminants (Kumar et al., 2018; Truskewycz et al., 2019). Microbes, on the other hand, produce biosurfactants that have been found to significantly aid hydrocarbon bioremediation, metal detoxification and removal and soil cleaning technologies (Patowary et al., 2022a, b). Biosurfactants also aid in the formation of root cells and nutrient absorption. Furthermore, biosurfactants produced by root-associated bacteria promote metal and micronutrient dispersion in the soil, boosting nutrient availability and uptake and thereby driving plant growth (Fenibo et al., 2019; Rasheed et al., 2020).

In the last few decades, many research projects have explored the role of beneficial plants and microbes in bioremediation practices (Frenzel et al., 2010; Mazzeo et al., 2010; Bacosa and Inoue, 2015). Despite positive findings in the greenhouse, the practice of in situ bioremediation is limited. There are numerous causes for this, including low soil temperature, changing pH levels, a lack of synergy between plant-microbe associations, the type of contaminates and their concentration, soil qualities and so on (Ma et al., 2018). Understanding the nature, composition, properties, sources of pollution, type of environment, fate, transport, and distribution of pollutants, mechanism of degradation, interaction and relationships with microorganisms, and intrinsic and extrinsic factors affecting remediation are critical for selecting the best treatment option (Ossai et al., 2020).

Although many technical issues exist with this phytotechnology, many reports have been published currently showing that by using proper high-end technology, degraded land mass can be made fertile again, which will boost the bio-economy of a nation (Sarma et al., 2017, 2021). This study aims to provide an inventory of some plant-microbe consortia that can potentially be used for in situ bioremediation of crude oil-contaminated soil in the tropical environment of Assam. Emphasis has also been made on developing efficient methods accessible to economically marginalized small-scale tea growers and paddy farmers in the state. Table 1 details the recent work appraising the phytomanagement of polycyclic aromatic hydrocarbon (PAH) and heavy metal contaminated sites in different parts of the world.

The majority of the rural people live on less than US $1 a day in Asia (World Bank, 2001, 2004; FAO, 2005; CIFOR, 2002). These people sustain their livelihoods from the forests

TABLE 1   Aspects and appraisal of phytomanagement of polycyclic aromatic hydrocarbon and heavy metal contaminated sites.

Salient research finding/observation	Author(s)	Year
Review of polycyclic aromatic hydrocarbons, including their source, effects on the environment, human health effects, and remedies	Abdel-Shafy et al.	2016
Classification of bioremediation techniques according to the place of application: principles, benefits, drawbacks, and prospects	Azubuike et al.	2016
Implications for the bio-economy of the localization of polycyclic aromatic hydrocarbons and heavy metals in the topsoil of Asia's oldest oil and gas drilling site in Assam, northeast India	Sarma et al.	2016
Perspective on petroleum refining industry from clay adsorption	Abdelwahab Emam	2018
principles of environmental petroleum hydrocarbon microbial degradation	Al-Hawash et al.	2018
Site evaluation and cleanup for petroleum-contaminated areas	Cole	2018
A evaluation of treatment options for PAH-contaminated locations	Gitipour et al	2018
Treatment and disposal of petroleum sludge: a review	Johnson et al	2018
An eco-sustainable method for restoring contaminated sites	Kumar et al.	2018
Fast pyrolysis for the treatment of oil-contaminated soil and simultaneous oil recovery	Li et al.	2018
Review of the toxicity, microbial degradation and risk-based remediation techniques for petroleum hydrocarbons (PH) in groundwater aquifers	Logeshwaran et al.	2018
remediation of soil contaminated with heavy metals and hydrocarbons using electrokinetics and biostimulation	Ma et al.	2018
Constructed wetlands to treat petroleum wastewater	Mustapha et al.	2018
Wetlands were built to treat petroleum wastewater Bioaugmentation mediated by *Pseudomonas* spp. has an effect on bioremediation, increasing the bioavailability of weathered hydrocarbons in motor oil-contaminated soil	Ramadass et al.	2018
Reaction mechanisms, soil changes and implications for treated soil fertility in pyrolytic remediation of oil-contaminated soils	Sharma et al.	2018
Pyrolytic remediation of oil-contaminated soils: reaction mechanisms, soil changes and implications for treated soil fertility	Vidonish et al.	2018
Petroleum hydrocarbon-degrading bacteria for the remediation of oil pollution under aerobic conditions	Xu et al.	2018

VII. Case studies

TABLE 1   Aspects and appraisal of phytomanagement of polycyclic aromatic hydrocarbon and heavy metal contaminated sites—cont'd

Salient research finding/observation	Author(s)	Year
Monitoring of volatile organic compounds (VOCs) from an oil and gas station in northwest China for a year to identify petroleum hydrocarbon-degrading bacteria for oil pollution remediation under aerobic conditions	Zheng et al.	2018
Identification of petroleum hydrocarbons biodegradation bacteria that were isolated from oil-polluted soil in Dhahran, Saudi Arabia, using morphological, biochemical and molecular methods	Al-Dhabaan	2019
A study of methods and possibilities for decommissioning oil and gas platforms globally	Bull and Love	2019
Biochar made from activated petroleum waste sludge for effective catalytic ozonation of effluent from refineries	Chen et al.	2019
Wetland loss and restoration's effects on ecosystem services: a Danube Delta ecological evaluation from 1960 to 2010	Gómez-Baggethun et al.	2019
A comprehensive review on the role of plants and microbes in the bioremediation of petroleum	Hossain et al.	2019
A review on the application of chemical surfactant and surfactant foam for remediation of petroleum oil contaminated soil	Ite and Ibok	2019
Evidence review on the relationship between upstream oil extraction and environmental public health	Johnston et al.	2019
	Karthick et al.	2019
Solar-heated carbon absorbent inspired by biotechnology for effective cleaning of highly viscous crude oil	Kuang et al.	2019
Review of the role of sustainable remediation methods in relation to the contamination of vegetables and the food chain with hazardous heavy metals	Kumar et al.	2019
PAHs in the urban soils of two Florida cities Polycyclic aromatic hydrocarbons: treatment of soil pollution	Liu et al.	2019
Polycyclic aromatic hydrocarbons: soil pollution and remediation	Sakshi et al.	2019
Plant-microbiome assisted and biochar-amended remediation of heavy metals and polyaromatic compounds—a microcosmic study	Sarma et al.	2019
Environmental issues, field studies, sustainability concerns and future prospects in phytoremediation of heavy metal-contaminated sites	Saxena et al.	2019
Benefits and restrictions of in situ bioremediation techniques	Sharma	2019
Fate and microbial responses to petroleum hydrocarbon contamination in terrestrial ecosystems	Truskewycz et al.	2019
Abiotic variables, natural attenuation, bioaugmentation and nutrient supplementation have an impact on the bioremediation of agricultural soil contaminated by petroleum crude	Varjani et al.	2019
Phenanthrene bioelimination utilizing bacteria isolated from petroleum soil: a secure method	Abdel-Razek et al.	2020

Continued

VII. Case studies

TABLE 1   Aspects and appraisal of phytomanagement of polycyclic aromatic hydrocarbon and heavy metal contaminated sites—cont'd

Salient research finding/observation	Author(s)	Year
Hydrocarbon contamination in water and soil cleanup using plants: Application, methods, difficulties and potential	Abdullah et al.	2020
Bioremediation of soils saturated with spilled crude oil	Ali et al.	2020
A discussion of remediation methods for removing heavy metals from soil that has been contaminated by various sources	Dhaliwal et al.	2020
Review of the removal of petroleum hydrocarbons from wastewaters	Mohammadi et al.	2020
Ahead of the curve in bioremediation technology for petroleum hydrocarbon elimination	Naeem and Qazi	2020
An overview of soil and water remediation for petroleum hydrocarbon contamination	Ossai et al.	2020
Remediation using surfactants as a successful strategy for removing environmental pollutants a thorough analysis of PAH-metabolic degradation's and genetic features	Rasheed et al	2020
Technologies for sulfur characterization, identification and eradication from petroleum: toward cleaner fuel and a secure environment	Sakshi and Haritash	2020
Characterization, determination and elimination technologies for sulfur from petroleum: toward cleaner fuel and a safe environment	Saleh	2020
Using pig droppings and bone char to bioremediate crude oil-contaminated soil	Ugwoha et al.	2020
Using a unique strain of *Pseudomonas aeruginosa* and phytotoxicity of petroleum hydrocarbons for seed germination, oily sludge-polluted soil can be bioremediated	Varjani et al.	2020
Utilizing a metagenomics approach, oil-contaminated soil was chosen for bacteria that produce biosurfactants. Environmental sustainability and biotechnology	Islam et al.	2021
Biotechnology for sustainable environment	Joshi et al.	2021
Australian offshore oil and gas infrastructure decommissioning research needs	Melbourne-Thomas et al.	2021
Potential of biochar to stimulate plant growth and clean up soil contaminated by crude oil	Saeed et al.	2021
Polycyclic aromatic hydrocarbons in foods: biological effects, legislation, occurrence, analytical methods and strategies to reduce their formation	Sampaio et al.	2021
Enhancing the use of CRISPR-Cas9genome engineering technology in the phytoremediation of dangerous metal(loid)s	Sarma et al.	2021
Gaining knowledge of plant-microbe remediation strategies that remove radionuclides and heavy metals from soil holistically	Thakare et al.	2021
Adding biotechnology to *P. aeruginosa*, a new oily waste degrader for treating petroleum hydrocarbons is NCIM 5514	Varjani et al.	2021
Novel nanomaterials for nanobioremediation of polyaromatic hydrocarbons	Borah et al.	2022
Using a brand-new, biostimulated strain of *Enterobacter hormaechei* ODB H32, an environmentally acceptable bioremediation method for soils contaminated by crude oil	El-Liethy et al.	2022

TABLE 1 Aspects and appraisal of phytomanagement of polycyclic aromatic hydrocarbon and heavy metal contaminated sites—cont'd

Salient research finding/observation	Author(s)	Year
For the sustainable cleanup of developing pollutants, assisted and modified technology	Ghahari et al.	2022
Phytoremediation with help from biosurfactants for a sustainable future	Islam et al.	2022
Petroleum production activities and depletion of biodiversity: a case of oil spillage in the Niger Delta	Nnadi et al.	2022
An example of oil spill in the Niger Delta and the destruction of biodiversity caused by petroleum production activities	Rajendran et al.	2022
Recent developments in the cleanup of petroleum-contaminated soil and groundwater using in situ chemical oxidation	Wei et al.	2022

through land-extensive cultivation, logging and exploitation of non-wood products and sometimes these activities degrade the natural environment (Sarma and Sarma, 2008). It has been estimated that two-thirds of the world's ecosystems are shrinking and are considered degraded as a result of indiscriminate use of renewable and non-renewable resources, mismanagement, contamination and failure to look after these resources (Gómez-Baggethun et al., 2019). India's recognition as one of the four "mega-diversity" countries of Asia is derived largely from three of its most important biodiversity "hotspots": the Himalayas, Indo Burma and Western Ghats. Assam is one of the critical priority biodiversity hotspots of India under Himalayan and Indo-Burma hotspots. The major forest types of Assam have received international conservation efforts based on the levels of endemism, medicinal plant diversity, and human threat. The total wildlife protected areas (PAs) is $3925 km^2$ out of the total $28,748 km^2$ forest area of the state and comprises 18 Wildlife Sanctuaries and five National Parks (Sarma et al., 2011). The northeastern region of India forms a unique biogeographic province comprising major biomes recognized in the world and harbors about 50% (± 8500 sp) of the floristic wealth of India and 40% of them are endemic (Mao et al., 2009). This region is regarded as the place of origin of progenitors of many cultivated crops.

The upper Brahmaputra valley agro climatic zone is a land rich in both bioresources and natural resources, especially petroleum, natural gases and coal, tea, and rice. Crude oil drilling, mining of coal, deforestation, flood, etc., have severely degraded the land ecosystem in this zone and have had a negative impact on bioresources, i.e., tea plantation and rice cultivation. It has been recorded that coal mining and crude oil drilling are the major contributors which transform fertile, cultivable land into wasteland, as crude oil exploration and mining activity generate a vast quantity of solid waste (Johnston et al., 2019; Joshi et al., 2021) which is deposited on the surface and occupies huge areas of land (Table 2). In upper Assam, while many hectares of land have been transformed into a wasteland due to drilling of crude oil, coal mining, there is still much potential for agriculture as evidenced by the number of tea gardens and rice fields. The available data show that the annual average rice production in this zone is estimated as $1637 kg/ha$ (Ahmed et al., 2014). Recently, in rural Assam, farmers have also started planting tea in unutilized and underutilized uplands and thus brought

TABLE 2 Land utilization statistics (area in hectare) of upper Brahmaputra valley agro climatic zone.

District	Geographical area	Forest	Area not available for cultivation	Permanent pastures and other grazing land	Land under miscellaneous trees groves, etc.	Cultivable waste land	Fallow land	Net cropped area	Gross cropped area	Area sown more than once	Cropping intensity
Jorhat	851,000	21,904	110,567	4406	0024	6686	12,273	120,240	173,020	52,780	144%
Golaghat	350,200	156,905	51,232	8314	3217	5801	4555	119,046	180,097	61,051	151%
Sivasagar	266,800	30,465	56,151	7330	20,061	1820	7164	136,822	155,062	18,220	113%
Tinsukia	379,000	131,595	114,883	3560	19,786	1586	2876	104,714	147,936	43,222	141%
Dibrugarh	338,100	23,341	142,488	6170	16,883	7126	3276	139,498	158,917	19,419	114%

Source: Statistical Hand Book, Assam, 2011.

about huge socio-economic changes in Assam. In the last two decades, the number of small tea growers has swelled to a resounding number of 65,000 (http://aastga.org). A small tea grower is one whose land under tea cultivation does not exceed 10 ha. Almost 90,00,000 people are engaged directly and indirectly with tea cultivation. Around 2,50,000 ha of land have been covered by small-scale tea plantations. The contribution of small tea growers is about 29% of the total tea produced by Assam, which is approximately 14% of the total tea production of India (http://aastga.org). Most of the farmers have typical homesteads where they cultivate vegetables, horticulture crops, tree, and bamboo. This zone is also suitable for the cultivation of other plantation crops such as arecanut, coconut, banana, and lemon. The soil type In this zone is mostly sandy loam. The total number of farmers and farm families living in this zone is around 581,014 as per the Agriculture Census 2005–06 (Tables 3 and 4).

The sites where accidental seepage occurs become desert-like in nature and disturb the rice ecosystem (Fig. 2). These activities directly lead to fragmented landscapes, loss of cultivated land, imbalances in beneficial soil microbes to finally result in the overall loss of crop production and economic loss. The indirect effects can be loss of the water-holding capacity of soil, imbalances in physiochemical properties, discharge of carcinogenic pollutants to ambient air

TABLE 3 Crop profile of the upper Brahmaputra valley zone.

Agro climatic zone	District	Major field crops cultivated	Horticulture crops-fruits	Plantation crops
Upper Brahmaputra valley zone	Jorhat	Winter rice, blackgram	Banana	Arecanut, tea
	Sivasagar	Winter rice, autumn rice, summer rice	Banana, orange, pineapple	Tea, arecanut, coconut
	Golaghat	Terrace rice cultivation, wet rice cultivation, maize	Banana, lemon, pineapple	Tea, arecanut, coconut
	Dibrugarh	Winter paddy, autumn paddy	Arecanut, banana, assam lemon	Tea, black pepper
	Tinsukia	Paddy, maize, blackgram	Banana, Khasi Mandarin, Pineapple	Tea, arecanut, coconut

TABLE 4 District wise breakup small tea gardens, areas under paddy cultivation, and farmers/farming families.

District	Small tea gardens	Area under paddy ('000 ha)	Farmers/farming families
Jorhat	5890	83.10	194,423
Sivasagar	9592	107.18	113,440
Golaghat	11,287	11.036	78,189
Dibrugarh	19,160	77.462	93,605
Tinsukia	18,595	68.434	101,357

Source: Statistical Hand Book, Assam, 2011.

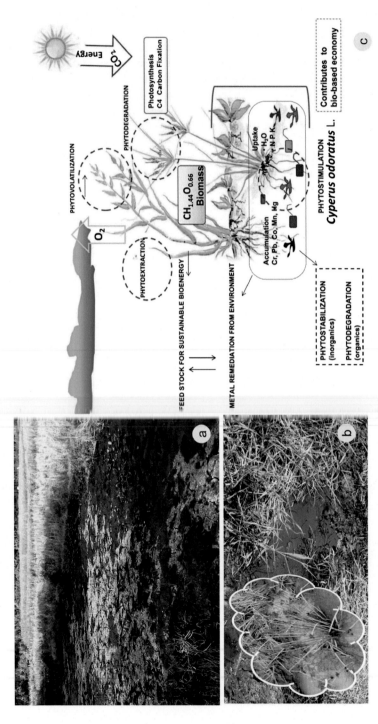

FIG. 2 (A and B) *Cyperus rotundus* (sedge, Cyperaceae), a weed in paddy field. (C) Its association is beneficial for biodegradation of PAH associated with microbial consortia. Sarma, H., Islam, N.F., Prasad, M.N.V., 2017. Plant-microbial association in petroleum and gas exploration sites in the state of Assam, North-East India—significance for bioremediation. Environ. Sci. Pollut. Res. 2, 119–127 with permission.

and perennial water sources, loss of biodiversity and ultimately, loss of economic wealth (Borah et al., 2022).

In a tropical country like India, beneficial plant-microbe technology to decontaminate PAHs infiltrated soil is feasible because o thus in situ bioremediation can potentially be an appropriate technology for this region (Sarma et al., 2018). This is especially true for Assam, which has substantial bioresources, but where remediation in crude oil-contaminated areas (of little economic value) cannot be implemented with more expensive methods.

2 Oil spillage—Causes and concerns

Oil exploration globally results in the accidental release of crude oil into the environment, which calls for the need for efficient and affordable remediation of PCS (Li et al., 2018). Crude oil exploration by the petroleum industry and its transport contaminates the environment with polyaromatic hydrocarbons and heavy metals due to oil spillage and leakage from pipelines, refining processes, oil field installations, petroleum plants, liquid fuel distribution and storage devices, transportation equipment for petroleum products, airports, and illegal drillings in pipelines (Nnadi et al., 2022; Tiwari et al., 2011; Auta et al., 2014). These spills, leaks and other releases of petroleum often result in large-scale contamination of soil and groundwater (Wei et al., 2022; Borah et al., 2022; Naeem and Qazi, 2020) (Fig. 2). These wastewaters contain contaminants that can cause mutations, cancer, and damage to nerve cells (Mustapha and Lens, 2018; Sakshi and Haritash, 2020). Petroleum hydrocarbons, such as benzene, toluene, ethylbenzene, and xylene, occur naturally in crude oil and gasoline (Mohammadi et al., 2020; Melbourne-Thomas et al., 2021). These components are released into the environment as a result of an accidental leak. Reduced agricultural output is one of the most evident results, particularly in north-east India, where multiple incidents of accidental spill have been reported. Over time, degraded soil becomes dry and unsuitable for cereal crop growth (Sharma et al., 2018).

Crude oil is a naturally occurring composite organic liquid with natural gas in underground reserves formed millions of years ago (Joshi et al., 2021). Chemically it consists of saturated non-cyclic hydrocarbons, cyclic hydrocarbons, alkenes, aromatics, sulfur compounds, nitrogen–oxygen compounds, and heavy metals (Sarma et al., 2016), and the refined products of crude oil such as petrol, diesel, etc., are major sources of energy. The refining process of crude oil generates huge quantities of toxic and persistent biodegradable pollutants like PAHs and non-biodegradable pollutants like heavy metals (chromium, lead, nickel, cadmium, cobalt, etc.). The refining process also discharges other hazardous compounds like phenols (cresols and xylenols), sulfides, ammonia, suspended solids, cyanides, and nitrogen compounds which contaminate the ecosystem (Tiwari et al., 2011; Cote, 1976). Of all these contaminants, the presence of PAHs and heavy metals is most disturbing due to their toxic effects on living organisms, as they contain mutagenic, and carcinogenic properties (Ghahari et al., 2022). Heavy metals can also act as endocrine disruptors in addition to causing neurological and developmental problems (Saxena et al., 2019). These pollutants have considerable environmental impacts and present substantial hazards to all living being, and destroy soil health, which may take years or even decades to restore (Bull and Love, 2019).

Several researches have been done on ways to reduce ecological threats of crude oil contamination and its risks to living organisms and ecological systems. Physical absorbers, skimmers, and vacuum technologies employed in conventional oil remediation approaches have either low absorption performance or very limited application possibilities (Kuang et al., 2019). Development of low-cost, high-performance catalysts for the catalytic ozonation process (COP) to treat petroleum refinery wastewater (PRW) remains a significant challenge (Chen et al., 2019). Biodegradation of hydrocarbons by microorganisms has been recognized as one leading, effective, and environment-friendly method by which crude oil is eliminated from contaminated sites (Azubuike et al., 2016; El-Liethy et al., 2022). Several factors and variables affect the removal rate, including the distribution and concentration of pollutants and co-contaminants, local microbial populations and reaction kinetics, and variables such as pH, moisture content, nutrition availability, and temperature (Sharma, 2019). This is acceptable when the contamination occurs in non-cultivated land, but it becomes a serious issue when contaminates spread to net cropped areas. There are several instances where crude oil influx in tea gardens and paddy fields has caused considerable damage. In general, crude oil contamination reduces plant growth in various ways and also affects their biochemical and physiological parameters (Saeed et al., 2021). These damaged areas become unsuitable for cultivation for long periods unless they are remediated through proper methods, as seen in the course of our visit (Fig. 3).

There is a potential for significant emissions of volatile organic compounds in oil and gas areas during activities such as exploration, drilling, transportation, and processing of raw materials (Zheng et al., 2018). It is evident that during the transportation process of crude oil from

FIG. 3   (A–C) Paddy fields in petroleum and natural gas explored areas (A–D) in Assam, India. Crops in petroleum and natural gas explored area (A–D) in Assam, India. (E and F) Oil driling site of Borholla GGS.

drilling sites, high pressure generated in the pipeline causes leakage; this seepage into the adjacent soil and water ecosystem is detrimental to the surrounding crops, especially tea and rice. Furthermore, contamination might also take place due to seepage of crude oil from effluent pits and Group Gathering Stations/Oil Collecting Stations where crude oil is stored for refining purposes.

During the course of the visit, it was observed that almost invariably, the tea growers and rice cultivators have not been keeping any records about their economic loss due to crude oil contamination in the periphery of oil exploration sites. Besides, most of the farmers are also not aware of the importance of soil testing as they have very little knowledge of the scientific practices of growing tea. In addition to that, on quite a few occasions, it has also been noticed that farmers have resorted to indiscriminate use of pesticides either in sub-lethal or in over-dose (Ajeigbe et al., 2012). This is a major concern for the bio-economy in this zone and in order to mitigate the aforesaid shortcoming agricultural extension activities and awareness programs are urgently needed. In fact, the issue of persistent organic pollutant residues is very dynamic in nature and hence tea and rice growers need to be trained periodically about safe cultivation practices including health hazards of pesticides. Those chemicals containing sulfur are the ones found in crude oil in the greatest abundance. The combustion of liquid fuel oil that contains sulfur results in the emission of sulfur oxides and sulphate particulate matter, which poses a risk not only to human health but also to the property of the surrounding community. Furthermore, the corrosion that results from these emissions shortens the useful life of engines and catalysts (Saleh, 2020). Otherwise, the tea produced in this region will fail quality-control tests and remain unsold in international markets, which will cause significant economic loss. In this situation, the combined application of native plants and microbes has a tremendous opportunity to decontaminate the cultivated areas. This is all the more so because bacterial strains able to degrade crude oil are known to be ubiquitous in nature since they use these contaminants as a carbon source, and the process is called microbial catabolism (Islam and Sarma, 2021). For this reason, the biodegradation of crude oil by bacteria is considered a useful tool in the reclamation of oil spill areas (Al-Dhabaan, 2019; Masakorala et al., 2013). Hydrocarbons were reduced by $93.67 \pm 1.80\%$ after 45 days of testing using bioaugmentation with the Hydrocarbon Utilizing Bacterial Consortium (HUBC) and biostimulation (with nutrient inputs) in the soil. The organic carbon content of the bioaugmented and biostimulated soil microcosm reduced from $3.49 \pm 0.08\%$ to $0.62 \pm 0.11\%$, while other nutrients also reduced (Varjani and Upasani, 2019). After 56 days, *P. aeruginosa* NCIM 5514 reduced oily waste by $76.14 \pm 0.85\%$. (Varjani and Upasani, 2021). These investigations will provide a clear explanation of the involvement of hydrocarbon-degrading microorganisms in the removal of oily waste from the environment.

3 Beneficial plant-microbe interaction in biodegradation

The US Environmental Protection Agency has prioritized 16 PAHs because they are physically, chemically, and biologically dangerous to soil and the environment (Ugwoha et al., 2020). Organic contaminates, e.g., PAHs can be removed from the soil through adsorption, volatilization, photolysis and chemical degradation, but these are expensive processes (Abdel-Shafy and Mansour, 2016; Lu et al., 2014). On the other hand, plant–microbial degradation is considered one of the prospective low-cost removal procedures (Islam et al., 2022). Temperature, oxygen, pH, and nutrition can all have an effect on microbial activity (Al-Hawash et al., 2018). PAHs are frequent soil pollutants that affect the grain size, porosity, and water-holding capacity of the soil, as well as the diversity and population of microbes (Sakshi et al., 2019). Plants have a range of cellular mechanisms that make them capable of resisting high concentrations of organic pollutants without displaying their toxic effects; such plants often accumulate and convert these organics into less toxic metabolites. Furthermore, the release of root exudates and enzymes in the rhizosphere in the presence of organic carbon in the soil stimulates faster degradation of organic contaminants. Several bacteria in nature may use petroleum hydrocarbons as carbon and energy sources (Patowary et al., 2022a, b). Bacteria having these abilities are commonly used in the bioremediation of petroleum-contaminated environments (Xu et al., 2018). Among the discovered bacterial species, those belonging to the genera Nocardioides (particularly *Nannocyrtopogon deserti*), Dietzia (particularly *Dietzia papillomatosis*), Microbacterium, Micrococcus, Arthrobacter, Pseudomonas, Cellulomonas, Gordonia, and others contributed the most to oil consumption (Ali et al., 2020). Endophytic bacteria commonly produce a wide range of organic molecules that are comparable to several xenobiotics found in polluted soil, hence facilitating the microbial breakdown of organic contaminants in soils (i.e., phytoremediation) (Ite and Ibok, 2019). On the other hand, for remediation of metal contaminants, some plants have shown the potential role for phytoextraction (contaminants uptake and accumulation into aboveground biomass through xylem conducting system), rhizofiltration (metals filtering from water through root systems), phytostabilization (stabilizing the contaminants by erosion control). *Axonopus compressus* (Sw.) P. Beauv (Bordoloi et al., 2012), *Cyperus brevifolius* (Rottb.) Hassk (Basumatary et al., 2012), and *Cyperus rotundus* Linn. (Basumatary et al., 2013) have already been identified as model plants showing optimum performance when planted in the hydrocarbon-contaminated soil of upper Assam. In the figure (Fig. 4), a hypothetical representation of a hyperaccumulator (*Cyperus odoratus* L.) along with an illustration depicting the process of phytoremediation by it has been provided to show the possibility of planting such hyperaccumulators in contaminated sites to reduce the amount of PAH and heavy metals from soil (Fig. 5).

The potential for adverse effects on humans, animals, and the environment exists whenever soil is subjected to pollutants that are produced after the use of petroleum oil. Used motor oil and PAHs are a few examples of the contaminants that can be found in the environment (Karthick et al., 2019) Researchers have looked into more efficient ways of bioremediation though the results attained to date leave much to be desired, few PAHs are absorbed by plants due to their lesser bioavailability. Many researchers have therefore explored microbe-based technologies and have been able to attain scientific achievements using microbial strains in PAHs degradation studies (Shin et al., 1999; Juhasz and Naidu, 2000). Both beneficial plants and microbes capable of degrading PAHs were isolated and characterized from the natural environment in recent days (Table 5).

4 Potential microbes boosting bio-economy in rice and tea

The tea industry is the largest agro industry in Assam which plays a predominant role in its economy, cultural heritage, and global recognition. In this zone the importance of tea industry can be seen from the fact that Assam alone produces more than 50% of country's tea. Assam

FIG. 4 *Areca catechu* plantations with *Oryza sativa* (A and D), *Musa* sp. plantation (B), *Aquilaria malaccensis* with *Cameli sin* (C, E, and F). Visible oil spills in rice paddies. Rice (*O. sativa*) being a grass, its rhizosphere is involved in the degradation of PAH.

FIG. 5 *Aquilaria malaccensis*—a traditional agro-based industry in petroleum and natural gas explored areas of Assam. Multiple uses of this high-value crops: (A) It is the most common shed tree in tea plantation of Assam, preferred by small-scale tea growers for dual benefits. (B) The fungi-infected clump is used for agar oil extractions. (C) Agar wood being ready for the use of oil extraction. (D) Hydro-distillation process of agar oil extractions. (E) Farmers shape agar wood as incense sticks that have a high demand in Middle East Asia for their traditional beliefs, (F–H) agar oil produced from its extraction through hydro-distillation for commercial production of perfumes.

TABLE 5 Beneficial plant-microbe interactions for degradation of PAHs and heavy metals.

Plants	Bacterial strain	Names of PAHs/HM	References
1. *Lolium perenne, Bouteloua gracilis Artemisia frigida, Banksia integrifolia, Lupinus albus, Lupinus luteus, Bidens pilosa, Alternanthera ficoidea, Taraxacum officinale, Amaranthus deflexu, Phaseolus aureus, Triticum durum, Camellia sinensis, Morus rubra,* and *Avena barbata*	*Pseudomonas* sp., *Rhizobium* sp., *Arthrobacter* sp., *Nocardia* sp., *Streptomyces* sp., *Burkholderia* sp.	Phenanthrene and α-pyrene	Chaudhry et al. (2005)
2. *Viola baoshanensis, Sedum alfredii, Rumex crispus, Helianthus annus, Anthyllis vulneraria, Festuca arvernensis, Koeleria vallesiana Armeria arenaria,* and *Lupinus albus*	*Pseudomonas putida, Pseudomonas fluorescens, Chlorella vulgaris, Methylobacterium oryzae, Berknolderia* sp., *Pseudomonas aeruginosa Citrobacter* sp., *Zooglea* sp., *Arthrobacter* sp., *Ochrobactrum* sp., *Serratia* sp., *Bacillus subtilis*	Naphthalene, anthracene, fluoranthene, pyrene, benzo(*a*)pyrene, As, Zn, Cu, Ni, Cr, Cd, Hg, and Pb	Shukla et al. (2010)
3. *Bouteloua gracilis, Cyanodon dactylon, Elymus canadensis, Festuca arundinacea, Festuca rubra,* and *Melilotus officinalis*	*Mycobacterium* sp., *Haemophilus* sp., *Rhodococcus* sp., *Paenibacillus* sp., *Sphingomonas paucimobilis, Agrobacterium* sp., *Burkholderia* sp., *Rhodococcus* sp., and *Mycobacterium* sp.	Benzo(*a*)pyrene, benzo[*k*] fluoranthene, anthracene, benzo[*b*]fluoranthene, benzo(*e*)pyrene, fluoranthene, naphthalene, phenanthrene, and benzo [ghi]pyrene	Haritash and Kaushik (2009)
1. *Bromus hordeaceous, Festuca arundinacea, Trifolium fragiferum, Trifolium hirtum* and *Vulpia microstachys, Bromus carinatus, Elymus glaucus, Festuca ruba, Hordeum californicum, Leymus triticoides,* and *Nassella pulchra*	*Pseudomonas oleovorans, Aquaspirillum* sp., *Flavobacterium indologenes, Pseudomonas* sp., and *Burkholderia* sp.	Hexadecane, Naphthalene, and Phenanthrene	Siciliano et al. (2003)
5. *Cordia subcordata, Thespesia populnea, Prosopis pallida, Scaevola serica,* and *Medicago sativa*	*Pseudomonas fluorescens, P. aeruginosa, Bacillus subtilis, Bacillus* sp., *Alcaligenes* sp., *Acinetobacter lwoffi,* and *Flavobacterium* sp.	Naphthalene	Das and Chandran (2011)
6. *Sinapis alba, Lepidium sativum,* and *Sorghum saccharatum*	*Acinetobacter* sp., *Acinetobacter iwoffii, Actynomices* sp., *Actynomices viscosus, Agrobacterium radiobacter,* and *Alcaligenes faecalis*	Benzo(*a*)pyrene	Coccia et al. (2009)

TABLE 5 Beneficial plant-microbe interactions for degradation of PAHs and heavy metals—cont'd

Plants	Bacterial strain	Names of PAHs/HM	References
7. *Nymphaea pubescens, Typha* sp., *Juncus effusus, Phragmites australis,* and *Schoenoplectus validus*	*Acinetobacter* sp., *Alcaligens* sp., *Listeria* sp., *Staphylococcus* sp., *Acinetobacter* sp., *Alcaligens* sp., and *Listeria* sp.	Cu, Zn, Pb, Cd, and Fe	Kabeer et al. (2014)
8. *Brassica napus*	*Micromonospora* sp., *Bacillus* sp., *Arthrobacter* sp., *Leifsonia* sp., and *Staphylococcus* sp.	Na, Mg, K, Fe, Cu, Zn, Cd, and Pb	Croes et al. (2013)

tea is considered to be the finest tea which has high demand in global market. The upper Brahmaputra valley produces 15% of the global tea which is higher than any other tea-producing countries like Kenya and Sri Lanka. Again, crude oil production in Assam, India, is a century-old process; therefore, the environmental contamination due to crude oil exploration is highly alarming. The petroleum industry leaves behind a solid residue that is potentially harmful, called oily sludge. It is of utmost importance to develop remedial methods for agricultural regions that have been contaminated by oily sludge (Varjani et al., 2020). Juxtaposition of crude oil exploration sites with rice fields or tea plantations is a predominant feature in upper Assam. Therefore, efforts have been made by various R&D laboratories in recent times to develop a feasible bioremediation technique for eco-restoration of crude oil-contaminated areas of Assam. Recently, the Northeast Institute of Science and Technology, Jorhat, which is a research laboratory of India's premier national R&D organization, the Council of Scientific & Industrial Research (CSIR), has made tremendous contribution toward eco-restoration of degraded land through the application of various bio-formulations consisting of native strains of bacteria along with plantation of hyperaccumulating species (Borah et al., 2022). Despite these efforts many areas are still contaminated, and there is a need to develop proper policy-making by the government and generate awareness among the farmers (Vidonish et al., 2018). Specific efficient plant species have already been used in trial studies for remediation of land degraded by crude oil in many parts of upper Assam's tea garden and rice field which absorb, accumulate and detoxify PAHs and heavy metals (Borah et al., 2022). It is evident that certain plants have capacities to stabilize hydrocarbon-polluted soils and to stimulate soil microbes/microbial consortia in the rhizosphere by releasing root exudates (Table 6). For boosting bio-economy and withstanding the toxic effects of hydrocarbons, particularly for rice and tea grown in crude oil exploration sites, it becomes necessary to isolate some potential microbes for pilot scale bioremediation application.

The data available particularly in the oil exploration sites, tea gardens, and rice fields (Lakuwa, Geleky, Amguri, and Borhola) show that as many as 39 native crude oil-degrading bacteria have been isolated by Yenn et al. (2014), which have the capacity to degrade polyaromatic hydrocarbons. Furthermore, the researchers confirmed that soil quality was also improved through earthworm mortality bioassay and plant tests on rice (*Oryza sativa*) and mung (*Vigna radiata*). These findings have the potential for tea and rice farmers, and we can conclude that the combined use of crude oil degrading bacteria along with nutrient supplements could revive crude oil contaminated soil effectively on a large scale, which is not

TABLE 6 Potential PAH degrading bacterial strains.

Location	Bacterial strain	Names of PAHs	References
1. Shenfu Irrigation Area, Liaoning Province, China	*Mycobacterium* sp., *Pseudomonas* sp., *Sphingomonas* sp., and *Rhodococcus* sp.	Anthracene, fluoranthene, benz(*a*)anthracene, phenanthrene, and pyrene	Li et al. (2007)
2. Tehran Oil Refinery Site, Persian Gulf coasts	*Pseudomonas pudita, Pseudomonas fluorescence, Serratia liquefaciens,* and *Micrococcus strains*	Phenanthrene, benzo[*a*]pyrene, benzo[*a*]anthracene, and chrysene	Mohsen et al. (2009)
3. San Diego Bay, California and Central Pacific ocean	*Pseudomonas* sp., *Rhodococcus* sp., *Mycobacterium* sp., *Burkholderia cocovenenas, Sphingomonas paucimobilis, Pseudomonas fredrikbergensis,* and *Pseudomonas fluorescens*	BTEX, benzo[*a*]pyrene, phenanthrene, naphthalene, fluoranthene, and benzo[*a*] anthracene	Bamforth and Singleton (2005)
4. Jubany station, Argentina	*Pseudomonas aeruginosa*	Phenanthrene	Ruberto et al. (2006)
5. Haldia Refinery site of India	*Bacillus weihenstephansis, Bacillus anthracis, Bacillus mycoides,* and *Bacillus thuringiensis*	Phenanthrene, anthracene, benzo(*a*)pyrene and fluoranthene	Maiti and Bhattacharyya (2012)
6. Beijing Coking Plant	*Acinetobacter* sp. and *Rhodococcus ruber*	Naphthalene, acenaphtylene, acenaphthene, fluorene, phenanthrene, anthracene, fluoranthene, pyrene, benz[*a*] anthracene, chrysene, benzo[*b*] fluoranthene, benzo[*k*] fluoranthene, benzo[*a*]pyrene, dibenzo[*a,h*]anthracene, benzo [*ghi*]perylene, and indeno[1,2,3-*cd*]pyrene	Sun et al. (2012)
7. Gas plant site in Australia	*Alcaligenes* sp. and *Paenibacillus* sp., *Escherichia coli, Pseudomonas* sp., *Pandorea* sp., *Pseudomonas putida, Burkholderia* sp., *Burkholderia cepaciadegrades, Mycobacterium* sp., and *Sphingomonas maltophilia*	Anthracene, phenanthrene, fluoranthene, pyrene, benzo(*a*) pyrene, chrysene, benzo[*b*] fluoranthene, benzo[*k*] fluoranthene, benzo[*a*]pyrene, chrysene, acenapthene, acenapthylene, indeno[1, 2, 3-*cd*] pyrene, dibenz[*ah*]anthracene, and benzo[*ghi*]perylene	Palanisami et al. (2012)
8. Southern Illinois, USA	*Mycobacterium* sp. and *Xanthomonas ampelina*	Pyrene, benzo[*a*]pyrene, anthracene, and benz[a] anthracene	Grosser et al. (1991)
9. Oil and Natural Gas Commission of India (ONGC) oil field, Assam	*Bacillus subtilis, Pseudomonas aeruginosa*	Pyrene, asphaltene, phenanthrene (PHE), and anthracene	Das and Mukherjee (2007)

VII. Case studies

only a cheaper but also an environment-friendly process. Soil contaminated by crude oil is a good habitat for potent hydrocarbon degraders of the genus *Lysinibacillus*, *Brevibacillus*, *Bacillus*, *Paenibacillus*, *Stenotrophomonas*, *Alcaligenes*, *Achromobacter*, and *Pseudomonas* strain. These bacteria singly and in consortia might have contributed to improve the quality of hydrocarbon-contaminated soil, which is supported by studies conducted on this aspect.

The scientists of the Northeast Institute of Science and Technology, Jorhat, and the Institute of Advanced Study in Science and Technology (IASST), Guwahati, an autonomous institute under the Department of Science and Technology, Government of India, have been actively engaged in bioremediation of crude oil contaminated sites in upper Assam oil fields owned by the Oil and Natural Gas Corporation Limited (ONGC) through a CSIR-ONGC joint venture project. Petroleum sludge from refineries was found to include high concentrations of Zn (1299 mg/kg), Fe (60,200 mg/kg), Cu (500 mg/kg), Cr (480 mg/kg), Ni (480 mg/kg), and Pb (565 mg/kg) (Johnson and Affam, 2018). The level of heavy metals in crude oil contaminated soil in some areas of Sivasagar district, Assam, have attained alarming concentrations, such as Arsenic (2.43 ppm), cadmium (4.75 ppm), chromium (7.72), mercury (10.63) and lead (7.98), and these elevated concentrations have been successfully mitigated by the application of bio-formulations developed by NEIST Jorhat (Borah et al., 2022). It further reported that the number of beneficial microbes in soil (e.g., phosphorous solubilizers, sulfur oxidizers, cellulose degraders) in crude oil contaminated soil, which showed a decrease due to contamination, showed an increase during and after remediation. A highly alkaline condition of the soil (pH 10.0–10.5) was reported in their study, which might be due to the low presence of these beneficial microbes. NEIST-Jorhat has successfully reclaimed six sites in ONGC oil fields in upper Assam using technology developed by this institute for eco-restoration of crude oil contaminated sites (Borah et al., 2022). Several crude oil drilling sites, oil wells located in and around tea plantation areas, and paddy fields are common features in the landscape of Assam. Understandably, crude oil contamination in such tea gardens and fields is very high due to spillage, pipeline leakages, etc., and these affect the soil's physical and biological properties. This reduces the growth and resistance of the plants to biotic and abiotic factors, making them more vulnerable to pathogen infestation (Udo and Fayemi, 1995; De Jong, 1980; Schutzendubel and Polle, 2002). A few consortia are mentioned below (Table 7) which can be used in agricultural fields as well as in tea gardens to fight crude oil contamination.

Many Plant microbe consortia have been inventoried, and their bio-economic significance has been outlined. In this article, many of them have been presented to show the tremendous bioremediation capacities of these consortia with respect to PAHs and Heavy metals in laboratory as well as field applications for phytomanagement of moderate to heavily contaminated sites. This is definitely of tremendous bioeconomic significance although, at the same time, it has many limitations, as has already been discussed above. The small-scale tea growers and rice farmers of this region should use this technology to enhance their crop productivity. However, it is to be emphasized that, although many consortia have been developed, only those plant microbes which are native strains will show maximum bioremediation capacities due to their easy adaptability as well as acclimatization in contaminated habitats of this region. Therefore, those interested farmers who would like to apply these consortia could consult with R&D laboratory that has come up with this innovative technology. Concurrently, the R&D lab should welcome not only oil exploration companies but should extend their expertise also to economically marginalized farmers for the greater good of the economy.

VII. Case studies

TABLE 7 List of some consortia which can be used in agricultural fields & tea gardens to fight crude oil contamination.

Consortia	Type of contaminates	References
Cyanobacterium sp., *Synechococcus elongatus*, *Methanocaldococcus indiensis*, *Corynebacterium* sp., *Nocardioides* sp., *Gordonia* sp., *Sinorhizobium* sp., *Rhizobium* sp., *Agrobacterium* sp. *Chelatococcus* sp., *Methylobacterium* sp., *Ochrobacter* sp., *Skermanella* sp., *Corynebacterium* sp., *Pseudomonas* sp., *Sinorhizobium* sp., *Brevibacillus* sp., *Rhizobium* sp., *Agrobacterium* sp.,*Streptomyces* sp., *Actinomycetales* sp., *Bacillales* sp., and *Rhizobiales* sp.	Pyrene metabolizing microbial consortia from the plant rhizoplane	Balcom and Crowley (2010)
Betaproteobacteria, Gammaproteobacteria, Bacteroidetes, Alpha proteobacteria, Stenotrophomonas maltophilia, Acidovorax avenae, Lysinibacillus sphaericus, Stenotrophomonas maltophilia, Achromobacter xylosoxidans and *Serratia marcescens, Lysinibacillus, Bacillus subtilis, Caulobacter* sp., *Bacillus pumilus, Bacillus sp, Stenotrophomonas maltophilia, Erythromicrobium ramosum, Acidovorax avenae, Labrys* sp., and *Burkholderia* sp.	Consortia of PAH-degrading bacteria from crops rhizosphere	Ma et al. (2010)
Sphingobacteria, Mesorhizobium sp., *Alcaligenes* sp. *Bacillus* sp., *Pedobacter* sp., *Paenibacillus* sp., and *Caulobacter* sp.	Consortia for remediation of PAH contaminated soil	Mao et al. (2012)
Klebsiella oxytoca, Bacillus subtilis, Streptococcus sp., *Pseudomonas aeruginosa, Bacillus megaterium, Staphylococcus epidermidis, Enterobacter aerogenes, Escherichia coli, Arthrobacter* sp., *Nocardia* sp. *Corynebacterium* sp., *Aspergillus versicolor, Aspergillus niger, Mucor mucedo, Penicillium chrysogenum, Penicillium* sp., *Acinetobacter iwoffii, Nocardia* sp., *Enterobacter agglomerans, Paenibacillus* sp., *Bacillus thuringeinsis, Alcaligenes* sp., *Agrobacterium* sp., *Chromobacterium* sp., *Enterobacter, Proteus* sp., *Rhizobium* sp., *Brevibacterium* sp., *Corynebacterium* sp., and *Micrococcus* sp.	Consortia for remediation of heavy metals and hydrocarbon contaminated soil	Jaboro et al. (2012)

5 Conclusion

Current research has shown that among the different microbes, use of plant growth-promoting rhizobacteria (PGPR) for bioremediation activity is gaining importance due to their differential abilities to degrade and detoxify contaminants and their positive effects on plant growth promotion (Glick, 2010). Some PGPR strains have the ability to produce IAA, siderophore, phosphorus solubilization, HCN production, phosphorus solubilization and in vitro anti-fungal activity, which enhance tea and rice production and, on the other hand, remove contaminants from the soil. Sometimes the concentration of contaminants can be so high that the environment becomes toxic to microbial populations. Therefore, the technologist must use advanced bioremediation techniques to modify the environment to make it more habitable for plant microbes.

Acknowledgments

The authors would like to express their gratitude to Bodoland University in India, for facilitating and supporting this research. The authors would also like to thank Bhoirob Gogoi and Suprity Shyam for their technical assistance.

References

Abdel-Razek, A.S., El-Sheikh, H.H., Suleiman, W.B., Taha, T.H., Mohamed, M.K., 2020. Bioelimination of phenanthrene using degrading bacteria isolated from petroleum soil: safe approach. Desalin. Water Treat. https://doi.org/10.5004/dwt.2020.25109.

Abdel-Shafy, H.I., Mansour, M.S.M., 2016. A review on polycyclic aromatic hydrocarbons: source, environmental impact, effect on human health and remediation. Egypt. J. Pet. 25 (1), 107–123. Egyptian Petroleum Research Institute https://doi.org/10.1016/j.ejpe.2015.03.011.

Abdelwahab Emam, E., 2018. Clay adsorption perspective on petroleum refining industry. Ind. Eng. 2 (1), 19–25.

Abdullah, S.R.S., Al-Baldawi, I.A., Almansoory, A.F., Purwanti, I.F., Al-Sbani, N.H., Sharuddin, S.S.N., 2020. Plant-assisted remediation of hydrocarbons in water and soil: application, mechanisms, challenges and opportunities. Chemosphere 247, 125932.

Ahmed, T., Chetia, S.K., Chowdhury, R., Ali, S., 2014. Status Paper on Rice in Assam. Rice Knowledge Management Portal, pp. 1–49. http://www.rkmp.co.in.

Ajeigbe, H.A., Adamu, R.S., Singh, B., 2012. Yield performance of cowpea as influenced by insecticide types and their combinations in the dry savannas of Nigeria. Afr. J. Agric. Res. 7 (44), 5930–5938.

Al-Dhabaan, F.A., 2019. Morphological, biochemical and molecular identification of petroleum hydrocarbons biodegradation bacteria isolated from oil polluted soil in Dhahran, Saud Arabia. Saudi J. Biol. Sci. 26 (6), 1247–1252.

Al-Hawash, A.B., Dragh, M.A., Li, S., Alhujaily, A., Abbood, H.A., Zhang, X., Ma, F., 2018. Principles of microbial degradation of petroleum hydrocarbons in the environment. Egypt. J. Aquat. Res. 44 (2), 71–76. https://doi.org/10.1016/J.EJAR.2018.06.001.

Ali, N., Dashti, N., Khanafer, M., Al-Awadhi, H., Radwan, S., 2020. Bioremediation of soils saturated with spilled crude oil. Sci. Rep. 10 (1). https://doi.org/10.1038/s41598-019-57224-x.

Auta, H.S., Ijah, U.J., Mojuetan, M.A., 2014. Bioaugumentation of crude oil contaminated soil using bacterial consortium. Adv. Sci. Focus 2 (1), 26–33.

Azubuike, C.C., Chikere, C.B., Okpokwasili, G.C., 2016. Bioremediation techniques—classification based on site of application: principles, advantages, limitations and prospects. World J. Microbiol. Biotechnol. 32 (11), 1–18. Springer Netherlands.

Bacosa, P.H., Inoue, C., 2015. Polycyclic aromatic hydrocarbons (PAHs) biodegradation potential and diversity of microbial consortia enriched from tsunami sediments in Miyagi, Japan. J. Hazard. Mater. 283, 689–697.

Balcom, N.I., Crowley, E.D., 2010. Isolation and characterization of pyrene metabolizing microbial consortia from the plant rhizoplane. Int. J. Phytoremediation 12, 599–615.

Bamforth, M.S., Singleton, I., 2005. Review bioremediation of polycyclic aromatic hydrocarbons: current knowledge and future directions. J. Chem. Technol. Biotechnol. 80, 723–736.

Basumatary, B., Bordoloi, S., Sarma, H.P., 2012. Crude oil-contaminated soil phytoremediation by using *Cyperus brevifolius* (Rottb.) Hassk. Water Air Soil Pollut. 223 (6), 3373–3383.

Basumatary, B., Saikia, R., Das, C.H., Bordoloi, S., 2013. Field note: phytoremediation of petroleum sludge contaminated field using sedge species, *Cyperus rotundus* (Linn.) and *Cyperus brevifolius* (Rottb.) Hassk. Int. J. Phytoremediation 15 (9), 877–888.

Borah, S.N., Koch, N., Sen, S., Prasad, R., Sarma, H., 2022. Novel nanomaterials for nanobioremediation of polyaromatic hydrocarbons. In: Emerging Contaminants in the Environment. Elsevier, pp. 643–667.

Bordoloi, S., Basumatary, B., Saikia, R., Das, C.H., 2012. *Axonopus compressus* (Sw.) P. Beauv. A native grass species for phytoremediation of hydrocarbon-contaminated soil in Assam, India. J. Chem. Technol. Biotechnol. 87 (9), 1335–1341.

Bull, A.S., Love, M.S., 2019. Worldwide oil and gas platform decommissioning: a review of practices and reefing options. Ocean Coast. Manag. 168, 274–306.

Chaudhry, Q., Zandstra, B.M., Satish Gupta, S., Joner, J.E., 2005. Utilising the synergy between plants and rhizosphere microorganisms to enhance breakdown of organic pollutants in the environment. Environ. Sci. Pollut. Res. 12 (1), 34–48.

Chen, C., Yan, X., Xu, Y., Yoza, B.A., Wang, X., Kou, Y., Ye, H., Wang, Q., Li, Q.X., 2019. Activated petroleum waste sludge biochar for efficient catalytic ozonation of refinery wastewater. Sci. Total Environ. 651, 2631–2640.

CIFOR, 2002. Center for International Forestry Research. Woodlands and rural livelihoods in dryland Africa, Indonesia. CIFOR Info Brief. 4, 1–4.

Coccia, M.A., Gucci, B.M.P., Lacchetti, I., Beccaloni, E., Paradiso, R., Beccaloni, M., Musmeci, L., 2009. Hydrocarbon contaminated soil treated by bioremediation technology: microbiological and toxicological preliminary findings. Environ. Biotechnol. 5 (2), 61–72.

Cole, G.M., 2018. Assessment and Remediation of Petroleum Contaminated Sites. CRC Press.

Cote, R.P., 1976. The Effects of Petroleum Refinery Liquid Wastes on Aquatic Life, with Special Emphasis on the Canadian Environment. Vol. 77 National Research Council of Canada, NRC Associate Committee on Scientific Criteria for Environmental Quality, Ottawa, Ontario, Canada.

Croes, S., Weyens, N., Janssen, J., Vercampt, H., Colpaert, V.J., Carleer, R., Vangronsveld, J., 2013. Bacterial communities associated with *Brassica napus* L. grown on trace element-contaminated and non-contaminated fields: a genotypic and phenotypic comparison. Microb. Biotechnol. 6 (4), 371–384.

Das, N., Chandran, P., 2011. Microbial degradation of petroleum hydrocarbon contaminants: an overview. Biotechnol. Res. Int. https://doi.org/10.4061/2011/941810.

Das, K., Mukherjee, A.K., 2007. Crude petroleum-oil biodegradation efficiency of *Bacillus subtilis* and *Pseudomonas aeruginosa* strains isolated from a petroleum-oil contaminated soil from north-East India. Bioresour. Technol. 98 (7), 1339–1345.

De Jong, E., 1980. Effect of a crude oil spill on cereals. Environ. Pollut. 22, 187–307.

Dhaliwal, S.S., Singh, J., Taneja, P.K., Mandal, A., 2020. Remediation techniques for removal of heavy metals from the soil contaminated through different sources: a review. Environ. Sci. Pollut. Res. 27 (2), 1319–1333.

El-Liethy, M.A., El-Noubi, M.M., Abia, A.L.K., El-Malky, M.G., Hashem, A.I., El-Taweel, G.E., 2022. Eco-friendly bioremediation approach for crude oil-polluted soils using a novel and biostimulated *Enterobacter hormaechei* ODB H32 strain. Int. J. Environ. Sci. Technol. 19 (11), 10577–10588.

FAO, 2005. The State of Food Insecurity in the World: Eradicating World Hunger-Key to Achieving the Millennium Development Goals. FAO, Rome, Italy.

Fenibo, E.O., Ijoma, G.N., Selvarajan, R., Chikere, C.B., 2019. Microbial surfactants: the next generation multifunctional biomolecules for applications in the petroleum industry and its associated environmental remediation. Microorganisms 7 (11), 581.

Frenzel, M., Scarlett, A., Rowland, S.J., Galloway, T.S., Burton, S.K., Lappin-Scott, H.M., Booth, A.M., 2010. Complications with remediation strategies involving the biodegradation and detoxification of recalcitrant contaminant aromatic hydrocarbons. Sci. Total Environ. 408, 4093–4101.

Ghahari, S., Ghahari, S., Ghahari, S., Nematzadeh, G.A., Saikia, R.R., Islam, N.F., Sarma, H., 2022. Assisted and amended technology for the sustainable remediation of emerging contaminants. In: Emerging Contaminants in the Environment. Elsevier, pp. 547–577.

Gitipour, S., Sorial, G.A., Ghasemi, S., Bazyari, M., 2018. Treatment technologies for PAH-contaminated sites: a critical review. Environ. Monit. Assess. 190 (9), 546.

Glick, B.R., 2010. Using soil bacteria to facilitate phytoremediation. Biotechnol. Adv. 28, 367–374.

Gómez-Baggethun, E., Tudor, M., Doroftei, M., Covaliov, S., Năstase, A., Onără, D.-F., Mierlă, M., Marinov, M., Doroșencu, A.-C., Lupu, G., Teodorof, L., Tudor, I.-M., Köhler, B., Museth, J., Aronsen, E., Ivar Johnsen, S., Ibram, O., Marin, E., Crăciun, A., Cioacă, E., 2019. Changes in ecosystem services from wetland loss and restoration: an ecosystem assessment of the Danube Delta (1960–2010). Ecosyst. Serv. 39, 100965.

Grosser, R.J., Warshawsky, D., Vestal, J.R., 1991. Indigenous and enhanced mineralization of pyrene, benzo[a]pyrene, and carbazole in soils. Appl. Environ. Microbiol. 57 (12), 3462–3469.

Haritash, A.K., Kaushik, C.P., 2009. Biodegradation aspects of polycyclic aromatic hydrocarbons (PAHs): a review. J. Hazard. Mater. 169, 1–15.

Islam, N.F., Sarma, H., 2021. Metagenomics approach for selection of biosurfactant producing Bacteria from oil contaminated soil. In: Biosurfactants for a Sustainable Future. John Wiley & Sons, Ltd, pp. 43–58.

Islam, N.F., Patowary, R., Sarma, H., 2022. Biosurfactant-assisted phytoremediation for a sustainable future. In: Assisted Phytoremediation. Elsevier, pp. 399–414.

Ite, A.E., Ibok, U.J., 2019. Role of plants and microbes in bioremediation of petroleum hydrocarbons contaminated soils. Int. J. Environ. Bioremediat. Biodegrad. 7 (1), 1–19.

Jaboro, A.G., Akortha, E., Obayagbona, O.N., 2012. Susceptibility to heavy metals and hydrocarbonclastic attributes of soil microbiota. Int. J. Agric. Biosci. 2 (5), 206–212.

Johnson, O.A., Affam, A.C., 2018. Petroleum sludge treatment and disposal: a review. Environ. Eng. Res. 24 (2), 191–201.

Johnston, J.E., Lim, E., Roh, H., 2019. Impact of upstream oil extraction and environmental public health: a review of the evidence. Sci. Total Environ. 657, 187–199.

Joshi, S.J., Deshmukh, A., Sarma, H. (Eds.), 2021. Biotechnology for Sustainable Environment. Springer.

Juhasz, A.L., Naidu, R., 2000. Bioremediation of high molecular weight polycyclic aromatic hydrocarbons: a review of the microbial degradation of benzo[a]pyrene. Int. Biodeterior. Biodegrad. 45, 57–58.

Kabeer, R., Varghese, R., Kannan, V.M., Thomas, J.R., Poulose, S.V., 2014. Rhizosphere bacterial diversity and heavy metal accumulation in Nymphaea pubescens in aid of phytoremediation potential. J. BioSci. Biotechnol. 3 (1), 89–95.

Karthick, A., Roy, B., Chattopadhyay, P., 2019. A review on the application of chemical surfactant and surfactant foam for remediation of petroleum oil contaminated soil. J. Environ. Manag. 243, 187–205.

Kuang, Y., Chen, C., Chen, G., Pei, Y., Pastel, G., Jia, C., Song, J., Mi, R., Yang, B., Das, S., Hu, L., 2019. Bioinspired solar-heated carbon absorbent for efficient cleanup of highly viscous crude oil. Adv. Funct. Mater. 29 (16), 1900162.

Kumar, V., Shahi, S.K., Singh, S., 2018. Bioremediation: an eco-sustainable approach for restoration of contaminated sites. In: Microbial Bioprospecting for Sustainable Development. Springer Singapore.

Li, X., Li, P., Lin, X., Zhang, C., Lia, Q., Gong, Z., 2007. Biodegradation of aged polycyclic aromatic hydrocarbons (PAHs) by microbial consortia in soil and slurry phases. J. Hazard. Mater. 150 (1), 21–26.

Li, D.-C., Xu, W.-F., Mu, Y., Yu, H.-Q., Jiang, H., Crittenden, J.C., 2018. Remediation of petroleum-contaminated soil and simultaneous recovery of oil by fast pyrolysis. Environ. Sci. Technol. 52 (9), 5330–5338.

Liu, Y., Gao, P., Su, J., da Silva, E.B., de Oliveira, L.M., Townsend, T., Xiang, P., Ma, L.Q., 2019. PAHs in urban soils of two Florida cities: background concentrations, distribution, and sources. Chemosphere 214, 220–227.

Logeshwaran, P., Megharaj, M., Chadalavada, S., Bowman, M., Naidu, R., 2018. Petroleum hydrocarbons (PH) in groundwater aquifers: an overview of environmental fate, toxicity, microbial degradation and risk-based remediation approaches. Environ. Technol. Innov. 10. https://doi.org/10.1016/j.eti.2018.02.001.

Lu, J., Guo, C., Zhang, M., Lu, G., Dang, Z., 2014. Biodegradation of single pyrene and mixtures of pyrene by a fusant bacterial strain F14. Int. Biodeterior. Biodegrad. 87, 75–80.

Ma, B., Chen, H., He, Y., Xu, J., 2010. Isolations and consortia of PAH-degrading bacteria from the rhizosphere of four crops in PAH-contaminated field. In: 19th World Congress of Soil Science, Soil Solutions for a Changing World 1–6 August 2010.

Ma, Y., Li, X., Mao, H., Wang, B., Wang, P., 2018. Remediation of hydrocarbon–heavy metal co-contaminated soil by electrokinetics combined with biostimulation. Chem. Eng. J. 353, 410–418.

Maiti, A., Bhattacharyya, N., 2012. Biochemical characteristics of a polycyclic aromatic hydrocarbon degrading bacterium isolated from an oil refinery site of West Bengal, India. Adv. Life Sci. Appl. 1 (3), 48–53.

Mao, A.A., Hynniewta, T.M., Sanjappa, M., 2009. Plant wealth of Northeast India with reference to ethnobotany. Indian J. Tradit. Knowl. 8 (1), 96–103.

Mao, J., Luo, Y., Teng, Y., Li, Z., 2012. Bioremediation of polycyclic aromatic hydrocarbon-contaminated soil by a bacterial consortium and associated microbial community changes. Int. Biodeterior. Biodegrad. 70, 141–147.

Masakorala, K., Yao, J., Cai, M., Chandankere, R., Yuan, H., Chen, H., 2013. Isolation and characterization of a novel phenanthrene (PHE) degrading strain Pseudomonas sp. USTB-RU from petroleum contaminated soil. J. Hazard. Mater. 263 (2), 493–500.

Mazzeo, D.E.C., Levy, C.E., De Angelis, D.D.F., Marin-Morales, M.A., 2010. BTEX biodegradation by bacteria from effluents of petroleum refinery. Sci. Total Environ. 408, 4334–4340.

Melbourne-Thomas, J., Hayes, K.R., Hobday, A.J., Little, L.R., Strzelecki, J., Thomson, D.P., van Putten, I., Hook, S.E., 2021. Decommissioning research needs for offshore oil and gas infrastructure in Australia. Front. Mar. Sci. 8, 1–19.

Mohammadi, L., Rahdar, A., Bazrafshan, E., Dahmardeh, H., Susan, M.A.B.H., Kyzas, G.Z., 2020. Petroleum hydrocarbon removal from wastewaters: a review. Processes 8 (4), 447.

VII. Case studies

Mohsen, A., Simin, N., Chimezie, A., 2009. Biodegradation of polycyclic aromatic hydrocarbons (PAHs) in petroleum contaminated soils. Iran. J. Chem. Chem. Eng. 28 (3), 53–59.

Mustapha, H.I., Lens, P.N.L., 2018. Constructed wetlands to treat petroleum wastewater. In: Approaches in Bioremediation. Springer.

Naeem, U., Qazi, M.A., 2020. Leading edges in bioremediation technologies for removal of petroleum hydrocarbons. Environ. Sci. Pollut. Res. 27 (22), 27370–27382.

Nnadi, V.E., Udokporo, E.L., Okolo, O.J., 2022. Petroleum production activities and depletion of biodiversity: a case of oil spillage in the Niger Delta. In: Handbook of Environmentally Conscious Manufacturing. Springer International Publishing, pp. 95–111.

Ossai, I.C., Ahmed, A., Hassan, A., Hamid, F.S., 2020. Remediation of soil and water contaminated with petroleum hydrocarbon: a review. Environ. Technol. Innov. 17, 100526.

Palanisami, T., Mallavarapu, M., Ravi, N., 2012. Bioremediation of high molecular weight polyaromatic hydrocarbons co-contaminated with metals in liquid and soil slurries by metal tolerant PAHs degrading bacterial consortium. Biodegradation 23 (6), 823–835.

Patowary, R., Patowary, K., Kalita, M.C., Deka, S., Borah, J.M., Joshi, S.J., Zhang, M., Peng, W., Sharma, G., Rinklebe, J., Sarma, H., 2022a. Biodegradation of hazardous naphthalene and cleaner production of rhamnolipids—green approaches of pollution mitigation. Environ. Res. 209, 112875.

Patowary, R., Patowary, K., Kalita, M.C., Deka, S., Lam, S.S., Sarma, H., 2022b. Green production of noncytotoxic rhamnolipids from jackfruit waste: process and prospects. Biomass Convers. Biorefin. 12, 1–14.

Rajendran, S., Priya, T.A.K., Khoo, K.S., Hoang, T.K.A., Ng, H.-S., Munawaroh, H.S.H., Karaman, C., Orooji, Y., Show, P.L., 2022. A critical review on various remediation approaches for heavy metal contaminants removal from contaminated soils. Chemosphere 287, 132269.

Rasheed, T., Shafi, S., Bilal, M., Hussain, T., Sher, F., Rizwan, K., 2020. Surfactants-based remediation as an effective approach for removal of environmental pollutants—a review. J. Mol. Liq. 318, 113960.

Ruberto, M.A.L., Lucas, A.M.R., Vazquez, C.S., Curtosi, A., Mestre, C.M., Pelletier, E., Cormack, M.P.W., 2006. Phenanthrene biodegradation in soils using an Antarctic bacterial consortium. Bioremediat. J. 10 (4), 191–201.

Saeed, M., Ilyas, N., Jayachandran, K., Gaffar, S., Arshad, M., Sheeraz Ahmad, M., Bibi, F., Jeddi, K., Hessini, K., 2021. Biostimulation potential of biochar for remediating the crude oil contaminated soil and plant growth. Saudi J. Biol. Sci. 28 (5), 2667–2676.

Sakshi, Haritash, A.K., 2020. A comprehensive review of metabolic and genomic aspects of PAH-degradation. Arch. Microbiol. 202 (8), 2033–2058.

Sakshi, Singh, S.K., Haritash, A.K., 2019. Polycyclic aromatic hydrocarbons: soil pollution and remediation. Int. J. Environ. Sci. Technol. 16 (10), 6489–6512.

Saleh, T.A., 2020. Characterization, determination and elimination technologies for sulfur from petroleum: toward cleaner fuel and a safe environment. Trends Environ. Anal. Chem. 25, e00080.

Sarma, H., Sarma, C.M., 2008. Alien traditionally used plant species of Manas Biosphere Reserve, Indo-Burma hotspot. Z. Arznei Gewurzpfla. 13 (3), 117–120.

Sarma, H., Tripathi, K.A., Borah, S., Kumar, D., 2011. Updated estimates of wild edible and threatened plants of Assam: a meta-analysis. Int. J. Bot. 6, 414–423.

Sarma, H., Islam, N.F., Borgohain, P., Sarma, A., Prasad, M.N.V., 2016. Localization of polycyclic aromatic hydrocarbons and heavy metals in surface soil of Asia's oldest oil and gas drilling site in Assam, north-east India: implications for the bio-economy. Emerg. Contam. 2 (3), 119–127.

Sarma, H., Islam, N.F., Prasad, M.N.V., 2017. Plant-microbial association in petroleum and gas exploration sites in the state of Assam, North-East India—significance for bioremediation. Environ. Sci. Pollut. Res. 2, 119–127.

Sarma, H., Nava, A.R., Prasad, M.N.V., 2018. Mechanistic understanding and future prospect of microbe-enhanced phytoremediation of polycyclic aromatic hydrocarbons in soil Environmental Technology & Innovation Mechanistic understanding and future prospect of microbe-enhanced phytoremediation of poly. Environ. Technol. Innov. 13, 318–330.

Sarma, H., Sonowal, S., Prasad, M.N.V., 2019. Plant-microbiome assisted and biochar-amended remediation of heavy metals and polyaromatic compounds—a microcosmic study. Ecotoxicol. Environ. Saf. 176, 288–299.

Sarma, H., Islam, N.F., Prasad, R., Prasad, M.N.V., Ma, L.Q., Rinklebe, J., 2021. Enhancing phytoremediation of hazardous metal(loid)s using genome engineering CRISPR-Cas9 technology. J. Hazard. Mater. 414, 125493.

Saxena, G., Purchase, D., Mulla, S.I., Saratale, G.D., Bharagava, R.N., 2019. Phytoremediation of heavy metal-contaminated sites: eco-environmental concerns, field studies, sustainability issues, and future prospects. Rev. Environ. Contam. Toxicol. 249, 71–131.

Schutzendubel, A., Polle, A., 2002. Plant responses to abiotic stresses: heavy metal-induced oxidative stress and protection by mycorrhization. J. Exp. Bot. 53 (372), 1351–1365.

Sharma, J., 2019. Advantages and limitations of in situ methods of bioremediation. Recent Adv. Biol. Med. 5.

Sharma, D., Sarma, H., Hazarika, S., Islam, N.F., Prasad, M.N.V., 2018. Agro-Ecosystem Diversity in Petroleum and Natural Gas Explored Sites in Assam State, North-Eastern India: Socio-Economic Perspectives.

Shin, S.K., Oh, Y.S., Kim, S.J., 1999. Biodegradation of phenanthrene by *Sphingomonas* sp. strain KH3-2. J. Microbiol. 37, 185–192.

Shukla, P.K., Singh, K.N., Sharma, S., 2010. Bioremediation: developments, current practices and perspectives. Genet. Eng. Biotechnol. J. 3, 1–20.

Siciliano, D.S., Germida, J.J., Banks, K., Greer, W.C., 2003. Changes in microbial community composition and function during a polyaromatic hydrocarbon phytoremediation field trial. Appl. Environ. Microbiol. 69, 483–489.

Sun, D.G., Xu, Y., Jin, H.J., Zhong, P.Z., Liu, Y., Luo, M., Liu, P.Z., 2012. Pilot scale ex-situ bioremediation of heavily PAHs-contaminated soil by indigenous microorganisms and bioaugmentation by a PAHs-degrading and bioemulsifier-producing strain. J. Hazard. Mater. 233–234, 72–78.

Thakare, M., Sarma, H., Datar, S., Roy, A., Pawar, P., Gupta, K., Pandit, S., Prasad, R., 2021. Understanding the holistic approach to plant-microbe remediation technologies for removing heavy metals and radionuclides from soil. Curr. Res. Biotechnol. 3, 84–98.

Tiwari, J.N., Chaturvedi, P., Ansari, N.G., Patel, D.K., Jain, S.K., Murthy, R.C., 2011. Assessment of polycyclic aromatic hydrocarbons (PAH) and heavy metals in the vicinity of an oil refinery in India. Soil Sediment Contam. Int. J. 20 (3), 315–328.

Truskewycz, A., Gundry, T.D., Khudur, L.S., Kolobaric, A., Taha, M., Aburto-Medina, A., Ball, A.S., Shahsavari, E., 2019. Petroleum hydrocarbon contamination in terrestrial ecosystems—fate and microbial responses. Molecules 24 (18), 3400.

Udo, E.J., Fayemi, A.A.A., 1995. The effect of oil pollution on soil germination, growth and nutrient uptake of corn. J. Environ. Qual. 4, 537–540.

Ugwoha, E., Amah, V.E., Oweh, G.O., 2020. Bioremediation of crude oil contaminated soil using pig droppings and bone char. J. Adv. Biol. Biotechnol. 23, 13–24.

Varjani, S., Upasani, V.N., 2019. Influence of abiotic factors, natural attenuation, bioaugmentation and nutrient supplementation on bioremediation of petroleum crude contaminated agricultural soil. J. Environ. Manag. 245, 358–366.

Varjani, S., Upasani, V.N., 2021. Bioaugmentation of *Pseudomonas aeruginosa* NCIM 5514—a novel oily waste degrader for treatment of petroleum hydrocarbons. Bioresour. Technol. 319, 124240.

Varjani, S., Upasani, V.N., Pandey, A., 2020. Bioremediation of oily sludge polluted soil employing a novel strain of *Pseudomonas aeruginosa* and phytotoxicity of petroleum hydrocarbons for seed germination. Sci. Total Environ. 737, 139766.

Vidonish, J.E., Alvarez, P.J.J., Zygourakis, K., 2018. Pyrolytic remediation of oil-contaminated soils: reaction mechanisms, soil changes, and implications for treated soil fertility. Ind. Eng. Chem. Res. 57 (10). https://doi.org/10.1021/acs.iecr.7b04651.

Wei, K.-H., Ma, J., Xi, B.-D., Yu, M.-D., Cui, J., Chen, B.-L., Li, Y., Gu, Q.-B., He, X.-S., 2022. Recent progress on in-situ chemical oxidation for the remediation of petroleum contaminated soil and groundwater. J. Hazard. Mater. 432, 128738.

World Bank, 2001. World Development Report 2000/2001: Attacking Poverty. Oxford University Press, Oxford, pp. 1–200.

World Bank, 2004. Sustaining Forests: A Development Strategy. World Bank, Washington, DC.

Xu, X., Liu, W., Tian, S., Wang, W., Qi, Q., Jiang, P., Gao, X., Li, F., Li, H., Yu, H., 2018. Petroleum hydrocarbon-degrading bacteria for the remediation of oil pollution under aerobic conditions: a perspective analysis. Front. Microbiol. 9, 2885.

Yenn, R., Borah, M., Boruah, H.P.D., Roy, A.S., Baruah, R., Saikia, N., Sahu, O.P., Tamuli, A.K., 2014. Phytoremediation of abandoned crude oil contaminated drill sites of Assam with the aid of a hydrocarbon-degrading bacterial formulation. Int. J. Phytoremediation 16 (9), 909–925.

Zheng, H., Kong, S., Xing, X., Mao, Y., Hu, T., Ding, Y., Li, G., Liu, D., Li, S., Qi, S., 2018. Monitoring of volatile organic compounds (VOCs) from an oil and gas station in Northwest China for 1 year. Atmos. Chem. Phys. 18 (7), 4567–4595.

New biology

From preservation of aquatic ecosystems to ecocatalysis®

Claude Grison[a], Pierre-Alexandre Deyris[a], Cyril Poullain[a], and Tomasz K. Olszewski[b]

[a]Laboratory of Bioinspired Chemistry and Ecological Innovations, UMR 5021, CNRS-University of Montpellier, Grabels, France [b]Department of Physical and Quantum Chemistry, Wroclaw University of Science and Technology, Wroclaw, Poland

1 Context

1.1 Preserving water quality: A major challenge

Everyone agrees that water is a vital resource and a precious common good. No priority can precede access to water. Water is sometimes referred to as "blue gold." On every 22nd March (World Water Day), the United Nations reminds us all that "access to clean water" is a crisis as serious as climate change (UNESCO, 2020). Although sustainable development goal #6, adopted by UN member states, requires providing access to safe drinking water for all, we already know that this objective will not be achieved by 2030.

For Alain Boinet, founder of Solidarités International, access to clean water is a vital humanitarian struggle. Nearly 2.2 billion people still do not have access to safely managed domestic drinking water supply services (Solidarités International, n.d.). Over the past 100 years, global water use has increased sixfold due to population pressure, economic development and overconsumption. Increasingly rare, this blue gold is also increasingly polluted and thus finds itself at the heart of a disruption of aquatic ecosystems that have been providing natural water quality treatment for millennia.

1.2 Ecosystem-based disruption of water resources

The world's freshwater resources are increasingly polluted by household waste (Brausch and Rand, 2011) (e.g., plastics (Blettler and Wantzen, 2019), pathogens (Schmeller et al., 2018), pesticides (Boxall, 2012), metallic elements (The European Parliament and the Council of the European Union, 2000) and emerging pollutants such as endocrine disruptors (Vilela et al., 2018), poly- and perfluorinated compounds (Cousins et al., 2022), solar filters (Emmanouil et al., 2019), and organic and mineral nanoparticles). Polluted urban stormwater runoff, effluent from mining activities, industrial spills, polluted sediments, and water transport of waste also have direct impacts on surface and groundwater quality. Even today, more than 80% of wastewater from domestic activities, urban areas, industry, and agriculture is discharged into nature without remediation action.

Wetlands are also affected. While wetlands are the largest reservoirs of carbonaceous organic matter, they contribute to the mitigation of extreme weather events (floods and droughts), water purification, and the preservation of biodiversity. However, 85% of wetlands have been lost in 300 years; they continue to disappear three times faster than forests (Davidson, 2014). Since the 1970s, an additional problem has been highlighted: the number of invasive alien species (IAS) in wetlands and aquatic environments (e.g., water hyacinth, water lettuce, Asian knotweed, Asian carp, American crayfish, etc.) has increased by 70% (Blottiere and Sarat, 2019; IPBES, 2019). These IAS pose a threat to water quality and availability. Together with global change, they represent one of the five factors responsible for declining biodiversity, animal and plant health, ecosystem services, the environment and the economy.

Thus, water depletion and pollution are the main causes of biodiversity loss and degradation of these ecosystems, which no longer provide their regulatory action, thus reducing the resilience of neighboring ecosystems in a worrying chain of events.

More vulnerable societies are increasingly exposed to climate and non-climate risks (UNESCO, 2018). However, the threat of access to drinking water resources not only concerns countries with dry climates and poorly developed infrastructure, but Europe is also concerned. European directive 2000/60/EC established a framework for improving water quality (The European Parliament and the Council of the European Union, 2000). It presents a strategy for reducing water pollution which comes from substances of greatest concern. Metals are on the priority list.

1.3 The ecological challenge of metals: As essential as they are worrying

Because of their interesting physical properties (electrical and thermal conduction, hardness, malleability, potential alloy, catalytic properties, etc.), metals are very widely used. Metal deposits have been exploited for a long time. Past and current mining activity has resulted in a dispersion of metals in the environment. The aquatic environment is contaminated by runoff from mineral storage sites. Agricultural activities also cause environmental contamination by metals. For example, copper sulfate is still used to treat vines and fruit trees. Fertilizers, fungicides, industrial sludge, composts, or slurry used to amend agricultural land are also susceptible to metal contamination. The problem is greatly amplified during floods. The overflow of settling ponds and the leaching of mining waste lead to the formation of runoff water that permanently disperses metal elements in the environment. Catastrophic breakages of mine waste embankments in Aznalcóllar, Spain (Aguilar et al., 2000), and in Bahia Mare, Romania (Cunningham, 2005), dam failures in the Brazilian municipality of Brumadinho (Siqueira et al., 2022) or the city of Jagersfontein, South Africa (Brown and van Wyk, 2022), quickly recall the inherent dangers. A major problem is also that of mines that are no longer in operation and have been abandoned. In southern Europe, the mines of the Cévennes (France) (Lefèvre and Richard, 2022) or Montevecchio (Sardinia) (Cidu and Fanfani, 2002) are perfect illustrations of the situation: Abandonment of mines and violent rainfall episodes represent a very worrying context, leading to the contamination of watercourses by zinc, cadmium, and lead. In the United Kingdom, the Environmental Protection Agency has identified 150 unstable or highly polluted sites out of tens of thousands of former mines. All 653 rivers are threatened, 90% of the alluvial plains of northern England are polluted with heavy metals, and it is estimated that the same is true in Wales and parts of southern England. In Italy, the Ministry of Ecological Transition updates an inventory of about a hundred pages every 3 years. The Environmental Protection Agency describes more than 300 sites with a high or very high risk to the environment and health (Izoard, 2022).

In France, the Aude department, and more particularly the Orbiel valley, perfectly illustrates the ecological and health emergency resulting from past mining activities, with the dispersion of toxic metal elements, such as arsenic, in waterways. The largest gold mine in Europe from 1870 to 2004, the Salsigne mine (Aude, France), was the world's leading arsenic production site (10% of production). In October 2018, exceptionally heavy rains in the Cévennes caused runoff under and on arsenic storage sites (about 6 million tons spread over three sites). The polluted water poured into the Orbiel river and its tributaries before reaching groundwater, gardens, schools, homes, wine-growing areas, etc. Similarly, in June 2008, Chinese companies dumped arsenic-containing chemical waste into Yangzonghai Lake, making it impossible to use the lake's water for fishing, agricultural irrigation, or human consumption (Chen et al., 2015). In Africa, too, arsenic contamination is a real problem. Although some of it is of geological origin, anthropogenic practices such as mining, agriculture, waste incineration are at the root of the contamination of water resources (Ahoulé et al., 2015).

Around the world, some 108 countries are affected by arsenic pollution of their groundwater, whether of geological or industrial origin (Shaji et al., 2021).

All these dramatic facts remind us of an essential point: water can become a vector of pollution (Fig. 1).

FIG. 1   (A) River downstream of the Montevecchio mine. (B) River downstream of the Salsigne mine (©Claude Grison).

FIG. 2   Sludge generated by the treatment of
pyrite quarry effluents (©Claude Grison).

Other lesser-known situations are also of concern. For example, some quarries have traces of problematic minerals such as pyrite (Fig. 2), where the gangue of iron disulfide and manganese is transformed by rainwater under aerobic conditions into sulfuric acid, Fe, and Mn sulfates, thus leading to highly acidic effluents (pH$=3$). The bottom water of the pits and treatment basins of these quarries must be treated.

The scenario is analogous to acid mine drainage. This is the case in South Africa in a former mine that ceased operation in the 1980s (Akcil and Koldas, 2006), but also in Ireland where many homes have been built with pyrite-rich materials, causing widespread damage to homes and the environment (O'Connell, 2012).

The solutions implemented transfer the problem since they generate toxic sludge stored on sites with significant management difficulties (Alastuey et al., 1999). Although stabilized, the alternation of hot seasons and rainy events weakens this sludge which erodes and finally flows into nearby rivers, as was the case in Spain in 2008.

Developing new approaches to anticipate these situations by preventing sources of pollution, treating polluted water upstream, and implementing innovative treatment processes using sustainable technologies have thus become a priority.

2   A triple ecological solution: Management of invasive alien species, effluent depollution, and sustainable chemistry recovery

Water comes from nature, which is why "preserving the natural environment is necessarily the best way to preserve the resource" (Orsenna, 2008). According to the 2018 World Water Development Report, "working with nature, not against it, would preserve natural capital and support a circular economy," using resources efficiently but sustainably. Nature-based solutions can be viable and provide environmental, social, and economic benefits. These combined benefits are critical to achieving the 2030 Agenda as defined by the UN.

FIG. 3 The 4 of the 17 sustainable development goals.

The Bio-inspired Chemistry and Ecological Innovations laboratory and the company Bioinspir joined forces at the beginning of 2020 to set up an ambitious and innovative response in the short to medium term to actively contribute to the depollution of aquatic environments.

The proposed process is based on breakthrough innovations reconciling ecological solutions and valorization in sustainable chemistry, in the spirit of Sustainable Development Goals #6 (Clean water and sanitation), #9 (Industry, innovation, and infrastructure), #13 (Climate action), and #14 (life below water) (Fig. 3).

2.1 Green solutions

2.1.1 Objectives

Faced with the findings of pollution of aquatic environments described above, ecological solutions must be efficient and selective. The principles of rhizofiltration and biosorption described in this chapter meet these expectations. Indeed, in contact with polluted environments, several aquatic plants have developed a tolerance and even affinities for certain metal pollutants. The ability to retain metals in their root parts makes these plants prime candidates for these ecological solutions. In the context of using these plants in a sustainable chemistry recovery process, it is immediately necessary to comply with the REACH regulation; this is why only certain categories of metal elements are targeted. The first are strategic metals, namely, platinoids and rare-earth metals. The second are the more common transition metals, called primary ores (Zn, Mn, Cu, Fe, etc.).

2.1.2 Rhizofiltration

Rhizofiltration is an effective method used in phytoremediation. The roots of the plants are used to absorb and sequester metals dissolved in polluted water. The first experiments were conducted in the early 1970s by Sutton and Blackburn on the accumulation potential of copper by *Eichhornia crassipes* and *Hydrilla verticillata* (Sutton and Blackburn, 1971a, b). Subsequently, nearly 40 aquatic and non-aquatic plants (Tiwari et al., 2019) were studied for their ability to bioaccumulate several dissolved metals (primary, strategic, or toxic to the environment) in their root systems. Of these, some have particular affinities for strategic and primary metallic elements (Table 1).

Among these results, an in-depth study of five plant species, *Bacopa monnieri, Brassica juncea, E. crassipes, Lolium multiflorum*, and *Pistia stratiotes*, was carried out to deepen the mechanisms of bioconcentration of key metallic elements, Cu, Mn, Pd, Pt, and Rh by rhizofiltration (Fig. 4) (Bihanic et al., 2021; Clavé et al., 2016a; Garel et al., 2015; Grison et al., 2015).

TABLE 1   Review of the works that study the ability of plants for rhizofiltration of some recoverable metallic elements.

Reference	Plant species	Contaminants
Sweta et al. (2015)	*Trapa natans*	Cu
Matache et al. (2013)	*Ceratophyllum demersum, Potamogeton pectinatus, Potamogeton perfoliatus, Potamogeton lucens*	Cu and Zn
Pajević et al. (2008)	*Potamogeton pectinatus, P. lucens, P. perfoliatus*	Cu, Mn, and Fe
Pavlovic et al. (2005)	*Lemna minor, L. gibba, Mentha aquatica, Myriophyllum spicatum*	Fe, Mn, and Cu
Milošković et al. (2013)	*Iris pseudacorus, Lemna gibba, Myriophyllum spicatum, Polygonum amphibium, Typha angustifolia*	Fe, Cu, and Mn
Baldantoni et al. (2004)	*Najas marina, Phragmites communis*	Cu, Mn, Fe, Ni, and Zn
Dushenkov et al. (1995)	*Brassica juncea, Helianthus annuus*	Cu, Ni, and Zn
Clavé et al. (2016a), Garel et al. (2015), Grison et al. (2015), and Muramoto and Oki (1983)	*Eichhornia crassipes, Brassica juncea, Lolium multiflorum*	Cu, Ni, Zn, Pd, Pt, and Rh
Kumar et al. (2012) and Sinha and Chandra (1990)	*Bacopa monnieri, Hydrilla verticillata, Iris aquatic, Marsilea minuta*	Cu and Ni
Bihanic et al. (2021)	*Pistia stratiotes*	Mn

FIG. 4   Plant species of interest in rhizofiltration (©Claude Grison).

The results obtained are collected in Table 2. In general, it is observed that the vast majority of plants presented have a concentration of metal elements studied of more than $1\,g\,kg^{-1}$ in their root system. Of the three plants studied in palladium rhizofiltration, *E. crassipes* has the highest concentration of palladium in its roots (Table 2, entries 1–3). *L. multiflorum* is more effective than *B. juncea* for platinum adsorption, while the reverse is observed for rhodium (Table 2, entries 4–7). *E. crassipes* is the aquatic species with the highest concentration of

TABLE 2 Concentration in roots of plants after rhizofiltration of Pd-, Pt-, Rh-, Cu-, and Mn-containing solutions; TME (trace metallic elements).

Entry	Plant species	TME	Concentration in roots (mg kg^{-1})	Reference
1	*Brassica juncea*	Pd	21,150	Grison et al. (2015)
2	*Lolium multiflorum*	Pd	4400	Grison et al. (2015)
3	*Eichhornia crassipes*	Pd	89,000	Clavé et al. (2017)
4	*Brassica juncea*	Pt	1938	Grison et al. (2015)
5	*Lolium multiflorum*	Pt	3781	Grison et al. (2015)
6	*Brassica juncea*	Rh	1512	Grison et al. (2015)
7	*Lolium multiflorum*	Rh	509	Grison et al. (2015)
8	*Bacopa monnieri*	Cu	13,400	Clavé et al. (2016a)
9	*Lolium multiflorum*	Cu	7100	Clavé et al. (2016a)
10	*Eichhornia crassipes*	Cu	25,500	Clavé et al. (2016a)
11	*Pistia stratiotes*	Mn	52,000	Bihanic et al. (2021)

Cu in roots after *B. monnieri* and *L. multiflorum* (Table 2, entries 8–10). Finally, *P. stratiotes* phytoaccumulates manganese strongly (Table 2, entry 11).

Rhizofiltration is an efficient phytotechnology to phytoaccumulate valuable metal elements. It is based on using unprotected, abundant plant species with easy access (easy crops or IAS). However, a limit is necessary for large-scale transposition: it requires substantial facilities such as large cultivation basins and then treatment of polluted water. Therefore, other natural solutions have been studied to facilitate scaling up and offer real industrial prospects.

2.1.3 Biosorption

Surprisingly, the Grison group has shown that the treatment of water polluted by the root biomass of the aforementioned plants is just as effective whether the plants are dead or alive (Grison et al., 2018). Dehydrated and powdered roots are ideal! It has been shown that the mechanism of concentration of metallic elements in the root system of aquatic and semi-aquatic plants is mainly an adsorption-like mechanism (Grison et al., 2018; Stanovych et al., 2019). The treatment of a polluted aquatic system could therefore be extended to biosorption. This technology is based on the development of an innovative filter system consisting exclusively of plant materials packaged in the form of vegetable powder (Grison et al., 2018). The selected plant species are either aquatic (or hydrophytes) or wetland plants (or hygrophytes). Their origin is twofold:

— The first are native plants led by the Water mint (Fig. 5). It has all the required assets: native plant, important root system capable of depolluting large volumes of effluents, and remarkable ability to bioconcentrate trace metal elements (TME). It is as efficient dead or

FIG. 5   *Mentha aquatica*: a plant species with a root system of interest for the ecological recycling of metal elements by biosorption (©Claude Grison).

   alive: the dehydrated and powdered roots retain their extractive properties. They are
   grown hydroponically to preserve natural resources.
 — The second are IAS that develop uncontrolled in aquatic environments (rivers, bodies of
   water, wetlands, etc.) (Fig. 6). They pose a real threat to aquatic ecosystems.

FIG. 6   (A) *Eichhornia crassipes*, (B) *Pistia stratiotes*, (C) *Ludwigia peploides*, (D) *Fallopia japonica*, are examples of IAS
which are used to decontaminate industrial effluents by biosorption (©Claude Grison).

As part of this program, the challenges are twofold: to clean up effluents using filters from abundant plant raw materials and to stimulate efforts to control the proliferation of these invasive species. Finding an economic outlet for harvested invasive plants becomes a condition of sustainability for such environmental actions.

The root system of these plants has a special chemical structure. It is rich in hemicellulose and in particular in carboxylate functions, which have a high affinity for TME. Since hemicellulose is a structural (not metabolic) component of the adult plant, the complexing properties of TME survive the death of the plant. An ATR FT-IR-fitting (attenuated total reflectance-Fourier transform infrared spectroscopy) method was developed to rationalize and then predict the ability of new aquatic species to phytoaccumulate TME (Stanovych et al., 2019). Direct correlations between the number of carboxylate functions and the percentage of biosorption of TME have been established. Indeed, for each of these materials, it is possible to determine the ratio of the intensity of the vibration band of the C=O bond of the carboxylate group and the intensity of each of the vibration bands of the aromatic ring (assimilated to lignin, another biopolymer constituting the materials). This ratio corresponds to the ratio of absorbances of the respective bands (Table 3). Two groups of biomaterials are then observed:

- Group 1, rich in carboxylate, corresponds to a ratio C=O/Ar > 1.
- Group 2, rich in lignin, corresponds to a ratio C=O/Ar < 1.

It is thus possible to extract polluting metallic elements of different nature and in different contexts if the biomaterial comes from well-chosen aquatic and semi-aquatic plant species. The developed technology enables responses to very different scenarios of metal pollution: strategic metals such as platinoids (Pd, Rh, Pt) (Garcia et al., 2021), rare earth (Eu, Yb, Ce) and similar (Sc) (Grison et al., 2018), primary metals (Fe, Zn, Ni, Mn, Cu) (Cases et al., 2021; Grison et al., 2018; Olszewski et al., 2019; Richards et al., 2022), and toxic metals (Cd, Pb) (Stanovych et al., 2019).

Direct biosorption

Maximum biosorption capacities have been established with different metallic elements frequently encountered in industrial effluents.

TABLE 3   Categories of biomaterials according to their C=O/R ratio.

Plant species		C=O/Ar ratio	Biosorbed TME
Group 1	*Eichhornia crassipes*	2.02	Fe, Zn, Mn, Ni, Cu, Co, Ce, Eu, Yb, Sc, Pd, Rh, Pt
	Mentha aquatica	1.65	
	Pistia stratiotes	1.56	
	Ludwigia peploides	1.15	
Group 2	*Fallopia japonica*	0.72	Fe, Ni, Pd
	Pine cones	0.53	

The different plants are compared to pine cones, which are mostly made of lignin.

Biosorption of iron and zinc While Fe(II) can be biosorbed by common lignin-rich bio-materials, the best results have been obtained with carboxylate-rich biosorbents derived from aquatic plants. Table 4 shows the equilibrium adsorption capacity of Fe(II) by several biosorbents [mg of Fe(II) per g of biosorbent] and illustrates this conclusion (Richards et al., 2022).

For example, materials rich in lignin (group 2) allow biosorption from 2.8 to 7.2 mg of Fe(II) per gram of biosorbent from an initial solution of iron sulfate concentrated at $1.2\,g\,L^{-1}$. However, aquatic plant species rich in carboxylates provide biosorption capacities ranging from 12.5 to $58.07\,mg\,g^{-1}$. The most spectacular example is *P. stratiotes* at $58.07\,mg\,g^{-1}$ (Richards et al., 2022).

The different biosorption capacities of these biomaterials were exploited to carry out the remediation of a mining effluent rich in Fe(II) and Zn (Table 5) (Stanovych et al., 2019).

The results are consistent with the known high affinity of iron for phenol groups of lignin-based materials (pine cones) (Merdy et al., 2003) and phenolic acids (*E. crassipes, Mentha aquatica*). In contrast, the ability of materials to remove Zn is more specific. For example, by using 3 g of root biomass of *E. crassipes*, the mine effluent containing Zn at $11.7\,mg\,L^{-1}$ could be purified to a Zn concentration of $0.9\,mg\,L^{-1}$ after biosorption is complete (Table 5). This is particularly interesting because this Zn concentration is below the permitted standards ($2\,mg\,L^{-1}$) (French Republic and INERIS, 2019). Importantly, no desorption was observed. Plant materials derived from pine cones are generally effective but do not achieve acceptable concentrations of Zn after biosorption. For example, with $3\,g\,L^{-1}$ of pine cones, the Zn concentration of an effluent initially containing $12.3\,mg\,L^{-1}$ could be reduced to $3.7\,mg\,L^{-1}$. Thus, the interest in using a pinecone material for biosorption is limited to immediate and highly effective retention of Fe (Table 5). Regarding Zn, the results obtained with

TABLE 4 Equilibrium adsorption capacity of Fe(II) by biosorbents derived from biomaterials rich in carboxylate functions (Group 1) or rich in lignin (Group 2).

Biosorbents	Group 1 (biomaterials rich in carboxylates)				Group 2 (biomaterials rich in lignin)	
	P. stratiotes	*M. aquatica*	*L. peploides*	*E. crassipes*	*F. japonica*	Pine cones
q_e $(mg\,g^{-1})$	58.07	16.70	15.74	12.50	7.18	2.78

TABLE 5 Biosorption of mining effluents rich in Fe and Zn using different biosorbents in batch mode.

Metal element		Fe $(mg\,L^{-1})$	Zn $(mg\,L^{-1})$
Initial concentration of mining effluent		5.5–5.9	11.7–12.4
Eichhornia crassipes roots	$1\,g\,L^{-1}$	0.2	3.6
	$3\,g\,L^{-1}$	0.3	0.9
Mentha aquatica roots	$1\,g\,L^{-1}$	0.2	2.4
	$3\,g\,L^{-1}$	0.2	1.0
Pine cones	$1\,g\,L^{-1}$	<0.1	5.5
	$3\,g\,L^{-1}$	<0.1	3.7

M. aquatica and *E. crassipes* are very similar (Table 5). With $3\,g\,L^{-1}$ of these biomaterials, the Zn concentration of an effluent initially containing $12.4\,mg\,L^{-1}$ could be reduced to 0.9 and $1.0\,mg\,L^{-1}$ of Zn.

Biosorption processes in gravity column mode were carried out to approach an automated treatment of industrial effluent. Trials focused on pine cones and *E. crassipes* (Table 6) (Stanovych et al., 2019).

These results remain consistent with the structural properties of plant-based biomaterials, as shown by ATR FT-IR analysis. Pine cones, rich in lignin and low in carboxylate groups, are ideal for the biosorption of Fe, while the biomass of *E. crassipes*, rich in carboxylate groups, is particularly suitable for the biosorption of Fe and Zn.

Biosorption of platinoids The differences between biomaterials are more nuanced with palladium biosorption (Garcia et al., 2021). Table 7 shows the biosorption results of a solution containing $14\,mg\,L^{-1}$ of palladium by different biomaterials at $1\,g\,L^{-1}$. These results show that different biomaterials, regardless of their origins, have the ability to biosorb palladium effectively. For biomaterials from carboxylate-rich plants, the best results were obtained by *P. stratiotes* and *E. crassipes* with 0.4 and $1.3\,mg\,L^{-1}$, respectively, of palladium remaining, i.e., biosorption yields of 97% and 91%. Pine cones, rich in lignin, leave a concentration of $0.6\,mg\,L^{-1}$ palladium solution (96% biosorption).

It should be noted that these results can be extended to other platinoids of interest in catalytic chemistry. Indeed, biosorption experiments on solutions containing different platinoids were performed with powdered roots of *E. crassipes* ($1\,g\,L^{-1}$) to determine the maximum individual adsorption capacity (Garcia et al., 2021). The results are listed in Table 8.

TABLE 6 MP-AES analysis (microwave plasma atomic emission spectroscopy) of effluent after biosorption in gravity column mode.

Element	Initial concentration $(mg\,L^{-1})$	After biosorption with pine cones at $1\,g\,L^{-1}$ $(mg\,L^{-1})$	After biosorption with *Eichhornia crassipes* at $3\,g\,L^{-1}$ $(mg\,L^{-1})$
Fe	5.6	<0.1	<0.1
Zn	12.4	9.4	1.1

TABLE 7 Comparison of biosorption on a synthetic effluent based on the content of Pd at $14\,mg\,L^{-1}$ with different biomaterials.

Biosorbent	Palladium concentration after biosorption at $1\,g\,L^{-1}$ $(mg\,L^{-1})$	Biosorption yield
Eichhornia crassipes	1.3	91%
Pistia stratiotes	0.4	97%
Ludwigia peploides	2.5	82%
Mentha aquatica	2.7	81%
Pine cones	0.6	96%

TABLE 8 Biosorption yields of several platinoids in solution at $14\,mmol\,L^{-1}$ by powdered dried roots of *Eichhornia crassipes* ($1\,g\,L^{-1}$).

Platinoid	Pd	Rh	Pt	Ru	Ir
Yield	89%	68%	64%	11%	4%

TABLE 9 Competitive biosorption of Pd, Pt, and Rh solution at $0.132\,mmol\,L^{-1}$ each by powdered roots of *Eichhornia crassipes* and selectivity.

Platinoid	Pd	Rh	Pt
Biosorption yield after *E. crassipes* at $1\,g\,L^{-1}$	58%	22%	7%
Selectivity	**Pd/Rh**	**Pd/Pt**	
	4.9	18.1	

While Ru and Ir are biosorbed only in negligible quantities, Pd, Rh, and Pt each have an interesting and different affinity for vegetable powder (Table 8). Taken individually, the biosorption capacity of Pd (89%) is higher than that of Rh (68%) and Pt (64%). These preliminary results show that the higher affinity of Pd on Rh and Pt for powdered roots of *E. crassipes* is interesting in terms of biosorption selectivity. A competitive biosorption experiment was therefore carried out (an equimolar ratio of $0.132\,mmol\,L^{-1}$ for each platinoid) with *E. crassipes* root powder (at $1\,g\,L^{-1}$) as biosorbent (Table 9).

Under competitive conditions, the biosorption capacity of platinoids with vegetable powder from *E. crassipes* can be classified as follows: Pd > Rh > Pt. Under selected experimental conditions, separation selectivity was calculated for Pd with respect to Rh and Pt (Table 9). It is observed that contrary to the satisfactory selectivity close to 5 for Pd/Rh, that calculated for the Pd/Pt is 18. A selectivity of 18 (Pd/Pt) in a competitive experiment is very interesting for metals often present together in the form of ore during extraction from the natural environment. A higher degree of selectivity was observed between Pd and Pt (18.1) than between Pd and Rh (4.9), meaning that the affinities of Pd and Rh for *E. crassipes* are closer. These results are consistent with those obtained for *Terminalia catappa* leaf biomass (Zhang et al., 2019) and could be a bio-based alternative to ion exchange resins (Lee et al., 2020) and chemical extraction treatments (Wang et al., 2021) that are currently in use.

Biosorption of nickel Biosorption of nickel from an aqueous solution has been studied in batches using three biomaterials with different chemical properties and which are proven in the biosorption of other metals: the roots of *E. crassipes*, an aquatic plant rich in carboxylate functions, pine cones rich in lignin, and coffee grounds rich in tannins (Cases et al., 2021). Biomaterials were dried, milled, and added to the solutions containing increasing concentrations of nickel sulfate (12 and $40\,mg\,L^{-1}$) (Table 10). Although exhibiting a high biosorption efficacy of $12\,mg\,L^{-1}$ up to 62% (Table 10, entries 1–3), these biomaterials quickly showed a limited biosorption efficacy of the order of 36% for a $40\,mg\,L^{-1}$ solution of $NiSO_4$ (Table 10, entries 4–6).

TABLE 10   Nickel biosorption by *Eichhornia crassipes* and pine cones.

Entry	Biomaterial at $1\,g\,L^{-1}$	$NiSO_4$ solution in water $(mg\,L^{-1})$	Biosorption efficiency
1	*E. crassipes*	12	62%
2	Pine cone		62%
3	Coffee grounds		30%
4	*E. crassipes*	40	36%
5	Pine cone		16%
6	Coffee grounds		17%

These results illustrate the interest of functionalizing biosorbents to increase their adsorption capacity of metallic elements that have a low affinity for these biomaterials.

Biosorption after functionalization

Biosorption of nickel   The described functionalization methodology (Grison et al., 2018) is inspired by the natural mechanism of nickel storage in the leaves of hyperaccumulating plants. The metallic elements are stored as metallic carboxylate, such as citrate, malate, and malonate (Grison et al., 2013).

Different modes of functionalization have been described: a possible strategy is based on a carboxylation reaction of hydroxyl functions to introduce a functionalized or non-functionalized ester bond. Several works in the literature describe the enrichment of materials of plant origin by the use of citric acid. For example, the process described by Zhu et al. consists of impregnating the material in water, heating to 60°C, then at 100°C until concentrating the medium, then heating at 120°C to cause esterification of the hydroxyl functions of the material via the intermediate formation of citric anhydride (Zhu et al., 2008). The process, however, suffers from several limitations:

— The removal of water is difficult and unfavorable to the formation of an anhydride;
— The decomposition of citric acid takes place at a higher temperature (150–170°C) (Wyrzykowski et al., 2011).
— The formation of citric anhydride is a minority product of the thermal decomposition of citric acid (Noordover et al., 2007).

In order to overcome these limitations, authors (Cases et al., 2021; Grison et al., 2018) described two different reaction strategies:

— Esterification using carboxylic anhydride (Strategy A).
— Direct esterification by autocatalysis without passing through an intermediate anhydride (Strategy B).

It is then possible to replace the water with polar green solvents that are easier to remove and are controlled in order to limit the formation of water-soluble fractions that are of no interest in depolluting the water. The first method (Strategy A) uses a carboxylic anhydride, allowing a high degree of functionalization of materials of plant origin (carboxylic, sulfonic

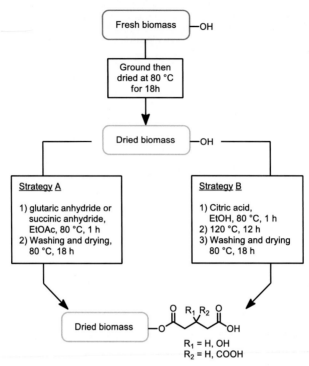

SCHEME 1 Presentation of the two mechanisms of functionalization of biomass. Fresh biomass = *E. crassipes*, pinecones, coffee grounds.

acids, phosphonic, carboxylic esters, or amides). The second methodology (strategy B) is carried out by autocatalysis using different polyacids such as citric acid. These functionalization strategies are illustrated in Scheme 1.

Strategy A has been described with succinic and glutaric succinic anhydride. Ethyl acetate was the solvent chosen for this method because it does not break down anhydrides, unlike a nucleophilic protic solvent. Strategy B is mainly based on the self-catalyzed esterification of hydroxyl groups by citric acid. Unlike strategy A, ethyl acetate could not be used due to the low citric acid solubility in organic solvents. Because water would rapidly degrade the anhydride, ethanol was thus chosen as the solvent. Both strategies were conducted under green conditions, using bio-based reagents (biomass and carboxylic acids or anhydrides) in a green solvent (ethyl acetate or ethanol).

After the functionalization reaction, the functionalized biomaterials were purified by water washing to remove excess carboxylic acids and were characterized by infrared spectroscopy. Infrared analyses showed new bands of higher transmission percentage in the $1720–1740\,cm^{-1}$ region corresponding to the stretching vibration of the ester group, which formed during the functionalization process (Cases et al., 2021).

The effectiveness of functionalization strategies was compared by titrating the carboxylate units introduced using sodium hydroxide solution (Table 11, entries 1–7). Both strategies A and B led to similar molar amounts of carboxylates per gram of biomaterial, except when

TABLE 11 Nickel biosorption efficiency of the different biosorbents, natural or functionalized.

Entry	Biomaterial after functionalization	Strategy/ functionalization reagent	Carboxylate ($mmol\,g^{-1}$ of material)	$NiSO_4$ solution ($mg\,L^{-1}$)	Biosorption efficiency (%)
1	Ec-SA	Strategy A Succinic anhydride	1.4	40	99
2	Pn-SA		1.1		78
3	Cg-SA		1.4		80
4	Ec-GA	Strategy A Glutaric anhydride	0.7		64
5	Ec-CA	Strategy B Citric acid	1.4		89
6	Pn-CA		1.6		70
7	Cg-CA		1.8		99
8	Ec-SA	Strategy A Succinic anhydride	1.4	1024	$51\,mg\,g^{-1}$

Ec = E. crassipes *roots, Pn = pine cones, Cg = coffee grounds, SA = succinic anhydride, GA = glutaric anhydride, CA = citric acid.*

glutaric anhydride is used, where functionalization is significantly less efficient (Table 11, entry 4).

The functionalized biomaterials were then used to test whether functionalization had improved their ability of nickel biosorption in batch mode (Table 11, entries 1–7). They were subjected to $40\,mg\,L^{-1}$ nickel sulfate solutions, the highest concentration at which the unfunctionalized biomass was saturated.

In general, functionalized biomass increased biosorption capacity by one-third to twice as much as unmodified biomass (Table 11, entries 5–7), except for Ec-GA (Table 11, entry 4). Indeed, the biosorption capacity of Ec-GA is similar to that of Ec, which may be related to the low functionalization efficiency using glutaric acid and can be considered a negative control of this experiment. It should be noted that Ec-SA and Cg-CA removed all Ni from the concentrated aqueous solution. The maximum biosorption capacity of Ec-SA was therefore tested using a saturation solution of $1024\,mg\,L^{-1}$ nickel sulfate (Table 11, entry 8). The Ec-SA was able to adsorb up to $51\,mg\,g^{-1}$ Ni, which is higher than unmodified natural materials (Akram Husain et al., 2013; Gupta et al., 2019; Reddy et al., 2011; Sarı et al., 2007; Villen-Guzman et al., 2019) and in the same range as materials that have been modified under environmentally unfriendly conditions (Deng and Ting, 2005; Feng et al., 2011). These new biomaterials are innovative, efficient, and sustainable solutions for treating effluents polluted by metallic elements.

Palladium biosorption   Similar results were obtained during the biosorption of Pd (Grison et al., 2018). The maximum biosorption capacity of *E. crassipes* increased from 14 to $43\,mg\,g^{-1}$ after functionalization with citric acid. No desorption was observed (Table 12).

This advantage is transposable to various transition metals such as Mn, Cu, toxic metals (Pb, Cd) as well as rare earth metals such as Ce, Yb, and Eu and similar types of Sc (Grison et al., 2018). These values are reported in Table 13.

TABLE 12   Comparison of palladium biosorption capacity between unmodified *Eichhornia crassipes* roots and modified by citric acid.

Biosorbent (1 g L^{-1})	Initial palladium concentration (mg L^{-1})	After biosorption
E. crassipes roots	14	1.6
E. crassipes roots functionalized by citric acid (Ec-CA)	14	0
	30	0
	44	1

TABLE 13   Maximum biosorption capacities of different metals by functionalized biosorbents.

Metal element	Mn	Cu	Pb	Sc	Ce
Functionalized biosorbent	Ec-CA	Cg-CA	Cg-CA	Ec-CA	Ec-CA
Maximum biosorption capacity (mg g^{-1})	23	47	56	21	31

Ec = E. crassipes *roots, Cg* = Coffee grounds, CA = Citric acid.

This functionalization has therefore allowed the ecological recycling of primary metals but also platinoids and rare earth metals, whose properties allow the development of a constantly diversified range of industrial applications, including electronics, green energy, aerospace, automotive, chemicals, and defense (Charalampides et al., 2015).

Competitive biosorptions

Competitive biosorption studies have been performed (Grison et al., 2018). Herein, we will focus on the biosorption of Sc in the presence of Ni. This situation represents a real case, as scandium is often extracted from other minerals from which it must be separated, including Ni (Wang et al., 2011).

To investigate the ability of *E. crassipes* functionalized with citric acid to selectively retain scandium relative to nickel, a synthetic effluent with concentrations of Sc = 25 mg L^{-1} and Ni = 72 mg L^{-1} was prepared. The separation was carried out in batches (Table 14). The

TABLE 14   Comparison of batch/column processes for the biosorption of a nickel-scandium mixture.

Metal element		Nickel (mg L^{-1})	Scandium (mg L^{-1})
Batch process	Initial conc.	72	25
	After 2 h	51	4
Column process	Initial conc.	112	2,3
	Run 1	68	0
	Run 2	64	0
	Run 3	64	0

functionalized biosorbent selectively extracted scandium over nickel, even if the latter was in excess. Compared to the biosorption of only scandium, where the maximum was $21\,mg\,g^{-1}$ (Table 13), the functionalized biosorbent appears to have biosorbed 20 mg of scandium, and the nickel was fixed to the remaining complexing sites.

To mimic the purification of an effluent at the industrial level, purification on a gravity column filled with carboxylated biosorbent was considered (Grison et al., 2018). Since the concentration of scandium in existing ores is very low, a solution of nickel and scandium was prepared with the ratio of $112\,mg\,L^{-1}$ of Ni for $2.3\,mg\,L^{-1}$ of Sc. The aim was to see if the remaining scandium fixed on the column in the presence of a large excess of nickel in order to enrich the biosorbent with scandium after several passages of effluent with a low concentration of scandium.

The synthetic effluent containing $112\,mg\,L^{-1}$ of Ni and $2.3\,mg\,L^{-1}$ of scandium was passed three times through a column containing citric acid-functionalized *E. crassipes* powder (Table 14). The residency time was 10 min.

As predicted by the batch results, industrial effluent loaded with nickel and scandium can be treated as preferentially extracting scandium. In addition, runs 2 and 3 show that there is no desorption, the two remaining metals are fixed on the biosorbent. The gravity column process is therefore robust, efficient and can be operated on a larger scale.

All these examples show the relevance of these biosorbents in cleaning up water contaminated by cationic elements. The problem is very different if the pollutant is present in anionic form. The most common example is arsenic, which occurs mainly as arsenate in aquatic systems. The right technology must be implemented. It is based on a technology of successive biosorptions.

Successive biosorptions

An innovative methodology has been implemented to remove arsenic from an aqueous solution by developing successive biosorption experiments (Richards et al., 2022). The principle relies on the biosorption of a first effluent loaded with iron followed by the biosorption of a second effluent, this time loaded with arsenic, using the same biosorbent. Mining effluent from the Malines mine (Gard, France) with a high Fe(II) content was a prime candidate for the first biosorption. As shown in Table 4, *P. stratiotes* has an extremely high affinity for Fe(II) and was chosen for this experiment. The high affinity of arseniate ions for Fe(III) (Grafe et al., 2001; Roberts et al., 2004) required the transformation of biosorbed Fe(II) to Fe(III). After oxidation of Fe(II) to Fe(III) by heat treatment, the plant filter was subjected to a contaminated synthetic effluent of $0.5\text{--}20\,mg\,L^{-1}$ of arsenic (Fig. 7).

The experiment gave a removal efficiency of 92%, determined by GF-AAS (graphite furnace atomic absorption spectrometry) and a q_{max} (maximum adsorption capacity) of $5.1\,mg\,g^{-1}$. This result can be advantageously compared to data from the literature, especially compared to other biomaterials. Indeed, other bio-based or natural materials enriched with iron, such as iron-coated sand (Kumar et al., 2008), iron-coated cellulose, iron-coated cork aggregate (Pintor et al., 2018), iron-coated wheat straw (Tian et al., 2011), iron-coated sawdust (Arshad and Imran, 2020), iron-treated coconut shell (Emahi et al., 2019), and iron-chitosan composites (Gupta et al., 2009), were tested for removal of As(V) and yielded a q_{max} range of 1.6 to $8.1\,mg\,g^{-1}$. Higher maximum As(V) biosorption capacities have been achieved with other materials, but the nature, abundance, and method of preparation of

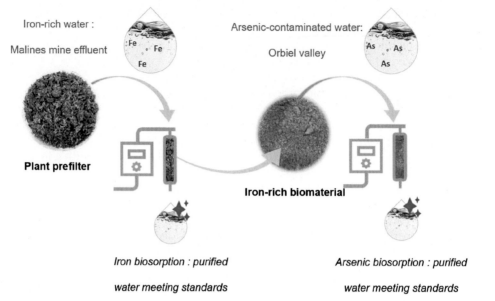

FIG. 7 Successive biosorption: depollution of an iron-enriched effluent followed by the depollution of an arsenic-rich effluent (©Claude Grison).

the materials used in adsorption must be considered in addition to the biosorption performance.

The biosorbent from Fe(III)-filled vegetable powder was characterized by Mössbauer spectroscopy. The Fe(III) is present as iron hydroxide oxide, HO(O)Fe, a ferric form known for its affinity with arsenates. The mechanisms of arsenic adsorption have been studied using theoretical models. Langmuir isotherms and pseudo-second-order kinetic models revealed excellent linearity and demonstrated the robustness of the method.

The methodology developed could be adapted to a real case, the Russec River (Orbiel Valley, France), which receives water from a former arsenic mine. Prior to biosorption, speciation analysis of the Russec River water was performed by anionic column separation (PRP-X100) and quantification by ICP-MS (Inductively Coupled Plasma Mass Spectrometry). These analyses were able to demonstrate that, as expected from surface water, 96.9% of arsenic is in its oxidized As(V) form. The total arsenic concentration was $62\,\mu g\,L^{-1}$. A pilot decontamination device was designed for this study (Fig. 8). It was a mobile unit whose characteristics have been chosen to make it usable in different specific cases. This filtration device consisted of two adsorption columns (1 L), usable in series or in parallel, allowing the removal of different metal pollutants depending on the filter media loaded inside. A pump allowed the circulation of polluted effluent, up to $200\,L\,h^{-1}$ and 10 bars pressure, through the columns (50 L inlet and outlet tank). Sensors measure pressure and flow, and a control box enables regulation of the pump, flow, and pressure with a real-time display and data collection. This pilot was tested on-site using 100 g of Fe(III) biosorbent from *P. stratiotes* and pumping water from the Russec River at a rate of $30\,L\,h^{-1}$ for 3 h.

FIG. 8 In natura tests for the remediation of the arsenic-polluted Russec River (©Claude Grison).

Samples were taken every 10 min and analyzed by GF-AAS (Fig. 9). The results were very encouraging since the pilot was able to remove 67% of As(V) from wastewater and fell below the European standard for industrial effluents of $25 \mu g L^{-1}$ (French Republic and INERIS, 2019). In this experiment, 90 L of wastewater was decontaminated, and the biosorbent was not saturated after 3 h and could be used longer (Richards et al., 2022).

This result reveals that these natural and inexpensive materials can follow mathematical models remarkably well, with good repeatability, and proves the interest of the method.

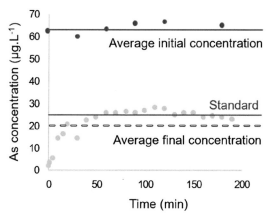

FIG. 9 Presentation of the mobile biosorption pilot.

VIII. New biology

2.1.4 Conclusions

A new system for the depollution of effluents contaminated by metallic elements has been developed (Fig. 10). Aquatic invasive plants were harvested, and their roots were collected and milled. The resulting root powder can be functionalized by a bio-inspired process and then used to biosorb metallic species. This system is currently being industrially developed by the company Bioinspir.

The new pilot device allowed the scaling-up of the laboratory's research results to a demonstration pilot process in the context of Fe, Zn, Cd and Pb pollution. The general process was tested on a polluted site, which suffers of alternating periods of drought and floods: the large mining site at Les Malines (Saint-Laurent-Le-Minier, Gard, France). The experiments were then extended *in natura* in the Orbiel valley (Salsigne, Aude, France), proving the capacity to decontaminate arsenic polluted waters.

Biosorption technology is remarkably simple to implement, fast, ecological, effective and robust. The optimal ratios, nature of the initial plant species/amount of plant filter/biomimetic functionalization/metal element loads/water flows to be treated, were established on industrial effluents with various compositions.

The possible variability of natural raw materials has been evaluated and controlled. The physicochemical parameters of the adsorption revealed a robustness in the process equivalent to that of synthetic materials. The kinetics of biosorption are very fast (total biosorption in 60 min, residence time less than 30 s) and fits faithfully to a pseudo-second order. The

FIG. 10 Environmentally friendly treatment of industrial effluents by biomass from aquatic invasive species (©Claude Grison and ©Pierre-Alexandre Deyris).

biosorption isotherms correspond in all respects to a Langmuir model, i.e., a monolayer filter media and homogeneous adsorption sites (Richards et al., 2022). The maximum adsorption capacities are very satisfactory.

These experiments provided a detailed study of sizing and modeling that led to the design and construction of flow devices on fixed or mobile beds, and in batch. The presented approach allows to activate three levers to develop the circular economy and the local industrial fabric at the input/output of the process:

- Harvesting and valorization of aquatic invasive plants as part of the management of their development, or the production of indigenous aquatic plants.
- Early depollution of industrial effluents using plant filters.
- Valorization of the plant filter in a sustainable cutting-edge chemistry which is capable of providing answers to a strong societal demand: the creation of 100% biosourced molecules without an environmental footprint. This process is based on the concept of ecocatalysis® presented hereunder.

2.2 Recovery of plant waste enriched with metallic elements in sustainable chemistry: Ecocatalysis®

A coherent ecological solution cannot lead to the generation of new waste. Thus, for the past 10 years, the Bio-inspired Chemistry and Ecological Innovations laboratory has been studying and developing an unprecedented valorization of remediation phytotechnologies (phytoextraction, rhizofiltration, biosorption).

Taking advantage of the remarkable capacity of roots and plant filters to bioconcentrate transition metals, it was possible to transform vegetal waste rich in metallic elements into bio-based metal catalysts for organic synthesis (Fig. 11). This original approach is a unique example of a chemical catalyst based on phytotechnologies. It offers the valorization of this specific biomass and has initiated a new branch of green and sustainable chemistry: ecocatalysis® (Deyris and Grison, 2018; Grison and Lock Toy Ki, 2021).

Ecocatalysis® has created a paradigm shift: the plant filters full of transition metals are not contaminated waste, but a natural depollution system with a high added value. It is a reservoir of transition metals or precious rare earths in organic synthesis. In other words, an ecological remediation solution leads to the creation of useful, innovative and motivating chemical objects. The originality of ecocatalysts can be summarized in five points (Fig. 12).

1. A crucial feature of ecological catalysts is their polymetallic composition resulting from the combination of transition elements at very high concentrations (e.g., $Zn^{2+}, Ni^{2+}, Mn^{2+}, Cu^{2+}, Fe^{2+}, Pd^2, Rh^{3+}, Pt^{2+}, Sc^{3+}$, and Ce^{2+}) with classical elements generally necessary for plant development (e.g., $Na^+, K^+, Ca^{2+}, Mg^{2+}$, and Fe^{3+}). The phenomenon of biosorption provides the development of ecocatalysts with a very high mass percentage of transition metal or rare earth metals (15%–40%). These ecocatalysts are also adjustable according to the selected plant species and biosorption parameters (residence time, bio-sorbent load, flow rates, column sizing, initial concentrations, etc.).

Ecocatalysts are composed of a variety of different metallic species which offers a combination of well-defined active sites. Therefore, sequences composed of original reaction steps enable access to unique selectivities. Indeed, a classical catalyst could be limited to influencing only some of the steps during the reaction process, thus limiting opportunities in organic synthesis. Here, the richness of the different interactions between the present species leads to unusual metal/reagent interactions in solution.

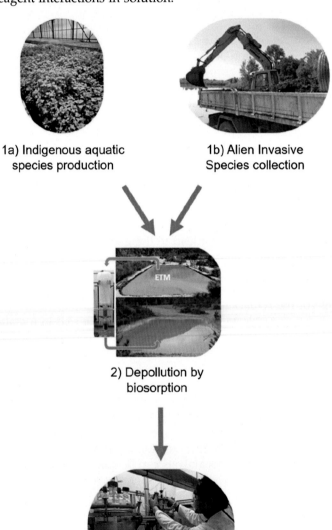

FIG. 11	A green sector based on the harvesting and valorization of aquatic species in ecocatalysis® (©Claude Grison).

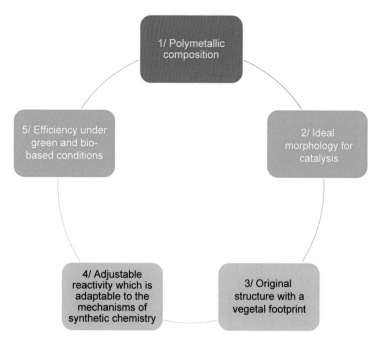

FIG. 12   Ecocatalysts' key points.

2. This polymetallic structure induces an unusual morphology. The presence of a mineral matrix consisting of physiological cationic elements allows dispersion and stabilization of the nanoparticles (2 4 nm) formed by the transition metals (Garel et al., 2015). Their aggregation is thus avoided, and leads to reactions which are catalyzed with very low catalyst loadings. This is particularly advantageous in the case of platinoids (Fig. 13). In scanning electronic microscopy (SEM) (Fig. 13, left), elements with a high molecular weight appear lighter while in transmission electronic microscopy (TEM) (Fig. 13, right), these are darker. In both cases, it is perfectly clear that palladium is not in the form of agglomerates but well dispersed within the plant mineral matrix.

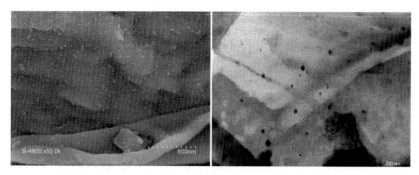

FIG. 13   Dispersion of Pd particles in Eco-Pd ecocatalyst on SEM (left) and TEM (right) images (©Claude Grison).

3. Thanks to a global approach to materials chemistry and molecular modeling, it has recently been proven that ecocatalysts have novel compositions and microstructures resulting from a plant footprint (Grison and Lock Toy Ki, 2022). Transition metals and alkaline earth metal cations form mixed salts, which are characteristic of the plant species they stem from and of the treatment they have underwent. Thus, for example, XRPD (X-ray powder diffraction) and HRTEM/EDS (high-resolution transmission electron microscopy-energy-dispersive X-ray spectroscopy) studies have demonstrated the presence of uncommon salts such as $Mg_{0.88}Zn_{0.12}O$, $Ca_2Mn_3O_4$, or $K_2Ca(CO_3)_2$ (Fig. 14) (Cybulska et al., 2022; Bihanic et al., 2021; Lock Toy Ki et al., 2022). Other examples include unusual oxidation degrees which were established by XPS (X-ray photoelectron spectroscopy) assays such as Pd(IV) in the form of K_2PdCl_6 (Grison et al., 2021).
4. Beyond their original composition, ecocatalysts are not simple catalytic tools for the synthesis of molecules, but innovative, structured and bio-based materials. These materials can be easily tuned by choice of the biosorbent or the treatment they underwent in order to obtain an ecocatalyst specialized in a specific and sensitive chemical reaction.

For example, among the various ecocatalysts, manganese-based ecocatalysts (Eco-Mn) revealed a strong potential in organic synthesis. Different Eco-Mn have been developed and studied. Many results have been accumulated in the field of green oxidations such as epoxidation reactions whose final products are known for their low stabilities (Bihanic et al., 2021). The Eco-CaMnOx clearly differentiates itself from conventional catalysts by allowing quantitative epoxidation of fragile substrates such as β-pinene (Scheme 2). In addition to the epoxidation step that can lead to more than nine different products (Neuenschwander et al., 2011), β-pinene oxide is very sensitive and can also be transformed

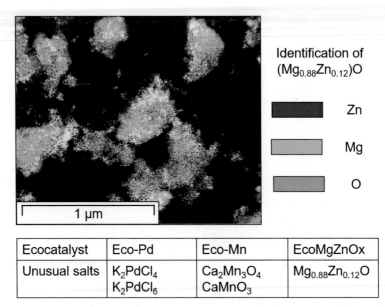

Ecocatalyst	Eco-Pd	Eco-Mn	EcoMgZnOx
Unusual salts	K_2PdCl_4 K_2PdCl_6	$Ca_2Mn_3O_4$ $CaMnO_3$	$Mg_{0.88}Zn_{0.12}O$

FIG. 14   Identification of $(Mg_{0.88}Zn_{0.12})O$ salt in Eco-MgZnOx ecocatalysts and some unusual salts composing different ecocatalysts.

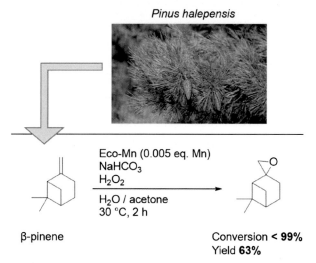

SCHEME 2 Epoxidation of β-pinene to β-pinene oxide using Eco-Mn stem from *Pistia stratiotes* roots.

into a dozen products (da Silva et al., 2013). The use of Eco-Mn in very mild catalytic conditions enables obtention of the desired product with a 63% yield (isolated product) (Bihanic et al., 2021).

These results showed that ecocatalysts could have much higher performance and selectivity than conventional catalysts. Cerium-based ecocatalysts proved their efficiency in multicomponent reactions, such as the Biginelli reaction (Grison et al., 2018), but other examples using Eco-Pd ecocatalysts were also relevant (Fig. 15). Indeed, the double-bond of an enone group was reduced chemoselectively without impacting the carbonyl group. Aromatic aldehydes could be reduced into alcohols while sensitive chemical groups such as cyano, bromo or nitro groups remained untouched. In other mild conditions, nitro groups could

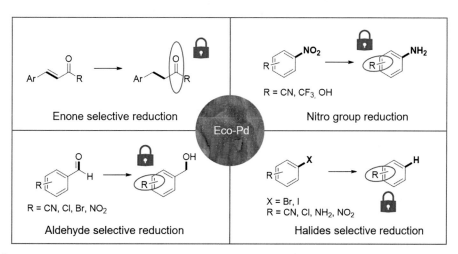

FIG. 15 Selective chemical reactions performed by Eco-Pd ecocatalysts.

VIII. New biology

be selectively transformed into amines without reducing cyano, trifluoromethyl or hydroxyl groups. In the same way, it is also possible to selectively reduce aromatic halides such as bromo and iodoarenes (Grison et al., 2021).

The concept of ecocatalysis® enabled the preparation of a wide variety of high-performance and modular reactivity ecological catalysts, while respecting the principles of sustainable chemistry. All the major mechanisms of organic synthesis were revisited with success by ecocatalysis® and were summarized in Fig. 16. Eco-Zn, Eco-Ce, Eco-Fe, and Eco-Cu promoted Lewis acid-catalyzed reactions (Clavé et al., 2016b; Grison et al., 2018). Isomerization and green reductions were performed by Eco-Pd ecocatalysts (Grison et al., 2018, 2021), while green oxidation reactions were realized by Eco-Mn and Eco-Fe (Bihanic et al., 2021; Grison et al., 2018). Of course, cross-coupling reactions were ecocatalyzed by Eco-Pd, Eco-Ni, and Eco-Cu (Adler et al., 2021; Cases et al., 2021; Clavé et al., 2016a, 2017; Garel et al., 2015; Grison et al., 2018).

5. Despite the emergence of new green solvents, such as 2-methyltetrahydrofuran, γ-valerolactone, or cyrene, most chemical reactions are currently carried out in organic solvents whose energy-intensive production comes mainly from petrochemicals and was estimated at 20 million tons in 2015 (Clarke et al., 2018). The use of alternative solvents from eco-compatible and bio-based processes is therefore a priority. Ecocatalyzed reactions were designed to be carried out without supplementary ligands or additives in green solvents, including water (Grison and Lock Toy Ki, 2021). To illustrate this point, three examples are given in Table 15. The coupling reactions of Sonogashira and Suzuki are prime examples of this approach. Under conventional conditions, these reactions are carried out in solvents such as toluene or dimethylformamide, which are considered to be

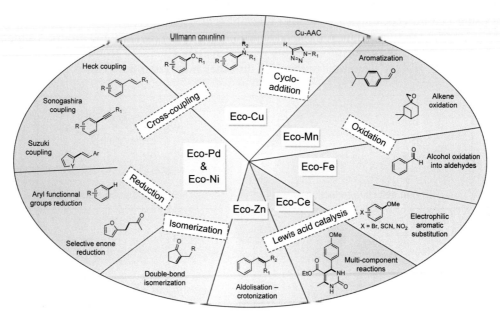

FIG. 16   Review of the reactions performed by ecocatalysis® with the use of catalysts issued from biomass from rhizofiltration/biosorption processes.

TABLE 15 Improvement of the reaction conditions towards eco-friendly and greener conditions.

Entry	Reaction name	Classical catalytic conditions	Ecocatalysis® conditions
1	Sonogashira cross-coupling	Pd(OAc)$_2$ (0.01 eq.) Cs$_2$CO$_3$ (2.0 eq.) Gallic acid (0.01 eq.) EtOH Sarmah et al. (2017)	Eco-Pd (0.001 eq. Pd) Glycerol/n-BuOH Eco-CaCO$_3$ (derived from oyster shells, 2 eq.) Adler et al. (2021)
2	Suzuki cross-coupling	PdCl$_2$ (0.05 eq.) K$_3$PO$_4$ (3.0 eq.) Toluene Pan et al. (2008)	Eco-Pd (0.01 eq. Pd) K$_2$CO$_3$ (2.0 eq.) H$_2$O Grison et al. (2018)
3	Alkene epoxidation	m-CPBA (1.02 eq.) NaHCO$_3$ (1.3 eq.) DCM Constantino et al. (2007)	Eco-MnOx-Ps (0.005 eq. Mn) NaHCO$_3$ 5 eq. H$_2$O$_2$ 5 eq. H$_2$O/acetone Bihanic et al. (2021)

of concern by REACH. As a replacement, ecocatalyzed reactions are carried out in bio-based polar solvents such as a glycerol/n-butanol mixture and even water while significantly reducing the catalytic loading (Table 15, entries 1 and 2). The use of a water/acetone mixture as a replacement for dichloromethane (DCM) in the alkene epoxidation reaction is an equally significant example (Table 15, entry 3).

In some cases, the joint use of ecocatalysts and the activation techniques of modern green chemistry allows them to perform neat chemical reactions and to optimize their E-factor. These include Michael's additions in mechanochemistry (Lock Toy Ki et al., 2022) and microwave-activated transesterification reactions (Grison and Lock Toy Ki, 2022).

Thanks to the work on biosorption, it is possible to recycle and re-use ecocatalysts, including in the homogeneous phase. This unusual possibility was tested with the Suzuki (Clavé et al., 2017) and Sonogashira (Adler et al., 2021) cross-coupling reactions and was reported in Fig. 17. Two recycling methods have been described. The first approach was based on an effluent treatment to solubilize all the palladium engaged in the reaction before biosorption (Clavé et al., 2017), while the second approach realized a direct biosorption in the reaction organic effluent (Adler et al., 2021). For the study of the first recycling method, the effluent was treated with aqua regia to solubilize palladium and was then placed in the presence of *E. crassipes* milled roots for the biosorption process. The biosorbent was then heat treated and activated to give back a new Eco-Pd suitable for a coupling reaction. It is easy to observe that on three successive recycling processes, the conversion of the reaction was total, and the ecocatalyst preserved its original activity (Fig. 17, left). In the second approach, biosorption was carried out by *Ludwigia peploides* milled roots directly in the effluent in order to avoid the preliminary acidification step. The biosorbent was also heat treated and activated to regenerate the ecocatalyst. It has to be noted that the conversion was slightly reduced to an average of 80% for the three successive replacements (Fig. 17, right) but remains constant.

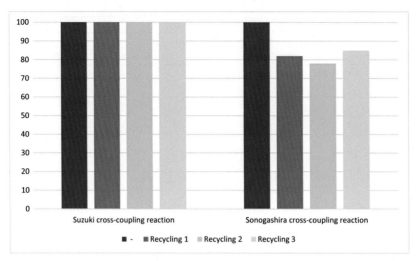

FIG. 17   Conversions of Suzuki and Sonogashira cross-coupling reactions during the recycling process.

This can be explained by the presence of the Eco-CaCO$_3$ base used in the reaction that changes the initial composition of the recycled Eco-Pd (Table 15, entry 1).

Both recycling methods are effective from the point of view of palladium recovery but also at the level of catalytic activity. The recycling of palladium after homogeneous catalysis reactions has always been a challenge that the ecocatalysis® process solves in an efficient and eco-compatible way.

A qualitative life cycle analysis illustrates the interest of the entire ecocatalysis® sector compared to conventional catalysis: from the preservation of aquatic systems to the development of sustainable chemistry for decarbonization (Fig. 18).

Most of the time, excavating metal-containing ore is made in open-pit mines. Although this method quickly extracts metals, the ecological consequences are catastrophic. In addition to the loss of biodiversity and carbon depletion, the mine waste products which contain TMEs are often stored in the open air and are thus subject to rainfall. This implies early soil erosion and pollution of the aquatic system by the TMEs. After the excavating process, the extraction of metals and their purification involve using huge quantities of chemical reagents, which produce more waste and increase the environmental footprint. Conventional catalytic reactions which use purified metal catalysts are performed in non-green solvents, and the catalysts are barely recycled. Knowing that some metal resource levels are critical (e.g., palladium, copper, etc.), an alternative to industrial mining had to be found quickly.

In parallel, the ecocatalysis® process works on several levels. It plays a role in the sustainable management of IAS. Plant roots are transformed into a plant filter which is able to depollute industrial effluent. By their action, these plant filters help the preservation of water resources and aquatic life. After treatment, the polluted plant filters, which are considered as one more waste, become an ecocatalyst with high-added value. These ecocatalysts are recyclable and are used in green and automated chemical reactions.

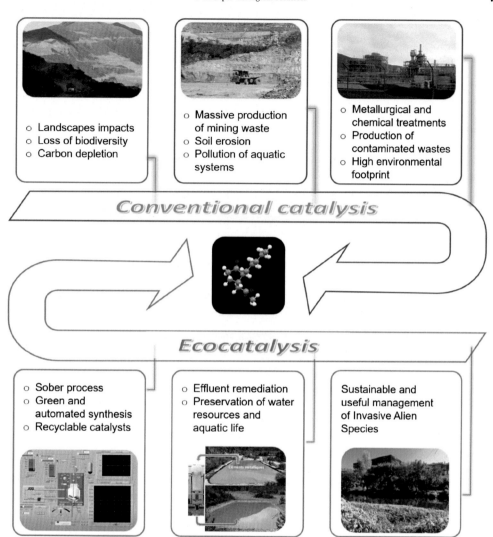

FIG. 18   Comparative life cycle analysis (©Claude Grison).

In conclusion, ecocatalysis® has broken down all the limitations of existing methods through a new generation of stable and recyclable ecological functional materials. There are several reactive interfaces whose properties can be controlled with regard to the intensity and nature of the metal/substrate interaction sought. These are linked to the biodiversity of plant species used in biosorption. For the first time, it is possible to prepare bio-based metal catalysts. Appropriate reagents and solvents allow the synthesis of 100% biobased products thanks to innovative processes. Ecocatalysts are therefore not just simple substitutes for catalysts from metallurgy but new tools which integrate a triple vision: chemistry, ecology, and environment.

Acknowledgments

The authors thank the CNRS, the Occitanie Region and all the partners of the ERDF "Eaux végétales" program (ChimEco, CNR, ETPB Gardons, SM Ganges-Le Vigan, Klorane Botanical Foundation, Suez) for their contribution or financial assistance.

References

Adler, P., Dumas, T., Deyris, P.-A., Petit, E., Diliberto, S., Boulanger, C., Grison, C., 2021. II—from ecological recycling of Pd to greener sonogashira cross-coupling reactions. J. Clean. Prod. 293, 126164. https://doi.org/10.1016/j.jclepro.2021.126164.

Aguilar, J., Dorronsoro, C., Fernández, E., Fernández, J., García, I., Martín, F., Ortiz, I., Simón, M., 2000. El desastre ecológico de Aznalcóllar.

Ahoulé, D.G., Lalanne, F., Mendret, J., Brosillon, S., Maïga, A.H., 2015. Arsenic in African waters: a review. Water Air Soil Pollut. 226, 302. https://doi.org/10.1007/s11270-015-2558-4.

Akcil, A., Koldas, S., 2006. Acid mine drainage (AMD): causes, treatment and case studies. J. Clean. Prod. 14, 1139–1145. https://doi.org/10.1016/j.jclepro.2004.09.006.

Akram Husain, R.S., Thatheyus, A.J., Ramya, D., 2013. Bioremoval of nickel using *Pseudomonas fluorescens*. Am. J. Microbiol. Res. 1, 48–52. https://doi.org/10.12691/ajmr-1-3-3.

Alastuey, A., García-Sánchez, A., López, F., Querol, X., 1999. Evolution of pyrite mud weathering and mobility of heavy metals in the Guadiamar valley after the Aznalcóllar spill, South-West Spain. Sci. Total Environ. 242, 41–55. https://doi.org/10.1016/S0048-9697(99)00375-7.

Arshad, N., Imran, S., 2020. Indigenous waste plant materials: an easy and cost-effective approach for the removal of heavy metals from water. Curr. Res. Green Sustain. Chem. 3, 100040. https://doi.org/10.1016/j.crgsc.2020.100040.

Baldantoni, D., Alfani, A., Di Tommasi, P., Bartoli, G., De Santo, A.V., 2004. Assessment of macro and microelement accumulation capability of two aquatic plants. Environ. Pollut. 130, 149–156. https://doi.org/10.1016/j.envpol.2003.12.015.

Bihanic, C., Lasbleiz, A., Regnier, M., Petit, E., Le Blainvaux, P., Grison, C., 2021. New sustainable synthetic routes to cyclic oxyterpenes using the ecocatalyst toolbox. Molecules 26, 7194. https://doi.org/10.3390/molecules26237194.

Blettler, M.C.M., Wantzen, K.M., 2019. Threats underestimated in freshwater plastic pollution: mini-review. Water Air Soil Pollut. 230, 174. https://doi.org/10.1007/s11270-019-4220-z.

Blottiere, D., Sarat, E., 2019. IPBES Report: The Invasive Alien Species are among the Five Factors of the Biodiversity Decline.

Boxall, A.B.A., 2012. New and Emerging Water Pollutants Arising from Agriculture. OECD 49.

Brausch, J.M., Rand, G.M., 2011. A review of personal care products in the aquatic environment: environmental concentrations and toxicity. Chemosphere 82, 1518–1532. https://doi.org/10.1016/j.chemosphere.2010.11.018.

Brown, M., van Wyk, D., 2022. Report on the Jagersfontein Tailings Disaster (No. NPO 048-041).

Cases, L., Adler, P., Pelissier, F., Diliberto, S., Boulanger, C., Grison, C., 2021. New biomaterials for Ni biosorption turned into catalysts for Suzuki-Miyaura cross coupling of aryl iodides in green conditions. RSC Adv. 11, 28085–28091. https://doi.org/10.1039/D1RA04478H.

Charalampides, G., Vatalis, K.I., Apostoplos, B., Ploutarch-Nikolas, B., 2015. Rare earth elements: industrial applications and economic dependency of Europe. Procedia Econ. Finance 24, 126–135. https://doi.org/10.1016/S2212-5671(15)00630-9.

Chen, J., Wang, S., Zhang, S., Yang, X., Huang, Z., Wang, C., Wei, Q., Zhang, G., Xiao, J., Jiang, F., Chang, J., Xiang, X., Wang, J., 2015. Arsenic pollution and its treatment in Yangzonghai lake in China: in situ remediation. Ecotoxicol. Environ. Saf. 122, 178–185. https://doi.org/10.1016/j.ecoenv.2015.07.032.

Cidu, R., Fanfani, L., 2002. Overview of the environmental geochemistry of mining districts in southwestern Sardinia, Italy. Geochem. Explor. Environ. Anal. 2, 243–251. https://doi.org/10.1144/1467-787302-028.

Clarke, C.J., Tu, W.-C., Levers, O., Bröhl, A., Hallett, J.P., 2018. Green and sustainable solvents in chemical processes. Chem. Rev. 118, 747–800. https://doi.org/10.1021/acs.chemrev.7b00571.

Clavé, G., Garel, C., Poullain, C., Renard, B.-L., Olszewski, T.K., Lange, B., Shutcha, M., Faucon, M.-P., Grison, C., 2016a. Ullmann reaction through ecocatalysis®: insights from bioresource and synthetic potential. RSC Adv. 6, 59550–59564. https://doi.org/10.1039/C6RA08664K.

Clavé, G., Garoux, L., Boulanger, C., Hesemann, P., Grison, C., 2016b. Ecological recycling of a bio-based catalyst for Cu click reaction: a new strategy for a greener sustainable catalysis. ChemistrySelect 1, 1410–1416. https://doi.org/10.1002/slct.201600430.

Clavé, G., Pelissier, F., Campidelli, S., Grison, C., 2017. Ecocatalyzed Suzuki cross coupling of heteroaryl compounds. Green Chem. 19, 4093–4103. https://doi.org/10.1039/C7GC01672G.

Constantino, M.G., Júnior, V.L., Invernize, P.R., Filho, L.C.d.S., José da Silva, G.V., 2007. Opening of epoxide rings catalyzed by niobium pentachloride. Synth. Commun. 37, 3529–3539. https://doi.org/10.1080/00397910701555790.

Cousins, I.T., Johansson, J.H., Salter, M.E., Sha, B., Scheringer, M., 2022. Outside the safe operating space of a new planetary boundary for per- and polyfluoroalkyl substances (PFAS). Environ. Sci. Technol. 56, 11172–11179. https://doi.org/10.1021/acs.est.2c02765.

Cunningham, S.A., 2005. Incident, accident, catastrophe: cyanide on the Danube: incident. Disasters 29, 99–128. https://doi.org/10.1111/j.0361-3666.2005.00276.x.

Cybulska, P., Legrand, Y.-M., Babst-Kostecka, A., Diliberto, S., Leśniewicz, A., Oliviero, E., Bert, V., Boulanger, C., Grison, C., Olszewski, T.K., 2022. Green and effective preparation of α-hydroxyphosphonates by Ecocatalysis®. Molecules 27, 3075. https://doi.org/10.3390/molecules27103075.

da Silva, M.J., Vieira, L.M.M., Oliveira, A.A., Ribeiro, M.C., 2013. Novel effect of palladium catalysts on chemoselective oxidation of β-pinene by hydrogen peroxide. Monatsh. Chem. 144, 321–326. https://doi.org/10.1007/s00706-012-0875-5.

Davidson, N.C., 2014. How much wetland has the world lost? Long-term and recent trends in global wetland area. Mar. Freshw. Res. 65, 934. https://doi.org/10.1071/MF14173.

Deng, S., Ting, Y.-P., 2005. Characterization of PEI-modified biomass and biosorption of Cu(II), Pb(II) and Ni(II). Water Res. 39, 2167–2177. https://doi.org/10.1016/j.watres.2005.03.033.

Deyris, P.-A., Grison, C., 2018. Nature, ecology and chemistry: an unusual combination for a new green catalysis, ecocatalysis®. Curr. Opin. Green Sustain. Chem. 10, 6–10. https://doi.org/10.1016/j.cogsc.2018.02.002.

Dushenkov, V., Kumar, P.B., Motto, H., Raskin, I., 1995. Rhizofiltration: the use of plants to remove heavy metals from aqueous streams. Environ. Sci. Technol. 29, 1239–1245. https://doi.org/10.1021/es00005a015.

Emahi, I., Sakyi, P.O., Bruce-Vanderpuije, P., Issifu, A.R., 2019. Effectiveness of raw versus activated coconut shells for removing arsenic and mercury from water. Environ. Nat. Resour. Res. 9, 127. https://doi.org/10.5539/enrr.v9n3p127.

Emmanouil, C., Bekyrou, M., Psomopoulos, C., Kungolos, A., 2019. An insight into ingredients of toxicological interest in personal care products and a small-scale sampling survey of the Greek market: delineating a potential contamination source for water resources. Water 11, 2501. https://doi.org/10.3390/w11122501.

Feng, N., Guo, X., Liang, S., Zhu, Y., Liu, J., 2011. Biosorption of heavy metals from aqueous solutions by chemically modified orange peel. J. Hazard. Mater. 185, 49–54. https://doi.org/10.1016/j.jhazmat.2010.08.114.

French Republic, INERIS, 2019. NOR: ATEP9870017A.

Garcia, A., Deyris, P.-A., Adler, P., Pelissier, F., Dumas, T., Legrand, Y.-M., Grison, C., 2021. I—ecologically responsible and efficient recycling of Pd from aqueous effluents using biosorption on biomass feedstock. J. Clean. Prod. 299, 126895. https://doi.org/10.1016/j.jclepro.2021.126895.

Garel, C., Renard, B.-L., Escande, V., Galtayries, A., Hesemann, P., Grison, C., 2015. CC bond formation strategy through ecocatalysis®: insights from structural studies and synthetic potential. Appl. Catal. A Gen. 504, 272–286. https://doi.org/10.1016/j.apcata.2015.01.021.

Grafe, M., Eick, M.J., Grossl, P.R., 2001. Adsorption of arsenate (V) and arsenite (III) on goethite in the presence and absence of dissolved organic carbon. Soil Sci. Soc. Am. J. 65, 1680–1687. https://doi.org/10.2136/sssaj2001.1680.

Grison, C., Adler, P., Deyris, P.-A., Diliberto, S., Boulanger, C., 2021. A green approach for the reduction of representative aryl functional groups using palladium ecocatalysts. Green Chem. Lett. Rev. 14, 234–245. https://doi.org/10.1080/17518253.2021.1898682.

Grison, C., Carrasco, D., Stanovych, A., 2018. Method for the Production of a Material of Plant Origin that is Rich in Phenolic Acids, Comprising at Least One Metal, for Carrying out Organic Synthesis Reactions. WO2018178374A1.

Grison, C., Escande, V., Bes, C., Renard, B.-L., 2015. Uses of Certain Platinoid Accumulating Plants for Use in Organic Chemical Reactions. WO2015007990.

Grison, C., Lock Toy Ki, Y., 2022. Novel Compositions for the Sustainable Catalysis of Organic Synthesis Reactions. EP22305074.1.

Grison, C., Lock Toy Ki, Y., 2021. Ecocatalysis®, a new vision of green and sustainable chemistry. Curr. Opin. Green Sustain. Chem. 29, 100461. https://doi.org/10.1016/j.cogsc.2021.100461.

Grison, C., Escande, V., Petit, E., Garoux, L., Boulanger, C., Grison, C., 2013. Psychotriadouarrei and *Geissois pruinosa*, novel resources for the plant-based catalytic chemistry. RSC Adv. 3, 22340. https://doi.org/10.1039/c3ra43995j.

Gupta, A., Chauhan, V.S., Sankararamakrishnan, N., 2009. Preparation and evaluation of iron-chitosan composites for removal of As(III) and As(V) from arsenic contaminated real life groundwater. Water Res. 43, 3862–3870. https://doi.org/10.1016/j.watres.2009.05.040.

Gupta, S., Sharma, S.K., Kumar, A., 2019. Biosorption of Ni(II) ions from aqueous solution using modified *Aloe barbadensis* miller leaf powder. Water Sci. Eng. 12, 27–36. https://doi.org/10.1016/j.wse.2019.04.003.

IPBES, 2019. Global Assessment Report on Biodiversity and Ecosystem Services of the Intergovernmental Science-Policy Platform on Biodiversity and Ecosystem Services. Zendo, https://doi.org/10.5281/ZENODO.3831673.

Izoard, C., 2022. Exclusif: la liste des sites miniers empoisonnés que l'État dissimule.

Kumar, A., Gurian, P.L., Bucciarelli-Tieger, R.H., Mitchell-Blackwood, J., 2008. Iron oxide-coated fibrous sorbents for arsenic removal. J. Am. Water Works Assoc. 100, 151. https://doi.org/10.1002/j.1551-8833.2008.tb09611.x.

Kumar, N., Bauddh, K., Dwivedi, N., Barman, S.C., Singh, D.P., 2012. Accumulation of metals in selected macrophytes grown in mixture of drain water and tannery effluent and their phytoremediation potential. J. Environ. Biol. 33, 923–927.

Lee, J., Kurniawan, Hong, H.-J., Chung, K.W., Kim, S., 2020. Separation of platinum, palladium and rhodium from aqueous solutions using ion exchange resin: a review. Sep. Purif. Technol. 246, 116896. https://doi.org/10.1016/j.seppur.2020.116896.

Lefèvre, S., Richard, E., 2022. Cancers et métaux lourds: une fois la mine fermée, la vie empoisonnée.

Lock Toy Ki, Y., Garcia, A., Pelissier, F., Olszewski, T.K., Babst-Kostecka, A., Legrand, Y.-M., Grison, C., 2022. Mechanochemistry and eco-bases for sustainable Michael addition reactions. Molecules 27, 3306. https://doi.org/10.3390/molecules27103306.

Matache, M.L., Marin, C., Rozylowicz, L., Tudorache, A., 2013. Plants accumulating heavy metals in the Danube River wetlands. J. Environ. Health Sci. Eng. 11, 39. https://doi.org/10.1186/2052-336X-11-39.

Merdy, P., Guillon, E., Frapart, Y.-M., Aplincourt, M., 2003. Iron and manganese surface complex formation with extracted lignin. Part 2. Characterisation of magnetic interaction between transition metal and quinonic radical by EPR microwave power saturation experiments. New J. Chem. 27, 577–582. https://doi.org/10.1039/b209665j.

Miloškovic, A., Branković, S., Simić, V., Kovačević, S., Čirković, M., Manojlović, D., 2013. The accumulation and distribution of metals in water, sediment, aquatic macrophytes and fishes of the Gruža Reservoir, Serbia. Bull. Environ. Contam. Toxicol. 90, 563–569. https://doi.org/10.1007/s00128-013-0969-8.

Muramoto, S., Oki, Y., 1983. Removal of some heavy metals from polluted water by water hyacinth (*Eichhornia crassipes*). Bull. Environ. Contam. Toxicol. 30, 170–177. https://doi.org/10.1007/BF01610117.

Neuenschwander, U., Meier, E., Hermans, I., 2011. Peculiarities of β-pinene autoxidation. ChemSusChem 4, 1613–1621. https://doi.org/10.1002/cssc.201100266.

Noordover, B.A.J., Duchateau, R., van Benthem, R.A.T.M., Ming, W., Koning, C.E., 2007. Enhancing the functionality of biobased polyester coating resins through modification with citric acid. Biomacromolecules 8, 3860–3870. https://doi.org/10.1021/bm700775e.

O'Connell, H., 2012. HomeBond "Snub" over Pyrite "a Matter of Serious Public Concern"—Committee.

Olszewski, T.K., Adler, P., Grison, C., 2019. Bio-based catalysts from biomass issued after decontamination of effluents rich in copper—an innovative approach towards greener copper-based catalysis. Catalysts 9, 214. https://doi.org/10.3390/catal9030214.

Orsenna, E., 2008. L'avenir de l'eau: petit précis de mondialisation II. Fayard, Paris.

Pajević, S., Borišev, M., Rončević, S., Vukov, D., Igić, R., 2008. Heavy metal accumulation of Danube river aquatic plants—indication of chemical contamination. Open Life Sci. 3, 285–294. https://doi.org/10.2478/s11535-008-0017-6.

Pan, C., Liu, M., Zhang, L., Wu, H., Ding, J., Cheng, J., 2008. Palladium catalyzed ligand-free Suzuki cross-coupling reaction. Catal. Commun. 9, 508–510. https://doi.org/10.1016/j.catcom.2007.06.022.

Pavlovic, S., Pavlovic, D., Topuzovic, M., 2005. Comparative analysis of heavy metal content in aquatic macrophytes in the reservoirs Gruža, Bubanj and Memorial Park. Kragujev. J. Sci. 27, 147–156.

Pintor, A.M.A., Vieira, B.R.C., Santos, S.C.R., Boaventura, R.A.R., Botelho, C.M.S., 2018. Arsenate and arsenite adsorption onto iron-coated cork granulates. Sci. Total Environ. 642, 1075–1089. https://doi.org/10.1016/j. scitotenv.2018.06.170.

Reddy, D.H.K., Ramana, D.K.V., Seshaiah, K., Reddy, A.V.R., 2011. Biosorption of Ni(II) from aqueous phase by *Moringa oleifera* bark, a low cost biosorbent. Desalination 268, 150–157. https://doi.org/10.1016/j. desal.2010.10.011.

Richards, K., Garçia, A., Legrand, Y.-M., Grison, C., 2022. A two-step biosorption methodology for efficient and rapid removal of Fe(II) following As(V) from aqueous solution using abundant biomaterials. Int. J. Environ. Sci. Technol. https://doi.org/10.1007/s13762-022-04584-z.

Roberts, L.C., Hug, S.J., Ruettimann, T., Billah, M.M., Khan, A.W., Rahman, M.T., 2004. Arsenic removal with iron(II) and iron(III) in waters with high silicate and phosphate concentrations. Environ. Sci. Technol. 38, 307–315. https://doi.org/10.1021/es0343205.

Sarı, A., Tuzen, M., Uluözlü, Ö.D., Soylak, M., 2007. Biosorption of Pb(II) and Ni(II) from aqueous solution by lichen (*Cladonia furcata*) biomass. Biochem. Eng. J. 37, 151–158. https://doi.org/10.1016/j.bej.2007.04.007.

Sarmah, M., Mondal, M., Gohain, S.B., Bora, U., 2017. Gallic acid-derived palladium(0) nanoparticles as in situ-formed catalyst for Sonogashira cross-coupling reaction in ethanol under open air. Catal. Commun. 90, 31–34. https://doi.org/10.1016/j.catcom.2016.10.034.

Schmeller, D.S., Loyau, A., Bao, K., Brack, W., Chatzinotas, A., De Vleeschouwer, F., Friesen, J., Gandois, L., Hansson, S.V., Haver, M., Le Roux, G., Shen, J., Teisserenc, R., Vredenburg, V.T., 2018. People, pollution and pathogens—global change impacts in mountain freshwater ecosystems. Sci. Total Environ. 622–623, 756–763. https://doi.org/10.1016/j.scitotenv.2017.12.006.

Shaji, E., Santosh, M., Sarath, K.V., Prakash, P., Deepchand, V., Divya, B.V., 2021. Arsenic contamination of groundwater: a global synopsis with focus on the Indian Peninsula. Geosci. Front. 12, 101079. https://doi.org/10.1016/j.gsf.2020.08.015.

Sinha, S., Chandra, P., 1990. Removal of Cu and Cd from water by *Bacopa monnieri* L. Water Air Soil Pollut. 51, 271–276. https://doi.org/10.1007/BF00158224.

Siqueira, D., Cesar, R., Lourenço, R., Salomão, A., Marques, M., Polivanov, H., Teixeira, M., Vezzone, M., Santos, D., Koifman, G., Fernandes, Y., Rodrigues, A.P., Alexandre, K., Carneiro, M., Bertolino, L.C., Fernandes, N., Domingos, L., Castilhos, Z.C., 2022. Terrestrial and aquatic ecotoxicity of iron ore tailings after the failure of VALE S.A. mining dam in Brumadinho (Brazil). J. Geochem. Explor. 235, 106954. https://doi.org/10.1016/j.gexplo.2022.106954.

Solidarités International, n.d.. https://www.solidarites.org/en/.

Stanovych, A., Balloy, M., Olszewski, T.K., Petit, E., Grison, C., 2019. Depollution of mining effluents: innovative mobilization of plant resources. Environ. Sci. Pollut. Res. 26, 19327–19334. https://doi.org/10.1007/s11356-019-05027-y.

Sutton, D.L., Blackburn, R.D., 1971a. Uptake of copper by water hyacinth. Hyacinth Contr. J. 9, 18–20.

Sutton, D.L., Blackburn, R.D., 1971b. Uptake of copper in hydrilla. Weed Res. 11, 47–53. https://doi.org/10.1111/j.1365-3180.1971.tb00975.x.

Sweta, Bauddh, K., Singh, R., Singh, R.P., 2015. The suitability of *Trapa natans* for phytoremediation of inorganic contaminants from the aquatic ecosystems. Ecol. Eng. 83, 39–42. https://doi.org/10.1016/j.ecoleng.2015.06.003.

The European Parliament and the Council of the European Union, 2000. Directive 2000/60/EC of the European Parliament and of the Council of 23 October 2000 Establishing a Framework for Community Action in the Field of Water Policy.

Tian, Y., Wu, M., Lin, X., Huang, P., Huang, Y., 2011. Synthesis of magnetic wheat straw for arsenic adsorption. J. Hazard. Mater. 193, 10–16. https://doi.org/10.1016/j.jhazmat.2011.04.093.

Tiwari, J., Ankit, Sweta, Kumar, S., Korstad, J., Bauddh, K., 2019. Ecorestoration of polluted aquatic ecosystems through rhizofiltration. In: Phytomanagement of Polluted Sites. Elsevier, pp. 179–201, https://doi.org/10.1016/B978-0-12-813912-7.00005-3.

UNESCO, 2018. The United Nations World Water Development Report 2018: Nature-Based Solutions for Water.

UNESCO, W.W.A.P, 2020. The United Nations World Water Development Report 2020: Water and Climate Change.

Vilela, C.L.S., Bassin, J.P., Peixoto, R.S., 2018. Water contamination by endocrine disruptors: impacts, microbiological aspects and trends for environmental protection. Environ. Pollut. 235, 546–559. https://doi.org/10.1016/j.envpol.2017.12.098.

VIII. New biology

Villen-Guzman, M., Gutierrez-Pinilla, D., Gomez-Lahoz, C., Vereda-Alonso, C., Rodriguez-Maroto, J.M., Arhoun, B., 2019. Optimization of Ni(II) biosorption from aqueous solution on modified lemon peel. Environ. Res. 179, 108849. https://doi.org/10.1016/j.envres.2019.108849.

Wang, J., Xu, W., Liu, H., Yu, F., Wang, H., 2021. Extractant structures and their performance for palladium extraction and separation from chloride media: a review. Miner. Eng. 163, 106798. https://doi.org/10.1016/j.mineng.2021.106798.

Wang, W., Pranolo, Y., Cheng, C.Y., 2011. Metallurgical processes for scandium recovery from various resources: a review. Hydrometallurgy 108, 100–108. https://doi.org/10.1016/j.hydromet.2011.03.001.

Wyrzykowski, D., Hebanowska, E., Nowak-Wiczk, G., Makowski, M., Chmurzyński, L., 2011. Thermal behaviour of citric acid and isomeric aconitic acids. J. Therm. Anal. Calorim. 104, 731–735. https://doi.org/10.1007/s10973-010-1015-2.

Zhang, L., Song, Q., Liu, Y., Xu, Z., 2019. Novel approach for recovery of palladium in spent catalyst from automobile by a capture technology of eutectic copper. J. Clean. Prod. 239, 118093. https://doi.org/10.1016/j.jclepro.2019.118093.

Zhu, B., Fan, T., Zhang, D., 2008. Adsorption of copper ions from aqueous solution by citric acid modified soybean straw. J. Hazard. Mater. 153, 300–308. https://doi.org/10.1016/j.jhazmat.2007.08.050.

Synthetic biology: An emerging field for developing economies

Boda Ravi Kiran[a], Majeti Narasimha Vara Prasad[b], and S. Venkata Mohan[a]

[a]Bioengineering and Environmental Sciences Lab, Department of Energy and Environmental Engineering (DEE), CSIR-Indian Institute of Chemical Technology (CSIR-IICT), Hyderabad, Telangana, India [b]Department of Plant Sciences, School of Life Sciences, University of Hyderabad (An Institution of Eminence), Hyderabad, Telangana, India

1 Introduction

Synthetic biology (SynBio), an emerging interdisciplinary scientific field, is known for designing and building new artificial biological pathways, creatures, or devices, as well as reworking existing natural biological systems. It is therefore unsurpassed as "an umbrella phrase… that encompasses activity spanning from the basic biology to creative technologies, instead of being a new conceptual framework" (Li et al., 2021). SynBio became popular in the late 2000s when DNA sequencing and synthesis became affordable and fast enough which revolutionized molecular sciences. The U.S. Department of Defense recognized synthetic biology as one of the top six disruptive technologies for development in the 21st century (U.S. Department of Defense, 2014). The development of advanced technologies for reading and writing DNA has resulted in groundbreaking progress in the design, assembly, manipulation of genes, materials, circuits, and metabolic pathways. These progressions have enabled scientists to manipulate biological systems and organisms to a greater extent than ever before (Fig. 1). Examples of this include the production of compounds like leghemoglobin, sitagliptin, and diamines by engineered cells, which were awarded the 2018 Nobel Prize. Meanwhile, the 2020 Nobel Prize was awarded for the engineering of modified cells themselves, such as bacteria, CAR-Ts, and genome-edited soy (Cumbers, 2020; Voigt, 2020). Recent developments in information technology (bioinformatics and design tools) and biotechnology (genome sequencing, genome editing, gene synthesis, and biofoundries) concomitantly

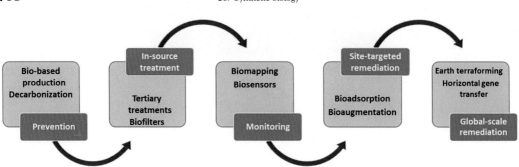

FIG. 1 Advancements in bio-based approaches to tackle environmental pollution. The direction of the arrow indicates the increasing complexity of the technologies.

artificial intelligence (AI) (machine learning) accelerated the discovery and optimization of metabolic pathways through the Design-Build-Test-Learn cycle.

Next-generation sequencing (NGS) and DNA synthesis techniques are the key technologies driving synthetic biology. They have the potential to bring the predictability and reliability of engineered biological systems that chemical engineering has brought to chemical systems. These technologies have led to a rise in DNA sequence libraries and advancements in bioinformatics methods and software, which allow the mining of genetic information from a diverse range of organisms. The genetic information can then be used to build new genetic constructs that can alter the function of living organisms. This may involve genetically modified organisms, extending the uses of genetics and genomics beyond systems biology and into synthetic life. People's comprehension of the biological components' mechanisms of action and the intricate regulatory network in organisms is improved by combining synthetic biology and engineering (Bashor and Collins, 2018; Ozdemir et al., 2018). By 2030, it is likely that you will have consumed, used, worn, or been treated with a product from synthetic biology as they are pervading the society rapidly. Genetically modified living organisms (including diseases) have made tremendous strides, but this has also raised the possibility of bio-risks in terms of biosafety, biosecurity, and cyber-biosecurity (World Health Organization, 2018). Using the terms "synthetic biology" as a database search in indexed journals (Web of Science—WoS), the publication number across the world (586–38,298) is rising rapidly, illuminating the scientific community's growing interest (Fig. 2A and B).

2 Recent trends and approaches in phytoremediation

Plants are a plethora of value-added compounds, including oils, medications, flavorings, and more. Its use as metabolite producers, however, comes with limitations, including reliance on arable land and water, photoperiod and seasonality (Li et al., 2018a, b; Moses et al., 2017; Pouvreau et al., 2018). Phytoremediation technology is an effective method for developing agricultural and industrial integration. It uses plants that accumulate metals and enrich themselves with potentially hazardous substances to restore the contaminated soil (Lorenzo et al., 2018; Zhang et al., 2022a, b). Biological processes can offer moderate advantages in terms of environmental friendliness and economic benefits, particularly when it comes to soil

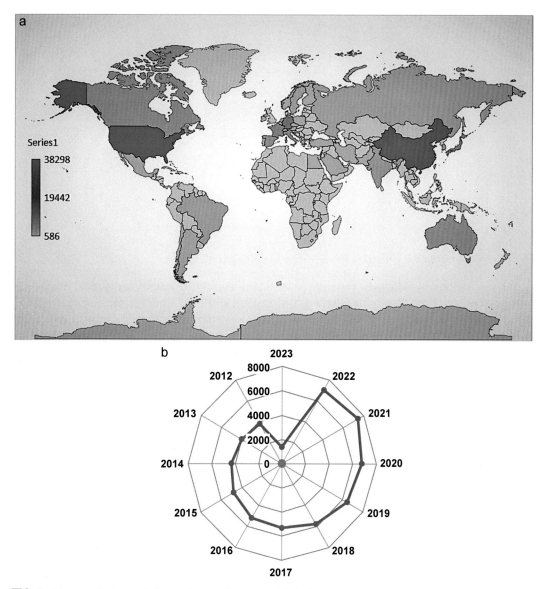

FIG. 2 The cumulative total of annual research papers published on Synthetic biology (A) SynBio research network on a global scale in different countries. (*The color intensity indicates synthetic biology progressive development in emerging nations.) (B) Publication from 2012 to 2023 (keyword used = synthetic biology. Data source: http://www.webofscience.com).

remediation efficacy when compared to chemical approaches (Kumar et al., 2016; Sabreena et al., 2022; Ubando et al., 2021). The WoS tool is used to assess the network and econometric analyses by using "Phytoremediation" as the keyword to analyze the literature published between 2012 and 2023 (Fig. 3).

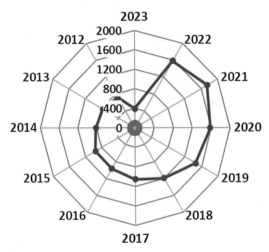

FIG. 3   Phytoremediation of potentially toxic elements in soil from 2012 to 2023.

Systems biology investigates the interactions between living cells, plants, and microbial communities, and how they relate to the physiological and biochemical processes of the ecosystem. Recent research has revealed the possibility of enhancing plant-microbe interactions through genetic manipulation of plants and microbes (Hassan et al., 2022). Similarly, exogenous chemicals have also been found to produce positive effects on these interactions. The efficacy of phytoremediation can be enhanced by using gene editing and manipulation technologies. For example, HhSSB protein transformation in *Arabidopsis* can boost the plant's stress tolerance, resulting in an increase in Cd accumulation (Basu et al., 2018; Sanz-Fernández et al., 2017). Metagenomics, metatranscriptomics, and metabolomics are techniques that can aid in discovering novel genes for improving resource utilization through transgenic technologies. A recent computational approach, known as Mergeomics, has been developed to identify disease-related processes using the vast datasets generated from plant-microbe interaction studies in the omics field (Hou et al., 2019; Malla et al., 2018; Zhang et al., 2022a, b). Model-guided design will become increasingly critical as SynBio advances toward developing larger systems that integrate gene circuits with other biological systems, such as metabolic and signaling networks (Mansoori, 2018; Zhang et al., 2022a, b). Designers have the ability to experiment with system models to verify the effectiveness of system structures, various parameter combinations, and performance under different conditions. For instance, dynamic metabolic engineering models have proven how managing system-level features, such as metabolic heterogeneity, can be utilized for biosynthesis. Advanced algorithms based on AI and machine learning, coupled with the rapidly expanding volume of big data in the field of biotechnology, will be essential in developing microbial cell factories and optimizing bioprocesses. By leveraging these innovative modeling and simulation techniques, it is possible to overcome the significant hurdles of scaling up and improving the speed and predictability of large bioprocesses, which have contributed to the chemical industry's continuing reservations about bio-based manufacturing (Gao et al., 2022). These cutting-edge methods can facilitate the identification of new metabolic engineering targets, optimize production conditions, and accurately predict the behavior of cells in response to changing environmental conditions, ultimately driving the adoption of bio-based manufacturing on a larger scale and promoting a more sustainable future (Cazzolla Gatti et al., 2022).

3 Synthetic biology in agriculture

The field of synthetic biology, which is rapidly growing, has mainly concentrated on microbial sciences and human health. To date, emerging molecular technologies and genetics are thought to have contributed around 1%–3% raise in crop and livestock productivity annually (Awan et al., 2016; Bueso and Tangney, 2017; Goold et al., 2018; Kiran et al., 2023). This industry has a well-established trend of rapidly adopting groundbreaking innovations. For instance, from 1996 to 2013, genetic technologies were credited with generating an additional 110 million tonnes of soybean production and 195 million tonnes of maize production due to their positive yield effects. In the development of many agronomically significant characteristics in various crops, traditional breeding techniques have played a vital role. Applying contemporary biotechnology methods has allowed for advancements such as longer flowering times, increased water use efficiency (WUE), and disease and herbicide resistance that are beyond the capability or competence of conventional breeding. Notwithstanding these improvements over time, pests, weeds, nutrient absorption, and abiotic challenges still pose a threat to agricultural yields. It is essential to incorporate new characteristics and genetic variability in collections of crop germplasm (Dwivedi et al., 2017; Sharwood et al., 2022). Synthetic biology (SynBio) refers to techniques used to create new biological components (such as cells, genetic circuits, and enzymes) or to redesign already-existing biological systems to provide novel and enhanced functionality (Table 1). For instance, five genes from the NADP-ME biochemical subtype were introduced during the early stages of C4 rice development. If conventional genetic engineering techniques were used, which entail introducing single genes and then stacking them through multiple cycles, this transformation would have required several years to accomplish. This complicated metamorphosis was made possible by SynBio methods (such as Golden Gate cloning) and took place within 6–12months (Ermakova et al., 2020). Metabolic engineering can enhance the fatty acid nutritional quality of traditional seed crops like canola by integrating seven enzymes from five diverse organisms (*Lanchancea kluyveri*, *Pichia pastoris*, *Micromonas pusilla*, *Pyramimonas cordata*, and *Pavlova salina*).

Agriculture depends on photosynthesis, which is the only characteristic that distinguishes and unites green lineage species. Yet, due to its inherent inefficiency—which seldom exceeds a few percent and has a theoretical maximum efficiency of about 11%—synthetic biology has a wealth of possible targets for improving results. Gene editing and gene drives have been proposed as potential tools for mitigating atmospheric levels of CO_2 and greenhouse gases (GHGs). For instance, it could involve introducing non-photosynthetic CO_2 capture pathways, such as artificial pathways like the reverse tricarboxylic acid cycle, the Woods-Ljungdahl cycle, the hydroxybutyrate cycle, or new-to-nature pathways into microbes or plants (Chen et al., 2021; Sargent et al., 2022). CRISPR/Cas9 gene drives could be used to introduce these metabolic pathways into wild populations or ecosystems to reduce the buildup of GHGs like CO_2. For instance, inserting cyanobacterial carboxysomes into the chloroplast may be able to compensate for the CO_2-fixing enzyme RuBisCO's inherently poor activity

TABLE 1 Synthetic biology (SynBio) technological advances for improved crops and sustainable agriculture.

Technologies	Description	References
Plant breeding	Genetic improvement of the crop to create desired plant type. They are classified into two categories: 1. Conventional method: Selection, hybridization, polyploidy and mutation breeding 2. Non-conventional method: Marker-assisted & Biotechnological approaches	Anand et al. (2022) and Ahmad (2023)
Multi-omic analysis	Proteomics—to identify changes in protein expression in response to abiotic stress factors Metabolomics—to quantify all endogenous and exogenous low-molecular-weight (<1kDa) small molecules/metabolites in a biological system Transcriptomics—to study the expression levels of mRNAs in a given cell population	Sharwood et al. (2022), Ermakova et al. (2021), Cousins et al. (2020), and Rabara et al. (2017)
Gene editing tools	CRISPR-Cas9—gene editing (silencing, changing, or enhancing specific genes or integrating transgenes) technology for crop improvement CRMAGE—gene editing in *Escherichia coli* that combines CRISPR-Cas9 and Lambda (Λ) Red recombineering with the MAGE technique Golden Gate—in vitro assembly of multiple DNA fragments into a single construct RNAi—targeted gene silencing by RNA interference (RNAi) Multiplex genome editing by natural transformation (MuGENT)—generates complex population of mutants Gene drives—promoting inheritance of deleterious alleles (i.e., lethal or sterile alleles in insect pests) RNA-guided endonucleases(RGENs)—ribonucleoproteins comprising Cas9 and guide RNA derived from *Streptococcus pyogenes* Transcription activator-like effector nucleases (TALENs)—non-specific exonucleases fused to a DNA-binding domain Zinc-finger nucleases (ZFNs)—genetically engineered DNA-binding proteins that enable genome editing at targeted sites	Ahmad (2023), Bier (2022) Degen et al. (2021), Niehl et al. (2018), Giessen and Silver (2017), www.genscript.com

	Regulated and artificial promoters—control gene expression by activating or deactivating downstream genes under specific conditions such as environmental stress or phenological development Riboswitches and Ribozymes-regulate the expression of mRNA by binding to the aptamer domain	Basu et al. (2018), Chen et al. (2021), Ermakova et al. (2020), Hung and Slotkin (2021), and Sargent et al. (2022)
Bioinformatics and web-based tools	Constraint-based modeling (CBM)—combines genomic, biochemical, and genetic information to explain the physiology of metabolism Flux balance analysis (FBA)—simulate metabolic pathways on a genome-wide scale KeyPathwayMinerWeb—enables pathway enrichment analysis (de novo) directly in the browser (https://keypathwayminer.compbio.sdu.dk/keypathwayminer/) Omics—integrative genomics analysis (https://pypi.python.org/pypi/omics) Omictools—for microarray, NGS, PCR, mass spectrometry(MS), and NMR techniques (http://omictools.com/) Optknock—identify overproduced genes in a biochemical pathway	

(Degen et al., 2021; Giessen and Silver, 2017). The SynBio toolkit will be critical for the efficient transfer of large gene constructs with desired expression patterns to improve photosynthetic pathways. Examples include boosting photosynthetic enzymes, increasing WUE through aquaporins, and varying cells anatomy to upsurge mesophyll conductivity (Cousins et al., 2020; Ermakova et al., 2021; Sharwood et al., 2022; Kiran et al., 2023). Plants' capacity and effectiveness in fixing atmospheric carbon could be increased, which would have obvious implications for agricultural productivity and resource management.

SynBio technologies offer a variety of ways to deal with insect infestations. In the past, toxin-producing genes have been introduced into certain crop tissues with considerable success. An excellent example of utilizing "traditional" genetic engineering techniques to produce insect pest resistance is the integration of the Cry protein genes from *Bacillus thuringiensis* into cotton and maize to produce Bt varieties (Downes et al., 2016). Comparing data from 1986 to 1995 to that of 1996–2015 in the USA, it is evident that the introduction of Bt cotton resulted in a substantial reduction in both insecticide use and damage losses caused by insects (Williams, 2015). The implementation of Bt cotton varieties in Australia since 1992 has led to an increase of approximately AUD$180 per hectare in total farm income and a reduction of approximately 97% in the usage of insecticides (Cotton Australia, 2021). Biologically active compounds and vectors may avoid risk assessment and regulatory evaluation due to their frequent exclusion from the hazardous chemical category and their explicit exclusion from the category of "genetically modified agents." This new oversight gap may result in allocations for dual purposes or inadvertent harm to the environment or human health (Heinemann and Walker, 2019; Hung and Slotkin, 2021). Synthetic biology's field of biosensors, or genetically encoded sensors, has the potential to revolutionize agriculture (Williams et al., 2017). To produce elicit results, a biosensor can be coupled with a variety of genetically encoded components, such as new receptors, deactivated Cas9 and derivatives (capable of binding but not cleaving DNA), and transcription factors (Gander et al., 2017; Kim et al., 2018). For instance, plants can be designed to stimuli to various environmental pollutants, nutrients, and abiotic stress factors (Pouvreau et al., 2018). The use of genetically modified organisms may have a negative impact on biodiversity globally. Genetically engineered crops can spread alien genes to non-modified varieties and crop wild cousins via wind pollination. This could change plant communities by favoring one species over another and upsetting the vertebrate and invertebrate feeding webs in which they are entrenched. According to some studies, genetically modified crops may also change the composition of soil microbial communities that can break down the root exudates of GM plants. Stronger biocontainment techniques can reduce the hazards posed by synthetic biology. For example, the insertion of foreign genes into the genomes of organelles; the use of "kill-switches" to immediately remove engineered organisms from the environment; and the use of localized (rather than global) gene drives that gradually lose their effectiveness over time (Collins, 2018; Stirling et al., 2017).

4 Synthetic biology for handling biowaste

The increase in urbanization, fast-paced economic growth, and unchecked population expansion have resulted in a significant rise in food consumption, leading to a substantial increase in annual waste generation (Zhou et al., 2022). FAO approximations suggest that approximately 1.3 billion tons of food go to waste annually, which is equivalent to one-third of the world's total food production and represents 28% of the 1.4 billion hectares of arable land worldwide (Karthikeyan et al., 2018; Paritosh et al., 2017). In the next 25 years, it is predicted that the food waste generation would rise in Asian nations due to economic and population expansion. By 2025, the urban food waste is anticipated to increase by 138 million tonnes compared to 2005. According to estimates, the European Union produces 2.5 billion tonnes of F.W. each year outside the food supply system (Li et al., 2017). Food waste comprises discharges from various sources, such as households, hospitality sector, food processing industries, slaughterhouses, commercial kitchens, and agricultural waste (Karthikeyan et al., 2018; Kiran et al., 2014; Ning et al., 2018). These sources are responsible for emitting approximately 3.3 billion tonnes of CO_2 annually, which contribute to the increase in GHG emissions (Adhikari et al., 2018; Paritosh et al., 2017). Effective waste treatment can reduce global GHG emissions by up to 21% (Thakur et al., 2022). Socioeconomic factors play a significant role in the generation and management of waste, with high-income countries' 16% of the population responsible for generating 34% of the total waste produced. Conversely, low-income countries generate only 5% of the waste, and a mere 39% of the generated waste is collected, with the rest ending up in landfills and dump yards. This is mainly attributed to the high costs of waste management, which range between $35 and $100 per tonne (Ahmed et al., 2018; Awasthi et al., 2022; FCCC, 2017; Zhang et al., 2022a, b). Valorization is gaining popularity because the conventional disposal or incineration methods for waste do not relieve the economic and environmental burden. Biogas, hydrogen, ethanol, biodiesel, butanol, and methane are just a few of the energy molecules or biofuels derived from food waste through Acidogenesis/dark fermentation. Volatile fatty acids (VFAs) derived from acidogenesis can be used as platform chemicals to produce alcohols, carbonyl compounds, and esters. With the use of hydrothermal liquefaction (HTL) technology, food waste that has been received in a wet state can be efficiently converted into high-heating value bio-oil/bio-crude, biochar, and other important gaseous products (Katakojwala et al., 2021; Otieno et al., 2022). Agricultural waste is composed of organic materials such as straw, bagasse, molasses, spent grains, husk (rice, maize, and wheat), shell (walnut, coconut, and groundnut), skin (banana, avocado), crop stalks (cotton), plant debris, and animal and bird dung (Dai et al., 2018; Gong et al., 2022). According to a report by the FAO, roughly 250 million tonnes of non-edible plant waste are produced annually as agro-industrial waste from various agricultural processes (Heredia-Guerrero et al., 2017). In 2013, China, the world's top grain producer, generated 1.75×10^9 t of agricultural waste, with crop straw accounting for 9.93×10^8 t, animal and poultry manure contributing 4.52×10^8 t, and forest leftovers making up 3.03×10^8 t (Dai et al., 2018). India, on the other hand, produces more than 350 million tonnes of agro-industrial waste annually from various sources. The utilization of renewable resources for the bio-based production of chemicals and materials has become increasingly critical in the face of ongoing climate change and the depletion of fossil fuels. The progress of modern biotechnology has yielded not only innovative approaches to manufacturing molecules and materials but also novel techniques and processes. Additionally, biotechnology has facilitated the development of new strategies for the management, monitoring, and remediation of pollutants, as well as the conversion of waste into valuable energy or useful molecules. Biomaterials such as polysaccharides, cellulose, protein, lipids, silk, and other polyhydroxyalkanoates/butyrates (PHA/PHBs) have all been successfully produced through biology-by-design (Tai et al., 2016; Nduko and Taguchi, 2021; Zhou et al., 2022). Reducing and preventing dangerous chemical waste emissions at the site of manufacture is a second strategy. The potential of microorganisms to decompose harmful compounds has significantly increased as a result of directed evolution, genetic engineering, or a fusion of both approaches. The resulting biocatalysts can be integrated with industrial pipelines that emit no pollution. For instance, microbes have been effectively manipulated to make PHAs, a family of varied biopolyesters used in medicine, smart materials, and ecologically friendly packaging. In addition to PHA production, metabolic engineers have created microbial strains that can also produce artificial materials, like the FDA-approved biomedical polymer poly(lactate-*co*-glycolate). Numerous microorganisms capable of fixing CO_2, beyond just cyanobacteria, can be enhanced through genetic modification or even entirely redesigned to achieve superior performance in capturing carbon in either mineral or organic forms. This, in turn, enables the conversion of captured carbon into value-added products such as sugars and polymers. These microorganisms are capable of fermenting complex municipal, commercial, sludge, or agricultural biowaste (Kumari et al., 2022; Tan et al., 2021). Solid waste contains 12% plastic, which poses a significant threat to the survival of ecosystems as they have infiltrated the bloodstream of humans and other aquatic organisms (Kiran et al., 2022). Microbial enzymes such as laccase, peroxidase, and oxidoreductase can effectively break down synthetic plastics. These enzymes break down synthetic plastic polymers into shorter chains of monomers, dimers, and oligomers that can easily penetrate cellular membranes and serve as an energy source for microbiota (Reshmy et al., 2022; Yang et al., 2021). Modern biotechnology also offers practical solutions by introducing novel metabolic pathways, derivative enzymes, or designer microorganisms that upcycle plastic waste into valuable products (Urbina et al., 2015). Omics techniques such as transcriptomics, metabolomics, proteomics, gene editing systems like clustered regularly interspaced short palindromic repeats-Cas9 (CRISPRCas9), and others can also help address the challenges of converting plastic waste into novel products (Kato et al., 2022; Lorenzo et al., 2018).

4.1 Biorecovery/recycling of metals from waste

Metals are a limited resource, and there is a growing global demand for them to be used in current technology. Despite a growing need for their use in present and future technologies, rare earth elements (REE) are not abundant and are therefore difficult to extract effectively (Capeness and Horsfall, 2020; Gumulya et al., 2018). Water and soil are contaminated during the mining and extraction of natural resources because the material is lost throughout the whole lifecycle of metals. The metal environs contain harmful substances and are extremely hot, osmolar, and acidic or basic (Kaksonen et al., 2017). To access metal resources that are trapped in contaminated land or lost in extraction and manufacturing processes, industries are increasingly turning to newer technologies. A promising approach to achieve this involves the use of biological methods, which not only extract resources but also clean up the environment and reduce waste (Urbina et al., 2019). Synthetic biology is going to resurrect the use of biology to discover and recover metals, which is thought to be a more "greener" alternative to chemical processes. In place of the typical laboratory strains, this has made it possible to use non-model organisms and to adapt genetic material derived from the environment to standardized components and procedures.

Once the high-grade ore has been extracted using physical and chemical methods, the remaining waste material is considered too low in metal concentration for further processing.

However, this low-grade waste ore can still be valuable if treated biologically. For example, bio-oxidation can be used to recover gold and silver, while bioleaching can extract copper, zinc, cobalt, and uranium from the ore, depending on the metal content (Kaksonen et al., 2017). Sulfuric acids are used in both instances to enhance the solubility of metal ions, which are then assimilated or bioaccumulated by naturally occurring or introduced single or mixed cultures of bacteria and archaea. The deposits are cleaned to extract the metal-containing bacteria for further processing. The majority of microorganisms utilized in this process are acidophilic bacteria or archaea that possess high heat resistance and the ability to oxidize compounds containing iron and sulfur. One such microbe commonly used in biomining cultures is *Acidithiobacillus* sp., either alone or in combination with other bacteria such as *Acidiphilium* sp., *Leptospirillum* sp., and *Sulfobacillus* sp. (García-García et al., 2016; Gumulya et al., 2018). These microorganisms' metabolism, reduction capacity, and frequent requirement for electron shuttle-mediators such as hydrogenases or cytochromes all contribute to bioremediation. The cell also uses ATP as a terminal electron acceptor, which increases its ability to reduce. Microalgae are another key player in the treatment of wastewater, in addition to bacteria and fungi. Distinctive microalgal genes and proteins were investigated by Sattayawat et al. for their biological functions in eradicating heavy metals. According to their findings, the genetic elements in their library could have potential applications in synthetic biology, which could pave the way for the use of microalgae as heavy metal bioremediators on a global scale (Diep et al., 2018; Jiang et al., 2022). To increase the effectiveness of metal apprehension and exclusion, candid cultures are utilized in the bioremediation of metals employing a variety of methodological approaches (biosensors, adsorption/chelation, and bioconversion). Although there are toolkits available for various organisms, there is a lack of research on the bacteria and archaea employed in biomining and wastewater treatment.

Biosensors: Biosensors offer a promising alternative for on-site analysis that is environmentally friendly and cost-effective. Organisms that thrive in high metal concentration environments contain valuable genetic material for biological detection and response mechanisms. Synthetic biology allows for the extraction of this genetic material and its insertion into non-heterologous hosts. Both prokaryotes and eukaryotes have sensing/regulatory systems such as CusSR (copper), ArsR (arsenic), MerR (mercury), NikR (nickel), and the extensively researched Fur system (iron), which are crucial in responding to excess metal concentration within cells (Bernard and Wang, 2017). Currently, biosensors are used for in situ detection of arsenic in drinking water, mercury and cadmium in anaerobic environments, and gold, copper, silver, zinc, and lead in soils, among others, providing field study platforms without the need for laboratory analysis (Martinez et al., 2019; Stenzler et al., 2018).

Adsorption/chelation: Adsorption is a process in which metal ions are bound to extracellular polysaccharides and biofilm matrix without causing oxidative stress or mismetalation, which can interfere with cellular physiology. Many species, including bacteria, algae, and fungi, possess metallothioneins and phytochelatins that bind metal ions through thiolate linkages (García-García et al., 2016; Muthu et al., 2017; Sun et al., 2020). For example, plant-derived phytochelatins, which are rich in cysteine and glutamine residues, are well known for their ability to enhance heavy-metal tolerance to cadmium, lead, and zinc by sequestering the metal ions and preventing them from disrupting crucial cellular activities, both intracellular and extracellular (Dennis et al., 2019; El-Amier et al., 2019).

Bioconversion: Bioconversion involves converting the metal ions themselves, while chelation and compartmentalization involve sequestering the ions to mitigate their harmful effects on cell viability (Cao et al., 2019; Cook et al., 2018). Bio-precipitation, which involves removing metal ions from the cytosol or surrounding medium as a defense mechanism against metal-ion stress, can occur independently of metabolism (Li et al., 2018a, b).

5 Synthetic biology—Medical research

Advancements in synthetic biology have allowed for integrating artificial metabolic networks into engineered strains of organisms, including bacteria, yeast, and mammalian cells, resulting in a better understanding of their metabolic processes. *Saccharomyces cerevisiae* is an often used industrial model organism because it was the first eukaryote with full genome sequencing (Wang et al., 2015). Engineered *S. cerevisiae* may be utilized to manufacture a range of compounds with the help of cutting-edge gene editing techniques like CRISPR/Cas. Manufacturing artemisinin acid and opioids in yeast are an example of a successful venture (Galanie et al., 2015; Heidari et al., 2017). This technique can take the place of extensive medicinal plant cultivation, making it a significant application of synthetic biology in chemical manufacturing (Zhao et al., 2018). Focused interest has been drawn for CRISPR/Cas-modified cells and species on living beings and governance (DiEuliis and Giordano, 2017). Simplification of the process for editing sequence-specific genes has opened up possibilities for the development of disease-specific genetic models and the exploration of therapeutic options for human genetic disorders. Moreover, gene-editing tools and synthetic biology can be utilized to enhance the production of conventional or novel infectious agents or neurotoxins in vivo or in vitro and to modify existing agents that target the nervous system and brain, thereby influencing neural phenotypes that impact cognition, emotion, and behavior. Any early clinical applications of this gene-editing technology, which has the capacity to genetically modify humans and embryos, could have unexpected repercussions. These techniques are thought to be a potential means of delivering biological weapons that might cause great destruction (Servick, 2016). It could potentially impact the course of neuro-weapons (DiEuliis and Giordano, 2017).

The comprehensive merger of synthetic biology and medical domains has led to the development of a number of diagnostic and therapeutic techniques for the prompt treatment of diseases and has also opened up new avenues for research (Danino et al., 2015; Xie et al., 2016). For identification and eradication of antibiotic-resistant infections, cells or bacteria are engineered (Peplow, 2016; Yehl et al., 2019). *Salmonella*, which possesses genes for creating synthetic anti-cancer drugs, is being utilized in innovative cancer therapy. The bacterium is programmed to detect the hypoxic microenvironment of tumors and gradually release the medication, leading to the inhibition of tumor growth, as described by Din et al. (2016). By incorporating artificial circuits that serve as biosensors and regulators into host cells, it is possible to alter the development of illness or cellular homeostasis (Bai et al., 2016; Saxena et al., 2016). For example, leukemia could be treated using synthetic biology approaches to apply modified T cells, such as chimeric antigen receptor T-cell immunotherapy (CAR-T) (Fraietta et al., 2018). The fusion of synthetic biology and computer science has led to the successful realization of manipulating artificially created cells using an external electronic system, as demonstrated by Shao et al. (2017). The global impact of the COVID-19 pandemic, which rapidly spread across the world, has been significant. Various genesis theories have been proposed, including the possibility of the novel coronavirus SAR-CoV-2 being created in a lab and subsequently released, causing widespread concern. Despite extensive research efforts by experts, the origin of SARS-CoV-2 remains unknown (Centers for Disease Control and Prevention, 2020; Paraskevis et al., 2020; Wan et al., 2020). Most vaccines currently under development or production, such as inactivated, subunit, and viral vector-based vaccines, are based on synthetic biology (Wang et al., 2021). In their review article, "The Proximal Origin of SARS-CoV-2" published in Nature Medicine, viral evolution experts Kristian Andersen and Andrew Rambaut provided evidence to support the theory that the virus originated from animals (Andersen et al., 2020). The use of CRISPR-Cas technology for pathogen identification has significantly reduced the cost of detecting SARS-CoV-2, with a much lower total cost compared to traditional RT-PCR assays (Rahimi et al., 2021).

6 Synthetic biology for sustainable development

The synthetic biology community and life sciences are being challenged by the United Nations' sustainable development goals (SDGs) to develop transformative technologies that can safeguard and enhance the habitability of our planet. The current industrial processes, consumption patterns, and intensive agricultural practices pose a threat to the stability of both human societies and ecosystems around the world (French, 2019). In 2015, the United Nations formulated 17 SDGs with the aim of promoting a more sustainable society by 2030 (https://www.un.org/sustainabledevelopment/). The UN SDGs highlight the importance of preserving and safeguarding Earth's natural resources, as well as promoting the development of innovative technologies. Synthetic biology has the potential to address eight SDGs, including issues such as hunger, lack of access to clean water, sanitation, sterility, pollution, environmental hazards, and climate change, among others. SynBio's impact could be significant in four areas, namely (1) reducing the use of harmful industrial chemicals by developing biologically based alternatives, (2) eliminating environmental pollutants, (3) enhancing crop productivity and soil health, and (4) substituting non-renewable synthetic materials with biologically derived ones. These contributions align with objectives 6, 9, 14, and 15 (Table 2).

Significant advancements in DNA synthesis and sequencing, as well as systems and synthetic biology, and CRISPR-based genome editing tools have accelerated the revolutionary impact of modern, science-based biotechnology, which first emerged in the late 1970s. The development of synthetic biology has been greatly aided by groundbreaking innovations in genetic engineering and genome editing, including the revolutionary CRISPR-Cas9 gene-editing tool, which originates from the bacterial immune system and has the potential to address various problems related to the SDGs. Rich natural resources exist in many developing nations, and with the right assistance, they may be directed toward biotech businesses. SynBio may be utilized to boost the output of organic plant products, lessen the vulnerability of crops to infections, or create new goods that cater to local demands. Ben-Erik Van Wyk and Nigel Gericke identified more than 700 plants in South Africa that could be economically valuable, providing the country with the necessary tools to establish a bio-based economy

TABLE 2 Synthetic biology's contribution to the sustainable development goals (SDGs).

Sustainable development goals (SDGs)	Changes needed in synthetic biology practices	Innovations/scenarios
6 CLEAN WATER AND SANITATION	Upscale lab experiments to real-field applications; join forces with non-governmental organizations (NGOs) and socially funded groups to identify key toxins	Benzene is removed from contaminated wells using a water filter made of wheat proteins and potato starch
8 DECENT WORK AND ECONOMIC GROWTH	Collaborate with small-scale farmers and industries to provide education and training; focus industrial investment on assisting new businesses	A start-up in Kenya produces value-added products from *Chlamydomonas reinhardtii* using synthetic biology
9 INDUSTRY, INNOVATION AND INFRASTRUCTURE	Reduce jeopardy; enumerate important industrial substances and colors that could be replaced with synthetic alternatives	Dyes produced from silica diatom frustules exploit structural color to replace Azo dyes used in the textile industry
10 REDUCED INEQUALITIES	Encourage non-Western nations to have access to the resources (such as vectors, gene synthesis, and lab equipment) required for synthetic biology; collaborating with decision-makers to create new regulations for the use of organisms and genetic sequences in synthetic biology applications	Gene synthesis companies automatically apply designated benefit-sharing charge for the use of certain genetic materials
11 SUSTAINABLE CITIES AND COMMUNITIES	Create and distribute materials to increase awareness and literacy about environmental health risks. Educate the general population about the social benefits of synthetic biology	Products made using synthetic biology are found in everyday life
12 RESPONSIBLE CONSUMPTION AND PRODUCTION	Minimize plastic usage and antibiotics; shift to using waste for industrial production of recombinant proteins	Industrial chasses use biosynthetic pathways for waste valorization as a source of energy
14 LIFE BELOW WATER	Create innovative technologies that will allow bioengineered organisms to be used safely in the environment to remove pollutants, restore landscapes that have been damaged, and increase agricultural production	Paper-based biosensors based on glutamate sensors from *Bacillus subtilus* detect algal neurotoxins in fish and shellfish to prevent human deaths
15 LIFE ON LAND	(All the above)	Bacteria and plants are engineered to remove polyaromatic hydrocarbons and polychlorinated biphenyls from polluted atmosphere

(Gericke, 2018). The stevia plant, a sweetener used by the indigenous Guarani people of Brazil, was documented by an ethnographer in the early 20th century. It was then mass-produced in Japan in the 1970s and became a substitute for sugar. There are currently around 20 patents by significant food firms on chemicals made from plants and their synthetic versions, as the demand for sweeteners grows (Olsson et al., 2016). Moreover, synthetic biology might help university labs and current biotech production processes produce less waste (Goal 12). A study conducted by the University of Exeter revealed that in 2015, laboratories around the world generated 5.5 million tonnes of plastic waste, with 20,500 labs contributing to this amount. Biomanufacturing could become more sustainable by expanding the availability of biodegradable plastics and suitable disposal techniques (such as landfills and anaerobic bioreactors). The convergence of these disciplines and methodologies with other key technologies of the current fourth industrial revolution, including AI, robotics, big data, and information technologies, is expected to give rise to a novel society, economy, and industry that differ significantly from the ones we currently know. WoS data show that out of the 13,050 papers on synthetic biology published globally since 1980, the majority (42%) are from the United States, followed by England (10.5%), Germany (9.4%), and the People's Republic of China (8.7%). Non-Western nations, on the other hand, are significantly underrepresented, with only 0.19% of papers coming from 17 African countries. For synthetic biology to be more widely adopted in non-Western nations, national governments must commit to supporting this research. To promote the growth of synthetic biology in non-Western countries, it is essential for governments to allocate funding for this research. Furthermore, philanthropic organizations such as the Gates Foundation could play a vital role in supporting the development of synthetic biology capabilities in underdeveloped nations, particularly by funding projects related to creating new vaccines. Better knowledge-sharing mechanisms could also facilitate the emergence of bio-based industries in less affluent countries.

7 Biosafety governance—Laws and regulations

SynBio is "double-edged." It may increase our understanding of pathogens and diseases for the good of all people. If unintentionally unleashed, genetically modified creatures could have catastrophic repercussions on ecosystems. Gene drive technologies and synthetic commodities have the potential to create new diseases and harm normally found bacteria (World Health Organization, 2018). The National Science Advisory Board for Biosecurity (NSABB) has highlighted in a report that synthetic biology could lead to biosecurity risks arising from the intentional misuse of the technology for bioterrorism or warfare or from accidental proclamations (Johns Hopkins Center for Health Security, 2019). To address bio-risk concerns, China has enacted laws and regulations governing biosafety and biosecurity in synthetic biology laboratory practices. The Biosafety Law, which was published in 2020 by the Standing Committee of the National People's Congress, aims to ensure biosafety and biosecurity. The majority of synthetic biology research falls under the category of genetic engineering and is subject to the current GMO laws in the E.U. These laws regulate the use and marketing of genetically modified organisms and their products (Keiper and Atanassova, 2020; Dunlap and Pauwels, 2017a, b). The genetic modification of microorganisms, as well as their cultivation, storage, transport, destruction, and disposal, are regulated by the European Union's GMO directives, including Directive 90/219/EC and Directive 2001/18/EC, which oversee the intentional release of GMOs. In laboratory management, the US Centers for Disease Control and Prevention (CDC) and the National Institutes of Health (NIH) have issued "Biosafety in Microbiology and Biomedical Laboratories," a manual providing guidelines for physical containment of pathogens. This has been documented in various studies, such as Eriksson et al. (2018, 2020) and Bratlie et al. (2019). The United Nations' 2030 Agenda for Sustainable Development places people and environment at its core, seeking to achieve a balanced approach to socioeconomic and environmental advancement. To further this goal, possible modifications may involve forbidding the production of substances obtained from certain rare or noteworthy organisms, along with implementing a particular "levy" on the utilization of genes derived from a selection of priority or safeguarded organisms.

8 Future perspectives

The progress in computational biology is crucial for accurately forecasting the behavior of engineered biological systems through in silico design. The integration of novel biotechnology and information technology could pose significant risks as life sciences become more automated. The combination of intelligence and biolabs could lead to malicious exploitation of engineering or modifying biological agents or toxins in the fourth industrial revolution. Synthetic biology presents potential bio-risks that could result from both intended and unintended consequences caused by cyber-overlap and automation, as a consequence of the fusion of synthetic biology with AI and automation. With the advancement of DNA synthesis techniques and the introduction of robotic cloud laboratories, there may be ways to surpass existing governance barriers, resulting in an increase in the virulence and transmissibility of pathogens.

References

Adhikari, B.B., Chae, M., Bressler, D.C., 2018. Utilization of slaughterhouse waste in value-added applications: recent advances in the development of wood adhesives. Polymers 10 (2), 176.

Ahmad, M., 2023. Plant breeding advancements with "CRISPR-Cas" genome editing technologies will assist future food security. Front. Plant Sci. 14, 1133036.

Ahmed, T., Shahid, M., Azeem, F., Rasul, I., Shah, A.A., Noman, M., Hameed, A., Manzoor, N., Manzoor, I., Muhammad, S., 2018. Biodegradation of plastics: current scenario and future prospects for environmental safety. Environ. Sci. Pollut. Res. 25, 7287–7298.

Anand, A., Subramanian, M., Kar, D., 2022. Breeding techniques to dispense higher genetic gains. Front. Plant Sci. 13, 1076094.

Andersen, K.G., Rambaut, A., Lipkin, W.I., Holmes, E.C., Garry, R.F., 2020. The proximal origin of SARS-CoV-2. Nat. Med. 26 (4), 450–452.

Awan, A.R., Shaw, W.M., Ellis, T., 2016. Biosynthesis of therapeutic natural products using synthetic biology. Adv. Drug Deliv. Rev. 105, 96–106.

Awasthi, S.K., Sarsaiya, S., Kumar, V., Chaturvedi, P., Sindhu, R., Binod, P., Zhang, Z., Pandey, A., Awasthi, M.K., 2022. Processing of municipal solid waste resources for a circular economy in China: an overview. Fuel 317, 123478.

Bai, P., Ye, H., Xie, M., Saxena, P., Zulewski, H., Charpin-El Hamri, G., Djonov, V., Fussenegger, M., 2016. A synthetic biology-based device prevents liver injury in mice. J. Hepatol. 65 (1), 84–94.

Bashor, C.J., Collins, J.J., 2018. Understanding biological regulation through synthetic biology. Annu. Rev. Biophys. 47, 399–423.

Basu, S., Rabara, R.C., Negi, S., Shukla, P., 2018. Engineering PGPMOs through gene editing and systems biology: a solution for phytoremediation? Trends Biotechnol. 36 (5), 499–510.

Bernard, E., Wang, B., 2017. Synthetic cell-based sensors with programmed selectivity and sensitivity. In: Biosensors and Biodetection: Methods and Protocols, Volume 2: Electrochemical, Bioelectronic, Piezoelectric, Cellular and Molecular Biosensors. Springer, pp. 349–363.

Bier, E., 2022. Gene drives gaining speed. Nat. Rev. Genet. 23, 5–22.

Bratlie, S., Halvorsen, K., Myskja, B.K., Mellegård, H., Bjorvatn, C., Frost, P., Heiene, G., Hofmann, B., Holst-Jensen, A., Holst-Larsen, T., Malnes, R.S., 2019. A novel governance framework for GMO: a tiered, more flexible regulation for GMO s would help to stimulate innovation and public debate. EMBO Rep. 20 (5), e47812.

Bueso, Y.F., Tangney, M., 2017. Synthetic biology in the driving seat of the bioeconomy. Trends Biotechnol. 35 (5), 373–378.

Cao, Y., Song, M., Li, F., Li, C., Lin, X., Chen, Y., Chen, Y., Xu, J., Ding, Q., Song, H., 2019. A synthetic plasmid toolkit for *Shewanella oneidensis* MR-1. Front. Microbiol. 10, 410.

Capeness, M.J., Horsfall, L.E., 2020. Synthetic biology approaches towards the recycling of metals from the environment. Biochem. Soc. Trans. 48 (4), 1367–1378.

Cazzolla Gatti, R., Reich, P.B., Gamarra, J.G., Crowther, T., Hui, C., Morera, A., Bastin, J.F., De-Miguel, S., Nabuurs, G.-J., Svenning, J.C., Serra-Diaz, J.M., 2022. The number of tree species on Earth. Proc. Natl. Acad. Sci. 119 (6), e2115329119.

Centers for Disease Control and Prevention, 2020. CDC COVID Data Tracker. Available from: https://www.cdc.gov/coronavirus/2019-nCoV/summary.html. (Accessed 8 February 2020).

Chen, Q.L., Hu, H.W., He, Z.Y., Cui, L., Zhu, Y.G., He, J.Z., 2021. Potential of indigenous crop microbiomes for sustainable agriculture. Nat. Food 2 (4), 233–240.

Collins, J.P., 2018. Gene drives in our future: challenges of and opportunities for using a self-sustaining technology in pest and vector management. BMC Proc. 12 (8), 37–41 (BioMed Central).

Cook, T.B., Rand, J.M., Nurani, W., Courtney, D.K., Liu, S.A., Pfleger, B.F., 2018. Genetic tools for reliable gene expression and recombineering in *Pseudomonas putida*. J. Ind. Microbiol. Biotechnol. 45 (7), 517–527.

Cotton Australia. Biotechnology and Cotton (2021) https://cottonaustralia.com.au/fact-sheet (online). Accessed 16 March 2021.

Cousins, A.B., Mullendore, D.L., Sonawane, B.V., 2020. Recent developments in mesophyll conductance in C3, C4, and crassulacean acid metabolism plants. Plant J. 101 (4), 816–830.

Cumbers, J., 2020. Synthetic Biology Startups Raised $3 Billion in the First Half of 2020. Forbes (online), 9 September 2020).

Dai, Y., Sun, Q., Wang, W., Lu, L., Liu, M., Li, J., Yang, S., Sun, Y., Zhang, K., Xu, J., Zheng, W., 2018. Utilizations of agricultural waste as adsorbent for the removal of contaminants: a review. Chemosphere 211, 235–253.

Danino, T., Prindle, A., Kwong, G.A., Skalak, M., Li, H., Allen, K., Hasty, J., Bhatia, S.N., 2015. Programmable probiotics for detection of cancer in urine. Sci. Transl. Med. 7 (289) (289ra84).

Degen, G.E., Orr, D.J., Carmo-Silva, E., 2021. Heat-induced changes in the abundance of wheat Rubisco activase isoforms. New Phytol. 229 (3), 1298–1311.

Dennis, K.K., Uppal, K., Liu, K.H., Ma, C., Liang, B., Go, Y.M., Jones, D.P., 2019. Phytochelatin database: a resource for phytochelatin complexes of nutritional and environmental metals. Database 2019, baz083. https://doi.org/10.1093/database/baz083.

Diep, P., Mahadevan, R., Yakunin, A.F., 2018. Heavy metal removal by bioaccumulation using genetically engineered microorganisms. Front. Bioeng. Biotechnol. 6, 157.

DiEuliis, D., Giordano, J., 2017. Why gene editors like CRISPR/Cas may be a game-changer for neuroweapons. Health Secur. 15 (3), 296–302.

Din, M.O., Danino, T., Prindle, A., Skalak, M., Selimkhanov, J., Allen, K., Julio, E., Atolia, E., Tsimring, L.S., Bhatia, S.-N., Hasty, J., 2016. Synchronized cycles of bacterial lysis for in vivo delivery. Nature 536 (7614), 81–85.

Downes, S., Walsh, T., Tay, W.T., 2016. Bt resistance in Australian insect pest species. Curr. Opin. Insect Sci. 15, 78–83.

Dunlap, G., Pauwels, E., 2017a. The Intelligent and Connected Bio-Labs of the Future: Promise and Peril in the Fourth Industrial Revolution. Wilson Center Policy Briefs, pp. 4–6.

Dunlap, G., Pauwels, E., 2017b. Available from: https://www.weforum.org/agenda/2017/10/intelligent-and-connected-bio-labs-promiseand-peril-inthe-fourth-industrial-revolution. (Accessed 6 February 2021).

Dwivedi, S.L., Scheben, A., Edwards, D., Spillane, C., Ortiz, R., 2017. Assessing and exploiting functional diversity in germplasm pools to enhance abiotic stress adaptation and yield in cereals and food legumes. Front. Plant Sci. 8, 1461.

El-Amier, Y., Elhindi, K., El-Hendawy, S., Al-Rashed, S., Abd-ElGawad, A., 2019. Antioxidant system and biomolecules alteration in *Pisum sativum* under heavy metal stress and possible alleviation by 5-aminolevulinic acid. Molecules 24 (22), 4194.

Eriksson, D., Harwood, W., Hofvander, P., Jones, H., Rogowsky, P., Stöger, E., Visser, R.G., 2018. A welcome proposal to amend the GMO legislation of the EU. Trends Biotechnol. 36 (11), 1100–1103.

Eriksson, D., Custers, R., Björnberg, K.E., Hansson, S.O., Purnhagen, K., Qaim, M., Romeis, J., Schiemann, J., Schleissing, S., Tosun, J., Visser, R.G., 2020. Options to reform the European Union legislation on GMOs: scope and definitions. Trends Biotechnol. 38 (3), 231–234.

Ermakova, M., Danila, F.R., Furbank, R.T., von Caemmerer, S., 2020. On the road to C4 rice: advances and perspectives. Plant J. 101 (4), 940–950.

Ermakova, M., Osborn, H., Groszmann, M., Bala, S., Bowerman, A., McGaughey, S., Byrt, C., Alonso-Cantabrana, H., Tyerman, S., Furbank, R.T., Sharwood, R.E., 2021. Expression of a CO_2-permeable aquaporin enhances mesophyll conductance in the C4 species *Setaria viridis*. elife 10, e70095.

FCCC, 2017. National Greenhouse Gas Inventory Data for the Period 1990–2015. Framework Convention on Climate Change 16479, pp. 1–5 (September).

Fraietta, J.A., Lacey, S.F., Orlando, E.J., Pruteanu-Malinici, I., Gohil, M., Lundh, S., Boesteanu, A.C., Wang, Y., O'Connor, R.S., Hwang, W.T., Pequignot, E., 2018. Determinants of response and resistance to CD19 chimeric antigen receptor (CAR) T cell therapy of chronic lymphocytic leukemia. Nat. Med. 24 (5), 563–571.

French, K.E., 2019. Harnessing synthetic biology for sustainable development. Nat. Sustain. 2 (4), 250–252.

Galanie, S., Thodey, K., Trenchard, I.J., Filsinger Interrante, M., Smolke, C.D., 2015. Complete biosynthesis of opioids in yeast. Science 349 (6252), 1095–1100.

Gander, M.W., Vrana, J.D., Voje, W.E., Carothers, J.M., Klavins, E., 2017. Digital logic circuits in yeast with CRISPR-dCas9 NOR gates. Nat. Commun. 8 (1), 15459.

Gao, J., Faheem, M., Yu, X., 2022. Global research on contaminated soil remediation: a bibliometric network analysis. Land 11 (9), 1581.

García-García, J.D., Sánchez-Thomas, R., Moreno-Sánchez, R., 2016. Bio-recovery of non-essential heavy metals by intra-and extracellular mechanisms in free-living microorganisms. Biotechnol. Adv. 34 (5), 859–873.

Gericke, N., 2018. People's Plants: A Guide to Useful Plants of Southern Africa. Briza Publications.

Giessen, T.W., Silver, P.A., 2017. Engineering carbon fixation with artificial protein organelles. Curr. Opin. Biotechnol. 46, 42–50.

Gong, C., Cao, L., Fang, D., Zhang, J., Awasthi, M.K., Xue, D., 2022. Genetic manipulation strategies for ethanol production from bioconversion of lignocellulose waste. Bioresour. Technol., 127105.

Goold, H.D., Wright, P., Hailstones, D., 2018. Emerging opportunities for synthetic biology in agriculture. Genes 9 (7), 341.

Gumulya, Y., Boxall, N.J., Khaleque, H.N., Santala, V., Carlson, R.P., Kaksonen, A.H., 2018. In a quest for engineering acidophiles for biomining applications: challenges and opportunities. Genes 9 (2), 116.

Hassan, S., Bhat, S.A., Kumar, V., Ganai, B.A., Ameen, F., 2022. Phytoremediation of heavy metals: an indispensable contrivance in green remediation technology. Plan. Theory 11 (9), 1255.

Heidari, R., Shaw, D.M., Elger, B.S., 2017. CRISPR and the rebirth of synthetic biology. Sci. Eng. Ethics 23, 351–363.

Heinemann, J.A., Walker, S., 2019. Environmentally applied nucleic acids and proteins for purposes of engineering changes to genes and other genetic material. Biosaf. Health 1 (3), 113–123.

Heredia-Guerrero, J.A., Heredia, A., Domínguez, E., Cingolani, R., Bayer, I.S., Athanassiou, A., Benítez, J.J., 2017. Cutin from agro-waste as a raw material for the production of bioplastics. J. Exp. Bot. 68 (19), 5401–5410.

Hou, J., Lin, D., White, J.C., Gardea-Torresdey, J.L., Xing, B., 2019. Joint nanotoxicology assessment provides a new strategy for developing nanoenabled bioremediation technologies. Environ. Sci. Technol. 53, 7927–7929.

Hung, Y.H., Slotkin, R.K., 2021. The initiation of RNA interference (RNAi) in plants. Curr. Opin. Plant Biol. 61, 102014.

Jiang, S., Tang, J., Rahimi, S., Mijakovic, I., Wei, Y., 2022. Efficient treatment of industrial wastewater with microbiome and synthetic biology. Front. Environ. Sci. 10, 342. https://doi.org/10.3389/fenvs.2022.902926.

Johns Hopkins Center for Health Security, 2019. Biosafety and Biosecurity in the Era of Synthetic Biology: Perspectives from the United States and China. Available from: https://www.centerforhealthsecurity.org/our-work/pubs_archive/pubs-pdfs/2019/190916-ChinaUSmtgReport.pdf. (Accessed 6 February 2021).

Kaksonen, A.H., Boxall, N.J., Usher, K.M., Ucar, D., Sahinkaya, E., 2017. Biosolubilisation of metals and metalloids. In: Sustainable Heavy Metal Remediation: Volume 1: Principles and Processes. Springer, pp. 233–283.

Karthikeyan, O.P., Trably, E., Mehariya, S., Bernet, N., Wong, J.W., Carrere, H., 2018. Pretreatment of food waste for methane and hydrogen recovery: a review. Bioresour. Technol. 249, 1025–1039.

Katakojwala, R., Kopperi, H.S., Avanthi, A., Mohan, S.V., 2021. Hydrothermal liquefaction of food waste: a potential resource recovery strategy. In: Sustainable Resource Management: Technologies for Recovery and Reuse of Energy and Waste Materials. John Wiley & Sons, Inc.

Kato, Y., Inabe, K., Hidese, R., Kondo, A., Hasunuma, T., 2022. Metabolomics-based engineering for biofuel and bio-based chemical production in microalgae and cyanobacteria: a review. Bioresour. Technol. 344, 126196.

Keiper, F., Atanassova, A., 2020. Regulation of synthetic biology: developments under the convention on biological diversity and its protocols. Front. Bioeng. Biotechnol. 8, 310.

Kim, H.J., Jeong, H., Lee, S.J., 2018. Synthetic biology for microbial heavy metal biosensors. Anal. Bioanal. Chem. 410, 1191–1203.

Kiran, E.U., Trzcinski, A.P., Ng, W.J., Liu, Y., 2014. Bioconversion of food waste to energy: a review. Fuel 134, 389–399.

Kiran, B.R., Kopperi, H., Venkata Mohan, S., 2022. Micro/nano-plastics occurrence, identification, risk analysis and mitigation: challenges and perspectives. Rev. Environ. Sci. Biotechnol. 21 (1), 169–203.

Kiran, B.R., Prasad, M.N.V., Venkata Mohan, S., 2023. Farm to fork: sustainable agrifood systems. In: Sustainable and Circular Management of Resources and Waste Towards a Green Deal. Elsevier, pp. 25–38.

Kumar, V., Baweja, M., Singh, P.K., Shukla, P., 2016. Recent developments in systems biology and metabolic engineering of plant-microbe interactions. Front. Plant Sci. 7, 1421.

Kumari, P., Kiran, B.R., Mohan, S.V., 2022. Polyhydroxybutyrate production by *Chlorella sorokiniana* SVMIICT8 under nutrient-deprived mixotrophy. Bioresour. Technol. 354, 127135.

Li, P., Zeng, Y., Xie, Y., Li, X., Kang, Y., Wang, Y., Xie, T., Zhang, Y., 2017. Effect of pretreatment on the enzymatic hydrolysis of kitchen waste for xanthan production. Bioresour. Technol. 223, 84–90.

Li, S., Li, Y., Smolke, C.D., 2018a. Strategies for microbial synthesis of high-value phytochemicals. Nat. Chem. 10 (4), 395–404.

Li, X., Lan, S.M., Zhu, Z.P., Zhang, C., Zeng, G.M., Liu, Y.G., Cao, W.C., Song, B., Yang, H., Wang, S.F., Wu, S.H., 2018b. The bioenergetics mechanisms and applications of sulfate-reducing bacteria in remediation of pollutants in drainage: a review. Ecotoxicol. Environ. Saf. 158, 162–170.

Li, J., Zhao, H., Zheng, L., An, W., 2021. Advances in synthetic biology and biosafety governance. Front. Bioeng. Biotechnol. 9, 598087.

Lorenzo, V.D., Prather, K.L., Chen, G.Q., O'Day, E., Kameke, C.V., Oyarzún, D.A., Hosta-Rigau, L., Alsafar, H., Cao, C., Ji, W., Okano, H., 2018. The power of synthetic biology for bioproduction, remediation and pollution control. EMBO Rep. 19 (4), 4–9.

Malla, M.A., Dubey, A., Yadav, S., Kumar, A., Hashem, A., Abd Allah, E.F., 2018. Understanding and designing the strategies for the microbe-mediated remediation of environmental contaminants using omics approaches. Front. Microbiol. 9, 1132.

Mansoori, P., 2018. 50 years of Iranian clinical, biomedical, and public health research: a bibliometric analysis of the Web of Science Core Collection (1965–2014). J. Glob. Health 8 (2), 020701.

Martinez, A.R., Heil, J.R., Charles, T.C., 2019. An engineered GFP fluorescent bacterial biosensor for detecting and quantifying silver and copper ions. Biometals 32 (2), 265–272.

Moses, T., Mehrshahi, P., Smith, A.G., Goossens, A., 2017. Synthetic biology approaches for the production of plant metabolites in unicellular organisms. J. Exp. Bot. 68 (15), 4057–4074.

Muthu, M., Wu, H.F., Gopal, J., Sivanesan, I., Chun, S., 2017. Exploiting microbial polysaccharides for biosorption of trace elements in aqueous environments—scope for expansion via nanomaterial intervention. Polymers 9 (12), 721.

Nduko, J.M., Taguchi, S., 2021. Microbial production of biodegradable lactate-based polymers and oligomeric building blocks from renewable and waste resources. Front. Bioeng. Biotechnol. 8, 618077.

Niehl, A., Soininen, M., Poranen, M.M., Heinlein, M., 2018. Synthetic biology approach for plant protection using ds RNA. Plant Biotechnol. J. 16 (9), 1679–1687.

Ning, Z., Zhang, H., Li, W., Zhang, R., Liu, G., Chen, C., 2018. Anaerobic digestion of lipid-rich swine slaughterhouse waste: methane production performance, long-chain fatty acids profile and predominant microorganisms. Bioresour. Technol. 269, 426–443. https://doi.org/10.1016/j.biortech.2018.08.001.

VIII. New biology

Olsson, K., Carlsen, S., Semmler, A., Simón, E., Mikkelsen, M.D., Møller, B.L., 2016. Microbial production of next-generation stevia sweeteners. Microb. Cell Factories 15, 1–14.

Otieno, O.D., Mulaa, F.J., Obiero, G., Midiwo, J., 2022. Utilization of fruit waste substrates in mushroom production and manipulation of chemical composition. Biocatal. Agric. Biotechnol. 39, 102250.

Ozdemir, T., Fedorec, A.J., Danino, T., Barnes, C.P., 2018. Synthetic biology and engineered live biotherapeutics: toward increasing system complexity. Cell Syst. 7 (1), 5–16.

Paraskevis, D., Kostaki, E.G., Magiorkinis, G., Panayiotakopoulos, G., Sourvinos, G., Tsiodras, A.S., 2020. Full-genome evolutionary analysis of the novel corona virus (2019-nCoV) rejects the hypothesis of emergence as a result of a recent recombination event. Infect. Genet. Evol. 79, 104212.

Paritosh, K., Kushwaha, S.K., Yadav, M., Pareek, N., Chawade, A., Vivekanand, V., 2017. Food waste to energy: an overview of sustainable approaches for food waste management and nutrient recycling. Biomed. Res. Int. 2017, 2370927.

Peplow, M., 2016. Synthetic malaria drug meets market resistance: first commercial deployment of synthetic biology for medicine has modest impact. Nature 530 (7591), 389–391.

Pouvreau, B., Vanhercke, T., Singh, S., 2018. From plant metabolic engineering to plant synthetic biology: the evolution of the design/build/test/learn cycle. Plant Sci. 273, 3–12.

Rabara, R.C., Tripathi, P., Rushton, P.J., 2017. Comparative metabolome profile between tobacco and soybean grown under water-stressed conditions. Biomed. Res. Int. 2017, 3065251.

Rahimi, H., Salehiabar, M., Barsbay, M., Ghaffarlou, M., Kavetskyy, T., Sharafi, A., Davaran, S., Chauhan, S.C., Danafar, H., Kaboli, S., Nosrati, H., 2021. CRISPR systems for COVID-19 diagnosis. ACS Sensors 6 (4), 1430–1445.

Reshmy, R., Philip, E., Madhavan, A., Tarfdar, A., Sindhu, R., Binod, P., Sirohi, R., Awasthi, M.K., Pandey, A., 2022. Biorefinery aspects for cost-effective production of nanocellulose and high value-added biocomposites. Fuel 311, 122575.

Sabreena, Hassan, S., Bhat, S.A., Kumar, V., Ganai, B.A., Fuad, A., 2022. Phytoremediation of heavy metals: an indispensable contrivance in green remediation technology. Plants 11 (9), 1255. https://doi.org/10.3390/plants11091255.

Sanz-Fernández, M., Rodríguez-Serrano, M., Sevilla-Perea, A., Pena, L., Mingorance, M.D., Sandalio, L.M., Romero-Puertas, M.C., 2017. Screening *Arabidopsis* mutants in genes useful for phytoremediation. J. Hazard. Mater. 335, 143–151.

Sargent, D., Conaty, W.C., Tissue, D.T., Sharwood, R.E., 2022. Synthetic biology and opportunities within agricultural crops. J. Sustain. Agric. Environ. 1 (2), 89–107. https://doi.org/10.1002/sae2.12014.

Saxena, P., Charpin-El Hamri, G., Folcher, M., Zulewski, H., Fussenegger, M., 2016. Synthetic gene network restoring endogenous pituitary-thyroid feedback control in experimental graves' disease. Proc. Natl. Acad. Sci. 113 (5), 1244–1249.

Servick, K., 2016. CRISPR—A Weapon of Mass Destruction? Available from: https://www.sciencemag.org/news/2016/02/crispr-weapon-mass-destruction. (Accessed 6 February 2021).

Shao, J., Xue, S., Yu, G., Yu, Y., Yang, X., Bai, Y., Zhu, S., Yang, L., Yin, J., Wang, Y., Liao, S., 2017. Smartphone-controlled optogenetically engineered cells enable semiautomatic glucose homeostasis in diabetic mice. Sci. Transl. Med. 9 (387), eaal2298.

Sharwood, R.E., Quick, W.P., Sargent, D., Estavillo, G.M., Silva-Perez, V., Furbank, R.T., 2022. Mining for allelic gold: finding genetic variation in photosynthetic traits in crops and wild relatives. J. Exp. Bot. 73 (10), 3085–3108.

Stenzler, B.R., Gaudet, J., Poulain, A.J., 2018. An anaerobic biosensor assay for the detection of mercury and cadmium. J. Vis. Exp. 142, e58324.

Stirling, F., Bitzan, L., O'Keefe, S., Redfield, E., Oliver, J.W., Way, J., Silver, P.A., 2017. Rational design of evolutionarily stable microbial kill switches. Mol. Cell 68 (4), 686–697.

Sun, G.L., Reynolds, E.E., Belcher, A.M., 2020. Using yeast to sustainably remediate and extract heavy metals from waste waters. Nat. Sustain. 3 (4), 303–311.

Tai, Y.S., Xiong, M., Jambunathan, P., Wang, J., Wang, J., Stapleton, C., Zhang, K., 2016. Engineering nonphosphorylative metabolism to generate lignocellulose-derived products. Nat. Chem. Biol. 12 (4), 247–253.

Tan, D., Wang, Y., Tong, Y., Chen, G.Q., 2021. Grand challenges for industrializing polyhydroxyalkanoates (PHAs). Trends Biotechnol. 39 (9), 953–963.

Thakur, S., Chaudhary, J., Singh, P., Alsanie, W.F., Grammatikos, S.A., Thakur, V.K., 2022. Synthesis of bio-based monomers and polymers using microbes for a sustainable bioeconomy. Bioresour. Technol. 344, 126156.

U.S. Department of Defense, 2014. DoD Science&Technology Priorities. Available from: https://community.apan. org/wg/afosr/m/alea_stewart/135113. (Accessed 2 June 2020).

Ubando, A.T., Africa, A.D.M., Maniquiz-Redillas, M.C., Culaba, A.B., Chen, W.H., Chang, J.S., 2021. Microalgal biosorption of heavy metals: a comprehensive bibliometric review. J. Hazard. Mater. 402, 123431.

Urbina, M.A., Watts, A.J., Reardon, E.E., 2015. Labs should cut plastic waste too. Nature 528 (7583), 479.

Urbina, J., Patil, A., Fujishima, K., Paulino-Lima, I.G., Saltikov, C., Rothschild, L.J., 2019. A new approach to biomining: bioengineering surfaces for metal recovery from aqueous solutions. Sci. Rep. 9 (1), 1–11.

Voigt, C.A., 2020. Synthetic biology 2020–2030: six commercially-available products that are changing our world. Nat. Commun. 11 (1), 6379.

Wan, Y., Shang, J., Graham, R., Baric, R.S., Li, F., 2020. Receptor recognition by the novel coronavirus from Wuhan: an analysis based on decade-long structural studies of SARS coronavirus. J. Virol. 94 (7), e00127-20.

Wang, P., Wei, Y., Fan, Y., Liu, Q., Wei, W., Yang, C., Zhang, L., Zhao, G., Yue, J., Yan, X., Zhou, Z., 2015. Production of bioactive ginsenosides Rh2 and Rg3 by metabolically engineered yeasts. Metab. Eng. 29, 97–105.

Wang, Y., Zhang, Y., Chen, J., Wang, M., Zhang, T., Luo, W., Li, Y., Wu, Y., Zeng, B., Zhang, K., Deng, R., 2021. Detection of SARS-CoV-2 and its mutated variants via CRISPR-Cas13-based transcription amplification. Anal. Chem. 93 (7), 3393–3402.

Williams, M., 2015. Cotton Insect Losses (online) https://www.entomology.msstate.edu/resources/cottoncrop.asp. (Accessed 20 April 2016).

Williams, T.C., Xu, X., Ostrowski, M., Pretorius, I.S., Paulsen, I.T., 2017. Positive-feedback, ratiometric biosensor expression improves high-throughput metabolite-producer screening efficiency in yeast. Synth. Biol. 2 (1), ysw002.

World Health Organization [WHO], 2018. Biosafety and Biosecurity. WHO, Geneva.

Xie, M., Ye, H., Wang, H., Charpin-El Hamri, G., Lormeau, C., Saxena, P., Stelling, J., Fussenegger, M., 2016. β-cell–mimetic designer cells provide closed-loop glycemic control. Science 354 (6317), 1296–1301.

Yang, Z., Lü, F., Zhang, H., Wang, W., Shao, L., Ye, J., He, P., 2021. Is incineration the terminator of plastics and microplastics? J. Hazard. Mater. 401, 123429.

Yehl, K., Lemire, S., Yang, A.C., Ando, H., Mimee, M., Torres, M.D.T., de la Fuente-Nunez, C., Lu, T.K., 2019. Engineering phage host-range and suppressing bacterial resistance through phage tail fiber mutagenesis. Cell 179 (2), 459–469.

Zhang, X., Ma, D., Lv, J., Feng, Q., Liang, Z., Chen, H., Feng, J., 2022a. Food waste composting based on patented compost bins: carbon dioxide and nitrous oxide emissions and the denitrifying community analysis. Bioresour. Technol. 346, 126643.

Zhang, Y.L., He, G.D., He, Y.Q., He, T.B., 2022b. Bibliometrics-based: trends in phytoremediation of potentially toxic elements in soil. Land 11 (11), 2030.

Zhao, X., Park, S.Y., Yang, D., Lee, S.Y., 2018. Synthetic biology for natural compounds. Biochemistry 58 (11), 1454–1456.

Zhou, Y., Kumar, V., Harirchi, S., Vigneswaran, V.S., Rajendran, K., Sharma, P., Tong, Y.W., Binod, P., Sindhu, R., Sarsaiya, S., Balakrishnan, D., 2022. Recovery of value-added products from biowaste: a review. Bioresour. Technol. 360, 127565.

Index

Note: Page numbers followed by *f* indicate figures, *t* indicate tables, and *b* indicate boxes.

Printed in the United States
by Baker & Taylor Publisher Services